BIG IDEAS
MATH.
INTEGRATED MATHEMATICS II

TEACHING EDITION

Ron Larson
Laurie Boswell

BIG IDEAS LEARNING

Erie, Pennsylvania
BigIdeasLearning.com

Big Ideas Learning, LLC
1762 Norcross Road
Erie, PA 16510-3838
USA

For product information and customer support, contact Big Ideas Learning
at **1-877-552-7766** or visit us at ***BigIdeasLearning.com***.

Cover Image
Mopic/Shutterstock.com, © Abidal | Dreamstime.com

Copyright © 2016 by Big Ideas Learning, LLC. All rights reserved.

No part of this work may be reproduced or transmitted in any form or by any means, electronic or mechanical, including, but not limited to, photocopying and recording, or by any information storage or retrieval system, without prior written permission of Big Ideas Learning, LLC unless such copying is expressly permitted by copyright law. Address inquiries to Permissions, Big Ideas Learning, LLC, 1762 Norcross Road, Erie, PA 16510.

Big Ideas Learning and *Big Ideas Math* are registered trademarks of Larson Texts, Inc.

Printed in the U.S.A.

ISBN 13: 978-1-68033-070-0
ISBN 10: 1-68033-070-5

2 3 4 5 6 7 8 9 10 WEB 19 18 17 16 15

Authors

Ron Larson, Ph.D., is well known as the lead author of a comprehensive program for mathematics that spans middle school, high school, and college courses. He holds the distinction of Professor Emeritus from Penn State Erie, The Behrend College, where he taught for nearly 40 years. He received his Ph.D. in mathematics from the University of Colorado. Dr. Larson's numerous professional activities keep him actively involved in the mathematics education community and allow him to fully understand the needs of students, teachers, supervisors, and administrators.

Ron Larson

Laurie Boswell, Ed.D., is the Head of School and a mathematics teacher at the Riverside School in Lyndonville, Vermont. Dr. Boswell is a recipient of the Presidential Award for Excellence in Mathematics Teaching and has taught mathematics to students at all levels, from elementary through college. Dr. Boswell was a Tandy Technology Scholar and served on the NCTM Board of Directors from 2002 to 2005. She currently serves on the board of NCSM and is a popular national speaker.

Laurie Boswell

Dr. Ron Larson and **Dr. Laurie Boswell** began writing together in 1992. Since that time, they have authored over two dozen textbooks. In their collaboration, Ron is primarily responsible for the student edition while Laurie is primarily responsible for the teaching edition.

For the Student

Welcome to *Big Ideas Math Integrated Mathematics II*. From start to finish, this program was designed with you, the learner, in mind.

As you work through the chapters in your Integrated Mathematics II course, you will be encouraged to think and to make conjectures while you persevere through challenging problems and exercises. You will make errors—and that is ok! Learning and understanding occur when you make errors and push through mental roadblocks to comprehend and solve new and challenging problems.

In this program, you will also be required to explain your thinking and your analysis of diverse problems and exercises. You will master content through engaging explorations that will provide deeper understanding, concise stepped-out examples and rich thought-provoking exercises. Being actively involved in learning will help you develop mathematical reasoning and use it to solve math problems and work through other everyday challenges. We wish you the best of luck as you explore Integrated Mathematics II. We are excited to be a part of your preparation for the challenges you will face in the remainder of your high school career and beyond.

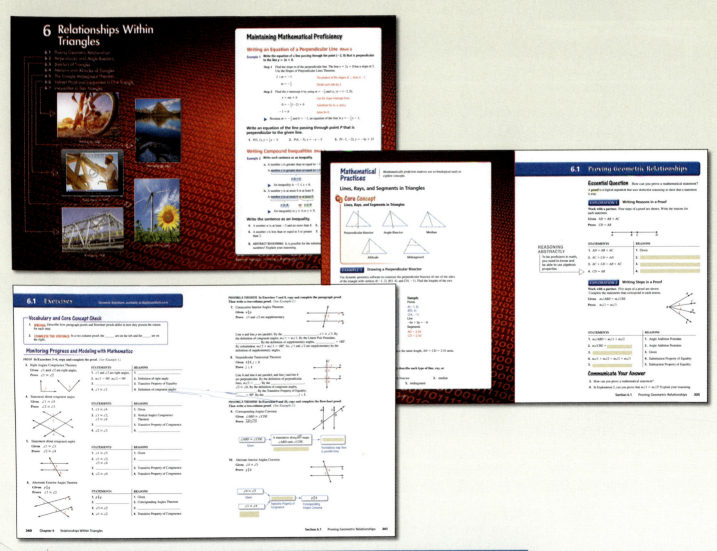

Big Ideas Math High School Research

Big Ideas Math Integrated Mathematics I, II, and *III* is a research-based program providing a rigorous, focused, and coherent curriculum for high school students. Ron Larson and Laurie Boswell utilized their expertise as well as the body of knowledge collected by additional expert mathematicians and researchers to develop each course.

The pedagogical approach to this program follows the best practices outlined in the most prominent and widely-accepted educational research and standards, including:

Achieve, ACT, and The College Board

Adding It Up: Helping Children Learn Mathematics
National Research Council ©2001

Common Core State Standards for Mathematics
National Governors Association Center for Best Practices and the Council of Chief State School Officers ©2010

Curriculum Focal Points and the *Principles and Standards for School Mathematics* ©2000
National Council of Teachers of Mathematics (NCTM)

Project Based Learning
The Buck Institute

Rigor/Relevance Framework™
International Center for Leadership in Education

Universal Design for Learning Guidelines
CAST ©2011

Big Ideas Math would like to express our gratitude to the mathematics education and instruction experts who served as consultants during the writing of *Big Ideas Math Integrated Mathematics I, II,* and *III*. Their input was an invaluable asset during the development of this program.

Kristen Karbon
Curriculum and Assessment Coordinator
Troy School District
Troy, Michigan

Jean Carwin
Math Specialist/TOSA
Snohomish School District
Snohomish, Washington

Carolyn Briles
Performance Tasks Consultant
Mathematics Teacher, Loudoun County Public Schools
Leesburg, Virginia

Bonnie Spence
Differentiated Instruction Consultant
Mathematics Lecturer, The University of Montana
Missoula, Montana

Connie Schrock, Ph.D.
Performance Tasks Consultant
Mathematics Professor, Emporia State University
Emporia, Kansas

We would also like to thank all of our reviewers who took the time to provide feedback during the final development phases. For a complete list of the *Big Ideas Math* program reviewers, please visit *www.BigIdeasLearning.com*.

Mathematical Practices

Make sense of problems and persevere in solving them.
- *Essential Questions* help students focus on core concepts as they analyze and work through each *Exploration*.
- Section opening *Explorations* allow students to struggle with new mathematical concepts and explain their reasoning in the *Communicate Your Answer* questions.

Reason abstractly and quantitatively.
- *Reasoning*, *Critical Thinking*, *Abstract Reasoning*, and *Problem Solving* exercises challenge students to apply their acquired knowledge and reasoning skills to solve each problem.
- *Thought Provoking* exercises test the reasoning skills of students as they analyze and interpret perplexing scenarios.

Construct viable arguments and critique the reasoning of others.
- Students must justify their responses to each *Essential Question* in the *Communicate Your Answer* questions at the end of each *Exploration* set.
- Students are asked to construct arguments and critique the reasoning of others in specialized exercises, including *Making an Argument*, *How Do You See It?*, *Drawing Conclusions*, *Reasoning*, *Error Analysis*, *Problem Solving*, and *Writing*.

Model with mathematics.
- Real-life scenarios are utilized in *Explorations*, *Examples*, *Exercises*, and *Assessments* so students have opportunities to apply the mathematical concepts they have learned to realistic situations.
- *Modeling with Mathematics* exercises allow students to interpret a problem in the context of a real-life situation, often utilizing tables, graphs, visual representations, and formulas.

Use appropriate tools strategically.
- Students are provided opportunities for selecting and utilizing the appropriate mathematical tool in *Using Tools* exercises. Students work with graphing calculators, dynamic geometry software, models, and more.
- A variety of tool papers and manipulatives are available for students to use in problems as strategically appropriate.

Attend to precision.
- *Vocabulary and Core Concept Check* exercises require students to use clear, precise mathematical language in their solutions and explanations.
- The many opportunities for cooperative learning in this program, including working with partners for each *Exploration*, support precise, explicit mathematical communication.

Look for and make use of structure.
- *Using Structure* exercises provide students with the opportunity to explore patterns and structure in mathematics.
- Students analyze structure in problems through *Justifying Steps* and *Analyzing Equations* exercises.

Look for and express regularity in repeated reasoning.
- Students are continually encouraged to evaluate the reasonableness of their solutions and their steps in the problem-solving process.
- Stepped-out *Examples* encourage students to maintain oversight of their problem-solving process and pay attention to the relevant details in each step.

Course Overview

Big Ideas Math Integrated Mathematics I, II, and *III* was developed using the consistent, dependable learning and instructional theory that have become synonymous with *Big Ideas Math*. Students will gain a deeper understanding of mathematics by narrowing their focus to fewer topics at each grade level. They will also master content through inductive reasoning opportunities, engaging explorations, concise stepped-out examples, and rich thought-provoking exercises.

The research-based curriculum features a continual development of concepts that have been previously taught while integrating algebra, geometry, probability, and statistics topics throughout each course.

In *Integrated Mathematics I*, students will study linear and exponential equations and functions. Students will use linear regression and perform data analysis. They will also learn about geometry topics such as simple proofs, congruence, and transformations.

Integrated Mathematics II expands into quadratic, absolute value, and other functions. Students will also explore polynomial equations and factoring, and probability and its applications. Coverage of geometry topics extends to polygon relationships, proofs, similarity, trigonometry, circles, and three-dimensional figures.

In *Integrated Mathematics III*, students will expand their understanding of area and volume with geometric modeling, which students will apply throughout the course as they learn new types of functions. Students will study polynomial, radical, logarithmic, rational, and trigonometric functions. They will also learn how visual displays and statistics relate to different types of data and probability distributions.

1 Functions and Exponents

Laurie's Notes

Chapter Summary	T-xxxii
1.1 Overview	T-2
1.2 Overview	T-10
1.3 Overview	T-18
1.4 Overview	T-26
1.5 Overview	T-34
1.6 Overview	T-40

Maintaining Mathematical Proficiency 1
Mathematical Practices 2

1.1 Absolute Value Functions
Exploration 3
Lesson 4

1.2 Piecewise Functions
Explorations 11
Lesson 12

1.3 Inverse of a Function
Explorations 19
Lesson 20

Study Skills: Making Note Cards 25
1.1–1.3 Quiz 26

1.4 Properties of Exponents
Exploration 27
Lesson 28

1.5 Radicals and Rational Exponents
Explorations 35
Lesson 36

1.6 Exponential Functions
Explorations 41
Lesson 42

Performance Task: Pool Shots 51
Chapter Review 52
Chapter Test 55
Cumulative Assessment 56

See the Big Idea
Learn how to make bank shots on a pool table using the reflection of a line.

2 Polynomial Equations and Factoring

Laurie's Notes

Chapter Summary	T-58
2.1 Overview	T-60
2.2 Overview	T-68
2.3 Overview	T-74
2.4 Overview	T-80
2.5 Overview	T-88
2.6 Overview	T-94
2.7 Overview	T-100
2.8 Overview	T-106

	Maintaining Mathematical Proficiency	59
	Mathematical Practices	60
2.1	**Adding and Subtracting Polynomials**	
	Explorations	61
	Lesson	62
2.2	**Multiplying Polynomials**	
	Explorations	69
	Lesson	70
2.3	**Special Products of Polynomials**	
	Explorations	75
	Lesson	76
2.4	**Solving Polynomial Equations in Factored Form**	
	Explorations	81
	Lesson	82
	Study Skills: Preparing for a Test	87
	2.1–2.4 Quiz	88
2.5	**Factoring $x^2 + bx + c$**	
	Exploration	89
	Lesson	90
2.6	**Factoring $ax^2 + bx + c$**	
	Exploration	95
	Lesson	96
2.7	**Factoring Special Products**	
	Explorations	101
	Lesson	102
2.8	**Factoring Polynomials Completely**	
	Explorations	107
	Lesson	108
	Performance Task: Flight Path of a Bird	113
	Chapter Review	114
	Chapter Test	117
	Cumulative Assessment	118

See the Big Idea
Explore whether seagulls and crows use the optimal height while dropping hard-shelled food to crack it open.

3 Graphing Quadratic Functions

Laurie's Notes

Chapter Summary	T-120
3.1 Overview	T-122
3.2 Overview	T-128
3.3 Overview	T-134
3.4 Overview	T-144
3.5 Overview	T-152
3.6 Overview	T-160
3.7 Overview	T-168

	Maintaining Mathematical Proficiency	121
	Mathematical Practices	122
3.1	**Graphing $f(x) = ax^2$**	
	Exploration	123
	Lesson	124
3.2	**Graphing $f(x) = ax^2 + c$**	
	Explorations	129
	Lesson	130
3.3	**Graphing $f(x) = ax^2 + bx + c$**	
	Explorations	135
	Lesson	136
	Study Skills: Learning Visually	143
	3.1–3.3 Quiz	144
3.4	**Graphing $f(x) = a(x - h)^2 + k$**	
	Explorations	145
	Lesson	146
3.5	**Graphing $f(x) = a(x - p)(x - q)$**	
	Exploration	153
	Lesson	154
3.6	**Focus of a Parabola**	
	Explorations	161
	Lesson	162
3.7	**Comparing Linear, Exponential, and Quadratic Functions**	
	Explorations	169
	Lesson	170
	Performance Task: Solar Energy	179
	Chapter Review	180
	Chapter Test	185
	Cumulative Assessment	186

See the Big Idea
Learn how to build your own parabolic mirror that uses sunlight to generate electricity.

4 Solving Quadratic Equations

Laurie's Notes

Chapter Summary	T-188
4.1 Overview	T-190
4.2 Overview	T-200
4.3 Overview	T-208
4.4 Overview	T-214
4.5 Overview	T-224
4.6 Overview	T-236
4.7 Overview	T-244
4.8 Overview	T-250
4.9 Overview	T-258

Maintaining Mathematical Proficiency 189
Mathematical Practices 190

4.1 Properties of Radicals
Explorations 191
Lesson 192

4.2 Solving Quadratic Equations by Graphing
Explorations 201
Lesson 202

4.3 Solving Quadratic Equations Using Square Roots
Explorations 209
Lesson 210

4.4 Solving Quadratic Equations by Completing the Square
Explorations 215
Lesson 216

4.5 Solving Quadratic Equations Using the Quadratic Formula
Explorations 225
Lesson 226

Study Skills: Keeping a Positive Attitude 235
4.1–4.5 Quiz 236

4.6 Complex Numbers
Explorations 237
Lesson 238

4.7 Solving Quadratic Equations With Complex Solutions
Explorations 245
Lesson 246

4.8 Solving Nonlinear Systems of Equations
Explorations 251
Lesson 252

4.9 Quadratic Inequalities
Explorations 259
Lesson 260

Performance Task: The Golden Ratio 267
Chapter Review 268
Chapter Test 273
Cumulative Assessment 274

See the Big Idea
Explore the Parthenon and investigate how the use of the golden rectangle has evolved since its discovery.

5 Probability

Laurie's Notes

Chapter Summary	T-276
5.1 Overview	T-278
5.2 Overview	T-286
5.3 Overview	T-294
5.4 Overview	T-304
5.5 Overview	T-310
5.6 Overview	T-318

Maintaining Mathematical Proficiency ... 277
Mathematical Practices ... 278

5.1 Sample Spaces and Probability
Explorations ... 279
Lesson ... 280

5.2 Independent and Dependent Events
Explorations ... 287
Lesson ... 288

5.3 Two-Way Tables and Probability
Explorations ... 295
Lesson ... 296
Study Skills: Making a Mental Cheat Sheet ... 303
5.1–5.3 Quiz ... 304

5.4 Probability of Disjoint and Overlapping Events
Explorations ... 305
Lesson ... 306

5.5 Permutations and Combinations
Explorations ... 311
Lesson ... 312

5.6 Binomial Distributions
Explorations ... 319
Lesson ... 320

Performance Task: Risk Analysis ... 325
Chapter Review ... 326
Chapter Test ... 329
Cumulative Assessment ... 330

See the Big Idea
Learn about caring for trees at an arboretum.

6 Relationships Within Triangles

Laurie's Notes

Chapter Summary	T-332
6.1 Overview	T-334
6.2 Overview	T-342
6.3 Overview	T-350
6.4 Overview	T-360
6.5 Overview	T-370
6.6 Overview	T-376
6.7 Overview	T-384

Maintaining Mathematical Proficiency 333
Mathematical Practices 334

6.1 Proving Geometric Relationships
Explorations 335
Lesson 336

6.2 Perpendicular and Angle Bisectors
Explorations 343
Lesson 344

6.3 Bisectors of Triangles
Explorations 351
Lesson 352

6.4 Medians and Altitudes of Triangles
Explorations 361
Lesson 362

Study Skills: Rework Your Notes 369
6.1–6.4 Quiz 370

6.5 The Triangle Midsegment Theorem
Explorations 371
Lesson 372

6.6 Indirect Proof and Inequalities in One Triangle
Exploration 377
Lesson 378

6.7 Inequalities in Two Triangles
Exploration 385
Lesson 386

Performance Task: Building a Roof Truss 391
Chapter Review 392
Chapter Test 397
Cumulative Assessment 398

See the Big Idea
Discover why triangles are used in building for strength.

xiii

7 Quadrilaterals and Other Polygons

Laurie's Notes

Chapter Summary	T-400
7.1 Overview	T-402
7.2 Overview	T-410
7.3 Overview	T-418
7.4 Overview	T-430
7.5 Overview	T-440

Maintaining Mathematical Proficiency 401
Mathematical Practices 402

7.1 Angles of Polygons
Explorations 403
Lesson 404

7.2 Properties of Parallelograms
Explorations 411
Lesson 412

7.3 Proving That a Quadrilateral is a Parallelogram
Explorations 419
Lesson 420

Study Skills: Keeping Your Mind Focused during Class 429
7.1–7.3 Quiz 430

7.4 Properties of Special Parallelograms
Explorations 431
Lesson 432

7.5 Properties of Trapezoids and Kites
Explorations 441
Lesson 442

Performance Task: Diamonds 451
Chapter Review 452
Chapter Test 455
Cumulative Assessment 456

See the Big Idea
Explore what the refractive index, reflected light, and light dispersion have to do with diamonds.

xiv

8 Similarity

Laurie's Notes

Chapter Summary	T-458
8.1 Overview	T-460
8.2 Overview	T-468
8.3 Overview	T-474
8.4 Overview	T-484
8.5 Overview	T-492
8.6 Overview	T-498

Maintaining Mathematical Proficiency 459
Mathematical Practices 460

8.1 Dilations
Explorations 461
Lesson 462

8.2 Similarity and Transformations
Explorations 469
Lesson 470

8.3 Similar Polygons
Explorations 475
Lesson 476

8.4 Proving Triangle Similarity by AA
Exploration 485
Lesson 486

Study Skills: Analysing Your Errors 491
8.1–8.4 Quiz 492

8.5 Proving Triangle Similarity by SSS and SAS
Explorations 493
Lesson 494

8.6 Proportionality Theorems
Explorations 499
Lesson 500

Performance Task: Pool Maintenance 507
Chapter Review 508
Chapter Test 511
Cumulative Assessment 512

See the Big Idea
Discover how many different ways you can scale a model.

xv

9 Right Triangles and Trigonometry

Laurie's Notes

Chapter Summary	T-514
9.1 Overview	T-516
9.2 Overview	T-524
9.3 Overview	T-530
9.4 Overview	T-540
9.5 Overview	T-546
9.6 Overview	T-556

Maintaining Mathematical Proficiency 515
Mathematical Practices 516

9.1 The Pythagorean Theorem
Explorations 517
Lesson 518

9.2 Special Right Triangles
Explorations 525
Lesson 526

9.3 Similar Right Triangles
Explorations 531
Lesson 532

Study Skills: Form a Weekly Study Group, Set Up Rules 539
9.1–9.3 Quiz 540

9.4 The Tangent Ratio
Explorations 541
Lesson 542

9.5 The Sine and Cosine Ratios
Exploration 547
Lesson 548

9.6 Solving Right Triangles
Explorations 557
Lesson 558

Performance Task: Challenging the Rock Wall 563
Chapter Review 564
Chapter Test 567
Cumulative Assessment 568

See the Big Idea
Test the accuracy of two measurement methods and discover which one prevails.

10 Circles

Laurie's Notes

Chapter Summary	T-570
10.1 Overview	T-572
10.2 Overview	T-580
10.3 Overview	T-588
10.4 Overview	T-596
10.5 Overview	T-604
10.6 Overview	T-612
10.7 Overview	T-618

Maintaining Mathematical Proficiency 571
Mathematical Practices 572

10.1 Lines and Segments That Intersect Circles
Explorations 573
Lesson 574

10.2 Finding Arc Measures
Exploration 581
Lesson 582

10.3 Using Chords
Explorations 589
Lesson 590
Study Skills: Keeping Your Mind Focused While Completing Homework 595
10.1–10.3 Quiz 596

10.4 Inscribed Angles and Polygons
Explorations 597
Lesson 598

10.5 Angle Relationships in Circles
Explorations 605
Lesson 606

10.6 Segment Relationships in Circles
Explorations 613
Lesson 614

10.7 Circles in the Coordinate Plane
Explorations 619
Lesson 620

Performance Task: Finding Locations 627
Chapter Review 628
Chapter Test 633
Cumulative Assessment 634

See the Big Idea
Utilize trilateration to find the epicenters of historical earthquakes and discover where they lie on known fault lines.

xvii

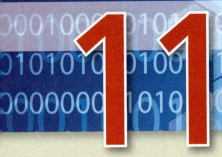

11 Circumference, Area, and Volume

Laurie's Notes

Chapter Summary	T-636
11.1 Overview	T-638
11.2 Overview	T-646
11.3 Overview	T-654
11.4 Overview	T-664
11.5 Overview	T-672
11.6 Overview	T-678
11.7 Overview	T-684

	Maintaining Mathematical Proficiency	637
	Mathematical Practices	638
11.1	**Circumference and Arc Length**	
	Explorations	639
	Lesson	640
11.2	**Areas of Circles and Sectors**	
	Explorations	647
	Lesson	648
11.3	**Areas of Polygons**	
	Explorations	655
	Lesson	656
	Study Skills: Kinesthetic Learners	663
	11.1–11.3 Quiz	664
11.4	**Volumes of Prisms and Cylinders**	
	Explorations	665
	Lesson	666
11.5	**Volumes of Pyramids**	
	Explorations	673
	Lesson	674
11.6	**Surface Areas and Volumes of Cones**	
	Explorations	679
	Lesson	680
11.7	**Surface Areas and Volumes of Spheres**	
	Explorations	685
	Lesson	686
	Performance Task: Tabletop Tiling	693
	Chapter Review	694
	Chapter Test	699
	Cumulative Assessment	700
	Additional Answers	A1
	English-Spanish Glossary	A67
	Index	A81
	Postulates and Theorems	A97
	Reference	A103

See the Big Idea
Learn how to cover a decagon-shaped tabletop using Penrose tiling.

How to Use Your Math Book

Get ready for each chapter by **Maintaining Mathematical Proficiency** and reviewing the **Mathematical Practices**. Begin each section by working through the **EXPLORATIONS** to **Communicate Your Answer** to the **Essential Question**. Each **Lesson** will explain **What You Will Learn** through **EXAMPLES**, **Core Concepts**, and **Core Vocabulary**. Answer the **Monitoring Progress** questions as you work through each lesson. Look for **STUDY TIPS**, **COMMON ERRORS**, and suggestions for looking at a problem **ANOTHER WAY** throughout the lessons. Take note of **CONNECTIONS TO ALGEBRA** and **CONNECTIONS TO GEOMETRY** which will inform you that the current concept can be applied to other topics throughout the program. We will also provide you with guidance for accurate mathematical **READING** and concept details you should **REMEMBER**.

Sharpen your newly acquired skills with **Exercises** at the end of every section. Halfway through each chapter you will be asked **What Did You Learn?** and you can use the Mid-Chapter **Quiz** to check your progress. You can also use the **Chapter Review** and **Chapter Test** to review and assess yourself after you have completed a chapter.

Apply what you learned in each chapter to a **Performance Task** and build your confidence for taking standardized tests with each chapter's **Cumulative Assessment**. For extra practice in any chapter, use your *Online Resources*, *Skills Review Handbook*, or your *Student Journal*.

Program Overview

Program Philosophy: Rigor and Balance with Real-Life Applications

The *Big Ideas Math*® program balances conceptual understanding with procedural fluency. Real-life applications help turn mathematical learning into an engaging and meaningful way to see and explore the real world.

Essential Question How can you prove a mathematical statement?

A **proof** is a logical argument that uses deductive reasoning to show that a statement is true.

EXPLORATION 1   Writing Reasons in a Proof

Work with a partner. Four steps of a proof are shown. Write the reasons for each statement.

Given $AD = AB + AC$

Prove $CD = AB$

STATEMENTS	REASONS
1. $AD = AB + AC$	1. Given
2. $AC + CD = AD$	2.
3. $AC + CD = AB + AC$	3.
4. $CD = AB$	4.

> Explorations and guiding *Essential Questions* encourage **conceptual understanding**.

6.1 Lesson

Core Vocabulary
proof, *p. 336*
two-column proof, *p. 336*
paragraph proof, *p. 337*
flowchart proof, or flow proof, *p. 338*
coordinate proof, *p. 339*

What You Will Learn
▶ Write two-column proofs to prove geometric relationships.
▶ Write paragraph proofs to prove geometric relationships.
▶ Write flowchart proofs to prove geometric relationships.
▶ Write coordinate proofs to prove geometric relationships.

Writing Two-Column Proofs

A **proof** is a logical argument that uses deductive reasoning to show that a statement is true. There are several formats for proofs. A **two-column proof** has numbered statements and corresponding reasons that show an argument in a logical order.

In a two-column proof, each statement in the left-hand column is either given information or the result of applying a known property or fact to statements already made. Each reason in the right-hand column is the explanation for the corresponding statement.

EXAMPLE 1   Writing a Two-Column Proof

Write a two-column proof of the Vertical Angles Congruence Theorem.

Given $\angle 1$ and $\angle 3$ are vertical angles.

Prove $\angle 1 \cong \angle 3$

> Direct instruction lessons allow for **procedural fluency** and provide the opportunity to use clear, precise mathematical language.

EXAMPLE 4   Solving a Real-Life Problem

A soccer goalie's position relative to the ball and goalposts forms congruent angles, as shown. Will the goalie have to move farther to block a shot toward the right goalpost R or the left goalpost L?

> Real-life applications provide students with opportunities to connect classroom lessons to realistic scenarios.

SOLUTION
The congruent angles tell you that the goalie is on the bisector of $\angle LBR$. By the Angle Bisector Theorem, the goalie is equidistant from $\overrightarrow{BR}$ and $\overrightarrow{BL}$.

▶ So, the goalie must move the same distance to block either shot.

Chapter openers focused on **Maintaining Mathematical Proficiency** promote the development of the habits of mind mathematically proficient students demonstrate.

Maintaining Mathematical Proficiency

Writing an Equation of a Perpendicular Line (Math I)

Example 1 Write the equation of a line passing through the point $(-2, 0)$ that is perpendicular to the line $y = 2x + 8$.

Step 1 Find the slope m of the perpendicular line. The line $y = 2x + 8$ has a slope of 2. Use the Slopes of Perpendicular Lines Theorem.

$2 \cdot m = -1$ The product of the slopes of ⊥ lines is -1.

$m = -\frac{1}{2}$ Divide each side by 2.

Step 2 Find the y-intercept b by using $m = -\frac{1}{2}$ and $(x, y) = (-2, 0)$.

$y = mx + b$ Use the slope-intercept form.

$0 = -\frac{1}{2}(-2) + b$ Substitute for m, x, and y.

$-1 = b$ Solve for b.

▶ Because $m = -\frac{1}{2}$ and $b = -1$, an equation of the line is $y = -\frac{1}{2}x - 1$.

Write an equation of the line passing through point P that is perpendicular to the given line.

1. $P(3, 1), y = \frac{1}{3}x - 5$ 2. $P(4, -3), y = -x - 5$ 3. $P(-1, -2), y = -4x + 13$

Writing Compound Inequalities (Math I)

Example 2 Write each sentence as an inequality.

a. A number x is greater than or equal to -1 and less than 6.

A number x is greater than or equal to -1 and less than 6.

$x \geq -1$ and $x < 6$

▶ An inequality is $-1 \leq x < 6$.

b. A number y is at most 4 or at least 9.

A number y is at most 4 or at least 9.

$y \leq 4$ or $y \geq 9$

▶ An inequality is $y \leq 4$ or $y \geq 9$.

the sentence as an inequality.

umber w is at least -3 and no more than 8.   5. A number m is more than 0 and less than 11.

umber s is less than or equal to 5 or greater   7. A number d is fewer than 12 or no less than -7.

TRACT REASONING Is it possible for the solution of a compound inequality to be all real bers? Explain your reasoning.

ns available at *BigIdeasMath.com*

333

Mathematical practices are woven into every chapter, including a full page dedicated to mastering one of the **Mathematical Practices**.

Mathematical Practices

Mathematically proficient students use technological tools to explore concepts.

Lines, Rays, and Segments in Triangles

Core Concept

Lines, Rays, and Segments in Triangles

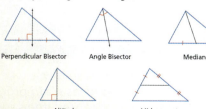

Perpendicular Bisector   Angle Bisector   Median

Altitude   Midsegment

EXAMPLE 1   Drawing a Perpendicular Bisector

Use dynamic geometry software to construct the perpendicular bisector of one of the sides of the triangle with vertices $A(-1, 2)$, $B(5, 4)$, and $C(4, -1)$. Find the lengths of the two segments of the bisected side.

SOLUTION

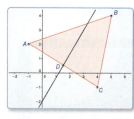

Sample
Points
$A(-1, 2)$
$B(5, 4)$
$C(4, -1)$
Line
$-5x + 3y = -6$
Segments
$AD = 2.92$
$CD = 2.92$

▶ The two segments of the bisected side have the same length, $AD = CD = 2.92$ units.

Monitoring Progress

Refer to the figures at the top of the page to describe each type of line, ray, or segment in a triangle.

1. perpendicular bisector   2. angle bisector   3. median
4. altitude   5. midsegment

334   Chapter 6   Relationships Within Triangles

Monitoring Progress problems allow students to practice and sharpen their skills as they work toward mathematical understanding.

xxi

Personalized Learning

The *Big Ideas Math* program offers teachers and students many ways to personalize and enrich the learning experience of all levels of learners.

Dynamic Student Edition

This unique tool, available online or as an eBook App, provides students with embedded 21st century learning resources. Students have the opportunity to interact with the underlying mathematics in a number of ways, including engaging tutorials, interactive manipulatives, flashcards, vocabulary support, and games that enhance the learning experience and promote mathematical understanding.

Dynamic Assessment and Progress Monitoring Tool

This tool allows teachers to provide customizable homework directly related to the *Big Ideas Math* program. Question types include technology enhanced items such as drag and drop, graphing, point plotting, multiple select, and fill in the blank using math expressions. Assignments are automatically scored and students have access to immediate remediation on homework questions.

Online Lesson Tutorials

Two- to three-minute lesson tutorial videos provide video and audio support for every example in the textbook. These are valuable for students who miss a class, need a second explanation, or need extra assistance with a homework assignment. Parents can also utilize the tutorials to stay connected or to provide additional help at home.

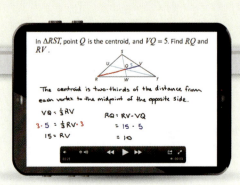

Differentiated Instruction

Through print and digital resources, the *Big Ideas Math* program completely supports the 3-Tier Response to Intervention model. Using research-based strategies, teachers can reach, challenge, and motivate each student with high-quality instruction targeted to individual needs.

Customized Learning Intervention

- *Big Ideas Math* Middle School program at BigIdeasMath.com

Tier 3

Strategic Intervention

- Lesson Tutorials
- Game Closet
- Skills Review Handbook
- Differentiated Instruction
- Dynamic Assesssment and Progress Monitoring Tool

Tier 2

Daily Intervention

- Student Journal
- Vocabulary Support
- Lesson Tutorials
- Communicate Your Answer
- Monitoring Progress
- Maintaining Mathematical Proficiency
- Dynamic Assessment and Progress Monitoring Tool

Tier 1

Big Ideas Math Dynamic Technology

The Dynamic Assessment and Progress Monitoring Tool

The Dynamic Assessment and Progress Monitoring Tool allows teachers to track and evaluate their students' advancement through the curriculum. Developed exclusively for *Big Ideas Math*, this technology provides teachers and students an intuitive and state-of-the-art tool to help students effectively learn mathematics. Built for ease of use, the tool is available on a wide range of devices.

Homework and Assessment

- Includes multiple assignments available for each chapter that are customizable
- Allows you to assign homework and assessments for the entire class or a select group of students
- Offers progress monitoring assessments for an adaptive testing experience

Direct Ties to Remediation

- Includes direct links to Lesson Tutorial Videos and relevant lesson sections

All-In-One Reporting

- Offers real time reporting at both the class and student level
- Tracks progress through Item Analysis, Standard Analysis, and Remediation reports

Assessment Delivery

- Provides embedded tools for students
- Includes auto-scored technology enhanced items such as drag and drop, graphing, point plotting, multiple select and fill in the blank using math expressions
- Allows you to include reminders or notes to students

Intuitive Design

- Operates on a wide range of devices with large and clear icons for visibility
- Allows for multiple reporting views through toggle options
- Includes intelligent presets and easy navigation

Dynamic Student Edition

Through the **Dynamic Student Edition eBook App,** students not only have access to the complete textbook, but they can also explore robust interactive digital resources embedded within each lesson. There is audio support for the text and Lesson Tutorial Videos in both English and Spanish. Interactive investigations, direct links to remediation, and additional resources are linked right to the lesson.

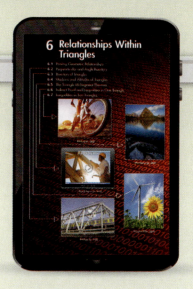

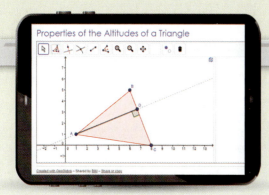

Dynamic Investigations

Dynamic Investigations in the *Big Ideas Math* program are powered by Desmos® and GeoGebra®. Teachers and students can integrate these investigations into their discovery learning to interact with the Explorations in the Student Edition.

Real-Life STEM Videos

Science - Technology - Engineering - Mathematics

Every chapter in the *Big Ideas Math* program contains a Real-Life STEM Video allowing students to further engage with mathematical concepts. Students learn about the speed of light, natural disasters, solar power, and more!

Teaching Support

Lesson Planning Support

Online Lesson Plans

Complete, editable Lesson Plans are included for every lesson in the program.

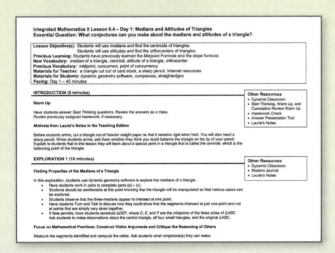

Laurie's Notes

Laurie Boswell provides comprehensive teaching support for every student page in the Teaching Edition.

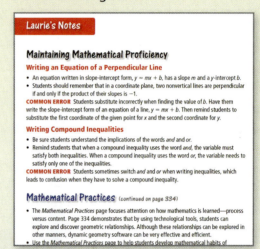

Lesson Presentation Support

Dynamic Classroom

This online lesson presentation tool includes point-of-use resources that can be used with a projector or interactive whiteboard.

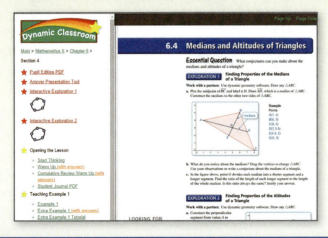

Interactive Whiteboard Lesson Library

Customizable SMART®, Promethean®, and Mimio® lessons are provided for every section in the program.

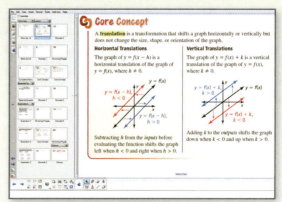

Program Resources

Print

Student Edition

Teaching Edition
- Laurie's Notes

Student Journal
Available in English and Spanish
- Maintaining Mathematical Proficiency
- Exploration Journal
- Notetaking with Vocabulary
- Extra Practice

Resources by Chapter
- Start Thinking
- Warm-Up
- Cumulative Review Warm-Up
- Practice A and B
- Enrichment and Extension
- Puzzle Time
- Cumulative Review
- Family Communication Letters
  Available in English and Spanish

Assessment Book
- Performance Tasks
- Prerequisite Skills Tests with Item Analysis
- Cumulative Tests
- Quizzes
- Chapter Tests
- Alternative Assessments with Scoring Rubrics
- Pre-Course Test with Item Analysis
- Post-Course Test with Item Analysis

Technology

Student Edition
With complete English and Spanish audio
- Dynamic eBook App
  - Lesson Tutorial Videos
  - Dynamic Investigations
- Online Home Edition
  - Multi-Language Glossary
  - Skills Review Handbook
- Dynamic Solutions Tool

Dynamic Classroom
- Interactive Manipulatives
- Answer Presentation Tool
- Extra Examples
- Mini Assessment

Dynamic Teaching Tools
- Interactive Whiteboard Lesson Library
  - Includes standard and customizable lessons.
  - Compatible with SMART®, Promethean®, and Mimio® technology
- **Exam**View® Assessment Suite
- Real-Life STEM Videos
- Editable Online Resources
  - Lesson Plans
  - Pacing Guides
  - Assessment Book
  - Resources by Chapter
- Answer Presentation Tool

Dynamic Assessment and Progress Monitoring Tool
- Homework and Assessment Creation
- Progress Monitoring
- Direct Ties to Remediation
- All-In-One Reporting
- Online Chat Tutor

Learning Progression

Integrated Mathematics I, II, and III

Domain	Integrated Mathematics I	Integrated Mathematics II	Integrated Mathematics III
Number and Quantity			
Quantities	Reason quantitatively and use units. *Chapters 1, 3–6*		
Real Numbers		Use properties of rational exponents, rational numbers, and irrational numbers. *Chapters 1, 4*	
Complex Numbers		Perform arithmetic with complex numbers and use complex numbers in quadratic identities and equations. *Chapter 4*	Use complex numbers in polynomial identities and equations. *Chapter 3*
Algebra			
Structure	Interpret the structure of linear and exponential expressions. *Chapters 1, 3, 4, 6*	Interpret the structure of quadratic and exponential expressions. *Chapters 1–4*	Interpret the structure of polynomial and rational expressions. *Chapters 2–7, 9*
Arithmetic with Polynomials and Rational Expressions		Perform arithmetic on polynomials that simplify to quadratics. *Chapter 2*	Perform arithmetic on polynomials beyond quadratics. *Chapter 3* Rewrite rational expressions. *Chapters 3, 6*
Creating Equations	Create linear and exponential equations with integer inputs. *Chapters 1, 6*	Create quadratic equations. *Chapters 3, 4*	Create equations of all available types. *Chapters 1–9*
Reasoning with Equations and Inequalities	Solve linear equations and inequalities, and exponential equations. *Chapters 1, 2, 6* Solve systems of linear equations. *Chapter 5*	Solve quadratic equations and inequalities. *Chapters 2–4, 9–11* Solve systems of linear and quadratic equations. *Chapter 4*	Solve polynomial, radical, logarithmic, and rational equations. *Chapters 3–6*

Domain	Integrated Mathematics I	Integrated Mathematics II	Integrated Mathematics III
Functions			
Interpreting Functions	Interpret and analyze linear and exponential functions. *Chapters 3, 6*	Interpret and analyze quadratic, absolute value, step, and piecewise functions. *Chapters 1, 3, 4*	Interpret and analyze polynomial, radical, logarithmic, and rational functions. *Chapters 3–6*
Building Functions	Build linear and exponential models. *Chapters 4, 6* Build linear and exponential functions from existing functions. *Chapters 3, 6*	Build quadratic and exponential models. *Chapters 3, 4* Build quadratic and absolute value functions from existing functions. *Chapters 1, 3*	Build models from all types of functions studied. *Chapters 4–6, 9* Build polynomial, radical, rational, exponential, and logarithmic functions from existing functions. *Chapters 3–6*
Trigonometric Functions		Prove and use trigonometric identities. *Chapter 9*	Extend the domain of trigonometric functions. Model with trigonometric functions. *Chapter 8*
Geometry			
Congruence	Understand transformations and congruence. *Chapters 11–12* Make constructions. *Chapters 8, 10, 12*	Prove geometric theorems. *Chapters 6, 7*	
Similarity and Trigonometry		Understand similarity and prove theorems involving similarity. *Chapters 8, 9* Use trigonometry with right triangles. *Chapter 9*	Use trigonometry with general triangles. *Chapter 9*
Circles		Use theorems about circles. Find arc lengths and areas of sectors. *Chapters 9, 10*	
Measurement		Understand and use volume formulas. *Chapter 11*	Relate two-dimensional and three-dimensional objects. *Chapter 1*
Statistics and Probability			
Interpreting Data	Summarize, represent, and interpret data using measures of center and spread. *Chapter 7*		Fit data to a normal distribution using the mean and standard deviation. *Chapter 10*
Making Inferences			Make inferences and justify conclusions. *Chapter 10*
Probability		Use conditional probability and find the probabilities of compound events. *Chapter 5*	Use probability in complex situations. *Chapter 10*

Pacing Guide

Chapters 1–11 158 Days

Chapter 1 (13 Days)

Chapter Opener/Mathematical Practices	0.5 Day
Section 1	1.5 Days
Section 2	2 Days
Section 3	1 Day
Quiz	0.5 Day
Section 4	1.5 Days
Section 5	2 Days
Section 6	2 Days
Chapter Review/Chapter Tests	2 Days
Year-to-Date	**13 Days**

Chapter 2 (18 Days)

Chapter Opener/Mathematical Practices	0.5 Day
Section 1	1.5 Days
Section 2	2 Days
Section 3	2 Days
Section 4	2 Days
Quiz	0.5 Day
Section 5	1.5 Days
Section 6	2 Days
Section 7	2 Days
Section 8	2 Days
Chapter Review/Chapter Tests	2 Days
Year-to-Date	**31 Days**

Chapter 3 (16 Days)

Chapter Opener/Mathematical Practices	0.5 Day
Section 1	1.5 Days
Section 2	2 Days
Section 3	2 Days
Quiz	0.5 Day
Section 4	1.5 Days
Section 5	2 Days
Section 6	2 Days
Section 7	2 Days
Chapter Review/Chapter Tests	2 Days
Year-to-Date	**47 Days**

Chapter 4 (19 Days)

Chapter Opener/Mathematical Practices	0.5 Day
Section 1	1.5 Days
Section 2	2 Days
Section 3	2 Days
Section 4	2 Days
Section 5	2 Days
Quiz	0.5 Day
Section 6	1.5 Days
Section 7	1 Day
Section 8	2 Days
Section 9	2 Days
Chapter Review/Chapter Tests	2 Days
Year-to-Date	**66 Days**

Chapter 5 (14 Days)

Chapter Opener/Mathematical Practices	0.5 Day
Section 1	1.5 Days
Section 2	2 Days
Section 3	2 Days
Quiz	0.5 Day
Section 4	1.5 Days
Section 5	2 Days
Section 6	2 Days
Chapter Review/Chapter Tests	2 Days
Year-to-Date	**80 Days**

Chapter 6 (15 Days)

Chapter Opener/Mathematical Practices	0.5 Day
Section 1	1.5 Days
Section 2	2 Days
Section 3	2 Days
Section 4	2 Days
Quiz	0.5 Day
Section 5	1.5 Days
Section 6	2 Days
Section 7	1 Day
Chapter Review/Chapter Tests	2 Days
Year-to-Date	**95 Days**

Chapter 7 (12 Days)

Chapter Opener/Mathematical Practices	0.5 Day
Section 1	1.5 Days
Section 2	2 Days
Section 3	2 Days
Quiz	0.5 Day
Section 4	1.5 Days
Section 5	2 Days
Chapter Review/Chapter Tests	2 Days
Year-to-Date	**107 Days**

Chapter 8 (11 Days)

Chapter Opener/Mathematical Practices	0.5 Day
Section 1	1.5 Days
Section 2	1 Day
Section 3	2 Days
Section 4	1 Day
Quiz	0.5 Day
Section 5	1.5 Days
Section 6	1 Day
Chapter Review/Chapter Tests	2 Days
Year-to-Date	**118 Days**

Chapter 9 (13 Days)

Chapter Opener/Mathematical Practices	0.5 Day
Section 1	1.5 Days
Section 2	1 Day
Section 3	2 Days
Quiz	0.5 Day
Section 4	1.5 Days
Section 5	2 Days
Section 6	2 Days
Chapter Review/Chapter Tests	2 Days
Year-to-Date	**131 Days**

Chapter 10 (12 Days)

Chapter Opener/Mathematical Practices	0.5 Day
Section 1	1.5 Days
Section 2	1 Day
Section 3	1 Day
Quiz	0.5 Day
Section 4	1.5 Days
Section 5	2 Days
Section 6	1 Day
Section 7	1 Day
Chapter Review/Chapter Tests	2 Days
Year-to-Date	**143 Days**

Chapter 11 (15 Days)

Chapter Opener/Mathematical Practices	0.5 Day
Section 1	1.5 Days
Section 2	1 Day
Section 3	2 Days
Quiz	0.5 Day
Section 4	1.5 Days
Section 5	2 Days
Section 6	2 Days
Section 7	2 Days
Chapter Review/Chapter Tests	2 Days
Year-to-Date	**158 Days**

Chapter 1 Pacing Guide	
Chapter Opener/Mathematical Practices	0.5 Day
Section 1	1.5 Days
Section 2	2 Days
Section 3	1 Day
Quiz	0.5 Day
Section 4	1.5 Days
Section 5	2 Days
Section 6	2 Days
Chapter Review/Chapter Tests	2 Days
Total Chapter 1	13 Days
Year-to-Date	13 Days

1 Functions and Exponents

1.1 Absolute Value Functions
1.2 Piecewise Functions
1.3 Inverse of a Function
1.4 Properties of Exponents
1.5 Radicals and Rational Exponents
1.6 Exponential Functions

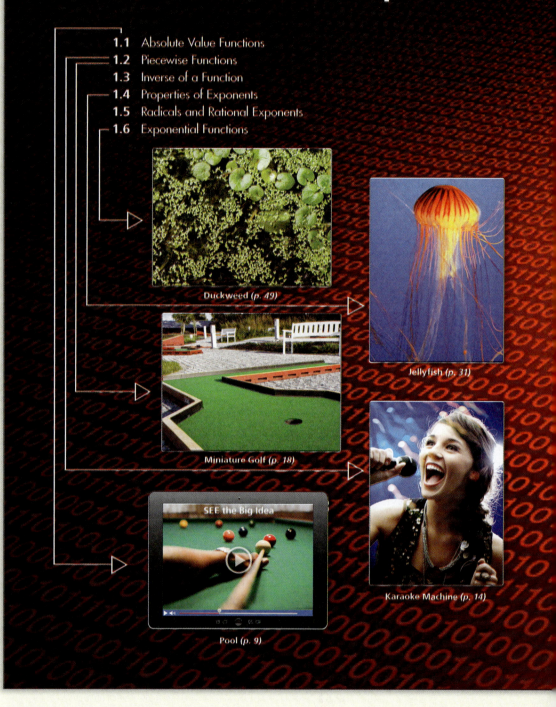

Duckweed (p. 49)
Jellyfish (p. 31)
Miniature Golf (p. 18)
Karaoke Machine (p. 14)
Pool (p. 9)

Laurie's Notes

Chapter Summary

Welcome to a new school year, and for many students, a new school. There is always great excitement, and students are anxious to start anew. As teachers, we need to capitalize on the opportunity, establishing norms and routines for student discourse and classroom climate.

- In this book, students are expected to work together on explorations, to make conjectures, to construct viable arguments, and to critique the reasoning of others. Take time in this first chapter to make explicit what classroom productive dialogue sounds like. Listen for students explaining their thinking, not just their process.
- Chapter 1 presents new functions, building on work done with linear and exponential functions in Math I. Do students still have their notebooks from last year? Take time to discuss expectations you have for students with regard to using resources (notes and text) and taking personal responsibility for seeking support when necessary.
- In the first lesson, students will apply their knowledge about transformations of functions to the absolute value function. Next they will study piecewise functions, which allow you to integrate a review of linear functions. Finding and graphing inverses of functions is the topic of Lesson 3.
- The last three lessons deal with exponents, radicals, rational exponents, and exponential functions. Though exponential functions were studied in Math I, the domain was limited to integer values. Once rational exponents are defined, the domain of exponential functions is extended.

What Your Students Have Learned

Middle School
- Understand the absolute value of a rational number as its distance from 0 on the number line.
- Represent functions as equations, input-output tables, and graphs.
- Write and evaluate numerical expressions involving integer exponents.
- Use properties of exponents and Order of Operations to evaluate expressions.
- Compare linear and nonlinear functions represented by tables, graphs, and equations.

Math I
- Use function notation to evaluate, interpret, solve, and graph functions.
- Determine whether relations are functions.
- Identify exponential growth and exponential decay functions and the percent rate of change.
- Graph exponential functions, including transformations, and compare the graphs of exponential functions.
- Write terms and rules of recursively defined sequences and translate between recursive rules and explicit rules.

What Your Students Will Learn

Math II
- Evaluate, graph, and write piecewise functions, including step and absolute value functions.
- Find inverses of relations and linear functions.
- Evaluate and simplify expressions with exponents, including rational exponents.
- Graph exponential growth and decay functions and identify growth and decay factors.
- Write exponential growth and decay models and recursive rules for exponential functions.
- Rewrite exponential functions to identify the percent rate of change.

Dynamic Teaching Tools
- Dynamic Assessment & Progress Monitoring Tool
- Lesson Planning Tool
- Interactive Whiteboard Lesson Library
- Dynamic Classroom with Dynamic Investigations
- Real-Life STEM Videos

Scaffolding in the Classroom

Graphic Organizers: Information Frame

An Information Frame can be used to help students organize and remember concepts. Students write the topic in the middle rectangle. Then students write related concepts in the spaces around the rectangle. Related concepts can include *Words, Numbers, Algebra, Definition, Example, Non-Example, Visual, Procedure, Details,* and *Vocabulary*. Students can place their Information Frames on note cards to use as a quick study reference.

Questioning in the Classroom
Be open to multiple answers.
Questions can frequently be interpreted differently. Allow students time to discuss and explain their point of view.

Laurie's Notes

Maintaining Mathematical Proficiency

Transforming Graphs of Linear Functions
- Note the color coding in Example 1 to connect the transformation and its graph.
- For the graph of $y = f(x - h)$, a horizontal translation is to the right when $h > 0$ and to the left when $h < 0$.
- For the graph of $y = f(x) + k$, a vertical translation is up when $k > 0$ and down when $k < 0$.
- Students need to decide when they are performing the transformation to the input or the output. $y = -f(x)$ is affecting the output (sign change), so the reflection is over the x-axis. $y = f(-x)$ is affecting the input (sign change), so the reflection is over the y-axis.

COMMON ERROR In Exercise 1, students may think that the graph is a horizontal translation 4 units left. Remind them that $h = 4 > 0$, so this is a translation to the right.

Reflecting Figures in the Line $y = x$
- One strategy students use to reflect in the line $y = x$ is to rotate their graph paper so that the line of reflection is horizontal. It is easier for many students to then reflect the endpoints of $\overline{PQ}$ in a vertical direction.
- Summarize the rules for four common reflections.
  - If (a, b) is reflected in the x-axis, then its image is the point $(a, -b)$.
  - If (a, b) is reflected in the y-axis, then its image is the point $(-a, b)$.
  - If (a, b) is reflected in the line $y = x$, then its image is the point (b, a).
  - If (a, b) is reflected in the line $y = -x$, then its image is the point $(-b, -a)$.

COMMON ERROR Students may incorrectly plot the image. Have them check by connecting corresponding points of the preimage and image. The connecting line segments should be perpendicular to the line of reflection.

Mathematical Practices (continued on page 2)
- The *Mathematical Practices* page focuses attention on how mathematics is learned—process versus content. This page demonstrates how to graph a function with a restricted domain on a graphing calculator.
- Use the *Mathematical Practices* page to help students develop mathematical habits of mind—how mathematics can be explored and how mathematics is thought about.

If students need help...	If students got it...
Student Journal • Maintaining Mathematical Proficiency	Game Closet at *BigIdeasMath.com*
Lesson Tutorials	Start the *next* Section
Skills Review Handbook	

Maintaining Mathematical Proficiency

Transforming Graphs of Linear Functions

Example 1 Graph $f(x) = x$ and $g(x) = -3x - 2$. Describe the transformations from the graph of f to the graph of g.

Note that you can rewrite g as $g(x) = -3f(x) - 2$.

Step 1 There is no horizontal translation from the graph of f to the graph of g.

Step 2 Stretch the graph of f vertically by a factor of 3 to get the graph of $h(x) = 3x$.

Step 3 Reflect the graph of h in the x-axis to get the graph of $r(x) = -3x$.

Step 4 Translate the graph of r vertically 2 units down to get the graph of $g(x) = -3x - 2$.

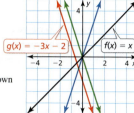

Let $f(x) = x + 1$. Graph f and g. Describe the transformation from the graph of f to the graph of g.

1. $g(x) = f(x - 4)$
2. $g(x) = f(x) + 5$
3. $g(x) = f(2x)$
4. $g(x) = -\frac{1}{2}f(x)$

Reflecting Figures in the Line $y = x$

Example 2 Graph $\overline{PQ}$ with endpoints $P(-3, 1)$ and $Q(2, 4)$ and its image after a reflection in the line $y = x$.

Graph $\overline{PQ}$ and the line $y = x$. Use the coordinate rule for reflecting in the line $y = x$ to find the coordinates of the endpoints of the image. Then graph the image $\overline{P'Q'}$.

$(a, b) \to (b, a)$

$P(-3, 1) \to P'(1, -3)$

$Q(2, 4) \to Q'(4, 2)$

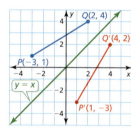

Graph the figure and its image after a reflection in the line $y = x$.

5. $\overline{AB}$ with endpoints $A(0, 0)$ and $B(0, -3.5)$
6. $\overline{CD}$ with endpoints $C(1, -2)$ and $D(-4, -1)$
7. $\triangle LMN$ with vertices $L(5, 3)$, $M(5, -3)$, and $N(0, -3)$
8. $\square STUV$ with vertices $S(-4, 6)$, $T(-1, 7)$, $U(-1, -4)$, and $V(-4, -5)$
9. Give an example of a line segment that when reflected in the line $y = x$, has the same endpoints as the original figure.
10. **ABSTRACT REASONING** Describe the relationship between the possible input and output values of the graph of $f(x) = a$ and its reflection in the line $y = x$.

Dynamic Solutions available at *BigIdeasMath.com*

Vocabulary Review

Have students make an Information Frame for each topic.

- Transforming the graph of $g(x) = a \cdot f(x - h) + k$
- Reflecting figures in the line $y = x$

What Your Students Have Learned

- Translate, reflect, shrink, and stretch graphs of linear functions, and combine transformations of graphs of linear functions.
- Perform reflections in horizontal lines, vertical lines, the line $y = x$, and the line $y = -x$.

ANSWERS

1.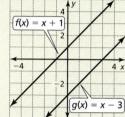

   The graph of g is a horizontal translation 4 units right of the graph of f.

2.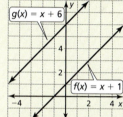

   The graph of g is a vertical translation 5 units up of the graph of f.

3.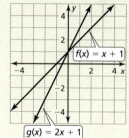

   The graph of g is a horizontal shrink of the graph of f by a factor of $\frac{1}{2}$.

4.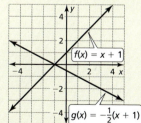

   The graph of g is a vertical shrink by a factor of $\frac{1}{2}$ and a reflection in the x-axis of the graph of f.

5–10. See Additional Answers.

Chapter 1

MONITORING PROGRESS ANSWERS

1.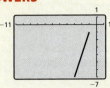

range: $y \leq -1$

2.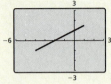

range: $y < -4$

3.

range: $-1 \leq y \leq \frac{3}{2}$

Mathematical Practices

Mathematically proficient students use technological tools to explore concepts.

Using a Graphing Calculator

Core Concept

Restricting the Domain of a Function

You can use a graphing calculator to graph a function with a restricted domain.

1. Enter the function, in parentheses, into a graphing calculator.
2. Press the division symbol key.
3. Enter the specified domain, in parentheses.
4. Graph the function in an appropriate viewing window.

EXAMPLE 1 Using a Graphing Calculator

Use a graphing calculator to graph (a) $y = -2x + 7$, $x \geq 2$, and (b) $y = \frac{3}{4}x - 1$, $-3 < x < 4$.

SOLUTION

a. Enter the function $y = -2x + 7$ and the domain $x \geq 2$ into a graphing calculator, as shown. Then graph the function in an appropriate viewing window.

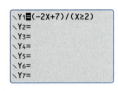 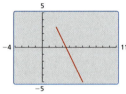

b. Enter the function $y = \frac{3}{4}x - 1$ and the domain $-3 < x < 4$ into a graphing calculator. In this case, you must enter the inequality using "and," as shown. Then graph the function in an appropriate viewing window.

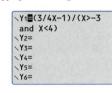 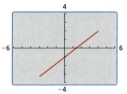

Notice that the graphing calculator does not distinguish between domains such as $x \geq 2$ and $x > 2$. You must distinguish between these yourself, based on the inequality symbol in the problem statement.

Monitoring Progress

Use a graphing calculator to graph the function with the specified domain. Describe the range of the function.

1. $y = 3x + 2$, $x \leq -1$ **2.** $y = -x - 4$, $x > 0$ **3.** $-\frac{1}{2}x + y = 1$, $-4 \leq x \leq 1$

Laurie's Notes   Mathematical Practices (continued from page T-1)

- Discuss the *Core Concept* and have students practice entering a function with a restricted domain as shown in Example 1. Be sure that students understand the syntax used. It would not make sense to divide an expression $(-2x + 7)$ by an inequality $(x \geq 2)$. The division symbol in this context is interpreted as saying that for the function entered $(-2x + 7)$, only display the graph when $x \geq 2$.
- **Use Appropriate Tools Strategically:** The example shown demonstrates how a graphing calculator is an appropriate tool for graphing functions with restricted domains.
- Give time for students to work through the questions in the *Monitoring Progress* and then discuss as a class.

Laurie's Notes

Overview of Section 1.1

Introduction
- The graph of the absolute value function will be new to students. I think it is important for them to discover the V-shape of the graph by making a simple table of values versus using the calculator. This can be done as the *Motivate* to begin the lesson before engaging in the explorations.

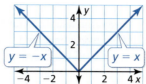

- The geometric definition that students have learned for absolute value is that the absolute value of a number is the distance the number is from 0. The algebraic definition is a split-rule function:

$$|x| = \begin{array}{l} x, \text{ for } x \geq 0 \\ -x, \text{ for } x < 0 \end{array}.$$

- When students make a quick table of values, the domain and range of the parent function are obvious.
- All of the transformations of the graphs of linear functions apply to absolute value functions. Transformations of absolute value functions are much easier for students to see and understand. Whether you reflect $f(x) = x$ vertically or horizontally, it looks the same. That is not the case with $f(x) = |x|$.

Resources
- The graphing calculator is used in the exploration as a way to verify that the matches are correct. Ensure that students are making a reasoned attempt to do the matching before using the graphing calculator. The graphing calculator can also be a helpful teaching tool in the formal lesson.

Formative Assessment Tips
- **Turn and Talk:** This technique allows all students in the class to have a voice. Using a 3-foot voice, students turn and talk to their partners about a problem or to discuss a question. There may be different roles I ask partners to assume, so I refer to partner A and partner B. In discussing a procedure or explaining an answer, I might ask partner A to talk uninterrupted for a fixed period of time. Then partner B might be asked to repeat back what he/she heard or to ask a question about what has been shared.
  Example: Turn and Talk, so that partner A explains the effect of the constant 2 on the parent function in the graph of $f(x) = 2|x|$.
- It is important to establish norms: Three-foot voices should be expected when students are doing partner work. Discuss with students the difference between *authentic listening* and being quiet while your partners are speaking.

Pacing Suggestion
- The formal lesson is long. Important discussion and understanding can occur during the exploration. Take time for students to make sense of each problem.

Dynamic Teaching Tools
Dynamic Assessment & Progress Monitoring Tool
Lesson Planning Tool
Interactive Whiteboard Lesson Library
Dynamic Classroom with Dynamic Investigations

What Your Students Will Learn

- Translate graphs of absolute value functions in the coordinate plane.
- Stretch, shrink, and reflect graphs of absolute value functions in the coordinate plane.
- Combine transformations of graphs of absolute value functions in the coordinate plane.

Laurie's Notes

Exploration

Motivate

- Distribute whiteboards to seven or nine volunteers. Tell the volunteers that they represent the domain values of a function. For seven volunteers they represent the integers -3 to 3. Orientation is important. They should stand at the front of the room facing their classmates with the left-most person representing $x = -3$ and the right-most person representing $x = 3$.
- Explain that you are going to say a function and the volunteers are to evaluate the function for the x-value they represent. They write the outputs on their whiteboards and hold the whiteboards in front of them. Students seated in the class will see the range values for the function. In other words, students will see the outputs for a table of values.
- Begin with a few linear functions: $y = x$; $y = x + 3$; $y = -x - 2$. Any miscalculations are obvious to spot because the slopes are 1 or -1. You could also ask about y-intercepts.
- Now ask about $y = |x|$.
- ? "What do you notice about the outputs?" Listen for students to observe that the outputs are all positive and that there is a symmetry in the outputs.
- ? **Turn and Talk:** "What will $y = |x|$ look like when graphed?" It is not uncommon for students to make a V-shape with their fingers, uncertain what to call it!
- Describe the parent function $f(x) = |x|$, and explain to students that in this lesson they will explore transformations of the parent function graph. Give students vocabulary to use in discussing the graph of the absolute value function. The *vertex* is the point where the graph changes direction, and the graph is symmetric about a vertical line through the vertex.

Exploration 1

- The goal is for students to use their understanding of transformations from the last lesson to match the equations and graphs.
- As you circulate, you should hear students make observations about three of the graphs being upward V's and three graphs being upside down V's. Students quickly observe that three of the functions have a negative sign outside the absolute value function notation.
- Students will need to make a table of values to help them do the matching.
- **Look For and Make Use of Structure:** Mathematically proficient students will pay attention to whether the transformation is done before or after the absolute value function is evaluated.
- **Selective Responses:** As you circulate, record the different methods and reasoning students use to match functions with graphs. When it is time for a whole class discussion, refer to your notes. Begin with a student who used little reasoning—he or she made a table of values. Work toward the student who used more reasoning. This is a teaching strategy that is referred to as *Selective Responses*. Carefully controlling how responses are shared with the whole class focuses students' attention on the learning outcome, namely how the values of a, h, and k affect the graph of the absolute value function $g(x) = a|x - h| + k$. See Section 4.9 for a more detailed description of *Selective Responses*.
- The graphing calculator only verifies students' answers. It should not be used in advance of thorough discussions by partners about the process.

Communicate Your Answer

- Students should describe the effect of a, h, and k on the graph of the absolute value function.
- **Use Appropriate Tools Strategically:** The graphing calculator is a useful tool in checking students' equations for Question 3.

Connecting to Next Step

- Discussion of the reasoning used by students to do the matching should be followed by Example 1 in the formal lesson.

1.1 Absolute Value Functions

Essential Question How do the values of a, h, and k affect the graph of the absolute value function $g(x) = a|x - h| + k$?

The parent absolute value function is

$f(x) = |x|$.   Parent absolute value function

The graph of f is V-shaped.

EXPLORATION 1 Identifying Graphs of Absolute Value Functions

Work with a partner. Match each absolute value function with its graph. Then use a graphing calculator to verify your answers.

a. $g(x) = -|x - 2|$
b. $g(x) = |x - 2| + 2$
c. $g(x) = -|x + 2| - 2$
d. $g(x) = |x - 2| - 2$
e. $g(x) = 2|x - 2|$
f. $g(x) = -|x + 2| + 2$

A.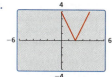

B.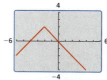

C.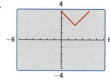

D.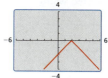

E.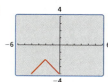

F.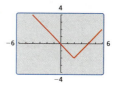

LOOKING FOR STRUCTURE
To be proficient in math, you need to look closely to discern a pattern or structure.

Communicate Your Answer

2. How do the values of a, h, and k affect the graph of the absolute value function $g(x) = a|x - h| + k$?

3. Write the equation of the absolute value function whose graph is shown. Use a graphing calculator to verify your equation.

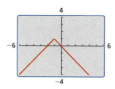

Section 1.1   Absolute Value Functions   3

Dynamic Teaching Tools
- Dynamic Assessment & Progress Monitoring Tool
- Lesson Planning Tool
- Interactive Whiteboard Lesson Library
- Dynamic Classroom with Dynamic Investigations

ANSWERS

1. a. D
   b. C
   c. E
   d. F
   e. A
   f. B

2. a stretches or shrinks the graph and determines whether the graph opens up or down; h translates the graph horizontally; k translates the graph vertically

3. $g(x) = -|x + 1| + 1$

English Language Learners

Build on Past Knowledge
Some students may know the definition of a vertex as it applies to angles and other geometric figures. Connect their understanding of the vertex of an angle to the vertex of the graph of an absolute value equation.

Extra Example 1
Graph each function. Compare each graph to the graph of $f(x) = |x|$. Describe the domain and range.

a. $g(x) = |x| + 2$

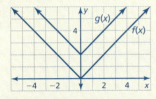

The graph of g is a vertical translation 2 units up of the graph of f. The domain is all real numbers. The range is $y \geq 2$.

b. $m(x) = |x + 3|$

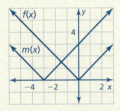

The graph of m is a horizontal translation 3 units left of the graph of f. The domain is all real numbers. The range is $y \geq 0$.

MONITORING PROGRESS ANSWERS
1–2. See Additional Answers.

1.1 Lesson

Core Vocabulary
absolute value function, p. 4
vertex, p. 4
vertex form, p. 6
Previous
domain
range

What You Will Learn
- Translate graphs of absolute value functions.
- Stretch, shrink, and reflect graphs of absolute value functions.
- Combine transformations of graphs of absolute value functions.

Translating Graphs of Absolute Value Functions

> **Core Concept**
>
> **Absolute Value Function**
> An **absolute value function** is a function that contains an absolute value expression. The parent absolute value function is $f(x) = |x|$. The graph of $f(x) = |x|$ is V-shaped and symmetric about the y-axis. The **vertex** is the point where the graph changes direction. The vertex of the graph of $f(x) = |x|$ is (0, 0).
>
> The domain of $f(x) = |x|$ is all real numbers. The range is $y \geq 0$.

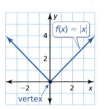

The graphs of all other absolute value functions are transformations of the graph of the parent function $f(x) = |x|$. The transformations of graphs of linear functions that you learned in a previous course also apply to absolute value functions.

REMEMBER
The graph of $y = f(x - h)$ is a horizontal translation and $y = f(x) + k$ is a vertical translation of the graph of $y = f(x)$, where $h, k \neq 0$.

EXAMPLE 1 Graphing $g(x) = |x| + k$ and $g(x) = |x - h|$

Graph each function. Compare each graph to the graph of $f(x) = |x|$. Describe the domain and range.

a. $g(x) = |x| + 3$ b. $m(x) = |x - 2|$

SOLUTION

a. **Step 1** Make a table of values.

x	−2	−1	0	1	2
g(x)	5	4	3	4	5

Step 2 Plot the ordered pairs.
Step 3 Draw the V-shaped graph.

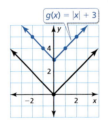

▶ The function g is of the form $y = f(x) + k$, where $k = 3$. So, the graph of g is a vertical translation 3 units up of the graph of f. The domain is all real numbers. The range is $y \geq 3$.

b. **Step 1** Make a table of values.

x	0	1	2	3	4
m(x)	2	1	0	1	2

Step 2 Plot the ordered pairs.
Step 3 Draw the V-shaped graph.

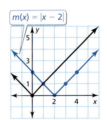

▶ The function m is of the form $y = f(x - h)$, where $h = 2$. So, the graph of m is a horizontal translation 2 units right of the graph of f. The domain is all real numbers. The range is $y \geq 0$.

Monitoring Progress 🔊 Help in English and Spanish at *BigIdeasMath.com*

Graph the function. Compare the graph to the graph of $f(x) = |x|$. Describe the domain and range.

1. $h(x) = |x| - 1$ 2. $n(x) = |x + 4|$

4   Chapter 1   Functions and Exponents

Laurie's Notes — Teacher Actions

- Students who have worked through the exploration should have the *Core Concept* in their notes and be ready for the first example.
- Making the table of values helps students focus on the vertex and where the symmetry occurs in the y-values.

? Advancing Question: "You can subtract (or add) a number before or after the function is evaluated. In which direction will the transformation occur?" *Sample answer:* before: $f(x) = |x - 2|$, right 2 units; after: $f(x) = |x| - 2$, down 2 units

- Subtracting 2 before evaluating the absolute value function affects the x-value. The misconception for students is that they think the graph should shift 2 units left. Looking at the table of values helps students see that when $x = 2$, you are evaluating $|2 - 2| = |0|$, which is the x-value of the vertex of the parent function.

? "How does the graph of $f(x) = |x| + k$ compare to $f(x) = |x|$?" It is a translation up ($k > 0$) or down ($k < 0$). "Does the range change?" yes; $y \geq k$

? "How does the graph of $f(x) = |x - h|$ compare to $f(x) = |x|$?" It is a translation right ($h > 0$) or left ($h < 0$). "Does the range change?" no

4   Chapter 1

Stretching, Shrinking, and Reflecting

EXAMPLE 2 Graphing $g(x) = a|x|$

> **REMEMBER**
> Recall the following transformations of the graph of $y = f(x)$.
> $y = -f(x)$: reflection in the x-axis
> $y = f(-x)$: reflection in the y-axis
> $y = f(ax)$: horizontal stretch or shrink by a factor of $\frac{1}{a}$, where $a > 0$ and $a \neq 1$
> $y = a \cdot f(x)$: vertical stretch or shrink by a factor of a, where $a > 0$ and $a \neq 1$

Graph each function. Compare each graph to the graph of $f(x) = |x|$. Describe the domain and range.

a. $q(x) = 2|x|$ **b.** $p(x) = -\frac{1}{2}|x|$

SOLUTION

a. Step 1 Make a table of values.

x	−2	−1	0	1	2
q(x)	4	2	0	2	4

Step 2 Plot the ordered pairs.

Step 3 Draw the V-shaped graph.

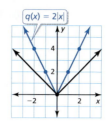

> **STUDY TIP**
> A vertical stretch of the graph of $f(x) = |x|$ is narrower than the graph of $f(x) = |x|$.

▶ The function q is of the form $y = a \cdot f(x)$, where $a = 2$. So, the graph of q is a vertical stretch of the graph of f by a factor of 2. The domain is all real numbers. The range is $y \geq 0$.

b. Step 1 Make a table of values.

x	−2	−1	0	1	2
p(x)	−1	−$\frac{1}{2}$	0	−$\frac{1}{2}$	−1

Step 2 Plot the ordered pairs.

Step 3 Draw the V-shaped graph.

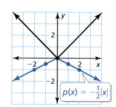

> **STUDY TIP**
> A vertical shrink of the graph of $f(x) = |x|$ is wider than the graph of $f(x) = |x|$.

 The function p is of the form $y = -a \cdot f(x)$, where $a = \frac{1}{2}$. So, the graph of p is a vertical shrink of the graph of f by a factor of $\frac{1}{2}$ and a reflection in the x-axis. The domain is all real numbers. The range is $y \leq 0$.

Monitoring Progress
Help in English and Spanish at BigIdeasMath.com

Graph the function. Compare the graph to the graph of $f(x) = |x|$. Describe the domain and range.

3. $t(x) = -3|x|$ **4.** $v(x) = \frac{1}{4}|x|$

Section 1.1   Absolute Value Functions   5

Differentiated Instruction

Kinesthetic/Visual
Some students will benefit by having access to graphing calculators or the graphing software programs available on many computer operating systems. Students can enter the absolute value functions, view and compare their graphs, and see the domain and range.

Extra Example 2
Graph each function. Compare each graph to the graph of $f(x) = |x|$. Describe the domain and range.

a. $s(x) = \frac{1}{2}|x|$

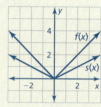

The graph of s is a vertical shrink of the graph of f by a factor of $\frac{1}{2}$. The domain is all real numbers. The range is $y \geq 0$.

b. $r(x) = -2|x|$

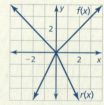

The graph of r is a vertical stretch of the graph of f by a factor of 2 and a reflection in the x-axis. The domain is all real numbers. The range is $y \leq 0$.

MONITORING PROGRESS ANSWERS
3–4. See Additional Answers.

Laurie's Notes — Teacher Actions

- **Turn and Talk:** "Write the function $q(x) = 2|x|$. What effect do you think the 2 has on the graph of $f(x) = |x|$?" Give ample time for students to discuss before soliciting answers. *Sample answer:* The graph stretches upward.
- **Teaching Tip:** Graph the parent function and the transformed function so that students can compare and see the effect of the transformation.
- "When you graph $y = x$ and $y = mx$ on the same graph, what does m do to the graph of $y = x$?" It changes the steepness of the line.
- **Connection:** Discuss the transformations of the graph of $f(x) = a|x|$. In part (a), $a = 2$. Because $|a| > 1$, it makes the V-shape steeper than the parent function. This is a stretch. In part (b), $a = -\frac{1}{2}$. Because $|a| < 1$, it makes the V-shape wider than the parent function, and the graph opens down because $a < 0$. This is a shrink and a reflection over the x-axis.
- Discuss the range of each function.

Extra Example 3

Graph $f(x) = |x - 3| + 1$ and $g(x) = |3x - 3| + 1$. Compare the graph of g to the graph of f.

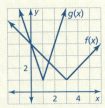

The graph of g is a horizontal shrink of the graph of f by a factor of $\frac{1}{3}$. The y-intercept is the same for both graphs and the points on the graph of f move $\frac{2}{3}$ closer to the y-axis, resulting in the graph of g. When the input values of f are 3 times the input values of g, the outputs are the same.

MONITORING PROGRESS ANSWERS

5.

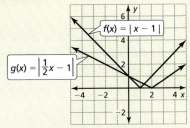

The graph of g is a horizontal stretch of the graph of f by a factor of 2.

6.

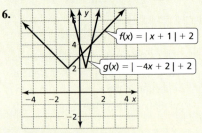

The graph of g is a horizontal shrink of the graph of f by a factor of $\frac{1}{4}$ and a reflection in the y-axis.

Core Concept

Vertex Form of an Absolute Value Function

An absolute value function written in the form $g(x) = a|x - h| + k$, where $a \ne 0$, is in **vertex form**. The vertex of the graph of g is (h, k).

Any absolute value function can be written in vertex form, and its graph is symmetric about the line $x = h$.

STUDY TIP
The function g is *not* in vertex form because the x variable does not have a coefficient of 1.

EXAMPLE 3 — Graphing $f(x) = |x - h| + k$ and $g(x) = f(ax)$

Graph $f(x) = |x + 2| - 3$ and $g(x) = |2x + 2| - 3$. Compare the graph of g to the graph of f.

SOLUTION

Step 1 Make a table of values for each function.

x	−4	−3	−2	−1	0	1	2
f(x)	−1	−2	−3	−2	−1	0	1

x	−2	−1.5	−1	−0.5	0	0.5	1
g(x)	−1	−2	−3	−2	−1	0	1

Step 2 Plot the ordered pairs.

Step 3 Draw the V-shaped graph of each function. Notice that the vertex of the graph of f is $(-2, -3)$ and the graph is symmetric about $x = -2$.

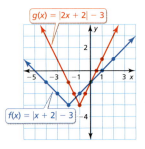

Note that you can rewrite g as $g(x) = f(2x)$, which is of the form $y = f(ax)$, where $a = 2$. So, the graph of g is a horizontal shrink of the graph of f by a factor of $\frac{1}{2}$. The y-intercept is the same for both graphs. The points on the graph of f move halfway closer to the y-axis, resulting in the graph of g. When the input values of f are 2 times the input values of g, the output values of f and g are the same.

Monitoring Progress Help in English and Spanish at BigIdeasMath.com

5. Graph $f(x) = |x - 1|$ and $g(x) = \left|\frac{1}{2}x - 1\right|$. Compare the graph of g to the graph of f.

6. Graph $f(x) = |x + 2| + 2$ and $g(x) = |-4x + 2| + 2$. Compare the graph of g to the graph of f.

Laurie's Notes — Teacher Actions

- **? Use Appropriate Tools Strategically:** Have students use a graphing calculator to graph $f(x) = 2|x - 3| + 1$. "What is the vertex?" $(3, 1)$
- **? Reason Abstractly and Quantitatively:** "Could you predict what the vertex of $f(x) = \frac{1}{2}|x + 5| - 4$ would be?" Give time for students to talk with neighbors. Ask volunteers to share their reasoning. $(-5, -4)$
- Write the *Core Concept*. Make connections to the transformations of graphs of linear functions. Note that the coefficient of x is 1.
- Write the two functions in Example 3 and give students time to work with their partners.
- **Discuss:** $g(x)$ is a horizontal shrink of the graph of f by a factor of $\frac{1}{2}$.

Combining Transformations

EXAMPLE 4 Graphing $g(x) = a|x - h| + k$

Let $g(x) = -2|x - 1| + 3$. (a) Describe the transformations from the graph of $f(x) = |x|$ to the graph of g. (b) Graph g.

SOLUTION

a. Step 1 Translate the graph of f horizontally 1 unit right to get the graph of $t(x) = |x - 1|$.

Step 2 Stretch the graph of t vertically by a factor of 2 to get the graph of $h(x) = 2|x - 1|$.

Step 3 Reflect the graph of h in the x-axis to get the graph of $r(x) = -2|x - 1|$.

Step 4 Translate the graph of r vertically 3 units up to get the graph of $g(x) = -2|x - 1| + 3$.

b. Method 1

Step 1 Make a table of values.

x	−1	0	1	2	3
g(x)	−1	1	3	1	−1

Step 2 Plot the ordered pairs.

Step 3 Draw the V-shaped graph.

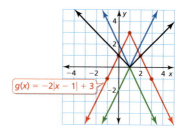

Method 2

Step 1 Identify and plot the vertex. $(h, k) = (1, 3)$

Step 2 Plot another point on the graph, such as (2, 1). Because the graph is symmetric about the line $x = 1$, you can use symmetry to plot a third point, (0, 1).

Step 3 Draw the V-shaped graph.

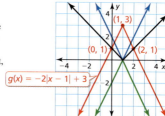

Monitoring Progress Help in English and Spanish at BigIdeasMath.com

7. Let $g(x) = \left|-\frac{1}{2}x + 2\right| + 1$. (a) Describe the transformations from the graph of $f(x) = |x|$ to the graph of g. (b) Graph g.

Section 1.1 Absolute Value Functions 7

Extra Example 4

Let $g(x) = -|x + 1| + 2$.

a. Describe the transformations from the graph of $f(x) = |x|$ to the graph of g. Translate the graph of f horizontally 1 unit left to get the graph of $h(x) = |x + 1|$. Reflect the graph of h in the x-axis to get the graph of $r(x) = -|x + 1|$. Translate the graph of r vertically 2 units up to get the graph of $g(x) = -|x + 1| + 2$.

b. Graph g.

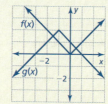

MONITORING PROGRESS ANSWER

7. a. The transformations are a horizontal translation 2 units left, then a horizontal stretch by a factor of 2, then a reflection in the y-axis, then a vertical translation 1 unit up.

b.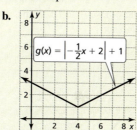

Laurie's Notes — Teacher Actions

- Like linear functions, all of the transformations can be performed to the graph of an absolute value function. The vertex form showed this.
- **Turn and Talk:** "Describe the effect of the −2, 1, and 3 on the parent function $f(x) = |x|$." The 1 translates the graph 1 unit right. The −2 reflects the graph in the x-axis and stretches the graph by a factor of 2. The 3 translates the graph 3 units up.
- **Big Idea:** Students often ask what values they should use in the table of values to be sure it includes the vertex and a few values on either side of the vertex. In the parent function, the vertex occurs at $x = 0$. In the transformed function, the vertex occurs when the expression inside the absolute value symbols equals 0. In other words, when does $x - h = 0$?

Closure

- Have students summarize the learning outcome for Section 1.1 by first writing it in their own notebooks and then sharing their summaries with partners. Encourage students to then look back at the *What You Will Learn* on page 4.

Assignment Guide and Homework Check

ASSIGNMENT

Basic: 1–4, 5–13 odd, 19, 23, 27–41 odd, 45, 50, 62, 64–69

Average: 1–4, 6–24 even, 28–46 even, 50–56 even, 62, 64–69

Advanced: 1–4, 8–26 even, 30, 32, 38–62 even, 64–69

HOMEWORK CHECK

Basic: 5, 9, 27, 35

Average: 6, 12, 28, 36

Advanced: 8, 12, 30, 38

ANSWERS

1. vertex
2. If $|a| > 1$, the graph is a vertical stretch. If $0 < |a| < 1$, the graph is a vertical shrink.
3. Sample answer: $g(x) = a|x|$ is a vertical stretch or shrink, $g(x) = |x - h|$ is a horizontal translation, and $g(x) = |x| + k$ is a vertical translation.
4. $g(x) = |3x - 2| + 5$; g is a horizontal shrink of f, so it will have the same y-intercept.

5.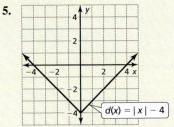

The graph of d is a vertical translation 4 units down of the graph of f; domain: all real numbers; range: $y \geq -4$

6.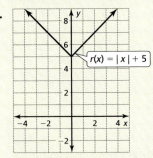

The graph of r is a vertical translation 5 units up of the graph of f; domain: all real numbers; range: $y \geq 5$

1.1 Exercises

Dynamic Solutions available at BigIdeasMath.com

Vocabulary and Core Concept Check

1. **COMPLETE THE SENTENCE** The point $(1, -4)$ is the _____ of the graph of $f(x) = -3|x - 1| - 4$.

2. **USING STRUCTURE** How do you know whether the graph of $f(x) = a|x - h| + k$ is a vertical stretch or a vertical shrink of the graph of $f(x) = |x|$?

3. **WRITING** Describe three different types of transformations of the graph of an absolute value function.

4. **REASONING** The graph of which function has the same y-intercept as the graph of $f(x) = |x - 2| + 5$? Explain.

   $g(x) = |3x - 2| + 5$ $h(x) = 3|x - 2| + 5$

Monitoring Progress and Modeling with Mathematics

In Exercises 5–12, graph the function. Compare the graph to the graph of $f(x) = |x|$. Describe the domain and range. *(See Examples 1 and 2.)*

5. $d(x) = |x| - 4$
6. $r(x) = |x| + 5$
7. $m(x) = |x + 1|$
8. $v(x) = |x - 3|$
9. $p(x) = \frac{1}{3}|x|$
10. $j(x) = 3|x|$
11. $a(x) = -5|x|$
12. $q(x) = -\frac{3}{2}|x|$

In Exercises 13–16, graph the function. Compare the graph to the graph of $f(x) = |x - 6|$.

13. $h(x) = |x - 6| + 2$
14. $n(x) = \frac{1}{2}|x - 6|$
15. $k(x) = -3|x - 6|$
16. $g(x) = |x - 1|$

In Exercises 17 and 18, graph the function. Compare the graph to the graph of $f(x) = |x + 3| - 2$.

17. $y(x) = |x + 4| - 2$
18. $b(x) = |x + 3| + 3$

In Exercises 19–22, compare the graphs. Find the value of h, k, or a.

19.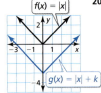
20.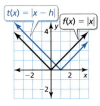
21.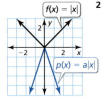
22.

In Exercises 23–26, write an equation that represents the given transformation(s) of the graph of $g(x) = |x|$.

23. vertical translation 7 units down
24. horizontal translation 10 units left
25. vertical shrink by a factor of $\frac{1}{4}$
26. vertical stretch by a factor of 3 and a reflection in the x-axis

In Exercises 27–32, graph and compare the two functions. *(See Example 3.)*

27. $f(x) = |x - 4|$; $g(x) = |3x - 4|$
28. $h(x) = |x + 5|$; $t(x) = |2x + 5|$
29. $p(x) = |x + 1| - 2$; $q(x) = \left|\frac{1}{4}x + 1\right| - 2$
30. $w(x) = |x - 3| + 4$; $y(x) = |5x - 3| + 4$
31. $a(x) = |x + 2| + 3$; $b(x) = |-4x + 2| + 3$
32. $u(x) = |x - 1| + 2$; $v(x) = \left|-\frac{1}{2}x - 1\right| + 2$

8   Chapter 1   Functions and Exponents

7.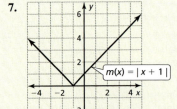

The graph of m is a horizontal translation 1 unit left of the graph of f; domain: all real numbers; range: $y \geq 0$

8–32. See Additional Answers.

In Exercises 33–40, describe the transformations from the graph of $f(x) = |x|$ to the graph of the given function. Then graph the given function. *(See Example 4.)*

33. $r(x) = |x + 2| - 6$ 34. $c(x) = |x + 4| + 4$

35. $d(x) = -|x - 3| + 5$ 36. $v(x) = -3|x + 1| + 4$

37. $m(x) = \frac{1}{2}|x + 4| - 1$ 38. $s(x) = |2x - 2| - 3$

39. $j(x) = |-x + 1| - 5$ 40. $n(x) = \left|-\frac{1}{3}x + 1\right| + 2$

41. **MODELING WITH MATHEMATICS** The number of pairs of shoes sold s (in thousands) increases and then decreases as described by the function $s(t) = -2|t - 15| + 50$, where t is the time (in weeks).

a. Graph the function.

b. What is the greatest number of pairs of shoes sold in 1 week?

42. **MODELING WITH MATHEMATICS** On the pool table shown, you bank the five ball off the side represented by the x-axis. The path of the ball is described by the function $p(x) = \frac{4}{3}\left|x - \frac{5}{4}\right|$.

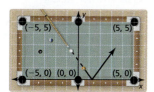

a. At what point does the five ball bank off the side?

b. Do you make the shot? Explain your reasoning.

43. **USING TRANSFORMATIONS** The points $A\left(-\frac{1}{2}, 3\right)$, $B(1, 0)$, and $C(-4, -2)$ lie on the graph of the absolute value function f. Find the coordinates of the points corresponding to A, B, and C on the graph of each function.

a. $g(x) = f(x) - 5$ b. $h(x) = f(x - 3)$
c. $j(x) = -f(x)$ d. $k(x) = 4f(x)$

44. **USING STRUCTURE** Explain how the graph of each function compares to the graph of $y = |x|$ for positive and negative values of k, h, and a.

a. $y = |x| + k$
b. $y = |x - h|$
c. $y = a|x|$
d. $y = |ax|$

ERROR ANALYSIS In Exercises 45 and 46, describe and correct the error in graphing the function.

45.
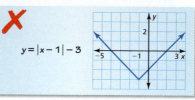

46.
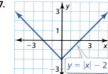

MATHEMATICAL CONNECTIONS In Exercises 47 and 48, write an absolute value function whose graph forms a square with the given graph.

47.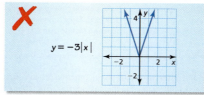

48.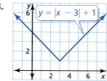

49. **WRITING** Compare the graphs of $p(x) = |x - 6|$ and $q(x) = |x| - 6$.

Section 1.1 Absolute Value Functions 9

ANSWERS

33. The transformations are a horizontal translation 2 units left then a vertical translation 6 units down.

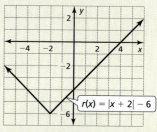

34. The transformations are a horizontal translation 4 units left then a vertical translation 4 units up.

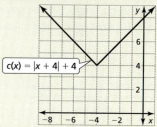

35. The transformations are a horizontal translation 3 units right, then a reflection in the x-axis, then a vertical translation 5 units up.

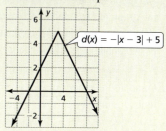

36. The transformations are a horizontal translation 1 unit left, then a vertical stretch by a factor of 3, then a reflection in the x-axis, then a vertical translation 4 units up.

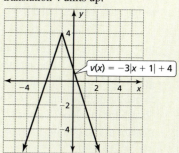

37. The transformations are a horizontal translation 4 units left, then a vertical shrink by a factor of $\frac{1}{2}$, then a vertical translation 1 unit down.

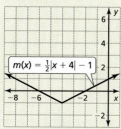

38–49. See Additional Answers.

Section 1.1 9

ANSWERS

50. a. (7, 1)

b. Write an equation in vertex form with (7, 1) as the vertex; $f(x) = |x - 7| + 1$

51–69. See Additional Answers.

Mini-Assessment

Graph the function. Compare the graph to the graph of $f(x) = |x|$. Describe the domain and range.

1. $g(x) = |x| + 1$

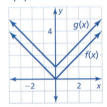

The graph of g is a vertical translation 1 unit up of the graph of f. The domain is all real numbers. The range is $y \geq 1$.

2. $d(x) = -\frac{1}{3}|x|$

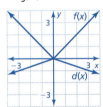

The graph of d is a vertical shrink of the graph of f by a factor of $\frac{1}{3}$ and a reflection in the x-axis. The domain is all real numbers. The range is $y \leq 0$.

3. Compare the graphs of the two functions $m(x) = |x + 2| - 1$ and $t(x) = |2x + 2| - 1$. The graph of t is a horizontal shrink of the graph of m by a factor of $\frac{1}{2}$. The y-intercept is the same for both graphs and the points on the graph of m move halfway closer to the y-axis, resulting in the graph of t.

4. Describe the transformations from the graph of $f(x) = |x|$ to the graph of $g(x) = 2|x + 4|$. Translate the graph of f horizontally 4 units left to get the graph of $h(x) = |x + 4|$. Stretch the graph of h vertically by a factor of 2 to get the graph of $g(x) = 2|x + 4|$.

---

50. HOW DO YOU SEE IT? The object of a computer game is to break bricks by deflecting a ball toward them using a paddle. The graph shows the current path of the ball and the location of the last brick.

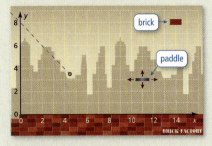

a. You can move the paddle up, down, left, and right. At what coordinates should you place the paddle to break the last brick? Assume the ball deflects at a right angle.

b. You move the paddle to the coordinates in part (a), and the ball is deflected. How can you write an absolute value function that describes the path of the ball?

In Exercises 51–54, graph the function. Then rewrite the absolute value function as two linear functions, one that has the domain $x < 0$ and one that has the domain $x \geq 0$.

51. $y = |x|$

52. $y = |x| - 3$

53. $y = -|x| + 9$

54. $y = -4|x|$

In Exercises 55–58, graph and compare the two functions.

55. $f(x) = |x - 1| + 2$; $g(x) = 4|x - 1| + 8$

56. $s(x) = |2x - 5| - 6$; $t(x) = \frac{1}{2}|2x - 5| - 3$

57. $v(x) = -2|3x + 1| + 4$; $w(x) = 3|3x + 1| - 6$

58. $c(x) = 4|x + 3| - 1$; $d(x) = -\frac{4}{3}|x + 3| + \frac{1}{3}$

59. REASONING Describe the transformations from the graph of $g(x) = -2|x + 1| + 4$ to the graph of $h(x) = |x|$. Explain your reasoning.

60. THOUGHT PROVOKING Graph an absolute value function f that represents the route a wide receiver runs in a football game. Let the x-axis represent distance (in yards) across the field horizontally. Let the y-axis represent distance (in yards) down the field. Be sure to limit the domain so the route is realistic.

61. SOLVING BY GRAPHING Graph $y = 2|x + 2| - 6$ and $y = -2$ in the same coordinate plane. Use the graph to solve the equation $2|x + 2| - 6 = -2$. Check your solutions.

62. MAKING AN ARGUMENT Let p be a positive constant. Your friend says that because the graph of $y = |x| + p$ is a *positive* vertical translation of the graph of $y = |x|$, the graph of $y = |x + p|$ is a *positive* horizontal translation of the graph of $y = |x|$. Is your friend correct? Explain.

63. ABSTRACT REASONING Write the vertex of the absolute value function $f(x) = |ax - h| + k$ in terms of a, h, and k.

Maintaining Mathematical Proficiency
Reviewing what you learned in previous grades and lessons

Solve the formula for the indicated variable. *(Skills Review Handbook)*

64. Solve for h.

$V = \pi r^2 h$

65. Solve for w.

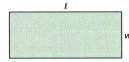

$P = 2\ell + 2w$

Write an equation of the line that passes through the points. *(Skills Review Handbook)*

66. (1, −4), (2, −1)

67. (−3, 12), (0, 6)

68. (3, 1), (−5, 5)

69. (2, −1), (0, 1)

If students need help...	If students got it...
Resources by Chapter • Practice A and Practice B • Puzzle Time	Resources by Chapter • Enrichment and Extension • Cumulative Review
Student Journal • Practice	Start the *next* Section
Differentiating the Lesson Skills Review Handbook	

Laurie's Notes

Overview of Section 1.2

Introduction
- In this section, students will apply skills learned in Math I when they evaluate, graph, and write *piecewise functions*. Another name for piecewise function is split-rule function. Two special types of piecewise functions, the step function and absolute value functions, are presented at the end of the lesson.

Teaching Strategy
- Piecewise functions can be graphed on a calculator using Boolean logic.
- To graph the function

$$f(x) = \begin{cases} x-4, & \text{if } x < 0 \\ -2, & \text{if } 0 < x < 2, \\ -x+6, & \text{if } x > 2 \end{cases}$$

enter the following in the equation editor.
$(x-4)(x<0) + (-2)(x>0 \text{ and } x<2) + (-x+6)(x>2)$

- Each of the three expressions (or rules) is entered in parentheses and is followed by a specified domain. The domain is the Boolean operator. When the value of *x* is in that domain, it returns a Boolean value of 1 (true) and multiplies the expression by 1. When the value of *x* is *not* in that domain, it returns a Boolean value of 0 (false) and multiplies the expression by 0. By entering the three expressions with addition signs between them, only one of the three xpressions will be true at a time and the other two expressions will be 0.

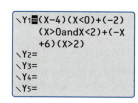 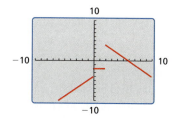

Another Way
- To sketch a piecewise function, lightly graph each line using the slope and *y*-intercept. Then use the domains to decide what portion of each line to keep and what to erase. This method is sometimes easier for students when the change between the expressions does not occur at $x = 0$ and when there are more than two expressions.

Pacing Suggestion
- Once students have worked the explorations, continue with the formal lesson.

What Your Students Will Learn

- Evaluate piecewise functions.
- Graph and write piecewise functions and describe the domains and ranges.
- Graph and write step functions.
- Write absolute value functions as piecewise functions.

Laurie's Notes

Exploration

Motivate
- Ask students whether they have seen a utility bill where the cost per kilowatt-hour depended on the amount used. For instance, the first 5000 kilowatt-hours might be $0.1782 per kilowatt-hour, then $0.1219 per kilowatt-hour for the next 3000 kilowatt-hours, and so on.
- Explain to students that this type of billing scheme is an example of the type of function they will learn about today.

Discuss
- Review with students how to determine whether a graph is a function. If you have described the "vertical line test," then bring this into the discussion. Review the language associated with functions: domain, range, independent and dependent variables. Finally, mention the notation used for functions and how it is read. *Example:* $f(4) = -6$ is read, "f of 4 is equal to -6" or "the value of the function f at 4 is -6."
- You may also need to review compound inequalities, including how they are read and interpreted.

Exploration 1
- **Connection:** When solutions are plotted on a number line, the open and closed circles are used to represent when the value of x is or is not included in the solution set. When graphing piecewise-defined functions, also known as split-rule functions, the open and closed circles are used to represent when the value of x is included as a solution for each piece of the function.
- The first two questions help focus attention on where the graph changes from one rule to a new rule.
- **?** "What information is needed to write the equation of either piece of the graph?" You need the slope and y-intercept.
- What can be confusing for students is the notation of specifying the domain for x (if $x < 0$). Explain that each rule applies to a different part of the domain.

Exploration 2
- In this exploration, students should recognize a series of constant functions. It is very common for students to describe the graph as looking like steps!
- **?** **Assessing Question:** When students have finished, ask, "How do you know the graph represents a function?" Answers will vary. Listen for understanding that each value of the domain is associated with only one range value.
- **Extension:** Ask students whether this pattern, meaning the steps, could continue. You could also ask how the function would change if the segments were an open circle on the left endpoint and a closed circle on the right endpoint.

Communicate Your Answer
- In Question 4, check that students have only one function rule defined when $x = 0$.

Connecting to Next Step
- The short explorations provide an introduction to piecewise functions. You might consider omitting Example 3 in the formal lesson and checking for understanding with a few *Monitoring Progress* questions.

1.2 Piecewise Functions

Essential Question How can you describe a function that is represented by more than one equation?

EXPLORATION 1 Writing Equations for a Function

Work with a partner.

CONSTRUCTING VIABLE ARGUMENTS
To be proficient in math, you need to justify your conclusions and communicate them to others.

a. Does the graph represent y as a function of x? Justify your conclusion.

b. What is the value of the function when $x = 0$? How can you tell?

c. Write an equation that represents the values of the function when $x \leq 0$.

$$f(x) = \underline{}, \text{ if } x \leq 0$$

d. Write an equation that represents the values of the function when $x > 0$.

$$f(x) = \underline{}, \text{ if } x > 0$$

e. Combine the results of parts (c) and (d) to write a single description of the function.

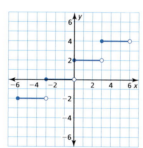

$$f(x) = \begin{cases} \underline{}, & \text{if } x \leq 0 \\ \underline{}, & \text{if } x > 0 \end{cases}$$

EXPLORATION 2 Writing Equations for a Function

Work with a partner.

a. Does the graph represent y as a function of x? Justify your conclusion.

b. Describe the values of the function for the following intervals.

$$f(x) = \begin{cases} \underline{}, & \text{if } -6 \leq x < -3 \\ \underline{}, & \text{if } -3 \leq x < 0 \\ \underline{}, & \text{if } 0 \leq x < 3 \\ \underline{}, & \text{if } 3 \leq x < 6 \end{cases}$$

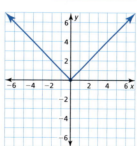

Communicate Your Answer

3. How can you describe a function that is represented by more than one equation?

4. Use two equations to describe the function represented by the graph.

ANSWERS

1. a. yes; No vertical line can be drawn through more than one point on the graph.
   b. 0; The point (0, 0) is plotted.
   c. $-x$
   d. 2
   e. $-x$; 2

2. a. yes; No vertical line can be drawn through more than one point on the graph.
   b. -2; 0; 2; 4

3. Write the expression for each part of the function along with the part of the domain to which the expression applies.

4. $f(x) = \begin{cases} -x, & \text{if } x \leq 0 \\ x, & \text{if } x > 0 \end{cases}$ or

   $f(x) = \begin{cases} -x, & \text{if } x < 0 \\ x, & \text{if } x \geq 0 \end{cases}$

Differentiated Instruction

Inclusion
Some students may have difficulty associating piecewise functions with parts of a graph. Remind them that they can use two points from each piece of the graph to find the slope for each piece. They can also locate the *y*-intercept for each piece. Have them use the slope and *y*-intercept to verify the function given for each piece.

Extra Example 1
Evaluate the function.
$$f(x) = \begin{cases} x - 2, & \text{if } x \leq 0 \\ 2x + 1, & \text{if } x > 0 \end{cases}$$

a. when $x = -1$
   The value of *f* is -3 when $x = -1$.
b. when $x = 1$
   The value of *f* is 3 when $x = 1$.

MONITORING PROGRESS ANSWERS
1. 3
2. 0
3. 2
4. 5
5. 7
6. 40

1.2 Lesson

What You Will Learn
- Evaluate piecewise functions.
- Graph and write piecewise functions.
- Graph and write step functions.
- Write absolute value functions.

Core Vocabulary
piecewise function, *p. 12*
step function, *p. 14*
Previous
absolute value function
vertex form
vertex

Evaluating Piecewise Functions

Core Concept

Piecewise Function
A **piecewise function** is a function defined by two or more equations. Each "piece" of the function applies to a different part of its domain. An example is shown below.

$$f(x) = \begin{cases} x - 2, & \text{if } x \leq 0 \\ 2x + 1, & \text{if } x > 0 \end{cases}$$

- The expression $x - 2$ represents the value of *f* when *x* is less than or equal to 0.
- The expression $2x + 1$ represents the value of *f* when *x* is greater than 0.

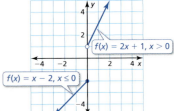

EXAMPLE 1   Evaluating a Piecewise Function

Evaluate the function *f* above when (a) $x = 0$ and (b) $x = 4$.

SOLUTION

a. $f(x) = x - 2$ Because $0 \leq 0$, use the first equation.
   $f(0) = 0 - 2$ Substitute 0 for *x*.
   $f(0) = -2$ Simplify.

  ▶ The value of *f* is -2 when $x = 0$.

b. $f(x) = 2x + 1$ Because $4 > 0$, use the second equation.
   $f(4) = 2(4) + 1$ Substitute 4 for *x*.
   $f(4) = 9$ Simplify.

  ▶ The value of *f* is 9 when $x = 4$.

Monitoring Progress Help in English and Spanish at *BigIdeasMath.com*

Evaluate the function.

$$f(x) = \begin{cases} 3, & \text{if } x < -2 \\ x + 2, & \text{if } -2 \leq x \leq 5 \\ 4x, & \text{if } x > 5 \end{cases}$$

1. $f(-8)$ 2. $f(-2)$
3. $f(0)$ 4. $f(3)$
5. $f(5)$ 6. $f(10)$

Laurie's Notes | Teacher Actions

- Define a *piecewise function* and connect to the explorations.
- **Teaching Tip:** Graph or project a piecewise function. Cover the right portion of the function and ask students to describe the left portion. Switch and describe the right portion. Write the two descriptions together with the appropriate domains.
- **Monitoring Progress:** These questions are a quick check for understanding of how to evaluate a piecewise function.

Graphing and Writing Piecewise Functions

EXAMPLE 2 Graphing a Piecewise Function

Graph $y = \begin{cases} -x - 4, & \text{if } x < 0 \\ x, & \text{if } x \geq 0 \end{cases}$. Describe the domain and range.

SOLUTION

Step 1 Graph $y = -x - 4$ for $x < 0$. Because x is not equal to 0, use an open circle at $(0, -4)$.

Step 2 Graph $y = x$ for $x \geq 0$. Because x is greater than or equal to 0, use a closed circle at $(0, 0)$.

▶ The domain is all real numbers. The range is $y > -4$.

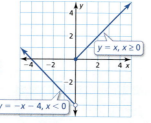

Monitoring Progress Help in English and Spanish at BigIdeasMath.com

Graph the function. Describe the domain and range.

7. $y = \begin{cases} x + 1, & \text{if } x \leq 0 \\ -x, & \text{if } x > 0 \end{cases}$
8. $y = \begin{cases} x - 2, & \text{if } x < 0 \\ 4x, & \text{if } x \geq 0 \end{cases}$

EXAMPLE 3 Writing a Piecewise Function

Write a piecewise function for the graph.

SOLUTION

Each "piece" of the function is linear.

Left Piece When $x < 0$, the graph is the line given by $y = x + 3$.

Right Piece When $x \geq 0$, the graph is the line given by $y = 2x - 1$.

▶ So, a piecewise function for the graph is

$f(x) = \begin{cases} x + 3, & \text{if } x < 0 \\ 2x - 1, & \text{if } x \geq 0 \end{cases}$.

Monitoring Progress Help in English and Spanish at BigIdeasMath.com

Write a piecewise function for the graph.

9.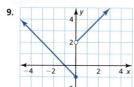
10.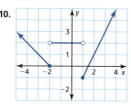

Extra Example 2

Graph $y = \begin{cases} -x + 2, & \text{if } x \leq 0 \\ 2x, & \text{if } x > 0 \end{cases}$.

Describe the domain and range.

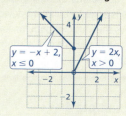

The domain is all real numbers. The range is $y > 0$.

Extra Example 3

Write a piecewise function for the graph.

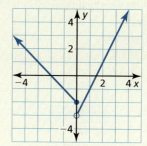

$y = \begin{cases} -x - 2, & \text{if } x \leq 0 \\ 2x - 3, & \text{if } x > 0 \end{cases}$

MONITORING PROGRESS ANSWERS

7.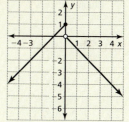

domain: all real numbers;
range: $y \leq 1$

8.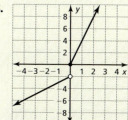

domain: all real numbers;
range: $y < -2$ or $y \geq 0$

9. $f(x) = \begin{cases} -x - 1, & \text{if } x \leq 0 \\ x + 2, & \text{if } x > 0 \end{cases}$

10. $f(x) = \begin{cases} -x - 2, & \text{if } x \leq -2 \\ 2, & \text{if } -2 < x < 1 \\ 2x - 3, & \text{if } x \geq 1 \end{cases}$

Laurie's Notes — Teacher Actions

- **Use Appropriate Tools Strategically:** It is possible to graph piecewise functions with a graphing calculator. See the *Teaching Strategy*.
- In stating the range for Example 2, students should notice that all of the y-values are greater than -4. Sometimes students will find this difficult to answer until I move my hand in an upward motion starting at $y = -4$.
- If you have access to a document camera or similar visual display, have students work with partners on Examples 2 and 3. Then display results and discuss as a class.

English Language Learners

Culture
Explain to students that a karaoke machine is an entertainment device that plays the music of popular songs. People sing the words of the songs as the music plays.

Extra Example 4
You rent a bicycle for 4 days. The bike store charges $20 for the first day and $15 for each additional day. Write and graph a step function that represents the relationship between the number x of days and the total cost y (in dollars) of renting the bicycle.

$$f(x) = \begin{cases} 20, & \text{if } 0 < x \leq 1 \\ 35, & \text{if } 1 < x \leq 2 \\ 50, & \text{if } 2 < x \leq 3 \\ 65, & \text{if } 3 < x \leq 4 \end{cases}$$

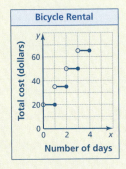

MONITORING PROGRESS ANSWER

11. $f(x) = \begin{cases} 100, & \text{if } 0 < x \leq 1 \\ 150, & \text{if } 1 < x \leq 2 \\ 200, & \text{if } 2 < x \leq 3 \\ 250, & \text{if } 3 < x \leq 4 \end{cases}$

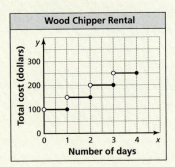

STUDY TIP
The graph of a step function looks like a staircase.

Graphing and Writing Step Functions

A **step function** is a piecewise function defined by a constant value over each part of its domain. The graph of a step function consists of a series of line segments.

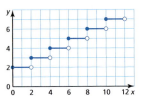

$$f(x) = \begin{cases} 2, & \text{if } 0 \leq x < 2 \\ 3, & \text{if } 2 \leq x < 4 \\ 4, & \text{if } 4 \leq x < 6 \\ 5, & \text{if } 6 \leq x < 8 \\ 6, & \text{if } 8 \leq x < 10 \\ 7, & \text{if } 10 \leq x < 12 \end{cases}$$

EXAMPLE 4 Graphing and Writing a Step Function

You rent a karaoke machine for 5 days. The rental company charges $50 for the first day and $25 for each additional day. Write and graph a step function that represents the relationship between the number x of days and the total cost y (in dollars) of renting the karaoke machine.

SOLUTION

Step 1 Use a table to organize the information.

Number of days	Total cost (dollars)
$0 < x \leq 1$	50
$1 < x \leq 2$	75
$2 < x \leq 3$	100
$3 < x \leq 4$	125
$4 < x \leq 5$	150

Step 2 Write the step function.

$$f(x) = \begin{cases} 50, & \text{if } 0 < x \leq 1 \\ 75, & \text{if } 1 < x \leq 2 \\ 100, & \text{if } 2 < x \leq 3 \\ 125, & \text{if } 3 < x \leq 4 \\ 150, & \text{if } 4 < x \leq 5 \end{cases}$$

Step 3 Graph the step function.

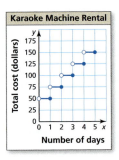

Monitoring Progress Help in English and Spanish at *BigIdeasMath.com*

11. A landscaper rents a wood chipper for 4 days. The rental company charges $100 for the first day and $50 for each additional day. Write and graph a step function that represents the relationship between the number x of days and the total cost y (in dollars) of renting the chipper.

14 Chapter 1 Functions and Exponents

Laurie's Notes — Teacher Actions

- **Make Sense of Problems and Persevere in Solving Them** and **Model with Mathematics:** The table helps organize the data. Compound inequalities are used to define the days. Because there are five days, the step function will have five expressions, each being a constant.
- ❓ "Why are the steps open at the left and closed at the right?" Once you go beyond a whole day, you pay for the next full day.
- **Think-Pair-Share:** Have students answer the *Monitoring Progress* question and then share and discuss as a class. (*Note*: This *Formative Assessment Tip* is described on page T-18.)

Writing Absolute Value Functions

The absolute value function $f(x) = |x|$ can be written as a piecewise function.

$$f(x) = \begin{cases} -x, & \text{if } x < 0 \\ x, & \text{if } x \geq 0 \end{cases}$$

Similarly, the vertex form of an absolute value function $g(x) = a|x - h| + k$ can be written as a piecewise function.

$$g(x) = \begin{cases} a[-(x - h)] + k, & \text{if } x - h < 0 \\ a(x - h) + k, & \text{if } x - h \geq 0 \end{cases}$$

EXAMPLE 5 Writing an Absolute Value Function

In holography, light from a laser beam is split into two beams, a reference beam and an object beam. Light from the object beam reflects off an object and is recombined with the reference beam to form images on film that can be used to create three-dimensional images.

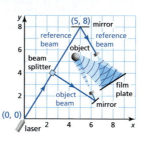

a. Write an absolute value function that represents the path of the reference beam.

b. Write the function in part (a) as a piecewise function.

SOLUTION

a. The vertex of the path of the reference beam is (5, 8). So, the function has the form $g(x) = a|x - 5| + 8$. Substitute the coordinates of the point (0, 0) into the equation and solve for a.

$g(x) = a|x - 5| + 8$ Vertex form of the function
$0 = a|0 - 5| + 8$ Substitute 0 for x and 0 for $g(x)$.
$-1.6 = a$ Solve for a.

▶ So, the function $g(x) = -1.6|x - 5| + 8$ represents the path of the reference beam.

STUDY TIP
The graph of an absolute value function is symmetric about the line $x = h$. So, it makes sense that the piecewise definition "splits" the function at $x = 5$.

b. Write $g(x) = -1.6|x - 5| + 8$ as a piecewise function.

$$g(x) = \begin{cases} -1.6[-(x - 5)] + 8, & \text{if } x - 5 < 0 \\ -1.6(x - 5) + 8, & \text{if } x - 5 \geq 0 \end{cases}$$

Simplify each expression and solve the inequalities.

▶ So, a piecewise function for $g(x) = -1.6|x - 5| + 8$ is

$$g(x) = \begin{cases} 1.6x, & \text{if } x < 5 \\ -1.6x + 16, & \text{if } x \geq 5 \end{cases}$$

Monitoring Progress Help in English and Spanish at BigIdeasMath.com

12. **WHAT IF?** The reference beam originates at (3, 0) and reflects off a mirror at (5, 4).

   a. Write an absolute value function that represents the path of the reference beam.

   b. Write the function in part (a) as a piecewise function.

Extra Example 5

The reference beam in Example 5 originates at (1, 0) and reflects off a mirror at (5, 6).

a. Write an absolute value function that represents the path of the reference beam.

$g(x) = -1.5|x - 5| + 6$

b. Write the function in part (a) as a piecewise function.

$$g(x) = \begin{cases} 1.5x - 1.5, & \text{if } x < 5 \\ -1.5x + 13.5, & \text{if } x \geq 5 \end{cases}$$

MONITORING PROGRESS ANSWER

12. **a.** $g(x) = -2|x - 5| + 4$

   b. $g(x) = \begin{cases} 2x - 6, & \text{if } x < 5 \\ -2x + 14, & \text{if } x \geq 5 \end{cases}$

Laurie's Notes   Teacher Actions

- Discuss how to write the absolute value function as a piecewise function. Have students think about the equation for each part. Discuss the vertex form and how it can be written in two parts.
- **Turn and Talk:** Pose Example 5 and have partners discuss ideas for writing the functions for each part.

Closure

- Ask students to reflect on the lesson and identify the point or part of the lesson that most aided their learning. This feedback can be shared aloud or on a paper to be collected.

Assignment Guide and Homework Check

ASSIGNMENT

Basic: 1, 2, 3–17 odd, 21–31 odd, 35–41 odd, 47, 52, 55–61

Average: 1, 2–18 even, 22–48 even, 49, 52, 55–61

Advanced: 1, 2, 10–14 even, 18–22 even, 28–36 even, 46, 48–61

HOMEWORK CHECK

Basic: 3, 15, 23, 35, 47

Average: 4, 16, 24, 36, 48

Advanced: 14, 18, 28, 36, 48

ANSWERS

1. A piecewise function is a function defined by two or more equations. A step function is a special piecewise function defined by a constant value over each part of its domain.

2. It can be written as a piecewise function by writing a linear function for each of the two lines that form the graph.

3. -16 4. 1
5. 3 6. 8
7. 8 8. 5
9. 3 10. 3
11. -1 12. 5
13. 240 mi   14. $430.80

15.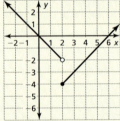
domain: all real numbers; range: $y \geq -4$

16.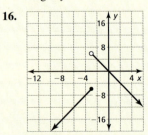
domain: all real numbers; range: $y < 6$

1.2 Exercises

Dynamic Solutions available at *BigIdeasMath.com*

Vocabulary and Core Concept Check

1. **VOCABULARY** Compare piecewise functions and step functions.

2. **WRITING** Use a graph to explain why you can write the absolute value function $y = |x|$ as a piecewise function.

Monitoring Progress and Modeling with Mathematics

In Exercises 3–12, evaluate the function. *(See Example 1.)*

$$f(x) = \begin{cases} 5x - 1, & \text{if } x < -2 \\ x + 3, & \text{if } x \geq -2 \end{cases}$$

$$g(x) = \begin{cases} -x + 4, & \text{if } x \leq -1 \\ 3, & \text{if } -1 < x < 2 \\ 2x - 5, & \text{if } x \geq 2 \end{cases}$$

3. $f(-3)$ 4. $f(-2)$
5. $f(0)$ 6. $f(5)$
7. $g(-4)$ 8. $g(-1)$
9. $g(0)$ 10. $g(1)$
11. $g(2)$ 12. $g(5)$

13. **MODELING WITH MATHEMATICS** On a trip, the total distance (in miles) you travel in x hours is represented by the piecewise function

$$d(x) = \begin{cases} 55x, & \text{if } 0 \leq x \leq 2 \\ 65x - 20, & \text{if } 2 < x \leq 5 \end{cases}.$$

How far do you travel in 4 hours?

14. **MODELING WITH MATHEMATICS** The total cost (in dollars) of ordering x custom shirts is represented by the piecewise function

$$c(x) = \begin{cases} 17x + 20, & \text{if } 0 \leq x < 25 \\ 15.80x + 20, & \text{if } 25 \leq x < 50 \\ 14x + 20, & \text{if } x \geq 50 \end{cases}.$$

Determine the total cost of ordering 26 shirts.

In Exercises 15–20, graph the function. Describe the domain and range. *(See Example 2.)*

15. $y = \begin{cases} -x, & \text{if } x < 2 \\ x - 6, & \text{if } x \geq 2 \end{cases}$

16. $y = \begin{cases} 2x, & \text{if } x \leq -3 \\ -2x, & \text{if } x > -3 \end{cases}$

17. $y = \begin{cases} -3x - 2, & \text{if } x \leq -1 \\ x + 2, & \text{if } x > -1 \end{cases}$

18. $y = \begin{cases} x + 8, & \text{if } x < 4 \\ 4x - 4, & \text{if } x \geq 4 \end{cases}$

19. $y = \begin{cases} 1, & \text{if } x < -3 \\ x - 1, & \text{if } -3 \leq x \leq 3 \\ -2x + 4, & \text{if } x > 3 \end{cases}$

20. $y = \begin{cases} 2x + 1, & \text{if } x \leq -1 \\ -x + 2, & \text{if } -1 < x < 2 \\ -3, & \text{if } x \geq 2 \end{cases}$

21. **ERROR ANALYSIS** Describe and correct the error in finding $f(5)$ when $f(x) = \begin{cases} 2x - 3, & \text{if } x < 5 \\ x + 8, & \text{if } x \geq 5 \end{cases}$.

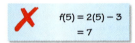

22. **ERROR ANALYSIS** Describe and correct the error in graphing $y = \begin{cases} x + 6, & \text{if } x \leq -2 \\ 1, & \text{if } x > -2 \end{cases}$.

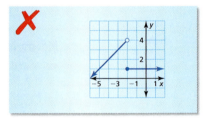

16   Chapter 1   Functions and Exponents

17. domain: all real numbers; range: $y \geq 1$

18. domain: all real numbers; range: all real numbers

19–22. See Additional Answers.

In Exercises 23–30, write a piecewise function for the graph. (See Example 3.)

23.
24.
25.
26.
27.
28.
29.
30.

In Exercises 31–34, graph the step function. Describe the domain and range.

31. $f(x) = \begin{cases} 3, & \text{if } 0 \le x < 2 \\ 4, & \text{if } 2 \le x < 4 \\ 5, & \text{if } 4 \le x < 6 \\ 6, & \text{if } 6 \le x < 8 \end{cases}$

32. $f(x) = \begin{cases} -4, & \text{if } 1 < x \le 2 \\ -6, & \text{if } 2 < x \le 3 \\ -8, & \text{if } 3 < x \le 4 \\ -10, & \text{if } 4 < x \le 5 \end{cases}$

33. $f(x) = \begin{cases} 9, & \text{if } 1 < x \le 2 \\ 6, & \text{if } 2 < x \le 4 \\ 5, & \text{if } 4 < x \le 9 \\ 1, & \text{if } 9 < x \le 12 \end{cases}$

34. $f(x) = \begin{cases} -2, & \text{if } -6 \le x < -5 \\ -1, & \text{if } -5 \le x < -3 \\ 0, & \text{if } -3 \le x < -2 \\ 1, & \text{if } -2 \le x < 0 \end{cases}$

35. **MODELING WITH MATHEMATICS** The cost to join an intramural sports league is $180 per team and includes the first five team members. For each additional team member, there is a $30 fee. You plan to have nine people on your team. Write and graph a step function that represents the relationship between the number p of people on your team and the total cost of joining the league. (See Example 4.)

36. **MODELING WITH MATHEMATICS** The rates for a parking garage are shown. Write and graph a step function that represents the relationship between the number x of hours a car is parked in the garage and the total cost of parking in the garage for 1 day.

Daily Parking Garage Rates
$4 per hour
$15 daily maximum

In Exercises 37–46, write the absolute value function as a piecewise function.

37. $y = |x| + 1$
38. $y = |x| - 3$
39. $y = |x - 2|$
40. $y = |x + 5|$
41. $y = 2|x + 3|$
42. $y = 4|x - 1|$
43. $y = -5|x - 8|$
44. $y = -3|x + 6|$
45. $y = -|x - 3| + 2$
46. $y = 7|x + 1| - 5$

47. **MODELING WITH MATHEMATICS** You are sitting on a boat on a lake. You can get a sunburn from the sunlight that hits you directly and also from the sunlight that reflects off the water. (See Example 5.)

a. Write an absolute value function that represents the path of the sunlight that reflects off the water.
b. Write the function in part (a) as a piecewise function.

Section 1.2   Piecewise Functions   17

Dynamic Teaching Tools
Dynamic Assessment & Progress Monitoring Tool
Interactive Whiteboard Lesson Library
Dynamic Classroom with Dynamic Investigations

ANSWERS

23. $f(x) = \begin{cases} x + 2, & \text{if } x < 0 \\ 2, & \text{if } x \ge 0 \end{cases}$

24. $f(x) = \begin{cases} -3, & \text{if } x \le 0 \\ -3x + 3, & \text{if } x > 0 \end{cases}$

25. $f(x) = \begin{cases} -x, & \text{if } x < 4 \\ -x + 1, & \text{if } x \ge 4 \end{cases}$

26. $f(x) = \begin{cases} 2x + 2, & \text{if } x \le -2 \\ \frac{1}{2}x - 1, & \text{if } x > -2 \end{cases}$

27. $f(x) = \begin{cases} 1, & \text{if } x \le -2 \\ 2x, & \text{if } -2 < x \le 0 \\ -\frac{1}{2}x + 2, & \text{if } x > 0 \end{cases}$

28. $f(x) = \begin{cases} x + 4, & \text{if } x \le -1 \\ -\frac{1}{4}x - \frac{1}{4}, & \text{if } -1 < x < 3 \\ -3, & \text{if } x \ge 3 \end{cases}$

29. $f(x) = \begin{cases} -5, & \text{if } -5 \le x < -3 \\ -3, & \text{if } -3 \le x < -1 \\ -1, & \text{if } -1 \le x < 1 \end{cases}$

30. $f(x) = \begin{cases} 4, & \text{if } 0 < x \le 1 \\ 3, & \text{if } 1 < x \le 2 \\ 2, & \text{if } 2 < x \le 3 \\ 1, & \text{if } 3 < x \le 4 \end{cases}$

31.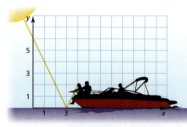
domain: $0 \le x < 8$; range: 3, 4, 5, 6

32.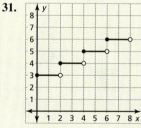
domain: $1 < x \le 5$;
range: $-10, -8, -6, -4$

33.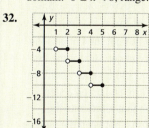
domain: $1 < x \le 12$; range: 1, 5, 6, 9

34–36. See Additional Answers.

37. $y = \begin{cases} -x + 1, & \text{if } x < 0 \\ x + 1, & \text{if } x \ge 0 \end{cases}$

38. $y = \begin{cases} -x - 3, & \text{if } x < 0 \\ x - 3, & \text{if } x \ge 0 \end{cases}$

39. $y = \begin{cases} -x + 2, & \text{if } x < 2 \\ x - 2, & \text{if } x \ge 2 \end{cases}$

40. $y = \begin{cases} -x - 5, & \text{if } x < -5 \\ x + 5, & \text{if } x \ge -5 \end{cases}$

41. $y = \begin{cases} -2x - 6, & \text{if } x < -3 \\ 2x + 6, & \text{if } x \ge -3 \end{cases}$

42. $y = \begin{cases} -4x + 4, & \text{if } x < 1 \\ 4x - 4, & \text{if } x \ge 1 \end{cases}$

43. $y = \begin{cases} 5x - 40, & \text{if } x < 8 \\ -5x + 40, & \text{if } x \ge 8 \end{cases}$

44. $y = \begin{cases} 3x + 18, & \text{if } x < -6 \\ -3x - 18, & \text{if } x \ge -6 \end{cases}$

45. $y = \begin{cases} x - 1, & \text{if } x < 3 \\ -x + 5, & \text{if } x \ge 3 \end{cases}$

46. $y = \begin{cases} -7x - 12, & \text{if } x < -1 \\ 7x + 2, & \text{if } x \ge -1 \end{cases}$

47. a. $f(x) = 2|x - 3|$
    b. $f(x) = \begin{cases} -2x + 6, & \text{if } x < 3 \\ 2x - 6, & \text{if } x \ge 3 \end{cases}$

Section 1.2   17

ANSWERS
48–61. See Additional Answers.

Mini-Assessment

1. Evaluate the function
$$f(x) = \begin{cases} -2x + 1, & \text{if } x \leq 0 \\ x - 3, & \text{if } x > 0 \end{cases}$$
when $x = 1$. $f(1) = -2$

2. Graph $y = \begin{cases} -x - 3, & \text{if } x \leq 0 \\ 2x - 1, & \text{if } x > 0 \end{cases}$.

Describe the domain and range.

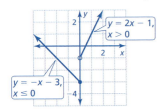

The domain is all real numbers. The range is $y \geq -3$.

3. Write a piecewise function for the graph.

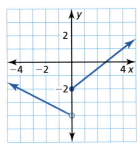

$f(x) = \begin{cases} -0.5x - 4, & \text{if } x < 0 \\ 0.75x - 2, & \text{if } x \geq 0 \end{cases}$

4. You rent a boat for 4 hours. The marina charges $60 for the first hour and $40 for each additional hour. Write and graph a step function that represents the relationship between the number x of hours and the total cost y (in dollars) of renting the boat.
See Additional Answers.

5. The reference beam in Example 5 on page 15 originates at (4, 0) and reflects off a mirror at (5, 2). Write an absolute value function that represents the path of the reference beam.

$g(x) = -2|x - 5| + 2$

18 Chapter 1

---

48. **MODELING WITH MATHEMATICS** You are trying to make a hole in one on the miniature golf green.

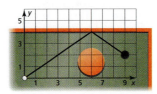

   a. Write an absolute value function that represents the path of the golf ball.
   b. Write the function in part (a) as a piecewise function.

49. **REASONING** The piecewise function f consists of two linear "pieces." The graph of f is shown.

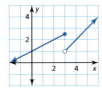

   a. What is the value of $f(-10)$?
   b. What is the value of $f(8)$?

50. **CRITICAL THINKING** Describe how the graph of each piecewise function changes when < is replaced with ≤ and ≥ is replaced with >. Do the domain and range change? Explain.

   a. $f(x) = \begin{cases} x + 2, & \text{if } x < 2 \\ -x - 1, & \text{if } x \geq 2 \end{cases}$

   b. $f(x) = \begin{cases} \frac{1}{2}x + \frac{3}{2}, & \text{if } x < 1 \\ -x + 3, & \text{if } x \geq 1 \end{cases}$

51. **USING STRUCTURE** Graph
$$y = \begin{cases} -x + 2, & \text{if } x \leq -2 \\ |x|, & \text{if } x > -2 \end{cases}.$$
Describe the domain and range.

Maintaining Mathematical Proficiency
Reviewing what you learned in previous grades and lessons

Write the sentence as an inequality. Graph the inequality. *(Skills Review Handbook)*

56. A number r is greater than -12 and no more than 13.

57. A number t is less than or equal to 4 or no less than 18.

Describe the transformation(s) from the graph of $f(x) = |x|$ to the graph of the given function. *(Section 1.1)*

58. $g(x) = |x - 5|$ 59. $h(x) = -2.5|x|$ 60. $p(x) = |x + 1| + 6$ 61. $v(x) = 4|x + 3|$

18 Chapter 1 Functions and Exponents

---

52. **HOW DO YOU SEE IT?** The graph shows the total cost C of making x photocopies at a copy shop.

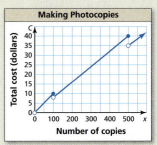

   a. Does it cost more money to make 100 photocopies or 101 photocopies? Explain.
   b. You have $40 to make photocopies. Can you buy more than 500 photocopies? Explain.

53. **USING STRUCTURE** The output y of the *greatest integer function* is the greatest integer less than or equal to the input value x. This function is written as $f(x) = [\![x]\!]$. Graph the function for $-4 \leq x < 4$. Is it a piecewise function? a step function? Explain.

54. **THOUGHT PROVOKING** Explain why
$$y = \begin{cases} 2x - 2, & \text{if } x \leq 3 \\ -3, & \text{if } x \geq 3 \end{cases}$$
does not represent a function. How can you redefine y so that it does represent a function?

55. **MAKING AN ARGUMENT** During a 9-hour snowstorm, it snows at a rate of 1 inch per hour for the first 2 hours, 2 inches per hour for the next 6 hours, and 1 inch per hour for the final hour.

   a. Write and graph a piecewise function that represents the depth of the snow during the snowstorm.
   b. Your friend says 12 inches of snow accumulated during the storm. Is your friend correct? Explain.

If students need help...	If students got it...
Resources by Chapter • Practice A and Practice B • Puzzle Time	**Resources by Chapter** • Enrichment and Extension • Cumulative Review
Student Journal • Practice	Start the *next* Section
Differentiating the Lesson Skills Review Handbook	

Laurie's Notes

Overview of Section 1.3

Introduction
- This lesson presents inverse relations and inverse functions.
- The inverse of a linear equation is a linear equation.
- The inverse of a function can be found algebraically by solving $y = f(x)$ for x. When graphed, inverse functions are symmetric about the line $y = x$.

Formative Assessment Tips
- **Think-Pair-Share:** This technique allows students to share their thinking about a problem with their partners after they have had time to consider the problem alone. Once partners have discussed the problem, small groups or the whole class should discuss the problem.
- The initial time working alone is important for students to develop their own understanding of the mathematics. "Private think time" is what I call it. Once students have engaged in the problem, sharing with partners helps confirm their understanding or perhaps the need to modify their thinking. Sharing their thinking with the whole class is more comfortable for students when they have had the chance to discuss their thinking with partners.
- **Construct Viable Arguments and Critique the Reasoning of Others:** Using this formative assessment technique allows you to check students' conceptual knowledge, and their ability to construct a viable argument.

Pacing Suggestion
- The explorations are straightforward and should be relatively brief in terms of time. Transition to the formal lesson when students finish.

Dynamic Teaching Tools
Dynamic Assessment & Progress Monitoring Tool
Lesson Planning Tool
Interactive Whiteboard Lesson Library
Dynamic Classroom with Dynamic Investigations

What Your Students Will Learn

- Find inverses of relations.
- Explore inverses of linear functions using equations, tables, and graphs.
- Find inverses of linear functions algebraically.

Laurie's Notes

Exploration

Motivate
- ❓ "What was the average daily temperature yesterday?" Answers will vary. Find this information in advance.
- Write a table on the board listing the date (input) and average daily temperature (output) for a specific location. Feel free to make up the data. Make sure that at least two of the temperatures are the same.
- ❓ "Does this table represent a function? Explain." yes; Every date is associated with exactly one average daily temperature.
- ❓ "If we switch the inputs and outputs, would it represent a function? Explain." no; An average daily temperature would be associated with more than one date.

Exploration 1
- The introduction mentions that f and g are inverses, but inverse functions are not defined. From looking at the two tables, students will quickly and imprecisely state, "The x's and y's are switched."
- Probe students to see whether there is alternate language they could use to describe the inverse functions. When the ordered pair (a, b) is a solution of one function, (b, a) is a solution of the inverse function.

Exploration 2
- Students should have little difficulty plotting the two functions.
- ❓ "What observation do you have about the two graphs?" Answers will vary. Some students may observe the symmetry about the line $y = x$, but if they don't, it will be discussed in the lesson.
- The ordered pairs that have integer coordinates should give students enough information to write the equation for each function.

Communicate Your Answer
- Listen for student understanding of how inverse functions are related.
- In Question 4, students should recognize that $f(x) = x + 1$, and they will quickly guess that the inverse of "adding 1" is "subtracting 1." So the inverse will be $g(x) = x - 1$.

Connecting to Next Step
- The explorations will help students develop a beginning understanding of inverse functions. Inverse relations and inverse functions are discussed in the lesson.

1.3 Inverse of a Function

Essential Question How are a function and its inverse related?

EXPLORATION 1 Exploring Inverse Functions

Work with a partner. The functions f and g are *inverses* of each other. Compare the tables of values of the two functions. How are the functions related?

x	0	0.5	1	1.5	2	2.5	3	3.5
f(x)	−2	−0.75	0.5	1.75	3	4.25	5.5	6.75

x	−2	−0.75	0.5	1.75	3	4.25	5.5	6.75
g(x)	0	0.5	1	1.5	2	2.5	3	3.5

EXPLORATION 2 Exploring Inverse Functions

Work with a partner.

a. Use the coordinate plane below to plot the two sets of points represented by the tables in Exploration 1. Then draw a line through each set of points.

b. Describe the relationship between the two graphs.

c. Write an equation for each function.

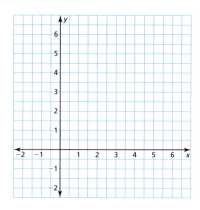

Communicate Your Answer

3. How are a function and its inverse related?

4. A table of values for a function f is given. Create a table of values for a function g, the inverse of f.

x	0	1	2	3	4	5	6	7
f(x)	1	2	3	4	5	6	7	8

5. Sketch the graphs of $f(x) = x + 4$ and its inverse in the same coordinate plane. Then write an equation of the inverse of f. Explain your reasoning.

ATTENDING TO PRECISION
To be proficient in math, you need to communicate precisely with others.

Dynamic Teaching Tools
- Dynamic Assessment & Progress Monitoring Tool
- Lesson Planning Tool
- Interactive Whiteboard Lesson Library
- Dynamic Classroom with Dynamic Investigations

ANSWERS

1. The x-values in each table are the same as the function values in the other table.

2. a.

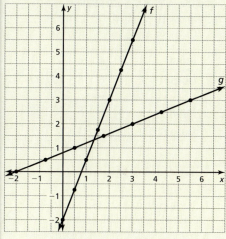

b. The graphs are reflections of each other in the line $y = x$.

c. $f(x) = \frac{5}{2}x - 2$; $g(x) = \frac{2}{5}x + \frac{4}{5}$

3. The x-coordinates of a function are the y-coordinates of its inverse function, and the y-coordinates of the function are the x-coordinates of its inverse function.

4.

x	1	2	3	4	5	6	7	8
g(x)	0	1	2	3	4	5	6	7

5.

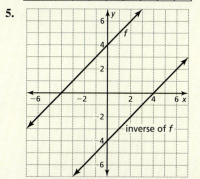

$y = x - 4$; This is an equation of the line containing the points obtained by switching the inputs and outputs of f to get the inverse of f.

English Language Learners

Notebook Development
Have students create an inverse relations and functions page in their notebooks. Students should write the definitions of *inverse relation* and *inverse function* and include examples with each definition.

Extra Example 1
Find the inverse of each relation.

a. (−2, 6), (1, 5), (4, 4), (7, 3), (10, 2)
   (6, −2), (5, 1), (4, 4), (3, 7), (2, 10)

b.
Input	0	2	4	4	2	0
Output	2	4	6	8	10	12

Input	2	4	6	8	10	12
Output	0	2	4	4	2	0

MONITORING PROGRESS ANSWERS

1. (−4, −3), (0, −2), (4, −1), (8, 0), (12, 1), (16, 2), (20, 3)

2.
Input	4	1	0	1	4
Output	−2	−1	0	1	2

1.3 Lesson

What You Will Learn
▶ Find inverses of relations.
▶ Explore inverses of functions.
▶ Find inverses of functions algebraically.

Core Vocabulary
inverse relation, p. 20
inverse function, p. 21
Previous
input
output
inverse operations
reflection
line of reflection

Finding Inverses of Relations

Recall that a relation pairs inputs with outputs. An **inverse relation** switches the input and output values of the original relation.

> **Core Concept**
>
> **Inverse Relation**
> When a relation contains (a, b), the inverse relation contains (b, a).

EXAMPLE 1 Finding Inverses of Relations

Find the inverse of each relation.

a. 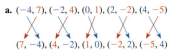   Switch the coordinates of each ordered pair.

   (7, −4), (4, −2), (1, 0), (−2, 2), (−5, 4)   Inverse relation

b.
Input	−1	0	1	2	3	4
Output	5	10	15	20	25	30

Inverse relation:

Input	5	10	15	20	25	30
Output	−1	0	1	2	3	4

Switch the inputs and outputs.

Monitoring Progress Help in English and Spanish at *BigIdeasMath.com*

Find the inverse of the relation.

1. (−3, −4), (−2, 0), (−1, 4), (0, 8), (1, 12), (2, 16), (3, 20)

2.
Input	−2	−1	0	1	2
Output	4	1	0	1	4

Exploring Inverses of Functions

You have previously used given inputs to find corresponding outputs of $y = f(x)$ for various types of functions. You have also used given outputs to find corresponding inputs. Now you will solve equations of the form $y = f(x)$ for x to obtain a formula for finding the input given a specific output of the function f.

Laurie's Notes — Teacher Actions

- Students should recall relations, which were introduced in Math I. When the input and output of a relation are switched, an inverse relation is created. Color coding is used to draw attention to the switch of the input and output.
- **Think-Pair-Share:** Have students answer Questions 1 and 2, and then share and discuss as a class.
- **Turn and Talk:** "If you knew the equation (rule) for a relation, how could you find the equation (rule) for the inverse relation?"

EXAMPLE 2 Writing a Formula for the Input of a Function

Let $f(x) = 2x + 1$. Solve $y = f(x)$ for x. Then find the input when the output is -3.

SOLUTION

$y = 2x + 1$	Set y equal to $f(x)$.
$y - 1 = 2x$	Subtract 1 from each side.
$\dfrac{y-1}{2} = x$	Divide each side by 2.

Find the input when $y = -3$.

$x = \dfrac{-3 - 1}{2}$	Substitute -3 for y.
$= \dfrac{-4}{2}$	Subtract.
$= -2$	Divide.

Check
$f(-2) = 2(-2) + 1$
$= -4 + 1$
$= -3$ ✓

▶ So, the input is -2 when the output is -3.

Monitoring Progress Help in English and Spanish at BigIdeasMath.com

Solve $y = f(x)$ for x. Then find the input when the output is 4.

3. $f(x) = x - 6$ **4.** $f(x) = \tfrac{1}{2}x + 3$ **5.** $f(x) = 10 - x$

UNDERSTANDING MATHEMATICAL TERMS
The term *inverse functions* does not refer to a new type of function. Rather, it describes any pair of functions that are inverses.

In Example 2, notice the steps involved after substituting for x in $y = 2x + 1$ and after substituting for y in $x = \dfrac{y-1}{2}$.

$y = 2x + 1$ $x = \dfrac{y-1}{2}$

Step 1 Multiply by 2.   Step 1 Subtract 1.
Step 2 Add 1.   *inverse operations in the reverse order*   Step 2 Divide by 2.

Notice that these steps *undo* each other. **Inverse functions** are functions that undo each other. In Example 2, you can use the equation solved for x to write the inverse of f by switching the roles of x and y.

$f(x) = 2x + 1$ original function $g(x) = \dfrac{x-1}{2}$ inverse function

Because an inverse function interchanges the input and output values of the original function, the domain and range are also interchanged.

LOOKING FOR A PATTERN
Notice that the graph of the inverse function g is a reflection of the graph of the original function f. The line of reflection is $y = x$.

Original function: $f(x) = 2x + 1$

x	-2	-1	0	1	2
y	-3	-1	1	3	5

Inverse function: $g(x) = \dfrac{x-1}{2}$

x	-3	-1	1	3	5
y	-2	-1	0	1	2

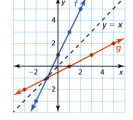

Extra Example 2
Let $f(x) = 3x - 5$. Solve $y = f(x)$ for x. Then find the input when the output is 4.
$x = \dfrac{y + 5}{3}$; The input is 3 when the output is 4.

MONITORING PROGRESS ANSWERS
3. $x = y + 6$; 10
4. $x = 2y - 6$; 2
5. $x = -y + 10$; 6

Laurie's Notes   Teacher Actions

- Pose Example 2. This question is asking students to solve the function for x, not simply find the input for a given output.
- Summarize by saying, "For the function $f(x) = 2x + 1$, the formula to find the input is $\dfrac{y-1}{2} = x$."
- Ask whether students observe a pattern. Most students make the connection that $f(x)$ multiplies the input by 2 and then adds 1. The inverse relation subtracts 1 and divides by 2.
- Discuss inverse operations in the reverse order.
- ❓ "If the ordered pair $(1, -3)$ satisfies the original function, what ordered pair satisfies the inverse function?" $(-3, 1)$

Differentiated Instruction

Organization

Some students may benefit by writing down the steps for finding the inverses of functions algebraically. Have them copy into their notebooks the steps shown in the *Core Concept*. Students can use the steps as a reference as they solve problems like Example 3.

Extra Example 3

Find the inverse of $f(x) = 3x + 4$. The inverse of f is $g(x) = \dfrac{x-4}{3}$, or $g(x) = \dfrac{1}{3}x - \dfrac{4}{3}$.

MONITORING PROGRESS ANSWERS

6. $g(x) = \dfrac{1}{6}x$

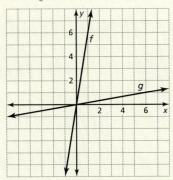

7. $g(x) = -x + 5$

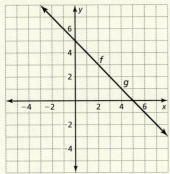

8. $g(x) = 4x + 4$

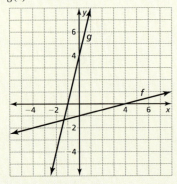

Finding Inverses of Functions Algebraically

Core Concept

Finding Inverses of Functions Algebraically

Step 1 Set y equal to $f(x)$.
Step 2 Switch x and y in the equation.
Step 3 Solve the equation for y.

STUDY TIP
On the previous page, you solved a function for x and switched the roles of x and y to find the inverse function. You can also find the inverse function by switching x and y first, and then solving for y.

EXAMPLE 3 Finding the Inverse of a Function

Find the inverse of $f(x) = 4x - 9$.

SOLUTION

Method 1 Use the method above.

Step 1	$f(x) = 4x - 9$	Write the function.
	$y = 4x - 9$	Set y equal to $f(x)$.
Step 2	$x = 4y - 9$	Switch x and y in the equation.
Step 3	$x + 9 = 4y$	Add 9 to each side.
	$\dfrac{x+9}{4} = y$	Divide each side by 4.

▶ The inverse of f is $g(x) = \dfrac{x+9}{4}$, or $g(x) = \dfrac{1}{4}x + \dfrac{9}{4}$.

Method 2 Use inverse operations in the reverse order.

$f(x) = 4x - 9$ Multiply the input x by 4 and then subtract 9.

To find the inverse, apply inverse operations in the reverse order.

$g(x) = \dfrac{x+9}{4}$ Add 9 to the input x and then divide by 4.

▶ The inverse of f is $g(x) = \dfrac{x+9}{4}$, or $g(x) = \dfrac{1}{4}x + \dfrac{9}{4}$.

Check

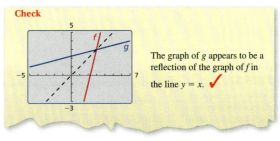

The graph of g appears to be a reflection of the graph of f in the line $y = x$.

Monitoring Progress   Help in English and Spanish at BigIdeasMath.com

Find the inverse of the function. Then graph the function and its inverse.

6. $f(x) = 6x$
7. $f(x) = -x + 5$
8. $f(x) = \dfrac{1}{4}x - 1$

Laurie's Notes — Teacher Actions

- Write the *Core Concept*. Work through Example 3, showing both methods.
- Group students (2–4 in a group) and ask, "Does every function have an inverse? Explain your reasoning." Give groups sufficient time to consider the question and to consider their reasoning. Probe with additional questions as students consider how to investigate their prediction. Some may think about the absolute value function and what its reflection would be in the line $y = x$. Students may need to revise their thinking!

Closure

- **Exit Ticket:** Explain the steps you would take to find the inverse of $f(x) = 3x - 4$. Then find the inverse. *Replace $f(x)$ with y, switch x and y, solve for y, and then replace y with $g(x)$ to obtain $g(x) = \dfrac{1}{3}x + \dfrac{4}{3}$.*

1.3 Exercises

Dynamic Solutions available at *BigIdeasMath.com*

Vocabulary and Core Concept Check

1. **COMPLETE THE SENTENCE** A relation contains the point $(-3, 10)$. The _____ contains the point $(10, -3)$.

2. **DIFFERENT WORDS, SAME QUESTION** Consider the function f represented by the graph. Which is different? Find "both" answers.

   Graph the inverse of the function.

   Reflect the graph of the function in the x-axis.

   Reflect the graph of the function in the line $y = x$.

   Switch the inputs and outputs of the function and graph the resulting function.

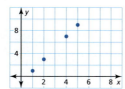

Monitoring Progress and Modeling with Mathematics

In Exercises 3–8, find the inverse of the relation. *(See Example 1.)*

3. $(1, 0), (3, -8), (4, -3), (7, -5), (9, -1)$

4. $(2, 1), (4, -3), (6, 7), (8, 1), (10, -4)$

5.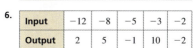

6.
Input	-12	-8	-5	-3	-2
Output	2	5	-1	10	-2

7. Input / Output

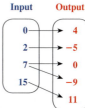

8. Input / Output

   0 → 4
   2 → -5
   7 → 0
   15 → -9
        11

In Exercises 9–12, solve $y = f(x)$ for x. Then find the input when the output is 2. *(See Example 2.)*

9. $f(x) = x + 5$

10. $f(x) = 2x - 3$

11. $f(x) = \frac{1}{4}x - 1$

12. $f(x) = \frac{2}{3}x + 4$

In Exercises 13 and 14, graph the inverse of the function by reflecting the graph in the line $y = x$. Describe the domain and range of the inverse.

13.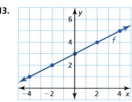

14.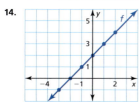

In Exercises 15–22, find the inverse of the function. Then graph the function and its inverse. *(See Example 3.)*

15. $f(x) = 3x$

16. $f(x) = -5x$

17. $f(x) = 4x - 1$

18. $f(x) = -2x + 5$

19. $f(x) = -3x - 2$

20. $f(x) = 2x + 3$

21. $f(x) = \frac{1}{3}x + 8$

22. $f(x) = -\frac{3}{2}x + \frac{7}{2}$

Assignment Guide and Homework Check

ASSIGNMENT
Basic: 1, 2, 3–25 odd, 28, 32–39
Average: 1, 2–26 even, 27, 28, 32–39
Advanced: 1, 2, 8, 12, 14, 22, 24, 27–39

HOMEWORK CHECK
Basic: 3, 9, 13, 17
Average: 6, 10, 14, 18
Advanced: 8, 12, 14, 22

ANSWERS

1. inverse relation

2. Reflect the graph of the function in the x-axis; This is the only statement that does not describe graphing the inverse of the function;

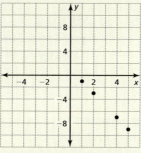

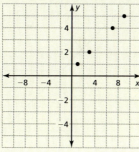

3. $(0, 1), (-8, 3), (-3, 4), (-5, 7), (-1, 9)$

4. $(1, 2), (-3, 4), (7, 6), (1, 8), (-4, 10)$

5.
Input	8	6	0	6	8
Output	-5	-5	0	5	10

6.
Input	2	5	-1	10	-2
Output	-12	-8	-5	-3	-2

7. Input / Output

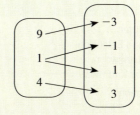

8. Input / Output

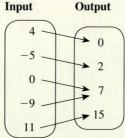

9. $x = y - 5; -3$

10. $x = \dfrac{y + 3}{2}; \dfrac{5}{2}$

11–22. See Additional Answers.

Dynamic Teaching Tools
- Dynamic Assessment & Progress Monitoring Tool
- Interactive Whiteboard Lesson Library
- Dynamic Classroom with Dynamic Investigations

ANSWERS

23. x and y were not switched in the equation; The inverse of f is $g(x) = \dfrac{x-5}{3}$ or $g(x) = \dfrac{x}{3} - \dfrac{5}{3}$.

24. The graph of g should be a reflection in the line $y = x$ instead of the x-axis.

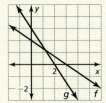

25. $D = 1.33899E$; about 334.75 U.S. dollars

26. a. $w = 0.5x + 11$
    b. $x = 2w - 22$; w represents your hourly wage and x represents how many units you produce each hour.
    c. $15
    d. 38

27. Sample answer: $f(x) = \frac{1}{4}x$

28. A, E; B, C; D, F

29. a.

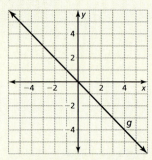

For each point $(a, -a)$ that is on the graph of $g(x) = -x$, the point $(-a, a)$ is also on the graph.

29b–39. See Additional Answers.

Mini-Assessment

1. Find the inverse of the relation.
   (6, 0), (8, 1), (11, 2), (15, 3), (20, 4)
   (0, 6), (1, 8), (2, 11), (3, 15), (4, 20)

2. Let $f(x) = -x + 2$. Solve $y = f(x)$ for x. Then find the input when the output is -5. $x = 2 - y$; input: 7

3. Find the inverse of $f(x) = 4x - 3$.
   $g(x) = \dfrac{x+3}{4}$, or $g(x) = \dfrac{1}{4}x + \dfrac{3}{4}$

24 Chapter 1

---

23. **ERROR ANALYSIS** Describe and correct the error in finding the inverse of the function $f(x) = 3x + 5$.

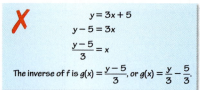

24. **ERROR ANALYSIS** Describe and correct the error in graphing the inverse g of the function f.

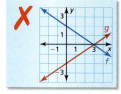

25. **MODELING WITH MATHEMATICS** The euro is the unit of currency for the European Union. On a certain day, the number E of euros that could be obtained for D U.S. dollars was represented by the formula shown.

$$E = 0.74683D$$

Solve the formula for D. Then find the number of U.S. dollars that could be obtained for 250 euros on that day.

26. **MODELING WITH MATHEMATICS** You work at a manufacturing plant and earn $11 per hour plus $0.50 for each unit you produce per hour.
    a. Write a linear model that represents your hourly wage as a function of the number of units you produce per hour.
    b. Find the inverse of the function you wrote in part (a). What does each variable in the inverse function represent?
    c. What is your hourly wage when you produce 8 units per hour?
    d. How many units must you produce per hour to double your hourly wage in part (c)?

27. **OPEN-ENDED** Write a function such that the graph of its inverse is a line with a slope of 4.

Maintaining Mathematical Proficiency
Reviewing what you learned in previous grades and lessons

Evaluate the expression. *(Skills Review Handbook)*

32. 12^5 33. -11^3 34. $\left(\dfrac{5}{6}\right)^4$ 35. $-\left(\dfrac{1}{3}\right)^6$

Write the number in scientific notation. *(Skills Review Handbook)*

36. 84,375 37. 0.00944 38. 400,700,000 39. 0.000000082

24 Chapter 1 Functions and Exponents

---

28. **HOW DO YOU SEE IT?** Pair the graph of each function with the graph of its inverse.

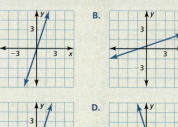

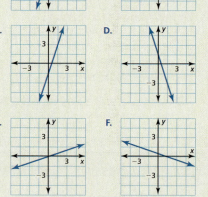

29. **CRITICAL THINKING** Consider the function $g(x) = -x$.
    a. Graph $g(x) = -x$ and explain why it is its own inverse.
    b. Graph other linear functions that are their own inverses. Write equations of the lines you graph.
    c. Use your results from part (b) to write a general equation that describes the family of linear functions that are their own inverses.

30. **THOUGHT PROVOKING** Find a formula (for instance, from geometry or physics) that has an inverse function. Explain what the variables in your formula represent.

31. **REASONING** Show that the inverse of any linear function $f(x) = mx + b$, where $m \neq 0$, is also a linear function. Write the slope and y-intercept of the graph of the inverse in terms of m and b.

If students need help...	If students got it...
Resources by Chapter • Practice A and Practice B • Puzzle Time	Resources by Chapter • Enrichment and Extension • Cumulative Review
Student Journal • Practice	Start the *next* Section
Differentiating the Lesson Skills Review Handbook	

1.1–1.3 What Did You Learn?

Dynamic Teaching Tools
Dynamic Assessment & Progress Monitoring Tool
Interactive Whiteboard Lesson Library
Dynamic Classroom with Dynamic Investigations

Core Vocabulary

absolute value function, *p. 4*
vertex, *p. 4*
vertex form, *p. 6*
piecewise function, *p. 12*
step function, *p. 14*
inverse relation, *p. 20*
inverse function, *p. 21*

Core Concepts

Section 1.1
Absolute Value Function, *p. 4*
Vertex Form of an Absolute Value Function, *p. 6*

Section 1.2
Piecewise Function, *p. 12*
Step Function, *p. 14*
Writing Absolute Value Functions, *p. 15*

Section 1.3
Inverse Relation, *p. 20*
Finding Inverses of Functions Algebraically, *p. 22*

Mathematical Practices

1. Describe the definitions you used when you explained your answer in Exercise 53 on page 18.
2. What external resources could you use to check the reasonableness of your answer in Exercise 25 on page 24?

ANSWERS
1. The definition of piecewise function, which is a function defined by two or more equations, and the definition of step function, which is a special piecewise function defined by a constant value over each part of its domain.
2. *Sample answer:* The Internet or a smart phone application that performs real-time currency conversions.

Making Note Cards

Invest in three different colors of note cards. Use one color for each of the following: vocabulary words, rules, and calculator keystrokes.

- Using the first color of note cards, write a vocabulary word on one side of a card. On the other side, write the definition and an example. If possible, put the definition in your own words.
- Using the second color of note cards, write a rule on one side of a card. On the other side, write an explanation and an example.
- Using the third color of note cards, write a calculation on one side of a card. On the other side, write the keystrokes required to perform the calculation.

Use the note cards as references while completing your homework. Quiz yourself once a day.

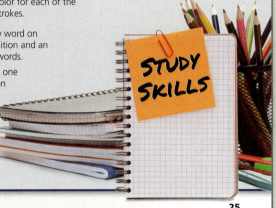

ANSWERS

1. The transformations are a horizontal translation 5 units left then a vertical translation 4 units up.

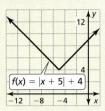

2. The transformations are a horizontal translation 3 units right, then a vertical shrink by a factor of $\frac{1}{2}$, then a reflection in the x-axis, then a vertical translation 6 units down.

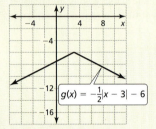

3. The transformations are a horizontal translation 2 units left, then a reflection in the y-axis, then a vertical translation 4 units down.

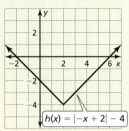

4. $g(x) = |x - 9|$
5. $g(x) = |x| + 4.5$
6. $g(x) = \left|\frac{1}{5}x\right|$
7. $g(x) = -\frac{1}{4}|x|$
8.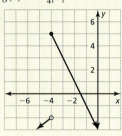

domain: all real numbers; range: $y \leq 5$

9.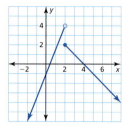

domain: $-4 \leq x < 2$; range: $-3, -2, -1, 0$

10. $y = \begin{cases} \frac{5}{2}x - 1, & \text{if } x < 2 \\ -x + 4, & \text{if } x \geq 2 \end{cases}$

11. $y = \begin{cases} 4x + 4, & \text{if } x < -1 \\ -4x - 4, & \text{if } x \geq 1 \end{cases}$

12.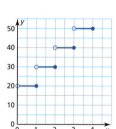

13–18. See Additional Answers.

1.1–1.3 Quiz

Describe the transformations from the graph of $f(x) = |x|$ to the graph of the given function. Then graph the given function. *(Section 1.1)*

1. $f(x) = |x + 5| + 4$
2. $g(x) = -\frac{1}{2}|x - 3| - 6$
3. $h(x) = |-x + 2| - 4$

Write an equation that represents the given transformation(s) of the graph of $f(x) = |x|$. *(Section 1.1)*

4. horizontal translation 9 units right
5. vertical translation 4.5 units up
6. horizontal stretch by a factor of 5
7. vertical shrink by a factor of $\frac{1}{4}$ and a reflection in the x-axis

Graph the function. Describe the domain and range. *(Section 1.2)*

8. $y = \begin{cases} \frac{3}{4}x + 1, & \text{if } x < -4 \\ -2x - 3, & \text{if } x \geq -4 \end{cases}$

9. $y = \begin{cases} -3, & \text{if } -4 \leq x < -2 \\ -2, & \text{if } -2 \leq x < -1 \\ -1, & \text{if } -1 \leq x < 1 \\ 0, & \text{if } 1 \leq x < 2 \end{cases}$

10. Write a piecewise function represented by the graph shown. *(Section 1.2)*

11. Write the function $y = -4|x + 1|$ as a piecewise function. *(Section 1.2)*

12. Find the inverse of the relation shown in the table. *(Section 1.3)*

Input	−4	−2	0	2	4
Output	6	3	0	−3	−6

Find the inverse of the function. Then graph the function and its inverse. *(Section 1.3)*

13. $k(x) = -\frac{1}{2}x$
14. $p(x) = 3x - 5$
15. $r(x) = 6 - 0.2x$

16. Write two absolute value functions whose graphs form a square in a coordinate plane. *(Section 1.1)*

17. Describe a real-life situation that can be represented by the graph shown. *(Section 1.2)*

18. You are ordering customized bracelets for a fundraiser. The total cost (in dollars) of ordering x bracelets is represented by the function below. *(Section 1.2)*

$$f(x) = \begin{cases} 0.9x + 25, & \text{if } 0 \leq x < 100 \\ 0.65x + 25, & \text{if } 100 \leq x < 500 \\ 0.4x + 25, & \text{if } 500 \leq x < 1000 \\ 0.1x + 25, & \text{if } x \geq 1000 \end{cases}$$

a. Find and interpret $f(35)$.
b. Determine the costs of ordering 100, 500, and 2500 bracelets.
c. What do you think the 25s in the function represent?

Laurie's Notes

Overview of Section 1.4

Introduction
- Students have worked with exponents previously, and the properties presented in this lesson should be familiar from Grade 8 mathematics.
- Seven properties of exponents are presented in this lesson. Each of the properties is practiced independently to simplify expressions. Applying more than one property to the simplifying of an expression is where students can become challenged.

Formative Assessment Tips
- **Whiteboarding:** Whiteboards can be used to provide individual responses, or used with small groups to encourage student collaboration and consensus on a problem or solution method. Whiteboards can be used at the beginning of class for warm-ups or throughout the lesson to elicit student responses. Unlike writing on scrap paper (individual response) or chart paper (group response), responses can be erased and modified easily. As understanding progresses, responses can reflect this growth.
- Use boards for more than quick responses. Sizable boards can be used to communicate thinking, providing evidence of how a problem was solved. When a student displays their whiteboard in front of the room, classmates can critique their reasoning or method of solution.

Pacing Suggestion
- The explorations provide an opportunity for students to develop a conceptual understanding of the properties of exponents. Complete all of the explorations and then transition to the formal lesson.

Dynamic Teaching Tools
Dynamic Assessment & Progress Monitoring Tool
Lesson Planning Tool
Interactive Whiteboard Lesson Library
Dynamic Classroom with Dynamic Investigations

What Your Students Will Learn

- Evaluate and simplify expressions involving zero and negative exponents.
- Use the properties of exponents to rewrite expressions using only positive exponents.
- Solve real-life problems involving exponents.

Laurie's Notes

Exploration

Motivate
- The $10,000 bill, which is no longer in circulation, would be much easier to carry than the same amount in pennies.
- **?** Ask a few questions about money.
  - "How many pennies equal $10,000?" $100 \times 10,000 = 1,000,000$, or 10^6
  - "How many dimes equal $10,000?" $10 \times 10,000 = 100,000$, or 10^5
  - "How many $10 bills equal $10,000?" $\frac{1}{10} \times 10,000 = 1000$, or 10^3
  - "How many $100 bills equal $10,000?" $\frac{1}{100} \times 10,000 = 100$, or 10^2

Discuss
- Inductive reasoning is the use of specific examples to make a general statement. Encourage students to focus on the directions for each part and the language that is used. Each part begins with a question and ends with a statement about writing a general rule that answers the question that was posed. In writing the general rule, students may notice a pattern with the exponents, but they may need help in writing the rule using variables for the exponents. It may be necessary to model how to write the first rule after students have recognized the pattern.

Exploration 1
- For each of the problems in each part, students may need to expand the expression written and then simplify to see a pattern. For example:
  $(2^2)(2^3) = (2 \cdot 2)(2 \cdot 2 \cdot 2) = 2^5$
  $\frac{4^3}{4^2} = \frac{4 \cdot \cancel{4} \cdot \cancel{4}}{\cancel{4} \cdot \cancel{4}} = 4$
  $(2^2)^4 = (2^2)(2^2)(2^2)(2^2) = (2 \cdot 2)(2 \cdot 2)(2 \cdot 2)(2 \cdot 2) = 2^8$
- Some students may not need to write the expanded version, as they will recognize a pattern, or understand the structure, and will be able to write the general rule.
- **FYI:** All the quotients in the exploration result in nonnegative exponents. The formal lesson includes quotients that result in negative exponents.

Communicate Your Answer
- **Look For and Express Regularity in Repeated Reasoning:** Both questions involve repeated reasoning.

Connecting to Next Step
- Each part of the exploration is related to a property of exponents which will be stated in the formal lesson.

1.4 Properties of Exponents

Essential Question How can you write general rules involving properties of exponents?

EXPLORATION 1 Writing Rules for Properties of Exponents

Work with a partner.

WRITING GENERAL RULES
To be proficient in math, you need to understand and use stated assumptions, definitions, and previously established results in writing general rules.

a. What happens when you multiply two powers with the same base? Write the product of the two powers as a single power. Then write a *general rule* for finding the product of two powers with the same base.

  i. $(2^2)(2^3) = $ _____ ii. $(4^1)(4^5) = $ _____
  iii. $(5^3)(5^5) = $ _____ iv. $(x^2)(x^6) = $ _____

b. What happens when you divide two powers with the same base? Write the quotient of the two powers as a single power. Then write a *general rule* for finding the quotient of two powers with the same base.

  i. $\dfrac{4^3}{4^2} = $ _____ ii. $\dfrac{2^5}{2^2} = $ _____
  iii. $\dfrac{x^6}{x^3} = $ _____ iv. $\dfrac{3^4}{3^4} = $ _____

c. What happens when you find a power of a power? Write the expression as a single power. Then write a *general rule* for finding a power of a power.

  i. $(2^2)^4 = $ _____ ii. $(7^3)^2 = $ _____
  iii. $(y^3)^3 = $ _____ iv. $(x^4)^2 = $ _____

d. What happens when you find a power of a product? Write the expression as the product of two powers. Then write a *general rule* for finding a power of a product.

  i. $(2 \cdot 5)^2 = $ _____ ii. $(5 \cdot 4)^3 = $ _____
  iii. $(6a)^2 = $ _____ iv. $(3x)^2 = $ _____

e. What happens when you find a power of a quotient? Write the expression as the quotient of two powers. Then write a *general rule* for finding a power of a quotient.

  i. $\left(\dfrac{2}{3}\right)^2 = $ _____ ii. $\left(\dfrac{4}{3}\right)^3 = $ _____
  iii. $\left(\dfrac{x}{2}\right)^3 = $ _____ iv. $\left(\dfrac{a}{b}\right)^4 = $ _____

Communicate Your Answer

2. How can you write general rules involving properties of exponents?

3. There are 3^3 small cubes in the cube below. Write an expression for the number of small cubes in the large cube at the right.

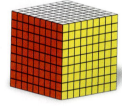

Dynamic Teaching Tools
- Dynamic Assessment & Progress Monitoring Tool
- Lesson Planning Tool
- Interactive Whiteboard Lesson Library
- Dynamic Classroom with Dynamic Investigations

ANSWERS

1. **a.** $a^m a^n = a^{m+n}$
   i. 2^5
   ii. 4^6
   iii. 5^8
   iv. x^8

   b. $\dfrac{a^m}{a^n} = a^{m-n}$
   i. 4^1
   ii. 2^3
   iii. x^3
   iv. 3^0

   c. $(a^m)^n = a^{mn}$
   i. 2^8
   ii. 7^6
   iii. y^9
   iv. x^8

   d. $(ab)^m = a^m b^m$
   i. $2^2 \cdot 5^2$
   ii. $5^3 \cdot 4^3$
   iii. $6^2 a^2$
   iv. $3^2 x^2$

   e. $\left(\dfrac{a}{b}\right)^m = \dfrac{a^m}{b^m}$
   i. $\dfrac{2^2}{3^2}$
   ii. $\dfrac{4^3}{3^3}$
   iii. $\dfrac{x^3}{2^3}$
   iv. $\dfrac{a^4}{b^4}$

2. Try several examples to find a pattern, then express the pattern using variables.

3. 9^3

Differentiated Instruction

Inclusion
Challenge students to write expressions that are equivalent to a given simplified expression. For example, give students the expression x^2y. Have them write two different expressions that can be simplified to x^2y.

Sample answer: $\dfrac{x^3}{xy^{-1}}$, $\dfrac{x^{-1}y^2}{x^{-3}y}$

Extra Example 1
Evaluate each expression.

a. 2.5^0 1

b. $(-3)^{-2}$ $\dfrac{1}{9}$

Extra Example 2
Simplify the expression $\dfrac{2^0 x^2}{y^{-2}}$. Write your answer using only positive exponents. x^2y^2

MONITORING PROGRESS ANSWERS
1. 1
2. $\dfrac{1}{27}$
3. -4
4. $\dfrac{1}{9x^5}$

1.4 Lesson

What You Will Learn
- Use zero and negative exponents.
- Use the properties of exponents.
- Solve real-life problems involving exponents.

Core Vocabulary
Previous
power
exponent
base
scientific notation

Using Zero and Negative Exponents

Core Concept

Zero Exponent

Words For any nonzero number a, $a^0 = 1$. The power 0^0 is undefined.

Numbers $4^0 = 1$ **Algebra** $a^0 = 1$, where $a \neq 0$

Negative Exponents

Words For any integer n and any nonzero number a, a^{-n} is the reciprocal of a^n.

Numbers $4^{-2} = \dfrac{1}{4^2}$ **Algebra** $a^{-n} = \dfrac{1}{a^n}$, where $a \neq 0$

EXAMPLE 1 Using Zero and Negative Exponents

Evaluate each expression.

a. 6.7^0 b. $(-2)^{-4}$

SOLUTION

a. $6.7^0 = 1$ Definition of zero exponent

b. $(-2)^{-4} = \dfrac{1}{(-2)^4}$ Definition of negative exponent

$= \dfrac{1}{16}$ Simplify.

EXAMPLE 2 Simplifying an Expression

Simplify the expression $\dfrac{4x^0}{y^{-3}}$. Write your answer using only positive exponents.

SOLUTION

$\dfrac{4x^0}{y^{-3}} = 4x^0y^3$ Definition of negative exponent

$= 4y^3$ Definition of zero exponent

Monitoring Progress Help in English and Spanish at *BigIdeasMath.com*

Evaluate the expression.

1. $(-9)^0$ 2. 3^{-3} 3. $\dfrac{-5^0}{2^{-2}}$

4. Simplify the expression $\dfrac{3^{-2}x^{-5}}{y^0}$. Write your answer using only positive exponents.

28   Chapter 1   Functions and Exponents

Laurie's Notes Teacher Actions

- Write and discuss the two properties. Zero and negative exponents can be the most challenging for students to make sense of.
- Take time to discuss negative exponents in the denominator. A corollary would be $\dfrac{1}{a^{-n}} = a^n$. This situation shows up in Example 2.
- **Think-Pair-Share:** Have students answer Questions 1–4, and then share and discuss as a class.
- Students may ask what order to apply the properties. Different orders are possible. Model this with Question 4.

Using the Properties of Exponents

REMEMBER
The expression x^3 is called a *power*. The *base*, x, is used as a factor 3 times because the *exponent* is 3.

Core Concept

Product of Powers Property

Let a be a real number, and let m and n be integers.

Words To multiply powers with the same base, add their exponents.

Numbers $4^6 \cdot 4^3 = 4^{6+3} = 4^9$ **Algebra** $a^m \cdot a^n = a^{m+n}$

Quotient of Powers Property

Let a be a nonzero real number, and let m and n be integers.

Words To divide powers with the same base, subtract their exponents.

Numbers $\dfrac{4^6}{4^3} = 4^{6-3} = 4^3$ **Algebra** $\dfrac{a^m}{a^n} = a^{m-n}$, where $a \neq 0$

Power of a Power Property

Let a be a real number, and let m and n be integers.

Words To find a power of a power, multiply the exponents.

Numbers $(4^6)^3 = 4^{6 \cdot 3} = 4^{18}$ **Algebra** $(a^m)^n = a^{mn}$

EXAMPLE 3 Using Properties of Exponents

Simplify each expression. Write your answer using only positive exponents.

a. $3^2 \cdot 3^6$ b. $\dfrac{(-4)^2}{(-4)^7}$ c. $(z^4)^{-3}$

SOLUTION

a. $3^2 \cdot 3^6 = 3^{2+6}$ Product of Powers Property
   $= 3^8 = 6561$ Simplify.

b. $\dfrac{(-4)^2}{(-4)^7} = (-4)^{2-7}$ Quotient of Powers Property
   $= (-4)^{-5}$ Simplify.
   $= \dfrac{1}{(-4)^5} = -\dfrac{1}{1024}$ Definition of negative exponent

c. $(z^4)^{-3} = z^{4 \cdot (-3)}$ Power of a Power Property
   $= z^{-12}$ Simplify.
   $= \dfrac{1}{z^{12}}$ Definition of negative exponent

Monitoring Progress Help in English and Spanish at *BigIdeasMath.com*

Simplify the expression. Write your answer using only positive exponents.

5. $10^4 \cdot 10^{-6}$ 6. $x^9 \cdot x^{-9}$ 7. $\dfrac{-5^8}{-5^4}$

8. $\dfrac{y^6}{y^7}$ 9. $(6^{-2})^{-1}$ 10. $(w^{12})^5$

English Language Learners

Build on Past Knowledge
Remind students from their study of exponents that the expressions $(-4)^2$ and -4^2 are not equivalent. The expression $(-4)^2 = (-4) \cdot (-4) = 16$ while the expression $-4^2 = -(4 \cdot 4) = -16$. Remembering this expanded form will help students as they evaluate expressions similar to the ones in Example 3b and Exercise 7 in *Monitoring Progress*.

Extra Example 3

Simplify each expression. Write your answer using only positive exponents.

a. $2^3 \cdot 2^5$ $2^8 = 256$

b. $\dfrac{(-7)^2}{(-7)^4}$ $\dfrac{1}{(-7)^2} = \dfrac{1}{49}$

c. $(d^{-3})^5$ $d^{-15} = \dfrac{1}{d^{15}}$

MONITORING PROGRESS ANSWERS

5. $\dfrac{1}{100}$
6. 1
7. 625
8. $\dfrac{1}{y}$
9. 36
10. w^{60}

Laurie's Notes — Teacher Actions

- These three properties are grouped together because they are about an operation being done to one or more powers.
- The expression in part (b) is different from the examples in the exploration because the exponent in the denominator is greater than the exponent in the numerator.

COMMON ERROR Students believe $(-4)^{-5}$ is the final answer. Remind them that answers must have positive exponents.

- **Whiteboarding:** Have students answer Questions 5–7 on their whiteboards and then reveal their answers. Repeat for Questions 8–10. This feedback will help you determine whether students are ready to move on.

Extra Example 4

Simplify each expression. Write your answer using only positive exponents.

a. $(-3x)^3 \quad -27x^3$

b. $\left(\dfrac{b}{-3}\right)^4 \quad \dfrac{b^4}{81}$

c. $\left(\dfrac{2x}{5}\right)^3 \quad \dfrac{8x^3}{125}$

d. $\left(\dfrac{3c}{4}\right)^{-2} \quad \dfrac{16}{9c^2}$

MONITORING PROGRESS ANSWERS

11. $\dfrac{1}{1000y^3}$

12. $\dfrac{-1024}{n^5}$

13. $\dfrac{1}{32k^{10}}$

14. $\dfrac{49}{36c^2}$

Core Concept

Power of a Product Property

Let a and b be real numbers, and let m be an integer.

Words To find a power of a product, find the power of each factor and multiply.

Numbers $(3 \cdot 2)^5 = 3^5 \cdot 2^5$ **Algebra** $(ab)^m = a^m b^m$

Power of a Quotient Property

Let a and b be real numbers with $b \neq 0$, and let m be an integer.

Words To find the power of a quotient, find the power of the numerator and the power of the denominator and divide.

Numbers $\left(\dfrac{3}{2}\right)^5 = \dfrac{3^5}{2^5}$ **Algebra** $\left(\dfrac{a}{b}\right)^m = \dfrac{a^m}{b^m}$, where $b \neq 0$

EXAMPLE 4   Using Properties of Exponents

Simplify each expression. Write your answer using only positive exponents.

a. $(-1.5y)^2$ b. $\left(\dfrac{a}{-10}\right)^3$ c. $\left(\dfrac{3d}{2}\right)^4$ d. $\left(\dfrac{2x}{3}\right)^{-5}$

SOLUTION

a. $(-1.5y)^2 = (-1.5)^2 \cdot y^2$ Power of a Product Property

            $= 2.25y^2$ Simplify.

b. $\left(\dfrac{a}{-10}\right)^3 = \dfrac{a^3}{(-10)^3}$ Power of a Quotient Property

            $= -\dfrac{a^3}{1000}$ Simplify.

ANOTHER WAY

Because the exponent is negative, you could find the reciprocal of the base first. Then simplify.

$\left(\dfrac{2x}{3}\right)^{-5} = \left(\dfrac{3}{2x}\right)^5 = \dfrac{243}{32x^5}$

c. $\left(\dfrac{3d}{2}\right)^4 = \dfrac{(3d)^4}{2^4}$ Power of a Quotient Property

            $= \dfrac{3^4 d^4}{2^4}$ Power of a Product Property

            $= \dfrac{81d^4}{16}$ Simplify.

d. $\left(\dfrac{2x}{3}\right)^{-5} = \dfrac{(2x)^{-5}}{3^{-5}}$ Power of a Quotient Property

            $= \dfrac{3^5}{(2x)^5}$ Definition of negative exponent

            $= \dfrac{3^5}{2^5 x^5}$ Power of a Product Property

            $= \dfrac{243}{32x^5}$ Simplify.

Monitoring Progress Help in English and Spanish at *BigIdeasMath.com*

Simplify the expression. Write your answer using only positive exponents.

11. $(10y)^{-3}$ 12. $\left(-\dfrac{4}{n}\right)^5$ 13. $\left(\dfrac{1}{2k^2}\right)^5$ 14. $\left(\dfrac{6c}{7}\right)^{-2}$

Laurie's Notes   Teacher Actions

- **Common Misconception:** Students often refer to both of these properties as "distributing" the exponent. This is not!
- Work through each part as shown, pausing to ask questions. There are numerous places where common errors occur, such as when raising the product $(3d)^4$—students raise d to the fourth but do not raise 3 to the fourth.
- Students may suggest or ask about alternate solution methods. Try *Another Way* for the expression in part (d).
- **Whiteboarding:** Have students answer Questions 11–14 on their whiteboards and then reveal their answers.

Solving Real-Life Problems

EXAMPLE 5 Simplifying a Real-Life Expression

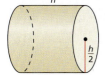
Volume = ?

$2\pi r^3 \quad \pi h^3 2^{-2} \quad \pi h 4^{-1}$

$\dfrac{\pi h^2}{4} \quad \dfrac{\pi h^3}{4} \quad \dfrac{\pi h^3}{2}$

Which of the expressions shown represent the volume of the cylinder, where r is the radius and h is the height?

SOLUTION

$V = \pi r^2 h$ Formula for the volume of a cylinder

$= \pi \left(\dfrac{h}{2}\right)^2 (h)$ Substitute $\dfrac{h}{2}$ for r.

$= \pi \left(\dfrac{h^2}{2^2}\right)(h)$ Power of a Quotient Property

$= \dfrac{\pi h^3}{4}$ Simplify.

Any expression equivalent to $\dfrac{\pi h^3}{4}$ represents the volume of the cylinder.

- You can use the properties of exponents to write $\pi h^3 2^{-2}$ as $\dfrac{\pi h^3}{4}$.
- Note $h = 2r$. When you substitute $2r$ for h in $\dfrac{\pi h^3}{4}$, you can write $\dfrac{\pi(2r)^3}{4}$ as $2\pi r^3$.
- None of the other expressions are equivalent to $\dfrac{\pi h^3}{4}$.

▶ The expressions $2\pi r^3$, $\pi h^3 2^{-2}$, and $\dfrac{\pi h^3}{4}$ represent the volume of the cylinder.

REMEMBER
A number is written in scientific notation when it is of the form $a \times 10^b$, where $1 \le a < 10$ and b is an integer.

EXAMPLE 6 Solving a Real-Life Problem

A jellyfish emits about 1.25×10^8 particles of light, or photons, in 6.25×10^{-4} second. How many photons does the jellyfish emit each second? Write your answer in scientific notation and in standard form.

SOLUTION

Divide to find the unit rate.

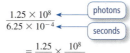

$\dfrac{1.25 \times 10^8}{6.25 \times 10^{-4}}$ Write the rate.

$= \dfrac{1.25}{6.25} \times \dfrac{10^8}{10^{-4}}$ Rewrite.

$= 0.2 \times 10^{12}$ Simplify.

$= 2 \times 10^{11}$ Write in scientific notation.

▶ The jellyfish emits 2×10^{11}, or 200,000,000,000 photons per second.

Monitoring Progress Help in English and Spanish at BigIdeasMath.com

15. Write two expressions that represent the area of a base of the cylinder in Example 5.

16. It takes the Sun about 2.3×10^8 years to orbit the center of the Milky Way. It takes Pluto about 2.5×10^2 years to orbit the Sun. How many times does Pluto orbit the Sun while the Sun completes one orbit around the center of the Milky Way? Write your answer in scientific notation.

Section 1.4 Properties of Exponents 31

Extra Example 5

Which of the expressions shown represent the volume of the cone, where r is the radius and h is the height?

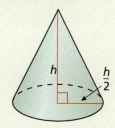

$\dfrac{\pi h^3}{12} \quad \dfrac{\pi r^3}{3} \quad \dfrac{\pi h^2}{6}$

$\dfrac{\pi h^2}{12} \quad \dfrac{2\pi r^3}{3} \quad 8\pi r^3 \quad \dfrac{\pi h^3}{12}, \dfrac{2\pi r^3}{3}$

Extra Example 6

A laptop computer uses approximately 8.4×10^{-3} kilowatt-hours of energy in 3.5×10^{-1} hours. How many kilowatt-hours of energy does the laptop use per hour? Write your answer in scientific notation and in standard form. The laptop uses 2.4×10^{-2}, or 0.024 kilowatt-hour of energy per hour.

MONITORING PROGRESS ANSWERS

15. πr^2; $\dfrac{\pi h^2}{4}$

16. 9.2×10^5

Laurie's Notes Teacher Actions

? "How do you find the volume of a cylinder?" area of the base times the height Have students work with their partners to simplify the six expressions to determine which are equivalent to $\dfrac{\pi h^3}{4}$.

- **Look For and Make Use of Structure:** In Example 6, help students see this problem as simplifying $\dfrac{a \cdot b^n}{c \cdot d^m}$, which is equivalent to $\dfrac{a}{c} \cdot \dfrac{b^n}{d^m}$.

Closure

- Have students write a statement about their understanding of this lesson. They should begin with "I feel confident about …" or "I am still a bit confused about …." Collect their statements.

Assignment Guide and Homework Check

ASSIGNMENT

Basic: 1–4, 5–45 odd, 55, 64, 67, 70–75

Average: 1–3, 4–44 even, 45, 46–58 even, 64, 67, 70–75

Advanced: 1–4, 10–24 even, 32–44 even, 45, 46–64 even, 65–75

HOMEWORK CHECK

Basic: 5, 13, 23, 37, 55

Average: 8, 16, 24, 38, 56

Advanced: 10, 18, 32, 40, 56

ANSWERS

1. Product of Powers Property, Power of a Power Property, definition of negative exponents; Use the Product of Powers Property to simplify the expression inside the parentheses to 4^4. Then, use the Power of a Power Property to simplify the entire expression to 4^{-8}. Then, use the definition of negative exponents to produce the final answer, $\frac{1}{4^8} = \frac{1}{65{,}536}$.

2. The Power of a Product Property is used when finding a power of a product by finding the power of each factor and multiplying.

3. The Quotient of Powers Property is used when dividing powers that have the same base. The answer is the common base raised to the difference of the exponents of the numerator and denominator.

4. Simplify $3^6 \cdot 3$; 3^{18}; 3^9

5. 1
6. 1
7. $\frac{1}{625}$
8. $-\frac{1}{32}$
9. $\frac{1}{16}$
10. $-\frac{1}{5}$
11. $-\frac{4}{3}$
12. $\frac{81}{64}$
13. $\frac{1}{x^7}$
14. 1
15. $\frac{9}{y^3}$
16. $\frac{15}{c^8}$

1.4 Exercises

Dynamic Solutions available at BigIdeasMath.com

Vocabulary and Core Concept Check

1. **VOCABULARY** Which definitions or properties would you use to simplify the expression $(4^8 \cdot 4^{-4})^{-2}$? Explain.

2. **WRITING** Explain when and how to use the Power of a Product Property.

3. **WRITING** Explain when and how to use the Quotient of Powers Property.

4. **DIFFERENT WORDS, SAME QUESTION** Which is different? Find "both" answers.

   Simplify $3^3 \cdot 3^6$. | Simplify 3^{3+6}. | Simplify $3^6 \cdot 3$. | Simplify $3^6 \cdot 3^3$.

Monitoring Progress and Modeling with Mathematics

In Exercises 5–12, evaluate the expression. (See Example 1.)

5. $(-7)^0$
6. 4^0
7. 5^{-4}
8. $(-2)^{-5}$
9. $\frac{2^{-4}}{4^0}$
10. $\frac{5^{-1}}{-9^0}$
11. $\frac{-3^{-3}}{6^{-2}}$
12. $\frac{(-8)^{-2}}{3^{-4}}$

In Exercises 13–22, simplify the expression. Write your answer using only positive exponents. (See Example 2.)

13. x^{-7}
14. y^0
15. $9x^0y^{-3}$
16. $15c^{-8}d^0$
17. $\frac{2^{-2}m^{-3}}{n^0}$
18. $\frac{10^0 r^{-11} s}{3^2}$
19. $\frac{4^{-3}a^0}{b^{-7}}$
20. $\frac{p^{-8}}{7^{-2}q^{-9}}$
21. $\frac{2^2 y^{-6}}{8^{-1} z^0 x^{-7}}$
22. $\frac{13x^{-5}y^0}{5^{-3}z^{-10}}$

In Exercises 23–32, simplify the expression. Write your answer using only positive exponents. (See Example 3.)

23. $\frac{5^6}{5^2}$
24. $\frac{(-6)^8}{(-6)^5}$
25. $(-9)^2 \cdot (-9)^2$
26. $4^{-5} \cdot 4^5$
27. $(p^6)^4$
28. $(s^{-5})^3$
29. $6^{-8} \cdot 6^5$
30. $-7 \cdot (-7)^{-4}$
31. $\frac{x^5}{x^4} \cdot x$
32. $\frac{z^8 \cdot z^2}{z^5}$

33. **USING PROPERTIES** A microscope magnifies an object 10^5 times. The length of an object is 10^{-7} meter. What is its magnified length?

34. **USING PROPERTIES** The area of the rectangular computer chip is $112a^3b^2$ square microns. What is the length?

width = $8ab$ microns

ERROR ANALYSIS In Exercises 35 and 36, describe and correct the error in simplifying the expression.

35.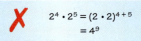

36.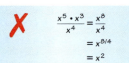

32   Chapter 1   Functions and Exponents

17. $\frac{1}{4m^3}$
18. $\frac{s}{9r^{11}}$
19. $\frac{b^7}{64}$
20. $\frac{49q^9}{p^8}$
21. $\frac{32x^7}{y^6}$
22. $\frac{1625z^{10}}{x^5}$

23. 625
24. -216
25. 6561
26. 1
27. p^{24}
28. $\frac{1}{s^{15}}$
29. $\frac{1}{216}$
30. $-\frac{1}{343}$
31. x^2
32. z^5

33. 10^{-2} m
34. $14a^2b$ microns
35. The product has a base of 2, not $2 \cdot 2$; $2^4 \cdot 2^5 = 2^9$
36. The exponent of the quotient should be the difference of the exponents, not the quotient of the exponents; $\frac{x^8}{x^4} = x^4$

In Exercises 37–44, simplify the expression. Write your answer using only positive exponents. (See Example 4.)

37. $(-5z)^3$
38. $(4x)^{-4}$
39. $\left(\dfrac{6}{n}\right)^{-2}$
40. $\left(\dfrac{-t}{3}\right)^2$
41. $(3s^8)^{-5}$
42. $(-5p^3)^3$
43. $\left(-\dfrac{w^3}{6}\right)^{-2}$
44. $\left(\dfrac{1}{2r^6}\right)^{-6}$

45. **USING PROPERTIES** Which of the expressions represent the volume of the sphere? Explain. (See Example 5.)

Ⓐ $\left(\dfrac{3s^2}{2^4\pi s^8}\right)^{-1}$
Ⓑ $(2^5\pi s^6)(3^{-1})$
Ⓒ $\dfrac{32\pi s^6}{3}$
Ⓓ $(2s)^5 \cdot \dfrac{\pi s}{3}$
Ⓔ $\left(\dfrac{3\pi s^6}{32}\right)^{-1}$
Ⓕ $\dfrac{32}{3}\pi s^5$

46. **MODELING WITH MATHEMATICS** Diffusion is the movement of molecules from one location to another. The time t (in seconds) it takes molecules to diffuse a distance of x centimeters is given by $t = \dfrac{x^2}{2D}$, where D is the diffusion coefficient. The diffusion coefficient for a drop of ink in water is about 10^{-5} square centimeters per second. How long will it take the ink to diffuse 1 micrometer (10^{-4} centimeter)?

In Exercises 47–50, simplify the expression. Write your answer using only positive exponents.

47. $\left(\dfrac{2x^{-2}y^3}{3xy^{-4}}\right)^4$
48. $\left(\dfrac{4s^5t^{-7}}{-2s^{-2}t^4}\right)^3$
49. $\left(\dfrac{3m^{-5}n^2}{4m^{-2}n^0}\right)^2 \cdot \left(\dfrac{mn^4}{9n}\right)^2$
50. $\left(\dfrac{3x^3y^0}{x^{-2}}\right)^4 \cdot \left(\dfrac{y^2x^{-4}}{5xy^{-8}}\right)^3$

In Exercises 51–54, evaluate the expression. Write your answer in scientific notation and standard form.

51. $(3 \times 10^2)(1.5 \times 10^{-5})$
52. $(6.1 \times 10^{-3})(8 \times 10^9)$
53. $\dfrac{(6.4 \times 10^7)}{(1.6 \times 10^5)}$
54. $\dfrac{(3.9 \times 10^{-5})}{(7.8 \times 10^{-8})}$

55. **PROBLEM SOLVING** In 2012, on average, about 9.46×10^{-1} pound of potatoes was produced for every 2.3×10^{-5} acre harvested. How many pounds of potatoes on average were produced for each acre harvested? Write your answer in scientific notation and in standard form. (See Example 6.)

56. **PROBLEM SOLVING** The speed of light is approximately 3×10^5 kilometers per second. How long does it take sunlight to reach Jupiter? Write your answer in scientific notation and in standard form.

57. **MATHEMATICAL CONNECTIONS** Consider Cube A and Cube B.

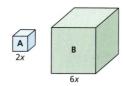

a. Which property of exponents should you use to simplify an expression for the volume of each cube?

b. How can you use the Power of a Quotient Property to find how many times greater the volume of Cube B is than the volume of Cube A?

58. **PROBLEM SOLVING** A byte is a unit used to measure a computer's memory. The table shows the numbers of bytes in several units of measure.

Unit	kilobyte	megabyte	gigabyte	terabyte
Number of bytes	2^{10}	2^{20}	2^{30}	2^{40}

a. How many kilobytes are in 1 terabyte?

b. How many megabytes are in 16 gigabytes?

c. Another unit used to measure a computer's memory is a bit. There are 8 bits in a byte. How can you convert the number of bytes in each unit of measure given in the table to bits? Can you still use a base of 2? Explain.

ANSWERS

37. $-125z^3$
38. $\dfrac{1}{256x^4}$
39. $\dfrac{n^2}{36}$
40. $\dfrac{t^2}{9}$
41. $\dfrac{1}{243s^{40}}$
42. $-125p^9$
43. $\dfrac{36}{w^6}$
44. $64r^{36}$
45. B, C, D; These expressions simplify to be the volume of the sphere, which is $\dfrac{32\pi s^6}{3}$.
46. 5×10^{-4} sec, or 0.0005 sec
47. $\dfrac{16y^{28}}{81x^{12}}$
48. $\dfrac{-8s^{21}}{t^{33}}$
49. $\dfrac{n^{10}}{144m^4}$
50. $\dfrac{81x^5y^{30}}{125}$
51. 4.5×10^{-3}; 0.0045
52. 4.88×10^7; 48,800,000
53. 4×10^2; 400
54. 5×10^2; 500
55. about 4.113×10^4 lb/acre; about 41,130 lb/acre
56. 2.6×10^3 sec; 2600 sec
57. a. Power of a Product Property

 b. Express $\dfrac{(6x)^3}{(2x)^3}$ as $\left(\dfrac{6x}{2x}\right)^3$. Simplify the expression inside the parentheses to produce $(3)^3$, so the volume is 27 times greater.

58. a. 2^{30} kilobytes

b. 16,384 megabytes

c. Multiply each number in the table by 8; yes; The number 8 can be expressed as 2^3, so multiply each number in the table by 2^3. Because the values have a common base of 2, they can be simplified using the Product of Powers Property.

ANSWERS

59. $(2ab)^3$
60. $(4rs)^2$
61. $(2w^3z^2)^6$
62. $(9x^2y^4)^2$ or $(3xy^2)^4$
63. a. $\left(\frac{1}{6}\right)^n$
    b. $\frac{1}{1296}$
    c. $\frac{1}{32}$; The probability of flipping heads once is $\frac{1}{2}$, and $\left(\frac{1}{2}\right)^5$ is $\frac{1}{32}$.
64. a. $\frac{1}{2}; \frac{1}{4}; \frac{1}{8}; \frac{1}{16}$
    b. $2^{-1}; 2^{-2}; 2^{-3}; 2^{-4}$
65. $x = 8, y = -1$; Using the Quotient of Powers Property, you can conclude from the first equation that $x - y = 9$. Using the Product of Powers Property and the Quotient of Powers Property, you can conclude from the second equation that $x + 2 - 3y = 13$. Use these equations to solve a system of linear equations.
66. Sample answer: $r = x^2, h = 81x^4$
67. no; The mass of the seed from the double coconut palm is 10 kilograms.
68. a. 2^{13}
    b. 2^{23}
    c. They become 3^{13} and 3^{23}.
69–75. See Additional Answers.

Mini-Assessment

1. Evaluate the expression $\frac{6^0}{7^{-2}}$. 49

Simplify the expression. Write your answer using only positive exponents.

2. $(x^{-4})^3$ $x^{-12} = \frac{1}{x^{12}}$

3. $\left(\frac{5n}{-3}\right)^{-3}$ $-\frac{27}{125n^3}$

4. A DVD player uses approximately 1.35×10^{-3} kilowatt-hours of energy in 4.5×10^{-2} hours. How many kilowatt-hours of energy does the DVD player use per hour? Write your answer in scientific notation and in standard form.

    3.0×10^{-2}, or 0.03 kilowatt-hour of energy per hour

34   Chapter 1

REWRITING EXPRESSIONS In Exercises 59–62, rewrite the expression as a power of a product.

59. $8a^3b^3$
60. $16r^2s^2$
61. $64w^{18}z^{12}$
62. $81x^4y^8$

63. **USING STRUCTURE** The probability of rolling a 6 on a number cube is $\frac{1}{6}$. The probability of rolling a 6 twice in a row is $\left(\frac{1}{6}\right)^2 = \frac{1}{36}$.
    a. Write an expression that represents the probability of rolling a 6 n times in a row.
    b. What is the probability of rolling a 6 four times in a row?
    c. What is the probability of flipping heads on a coin five times in a row? Explain.

64. **HOW DO YOU SEE IT?** The shaded part of Figure n represents the portion of a piece of paper visible after folding the paper in half n times.

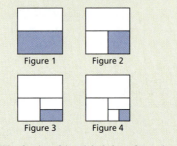

Figure 1   Figure 2

Figure 3   Figure 4

   a. What fraction of the original piece of paper is each shaded part?
   b. Rewrite each fraction from part (a) in the form 2^x.

65. **REASONING** Find x and y when $\frac{b^x}{b^y} = b^9$ and $\frac{b^x \cdot b^2}{b^{3y}} = b^{13}$. Explain how you found your answer.

Maintaining Mathematical Proficiency Reviewing what you learned in previous grades and lessons

Find the square root(s). *(Skills Review Handbook)*

70. $\sqrt{25}$
71. $-\sqrt{100}$
72. $\pm\sqrt{\frac{1}{64}}$

Classify the real number in as many ways as possible. *(Skills Review Handbook)*

73. 12
74. $\frac{65}{9}$
75. $\frac{\pi}{4}$

34   Chapter 1   Functions and Exponents

---

66. **THOUGHT PROVOKING** Write expressions for r and h so that the volume of the cone can be represented by the expression $27\pi x^8$. Find r and h.

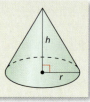

67. **MAKING AN ARGUMENT** One of the smallest plant seeds comes from an orchid, and one of the largest plant seeds comes from a double coconut palm. A seed from an orchid has a mass of 10^{-6} gram. The mass of a seed from a double coconut palm is 10^{10} times the mass of the seed from the orchid. Your friend says that the seed from the double coconut palm has a mass of about 1 kilogram. Is your friend correct? Explain.

68. **CRITICAL THINKING** Your school is conducting a survey. Students can answer the questions in either part with "agree" or "disagree."

    | Part 1: 13 questions | | |
    | Part 2: 10 questions | | |
    | Part 1: Classroom | Agree | Disagree |
    | 1. I come prepared for class. | ○ | ○ |
    | 2. I enjoy my assignments. | ○ | ○ |

   a. What power of 2 represents the number of different ways that a student can answer all the questions in Part 1?
   b. What power of 2 represents the number of different ways that a student can answer all the questions on the entire survey?
   c. The survey changes, and students can now answer "agree," "disagree," or "no opinion." How does this affect your answers in parts (a) and (b)?

69. **ABSTRACT REASONING** Compare the values of a^n and a^{-n} when $n < 0$, when $n = 0$, and when $n > 0$ for (a) $a > 1$ and (b) $0 < a < 1$. Explain your reasoning.

If students need help...

Resources by Chapter
• Practice A and Practice B
• Puzzle Time

Student Journal
• Practice

Differentiating the Lesson
Skills Review Handbook

If students got it...

Resources by Chapter
• Enrichment and Extension
• Cumulative Review

Start the *next* Section

Laurie's Notes

Overview of Section 1.5

Introduction
- This lesson introduces the language and notation of *n*th roots and how to evaluate expressions involving *n*th roots. The examples presented are numeric so that it is possible to evaluate them. Simplifying algebraic expressions is presented in Chapter 4.
- Real-life applications are presented at the end of the lesson.

Resources
- Calculators will be necessary to evaluate *n*th roots. Students may need instruction on how to evaluate expressions, both in radical form and rational exponent form.

Formative Assessment Tips
- **Thumbs Up:** This technique asks students to indicate the extent to which they understand a concept, procedure, or even the directions for an activity.

I get it.

I don't get it.

I'm not sure.

- Use this technique to assess students' understanding of directions for an exploration, getting started with a problem, or in response to a peer's explanation of a problem.

Another Way
- Consider the cube root of 8, or $8^{1/3}$, rewritten as $(2 \cdot 2 \cdot 2)^{1/3}$. The index 1/3 indicates that the answer is 1 of 3 equal factors of 8, or 2.
- This helps students when they get to $8^{2/3}$, rewritten as $(2 \cdot 2 \cdot 2)^{2/3}$. Here the answer is 2 of the 3 equal factors of 8, or $2 \cdot 2 = 4$.
- This will also work with negatives, bases with variables, and the like.

Pacing Suggestion
- Complete the explorations and then start the formal lesson by defining the *n*th root of a number.

What Your Students Will Learn

- Find nth roots and write the nth root of a as a power of a.
- Evaluate expressions with rational exponents.
- Solve real-life problems involving rational exponents.

Laurie's Notes

Exploration

Motivate
- "What do you think 8 yards of gravel looks like?" Students may not know that gravel is generally sold by the cubic yard, and the word *cubic* is rarely stated. So, 8 yards refers to 8 cubic yards.
- ? "If the gravel were dumped inside a cube-shaped container, what would the dimensions of the cube be?" 2 yards on each edge
- ? "What would the dimensions be in feet?" 6 feet on each edge
- ? "What would the volume be in cubic feet?" 216 ft³
- Make the connection that 8 yd³ = 216 ft³.

Discuss
- Introduce the cube root symbol.

Exploration 1
- Students will find a calculator helpful in testing their guesses for side length. For instance, they probably do not know that $15^3 = 3375$, although the 5 in the ones place should be an important clue!
- **? Reason Abstractly and Quantitatively:** When students have finished, spend time discussing the two additional questions posed. "How did you determine which cube was the largest?" Listen for something about converting units and that 1 meter is longer than 1 yard.
- After finishing this exploration, students should be comfortable with the idea that cubing and finding a cube root are inverse operations. Have students write the results for each cube and then substitute to write an equivalent statement involving the cube. For example, in part (a), students should write $\sqrt[3]{27} = 3 \rightarrow \sqrt[3]{3^3} = 3$.

Exploration 2
- Write: "When you raise the nth root of a number to the nth power, you get the original number." Relate this to the solution statements in Exploration 1, and note the distinction. In Exploration 1, the cube root of a cubed number is the number. In the statement above, the cube of the cube root of a number is the number.

    Exploration 1: $\sqrt[3]{8^3} = 8$ Exploration 2: $\left(\sqrt[3]{8}\right)^3 = 8$

- Use this exploration to talk about nth roots for values of n other than 3.
- **Reason Abstractly and Quantitatively:** Discuss the justifications for part (a). Listen for student reasoning, such as $\sqrt[4]{25}$ is between 2 and 3 because $2^4 = 16$ and $3^4 = 81$. Also, $\sqrt[4]{25}$ is closer to 2 than 3.
- **Construct Viable Arguments and Critique the Reasoning of Others:** Ask volunteers to share their answers, explaining how they made their matches.

Communicate Your Answer

- ? "Could you lift 4000 kilograms? About how many pounds is 4000 kilograms?" no; 8800 lb
- ? "Where is the femur on a human? Do you think the size of your femur is greater than that of a dinosaur?" thighbone; yes, for some dinosaurs

Connecting to Next Step
- The explorations give students an opportunity to explore nth roots, particularly cube roots. In the formal lesson, the language and notation of nth roots are presented.

1.5 Radicals and Rational Exponents

Essential Question How can you write and evaluate an *n*th root of a number?

Recall that you cube a number as follows.

3rd power
$2^3 = 2 \cdot 2 \cdot 2 = 8$ 2 cubed is 8.

To "undo" cubing a number, take the cube root of the number.

Symbol for cube root is $\sqrt[3]{}$.
$\sqrt[3]{8} = \sqrt[3]{2^3} = 2$ The cube root of 8 is 2.

EXPLORATION 1   Finding Cube Roots

Work with a partner. Use a cube root symbol to write the side length of each cube. Then find the cube root. Check your answers by multiplying. Which cube is the largest? Which two cubes are the same size? Explain your reasoning.

JUSTIFYING CONCLUSIONS
To be proficient in math, you need to justify your conclusions and communicate them to others.

a. Volume = 27 ft³
b. Volume = 125 cm³
c. Volume = 3375 in.³
d. Volume = 3.375 m³
e. Volume = 1 yd³
f. Volume = $\frac{125}{8}$ mm³

EXPLORATION 2   Estimating *n*th Roots

Work with a partner. Estimate each positive *n*th root. Then match each *n*th root with the point on the number line. Justify your answers.

a. $\sqrt[4]{25}$
b. $\sqrt{0.5}$
c. $\sqrt[5]{2.5}$
d. $\sqrt[3]{65}$
e. $\sqrt[3]{55}$
f. $\sqrt[6]{20{,}000}$

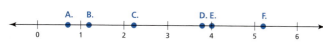

Communicate Your Answer

3. How can you write and evaluate an *n*th root of a number?

4. The body mass *m* (in kilograms) of a dinosaur that walked on two feet can be modeled by

   $m = (0.00016)C^{2.73}$

   where *C* is the circumference (in millimeters) of the dinosaur's femur. The mass of a *Tyrannosaurus rex* was 4000 kilograms. Use a calculator to approximate the circumference of its femur.

Section 1.5   Radicals and Rational Exponents   35

Dynamic Teaching Tools

Dynamic Assessment & Progress Monitoring Tool
Lesson Planning Tool
Interactive Whiteboard Lesson Library
Dynamic Classroom with Dynamic Investigations

ANSWERS

1. a. $\sqrt[3]{27}$ ft; 3 ft
   b. $\sqrt[3]{125}$ cm; 5 cm
   c. $\sqrt[3]{3375}$ in.; 15 in.
   d. $\sqrt[3]{3.375}$ m; 1.5 m
   e. $\sqrt[3]{1}$ yd; 1 yd
   f. $\sqrt[3]{\frac{125}{8}}$; 2.5 mm

   The cube in part (d) has the largest side length of 1.5 meters. The cubes in parts (a) and (e) have equal side lengths because 3 feet = 1 yard.

2. a. *Sample answer:* 2.2; C; 25 is between $2^4 = 16$ and $3^4 = 81$, and C is the only point on the graph between 2 and 3.
   b. *Sample answer:* 0.7; A; 0.5 is between $0^2 = 0$ and $1^2 = 1$, and A is the only point on the graph between 0 and 1.
   c. *Sample answer:* 1.2; B; 2.5 is between $1^5 = 1$ and $2^5 = 32$, and B is the only point on the graph between 1 and 2.
   d. *Sample answer:* 4.0; E; 65 is between $4^3 = 64$ and $5^3 = 125$, and E is the only point on the graph between 4 and 5.
   e. *Sample answer:* 3.8; D; 55 is between $3^3 = 27$ and $4^3 = 64$, and D is the only point on the graph between 3 and 4.
   f. *Sample answer:* 5.2; F; 20,000 is between $5^6 = 15{,}625$ and $6^6 = 46{,}656$, and F is the only point on the graph between 5 and 6.

3. Find what real number multiplied by itself *n* times gives you that number. If that is not possible, determine which *n*th powers the number is between and estimate the decimal part.

4. about 512.7 mm

Section 1.5   35

Extra Example 1

Find the indicated real nth root(s) of a.

a. $n = 3, a = -64$

$\sqrt[3]{-64} = -4$, or $(-64)^{1/3} = -4$

b. $n = 2, a = 81$

$\pm\sqrt{81} = \pm 9$, or $\pm(81)^{1/2} = \pm 9$

MONITORING PROGRESS ANSWERS

1. -5
2. ± 2

1.5 Lesson

What You Will Learn

▶ Find nth roots.
▶ Evaluate expressions with rational exponents.
▶ Solve real-life problems involving rational exponents.

Core Vocabulary

nth root of a, p. 36
radical, p. 36
index of a radical, p. 36

Previous
square root

Finding nth Roots

You can extend the concept of a square root to other types of roots. For example, 2 is a cube root of 8 because $2^3 = 8$, and 3 is a fourth root of 81 because $3^4 = 81$. In general, for an integer n greater than 1, if $b^n = a$, then b is an **nth root of a**. An nth root of a is written as $\sqrt[n]{a}$, where the expression $\sqrt[n]{a}$ is called a **radical** and n is the **index** of the radical.

You can also write an nth root of a as a power of a. If you assume the Power of a Power Property applies to rational exponents, then the following is true.

$$(a^{1/2})^2 = a^{(1/2) \cdot 2} = a^1 = a$$
$$(a^{1/3})^3 = a^{(1/3) \cdot 3} = a^1 = a$$
$$(a^{1/4})^4 = a^{(1/4) \cdot 4} = a^1 = a$$

Because $a^{1/2}$ is a number whose square is a, you can write $\sqrt{a} = a^{1/2}$. Similarly, $\sqrt[3]{a} = a^{1/3}$ and $\sqrt[4]{a} = a^{1/4}$. In general, $\sqrt[n]{a} = a^{1/n}$ for any integer n greater than 1.

READING

$\pm\sqrt[n]{a}$ represents both the positive and negative nth roots of a.

Core Concept

Real nth Roots of a

Let n be an integer greater than 1, and let a be a real number.

- If n is odd, then a has one real nth root: $\sqrt[n]{a} = a^{1/n}$
- If n is even and $a > 0$, then a has two real nth roots: $\pm\sqrt[n]{a} = \pm a^{1/n}$
- If n is even and $a = 0$, then a has one real nth root: $\sqrt[n]{0} = 0$
- If n is even and $a < 0$, then a has no real nth roots.

The nth roots of a number may be real numbers or *imaginary numbers*. You will study imaginary numbers in Chapter 4.

EXAMPLE 1 Finding nth Roots

Find the indicated real nth root(s) of a.

a. $n = 3, a = -27$ b. $n = 4, a = 16$

SOLUTION

a. The index $n = 3$ is odd, so -27 has one real cube root. Because $(-3)^3 = -27$, the cube root of -27 is $\sqrt[3]{-27} = -3$, or $(-27)^{1/3} = -3$.

b. The index $n = 4$ is even, and $a > 0$. So, 16 has two real fourth roots. Because $2^4 = 16$ and $(-2)^4 = 16$, the fourth roots of 16 are $\pm\sqrt[4]{16} = \pm 2$, or $\pm 16^{1/4} = \pm 2$.

Monitoring Progress Help in English and Spanish at BigIdeasMath.com

Find the indicated real nth root(s) of a.

1. $n = 3, a = -125$
2. $n = 6, a = 64$

Laurie's Notes | Teacher Actions

- Introduce the vocabulary and notation associated with nth roots. Connect the radical form with the rational exponent form.
- **? Probing Question** and **Turn and Talk:** "Will the nth root of 64 always be a real number? yes Will the nth root of -64 always be a real number?" no You want students to activate their thinking about the various cases of $\sqrt[n]{a}$ (n is odd or even; sign of a) before writing the Core Concept.
- Students will say that $\sqrt{-64}$ is impossible or cannot be done. Explain that imaginary numbers will be introduced in Chapter 4.

Evaluating Expressions with Rational Exponents

Recall that the radical $\sqrt{a}$ indicates the positive square root of a. Similarly, an nth root of a, $\sqrt[n]{a}$, with an *even* index indicates the positive nth root of a.

REMEMBER
The expression under the radical sign is the radicand.

EXAMPLE 2 Evaluating nth Root Expressions

Evaluate each expression.

a. $\sqrt[3]{-8}$ b. $-\sqrt[3]{8}$ c. $16^{1/4}$ d. $(-16)^{1/4}$

SOLUTION

a. $\sqrt[3]{-8} = \sqrt[3]{(-2) \cdot (-2) \cdot (-2)}$ Rewrite the expression showing factors.
   $= -2$ Evaluate the cube root.

b. $-\sqrt[3]{8} = -(\sqrt[3]{2 \cdot 2 \cdot 2})$ Rewrite the expression showing factors.
   $= -(2)$ Evaluate the cube root.
   $= -2$ Simplify.

c. $16^{1/4} = \sqrt[4]{16}$ Rewrite the expression in radical form.
   $= \sqrt[4]{2 \cdot 2 \cdot 2 \cdot 2}$ Rewrite the expression showing factors.
   $= 2$ Evaluate the fourth root.

d. $(-16)^{1/4}$ is not a real number because there is no real number that can be multiplied by itself four times to produce -16.

A rational exponent does not have to be of the form $1/n$. Other rational numbers such as $3/2$ can also be used as exponents. You can use the properties of exponents to evaluate or simplify expressions involving rational exponents.

STUDY TIP
You can rewrite $27^{2/3}$ as $27^{(1/3) \cdot 2}$ and then use the Power of a Power Property to show that
$27^{(1/3) \cdot 2} = (27^{1/3})^2$.

Core Concept

Rational Exponents
Let $a^{1/n}$ be an nth root of a, and let m be a positive integer.

Algebra $a^{m/n} = (a^{1/n})^m = (\sqrt[n]{a})^m$

Numbers $27^{2/3} = (27^{1/3})^2 = (\sqrt[3]{27})^2$

EXAMPLE 3 Evaluating Expressions with Rational Exponents

Evaluate (a) $16^{3/4}$ and (b) $27^{4/3}$.

SOLUTION

a. $16^{3/4} = (16^{1/4})^3$ Rational exponents   b. $27^{4/3} = (27^{1/3})^4$
   $= 2^3$ Evaluate the nth root. $= 3^4$
   $= 8$ Evaluate the power. $= 81$

Monitoring Progress Help in English and Spanish at *BigIdeasMath.com*

Evaluate the expression.

3. $\sqrt[3]{-125}$ 4. $(-64)^{2/3}$ 5. $9^{5/2}$ 6. $256^{3/4}$

Differentiated Instruction

Kinesthetic
Some students may benefit by having access to a scientific or graphing calculator. After evaluating the expressions, students can enter the expressions in the calculator, find the values, and compare the calculated values to their answers.

Extra Example 2
Evaluate each expression.
a. $\sqrt[4]{81}$ 3
b. $-\sqrt[4]{81}$ -3
c. $25^{1/2}$ 5
d. $(-25)^{1/2}$ $(-25)^{1/2}$ is not a real number because there is no real number that can be multiplied by itself to produce -25.

Extra Example 3
a. Evaluate $243^{2/5}$. 9
b. Evaluate $16^{3/2}$. 64

MONITORING PROGRESS ANSWERS
3. -5
4. 16
5. 243
6. 64

Laurie's Notes — Teacher Actions

- Work through each part of Example 2.
- **? Attend to Precision:** Make sure students read $\sqrt[3]{-8}$ and $-\sqrt[3]{8}$ correctly and understand why the solutions are the same. Ask, "Are $\sqrt[4]{-16}$ and $-\sqrt[4]{16}$ also equivalent?" no; $\sqrt[4]{-16}$ has no real roots, and $-\sqrt[4]{16} = -2$.
- Write the expression in Example 3(a). Explain that you can rewrite the exponent as $3 \cdot \frac{1}{4}$, but it is usually easier to evaluate the root first.
- **Thumbs Up:** Have students answer Questions 3–6 and then give a *Thumbs Up* self-assessment to indicate their confidence in evaluating expressions with rational exponents.

Extra Example 4

The radius r of a sphere is given by the equation $r = \left(\dfrac{3V}{4\pi}\right)^{1/3}$, where V is the volume of the sphere.

Volume = 463 cubic inches

Find the radius of the basketball to the nearest inch. Use 3.14 for π. *The radius of the basketball is about 5 inches.*

Extra Example 5

To calculate the annual inflation rate r (in decimal form) of an item that increases in value from P to F over a period of n years, you can use the equation $r = \left(\dfrac{F}{P}\right)^{1/n} - 1$. Find the annual inflation rate to the nearest tenth of a percent of a house that increases in value from \$300,000 to \$360,000 over a period of 8 years. *The annual inflation rate is about 2.3%.*

MONITORING PROGRESS ANSWERS

7. about 16 in.
8. about 6.0%

Solving Real-Life Problems

EXAMPLE 4 Solving a Real-Life Problem

Volume = 113 cubic feet

The radius r of a sphere is given by the equation $r = \left(\dfrac{3V}{4\pi}\right)^{1/3}$, where V is the volume of the sphere. Find the radius of the beach ball to the nearest foot. Use 3.14 for π.

SOLUTION

1. **Understand the Problem** You know the equation that represents the radius of a sphere in terms of its volume. You are asked to find the radius for a given volume.
2. **Make a Plan** Substitute the given volume into the equation. Then evaluate to find the radius.
3. **Solve the Problem**

$$r = \left(\dfrac{3V}{4\pi}\right)^{1/3} \qquad \text{Write the equation.}$$
$$= \left(\dfrac{3(113)}{4(3.14)}\right)^{1/3} \qquad \text{Substitute 113 for } V \text{ and 3.14 for } \pi.$$
$$= \left(\dfrac{339}{12.56}\right)^{1/3} \qquad \text{Multiply.}$$
$$\approx 3 \qquad \text{Use a calculator.}$$

▶ The radius of the beach ball is about 3 feet.

4. **Look Back** To check that your answer is reasonable, compare the size of the ball to the size of the woman pushing the ball. The ball appears to be slightly taller than the woman. The average height of a woman is between 5 and 6 feet. So, a radius of 3 feet, or height of 6 feet, seems reasonable for the beach ball.

EXAMPLE 5 Solving a Real-Life Problem

To calculate the annual inflation rate r (in decimal form) of an item that increases in value from P to F over a period of n years, you can use the equation $r = \left(\dfrac{F}{P}\right)^{1/n} - 1$. Find the annual inflation rate to the nearest tenth of a percent of a house that increases in value from \$200,000 to \$235,000 over a period of 5 years.

SOLUTION

$$r = \left(\dfrac{F}{P}\right)^{1/n} - 1 \qquad \text{Write the equation.}$$
$$= \left(\dfrac{235{,}000}{200{,}000}\right)^{1/5} - 1 \qquad \text{Substitute 235,000 for } F, 200{,}000 \text{ for } P, \text{ and 5 for } n.$$
$$= 1.175^{1/5} - 1 \qquad \text{Divide.}$$
$$\approx 0.03278 \qquad \text{Use a calculator.}$$

REMEMBER
To write a decimal as a percent, move the decimal point two places to the right. Then add a percent symbol.

▶ The annual inflation rate is about 3.3%.

Monitoring Progress Help in English and Spanish at BigIdeasMath.com

7. **WHAT IF?** In Example 4, the volume of the beach ball is 17,000 cubic inches. Find the radius to the nearest inch. Use 3.14 for π.
8. The average cost of college tuition increases from \$8500 to \$13,500 over a period of 8 years. Find the annual inflation rate to the nearest tenth of a percent.

Laurie's Notes — Teacher Actions

- **? Make Sense of Problems and Persevere in Solving Them:** "Can you use the photo in Example 4 to estimate the radius?" yes; about 2 to 3 feet Discuss and let students work the problem with partners.
- The solution calls for a calculator, but without a calculator you could simplify $339 \div 12.56 \approx 26.99$ and estimate that $\sqrt[3]{26.99} \approx 3.0$.
- **Thumbs Up:** Discuss inflation rate. Have students work Example 5 and share as a class. Ask for a *Thumbs Up* self-assessment at this point.

Closure

- **Whiteboarding:** Evaluate $32^{2/5}$ and $-27^{2/3}$. 4, -9

1.5 Exercises

Dynamic Solutions available at *BigIdeasMath.com*

Vocabulary and Core Concept Check

1. **WRITING** Explain how to evaluate $81^{1/4}$.

2. **WHICH ONE DOESN'T BELONG?** Which expression does *not* belong with the other three? Explain your reasoning.

$$(\sqrt[3]{27})^2 \qquad 27^{2/3} \qquad 3^2 \qquad (\sqrt[2]{27})^3$$

Monitoring Progress and Modeling with Mathematics

In Exercises 3 and 4, rewrite the expression in rational exponent form.

3. $\sqrt{10}$
4. $\sqrt[5]{34}$

In Exercises 5 and 6, rewrite the expression in radical form.

5. $15^{1/3}$
6. $140^{1/8}$

In Exercises 7–10, find the indicated real *n*th root(s) of *a*. *(See Example 1.)*

7. $n = 2, a = 36$
8. $n = 4, a = 81$
9. $n = 3, a = 1000$
10. $n = 9, a = -512$

MATHEMATICAL CONNECTIONS In Exercises 11 and 12, find the dimensions of the cube. Check your answer.

11. Volume = 64 in.³
12. Volume = 216 cm³

In Exercises 13–18, evaluate the expression. *(See Example 2.)*

13. $\sqrt[4]{256}$
14. $\sqrt[3]{-216}$
15. $\sqrt[3]{-343}$
16. $-\sqrt[5]{1024}$
17. $128^{1/7}$
18. $(-64)^{1/2}$

In Exercises 19 and 20, rewrite the expression in rational exponent form.

19. $(\sqrt[5]{8})^4$
20. $(\sqrt[5]{-21})^6$

In Exercises 21 and 22, rewrite the expression in radical form.

21. $(-4)^{2/7}$
22. $9^{5/2}$

In Exercises 23–28, evaluate the expression. *(See Example 3.)*

23. $32^{3/5}$
24. $125^{2/3}$
25. $(-36)^{3/2}$
26. $(-243)^{2/5}$
27. $(-128)^{5/7}$
28. $343^{4/3}$

29. **ERROR ANALYSIS** Describe and correct the error in rewriting the expression in rational exponent form.

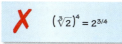

30. **ERROR ANALYSIS** Describe and correct the error in evaluating the expression.

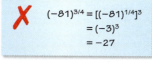

In Exercises 31–34, evaluate the expression.

31. $\left(\dfrac{1}{1000}\right)^{1/3}$
32. $\left(\dfrac{1}{64}\right)^{1/6}$
33. $(27)^{-2/3}$
34. $(9)^{-5/2}$

Section 1.5 Radicals and Rational Exponents 39

31. $\dfrac{1}{10}$
32. $\dfrac{1}{2}$
33. $\dfrac{1}{9}$
34. $\dfrac{1}{243}$

Assignment Guide and Homework Check

ASSIGNMENT

Basic: 1, 2, 3–29 odd, 37, 40, 41, 44, 57–60

Average: 1, 2–44 even, 57–60

Advanced: 1, 2, 8–12 even, 16–44 even, 45–60

HOMEWORK CHECK

Basic: 7, 13, 23, 37, 41

Average: 8, 14, 26, 38, 42

Advanced: 10, 16, 26, 38, 42

ANSWERS

1. Find the fourth root of 81, or what real number multiplied by itself four times produces 81.
2. $(\sqrt[2]{27})^3$; It is the only one that is not equivalent to 9.
3. $10^{1/2}$
4. $34^{1/5}$
5. $\sqrt[3]{15}$
6. $\sqrt[8]{140}$
7. ± 6
8. ± 3
9. 10
10. -2
11. $s = 4$ in.
12. $s = 6$ cm
13. 4
14. -6
15. -7
16. -4
17. 2
18. not a real number
19. $8^{4/5}$
20. $-21^{6/5}$
21. $(\sqrt[7]{-4})^2$
22. $(\sqrt{9})^5$
23. 8
24. 25
25. not a real number
26. 9
27. -32
28. 2401
29. The numerator and denominator are reversed; $(\sqrt[3]{2})^4 = 2^{4/3}$
30. The number -81 does not have a real fourth root; not a real number

Dynamic Teaching Tools

Dynamic Assessment & Progress Monitoring Tool

Interactive Whiteboard Lesson Library

Dynamic Classroom with Dynamic Investigations

ANSWERS

35. 6 ft^2
36. 243 mm
37. about 1 in.
38. about 44 m^3
39. Write the radicand, a, as the base and write the exponent as a fraction with the power, m, as the numerator and the index, n, as the denominator.
40. $x^{1/2}$ in.
41. about 5.5%
42. about 9.2%
43. -1, 0, and 1
44. no; If n is odd and a is negative, $\sqrt[n]{a}$ will be negative, and $-\sqrt[n]{a}$ will be positive.
45. $(xy)^{1/2}$
46. y^2
47. $2xy^2$
48. x^3y^5
49. about 1.38 ft
50. The geometric mean is always less than or equal to the arithmetic mean.
51. always; Power of a Power Property
52. sometimes; true when $x = 1$, otherwise false
53–60. See Additional Answers.

Mini-Assessment

1. Find the indicated real nth roots of a.
   $n = 4$, $a = 625$ $\pm\sqrt[4]{625} = \pm 5$, or $\pm(625)^{1/4} = \pm 5$

Evaluate the expression.

2. $-\sqrt[4]{1296}$ -6
3. $64^{4/3}$ 256
4. Use the equation $r = \left(\dfrac{F}{P}\right)^{1/n} - 1$ to calculate the annual inflation rate r (in decimal form) of an item that increases in value from P to F over n years. The cost of a gallon of milk increased from $2.81 to $3.48 over 10 years. Find the annual inflation rate to the nearest tenth of a percent. about 2.2%

40   Chapter 1

---

35. **PROBLEM SOLVING** A math club is having a bake sale. Find the area of the bake sale sign.

$4^{1/2}$ ft

$\sqrt[6]{729}$ ft

36. **PROBLEM SOLVING** The volume of a cube-shaped box is 27^5 cubic millimeters. Find the length of one side of the box.

37. **MODELING WITH MATHEMATICS** The radius r of the base of a cone is given by the equation

$$r = \left(\dfrac{3V}{\pi h}\right)^{1/2}$$

where V is the volume of the cone and h is the height of the cone. Find the radius of the paper cup to the nearest inch. Use 3.14 for π. (See Example 4.)

4 in.

Volume = 5 in.^{3}

38. **MODELING WITH MATHEMATICS** The volume of a sphere is given by the equation $V = \dfrac{1}{6\sqrt{\pi}} S^{3/2}$, where S is the surface area of the sphere. Find the volume of a sphere, to the nearest cubic meter, that has a surface area of 60 square meters. Use 3.14 for π.

39. **WRITING** Explain how to write $(\sqrt[n]{a})^m$ in rational exponent form.

40. **HOW DO YOU SEE IT?** Write an expression in rational exponent form that represents the side length of the square.

Area = x in.^{2}

In Exercises 41 and 42, use the formula $r = \left(\dfrac{F}{P}\right)^{1/n} - 1$ to find the annual inflation rate to the nearest tenth of a percent. (See Example 5.)

41. A farm increases in value from $800,000 to $1,100,000 over a period of 6 years.

42. The cost of a gallon of gas increases from $1.46 to $3.53 over a period of 10 years.

43. **REASONING** For what values of x is $x = x^{1/5}$?

44. **MAKING AN ARGUMENT** Your friend says that for a real number a and a positive integer n, the value of $\sqrt[n]{a}$ is always positive and the value of $-\sqrt[n]{a}$ is always negative. Is your friend correct? Explain.

In Exercises 45–48, simplify the expression.

45. $(y^{1/6})^3 \cdot \sqrt{x}$
46. $(y \cdot y^{1/3})^{3/2}$
47. $x \cdot \sqrt[3]{y^6} + y^2 \cdot \sqrt[3]{x^3}$
48. $(x^{1/3} \cdot y^{1/2})^9 \cdot \sqrt{y}$

49. **PROBLEM SOLVING** The formula for the volume of a regular dodecahedron is $V \approx 7.66\ell^3$, where ℓ is the length of an edge. The volume of the dodecahedron is 20 cubic feet. Estimate the edge length.

50. **THOUGHT PROVOKING** To find the arithmetic mean of n numbers, divide the sum of the numbers by n. To find the geometric mean of n numbers $a_1, a_2, a_3, \ldots, a_n$, take the nth root of the product of the numbers.

geometric mean = $\sqrt[n]{a_1 \cdot a_2 \cdot a_3 \cdot \ldots \cdot a_n}$

Compare the arithmetic mean to the geometric mean of n numbers.

ABSTRACT REASONING In Exercises 51–56, let x be a nonnegative real number. Determine whether the statement is *always*, *sometimes*, or *never* true. Justify your answer.

51. $(x^{1/3})^3 = x$
52. $x^{1/3} = x^{-3}$
53. $x^{1/3} = \sqrt[3]{x}$
54. $x^{1/3} = x^3$
55. $\dfrac{x^{2/3}}{x^{1/3}} = \sqrt[3]{x}$
56. $x = x^{1/3} \cdot x^3$

Maintaining Mathematical Proficiency — Reviewing what you learned in previous grades and lessons

Evaluate the function when $x = -3$, 0, and 8. (*Skills Review Handbook*)

57. $f(x) = 2x - 10$
58. $w(x) = -5x - 1$
59. $h(x) = 13 - x$
60. $g(x) = 8x + 16$

40   Chapter 1   Functions and Exponents

If students need help...	If students got it...
Resources by Chapter • Practice A and Practice B • Puzzle Time	Resources by Chapter • Enrichment and Extension • Cumulative Review
Student Journal • Practice	Start the *next* Section
Differentiating the Lesson Skills Review Handbook	

Laurie's Notes

Overview of Section 1.6

Introduction
- Students studied exponential functions in Math I, where the domain was limited to integer values. Now that students have evaluated expressions with rational exponents, they should be able to evaluate and graph $y = ab^x$.
- In this lesson, students will work with applications of both exponential growth and decay.

Formative Assessment Tips
- **Popsicle Sticks:** This technique ensures that *any* student can be called on during questioning time in class. Write the name of each student on a Popsicle stick. Place the sticks in a cup (or can). When questions are posed in class during *No-Hands Questioning**, each student should think and be prepared to answer. If you only call on students who raise their hands, students can then opt out of being engaged. All students think they have an equal chance of being called on when a stick is pulled randomly, so they engage more in the lesson.

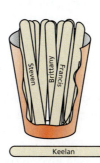

- If there are certain students that you want to hear from, the cup could have an inner cylinder where *select* sticks are placed. It will appear that the process is still random, and the voices you have not heard much from will still have the opportunity to prepare and formulate their answer.

* See Section 2.4 for a description of *No-Hands Questioning*.

Pacing Suggestion
- Once students have worked the explorations, continue with the formal lesson.

Dynamic Teaching Tools
Dynamic Assessment & Progress Monitoring Tool
Lesson Planning Tool
Interactive Whiteboard Lesson Library
Dynamic Classroom with Dynamic Investigations

What Your Students Will Learn

- Graph exponential growth and decay functions and identify growth and decay factors.
- Write exponential growth and decay models.
- Write recursive rules for exponential functions.
- Rewrite exponential functions to identify the percent rate of change.

Laurie's Notes

Exploration

Motivate
- **?** Ask, "What do the words *appreciate* and *depreciate* mean with respect to the value of an automobile?" Answers will vary.
- Discuss various types of cars that have increased in value (i.e. 1960 Mercedes-Benz 190SL or 1970 Corvette) and the vast majority which decrease in value.
- **?** "If you purchased a new car for $30,000 three years ago that is now worth $18,000, do you think the value of the car decreased $4000 each year?" Answers will vary.
- Explain that in this lesson they will look at a model for how the value of a car decreases over time.

Exploration Note
- Students worked with exponential functions in Math I, so it should not be necessary to give directions for evaluating an exponential function or graphing a simple exponential function. Students should be familiar with the terms *growth* and *decay*.

Exploration 1
- **Reason Abstractly and Quantitatively:** Consider the type of reasoning students use to match the equations and graphs. Three of the graphs are increasing, and three are decreasing. Three of the equations involve a whole number raised to an exponent, and three involve a fraction between 0 and 1 raised to an exponent.
- **COMMON ERROR** Students may refer to three of the graphs as having a positive slope and three of the graphs as having a negative slope. These are not graphs of linear functions, so none of them have a slope. Certainly, what students are suggesting is that three of the graphs are rising from left to right and three of the graphs are falling from left to right.
- When students have finished, solicit responses and explanations as to how the matching was done.

Exploration 2
- **Construct Viable Arguments and Critique the Reasoning of Others:** This exploration asks students to make conjectures about the attributes of an exponential graph. Their conjectures may be based on recall from Math I or from observation and partner discussion.

Communicate Your Answer
- In describing the characteristics of the graph of an exponential function, students may describe the graph as not touching the *x*-axis, thus having no *x*-intercept.

Connecting to Next Step
- The explorations should be a review of the graphs of exponential functions. Quickly transition to the formal lesson.

1.6 Exponential Functions

Essential Question What are some of the characteristics of the graph of an exponential function?

You can use a graphing calculator to evaluate an exponential function. For example, consider the exponential function $f(x) = 2^x$.

Function Value	Graphing Calculator Keystrokes	Display
$f(-3.1) = 2^{-3.1}$	2 ^ (−) 3.1 ENTER	0.1166291
$f\left(\frac{2}{3}\right) = 2^{2/3}$	2 ^ (2 ÷ 3) ENTER	1.5874011

EXPLORATION 1 Identifying Graphs of Exponential Functions

Work with a partner. Match each exponential function with its graph. Use a table of values to sketch the graph of the function, if necessary.

a. $f(x) = 2^x$
b. $f(x) = 3^x$
c. $f(x) = 4^x$
d. $f(x) = \left(\frac{1}{2}\right)^x$
e. $f(x) = \left(\frac{1}{3}\right)^x$
f. $f(x) = \left(\frac{1}{4}\right)^x$

A.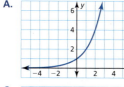
B.
C.
D.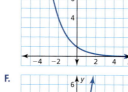
E.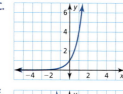
F.

EXPLORATION 2 Characteristics of Graphs of Exponential Functions

CONSTRUCTING VIABLE ARGUMENTS

To be proficient in math, you need to justify your conclusions and communicate them to others.

Work with a partner. Use the graphs in Exploration 1 to determine the domain, range, and y-intercept of the graph of $f(x) = b^x$, where b is a positive real number other than 1. Explain your reasoning.

Communicate Your Answer

3. What are some of the characteristics of the graph of an exponential function?
4. In Exploration 2, is it possible for the graph of $f(x) = b^x$ to have an x-intercept? Explain your reasoning.

Differentiated Instruction

Organization

Students have studied many function families. Have students make a chart of the function families they know. Include the parent function and a sketch of the graph for each function family. Make sure students add exponential growth and exponential decay functions to the chart. The chart below shows a sample row.

Family	Parent Function	Graph
Linear	$f(x) = mx + b$	

1.6 Lesson

Core Vocabulary

exponential function, p. 42
exponential growth function, p. 42
growth factor, p. 42
asymptote, p. 42
exponential decay function, p. 42
decay factor, p. 42
recursive rule for an exponential function, p. 44

Previous
sequences
properties of exponents

What You Will Learn

▶ Graph exponential growth and decay functions.
▶ Write exponential models and recursive rules.
▶ Rewrite exponential functions.

Exponential Growth and Decay Functions

An **exponential function** has the form $y = ab^x$, where $a \neq 0$ and the base b is a positive real number other than 1. If $a > 0$ and $b > 1$, then $y = ab^x$ is an **exponential growth function**, and b is called the **growth factor**. The simplest type of exponential growth function has the form $y = b^x$.

🌀 Core Concept

Parent Function for Exponential Growth Functions

The function $f(x) = b^x$, where $b > 1$, is the parent function for the family of exponential growth functions with base b. The graph shows the general shape of an exponential growth function.

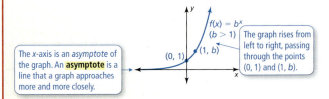

The x-axis is an *asymptote* of the graph. An **asymptote** is a line that a graph approaches more and more closely.

$f(x) = b^x$ $(b > 1)$
The graph rises from left to right, passing through the points $(0, 1)$ and $(1, b)$.

The domain of $f(x) = b^x$ is all real numbers. The range is $y > 0$.

If $a > 0$ and $0 < b < 1$, then $y = ab^x$ is an **exponential decay function**, and b is called the **decay factor**.

🌀 Core Concept

Parent Function for Exponential Decay Functions

The function $f(x) = b^x$, where $0 < b < 1$, is the parent function for the family of exponential decay functions with base b. The graph shows the general shape of an exponential decay function.

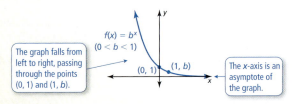

$f(x) = b^x$ $(0 < b < 1)$
The graph falls from left to right, passing through the points $(0, 1)$ and $(1, b)$.

The x-axis is an asymptote of the graph.

The domain of $f(x) = b^x$ is all real numbers. The range is $y > 0$.

Laurie's Notes   Teacher Actions

- **Reason Abstractly and Quantitatively** and **Attend to Precision:** Refer to the definition of an *exponential function* and ask a series of *Assessing Questions*: "Why must $a \neq 0$? Why must b be a positive number? Why can't $b = 1$? What is the domain of $y = ab^x$? What is the range of $y = ab^x$ if $a > 0$ and $b > 1$?" Define *asymptote*.
- ❓ Say, "If $b > 1$, then $y = b^x$ is exponential growth. Explain why. If $0 < b < 1$, then $y = b^x$ is exponential decay. Explain why." *Answers will vary. Students should convey understanding of raising a base to an exponent that is increasing.*
- ❓ "What two ordered pairs are on every graph of $y = b^x$?" *$(0, 1)$ and $(1, b)$*

STUDY TIP

When graphing exponential functions, you connect the points with a smooth curve because $y = ab^x$ is defined for rational x-values. This should make sense from your study of rational exponents. For instance, in Example 1(a), when $x = \frac{1}{2}$,

$y = 2^{1/2} = \sqrt{2} \approx 1.4.$

EXAMPLE 1 Graphing Exponential Growth and Decay Functions

Determine whether each function represents *exponential growth* or *exponential decay*. Then graph the function.

a. $y = 2^x$
b. $y = \left(\frac{1}{2}\right)^x$

SOLUTION

a. **Step 1** Identify the value of the base. The base, 2, is greater than 1, so the function represents exponential growth.

Step 2 Make a table of values.

x	−2	−1	0	1	2	3
y	$\frac{1}{4}$	$\frac{1}{2}$	1	2	4	8

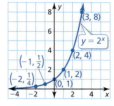

Step 3 Plot the points from the table.

Step 4 Draw, from *left to right*, a smooth curve that begins just above the x-axis, passes through the plotted points, and moves up to the right.

b. **Step 1** Identify the value of the base. The base, $\frac{1}{2}$, is greater than 0 and less than 1, so the function represents exponential decay.

Step 2 Make a table of values.

x	−3	−2	−1	0	1	2
y	8	4	2	1	$\frac{1}{2}$	$\frac{1}{4}$

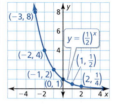

Step 3 Plot the points from the table.

Step 4 Draw, from *right to left*, a smooth curve that begins just above the x-axis, passes through the plotted points, and moves up to the left.

Monitoring Progress Help in English and Spanish at BigIdeasMath.com

Determine whether the function represents *exponential growth* or *exponential decay*. Then graph the function.

1. $y = 4^x$
2. $y = \left(\frac{2}{3}\right)^x$
3. $f(x) = (0.25)^x$
4. $f(x) = (1.5)^x$

Writing Exponential Models and Recursive Rules

Some real-life quantities increase or decrease by a fixed percent each year (or some other time period). The amount y of such a quantity after t years can be modeled by one of these equations.

Exponential Growth Model   **Exponential Decay Model**

$y = a(1 + r)^t$ $y = a(1 − r)^t$

Note that a is the initial amount and r is the percent increase or decrease written as a decimal. The quantity $1 + r$ is the growth factor, and $1 − r$ is the decay factor.

Section 1.6 Exponential Functions 43

English Language Learners

Labels

Have students write and label the exponential growth model.

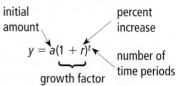

Remind students that "growth factor = 1 + percent increase." Have students write and label a similar diagram for the exponential decay model. Then remind students that "decay factor = 1 − percent decrease."

Extra Example 1

Tell whether each function represents *exponential growth* or *exponential decay*. Then graph the function.

a. $y = (1.3)^x$

The base, 1.3, is greater than 1, so the function represents exponential growth.

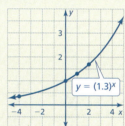

b. $y = \left(\frac{1}{3}\right)^x$

The base, $\frac{1}{3}$, is greater than 0 and less than 1, so the function represents exponential decay.

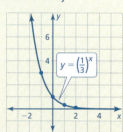

MONITORING PROGRESS ANSWERS

1–4. See Additional Answers.

Laurie's Notes — Teacher Actions

- **Turn and Talk** and **Whiteboarding:** Pose the two functions in Example 1 and have partners work together. Have partners use *whiteboards* to share their graphs.
- **Think-Pair-Share:** Have students answer Questions 1–4, and then share and discuss as a class.
- Introduce exponential models. Explain that $(1 + r)$ is the base b and the independent variable is t. The model $y = a(1 + r)^t$ has the same form as $y = ab^x$.

Extra Example 2

The population of a city was about 1.04 million on January 1, 2000. The population at the beginning of each subsequent year increased by about 2.05%.

a. Write an exponential growth model that represents the population y (in millions) t years after January 1, 2000. Find and interpret the y-value when $t = 8.5$.

$y = 1.04(1.0205)^t$; Using this model, you can estimate the midyear population in 2008 ($t = 8.5$) to be $y = 1.04(1.0205)^{8.5} \approx 1.24$ million.

b. Estimate when the population was 1.3 million. The city's population was about 1.3 million at the beginning of 2011.

MONITORING PROGRESS ANSWERS

5. $y = 6.09(1.015)^t$; near the middle of 2009

EXAMPLE 2 Writing an Exponential Model

The population of a country was about 6.09 million on January 1, 2000. The population at the beginning of each subsequent year increased by about 1.18%.

a. Write an exponential growth model that represents the population y (in millions) t years after January 1, 2000. Find and interpret the y-value when $t = 7.5$.

b. Estimate when the population was 7 million.

SOLUTION

a. The initial amount is $a = 6.09$, and the percent increase is $r = 0.0118$. So, the exponential growth model is

$y = a(1 + r)^t$ Write exponential growth model.
$= 6.09(1 + 0.0118)^t$ Substitute 6.09 for a and 0.0118 for r.
$= 6.09(1.0118)^t$.   Simplify.

Using this model, you can estimate the *midyear* population in 2007 ($t = 7.5$) to be $y = 6.09(1.0118)^{7.5} \approx 6.65$ million.

b. Use the *table* feature of a graphing calculator to determine that $y \approx 7$ when $t \approx 11.9$. So, the population was about 7 million near the end of 2011.

Monitoring Progress Help in English and Spanish at *BigIdeasMath.com*

5. **WHAT IF?** Assume the population increased by 1.5% each year. Write an equation to model this situation. Estimate when the population was 7 million.

In real-life situations, you can also show exponential relationships using *recursive rules*.

REMEMBER
Recall that for a sequence, a recursive rule gives the beginning term(s) of the sequence and a recursive equation that tells how a_n is related to one or more preceding terms.

A **recursive rule** for an exponential function gives the initial value of the function $f(0)$, and a recursive equation that tells how a value $f(n)$ is related to a preceding value $f(n-1)$.

Core Concept

Writing Recursive Rules for Exponential Functions

An exponential function of the form $f(x) = ab^x$ is written using a recursive rule as follows.

Recursive Rule $f(0) = a,\ f(n) = r \cdot f(n-1)$

where $a \neq 0$, r is the common ratio, and n is a natural number

Example $y = 6(3)^x$ can be written as $f(0) = 6$, $f(n) = 3 \cdot f(n-1)$

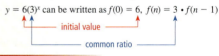

Notice that the base b of the exponential function is the common ratio r in the recursive equation. Also, notice the value of a in the exponential function is the initial value of the recursive rule.

44   Chapter 1   Functions and Exponents

Laurie's Notes   **Teacher Actions**

- **Think-Pair-Share:** Pose Example 2. Give time for students to think through the problem and then have partners share their problem-solving processes and solutions with one another. When students have finished the example, use *Popsicle Sticks* for responses.
- Recursive rules were introduced in Math I. The recursive equation for a geometric sequence is $a_n = r \cdot a_{n-1}$, where r is the common ratio.
- Write the *Core Concept* and relate it to geometric sequences. Discuss the example shown.

STUDY TIP

Notice that the domain consists of the natural numbers when written recursively.

EXAMPLE 3 Writing a Recursive Rule for an Exponential Function

Write a recursive rule for the function you wrote in Example 2.

SOLUTION

The function $y = 6.09(1.0118)^t$ is exponential with initial value $f(0) = 6.09$ and common ratio $r = 1.0118$. So, a recursive equation is

$f(n) = r \cdot f(n - 1)$ Recursive equation for exponential functions

$= 1.0118 \cdot f(n - 1)$.   Substitute 1.0118 for r.

▶ A recursive rule for the exponential function is $f(0) = 6.09$, $f(n) = 1.0118 \cdot f(n - 1)$.

Monitoring Progress Help in English and Spanish at BigIdeasMath.com

Write an recursive rule for the exponential function.

6. $f(x) = 4(7)^x$ 7. $y = 9\left(\frac{1}{3}\right)^t$

Rewriting Exponential Functions

EXAMPLE 4 Rewriting Exponential Functions

Rewrite each function to determine whether it represents *exponential growth* or *exponential decay*. Then identify the percent rate of change.

a. $y = 100(0.96)^{t/4}$ **b.** $f(t) = (1.1)^{t-3}$

STUDY TIP

You can rewrite exponential expressions and functions using the properties of exponents. Changing the form of an exponential function can reveal important attributes of the function.

SOLUTION

a. $y = 100(0.96)^{t/4}$ Write the function.

$= 100(0.96^{1/4})^t$ Power of a Power Property

$\approx 100(0.99)^t$ Evaluate the power.

▶ So, the function represents exponential decay. Use the decay factor $1 - r \approx 0.99$ to find the rate of decay $r \approx 0.01$, or about 1%.

b. $f(t) = (1.1)^{t-3}$ Write the function.

$= \dfrac{(1.1)^t}{(1.1)^3}$ Quotient of Powers Property

$\approx 0.75(1.1)^t$ Evaluate the power and simplify.

▶ So, the function represents exponential growth. Use the growth factor $1 + r = 1.1$ to find the rate of growth $r = 0.1$, or 10%.

Monitoring Progress Help in English and Spanish at BigIdeasMath.com

Rewrite the function to determine whether it represents *exponential growth* or *exponential decay*. Then identify the percent rate of change.

8. $f(t) = 3(1.02)^{10t}$ 9. $y = (0.95)^{t+2}$

Section 1.6   Exponential Functions   45

Extra Example 3
Write a recursive rule for the function you wrote in Extra Example 2.
$f(0) = 1.04$, $f(n) = 1.0205 \cdot f(n - 1)$

Extra Example 4
Rewrite each function to determine whether it represents *exponential growth* or *exponential decay*. Then identify the percent rate of change.

a. $y = 20(0.81)^{t/2}$ $y = 20(0.9)^t$; The function represents exponential decay; 10%.

b. $f(t) = (2.5)^{t-2}$ $f(t) = 0.16(2.5)^t$; The function represents exponential growth; 150%

MONITORING PROGRESS ANSWERS

6. $f(0) = 4, f(n) = 7 \cdot f(n - 1)$
7. $f(0) = 9, f(n) = \frac{1}{3} \cdot f(n - 1)$
8. $f(t) \approx 3(1.22)^t$; exponential growth; about 22%
9. $y \approx 0.9(0.95)^t$; exponential decay; 5%

Laurie's Notes   Teacher Actions

- **Think-Pair-Share:** Pose Example 3 and ask partners to think independently first and then work to write Example 2 as a recursive function.
- **?** Ask students whether the following is *always true*, *sometimes true*, or *never true*. They need to explain their reasoning. "Given the equation $y = ab^x$, when $b > 1$, the equation represents exponential growth." always true; When the base of the exponential is greater than 1, the y-values will increase and you have exponential growth.
- **Popsicle Sticks:** Give students time to work on Example 4 with their partners. Use *Popsicle Sticks* to solicit responses.
- **Look For and Make Use of Structure:** Exponential equations are not always in the form $y = ab^x$. Rewriting equations requires applying properties of exponents.

Extra Example 5

The value of a car is $15,000. It loses 10% of its value every year.

a. Write a function that represents the value y (in dollars) of the car after t years.
  The value of the car can be represented by $y = 15,000(0.9)^t$.

b. Find the approximate monthly percent decrease in value.
  The monthly percent decrease is about 0.9%.

c. Graph the function from part (a). Use the graph to estimate the value of the car after 6 years.

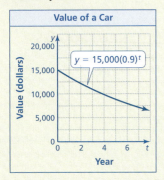

The value of the car is about $8,000 after 6 years.

MONITORING PROGRESS ANSWER

10. a. $y = 21,500(0.91)^t$
    b. about 0.8%
    c.
    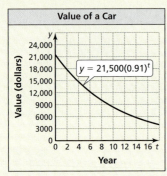
    about $7000

EXAMPLE 5 Solving a Real-Life Problem

The value of a car is $21,500. It loses 12% of its value every year. (a) Write a function that represents the value y (in dollars) of the car after t years. (b) Find the approximate monthly percent decrease in value. (c) Graph the function from part (a). Use the graph to estimate the value of the car after 6 years.

SOLUTION

1. **Understand the Problem** You know the value of the car and its annual percent decrease in value. You are asked to write a function that represents the value of the car over time and approximate the monthly percent decrease in value. Then graph the function and use the graph to estimate the value of the car in the future.

2. **Make a Plan** Use the initial amount and the annual percent decrease in value to write an exponential decay function. Note that the annual percent decrease represents the rate of decay. Rewrite the function using the properties of exponents to approximate the monthly percent decrease (rate of decay). Then graph the original function and use the graph to estimate the y-value when the t-value is 6.

3. **Solve the Problem**

   a. The initial value is $21,500, and the rate of decay is 12%, or 0.12.

$y = a(1 - r)^t$	Write exponential decay model.
$= 21,500(1 - 0.12)^t$	Substitute 21,500 for a and 0.12 for r.
$= 21,500(0.88)^t$	Simplify.

   ▶ The value of the car can be represented by $y = 21,500(0.88)^t$.

   b. Use the fact that $t = \frac{1}{12}(12t)$ and the properties of exponents to rewrite the function in a form that reveals the monthly rate of decay.

$y = 21,500(0.88)^t$	Write the original function.
$= 21,500(0.88)^{(1/12)(12t)}$	Rewrite the exponent.
$= 21,500(0.88^{1/12})^{12t}$	Power of a Power Property
$\approx 21,500(0.989)^{12t}$	Evaluate the power.

   Use the decay factor $1 - r \approx 0.989$ to find the rate of decay $r \approx 0.011$.

   ▶ So, the monthly percent decrease is about 1.1%.

   c. From the graph, you can see that the y-value is about 10,000 when $t = 6$.

   ▶ So, the value of the car is about $10,000 after 6 years.

4. **Look Back** To check that the monthly percent decrease is reasonable, multiply it by 12 to see if it is close in value to the annual percent decrease of 12%.

   $1.1\% \times 12 = 13.2\%$ 13.2% is close to 12%, so 1.1% is reasonable.

   When you evaluate $y = 21,500(0.88)^t$ for $t = 6$, you get about $9985. So, $10,000 is a reasonable estimation.

REASONING QUANTITATIVELY

The decay factor, 0.88, tells you what fraction of the car's value *remains* each year. The rate of decay, 12%, tells you how much value the car *loses* each year. In real life, the percent decrease in value of an asset is the *depreciation rate*.

Monitoring Progress Help in English and Spanish at *BigIdeasMath.com*

10. **WHAT IF?** The car loses 9% of its value every year. (a) Write a function that represents the value y (in dollars) of the car after t years. (b) Find the approximate monthly percent decrease in value. (c) Graph the function from part (a). Use the graph to estimate the value of the car after 12 years. Round your answer to the nearest thousand.

Laurie's Notes — Teacher Actions

- Discuss depreciation of a car. Help students write the depreciation equation from the given information.
- **?** In part (b), you are asked to find the monthly percent decrease in value. "What is $\frac{1}{12}(12t)$?" t Identify this substitution in the problem.
- **?** "Is the amount the car loses in value the same each year? Explain." no; It loses more the first year than any other year. The percent of the value that it loses is the same.

Closure

- **Thumbs Up:** Ask students to discuss the connection between exponential functions and recursive rules with their partners. Use a *Thumbs Up* to indicate how secure their understanding is.

1.6 Exercises

Dynamic Solutions available at *BigIdeasMath.com*

Vocabulary and Core Concept Check

1. **VOCABULARY** In the exponential growth model $y = 2.4(1.5)^x$, identify the initial amount, the growth factor, and the percent increase.

2. **WHICH ONE DOESN'T BELONG?** Which characteristic of an exponential decay function does *not* belong with the other three? Explain your reasoning.

base of 0.8	decay factor of 0.8
decay rate of 20%	80% decrease

Monitoring Progress and Modeling with Mathematics

In Exercises 3–8, evaluate the expression for (a) $x = -2$ and (b) $x = 3$.

3. 2^x
4. 4^x
5. $8 \cdot 3^x$
6. $6 \cdot 2^x$
7. $5 + 3^x$
8. $2^x - 2$

In Exercises 9–18, determine whether the function represents *exponential growth* or *exponential decay*. Then graph the function. *(See Example 1.)*

9. $y = 6^x$
10. $y = 7^x$
11. $y = \left(\frac{1}{6}\right)^x$
12. $y = \left(\frac{1}{8}\right)^x$
13. $y = \left(\frac{4}{3}\right)^x$
14. $y = \left(\frac{2}{5}\right)^x$
15. $y = (1.2)^x$
16. $y = (0.75)^x$
17. $y = (0.6)^x$
18. $y = (1.8)^x$

ANALYZING RELATIONSHIPS In Exercises 19 and 20, use the graph of $f(x) = b^x$ to identify the value of the base b.

19.
20.

21. **MODELING WITH MATHEMATICS** The population of Austin, Texas, was about 494,000 at the beginning of a decade. The population increased by about 3% each year. *(See Example 2.)*

    a. Write an exponential growth model that represents the population y (in thousands) t years after the beginning of the decade. Find and interpret the y-value when $t = 10$.

    b. Estimate when the population was about 590,000.

22. **MODELING WITH MATHEMATICS** You take a 325 milligram dosage of ibuprofen. During each subsequent hour, the amount of medication in your bloodstream decreases by about 29% each hour.

    a. Write an exponential decay model giving the amount y (in milligrams) of ibuprofen in your bloodstream t hours after the initial dose. Find and interpret the y-value when $t = 1.5$.

    b. Estimate how long it takes for you to have 100 milligrams of ibuprofen in your bloodstream.

23. **ERROR ANALYSIS** You invest $500 in the stock of a company. The value of the stock decreases 2% each year. Describe and correct the error in writing a model for the value of the stock after t years.

    ✗ $y = \begin{pmatrix}\text{Initial}\\\text{amount}\end{pmatrix}\begin{pmatrix}\text{Decay}\\\text{factor}\end{pmatrix}^t$

    $y = 500(0.02)^t$

Section 1.6 Exponential Functions 47

Assignment Guide and Homework Check

ASSIGNMENT

Basic: 1, 2, 7–23 odd, 24, 25–31 odd, 41–55 odd, 67, 68, 75–82

Average: 1, 2, 10–18 even, 22, 23, 26–46 even, 50–66 even, 67, 68, 75–82

Advanced: 1, 2, 18, 22, 23, 32–40 even, 50–62 even, 63, 64, 67–82

HOMEWORK CHECK

Basic: 9, 21, 25, 41, 49

Average: 10, 22, 26, 34, 50

Advanced: 18, 22, 32, 54, 62

ANSWERS

1. The initial amount is 2.4, the growth factor is 1.5, and the percent increase is 0.5 or 50%.

2. 80% decrease; It is the only one that, in $y = ab^x$, b is not 0.8.

3. a. $\frac{1}{4}$
   b. 8

4. a. $\frac{1}{16}$
   b. 64

5. a. $\frac{8}{9}$
   b. 216

6. a. $\frac{3}{2}$
   b. 48

7. a. $\frac{46}{9}$
   b. 32

8. a. $-\frac{7}{4}$
   b. 6

9. exponential growth
   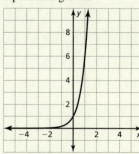

10. exponential growth
    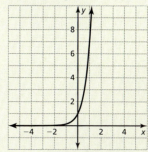

11. exponential decay
    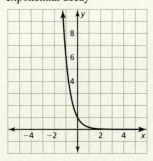

12. exponential decay
    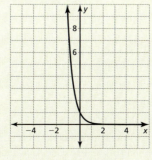

13–23. See Additional Answers.

Section 1.6 47

Dynamic Teaching Tools
- Dynamic Assessment & Progress Monitoring Tool
- Interactive Whiteboard Lesson Library
- Dynamic Classroom with Dynamic Investigations

ANSWERS

24.

n	0	1	2	3	4	5
f(n)	4	12	36	108	324	972

25. $f(0) = 3, f(n) = 7 \cdot f(n-1)$
26. $f(0) = 25, f(n) = 0.2 \cdot f(n-1)$
27. $f(0) = 12, f(n) = 0.5 \cdot f(n-1)$
28. $f(0) = 19, f(n) = 4 \cdot f(n-1)$
29. $f(0) = 0.5, f(n) = 3 \cdot f(n-1)$
30. $f(0) = \frac{1}{3}, f(n) = 2 \cdot f(n-1)$
31. $f(0) = 4, f(n) = \frac{1}{6} \cdot f(n-1)$
32. $f(0) = 0.25, f(n) = \frac{2}{3} \cdot f(n-1)$
33. The inputs are equally spaced and the common ratio is 2; $f(0) = 0.75, f(n) = 2 \cdot f(n-1)$
34. The inputs are equally spaced and the common ratio is 4; $f(0) = 2, f(n) = 4 \cdot f(n-1)$
35. The inputs are equally spaced and the common ratio is $\frac{1}{2}$; $f(0) = 96, f(n) = \frac{1}{2} \cdot f(n-1)$
36. The inputs are equally spaced and the common ratio is $\frac{1}{3}$; $f(0) = 162, f(n) = \frac{1}{3} \cdot f(n-1)$
37. $f(x) = 24(0.1)^x$
38. $f(x) = \frac{1}{2}\left(\frac{5}{2}\right)^x$
39. Sample answer: The number of bacteria in a dish after n hours; $f(0) = 1, f(n) = 2 \cdot f(n-1)$; 64; There are 64 bacteria after 6 hours.
40. a. Tree A: $f(0) = 120$, $f(t) = 1.1 \cdot f(t-1)$; Tree B: $f(0) = 154$, $f(t) = 1.06 \cdot f(t-1)$
    b. Tree B; Tree A; $154 > 120$ and $1.1 > 1.06$
    c.

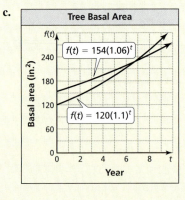
Tree Basal Area

---

24. USING EQUATIONS Complete a table of values for $0 \le n \le 5$ using the given recursive rule of an exponential function.

$$f(0) = 4, f(n) = 3 \cdot f(n-1)$$

In Exercises 25–32, write a recursive rule for the exponential function. *(See Example 3.)*

25. $y = 3(7)^x$
26. $y = 25(0.2)^x$
27. $y = 12(0.5)^t$
28. $y = 19(4)^t$
29. $g(x) = 0.5(3)^x$
30. $m(t) = \frac{1}{3}(2)^t$
31. $f(x) = 4\left(\frac{1}{6}\right)^x$
32. $f(t) = 0.25\left(\frac{2}{3}\right)^t$

In Exercises 33–36, show that an exponential model fits the data. Then write a recursive rule that models the data.

33.

n	0	1	2	3	4	5
f(n)	0.75	1.5	3	6	12	24

34.

n	0	1	2	3	4
f(n)	2	8	32	128	512

35.

n	0	1	2	3	4	5
f(n)	96	48	24	12	6	3

36.

n	0	1	2	3	4	5
f(n)	162	54	18	6	2	$\frac{2}{3}$

In Exercises 37 and 38, write an exponential function for the recursive rule.

37. $f(0) = 24, f(n) = 0.1 \cdot f(n-1)$
38. $f(0) = \frac{1}{2}, f(n) = \frac{5}{2} \cdot f(n-1)$

39. **PROBLEM SOLVING** Describe a real-life situation that can be represented by the graph. Write a recursive rule that models the data. Then find and interpret $f(6)$.

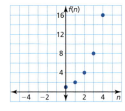

48   Chapter 1   Functions and Exponents

---

40. **PROBLEM SOLVING** The cross-sectional area of a tree 4.5 feet from the ground is called its *basal area*. The table shows the basal areas (in square inches) of Tree A over time.

Year	0	1	2	3	4
Basal area	120	132	145.2	159.7	175.7

Tree B
Growth rate: 6%
Initial basal area: 154 in.^{2}

a. Write recursive rules that represent the basal areas of the trees after t years.
b. Which tree has a greater initial basal area? a greater basal area growth rate? Use the recursive rules you wrote in part (a) to justify your answers.
c. Use a graph to represent the growth of the trees over the interval $0 \le t \le 10$. Does the graph support your answers in part (b)? Explain. Then make an additional observation from the graph.

In Exercises 41–44, determine whether the function represents *exponential growth* or *exponential decay*. Then identify the percent rate of change.

41. $y = 5(0.6)^t$
42. $y = 10(1.07)^t$
43. $f(t) = 200\left(\frac{4}{3}\right)^t$
44. $f(t) = 0.8\left(\frac{1}{4}\right)^t$

45. **PROBLEM SOLVING** A website recorded the number y of referrals it received from social media websites over a 10-year period. The results can be modeled by $y = 2500(1.50)^t$, where t is the year and $0 \le t \le 9$.

a. Describe the real-life meaning of 2500 and 1.50 in the model.
b. What is the annual percent increase? Explain.

46. **PROBLEM SOLVING** The population p of a small town after x years can be modeled by the function $p = 6850(0.97)^x$.

a. What is the annual percent decrease? Explain.
b. What is the average rate of change in the population over the first 6 years? Justify your answer.

---

yes; The graph for Tree B has a greater y-intercept, but the graph for Tree A increases at a greater rate; Sample answer: The difference between the basal areas is decreasing.

41. exponential decay; 40%
42. exponential growth; 7%
43. exponential growth; $33\frac{1}{3}$%
44. exponential decay; 75%
45. a. The website had 2500 referrals at the beginning of the 10-year period and the number of referrals increased by a factor of 1.50 each year.
    b. 50%; $1 + r = 1.50$, so $r = 0.50$ or 50%
46. a. 3%; $1 - r = 0.97$, so $r = 0.03$ or 3%
    b. about -191 people per year; When $x = 0, y = 6850$, and when $x = 6$, $y \approx 5706$, so the average rate of change is $\frac{5706 - 6850}{6 - 0} \approx -191$.

JUSTIFYING STEPS In Exercises 47 and 48, justify each step in rewriting the exponential function.

47. $y = a(3)^{t/14}$ Write original function.
   $= a[(3)^{1/14}]^t$
   $\approx a(1.0816)^t$
   $= a(1 + 0.0816)^t$

48. $y = a(0.1)^{t/3}$ Write original function.
   $= a[(0.1)^{1/3}]^t$
   $\approx a(0.4642)^t$
   $= a(1 - 0.5358)^t$

In Exercises 49–56, rewrite the function to determine whether it represents *exponential growth* or *exponential decay*. Then identify the percent rate of change. *(See Example 4.)*

49. $y = (0.9)^{t-4}$
50. $y = (1.4)^{t+8}$
51. $y = 2(1.06)^{9t}$
52. $y = 5(0.82)^{t/5}$
53. $x(t) = (1.45)^{t/2}$
54. $f(t) = 0.4(1.16)^{t-1}$
55. $b(t) = 4(0.55)^{t+3}$
56. $r(t) = (0.88)^{4t}$

In Exercises 57–62, rewrite the function in the form $y = a(1 + r)^t$ or $y = a(1 - r)^t$. Then state the growth or decay rate.

57. $y = a(2)^{t/3}$
58. $y = a(0.5)^{t/12}$
59. $y = a\left(\frac{2}{3}\right)^{t/10}$
60. $y = a\left(\frac{5}{4}\right)^{t/22}$
61. $y = a(2)^{8t}$
62. $y = a\left(\frac{1}{3}\right)^{3t}$

63. **PROBLEM SOLVING** When a plant or animal dies, it stops acquiring carbon-14 from the atmosphere. The amount y (in grams) of carbon-14 in the body of an organism after t years is $y = a(0.5)^{t/5730}$, where a is the initial amount (in grams). What percent of the carbon-14 is released each year?

64. **PROBLEM SOLVING** The number y of duckweed fronds in a pond after t days is $y = a(1230.25)^{t/16}$, where a is the initial number of fronds. By what percent does the duckweed increase each day?

65. **PROBLEM SOLVING** A city has a population of 25,000. The population is expected to increase by 5.5% annually for the next decade. *(See Example 5.)*

a. Write a function that represents the population y after t years.
b. Find the approximate monthly percent increase in population.
c. Graph the function from part (a). Use the graph to estimate the population after 4 years.

66. **PROBLEM SOLVING** Plutonium-238 is a material that generates steady heat due to decay and is used in power systems for some spacecraft. The function $y = a(0.5)^{t/x}$ represents the amount y of a substance remaining after t years, where a is the initial amount and x is the length of the half-life (in years).

Plutonium-238
Half-life ≈ 88 years

a. A scientist is studying a 3-gram sample. Write a function that represents the amount y of plutonium-238 after t years.
b. What is the yearly percent decrease of plutonium-238?
c. Graph the function from part (a). Use the graph to estimate the amount remaining after 12 years.

67. **MAKING AN ARGUMENT** Your friend says the graph of $f(x) = 2^x$ increases at a faster rate than the graph of $g(x) = 4x$ when $x \geq 0$. Is your friend correct? Explain your reasoning.

67. no; $f(x) = 2^x$ eventually increases at a faster rate than $g(x) = 4x$, but not for all $x \geq 0$.

ANSWERS

47. Power of a Power Property; Evaluate power; Rewrite in form $y = a(1 + r)^t$.
48. Power of a Power Property; Evaluate power; Rewrite in form $y = a(1 - r)^t$.
49. $y \approx 1.52(0.9)^t$; exponential decay; 10%
50. $y \approx 14.8(1.4)^t$; exponential growth; 40%
51. $y \approx 2(1.69)^t$; exponential growth; about 69%
52. $y \approx 5(0.96)^t$; exponential decay; about 4%
53. $x(t) \approx (1.20)^t$; exponential growth; about 20%
54. $f(t) \approx 0.34(1.16)^t$; exponential growth; 16%
55. $b(t) \approx 0.67(0.55)^t$; exponential decay; 45%
56. $r(t) \approx (0.60)^t$; exponential decay; about 40%
57. $y = a(1 + 0.26)^t$; 26% growth
58. $y = a(1 - 0.06)^t$; 6% decay
59. $y = a(1 - 0.04)^t$; 4% decay
60. $y = a(1 + 0.01)^t$; 1% growth
61. $y = a(1 + 255)^t$; 25,500% growth
62. $y = a(1 - 0.96)^t$; 96% decay
63. about 0.01%
64. about 56%
65. a. $y = 25,000(1.055)^t$
    b. about 0.45%
    c.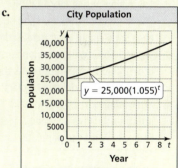

    City Population, $y = 25,000(1.055)^t$

    about 30,971 people
66. a. $y = 3(0.5)^{t/88}$
    b. about 0.8%
    c. Plutonium-238 Decay, $y = 3(0.5)^{t/88}$

    about 2.7 g

ANSWERS
68. a. exponential decay
b. domain: all real numbers, range: $y > 0$; It is an exponential function and any real number can be used as an exponent. $f(x) \to 0$ as $x \to \infty$ and $f(x) \to \infty$ as $x \to -\infty$.

69–82. See Additional Answers.

Mini-Assessment

1. Tell whether the function $y = \left(\frac{5}{4}\right)^x$ represents *exponential growth* or *exponential decay*. Then graph the function. **exponential growth**

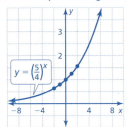

$y = \left(\frac{5}{4}\right)^x$

2. The population of a city was about 1.6 million on January 1, 2000. The population at the beginning of each subsequent year increased by about 1.76%. Write an exponential growth model that represents the population y (in millions) t years after January 1, 2000. Find and interpret the y-value when $t = 9.5$. Estimate when the population was 1.8 million.
$y = 1.6(1.0176)^t$; The midyear population in 2009 was about 1.9 million; near the end of 2006

3. Write a recursive rule for the exponential function $f(x) = 1.2(0.4)^x$.
$f(0) = 1.2, f(n) = 0.4 \cdot f(n-1)$

4. Rewrite the function $f(x) = 300(0.7)^{x/4}$ in the form $f(x) = ab^x$ to determine whether it represents *exponential growth* or *exponential decay*. Then identify the percent rate of change.
$f(x) \approx 300(0.91)^x$; exponential decay; about 9%

---

68. HOW DO YOU SEE IT? Consider the graph of an exponential function of the form $f(x) = ab^x$.

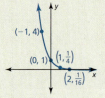

a. Determine whether the graph of f represents exponential growth or exponential decay.

b. What are the domain and range of the function? Explain.

69. COMPARING FUNCTIONS The two given functions describe the amount y of ibuprofen (in milligrams) in a person's bloodstream t hours after taking the dosage.

$$y \approx 325(0.9943)^{60t} \qquad y \approx 325(0.843)^{2t}$$

a. Show that these models are approximately equivalent to the model you wrote in Exercise 22.

b. Describe the information given by each of the models above.

70. DRAWING CONCLUSIONS The amount A in an account after t years with principal P, annual interest rate r (expressed as a decimal), and compounded n times per year is given by

$$A = P\left(1 + \frac{r}{n}\right)^{nt}.$$

You deposit $1000 into three separate bank accounts that each pay 3% annual interest. For each account, evaluate $\left(1 + \frac{r}{n}\right)^n$. Interpret this quantity in the context of the problem. Then complete the table.

Account	Compounding	Balance after 1 year
1	quarterly	
2	monthly	
3	daily	

Maintaining Mathematical Proficiency — Reviewing what you learned in previous grades and lessons

Simplify the expression. *(Skills Review Handbook)*

75. $x + 3x$ **76.** $8y - 21y$ **77.** $13z + 9 - 8z$ **78.** $-9w + w - 5$

Simplify the expression. Write your answer using only positive exponents. *(Section 1.4)*

79. $x^{-9} \cdot x^2$ **80.** $\dfrac{x^4}{x^3}$ **81.** $(-6x)^2$ **82.** $\left(\dfrac{4x^8}{2x^6}\right)^4$

50   Chapter 1   Functions and Exponents

---

71. REASONING Consider the exponential function $f(x) = ab^x$.

a. Show that $\dfrac{f(x+1)}{f(x)} = b$.

b. Use the equation in part (a) to explain why there is no exponential function of the form $f(x) = ab^x$ whose graph passes through the points in the table below.

x	0	1	2	3	4
y	4	4	8	24	72

72. THOUGHT PROVOKING The function $f(x) = b^x$ represents an exponential decay function. Write a second exponential decay function in terms of b and x.

73. PROBLEM SOLVING The number E of eggs a Leghorn chicken produces per year can be modeled by the equation $E = 179.2(0.89)^{w/52}$, where w is the age (in weeks) of the chicken and $w \geq 22$.

a. Identify the decay factor and the percent decrease.

b. Graph the model.

c. Estimate the egg production of a chicken that is 2.5 years old.

d. Explain how you can rewrite the model so that time is measured in years rather than in weeks.

74. CRITICAL THINKING You buy a new stereo for $1300 and are able to sell it 4 years later for $275. Assume that the resale value of the stereo decays exponentially with time. Write an equation giving the resale value V (in dollars) of the stereo as a function of the time t (in years) since you bought it.

If students need help...	If students got it...
Resources by Chapter • Practice A and Practice B • Puzzle Time	**Resources by Chapter** • Enrichment and Extension • Cumulative Review
Student Journal • Practice	Start the *next* Section
Differentiating the Lesson Skills Review Handbook	

1.4–1.6 What Did You Learn?

Core Vocabulary

nth root of a, p. 36
radical, p. 36
index of a radical, p. 36
exponential function, p. 42

exponential growth function, p. 42
growth factor, p. 42
asymptote, p. 42
exponential decay function, p. 42

decay factor, p. 42
recursive rule for an exponential function, p. 44

Core Concepts

Section 1.4
Zero Exponent, p. 28
Negative Exponents, p. 28
Product of Powers Property, p. 29
Quotient of Powers Property, p. 29

Power of a Power Property, p. 29
Power of a Product Property, p. 30
Power of a Quotient Property, p. 30

Section 1.5
Real nth Roots of a, p. 36

Rational Exponents, p. 37

Section 1.6
Parent Function for Exponential Growth Functions, p. 42
Parent Function for Exponential Decay Functions, p. 42

Exponential Growth and Decay Models, p. 43
Writing Recursive Rules for Exponential Functions, p. 44

Mathematical Practices

1. How did you apply what you know to simplify the complicated situation in Exercise 56 on page 33?
2. How can you use previously established results to construct an argument in Exercise 44 on page 40?

Performance Task:

Pool Shots

How can mathematics help you become a better pool player? What type of function could you use to sink a pool ball? What aspects of the function will help the shot be successful?

To explore the answers to these questions and more, check out the Performance Task and Real-Life STEM video at *BigIdeasMath.com*.

ANSWERS

1.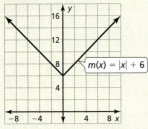

The graph of m is a vertical translation 6 units up of the graph of f; domain: all real numbers; range: $y \geq 6$

2.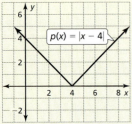

The graph of p is a horizontal translation 4 units right of the graph of f; domain: all real numbers; range: $y \geq 0$

3.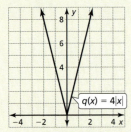

The graph of q is a vertical stretch of the graph of f by a factor of 4; domain: all real numbers; range: $y \geq 0$

4.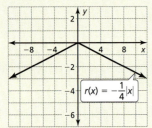

The graph of r is a vertical shrink of the graph of f by a factor of $\frac{1}{4}$ and a reflection in the x-axis; domain: all real numbers; range: $y \leq 0$

5.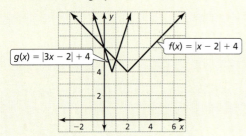

The graph of g is a horizontal shrink of the graph of f by a factor of $\frac{1}{3}$.

6. Sample answer: Identify and plot the vertex $(-1, 2)$. Plot another point such as $(0, -1)$ and use symmetry to plot a third point, $(-2, -1)$. Then draw the V-shaped graph.

7. a. The transformations are a horizontal translation 1 unit right, then a vertical shrink by a factor of $\frac{1}{3}$, then a vertical translation 2 units down.

b.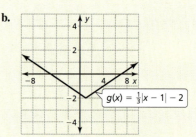

8. a. 3
b. -10

---

1 Chapter Review

Dynamic Solutions available at BigIdeasMath.com

1.1 Absolute Value Functions (pp. 3–10)

Let $g(x) = -3|x + 1| + 2$. (a) Describe the transformations from the graph of $f(x) = |x|$ to the graph of g. (b) Graph g.

a. Step 1 Translate the graph of f horizontally 1 unit left to get the graph of $t(x) = |x + 1|$.

Step 2 Stretch the graph of t vertically by a factor of 3 to get the graph of $h(x) = 3|x + 1|$.

Step 3 Reflect the graph of h in the x-axis to get the graph of $r(x) = -3|x + 1|$.

Step 4 Translate the graph of r vertically 2 units up to get the graph of $g(x) = -3|x + 1| + 2$.

b. Step 1 Make a table of values.

x	−3	−2	−1	0	1
g(x)	−4	−1	2	−1	−4

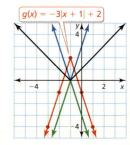

Step 2 Plot the ordered pairs.

Step 3 Draw the V-shaped graph.

Graph the function. Compare the graph to the graph of $f(x) = |x|$. Describe the domain and range.

1. $m(x) = |x| + 6$
2. $p(x) = |x - 4|$
3. $q(x) = 4|x|$
4. $r(x) = -\frac{1}{4}|x|$

5. Graph $f(x) = |x - 2| + 4$ and $g(x) = |3x - 2| + 4$. Compare the graph of g to the graph of f.

6. Describe another method you can use to graph g in the example above.

7. Let $g(x) = \frac{1}{3}|x - 1| - 2$. (a) Describe the transformations from the graph of $f(x) = |x|$ to the graph of g. (b) Graph g.

1.2 Piecewise Functions (pp. 11–18)

Graph $y = \begin{cases} \frac{3}{2}x + 3, & \text{if } x \leq 0 \\ -2x, & \text{if } x > 0 \end{cases}$. Describe the domain and range.

Step 1 Graph $y = \frac{3}{2}x + 3$ for $x \leq 0$. Because x is less than or equal to 0, use a closed circle at $(0, 3)$.

Step 2 Graph $y = -2x$ for $x > 0$. Because x is not equal to 0, use an open circle at $(0, 0)$.

▶ The domain is all real numbers. The range is $y \leq 3$.

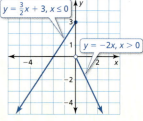

8. Evaluate the function in the example when (a) $x = 0$ and (b) $x = 5$.

Graph the function. Describe the domain and range.

9. $y = \begin{cases} x + 6, & \text{if } x \leq 0 \\ -3x, & \text{if } x > 0 \end{cases}$

10. $y = \begin{cases} 4x + 2, & \text{if } x < -4 \\ 2x - 6, & \text{if } x \geq -4 \end{cases}$

Write the absolute value function as a piecewise function.

11. $y = |x| + 15$
12. $y = 4|x + 5|$
13. $y = 2|x + 2| - 3$

14. You are organizing a school fair and rent a popcorn machine for 3 days. The rental company charges $65 for the first day and $35 for each additional day. Write and graph a step function that represents the relationship between the number x of days and the total cost y (in dollars) of renting the popcorn machine.

1.3 Inverse of a Function (pp. 19–24)

Find the inverse of $f(x) = \frac{1}{3}x - 2$. Then graph the function and its inverse.

$y = \frac{1}{3}x - 2$ Set y equal to $f(x)$.

$x = \frac{1}{3}y - 2$ Switch x and y in the equation.

$x + 2 = \frac{1}{3}y$ Add 2 to each side.

$3x + 6 = y$ Multiply each side by 3.

▶ The inverse of f is $g(x) = 3x + 6$.

15. Find the inverse of the relation: $(1, -10), (3, -4), (5, 4), (7, 14), (9, 26)$.

Find the inverse of the function. Then graph the function and its inverse.

16. $f(x) = \frac{3}{4}x$
17. $f(x) = -5x + 10$
18. $f(x) = -\frac{2}{5}x + 6$

19. In bowling, a handicap is an adjustment to a bowler's score to even out differences in ability levels. In a particular league, you can find a bowler's handicap h by using the formula $h = 0.8(210 - a)$, where a is the bowler's average. Solve the formula for a. Then find a bowler's average when the bowler's handicap is 28.

1.4 Properties of Exponents (pp. 27–34)

Simplify $\left(\dfrac{x}{4}\right)^{-4}$. Write your answer using only positive exponents.

$\left(\dfrac{x}{4}\right)^{-4} = \dfrac{x^{-4}}{4^{-4}}$ Power of a Quotient Property

$= \dfrac{4^4}{x^4}$ Definition of negative exponent

$= \dfrac{256}{x^4}$ Simplify.

Simplify the expression. Write your answer using only positive exponents.

20. $y^3 \cdot y^{-5}$
21. $\dfrac{x^4}{x^7}$
22. $(x^0 y^2)^3$
23. $\left(\dfrac{2x^2}{5y^4}\right)^{-2}$

17. $g(x) = \dfrac{10 - x}{5}$

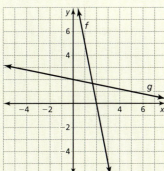

18. $g(x) = -\dfrac{5}{2}x + 15$

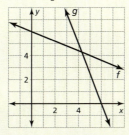

19–23. See Additional Answers.

ANSWERS

9.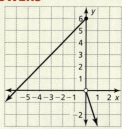

domain: all real numbers; range: $y \leq 6$

10.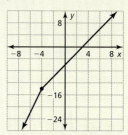

domain: all real numbers; range: all real numbers

11. $y = \begin{cases} -x + 15, & \text{if } x < 0 \\ x + 15, & \text{if } x \geq 0 \end{cases}$

12. $y = \begin{cases} -4x - 20, & \text{if } x < -5 \\ 4x + 20, & \text{if } x \geq -5 \end{cases}$

13. $y = \begin{cases} -2x - 7, & \text{if } x < -2 \\ 2x + 1, & \text{if } x \geq -2 \end{cases}$

14. $f(x) = \begin{cases} 65, & \text{if } 0 < x \leq 1 \\ 100, & \text{if } 1 < x \leq 2 \\ 135, & \text{if } 2 < x \leq 3 \end{cases}$

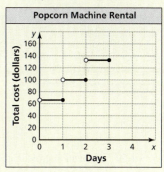

15. $(-10, 1), (-4, 3), (4, 5), (14, 7), (26, 9)$

16. $g(x) = \dfrac{4}{3}x$

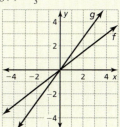

ANSWERS

24. 2
25. −3
26. 125
27. not a real number
28. exponential decay; 66.67%

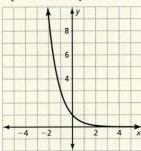

29. exponential growth; 400%

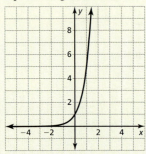

30. exponential decay; 80%

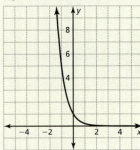

31. $f(t) \approx 7.81(1.25)^t$, exponential growth; 25%
32. $y \approx 1.59^t$, exponential growth, about 59%
33. $f(t) \approx 12.05(0.84)^t$, exponential decay; 16%

1.5 Radicals and Rational Exponents (pp. 35–40)

Evaluate $512^{1/3}$.

$512^{1/3} = \sqrt[3]{512}$ Rewrite the expression in radical form.
$= \sqrt[3]{8 \cdot 8 \cdot 8}$ Rewrite the expression showing factors.
$= 8$ Evaluate the cube root.

Evaluate the expression.

24. $\sqrt[3]{8}$
25. $\sqrt[5]{-243}$
26. $625^{3/4}$
27. $(-25)^{1/2}$

1.6 Exponential Functions (pp. 41–50)

a. Determine whether the function $y = 3^x$ represents *exponential growth* or *exponential decay*. Then graph the function.

Step 1 Identify the value of the base. The base, 3, is greater than 1, so the function represents exponential growth.

Step 2 Make a table of values.

x	−2	−1	0	1	2
y	$\frac{1}{9}$	$\frac{1}{3}$	1	3	9

Step 3 Plot the points from the table.

Step 4 Draw, from *left to right*, a smooth curve that begins just above the x-axis, passes through the plotted points, and moves up to the right.

b. Rewrite the function $y = 10(0.65)^{t/8}$ to determine whether it represents *exponential growth* or *exponential decay*. Then identify the percent rate of change.

$y = 10(0.65)^{t/8}$ Write the function.
$= 10(0.65^{1/8})^t$ Power of a Power Property
$\approx 10(0.95)^t$ Evaluate the power.

▶ So, the function represents exponential decay. Use the decay factor $1 − r \approx 0.95$ to find the rate of decay $r \approx 0.05$, or about 5%.

Determine whether the function represents exponential growth or exponential decay. Identify the percent rate of change. Then graph the function.

28. $f(x) = \left(\frac{1}{3}\right)^x$
29. $y = 5^x$
30. $f(x) = (0.2)^x$

Rewrite the function to determine whether it represents *exponential growth* or *exponential decay*. Then identify the percent rate of change.

31. $f(t) = 4(1.25)^{t+3}$
32. $y = (1.06)^{8t}$
33. $f(t) = 6(0.84)^{t-4}$

1 Chapter Test

Evaluate the expression.

1. $-\sqrt[4]{16}$
2. $729^{1/6}$
3. $(-32)^{7/5}$

Simplify the expression. Write your answer using only positive exponents.

4. $z^{-2} \cdot z^4$
5. $\dfrac{b^{-5}}{a^0 b^{-8}}$
6. $\left(\dfrac{2c^4}{5}\right)^{-3}$

Graph the function. Describe the domain and range.

7. $h(x) = -|x - 1| - 4$
8. $p(x) = \left|\dfrac{1}{2}x + 3\right| + 2$
9. $y = \begin{cases} 2x + 4, & \text{if } x \le -1 \\ \dfrac{1}{3}x - 1, & \text{if } x > -1 \end{cases}$
10. $y = \begin{cases} 1, & \text{if } 0 \le x < 3 \\ 0, & \text{if } 3 \le x < 6 \\ -1, & \text{if } 6 \le x < 9 \\ -2, & \text{if } 9 \le x < 12 \end{cases}$

Find the inverse of the function. Then graph the function and its inverse.

11. $g(x) = -4x$
12. $r(x) = 3x + 5$
13. $v(x) = -\dfrac{3}{4}x - \dfrac{1}{3}$

Use the equation to complete the statement "$a \;\square\; b$" with the symbol $<$, $>$, or $=$. Do not attempt to solve the equation.

14. $\dfrac{5^a}{5^b} = 5^{-3}$
15. $9^a \cdot 9^{-b} = 1$

16. Write a piecewise function defined by three equations that has a domain of all real numbers and a range of $-3 < y \le 1$.

17. A rock band releases a new single. Weekly sales s (in thousands of dollars) increase and then decrease as described by the function $s(t) = -2|t - 20| + 40$, where t is the time (in weeks).
    a. Graph s. Describe the transformations from the graph of $f(x) = |x|$ to the graph of s.
    b. Identify and interpret the vertex of the graph of s in the context of the problem.

18. The speed of light is approximately 3×10^5 kilometers per second. The average distance from the Sun to Neptune is about 4.5 billion kilometers. How long does it take sunlight to reach Neptune? Write your answer in scientific notation and in standard form.

19. The value of a mountain bike y (in dollars) can be approximated by the model $y = 200(0.75)^t$, where t is the number of years since the bike was new.
    a. Determine whether the model represents exponential growth or exponential decay.
    b. Identify the annual percent increase or decrease in the value of the bike.
    c. Estimate when the value of the bike will be $50.

20. The amount y (in grams) of the radioactive isotope iodine-123 remaining after t hours is $y = a(0.5)^{t/13}$, where a is the initial amount (in grams). What percent of the iodine-123 decays each hour?

ANSWERS

1. -2
2. 3
3. -128
4. z^2
5. b^3
6. $\dfrac{125}{8c^{12}}$
7.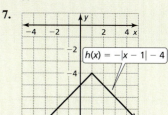

   domain: all real numbers; range: $y \le -4$

8.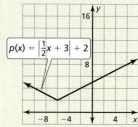

   domain: all real numbers; range: $y \ge 2$

9.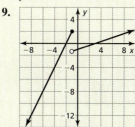

   domain: all real numbers; range: all real numbers

10.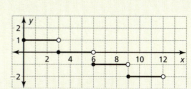

    domain: $0 \le x < 12$; range: $-2, -1, 0, 1$

11. $h(x) = -\dfrac{1}{4}x$

    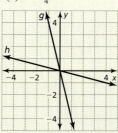

12–20. See Additional Answers.

1 Cumulative Assessment

ANSWERS
1. 3
2. $+$; $<$; $-$; $\geq$
3. B
4. *Sample answer:* D, F, A
5. $2\frac{1}{2}, 3, 3\frac{1}{2}, 4$

1. Fill in the exponent of x with a number to simplify the expression.

   $$\frac{x^{5/3} \cdot x^{-1} \cdot \sqrt[3]{x}}{x^{-2} \cdot x^0} = x^{\boxed{}}$$

2. Fill in the piecewise function with $-$, $+$, $<$, $\leq$, $>$, or $\geq$ so that the function is represented by the graph.

   $$y = \begin{cases} 2x \,\boxed{}\, 3, & \text{if } x \,\boxed{}\, 0 \\ 2x \,\boxed{}\, 3, & \text{if } x \,\boxed{}\, 0 \end{cases}$$

   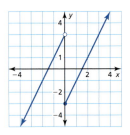

3. Which graph represents the inverse of the function $f(x) = 2x + 4$?

   (A)   (B)

   (C)   (D)

4. Let $f(x) = |x|$ and $g(x) = -4|x + 11|$. Select the possible transformations of the graph of f represented by the function g.

   (A) reflection in the x-axis
   (B) reflection in the y-axis
   (C) horizontal translation 11 units right
   (D) horizontal translation 11 units left
   (E) horizontal shrink by a factor of $\frac{1}{4}$
   (F) vertical stretch by a factor of 4

5. Select all the numbers that are in the range of the function shown.

   $$y = \begin{cases} |x + 2| + 3, & \text{if } x \leq -2 \\ \frac{1}{2}x + 3, & \text{if } x > -2 \end{cases}$$

   $\boxed{0}$ $\boxed{\tfrac{1}{2}}$ $\boxed{1}$ $\boxed{1\tfrac{1}{2}}$ $\boxed{2}$ $\boxed{2\tfrac{1}{2}}$ $\boxed{3}$ $\boxed{3\tfrac{1}{2}}$ $\boxed{4}$

6. Select every value of b for the equation $y = b^x$ that could result in the graph shown.

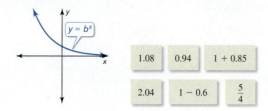

| 1.08 | 0.94 | $1 + 0.85$ |
| 2.04 | $1 - 0.6$ | $\frac{5}{4}$ |

7. Pair each function with its inverse.

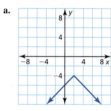 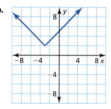

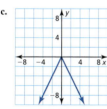 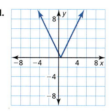

8. Describe the transformation of the graph of $f(x) = |x|$ represented in each graph.

   a.
   b.
   c.
   d.

9. Which of the expressions are equivalent?

Chapter 1   Cumulative Assessment   57

Chapter 2 Pacing Guide

Chapter Opener/Mathematical Practices	0.5 Day
Section 1	1.5 Days
Section 2	2 Days
Section 3	2 Days
Section 4	2 Days
Quiz	0.5 Day
Section 5	1.5 Days
Section 6	2 Days
Section 7	2 Days
Section 8	2 Days
Chapter Review/Chapter Tests	2 Days
Total Chapter 2	18 Days
Year-to-Date	31 Days

2 Polynomial Equations and Factoring

- **2.1** Adding and Subtracting Polynomials
- **2.2** Multiplying Polynomials
- **2.3** Special Products of Polynomials
- **2.4** Solving Polynomial Equations in Factored Form
- **2.5** Factoring $x^2 + bx + c$
- **2.6** Factoring $ax^2 + bx + c$
- **2.7** Factoring Special Products
- **2.8** Factoring Polynomials Completely

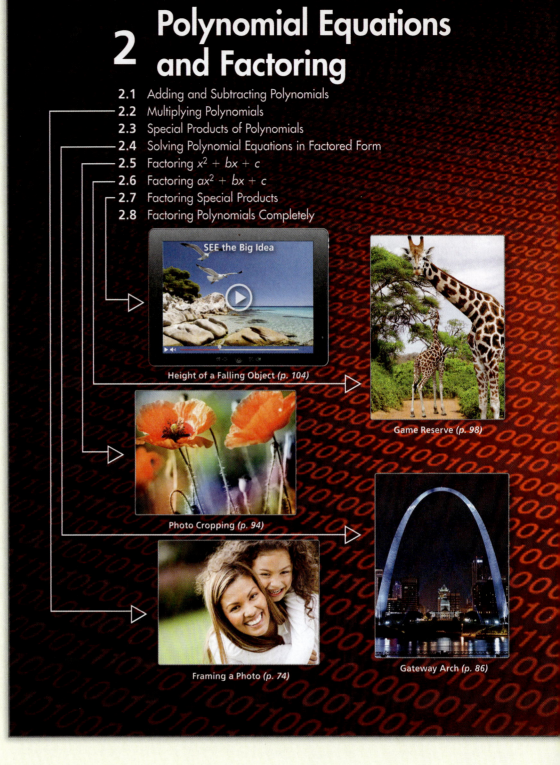

SEE the Big Idea

Height of a Falling Object (p. 104)

Game Reserve (p. 98)

Photo Cropping (p. 94)

Framing a Photo (p. 74)

Gateway Arch (p. 86)

Laurie's Notes

Chapter Summary

- This is a long chapter about polynomial equations and factoring. It is positioned here in the book in preparation for upcoming work with quadratics.
- In the first few lessons, the vocabulary and representation of polynomials is introduced, along with operations with polynomials. Operations of addition, subtraction, and multiplication are presented.
- The remainder of the chapter is on solving polynomial equations, which can be done when the polynomial is written in factored form. Students will use the Zero-Product Property to solve polynomial equations in factored form. Students will learn a series of techniques for factoring polynomials, aided by visual explorations using algebra tiles.

What Your Students Have Learned

Middle School
- Solve linear equations using the Distributive Property and combining like terms.
- Find the greatest common factor (GCF) of two whole numbers.
- Add, subtract, factor, and expand linear expressions with rational coefficients.
- Know and apply the properties of integer exponents to generate equivalent numerical expressions.

Math I
- Solve multi-step linear equations in one variable that have variables on both sides using inverse operations.

What Your Students Will Learn

Math II
- Classify, add, and subtract polynomials.
- Multiply two binomials using the FOIL Method, the square of a binomial pattern, and the sum and difference pattern.
- Multiply binomials and trinomials.
- Solve polynomial equations by using the Zero-Product Property and by factoring out the greatest common factor (GCF).
- Factor $x^2 + bx + c$, $ax^2 + bx + c$, and perfect square trinomials.
- Factor polynomials as the difference of two squares and by grouping.
- Use factoring to solve real-life problems involving polynomial equations.

Dynamic Teaching Tools
Dynamic Assessment & Progress Monitoring Tool
Lesson Planning Tool
Interactive Whiteboard Lesson Library
Dynamic Classroom with Dynamic Investigations
Real-Life STEM Videos

Scaffolding in the Classroom
Graphic Organizers: Concept Circle
A Concept Circle can be used to organize information about a concept. Students write the concept above the circle. Then students write associated information in the sectors of the circle. Associated information can include (an explanation of) *Concept, Apply, Solve, Check, Example,* and *Justify*. Concept Circles can have any number of sectors. Students can place their concept circles on note cards to use as a quick study reference.

Questioning in the Classroom

Share ideas.

Have students discuss closure questions or start thinking questions in small groups and then share their answers using a document camera, chart paper, etc. Have different students from each group support their answers.

Laurie's Notes

Maintaining Mathematical Proficiency

Simplifying Algebraic Expressions

- Remind students that when they use the Distributive Property to write a product such as $-8(y-3)$ as a sum or difference, they must multiply each number inside the parentheses by the factor outside the parentheses.

  COMMON ERROR Students may not change the sign of the values inside the parentheses when multiplying by a negative factor. For example, students may rewrite $2m - 7(3 - m)$ in Exercise 5 as $2m - 21 - 7m$ instead of $2m - 21 + 7m$.

Finding the Greatest Common Factor

- Remind students to find the prime factorization of each number to determine the GCF. If the numbers have two or more common prime factors, the GCF is the product of those common prime factors.

  COMMON ERROR Students may think that they need to identify only the greatest common prime factor and not all the common prime factors to find the GCF.

Mathematical Practices (continued on page 60)

- The *Mathematical Practices* page focuses attention on how mathematics is learned—process versus content. This page demonstrates that algebra tiles are an appropriate tool for modeling algebraic expressions. Algebra tiles will be used extensively in the lesson explorations in this chapter.
- Use the *Mathematical Practices* page to help students develop mathematical habits of mind—how mathematics can be explored and how mathematics is thought about.

If students need help...	If students got it...
Student Journal • Maintaining Mathematical Proficiency	Game Closet at *BigIdeasMath.com*
Lesson Tutorials	Start the *next* Section
Skills Review Handbook	

Maintaining Mathematical Proficiency

Simplifying Algebraic Expressions

Example 1 Simplify $6x + 5 - 3x - 4$.

$$6x + 5 - 3x - 4 = 6x - 3x + 5 - 4 \quad \text{Commutative Property of Addition}$$
$$= (6 - 3)x + 5 - 4 \quad \text{Distributive Property}$$
$$= 3x + 1 \quad \text{Simplify.}$$

Example 2 Simplify $-8(y - 3) + 2y$.

$$-8(y - 3) + 2y = -8(y) - (-8)(3) + 2y \quad \text{Distributive Property}$$
$$= -8y + 24 + 2y \quad \text{Multiply.}$$
$$= -8y + 2y + 24 \quad \text{Commutative Property of Addition}$$
$$= (-8 + 2)y + 24 \quad \text{Distributive Property}$$
$$= -6y + 24 \quad \text{Simplify.}$$

Simplify the expression.

1. $3x - 7 + 2x$
2. $4r + 6 - 9r - 1$
3. $-5t + 3 - t - 4 + 8t$
4. $3(s - 1) + 5$
5. $2m - 7(3 - m)$
6. $4(h + 6) - (h - 2)$

Finding the Greatest Common Factor

Example 3 Find the greatest common factor (GCF) of 42 and 70.

To find the GCF of two numbers, first write the prime factorization of each number. Then find the product of the common prime factors.

$42 = \boxed{2} \cdot 3 \cdot \boxed{7}$
$70 = \boxed{2} \cdot 5 \cdot \boxed{7}$

▶ The GCF of 42 and 70 is $2 \cdot 7 = 14$.

Find the greatest common factor.

7. 20, 36
8. 42, 63
9. 54, 81
10. 72, 84
11. 28, 64
12. 30, 77

13. **ABSTRACT REASONING** Is it possible for two integers to have no common factors? Explain your reasoning.

Dynamic Solutions available at *BigIdeasMath.com*

What Your Students Have Learned

- Simplify algebraic expressions using the Distributive Property and combining like terms.
- Find the greatest common factor (GCF) of two or more numbers by first writing the prime factorizations of the numbers.

ANSWERS

1. $5x - 7$
2. $-5r + 5$
3. $2t - 1$
4. $3s + 2$
5. $9m - 21$
6. $3h + 26$
7. 4
8. 21
9. 27
10. 12
11. 4
12. 1
13. no; The number 1 is a factor of every integer.

Vocabulary Review

Have students make Idea and Examples Charts for the following terms.
- Commutative Property of Addition
- Distributive Property
- Greatest common factor

MONITORING PROGRESS ANSWERS

1. $3x^2 - x + 1$
2. $-x^2 + x$
3. $-x^2 - 2x$
4. $x^2 + x - 1$
5. $x^2 + 2$
6. $x - 6$
7. $-x^2 + 2x$
8. 9
9. $2x^2$

Mathematical Practices

Mathematically proficient students consider concrete models when solving a mathematics problem.

Using Models

Core Concept

Using Algebra Tiles

When solving a problem, it can be helpful to use a model. For instance, you can use algebra tiles to model algebraic expressions and operations with algebraic expressions.

EXAMPLE 1 Writing Expressions Modeled by Algebra Tiles

Write the algebraic expression modeled by the algebra tiles.

a.   b.   c.

SOLUTION

a. The algebraic expression is x^2.
b. The algebraic expression is $3x + 4$.
c. The algebraic expression is $x^2 - x + 2$.

Monitoring Progress

Write the algebraic expression modeled by the algebra tiles.

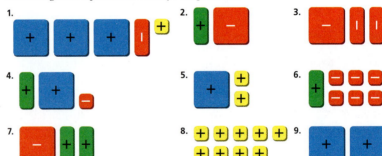

60   Chapter 2   Polynomial Equations and Factoring

Laurie's Notes Mathematical Practices (continued from page T-59)

- Throughout this chapter, algebra tiles will be used to model operations with algebraic expressions. The tiles can be used to help students understand why only like terms can be added or subtracted. The tiles also help students see a visual representation of two binomials being multiplied. Factoring trinomials can also be modeled with algebra tiles.
- It is assumed that students have worked with algebra tiles before, when learning about integer operations and when solving equations in one variable, as students did last year.
- Give time for students to work through the questions in the *Monitoring Progress,* and then discuss as a class.

Laurie's Notes

Overview of Section 2.1

Introduction
- This is the first chapter of three about polynomials and quadratics. In this first lesson, the language of polynomials is introduced. This language will help facilitate conversation for the remainder of the year and beyond.
- The second part of the lesson is on addition and subtraction of polynomials. The explorations help students recognize that only like terms can be combined by addition and subtraction.
- **Connection:** Students should recall that when you add or subtract fractions you need a common denominator (like terms). When you add or subtract whole numbers, you add or subtract like place values (like terms).

Formative Assessment Tips
- **Exit Ticket:** This technique asks students to respond to a question at the end of the lesson, activity, or learning experience. The *Exit Ticket* allows you to collect evidence of student learning. I cut scrap paper into smaller pieces so that "exit tickets" can be distributed quickly to students.
- The *Exit Ticket* is helpful in planning instruction. During the class, there may be students that you have not heard from. They may not have raised their hands, or they may have been less vocal when working with partners. The *Exit Ticket* helps you gauge the ability of all students to answer a particular type of question.
- Students write their names on the *Exit Tickets*, which are then collected.
- When a subset of students has difficulty with the skill addressed by the *Exit Ticket*, instruction for the following day should address this problem.

Applications
- Example 6 uses the vertical motion model (the penny) and the falling object model (the paintbrush). These two models are commonly used in application questions in algebra.

Pacing Suggestion
- The explorations will not take long for students to complete. Transition to the formal lesson when they have finished.

Dynamic Teaching Tools
- Dynamic Assessment & Progress Monitoring Tool
- Lesson Planning Tool
- Interactive Whiteboard Lesson Library
- Dynamic Classroom with Dynamic Investigations

What Your Students Will Learn

- Find the degrees of monomials.
- Classify a polynomial by the number of terms, and identify the degree and leading coefficients.
- Add and subtract polynomials by combining like terms.
- Solve real-life problems involving the addition and subtraction of polynomials.

Laurie's Notes

Exploration

Motivate
- Share a bit of "Did You Know?" with students today.
- The modern addition and subtraction symbols + and − were first used in Germany during the fifteenth century. They have been in use for less than 600 years.
- Ancient Greeks, such as Aristotle, used letters to represent numbers. So, variables were being used long before addition and subtraction symbols.

Discuss
- **Look For and Make Use of Structure:** The shapes of algebra tiles are a visual clue that when you add $2x^2$ and $3x$, you do not get $5x^3$. The shapes are different, so you cannot combine them by addition or subtraction. Mathematically proficient students see that $2x^2 + 3x$ means you have 2 of something (x^2-term) being added to 3 of something different (x-term).

Exploration 1
- Students should be comfortable with integer operations and with using algebra tiles to represent polynomials. You may want to review the addition and subtraction of integers with algebra tiles.
- Once the expressions are added (pushed together), like terms (like tiles) are grouped and zero pairs are removed.
- **Attend to Precision:** Do not allow students to use the phrase "The positive and negative cancel each other, or cancel out." Encourage them to refer to this as the Additive Inverse Property, $a + (-a) = 0$.
- Write the problem and solution: $(3x + 2) + (x - 5) = 4x - 3$.
- ? "Do you think the sum of two binomials is always a binomial?" Students should be able to give examples for which the sum is not a binomial.

Exploration 2
- ? "By using algebra tiles to subtract integers, we discovered that subtraction was the same as what?" adding the opposite
- Explain that Step 2 shows how to add the opposite. Change the subtraction symbol to an addition symbol and flip over each of the tiles being subtracted (or swap them out if the tiles are not two-sided).
- Write the problem and solution: $(x^2 + 2x + 2) - (x - 1) = x^2 + x + 3$.

Communicate Your Answer
- You are listening for students to say that you can only add or subtract tiles that are the same size, meaning like terms.

Connecting to Next Step
- These brief explorations will prepare students for the second part of the formal lesson in which students will be adding and subtracting polynomials.

2.1 Adding and Subtracting Polynomials

Essential Question How can you add and subtract polynomials?

EXPLORATION 1 Adding Polynomials

Work with a partner. Write the expression modeled by the algebra tiles in each step.

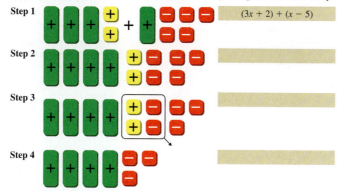

$(3x + 2) + (x - 5)$

EXPLORATION 2 Subtracting Polynomials

Work with a partner. Write the expression modeled by the algebra tiles in each step.

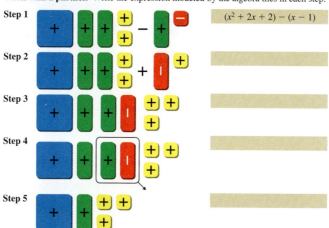

$(x^2 + 2x + 2) - (x - 1)$

REASONING ABSTRACTLY
To be proficient in math, you need to represent a given situation using symbols.

Communicate Your Answer

3. How can you add and subtract polynomials?

4. Use your methods in Question 3 to find each sum or difference.

   a. $(x^2 + 2x - 1) + (2x^2 - 2x + 1)$ b. $(4x + 3) + (x - 2)$
   c. $(x^2 + 2) - (3x^2 + 2x + 5)$ d. $(2x - 3x) - (x^2 - 2x + 4)$

Section 2.1 Adding and Subtracting Polynomials 61

Dynamic Teaching Tools

- Dynamic Assessment & Progress Monitoring Tool
- Lesson Planning Tool
- Interactive Whiteboard Lesson Library
- Dynamic Classroom with Dynamic Investigations

ANSWERS

1. $4x + 2 - 5$; $4x + (2 - 2) - 3$; $4x - 3$

2. $(x^2 + 2x + 2) + (-x + 1)$; $x^2 + 2x - x + 3$; $x^2 + x + (x - x) + 3$; $x^2 + x + 3$

3. Add or subtract like terms.

4. a. $3x^2$
   b. $5x + 1$
   c. $-2x^2 - 2x - 3$
   d. $-x^2 + x - 4$

Extra Example 1

Find the degree of each monomial.

a. $2x$

   The degree of the monomial is 1.

b. $\frac{3}{4}x^2y^2$

   The degree of the monomial is 4.

c. $5x^2y$

   The degree of the monomial is 3.

d. -15

   The degree of the monomial is 0.

MONITORING PROGRESS ANSWERS

1. 4
2. 5
3. 1
4. 0

2.1 Lesson

Core Vocabulary

monomial, p. 62
degree of a monomial, p. 62
polynomial, p. 63
binomial, p. 63
trinomial, p. 63
degree of a polynomial, p. 63
standard form, p. 63
leading coefficient, p. 63
closed, p. 64

What You Will Learn

▶ Find the degrees of monomials.
▶ Classify polynomials.
▶ Add and subtract polynomials.
▶ Solve real-life problems.

Finding the Degrees of Monomials

A **monomial** is a number, a variable, or the product of a number and one or more variables with whole number exponents.

The **degree of a monomial** is the sum of the exponents of the variables in the monomial. The degree of a nonzero constant term is 0. The constant 0 does not have a degree.

Monomial	Degree	Not a monomial	Reason
10	0	$5 + x$	A sum is not a monomial.
$3x$	1	$\frac{2}{n}$	A monomial cannot have a variable in the denominator.
$\frac{1}{2}ab^2$	$1 + 2 = 3$	4^a	A monomial cannot have a variable exponent.
$-1.8m^5$	5	x^{-1}	The variable must have a whole number exponent.

EXAMPLE 1 Finding the Degrees of Monomials

Find the degree of each monomial.

a. $5x^2$ b. $-\frac{1}{2}xy^3$ c. $8x^3y^3$ d. -3

SOLUTION

a. The exponent of x is 2.

   ▶ So, the degree of the monomial is 2.

b. The exponent of x is 1, and the exponent of y is 3.

   ▶ So, the degree of the monomial is $1 + 3$, or 4.

c. The exponent of x is 3, and the exponent of y is 3.

   ▶ So, the degree of the monomial is $3 + 3$, or 6.

d. You can rewrite -3 as $-3x^0$.

   ▶ So, the degree of the monomial is 0.

Monitoring Progress Help in English and Spanish at *BigIdeasMath.com*

Find the degree of the monomial.

1. $-3x^4$ 2. $7c^3d^2$ 3. $\frac{5}{3}y$ 4. -20.5

Laurie's Notes — Teacher Actions

- **Teaching Tip:** Make two lists: *Monomials* and *Not Monomials*. Give partners time to write a definition for monomials and degree of a monomial. Ask partners to write other examples of monomials. Use whiteboards to share and discuss results.
- **Turn and Talk:** Have partners discuss the degree of each monomial in Example 1. Use *Popsicle Sticks* to solicit answers.
- **Think-Pair-Share:** Have students answer Questions 1–4, and then share and discuss as a class.

Classifying Polynomials

Core Concept

Polynomials

A **polynomial** is a monomial or a sum of monomials. Each monomial is called a *term* of the polynomial. A polynomial with two terms is a **binomial**. A polynomial with three terms is a **trinomial**.

Binomial	Trinomial
$5x + 2$	$x^2 + 5x + 2$

The **degree of a polynomial** is the greatest degree of its terms. A polynomial in one variable is in **standard form** when the exponents of the terms decrease from left to right. When you write a polynomial in standard form, the coefficient of the first term is the **leading coefficient**.

leading coefficient → degree → constant term
$2x^3 + x^2 - 5x + 12$

EXAMPLE 2 Writing a Polynomial in Standard Form

Write $15x - x^3 + 3$ in standard form. Identify the degree and leading coefficient of the polynomial.

SOLUTION

Consider the degree of each term of the polynomial.

Degree is 1. → $15x - x^3 + 3$ ← Degree is 0.
Degree is 3.

▶ You can write the polynomial in standard form as $-x^3 + 15x + 3$. The greatest degree is 3, so the degree of the polynomial is 3, and the leading coefficient is -1.

EXAMPLE 3 Classifying Polynomials

Write each polynomial in standard form. Identify the degree and classify each polynomial by the number of terms.

a. $-3z^4$ b. $4 + 5x^2 - x$ c. $8q + q^5$

SOLUTION

Polynomial	Standard Form	Degree	Type of Polynomial
a. $-3z^4$	$-3z^4$	4	monomial
b. $4 + 5x^2 - x$	$5x^2 - x + 4$	2	trinomial
c. $8q + q^5$	$q^5 + 8q$	5	binomial

Monitoring Progress Help in English and Spanish at BigIdeasMath.com

Write the polynomial in standard form. Identify the degree and leading coefficient of the polynomial. Then classify the polynomial by the number of terms.

5. $4 - 9z$ 6. $t^2 - t^3 - 10t$ 7. $2.8x + x^3$

English Language Learners

Notebook Development
Have students create in their notebooks a polynomial page that includes all the vocabulary words on pages 62 and 63. For each word, have students include the definition and an example.

Extra Example 2
Write $-8 + 9x^2 - 2x$ in standard form. Identify the degree and leading coefficient of the polynomial.
standard form: $9x^2 - 2x - 8$; degree of polynomial: 2; leading coefficient: 9

Extra Example 3
Write each polynomial in standard form. Identify the degree and classify each polynomial by the number of terms.

a. $3 + 7a^2$
standard form: $7a^2 + 3$; degree: 2; type: binomial

b. $-b^2 + 2b^4 + 6b$
standard form: $2b^4 - b^2 + 6b$; degree: 4; type: trinomial

c. $12c^5$
standard form: $12c^5$; degree: 5; type: monomial

MONITORING PROGRESS ANSWERS

5. $-9z + 4$; 1; -9; binomial
6. $-t^3 + t^2 - 10t$; 3; -1; trinomial
7. $x^3 + 2.8x$; 3; 1; binomial

Laurie's Notes — Teacher Actions

- Have students read and write key information from the *Core Concept*. You might have students write several examples of each.
- Write Example 2 and have students share all that they know: not in standard form; degree = 3; trinomial; leading coefficient = -1; constant term = 3. As a follow-up, give students information about a polynomial and have them write the polynomial.
- Students can be distracted by the use of different variables. Remind students that polynomials can be written with any variable.

Differentiated Instruction

Auditory

Have students work in pairs. Ask one student to find a sum or difference of polynomials using the vertical format and the other student to find the same sum or difference using the horizontal format. Have students compare their answers. Partners should alternate methods as they continue to solve problems. After they solve several problems, have students tell which method they prefer and why.

Extra Example 4

Find the sum.

a. $(3x^2 + 4x + 3) + (2x + 5x^2 + 1)$
   $8x^2 + 6x + 4$

b. $(-5a^2 + a + 2) + (2a^2 - a - 9)$
   $-3a^2 - 7$

Extra Example 5

Find the difference.

a. $(3y^2 + 8) - (6y^2 + 4y - 2)$
   $-3y^2 - 4y + 10$

b. $(12t^2 + 5t - 7) - (-4t^2 + 3t - 1)$
   $16t^2 + 2t - 6$

Adding and Subtracting Polynomials

A set of numbers is **closed** under an operation when the operation performed on any two numbers in the set results in a number that is also in the set. For example, the set of integers is closed under addition, subtraction, and multiplication. This means that if a and b are two integers, then $a + b$, $a - b$, and ab are also integers.

The set of polynomials is closed under addition and subtraction. So, the sum or difference of any two polynomials is also a polynomial.

To add polynomials, add like terms. You can use a vertical or a horizontal format.

EXAMPLE 4 Adding Polynomials

Find the sum.

a. $(2x^3 - 5x^2 + x) + (2x^2 + x^3 - 1)$ b. $(3x^2 + x - 6) + (x^2 + 4x + 10)$

SOLUTION

STUDY TIP
When a power of the variable appears in one polynomial but not the other, leave a space in that column, or write the term with a coefficient of 0.

a. **Vertical format:** Align like terms vertically and add.

$$\begin{array}{r} 2x^3 - 5x^2 + x \\ +\ \ \ \ \ x^3 + 2x^2\ \ \ \ \ \ \ - 1 \\ \hline 3x^3 - 3x^2 + x - 1 \end{array}$$

▶ The sum is $3x^3 - 3x^2 + x - 1$.

b. **Horizontal format:** Group like terms and simplify.

$(3x^2 + x - 6) + (x^2 + 4x + 10) = (3x^2 + x^2) + (x + 4x) + (-6 + 10)$
$ = 4x^2 + 5x + 4$

▶ The sum is $4x^2 + 5x + 4$.

To subtract a polynomial, add its opposite. To find the opposite of a polynomial, multiply each of its terms by -1.

EXAMPLE 5 Subtracting Polynomials

Find the difference.

a. $(4n^2 + 5) - (-2n^2 + 2n - 4)$ b. $(4x^2 - 3x + 5) - (3x^2 - x - 8)$

COMMON ERROR
Remember to multiply *each* term of the polynomial by -1 when you write the subtraction as addition.

SOLUTION

a. **Vertical format:** Align like terms vertically and subtract.

$$\begin{array}{r} 4n^2\ \ \ \ \ \ \ \ + 5 \\ -\ (-2n^2 + 2n - 4) \end{array} \Rightarrow \begin{array}{r} 4n^2\ \ \ \ \ \ \ \ + 5 \\ +\ \ 2n^2 - 2n + 4 \\ \hline 6n^2 - 2n + 9 \end{array}$$

▶ The difference is $6n^2 - 2n + 9$.

b. **Horizontal format:** Group like terms and simplify.

$(4x^2 - 3x + 5) - (3x^2 - x - 8) = 4x^2 - 3x + 5 - 3x^2 + x + 8$
$ = (4x^2 - 3x^2) + (-3x + x) + (5 + 8)$
$ = x^2 - 2x + 13$

▶ The difference is $x^2 - 2x + 13$.

64   Chapter 2   Polynomial Equations and Factoring

Laurie's Notes   Teacher Actions

- Explain that when adding and subtracting polynomials, a vertical or horizontal format can be used, similar to whole-number operations.
- ❓ In Example 4(a) ask, "Why is a space left when writing $x^3 + 2x^2 - 1$?" so that like terms are lined up
- Subtracting polynomials is generally more challenging for students because they need to be careful with the signs of the terms.
- **Teaching Tip:** Before beginning Example 5(b), ask students to identify like terms.

Monitoring Progress Help in English and Spanish at *BigIdeasMath.com*

Find the sum or difference.

8. $(b - 10) + (4b - 3)$
9. $(x^2 - x - 2) + (7x^2 - x)$
10. $(p^2 + p + 3) - (-4p^2 - p + 3)$
11. $(-k + 5) - (3k^2 - 6)$

Solving Real-Life Problems

EXAMPLE 6 Solving a Real-Life Problem

A penny is thrown straight down from a height of 200 feet. At the same time, a paintbrush is dropped from a height of 100 feet. The polynomials represent the heights (in feet) of the objects after t seconds.

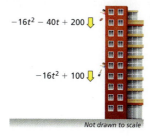

Not drawn to scale

a. Write a polynomial that represents the distance between the penny and the paintbrush after t seconds.

b. Interpret the coefficients of the polynomial in part (a).

SOLUTION

a. To find the distance between the objects after t seconds, subtract the polynomials.

Penny	$-16t^2 - 40t + 200$		$-16t^2 - 40t + 200$
Paintbrush	$-(-16t^2\qquad + 100)$	➡	$+\quad 16t^2\qquad\quad - 100$
			$-40t + 100$

▶ The polynomial $-40t + 100$ represents the distance between the objects after t seconds.

b. When $t = 0$, the distance between the objects is $-40(0) + 100 = 100$ feet. So, the constant term 100 represents the distance between the penny and the paintbrush when both objects begin to fall.

As the value of t increases by 1, the value of $-40t + 100$ decreases by 40. This means that the objects become 40 feet closer to each other each second. So, -40 represents the amount that the distance between the objects changes each second.

INTERPRETING EXPRESSIONS

Notice that each term of the resulting expression has special meaning in the context of the problem. Analyzing the terms helps you understand the problem in more depth.

Monitoring Progress Help in English and Spanish at *BigIdeasMath.com*

12. **WHAT IF?** The polynomial $-16t^2 - 25t + 200$ represents the height of the penny after t seconds.

    a. Write a polynomial that represents the distance between the penny and the paintbrush after t seconds.

    b. Interpret the coefficients of the polynomial in part (a).

Assignment Guide and Homework Check

ASSIGNMENT

Basic: 1–4, 5–41 odd, 53, 56, 57, 62–64

Average: 1–4, 6–52 even, 53–57, 62–64

Advanced: 1–4, 8, 12, 18–22 even, 26–52 even, 53–64

HOMEWORK CHECK

Basic: 5, 13, 23, 33, 53

Average: 10, 16, 24, 34, 53

Advanced: 12, 18, 28, 36, 54

ANSWERS

1. when the exponents of the terms decrease from left to right
2. Sample answer: $2x^5 - 3x + 7$
3. Determine if performing the operation on any two numbers in the set always results in a number that is also in the set.
4. $x^2 - 8^x$; It is the only one that is not a polynomial.
5. 1
6. 4
7. 2
8. 0
9. 9
10. 6
11. 11
12. 11
13. $2c^4 + 6c^2 - c$; 4; 2; trinomial
14. $-w^{12} + 4w^{11}$; 12; -1; binomial
15. $3p^2 + 7$; 2; 3; binomial
16. $-4d^3 + 8d - 2$; 3; -4; trinomial
17. $3t^8$; 8; 3; monomial
18. $3z^4 + 2z^3 + 5z$; 4; 3; trinomial
19. $-\frac{5}{7}r^8 + 2r^5 + \pi r^2$; 8; $-\frac{5}{7}$; trinomial
20. $\sqrt{7}n^4$; 4; $\sqrt{7}$; monomial
21. It is the product of a number, $\frac{4}{3}\pi$, and a variable with a whole number exponent, r^3; 3
22. binomial; 8
23. $3y + 10$
24. $x - 8$
25. $n^2 - 8n + 5$
26. $-4p^3 - 3p^2 - 17p$
27. $6g^2 - 9g + 4$
28. $12r^2 + r - 7$
29. $a^3 - 5a^2 + 4a + 5$
30. $-5s^3 + 2s^2 - s - 9$
31. $-2d - 8$
32. $-x + 8$
33. $-2y^2 + 2y + 18$
34. $7m^2 - 11m - 2$
35. $k^3 - k^2 - 7k + 14$
36. $4r^3 - r^2 - 8r - 10$

2.1 Exercises

Dynamic Solutions available at BigIdeasMath.com

Vocabulary and Core Concept Check

1. **VOCABULARY** When is a polynomial in one variable in standard form?
2. **OPEN-ENDED** Write a trinomial in one variable of degree 5 in standard form.
3. **VOCABULARY** How can you determine whether a set of numbers is closed under an operation?
4. **WHICH ONE DOESN'T BELONG?** Which expression does *not* belong with the other three? Explain your reasoning.

   $a^3 + 4a$ $\quad$ $x^2 - 8^x$ $\quad$ $b - 2^{-1}$ $\quad$ $-\frac{\pi}{3} + 6y^8z$

Monitoring Progress and Modeling with Mathematics

In Exercises 5–12, find the degree of the monomial. (See Example 1.)

5. $4g$
6. $23x^4$
7. $-1.75k^2$
8. $-\frac{4}{9}$
9. s^8t
10. $8m^2n^4$
11. $9xy^3z^7$
12. $-3q^4rs^6$

In Exercises 13–20, write the polynomial in standard form. Identify the degree and leading coefficient of the polynomial. Then classify the polynomial by the number of terms. (See Examples 2 and 3.)

13. $6c^2 + 2c^4 - c$
14. $4w^{11} - w^{12}$
15. $7 + 3p^2$
16. $8d - 2 - 4d^3$
17. $3t^8$
18. $5z + 2z^3 + 3z^4$
19. $\pi r^2 - \frac{5}{7}r^8 + 2r^5$
20. $\sqrt{7}n^4$

21. **MODELING WITH MATHEMATICS** The expression $\frac{4}{3}\pi r^3$ represents the volume of a sphere with radius r. Why is this expression a monomial? What is its degree?

22. **MODELING WITH MATHEMATICS** The amount of money you have after investing $400 for 8 years and $600 for 6 years at the same interest rate is represented by $400x^8 + 600x^6$, where x is the growth factor. Classify the polynomial by the number of terms. What is its degree?

In Exercises 23–30, find the sum. (See Example 4.)

23. $(5y + 4) + (-2y + 6)$
24. $(-8x - 12) + (9x + 4)$
25. $(2n^2 - 5n - 6) + (-n^2 - 3n + 11)$
26. $(-3p^3 + 5p^2 - 2p) + (-p^3 - 8p^2 - 15p)$
27. $(3g^2 - g) + (3g^2 - 8g + 4)$
28. $(9r^2 + 4r - 7) + (3r^2 - 3r)$
29. $(4a - a^3 - 3) + (2a^3 - 5a^2 + 8)$
30. $(s^3 - 2s - 9) + (2s^2 - 6s^3 + s)$

In Exercises 31–38, find the difference. (See Example 5.)

31. $(d - 9) - (3d - 1)$
32. $(6x + 9) - (7x + 1)$
33. $(y^2 - 4y + 9) - (3y^2 - 6y - 9)$
34. $(4m^2 - m + 2) - (-3m^2 + 10m + 4)$
35. $(k^3 - 7k + 2) - (k^2 - 12)$
36. $(-r - 10) - (-4r^3 + r^2 + 7r)$

66 Chapter 2 Polynomial Equations and Factoring

37. $(t^4 - t^2 + t) - (12 - 9t^2 - 7t)$

38. $(4d - 6d^3 + 3d^2) - (10d^3 + 7d - 2)$

ERROR ANALYSIS In Exercises 39 and 40, describe and correct the error in finding the sum or difference.

39.
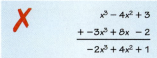

40.

41. **MODELING WITH MATHEMATICS** The cost (in dollars) of making b bracelets is represented by $4 + 5b$. The cost (in dollars) of making b necklaces is represented by $8b + 6$. Write a polynomial that represents how much more it costs to make b necklaces than b bracelets.

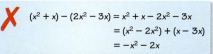

42. **MODELING WITH MATHEMATICS** The number of individual memberships at a fitness center in m months is represented by $142 + 12m$. The number of family memberships at the fitness center in m months is represented by $52 + 6m$. Write a polynomial that represents the total number of memberships at the fitness center.

In Exercises 43–46, find the sum or difference.

43. $(2s^2 - 5st - t^2) - (s^2 + 7st - t^2)$

44. $(a^2 - 3ab + 2b^2) + (-4a^2 + 5ab - b^2)$

45. $(c^2 - 6d^2) + (c^2 - 2cd + 2d^2)$

46. $(-x^2 + 9xy) - (x^2 + 6xy - 8y^2)$

REASONING In Exercises 47–50, complete the statement with *always*, *sometimes*, or *never*. Explain your reasoning.

47. The terms of a polynomial are _____ monomials.

48. The difference of two trinomials is _____ a trinomial.

49. A binomial is _____ a polynomial of degree 2.

50. The sum of two polynomials is _____ a polynomial.

MODELING WITH MATHEMATICS The polynomial $-16t^2 + v_0 t + s_0$ represents the height (in feet) of an object, where v_0 is the initial vertical velocity (in feet per second), s_0 is the initial height of the object (in feet), and t is the time (in seconds). In Exercises 51 and 52, write a polynomial that represents the height of the object. Then find the height of the object after 1 second.

51. You throw a water balloon from a building.
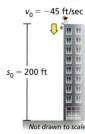
$v_0 = -45$ ft/sec
$s_0 = 200$ ft

52. You bounce a tennis ball on a racket.

$v_0 = 16$ ft/sec
$s_0 = 3$ ft

Not drawn to scale

53. **MODELING WITH MATHEMATICS** You drop a ball from a height of 98 feet. At the same time, your friend throws a ball upward. The polynomials represent the heights (in feet) of the balls after t seconds. (See Example 6.)

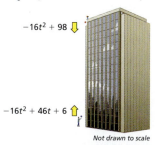

$-16t^2 + 98$

$-16t^2 + 46t + 6$

Not drawn to scale

a. Write a polynomial that represents the distance between your ball and your friend's ball after t seconds.

b. Interpret the coefficients of the polynomial in part (a).

ANSWERS

37. $t^4 + 8t^2 + 8t - 12$
38. $-16d^3 + 3d^2 - 3d + 2$
39. When writing the subtraction as addition, the last term of the polynomial was not multiplied by -1; $= (x^2 + x) + (-2x^2 + 3x) = (x^2 - 2x^2) + (x + 3x) = -x^2 + 4x$
40. $-4x^2$ and $8x$ are not like terms, so they cannot be added;
$$\begin{array}{r} x^3 - 4x^2 + 3 \\ + -3x^3 + 8x - 2 \\ \hline -2x^3 - 4x^2 + 8x + 1 \end{array}$$
41. $3b + 2$
42. $194 + 18m$
43. $s^2 - 12st$
44. $-3a^2 + 2ab + b^2$
45. $2c^2 - 2cd - 4d^2$
46. $-2x^2 + 3xy + 8y^2$
47. always; A polynomial is a monomial or a sum of monomials, and each monomial is a term of the polynomial.
48. sometimes; If like terms have the same coefficient, they will cancel when subtracted so the difference will have fewer than 3 terms. Or, if the terms in the trinomial are not all of the same degree, then the difference could have more than 3 terms.
49. sometimes; The two terms in the binomial can be of any degree.
50. always; Polynomials are closed under addition.
51. $-16t^2 - 45t + 200$; 139 ft
52. $-16t^2 + 16t + 3$; 3 ft
53. a. $-46t + 92$
    b. The constant term 92 indicates that the distance between the two balls is 92 feet when they begin. The coefficient of the linear term -46 indicates that the two balls become 46 feet closer to each other each second.

ANSWERS

54. a. $-0.028t^3 - 0.32t^2 + 1.6t + 59$
    b. $55.5 million
55. $12x - 3$
56. a. $-3x + 3$ or $3x - 3$
    b. when $x = 1$; This is the x-coordinate of the point of intersection of the two lines.
57. yes; Addition is commutative and associative, so you can add in any order.
58. $\frac{1}{2}x^2 + \frac{1}{2}, \frac{1}{2}x^2 - \frac{1}{2}$
59–64. See Additional Answers.

Mini-Assessment

1. Find the degree of the monomial $15d^3$. 3
2. Write the polynomial $-8 - x^2 + 2x$ in standard form. Identify the degree and classify the polynomial by the number of terms.
   standard form: $-x^2 + 2x - 8$;
   degree: 2; type: trinomial

Find the sum or difference.

3. $(6n^2 + 2n + 3) + (-5n^2 - 4n - 7)$
   $n^2 - 2n - 4$
4. $(5x^2 - 4x + 7) - (4x - 7)$
   $5x^2 - 8x + 14$
5. A ball is thrown straight down from a height of 300 feet. At the same time, a quarter is dropped from a height of 150 feet. The polynomials represent the heights (in feet) of the objects after t seconds.
   Ball: $-16t^2 - 50t + 300$
   Quarter: $-16t^2 + 150$
   a. Write a polynomial that represents the distance between the ball and the quarter after t seconds. $-50t + 150$
   b. Interpret the coefficients of the polynomial in part (a). The constant term 150 represents the distance between the ball and the quarter when both objects begin to fall. The coefficient -50 represents the amount that the distance between the objects changes each second.

68   Chapter 2

---

54. **MODELING WITH MATHEMATICS** During a 7-year period, the amounts (in millions of dollars) spent each year on buying new vehicles N and used vehicles U by United States residents are modeled by the equations

    $N = -0.028t^3 + 0.06t^2 + 0.1t + 17$
    $U = -0.38t^2 + 1.5t + 42$

    where $t = 1$ represents the first year in the 7-year period.

    a. Write a polynomial that represents the total amount spent each year on buying new and used vehicles in the 7-year period.
    b. How much is spent on buying new and used vehicles in the fifth year?

55. **MATHEMATICAL CONNECTIONS**
    Write the polynomial in standard form that represents the perimeter of the quadrilateral.

    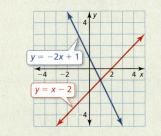

56. **HOW DO YOU SEE IT?** The right side of the equation of each line is a polynomial.

    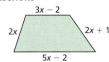

    a. The absolute value of the difference of the two polynomials represents the vertical distance between points on the lines with the same x-value. Write this expression.
    b. When does the expression in part (a) equal 0? How does this value relate to the graph?

57. **MAKING AN ARGUMENT** Your friend says that when adding polynomials, the order in which you add does not matter. Is your friend correct? Explain.

58. **THOUGHT PROVOKING** Write two polynomials whose sum is x^2 and whose difference is 1.

59. **REASONING** Determine whether the set is closed under the given operation. Explain.
    a. the set of negative integers; multiplication
    b. the set of whole numbers; addition

60. **PROBLEM SOLVING** You are building a multi-level deck.

    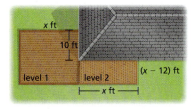

    a. For each level, write a polynomial in standard form that represents the area of that level. Then write the polynomial in standard form that represents the total area of the deck.
    b. What is the total area of the deck when $x = 20$?
    c. A gallon of deck sealant covers 400 square feet. How many gallons of sealant do you need to cover the deck in part (b) once? Explain.

61. **PROBLEM SOLVING** A hotel installs a new swimming pool and a new hot tub.

    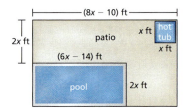

    a. Write the polynomial in standard form that represents the area of the patio.
    b. The patio will cost $10 per square foot. Determine the cost of the patio when $x = 9$.

Maintaining Mathematical Proficiency
Reviewing what you learned in previous grades and lessons

Simplify the expression. *(Skills Review Handbook)*

62. $2(x - 1) + 3(x + 2)$
63. $8(4y - 3) + 2(y - 5)$
64. $5(2r + 1) - 3(-4r + 2)$

68   Chapter 2   Polynomial Equations and Factoring

If students need help...	If students got it...
Resources by Chapter • Practice A and Practice B • Puzzle Time	Resources by Chapter • Enrichment and Extension • Cumulative Review
Student Journal • Practice	Start the *next* Section
Differentiating the Lesson Skills Review Handbook	

Laurie's Notes

Overview of Section 2.2

Introduction
- When students learned to multiply whole numbers, they likely began with an area model that was eventually represented using a box (table) model. This eventually led to an efficient algorithm. There is a connection between how whole numbers are multiplied and how binomials are multiplied.
- This lesson introduces multiplication of binomials, using three different approaches: the Distributive Property, a table, and FOIL (First-Outside-Inside-Last).
- The product of a binomial and a trinomial is also found in this lesson.

Teaching Strategy
- **Connection:** You want to be sure that students see the connection between multiplying two binomials and multiplying two 2-digit whole numbers. There are four multiplications that are performed.
- One early approach to multiplying whole numbers was to use an area model (base-ten blocks), which was later replaced with a simple table.
- A table model is shown in Example 2 for multiplying $(2x - 3)$ by $(x + 5)$.
- The table model can also be used to multiply a binomial by a trinomial. Example 4 is shown below.

	x^2	$-3x$	-2
x	x^3	$-3x^2$	$-2x$
5	$5x^2$	$-15x$	-10

Pacing Suggestion
- The explorations provide an opportunity for students to develop a conceptual understanding of what it means to square a binomial. Transition to the formal lesson as soon as students have discussed each exploration.

Dynamic Teaching Tools
- Dynamic Assessment & Progress Monitoring Tool
- Lesson Planning Tool
- Interactive Whiteboard Lesson Library
- Dynamic Classroom with Dynamic Investigations

What Your Students Will Learn

- Multiply binomials using the Distributive Property and using tables.
- Use the FOIL Method as a way to multiply two binomials.
- Multiply binomials and trinomials using the Distributive Property.

Laurie's Notes

Exploration

Motivate

- **Mental Math Challenge:** Ask students to find the following products using mental math. They should record only the answer.

    8×32 9×48 12×43

- Have students reveal each answer, and ask how they performed the computation. Listen for strategies that likely make use of the Distributive Property.

    $8 \times 32 = 8(30 + 2) = 240 + 16 = 256$
    $9 \times 48 = 9(50 - 2) = 450 - 18 = 432$
    $12 \times 43 = 12(40 + 3) = 480 + 36 = 516$

    OR $12 \times 43 = (10 + 2)(40 + 3)$
    $= 10 \times 40 + 10 \times 3 + 2 \times 40 + 2 \times 3 = 516$

- Make sure students recognize the use of the Distributive Property and that 12×43 can be done two different ways using the Distributive Property.

Exploration Note

- **Make Sense of Problems and Persevere in Solving Them** and **Look For and Make Use of Structure:** Working with algebra tiles in this exploration should be reminiscent of using base-ten blocks when performing multi-digit multiplication. The connection to whole-number multiplication, and how it is represented, should make sense to students. The underlying structure of the problem is the same.

Exploration 1

- In explaining their reasoning, you should expect students to write the problem using symbols, and then explain their answer. For instance, part (e) would be $(x)(-1) = -x$. The explanation might be that multiplying by -1 is the same as taking the opposite of the number.

Exploration 2

- In this exploration, the models for each factor and the product for parts (a) and (b) are shown. It is important to have a conversation about the product in part (a) before students begin. If students understand the area model, they should be able to fill in the product for parts (c) and (d).
- Make the connection to two-digit multiplication. Each binomial has two terms. There are four multiplications to perform, and the products are represented by the tiles in the lower right quadrant of the *t*-diagram.
- ? "In part (a), what are the factors in this product?" $(x + 3)$ and $(x - 2)$
- ? "What is the product of *x* and *x*?" x^2 "What is the product of *x* and -2?" $-2x$ "What is the product of 3 and *x*?" $3x$ "What is the product of 3 and -2?" -6
- Remind students to simplify expressions when writing a product.

Communicate Your Answer

- **Neighbor Check:** For Question 4, have students work independently and then have their neighbors check their work. Have students discuss any discrepancies.

Connecting to Next Step

- The explorations help students develop an understanding of the steps in squaring a binomial. When students see the symbolic approach in the formal lesson, they will have a visual model to connect it to.

2.2 Multiplying Polynomials

Essential Question How can you multiply two polynomials?

EXPLORATION 1 Multiplying Monomials Using Algebra Tiles

Work with a partner. Write each product. Explain your reasoning.

a. $(+) \cdot (+) =$ b. $(+) \cdot (-) =$
c. $(-) \cdot (-) =$ d. $(+) \cdot (+) =$
e. $(+) \cdot (-) =$ f. $(-) \cdot (+) =$
g. $(-) \cdot (-) =$ h. $(+) \cdot (+) =$
i. $(+) \cdot (-) =$ j. $(-) \cdot (-) =$

REASONING ABSTRACTLY

To be proficient in math, you need to reason abstractly and quantitatively. You need to pause as needed to recall the meanings of the symbols, operations, and quantities involved.

EXPLORATION 2 Multiplying Binomials Using Algebra Tiles

Work with a partner. Write the product of two binomials modeled by each rectangular array of algebra tiles. In parts (c) and (d), first draw the rectangular array of algebra tiles that models each product.

a. $(x + 3)(x - 2) =$ b. $(2x - 1)(2x + 1) =$

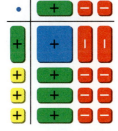 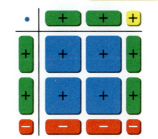

c. $(x + 2)(2x - 1) =$ d. $(-x - 2)(x - 3) =$

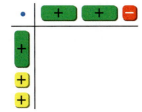 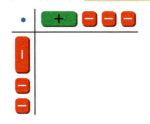

Communicate Your Answer

3. How can you multiply two polynomials?

4. Give another example of multiplying two binomials using algebra tiles that is similar to those in Exploration 2.

Section 2.2 Multiplying Polynomials 69

Dynamic Teaching Tools

Dynamic Assessment & Progress Monitoring Tool
Lesson Planning Tool
Interactive Whiteboard Lesson Library
Dynamic Classroom with Dynamic Investigations

ANSWERS

1. a. 1; The product of 1 and 1 is 1.
   b. -1; The product of 1 and -1 is -1.
   c. 1; The product of -1 and -1 is 1.
   d. x; The product of any number and 1 is that number.
   e. $-x$; The product of any number and -1 is the opposite of that number.
   f. $-x$; The product of any number and 1 is that number.
   g. x; The product of the opposite of a number and -1 is the number.
   h. x^2; The product of any number multiplied by itself is the number squared.
   i. $-x^2$; The product of any number and its opposite is the opposite of the number squared.
   j. x^2; The product of any number multiplied by itself is the number squared.

2. a. $x^2 + x - 6$
   b. $4x^2 - 1$
   c.
      $2x^2 + 3x - 2$
   d.
      $-x^2 + x + 6$

3. Multiply each term in one polynomial by each term in the other polynomial, then combine like terms.

4. *Sample answer:*
   $(x + 1)(x - 2) = x^2 - x - 2$

Section 2.2 69

Extra Example 1

a. Find $(x + 8)(x + 2)$.
   $x^2 + 10x + 16$

b. Find $(x - 5)(x + 1)$.
   $x^2 - 4x - 5$

Extra Example 2

Find $(2x + 3)(x - 3)$.
$2x^2 - 3x - 9$

MONITORING PROGRESS ANSWERS

1. $y^2 + 5y + 4$
2. $z^2 + 4z - 12$
3. $p^2 - 5p - 24$
4. $2r^2 - 11r + 5$

2.2 Lesson

Core Vocabulary
FOIL Method, *p. 71*
Previous
polynomial
closed
binomial
trinomial

What You Will Learn

- Multiply binomials.
- Use the FOIL Method.
- Multiply binomials and trinomials.

Multiplying Binomials

The product of two polynomials is always a polynomial. So, like the set of integers, the set of polynomials is closed under multiplication. You can use the Distributive Property to multiply two binomials.

EXAMPLE 1 Multiplying Binomials Using the Distributive Property

Find (a) $(x + 2)(x + 5)$ and (b) $(x + 3)(x - 4)$.

SOLUTION

a. Use the horizontal method.

$(x + 2)(x + 5) = x(x + 5) + 2(x + 5)$ Distribute $(x + 5)$ to each term of $(x + 2)$.
$= x(x) + x(5) + 2(x) + 2(5)$ Distributive Property
$= x^2 + 5x + 2x + 10$ Multiply.
$= x^2 + 7x + 10$ Combine like terms.

▶ The product is $x^2 + 7x + 10$.

b. Use the vertical method.

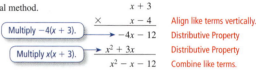

▶ The product is $x^2 - x - 12$.

EXAMPLE 2 Multiplying Binomials Using a Table

Find $(2x - 3)(x + 5)$.

SOLUTION

Step 1 Write each binomial as a sum of terms.

$(2x - 3)(x + 5) = [2x + (-3)](x + 5)$

Step 2 Make a table of products.

	$2x$	-3
x	$2x^2$	$-3x$
5	$10x$	-15

▶ The product is $2x^2 - 3x + 10x - 15$, or $2x^2 + 7x - 15$.

Monitoring Progress Help in English and Spanish at *BigIdeasMath.com*

Use the Distributive Property to find the product.

1. $(y + 4)(y + 1)$
2. $(z - 2)(z + 6)$

Use a table to find the product.

3. $(p + 3)(p - 8)$
4. $(r - 5)(2r - 1)$

70   Chapter 2   Polynomial Equations and Factoring

Laurie's Notes Teacher Actions

- Point out that multiplying $(x + 2)$ by $(x + 5)$ means that both terms of $(x + 2)$ must be multiplied by $(x + 5)$.
- The order in which the multiplications are performed is not important because of the Commutative Property of Addition. However, in order to make the connection to the FOIL Method, perform the operations in the order shown.
- The box or table model can be used to multiply binomials. See the *Teaching Strategy* on page T-68.

Using the FOIL Method

The **FOIL Method** is a shortcut for multiplying two binomials.

Core Concept

FOIL Method

To multiply two binomials using the FOIL Method, find the sum of the products of the

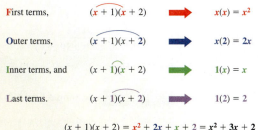

First terms, $(x + 1)(x + 2)$ ⟶ $x(x) = x^2$

Outer terms, $(x + 1)(x + 2)$ ⟶ $x(2) = 2x$

Inner terms, and $(x + 1)(x + 2)$ ⟶ $1(x) = x$

Last terms. $(x + 1)(x + 2)$ ⟶ $1(2) = 2$

$(x + 1)(x + 2) = x^2 + 2x + x + 2 = x^2 + 3x + 2$

EXAMPLE 3 Multiplying Binomials Using the FOIL Method

Find each product.

a. $(x - 3)(x - 6)$ **b.** $(2x + 1)(3x - 5)$

SOLUTION

a. Use the FOIL Method.

$$
\begin{aligned}
&\qquad\quad\text{First}\quad\text{Outer}\quad\text{Inner}\quad\text{Last}\\
(x - 3)(x - 6) &= x(x) + x(-6) + (-3)(x) + (-3)(-6) &&\text{FOIL Method}\\
&= x^2 + (-6x) + (-3x) + 18 &&\text{Multiply.}\\
&= x^2 - 9x + 18 &&\text{Combine like terms.}
\end{aligned}
$$

▶ The product is $x^2 - 9x + 18$.

b. Use the FOIL Method.

$$
\begin{aligned}
&\qquad\qquad\text{First}\quad\text{Outer}\quad\text{Inner}\quad\text{Last}\\
(2x + 1)(3x - 5) &= 2x(3x) + 2x(-5) + 1(3x) + 1(-5) &&\text{FOIL Method}\\
&= 6x^2 + (-10x) + 3x + (-5) &&\text{Multiply.}\\
&= 6x^2 - 7x - 5 &&\text{Combine like terms.}
\end{aligned}
$$

▶ The product is $6x^2 - 7x - 5$.

Monitoring Progress Help in English and Spanish at *BigIdeasMath.com*

Use the FOIL Method to find the product.

5. $(m - 3)(m - 7)$ **6.** $(x - 4)(x + 2)$

7. $\left(2u + \tfrac{1}{2}\right)\left(u - \tfrac{3}{2}\right)$ **8.** $(n + 2)(n^2 + 3)$

Differentiated Instruction

Auditory/Visual

Assign each group a different method for multiplying binominals, using the Distributive Property, a table, or FOIL. Have each student solve a different problem using the assigned method and share the solution within the group. Then have the groups solve the same problems again using the other two methods, and compare solutions.

Extra Example 3

Find each product.

a. $(x + 3)(x - 7)$
$\quad x^2 - 4x - 21$

b. $(2x - 1)(5x - 4)$
$\quad 10x^2 - 13x + 4$

MONITORING PROGRESS ANSWERS

5. $m^2 - 10m + 21$

6. $x^2 - 2x - 8$

7. $2u^2 - \tfrac{5}{2}u - \tfrac{3}{4}$

8. $n^3 + 2n^2 + 3n + 6$

Laurie's Notes   Teacher Actions

- Explain to students that the FOIL Method is simply an acronym to help them remember the four multiplications that need to be performed.
- Some students may be comfortable with saying the process aloud and simply writing the product, skipping the first step.
- **Teaching Tip:** Use your index fingers to point to the two terms being multiplied as you say, "first terms, outer terms, inner terms, and last terms."
- In part (b), students need to pay attention to the coefficients of the *x*-terms.

Extra Example 4
Find $(x - 3)(x^2 - 4x - 4)$.
$x^3 - 7x^2 + 8x + 12$

Extra Example 5
The purple piece of stained glass has a trapezoidal shape.

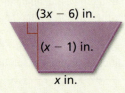

a. Write a polynomial that represents the area of the piece of glass.
$2x^2 - 5x + 3$

b. Find the area of the piece of glass when the shorter base is 7 inches.
66 square inches

MONITORING PROGRESS ANSWERS
9. $x^3 + 6x^2 + 13x + 8$
10. $n^3 - 5n^2 + 10n - 12$
11. It becomes $x^2 - \frac{3}{2}x - \frac{77}{2}$; The longer base becomes $x + 11$. Substituting this value in the formula for the area of a trapezoid along with the other unchanged values changes the linear and constant terms in the polynomial.

Multiplying Binomials and Trinomials

EXAMPLE 4 Multiplying a Binomial and a Trinomial

Find $(x + 5)(x^2 - 3x - 2)$.

SOLUTION

$$\begin{array}{r} x^2 - 3x - 2 \\ \times \quad x + 5 \\ \hline 5x^2 - 15x - 10 \\ x^3 - 3x^2 - 2x \\ \hline x^3 + 2x^2 - 17x - 10 \end{array}$$

Align like terms vertically.
Multiply $5(x^2 - 3x - 2)$. Distributive Property
Multiply $x(x^2 - 3x - 2)$. Distributive Property
Combine like terms.

▶ The product is $x^3 + 2x^2 - 17x - 10$.

EXAMPLE 5 Solving a Real-Life Problem

In hockey, a goalie behind the goal line can only play a puck in the trapezoidal region.

a. Write a polynomial that represents the area of the trapezoidal region.
b. Find the area of the trapezoidal region when the shorter base is 18 feet.

SOLUTION

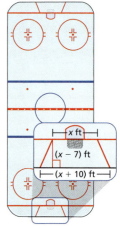

CONNECTIONS TO GEOMETRY
Recall that the formula for the area A of a trapezoid with height h and bases b_1 and b_2 is
$A = \frac{1}{2}h(b_1 + b_2)$.

a. $\frac{1}{2}h(b_1 + b_2) = \frac{1}{2}(x - 7)[x + (x + 10)]$
$= \frac{1}{2}(x - 7)(2x + 10)$.

 F O I L
$= \frac{1}{2}[2x^2 + 10x + (-14x) + (-70)]$
$= \frac{1}{2}(2x^2 - 4x - 70)$
$= x^2 - 2x - 35$

▶ A polynomial that represents the area of the trapezoidal region is $x^2 - 2x - 35$.

b. Find the value of $x^2 - 2x - 35$ when $x = 18$.

$x^2 - 2x - 35 = 18^2 - 2(18) - 35$ Substitute 18 for x.
$= 324 - 36 - 35$ Simplify.
$= 253$ Subtract.

▶ The area of the trapezoidal region is 253 square feet.

Monitoring Progress Help in English and Spanish at BigIdeasMath.com

Find the product.

9. $(x + 1)(x^2 + 5x + 8)$
10. $(n - 3)(n^2 - 2n + 4)$

11. **WHAT IF?** In Example 5(a), how does the polynomial change when the longer base is extended by 1 foot? Explain.

Laurie's Notes — Teacher Actions

- Solve Example 4 as shown, paying attention to like terms. See the *Teaching Strategy* on page T-68 for an alternate method.
- **Make Sense of Problems and Persevere in Solving Them** and **Thumbs Up:** Discuss the problem in Example 5. Do a *Thumbs Up* self-assessment to see whether students are secure in starting the problem independently.
- Have partners check their work on part (a) before beginning part (b).

Closure
- **Exit Ticket:** Find the products.
  a. $(x - 4)(x + 3)$
  $x^2 - x - 12$

  b. $(3x - 7)(x - 2)$
  $3x^2 - 13x + 14$

  c. $(4 + 2x)(x + 9)$
  $2x^2 + 22x + 36$

2.2 Exercises

Dynamic Solutions available at *BigIdeasMath.com*

Vocabulary and Core Concept Check

1. **VOCABULARY** Describe two ways to find the product of two binomials.

2. **WRITING** Explain how the letters of the word FOIL can help you to remember how to multiply two binomials.

Monitoring Progress and Modeling with Mathematics

In Exercises 3–10, use the Distributive Property to find the product. *(See Example 1.)*

3. $(x + 1)(x + 3)$
4. $(y + 6)(y + 4)$
5. $(z - 5)(z + 3)$
6. $(a + 8)(a - 3)$
7. $(g - 7)(g - 2)$
8. $(n - 6)(n - 4)$
9. $(3m + 1)(m + 9)$
10. $(5s + 6)(s - 2)$

In Exercises 11–18, use a table to find the product. *(See Example 2.)*

11. $(x + 3)(x + 2)$
12. $(y + 10)(y - 5)$
13. $(h - 8)(h - 9)$
14. $(c - 6)(c - 5)$
15. $(3k - 1)(4k + 9)$
16. $(5g + 3)(g + 8)$
17. $(-3 + 2j)(4j - 7)$
18. $(5d - 12)(-7 + 3d)$

ERROR ANALYSIS In Exercises 19 and 20, describe and correct the error in finding the product of the binomials.

19.
$$(t - 2)(t + 5) = t - 2(t + 5)$$
$$= t - 2t - 10$$
$$= -t - 10$$

20.
$(x - 5)(3x + 1)$

	$3x$	1
x	$3x^2$	x
5	$15x$	5

$(x - 5)(3x + 1) = 3x^2 + 16x + 5$

In Exercises 21–30, use the FOIL Method to find the product. *(See Example 3.)*

21. $(b + 3)(b + 7)$
22. $(w + 9)(w + 6)$
23. $(k + 5)(k - 1)$
24. $(x - 4)(x + 8)$
25. $\left(q - \frac{3}{4}\right)\left(q + \frac{1}{4}\right)$
26. $\left(z - \frac{5}{3}\right)\left(z - \frac{2}{3}\right)$
27. $(9 - r)(2 - 3r)$
28. $(8 - 4x)(2x + 6)$
29. $(w + 5)(w^2 + 3w)$
30. $(v - 3)(v^2 + 8v)$

MATHEMATICAL CONNECTIONS In Exercises 31–34, write a polynomial that represents the area of the shaded region.

31.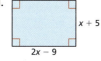

32.

33.

34.

In Exercises 35–42, find the product. *(See Example 4.)*

35. $(x + 4)(x^2 + 3x + 2)$
36. $(f + 1)(f^2 + 4f + 8)$
37. $(y + 3)(y^2 + 8y - 2)$
38. $(t - 2)(t^2 - 5t + 1)$
39. $(4 - b)(5b^2 + 5b - 4)$
40. $(d + 6)(2d^2 - d + 7)$
41. $(3e^2 - 5e + 7)(6e + 1)$
42. $(6v^2 + 2v - 9)(4 - 5v)$

Assignment Guide and Homework Check

ASSIGNMENT

Basic: 1, 2, 3–43 odd, 47, 48, 52–58

Average: 1, 2–44 even, 47, 48, 52–58

Advanced: 1, 2, 8, 10, 16–20 even, 28–34 even, 40–44 even, 45–58

HOMEWORK CHECK

Basic: 3, 13, 21, 35, 43

Average: 4, 14, 22, 38, 44

Advanced: 8, 16, 28, 40, 44

ANSWERS

1. *Sample answer:* Distribute one of the binomials over each term in the other binomial and simplify; Write each binomial as a sum of terms and make a table of products.

2. The letters stand for the sets of terms to multiply: first, outer, inner, and last.

3. $x^2 + 4x + 3$
4. $y^2 + 10y + 24$
5. $z^2 - 2z - 15$
6. $a^2 + 5a - 24$
7. $g^2 - 9g + 14$
8. $n^2 - 10n + 24$
9. $3m^2 + 28m + 9$
10. $5s^2 - 4s - 12$
11. $x^2 + 5x + 6$
12. $y^2 + 5y - 50$
13. $h^2 - 17h + 72$
14. $c^2 - 11c + 30$
15. $12k^2 + 23k - 9$
16. $5g^2 + 43g + 24$
17. $8j^2 - 26j + 21$
18. $15d^2 - 71d + 84$
19. t also should be multiplied by $t + 5$; $= t(t + 5) - 2(t + 5) = t^2 + 5t - 2t - 10 = t^2 + 3t - 10$
20. The 5 in the left column should be -5;

	$3x$	1
x	$3x^2$	x
-5	$-15x$	-5

$(x - 5)(3x + 1) = 3x^2 - 14x - 5$

21. $b^2 + 10b + 21$
22. $w^2 + 15w + 54$
23. $k^2 + 4k - 5$
24. $x^2 + 4x - 32$
25. $q^2 - \frac{1}{2}q - \frac{3}{16}$
26. $z^2 - \frac{7}{3}z + \frac{10}{9}$
27. $3r^2 - 29r + 18$
28. $-8x^2 - 8x + 48$
29. $w^3 + 8w^2 + 15w$
30. $v^3 + 5v^2 - 24v$
31. $2x^2 + x - 45$
32. $p^2 - 2p - 3$
33. $\frac{1}{2}x^2 + \frac{11}{2}x + 15$
34. $x^2 - 3x + 36$
35. $x^3 + 7x^2 + 14x + 8$
36. $f^3 + 5f^2 + 12f + 8$
37. $y^3 + 11y^2 + 22y - 6$
38. $t^3 - 7t^2 + 11t - 2$
39. $-5b^3 + 15b^2 + 24b - 16$
40. $2d^3 + 11d^2 + d + 42$
41. $18e^3 - 27e^2 + 37e + 7$
42. $-30v^3 + 14v^2 + 53v - 36$

Dynamic Teaching Tools
Dynamic Assessment & Progress Monitoring Tool
Interactive Whiteboard Lesson Library
Dynamic Classroom with Dynamic Investigations

ANSWERS
43. a. $(40x^2 + 240x + 200)$ ft^2
    b. 57,600 ft^2
44. a. $(4x^2 + 84x + 440)$ in.^{2}
    b. 840 in.^{2}
45. The degree of the product is the sum of the degrees of each binomial.
46. Sample answer: $(2x - 6)(x^2 + 3x + 4)$
47. no; FOIL would leave out the products that include the middle terms of the two trinomials.
48. a. $(-8x - 9)(-4x + 3)$
    b. a and c are positive, b and d are negative.
49. yes; You are both multiplying the same binomials, and neither the order in which you multiply nor the method used will make a difference.
50. $(4x^3 + 9x^2 - x - 6)$ ft^3
51. a. They have the same signs.
    b. They have opposite signs.
52. $y = \begin{cases} -x + 4, & \text{if } x < 0 \\ x + 4, & \text{if } x \geq 0 \end{cases}$
53. $y = \begin{cases} -6x + 18, & \text{if } x < 3 \\ 6x - 18, & \text{if } x \geq 3 \end{cases}$
54. $y = \begin{cases} 4x + 8, & \text{if } x < -2 \\ -4x - 8, & \text{if } x \geq -2 \end{cases}$
55. 10^{11}
56. $\dfrac{1}{x^2}$
57. $\dfrac{1}{27z^{18}}$
58. $\dfrac{1}{4y^2}$

---

43. **MODELING WITH MATHEMATICS** The football field is rectangular. *(See Example 5.)*

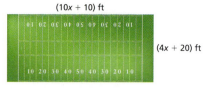
$(10x + 10)$ ft
$(4x + 20)$ ft

a. Write a polynomial that represents the area of the football field.
b. Find the area of the football field when the width is 160 feet.

44. **MODELING WITH MATHEMATICS** You design a frame to surround a rectangular photo. The width of the frame is the same on every side, as shown.

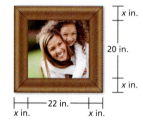
x in.
20 in.
x in.
x in. — 22 in. — x in.

a. Write a polynomial that represents the combined area of the photo and the frame.
b. Find the combined area of the photo and the frame when the width of the frame is 4 inches.

45. **WRITING** When multiplying two binomials, explain how the degree of the product is related to the degree of each binomial.

46. **THOUGHT PROVOKING** Write two polynomials that are not monomials whose product is a trinomial of degree 3.

47. **MAKING AN ARGUMENT** Your friend says the FOIL Method can be used to multiply two trinomials. Is your friend correct? Explain your reasoning.

48. **HOW DO YOU SEE IT?** The table shows one method of finding the product of two binomials.

	$-4x$	3
$-8x$	a	b
-9	c	d

a. Write the two binomials being multiplied.
b. Determine whether a, b, c, and d will be positive or negative when $x > 0$.

49. **COMPARING METHODS** You use the Distributive Property to multiply $(x + 3)(x - 5)$. Your friend uses the FOIL Method to multiply $(x - 5)(x + 3)$. Should your answers be equivalent? Justify your answer.

50. **USING STRUCTURE** The shipping container is a rectangular prism. Write a polynomial that represents the volume of the container.

$(x + 2)$ ft
$(4x - 3)$ ft
$(x + 1)$ ft

51. **ABSTRACT REASONING** The product of $(x + m)(x + n)$ is $x^2 + bx + c$.

a. What do you know about m and n when $c > 0$?
b. What do you know about m and n when $c < 0$?

Maintaining Mathematical Proficiency
Reviewing what you learned in previous grades and lessons

Write the absolute value function as a piecewise function. *(Section 1.2)*

52. $y = |x| + 4$
53. $y = 6|x - 3|$
54. $y = -4|x + 2|$

Simplify the expression. Write your answer using only positive exponents. *(Section 1.4)*

55. $10^2 \cdot 10^9$
56. $\dfrac{x^5 \cdot x}{x^8}$
57. $(3z^6)^{-3}$
58. $\left(\dfrac{2y^4}{y^3}\right)^{-2}$

Mini-Assessment

1. Use the Distributive Property to find $(x - 7)(x - 5)$.
   $x^2 - 12x + 35$
2. Use a table to find $(2a + 1)(a + 4)$.
   $2a^2 + 9a + 4$
3. Use the FOIL method to find $(3x - 1)(x + 5)$.
   $3x^2 + 14x - 5$
4. Find $(x - 2)(x^2 - 3x + 1)$.
   $x^3 - 5x^2 + 7x - 2$

If students need help...	If students got it...
Resources by Chapter • Practice A and Practice B • Puzzle Time	Resources by Chapter • Enrichment and Extension • Cumulative Review
Student Journal • Practice	Start the *next* Section
Differentiating the Lesson Skills Review Handbook	

Laurie's Notes

Overview of Section 2.3

Introduction
- Multiplication of binomials was presented in the last lesson. In this lesson, special products of binomials are computed: the square of a binomial pattern, and the sum and difference pattern.
- The lesson ends with a real-life application involving genetics.

Formative Assessment Tips
- **Point of Most Significance:** This technique asks students to identify the most significant idea, learning, or concept they gained in the lesson today. Students reflect on the lesson and are asked to identify the key example, problem, or point that was significant in their learning today. If the goal or learning intention was identified at the beginning of the class, then students assess what contributed to their attainment of the goal.
- It is important for teachers to know whether the lesson as designed helped move learning forward for students or whether the lesson needs to be modified. Share with students what you learned from their reflections. Students will take the reflection more seriously when they are valued and used.

Pacing Suggestion
- The explorations provide an opportunity for students to recognize the patterns when special binomials are squared. Transition to the formal lesson as soon as students have discussed each exploration.

Dynamic Teaching Tools
Dynamic Assessment & Progress Monitoring Tool
Lesson Planning Tool
Interactive Whiteboard Lesson Library
Dynamic Classroom with Dynamic Investigations

What Your Students Will Learn

- Use the square of a binomial pattern to multiply expressions having the form $(a + b)^2$ or $(a - b)^2$.
- Use the sum and difference pattern to multiply expressions having the form $(a + b)(a - b)$.
- Use special product patterns to solve real-life problems.

Laurie's Notes

Exploration

Motivate
- Challenge your students to a multiplication race! Have the students choose a two-digit number. Then you choose a "compatible" two-digit number so that you can use special product patterns to multiply.
- *Example:* If students choose 42, you choose 38 or 58.
  $42 \times 38 = (40 + 2)(40 - 2) = 40^2 - 2^2 = 1600 - 4 = 1596$
  $42 \times 58 = (50 - 8)(50 + 8) = 50^2 - 8^2 = 2500 - 64 = 2436$
- **FYI:** Note that the number of hundreds is 1 less than a perfect square, and the last two digits are the difference between 100 and a perfect square.
- After several problems, help students to see the patterns that you are using.

Exploration Note
- **Make Sense of Problems and Persevere in Solving Them** and **Use Appropriate Tools Strategically:** Working with algebra tiles in this exploration is a visual and tactile experience that should help students make sense of, and recognize, the patterns that appear when finding special products.

Exploration 1
- In this exploration, the models for each factor and the product are shown.
- **?** Ask the following questions.
  - "What are the factors in part (a)?" $(x + 2)$ and $(x - 2)$
  - "What is the product?" $x^2 - 4$
  - "What happened to the *x*-terms?" They added to zero.
  - "Did this also happen in part (b)?" yes
  - "How are the binomial factors different in each part?" The operation is different.

Exploration 2
- Students should be secure enough with algebra tiles at this point to be able to model the factors $(x + 2)$ in part (a) and $(2x - 1)$ in part (b).
- **Attend to Precision:** Part (a) is read, "*x* plus 2 quantity squared" not "*x* plus 2 squared."
- Students should recognize that there are two factors that are the same.
- **?** Ask the following questions.
  - "What are the factors in part (a)?" $(x + 2)$ and $(x + 2)$
  - "Are the factors the same in each part?" yes
  - "What is the product in part (b)?" $4x^2 - 4x + 1$
  - "Do you see a pattern in the *x*-terms of the products?" Students may not notice a pattern at this stage, although they may comment on the symmetry of the product along the diagonal.

Communicate Your Answer
- **Neighbor Check:** Have students work independently and then have their neighbors check their work. Have students discuss any discrepancies.

Connecting to Next Step
- The explorations help students recognize patterns in special products of binomials. When students see the symbolic approach in the formal lesson, they will have a visual model to connect it to.

2.3 Special Products of Polynomials

Essential Question What are the patterns in the special products $(a+b)(a-b)$, $(a+b)^2$, and $(a-b)^2$?

EXPLORATION 1 Finding a Sum and Difference Pattern

Work with a partner. Write the product of two binomials modeled by each rectangular array of algebra tiles.

a. $(x+2)(x-2) =$ _____ b. $(2x-1)(2x+1) =$ _____

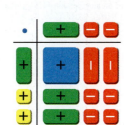 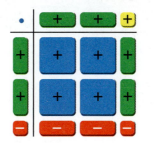

EXPLORATION 2 Finding the Square of a Binomial Pattern

Work with a partner. Draw the rectangular array of algebra tiles that models each product of two binomials. Write the product.

a. $(x+2)^2 =$ _____ b. $(2x-1)^2 =$ _____

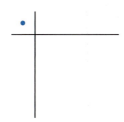 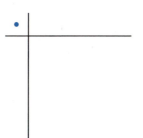

LOOKING FOR STRUCTURE

To be proficient in math, you need to look closely to discern a pattern or structure.

Communicate Your Answer

3. What are the patterns in the special products $(a+b)(a-b)$, $(a+b)^2$, and $(a-b)^2$?

4. Use the appropriate special product pattern to find each product. Check your answers using algebra tiles.

   a. $(x+3)(x-3)$ b. $(x-4)(x+4)$ c. $(3x+1)(3x-1)$
   d. $(x+3)^2$ e. $(x-2)^2$ f. $(3x+1)^2$

Section 2.3 Special Products of Polynomials 75

Dynamic Teaching Tools
- Dynamic Assessment & Progress Monitoring Tool
- Lesson Planning Tool
- Interactive Whiteboard Lesson Library
- Dynamic Classroom with Dynamic Investigations

ANSWERS

1. a. $x^2 - 4$
   b. $4x^2 - 1$

2. a.

   $x^2 + 4x + 4$

   b.

   $4x^2 - 4x + 1$

3. $(a+b)(a-b) = a^2 - b^2$;
   $(a+b)^2 = a^2 + 2ab + b^2$;
   $(a-b)^2 = a^2 - 2ab + b^2$

4. a. $x^2 - 9$
   b. $x^2 - 16$
   c. $9x^2 - 1$
   d. $x^2 + 6x + 9$
   e. $x^2 - 4x + 4$
   f. $9x^2 + 6x + 1$

Extra Example 1

Find each product.

a. $(2x + 6)^2$
   $4x^2 + 24x + 36$

b. $(3x - y)^2$
   $9x^2 - 6xy + y^2$

MONITORING PROGRESS ANSWERS

1. $x^2 + 14x + 49$
2. $49x^2 - 42x + 9$
3. $16x^2 - 8xy + y^2$
4. $9m^2 + 6mn + n^2$

2.3 Lesson

Core Vocabulary

Previous
binomial

What You Will Learn

▶ Use the square of a binomial pattern.
▶ Use the sum and difference pattern.
▶ Use special product patterns to solve real-life problems.

Using the Square of a Binomial Pattern

The diagram shows a square with a side length of $(a + b)$ units. You can see that the area of the square is

$$(a + b)^2 = a^2 + 2ab + b^2.$$

This is one version of a pattern called the square of a binomial. To find another version of this pattern, use algebra: replace b with $-b$.

$(a + (-b))^2 = a^2 + 2a(-b) + (-b)^2$ Replace b with $-b$ in the pattern above.
$(a - b)^2 = a^2 - 2ab + b^2$ Simplify.

🄲 Core Concept

Square of a Binomial Pattern

Algebra	Example
$(a + b)^2 = a^2 + 2ab + b^2$	$(x + 5)^2 = (x)^2 + 2(x)(5) + (5)^2$
	$= x^2 + 10x + 25$
$(a - b)^2 = a^2 - 2ab + b^2$	$(2x - 3)^2 = (2x)^2 - 2(2x)(3) + (3)^2$
	$= 4x^2 - 12x + 9$

LOOKING FOR STRUCTURE

When you use special product patterns, remember that a and b can be numbers, variables, or variable expressions.

EXAMPLE 1 Using the Square of a Binomial Pattern

Find each product.

a. $(3x + 4)^2$ b. $(5x - 2y)^2$

SOLUTION

a. $(3x + 4)^2 = (3x)^2 + 2(3x)(4) + 4^2$ Square of a binomial pattern
   $= 9x^2 + 24x + 16$ Simplify.

   ▶ The product is $9x^2 + 24x + 16$.

b. $(5x - 2y)^2 = (5x)^2 - 2(5x)(2y) + (2y)^2$ Square of a binomial pattern
   $= 25x^2 - 20xy + 4y^2$ Simplify.

   ▶ The product is $25x^2 - 20xy + 4y^2$.

Monitoring Progress Help in English and Spanish at *BigIdeasMath.com*

Find the product.

1. $(x + 7)^2$ 2. $(7x - 3)^2$ 3. $(4x - y)^2$ 4. $(3m + n)^2$

Chapter 2 Polynomial Equations and Factoring

Laurie's Notes — Teacher Actions

- Work through the examples as shown. You may want to use color coding to focus attention on the fact that the middle term of the product is twice the product of ab.
- **Make Sense of Problems and Persevere in Solving Them:** Some students will expand the square of a binomial and continue to use the FOIL Method. While this is not as efficient, it yields a correct solution and makes sense to the student.

Using the Sum and Difference Pattern

To find the product $(x + 2)(x - 2)$, you can multiply the two binomials using the FOIL Method.

$(x + 2)(x - 2) = x^2 - 2x + 2x - 4$ FOIL Method

$ = x^2 - 4$ Combine like terms.

This suggests a pattern for the product of the sum and difference of two terms.

Core Concept

Sum and Difference Pattern

Algebra	Example
$(a + b)(a - b) = a^2 - b^2$	$(x + 3)(x - 3) = x^2 - 9$

EXAMPLE 2 Using the Sum and Difference Pattern

Find each product.

a. $(t + 5)(t - 5)$ **b.** $(3x + y)(3x - y)$

SOLUTION

a. $(t + 5)(t - 5) = t^2 - 5^2$ Sum and difference pattern

$ = t^2 - 25$ Simplify.

▶ The product is $t^2 - 25$.

b. $(3x + y)(3x - y) = (3x)^2 - y^2$ Sum and difference pattern

$ = 9x^2 - y^2$ Simplify.

▶ The product is $9x^2 - y^2$.

The special product patterns can help you use mental math to find certain products of numbers.

EXAMPLE 3 Using Special Product Patterns and Mental Math

Use special product patterns to find the product $26 \cdot 34$.

SOLUTION

Notice that 26 is 4 less than 30, while 34 is 4 more than 30.

$26 \cdot 34 = (30 - 4)(30 + 4)$ Write as product of difference and sum.

$ = 30^2 - 4^2$ Sum and difference pattern

$ = 900 - 16$ Evaluate powers.

$ = 884$ Simplify.

▶ The product is 884.

Monitoring Progress   Help in English and Spanish at *BigIdeasMath.com*

Find the product.

5. $(x + 10)(x - 10)$ **6.** $(2x + 1)(2x - 1)$ **7.** $(x + 3y)(x - 3y)$

8. Describe how to use special product patterns to find 21^2.

Section 2.3   Special Products of Polynomials   77

Differentiated Instruction

Organization

Have students make a chart to show the patterns of special products. Have them include their own examples in the chart.

Patterns of Special Products
Square of a binomial (sum) $(a + b)^2 = a^2 + 2ab + b^2$
Example:
Square of a binomial (difference) $(a - b)^2 = a^2 - 2ab + b^2$
Example:
Sum and difference $(a + b)(a - b) = a^2 - b^2$
Example:

Extra Example 2

Find each product.

a. $(a + 8)(a - 8)$
$a^2 - 64$

b. $(4x + y)(4x - y)$
$16x^2 - y^2$

Extra Example 3

Use special product patterns to find the product $48 \cdot 52$.
$(50 - 2)(50 + 2) = 2500 - 4 = 2496$

MONITORING PROGRESS ANSWERS

5. $x^2 - 100$

6. $4x^2 - 1$

7. $x^2 - 9y^2$

8. Rewrite 21^2 as $(20 + 1)^2$, then use the square of a binomial pattern $(20 + 1)^2 = 20^2 + 2 \cdot 20 \cdot 1 + 1^2 = 400 + 40 + 1 = 441$.

Laurie's Notes   Teacher Actions

? Probing Question: "The product of two binomials is not always a trinomial. Why?"
Recalling the explorations, students should describe the sum and difference pattern. Write the Core Concept.

COMMON ERROR When substituting for *a* in Example 2(b), students may write $3x^2$ instead of $(3x)^2$. Remind them to use parentheses when the coefficient is not 1.

• Example 3 is related to the *Motivate*. Have students practice their mental math skills!

Extra Example 4

Each of two cats has one black patch gene (*B*) and one tan patch gene (*t*). Any gene combination with a *B* results in black patches in an offspring's fur. The Punnett square shows the possible gene combinations of the offspring and the resulting patch colors.

	B	t
B	BB	Bt
t	Bt	tt

a. What percent of the possible gene combinations result in tan patches?
   25%

b. Show how you could use a polynomial to model the possible gene combinations of the offspring.
   $0.25B^2 + 0.5Bt + 0.25t^2$; The coefficient of t^2 shows that 25% of the possible gene combinations result in tan patches.

MONITORING PROGRESS ANSWER

9. a. 25%
   b. $(0.5B + 0.5W)^2 = 0.25B^2 + 0.5BW + 0.25W^2$; The coefficients show 25% black, 50% gray, and 25% white.

Solving Real-Life Problems

EXAMPLE 4 Modeling with Mathematics

A combination of two genes determines the color of the dark patches of a border collie's coat. An offspring inherits one patch color gene from each parent. Each parent has two color genes, and the offspring has an equal chance of inheriting either one.

The gene *B* is for black patches, and the gene *r* is for red patches. Any gene combination with a *B* results in black patches. Suppose each parent has the same gene combination *Br*. The Punnett square shows the possible gene combinations of the offspring and the resulting patch colors.

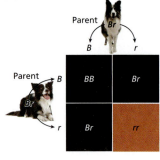

a. What percent of the possible gene combinations result in black patches?
b. Show how you could use a polynomial to model the possible gene combinations.

SOLUTION

a. Notice that the Punnett square shows four possible gene combinations of the offspring. Of these combinations, three result in black patches.

   ▶ So, 75% of the possible gene combinations result in black patches.

b. Model the gene from each parent with $0.5B + 0.5r$. There is an equal chance that the offspring inherits a black or a red gene from each parent.

   You can model the possible gene combinations of the offspring with $(0.5B + 0.5r)^2$. Notice that this product also represents the area of the Punnett square.

   Expand the product to find the possible patch colors of the offspring.

   $(0.5B + 0.5r)^2 = (0.5B)^2 + 2(0.5B)(0.5r) + (0.5r)^2$
   $= 0.25B^2 + 0.5Br + 0.25r^2$

   Consider the coefficients in the polynomial.

   $0.25B^2 + 0.5Br + 0.25r^2$

   - 25% *BB*, black patches
   - 50% *Br*, black patches
   - 25% *rr*, red patches

   The coefficients show that $25\% + 50\% = 75\%$ of the possible gene combinations result in black patches.

Monitoring Progress Help in English and Spanish at *BigIdeasMath.com*

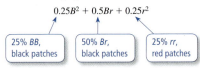
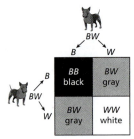

9. Each of two dogs has one black gene (*B*) and one white gene (*W*). The Punnett square shows the possible gene combinations of an offspring and the resulting colors.

   a. What percent of the possible gene combinations result in black?
   b. Show how you could use a polynomial to model the possible gene combinations of the offspring.

Chapter 2 Polynomial Equations and Factoring

Laurie's Notes   Teacher Actions

- Ask students whether they have studied genetics in science class. If they have, they will be familiar with the Punnett square, although they may not know it by that name.

COMMON ERROR Students may forget to square the decimal coefficients.

- **Think-Pair-Share:** Have students answer Question 9, and then share and discuss as a class.

Closure

- **Point of Most Significance:** Ask students to identify, aloud or on a paper to be collected, the most significant point (or part) in the lesson that aided their learning.

2.3 Exercises

Dynamic Solutions available at *BigIdeasMath.com*

Vocabulary and Core Concept Check

1. **WRITING** Explain how to use the square of a binomial pattern.
2. **WHICH ONE DOESN'T BELONG?** Which expression does *not* belong with the other three? Explain your reasoning.

$(x+1)(x-1)$ $(3x+2)(3x-2)$ $(x+2)(x-3)$ $(2x+5)(2x-5)$

Monitoring Progress and Modeling with Mathematics

In Exercises 3–10, find the product. (See Example 1.)

3. $(x+8)^2$
4. $(a-6)^2$
5. $(2f-1)^2$
6. $(5p+2)^2$
7. $(-7t+4)^2$
8. $(-12-n)^2$
9. $(2a+b)^2$
10. $(6x-3y)^2$

MATHEMATICAL CONNECTIONS In Exercises 11–14, write a polynomial that represents the area of the square.

11.
12.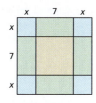
13.
14.

In Exercises 15–24, find the product. (See Example 2.)

15. $(t-7)(t+7)$
16. $(m+6)(m-6)$
17. $(4x+1)(4x-1)$
18. $(2k-4)(2k+4)$
19. $(8+3a)(8-3a)$
20. $\left(\frac{1}{2}-c\right)\left(\frac{1}{2}+c\right)$
21. $(p-10q)(p+10q)$
22. $(7m+8n)(7m-8n)$
23. $(-y+4)(-y-4)$
24. $(-5g-2h)(-5g+2h)$

In Exercises 25–30, use special product patterns to find the product. (See Example 3.)

25. $16 \cdot 24$
26. $33 \cdot 27$
27. 42^2
28. 29^2
29. 30.5^2
30. $10\frac{1}{3} \cdot 9\frac{2}{3}$

ERROR ANALYSIS In Exercises 31 and 32, describe and correct the error in finding the product.

31.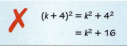
32.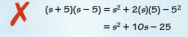

33. **MODELING WITH MATHEMATICS** A contractor extends a house on two sides.

a. The area of the house after the renovation is represented by $(x+50)^2$. Find this product.
b. Use the polynomial in part (a) to find the area when $x = 15$. What is the area of the extension?

Section 2.3 Special Products of Polynomials 79

31. The middle term in the square of a binomial pattern was not included;
$= k^2 + 2(k)(4) + 4^2 = k^2 + 8k + 16$

32. There is no middle term in the sum and difference pattern; $= s^2 - 5^2 = s^2 - 25$

33. a. $(x^2 + 100x + 2500)$ ft²
    b. 4225 ft²; 1725 ft²

Assignment Guide and Homework Check

ASSIGNMENT
Basic: 1, 2, 3–35 odd, 38, 43, 48–51
Average: 1, 2–34 even, 35–39, 43, 48–51
Advanced: 1, 2, 8–14 even, 22–34 even, 35–37, 38–42 even, 43–51

HOMEWORK CHECK
Basic: 3, 15, 17, 25, 35
Average: 4, 16, 18, 26, 35
Advanced: 10, 22, 24, 28, 35

ANSWERS

1. Substitute the first term of the binomial for *a* and the second term of the binomial for *b* in the square of a binomial pattern, then simplify.
2. $(x+2)(x-3)$; It is the only one that cannot be simplified using the sum and difference pattern.
3. $x^2 + 16x + 64$
4. $a^2 - 12a + 36$
5. $4f^2 - 4f + 1$
6. $25p^2 + 20p + 4$
7. $49t^2 - 56t + 16$
8. $n^2 + 24n + 144$
9. $4a^2 + 4ab + b^2$
10. $36x^2 - 36xy + 9y^2$
11. $x^2 + 8x + 16$
12. $4x^2 + 28x + 49$
13. $49n^2 - 70n + 25$
14. $16c^2 + 32cd + 16d^2$
15. $t^2 - 49$
16. $m^2 - 36$
17. $16x^2 - 1$
18. $4k^2 - 16$
19. $64 - 9a^2$
20. $\frac{1}{4} - c^2$
21. $p^2 - 100q^2$
22. $49m^2 - 64n^2$
23. $y^2 - 16$
24. $25g^2 - 4h^2$
25. 384
26. 891
27. 1764
28. 841
29. 930.25
30. $99\frac{8}{9}$

Section 2.3 79

Dynamic Teaching Tools
- Dynamic Assessment & Progress Monitoring Tool
- Interactive Whiteboard Lesson Library
- Dynamic Classroom with Dynamic Investigations

ANSWERS

34. a. $(10{,}000 - x^2)$ ft^2
    b. decrease; The original area is 10,000 ft^2, so the new area is the original area decreased by x^2.
    c. 9559 ft^2

35. a. 25%
    b. $(0.5N + 0.5a)^2 = 0.25N^2 + 0.5Na + 0.25a^2$

36–51. See Additional Answers.

Mini-Assessment

Find the product.

1. $(4x + 1)^2$ $16x^2 + 8x + 1$
2. $(3a - 5b)^2$ $9a^2 - 30ab + 25b^2$
3. $(2d + 3)(2d - 3)$ $4d^2 - 9$
4. Use special product patterns to find the product $43 \cdot 37$.
   $(40 + 3)(40 - 3) = 1600 - 9 = 1591$
5. Each of two pea plants has one tall gene (T) and one short gene (s). Any gene combination with a T results in a tall plant. The Punnett square shows the possible gene combinations of the offspring and the resulting plant heights.

	T	s
T	TT	Ts
s	Ts	ss

   a. What percent of the possible gene combinations result in tall plants? 75%
   b. Show how you could use a polynomial to model the possible gene combinations of the offspring.
   $0.25T^2 + 0.5Ts + 0.25s^2$; The coefficients show that $25\% + 50\% = 75\%$ of the possible gene combinations result in tall plants.

---

34. **MODELING WITH MATHEMATICS** A square-shaped parking lot with 100-foot sides is reduced by x feet on one side and extended by x feet on an adjacent side.
    a. The area of the new parking lot is represented by $(100 - x)(100 + x)$. Find this product.
    b. Does the area of the parking lot increase, decrease, or stay the same? Explain.
    c. Use the polynomial in part (a) to find the area of the new parking lot when $x = 21$.

35. **MODELING WITH MATHEMATICS** In deer, the gene N is for normal coloring and the gene a is for no coloring, or albino. Any gene combination with an N results in normal coloring. The Punnett square shows the possible gene combinations of an offspring and the resulting colors from parents that both have the gene combination Na. (See Example 4.)
    a. What percent of the possible gene combinations result in albino coloring?
    b. Show how you could use a polynomial to model the possible gene combinations of the offspring.

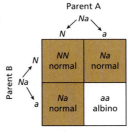

36. **MODELING WITH MATHEMATICS** Your iris controls the amount of light that enters your eye by changing the size of your pupil.

    a. Write a polynomial that represents the area of your pupil. Write your answer in terms of π.
    b. The width x of your iris decreases from 4 millimeters to 2 millimeters when you enter a dark room. How many times greater is the area of your pupil after entering the room than before entering the room? Explain.

37. **CRITICAL THINKING** Write two binomials that have the product $x^2 - 121$. Explain.

38. **HOW DO YOU SEE IT?** In pea plants, any gene combination with a green gene (G) results in a green pod. The Punnett square shows the possible gene combinations of the offspring of two Gy pea plants and the resulting pod colors.

    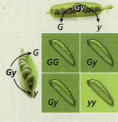

    G = Green gene
    y = Yellow gene

    A polynomial that models the possible gene combinations of the offspring is
    $(0.5G + 0.5y)^2 = 0.25G^2 + 0.5Gy + 0.25y^2$.
    Describe two ways to determine the percent of possible gene combinations that result in green pods.

In Exercises 39–42, find the product.

39. $(x^2 + 1)(x^2 - 1)$ 40. $(y^3 + 4)^2$
41. $(2m^2 - 5n^2)^2$ 42. $(r^3 - 6t^4)(r^3 + 6t^4)$

43. **MAKING AN ARGUMENT** Your friend claims to be able to use a special product pattern to determine that $\left(4\tfrac{1}{3}\right)^2$ is equal to $16\tfrac{1}{9}$. Is your friend correct? Explain.

44. **THOUGHT PROVOKING** Modify the dimensions of the original parking lot in Exercise 34 so that the area can be represented by two other types of special product patterns discussed in this section. Is there a positive x-value for which the three area expressions are equivalent? Explain.

45. **REASONING** Find k so that $9x^2 - 48x + k$ is the square of a binomial.

46. **REPEATED REASONING** Find $(x + 1)^3$ and $(x + 2)^3$. Find a pattern in the terms and use it to write a pattern for the cube of a binomial $(a + b)^3$.

47. **PROBLEM SOLVING** Find two numbers a and b such that $(a + b)(a - b) < (a - b)^2 < (a + b)^2$.

Maintaining Mathematical Proficiency *Reviewing what you learned in previous grades and lessons*

Factor the expression using the GCF. *(Skills Review Handbook)*

48. $12y - 18$ 49. $9r + 27$ 50. $49s + 35t$ 51. $15x - 10y$

80   Chapter 2   Polynomial Equations and Factoring

If students need help...	If students got it...
Resources by Chapter • Practice A and Practice B • Puzzle Time	**Resources by Chapter** • Enrichment and Extension • Cumulative Review
Student Journal • Practice	Start the *next* Section
Differentiating the Lesson Skills Review Handbook	

Laurie's Notes

Overview of Section 2.4

Introduction
- In this lesson, students will solve polynomial equations by using the Zero-Product Property. If the polynomial is already factored and set equal to zero, the equation can be solved by setting each of the factors equal to 0 and solving. If the equation is not in factored form, it can be rewritten by factoring out a common factor.
- Additional factoring techniques are presented in the next lesson.

Teaching Strategy
- In this lesson and in the remaining lessons in this chapter, students will be solving polynomials by factoring an expression and setting it equal to 0. Note that other techniques besides factoring will be presented in later chapters.
- When factors are set equal to 0, the Zero-Product Property is used to solve the polynomial. Many of the factors will be of the form $(ax + b)$, so students need to be proficient with solving $ax + b = 0$, where a and b are real numbers.
- Spend time solving a variety of equivalent forms so that students are efficient and accurate with their solutions, eventually using eyesight (inspection) to solve.

    Examples: $-5x + 7 = 0$ $8 - 3x = 0$ $6 + \frac{1}{2}x = 0$

    $-\frac{2}{3}x - 8 = 0$ $0.1x + 9 = 0$ $-5 - 4x = 0$

Pacing Suggestion
- Complete at least the first two explorations before beginning the formal lesson. The third exploration probes prior learning and connects to the Zero-Product Property.

Dynamic Teaching Tools
Dynamic Assessment & Progress Monitoring Tool
Lesson Planning Tool
Interactive Whiteboard Lesson Library
Dynamic Classroom with Dynamic Investigations

What Your Students Will Learn

- Use the Zero-Product Property to solve polynomial equations in factored form.
- Factor polynomials by finding the greatest common factor (GCF).
- Use the Zero-Product Property to solve real-life problems involving polynomial equations.

Laurie's Notes

Exploration

Motivate
- Write the following equations on the board and ask which one does not belong. Have students explain their reasoning.

    A. $3(x - 4) = 12$ **B.** $\frac{3}{4}x = 6$ **C.** $4 - 8x = 3$ **D.** $6x - 7 = 9 + 4x$

- Students should solve the equations and reason that C does not belong. The other three equations have the same solution, $x = 8$. These equations are equivalent.

Discuss
- Explain what equivalent polynomial equations are, connecting to the *Motivate*.
- Review the language associated with polynomials: factored form, standard form, and nonstandard form. Check the solutions in each form.

Exploration 1
- Students may approach the exploration in one of two ways. If they recognize the solutions from the factored form, then they may look at the other two forms and determine which forms are equivalent by substituting the solutions from the factored form. The alternate method would be to expand the factored form to see which expressions are equivalent in standard form.
- Give students time to work through the problems. Then ask for volunteers to share their findings and their method of solving.
- **?** "How did you determine which equations are equivalent?" Expect different methods.

Exploration 2
- For each part, students will substitute the given values and evaluate the polynomial.
- **Use Appropriate Tools Strategically:** A table is an appropriate tool to help organize what is known in this problem. The table might have seven columns. The first column lists the equation, and the next six columns are for the values of x being substituted in the equation. The value being substituted is either a solution or not a solution.
- **Big Idea:** Each polynomial has two solutions. When either or both of the solutions are substituted in the factored expression, the product is 0. If two numbers are substituted and neither is a solution, then the product is not equal to 0.

Exploration 3
- The numbers 0 and 1 have special properties. Students may not recall the formal names of the properties, but they should recall the essence of the properties.
- Remind students that they should check both 0 and 1 in each statement to determine whether the statement is true.
- When students are finished, ask for volunteers to share their answers and their reasoning.
- **Extension:** Ask whether any of the statements are true when $n = -1$.

Communicate Your Answer
- For Question 5, have students discuss which property they chose and why they chose it.

Connecting to Next Step
- Students have connected the solution of a quadratic in factored form with the Zero-Product Property, without necessarily knowing the name of the property. In the formal lesson, the approach is given a specific name.

2.4 Solving Polynomial Equations in Factored Form

Essential Question How can you solve a polynomial equation?

EXPLORATION 1: Matching Equivalent Forms of an Equation

Work with a partner. An equation is considered to be in *factored form* when the product of the factors is equal to 0. Match each factored form of the equation with its equivalent standard form and nonstandard form.

Factored Form		Standard Form		Nonstandard Form
a. $(x-1)(x-3)=0$	**A.**	$x^2-x-2=0$	**1.**	$x^2-5x=-6$
b. $(x-2)(x-3)=0$	**B.**	$x^2+x-2=0$	**2.**	$(x-1)^2=4$
c. $(x+1)(x-2)=0$	**C.**	$x^2-4x+3=0$	**3.**	$x^2-x=2$
d. $(x-1)(x+2)=0$	**D.**	$x^2-5x+6=0$	**4.**	$x(x+1)=2$
e. $(x+1)(x-3)=0$	**E.**	$x^2-2x-3=0$	**5.**	$x^2-4x=-3$

USING TOOLS STRATEGICALLY
To be proficient in math, you need to consider using tools such as a table or a spreadsheet to organize your results.

EXPLORATION 2: Writing a Conjecture

Work with a partner. Substitute 1, 2, 3, 4, 5, and 6 for x in each equation and determine whether the equation is true. Organize your results in a table. Write a conjecture describing what you discovered.

a. $(x-1)(x-2)=0$
b. $(x-2)(x-3)=0$
c. $(x-3)(x-4)=0$
d. $(x-4)(x-5)=0$
e. $(x-5)(x-6)=0$
f. $(x-6)(x-1)=0$

EXPLORATION 3: Special Properties of 0 and 1

Work with a partner. The numbers 0 and 1 have special properties that are shared by no other numbers. For each of the following, decide whether the property is true for 0, 1, both, or neither. Explain your reasoning.

a. When you add ▢ to a number n, you get n.

b. If the product of two numbers is ▢, then at least one of the numbers is 0.

c. The square of ▢ is equal to itself.

d. When you multiply a number n by ▢, you get n.

e. When you multiply a number n by ▢, you get 0.

f. The opposite of ▢ is equal to itself.

Communicate Your Answer

4. How can you solve a polynomial equation?

5. One of the properties in Exploration 3 is called the Zero-Product Property. It is one of the most important properties in all of algebra. Which property is it? Why do you think it is called the Zero-Product Property? Explain how it is used in algebra and why it is so important.

Section 2.4 Solving Polynomial Equations in Factored Form

Dynamic Teaching Tools

Dynamic Assessment & Progress Monitoring Tool
Lesson Planning Tool
Interactive Whiteboard Lesson Library
Dynamic Classroom with Dynamic Investigations

ANSWERS

1. **a.** C; 5
   b. D; 1
   c. A; 3
   d. B; 4
   e. E; 2
2. See Additional Answers.
3. **a.** 0; Adding 0 does not change a value, but adding 1 increases the value.
   b. 0; The product of 0 and any number is always 0, never 1.
   c. both; $0^2 = 0$ and $1^2 = 1$
   d. 1; The product of any number and 0 is 0, the product of any number and 1 is the number.
   e. 0; The product of any number and 0 is 0, the product of any number and 1 is the number.
   f. 0; Zero is neither positive nor negative, so it is its own opposite.
4. Set each polynomial factor equal to 0 and solve.
5. b; It describes what happens when you have a product that is equal to zero; It is used to solve polynomial equations by setting each polynomial factor equal to 0. It is important because it provides an easy way to solve polynomial equations.

Extra Example 1

Solve each equation.

a. $3x(x-6) = 0$

The roots are $x = 0$ and $x = 6$.

b. $(x+5)(x-4) = 0$

The roots are $x = -5$ and $x = 4$.

MONITORING PROGRESS ANSWERS

1. $x = 0, x = 1$
2. $t = 0, t = -2$
3. $z = 4, z = 6$

2.4 Lesson

What You Will Learn

- Use the Zero-Product Property.
- Factor polynomials using the GCF.
- Use the Zero-Product Property to solve real-life problems.

Core Vocabulary

factored form, *p. 82*
Zero-Product Property, *p. 82*
roots, *p. 82*
repeated roots, *p. 83*

Previous
polynomial
standard form
greatest common factor (GCF)
monomial

Using the Zero-Product Property

A polynomial is in **factored form** when it is written as a product of factors.

Standard form	Factored form
$x^2 + 2x$	$x(x+2)$
$x^2 + 5x - 24$	$(x-3)(x+8)$

When one side of an equation is a polynomial in factored form and the other side is 0, use the **Zero-Product Property** to solve the polynomial equation. The solutions of a polynomial equation are also called **roots**.

Core Concept

Zero-Product Property

Words If the product of two real numbers is 0, then at least one of the numbers is 0.

Algebra If a and b are real numbers and $ab = 0$, then $a = 0$ or $b = 0$.

EXAMPLE 1 Solving Polynomial Equations

Solve each equation.

a. $2x(x-4) = 0$ b. $(x-3)(x-9) = 0$

SOLUTION

Check

To check the solutions of Example 1(a), substitute each solution in the original equation.

$2(0)(0-4) \stackrel{?}{=} 0$
$0(-4) \stackrel{?}{=} 0$
$0 = 0$ ✓

$2(4)(4-4) \stackrel{?}{=} 0$
$8(0) \stackrel{?}{=} 0$
$0 = 0$ ✓

a. $2x(x-4) = 0$ Write equation.
   $2x = 0$ or $x - 4 = 0$ Zero-Product Property
   $x = 0$ or $x = 4$ Solve for *x*.

▶ The roots are $x = 0$ and $x = 4$.

b. $(x-3)(x-9) = 0$ Write equation.
   $x - 3 = 0$ or $x - 9 = 0$ Zero-Product Property
   $x = 3$ or $x = 9$ Solve for *x*.

▶ The roots are $x = 3$ and $x = 9$.

Monitoring Progress Help in English and Spanish at *BigIdeasMath.com*

Solve the equation. Check your solutions.

1. $x(x-1) = 0$
2. $3t(t+2) = 0$
3. $(z-4)(z-6) = 0$

82   Chapter 2   Polynomial Equations and Factoring

Laurie's Notes | Teacher Actions

- Explain that solutions of a polynomial equation are also called *roots*. Use "roots" and "solutions" interchangeably so students are comfortable with the vocabulary. Likewise, continue to use "factor" and "product" to focus attention on the operation involved.
- **? Teaching Tip:** In part (a), underline each of the factors and ask, "What are the factors that have a product of 0?" *2x and (x − 4)*
- **? Probing Question:** "If the product of two factors is 0, will there always be two roots? Explain." *no; The roots could be the same.*

When two or more roots of an equation are the same number, the equation has **repeated roots**.

EXAMPLE 2   Solving Polynomial Equations

Solve each equation.

a. $(2x + 7)(2x - 7) = 0$ b. $(x - 1)^2 = 0$ c. $(x + 1)(x - 3)(x - 2) = 0$

SOLUTION

a. $(2x + 7)(2x - 7) = 0$ Write equation.
   $2x + 7 = 0$ or $2x - 7 = 0$ Zero-Product Property
   $x = -\frac{7}{2}$ or $x = \frac{7}{2}$ Solve for x.

 The roots are $x = -\frac{7}{2}$ and $x = \frac{7}{2}$.

b. $(x - 1)^2 = 0$ Write equation.
    $(x - 1)(x - 1) = 0$ Expand equation.
    $x - 1 = 0$ or $x - 1 = 0$ Zero-Product Property
    $x = 1$ or $x = 1$ Solve for x.

▶ The equation has repeated roots of $x = 1$.

c. $(x + 1)(x - 3)(x - 2) = 0$ Write equation.
   $x + 1 = 0$ or $x - 3 = 0$ or $x - 2 = 0$ Zero-Product Property
   $x = -1$ or $x = 3$ or $x = 2$ Solve for x.

 The roots are $x = -1$, $x = 3$, and $x = 2$.

STUDY TIP
You can extend the Zero-Product Property to products of more than two real numbers.

Monitoring Progress   Help in English and Spanish at *BigIdeasMath.com*

Solve the equation. Check your solutions.

4. $(3s + 5)(5s + 8) = 0$ 5. $(b + 7)^2 = 0$ 6. $(d - 2)(d + 6)(d + 8) = 0$

Factoring Polynomials Using the GCF

To solve a polynomial equation using the Zero-Product Property, you may need to *factor* the polynomial, or write it as a product of other polynomials. Look for the *greatest common factor* (GCF) of the terms of the polynomial. This is a monomial that divides evenly into each term.

EXAMPLE 3   Finding the Greatest Common Monomial Factor

Factor out the greatest common monomial factor from $4x^4 + 24x^3$.

SOLUTION

The GCF of 4 and 24 is 4. The GCF of x^4 and x^3 is x^3. So, the greatest common monomial factor of the terms is $4x^3$.

▶ So, $4x^4 + 24x^3 = 4x^3(x + 6)$.

Monitoring Progress   Help in English and Spanish at *BigIdeasMath.com*

7. Factor out the greatest common monomial factor from $8y^2 - 24y$.

Section 2.4   Solving Polynomial Equations in Factored Form   83

Differentiated Instruction

Visual
When using the GCF to factor, some students may benefit from seeing the GCF in each term of the polynomial. For example, write $4x^4 + 24x^3$ in Example 3 as $4x^3(x) + 4x^3(6)$. This may help students see that $4x^3$ is a factor of each term and that $4x^4 + 24x^3 = 4x^3(x + 6)$.

Extra Example 2
Solve each equation.
a. $(4x + 5)(4x - 5) = 0$
   The roots are $x = -\frac{5}{4}$ and $x = \frac{5}{4}$.
b. $(c + 6)^2 = 0$ The equation has repeated roots of $c = -6$.
c. $(a + 5)(a - 2)(a - 7) = 0$ The roots are $a = -5$, $a = 2$, and $a = 7$.

Extra Example 3
Factor out the greatest common monomial factor from $12x^3 + 3x^2$. $3x^2(4x + 1)$

MONITORING PROGRESS ANSWERS

4. $s = -\frac{5}{3}, s = -\frac{8}{5}$
5. $b = -7$
6. $d = 2, d = -6, d = -8$
7. $8y(y - 3)$

Laurie's Notes   Teacher Actions

- In part (a), you want students to be fluent in solving $2x + 7 = 0$ and $2x - 7 = 0$ by inspection. See the *Teaching Strategy* on page T-80.
- Students benefit from initially seeing each factor set equal to 0. Remind students that this is the application of the *Zero-Product Property*.
- **Connection:** Work through Example 3 as shown. Students often fail to see that this is the Distributive Property. Instead of multiplying a factor times a binomial, they are factoring out a common factor from the binomial.

Extra Example 4

a. Solve $4x^2 + 12x = 0$. The roots are $x = 0$ and $x = -3$.
b. Solve $-10a^2 = 8a$. The roots are $a = 0$ and $a = -\frac{4}{5}$.

Extra Example 5

You can model the arch of an entrance to a train tunnel by using the equation $y = -\frac{5}{16}(x + 8)(x - 8)$, where x and y are measured in feet. The x-axis represents the ground. Find the width of the entrance at ground level. **16 feet**

MONITORING PROGRESS ANSWERS

8. $a = 0, a = -5$
9. $s = 0, s = 3$
10. $x = 0, x = \frac{1}{2}$
11. 8 ft

EXAMPLE 4 Solving Equations by Factoring

Solve (a) $2x^2 + 8x = 0$ and (b) $6n^2 = 15n$.

SOLUTION

a.
$2x^2 + 8x = 0$ Write equation.
$2x(x + 4) = 0$ Factor left side.
$2x = 0$ or $x + 4 = 0$ Zero-Product Property
$x = 0$ or $x = -4$ Solve for x.

▶ The roots are $x = 0$ and $x = -4$.

b.
$6n^2 = 15n$ Write equation.
$6n^2 - 15n = 0$ Subtract $15n$ from each side.
$3n(2n - 5) = 0$ Factor left side.
$3n = 0$ or $2n - 5 = 0$ Zero-Product Property
$n = 0$ or $n = \frac{5}{2}$ Solve for n.

▶ The roots are $n = 0$ and $n = \frac{5}{2}$.

Monitoring Progress Help in English and Spanish at BigIdeasMath.com

Solve the equation. Check your solutions.

8. $a^2 + 5a = 0$ 9. $3s^2 - 9s = 0$ 10. $4x^2 = 2x$

Solving Real-Life Problems

EXAMPLE 5 Modeling with Mathematics

You can model the arch of a fireplace using the equation $y = -\frac{1}{9}(x + 18)(x - 18)$, where x and y are measured in inches. The x-axis represents the floor. Find the width of the arch at floor level.

SOLUTION

Use the x-coordinates of the points where the arch meets the floor to find the width. At floor level, $y = 0$. So, substitute 0 for y and solve for x.

$y = -\frac{1}{9}(x + 18)(x - 18)$ Write equation.
$0 = -\frac{1}{9}(x + 18)(x - 18)$ Substitute 0 for y.
$0 = (x + 18)(x - 18)$ Multiply each side by -9.
$x + 18 = 0$ or $x - 18 = 0$ Zero-Product Property
$x = -18$ or $x = 18$ Solve for x.

The width is the distance between the x-coordinates, -18 and 18.

▶ So, the width of the arch at floor level is $|-18 - 18| = 36$ inches.

Monitoring Progress Help in English and Spanish at BigIdeasMath.com

11. You can model the entrance to a mine shaft using the equation

$y = -\frac{1}{2}(x + 4)(x - 4)$, where x and y are measured in feet. The x-axis represents the ground. Find the width of the entrance at ground level.

84   Chapter 2   Polynomial Equations and Factoring

Laurie's Notes   Teacher Actions

❓ "In part (a), how can you write $2x^2 + 8x$ as a product instead of a sum?" Factor out $2x$ so you have $2x(x + 4)$.

COMMON ERROR In Example 5, when multiplying each side by -9, students may think that all three factors on the right side must be multiplied by -9. To help students with this, ask them to find the product $3(4)(5)$. 60 Then ask them to find the product $2(3)(4)(5)$. 120

Closure

- **Exit Ticket:** Write a polynomial equation that has -6 and 4 as solutions. *Sample answer:* $(x + 6)(x - 4) = 0$

2.4 Exercises

Dynamic Solutions available at BigIdeasMath.com

Vocabulary and Core Concept Check

1. **WRITING** Explain how to use the Zero-Product Property to find the solutions of the equation $3x(x - 6) = 0$.

2. **DIFFERENT WORDS, SAME QUESTION** Which is different? Find *both* answers.

Solve the equation $(2k + 4)(k - 3) = 0$.	Find the values of k for which $2k + 4 = 0$ or $k - 3 = 0$.
Find the value of k for which $(2k + 4) + (k - 3) = 0$.	Find the roots of the equation $(2k + 4)(k - 3) = 0$.

Monitoring Progress and Modeling with Mathematics

In Exercises 3–8, solve the equation. (*See Example 1.*)

3. $x(x + 7) = 0$
4. $r(r - 10) = 0$
5. $12t(t - 5) = 0$
6. $-2v(v + 1) = 0$
7. $(s - 9)(s - 1) = 0$
8. $(y + 2)(y - 6) = 0$

In Exercises 9–20, solve the equation. (*See Example 2.*)

9. $(2a - 6)(3a + 15) = 0$
10. $(4q + 3)(q + 2) = 0$
11. $(5m + 4)^2 = 0$
12. $(h - 8)^2 = 0$
13. $(3 - 2g)(7 - g) = 0$
14. $(2 - 4d)(2 + 4d) = 0$
15. $z(z + 2)(z - 1) = 0$
16. $5p(2p - 3)(p + 7) = 0$
17. $(r - 4)^2(r + 8) = 0$
18. $w(w - 6)^2 = 0$
19. $(15 - 5c)(5c + 5)(-c + 6) = 0$
20. $(2 - n)\left(6 + \frac{2}{3}n\right)(n - 2) = 0$

In Exercises 21–24, find the *x*-coordinates of the points where the graph crosses the *x*-axis.

21.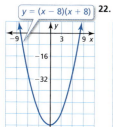
22.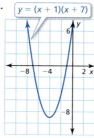

23.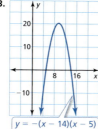
    $y = -(x - 14)(x - 5)$

24.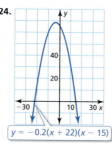
    $y = -0.2(x + 22)(x - 15)$

In Exercises 25–30, factor the polynomial. (*See Example 3.*)

25. $5z^2 + 45z$
26. $6d^2 - 21d$
27. $3y^3 - 9y^2$
28. $20x^3 + 30x^2$
29. $5n^6 + 2n^5$
30. $12a^4 + 8a$

In Exercises 31–36, solve the equation. (*See Example 4.*)

31. $4p^2 - p = 0$
32. $6m^2 + 12m = 0$
33. $25c + 10c^2 = 0$
34. $18q - 2q^2 = 0$
35. $3n^2 = 9n$
36. $-28r = 4r^2$

37. **ERROR ANALYSIS** Describe and correct the error in solving the equation.

    $6x(x + 5) = 0$
    $x + 5 = 0$
    $x = -5$
    The root is $x = -5$.

Section 2.4 Solving Polynomial Equations in Factored Form 85

---

34. $q = 0, q = 9$
35. $n = 0, n = 3$
36. $r = 0, r = -7$
37. also need to set $6x = 0$ and solve; $6x = 0$ or $x + 5 = 0$; $x = 0$ or $x = -5$; The roots are $x = 0$ and $x = -5$.

Assignment Guide and Homework Check

ASSIGNMENT

Basic: 1, 2, 3–39 odd, 42, 44, 49–52
Average: 1, 2–40 even, 41, 42, 44, 49–52
Advanced: 1, 2, 6, 8, 16–24 even, 28–40 even, 41–52

HOMEWORK CHECK

Basic: 3, 9, 25, 33, 39
Average: 4, 10, 26, 32, 40
Advanced: 8, 16, 28, 32, 40

ANSWERS

1. Set $3x = 0$ and $x - 6 = 0$, then solve both of the equations to get the solutions $x = 0$ and $x = 6$.
2. Find the value of k for which $(2k + 4) + (k - 3) = 0$; $k = -\frac{1}{3}$, $k = -2$, $k = 3$
3. $x = 0, x = -7$
4. $r = 0, r = 10$
5. $t = 0, t = 5$
6. $v = 0, v = -1$
7. $s = 9, s = 1$
8. $y = -2, y = 6$
9. $a = 3, a = -5$
10. $q = -\frac{3}{4}, q = -2$
11. $m = -\frac{4}{5}$
12. $h = 8$
13. $g = \frac{3}{2}, g = 7$
14. $d = \frac{1}{2}, d = -\frac{1}{2}$
15. $z = 0, z = -2, z = 1$
16. $p = 0, p = \frac{3}{2}, p = -7$
17. $r = 4, r = -8$
18. $w = 0, w = 6$
19. $c = 3, c = -1, c = 6$
20. $n = 2, n = -9$
21. $x = 8, x = -8$
22. $x = -1, x = -7$
23. $x = 14, x = 5$
24. $x = -22, x = 15$
25. $5z(z + 9)$
26. $3d(2d - 7)$
27. $3y^2(y - 3)$
28. $10x^2(2x + 3)$
29. $n^5(5n + 2)$
30. $4a(3a^3 + 2)$
31. $p = 0, p = \frac{1}{4}$
32. $m = 0, m = -2$
33. $c = 0, c = -\frac{5}{2}$

Section 2.4   85

Dynamic Teaching Tools
Dynamic Assessment & Progress Monitoring Tool
Interactive Whiteboard Lesson Library
Dynamic Classroom with Dynamic Investigations

ANSWERS

38. cannot divide both sides by y, because y could be 0 and division by 0 is undefined; $3y^2 - 21y = 0$; $3y(y - 7) = 0$; $3y = 0$ or $y - 7 = 0$; $y = 0$ or $y = 7$; The roots are $y = 0$ and $y = 7$.
39. 20 ft
40. a. 630 ft
    b. 630 ft
41. $x = 0$, $x = 0.3$ sec; The roots represent the times when the penguin is at water level. $x = 0$ is when it leaves the water, and $x = 0.3$ sec is when it returns to the water after the leap.
42. $-$; $+$; The point of intersection on the positive x-axis is farther from the origin than the point of intersection on the negative x-axis, so the positive solution to the equation will be the one with the greater absolute value.
43. 2; x-intercepts occur when $y = 0$ and the equation has 2 roots when $y = 0$.
44. no; If $a = b$, then there is a repeated root and the graph will only have one x-intercept.
45–52. See Additional Answers.

Mini-Assessment

Solve the equation.
1. $5x(x - 4) = 0$
   The roots are $x = 0$ and $x = 4$.
2. $(3x + 8)(3x - 8) = 0$
   The roots are $x = -\frac{8}{3}$ and $x = \frac{8}{3}$.
3. $3x^5 + 21x^4 = 0$ The roots are $x = 0$ and $x = -7$.
4. Factor out the greatest common monomial factor from $6x^4 + 8x^3$.
   $2x^3(3x + 4)$
5. You can model the arch of a building entrance by using the equation $y = -\frac{1}{3}(x + 6)(x - 6)$, where x and y are measured in feet. The x-axis represents the ground. Find the width of the arch at the ground. 12 feet

86   Chapter 2

---

38. **ERROR ANALYSIS** Describe and correct the error in solving the equation.

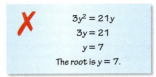

$$3y^2 = 21y$$
$$3y = 21$$
$$y = 7$$
The root is $y = 7$.

39. **MODELING WITH MATHEMATICS** The entrance of a tunnel can be modeled by $y = -\frac{11}{50}(x - 4)(x - 24)$, where x and y are measured in feet. The x-axis represents the ground. Find the width of the tunnel at ground level. *(See Example 5.)*

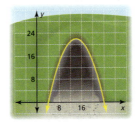

40. **MODELING WITH MATHEMATICS** The Gateway Arch in St. Louis can be modeled by $y = -\frac{2}{315}(x + 315)(x - 315)$, where x and y are measured in feet. The x-axis represents the ground.

a. Find the width of the arch at ground level.
b. How tall is the arch?

41. **MODELING WITH MATHEMATICS** A penguin leaps out of the water while swimming. This action is called porpoising. The height y (in feet) of a porpoising penguin can be modeled by $y = -16x^2 + 4.8x$, where x is the time (in seconds) since the penguin leaped out of the water. Find the roots of the equation when $y = 0$. Explain what the roots mean in this situation.

42. **HOW DO YOU SEE IT?** Use the graph to fill in each blank in the equation with the symbol $+$ or $-$. Explain your reasoning.

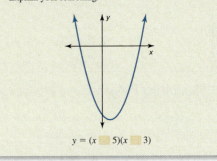

$y = (x \;\square\; 5)(x \;\square\; 3)$

43. **CRITICAL THINKING** How many x-intercepts does the graph of $y = (2x + 5)(x - 9)^2$ have? Explain.

44. **MAKING AN ARGUMENT** Your friend says that the graph of the equation $y = (x - a)(x - b)$ always has two x-intercepts for any values of a and b. Is your friend correct? Explain.

45. **CRITICAL THINKING** Does the equation $(x^2 + 3)(x^4 + 1) = 0$ have any real roots? Explain.

46. **THOUGHT PROVOKING** Write a polynomial equation of degree 4 whose only roots are $x = 1$, $x = 2$, and $x = 3$.

47. **REASONING** Find the values of x in terms of y that are solutions of each equation.
    a. $(x + y)(2x - y) = 0$
    b. $(x^2 - y^2)(4x + 16y) = 0$

48. **PROBLEM SOLVING** Solve the equation $(4^{x-5} - 16)(3^x - 81) = 0$.

Maintaining Mathematical Proficiency Reviewing what you learned in previous grades and lessons

List the factor pairs of the number. *(Skills Review Handbook)*
49. 10
50. 18
51. 30
52. 48

86   Chapter 2   Polynomial Equations and Factoring

If students need help...	If students got it...
Resources by Chapter • Practice A and Practice B • Puzzle Time	Resources by Chapter • Enrichment and Extension • Cumulative Review
Student Journal • Practice	Start the *next* Section
Differentiating the Lesson Skills Review Handbook	

2.1–2.4 What Did You Learn?

Core Vocabulary

monomial, *p. 62*
degree of a monomial, *p. 62*
polynomial, *p. 63*
binomial, *p. 63*
trinomial, *p. 63*
degree of a polynomial, *p. 63*
standard form, *p. 63*
leading coefficient, *p. 63*
closed, *p. 64*
FOIL Method, *p. 71*
factored form, *p. 82*
Zero-Product Property, *p. 82*
roots, *p. 82*
repeated roots, *p. 83*

Core Concepts

Section 2.1
Polynomials, *p. 63*
Adding Polynomials, *p. 64*
Subtracting Polynomials, *p. 64*

Section 2.2
Multiplying Binomials, *p. 70*
FOIL Method, *p. 71*
Multiplying Binomials and Trinomials, *p. 72*

Section 2.3
Square of a Binomial Pattern, *p. 76*
Sum and Difference Pattern, *p. 77*

Section 2.4
Zero-Product Property, *p. 82*
Factoring Polynomials Using the GCF, *p. 83*

Mathematical Practices

1. Explain how you wrote the polynomial in Exercise 11 on page 79. Is there another method you can use to write the same polynomial?
2. Find a shortcut for exercises like Exercise 7 on page 85 when the variable has a coefficient of 1. Does your shortcut work when the coefficient is *not* 1?

Preparing for a Test

- Review examples of each type of problem that could appear on the test. Use the tutorials at *BigIdeasMath.com* for additional help.
- Review the homework problems your teacher assigned.
- Take a practice test.

2.1–2.4 Quiz

Write the polynomial in standard form. Identify the degree and leading coefficient of the polynomial. Then classify the polynomial by the number of terms. *(Section 2.1)*

1. $-8q^3$
2. $9 + d^2 - 3d$
3. $\frac{2}{3}m^4 - \frac{5}{6}m^6$
4. $-1.3z + 3z^4 + 7.4z^2$

Find the sum or difference. *(Section 2.1)*

5. $(2x^2 + 5) + (-x^2 + 4)$
6. $(-3n^2 + n) - (2n^2 - 7)$
7. $(-p^2 + 4p) - (p^2 - 3p + 15)$
8. $(a^2 - 3ab + b^2) + (-a^2 + ab + b^2)$

Find the product. *(Section 2.2 and Section 2.3)*

9. $(w + 6)(w + 7)$
10. $(3 - 4d)(2d - 5)$
11. $(y + 9)(y^2 + 2y - 3)$
12. $(3z - 5)(3z + 5)$
13. $(t + 5)^2$
14. $(2q - 6)^2$

Solve the equation. *(Section 2.4)*

15. $5x^2 - 15x = 0$
16. $(8 - g)(8 - g) = 0$
17. $(3p + 7)(3p - 7)(p + 8) = 0$
18. $-3y(y - 8)(2y + 1) = 0$

19. You are making a blanket with a fringe border of equal width on each side. *(Section 2.1 and Section 2.2)*
    a. Write a polynomial that represents the perimeter of the blanket including the fringe.
    b. Write a polynomial that represents the area of the blanket including the fringe.
    c. Find the perimeter and the area of the blanket including the fringe when the width of the fringe is 4 inches.

20. You are saving money to buy an electric guitar. You deposit $1000 in an account that earns interest compounded annually. The expression $1000(1 + r)^2$ represents the balance after 2 years, where r is the annual interest rate in decimal form. *(Section 2.3)*
    a. Write the polynomial in standard form that represents the balance of your account after 2 years.
    b. The interest rate is 3%. What is the balance of your account after 2 years?
    c. The guitar costs $1100. Do you have enough money in your account *after 3 years*? Explain.

21. The front of a storage bunker can be modeled by $y = -\frac{5}{216}(x - 72)(x + 72)$, where x and y are measured in inches. The x-axis represents the ground. Find the width of the bunker at ground level. *(Section 2.4)*

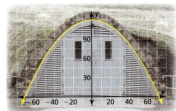

ANSWERS

1. $-8q^3$; 3; -8; monomial
2. $d^2 - 3d + 9$; 2; 1; trinomial
3. $-\frac{5}{6}m^6 + \frac{2}{3}m^4$; 6; $-\frac{5}{6}$; binomial
4. $3z^4 + 7.4z^2 - 1.3z$; 4; 3; trinomial
5. $x^2 + 9$
6. $-5n^2 + n + 7$
7. $-2p^2 + 7p - 15$
8. $-2ab + 2b^2$
9. $w^2 + 13w + 42$
10. $-8d^2 + 26d - 15$
11. $y^3 + 11y^2 + 15y - 27$
12. $9z^2 - 25$
13. $t^2 + 10t + 25$
14. $4q^2 - 24q + 36$
15. $x = 0, x = 3$
16. $g = 8$
17. $p = -\frac{7}{3}, p = \frac{7}{3}, p = -8$
18. $y = 0, y = 8, y = -\frac{1}{2}$
19. a. $(8x + 240)$ in.
    b. $(4x^2 + 240x + 3456)$ in.^{2}
    c. 272 in.; 4480 in.^{2}
20. a. $1000r^2 + 2000r + 1000$
    b. $1060.90
    c. no; The balance after 3 years is represented by the expression $1000(1 + 0.03)^3$, which is only about $1092.73.
21. 144 in.

Laurie's Notes

Overview of Section 2.5

Introduction
- In this lesson, students factor trinomials of the form $x^2 + bx + c$. The different cases of $c > 0$ and $c < 0$ are presented. The lesson ends with a real-life problem.
- Although factoring of trinomials is the focus of the lesson, the ultimate goal is for students to be able to solve polynomial equations and simplify expressions with common factors.

Teaching Strategy
- **Use Appropriate Tools Strategically:** Algebra tiles can be used to help develop a greater understanding of what it means to factor a trinomial, in the same way base-ten blocks were used when developing an understanding of writing a whole number as the product of two factors. *Example:* Given 12 unit squares, they can be arranged into a rectangle of dimensions 1×12, 2×6, or 3×4.
- The polynomial $x^2 + 8x + 12$ can be modeled using 1 x^2-tile, 8 x-tiles, and 12 unit tiles.
- To arrange the tiles into a rectangle means that you need to have the 12 unit pieces arranged as an array with dimensions 1×12, 2×6, or 3×4. If factorable, the possible rectangular array must be one of the following:

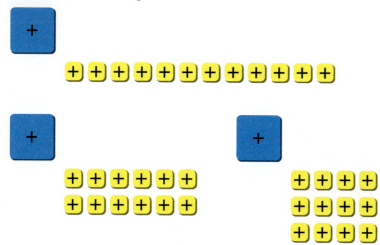

- The x-tiles now have to fill in the opposite regions of the array. The first array requires 13 x-tiles $(1 + 12)$, the second array requires 8 x-tiles $(2 + 6)$, and the third array requires 7 x-tiles $(3 + 4)$. Because there are 8 x-tiles, the second array is the correct arrangement. The trinomial $x^2 + 8x + 12$ equals $(x + 2)(x + 6)$.
- Consider the additional possibilities when the constant term is negative! Although this type of trinomial is not shown in the explorations, it can certainly be done.
- To factor $x^2 + 4x - 12$, make the same three arrangements as with the previous problem. You have 4 x-tiles. Two zero pairs of x-tiles are added to fill the rectangular array.

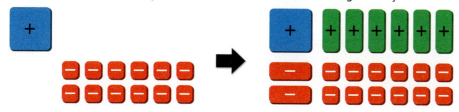

- So the trinomial $x^2 + 4x - 12$ equals $(x - 2)(x + 6)$.

Pacing Suggestion
- Complete the exploration and discuss students' observations. Transition to the formal lesson after the *Motivate*.

Dynamic Teaching Tools
Dynamic Assessment & Progress Monitoring Tool
Lesson Planning Tool
Interactive Whiteboard Lesson Library
Dynamic Classroom with Dynamic Investigations

What Your Students Will Learn

- Factor $x^2 + bx + c$ when b and c are positive, when b is negative and c is positive, and when c is negative.
- Use factoring to solve real-life problems involving polynomial equations.

Laurie's Notes

Exploration

Motivate
- Have students match the factored form with the correct standard form.
  1. $(x + 2)(x + 6)$ B
  2. $(x + 12)(x + 1)$ C
  3. $(x + 3)(x + 4)$ A

  A. $x^2 + 7x + 12$
  B. $x^2 + 8x + 12$
  C. $x^2 + 13x + 12$

- **?** "How are the problems alike? How are they different?" Answers will vary.
- Circle the constant terms in the binomials (2, 6, 12, 1, 3, and 4), the coefficients of the linear terms (7, 8, and 13), and the constant terms (12, 12, and 12).
- **?** "How are these numbers related?" The sum of each pair is the coefficient of the x-term, and the product of each pair is the constant term.

For Your Information
- Although students have used algebra tiles to multiply binomials, using tiles to factor trinomials can be challenging for them. See the *Teaching Strategy* on page T-88.

Exploration 1
- Model the sample problem with students. Ask students to model $x^2 + 5x + 6$ so that they form a rectangular array with the algebra tiles. There are two options for the six unit tiles: 1×6 or 2×3. If students try the 1×6, they will realize that they do not have enough x-tiles.
- **?** "What are the dimensions of the rectangle?" $(x + 2)$ and $(x + 3)$
- Write and say, "So $x^2 + 5x + 6 = (x + 2)(x + 3)$." Discuss that $x^2 + 5x + 6$ is the product, or area of the rectangle, and $(x + 2)$ and $(x + 3)$ are the factors, or dimensions of the rectangle.
- As students work the four parts with their partners, encourage them to think about the dimensions of the rectangles. Students also need to pay attention to the signs of their tiles.
- In parts (c) and (d), students should recognize that when the constant term is 12, this product could be the result of $(3)(4)$ or $(-3)(-4)$.
- Have students share their work at a document camera or overhead projector if possible.

Communicate Your Answer
- Students should recognize that to factor $x^2 + bx + c$, they need to find two numbers whose product is c and whose sum is b. Their language may not be precise at this point.

Connecting to Next Step
- The exploration helps students visualize and make sense of factoring a trinomial of the form $x^2 + bx + c$. Once explored, students will be better able to perform the algebraic manipulations.

2.5 Factoring $x^2 + bx + c$

Essential Question How can you use algebra tiles to factor the trinomial $x^2 + bx + c$ into the product of two binomials?

EXPLORATION 1 Finding Binomial Factors

Work with a partner. Use algebra tiles to write each polynomial as the product of two binomials. Check your answer by multiplying.

Sample $x^2 + 5x + 6$

Step 1 Arrange algebra tiles that model $x^2 + 5x + 6$ into a rectangular array.

Step 2 Use additional algebra tiles to model the dimensions of the rectangle.

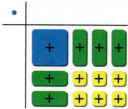   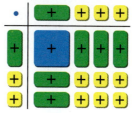

Step 3 Write the polynomial in factored form using the dimensions of the rectangle.

Area = $x^2 + 5x + 6 = (x + 2)(x + 3)$

a. $x^2 - 3x + 2 = $ ▭

b. $x^2 + 5x + 4 = $ ▭

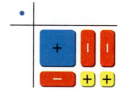   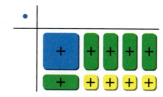

c. $x^2 - 7x + 12 = $ ▭

d. $x^2 + 7x + 12 = $ ▭

REASONING ABSTRACTLY
To be proficient in math, you need to understand a situation abstractly and represent it symbolically.

Communicate Your Answer

2. How can you use algebra tiles to factor the trinomial $x^2 + bx + c$ into the product of two binomials?

3. Describe a strategy for factoring the trinomial $x^2 + bx + c$ that does not use algebra tiles.

ANSWERS

1. a. $(x - 1)(x - 2)$

   b. $(x + 1)(x + 4)$

   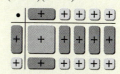

   c. $(x - 4)(x - 3)$

   d. $(x + 4)(x + 3)$

   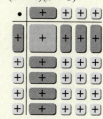

2. Arrange algebra tiles that model the trinomial into a rectangular array, use additional algebra tiles to model the dimensions of the rectangle, then write the polynomial in factored form using the dimensions of the rectangle.

3. Find two integer factors of c that have a sum of b, then write the binomial factors by adding each integer factor to x.

English Language Learners

Pair Activity

Have students work in pairs to factor polynomials. Each student factors a different polynomial. Then students take turns explaining their solutions to their partners.

Extra Example 1

Factor $x^2 + 9x + 14$.

$(x + 2)(x + 7)$

MONITORING PROGRESS ANSWERS

1. $(x + 1)(x + 6)$
2. $(x + 1)(x + 8)$

2.5 Lesson

Core Vocabulary

Previous
polynomial
FOIL Method
Zero-Product Property

What You Will Learn

▶ Factor $x^2 + bx + c$.
▶ Use factoring to solve real-life problems.

Factoring $x^2 + bx + c$

Writing a polynomial as a product of factors is called *factoring*. To factor $x^2 + bx + c$ as $(x + p)(x + q)$, you need to find p and q such that $p + q = b$ and $pq = c$.

$$(x + p)(x + q) = x^2 + px + qx + pq$$
$$= x^2 + (p + q)x + pq$$

Core Concept

Factoring $x^2 + bx + c$ When c Is Positive

Algebra $x^2 + bx + c = (x + p)(x + q)$ when $p + q = b$ and $pq = c$.

When c is positive, p and q have the same sign as b.

Examples $x^2 + 6x + 5 = (x + 1)(x + 5)$
$x^2 - 6x + 5 = (x - 1)(x - 5)$

EXAMPLE 1 Factoring $x^2 + bx + c$ When b and c Are Positive

Factor $x^2 + 10x + 16$.

SOLUTION

Notice that $b = 10$ and $c = 16$.

- Because c is positive, the factors p and q must have the same sign so that pq is positive.
- Because b is also positive, p and q must each be positive so that $p + q$ is positive.

Find two positive integer factors of 16 whose sum is 10.

Check

Use the FOIL Method.

$(x + 2)(x + 8)$
$= x^2 + 8x + 2x + 16$
$= x^2 + 10x + 16$ ✓

Factors of 16	Sum of factors
1, 16	17
2, 8	10
4, 4	8

The values of p and q are 2 and 8.

▶ So, $x^2 + 10x + 16 = (x + 2)(x + 8)$.

Monitoring Progress  Help in English and Spanish at BigIdeasMath.com

Factor the polynomial.

1. $x^2 + 7x + 6$
2. $x^2 + 9x + 8$

Laurie's Notes | Teacher Actions

- **Teaching Tip:** Write the example and say, "You need two numbers whose product is 16 and whose sum is 10."
- Some students may find it tedious to list the factor pairs. Share with them that in future problems the signs may not both be positive and the leading coefficient may be a number other than 1.
- Students should set up each problem before they factor. For example, $x^2 + 7x + 6 = (x + \_\_)(x + \_\_)$. "I need two numbers whose product is 6 and whose sum is 7."

EXAMPLE 2 — Factoring $x^2 + bx + c$ When b Is Negative and c Is Positive

Factor $x^2 - 8x + 12$.

SOLUTION

Notice that $b = -8$ and $c = 12$.

- Because c is positive, the factors p and q must have the same sign so that pq is positive.
- Because b is negative, p and q must each be negative so that $p + q$ is negative.

Find two negative integer factors of 12 whose sum is -8.

Factors of 12	$-1, -12$	$-2, -6$	$-3, -4$
Sum of factors	-13	-8	-7

The values of p and q are -2 and -6.

▶ So, $x^2 - 8x + 12 = (x - 2)(x - 6)$.

Check
Use the FOIL Method.
$(x - 2)(x - 6)$
$= x^2 - 6x - 2x + 12$
$= x^2 - 8x + 12$ ✓

Core Concept

Factoring $x^2 + bx + c$ When c Is Negative

Algebra $x^2 + bx + c = (x + p)(x + q)$ when $p + q = b$ and $pq = c$.

When c is negative, p and q have different signs.

Example $x^2 - 4x - 5 = (x + 1)(x - 5)$

EXAMPLE 3 — Factoring $x^2 + bx + c$ When c Is Negative

Factor $x^2 + 4x - 21$.

SOLUTION

Notice that $b = 4$ and $c = -21$. Because c is negative, the factors p and q must have different signs so that pq is negative.

Find two integer factors of -21 whose sum is 4.

Factors of -21	$-21, 1$	$-1, 21$	$-7, 3$	$-3, 7$
Sum of factors	-20	20	-4	4

The values of p and q are -3 and 7.

▶ So, $x^2 + 4x - 21 = (x - 3)(x + 7)$.

Check
Use the FOIL Method.
$(x - 3)(x + 7)$
$= x^2 + 7x - 3x - 21$
$= x^2 + 4x - 21$ ✓

Monitoring Progress Help in English and Spanish at *BigIdeasMath.com*

Factor the polynomial.

3. $w^2 - 4w + 3$
4. $n^2 - 12n + 35$
5. $x^2 - 14x + 24$
6. $x^2 + 2x - 15$
7. $y^2 + 13y - 30$
8. $v^2 - v - 42$

Extra Example 2
Factor $x^2 - 12x + 27$.
$(x - 3)(x - 9)$

Extra Example 3
Factor $x^2 + 7x - 8$.
$(x - 1)(x + 8)$

MONITORING PROGRESS ANSWERS

3. $(w - 1)(w - 3)$
4. $(n - 5)(n - 7)$
5. $(x - 2)(x - 12)$
6. $(x - 3)(x + 5)$
7. $(y - 2)(y + 15)$
8. $(v + 6)(v - 7)$

Laurie's Notes — Teacher Actions

? Write Example 2 and say, "You need two numbers whose product is 12 and whose sum is -8. What do you know about the two numbers?" *The numbers must both be negative.* Remind students to check their answers.

- **Turn and Talk:** In Example 3 say, "You need two numbers whose product is -21 and whose sum is 4. What are the possibilities?"
- **Think-Pair-Share:** Have students answer Questions 3–8, and then share and discuss as a class.

Extra Example 4

A gardener plants a rectangular lettuce patch in the southwest corner of a square garden. The area of the lettuce patch is 120 square feet. What is the area of the square garden?

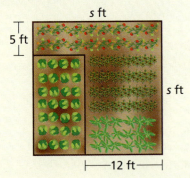

400 square feet

MONITORING PROGRESS ANSWER

9. 2500 m²

Solving Real-Life Problems

EXAMPLE 4 Solving a Real-Life Problem

A farmer plants a rectangular pumpkin patch in the northeast corner of a square plot of land. The area of the pumpkin patch is 600 square meters. What is the area of the square plot of land?

SOLUTION

1. **Understand the Problem** You are given the area of the pumpkin patch, the difference of the side length of the square plot and the length of the pumpkin patch, and the difference of the side length of the square plot and the width of the pumpkin patch.

2. **Make a Plan** The length of the pumpkin patch is $(s - 30)$ meters and the width is $(s - 40)$ meters. Write and solve an equation to find the side length s. Then use the solution to find the area of the square plot of land.

3. **Solve the Problem** Use the equation for the area of a rectangle to write and solve an equation to find the side length s of the square plot of land.

$600 = (s - 30)(s - 40)$	Write an equation.
$600 = s^2 - 70s + 1200$	Multiply.
$0 = s^2 - 70s + 600$	Subtract 600 from each side.
$0 = (s - 10)(s - 60)$	Factor the polynomial.
$s - 10 = 0$ or $s - 60 = 0$	Zero-Product Property
$s = 10$ or $s = 60$	Solve for s.

STUDY TIP The diagram shows that the side length is more than 40 meters, so a side length of 10 meters does not make sense in this situation. The side length is 60 meters.

▶ So, the area of the square plot of land is $60(60) = 3600$ square meters.

4. **Look Back** Use the diagram to check that you found the correct side length. Using $s = 60$, the length of the pumpkin patch is $60 - 30 = 30$ meters and the width is $60 - 40 = 20$ meters. So, the area of the pumpkin patch is 600 square meters. This matches the given information and confirms the side length is 60 meters, which gives an area of 3600 square meters.

Monitoring Progress Help in English and Spanish at BigIdeasMath.com

9. **WHAT IF?** The area of the pumpkin patch is 200 square meters. What is the area of the square plot of land?

Concept Summary

Factoring $x^2 + bx + c$ as $(x + p)(x + q)$

The diagram shows the relationships between the signs of b and c and the signs of p and q.

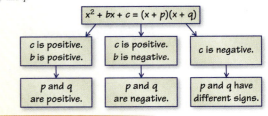

Laurie's Notes — Teacher Actions

- **Model with Mathematics:** In this problem, the diagram is a helpful model. When the problem has been solved, the results can be interpreted in the context of the problem, and again the diagram is helpful. It is clear that a side length of 10 is not possible.
- Students should recognize that in this problem they needed to subtract 600 from each side. The trinomial is set equal to 0 before factoring.

Closure

- Explain your strategy for factoring the following polynomials.
  a. $x^2 - 2x - 15$ $(x - 5)(x + 3)$
  b. $x^2 + 2x - 15$ $(x + 5)(x - 3)$
  c. $x^2 + 8x + 15$ $(x + 5)(x + 3)$
  d. $x^2 - 8x + 15$ $(x - 5)(x - 3)$

2.5 Exercises

Dynamic Solutions available at *BigIdeasMath.com*

Vocabulary and Core Concept Check

1. **WRITING** You are factoring $x^2 + 11x - 26$. What do the signs of the terms tell you about the factors? Explain.

2. **OPEN-ENDED** Write a trinomial that can be factored as $(x + p)(x + q)$, where p and q are positive.

Monitoring Progress and Modeling with Mathematics

In Exercises 3–8, factor the polynomial. (*See Example 1.*)

3. $x^2 + 8x + 7$
4. $z^2 + 10z + 21$
5. $n^2 + 9n + 20$
6. $s^2 + 11s + 30$
7. $h^2 + 11h + 18$
8. $y^2 + 13y + 40$

In Exercises 9–14, factor the polynomial. (*See Example 2.*)

9. $v^2 - 5v + 4$
10. $x^2 - 13x + 22$
11. $d^2 - 5d + 6$
12. $k^2 - 10k + 24$
13. $w^2 - 17w + 72$
14. $j^2 - 13j + 42$

In Exercises 15–24, factor the polynomial. (*See Example 3.*)

15. $x^2 + 3x - 4$
16. $z^2 + 7z - 18$
17. $n^2 + 4n - 12$
18. $s^2 + 3s - 40$
19. $y^2 + 2y - 48$
20. $h^2 + 6h - 27$
21. $x^2 - x - 20$
22. $m^2 - 6m - 7$
23. $-6t - 16 + t^2$
24. $-7y + y^2 - 30$

25. **MODELING WITH MATHEMATICS** A projector displays an image on a wall. The area (in square feet) of the projection is represented by $x^2 - 8x + 15$.

    a. Write a binomial that represents the height of the projection.
    b. Find the perimeter of the projection when the height of the wall is 8 feet.

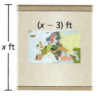

26. **MODELING WITH MATHEMATICS** A dentist's office and parking lot are on a rectangular piece of land. The area (in square meters) of the land is represented by $x^2 + x - 30$.

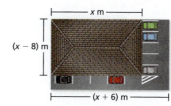

    a. Write a binomial that represents the width of the land.
    b. Find the area of the land when the length of the dentist's office is 20 meters.

ERROR ANALYSIS In Exercises 27 and 28, describe and correct the error in factoring the polynomial.

27.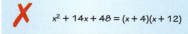
    ✗ $x^2 + 14x + 48 = (x + 4)(x + 12)$

28.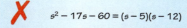
    ✗ $s^2 - 17s - 60 = (s - 5)(s - 12)$

In Exercises 29–38, solve the equation.

29. $m^2 + 3m + 2 = 0$
30. $n^2 - 9n + 18 = 0$
31. $x^2 + 5x - 14 = 0$
32. $v^2 + 11v - 26 = 0$
33. $t^2 + 15t = -36$
34. $n^2 - 5n = 24$
35. $a^2 + 5a - 20 = 30$
36. $y^2 - 2y - 8 = 7$
37. $m^2 + 10 = 15m - 34$
38. $b^2 + 5 = 8b - 10$

Section 2.5 Factoring $x^2 + bx + c$ 93

---

31. $x = 2, x = -7$
32. $v = 2, v = -13$
33. $t = -3, t = -12$
34. $n = -3, n = 8$
35. $a = 5, a = -10$
36. $y = -3, y = 5$
37. $m = 4, m = 11$
38. $b = 3, b = 5$

Assignment Guide and Homework Check

ASSIGNMENT

Basic: 1, 2, 3–29 odd, 39, 41, 46, 52–55

Average: 1, 2–40 even, 41, 46, 47, 52–55

Advanced: 1, 2, 8, 14, 22–28 even, 36–40 even, 41–55

HOMEWORK CHECK

Basic: 3, 9, 15, 39

Average: 4, 10, 16, 40

Advanced: 8, 14, 22, 40

ANSWERS

1. They have opposite signs; When factoring $x^2 + bx + c = (x + p)(x + q)$, if c is negative, p and q must have opposite signs.
2. Sample answer: $x^2 + 5x + 6$
3. $(x + 1)(x + 7)$
4. $(z + 3)(z + 7)$
5. $(n + 4)(n + 5)$
6. $(s + 5)(s + 6)$
7. $(h + 2)(h + 9)$
8. $(y + 5)(y + 8)$
9. $(v - 1)(v - 4)$
10. $(x - 2)(x - 11)$
11. $(d - 2)(d - 3)$
12. $(k - 4)(k - 6)$
13. $(w - 8)(w - 9)$
14. $(j - 6)(j - 7)$
15. $(x - 1)(x + 4)$
16. $(z - 2)(z + 9)$
17. $(n - 2)(n + 6)$
18. $(s - 5)(s + 8)$
19. $(y - 6)(y + 8)$
20. $(h - 3)(h + 9)$
21. $(x + 4)(x - 5)$
22. $(m + 1)(m - 7)$
23. $(t + 2)(t - 8)$
24. $(y + 3)(y - 10)$
25. a. $(x - 5)$ ft
    b. 16 ft
26. a. $(x - 5)$ m
    b. 390 m^2
27. $4 + 12$ is not 14; $= (x + 6)(x + 8)$
28. c is -60, so the signs of the factors must be opposite; $= (s + 3)(s - 20)$
29. $m = -1, m = -2$
30. $n = 3, n = 6$

Section 2.5 93

Dynamic Teaching Tools
- Dynamic Assessment & Progress Monitoring Tool
- Interactive Whiteboard Lesson Library
- Dynamic Classroom with Dynamic Investigations

ANSWERS
39. 100 in.^{2}
40. a. 6 in.
    b. length: 13 in., width: 8 in.
41. yes; p and q must be factors of -12 that have a sum of b, and -12 has 6 sets of integer factors, -1 and 12, -2 and 6, -3 and 4, -4 and 3, -6 and 2, and -12 and 1.
42. a.

$x^2 - x - 6 = (x + 2)(x - 3)$

b.

$x^2 + 2x - 8 = (x - 2)(x + 4)$

43. length: 11 ft, width: 4 ft
44. base: 10 m, height: 7 m
45–55. See Additional Answers.

Mini-Assessment
Factor the polynomial.
1. $x^2 + 13x + 12$ $(x + 1)(x + 12)$
2. $x^2 - 14x + 40$ $(x - 4)(x - 10)$
3. $x^2 + 3x - 54$ $(x - 6)(x + 9)$
4. A homeowner puts a patio in the northwest corner of a square back yard. The area of the patio is 750 square feet. What is the area of the backyard?

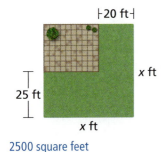

2500 square feet

94   Chapter 2

---

39. MODELING WITH MATHEMATICS You trimmed a large square picture so that you could fit it into a frame. The area of the cut picture is 20 square inches. What is the area of the original picture? *(See Example 4.)*

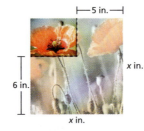

40. MODELING WITH MATHEMATICS A web browser is open on your computer screen.

a. The area of the browser window is 24 square inches. Find the length of the browser window x.
b. The browser covers $\frac{3}{13}$ of the screen. What are the dimensions of the screen?

41. MAKING AN ARGUMENT Your friend says there are six integer values of b for which the trinomial $x^2 + bx - 12$ has two binomial factors of the form $(x + p)$ and $(x + q)$. Is your friend correct? Explain.

42. THOUGHT PROVOKING Use algebra tiles to factor each polynomial modeled by the tiles. Show your work.

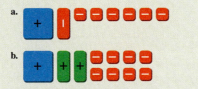

Maintaining Mathematical Proficiency — Reviewing what you learned in previous grades and lessons

Solve the equation. Check your solution. *(Skills Review Handbook)*

52. $p - 9 = 0$ 53. $z + 12 = -5$ 54. $6 = \dfrac{c}{-7}$ 55. $4k = 0$

94   Chapter 2   Polynomial Equations and Factoring

MATHEMATICAL CONNECTIONS In Exercises 43 and 44, find the dimensions of the polygon with the given area.

43. Area = 44 ft^2 44. Area = 35 m^2

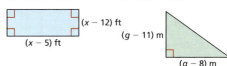

45. REASONING Write an equation of the form $x^2 + bx + c = 0$ that has the solutions $x = -4$ and $x = 6$. Explain how you found your answer.

46. HOW DO YOU SEE IT? The graph of $y = x^2 + x - 6$ is shown.

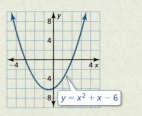

a. Explain how you can use the graph to factor the polynomial $x^2 + x - 6$.
b. Factor the polynomial.

47. PROBLEM SOLVING Road construction workers are paving the area shown.

a. Write an expression that represents the area being paved.
b. The area being paved is 280 square meters. Write and solve an equation to find the width of the road x.

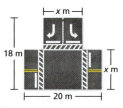

USING STRUCTURE In Exercises 48–51, factor the polynomial.

48. $x^2 + 6xy + 8y^2$ 49. $r^2 + 7rs + 12s^2$
50. $a^2 + 11ab - 26b^2$ 51. $x^2 - 2xy - 35y^2$

If students need help...	If students got it...
Resources by Chapter • Practice A and Practice B • Puzzle Time	Resources by Chapter • Enrichment and Extension • Cumulative Review
Student Journal • Practice	Start the *next* Section
Differentiating the Lesson Skills Review Handbook	

Laurie's Notes

Overview of Section 2.6

Introduction
- In this lesson, students factor trinomials of the form $ax^2 + bx + c$. In factoring trinomials, any common factor should be factored out first, leaving the resulting coefficients with fewer factors to work with.
- Students will recognize that the fewest combinations occur when both a and c are prime. When either a or c is composite, there are more combinations to check. When both a and c are composite, there are even more combinations to check.
- The lesson ends with a real-life problem where once the polynomial has been factored and set equal to zero, the Zero-Product Property is used to solve the equation.

Teaching Strategy
- Factoring trinomials in which the leading coefficient is not equal to 1 can be challenging for some students. They struggle to hold several computations in their heads. It is helpful to have a structure that allows students to test out different combinations quickly.
- *Example:* If $2x^2 + 13x + 6$ factors over the integers, then the binomial factors must be $(x + \_\_)(2x + \_\_)$. We also know that the factors of the constant term c are $\{1, 2, 3, 6\}$. The goal is to find the combination that gives a linear term of $13x$.
- We could write out the four possible combinations:

    $(x + 1)(2x + 6)$ ➡ $1x \cdot 6 + 1 \cdot 2x = 8x$
    $(x + 6)(2x + 1)$ ➡ $1x \cdot 1 + 6 \cdot 2x = 13x$
    $(x + 2)(2x + 3)$ ➡ $1x \cdot 3 + 2 \cdot 2x = 7x$
    $(x + 3)(2x + 2)$ ➡ $1x \cdot 2 + 3 \cdot 2x = 8x$

- There is a symbolic way of representing all four cases. For each set of brackets, sum the products along both solid lines; sum the products along the dotted lines. The sum that is 13 is highlighted. Write the binomials that give the same calculations when multiplied.

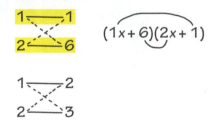

- The list of possible factor combinations can be written vertically for students to check. This helps students who need a structure for listing the possible combinations.

Pacing Suggestion
- Complete the exploration and discuss students' observations. Transition to the formal lesson.

What Your Students Will Learn

- Factor $ax^2 + bx + c$ when ac is positive, when ac is negative, and when a is negative.
- Use factoring to solve real-life problems involving polynomial equations.

Laurie's Notes

Exploration

Motivate
- Model $2x^2 + 3x + 1$ with tiles on the overhead. Have students do the same at their desks.
- ❓ "Can you arrange the algebra tiles so they form a rectangle with no holes or overlaps?" yes
- When students try to place all three x-tiles vertically, there is no place for the unit tile to fit. Model this when students do not make the attempt.
- ❓ "What are the dimensions of the rectangle?" $(x + 1)$ and $(2x + 1)$
- Label the dimensions. Explain that $2x^2 + 3x + 1 = (x + 1)(2x + 1)$.
- **Use Appropriate Tools Strategically:** This is similar to the visual model used when whole numbers are multiplied using base-ten blocks.

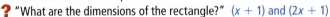

Exploration 1
- Ask students to form a rectangular array to model $2x^2 + 5x + 2$. The two x^2-tiles need to be placed side by side, either vertically or horizontally.
- Show the following three arrangements and explain that we use the one on the left because there is space for the two remaining unit tiles.

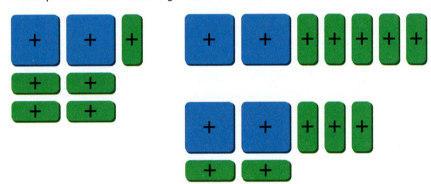

- ❓ "What are the dimensions of the rectangle?" $(x + 2)$ and $(2x + 1)$
- Label the dimensions. Explain that $2x^2 + 5x + 2 = (x + 2)(2x + 1)$.
- As students work part (a) with their partners, encourage them to think about the dimensions of the rectangle.
- Part (b) has been started so it eliminates the possibility of $(4x + \underline{})(x + \underline{})$.
- Part (c) has only one possibility for the x^2-tiles, though students need to pay attention to signs as they work the problem.

Communicate Your Answer
- **Use Appropriate Tools Strategically:** The algebra tiles provide a visual confirmation that the trinomial in Question 3 is not factorable.

Connecting to Next Step
- The exploration helps students visualize and make sense of factoring a trinomial of the form $ax^2 + bx + c$. Once explored, students will be better able to perform the algebraic manipulations in the formal lesson.

2.6 Factoring $ax^2 + bx + c$

Essential Question How can you use algebra tiles to factor the trinomial $ax^2 + bx + c$ into the product of two binomials?

EXPLORATION 1 Finding Binomial Factors

Work with a partner. Use algebra tiles to write each polynomial as the product of two binomials. Check your answer by multiplying.

Sample $2x^2 + 5x + 2$

Step 1 Arrange algebra tiles that model $2x^2 + 5x + 2$ into a rectangular array.

Step 2 Use additional algebra tiles to model the dimensions of the rectangle.

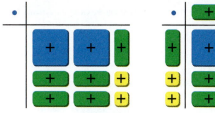

Step 3 Write the polynomial in factored form using the dimensions of the rectangle.

Area = $2x^2 + 5x + 2$ = $(x + 2)(2x + 1)$
  (width)   (length)

a. $3x^2 + 5x + 2 = $ _____

b. $4x^2 + 4x - 3 = $ _____   c. $2x^2 - 11x + 5 = $ _____

USING TOOLS STRATEGICALLY

To be proficient in math, you need to consider the available tools, including concrete models, when solving a mathematical problem.

Communicate Your Answer

2. How can you use algebra tiles to factor the trinomial $ax^2 + bx + c$ into the product of two binomials?

3. Is it possible to factor the trinomial $2x^2 + 2x + 1$? Explain your reasoning.

Dynamic Teaching Tools

Dynamic Assessment & Progress Monitoring Tool
Lesson Planning Tool
Interactive Whiteboard Lesson Library
Dynamic Classroom with Dynamic Investigations

ANSWERS

1. a. $(x + 1)(3x + 2)$

b. $(2x + 3)(2x - 1)$

c. $(x - 5)(2x - 1)$

2. Arrange algebra tiles that model the trinomial into a rectangular array, use additional algebra tiles to model the dimensions of the rectangle, then write the polynomial in factored form using the dimensions of the rectangle.

3. no; There is no way to model this expression as a rectangular array.

Extra Example 1

Factor $4x^2 + 32x + 60$.

$4(x + 3)(x + 5)$

Extra Example 2

Factor each polynomial.

a. $2x^2 + 7x + 6$ $(2x + 3)(x + 2)$

b. $4x^2 - 7x + 3$ $(4x - 3)(x - 1)$

2.6 Lesson

Core Vocabulary

Previous
polynomial
greatest common factor (GCF)
Zero-Product Property

What You Will Learn

▶ Factor $ax^2 + bx + c$.
▶ Use factoring to solve real-life problems.

Factoring $ax^2 + bx + c$

In Section 2.5, you factored polynomials of the form $ax^2 + bx + c$, where $a = 1$. To factor polynomials of the form $ax^2 + bx + c$, where $a \neq 1$, first look for the GCF of the terms of the polynomial and then factor further if possible.

EXAMPLE 1 Factoring Out the GCF

Factor $5x^2 + 15x + 10$.

SOLUTION

Notice that the GCF of the terms $5x^2$, $15x$, and 10 is 5.

$5x^2 + 15x + 10 = 5(x^2 + 3x + 2)$ Factor out GCF.

$\qquad\qquad\qquad\quad = 5(x + 1)(x + 2)$ Factor $x^2 + 3x + 2$.

▶ So, $5x^2 + 15x + 10 = 5(x + 1)(x + 2)$.

When there is no GCF, consider the possible factors of a and c.

EXAMPLE 2 Factoring $ax^2 + bx + c$ When ac Is Positive

Factor each polynomial.

a. $4x^2 + 13x + 3$ b. $3x^2 - 7x + 2$

STUDY TIP

You must consider the order of the factors of 3, because the middle terms formed by the possible factorizations are different.

SOLUTION

a. There is no GCF, so you need to consider the possible factors of a and c. Because b and c are both positive, the factors of c must be positive. Use a table to organize information about the factors of a and c.

Factors of 4	Factors of 3	Possible factorization	Middle term	
1, 4	1, 3	$(x + 1)(4x + 3)$	$3x + 4x = 7x$	✗
1, 4	3, 1	$(x + 3)(4x + 1)$	$x + 12x = 13x$	✓
2, 2	1, 3	$(2x + 1)(2x + 3)$	$6x + 2x = 8x$	✗

▶ So, $4x^2 + 13x + 3 = (x + 3)(4x + 1)$.

b. There is no GCF, so you need to consider the possible factors of a and c. Because b is negative and c is positive, both factors of c must be negative. Use a table to organize information about the factors of a and c.

Factors of 3	Factors of 2	Possible factorization	Middle term	
1, 3	−1, −2	$(x - 1)(3x - 2)$	$-2x - 3x = -5x$	✗
1, 3	−2, −1	$(x - 2)(3x - 1)$	$-x - 6x = -7x$	✓

▶ So, $3x^2 - 7x + 2 = (x - 2)(3x - 1)$.

Laurie's Notes Teacher Actions

- Students now understand that the ability to factor a trinomial is related to the values of a, b, and c. Factoring out the GCF will make sense!
- **Turn and Talk:** "When a and c are positive, what does the sign of b tell you?" Use *Popsicle Sticks* to solicit an explanation.
- The more factors a and c have, the more combinations must be checked. See the *Teaching Strategy* on page T-94 for a structure to list and check the various combinations.

EXAMPLE 3 Factoring $ax^2 + bx + c$ When ac Is Negative

Factor $2x^2 - 5x - 7$.

SOLUTION

There is no GCF, so you need to consider the possible factors of a and c. Because c is negative, the factors of c must have different signs. Use a table to organize information about the factors of a and c.

Factors of 2	Factors of −7	Possible factorization	Middle term	
1, 2	1, −7	$(x + 1)(2x - 7)$	$-7x + 2x = -5x$	✓
1, 2	7, −1	$(x + 7)(2x - 1)$	$-x + 14x = 13x$	✗
1, 2	−1, 7	$(x - 1)(2x + 7)$	$7x - 2x = 5x$	✗
1, 2	−7, 1	$(x - 7)(2x + 1)$	$x - 14x = -13x$	✗

▶ So, $2x^2 - 5x - 7 = (x + 1)(2x - 7)$.

STUDY TIP

When a is negative, factor -1 from each term of $ax^2 + bx + c$. Then factor the resulting trinomial as in the previous examples.

EXAMPLE 4 Factoring $ax^2 + bx + c$ When a Is Negative

Factor $-4x^2 - 8x + 5$.

SOLUTION

Step 1 Factor -1 from each term of the trinomial.

$$-4x^2 - 8x + 5 = -(4x^2 + 8x - 5)$$

Step 2 Factor the trinomial $4x^2 + 8x - 5$. Because c is negative, the factors of c must have different signs. Use a table to organize information about the factors of a and c.

Factors of 4	Factors of −5	Possible factorization	Middle term	
1, 4	1, −5	$(x + 1)(4x - 5)$	$-5x + 4x = -x$	✗
1, 4	5, −1	$(x + 5)(4x - 1)$	$-x + 20x = 19x$	✗
1, 4	−1, 5	$(x - 1)(4x + 5)$	$5x - 4x = x$	✗
1, 4	−5, 1	$(x - 5)(4x + 1)$	$x - 20x = -19x$	✗
2, 2	1, −5	$(2x + 1)(2x - 5)$	$-10x + 2x = -8x$	✗
2, 2	−1, 5	$(2x - 1)(2x + 5)$	$10x - 2x = 8x$	✓

▶ So, $-4x^2 - 8x + 5 = -(2x - 1)(2x + 5)$.

Monitoring Progress Help in English and Spanish at *BigIdeasMath.com*

Factor the polynomial.

1. $8x^2 - 56x + 48$
2. $14x^2 + 31x + 15$
3. $2x^2 - 7x + 5$
4. $3x^2 - 14x + 8$
5. $4x^2 - 19x - 5$
6. $6x^2 + x - 12$
7. $-2y^2 - 5y - 3$
8. $-5m^2 + 6m - 1$
9. $-3x^2 - x + 2$

Differentiated Instruction

Organization

Some students may benefit from completing a table to show the signs of the factors of c. Students can use their tables to check the signs of the factors of c in their factorizations.

$ax^2 + bx + c$

b	c	Factors of c
positive	positive	both positive
positive	negative	different signs
negative	positive	both negative
negative	negative	different signs

Extra Example 3

Factor $3x^2 - 7x - 6$.

$(x - 3)(3x + 2)$

Extra Example 4

Factor $-9x^2 - 3x + 2$.

$-(3x - 1)(3x + 2)$

MONITORING PROGRESS ANSWERS

1. $8(x - 1)(x - 6)$
2. $(7x + 5)(2x + 3)$
3. $(x - 1)(2x - 5)$
4. $(3x - 2)(x - 4)$
5. $(4x + 1)(x - 5)$
6. $(2x + 3)(3x - 4)$
7. $-(y + 1)(2y + 3)$
8. $-(m - 1)(5m - 1)$
9. $-(x + 1)(3x - 2)$

Laurie's Notes — Teacher Actions

? Write Example 3. "Because the constant term is -7, what do you know about the linear factors?" *They will have different signs.*

- Some students will have little difficulty using mental math to check the combinations. After modeling a problem, students must try problems independently. Watching someone factor trinomials cannot replicate the independent thought process that students need to experience.
- **Look For and Make Use of Structure:** In Example 4, factoring out a (-1) is not an intuitive step for students. Make the connection to the Distributive Property again.

Extra Example 5
The length of a rectangular state park is 2 miles longer than twice the width. The area of the park is 84 square miles. What is the width of the park? *The width of the park is 6 miles.*

MONITORING PROGRESS ANSWER

10. 8 mi

Solving Real-Life Problems

EXAMPLE 5 Solving a Real-Life Problem

The length of a rectangular game reserve is 1 mile longer than twice the width. The area of the reserve is 55 square miles. What is the width of the reserve?

SOLUTION

Use the formula for the area of a rectangle to write an equation for the area of the reserve. Let w represent the width. Then $2w + 1$ represents the length. Solve for w.

$w(2w + 1) = 55$ Area of the reserve

$2w^2 + w = 55$ Distributive Property

$2w^2 + w - 55 = 0$ Subtract 55 from each side.

Factor the left side of the equation. There is no GCF, so you need to consider the possible factors of a and c. Because c is negative, the factors of c must have different signs. Use a table to organize information about the factors of a and c.

Factors of 2	Factors of −55	Possible factorization	Middle term	
1, 2	1, −55	$(w + 1)(2w - 55)$	$-55w + 2w = -53w$	✗
1, 2	55, −1	$(w + 55)(2w - 1)$	$-w + 110w = 109w$	✗
1, 2	−1, 55	$(w - 1)(2w + 55)$	$55w - 2w = 53w$	✗
1, 2	−55, 1	$(w - 55)(2w + 1)$	$w - 110w = -109w$	✗
1, 2	5, −11	$(w + 5)(2w - 11)$	$-11w + 10w = -w$	✗
1, 2	11, −5	$(w + 11)(2w - 5)$	$-5w + 22w = 17w$	✗
1, 2	−5, 11	$(w - 5)(2w + 11)$	$11w - 10w = w$	✓
1, 2	−11, 5	$(w - 11)(2w + 5)$	$5w - 22w = -17w$	✗

So, you can rewrite $2w^2 + w - 55$ as $(w - 5)(2w + 11)$. Write the equation with the left side factored and continue solving for w.

$(w - 5)(2w + 11) = 0$ Rewrite equation with left side factored.

$w - 5 = 0$ or $2w + 11 = 0$ Zero-Product Property

$w = 5$ or $w = -\frac{11}{2}$ Solve for w.

A negative width does not make sense, so you should use the positive solution.

▶ So, the width of the reserve is 5 miles.

Check
Use mental math.
The width is 5 miles, so the length is $5(2) + 1 = 11$ miles and the area is $5(11) = 55$ square miles.

Monitoring Progress 🔊 Help in English and Spanish at *BigIdeasMath.com*

10. **WHAT IF?** The area of the reserve is 136 square miles. How wide is the reserve?

98   Chapter 2   Polynomial Equations and Factoring

Laurie's Notes   Teacher Actions

- **Make Sense of Problems and Persevere in Solving Them** and **Use Appropriate Tools Strategically:** Ask a student to read the problem, make a sketch, and label the dimensions of the rectangle.
- ❓ Once the trinomial is set equal to 0 ask, "To get a sum of $1w$ for the middle term, what factors of −55 might be reasonable to try first? Why?" *5 and 11, because the 5 can be multiplied by $2w$ first.*

Closure
- **Exit Ticket:** Factor $2x^2 - 7x + 3$. *$(2x - 1)(x - 3)$*

2.6 Exercises

Dynamic Solutions available at *BigIdeasMath.com*

Vocabulary and Core Concept Check

1. **REASONING** What is the greatest common factor of the terms of $3y^2 - 21y + 36$?
2. **WRITING** Compare factoring $6x^2 - x - 2$ with factoring $x^2 - x - 2$.

Monitoring Progress and Modeling with Mathematics

In Exercises 3–8, factor the polynomial. *(See Example 1.)*

3. $3x^2 + 3x - 6$
4. $8v^2 + 8v - 48$
5. $4k^2 + 28k + 48$
6. $6y^2 - 24y + 18$
7. $7b^2 - 63b + 140$
8. $9r^2 - 36r - 45$

In Exercises 9–16, factor the polynomial. *(See Examples 2 and 3.)*

9. $3h^2 + 11h + 6$
10. $8m^2 + 30m + 7$
11. $6x^2 - 5x + 1$
12. $10w^2 - 31w + 15$
13. $3n^2 + 5n - 2$
14. $4z^2 + 4z - 3$
15. $8g^2 - 10g - 12$
16. $18v^2 - 15v - 18$

In Exercises 17–22, factor the polynomial. *(See Example 4.)*

17. $-3t^2 + 11t - 6$
18. $-7v^2 - 25v - 12$
19. $-4c^2 + 19c + 5$
20. $-8h^2 - 13h + 6$
21. $-15w^2 - w + 28$
22. $-22d^2 + 29d - 9$

ERROR ANALYSIS In Exercises 23 and 24, describe and correct the error in factoring the polynomial.

23.
$$2x^2 - 2x - 24 = 2(x^2 - 2x - 24)$$
$$= 2(x - 6)(x + 4)$$

24.
$$6x^2 - 7x - 3 = (3x - 3)(2x + 1)$$

In Exercises 25–28, solve the equation.

25. $5x^2 - 5x - 30 = 0$
26. $2k^2 - 5k - 18 = 0$
27. $-12n^2 - 11n = -15$
28. $14b^2 - 2 = -3b$

In Exercises 29–32, find the x-coordinates of the points where the graph crosses the x-axis.

29.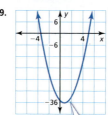
$y = 2x^2 - 3x - 35$

30.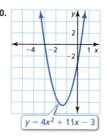
$y = 4x^2 + 11x - 3$

31.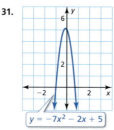
$y = -7x^2 - 2x + 5$

32.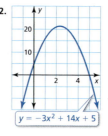
$y = -3x^2 + 14x + 5$

33. **MODELING WITH MATHEMATICS** The area (in square feet) of the school sign can be represented by $15x^2 - x - 2$.

   a. Write an expression that represents the length of the sign.

   b. Describe two ways to find the area of the sign when $x = 3$.

$(3x + 1)$ ft

Assignment Guide and Homework Check

ASSIGNMENT
Basic: 1, 2, 3–25 odd, 33, 35, 38, 40, 49–56
Average: 1, 2–40 even, 49–56
Advanced: 1, 2, 8, 12–36 even, 37–56

HOMEWORK CHECK
Basic: 3, 9, 13, 17, 35
Average: 4, 10, 14, 18, 34
Advanced: 8, 12, 16, 20, 36

ANSWERS

1. 3
2. Factoring $6x^2 - x - 2$ requires considering factors of 6 and -2 in different combinations until the combination is found that produces the correct middle term. Factoring $x^2 - x - 2$ only requires finding the factors of -2 that add up to -1.
3. $3(x - 1)(x + 2)$
4. $8(v - 2)(v + 3)$
5. $4(k + 3)(k + 4)$
6. $6(y - 1)(y - 3)$
7. $7(b - 4)(b - 5)$
8. $9(r + 1)(r - 5)$
9. $(3h + 2)(h + 3)$
10. $(2m + 7)(4m + 1)$
11. $(2x - 1)(3x - 1)$
12. $(2w - 5)(5w - 3)$
13. $(n + 2)(3n - 1)$
14. $(2z - 1)(2z + 3)$
15. $2(g - 2)(4g + 3)$
16. $3(2v - 3)(3v + 2)$
17. $-(t - 3)(3t - 2)$
18. $-(v + 3)(7v + 4)$
19. $-(c - 5)(4c + 1)$
20. $-(h + 2)(8h - 3)$
21. $-(3w - 4)(5w + 7)$
22. $-(2d - 1)(11d - 9)$
23. need to factor 2 out of every term; $= 2(x^2 - x - 12) = 2(x + 3)(x - 4)$
24. These factors do not give the correct middle term; $= (2x - 3)(3x + 1)$
25. $x = -2, x = 3$
26. $k = -2, k = \frac{9}{2}$
27. $n = -\frac{5}{3}, n = \frac{3}{4}$
28. $b = -\frac{1}{2}, b = \frac{2}{7}$
29. $x = -\frac{7}{2}, x = 5$

30. $x = -3, x = \frac{1}{4}$
31. $x = -1, x = \frac{5}{7}$
32. $x = -\frac{1}{3}, x = 5$
33. a. $(5x - 2)$ ft
   b. Substitute 3 for x into the expression for the area $15x^2 - x - 2$, then simplify; Substitute 3 for x into the expressions for the length $(5x - 2)$ and width $(3x + 1)$, simplify each, then multiply these two numbers.

Dynamic Teaching Tools
Dynamic Assessment & Progress Monitoring Tool
Interactive Whiteboard Lesson Library
Dynamic Classroom with Dynamic Investigations

ANSWERS
34. the initial height of the diver; 2.5 sec
35. length: 70 m, width: 31 m
36. yes; The length of the invitation is 5 inches, which is less than $5\frac{1}{8}$ inches. The width of the invitation is 3 inches, which is less than $3\frac{5}{8}$ inches.
37. Sample answer: $6x^2 + 3x$
38. The graph of k represents function g, and the graph of ℓ represents function h; Because c is positive, the constant terms in the factors must have the same sign. Because g has a positive value of b, the constant terms of the factors will both be positive, which results in negative roots, and k has two negative x-intercepts. Because h has a negative value of b, the constant terms of the factors will both be negative, which results in positive roots, and ℓ has two positive x-intercepts.
39. when no combination of factors of a and c produce the correct middle term; Sample answer: $2x^2 + x + 1$
40. no; To use the Zero-Product Property, one side of the equation needs to be 0. So, you must first subtract 2 from each side of the equation, then factor.
41. $\pm 9, \pm 12, \pm 21$
42–56. See Additional Answers.

---

34. **MODELING WITH MATHEMATICS** The height h (in feet) above the water of a cliff diver is modeled by $h = -16t^2 + 8t + 80$, where t is the time (in seconds). What does the constant term represent? How long is the diver in the air?

35. **MODELING WITH MATHEMATICS** The Parthenon in Athens, Greece, is an ancient structure that has a rectangular base. The length of the base of the Parthenon is 8 meters more than twice its width. The area of the base is about 2170 square meters. Find the length and width of the base. *(See Example 5.)*

36. **MODELING WITH MATHEMATICS** The length of a rectangular birthday party invitation is 1 inch less than twice its width. The area of the invitation is 15 square inches. Will the invitation fit in the envelope shown without being folded? Explain.

37. **OPEN-ENDED** Write a binomial whose terms have a GCF of $3x$.

38. **HOW DO YOU SEE IT?** Without factoring, determine which of the graphs represents the function $g(x) = 21x^2 + 37x + 12$ and which represents the function $h(x) = 21x^2 - 37x + 12$. Explain your reasoning.

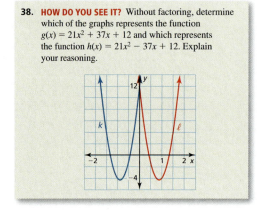

39. **REASONING** When is it not possible to factor $ax^2 + bx + c$, where $a \neq 1$? Give an example.

40. **MAKING AN ARGUMENT** Your friend says that to solve the equation $5x^2 + x - 4 = 2$, you should start by factoring the left side as $(5x - 4)(x + 1)$. Is your friend correct? Explain.

41. **REASONING** For what values of t can $2x^2 + tx + 10$ be written as the product of two binomials?

42. **THOUGHT PROVOKING** Use algebra tiles to factor each polynomial modeled by the tiles. Show your work.

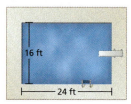

43. **MATHEMATICAL CONNECTIONS** The length of a rectangle is 1 inch more than twice its width. The value of the area of the rectangle (in square inches) is 5 more than the value of the perimeter of the rectangle (in inches). Find the width.

44. **PROBLEM SOLVING** A rectangular swimming pool is bordered by a concrete patio. The width of the patio is the same on every side. The area of the surface of the pool is equal to the area of the patio. What is the width of the patio?

In Exercises 45–48, factor the polynomial.

45. $4k^2 + 7jk - 2j^2$
46. $6x^2 + 5xy - 4y^2$
47. $-6a^2 + 19ab - 14b^2$
48. $18m^3 + 39m^2n - 15mn^2$

Maintaining Mathematical Proficiency — Reviewing what you learned in previous grades and lessons

Graph the function. Compare the graph to the graph of $f(x) = |x|$. Describe the domain and range. *(Section 1.1)*

49. $h(x) = 3|x|$
50. $v(x) = |x - 5|$
51. $g(x) = |x| + 1$
52. $r(x) = -2|x|$

Evaluate the expression. *(Section 1.5)*

53. $-\sqrt[3]{216}$
54. $\sqrt[5]{-32}$
55. $16^{5/4}$
56. $(-27)^{2/3}$

Mini-Assessment
Factor each polynomial.
1. $2x^2 + 14x + 24$ $2(x + 3)(x + 4)$
2. $3x^2 + 11x + 10$ $(3x + 5)(x + 2)$
3. $2x^2 - 7x - 4$ $(2x + 1)(x - 4)$
4. $-3x^2 - 4x + 7$ $-(3x + 7)(x - 1)$
5. The length of a rectangular garden is 4 yards less than twice the width. The area of the garden is 96 square yards. What is the width of the garden? The width of the garden is 8 yards.

If students need help...	If students got it...
Resources by Chapter • Practice A and Practice B • Puzzle Time	Resources by Chapter • Enrichment and Extension • Cumulative Review
Student Journal • Practice	Start the *next* Section
Differentiating the Lesson Skills Review Handbook	

Laurie's Notes

Overview of Section 2.7

Introduction
- This lesson on factoring special products is the inverse of finding special products earlier in the chapter. Instead of multiplying to find the product, $(x - a)(x + a) = x^2 - a^2$, in this lesson students will factor and write $x^2 - a^2 = (x - a)(x + a)$.
- In addition to factoring the difference of two squares and factoring a perfect square trinomial, factoring is used to solve a real-life problem, again using the Zero-Product Property.

Teaching Strategy
- This is frequently a challenging lesson for students, mainly because some students will say that they do not see the need for memorizing more "rules," the name they give for these patterns.
- This is particularly true for the perfect square trinomial. Given the trinomial $n^2 + 8n + 16$, students say, "I need two numbers whose product is 16 and whose sum is 8." They write $(n + 4)(n + 4)$. Certainly the result is the same.
- I do not require my students to memorize the factor pattern for the perfect square trinomials. I do suggest that they at least examine each trinomial and ask whether there are any perfect squares in the expression.

Another Way
- To solve the quadratic $x^2 - 25 = 0$, students write the quadratic in factored form: $(x + 5)(x - 5) = 0$.
- In Chapter 4, after square roots are introduced, $x^2 - 25 = 0$ will be solved another way, by square rooting.

$$x^2 - 25 = 0$$
$$x^2 = 25$$
$$x = \pm 5$$

Pacing Suggestion
- Complete the exploration and discuss students' observations. Transition to the formal lesson.

Dynamic Teaching Tools
Dynamic Assessment & Progress Monitoring Tool
Lesson Planning Tool
Interactive Whiteboard Lesson Library
Dynamic Classroom with Dynamic Investigations

What Your Students Will Learn

- Use the special product pattern to factor the difference of two squares.
- Use special product patterns to factor perfect square trinomials.
- Use factoring to solve real-life problems involving polynomial equations.

Laurie's Notes

Exploration

Motivate
- Copy the four equations for students. If you have sets of numeral tiles (0–9), they can be used for this activity, or students can write the digits.
- **Directions:** Use the digits 0–9 to make the following statements true. Use each digit only once.
  $(x + \underline{})^2 = x^2 + 6x + \underline{}$ 3; 9
  $(\underline{}x - \underline{})(2x + 1) = 4x^2 - 1$ 2; 1
  $(x + \underline{})^2 = x^2 + 16x + \underline{}\,\underline{}$ 8; 64
  $(\underline{}x - \underline{})^2 = 25x^2 - 7\underline{}x + 49$ 5; 7; 0
- Completing these problems will help students review the special products from Section 2.3.

Exploration 1
- Say, "In this exploration, you are to determine the factors of the polynomial shown in the rectangular array. Pay attention to the signs."
- The standard form of the polynomial is already given. Students only need write the dimensions (factors).
- ❓ "What do you recognize about arrays (a), (b), and (c)?" Listen for students to recognize that each array is a square, not just a rectangle. Each is a special product from Section 2.3.

Exploration 2
- Before students begin, you may wish to write the three special products from Section 2.3. Ask students to help you.
- Note that all the tiles will be used. Students need to determine the signs of the missing pieces.
- ❓ "What special shape is each of the rectangular arrays?" square

Communicate Your Answer
- **Neighbor Check:** For Question 4, have students work independently, and then have their neighbors check their work. Have students discuss any discrepancies.

Connecting to Next Step
- The explorations help students visualize and make sense of factoring a trinomial that is one of the special products from an earlier lesson. Once explored, students will be better able to perform the algebraic manipulations in the formal lesson.

2.7 Factoring Special Products

Essential Question How can you recognize and factor special products?

EXPLORATION 1 Factoring Special Products

Work with a partner. Use algebra tiles to write each polynomial as the product of two binomials. Check your answer by multiplying. State whether the product is a "special product" that you studied in Section 2.3.

LOOKING FOR STRUCTURE
To be proficient in math, you need to see complicated things as single objects or as being composed of several objects.

a. $4x^2 - 1 =$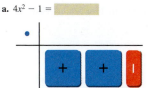

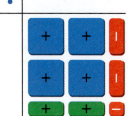

b. $4x^2 - 4x + 1 =$

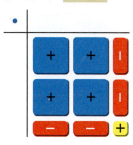

c. $4x^2 + 4x + 1 =$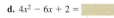

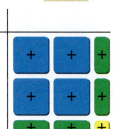

d. $4x^2 - 6x + 2 =$

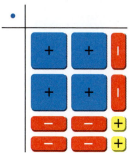

EXPLORATION 2 Factoring Special Products

Work with a partner. Use algebra tiles to complete the rectangular array at the left in three different ways, so that each way represents a different special product. Write each special product in standard form and in factored form.

Communicate Your Answer

3. How can you recognize and factor special products? Describe a strategy for recognizing which polynomials can be factored as special products.

4. Use the strategy you described in Question 3 to factor each polynomial.
   a. $25x^2 + 10x + 1$
   b. $25x^2 - 10x + 1$
   c. $25x^2 - 1$

Section 2.7   Factoring Special Products   101

Dynamic Teaching Tools
Dynamic Assessment & Progress Monitoring Tool
Lesson Planning Tool
Interactive Whiteboard Lesson Library
Dynamic Classroom with Dynamic Investigations

ANSWERS

1. a. $(2x + 1)(2x - 1)$; yes, sum and difference pattern

   b. $(2x - 1)^2$; yes, square of a binomial pattern

   c. $(2x + 1)^2$; yes, square of a binomial pattern

   d. $(2x - 2)(2x - 1)$; no, not a special pattern

2. See Additional Answers.

3. *Sample answer:* The algebra tiles for special patterns will always form a square array, and the x^2 tiles will also form a square array. Factor special products by using the special patterns studied in Lesson 2.3, but in reversed order; For the sum and difference pattern, the number of x and $-x$ tiles will be the same, and there will only be -1 tiles to complete the array. For the square of a binomial pattern, the x tiles will either be all positive or all negative and there will only be $+1$ tiles to complete the array.

4. a. $(5x + 1)^2$
   b. $(5x - 1)^2$
   c. $(5x + 1)(5x - 1)$

Extra Example 1
a. Factor $x^2 - 64$. $(x + 8)(x - 8)$
b. Factor $25b^2 - 36$. $(5b + 6)(5b - 6)$

Extra Example 2
Use a special product pattern to evaluate the expression $68^2 - 62^2$.
$(68 + 62)(68 - 62) = 130(6) = 780$

MONITORING PROGRESS ANSWERS
1. $(x + 6)(x - 6)$
2. $(10 + m)(10 - m)$
3. $(3n + 4)(3n - 4)$
4. $(4h + 7)(4h - 7)$
5. 140
6. 273
7. 525
8. 208

2.7 Lesson

Core Vocabulary
Previous
polynomial
trinomial

What You Will Learn
- Factor the difference of two squares.
- Factor perfect square trinomials.
- Use factoring to solve real-life problems.

Factoring the Difference of Two Squares
You can use special product patterns to factor polynomials.

Core Concept
Difference of Two Squares Pattern

Algebra	Example
$a^2 - b^2 = (a + b)(a - b)$	$x^2 - 9 = x^2 - 3^2 = (x + 3)(x - 3)$

EXAMPLE 1 Factoring the Difference of Two Squares

Factor (a) $x^2 - 25$ and (b) $4z^2 - 1$.

SOLUTION

a. $x^2 - 25 = x^2 - 5^2$ Write as $a^2 - b^2$.
   $ = (x + 5)(x - 5)$ Difference of two squares pattern

▶ So, $x^2 - 25 = (x + 5)(x - 5)$.

b. $4z^2 - 1 = (2z)^2 - 1^2$ Write as $a^2 - b^2$.
   $ = (2z + 1)(2z - 1)$ Difference of two squares pattern

▶ So, $4z^2 - 1 = (2z + 1)(2z - 1)$.

EXAMPLE 2 Evaluating a Numerical Expression

Use a special product pattern to evaluate the expression $54^2 - 48^2$.

SOLUTION

Notice that $54^2 - 48^2$ is a difference of two squares. So, you can rewrite the expression in a form that it is easier to evaluate using the difference of two squares pattern.

$54^2 - 48^2 = (54 + 48)(54 - 48)$ Difference of two squares pattern
$ = 102(6)$ Simplify.
$ = 612$ Multiply.

▶ So, $54^2 - 48^2 = 612$.

Monitoring Progress Help in English and Spanish at *BigIdeasMath.com*

Factor the polynomial.

1. $x^2 - 36$
2. $100 - m^2$
3. $9n^2 - 16$
4. $16h^2 - 49$

Use a special product pattern to evaluate the expression.

5. $36^2 - 34^2$
6. $47^2 - 44^2$
7. $55^2 - 50^2$
8. $28^2 - 24^2$

Laurie's Notes Teacher Actions

- Write the *Core Concept*. Point out to students that they used the reverse of this pattern to multiply binomials earlier in this chapter.
- Write the expression in part (a). Point out that x^2 and 25 are both perfect squares. Some students find it helpful to write $x^2 + 0x - 25$.
- ? In part (b) ask, "What do you notice about the numbers 4 and 1, and the expression z^2?"
  They are all perfect squares.
- **Think-Pair-Share:** Have students answer Questions 1–8, and then share and discuss as a class.

Factoring Perfect Square Trinomials

Core Concept

Perfect Square Trinomial Pattern

Algebra

$a^2 + 2ab + b^2 = (a + b)^2$

$a^2 - 2ab + b^2 = (a - b)^2$

Example

$x^2 + 6x + 9 = x^2 + 2(x)(3) + 3^2$
$= (x + 3)^2$

$x^2 - 6x + 9 = x^2 - 2(x)(3) + 3^2$
$= (x - 3)^2$

EXAMPLE 3 Factoring Perfect Square Trinomials

Factor each polynomial.

a. $n^2 + 8n + 16$ b. $4x^2 - 12x + 9$

SOLUTION

a. $n^2 + 8n + 16 = n^2 + 2(n)(4) + 4^2$ Write as $a^2 + 2ab + b^2$.
$= (n + 4)^2$ Perfect square trinomial pattern

▶ So, $n^2 + 8n + 16 = (n + 4)^2$.

b. $4x^2 - 12x + 9 = (2x)^2 - 2(2x)(3) + 3^2$ Write as $a^2 - 2ab + b^2$.
$= (2x - 3)^2$ Perfect square trinomial pattern

▶ So, $4x^2 - 12x + 9 = (2x - 3)^2$.

EXAMPLE 4 Solving a Polynomial Equation

Solve $x^2 + \frac{2}{3}x + \frac{1}{9} = 0$.

SOLUTION

LOOKING FOR STRUCTURE

Equations of the form $(x + a)^2 = 0$ always have repeated roots of $x = -a$.

$x^2 + \frac{2}{3}x + \frac{1}{9} = 0$	Write equation.
$9x^2 + 6x + 1 = 0$	Multiply each side by 9.
$(3x)^2 + 2(3x)(1) + 1^2 = 0$	Write left side as $a^2 + 2ab + b^2$.
$(3x + 1)^2 = 0$	Perfect square trinomial pattern
$3x + 1 = 0$	Zero-Product Property
$x = -\frac{1}{3}$	Solve for x.

▶ The solution is $x = -\frac{1}{3}$.

Monitoring Progress  Help in English and Spanish at BigIdeasMath.com

Factor the polynomial.

9. $m^2 - 2m + 1$ 10. $d^2 - 10d + 25$ 11. $9z^2 + 36z + 36$

Solve the equation.

12. $a^2 + 6a + 9 = 0$ 13. $w^2 - \frac{7}{3}w + \frac{49}{36} = 0$ 14. $n^2 - 81 = 0$

Section 2.7 Factoring Special Products 103

Differentiated Instruction

Inclusion

Some students may have difficulty writing a trinomial such as $4x^2 - 12x + 9$ in the form $a^2 - 2ab + b^2$ or $a^2 + 2ab + b^2$. Pair each of these students with one who understands the process. Have the students rewrite a trinomial together while discussing the process. Then provide a second trinomial for each student to rewrite, and have partners compare and justify their answers.

Extra Example 3

Factor each polynomial.

a. $x^2 + 26x + 169$ $(x + 13)^2$

b. $9x^2 - 24x + 16$ $(3x - 4)^2$

Extra Example 4

Solve $x^2 + 3x + \frac{9}{4} = 0$. $x = -\frac{3}{2}$

MONITORING PROGRESS ANSWERS

9. $(m - 1)^2$
10. $(d - 5)^2$
11. $9(z + 2)^2$
12. $a = -3$
13. $w = \frac{7}{6}$
14. $n = -9, n = 9$

Laurie's Notes Teacher Actions

- Write the *Core Concept*. Point out again that students used the reverse of this pattern to square binomials earlier in this chapter.
- ? Ask students whether the following is *always true*, *sometimes true*, or *never true*. They need to explain their reasoning. "If *a* and *c* are perfect squares in the trinomial $ax^2 + bx + c$, then you have a perfect square trinomial." sometimes true; You need to check the value of *b*.
- In Example 4, students may factor the trinomial as $\left(x + \frac{1}{3}\right)^2$ and the solution, when set equal to 0, is the same as the method shown.

Extra Example 5

Use the information in Example 5. Suppose the golf ball hits the top of a 17-foot oak tree. After how many seconds does the ball hit the tree? *The golf ball hits the tree after 2 seconds.*

MONITORING PROGRESS ANSWER

15. 2.25 sec

Solving Real-Life Problems

EXAMPLE 5   Modeling with Mathematics

A bird picks up a golf ball and drops it while flying. The function represents the height y (in feet) of the golf ball t seconds after it is dropped. The ball hits the top of a 32-foot-tall pine tree. After how many seconds does the ball hit the tree?

$y = 81 - 16t^2$

SOLUTION

1. **Understand the Problem** You are given the height of the golf ball as a function of the amount of time after it is dropped and the height of the tree that the golf ball hits. You are asked to determine how many seconds it takes for the ball to hit the tree.

2. **Make a Plan** Use the function for the height of the golf ball. Substitute the height of the tree for y and solve for the time t.

3. **Solve the Problem** Substitute 32 for y and solve for t.

$y = 81 - 16t^2$	Write equation.
$32 = 81 - 16t^2$	Substitute 32 for y.
$0 = 49 - 16t^2$	Subtract 32 from each side.
$0 = 7^2 - (4t)^2$	Write as $a^2 - b^2$.
$0 = (7 + 4t)(7 - 4t)$	Difference of two squares pattern
$7 + 4t = 0$ or $7 - 4t = 0$	Zero-Product Property
$t = -\frac{7}{4}$ or $t = \frac{7}{4}$	Solve for t.

   A negative time does not make sense in this situation.

   ▶ So, the golf ball hits the tree after $\frac{7}{4}$, or 1.75 seconds.

4. **Look Back** Check your solution, as shown, by substituting $t = \frac{7}{4}$ into the equation $32 = 81 - 16t^2$. Then verify that a time of $\frac{7}{4}$ seconds gives a height of 32 feet.

   Check
   $32 = 81 - 16t^2$
   $32 \stackrel{?}{=} 81 - 16\left(\frac{7}{4}\right)^2$
   $32 \stackrel{?}{=} 81 - 16\left(\frac{49}{16}\right)$
   $32 \stackrel{?}{=} 81 - 49$
   $32 = 32$ ✓

Monitoring Progress Help in English and Spanish at BigIdeasMath.com

15. **WHAT IF?** The golf ball does not hit the pine tree. After how many seconds does the ball hit the ground?

Laurie's Notes — Teacher Actions

- **FYI:** This is a falling object model where the polynomial is not written in standard form. The equation could have been written as $y = -16t^2 + 81$, but that does not fit the difference of two squares pattern.
- **Attend to Precision:** Because the coefficient of the linear term is 0, you could solve by square rooting. See *Another Way* on page T-100.

Closure

- **Exit Ticket:** Factor each polynomial.
  a. $81 - x^2$ $(9 - x)(9 + x)$
  b. $x^2 - 24x + 144$ $(x - 12)^2$

2.7 Exercises

Dynamic Solutions available at *BigIdeasMath.com*

Vocabulary and Core Concept Check

1. **REASONING** Can you use the perfect square trinomial pattern to factor $y^2 + 16y + 64$? Explain.

2. **WHICH ONE DOESN'T BELONG?** Which polynomial does *not* belong with the other three? Explain your reasoning.

   | $n^2 - 4$ | $g^2 - 6g + 9$ | $r^2 + 12r + 36$ | $k^2 + 25$ |

Monitoring Progress and Modeling with Mathematics

In Exercises 3–8, factor the polynomial. *(See Example 1.)*

3. $m^2 - 49$
4. $z^2 - 81$
5. $64 - 81d^2$
6. $25 - 4x^2$
7. $225a^2 - 36b^2$
8. $16x^2 - 169y^2$

In Exercises 9–14, use a special product pattern to evaluate the expression. *(See Example 2.)*

9. $12^2 - 9^2$
10. $19^2 - 11^2$
11. $78^2 - 72^2$
12. $54^2 - 52^2$
13. $53^2 - 47^2$
14. $39^2 - 36^2$

In Exercises 15–22, factor the polynomial. *(See Example 3.)*

15. $h^2 + 12h + 36$
16. $p^2 + 30p + 225$
17. $y^2 - 22y + 121$
18. $x^2 - 4x + 4$
19. $a^2 - 28a + 196$
20. $m^2 + 24m + 144$
21. $25n^2 + 20n + 4$
22. $49a^2 - 14a + 1$

ERROR ANALYSIS In Exercises 23 and 24, describe and correct the error in factoring the polynomial.

23.
$$n^2 - 64 = n^2 - 8^2$$
$$= (n - 8)^2$$

24.
$$y^2 - 6y + 9 = y^2 - 2(y)(3) + 3^2$$
$$= (y - 3)(y + 3)$$

25. **MODELING WITH MATHEMATICS** The area (in square centimeters) of a square coaster can be represented by $d^2 + 8d + 16$.

    a. Write an expression that represents the side length of the coaster.

    b. Write an expression for the perimeter of the coaster.

26. **MODELING WITH MATHEMATICS** The polynomial represents the area (in square feet) of the square playground.

    a. Write a polynomial that represents the side length of the playground.

    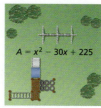

    b. Write an expression for the perimeter of the playground.

In Exercises 27–34, solve the equation. *(See Example 4.)*

27. $z^2 - 4 = 0$
28. $4x^2 = 49$
29. $k^2 - 16k + 64 = 0$
30. $s^2 + 20s + 100 = 0$
31. $n^2 + 9 = 6n$
32. $y^2 = 12y - 36$
33. $y^2 + \frac{1}{2}y = -\frac{1}{16}$
34. $-\frac{4}{3}x + \frac{4}{9} = -x^2$

In Exercises 35–40, factor the polynomial.

35. $3z^2 - 27$
36. $2m^2 - 50$
37. $4y^2 - 16y + 16$
38. $8k^2 + 80k + 200$
39. $50y^2 + 120y + 72$
40. $27m^2 - 36m + 12$

---

33. $y = -\frac{1}{4}$
34. $x = \frac{2}{3}$
35. $3(z + 3)(z - 3)$
36. $2(m + 5)(m - 5)$
37. $4(y - 2)^2$
38. $8(k + 5)^2$
39. $2(5y + 6)^2$
40. $3(3m - 2)^2$

Assignment Guide and Homework Check

ASSIGNMENT
Basic: 1, 2, 3–35 odd, 41, 46, 49–55
Average: 1, 2–42 even, 46, 47, 49–55
Advanced: 1, 2, 8, 14, 20–26 even, 32–42 even, 43–55

HOMEWORK CHECK
Basic: 3, 9, 15, 29, 41
Average: 4, 12, 16, 30, 42
Advanced: 8, 14, 22, 32, 42

ANSWERS

1. yes; The square roots of the first and last terms are y and 8, and the middle term is $2 \cdot y \cdot 8$, so it fits the pattern.
2. $k^2 + 25$; It is the only one that cannot be factored using a special pattern.
3. $(m + 7)(m - 7)$
4. $(z + 9)(z - 9)$
5. $(8 + 9d)(8 - 9d)$
6. $(5 + 2x)(5 - 2x)$
7. $9(5a + 2b)(5a - 2b)$
8. $(4x + 13y)(4x - 13y)$
9. 63
10. 240
11. 900
12. 212
13. 600
14. 225
15. $(h + 6)^2$
16. $(p + 15)^2$
17. $(y - 11)^2$
18. $(x - 2)^2$
19. $(a - 14)^2$
20. $(m + 12)^2$
21. $(5n + 2)^2$
22. $(7a - 1)^2$
23. should follow the difference of two squares pattern; $= (n + 8)(n - 8)$
24. should follow the perfect square trinomial pattern; $= (y - 3)^2$
25. a. $(d + 4)$ cm
    b. $(4d + 16)$ cm
26. a. $(x - 15)$ ft
    b. $(4x - 60)$ ft
27. $z = -2, z = 2$
28. $x = -\frac{7}{2}, x = \frac{7}{2}$
29. $k = 8$
30. $s = -10$
31. $n = 3$
32. $y = 6$

Dynamic Teaching Tools

- Dynamic Assessment & Progress Monitoring Tool
- Interactive Whiteboard Lesson Library
- Dynamic Classroom with Dynamic Investigations

ANSWERS

41. 1.25 sec

42. 0.25 sec

43. a. no; $w^2 + 18w + 81$
b. no; $y^2 - 10y + 25$

44. a.

$4x^2 - 1 = (2x+1)(2x-1)$

b.

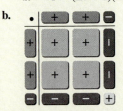

$4x^2 - 4x + 1 = (2x-1)^2$

45. Square each binomial, then combine like terms; Use the difference of two squares pattern with each binomial as one of the terms, then simplify; *Sample answer:* The difference of two squares pattern; You do not need to square any binomials.

46–55. See Additional Answers.

Mini-Assessment

Factor each polynomial.

1. $9x^2 - 25$ $(3x+5)(3x-5)$
2. $x^2 - 40x + 400$ $(x-20)^2$
3. Use a special product pattern to evaluate the expression $26^2 - 22^2$.
   $(26+22)(26-22) = 48(4) = 192$
4. Solve $16x^2 + 8x + 1 = 0$. $x = -\frac{1}{4}$
5. A bird picks up a stone and drops it while flying. The function $y = 49 - 16t^2$ represents the height y (in feet) of the stone t seconds after it is dropped. The stone hits the top of a 13-foot-tall building. After how many seconds does the stone hit the building? *The stone hits the building after $\frac{3}{2}$, or 1.5, seconds.*

106   Chapter 2

---

41. MODELING WITH MATHEMATICS While standing on a ladder, you drop a paintbrush. The function represents the height y (in feet) of the paintbrush t seconds after it is dropped. After how many seconds does the paintbrush land on the ground? (*See Example 5.*)

$y = 25 - 16t^2$

42. MODELING WITH MATHEMATICS The function represents the height y (in feet) of a grasshopper jumping straight up from the ground t seconds after the start of the jump. After how many seconds is the grasshopper 1 foot off the ground?

$y = -16t^2 + 8t$

43. REASONING Tell whether the polynomial can be factored. If not, change the constant term so that the polynomial is a perfect square trinomial.

a. $w^2 + 18w + 84$ **b.** $y^2 - 10y + 23$

44. THOUGHT PROVOKING Use algebra tiles to factor each polynomial modeled by the tiles. Show your work.

45. COMPARING METHODS Describe two methods you can use to simplify $(2x-5)^2 - (x-4)^2$. Which one would you use? Explain.

Maintaining Mathematical Proficiency
Reviewing what you learned in previous grades and lessons

Write the prime factorization of the number. (*Skills Review Handbook*)

49. 50   **50.** 44   **51.** 85   **52.** 96

Find the inverse of the function. Then graph the function and its inverse. (*Section 1.3*)

53. $f(x) = x - 3$ **54.** $f(x) = -2x + 5$ **55.** $f(x) = \frac{1}{2}x - 1$

106   Chapter 2   Polynomial Equations and Factoring

---

46. HOW DO YOU SEE IT? The figure shows a large square with an area of a^2 that contains a smaller square with an area of b^2.

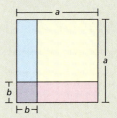

a. Describe the regions that represent $a^2 - b^2$. How can you rearrange these regions to show that $a^2 - b^2 = (a+b)(a-b)$?

b. How can you use the figure to show that $(a-b)^2 = a^2 - 2ab + b^2$?

47. PROBLEM SOLVING You hang nine identical square picture frames on a wall.

a. Write a polynomial that represents the area of the picture frames, not including the pictures.

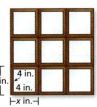

b. The area in part (a) is 81 square inches. What is the side length of one of the picture frames? Explain your reasoning.

48. MATHEMATICAL CONNECTIONS The composite solid is made up of a cube and a rectangular prism.

a. Write a polynomial that represents the volume of the composite solid.

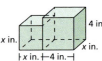

b. The volume of the composite solid is equal to $25x$. What is the value of x? Explain your reasoning.

If students need help...	If students got it...
Resources by Chapter • Practice A and Practice B • Puzzle Time	Resources by Chapter • Enrichment and Extension • Cumulative Review
Student Journal • Practice	Start the *next* Section
Differentiating the Lesson Skills Review Handbook	

Laurie's Notes

Overview of Section 2.8

Introduction
- This last lesson, on factoring polynomials completely, requires many prior skills. Students will learn the technique of factoring by grouping, which involves the Distributive Property. Then they will factor polynomials completely by first looking for the greatest common factor in each term.
- Although factoring of polynomials is the focus of the lesson, the ultimate goal is for students to be able to solve polynomial equations and simplify expressions with common factors.

Teaching Strategy
- **Look For and Make Use of Structure:** All the following expressions have the same structure. Each expression has two terms and can be simplified using the Distributive Property.
  $$ac + bc \qquad a(x+2) + b(x+2) \qquad a(x-4) - b(x-4)$$
- While factoring out a common binomial factor is no different from factoring out a common monomial factor, students do not always recognize the underlying structure. To help, use colors and underlining.
- Notice how the use of color and underlining helps in recognizing common factors:
  $$a\underline{c} + b\underline{c} = (a+b)\underline{c}$$
  $$a\underline{(x+2)} + b\underline{(x+2)} = (a+b)\underline{(x+2)}$$
  $$a\underline{(x-4)} - b\underline{(x-4)} = (a-b)\underline{(x-4)}$$
- The Distributive Property is often viewed as a right-pointing arrow versus a statement of equality. In each of the examples above, a common factor has been factored out using the Distributive Property.

Pacing Suggestion
- Complete the explorations, perhaps using a jigsaw approach in each exploration, and then transition to the formal lesson.

Dynamic Teaching Tools
Dynamic Assessment & Progress Monitoring Tool
Lesson Planning Tool
Interactive Whiteboard Lesson Library
Dynamic Classroom with Dynamic Investigations

What Your Students Will Learn

- Factor polynomials by grouping by finding the GCF of pairs of terms.
- Factor polynomials with integer coefficients completely.
- Use factoring to solve real-life problems involving polynomial equations.

Laurie's Notes

Exploration

Motivate
- Today's lesson involves factoring polynomials. A quick warm-up with factoring out greatest common factors (GCFs) would be helpful.
- Ask students to factor the GCF out of each of the following polynomials:
  $3x^2 - 15x = 3x(x - 5)$
  $-16x^3 - 8x^2 - 24x = -8x(2x^2 + x + 3)$
  $30x^2 + 5x = 5x(6x + 1)$

Exploration Note
- Most of the examples students have worked with in this chapter involve two factors and second-degree polynomials. In both explorations, students are working with three factors and mostly third-degree polynomials in preparation for this last lesson in the chapter on factoring polynomials completely.

Exploration 1
- Explain that this first exploration is a good review. Students will write the expression modeled by the algebra tiles and then find the product and express it in standard form. This is a skill taught in the early part of the chapter.
- **COMMON ERROR** When multiplying $(x + 1)(x + 1)(-2)$, students will sometimes multiply (-2) times each of the binomials and get $(-2x - 2)(-2x - 2)$ before multiplying the binomials. Remind them that when they were multiplying $(3)(4)(2)$, they would not multiply the factor of 2 by each of the other factors and get $(6)(8)$.
- If time is a factor, you might have half the students do parts (a), (c), and (e) and the rest of the students do parts (b), (d), and (f).

Exploration 2
- It is likely that most students will work with the factored form, multiply the expression, and look for their answer in the list of answers in standard form.
- Again, if time is a factor, you could split the class so they are doing every other problem. Half the class begins with A and half begins with B.
- **Popsicle Sticks:** Solicit responses when students have finished.
- ? Look back at the standard form expressions that are trinomials. "What do you notice about the terms in the trinomial?" Each of the terms has an x in it. "Could that x be factored out as a common factor?" yes
- ? Select one of the trinomial expressions, say part (d), and factor out an x. Write $x^3 - 4x^2 + 4x = x(x^2 - 4x + 4)$. "Can you factor the resulting trinomial?" yes; $(x^2 - 4x + 4) = (x - 2)^2$
- You could use some of the responses in the left-hand column (standard form) to review vocabulary associated with this chapter.
- **Extension:** Select a few of the factored form expressions and set them equal to 0 and ask what the solutions would be.

Communicate Your Answer
- **Neighbor Check:** For Question 4, have students work independently and then have their neighbors check their work. Have students discuss any discrepancies.

Connecting to Next Step
- The explorations help students visualize and make sense of factoring third-degree polynomials. In the formal lesson, students will work with factoring a polynomial completely.

2.8 Factoring Polynomials Completely

Essential Question How can you factor a polynomial completely?

EXPLORATION 1 Writing a Product of Linear Factors

Work with a partner. Write the product represented by the algebra tiles. Then multiply to write the polynomial in standard form.

a. (+ +)(+ +)(− −)
b. (+ + +)(+ +)(−)
c. (+ + + +)(+)(+ +)
d. (+ +)(+ −)(+)
e. (− +)(+ +)(−)
f. (− −)(+ +)(− −)

> **REASONING ABSTRACTLY**
> To be proficient in math, you need to know and flexibly use different properties of operations and objects.

EXPLORATION 2 Matching Standard and Factored Forms

Work with a partner. Match the standard form of the polynomial with the equivalent factored form. Explain your strategy.

a. $x^3 + x^2$ A. $x(x+1)(x-1)$
b. $x^3 - x$ B. $x(x-1)^2$
c. $x^3 + x^2 - 2x$ C. $x(x+1)^2$
d. $x^3 - 4x^2 + 4x$ D. $x(x+2)(x-1)$
e. $x^3 - 2x^2 - 3x$ E. $x(x-1)(x-2)$
f. $x^3 - 2x^2 + x$ F. $x(x+2)(x-2)$
g. $x^3 - 4x$ G. $x(x-2)^2$
h. $x^3 + 2x^2$ H. $x(x+2)^2$
i. $x^3 - x^2$ I. $x^2(x-1)$
j. $x^3 - 3x^2 + 2x$ J. $x^2(x+1)$
k. $x^3 + 2x^2 - 3x$ K. $x^2(x-2)$
l. $x^3 - 4x^2 + 3x$ L. $x^2(x+2)$
m. $x^3 - 2x^2$ M. $x(x+3)(x-1)$
n. $x^3 + 4x^2 + 4x$ N. $x(x+1)(x-3)$
o. $x^3 + 2x^2 + x$ O. $x(x-1)(x-3)$

Communicate Your Answer

3. How can you factor a polynomial completely?

4. Use your answer to Question 3 to factor each polynomial completely.

   a. $x^3 + 4x^2 + 3x$ b. $x^3 - 6x^2 + 9x$ c. $x^3 + 6x^2 + 9x$

Section 2.8 Factoring Polynomials Completely 107

Extra Example 1

Factor each polynomial by grouping.

a. $x^3 + 4x^2 + 2x + 8$ $(x^2 + 2)(x + 4)$
b. $x^2 + 4y + 2x + 2xy$ $(x + 2)(x + 2y)$

MONITORING PROGRESS ANSWERS

1. $(a^2 + 1)(a + 3)$
2. $(y + 2)(y + x)$

2.8 Lesson

What You Will Learn

- Factor polynomials by grouping.
- Factor polynomials completely.
- Use factoring to solve real-life problems.

Core Vocabulary

factoring by grouping, p. 108
factored completely, p. 108

Previous
polynomial
binomial

Factoring Polynomials by Grouping

You have used the Distributive Property to factor out a greatest common monomial from a polynomial. Sometimes, you can factor out a common binomial. You may be able to use the Distributive Property to factor polynomials with four terms, as described below.

> **Core Concept**
>
> **Factoring by Grouping**
>
> To factor a polynomial with four terms, group the terms into pairs. Factor the GCF out of each pair of terms. Look for and factor out the common binomial factor. This process is called **factoring by grouping**.

EXAMPLE 1   **Factoring by Grouping**

Factor each polynomial by grouping.

a. $x^3 + 3x^2 + 2x + 6$ b. $x^2 + y + x + xy$

SOLUTION

a. $x^3 + 3x^2 + 2x + 6 = (x^3 + 3x^2) + (2x + 6)$ Group terms with common factors.

   Common binomial factor is $x + 3$. $= x^2(x + 3) + 2(x + 3)$ Factor out GCF of each pair of terms.

   $= (x + 3)(x^2 + 2)$ Factor out $(x + 3)$.

▶ So, $x^3 + 3x^2 + 2x + 6 = (x + 3)(x^2 + 2)$.

b. $x^2 + y + x + xy = x^2 + x + xy + y$ Rewrite polynomial.

   $= (x^2 + x) + (xy + y)$ Group terms with common factors.

   Common binomial factor is $x + 1$. $= x(x + 1) + y(x + 1)$ Factor out GCF of each pair of terms.

   $= (x + 1)(x + y)$ Factor out $(x + 1)$.

▶ So, $x^2 + y + x + xy = (x + 1)(x + y)$.

Monitoring Progress Help in English and Spanish at *BigIdeasMath.com*

Factor the polynomial by grouping.

1. $a^3 + 3a^2 + a + 3$ 2. $y^2 + 2x + yx + 2y$

Factoring Polynomials Completely

You have seen that the polynomial $x^2 - 1$ can be factored as $(x + 1)(x - 1)$. This polynomial is factorable. Notice that the polynomial $x^2 + 1$ cannot be written as the product of polynomials with integer coefficients. This polynomial is unfactorable. A factorable polynomial with integer coefficients is **factored completely** when it is written as a product of unfactorable polynomials with integer coefficients.

Laurie's Notes Teacher Actions

- **Teaching Tip:** Use of color and underlining is helpful: $x^2\underline{(x + 3)} + 2\underline{(x + 3)}$. This is the point where it is important for students to view this expression as the sum of two terms that have a common factor. The common factor $\underline{(x + 3)}$ can be factored out just like a monomial.
- The polynomial in part (b) appears different to students because of the two variables and because terms with common factors are not adjacent. Group (put parentheses around) terms with common factors. Use the Distributive Property to factor.

Concept Summary

Guidelines for Factoring Polynomials Completely
To factor a polynomial completely, you should try each of these steps.

1. Factor out the greatest common monomial factor. $3x^2 + 6x = 3x(x + 2)$
2. Look for a difference of two squares or a perfect square trinomial. $x^2 + 4x + 4 = (x + 2)^2$
3. Factor a trinomial of the form $ax^2 + bx + c$ into a product of binomial factors. $3x^2 - 5x - 2 = (3x + 1)(x - 2)$
4. Factor a polynomial with four terms by grouping. $x^3 + x - 4x^2 - 4 = (x^2 + 1)(x - 4)$

EXAMPLE 2 Factoring Completely

Factor (a) $3x^3 + 6x^2 - 18x$ and (b) $7x^4 - 28x^2$.

SOLUTION

a. $3x^3 + 6x^2 - 18x = 3x(x^2 + 2x - 6)$ Factor out $3x$.

$x^2 + 2x - 6$ is unfactorable, so the polynomial is factored completely.

▶ So, $3x^3 + 6x^2 - 18x = 3x(x^2 + 2x - 6)$.

b. $7x^4 - 28x^2 = 7x^2(x^2 - 4)$ Factor out $7x^2$.
$= 7x^2(x^2 - 2^2)$ Write as $a^2 - b^2$.
$= 7x^2(x + 2)(x - 2)$ Difference of two squares pattern

▶ So, $7x^4 - 28x^2 = 7x^2(x + 2)(x - 2)$.

EXAMPLE 3 Solving an Equation by Factoring Completely

Solve $2x^3 + 8x^2 = 10x$.

SOLUTION

$2x^3 + 8x^2 = 10x$			Original equation
$2x^3 + 8x^2 - 10x = 0$			Subtract $10x$ from each side.
$2x(x^2 + 4x - 5) = 0$			Factor out $2x$.
$2x(x + 5)(x - 1) = 0$			Factor $x^2 + 4x - 5$.
$2x = 0$ or	$x + 5 = 0$ or	$x - 1 = 0$	Zero-Product Property
$x = 0$ or	$x = -5$ or	$x = 1$	Solve for x.

▶ The roots are $x = -5$, $x = 0$, and $x = 1$.

Monitoring Progress Help in English and Spanish at BigIdeasMath.com

Factor the polynomial completely.

3. $3x^3 - 12x$
4. $2y^3 - 12y^2 + 18y$
5. $m^3 - 2m^2 - 8m$

Solve the equation.

6. $w^3 - 8w^2 + 16w = 0$
7. $x^3 - 25x = 0$
8. $c^3 - 7c^2 + 12c = 0$

English Language Learners

Organization
Students will benefit by writing down the steps for factoring polynomials completely, as shown in the *Concept Summary*. Have students write the steps in their notebooks. Consider also displaying a poster with the same information in the classroom.

Extra Example 2
a. Factor $2x^3 + 6x^2 - 2x$.
   $2x(x^2 + 3x - 1)$
b. Factor $5x^4 - 45x^2$. $5x^2(x + 3)(x - 3)$

Extra Example 3
Solve $3x^3 + 6x^2 = 24x$.
The roots are $x = -4$, $x = 0$, and $x = 2$.

MONITORING PROGRESS ANSWERS

3. $3x(x + 2)(x - 2)$
4. $2y(y - 3)^2$
5. $m(m + 2)(m - 4)$
6. $w = 0$, $w = 4$
7. $x = 0$, $x = -5$, $x = 5$
8. $c = 0$, $c = 3$, $c = 4$

Laurie's Notes   Teacher Actions

- Explain what it means to factor completely. Discuss the guidelines and review the sample problems.

  COMMON ERROR Students may see a factor, but it may not be the greatest common factor. Always ask, "Are there any other factors?"

  ? Probing Question: Write $2x(x + 5)(x - 1) = 0$; $x(2x + 10)(x - 1) = 0$; $x(x + 5)(2x - 2) = 0$. Say, "All of these are equivalent. Explain why." They all have the same solutions, $x = -5$, $x = 0$, and $x = 1$.

Extra Example 4

A box in the shape of a rectangular prism has a volume of 96 cubic inches. The dimensions of the box in terms of its width are shown. Find the length, width, and height of the box.

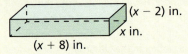

The width is 4 inches, the length is 12 inches, and the height is 2 inches.

MONITORING PROGRESS ANSWER

9. length: 3 ft, width: 2 ft, height: 12 ft

Solving Real-Life Problems

EXAMPLE 4 Modeling with Mathematics

A terrarium in the shape of a rectangular prism has a volume of 4608 cubic inches. Its length is more than 10 inches. The dimensions of the terrarium in terms of its width are shown. Find the length, width, and height of the terrarium.

SOLUTION

1. **Understand the Problem** You are given the volume of a terrarium in the shape of a rectangular prism and a description of the length. The dimensions are written in terms of its width. You are asked to find the length, width, and height of the terrarium.

2. **Make a Plan** Use the formula for the volume of a rectangular prism to write and solve an equation for the width of the terrarium. Then substitute that value in the expressions for the length and height of the terrarium.

3. **Solve the Problem**

Volume = length • width • height	Volume of a rectangular prism
$4608 = (36 - w)(w)(w + 4)$	Write equation.
$4608 = 32w^2 + 144w - w^3$	Multiply.
$0 = 32w^2 + 144w - w^3 - 4608$	Subtract 4608 from each side.
$0 = (-w^3 + 32w^2) + (144w - 4608)$	Group terms with common factors.
$0 = -w^2(w - 32) + 144(w - 32)$	Factor out GCF of each pair of terms.
$0 = (w - 32)(-w^2 + 144)$	Factor out $(w - 32)$.
$0 = -1(w - 32)(w^2 - 144)$	Factor -1 from $-w^2 + 144$.
$0 = -1(w - 32)(w - 12)(w + 12)$	Difference of two squares pattern
$w - 32 = 0$ or $w - 12 = 0$ or $w + 12 = 0$	Zero-Product Property
$w = 32$ or $w = 12$ or $w = -12$	Solve for w.

Disregard $w = -12$ because a negative width does not make sense. You know that the length is more than 10 inches. Test the solutions of the equation, 12 and 32, in the expression for the length.

length $= 36 - w = 36 - 12 = 24$ ✓ or length $= 36 - w = 36 - 32 = 4$ ✗

The solution 12 gives a length of 24 inches, so 12 is the correct value of w.

Use $w = 12$ to find the height, as shown.

height $= w + 4 = 12 + 4 = 16$

▶ The width is 12 inches, the length is 24 inches, and the height is 16 inches.

Check
$V = \ell wh$
$4608 \stackrel{?}{=} 24(12)(16)$
$4608 = 4608$

4. **Look Back** Check your solution. Substitute the values for the length, width, and height when the width is 12 inches into the formula for volume. The volume of the terrarium should be 4608 cubic inches.

Monitoring Progress Help in English and Spanish at BigIdeasMath.com

9. A box in the shape of a rectangular prism has a volume of 72 cubic feet. The box has a length of x feet, a width of $(x - 1)$ feet, and a height of $(x + 9)$ feet. Find the dimensions of the box.

Laurie's Notes — Teacher Actions

? Make Sense of Problems and Persevere in Solving Them and **Model with Mathematics:** Draw and label the diagram and ask, "How do you find the volume of this rectangular prism?" $V = \ell wh$, so by substituting, $4608 = (36 - w)(w)(w + 4)$.

- **Teaching Tip:** Pose Example 4 and discuss to make sure that students understand the problem. Then give time for students to work with their partner as you circulate. This is a problem where productive struggle is appropriate. Allow groups to check in with other groups to compare work in progress.

Closure

- **Exit Ticket:** Solve $3x^5 - 6x^4 - 45x^3 = 0$. $x = -3, x = 0, x = 5$

2.8 Exercises

Dynamic Solutions available at *BigIdeasMath.com*

Vocabulary and Core Concept Check

1. **VOCABULARY** What does it mean for a polynomial to be factored completely?
2. **WRITING** Explain how to choose which terms to group together when factoring by grouping.

Monitoring Progress and Modeling with Mathematics

In Exercises 3–10, factor the polynomial by grouping. (See Example 1.)

3. $x^3 + x^2 + 2x + 2$
4. $y^3 - 9y^2 + y - 9$
5. $3z^3 + 2z - 12z^2 - 8$
6. $2s^3 - 27 - 18s + 3s^2$
7. $x^2 + xy + 8x + 8y$
8. $q^2 + q + 5pq + 5p$
9. $m^2 - 3m + mn - 3n$
10. $2a^2 + 8ab - 3a - 12b$

In Exercises 11–22, factor the polynomial completely. (See Example 2.)

11. $2x^3 - 2x$
12. $36a^4 - 4a^2$
13. $2c^2 - 7c + 19$
14. $m^2 - 5m - 35$
15. $6g^3 - 24g^2 + 24g$
16. $-15d^3 + 21d^2 - 6d$
17. $3r^5 + 3r^4 - 90r^3$
18. $5w^4 - 40w^3 + 80w^2$
19. $-4c^4 + 8c^3 - 28c^2$
20. $8t^2 + 8t - 72$
21. $b^3 - 5b^2 - 4b + 20$
22. $h^3 + 4h^2 - 25h - 100$

In Exercises 23–28, solve the equation. (See Example 3.)

23. $5n^3 - 30n^2 + 40n = 0$
24. $k^4 - 100k^2 = 0$
25. $x^3 + x^2 = 4x + 4$
26. $2t^5 + 2t^4 - 144t^3 = 0$
27. $12s - 3s^3 = 0$
28. $4y^3 - 7y^2 + 28 = 16y$

In Exercises 29–32, find the x-coordinates of the points where the graph crosses the x-axis.

29.
$y = x^3 - 81x$

30.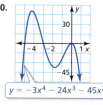
$y = -3x^4 - 24x^3 - 45x^2$

31.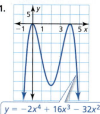
$y = -2x^4 + 16x^3 - 32x^2$

32.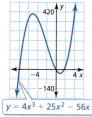
$y = 4x^3 + 25x^2 - 56x$

ERROR ANALYSIS In Exercises 33 and 34, describe and correct the error in factoring the polynomial completely.

33.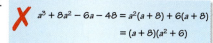
$a^3 + 8a^2 - 6a - 48 = a^2(a + 8) + 6(a + 8)$
$= (a + 8)(a^2 + 6)$

34. $x^3 - 6x^2 - 9x + 54 = x^2(x - 6) - 9(x - 6)$
$= (x - 6)(x^2 - 9)$

35. **MODELING WITH MATHEMATICS**
You are building a birdhouse in the shape of a rectangular prism that has a volume of 128 cubic inches. The dimensions of the birdhouse in terms of its width are shown. (See Example 4.)

$(w + 4)$ in.
w in., 4 in.

a. Write a polynomial that represents the volume of the birdhouse.
b. What are the dimensions of the birdhouse?

Section 2.8 Factoring Polynomials Completely

34. It is not factored completely because $x^2 - 9$ can be factored; $= (x - 6)(x + 3)(x - 3)$
35. a. $(4w^2 + 16w)$ in.³
    b. length: 4 in., width: 4 in., height: 8 in.

Dynamic Teaching Tools
Dynamic Assessment & Progress Monitoring Tool
Interactive Whiteboard Lesson Library
Dynamic Classroom with Dynamic Investigations

ANSWERS
36. length: 16 in., width: 6 in., height: 12 in.
37. $(x + 2y)(x + 1)(x - 1)$
38. $(2b - a)(2b + 3)(2b - 3)$
39. $(4s - 1)(s + 3t)$
40. $(m^2 - 2n)(6m + n)$
41. no; The factors of the polynomial are $x^2 + 3$ and $x + 2$. Using the Zero-Product Property, $x + 2 = 0$ will give 1 real solution, but $x^2 + 3 = 0$ has no real solutions.
42. The x-intercepts occur when $y = 0$, so set each factor equal to 0 and solve for x to get the x-coordinates of the x-intercepts.
43. a. Sample answer: $x^3 + x^2 + x + 2$
    b. Sample answer: $x^3 + x^2 + x + 1$
44. no; It is possible that the terms will have a common monomial factor.
45. $3z, 2z + 3, 2z - 3$
46–57. See Additional Answers.

Mini-Assessment
1. Factor $x^2 + 15y + 5x + 3xy$ by grouping. $(x + 5)(x + 3y)$
2. Factor $3x^5 - 48x^3$ completely. $3x^3(x + 4)(x - 4)$
3. Solve $6x^3 - 30x^2 = 36x$. The roots are $x = -1, x = 0,$ and $x = 6$.
4. A box in the shape of a rectangular prism has a volume of 180 cubic inches. The dimensions of the box in terms of its height are shown. Find the length, width, and height of the box.

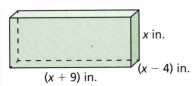

The length is 15 inches, the width is 2 inches, and the height is 6 inches.

112   Chapter 2

---

36. **MODELING WITH MATHEMATICS** A gift bag shaped like a rectangular prism has a volume of 1152 cubic inches. The dimensions of the gift bag in terms of its width are shown. The height is greater than the width. What are the dimensions of the gift bag?

In Exercises 37–40, factor the polynomial completely.

37. $x^3 + 2x^2y - x - 2y$ 38. $8b^3 - 4b^2a - 18b + 9a$

39. $4s^2 - s + 12st - 3t$

40. $6m^3 - 12mn + m^2n - 2n^2$

41. **WRITING** Is it possible to find three real solutions of the equation $x^3 + 2x^2 + 3x + 6 = 0$? Explain your reasoning.

42. **HOW DO YOU SEE IT?** How can you use the factored form of the polynomial $x^4 - 2x^3 - 9x^2 + 18x = x(x - 3)(x + 3)(x - 2)$ to find the x-intercepts of the graph of the function?

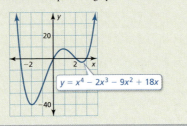

43. **OPEN-ENDED** Write a polynomial of degree 3 that satisfies each of the given conditions.
   a. is not factorable   b. can be factored by grouping

44. **MAKING AN ARGUMENT** Your friend says that if a trinomial cannot be factored as the product of two binomials, then the trinomial is factored completely. Is your friend correct? Explain.

45. **PROBLEM SOLVING** The volume (in cubic feet) of a room in the shape of a rectangular prism is represented by $12z^3 - 27z$. Find expressions that could represent the dimensions of the room.

46. **MATHEMATICAL CONNECTIONS** The width of a box in the shape of a rectangular prism is 4 inches more than the height h. The length is the difference of 9 inches and the height.
   a. Write a polynomial that represents the volume of the box in terms of its height (in inches).
   b. The volume of the box is 180 cubic inches. What are the possible dimensions of the box?
   c. Which dimensions result in a box with the least possible surface area? Explain your reasoning.

47. **MATHEMATICAL CONNECTIONS** The volume of a cylinder is given by $V = \pi r^2 h$, where r is the radius of the base of the cylinder and h is the height of the cylinder. Find the dimensions of the cylinder.

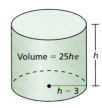

48. **THOUGHT PROVOKING** Factor the polynomial $x^5 - x^4 - 5x^3 + 5x^2 + 4x - 4$ completely.

49. **REASONING** Find a value for w so that the equation has (a) two solutions and (b) three solutions. Explain your reasoning.
$$5x^3 + wx^2 + 80x = 0$$

Maintaining Mathematical Proficiency
Reviewing what you learned in previous grades and lessons

Solve the system of linear equations by graphing. *(Skills Review Handbook)*

50. $y = x - 4$
    $y = -2x + 2$

51. $y = \frac{1}{2}x + 2$
    $y = 3x - 3$

52. $5x - y = 12$
    $\frac{1}{4}x + y = 9$

53. $x = 3y$
    $y - 10 = 2x$

Determine whether the function represents *exponential growth* or *exponential decay*. *(Section 1.6)*

54. $y = \left(\frac{4}{3}\right)^x$ 55. $y = 5(0.95)^x$ 56. $f(x) = (2.3)^{x-2}$ 57. $f(x) = 50(1.1)^{3x}$

112   Chapter 2   Polynomial Equations and Factoring

If students need help...	If students got it...
Resources by Chapter • Practice A and Practice B • Puzzle Time	Resources by Chapter • Enrichment and Extension • Cumulative Review
Student Journal • Practice	Start the *next* Section
Differentiating the Lesson Skills Review Handbook	

2.5–2.8 What Did You Learn?

Core Vocabulary

factoring by grouping, *p. 108*
factored completely, *p. 108*

Core Concepts

Section 2.5
Factoring $x^2 + bx + c$ When c Is Positive, *p. 90*
Factoring $x^2 + bx + c$ When c Is Negative, *p. 91*

Section 2.6
Factoring $ax^2 + bx + c$ When ac Is Positive, *p. 96*
Factoring $ax^2 + bx + c$ When ac Is Negative, *p. 97*

Section 2.7
Difference of Two Squares Pattern, *p. 102*
Perfect Square Trinomial Pattern, *p. 103*

Section 2.8
Factoring by Grouping, *p. 108*
Factoring Polynomials Completely, *p. 108*

Mathematical Practices

1. How are the solutions of Exercise 29 on page 93 related to the graph of $y = m^2 + 3m + 2$?
2. The equation in part (b) of Exercise 47 on page 94 has two solutions. Are both solutions of the equation reasonable in the context of the problem? Explain your reasoning.

Performance Task:

Flight Path of a Bird

Some birds, like parrots, have strong, large beaks to break open nuts and shells. But other birds, like crows, crack open their food by dropping it to the ground. How can a bird change its flight path to protect its falling food from other hungry birds?

To explore the answer to this question and more, check out the Performance Task and Real-Life STEM video at *BigIdeasMath.com*.

ANSWERS

1. $2x^2 + 6$; 2; 2; binomial
2. $5p^6 - 3p^3 - 4$; 6; 5; trinomial
3. $9x^7 + 13x^5 - 6x^2$; 7; 9; trinomial
4. $8y^3 - 12y$; 3; 8; binomial
5. $4a + 6$
6. $3x^2 + 6x + 10$
7. $-2y^2 + 6y + 4$
8. $-6p^2 - 12p + 7$
9. $x^2 + 2x - 24$
10. $3y^2 - 7y - 40$
11. $x^3 + 11x^2 + 28x$
12. $-12y^3 + 7y^2 + 20y - 7$
13. $x^2 - 81$
14. $4y^2 - 16$
15. $p^2 + 8p + 16$
16. $4d^2 - 4d + 1$

2 Chapter Review

Dynamic Solutions available at *BigIdeasMath.com*

2.1 Adding and Subtracting Polynomials (pp. 61–68)

Find $(2x^3 + 6x^2 - x) - (-3x^3 - 2x^2 - 9x)$.

$(2x^3 + 6x^2 - x) - (-3x^3 - 2x^2 - 9x) = (2x^3 + 6x^2 - x) + (3x^3 + 2x^2 + 9x)$
$= (2x^3 + 3x^3) + (6x^2 + 2x^2) + (-x + 9x)$
$= 5x^3 + 8x^2 + 8x$

Write the polynomial in standard form. Identify the degree and leading coefficient of the polynomial. Then classify the polynomial by the number of terms.

1. $6 + 2x^2$
2. $-3p^3 + 5p^6 - 4$
3. $9x^7 - 6x^2 + 13x^5$
4. $-12y + 8y^3$

Find the sum or difference.

5. $(3a + 7) + (a - 1)$
6. $(x^2 + 6x - 5) + (2x^2 + 15)$
7. $(-y^2 + y + 2) - (y^2 - 5y - 2)$
8. $(p + 7) - (6p^2 + 13p)$

2.2 Multiplying Polynomials (pp. 69–74)

Find $(x + 7)(x - 9)$.

$(x + 7)(x - 9) = x(x - 9) + 7(x - 9)$ Distribute $(x - 9)$ to each term of $(x + 7)$.
$= x(x) + x(-9) + 7(x) + 7(-9)$ Distributive Property
$= x^2 + (-9x) + 7x + (-63)$ Multiply.
$= x^2 - 2x - 63$ Combine like terms.

Find the product.

9. $(x + 6)(x - 4)$
10. $(y - 5)(3y + 8)$
11. $(x + 4)(x^2 + 7x)$
12. $(-3y + 1)(4y^2 - y - 7)$

2.3 Special Products of Polynomials (pp. 75–80)

Find each product.

a. $(6x + 4y)^2$

$(6x + 4y)^2 = (6x)^2 + 2(6x)(4y) + (4y)^2$ Square of a binomial pattern
$= 36x^2 + 48xy + 16y^2$ Simplify.

b. $(2x + 3y)(2x - 3y)$

$(2x + 3y)(2x - 3y) = (2x)^2 - (3y)^2$ Sum and difference pattern
$= 4x^2 - 9y^2$ Simplify.

Find the product.

13. $(x + 9)(x - 9)$
14. $(2y + 4)(2y - 4)$
15. $(p + 4)^2$
16. $(-1 + 2d)^2$

114   Chapter 2   Polynomial Equations and Factoring

2.4 Solving Polynomial Equations in Factored Form (pp. 81–86)

Solve $(x + 6)(x - 8) = 0$.

$(x + 6)(x - 8) = 0$ Write equation.

$x + 6 = 0$ or $x - 8 = 0$ Zero-Product Property

$x = -6$ or $x = 8$ Solve for x.

Solve the equation.

17. $x^2 + 5x = 0$ **18.** $(z + 3)(z - 7) = 0$ **19.** $(b + 13)^2 = 0$ **20.** $2y(y - 9)(y + 4) = 0$

2.5 Factoring $x^2 + bx + c$ (pp. 89–94)

Factor $x^2 + 6x - 27$.

Notice that $b = 6$ and $c = -27$. Because c is negative, the factors p and q must have different signs so that pq is negative.

Find two integer factors of -27 whose sum is 6.

Factors of −27	−27, 1	−1, 27	−9, 3	−3, 9
Sum of factors	−26	26	−6	6

The values of p and q are -3 and 9.

▶ So, $x^2 + 6x - 27 = (x - 3)(x + 9)$.

Factor the polynomial.

21. $p^2 + 2p - 35$ **22.** $b^2 + 18b + 80$ **23.** $z^2 - 4z - 21$ **24.** $x^2 - 11x + 28$

2.6 Factoring $ax^2 + bx + c$ (pp. 95–100)

Factor $5x^2 + 36x + 7$.

There is no GCF, so you need to consider the possible factors of a and c. Because b and c are both positive, the factors of c must be positive. Use a table to organize information about the factors of a and c.

Factors of 5	Factors of 7	Possible factorization	Middle term	
1, 5	1, 7	$(x + 1)(5x + 7)$	$7x + 5x = 12x$	✗
1, 5	7, 1	$(x + 7)(5x + 1)$	$x + 35x = 36x$	✓

▶ So, $5x^2 + 36x + 7 = (x + 7)(5x + 1)$.

Factor the polynomial.

25. $3t^2 + 16t - 12$ **26.** $-5y^2 - 22y - 8$ **27.** $6x^2 + 17x + 7$

28. $-2y^2 + 7y - 6$ **29.** $3z^2 + 26z - 9$ **30.** $10a^2 - 13a - 3$

ANSWERS

17. $x = 0, x = -5$
18. $z = -3, z = 7$
19. $b = -13$
20. $y = 0, y = 9, y = -4$
21. $(p + 7)(p - 5)$
22. $(b + 8)(b + 10)$
23. $(z + 3)(z - 7)$
24. $(x - 7)(x - 4)$
25. $(t + 6)(3t - 2)$
26. $-(y + 4)(5y + 2)$
27. $(2x + 1)(3x + 7)$
28. $-(y - 2)(2y - 3)$
29. $(z + 9)(3z - 1)$
30. $(2a - 3)(5a + 1)$

ANSWERS

31. $(x + 3)(x - 3)$
32. $(y + 10)(y - 10)$
33. $(z - 3)^2$
34. $(m + 8)^2$
35. $n(n + 3)(n - 3)$
36. $(x - 3)(x + 4a)$
37. $2x^2(x^2 + x - 10)$
38. $x = 0, x = 6, x = -3$
39. $x = -\frac{3}{2}, x = \frac{3}{2}$
40. $z = -3, z = 5, z = -5$
41. length: 12 ft, width: 4 ft, height: 2 ft

2.7 Factoring Special Products (pp. 101–106)

Factor each polynomial.

a. $x^2 - 16$

$\quad x^2 - 16 = x^2 - 4^2$ Write as $a^2 - b^2$.

$\qquad\qquad\ = (x + 4)(x - 4)$ Difference of two squares pattern

b. $25x^2 - 30x + 9$

$\quad 25x^2 - 30x + 9 = (5x)^2 - 2(5x)(3) + 3^2$ Write as $a^2 - 2ab + b^2$.

$\qquad\qquad\qquad\qquad\ = (5x - 3)^2$ Perfect square trinomial pattern

Factor the polynomial.

31. $x^2 - 9$ **32.** $y^2 - 100$ **33.** $z^2 - 6z + 9$ **34.** $m^2 + 16m + 64$

2.8 Factoring Polynomials Completely (pp. 107–112)

Factor each polynomial completely.

a. $x^3 + 4x^2 - 3x - 12$

$\quad x^3 + 4x^2 - 3x - 12 = (x^3 + 4x^2) + (-3x - 12)$ Group terms with common factors.

$\qquad\qquad\qquad\qquad\ = x^2(x + 4) + (-3)(x + 4)$ Factor out GCF of each pair of terms.

$\qquad\qquad\qquad\qquad\ = (x + 4)(x^2 - 3)$ Factor out $(x + 4)$.

b. $2x^4 - 8x^2$

$\quad 2x^4 - 8x^2 = 2x^2(x^2 - 4)$ Factor out $2x^2$.

$\qquad\qquad\ = 2x^2(x^2 - 2^2)$ Write as $a^2 - b^2$.

$\qquad\qquad\ = 2x^2(x + 2)(x - 2)$ Difference of two squares pattern

c. $2x^3 + 18x^2 - 72x$

$\quad 2x^3 + 18x^2 - 72x = 2x(x^2 + 9x - 36)$ Factor out $2x$.

$\qquad\qquad\qquad\qquad\ = 2x(x + 12)(x - 3)$ Factor $x^2 + 9x - 36$.

Factor the polynomial completely.

35. $n^3 - 9n$ **36.** $x^2 - 3x + 4ax - 12a$ **37.** $2x^4 + 2x^3 - 20x^2$

Solve the equation.

38. $3x^3 - 9x^2 - 54x = 0$ **39.** $16x^2 - 36 = 0$ **40.** $z^3 + 3z^2 - 25z - 75 = 0$

41. A box in the shape of a rectangular prism has a volume of 96 cubic feet. The box has a length of $(x + 8)$ feet, a width of x feet, and a height of $(x - 2)$ feet. Find the dimensions of the box.

2 Chapter Test

Find the sum or difference. Then identify the degree of the sum or difference and classify it by the number of terms.

1. $(-2p + 4) - (p^2 - 6p + 8)$
2. $(9c^6 - 5b^4) - (4c^6 - 5b^4)$
3. $(4s^4 + 2st + t) + (2s^4 - 2st - 4t)$

Find the product.

4. $(h - 5)(h - 8)$
5. $(2w - 3)(3w + 5)$
6. $(z + 11)(z - 11)$

7. Explain how you can determine whether a polynomial is a perfect square trinomial.

8. Is 18 a polynomial? Explain your reasoning.

Factor the polynomial completely.

9. $s^2 - 15s + 50$
10. $h^3 + 2h^2 - 9h - 18$
11. $-5k^2 - 22k + 15$

Solve the equation.

12. $(n - 1)(n + 6)(n + 5) = 0$
13. $d^2 + 14d + 49 = 0$
14. $6x^4 + 8x^2 = 26x^3$

15. The expression $\pi(r - 3)^2$ represents the area covered by the hour hand on a clock in one rotation, where r is the radius of the entire clock. Write a polynomial in standard form that represents the area covered by the hour hand in one rotation.

16. A magician's stage has a trapdoor.
    a. The total area (in square feet) of the stage can be represented by $x^2 + 27x + 176$. Write an expression for the width of the stage.
    b. Write an expression for the perimeter of the stage.
    c. The area of the trapdoor is 10 square feet. Find the value of x.
    d. The magician wishes to have the area of the stage be at least 20 times the area of the trapdoor. Does this stage satisfy his requirement? Explain.

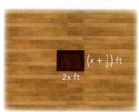
$(x + \frac{1}{2})$ ft
$2x$ ft
$(x + 16)$ ft

17. Write a polynomial equation in factored form that has three positive roots.

18. You are jumping on a trampoline. For one jump, your height y (in feet) above the trampoline after t seconds can be represented by $y = -16t^2 + 24t$. How many seconds are you in the air?

19. A cardboard box in the shape of a rectangular prism has the dimensions shown.
    a. Write a polynomial that represents the volume of the box.
    b. The volume of the box is 60 cubic inches. What are the length, width, and height of the box?

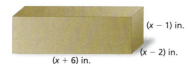
$(x - 1)$ in.
$(x - 2)$ in.
$(x + 6)$ in.

ANSWERS

1. $-p^2 + 4p - 4$; 2; trinomial
2. $5c^6$; 6; monomial
3. $6s^4 - 3t$; 4; binomial
4. $h^2 - 13h + 40$
5. $6w^2 + w - 15$
6. $z^2 - 121$
7. Check the first and last terms to verify they are perfect squares. If they are, find 2 times the product of their square roots. If this matches the middle term, or the opposite of the middle term, it is a perfect square trinomial.
8. yes; A monomial is a type of polynomial, and a number is a type of monomial.
9. $(s - 5)(s - 10)$
10. $(h + 2)(h + 3)(h - 3)$
11. $-(5k - 3)(k + 5)$
12. $n = 1, n = -6, n = -5$
13. $d = -7$
14. $x = 0, x = \frac{1}{3}, x = 4$
15. $\pi r^2 - 6\pi r + 9\pi$
16. a. $(x + 11)$ ft
    b. $(4x + 54)$ ft
    c. $x = 2$
    d. yes; The area of the stage is 234 square feet, and twenty times the area of the trap door is 200 square feet, so the area of the stage is greater than twenty times the area of the trap door.
17. Sample answer: $(x - 1)(x - 2)(x - 3) = 0$
18. 1.5 sec
19. a. $(x^3 + 3x^2 - 16x + 12)$ in.^{3}
    b. length: 10 in., width: 2 in., height: 3 in.

ANSWERS

1. a. monomial
   b. binomial
   c. trinomial
   d. binomial
   e. binomial
   f. trinomial
   c, a, d, b, f, e

2. D

3. $-4; 6$

4. a. $y = 65|x - 4|, x \geq 0$

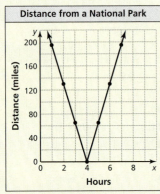

   b. after 4 hours; when $x = 4$, $y = 0$
   c. $y = \begin{cases} -65x + 260, & \text{if } 0 \leq x \leq 4 \\ 65x - 260, & \text{if } x > 4 \end{cases}$

5. f; The slope of f is 2 and the slope of the inverse of f is $\frac{1}{2}$.

2 Cumulative Assessment

1. Classify each polynomial by the number of terms. Then order the polynomials by degree from least to greatest.

   a. $-4x^3$
   b. $6y - 3y^5$
   c. $c^2 + 2 + c$
   d. $-10d^4 + 7d^2$
   e. $-5z^{11} + 8z^{12}$
   f. $3b^6 - 12b^8 + 4b^4$

2. Which exponential function has the greatest percent rate of change?

   Ⓐ $f(x) = 4(2.5)^{x/3}$

   Ⓑ
x	0	1	2	3	4
g(x)	8	12	18	27	40.5

   Ⓒ (graph of h)

   Ⓓ An exponential function j models a relationship in which the dependent variable is multiplied by 6 for every 1 unit the independent variable increases. The value of the function at 0 is 2.

3. Find the x-coordinates of the points where the graph of $f(x) = x^2 - 2x - 24$ crosses the x-axis.

   -24 -6 -4 -2 -1 0 1 2 4 6 24

4. The table shows your distances from a national park over time.

   a. Write and graph an absolute value function that represents the distance as a function of the number of hours.
   b. When do you reach the national park? Explain.
   c. Write the function in part (a) as a piecewise function.

Hours, x	Distance (miles), y
1	195
2	130
3	65
4	0
5	65
6	130
7	195

5. Which function has a steeper graph, $f(x) = 2x - 3$ or the inverse of f? Explain your reasoning.

6. Which expressions are equivalent to $-2x + 15x^2 - 8$?

$15x^2 - 2x - 8$	$(5x + 4)(3x + 2)$
$(5x - 4)(3x + 2)$	$15x^2 + 2x - 8$
$(3x - 2)(5x - 4)$	$(3x + 2)(5x - 4)$

7. The graph represents an exponential function.

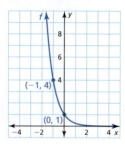

   a. Write the function.
   b. Find $f\left(\tfrac{1}{2}\right)$ and $f\left(-\tfrac{5}{2}\right)$.

8. Which polynomial represents the product of $3x + 1$ and $4x - 1$?

   Ⓐ $7x - 1$

   Ⓑ $12x^2 - 1$

   Ⓒ $12x^2 - x - 1$

   Ⓓ $12x^2 + x - 1$

9. You are playing miniature golf on the hole shown.
   a. Write a polynomial that represents the area of the golf hole.
   b. Write a polynomial that represents the perimeter of the golf hole.
   c. Find the perimeter of the golf hole when the area is 216 square feet.

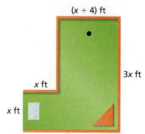

Chapter 3 Pacing Guide	
Chapter Opener/ Mathematical Practices	0.5 Day
Section 1	1.5 Days
Section 2	2 Days
Section 3	2 Days
Quiz	0.5 Day
Section 4	1.5 Days
Section 5	2 Days
Section 6	2 Days
Section 7	2 Days
Chapter Review/ Chapter Tests	2 Days
Total Chapter 3	16 Days
Year-to-Date	47 Days

3 Graphing Quadratic Functions

- **3.1** Graphing $f(x) = ax^2$
- **3.2** Graphing $f(x) = ax^2 + c$
- **3.3** Graphing $f(x) = ax^2 + bx + c$
- **3.4** Graphing $f(x) = a(x - h)^2 + k$
- **3.5** Graphing $f(x) = a(x - p)(x - q)$
- **3.6** Focus of a Parabola
- **3.7** Comparing Linear, Exponential, and Quadratic Functions

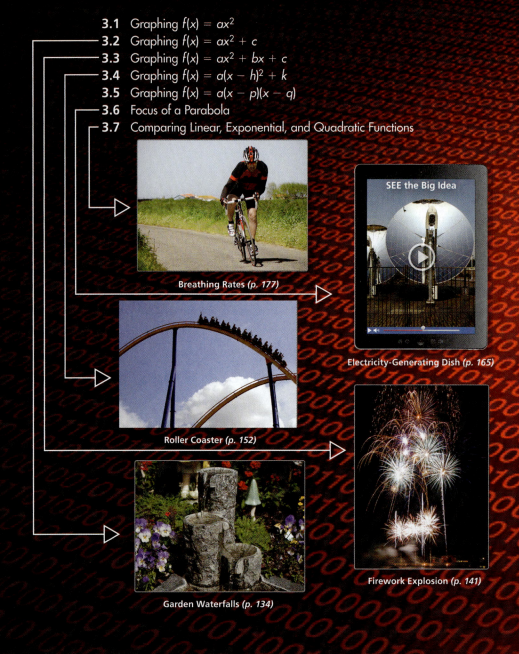

Breathing Rates (p. 177)

Electricity-Generating Dish (p. 165)

Roller Coaster (p. 152)

Firework Explosion (p. 141)

Garden Waterfalls (p. 134)

120   Chapter 3

Laurie's Notes

Chapter Summary

- This is the second of three chapters generally about polynomials and more specifically about quadratics. In the last chapter, students factored quadratics and used the Zero-Product Property to solve quadratics.
- The first five lessons of the chapter are about graphing quadratic functions. Connections to transformations of functions are made, just as students had seen with linear functions in Math I. In addition to graphing, the concept of zeros of functions is introduced and connected to the *x*-intercept of a graph.
- There are three forms of quadratic equations: standard, vertex, and intercept. Students should become familiar with what the parameters of each equation tell about the graph of the function.
- The graph of a quadratic function is a parabola which can be defined in terms of a fixed point, the focus, and a fixed line, the directrix. When students study conic sections in a later mathematics course, the parabola will be defined in this manner.
- The final lesson in the chapter compares the behavior of linear, exponential, and quadratic functions. This is done by looking at the data numerically and graphically, a theme emphasized throughout the chapter.

What Your Students Have Learned

Middle School
- Graph linear equations written in standard form and slope-intercept form.
- Write functions to model linear relationships, and determine and interpret rates of change.
- Compare linear and nonlinear functions represented by tables, graphs, and equations.

Math I
- Translate, reflect, shrink, and stretch graphs of linear functions, and combine transformations of graphs of linear functions.
- Identify, evaluate, and graph exponential functions and compare the graphs of exponential functions.
- Determine whether data can be modeled with linear or exponential functions and compare functions using average rates of change.
- Write terms and rules of recursively defined sequences and translate between recursive rules and explicit rules.

What Your Students Will Learn

Math II
- Graph, write, and use quadratic functions in standard form, vertex form, and intercept form.
- Find the minimum and maximum values of quadratic functions.
- Identify even and odd functions algebraically and graphically.
- Use the intercept form of quadratic functions to find the zeros of the functions.
- Define a parabola as the set of all points (*x*, *y*) in a plane that are equidistant from a fixed point called the focus and a fixed line called the directrix.
- Write equations of parabolas with a vertical axis of symmetry or with a horizontal axis of symmetry.
- Write quadratic functions to model data and write a recursive rule for a quadratic function.
- Compare linear, exponential, or quadratic functions using average rates of change.

Dynamic Teaching Tools
Dynamic Assessment & Progress Monitoring Tool
Lesson Planning Tool
Interactive Whiteboard Lesson Library
Dynamic Classroom with Dynamic Investigations
Real-Life STEM Videos

Scaffolding in the Classroom
How do you solve that?
Verbally state the thought processes you use to solve a problem. What questions do you ask yourself? Help the students develop their own set of questions.

Questioning in the Classroom
All students need to participate. When a small group of students continually answers the questions in your classroom, no one else has to think. Try asking a different student whether they agree with the answer given, and then follow up with why or why not.

Laurie's Notes

Maintaining Mathematical Proficiency

Graphing Linear Equations

- Remind students that the graph of a linear equation shows the relationship between two variables in a coordinate plane.
- Review how to use the given equation to make a table of values, then plot the points and draw a line through them.

COMMON ERROR Students may find the coordinates of only two points on the line, and if one contains an error, will graph the line incorrectly. Remind students to use at least three points.

Evaluating Expressions

- Remind students to substitute the given value of x in the expression first, and then simplify the expression.
- Review how to use the order of operations to simplify expressions involving exponents.

COMMON ERROR Students may use the order of operations incorrectly.

Mathematical Practices (continued on page 122)

- The *Mathematical Practices* page focuses attention on how mathematics is learned—process versus content. This page demonstrates the need for students to try simpler cases of a problem—in this case, try graphing the parent function first. Just as with linear and exponential functions, student will see that when graphing quadratics, they should consider how the parent function has been transformed.
- Use the *Mathematical Practices* page to help students develop mathematical habits of mind—how mathematics can be explored and how mathematics is thought about.

If students need help...	*If students got it...*
Student Journal • Maintaining Mathematical Proficiency	Game Closet at *BigIdeasMath.com*
Lesson Tutorials	Start the *next* Section
Skills Review Handbook	

Maintaining Mathematical Proficiency

Graphing Linear Equations

Example 1 Graph $y = -x - 1$.

Step 1 Make a table of values.

x	y = -x - 1	y	(x, y)
-1	y = -(-1) - 1	0	(-1, 0)
0	y = -(0) - 1	-1	(0, -1)
1	y = -(1) - 1	-2	(1, -2)
2	y = -(2) - 1	-3	(2, -3)

Step 2 Plot the ordered pairs.

Step 3 Draw a line through the points.

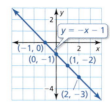

Graph the linear equation.

1. $y = 2x - 3$
2. $y = -3x + 4$
3. $y = -\frac{1}{2}x - 2$
4. $y = x + 5$

Evaluating Expressions

Example 2 Evaluate $2x^2 + 3x - 5$ when $x = -1$.

$2x^2 + 3x - 5 = 2(-1)^2 + 3(-1) - 5$ Substitute -1 for x.
$= 2(1) + 3(-1) - 5$ Evaluate the power.
$= 2 - 3 - 5$ Multiply.
$= -6$ Subtract.

Evaluate the expression when $x = -2$.

5. $5x^2 - 9$
6. $3x^2 + x - 2$
7. $-x^2 + 4x + 1$
8. $x^2 + 8x + 5$
9. $-2x^2 - 4x + 3$
10. $-4x^2 + 2x - 6$

11. **ABSTRACT REASONING** Complete the table. Find a pattern in the differences of consecutive y-values. Use the pattern to write an expression for y when $x = 6$.

x	1	2	3	4	5
y = ax²					

Dynamic Solutions available at *BigIdeasMath.com*

What Your Students Have Learned

- Graph linear equations written in slope-intercept form.
- Use Order of Operations to evaluate expressions, including those involving whole number exponents.

ANSWERS

1.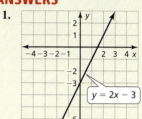

2.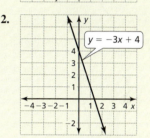

3.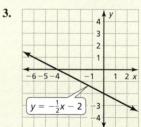

4.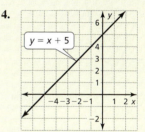

5. 11
6. 8
7. -11
8. -7
9. 3
10. -26
11. $a, 4a, 9a, 16a, 25a$; *Sample answer:* The coefficient of each difference is the next consecutive odd integer; $36a$

Vocabulary Review

Have students make Process Diagrams for the following topics.
- Graphing linear equations
- Evaluating algebraic expressions

MONITORING PROGRESS ANSWERS

1.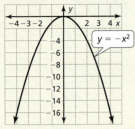

   The graph opens down and the highest point is at the origin.

2.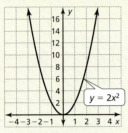

   The graph opens up and the lowest point is at the origin.

3.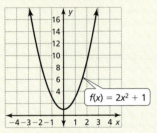

   The graph opens up and the lowest point is at (0, 1).

4.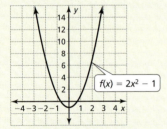

   The graph opens up and the lowest point is at (0, −1).

5.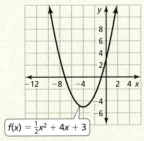

   The graph opens up and the lowest point is at (−4, −5).

6–9. See Additional Answers.

122   Chapter 3

Mathematical Practices

Mathematically proficient students try special cases of the original problem to gain insight into its solution.

Problem-Solving Strategies

Core Concept

Trying Special Cases

When solving a problem in mathematics, it can be helpful to try special cases of the original problem. For instance, in this chapter, you will learn to graph a quadratic function of the form $f(x) = ax^2 + bx + c$. The problem-solving strategy used is to first graph quadratic functions of the form $f(x) = ax^2$. From there, you progress to other forms of quadratic functions.

$f(x) = ax^2$	Section 3.1
$f(x) = ax^2 + c$	Section 3.2
$f(x) = ax^2 + bx + c$	Section 3.3
$f(x) = a(x - h)^2 + k$	Section 3.4
$f(x) = a(x - p)(x - q)$	Section 3.5

EXAMPLE 1   Graphing the Parent Quadratic Function

Graph the parent quadratic function $y = x^2$. Then describe its graph.

SOLUTION

The function is of the form $y = ax^2$, where $a = 1$. By plotting several points, you can see that the graph is U-shaped, as shown.

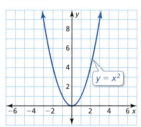

▶ The graph opens up, and the lowest point is at the origin.

Monitoring Progress

Graph the quadratic function. Then describe its graph.

1. $y = -x^2$
2. $y = 2x^2$
3. $f(x) = 2x^2 + 1$
4. $f(x) = 2x^2 - 1$
5. $f(x) = \frac{1}{2}x^2 + 4x + 3$
6. $f(x) = \frac{1}{2}x^2 - 4x + 3$
7. $y = -2(x + 1)^2 + 1$
8. $y = -2(x - 1)^2 + 1$

9. How are the graphs in Monitoring Progress Questions 1–8 similar? How are they different?

122   Chapter 3   Graphing Quadratic Functions

Laurie's Notes   Mathematical Practices (continued from page T-121)

- Transformations of functions will be a focus in this chapter. The basic quadratic $y = x^2$ will be shifted vertically and horizontally. It will be stretched and compressed vertically and horizontally.
- Give time for students to work through the questions in the *Monitoring Progress*, and then discuss as a class. Students should observe that the parent function has been shifted or reflected, but essentially students will describe seeing a U-shaped graph.

Laurie's Notes

Overview of Section 3.1

Introduction
- The lesson begins by describing the features of a *quadratic function*. The graph is called a *parabola*, and it has a *vertex* and an *axis of symmetry*. Unlike a linear function, which is increasing, decreasing, or constant, a quadratic function changes from a decreasing function to an increasing function or vice versa.
- Students are familiar with transformations from their work with linear functions. In this lesson, vertical stretches and shrinks of the parent function $y = x^2$ are performed, along with reflections in the *x*-axis. The connection for students is the effect of *m* on the graph of $y = mx$.

Resources
- Use dynamic geometry software to investigate the effect of *a* on the graph of $y = ax^2$. A slider can be used to generate values of *a*, $-5 < a < 5$.
- In lieu of creating your own, you can find online versions to download.

Teaching Strategy
- **Rule of 3:** Throughout this chapter, students will be graphing quadratic functions. Calculators can help students develop a stronger understanding of the behavior of quadratic functions, particularly understanding of transformations of these functions. This occurs by seeing the three representations of the function—symbolic, graphical, and numerical. The table of values can be particularly helpful in making sense of increasing and decreasing functions.
- In this example, the value of the coefficient is 2, meaning the parent function is stretched vertically. Graphically, you can see that the *y*-value of one is twice the *y*-value of the other. In the table, the values in the Y_2 column are twice the values in the Y_1 column for each value of *x*.

 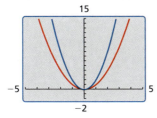 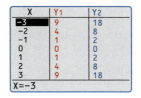

      Symbolic   Graphical   Numerical

- Beginning in this lesson and throughout this chapter, look for opportunities to integrate this *Rule of 3* in your teaching.

Pacing Suggestion
- The exploration helps students discover the effect of the value of *a* on the graph of $y = ax^2$. You may choose to omit a few examples and check for understanding with the *Monitoring Progress* questions.

Dynamic Teaching Tools
- Dynamic Assessment & Progress Monitoring Tool
- Lesson Planning Tool
- Interactive Whiteboard Lesson Library
- Dynamic Classroom with Dynamic Investigations

What Your Students Will Learn

- Identify characteristics of the graphs of quadratic functions such as the vertex, axis of symmetry, and behavior of the graph.
- Graph and use quadratic functions of the form $f(x) = ax^2$ when $a > 0$ and when $a < 0$.

Laurie's Notes

Exploration

Motivate
- Use a flashlight to cast a shadow that students can describe. If you do not have a flashlight, use something that has a parabolic cross section, such as a bulb in an overhead projector. Tell students that the graphs they will be sketching represent a common shape.

Exploration 1
- Having the graph of $y = x^2$ already on the grid may help students compare the graph of $y = ax^2$ to the graph of $y = x^2$.
- There are two features that the value of a affects, and you may need to prod students' thinking as you circulate. The obvious effect is whether the graph opens up or down. The feature that is more difficult for students to explain is the vertical shrink or stretch. Informally, this can be referred to as "more bowl-shaped" or "narrower."
- Students often say that when $a < 1$, the graphs become more bowl-shaped. Remind students that $-5 < 1$. What they really are trying to describe are rational numbers between -1 and 1, or $|a| < 1$.
- If a student fails to notice the shape of the graph, give the student additional functions to graph.
- **?** "Is the graph nonlinear when $a = 0$? Explain." no; When $a = 0$, the function becomes the constant function $y = 0$.
- Discuss as a class when all have finished.
- **?** "Describe the graph when $a < 0$." opens downward "Describe the graph when $|a| > 1$." narrower "Describe the graph when $|a| < 1$." more bowl-shaped
- If time allows, use a graphing calculator to check conjectures. You can also use the calculator to graph several functions and ask for reasonable equations for each graph.

Communicate Your Answer
- Have students summarize their conjectures in a table.

Connecting to Next Step
- The exploration allows students to discover the effect of the value of a on the graph of $f(x) = ax^2$. In the formal lesson, students will graph additional functions of this type.

3.1 Graphing $f(x) = ax^2$

Essential Question What are some of the characteristics of the graph of a quadratic function of the form $f(x) = ax^2$?

EXPLORATION 1 Graphing Quadratic Functions

Work with a partner. Graph each quadratic function. Compare each graph to the graph of $f(x) = x^2$.

a. $g(x) = 3x^2$

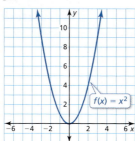

b. $g(x) = -5x^2$

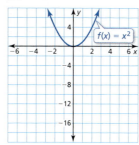

c. $g(x) = -0.2x^2$

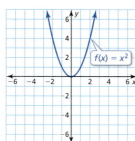

d. $g(x) = \frac{1}{10}x^2$

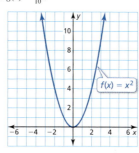

REASONING QUANTITATIVELY
To be proficient in math, you need to make sense of quantities and their relationships in problem situations.

Communicate Your Answer

2. What are some of the characteristics of the graph of a quadratic function of the form $f(x) = ax^2$?

3. How does the value of a affect the graph of $f(x) = ax^2$? Consider $0 < a < 1$, $a > 1$, $-1 < a < 0$, and $a < -1$. Use a graphing calculator to verify your answers.

4. The figure shows the graph of a quadratic function of the form $y = ax^2$. Which of the intervals in Question 3 describes the value of a? Explain your reasoning.

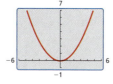

Section 3.1 Graphing $f(x) = ax^2$ 123

2. *Sample answer:* They are U-shaped and symmetric. They either open up with the lowest point at the origin, or they open down with the highest point at the origin.

3. *Sample answer:* When $0 < a < 1$ the graph of $f(x) = ax^2$ is a vertical shrink of the graph of $f(x) = x^2$, when $a > 1$ the graph of $f(x) = ax^2$ is a vertical stretch of the graph of $f(x) = x^2$, when $-1 < a < 0$ the graph of $f(x) = ax^2$ is a vertical shrink and a reflection in the x-axis of the graph of $f(x) = x^2$, and when $a < -1$ the graph of $f(x) = ax^2$ is a vertical stretch and a reflection in the x-axis of the graph of $f(x) = x^2$.

4. $0 < a < 1$; *Sample answer:* The graph is a vertical shrink of the graph of $f(x) = x^2$.

Dynamic Teaching Tools
Dynamic Assessment & Progress Monitoring Tool
Lesson Planning Tool
Interactive Whiteboard Lesson Library
Dynamic Classroom with Dynamic Investigations

ANSWERS

1. **a.**

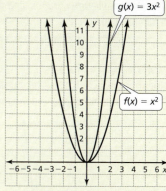

The graph of g is a vertical stretch by a factor of 3 of the graph of f.

b.

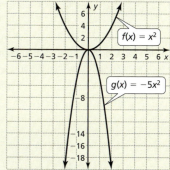

The graph of g is a vertical stretch by a factor of 5 and a reflection in the x-axis of the graph of f.

c.

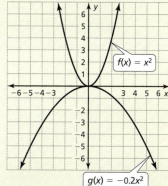

The graph of g is a vertical shrink by a factor of 0.2 and a reflection in the x-axis of the graph of f.

d. See Additional Answers.

Differentiated Instruction

Kinesthetic/Visual

Some students may benefit by having access to graphing calculators or graphing software programs available on many computers. Students can enter the quadratic function $f(x) = x^2$ and the given quadratic equation, view the graphs, and then compare the graphs.

Extra Example 1

Identify characteristics of the quadratic function and its graph.

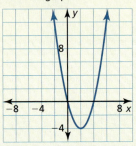

The vertex is $(2, -4)$, the axis of symmetry is $x = 2$, the domain is all real numbers, and the range is $y \geq -4$. When $x < 2$, y increases as x decreases, and when $x > 2$, y increases as x increases.

3.1 Lesson

What You Will Learn

- Identify characteristics of quadratic functions.
- Graph and use quadratic functions of the form $f(x) = ax^2$.

Core Vocabulary

quadratic function, p. 124
parabola, p. 124
vertex, p. 124
axis of symmetry, p. 124

Previous
domain
range
vertical shrink
vertical stretch
reflection

REMEMBER
The notation $f(x)$ is another name for y.

Identifying Characteristics of Quadratic Functions

A **quadratic function** is a nonlinear function that can be written in the standard form $y = ax^2 + bx + c$, where $a \neq 0$. The U-shaped graph of a quadratic function is called a **parabola**. In this lesson, you will graph quadratic functions, where b and c equal 0.

Core Concept

Characteristics of Quadratic Functions

The *parent quadratic function* is $f(x) = x^2$. The graphs of all other quadratic functions are *transformations* of the graph of the parent quadratic function.

The lowest point on a parabola that opens up or the highest point on a parabola that opens down is the **vertex**. The vertex of the graph of $f(x) = x^2$ is $(0, 0)$.

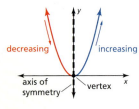

The vertical line that divides the parabola into two symmetric parts is the **axis of symmetry**. The axis of symmetry passes through the vertex. For the graph of $f(x) = x^2$, the axis of symmetry is the y-axis, or $x = 0$.

EXAMPLE 1 Identifying Characteristics of a Quadratic Function

Consider the graph of the quadratic function.

Using the graph, you can identify characteristics such as the vertex, axis of symmetry, and the behavior of the graph, as shown.

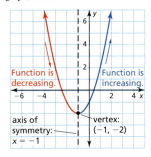

You can also determine the following:
- The domain is all real numbers.
- The range is all real numbers greater than or equal to -2.
- When $x < -1$, y increases as x decreases.
- When $x > -1$, y increases as x increases.

Laurie's Notes Teacher Actions

- **Look For and Make Use of Structure:** Define *quadratic function* in standard form. Explain that a, b, and c are parameters just as m and b are parameters in $y = mx + b$.
- Discuss the red and blue portions of the graph. Graphs are read left to right. So when the phrase "increasing function" is used, the values of the function (y-values) are increasing. When the phrase "decreasing function" is used, the values of the function are decreasing.
- In Example 1, help students understand that the range is $y \geq -2$ by moving your hand on the graph from $(0, -2)$ in an upward motion.

Monitoring Progress Help in English and Spanish at BigIdeasMath.com

Identify characteristics of the quadratic function and its graph.

1.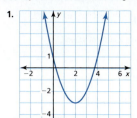

2.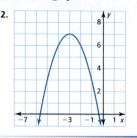

REMEMBER

The graph of $y = a \cdot f(x)$ is a vertical stretch or shrink by a factor of a of the graph of $y = f(x)$.

The graph of $y = -f(x)$ is a reflection in the x-axis of the graph of $y = f(x)$.

Graphing and Using $f(x) = ax^2$

Core Concept

Graphing $f(x) = ax^2$ When $a > 0$

- When $0 < a < 1$, the graph of $f(x) = ax^2$ is a vertical shrink of the graph of $f(x) = x^2$.
- When $a > 1$, the graph of $f(x) = ax^2$ is a vertical stretch of the graph of $f(x) = x^2$.

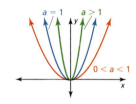

Graphing $f(x) = ax^2$ When $a < 0$

- When $-1 < a < 0$, the graph of $f(x) = ax^2$ is a vertical shrink with a reflection in the x-axis of the graph of $f(x) = x^2$.
- When $a < -1$, the graph of $f(x) = ax^2$ is a vertical stretch with a reflection in the x-axis of the graph of $f(x) = x^2$.

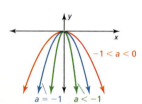

EXAMPLE 2 Graphing $y = ax^2$ When $a > 0$

Graph $g(x) = 2x^2$. Compare the graph to the graph of $f(x) = x^2$.

SOLUTION

Step 1 Make a table of values.

x	−2	−1	0	1	2
g(x)	8	2	0	2	8

Step 2 Plot the ordered pairs.

Step 3 Draw a smooth curve through the points.

▶ Both graphs open up and have the same vertex, (0, 0), and the same axis of symmetry, $x = 0$. The graph of g is narrower than the graph of f because the graph of g is a vertical stretch by a factor of 2 of the graph of f.

Section 3.1 Graphing $f(x) = ax^2$ 125

Extra Example 2

Graph $g(x) = 4x^2$. Compare the graph to the graph of $f(x) = x^2$.

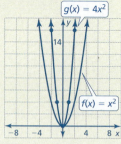

Both graphs open up and have the same vertex, (0, 0), and the same axis of symmetry, $x = 0$. The graph of g is a vertical stretch by a factor of 4 of the graph of f.

MONITORING PROGRESS ANSWERS

1. The vertex is $(2, -3)$. The axis of symmetry is $x = 2$. The domain is all real numbers. The range is $y \geq -3$. When $x < 2$, y increases as x decreases. When $x > 2$, y increases as x increases.

2. The vertex is $(-3, 7)$. The axis of symmetry is $x = -3$. The domain is all real numbers. The range is $y \leq 7$. When $x < -3$, y increases as x increases. When $x > -3$, y increases as x decreases.

Laurie's Notes — Teacher Actions

- Relate the *Core Concept* to the exploration. Use colors to distinguish $0 < |a| < 1$ and $|a| > 1$ from $|a| = 1$. Relate to transformations: the vertical stretch ($|a| > 1$) that students may call "narrower" and vertical shrink $0 < |a| < 1$ that they may call "more bowl-shaped."

COMMON ERROR Students multiply x by 2 and then square it instead of squaring x and then multiplying by 2. Recall order of operations!

- **Connection:** The y-values in the table of values decrease and then increase. Refer to the graph and point out the same feature.

Extra Example 3

Graph $h(x) = -\frac{2}{9}x^2$. Compare the graph to the graph of $f(x) = x^2$.

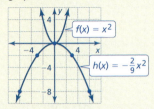

The graphs have the same vertex, (0, 0), and the same axis of symmetry, $x = 0$. The graph of h opens down and is wider than the graph of f. So, the graph of h is a vertical shrink by a factor of $\frac{2}{9}$ and a reflection in the x-axis of the graph of f.

Extra Example 4

The diagram shows the cross section of a parabolic mirror, where x and y are measured in feet. Find the width and the height of the mirror.

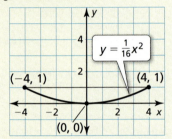

The mirror is 8 feet wide and 1 foot tall.

MONITORING PROGRESS ANSWERS

3–8. See Additional Answers.

9. 4 inches wide, 2 inches deep

STUDY TIP
To make the calculations easier, choose x-values that are multiples of 3.

EXAMPLE 3 Graphing $y = ax^2$ When $a < 0$

Graph $h(x) = -\frac{1}{3}x^2$. Compare the graph to the graph of $f(x) = x^2$.

SOLUTION

Step 1 Make a table of values.

x	−6	−3	0	3	6
h(x)	−12	−3	0	−3	−12

Step 2 Plot the ordered pairs.

Step 3 Draw a smooth curve through the points.

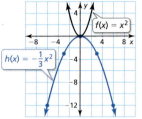

The graphs have the same vertex, (0, 0), and the same axis of symmetry, $x = 0$, but the graph of h opens down and is wider than the graph of f. So, the graph of h is a vertical shrink by a factor of $\frac{1}{3}$ and a reflection in the x-axis of the graph of f.

Monitoring Progress Help in English and Spanish at BigIdeasMath.com

Graph the function. Compare the graph to the graph of $f(x) = x^2$.

3. $g(x) = 5x^2$
4. $h(x) = \frac{1}{3}x^2$
5. $n(x) = \frac{3}{2}x^2$
6. $p(x) = -3x^2$
7. $q(x) = -0.1x^2$
8. $g(x) = -\frac{1}{4}x^2$

EXAMPLE 4 Solving a Real-Life Problem

The diagram at the left shows the cross section of a satellite dish, where x and y are measured in meters. Find the width and depth of the dish.

SOLUTION

Use the domain of the function to find the width of the dish. Use the range to find the depth.

The leftmost point on the graph is (−2, 1), and the rightmost point is (2, 1). So, the domain is $-2 \leq x \leq 2$, which represents 4 meters.

The lowest point on the graph is (0, 0), and the highest points on the graph are (−2, 1) and (2, 1). So, the range is $0 \leq y \leq 1$, which represents 1 meter.

So, the satellite dish is 4 meters wide and 1 meter deep.

Monitoring Progress Help in English and Spanish at BigIdeasMath.com

9. The cross section of a spotlight can be modeled by the graph of $y = 0.5x^2$, where x and y are measured in inches and $-2 \leq x \leq 2$. Find the width and depth of the spotlight.

Laurie's Notes — Teacher Actions

- ❓ "When $a = -\frac{1}{3}$, what do you know?" The graph opens down because $a < 0$. It is a vertical shrink (more bowl-shaped) because $0 < |a| < 1$.
- The parabolic shape is one that students have likely seen. The cross section of a satellite dish is parabolic.
- In Example 4, connect the domain and range to vertical and horizontal distances to finish the problem.

Closure

- **Exit Ticket:** Describe the differences between the graphs of $y = -3x^2$ and $y = \frac{1}{3}x^2$. The graph of $y = -3x^2$ opens down and the graph of $y = \frac{1}{3}x^2$ opens up. The graph of $y = -3x^2$ is narrower than the graph of $y = \frac{1}{3}x^2$. The range of $y = -3x^2$ is $y \leq 0$. The range of $y = \frac{1}{3}x^2$ is $y \geq 0$.

3.1 Exercises

Dynamic Solutions available at *BigIdeasMath.com*

Vocabulary and Core Concept Check

1. **VOCABULARY** What is the U-shaped graph of a quadratic function called?
2. **WRITING** When does the graph of a quadratic function open up? open down?

Monitoring Progress and Modeling with Mathematics

In Exercises 3 and 4, identify characteristics of the quadratic function and its graph. *(See Example 1.)*

3.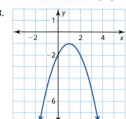

4.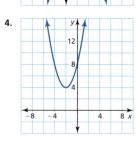

In Exercises 5–12, graph the function. Compare the graph to the graph of $f(x) = x^2$. *(See Examples 2 and 3.)*

5. $g(x) = 6x^2$
6. $b(x) = 2.5x^2$
7. $h(x) = \frac{1}{4}x^2$
8. $j(x) = 0.75x^2$
9. $m(x) = -2x^2$
10. $q(x) = -\frac{9}{2}x^2$
11. $k(x) = -0.2x^2$
12. $p(x) = -\frac{2}{3}x^2$

In Exercises 13–16, use a graphing calculator to graph the function. Compare the graph to the graph of $y = -4x^2$.

13. $y = 4x^2$
14. $y = -0.4x^2$
15. $y = -0.04x^2$
16. $y = -0.004x^2$

17. **ERROR ANALYSIS** Describe and correct the error in graphing and comparing $y = x^2$ and $y = 0.5x^2$.

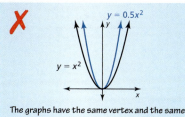

The graphs have the same vertex and the same axis of symmetry. The graph of $y = 0.5x^2$ is narrower than the graph of $y = x^2$.

18. **MODELING WITH MATHEMATICS** The arch support of a bridge can be modeled by $y = -0.0012x^2$, where x and y are measured in feet. Find the height and width of the arch. *(See Example 4.)*

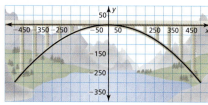

19. **PROBLEM SOLVING** The breaking strength z (in pounds) of a manila rope can be modeled by $z = 8900d^2$, where d is the diameter (in inches) of the rope.

    a. Describe the domain and range of the function.
    b. Graph the function using the domain in part (a).
    c. A manila rope has four times the breaking strength of another manila rope. Does the stronger rope have four times the diameter? Explain.

Section 3.1 Graphing $f(x) = ax^2$ 127

Assignment Guide and Homework Check

ASSIGNMENT

Basic: 1, 2, 3–17 odd, 18, 20–22, 31–35

Average: 1, 2–16 even, 17–23, 31–35

Advanced: 1, 2, 4, 8, 12–16 even, 17–23, 24–30 even, 31–35

HOMEWORK CHECK

Basic: 3, 5, 9, 18

Average: 4, 6, 12, 18

Advanced: 4, 8, 12, 18

ANSWERS

1. parabola
2. when $a > 0$; when $a < 0$
3. The vertex is $(1, -1)$. The axis of symmetry is $x = 1$. The domain is all real numbers. The range is $y \leq -1$. When $x < 1$, y increases as x increases. When $x > 1$, y increases as x decreases.
4. The vertex is $(-2, 4)$. The axis of symmetry is $x = -2$. The domain is all real numbers. The range is $y \geq 4$. When $x < -2$, y increases as x decreases. When $x > -2$, y increases as x increases.

5.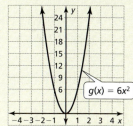

   The graph of g is a vertical stretch by a factor of 6 of the graph of f.

6.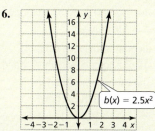

   The graph of b is a vertical stretch by a factor of 2.5 of the graph of f.

7.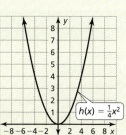

   The graph of h is a vertical shrink by a factor of $\frac{1}{4}$ of the graph of f.

8.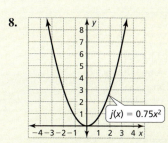

   The graph of j is a vertical shrink by a factor of 0.75 of the graph of f.

9–19. See Additional Answers.

Section 3.1 127

Dynamic Teaching Tools

Dynamic Assessment & Progress Monitoring Tool
Interactive Whiteboard Lesson Library
Dynamic Classroom with Dynamic Investigations

ANSWERS
20. a. $a > 1$
    b. $-1 < a < 0$
21. f is increasing when $x > 0$. g is increasing when $x < 0$.
22–35. See Additional Answers.

Mini-Assessment

1. Identify characteristics of the quadratic function and its graph.

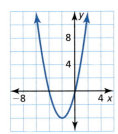

   See Additional Answers.

2. Graph $g(x) = 3.5x^2$. Compare the graph to the graph of $f(x) = x^2$.
   See Additional Answers.

3. Graph $n(x) = -\frac{3}{8}x^2$. Compare the graph to the graph of $f(x) = x^2$.

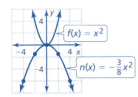

   The graph of n is a vertical shrink by a factor of $\frac{3}{8}$ and a reflection in the x-axis of the graph of f.

4. The diagram shows the cross section of a soup bowl, where x and y are measured in inches. Find the width and depth of the bowl.

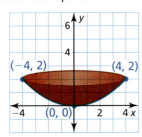

   8 inches wide, 2 inches deep

128   Chapter 3

---

20. **HOW DO YOU SEE IT?** Describe the possible values of a.

a.

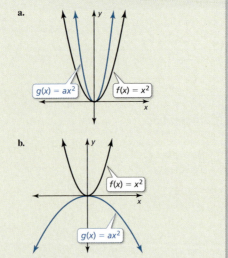

b.

ANALYZING GRAPHS In Exercises 21–23, use the graph.

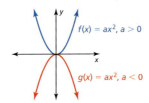

21. When is each function increasing?
22. When is each function decreasing?
23. Which function could include the point $(-2, 3)$? Find the value of a when the graph passes through $(-2, 3)$.
24. **REASONING** Is the x-intercept of the graph of $y = ax^2$ always 0? Justify your answer.
25. **REASONING** A parabola opens up and passes through $(-4, 2)$ and $(6, -3)$. How do you know that $(-4, 2)$ is not the vertex?

Maintaining Mathematical Proficiency *Reviewing what you learned in previous grades and lessons*

Evaluate the expression when $n = 3$ and $x = -2$. *(Skills Review Handbook)*

32. $n^2 + 5$ 33. $3x^2 - 9$ 34. $-4n^2 + 11$ 35. $n + 2x^2$

128   Chapter 3   Graphing Quadratic Functions

ABSTRACT REASONING In Exercises 26–29, determine whether the statement is *always*, *sometimes*, or *never* true. Explain your reasoning.

26. The graph of $f(x) = ax^2$ is narrower than the graph of $g(x) = x^2$ when $a > 0$.
27. The graph of $f(x) = ax^2$ is narrower than the graph of $g(x) = x^2$ when $|a| > 1$.
28. The graph of $f(x) = ax^2$ is wider than the graph of $g(x) = x^2$ when $0 < |a| < 1$.
29. The graph of $f(x) = ax^2$ is wider than the graph of $g(x) = dx^2$ when $|a| > |d|$.

30. **THOUGHT PROVOKING** Draw the isosceles triangle shown. Divide each leg into eight congruent segments. Connect the highest point of one leg with the lowest point of the other leg. Then connect the second highest point of one leg to the second lowest point of the other leg. Continue this process. Write a quadratic function whose graph models the shape that appears.

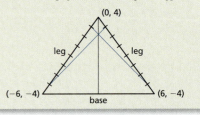

31. **MAKING AN ARGUMENT** The diagram shows the parabolic cross section of a swirling glass of water, where x and y are measured in centimeters.

   a. About how wide is the mouth of the glass?
   b. Your friend claims that the rotational speed of the water would have to increase for the cross section to be modeled by $y = 0.1x^2$. Is your friend correct? Explain your reasoning.

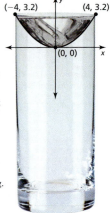

If students need help...	If students got it...
Resources by Chapter • Practice A and Practice B • Puzzle Time	Resources by Chapter • Enrichment and Extension • Cumulative Review
Student Journal • Practice	Start the *next* Section
Differentiating the Lesson Skills Review Handbook	

Laurie's Notes

Overview of Section 3.2

Introduction
- The lesson begins by describing the effect of c on the graph of $y = ax^2 + c$. The connection for students is the effect of b on the graph of $y = mx + b$.
- Students should recognize that vertical translations do not change the axis of symmetry, meaning $y = 3x^2$ and $y = 3x^2 - 4$ have the same axis of symmetry. Also, the vertex of the parabola remains on the y-axis.
- Also introduced in this lesson is the definition of a zero of a function. The zero of a function is where there is an x-intercept on the graph. The zeros of a function are found in a real-life application using the falling-objects model.

Resources
- Use dynamic geometry software to investigate the effect of c on the graph of $y = ax^2 + c$. A slider can be used to generate values of c, $-5 < c < 5$.
- In lieu of creating your own, you can find online versions to download.

Teaching Strategy
- **Rule of 3:** In this lesson, the *Rule of 3* strategy, connecting the symbolic, graphical, and numerical representation of functions, is still very important.
- In Example 3 in the lesson, students are given $f(x) = -0.5x^2 + 2$ and $g(x) = f(x) - 7$ and are asked to describe the transformation from the graph of f to the graph of g. In the equation editor of the calculator, it is possible to enter the functions as given. $Y_1 = f(x)$ and $Y_2 = g(x)$. Note in Figure 1 how the functions are entered.
- In Figure 2, the graph of Y_2 is a vertical shift of Y_1 down 7 units. In Figure 3, the table of values is shown, and students should recognize that subtracting 7 from each value in Y_1 gives the associated value in Y_2.

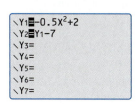

Figure 1

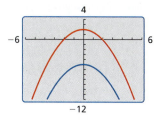

Figure 2

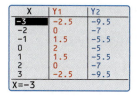

Figure 3

Pacing Suggestion
- The explorations help students discover the effect of the value of c on the graph of $y = ax^2 + c$. You may choose to omit a few examples and check for understanding with the *Monitoring Progress* questions.

Dynamic Teaching Tools
Dynamic Assessment & Progress Monitoring Tool
Lesson Planning Tool
Interactive Whiteboard Lesson Library
Dynamic Classroom with Dynamic Investigations

What Your Students Will Learn

- Graph quadratic functions of the form $f(x) = ax^2 + c$ by identifying the vertical translations of $|c|$ units up or down.
- Solve real-life problems involving the falling-objects model of the form $f(x) = ax^2 + c$.

Laurie's Notes

Exploration

Motivate

? "How many *y*-intercepts does a parabola have?" Students may say 1 or 0. A parabola always has one *y*-intercept, even if it cannot be seen in the standard viewing window of a graphing calculator.

? "How many *x*-intercepts does a parabola have?" Students will likely say 0, 1, or 2. Have them quickly sketch a parabola for each case.

For Your Information

- **Make Sense of Problems and Persevere in Solving Them:** Students should be comfortable with graphing $y = x^2$ and $y = ax^2$ and explaining the effect of *a* on the graph of $y = ax^2$. Today, they will make sense of another parameter when they graph $y = ax^2 + c$. Using tabular and graphical representations, they make sense of the effect of *c*.

Exploration 1

- Students should be comfortable graphing functions of the form $y = ax^2$ by making a table of values, plotting the ordered pairs, and drawing a smooth curve through the points.
- Students are graphing a pair of functions in the same coordinate plane.
- Discuss the characteristics of each graph, such as opens up or opens down and wider or narrower than $y = x^2$.
- Students may only notice that the vertex is no longer at (0, 0). They may not think of "*c*" as a vertical shift.
- **Connection:** Discuss the idea of the vertex being the *y*-intercept of the graph.

? "Could there be more than one *y*-intercept? Explain." no; It would not be a function.

COMMON ERROR Students do not always see that the pairs of graphs are similar, with one graph being a translation of the other. This is sometimes the case because different parts of each graph may be plotted. This may happen when using a graphing calculator with an inappropriate viewing window.

Exploration 2

? "What is an *x*-intercept?" It is where the graph crosses the *x*-axis.

? "Can a parabola have more than one *x*-intercept? Explain." yes; *Sample answer*: A parabola can open up and have a vertex below the *x*-axis, giving it two *x*-intercepts.

- Explain to students that in this exploration they will be looking for the *x*-intercept(s) of the graph.
- Again, students are graphing by first making a table of values. When making the table, students may identify an *x*-intercept.
- When students have finished both parts, ask for volunteers to share their graphs and explain how they found the *x*-intercepts. From the graph in part (a), they know that the roots occur between -3 and -2, and between 2 and 3. Students are familiar with the definition of square roots, so it should seem reasonable to them that solving $x^2 = 7$ will result in $x = \pm\sqrt{7}$.

Communicate Your Answer

- Make sure students are using appropriate viewing windows when verifying their conclusions.

Connecting to Next Step

- The exploration allows students to discover the effect of the value of *c* on the graph of $f(x) = ax^2 + c$. In the formal lesson, students will graph additional functions of this type.

3.2 Graphing $f(x) = ax^2 + c$

Essential Question How does the value of c affect the graph of $f(x) = ax^2 + c$?

EXPLORATION 1   Graphing $y = ax^2 + c$

Work with a partner. Sketch the graphs of the functions in the same coordinate plane. What do you notice?

a. $f(x) = x^2$ and $g(x) = x^2 + 2$

b. $f(x) = 2x^2$ and $g(x) = 2x^2 - 2$

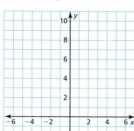
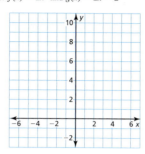

EXPLORATION 2   Finding x-Intercepts of Graphs

Work with a partner. Graph each function. Find the x-intercepts of the graph. Explain how you found the x-intercepts.

a. $y = x^2 - 7$

b. $y = -x^2 + 1$

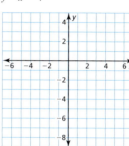
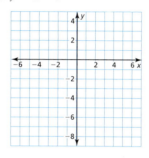

USING TOOLS STRATEGICALLY
To be proficient in math, you need to consider the available tools, such as a graphing calculator, when solving a mathematical problem.

Communicate Your Answer

3. How does the value of c affect the graph of $f(x) = ax^2 + c$?

4. Use a graphing calculator to verify your answers to Question 3.

5. The figure shows the graph of a quadratic function of the form $y = ax^2 + c$. Describe possible values of a and c. Explain your reasoning.

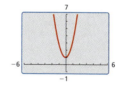

Section 3.2 Graphing $f(x) = ax^2 + c$ 129

Dynamic Teaching Tools

Dynamic Assessment & Progress Monitoring Tool
Lesson Planning Tool
Interactive Whiteboard Lesson Library
Dynamic Classroom with Dynamic Investigations

ANSWERS

1. a.

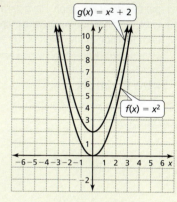

b.

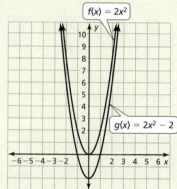

Sample answer: In $g(x) = ax^2 + c$, c causes a vertical shift in the graph of $f(x) = ax^2$.

2. a.

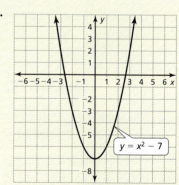

$x = \sqrt{7} \approx 2.6$, $x = -\sqrt{7} \approx -2.6$

2. b.

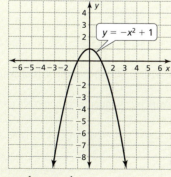

$x = 1, x = -1$

Sample answer: Estimate the points where the graph intersects the x-axis.

3. It causes a vertical shift of c units. When $c > 0$, the graph is translated up. When $c < 0$, the graph is translated down.

4. Check students' work.

5. *Sample answer:* $a > 1$, $c = 1$; The graph is narrower than $y = x^2$, so it is a vertical stretch. The vertex is 1 unit above the origin, so the graph is a vertical translation 1 unit up of the graph of $y = x^2$.

Section 3.2 129

English Language Learners

Pair Activity

Pair each English learner with an English speaker. Have the pairs graph quadratic functions and compare their graphs to the graph of the function $f(x) = x^2$. Have one student make a table of values and the other student use the ordered pairs to draw the graph. Have partners work together to compare the graph of each function to the graph of $f(x) = x^2$.

Extra Example 1

Graph $g(x) = x^2 - 4$. Compare the graph to the graph of $f(x) = x^2$.

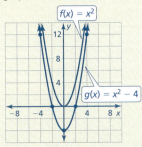

Both graphs open up and have the same axis of symmetry, $x = 0$. The vertex of the graph of g, $(0, -4)$, is below the vertex of the graph of f, $(0, 0)$, because the graph of g is a vertical translation 4 units down of the graph of f.

MONITORING PROGRESS ANSWERS

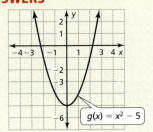

The graph of g is a vertical translation 5 units down of the graph of f.

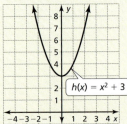

The graph of h is a vertical translation 3 units up of the graph of f.

130   Chapter 3

3.2 Lesson

What You Will Learn

▶ Graph quadratic functions of the form $f(x) = ax^2 + c$.
▶ Solve real-life problems involving functions of the form $f(x) = ax^2 + c$.

Core Vocabulary

zero of a function, *p. 132*
Previous
translation
vertex of a parabola
axis of symmetry
vertical stretch
vertical shrink

Graphing $f(x) = ax^2 + c$

Core Concept

Graphing $f(x) = ax^2 + c$

- When $c > 0$, the graph of $f(x) = ax^2 + c$ is a vertical translation c units up of the graph of $f(x) = ax^2$.
- When $c < 0$, the graph of $f(x) = ax^2 + c$ is a vertical translation $|c|$ units down of the graph of $f(x) = ax^2$.

The vertex of the graph of $f(x) = ax^2 + c$ is $(0, c)$, and the axis of symmetry is $x = 0$.

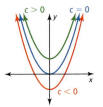

EXAMPLE 1   Graphing $y = x^2 + c$

Graph $g(x) = x^2 - 2$. Compare the graph to the graph of $f(x) = x^2$.

SOLUTION

Step 1 Make a table of values.

x	−2	−1	0	1	2
g(x)	2	−1	−2	−1	2

Step 2 Plot the ordered pairs.

Step 3 Draw a smooth curve through the points.

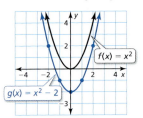

▶ Both graphs open up and have the same axis of symmetry, $x = 0$. The vertex of the graph of g, $(0, -2)$, is below the vertex of the graph of f, $(0, 0)$, because the graph of g is a vertical translation 2 units down of the graph of f.

Monitoring Progress   Help in English and Spanish at BigIdeasMath.com

Graph the function. Compare the graph to the graph of $f(x) = x^2$.

1. $g(x) = x^2 - 5$
2. $h(x) = x^2 + 3$

REMEMBER

The graph of $y = f(x) + k$ is a vertical translation, and the graph of $y = f(x - h)$ is a horizontal translation of the graph of f.

130   Chapter 3   Graphing Quadratic Functions

Laurie's Notes   Teacher Actions

- Write the *Core Concept*. Relate it to vertical translations of linear functions. Use different colors to connect the equations and the graphs.
- Explain that the shape of the graph is not changing. A vertical shift means the axis of symmetry is still $y = 0$ and the vertex is $(0, c)$.
- **Think-Pair-Share:** Pose Example 1. Have students work independently to complete a table of values and graph. Have them compare with their partners.
- ❓ "If $f(a) = 8$, what is $g(a)$? Explain." $g(a) = 6$; The function is shifted down 2 units, so $(a, 8)$ shifts down to $(a, 6)$.

EXAMPLE 2 Graphing $y = ax^2 + c$

Graph $g(x) = 4x^2 + 1$. Compare the graph to the graph of $f(x) = x^2$.

SOLUTION

Step 1 Make a table of values.

x	−2	−1	0	1	2
g(x)	17	5	1	5	17

Step 2 Plot the ordered pairs.

Step 3 Draw a smooth curve through the points.

▶ Both graphs open up and have the same axis of symmetry, $x = 0$. The graph of g is narrower, and its vertex, $(0, 1)$, is above the vertex of the graph of f, $(0, 0)$. So, the graph of g is a vertical stretch by a factor of 4 and a vertical translation 1 unit up of the graph of f.

EXAMPLE 3 Translating the Graph of $y = ax^2 + c$

Let $f(x) = -0.5x^2 + 2$ and $g(x) = f(x) - 7$.

a. Describe the transformation from the graph of f to the graph of g. Then graph f and g in the same coordinate plane.

b. Write an equation that represents g in terms of x.

SOLUTION

a. The function g is of the form $y = f(x) + k$, where $k = -7$. So, the graph of g is a vertical translation 7 units down of the graph of f.

x	−4	−2	0	2	4
f(x)	−6	0	2	0	−6
g(x)	−13	−7	−5	−7	−13

$-0.5x^2 + 2$ $f(x) - 7$

b. $g(x) = f(x) - 7$ Write the function g.
       $= -0.5x^2 + 2 - 7$ Substitute for $f(x)$.
       $= -0.5x^2 - 5$ Subtract.

▶ So, the equation $g(x) = -0.5x^2 - 5$ represents g in terms of x.

Monitoring Progress 🔊 Help in English and Spanish at *BigIdeasMath.com*

Graph the function. Compare the graph to the graph of $f(x) = x^2$.

3. $g(x) = 2x^2 - 5$

4. $h(x) = -\frac{1}{4}x^2 + 4$

5. Let $f(x) = 3x^2 - 1$ and $g(x) = f(x) + 3$.

   a. Describe the transformation from the graph of f to the graph of g. Then graph f and g in the same coordinate plane.

   b. Write an equation that represents g in terms of x.

Extra Example 2
Graph $g(x) = 2x^2 + 4$. Compare the graph to the graph of $f(x) = x^2$.

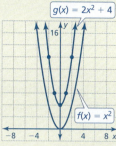

Both graphs open up and have the same axis of symmetry, $x = 0$. The graph of g is narrower, and its vertex, $(0, 4)$, is above the vertex of the graph of f, $(0, 0)$. So, the graph of g is a vertical stretch by a factor of 2 and a vertical translation 4 units up of the graph of f.

Extra Example 3
Let $f(x) = -0.75x^2 + 1$ and $g(x) = f(x) - 3$.

a. Describe the transformation from the graph of f to the graph of g. Then graph f and g in the same coordinate plane.

The graph of g is a vertical translation 3 units down of the graph of f.

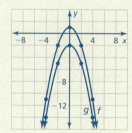

b. Write an equation that represents g in terms of x.
$g(x) = -0.75x^2 - 2$

MONITORING PROGRESS ANSWERS

3–5. See Additional Answers.

Laurie's Notes — Teacher Actions

- Example 2 is similar to Example 1, except now there is a leading coefficient of $a = 4$. The vertical shift of 1 unit is to the function $y = 4x^2$.
- **Connection:** Students should see the symmetry numerically in the table and graphically in the graph.
- **Attend to Precision:** In Example 3, help students make sense of the notation. One function is called $f(x)$ and is defined: $f(x) = -0.5x^2 + 2$. The second function is $g(x)$ and is defined: $g(x) = f(x) - 7$. The $g(x)$ function is the $f(x)$ function shifted down 7 units. See the *Teaching Strategy* on page T-128.

Extra Example 4

The function $f(t) = 4.9t^2$ represents the distance (in meters) a dropped object falls in t seconds. The function $g(t) = s_0$ represents the initial height (in meters) of the object.

a. Find and interpret $h(t) = -f(t) + g(t)$. The function $h(t) = -4.9t^2 + s_0$ represents the height (in meters) of a falling object t seconds after it is dropped from an initial height s_0 (in meters).

b. A coin is dropped from a height of 44.1 meters. After how many seconds does the coin hit the ground? The coin hits the ground 3 seconds after it is dropped.

c. Suppose the initial height is m meters lower. How will this affect part (b)? The coin will take less than 3 seconds to hit the ground.

MONITORING PROGRESS ANSWERS

6. *Sample answer:* Negative values of t would represent times before the egg was dropped, which have no meaning in the context of this problem.

7. 2.5 sec

Solving Real-Life Problems

A **zero of a function** f is an x-value for which $f(x) = 0$. A zero of a function is an x-intercept of the graph of the function.

EXAMPLE 4 Solving a Real-Life Problem

The function $f(t) = 16t^2$ represents the distance (in feet) a dropped object falls in t seconds. The function $g(t) = s_0$ represents the initial height (in feet) of the object.

a. Find and interpret $h(t) = -f(t) + g(t)$.

b. An egg is dropped from a height of 64 feet. After how many seconds does the egg hit the ground?

c. Suppose the initial height is adjusted by k feet. How will this affect part (b)?

SOLUTION

a. $h(t) = -f(t) + g(t)$ Write the function h.

$\quad\ = -(16t^2) + s_0$ Substitute for $f(t)$ and $g(t)$.

$\quad\ = -16t^2 + s_0$ Simplify.

▶ The function $h(t) = -16t^2 + s_0$ represents the height (in feet) of a falling object t seconds after it is dropped from an initial height s_0 (in feet).

b. The initial height is 64 feet. So, the function $h(t) = -16t^2 + 64$ represents the height of the egg t seconds after it is dropped. The egg hits the ground when $h(t) = 0$.

Step 1 Make a table of values and sketch the graph.

t	0	1	2
$h(t)$	64	48	0

COMMON ERROR

The graph in Step 1 shows the height of the object over time, not the path of the object.

Step 2 Find the positive zero of the function. When $t = 2$, $h(t) = 0$. So, the zero is 2.

▶ The egg hits the ground 2 seconds after it is dropped.

c. When the initial height is adjusted by k feet, the graph of h is translated up k units when $k > 0$ or down $|k|$ units when $k < 0$. So, the x-intercept of the graph of h will move right when $k > 0$ or left when $k < 0$.

▶ When $k > 0$, the egg will take more than 2 seconds to hit the ground. When $k < 0$, the egg will take less than 2 seconds to hit the ground.

Monitoring Progress Help in English and Spanish at BigIdeasMath.com

6. Explain why only nonnegative values of t are used in Example 4.

7. **WHAT IF?** The egg is dropped from a height of 100 feet. After how many seconds does the egg hit the ground?

Laurie's Notes Teacher Actions

- Define a *zero of a function*. Pose the egg drop problem. Introduce the falling-objects model.
- Ask questions to assess students' understanding of the problem, the model, and the known information. Give sufficient time for students to work with partners to graph the function and find the t-intercept.
- **? Probing Question:** "If the graph shifts up (down), what happens to the zero?" Listen for correct reasoning.

Closure

- **Exit Ticket:** Describe the graph of $y = -4x^2 + 12$. Do not graph it. *Sample answer:* The graph opens down and has $x = 0$ as its axis of symmetry. It is narrower than the graph of $y = -x^2$ and is a vertical stretch by a factor of 4. The graph is a vertical translation 12 units up and has its vertex at (0, 12).

3.2 Exercises

Dynamic Solutions available at BigIdeasMath.com

Vocabulary and Core Concept Check

1. **VOCABULARY** State the vertex and axis of symmetry of the graph of $y = ax^2 + c$.

2. **WRITING** How does the graph of $y = ax^2 + c$ compare to the graph of $y = ax^2$?

Monitoring Progress and Modeling with Mathematics

In Exercises 3–6, graph the function. Compare the graph to the graph of $f(x) = x^2$. (See Example 1.)

3. $g(x) = x^2 + 6$
4. $h(x) = x^2 + 8$
5. $p(x) = x^2 - 3$
6. $q(x) = x^2 - 1$

In Exercises 7–12, graph the function. Compare the graph to the graph of $f(x) = x^2$. (See Example 2.)

7. $g(x) = -x^2 + 3$
8. $h(x) = -x^2 - 7$
9. $s(x) = 2x^2 - 4$
10. $t(x) = -3x^2 + 1$
11. $p(x) = -\frac{1}{3}x^2 - 2$
12. $q(x) = \frac{1}{2}x^2 + 6$

In Exercises 13–16, describe the transformation from the graph of f to the graph of g. Then graph f and g in the same coordinate plane. Write an equation that represents g in terms of x. (See Example 3.)

13. $f(x) = 3x^2 + 4$
    $g(x) = f(x) + 2$
14. $f(x) = \frac{1}{2}x^2 + 1$
    $g(x) = f(x) - 4$
15. $f(x) = -\frac{1}{4}x^2 - 6$
    $g(x) = f(x) - 3$
16. $f(x) = 4x^2 - 5$
    $g(x) = f(x) + 7$

17. **ERROR ANALYSIS** Describe and correct the error in comparing the graphs.

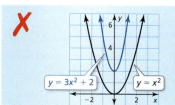

The graph of $y = 3x^2 + 2$ is a vertical shrink by a factor of 3 and a translation 2 units up of the graph of $y = x^2$.

18. **ERROR ANALYSIS** Describe and correct the error in graphing and comparing $f(x) = x^2$ and $g(x) = x^2 - 10$.

Both graphs open up and have the same axis of symmetry. However, the vertex of the graph of g, $(0, 10)$, is 10 units above the vertex of the graph of f, $(0, 0)$.

In Exercises 19–26, find the zeros of the function.

19. $y = x^2 - 1$
20. $y = x^2 - 36$
21. $f(x) = -x^2 + 25$
22. $f(x) = -x^2 + 49$
23. $f(x) = 4x^2 - 16$
24. $f(x) = 3x^2 - 27$
25. $f(x) = -12x^2 + 3$
26. $f(x) = -8x^2 + 98$

27. **MODELING WITH MATHEMATICS** A water balloon is dropped from a height of 144 feet. (See Example 4.)

   a. What does the function $h(t) = -16t^2 + 144$ represent in this situation?
   b. After how many seconds does the water balloon hit the ground?
   c. Suppose the initial height is adjusted by k feet. How does this affect part (b)?

28. **MODELING WITH MATHEMATICS** The function $y = -16x^2 + 36$ represents the height y (in feet) of an apple x seconds after falling from a tree. Find and interpret the x- and y-intercepts.

Section 3.2   Graphing $f(x) = ax^2 + c$ 133

Assignment Guide and Homework Check

ASSIGNMENT

Basic: 1, 2, 3–19 odd, 27, 29, 34, 36, 42–45

Average: 1, 2–22 even, 27–30, 34–38 even, 42–45

Advanced: 1, 2, 6, 10–26 even, 27, 28–32 even, 33–45

HOMEWORK CHECK

Basic: 3, 7, 13, 27

Average: 4, 10, 14, 27

Advanced: 6, 10, 16, 28

ANSWERS

1. $(0, c)$, $x = 0$

2. When $c > 0$, the graph of $f(x) = ax^2 + c$ is a vertical translation c units up of the graph of $f(x) = ax^2$. When $c < 0$, the graph of $f(x) = ax^2 + c$ is a vertical translation $|c|$ units down of the graph of $f(x) = ax^2$.

3.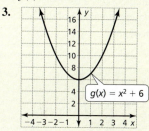

   The graph of g is a vertical translation 6 units up of the graph of f.

4.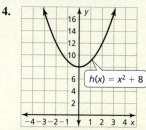

   The graph of h is a vertical translation 8 units up of the graph of f.

5.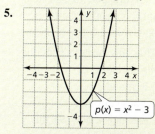

   The graph of p is a vertical translation 3 units down of the graph of f.

6.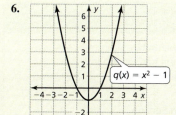

   The graph of q is a vertical translation 1 unit down of the graph of f.

7.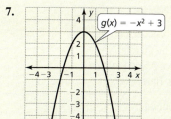

   The graph of g is a reflection in the x-axis, and a vertical translation 3 units up of the graph of f.

8–28. See Additional Answers.

Section 3.2   133

Dynamic Teaching Tools

Dynamic Assessment & Progress Monitoring Tool
Interactive Whiteboard Lesson Library
Dynamic Classroom with Dynamic Investigations

ANSWERS

29–45. See Additional Answers.

Mini-Assessment

Graph the function. Compare the graph to the graph of $f(x) = x^2$.

1. $g(x) = x^2 + 4$

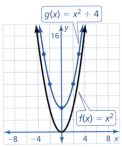

The graph of g is a vertical translation 4 units up of the graph of f.

2. $h(x) = 2x^2 - 6$

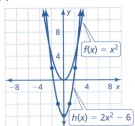

The graph of h is a vertical stretch by a factor of 2 and a vertical translation 6 units down of the graph of f.

3. The function $f(t) = -16t^2 + s_0$ represents the approximate height (in feet) of a falling object t seconds after it is dropped from an initial height s_0 (in feet). A marble is dropped from a height of 36 feet.

   a. After how many seconds does the marble hit the ground?

   1.5 seconds

   b. Suppose the initial height is k feet higher. How will this affect part (a)?

   The marble will take more than 1.5 seconds to hit the ground.

134 Chapter 3

In Exercises 29–32, sketch a parabola with the given characteristics.

29. The parabola opens up, and the vertex is (0, 3).

30. The vertex is (0, 4), and one of the x-intercepts is 2.

31. The related function is increasing when $x < 0$, and the zeros are -1 and 1.

32. The highest point on the parabola is (0, -5).

33. **DRAWING CONCLUSIONS** You and your friend both drop a ball at the same time. The function $h(x) = -16x^2 + 256$ represents the height (in feet) of your ball after x seconds. The function $g(x) = -16x^2 + 300$ represents the height (in feet) of your friend's ball after x seconds.

    a. Write the function $T(x) = h(x) - g(x)$. What does $T(x)$ represent?

    b. When your ball hits the ground, what is the height of your friend's ball? Use a graph to justify your answer.

34. **MAKING AN ARGUMENT** Your friend claims that in the equation $y = ax^2 + c$, the vertex changes when the value of a changes. Is your friend correct? Explain your reasoning.

35. **MATHEMATICAL CONNECTIONS** The area A (in square feet) of a square patio is represented by $A = x^2$, where x is the length of one side of the patio. You add 48 square feet to the patio, resulting in a total area of 192 square feet. What are the dimensions of the original patio? Use a graph to justify your answer.

36. **HOW DO YOU SEE IT?** The graph of $f(x) = ax^2 + c$ is shown. Points A and B are the same distance from the vertex of the graph of f. Which point is closer to the vertex of the graph of f as c increases?

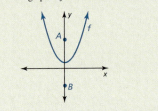

37. **REASONING** Describe two algebraic methods you can use to find the zeros of the function $f(t) = -16t^2 + 400$. Check your answer by graphing.

38. **PROBLEM SOLVING** The paths of water from three different garden waterfalls are given below. Each function gives the height h (in feet) and the horizontal distance d (in feet) of the water.

    Waterfall 1 $h = -3.1d^2 + 4.8$
    Waterfall 2 $h = -3.5d^2 + 1.9$
    Waterfall 3 $h = -1.1d^2 + 1.6$

    a. Which waterfall drops water from the highest point?

    b. Which waterfall follows the narrowest path?

    c. Which waterfall sends water the farthest?

39. **WRITING EQUATIONS** Two acorns fall to the ground from an oak tree. One falls 45 feet, while the other falls 32 feet.

    a. For each acorn, write an equation that represents the height h (in feet) as a function of the time t (in seconds).

    b. Describe how the graphs of the two equations are related.

40. **THOUGHT PROVOKING** One of two classic problems in calculus is to find the area under a curve. Approximate the area of the region bounded by the parabola and the x-axis. Show your work.

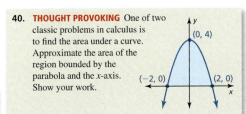

41. **CRITICAL THINKING** A cross section of the parabolic surface of the antenna shown can be modeled by $y = 0.012x^2$, where x and y are measured in feet. The antenna is moved up so that the outer edges of the dish are 25 feet above the x-axis. Where is the vertex of the cross section located? Explain.

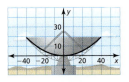

Maintaining Mathematical Proficiency Reviewing what you learned in previous grades and lessons

Evaluate the expression when $a = 4$ and $b = -3$. *(Skills Review Handbook)*

42. $\dfrac{a}{4b}$ 43. $-\dfrac{b}{2a}$ 44. $\dfrac{a-b}{3a+b}$ 45. $-\dfrac{b+2a}{ab}$

134   Chapter 3   Graphing Quadratic Functions

If students need help...	If students got it...
Resources by Chapter • Practice A and Practice B • Puzzle Time	Resources by Chapter • Enrichment and Extension • Cumulative Review
Student Journal • Practice	Start the *next* Section
Differentiating the Lesson Skills Review Handbook	

Laurie's Notes

Overview of Section 3.3

Introduction
- The first two lessons in the chapter looked at transformations of the parent function. In this lesson, the linear term, *bx*, is added to the equation. The lesson begins with the formula for the *x*-coordinate of the vertex of $y = ax^2 + bx + c$, which is also related to the equation of the axis of symmetry. This information is used to graph the quadratic function versus making a table of values that may or may not include the vertex.
- The second part of the lesson focuses on finding the maximum or minimum value of the function. This is done by evaluating the function using the *x*-coordinate of the vertex. The *y*-coordinate is the maximum or minimum value of the function.

Teaching Strategy
- In this lesson, students will use parameters from the equation to find the vertex of the parabola and also the maximum or minimum value of the function. Graphing technology could also be used.
- Example 5 in the lesson is about launching a water balloon. Enter the function $f(t) = -16t^2 + 80t + 5$, where *f* represents the height (in feet) of the water balloon *t* seconds after it is launched, in Y_1 of the equation editor, as shown in Figure 1.
- Go to the Calculate screen and choose "maximum," as shown in Figure 2. Follow the steps in the user's guide for your calculator to find and display the vertex and its value.
- Figure 3 shows the calculated value for the vertex.
- **Attend to Precision:** Note that the calculator generated a vertex of (2.4999994, 105). Calculated by hand, the vertex is (2.5, 105). Students need to know how to interpret results such as these and know that the vertex is indeed (2.5, 105).

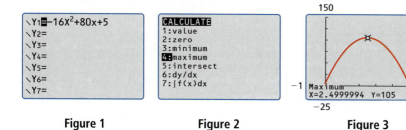

       Figure 1   **Figure 2**   **Figure 3**

Pacing Suggestion
- Once students have worked the explorations, continue with the formal lesson.

Dynamic Teaching Tools
Dynamic Assessment & Progress Monitoring Tool
Lesson Planning Tool
Interactive Whiteboard Lesson Library
Dynamic Classroom with Dynamic Investigations

What Your Students Will Learn

- Graph quadratic functions of the form $f(x) = ax^2 + bx + c$ by finding the axis of symmetry, plotting the vertex, the y-intercept, and additional points, and drawing a smooth curve through the points.
- Find the minimum and maximum values of quadratic functions by finding the y-coordinate of the vertex.

Laurie's Notes

Exploration

Motivate
- Discuss with students that, in a future course, they will study conic sections: ellipses, parabolas, and hyperbolas. A form of each of those words is a part of spoken English.
- Have students find the words *ellipsis*, *hyperbole*, and *parable* in a dictionary. Discuss the definitions.
- In rhetoric, "hyperbolic" speech goes beyond the facts, "elliptic" speech falls short of the facts, and a "parable" is a story that fits the facts exactly.

For Your Information
- **Look For and Make Use of Structure:** All of the quadratic functions that students have graphed have been symmetric about the y-axis. In the equation $y = ax^2 + c$, the values of a and c do not affect the graph in a horizontal direction. In this section, quadratic functions contain an x-term, which shifts the axis of symmetry away from the y-axis.

Exploration 1
- Students should use a table of values to graph the functions. In doing so, they gain a sense of the differences, numerically, for the two functions. Evaluating the functions when $x = 2$, the arithmetic is the same except for the second function where 6 is added.
- In both functions, when students use domain values of $-2, -1, 0, 1,$ and 2, the y-values are decreasing. When they evaluate either function for $x = 3$, the y-values start to increase. You want students to interpret this change. The vertex occurs when $x = 2$.
- The x-intercepts may have been found when generating the table of values. If not, students should substitute 0 for y and solve.
- Discuss the connections between the axis of symmetry, vertex, and x-intercepts.

Exploration 2
- Students may need guidance to factor $y = ax^2 + bx$. Suggest that they treat a and b as constants and focus on solving for x.
- When students complete the table, they are substituting the x-intercepts, 0 and $-\frac{b}{a}$, into the equation and solving. Both results should be $y = 0$.

Exploration 3
- Students need to reflect back on the two explorations they completed today.
- Students may describe the answer in words if they have difficulty finding an expression for x. For instance, in the first answer blank they may say that the vertex occurs halfway between 0 and $-\frac{b}{a}$, not realizing that they can express this as $x = -\frac{b}{2a}$. You may need to give examples and pose questions such as "What is halfway between 4 and 6? Halfway between 0 and a?"

Communicate Your Answer
- Listen for understanding of what students discovered in the explorations.

Connecting to Next Step
- The explorations help students make sense of why the x-coordinate of the vertex is $-\frac{b}{2a}$. In the formal lesson, this is stated at the beginning of the lesson.

3.3 Graphing $f(x) = ax^2 + bx + c$

Essential Question How can you find the vertex of the graph of $f(x) = ax^2 + bx + c$?

EXPLORATION 1 Comparing x-Intercepts with the Vertex

Work with a partner.

a. Sketch the graphs of $y = 2x^2 - 8x$ and $y = 2x^2 - 8x + 6$.

b. What do you notice about the x-coordinate of the vertex of each graph?

c. Use the graph of $y = 2x^2 - 8x$ to find its x-intercepts. Verify your answer by solving $0 = 2x^2 - 8x$.

d. Compare the value of the x-coordinate of the vertex with the values of the x-intercepts.

EXPLORATION 2 Finding x-Intercepts

Work with a partner.

a. Solve $0 = ax^2 + bx$ for x by factoring.

b. What are the x-intercepts of the graph of $y = ax^2 + bx$?

c. Copy and complete the table to verify your answer.

x	$y = ax^2 + bx$
0	
$-\dfrac{b}{a}$	

CONSTRUCTING VIABLE ARGUMENTS

To be proficient in math, you need to make conjectures and build a logical progression of statements.

EXPLORATION 3 Deductive Reasoning

Work with a partner. Complete the following logical argument.

The x-intercepts of the graph of $y = ax^2 + bx$ are 0 and $-\dfrac{b}{a}$.

The vertex of the graph of $y = ax^2 + bx$ occurs when $x = $ _____ .

The vertices of the graphs of $y = ax^2 + bx$ and $y = ax^2 + bx + c$ have the same x-coordinate.

The vertex of the graph of $y = ax^2 + bx + c$ occurs when $x = $ _____ .

Communicate Your Answer

4. How can you find the vertex of the graph of $f(x) = ax^2 + bx + c$?

5. Without graphing, find the vertex of the graph of $f(x) = x^2 - 4x + 3$. Check your result by graphing.

Section 3.3   Graphing $f(x) = ax^2 + bx + c$ 135

Dynamic Teaching Tools

Dynamic Assessment & Progress Monitoring Tool
Lesson Planning Tool
Interactive Whiteboard Lesson Library
Dynamic Classroom with Dynamic Investigations

ANSWERS

1. a.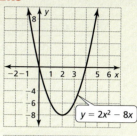

   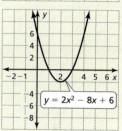

   b. They are the same.

   c. $x = 0, x = 4$

   d. *Sample answer:* The value of the x-coordinate of the vertex is the average of the values of the x-intercepts.

2. a. $x = 0, x = -\dfrac{b}{a}$

   b. $x = 0, x = -\dfrac{b}{a}$

   c. 0, 0

3. $-\dfrac{b}{2a}; -\dfrac{b}{2a}$

4. *Sample answer:* Find the x-coordinate using $x = -\dfrac{b}{2a}$, then use the function to find the y-coordinate.

5. $(2, -1)$

Section 3.3   135

Differentiated Instruction

Organization
Have students copy and complete the table below in their notebooks. Have students refer to the table to help them determine whether a function of the form $f(x) = ax^2 + bx + c$ has a maximum value or a minimum value.

$f(x) = ax^2 + bx + c$		
Value of a	$a > 0$	$a < 0$
Graph opens …	up	down
Maximum or minimum?		
Sketch		

Extra Example 1

a. Find the axis of symmetry of the graph of $f(x) = 2x^2 - 4x + 5$.
   The axis of symmetry is $x = 1$.

b. Find the vertex of the graph of $f(x) = 2x^2 - 4x + 5$.
   The vertex is $(1, 3)$.

MONITORING PROGRESS ANSWERS

1. a. $x = \frac{1}{3}$
   b. $\left(\frac{1}{3}, -\frac{1}{3}\right)$
2. a. $x = -3$
   b. $(-3, -4)$
3. a. $x = 7$
   b. $\left(7, \frac{41}{2}\right)$

3.3 Lesson

What You Will Learn

▶ Graph quadratic functions of the form $f(x) = ax^2 + bx + c$.
▶ Find maximum and minimum values of quadratic functions.

Core Vocabulary
maximum value, *p. 137*
minimum value, *p. 137*

Previous
independent variable
dependent variable

Graphing $f(x) = ax^2 + bx + c$

Core Concept

Graphing $f(x) = ax^2 + bx + c$

- The graph opens up when $a > 0$, and the graph opens down when $a < 0$.
- The y-intercept is c.
- The x-coordinate of the vertex is $-\frac{b}{2a}$.
- The axis of symmetry is $x = -\frac{b}{2a}$.

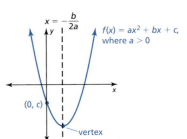

EXAMPLE 1 Finding the Axis of Symmetry and the Vertex

Find (a) the axis of symmetry and (b) the vertex of the graph of $f(x) = 2x^2 + 8x - 1$.

SOLUTION

a. Find the axis of symmetry when $a = 2$ and $b = 8$.

$x = -\dfrac{b}{2a}$ Write the equation for the axis of symmetry.

$x = -\dfrac{8}{2(2)}$ Substitute 2 for a and 8 for b.

$x = -2$ Simplify.

▶ The axis of symmetry is $x = -2$.

b. The axis of symmetry is $x = -2$, so the x-coordinate of the vertex is -2. Use the function to find the y-coordinate of the vertex.

$f(x) = 2x^2 + 8x - 1$ Write the function.

$f(-2) = 2(-2)^2 + 8(-2) - 1$ Substitute -2 for x.

$= -9$ Simplify.

▶ The vertex is $(-2, -9)$.

Check

Monitoring Progress Help in English and Spanish at BigIdeasMath.com

Find (a) the axis of symmetry and (b) the vertex of the graph of the function.

1. $f(x) = 3x^2 - 2x$
2. $g(x) = x^2 + 6x + 5$
3. $h(x) = -\frac{1}{2}x^2 + 7x - 4$

136   Chapter 3   Graphing Quadratic Functions

Laurie's Notes — Teacher Actions

- **Connection:** Write the *Core Concept*. The constant term in linear and quadratic functions is the y-intercept.
- **Thumbs Up:** Write the remaining information in the *Core Concept* about the vertex and line of symmetry. Have students self-assess with a *Thumbs Up* sign.
- **?** Pose Example 1. "How can you find the axis of symmetry?" *Listen as students describe using the formula $x = -\dfrac{b}{2a}$.*
- **?** "Now that you know the axis of symmetry, how do you find the vertex?" *Substitute -2 for x in the equation and simplify.*

COMMON ERROR
Be sure to include the negative sign before the fraction when finding the axis of symmetry.

EXAMPLE 2 Graphing $f(x) = ax^2 + bx + c$

Graph $f(x) = 3x^2 - 6x + 5$. Describe the domain and range.

SOLUTION

Step 1 Find and graph the axis of symmetry.

$$x = -\frac{b}{2a} = -\frac{(-6)}{2(3)} = 1 \qquad \text{Substitute and simplify.}$$

Step 2 Find and plot the vertex.

The axis of symmetry is $x = 1$, so the x-coordinate of the vertex is 1. Use the function to find the y-coordinate of the vertex.

$$f(x) = 3x^2 - 6x + 5 \qquad \text{Write the function.}$$
$$f(1) = 3(1)^2 - 6(1) + 5 \qquad \text{Substitute 1 for } x.$$
$$= 2 \qquad \text{Simplify.}$$

So, the vertex is (1, 2).

Step 3 Use the y-intercept to find two more points on the graph.

Because $c = 5$, the y-intercept is 5. So, (0, 5) lies on the graph. Because the axis of symmetry is $x = 1$, the point (2, 5) also lies on the graph.

Step 4 Draw a smooth curve through the points.

▶ The domain is all real numbers. The range is $y \geq 2$.

REMEMBER
The domain is the set of all possible input values of the independent variable x. The range is the set of all possible output values of the dependent variable y.

Monitoring Progress Help in English and Spanish at BigIdeasMath.com

Graph the function. Describe the domain and range.

4. $h(x) = 2x^2 + 4x + 1$ 5. $k(x) = x^2 - 8x + 7$ 6. $p(x) = -5x^2 - 10x - 2$

Finding Maximum and Minimum Values

Core Concept

Maximum and Minimum Values

The y-coordinate of the vertex of the graph of $f(x) = ax^2 + bx + c$ is the **maximum value** of the function when $a < 0$ or the **minimum value** of the function when $a > 0$.

$f(x) = ax^2 + bx + c, a < 0$ 	 $f(x) = ax^2 + bx + c, a > 0$

maximum 	 minimum

English Language Learners

Vocabulary

Use sketches of $f(x) = ax^2 + bx + c$ with positive and negative values of a to illustrate the maximum or minimum value of a quadratic function. Guide students to verbalize that "maximum" means "highest point" and "minimum" means "lowest point." If students are familiar with phrases such as "maximum speed" or "minimum wage," discuss the relationship between these phrases and the points on the graphs.

Extra Example 2

Graph $f(x) = x^2 - 2x - 3$. Describe the domain and range.

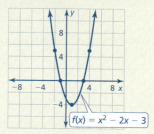

The domain is all real numbers. The range is $y \geq -4$.

MONITORING PROGRESS ANSWERS

4.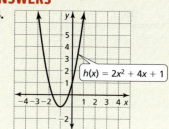

domain: all real numbers, range: $y \geq -1$

5.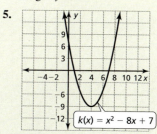

domain: all real numbers, range: $y \geq -9$

6. See Additional Answers.

Laurie's Notes Teacher Actions

- **Look For and Make Use of Structure:** Explain that besides a table of values and a calculator, information from the equation can be used to graph a quadratic. First, find the axis of symmetry, and then use the x-coordinate to find the vertex. To increase accuracy, use symmetry to find two additional points.
- **Think-Pair-Share:** Have students answer Questions 4–6, and then share and discuss as a class.
- **Misconception:** The vertex is not the maximum or minimum value. The y-coordinate of the vertex is the maximum or minimum value.

Extra Example 3
Tell whether the function $f(x) = 5x^2 + 15x - 10$ has a minimum value or a maximum value. Then find the value. The minimum value is -21.25.

Extra Example 4
A company plans to have coffee mugs produced with the company logo on them. The cost of each mug depends on the number of mugs that are ordered, and is modeled by $y = 0.0001x^2 - 0.1x + 30$, where x is the number of mugs ordered and y is the cost (in dollars). What is the minimum cost of each mug? The minimum cost of each mug is $5.00.

MONITORING PROGRESS ANSWERS
7. minimum value; 4
8. maximum value; 10
9. about 176 ft

EXAMPLE 3 Finding a Maximum or Minimum Value

Tell whether the function $f(x) = -4x^2 - 24x - 19$ has a minimum value or a maximum value. Then find the value.

SOLUTION

For $f(x) = -4x^2 - 24x - 19$, $a = -4$ and $-4 < 0$. So, the parabola opens down and the function has a maximum value. To find the maximum value, find the y-coordinate of the vertex.

First, find the x-coordinate of the vertex. Use $a = -4$ and $b = -24$.

$$x = -\frac{b}{2a} = -\frac{(-24)}{2(-4)} = -3 \qquad \text{Substitute and simplify.}$$

Then evaluate the function when $x = -3$ to find the y-coordinate of the vertex.

$$f(-3) = -4(-3)^2 - 24(-3) - 19 \qquad \text{Substitute } -3 \text{ for } x.$$
$$= 17 \qquad \text{Simplify.}$$

▶ The maximum value is 17.

EXAMPLE 4 Finding a Minimum Value

The suspension cables between the two towers of the Mackinac Bridge in Michigan form a parabola that can be modeled by $y = 0.000098x^2 - 0.37x + 552$, where x and y are measured in feet. What is the height of the cable above the water at its lowest point?

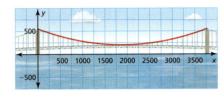

SOLUTION

The lowest point of the cable is at the vertex of the parabola. Find the x-coordinate of the vertex. Use $a = 0.000098$ and $b = -0.37$.

$$x = -\frac{b}{2a} = -\frac{(-0.37)}{2(0.000098)} \approx 1888 \qquad \text{Substitute and use a calculator.}$$

Substitute 1888 for x in the equation to find the y-coordinate of the vertex.

$$y = 0.000098(1888)^2 - 0.37(1888) + 552 \approx 203$$

▶ The cable is about 203 feet above the water at its lowest point.

Monitoring Progress Help in English and Spanish at *BigIdeasMath.com*

Tell whether the function has a minimum value or a maximum value. Then find the value.

7. $g(x) = 8x^2 - 8x + 6$
8. $h(x) = -\frac{1}{4}x^2 + 3x + 1$

9. The cables between the two towers of the Tacoma Narrows Bridge in Washington form a parabola that can be modeled by $y = 0.00016x^2 - 0.46x + 507$, where x and y are measured in feet. What is the height of the cable above the water at its lowest point?

138   Chapter 3   Graphing Quadratic Functions

Laurie's Notes   Teacher Actions

? Probing Question: "Do you need to graph the quadratic to know whether it has a maximum or minimum value? Explain." no; When $a > 0$, the quadratic opens up and has a minimum value. When $a < 0$, the quadratic opens down and has a maximum value.

- **Make Sense of Problems and Persevere in Solving Them:** Pose Example 4, and then give partners time to work on the problem. Do not rush in to solve this for them. Give sufficient time for the partners to work the problem.
- Students can use calculators to find the x-coordinate of the vertex and use this x-value to evaluate the function to find the minimum value.

MODELING WITH MATHEMATICS

Because time cannot be negative, use only nonnegative values of *t*.

EXAMPLE 5 Modeling with Mathematics

A group of friends is launching water balloons. The function $f(t) = -16t^2 + 80t + 5$ represents the height (in feet) of the first water balloon *t* seconds after it is launched. The height of the second water balloon *t* seconds after it is launched is shown in the graph. Which water balloon went higher?

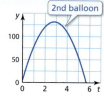

SOLUTION

1. **Understand the Problem** You are given a function that represents the height of the first water balloon. The height of the second water balloon is represented graphically. You need to find and compare the maximum heights of the water balloons.

2. **Make a Plan** To compare the maximum heights, represent both functions graphically. Use a graphing calculator to graph $f(t) = -16t^2 + 80t + 5$ in an appropriate viewing window. Then visually compare the heights of the water balloons.

3. **Solve the Problem** Enter the function $f(t) = -16t^2 + 80t + 5$ into your calculator and graph it. Compare the graphs to determine which function has a greater maximum value.

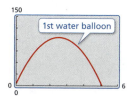 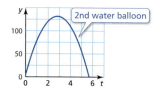

You can see that the second water balloon reaches a height of about 125 feet, while the first water balloon reaches a height of only about 100 feet.

▶ So, the second water balloon went higher.

4. **Look Back** Use the *maximum* feature to determine that the maximum value of $f(t) = -16t^2 + 80t + 5$ is 105. Use a straightedge to represent a height of 105 feet on the graph that represents the second water balloon to clearly see that the second water balloon went higher.

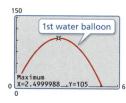 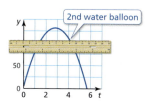

Monitoring Progress 🔊 Help in English and Spanish at *BigIdeasMath.com*

10. Which balloon is in the air longer? Explain your reasoning.
11. Which balloon reaches its maximum height faster? Explain your reasoning.

Section 3.3 Graphing $f(x) = ax^2 + bx + c$ 139

Extra Example 5

Two jugglers toss beanbags in the air. The function $h(t) = -5t^2 + 5t + 1$ represents the height (in meters) of a red beanbag after *t* seconds. The height of a blue beanbag *t* seconds after it is launched is shown in the graph. Which beanbag went higher?

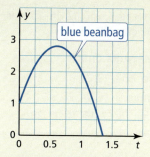

The blue beanbag went higher.

MONITORING PROGRESS ANSWERS

10. second water balloon; *Sample answer:* Comparing the graphs, the first water balloon has a height of 0 after about 5 seconds, and the second water balloon has a height of 0 after about 5.7 seconds.

11. first water balloon; *Sample answer:* Comparing the graphs, the first water balloon is at its maximum height after about 2.5 seconds, and the second water balloon is at its maximum height after about 2.8 seconds.

Laurie's Notes Teacher Actions

- **Model with Mathematics** and **Use Appropriate Tools Strategically:** Students are given two models represented differently—an equation and a graph. Have students work with partners to solve the problem. Students will likely graph both functions and use the maximum feature on their calculators. See the *Teaching Strategy* on page T-134.
- ❓ **Extension:** "When will the balloons hit the ground?" first water balloon: about 5 seconds, second water balloon: about 5.5 seconds

Closure

- Write an equation of a quadratic function that opens up, has a negative *y*-intercept, and is wider than the graph of $y = x^2$. *Sample answer:* $y = 0.25x^2 - 2$

Section 3.3 139

Assignment Guide and Homework Check

ASSIGNMENT

Basic: 1, 2, 3–29 odd, 32, 37, 42, 50–53

Average: 1, 2–36 even, 37, 38, 42, 45, 50–53

Advanced: 1, 2, 6, 12, 16–20 even, 24–38 even, 39–53

HOMEWORK CHECK

Basic: 7, 13, 23, 27, 37

Average: 8, 16, 24, 28, 37

Advanced: 12, 16, 24, 32, 38

ANSWERS

1. *Sample answer:* If the leading coefficient is positive, the graph has a minimum value. If the leading coefficient is negative, the graph has a maximum value.

2. What is the axis of symmetry of the graph of the function?; $x = 2$; 32

3. $(2, -1); x = 2; 1$
4. $(-3, 2); x = -3; -1$
5. $(-2, 0); x = -2; -3$
6. $(-1, 1); x = -1; 5$
7. a. $x = 1$
   b. $(1, -2)$
8. a. $x = -\frac{1}{3}$
   b. $\left(-\frac{1}{3}, -\frac{1}{3}\right)$
9. a. $x = -1$
   b. $(-1, 8)$
10. a. $x = 2$
    b. $(2, 4)$
11. a. $x = 5$
    b. $(5, 4)$
12. a. $x = 6$
    b. $(6, 9)$
13.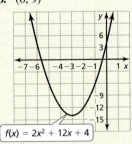
    domain: all real numbers, range: $y \geq -14$

3.3 Exercises

Dynamic Solutions available at BigIdeasMath.com

Vocabulary and Core Concept Check

1. **VOCABULARY** Explain how you can tell whether a quadratic function has a maximum value or a minimum value without graphing the function.

2. **DIFFERENT WORDS, SAME QUESTION** Consider the quadratic function $f(x) = -2x^2 + 8x + 24$. Which is different? Find "both" answers.

   - What is the maximum value of the function?
   - What is the greatest number in the range of the function?
   - What is the y-coordinate of the vertex of the graph of the function?
   - What is the axis of symmetry of the graph of the function?

Monitoring Progress and Modeling with Mathematics

In Exercises 3–6, find the vertex, the axis of symmetry, and the y-intercept of the graph.

3.
4.
5.
6.

In Exercises 7–12, find (a) the axis of symmetry and (b) the vertex of the graph of the function. *(See Example 1.)*

7. $f(x) = 2x^2 - 4x$
8. $y = 3x^2 + 2x$
9. $y = -9x^2 - 18x - 1$
10. $f(x) = -6x^2 + 24x - 20$
11. $f(x) = \frac{2}{5}x^2 - 4x + 14$
12. $y = -\frac{3}{4}x^2 + 9x - 18$

In Exercises 13–18, graph the function. Describe the domain and range. *(See Example 2.)*

13. $f(x) = 2x^2 + 12x + 4$
14. $y = 4x^2 + 24x + 13$
15. $y = -8x^2 - 16x - 9$
16. $f(x) = -5x^2 + 20x - 7$

17. $y = \frac{2}{3}x^2 - 6x + 5$
18. $f(x) = -\frac{1}{2}x^2 - 3x - 4$

19. **ERROR ANALYSIS** Describe and correct the error in finding the axis of symmetry of the graph of $y = 3x^2 - 12x + 11$.

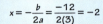
$$x = -\frac{b}{2a} = \frac{-12}{2(3)} = -2$$
The axis of symmetry is $x = -2$.

20. **ERROR ANALYSIS** Describe and correct the error in graphing the function $f(x) = x^2 + 4x + 3$.

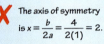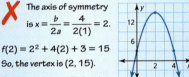
The axis of symmetry is $x = \frac{b}{2a} = \frac{4}{2(1)} = 2$.
$f(2) = 2^2 + 4(2) + 3 = 15$
So, the vertex is $(2, 15)$.
The y-intercept is 3. So, the points $(0, 3)$ and $(4, 3)$ lie on the graph.

In Exercises 21–26, tell whether the function has a minimum value or a maximum value. Then find the value. *(See Example 3.)*

21. $y = 3x^2 - 18x + 15$
22. $f(x) = -5x^2 + 10x + 7$
23. $f(x) = -4x^2 + 4x - 2$
24. $y = 2x^2 - 10x + 13$
25. $y = -\frac{1}{2}x^2 - 11x + 6$
26. $f(x) = \frac{1}{5}x^2 - 5x + 27$

140   Chapter 3   Graphing Quadratic Functions

14.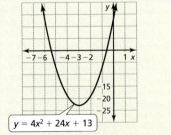
    domain: all real numbers, range: $y \geq -23$

15.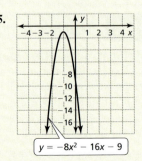
    domain: all real numbers, range: $y \leq -1$

16–26. See Additional Answers.

27. MODELING WITH MATHEMATICS The function shown represents the height h (in feet) of a firework t seconds after it is launched. The firework explodes at its highest point. *(See Example 4.)*

a. When does the firework explode?

b. At what height does the firework explode?

28. MODELING WITH MATHEMATICS The function $h(t) = -16t^2 + 16t$ represents the height (in feet) of a horse t seconds after it jumps during a steeplechase.

a. When does the horse reach its maximum height?

b. Can the horse clear a fence that is 3.5 feet tall? If so, by how much?

c. How long is the horse in the air?

29. MODELING WITH MATHEMATICS The cable between two towers of a suspension bridge can be modeled by the function shown, where x and y are measured in feet. The cable is at road level midway between the towers.

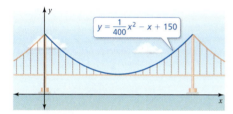

a. How far from each tower shown is the lowest point of the cable?

b. How high is the road above the water?

c. Describe the domain and range of the function shown.

30. REASONING Find the axis of symmetry of the graph of the equation $y = ax^2 + bx + c$ when $b = 0$. Can you find the axis of symmetry when $a = 0$? Explain.

31. ATTENDING TO PRECISION The vertex of a parabola is $(3, -1)$. One point on the parabola is $(6, 8)$. Find another point on the parabola. Justify your answer.

32. MAKING AN ARGUMENT Your friend claims that it is possible to draw a parabola through any two points with different x-coordinates. Is your friend correct? Explain.

USING TOOLS In Exercises 33–36, use the *minimum* or *maximum* feature of a graphing calculator to approximate the vertex of the graph of the function.

33. $y = 0.5x^2 + \sqrt{2}x - 3$

34. $y = -6.2x^2 + 4.8x - 1$

35. $y = -\pi x^2 + 3x$

36. $y = 0.25x^2 - 5^{2/3}x + 2$

37. MODELING WITH MATHEMATICS The opening of one aircraft hangar is a parabolic arch that can be modeled by the equation $y = -0.006x^2 + 1.5x$, where x and y are measured in feet. The opening of a second aircraft hangar is shown in the graph. *(See Example 5.)*

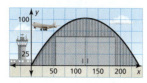

a. Which aircraft hangar is taller?

b. Which aircraft hangar is wider?

38. MODELING WITH MATHEMATICS An office supply store sells about 80 graphing calculators per month for $120 each. For each $6 decrease in price, the store expects to sell eight more calculators. The revenue from calculator sales is given by the function $R(n) = $ (unit price)(units sold), or $R(n) = (120 - 6n)(80 + 8n)$, where n is the number of $6 price decreases.

a. How much should the store charge to maximize monthly revenue?

b. Using a different revenue model, the store expects to sell five more calculators for each $4 decrease in price. Which revenue model results in a greater maximum monthly revenue? Explain.

Section 3.3   Graphing $f(x) = ax^2 + bx + c$ 141

ANSWERS

39. a. $x = 4.5$
    b. 20.25 in.^{2}
40. a. $x = 2$
    b. 32 ft^2
41. The graph of g is a reflection in the y-axis of the graph of h.
42. a. 1.5 m
    b. about 1.6 m
    c. about 90 m
43–53. See Additional Answers.

Mini-Assessment

1. Graph $f(x) = 3x^2 + 6x + 2$. Find the axis of symmetry and the vertex of the graph. Describe the domain and range.

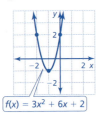

axis of symmetry: $x = -1$; vertex: $(-1, -1)$; domain: all real numbers; range: $y \geq -1$

2. Tell whether the function $f(x) = -6x^2 + 24x - 8$ has a minimum value or a maximum value. Then find the value. maximum value: 16

3. A group of friends is launching water balloons from the roof of a building. The function $f(t) = -16t^2 + 96t + 20$ represents the height (in feet) of the first balloon t seconds after it is launched. The height of the second balloon t seconds after it is launched is shown in the graph. Which balloon went higher?

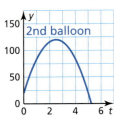

The first water balloon went higher.

142   Chapter 3

MATHEMATICAL CONNECTIONS In Exercises 39 and 40, (a) find the value of x that maximizes the area of the figure and (b) find the maximum area.

39.

40.

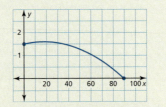

41. **WRITING** Compare the graph of $g(x) = x^2 + 4x + 1$ with the graph of $h(x) = x^2 - 4x + 1$.

42. **HOW DO YOU SEE IT?** During an archery competition, an archer shoots an arrow. The arrow follows the parabolic path shown, where x and y are measured in meters.

a. What is the initial height of the arrow?
b. Estimate the maximum height of the arrow.
c. How far does the arrow travel?

43. **USING TOOLS** The graph of a quadratic function passes through $(3, 2)$, $(4, 7)$, and $(9, 2)$. Does the graph open up or down? Explain your reasoning.

44. **REASONING** For a quadratic function f, what does $f\left(-\dfrac{b}{2a}\right)$ represent? Explain your reasoning.

45. **PROBLEM SOLVING** Write a function of the form $y = ax^2 + bx$ whose graph contains the points $(1, 6)$ and $(3, 6)$.

46. **CRITICAL THINKING** Parabolas A and B contain the points shown. Identify characteristics of each parabola, if possible. Explain your reasoning.

Parabola A		Parabola B	
x	y	x	y
2	3	1	4
6	4	3	−4
		5	4

47. **MODELING WITH MATHEMATICS** At a basketball game, an air cannon launches T-shirts into the crowd. The function $y = -\dfrac{1}{8}x^2 + 4x$ represents the path of a T-shirt. The function $3y = 2x - 14$ represents the height of the bleachers. In both functions, y represents vertical height (in feet) and x represents horizontal distance (in feet). At what height does the T-shirt land in the bleachers?

48. **THOUGHT PROVOKING** One of two classic problems in calculus is finding the slope of a *tangent line* to a curve. An example of a tangent line, which just touches the parabola at one point, is shown.

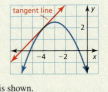

Approximate the slope of the tangent line to the graph of $y = x^2$ at the point $(1, 1)$. Explain your reasoning.

49. **PROBLEM SOLVING** The owners of a dog shelter want to enclose a rectangular play area on the side of their building. They have k feet of fencing. What is the maximum area of the outside enclosure in terms of k? (*Hint:* Find the y-coordinate of the vertex of the graph of the area function.)

Maintaining Mathematical Proficiency
Reviewing what you learned in previous grades and lessons

Describe the transformation(s) from the graph of $f(x) = |x|$ to the graph of the given function. *(Section 1.1)*

50. $q(x) = |x + 6|$ 51. $h(x) = -0.5|x|$ 52. $g(x) = |x - 2| + 5$ 53. $p(x) = 3|x + 1|$

If students need help...	If students got it...
Resources by Chapter • Practice A and Practice B • Puzzle Time	**Resources by Chapter** • Enrichment and Extension • Cumulative Review
Student Journal • Practice	Start the *next* Section
Differentiating the Lesson Skills Review Handbook	

3.1–3.3 What Did You Learn?

Core Vocabulary

quadratic function, *p. 124*
parabola, *p. 124*
vertex, *p. 124*
axis of symmetry, *p. 124*
zero of a function, *p. 132*
maximum value, *p. 137*
minimum value, *p. 137*

Core Concepts

Section 3.1
Characteristics of Quadratic Functions, *p. 124*
Graphing $f(x) = ax^2$ When $a > 0$, *p. 125*
Graphing $f(x) = ax^2$ When $a < 0$, *p. 125*

Section 3.2
Graphing $f(x) = ax^2 + c$, *p. 130*

Section 3.3
Graphing $f(x) = ax^2 + bx + c$, *p. 136*
Maximum and Minimum Values, *p. 137*

Mathematical Practices

1. Explain your plan for solving Exercise 18 on page 127.
2. How does graphing the function in Exercise 27 on page 133 help you answer the questions?
3. What definition and characteristics of the graph of a quadratic function did you use to answer Exercise 44 on page 142?

Learning Visually

- Draw a picture of a word problem before writing a verbal model. You do not have to be an artist.
- When making a review card for a word problem, include a picture. This will help you recall the information while taking a test.
- Make sure your notes are visually neat for easy recall.

Dynamic Teaching Tools

Dynamic Assessment & Progress Monitoring Tool
Interactive Whiteboard Lesson Library
Dynamic Classroom with Dynamic Investigations

ANSWERS

1. *Sample answer:* Because the highest point has a *y*-coordinate of 0, the height is the opposite of the *y*-coordinate of the lowest points. The width is the absolute value of the difference of the *x*-coordinates of the endpoints.
2. *Sample answer:* The *t*-intercept of the graph is the total time before the water balloon hits the ground.
3. *Sample answer:* the definition of maximum value/minimum value and the vertex

ANSWERS

1. The vertex is (1, 4). The axis of symmetry is $x = 1$. The domain is all real numbers. The range is $y \leq 4$. When $x < 1$, y increases as x increases. When $x > 1$, y increases as x decreases.

2. The vertex is (−2, 5). The axis of symmetry is $x = -2$. The domain is all real numbers. The range is $y \geq 5$. When $x < -2$, y increases as x decreases. When $x > -2$, y increases as x increases.

3.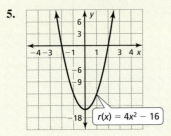

   The graph of h is a reflection in the x-axis of the graph of f.

4.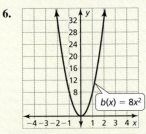

   The graph of p is a vertical stretch by a factor of 2 and a vertical translation 2 units up of the graph of f.

5.

   The graph of r is a vertical stretch by a factor of 4 and a vertical translation 16 units down of the graph of f.

6.

   The graph of b is a vertical stretch by a factor of 8 of the graph of f.

3.1–3.3 Quiz

Identify characteristics of the quadratic function and its graph. *(Section 3.1)*

1.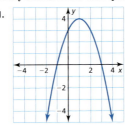
2.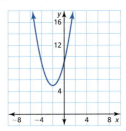

Graph the function. Compare the graph to the graph of $f(x) = x^2$. *(Section 3.1 and Section 3.2)*

3. $h(x) = -x^2$
4. $p(x) = 2x^2 + 2$
5. $r(x) = 4x^2 - 16$
6. $b(x) = 8x^2$
7. $g(x) = \frac{2}{5}x^2$
8. $m(x) = -\frac{1}{2}x^2 - 4$

Describe the transformation from the graph of f to the graph of g. Then graph f and g in the same coordinate plane. Write an equation that represents g in terms of x. *(Section 3.2)*

9. $f(x) = 2x^2 + 1$; $g(x) = f(x) + 2$
10. $f(x) = -3x^2 + 12$; $g(x) = f(x) - 9$
11. $f(x) = \frac{1}{2}x^2 - 2$; $g(x) = f(x) - 6$
12. $f(x) = 5x^2 - 3$; $g(x) = f(x) + 1$

Graph the function. Describe the domain and range. *(Section 3.3)*

13. $f(x) = -4x^2 - 4x + 7$
14. $f(x) = 2x^2 + 12x + 5$
15. $y = x^2 + 4x - 5$
16. $y = -3x^2 + 6x + 9$

Tell whether the function has a minimum value or a maximum value. Then find the value. *(Section 3.3)*

17. $f(x) = 5x^2 + 10x - 3$
18. $f(x) = -\frac{1}{2}x^2 + 2x + 16$
19. $y = -x^2 + 4x + 12$
20. $y = 2x^2 + 8x + 3$

21. The distance y (in feet) that a coconut falls after t seconds is given by the function $y = 16t^2$. Use a graph to determine how many seconds it takes for the coconut to fall 64 feet. *(Section 3.1)*

22. The function $y = -16t^2 + 25$ represents the height y (in feet) of a pinecone t seconds after falling from a tree. *(Section 3.2)*

    a. After how many seconds does the pinecone hit the ground?
    b. A second pinecone falls from a height of 36 feet. Which pinecone hits the ground in the least amount of time? Explain.

23. The function shown models the height (in feet) of a softball t seconds after it is pitched in an underhand motion. Describe the domain and range. Find the maximum height of the softball. *(Section 3.3)*

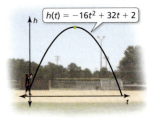

144   Chapter 3   Graphing Quadratic Functions

7.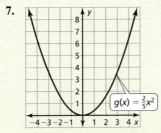

   The graph of g is a vertical shrink by a factor of $\frac{2}{5}$ of the graph of f.

8.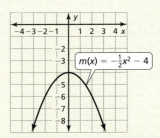

   The graph of m is a vertical shrink by a factor of $\frac{1}{2}$, a reflection in the x-axis, and a vertical translation 4 units down of the graph of f.

9–23. See Additional Answers.

Laurie's Notes

Overview of Section 3.4

Introduction
- In Math I, students were introduced to transformations of linear functions. In this chapter, students have performed several transformations on a quadratic function: translating vertically, stretching or shrinking vertically, and reflecting the function in the *x*-axis.
- In this lesson, the quadratic is translated horizontally, after the definitions of even and odd functions are given.
- The lesson develops the vertex form of a quadratic, $y = a(x - h)^2 + k$, from prior work with transformations. In the next chapter, when students learn the technique of completing the square, the vertex form could be derived from the standard form by completing the square.
- The lesson ends with a real-life example.

Resources
- Use dynamic geometry software to investigate the effect of *h* and *k* on the graph of $y = a(x - h)^2 + k$. Sliders can be used to generate values of *h* and *k*, $-5 < h < 5$ and $-5 < k < 5$.
- In lieu of creating your own, you can find online versions to download.

Formative Assessment Tips
- **Three-Minute Pause:** This technique is used when a lesson involves a particularly large amount of information or a lengthy process. The teacher takes a three-minute break after a period of instruction, and students work with partners or small groups to process the information they have just learned. They might ask one another questions or clarify what their understanding is, relative to the activity or instruction they have just experienced.
- The break in instruction allows students to resolve questions and ask for feedback from peers. Once the three minutes are up, the instruction continues until the next *Three-Minute Pause*.
- After the final *Three-Minute Pause*, any lingering questions can be written down and used by the teacher to clarify concepts taught in the lesson.

Pacing Suggestion
- The explorations prepare students to be thinking about horizontal translations before the vertex form of a quadratic is presented in the formal lesson. Have students work through the explorations, and then transition to the formal lesson.

Dynamic Teaching Tools
Dynamic Assessment & Progress Monitoring Tool
Lesson Planning Tool
Interactive Whiteboard Lesson Library
Dynamic Classroom with Dynamic Investigations

What Your Students Will Learn

- Identify even and odd functions algebraically and graphically.
- Graph quadratic functions of the form $f(x) = a(x - h)^2$ by identifying the horizontal translations of $|h|$ units right or left.
- Graph quadratic functions of the form $f(x) = a(x - h)^2 + k$ by identifying the vertex (h, k) and the axis of symmetry $x = h$.
- Model real-life problems using the vertex form of a quadratic function, $f(x) = a(x - h)^2 + k$.

Laurie's Notes

Exploration

Motivate
- Write two trinomials on the board for students to factor. Ask volunteers to share their answers.
  (a) $x^2 - 4x + 4$ $(x - 2)^2$ (b) $2x^2 + 4x + 2$ $2(x + 1)^2$
- Because $x^2 - 4x + 4 = (x - 2)^2$, the graphs of $y = x^2 - 4x + 4$ and $y = (x - 2)^2$ are the same.
- Today, students will graph quadratic functions in vertex form.

Discuss
- **Look For and Make Use of Structure:** All of the quadratic functions that students have graphed have been in standard form $y = ax^2 + bx + c$ with b and c sometimes equal to 0. In this lesson, students will investigate the vertex form of a quadratic function. Make the connection to linear functions and the different forms in which they are written. The structure of the equation gives information about the graph of the function.

Exploration 1
- There are several ways in which students will approach graphing $g(x) = (x - 2)^2$. They could make a table of values, expand the factored form to $x^2 - 4x + 4$ and find the vertex, and so on, or they may think about transformations and recall that there is a horizontal shift of 2 units right.
- **Make Sense of Problems and Persevere in Solving Them:** Take time to solicit different methods from students. In future problems, it will be helpful for students to recall a repertoire of approaches.
- In each part, students are graphing two functions. The goal is for students to recognize the translation.
- ? "How did the value of 2 affect the graph of $y = (x - 2)^2$?" The graph of $y = x^2$ was shifted 2 units right.

Exploration 2
- The approach in this exploration is the same as Exploration 1. There are different methods students may use, although they should be influenced by the results of the previous exploration!
- In each part, students are graphing two functions. The goal is for students to recognize the translation of the quadratic function that has been reflected in the x-axis.
- Be sure that students note that the general equation is written with the quantity $x - h$ in parentheses. Therefore, in the expression $(x + 2)$, $h = -2$.
- ? "How did the value of -2 affect the graph of $y = -(x + 2)^2$?" The graph of $y = -x^2$ was shifted 2 units left.

Communicate Your Answer
- **Turn and Talk:** Have partners discuss the problems together before sharing as a class.

Connecting to Next Step
- The explorations help students make sense of the horizontal translation for the graph of $y = (x - h)^2$. In the formal lesson, this is extended to the vertex form of a quadratic.

3.4 Graphing $f(x) = a(x - h)^2 + k$

Essential Question How can you describe the graph of $f(x) = a(x - h)^2$?

EXPLORATION 1 Graphing $y = a(x - h)^2$ When $h > 0$

Work with a partner. Sketch the graphs of the functions in the same coordinate plane. How does the value of h affect the graph of $y = a(x - h)^2$?

a. $f(x) = x^2$ and $g(x) = (x - 2)^2$

b. $f(x) = 2x^2$ and $g(x) = 2(x - 2)^2$

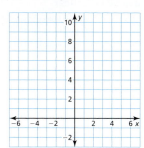
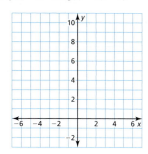

EXPLORATION 2 Graphing $y = a(x - h)^2$ When $h < 0$

Work with a partner. Sketch the graphs of the functions in the same coordinate plane. How does the value of h affect the graph of $y = a(x - h)^2$?

a. $f(x) = -x^2$ and $g(x) = -(x + 2)^2$

b. $f(x) = -2x^2$ and $g(x) = -2(x + 2)^2$

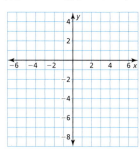
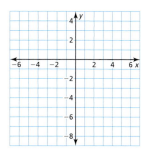

USING TOOLS STRATEGICALLY
To be proficient in math, you need to consider the available tools, such as a graphing calculator, when solving a mathematical problem.

Communicate Your Answer

3. How can you describe the graph of $f(x) = a(x - h)^2$?

4. Without graphing, describe the graph of each function. Use a graphing calculator to check your answer.

   a. $y = (x - 3)^2$
   b. $y = (x + 3)^2$
   c. $y = -(x - 3)^2$

Section 3.4 Graphing $f(x) = a(x - h)^2 + k$ 145

Dynamic Teaching Tools

Dynamic Assessment & Progress Monitoring Tool
Lesson Planning Tool
Interactive Whiteboard Lesson Library
Dynamic Classroom with Dynamic Investigations

ANSWERS

1. a.
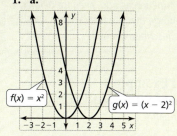

   b.
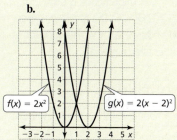

   Sample answer: The value of h causes a horizontal translation of the graph of $y = ax^2$.

2. a.
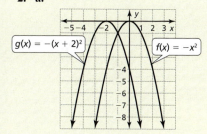

   b.
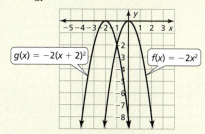

   Sample answer: The value of h causes a horizontal translation of the graph of $y = ax^2$.

3. When $h > 0$, the graph of $f(x) = a(x - h)^2$ is a horizontal translation h units to the right of the graph of $f(x) = ax^2$. When $h < 0$, the graph of $f(x) = a(x - h)^2$ is a horizontal translation $|h|$ units to the left of the graph of $f(x) = ax^2$.

4. a. The graph of $y = (x - 3)^2$ is a horizontal translation 3 units right of the graph of $y = x^2$.

   b. The graph of $y = (x + 3)^2$ is a horizontal translation 3 units left of the graph of $y = x^2$.

   c. The graph of $y = -(x - 3)^2$ is a horizontal translation 3 units right and a reflection in the x-axis of the graph of $y = x^2$.

Section 3.4 145

English Language Learners

Vocabulary
Students may be familiar with using the terms *odd* and *even* to describe numbers. Emphasize that, in this lesson, *odd* and *even* are used to describe functions. Sketch several functions and help students label each function as *odd*, *even*, or *neither*. Encourage them to refer to the sketches when needed.

Extra Example 1
Determine whether each function is *even*, *odd*, or *neither*.

a. $f(x) = 3x$ The function is odd.
b. $g(x) = 2x^2 - 6$ The function is even.
c. $h(x) = 3x^2 - 2x + 4$ The function is neither even nor odd.

MONITORING PROGRESS ANSWERS
1. odd
2. neither
3. even

3.4 Lesson

Core Vocabulary
even function, p. 146
odd function, p. 146
vertex form (of a quadratic function), p. 148
Previous
reflection

STUDY TIP
The graph of an odd function looks the same after a 180° rotation about the origin.

What You Will Learn
- Identify even and odd functions.
- Graph quadratic functions of the form $f(x) = a(x - h)^2$.
- Graph quadratic functions of the form $f(x) = a(x - h)^2 + k$.
- Model real-life problems using $f(x) = a(x - h)^2 + k$.

Identifying Even and Odd Functions

Core Concept

Even and Odd Functions
A function $y = f(x)$ is **even** when $f(-x) = f(x)$ for each x in the domain of f. The graph of an even function is symmetric about the y-axis.

A function $y = f(x)$ is **odd** when $f(-x) = -f(x)$ for each x in the domain of f. The graph of an odd function is symmetric about the origin. A graph is *symmetric about the origin* when it looks the same after reflections in the x-axis and then in the y-axis.

EXAMPLE 1 Identifying Even and Odd Functions

Determine whether each function is *even*, *odd*, or *neither*.

a. $f(x) = 2x$ b. $g(x) = x^2 - 2$ c. $h(x) = 2x^2 + x - 2$

SOLUTION

a. $f(x) = 2x$ — Write the original function.
   $f(-x) = 2(-x)$ — Substitute $-x$ for x.
   $= -2x$ — Simplify.
   $= -f(x)$ — Substitute $f(x)$ for $2x$.

▶ Because $f(-x) = -f(x)$, the function is odd.

b. $g(x) = x^2 - 2$ — Write the original function.
   $g(-x) = (-x)^2 - 2$ — Substitute $-x$ for x.
   $= x^2 - 2$ — Simplify.
   $= g(x)$ — Substitute $g(x)$ for $x^2 - 2$.

▶ Because $g(-x) = g(x)$, the function is even.

c. $h(x) = 2x^2 + x - 2$ — Write the original function.
   $h(-x) = 2(-x)^2 + (-x) - 2$ — Substitute $-x$ for x.
   $= 2x^2 - x - 2$ — Simplify.

▶ Because $h(x) = 2x^2 + x - 2$ and $-h(x) = -2x^2 - x + 2$, you can conclude that $h(-x) \ne h(x)$ and $h(-x) \ne -h(x)$. So, the function is neither even nor odd.

STUDY TIP
Most functions are neither even nor odd.

Monitoring Progress Help in English and Spanish at *BigIdeasMath.com*

Determine whether the function is *even*, *odd*, or *neither*.

1. $f(x) = 5x$ 2. $g(x) = 2^x$ 3. $h(x) = 2x^2 + 3$

146 Chapter 3 Graphing Quadratic Functions

Laurie's Notes Teacher Actions

- Define *even functions* and *odd functions*.
- The notation can be confusing, so be sure to use additional words. Say, "A function is even when you evaluate the function at x and at $-x$ and the answers (y-values) are the same."
- Similarly, "A function is odd when you evaluate the function at x and at $-x$ and the answers (y-values) are opposites."
- The *Study Tip* is helpful in recognizing an *odd function* from its graph.
- Work through each part of the example as shown. Discuss the *Study Tip*. Most functions are neither even nor odd.

Graphing $f(x) = a(x - h)^2$

Core Concept

Graphing $f(x) = a(x - h)^2$

- When $h > 0$, the graph of $f(x) = a(x - h)^2$ is a horizontal translation h units right of the graph of $f(x) = ax^2$.
- When $h < 0$, the graph of $f(x) = a(x - h)^2$ is a horizontal translation $|h|$ units left of the graph of $f(x) = ax^2$.

The vertex of the graph of $f(x) = a(x - h)^2$ is $(h, 0)$, and the axis of symmetry is $x = h$.

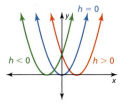

EXAMPLE 2 Graphing $y = a(x - h)^2$

Graph $g(x) = \frac{1}{2}(x - 4)^2$. Compare the graph to the graph of $f(x) = x^2$.

ANOTHER WAY
In Step 3, you could instead choose two x-values greater than the x-coordinate of the vertex.

SOLUTION

Step 1 Graph the axis of symmetry. Because $h = 4$, graph $x = 4$.

Step 2 Plot the vertex. Because $h = 4$, plot $(4, 0)$.

Step 3 Find and plot two more points on the graph. Choose two x-values less than the x-coordinate of the vertex. Then find $g(x)$ for each x-value.

When $x = 0$:
$g(0) = \frac{1}{2}(0 - 4)^2$
$= 8$

When $x = 2$:
$g(2) = \frac{1}{2}(2 - 4)^2$
$= 2$

So, plot $(0, 8)$ and $(2, 2)$.

Step 4 Reflect the points plotted in Step 3 in the axis of symmetry. So, plot $(8, 8)$ and $(6, 2)$.

Step 5 Draw a smooth curve through the points.

STUDY TIP
From the graph, you can see that $f(x) = x^2$ is an even function. However, $g(x) = \frac{1}{2}(x - 4)^2$ is neither even nor odd.

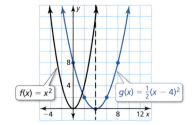

▶ Both graphs open up. The graph of g is wider than the graph of f. The axis of symmetry $x = 4$ and the vertex $(4, 0)$ of the graph of g are 4 units right of the axis of symmetry $x = 0$ and the vertex $(0, 0)$ of the graph of f. So, the graph of g is a translation 4 units right and a vertical shrink by a factor of $\frac{1}{2}$ of the graph of f.

Monitoring Progress Help in English and Spanish at *BigIdeasMath.com*

Graph the function. Compare the graph to the graph of $f(x) = x^2$.

4. $g(x) = 2(x + 5)^2$
5. $h(x) = -(x - 2)^2$

Section 3.4 Graphing $f(x) = a(x - h)^2 + k$ 147

Differentiated Instruction

Graphic Organizer
Have students make a table like this one for Example 2 to organize the attributes of a function they are comparing to $f(x) = x^2$. Encourage students to complete a table for each function and then use the information to write their comparison.

Function	$f(x) = x^2$	$f(x) = \frac{1}{2}(x - 4)^2$
Axis of symmetry	$x = 0$	$x = 4$
Vertex	$(0, 0)$	$(4, 0)$
Type of shrink or stretch (factor)	—	vertical shrink $\left(\frac{1}{2}\right)$
Translation	—	4 units right
Reflection	—	—

Extra Example 2
Graph $g(x) = \frac{1}{2}(x + 2)^2$. Compare the graph to the graph of $f(x) = x^2$.

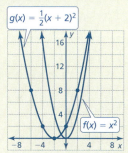

Both graphs open up. The graph of g is wider than the graph of f. The axis of symmetry, $x = -2$, and the vertex, $(-2, 0)$, of the graph of g are 2 units left of the axis of symmetry, $x = 0$, and the vertex, $(0, 0)$, of the graph of f. So, the graph of g is a translation 2 units left and a vertical shrink by a factor of $\frac{1}{2}$ of the graph of f.

MONITORING PROGRESS ANSWERS
4–5. See Additional Answers.

Laurie's Notes **Teacher Actions**

- Write the *Core Concept*, and connect it to the explorations.
- What is confusing for students is that when $h > 0$, the expression is $(x - h)$ and the shift is h units right. In Example 2, $h = 4$ and the transformation is 4 units right. In *Monitoring Progress* Question 4, where $h = -5$, the expression becomes $(x + 5)$ and the shift is 5 units left.
- Work through Example 2 as shown. Take time to compare graphs of $f(x)$ and $g(x)$, discussing the transformations.

Extra Example 3

Graph $g(x) = -3(x + 2)^2 + 2$. Compare the graph to the graph of $f(x) = x^2$.

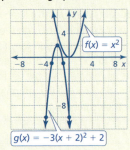

The graph of g opens down and is narrower than the graph of f. The vertex of the graph of g, $(-2, 2)$, is 2 units to the left and 2 units up from the vertex of the graph of f, $(0, 0)$. So, the graph of g is a vertical stretch by a factor of 3, a reflection in the x-axis, and a translation 2 units left and 2 units up of the graph of f.

Extra Example 4

Consider function g in Extra Example 3. Graph $f(x) = g(x - 4)$.

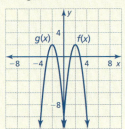

MONITORING PROGRESS ANSWERS

6.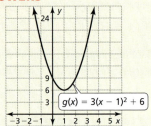

The graph of g is a vertical stretch by a factor of 3, and a translation 1 unit right and 6 units up of the graph of f.

7–8. See Additional Answers.

Graphing $f(x) = a(x - h)^2 + k$

Core Concept

Graphing $f(x) = a(x - h)^2 + k$

The **vertex form** of a quadratic function is $f(x) = a(x - h)^2 + k$, where $a \neq 0$. The graph of $f(x) = a(x - h)^2 + k$ is a translation h units horizontally and k units vertically of the graph of $f(x) = ax^2$.

The vertex of the graph of $f(x) = a(x - h)^2 + k$ is (h, k), and the axis of symmetry is $x = h$.

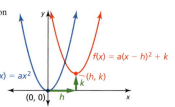

EXAMPLE 3 Graphing $y = a(x - h)^2 + k$

Graph $g(x) = -2(x + 2)^2 + 3$. Compare the graph to the graph of $f(x) = x^2$.

SOLUTION

Step 1 Graph the axis of symmetry. Because $h = -2$, graph $x = -2$.

Step 2 Plot the vertex. Because $h = -2$ and $k = 3$, plot $(-2, 3)$.

Step 3 Find and plot two more points on the graph. Choose two x-values less than the x-coordinate of the vertex. Then find $g(x)$ for each x-value. So, plot $(-4, -5)$ and $(-3, 1)$.

x	−4	−3
g(x)	−5	1

Step 4 Reflect the points plotted in Step 3 in the axis of symmetry. So, plot $(-1, 1)$ and $(0, -5)$.

Step 5 Draw a smooth curve through the points.

▶ The graph of g opens down and is narrower than the graph of f. The vertex of the graph of g, $(-2, 3)$, is 2 units left and 3 units up of the vertex of the graph of f, $(0, 0)$. So, the graph of g is a vertical stretch by a factor of 2, a reflection in the x-axis, and a translation 2 units left and 3 units up of the graph of f.

EXAMPLE 4 Transforming the Graph of $y = a(x - h)^2 + k$

Consider function g in Example 3. Graph $f(x) = g(x + 5)$.

SOLUTION

The function f is of the form $y = g(x - h)$, where $h = -5$. So, the graph of f is a horizontal translation 5 units left of the graph of g. To graph f, subtract 5 from the x-coordinates of the points on the graph of g.

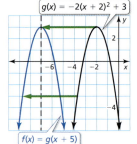

Monitoring Progress Help in English and Spanish at BigIdeasMath.com

Graph the function. Compare the graph to the graph of $f(x) = x^2$.

6. $g(x) = 3(x - 1)^2 + 6$
7. $h(x) = \frac{1}{2}(x + 4)^2 - 2$

8. Consider function g in Example 3. Graph $f(x) = g(x) - 3$.

148 Chapter 3 Graphing Quadratic Functions

Laurie's Notes   Teacher Actions

- **Three-Minute Pause:** For Example 3, pause three minutes for students to confer with partners or in small groups. Have them discuss what they know about transformations and the vertex form. There are four transformations (reflection, vertical stretch, horizontal shift, vertical shift). Does the order of the transformations matter? What is the vertex? Can they visualize the graph?
- Display the graph of Example 3 as you discuss Example 4. See the *Teaching Strategy* on page T-128 for graphing with technology.

Modeling Real-Life Problems

EXAMPLE 5 Modeling with Mathematics

Water fountains are usually designed to give a specific visual effect. For example, the water fountain shown consists of streams of water that are shaped like parabolas. Notice how the streams are designed to land on the underwater spotlights. Write and graph a quadratic function that models the path of a stream of water with a maximum height of 5 feet, represented by a vertex of (3, 5), landing on a spotlight 6 feet from the water jet, represented by (6, 0).

SOLUTION

1. **Understand the Problem** You know the vertex and another point on the graph that represents the parabolic path. You are asked to write and graph a quadratic function that models the path.

2. **Make a Plan** Use the given points and the vertex form to write a quadratic function. Then graph the function.

3. **Solve the Problem**
   Use the vertex form, vertex (3, 5), and point (6, 0) to find the value of a.

$f(x) = a(x - h)^2 + k$	Write the vertex form of a quadratic function.
$f(x) = a(x - 3)^2 + 5$	Substitute 3 for h and 5 for k.
$0 = a(6 - 3)^2 + 5$	Substitute 6 for x and 0 for $f(x)$.
$0 = 9a + 5$	Simplify.
$-\frac{5}{9} = a$	Solve for a.

   So, $f(x) = -\frac{5}{9}(x - 3)^2 + 5$ models the path of a stream of water. Now graph the function.

   Step 1 Graph the axis of symmetry. Because $h = 3$, graph $x = 3$.

   Step 2 Plot the vertex, (3, 5).

   Step 3 Find and plot two more points on the graph. Because the x-axis represents the water surface, the graph should only contain points with nonnegative values of $f(x)$. You know that (6, 0) is on the graph. To find another point, choose an x-value between $x = 3$ and $x = 6$. Then find the corresponding value of $f(x)$.

   $f(4.5) = -\frac{5}{9}(4.5 - 3)^2 + 5 = 3.75$

   So, plot (6, 0) and (4.5, 3.75).

   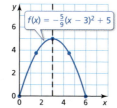

   Step 4 Reflect the points plotted in Step 3 in the axis of symmetry. So, plot (0, 0) and (1.5, 3.75).

   Step 5 Draw a smooth curve through the points.

4. **Look Back** Use a graphing calculator to graph $f(x) = -\frac{5}{9}(x - 3)^2 + 5$. Use the *maximum* feature to verify that the maximum value is 5. Then use the *zero* feature to verify that $x = 6$ is a zero of the function.

Monitoring Progress Help in English and Spanish at BigIdeasMath.com

9. **WHAT IF?** The vertex is (3, 6). Write and graph a quadratic function that models the path.

Section 3.4 Graphing $f(x) = a(x - h)^2 + k$ 149

Extra Example 5

Use the information in Example 5. Write and graph a quadratic function that models the path of a stream of water with a maximum height of 4 feet, represented by a vertex of (3, 4), landing on a spotlight 6 feet from the water jet, represented by (6, 0).
$f(x) = -\frac{4}{9}(x - 3)^2 + 4$

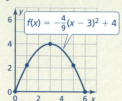

MONITORING PROGRESS ANSWER

9. $f(x) = -\frac{2}{3}(x - 3)^2 + 6$

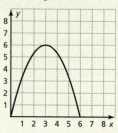

Laurie's Notes Teacher Actions

- **Make Sense of Problems and Persevere in Solving Them** and **Model with Mathematics:** A quick search will result in many online videos to motivate this example.
- **Three-Minute Pause:** Read the problem. Pause three minutes for students to confer with partners or in small groups. Discuss what is known and what they are trying to find. Ask probing questions without giving answers. Ask volunteers to share their progress, and then give additional work time. Trust that students can solve this problem.

Closure

- **Exit Ticket:** Given $f(x) = \frac{1}{2}(x + 8)^2 + 4$, tell what you know about the function and sketch its graph. *The graph opens up and is wider than the graph of $y = x^2$. The axis of symmetry is $x = -8$. The vertex is $(-8, 4)$. The domain is all real numbers. The range is $y \geq 4$. So, the graph of f is a vertical shrink by a factor of $\frac{1}{2}$ and a translation 8 units left and 4 units up of the graph of $y = x^2$. Check students' graphs.*

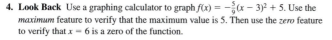

Assignment Guide and Homework Check

ASSIGNMENT

Basic: 1–4, 5–55 odd, 63, 70, 79–82
Average: 1–4, 6–66 even, 70, 79–82
Advanced: 1–4, 12–16 even, 22, 26–38 even, 42–48 even, 52–68 even, 69–82

HOMEWORK CHECK

Basic: 5, 23, 39, 49, 63
Average: 8, 24, 40, 50, 64
Advanced: 12, 26, 42, 52, 64

ANSWERS

1. Sample answer: The graph of an even function is symmetric about the y-axis. The graph of an odd function is symmetric about the origin.
2. Sample answer: $y = 5(x-1)^2 + 2$
3. The graph of g is a horizontal translation h units right if h is positive or $|h|$ units left if h is negative, and a vertical translation k units up if k is positive or $|k|$ units down if k is negative of the graph of f.
4. $f(x) = 2(x+0)^2$; It is the only function that is not a horizontal translation of the parent function $g(x) = x^2$.
5. neither
6. even
7. neither
8. neither
9. even
10. odd
11. neither
12. even
13. even
14. neither
15. neither
16. even
17. odd
18. neither
19. $(-1, 0); x = -1$
20. $(6, 0); x = 6$
21. $(4, 0); x = 4$
22. $(-9, 0); x = -9$

3.4 Exercises

Dynamic Solutions available at *BigIdeasMath.com*

Vocabulary and Core Concept Check

1. **VOCABULARY** Compare the graph of an even function with the graph of an odd function.
2. **OPEN-ENDED** Write a quadratic function whose graph has a vertex of (1, 2).
3. **WRITING** Describe the transformation from the graph of $f(x) = ax^2$ to the graph of $g(x) = a(x - h)^2 + k$.
4. **WHICH ONE DOESN'T BELONG?** Which function does *not* belong with the other three? Explain your reasoning.

   $f(x) = 8(x + 4)^2$ $f(x) = (x - 2)^2 + 4$ $f(x) = 2(x + 0)^2$ $f(x) = 3(x + 1)^2 + 1$

Monitoring Progress and Modeling with Mathematics

In Exercises 5–12, determine whether the function is *even*, *odd*, or *neither*. (See Example 1.)

5. $f(x) = 4x + 3$
6. $g(x) = 3x^2$
7. $h(x) = 5^x + 2$
8. $m(x) = 2x^2 - 7x$
9. $p(x) = -x^2 + 8$
10. $f(x) = -\frac{1}{2}x$
11. $n(x) = 2x^2 - 7x + 3$
12. $r(x) = -6x^2 + 5$

In Exercises 13–18, determine whether the function represented by the graph is *even*, *odd*, or *neither*.

13.
14.
15.
16.
17.
18.

In Exercises 19–22, find the vertex and the axis of symmetry of the graph of the function.

19. $f(x) = 3(x + 1)^2$
20. $f(x) = \frac{1}{4}(x - 6)^2$
21. $y = -\frac{1}{8}(x - 4)^2$
22. $y = -5(x + 9)^2$

In Exercises 23–28, graph the function. Compare the graph to the graph of $f(x) = x^2$. (See Example 2.)

23. $g(x) = 2(x + 3)^2$
24. $p(x) = 3(x - 1)^2$
25. $r(x) = \frac{1}{4}(x + 10)^2$
26. $n(x) = \frac{1}{3}(x - 6)^2$
27. $d(x) = \frac{1}{5}(x - 5)^2$
28. $q(x) = 6(x + 2)^2$

29. **ERROR ANALYSIS** Describe and correct the error in determining whether the function $f(x) = x^2 + 3$ is even, odd, or neither.

   $f(x) = x^2 + 3$
   $f(-x) = (-x)^2 + 3$
   $= x^2 + 3$
   $= f(x)$
   So, $f(x)$ is an odd function.

30. **ERROR ANALYSIS** Describe and correct the error in finding the vertex of the graph of the function.

   $y = -(x + 8)^2$
   Because $h = -8$, the vertex is $(0, -8)$.

23.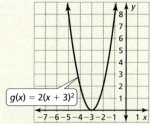

   The graph of g is a horizontal translation 3 units left and a vertical stretch by a factor of 2 of the graph of f.

24.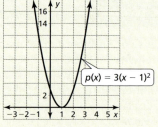

   The graph of p is a horizontal translation 1 unit right and a vertical stretch by a factor of 3 of the graph of f.

25–30. See Additional Answers.

In Exercises 31–34, find the vertex and the axis of symmetry of the graph of the function.

31. $y = -6(x + 4)^2 - 3$ **32.** $f(x) = 3(x - 3)^2 + 6$

33. $f(x) = -4(x + 3)^2 + 1$ **34.** $y = -(x - 6)^2 - 5$

In Exercises 35–38, match the function with its graph.

35. $y = -(x + 1)^2 - 3$ **36.** $y = -\frac{1}{2}(x - 1)^2 + 3$

37. $y = \frac{1}{3}(x - 1)^2 + 3$ **38.** $y = 2(x + 1)^2 - 3$

A.   **B.**

C.   **D.**

In Exercises 39–44, graph the function. Compare the graph to the graph of $f(x) = x^2$. (See Example 3.)

39. $h(x) = (x - 2)^2 + 4$ **40.** $g(x) = (x + 1)^2 - 7$

41. $r(x) = 4(x - 1)^2 - 5$ **42.** $n(x) = -(x + 4)^2 + 2$

43. $g(x) = -\frac{1}{3}(x + 3)^2 - 2$ **44.** $r(x) = \frac{1}{2}(x - 2)^2 - 4$

In Exercises 45–48, let $f(x) = (x - 2)^2 + 1$. Match the function with its graph.

45. $g(x) = f(x - 1)$ **46.** $r(x) = f(x + 2)$

47. $h(x) = f(x) + 2$ **48.** $p(x) = f(x) - 3$

A.   **B.**

C.   **D.**

In Exercises 49–54, graph g. (See Example 4.)

49. $f(x) = 2(x - 1)^2 + 1$; $g(x) = f(x + 3)$

50. $f(x) = -(x + 1)^2 + 2$; $g(x) = \frac{1}{2}f(x)$

51. $f(x) = -3(x + 5)^2 - 6$; $g(x) = 2f(x)$

52. $f(x) = 5(x - 3)^2 - 1$; $g(x) = f(x) - 6$

53. $f(x) = (x + 3)^2 + 5$; $g(x) = f(x - 4)$

54. $f(x) = -2(x - 4)^2 - 8$; $g(x) = -f(x)$

55. MODELING WITH MATHEMATICS The height (in meters) of a bird diving to catch a fish is represented by $h(t) = 5(t - 2.5)^2$, where t is the number of seconds after beginning the dive.

 a. Graph h.

 b. Another bird's dive is represented by $r(t) = 2h(t)$. Graph r.

 c. Compare the graphs. Which bird starts its dive from a greater height? Explain.

56. MODELING WITH MATHEMATICS A kicker punts a football. The height (in yards) of the football is represented by $f(x) = -\frac{1}{9}(x - 30)^2 + 25$, where x is the horizontal distance (in yards) from the kicker's goal line.

 a. Graph f. Describe the domain and range.

 b. On the next possession, the kicker punts the football. The height of the football is represented by $g(x) = f(x + 5)$. Graph g. Describe the domain and range.

 c. Compare the graphs. On which possession does the kicker punt closer to his goal line? Explain.

In Exercises 57–62, write a quadratic function in vertex form whose graph has the given vertex and passes through the given point.

57. vertex: $(1, 2)$; passes through $(3, 10)$

58. vertex: $(-3, 5)$; passes through $(0, -14)$

59. vertex: $(-2, -4)$; passes through $(-1, -6)$

60. vertex: $(1, 8)$; passes through $(3, 12)$

61. vertex: $(5, -2)$; passes through $(7, 0)$

62. vertex: $(-5, -1)$; passes through $(-2, 2)$

Section 3.4   Graphing $f(x) = a(x - h)^2 + k$ 151

Dynamic Teaching Tools

Dynamic Assessment & Progress Monitoring Tool

Interactive Whiteboard Lesson Library

Dynamic Classroom with Dynamic Investigations

ANSWERS

31. $(-4, -3)$; $x = -4$

32. $(3, 6)$; $x = 3$

33. $(-3, 1)$; $x = -3$

34. $(6, -5)$; $x = 6$

35. C   **36.** A

37. D   **38.** B

39.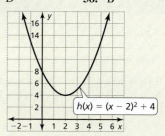

The graph of h is a translation 2 units right and 4 units up of the graph of f.

40.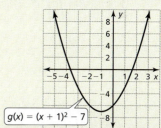

The graph of g is a translation 1 unit left and 7 units down of the graph of f.

41.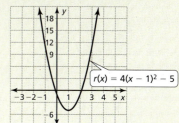

The graph of r is a vertical stretch by a factor of 4, and a translation 1 unit right and 5 units down of the graph of f.

42.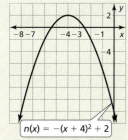

The graph of n is a reflection in the x-axis, and a translation 4 units left and 2 units up of the graph of f.

43–44. See Additional Answers.

45. A   **46.** C

47. B   **48.** D

49–56. See Additional Answers.

57. $f(x) = 2(x - 1)^2 + 2$

58. $f(x) = -\frac{19}{9}(x + 3)^2 + 5$

59. $f(x) = -2(x + 2)^2 - 4$

60. $f(x) = (x - 1)^2 + 8$

61. $f(x) = \frac{1}{2}(x - 5)^2 - 2$

62. $f(x) = \frac{1}{3}(x + 5)^2 - 1$

Section 3.4   151

ANSWERS

63–78. See Additional Answers.

79. no; *Sample answer:* If the absolute value function includes a horizontal translation, it will not be symmetric about the y-axis.

80. $x = 0, x = 1$

81. $x = -3, x = 8$

82. $x = 3, x = -3$

Mini-Assessment

1. Determine whether the function $f(x) = x^2 + 5$ is *even*, *odd*, or *neither*. **even**

2. Graph $g(x) = 2(x - 3)^2$. Compare the graph to the graph of $f(x) = x^2$.

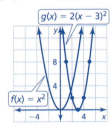

The graph of g is a translation 3 units right and a vertical stretch by a factor of 2 of the graph of f.

3. Graph $g(x) = -(x + 3)^2 - 1$. Compare the graph to the graph of $f(x) = x^2$.

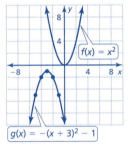

The graph of g is a reflection in the x-axis, and a translation 3 units left and 1 unit down of the graph of f.

4. A toy rocket is launched from the ground and travels in a path shaped like a parabola. Write and graph a quadratic equation that models the path of a rocket that lands 20 feet from where it was launched and reaches a maximum height of 36 feet, represented by the vertex (10, 36).

See Additional Answers.

152   Chapter 3

---

63. MODELING WITH MATHEMATICS A portion of a roller coaster track is in the shape of a parabola. Write and graph a quadratic function that models this portion of the roller coaster with a maximum height of 90 feet, represented by a vertex of (25, 90), passing through the point (50, 0). *(See Example 5.)*

64. MODELING WITH MATHEMATICS A flare is launched from a boat and travels in a parabolic path until reaching the water. Write and graph a quadratic function that models the path of the flare with a maximum height of 300 meters, represented by a vertex of (59, 300), landing in the water at the point (119, 0).

In Exercises 65–68, rewrite the quadratic function in vertex form.

65. $y = 2x^2 - 8x + 4$ **66.** $y = 3x^2 + 6x - 1$

67. $f(x) = -5x^2 + 10x + 3$

68. $f(x) = -x^2 - 4x + 2$

69. REASONING Can a function be symmetric about the x-axis? Explain.

70. HOW DO YOU SEE IT? The graph of a quadratic function is shown. Determine which symbols to use to complete the vertex form of the quadratic function. Explain your reasoning.

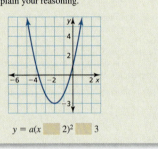

$y = a(x \;\square\; 2)^2 \;\square\; 3$

Maintaining Mathematical Proficiency
Reviewing what you learned in previous grades and lessons

Solve the equation. *(Section 2.4)*

80. $x(x - 1) = 0$ **81.** $(x + 3)(x - 8) = 0$ **82.** $(3x - 9)(4x + 12) = 0$

152   Chapter 3   Graphing Quadratic Functions

In Exercises 71–74, describe the transformation from the graph of f to the graph of h. Write an equation that represents h in terms of x.

71. $f(x) = -(x + 1)^2 - 2$ **72.** $f(x) = 2(x - 1)^2 + 1$
   $h(x) = f(x) + 4$ $h(x) = f(x - 5)$

73. $f(x) = 4(x - 2)^2 + 3$ **74.** $f(x) = -(x + 5)^2 - 6$
   $h(x) = 2f(x)$ $h(x) = \frac{1}{3}f(x)$

75. REASONING The graph of $y = x^2$ is translated 2 units right and 5 units down. Write an equation for the function in vertex form and in standard form. Describe advantages of writing the function in each form.

76. THOUGHT PROVOKING Which of the following are true? Justify your answers.

a. Any constant multiple of an even function is even.

b. Any constant multiple of an odd function is odd.

c. The sum or difference of two even functions is even.

d. The sum or difference of two odd functions is odd.

e. The sum or difference of an even function and an odd function is odd.

77. COMPARING FUNCTIONS A cross section of a birdbath can be modeled by $y = \frac{1}{81}(x - 18)^2 - 4$, where x and y are measured in inches. The graph shows the cross section of another birdbath.

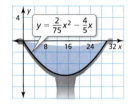

a. Which birdbath is deeper? Explain.

b. Which birdbath is wider? Explain.

78. REASONING Compare the graphs of $y = 2x^2 + 8x + 8$ and $y = x^2$ without graphing the functions. How can factoring help you compare the parabolas? Explain.

79. MAKING AN ARGUMENT Your friend says all absolute value functions are even because of their symmetry. Is your friend correct? Explain.

If students need help...	If students got it...
Resources by Chapter • Practice A and Practice B • Puzzle Time	Resources by Chapter • Enrichment and Extension • Cumulative Review
Student Journal • Practice	Start the *next* Section
Differentiating the Lesson Skills Review Handbook	

Laurie's Notes

Overview of Section 3.5

Introduction
- Earlier in the chapter, the standard form and vertex form of a quadratic were studied. In this lesson, the intercept form is presented and connections are made to the zeros of a function.
- The lesson begins with students graphing a quadratic written in intercept form. Then students are asked to use the intercepts to find the zeros of the function.
- The second half of the lesson focuses on the characteristics of quadratic functions so that the equations can be written and the functions graphed.

Formative Assessment Tips
- **Agree-Disagree Statement:** This technique has two parts. First, students are given a statement with which they may agree or disagree, or they could indicate that they need additional information in order to decide. They are then asked to explain why they agree or disagree, or why they might need additional information. The second part of the technique involves students describing how they could investigate, figure out, or test their thinking.
- This technique gives students the opportunity to think about their own understanding of a concept or process. Further, it helps students practice the skill of actively investigating their own thinking. In listening to their own words, and in listening to the thinking of others, students can solidify or modify their own beliefs.
- This technique can be used at the beginning of a class or unit of study to assess students' understanding of a concept. Student responses will help inform your own instruction, knowing what experiences students need to help guide their learning.
- *Note:* The statements made are not low cognitive demand questions such as true-false or yes-no questions. The goal is to uncover student conception about bigger ideas, such as whether all functions have an inverse or whether multiplying positive numbers gives a product greater than either factor.

Pacing Suggestion
- This is a long lesson with the exploration helping students to make the connection between the zeros of a function and the x-intercepts of its graph. Refer to the exploration as you work through the formal lesson.

Dynamic Teaching Tools

Dynamic Assessment & Progress Monitoring Tool
Lesson Planning Tool
Interactive Whiteboard Lesson Library
Dynamic Classroom with Dynamic Investigations

What Your Students Will Learn

- Graph quadratic functions of the form $f(x) = a(x − p)(x − q)$ by identifying the x-intercepts and axis of symmetry.
- Use the intercept form of a quadratic function to find the zeros of the function.
- Use characteristics to graph quadratic functions and to write quadratic functions in standard form or intercept form.

Laurie's Notes

Exploration

Motivate

- To motivate the explorations, share the graphs of two quadratic functions: $f(x) = (x − 3)^2$ and $g(x) = −(x + 5)^2$.
- Ask students to identify the vertex of each: $(3, 0)$ and $(−5, 0)$.

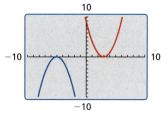

- ? "What are the zeros of each function?" 3 and −5, respectively
- Probe students to think of other ways in which the functions could be written, besides standard form.
- Write: $f(x) = (x − 3)^2 = $  . In other words, $f(x) = (x − 3)^2 = (x − 3)(x − 3)$. Say, "When $x = 3$, the function equals 0. So the zero of the function is the x-coordinate of the vertex."
- Ask students to represent $g(x)$ as the product of two factors. Interpret the vertex and zero of the function.

Exploration Note

- Before having students begin the exploration, be sure to complete the *Motivate* activity. This should be enough of an introduction for students to successfully explore the graphs shown.

Exploration 1

- Students can determine the x- and y-intercepts of each graph. Assure students that this is sufficient information in order to write the equations in the form $f(x) = (x − p)(x − q)$ or $f(x) = −(x − p)(x − q)$.
- Students should check their equations using a graphing calculator.
- **Reason Abstractly and Quantitatively:** Listen for correct reasoning as to why their equations make sense.

Communicate Your Answer

- **Use Appropriate Tools Strategically:** Have students explore the questions with their partners using a graphing calculator.

Connecting to Next Step

- The explorations help students make sense of the intercept form of a quadratic. In the formal lesson, this is stated along with how to find the axis of symmetry when the quadratic is written in the intercept form.

3.5 Graphing $f(x) = a(x - p)(x - q)$

Essential Question What are some of the characteristics of the graph of $f(x) = a(x - p)(x - q)$?

EXPLORATION 1 Using Zeros to Write Functions

Work with a partner. Each graph represents a function of the form $f(x) = (x - p)(x - q)$ or $f(x) = -(x - p)(x - q)$. Write the function represented by each graph. Explain your reasoning.

a.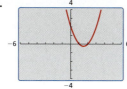

b.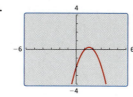

c.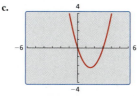

d.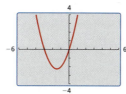

e.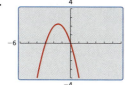

f.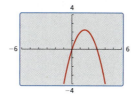

g.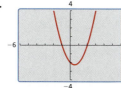

h.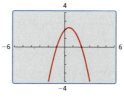

CONSTRUCTING VIABLE ARGUMENTS

To be proficient in math, you need to justify your conclusions and communicate them to others.

Communicate Your Answer

2. What are some of the characteristics of the graph of $f(x) = a(x - p)(x - q)$?

3. Consider the graph of $f(x) = a(x - p)(x - q)$.

   a. Does changing the sign of a change the x-intercepts? Does changing the sign of a change the y-intercept? Explain your reasoning.

   b. Does changing the value of p change the x-intercepts? Does changing the value of p change the y-intercept? Explain your reasoning.

Differentiated Instruction

Kinesthetic

Ask two students to assist you at the board to solve a problem similar to Example 1 or Example 2. Assign finding the *x*-intercepts, the axis of symmetry, and the vertex of a function to one student. Assign graphing the *x*-intercepts, axis of symmetry, the vertex, and the parabola to the second student. Have students discuss how the intercepts, axis, and vertex relate to the function and its graph.

Extra Example 1

Graph $f(x) = (x - 1)(x + 3)$. Describe the domain and range.

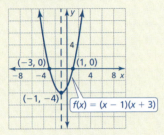

The domain is all real numbers. The range is $y \geq -4$.

3.5 Lesson

Core Vocabulary

intercept form, p. 154

What You Will Learn

- Graph quadratic functions of the form $f(x) = a(x - p)(x - q)$.
- Use intercept form to find zeros of functions.
- Use characteristics to graph and write quadratic functions.

Graphing $f(x) = a(x - p)(x - q)$

You have already graphed quadratic functions written in several different forms, such as $f(x) = ax^2 + bx + c$ (standard form) and $g(x) = a(x - h)^2 + k$ (vertex form). Quadratic functions can also be written in **intercept form**, $f(x) = a(x - p)(x - q)$, where $a \neq 0$. In this form, the polynomial that defines a function is in factored form and the *x*-intercepts of the graph can be easily determined.

Core Concept

Graphing $f(x) = a(x - p)(x - q)$

- The *x*-intercepts are *p* and *q*.
- The axis of symmetry is halfway between $(p, 0)$ and $(q, 0)$. So, the axis of symmetry is $x = \dfrac{p + q}{2}$.
- The graph opens up when $a > 0$, and the graph opens down when $a < 0$.

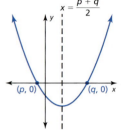

EXAMPLE 1 Graphing $f(x) = a(x - p)(x - q)$

Graph $f(x) = -(x + 1)(x - 5)$. Describe the domain and range.

SOLUTION

Step 1 Identify the *x*-intercepts. Because the *x*-intercepts are $p = -1$ and $q = 5$, plot $(-1, 0)$ and $(5, 0)$.

Step 2 Find and graph the axis of symmetry.
$$x = \frac{p + q}{2} = \frac{-1 + 5}{2} = 2$$

Step 3 Find and plot the vertex.

The *x*-coordinate of the vertex is 2. To find the *y*-coordinate of the vertex, substitute 2 for *x* and simplify.
$$f(2) = -(2 + 1)(2 - 5) = 9$$
So, the vertex is $(2, 9)$.

Step 4 Draw a parabola through the vertex and the points where the *x*-intercepts occur.

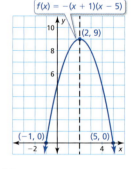

▶ The domain is all real numbers. The range is $y \leq 9$.

Laurie's Notes — Teacher Actions

? Agree-Disagree Statement: "All quadratic functions can be written $f(x) = (x - p)(x - q)$, where *p* and *q* are any real numbers. Agree/Disagree. Explain." Students should consider work from Chapter 2 and the exploration on page 153. What may concern students is the parent function and writing $f(x) = (x - 0)(x - 0)$. They may also wonder "What if the quadratic doesn't have *x*-intercepts?"

- Write the *Core Concept*. For quadratics that do have *x*-intercepts, discuss the equation, the axis of symmetry, and the effect of the constant *a*.

EXAMPLE 2 **Graphing a Quadratic Function**

Graph $f(x) = 2x^2 - 8$. Describe the domain and range.

SOLUTION

Step 1 Rewrite the quadratic function in intercept form.

$f(x) = 2x^2 - 8$ Write the function.

$= 2(x^2 - 4)$ Factor out common factor.

$= 2(x + 2)(x - 2)$ Difference of two squares pattern

Step 2 Identify the x-intercepts. Because the x-intercepts are $p = -2$ and $q = 2$, plot $(-2, 0)$ and $(2, 0)$.

Step 3 Find and graph the axis of symmetry.

$$x = \frac{p + q}{2} = \frac{-2 + 2}{2} = 0$$

Step 4 Find and plot the vertex.

The x-coordinate of the vertex is 0.
The y-coordinate of the vertex is

$f(0) = 2(0)^2 - 8 = -8$.

So, the vertex is $(0, -8)$.

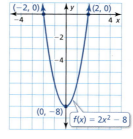

Step 5 Draw a parabola through the vertex and the points where the x-intercepts occur.

▶ The domain is all real numbers. The range is $y \geq -8$.

Monitoring Progress   Help in English and Spanish at *BigIdeasMath.com*

Graph the quadratic function. Label the vertex, axis of symmetry, and x-intercepts. Describe the domain and range of the function.

1. $f(x) = (x + 2)(x - 3)$ **2.** $g(x) = -2(x - 4)(x + 1)$ **3.** $h(x) = 4x^2 - 36$

REMEMBER
Functions have zeros, and graphs have x-intercepts.

Using Intercept Form to Find Zeros of Functions

In Section 3.2, you learned that a zero of a function is an x-value for which $f(x) = 0$. You can use the intercept form of a function to find the zeros of the function.

EXAMPLE 3 **Finding Zeros of a Function**

Find the zeros of $f(x) = (x - 1)(x + 2)$.

Check

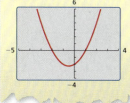

SOLUTION

To find the zeros, determine the x-values for which $f(x)$ is 0.

$f(x) = (x - 1)(x + 2)$ Write the function.

$0 = (x - 1)(x + 2)$ Substitute 0 for $f(x)$.

$x - 1 = 0$ or $x + 2 = 0$ Zero-Product Property

$x = 1$ or $x = -2$ Solve for x.

▶ So, the zeros of the function are -2 and 1.

Section 3.5   Graphing $f(x) = a(x - p)(x - q)$ 155

English Language Learners

Build on Past Knowledge
Remind students that a zero of a function $f(x)$ is a value of x that makes $f(x) = 0$, not the value $f(0)$.

Extra Example 2
Graph $f(x) = 3x^2 - 3$. Describe the domain and range.

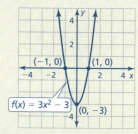

The domain is all real numbers. The range is $y \geq -3$.

Extra Example 3
Find the zeros of $f(x) = (x - 5)(x + 7)$.
The zeros of the function are -7 and 5.

MONITORING PROGRESS ANSWERS

1.

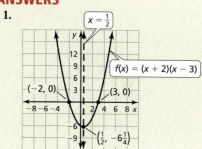

domain: all real numbers,
range: $y \geq -6\frac{1}{4}$

2.

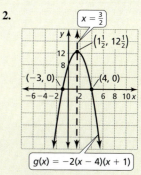

domain: all real numbers,
range: $y \leq 12\frac{1}{2}$

3. See Additional Answers.

Laurie's Notes   Teacher Actions

• **?** "What do you know about the graph of $f(x) = 2x^2 - 8$?" It opens upward, is stretched vertically, and is shifted down 8 units. The vertex is $(0, -8)$.
• **?** "Do you know the x-intercepts or could you determine them?" Factor $2x^2 - 8$ into $2(x - 2)(x + 2)$ and you have x-intercepts of $x = 2$ and $x = -2$.
• **Think-Pair-Share:** Have students answer Questions 1–3, and then share and discuss as a class.
• Say, "Recall the connection between zeros of a function and x-intercepts of a graph." Have students work independently on Example 3.

Extra Example 4

Find the zeros of each function.

a. $h(x) = x^2 - 25$

The zeros of the function are -5 and 5.

b. $f(x) = 2x^2 - 8x - 24$

The zeros of the function are -2 and 6.

Extra Example 5

Use zeros to graph $h(x) = x^2 + 3x - 4$.

The zeros of the function are -4 and 1.

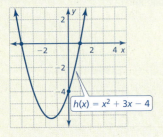

MONITORING PROGRESS ANSWERS

4. $1, 6$
5. $-1, 1$
6. 2

 **Core Concept**

Factors and Zeros

For any factor $x - n$ of a polynomial, n is a zero of the function defined by the polynomial.

EXAMPLE 4 Finding Zeros of Functions

Find the zeros of each function.

a. $h(x) = x^2 - 16$ b. $f(x) = -2x^2 - 10x - 12$

SOLUTION

Write each function in intercept form to identify the zeros.

a. $h(x) = x^2 - 16$ Write the function.

        $= (x + 4)(x - 4)$ Difference of two squares pattern

▶ So, the zeros of the function are -4 and 4.

b. $f(x) = -2x^2 - 10x - 12$ Write the function.

       $= -2(x^2 + 5x + 6)$ Factor out common factor.

       $= -2(x + 3)(x + 2)$ Factor the trinomial.

▶ So, the zeros of the function are -3 and -2.

Monitoring Progress Help in English and Spanish at *BigIdeasMath.com*

Find the zero(s) of the function.

4. $f(x) = (x - 6)(x - 1)$ 5. $h(x) = x^2 - 1$ 6. $g(x) = 3x^2 - 12x + 12$

Using Characteristics to Graph and Write Quadratic Functions

EXAMPLE 5 Graphing a Quadratic Function Using Zeros

Use zeros to graph $h(x) = x^2 - 2x - 3$.

SOLUTION

The function is in standard form. You know that the parabola opens up ($a > 0$) and the *y*-intercept is -3. So, begin by plotting $(0, -3)$.

ATTENDING TO PRECISION

To sketch a more precise graph, make a table of values and plot other points on the graph.

Notice that the polynomial that defines the function is factorable. So, write the function in intercept form and identify the zeros.

$h(x) = x^2 - 2x - 3$ Write the function.

$= (x + 1)(x - 3)$ Factor the trinomial.

The zeros of the function are -1 and 3. So, plot $(-1, 0)$ and $(3, 0)$. Draw a parabola through the points.

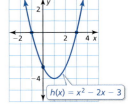

156   Chapter 3   Graphing Quadratic Functions

Laurie's Notes   Teacher Actions

- **Connection:** Students need to recall factoring polynomials in Chapter 2 to find the zeros of a function.
- **Whiteboarding:** Pose Example 4. Give sufficient time while students work with partners to solve. Use whiteboards for partners to record work.
- **? Agree-Disagree Statement:** "If you know the zeros of a quadratic, you can graph the function. Agree/Disagree. Explain." no; There are infinitely many quadratics that have the same zeros.
- In Example 5, the equation is known, so finding the zeros aids in graphing.

STUDY TIP

In part (a), many possible functions satisfy the given condition. The value a can be *any* nonzero number. To allow easier calculations, let $a = 1$. By letting $a = 2$, the resulting function would be $f(x) = 2x^2 + 12x + 22$.

EXAMPLE 6 Writing Quadratic Functions

Write a quadratic function in standard form whose graph satisfies the given condition(s).

a. vertex: $(-3, 4)$

b. passes through $(-9, 0)$, $(-2, 0)$, and $(-4, 20)$

SOLUTION

a. Because you know the vertex, use vertex form to write a function.

$f(x) = a(x - h)^2 + k$ Vertex form

$= 1(x + 3)^2 + 4$ Substitute for a, h, and k.

$= x^2 + 6x + 9 + 4$ Find the product $(x + 3)^2$.

$= x^2 + 6x + 13$ Combine like terms.

b. The given points indicate that the x-intercepts are -9 and -2. So, use intercept form to write a function.

$f(x) = a(x - p)(x - q)$ Intercept form

$= a(x + 9)(x + 2)$ Substitute for p and q.

Use the other given point, $(-4, 20)$, to find the value of a.

$20 = a(-4 + 9)(-4 + 2)$ Substitute -4 for x and 20 for $f(x)$.

$20 = a(5)(-2)$ Simplify.

$-2 = a$ Solve for a.

Use the value of a to write the function.

$f(x) = -2(x + 9)(x + 2)$ Substitute -2 for a.

$= -2x^2 - 22x - 36$ Simplify.

Check

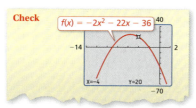

Monitoring Progress

Help in English and Spanish at *BigIdeasMath.com*

Use zeros to graph the function.

7. $f(x) = (x - 1)(x - 4)$

8. $g(x) = x^2 + x - 12$

Write a quadratic function in standard form whose graph satisfies the given condition(s).

9. x-intercepts: -1 and 1

10. vertex: $(8, 8)$

11. passes through $(0, 0)$, $(10, 0)$, and $(4, 12)$

12. passes through $(-5, 0)$, $(4, 0)$, and $(3, -16)$

Section 3.5   Graphing $f(x) = a(x - p)(x - q)$ 157

Extra Example 6

Write a quadratic function in standard form whose graph satisfies the given condition(s).

a. x-intercepts: -1 and 5

Sample answer: $f(x) = x^2 - 4x - 5$

b. vertex: $(-3, -2)$

Sample answer: $f(x) = x^2 + 6x + 7$

MONITORING PROGRESS ANSWERS

7.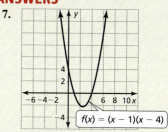

8.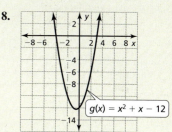

9. Sample answer: $f(x) = x^2 - 1$

10. Sample answer: $f(x) = x^2 - 16x + 72$

11. $f(x) = -\frac{1}{2}x^2 + 5x$

12. $f(x) = 2x^2 + 2x - 40$

Laurie's Notes   Teacher Actions

- Example 6 helps students focus on the three forms of a quadratic function and what information is known about the graph from each equation.
- **Three-Minute Pause:** Work through both parts of the example as shown. Give students time to discuss the example with partners and reflect on their understanding. Then have students self-assess and determine what equations are needed to answer Questions 9–12.

Closure

- **Exit Ticket:** Use zeros to sketch the graphs of $f(x) = -(x + 2)(x - 3)$ and $g(x) = x^2 + 7x + 10$. Check students' work.

Section 3.5   157

3.5 Exercises

Dynamic Solutions available at BigIdeasMath.com

Assignment Guide and Homework Check

ASSIGNMENT

Basic: 1, 2, 3–57 odd, 65, 72, 76, 81–83

Average: 1, 2–72 even, 76, 81–83

Advanced: 1, 2, 10–42 even, 50–70 even, 71–83

HOMEWORK CHECK

Basic: 7, 21, 43, 55

Average: 8, 24, 44, 56

Advanced: 10, 24, 50, 56

ANSWERS

1. x-intercepts

2. Sample answer: Find the x-coordinate of the vertex using $x = \dfrac{p+q}{2}$, then evaluate the function to determine the maximum value or minimum value.

3. $-3, 1; x = -1$

4. $2, 5; x = \dfrac{7}{2}$

5. $-7, 5; x = -1$

6. $-8, 0; x = -4$

7.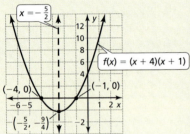

   domain: all real numbers, range: $y \geq -\dfrac{9}{4}$

8.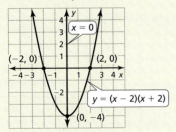

   domain: all real numbers, range: $y \geq -4$

Vocabulary and Core Concept Check

1. **COMPLETE THE SENTENCE** The values p and q are _____ of the graph of the function $f(x) = a(x - p)(x - q)$.

2. **WRITING** Explain how to find the maximum value or minimum value of a quadratic function when the function is given in intercept form.

Monitoring Progress and Modeling with Mathematics

In Exercises 3–6, find the x-intercepts and axis of symmetry of the graph of the function.

3.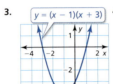

4.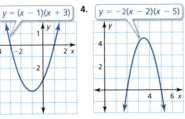

5. $f(x) = -5(x + 7)(x - 5)$ 6. $g(x) = \dfrac{2}{3}x(x + 8)$

In Exercises 7–12, graph the quadratic function. Label the vertex, axis of symmetry, and x-intercepts. Describe the domain and range of the function. (See Example 1.)

7. $f(x) = (x + 4)(x + 1)$ 8. $y = (x - 2)(x + 2)$

9. $y = -(x + 6)(x - 4)$ 10. $h(x) = -4(x - 7)(x - 3)$

11. $g(x) = 5(x + 1)(x + 2)$ 12. $y = -2(x - 3)(x + 4)$

In Exercises 13–20, graph the quadratic function. Label the vertex, axis of symmetry, and x-intercepts. Describe the domain and range of the function. (See Example 2.)

13. $y = x^2 - 9$ 14. $f(x) = x^2 - 8x$

15. $h(x) = -5x^2 + 5x$ 16. $y = 3x^2 - 48$

17. $q(x) = x^2 + 9x + 14$ 18. $p(x) = x^2 + 6x - 27$

19. $y = 4x^2 - 36x + 32$ 20. $y = -2x^2 - 4x + 30$

In Exercises 21–28, find the zero(s) of the function. (See Examples 3 and 4.)

21. $y = -2(x - 2)(x - 10)$ 22. $f(x) = \dfrac{1}{3}(x + 5)(x - 1)$

23. $f(x) = x^2 - 4$ 24. $h(x) = x^2 - 36$

25. $g(x) = x^2 + 5x - 24$ 26. $y = x^2 - 17x + 52$

27. $y = 3x^2 - 15x - 42$ 28. $g(x) = -4x^2 - 8x - 4$

In Exercises 29–34, match the function with its graph.

29. $y = (x + 5)(x + 3)$ 30. $y = (x + 5)(x - 3)$

31. $y = (x - 5)(x + 3)$ 32. $y = (x - 5)(x - 3)$

33. $y = (x + 5)(x - 5)$ 34. $y = (x + 3)(x - 3)$

A. 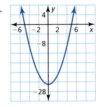   B.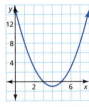

C. 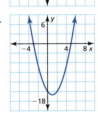   D.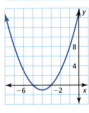

E. 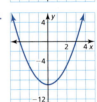   F.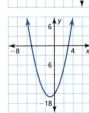

158   Chapter 3   Graphing Quadratic Functions

9.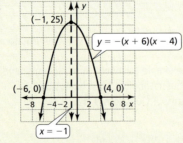

   domain: all real numbers, range: $y \leq 25$

10.

    domain: all real numbers, range: $y \leq 16$

11–34. See Additional Answers.

In Exercises 35–40, use zeros to graph the function.
(See Example 5.)

35. $f(x) = (x + 2)(x - 6)$ **36.** $g(x) = -3(x + 1)(x + 7)$

37. $y = x^2 - 11x + 18$ **38.** $y = x^2 - x - 30$

39. $y = -5x^2 - 10x + 40$ **40.** $h(x) = 8x^2 - 8$

ERROR ANALYSIS In Exercises 41 and 42, describe and correct the error in finding the zeros of the function.

41.

$y = 5(x + 3)(x - 2)$
The zeros of the function are 3 and −2.

42.
$y = x^2 - 9$
The zero of the function is 9.

In Exercises 43–54, write a quadratic function in standard form whose graph satisfies the given condition(s). (See Example 6.)

43. vertex: $(7, -3)$ **44.** vertex: $(4, 8)$

45. x-intercepts: 1 and 9 **46.** x-intercepts: −2 and −5

47. passes through $(-4, 0), (3, 0),$ and $(2, -18)$

48. passes through $(-5, 0), (-1, 0),$ and $(-4, 3)$

49. passes through $(7, 0)$

50. passes through $(0, 0)$ and $(6, 0)$

51. axis of symmetry: $x = -5$

52. y increases as x increases when $x < 4$; y decreases as x increases when $x > 4$.

53. range: $y \geq -3$ **54.** range: $y \leq 10$

In Exercises 55–58, write the quadratic function represented by the graph.

55. 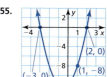 **56.**

57. 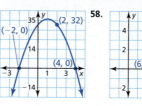 **58.**

In Exercises 59 and 60, all the zeros of a function are given. Use the zeros and the other point given to write a quadratic function represented by the table.

59.
x	y
0	0
2	30
7	0

60.
x	y
−3	0
1	−72
4	0

In Exercises 61–64, sketch a parabola that satisfies the given conditions.

61. x-intercepts: −4 and 2; range: $y \geq -3$

62. axis of symmetry: $x = 6$; passes through $(4, 15)$

63. range: $y \leq 5$; passes through $(0, 2)$

64. x-intercept: 6; y-intercept: 1; range: $y \geq -4$

65. MODELING WITH MATHEMATICS Satellite dishes are shaped like parabolas to optimally receive signals. The cross section of a satellite dish can be modeled by the function shown, where x and y are measured in feet. The x-axis represents the top of the opening of the dish.

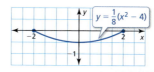

a. How wide is the satellite dish?

b. How deep is the satellite dish?

c. Write a quadratic function in standard form that models the cross section of a satellite dish that is 6 feet wide and 1.5 feet deep.

Section 3.5 Graphing $f(x) = a(x - p)(x - q)$ 159

ANSWERS

35.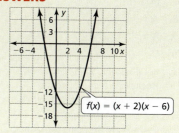

36.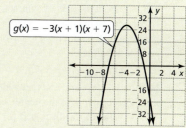

37.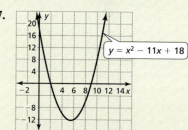

38.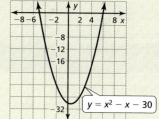

39.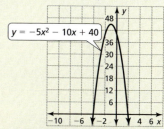

40.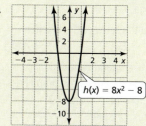

41. The factors need to be set equal to 0 and solved to find the zeros; $x + 3 = 0$ or $x - 2 = 0$; $x = -3$ or $x = 2$; The zeros of the function are −3 and 2.

42. $x^2 - 9$ needs to be factored; $y = (x + 3)(x - 3)$; The zeros of the function are −3 and 3.

43. Sample answer: $f(x) = x^2 - 14x + 46$

44. Sample answer: $f(x) = x^2 - 8x + 24$

45. Sample answer: $f(x) = x^2 - 10x + 9$

46. Sample answer: $f(x) = x^2 + 7x + 10$

47. $f(x) = 3x^2 + 3x - 36$

48. $f(x) = -x^2 - 6x - 5$

49. Sample answer: $f(x) = x^2 - 7x$

50. Sample answer: $f(x) = x^2 - 6x$

51. Sample answer: $f(x) = x^2 + 10x + 25$

52. Sample answer: $f(x) = -x^2 + 8x - 16$

53. Sample answer: $f(x) = x^2 - 3$

54. Sample answer: $f(x) = -x^2 + 10$

55. $f(x) = 2x^2 + 2x - 12$

56. $f(x) = -x^2 + 8x - 7$

57. $f(x) = -4x^2 + 8x + 32$

58. $f(x) = \frac{1}{2}x^2 - 8x + 30$

59. $f(x) = -3x^2 + 21x$

60. $f(x) = 6x^2 - 6x - 72$

61–65. See Additional Answers.

ANSWERS

66. a. yes; *Sample answer:* Using the function, when the horizontal distance of the ball is 3 feet, the vertical distance is 0, which is the location of the basket.

b. $y = -\frac{1}{15}x^2 + \frac{16}{15}x - \frac{13}{5}$

67. D

68. A

69. B

70. C

71. not possible; *Sample answer:* Because -5 and 1 are the x-intercepts, the axis of symmetry is $x = -2$. The points $(-3, 12)$ and $(-1, 4)$ are the same horizontal distance from the axis of symmetry, so for both of them to lie on the parabola they would have to have the same y-coordinate.

72. a. *Sample answer:* $p = 75$, $q = 425$

b. *Sample answer:* 350 feet wide, 61 feet high; Use the height as the value of $f(x)$ from part (a) along with the values of p and q estimated in part (a). Use the values of p and q to calculate the value of the x-coordinate of the vertex to substitute for x in the function in part (a), and solve the function for a.

73–83. See Additional Answers.

Mini-Assessment

1. Graph $f(x) = (x - 2)(x + 4)$. Describe the domain and range.

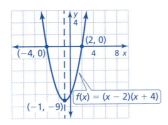

domain: all real numbers; range: $y \geq -9$

2. Find the zeros of the function $f(x) = 2x^2 + 2x - 60$. -6 and 5

3. Write a quadratic function in standard form whose graph has x-intercepts -4 and -3.
Sample answer:
$f(x) = x^2 + 7x + 12$

160   Chapter 3

---

66. MODELING WITH MATHEMATICS A professional basketball player's shot is modeled by the function shown, where x and y are measured in feet.

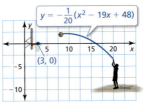

a. Does the player make the shot? Explain.

b. The basketball player releases another shot from the point $(13, 0)$ and makes the shot. The shot also passes through the point $(10, 1.4)$. Write a quadratic function in standard form that models the path of the shot.

USING STRUCTURE In Exercises 67–70, match the function with its graph.

67. $y = -x^2 + 5x$

68. $y = x^2 - x - 12$

69. $y = -2x^2 - 8$

70. $y = 3x^2 + 6$

71. CRITICAL THINKING Write a quadratic function represented by the table, if possible. If not, explain why.

x	-5	-3	-1	1
y	0	12	4	0

Maintaining Mathematical Proficiency — Reviewing what you learned in previous grades and lessons

Find the distance between the two points. *(Skills Review Handbook)*

81. $(2, 1)$ and $(3, 0)$

82. $(-1, 8)$ and $(-9, 2)$

83. $(-1, -6)$ and $(-3, -5)$

160   Chapter 3   Graphing Quadratic Functions

---

72. HOW DO YOU SEE IT? The graph shows the parabolic arch that supports the roof of a convention center, where x and y are measured in feet.

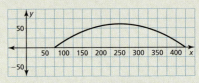

a. The arch can be represented by a function of the form $f(x) = a(x - p)(x - q)$. Estimate the values of p and q.

b. Estimate the width and height of the arch. Explain how you can use your height estimate to calculate a.

ANALYZING EQUATIONS In Exercises 73 and 74, (a) rewrite the quadratic function in intercept form and (b) graph the function using any method. Explain the method you used.

73. $f(x) = -3(x + 1)^2 + 27$

74. $g(x) = 2(x - 1)^2 - 2$

75. WRITING Can a quadratic function with exactly one real zero be written in intercept form? Explain.

76. MAKING AN ARGUMENT Your friend claims that any quadratic function can be written in standard form and in vertex form. Is your friend correct? Explain.

77. REASONING Let k be a constant. Find the zeros of the function $f(x) = kx^2 - k^2x - 2k^3$ in terms of k.

78. THOUGHT PROVOKING Sketch the graph of each function. Explain your procedure.

a. $f(x) = (x^2 - 1)(x^2 - 4)$

b. $g(x) = x(x^2 - 1)(x^2 - 4)$

PROBLEM SOLVING In Exercises 79 and 80, write two quadratic functions whose graphs intersect at the given points. Explain your reasoning.

79. $(-4, 0)$ and $(2, 0)$

80. $(3, 6)$ and $(7, 6)$

If students need help...	If students got it...
Resources by Chapter • Practice A and Practice B • Puzzle Time	Resources by Chapter • Enrichment and Extension • Cumulative Review
Student Journal • Practice	Start the *next* Section
Differentiating the Lesson Skills Review Handbook	

Laurie's Notes

Overview of Section 3.6

Introduction
- Students have likely seen parabolic satellite dishes of different sizes: large transmission satellites, small dishes on a rooftop, or handheld dishes on the sideline of a sporting event. The cross section of each of these satellite dishes is a parabola.
- The explorations allow students to discover properties of the focus of a parabola. By sketching the reflected rays, students quickly discover the focus. The explorations do not take long and are important for students to experience.
- The formal lesson begins with the definition of the *focus* and the *directrix*. Careful sketches are needed to demonstrate what is meant by "Any point on the parabola is equidistant from the focus and directrix."
- Symbolic manipulation skills are necessary in this lesson as students write the equations of a parabola or as equations are rewritten in a general form.

Resources
- Dynamic geometry software can be used to graph a parabola given its focus and directrix. Clicking and dragging on the focus or directrix demonstrates the change in position and shape of the graph.
- Search the Internet for "parabola + focus + applet" to find dynamic geometry software.

Teaching Strategy
- Dynamic geometry software can be used to graph a parabola given the focus and directrix.
- Have students explore dynamically the shape of the parabola by changing the focus and directrix.
- There are many applets that you'll find by doing an Internet search, or you could have students explore on their own.

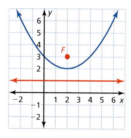

Pacing Suggestion
- Use the *Motivate* and the two explorations before beginning the formal lesson.

Dynamic Teaching Tools
Dynamic Assessment & Progress Monitoring Tool
Lesson Planning Tool
Interactive Whiteboard Lesson Library
Dynamic Classroom with Dynamic Investigations

What Your Students Will Learn

- Define a parabola as the set of all points (x, y) in a plane that are equidistant from a fixed point called the focus and a fixed line called the directrix.
- Write equations in standard form of parabolas with a vertical axis of symmetry or with a horizontal axis of symmetry.
- Solve real-life problems involving objects with parabolic shapes.

Laurie's Notes

Exploration

Motivate
- Display an image of a satellite dish. Have students share their knowledge of satellite dishes and how they work. The essential piece is that the satellite dish is an antenna that receives electromagnetic signals from an orbiting satellite. The shape of the satellite dish is parabolic.
- Discuss what it would mean to take a cross section of the satellite dish. The cross section is parabolic.
- Explain to students that in the exploration they will observe a new characteristic of the parabola.

Exploration 1
? "Have any of you played pool or billiards before?"
Answers will vary.
- Make a quick sketch of a pool table and its six pockets. Discuss what happens when a ball hits the edge of the table, namely the (angle of incidence) = (angle of reflection).
- Explain that a similar property is true for rays hitting a satellite dish.
- If students are accurate with estimating the outgoing angle, then the three rays they draw should be concurrent at the focus of the parabola.
- Students' reasoning of why this makes sense is connected to their knowledge of satellite dishes.

Exploration 2
- **Teaching Tip:** If you have a flashlight or overhead projector bulb, use it as a model for students to see as they begin this exploration. The bulb is located at the focus, and the rays are emanating from the bulb toward the parabolic surface.
- If students are accurate with estimating the outgoing angle, then the three rays they draw should be parallel.

Communicate Your Answer
- Students' understanding of a focus may be no more than knowing it is a point in the interior of the parabola. It may be necessary to ask a probing question for students to observe that the focus is on the line of symmetry. The properties of the focus are related to the observations made in the two explorations.

Connecting to Next Step
- The explorations should not take long to complete, and it is great for students to discover the properties of the focus. Begin the formal lesson with the definition of *focus* and *directrix*.

3.6 Focus of a Parabola

Essential Question What is the focus of a parabola?

EXPLORATION 1   Analyzing Satellite Dishes

Work with a partner. Vertical rays enter a satellite dish whose cross section is a parabola. When the rays hit the parabola, they reflect at the same angle at which they entered. (See Ray 1 in the figure.)

a. Draw the reflected rays so that they intersect the *y*-axis.

b. What do the reflected rays have in common?

c. The optimal location for the receiver of the satellite dish is at a point called the *focus* of the parabola. Determine the location of the focus. Explain why this makes sense in this situation.

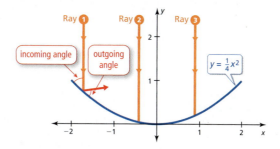

> **CONSTRUCTING VIABLE ARGUMENTS**
>
> To be proficient in math, you need to make conjectures and build logical progressions of statements to explore the truth of your conjectures.

EXPLORATION 2   Analyzing Spotlights

Work with a partner. Beams of light are coming from the bulb in a spotlight, located at the focus of the parabola. When the beams hit the parabola, they reflect at the same angle at which they hit. (See Beam 1 in the figure.) Draw the reflected beams. What do they have in common? Would you consider this to be the optimal result? Explain.

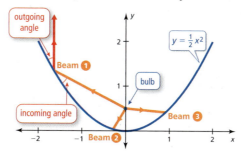

Communicate Your Answer

3. What is the focus of a parabola?

4. Describe some of the properties of the focus of a parabola.

Section 3.6   Focus of a Parabola   161

English Language Learners

Comprehension

Brainstorm with students the various meanings of the word *focus* in everyday life. Talk about how your eyes can focus on something, you focus a camera, or you must focus your attention on a task. Allow students to share how to pronounce the word in their native language. Then, explain that there is a mathematical meaning of the word *focus*. Guide students in making a list of the characteristics of the focus of a parabola.

Extra Example 1

Use the Distance Formula to write an equation of the parabola with focus $F(0, 4)$ and directrix $y = -4$. $y = \frac{1}{16}x^2$

MONITORING PROGRESS ANSWER

1. $y = -\frac{1}{12}x^2$

3.6 Lesson

Core Vocabulary

focus, p. 162
directrix, p. 162
Previous
perpendicular
Distance Formula
congruent

REMEMBER
The distance from a point to a line is defined as the length of the perpendicular segment from the point to the line.

REMEMBER
Squaring a positive number and finding a square root are inverse operations. So, you can square each side of the equation to eliminate the radicals.

What You Will Learn

▶ Explore the focus and the directrix of a parabola.
▶ Write equations of parabolas.
▶ Solve real-life problems.

Exploring the Focus and Directrix

Previously, you learned that the graph of a quadratic function is a parabola that opens up or down. A parabola can also be defined as the set of all points (x, y) in a plane that are equidistant from a fixed point called the **focus** and a fixed line called the **directrix**.

The **focus** is in the interior of the parabola and lies on the axis of symmetry.

The **vertex** lies halfway between the focus and the directrix.

The **directrix** is perpendicular to the axis of symmetry.

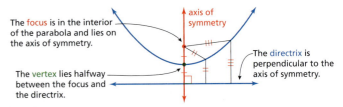

EXAMPLE 1 Using the Distance Formula to Write an Equation

Use the Distance Formula to write an equation of the parabola with focus $F(0, 2)$ and directrix $y = -2$.

SOLUTION

Notice the line segments drawn from point F to point P and from point P to point D. By the definition of a parabola, these line segments must be congruent.

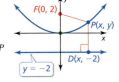

$$PD = PF \quad \text{Definition of a parabola}$$
$$\sqrt{(x - x_1)^2 + (y - y_1)^2} = \sqrt{(x - x_2)^2 + (y - y_2)^2} \quad \text{Distance Formula}$$
$$\sqrt{(x - x)^2 + (y - (-2))^2} = \sqrt{(x - 0)^2 + (y - 2)^2} \quad \text{Substitute for } x_1, y_1, x_2, \text{ and } y_2.$$
$$\sqrt{(y + 2)^2} = \sqrt{x^2 + (y - 2)^2} \quad \text{Simplify.}$$
$$(y + 2)^2 = x^2 + (y - 2)^2 \quad \text{Square each side.}$$
$$y^2 + 4y + 4 = x^2 + y^2 - 4y + 4 \quad \text{Expand.}$$
$$8y = x^2 \quad \text{Combine like terms.}$$
$$y = \frac{1}{8}x^2 \quad \text{Divide each side by 8.}$$

Monitoring Progress  Help in English and Spanish at *BigIdeasMath.com*

1. Use the Distance Formula to write an equation of the parabola with focus $F(0, -3)$ and directrix $y = 3$.

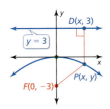

162 Chapter 3 Graphing Quadratic Functions

Laurie's Notes — Teacher Actions

- **Teaching Tip:** Demonstrate what is meant by the language "fixed point" and "fixed line" by drawing a line on the board and a point not on the line. "This line and this point define a set of points that we call a parabola."
- Discuss the sketch of the parabola, focus, and directrix. Demonstrate that any point on the parabola is equidistant to the focus and directrix. Mark the congruent segments. The focus and directrix are aids in graphing, but our sketches may not be exact; hence, marking the congruent segments is helpful.
- **Note:** The vertex is midway between the focus and directrix.
- The first *Remember* note reminds students of how to find the distance between a point and a line.
- Review the technique of squaring both sides of an equation if students have forgotten.
- **Think-Pair-Share:** Before students begin work on Question 1, ask how this question differs from Example 1. Students should note that the parabola opens downward and therefore $a < 0$.

You can derive the equation of a parabola that opens up or down with vertex (0, 0), focus (0, p), and directrix $y = -p$ using the procedure in Example 1.

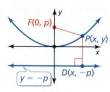

$$\sqrt{(x-x)^2 + (y-(-p))^2} = \sqrt{(x-0)^2 + (y-p)^2}$$
$$(y+p)^2 = x^2 + (y-p)^2$$
$$y^2 + 2py + p^2 = x^2 + y^2 - 2py + p^2$$
$$4py = x^2$$
$$y = \frac{1}{4p}x^2$$

The focus and directrix each lie $|p|$ units from the vertex. Parabolas can also open left or right, in which case the equation has the form $x = \frac{1}{4p}y^2$ when the vertex is (0, 0).

LOOKING FOR STRUCTURE

Notice that $y = \frac{1}{4p}x^2$ is of the form $y = ax^2$. So, changing the value of p vertically stretches or shrinks the parabola.

Core Concept

Standard Equations of a Parabola with Vertex at the Origin

Vertical axis of symmetry ($x = 0$)

Equation: $y = \frac{1}{4p}x^2$

Focus: $(0, p)$

Directrix: $y = -p$

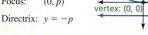

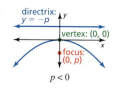

Horizontal axis of symmetry ($y = 0$)

Equation: $x = \frac{1}{4p}y^2$

Focus: $(p, 0)$

Directrix: $x = -p$

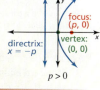

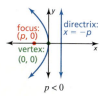

STUDY TIP

Notice that parabolas opening left or right do *not* represent functions.

EXAMPLE 2 Graphing an Equation of a Parabola

Identify the focus, directrix, and axis of symmetry of $-4x = y^2$. Graph the equation.

SOLUTION

Step 1 Rewrite the equation in standard form.

$-4x = y^2$ Write the original equation.

$x = -\frac{1}{4}y^2$ Divide each side by −4.

Step 2 Identify the focus, directrix, and axis of symmetry. The equation has the form $x = \frac{1}{4p}y^2$, where $p = -1$. The focus is $(p, 0)$, or $(-1, 0)$. The directrix is $x = -p$, or $x = 1$. Because y is squared, the axis of symmetry is the x-axis.

Step 3 Use a table of values to graph the equation. Notice that it is easier to substitute y-values and solve for x. Opposite y-values result in the same x-value.

y	0	±1	±2	±3	±4
x	0	−0.25	−1	−2.25	−4

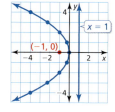

Section 3.6 Focus of a Parabola 163

Extra Example 3

Write an equation of the parabola shown.

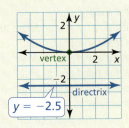

$y = 0.1x^2$

MONITORING PROGRESS ANSWERS

2. The focus is $\left(0, \frac{1}{2}\right)$, the directrix is $y = -\frac{1}{2}$, and the axis of symmetry is the y-axis.

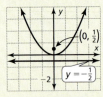

3. The focus is $\left(0, -\frac{1}{4}\right)$, the directrix is $y = \frac{1}{4}$, and the axis of symmetry is the y-axis.

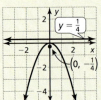

4. See Additional Answers.
5. $x = \frac{1}{12}y^2$
6. $x = -\frac{1}{8}y^2$
7. $y = \frac{1}{6}x^2$

Writing Equations of Parabolas

EXAMPLE 3 Writing an Equation of a Parabola

Write an equation of the parabola shown.

SOLUTION

Because the vertex is at the origin and the axis of symmetry is vertical, the equation has the form $y = \frac{1}{4p}x^2$. The directrix is $y = -p = 3$, so $p = -3$. Substitute -3 for p to write an equation of the parabola.

$$y = \frac{1}{4(-3)}x^2 = -\frac{1}{12}x^2$$

▶ So, an equation of the parabola is $y = -\frac{1}{12}x^2$.

Monitoring Progress Help in English and Spanish at BigIdeasMath.com

Identify the focus, directrix, and axis of symmetry of the parabola. Then graph the equation.

2. $y = 0.5x^2$ 3. $-y = x^2$ 4. $y^2 = 6x$

Write an equation of the parabola with vertex at (0, 0) and the given directrix or focus.

5. directrix: $x = -3$ 6. focus: $(-2, 0)$ 7. focus: $\left(0, \frac{3}{2}\right)$

The vertex of a parabola is not always at the origin. As in previous transformations, adding a value to the input or output of a function translates its graph.

Core Concept

Standard Equations of a Parabola with Vertex at (h, k)

Vertical axis of symmetry (x = h)

Equation: $y = \frac{1}{4p}(x - h)^2 + k$

Focus: $(h, k + p)$

Directrix: $y = k - p$

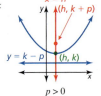 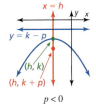

$p > 0$ $p < 0$

Horizontal axis of symmetry (y = k)

Equation: $x = \frac{1}{4p}(y - k)^2 + h$

Focus: $(h + p, k)$

Directrix: $x = h - p$

 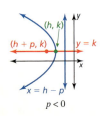

$p > 0$ $p < 0$

STUDY TIP

The standard form for a vertical axis of symmetry looks like vertex form. To remember the standard form for a horizontal axis of symmetry, switch x and y, and h and k.

Laurie's Notes — Teacher Actions

? "To write the equation of a parabola with its vertex at the origin, what information will you need?" *The value of p and which way the parabola opens, vertically or horizontally. This could be determined by knowing the directrix or the focus.*

- Work through the example as shown.
- Ask students to assess their level of understanding with a *Thumbs Up* signal. Have students work some or all of the *Monitoring Progress* questions based on their responses.

? Have students discuss the following question with their partners or in groups. "If the vertex of the parabola is translated to (h, k), what would the equations look like and what would happen to the directrix and focus?" *Listen to student conversations before writing the Core Concept that summarizes this.*

- **Note:** Be sure that students notice that with a horizontal axis of symmetry the general form is written in terms of "minus k" and "plus h," which is not what they are used to seeing.

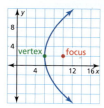

EXAMPLE 4 Writing an Equation of a Translated Parabola

Write an equation of the parabola shown.

SOLUTION

Because the vertex is not at the origin and the axis of symmetry is horizontal, the equation has the form $x = \frac{1}{4p}(y - k)^2 + h$. The vertex (h, k) is $(6, 2)$ and the focus $(h + p, k)$ is $(10, 2)$, so $h = 6$, $k = 2$, and $p = 4$. Substitute these values to write an equation of the parabola.

$$x = \frac{1}{4(4)}(y - 2)^2 + 6 = \frac{1}{16}(y - 2)^2 + 6$$

▶ So, an equation of the parabola is $x = \frac{1}{16}(y - 2)^2 + 6$.

Solving Real-Life Problems

Parabolic reflectors have cross sections that are parabolas. Incoming sound, light, or other energy that arrives at a parabolic reflector parallel to the axis of symmetry is directed to the focus (Diagram 1). Similarly, energy that is emitted from the focus of a parabolic reflector and then strikes the reflector is directed parallel to the axis of symmetry (Diagram 2).

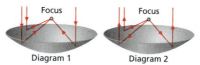

Diagram 1   Diagram 2

EXAMPLE 5 Solving a Real-Life Problem

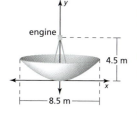

An electricity-generating dish uses a parabolic reflector to concentrate sunlight onto a high-frequency engine located at the focus of the reflector. The sunlight heats helium to 650°C to power the engine. Write an equation that represents the cross section of the dish shown with its vertex at $(0, 0)$. What is the depth of the dish?

SOLUTION

Because the vertex is at the origin, and the axis of symmetry is vertical, the equation has the form $y = \frac{1}{4p}x^2$. The engine is at the focus, which is 4.5 meters above the vertex. So, $p = 4.5$. Substitute 4.5 for p to write the equation.

$$y = \frac{1}{4(4.5)}x^2 = \frac{1}{18}x^2$$

The depth of the dish is the y-value at the dish's outside edge. The dish extends $\frac{8.5}{2} = 4.25$ meters to either side of the vertex $(0, 0)$, so find y when $x = 4.25$.

$$y = \frac{1}{18}(4.25)^2 \approx 1$$

▶ The depth of the dish is about 1 meter.

Monitoring Progress Help in English and Spanish at *BigIdeasMath.com*

8. Write an equation of a parabola with vertex $(-1, 4)$ and focus $(-1, 2)$.
9. A parabolic microwave antenna is 16 feet in diameter. Write an equation that represents the cross section of the antenna with its vertex at $(0, 0)$ and its focus 10 feet to the right of the vertex. What is the depth of the antenna?

Section 3.6 Focus of a Parabola 165

Extra Example 4
Write an equation of the parabola shown.

$x = \frac{1}{16}(y - 4)^2 - 2$

Extra Example 5
An archway in front of a school is in the shape of a parabola. The top of the arch is the vertex $(0, 0)$. The school seal is at the focus, 2.5 feet below the vertex, and the arch is 18 feet wide at the ground. Write an equation that represents a cross section of the arch. What is the height from the top of the arch to the ground?
$y = -0.1x^2$; 8.1 feet

MONITORING PROGRESS ANSWERS

8. $y = -\frac{1}{8}(x + 1)^2 + 4$
9. $x = \frac{1}{40}y^2$; 1.6 ft

Laurie's Notes Teacher Actions

? "What is the vertex of the parabola?" $(6, 2)$
- Have students identify h, k, and p, and let them work through the example with partners.
- **Popsicle Sticks:** Use *Popsicle Sticks* to identify the student who volunteers his/her equation.
- The real-life example connects to the explorations and the *Motivate*.

COMMON ERROR Students think the depth of the dish is the distance from the engine (focus) to the vertex. Explain that the depth is the y-value at the outside edge of the dish.

Closure
- **Exit Ticket:** Identify the focus, directrix, and axis of symmetry for $4y = -x^2$. Graph.
  focus: $(0, -1)$, directrix: $y = 1$, axis of symmetry: $x = 0$

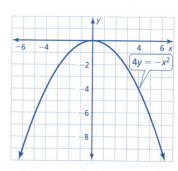

Section 3.6 165

Assignment Guide and Homework Check

ASSIGNMENT

Basic: 1–2, 3–9 odd, 13–21 odd, 25–39 odd, 50, 55–58

Average: 1–2, 4–10 even, 11, 14–22 even, 26–40 even, 50, 55–58

Advanced: 1–2, 10–12, 20–22, 26–54 even, 55–58

HOMEWORK CHECK

Basic: 3, 19, 25, 33, 39

Average: 4, 20, 26, 34, 40

Advanced: 20, 32, 34, 46, 48

ANSWERS

1. focus; directrix
2. Using the equation for directrix $y = -p$, the result is $p = -5$. Since the focus is $(0, p)$, the focus must be $(0, -5)$.
3. $y = \frac{1}{4}x^2$
4. $y = -\frac{1}{16}x^2$
5. $y = -\frac{1}{8}x^2$
6. $y = -\frac{1}{28}x^2$
7. $y = \frac{1}{24}x^2$
8. $y = \frac{1}{20}x^2$
9. $y = -\frac{1}{40}x^2$
10. $y = \frac{1}{36}x^2$
11. A, B and D; Each has a value for p that is negative. Substituting in a negative value for p in $y = \frac{1}{4p}x^2$ results in a parabola that has been reflected across the x-axis.
12. B, C, and E; Use the focus to create the equation $y = -\frac{1}{36}x^2$. Points B, C, and E are the fourth quadrant points that satisfy the equation.
13. The focus is $(0, 2)$. The directrix is $y = -2$. The axis of symmetry is the y-axis.

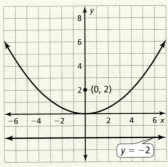

3.6 Exercises

Dynamic Solutions available at BigIdeasMath.com

Vocabulary and Core Concept Check

1. **COMPLETE THE SENTENCE** A parabola is the set of all points in a plane equidistant from a fixed point called the _____ and a fixed line called the _____ .

2. **WRITING** Explain how to find the coordinates of the focus of a parabola with vertex $(0, 0)$ and directrix $y = 5$.

Monitoring Progress and Modeling with Mathematics

In Exercises 3–10, use the Distance Formula to write an equation of the parabola. *(See Example 1.)*

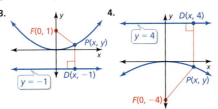

5. focus: $(0, -2)$
   directrix: $y = 2$

6. directrix: $y = 7$
   focus: $(0, -7)$

7. vertex: $(0, 0)$
   directrix: $y = -6$

8. vertex: $(0, 0)$
   focus: $(0, 5)$

9. vertex: $(0, 0)$
   focus: $(0, -10)$

10. vertex: $(0, 0)$
    directrix: $y = -9$

11. **ANALYZING RELATIONSHIPS** Which of the given characteristics describe parabolas that open down? Explain your reasoning.

    Ⓐ focus: $(0, -6)$
       directrix: $y = 6$
    Ⓑ focus: $(0, -2)$
       directrix: $y = 2$
    Ⓒ focus: $(0, 6)$
       directrix: $y = -6$
    Ⓓ focus: $(0, -1)$
       directrix: $y = 1$

12. **REASONING** Which of the following are possible coordinates of the point P in the graph shown? Explain.

    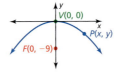

    Ⓐ $(-6, -1)$
    Ⓑ $\left(3, -\frac{1}{4}\right)$
    Ⓒ $\left(4, -\frac{4}{9}\right)$
    Ⓓ $\left(1, \frac{1}{36}\right)$
    Ⓔ $(6, -1)$
    Ⓕ $\left(2, -\frac{1}{18}\right)$

In Exercises 13–20, identify the focus, directrix, and axis of symmetry of the parabola. Graph the equation. *(See Example 2.)*

13. $y = \frac{1}{8}x^2$
14. $y = -\frac{1}{12}x^2$
15. $x = -\frac{1}{20}y^2$
16. $x = \frac{1}{24}y^2$
17. $y^2 = 16x$
18. $-x^2 = 48y$
19. $6x^2 + 3y = 0$
20. $8x^2 - y = 0$

ERROR ANALYSIS In Exercises 21 and 22, describe and correct the error in graphing the parabola.

21.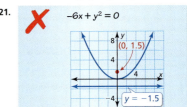

22.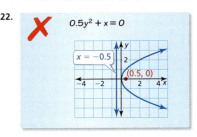

23. **ANALYZING EQUATIONS** The cross section (with units in inches) of a parabolic satellite dish can be modeled by the equation $y = \frac{1}{38}x^2$. How far is the receiver from the vertex of the cross section? Explain.

166   Chapter 3   Graphing Quadratic Functions

14. The focus is $(0, -3)$. The directrix is $y = 3$. The axis of symmetry is the y-axis.

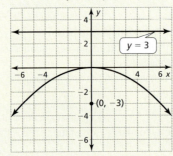

15. The focus is $(-5, 0)$. The directrix is $x = 5$. The axis of symmetry is the x-axis.

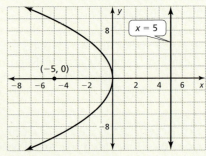

16–23. See Additional Answers.

24. ANALYZING EQUATIONS The cross section (with units in inches) of a parabolic spotlight can be modeled by the equation $x = \frac{1}{20}y^2$. How far is the bulb from the vertex of the cross section? Explain.

In Exercises 25–28, write an equation of the parabola shown. (See Example 3.)

25.

26. $y = \frac{3}{4}$, directrix, vertex

27. $x = \frac{5}{2}$, vertex, directrix

28. $x = -2$, vertex, directrix

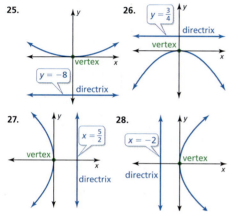

In Exercises 29–36, write an equation of the parabola with the given characteristics.

29. focus: $(3, 0)$
    directrix: $x = -3$

30. focus: $\left(\frac{2}{3}, 0\right)$
    directrix: $x = -\frac{2}{3}$

31. directrix: $x = -10$
    vertex: $(0, 0)$

32. directrix: $y = \frac{8}{3}$
    vertex: $(0, 0)$

33. focus: $\left(0, -\frac{5}{3}\right)$
    directrix: $y = \frac{5}{3}$

34. focus: $\left(0, \frac{5}{4}\right)$
    directrix: $y = -\frac{5}{4}$

35. focus: $\left(0, \frac{6}{7}\right)$
    vertex: $(0, 0)$

36. focus: $\left(-\frac{4}{5}, 0\right)$
    vertex: $(0, 0)$

In Exercises 37–40, write an equation of the parabola shown. (See Example 4.)

37.

38.

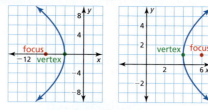

39.

40.

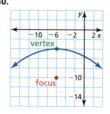

In Exercises 41–46, identify the vertex, focus, directrix, and axis of symmetry of the parabola. Describe the transformations of the graph of the standard equation with $p = 1$ and vertex $(0, 0)$.

41. $y = \frac{1}{8}(x - 3)^2 + 2$
42. $y = -\frac{1}{4}(x + 2)^2 + 1$
43. $x = \frac{1}{16}(y - 3)^2 + 1$
44. $y = (x + 3)^2 - 5$
45. $x = -3(y + 4)^2 + 2$
46. $x = 4(y + 5)^2 - 1$

47. MODELING WITH MATHEMATICS Scientists studying dolphin echolocation simulate the projection of a bottlenose dolphin's clicking sounds using computer models. The models originate the sounds at the focus of a parabolic reflector. The parabola in the graph shows the cross section of the reflector with focal length of 1.3 inches and aperture width of 8 inches. Write an equation to represent the cross section of the reflector. What is the depth of the reflector? (See Example 5.)

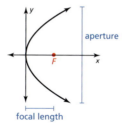

Section 3.6   Focus of a Parabola   167

Dynamic Teaching Tools

Dynamic Assessment & Progress Monitoring Tool
Interactive Whiteboard Lesson Library
Dynamic Classroom with Dynamic Investigations

ANSWERS

24. 5 in.; The bulb should be placed at the focus. The distance from the vertex to the focus is $p = \frac{20}{4} = 5$ in.
25. $y = \frac{1}{32}x^2$
26. $y = -\frac{1}{3}x^2$
27. $x = -\frac{1}{10}y^2$
28. $x = \frac{1}{8}y^2$
29. $x = \frac{1}{12}y^2$
30. $x = \frac{3}{8}y^2$
31. $x = \frac{1}{40}y^2$
32. $y = -\frac{3}{32}x^2$
33. $y = -\frac{3}{20}x^2$
34. $y = \frac{1}{5}x^2$
35. $y = \frac{7}{24}x^2$
36. $x = -\frac{5}{16}y^2$
37. $x = -\frac{1}{16}y^2 - 4$
38. $x = \frac{1}{8}(y - 1)^2 + 4$
39. $y = \frac{1}{6}x^2 + 1$
40. $y = -\frac{1}{24}(x + 6)^2 - 4$
41. The vertex is $(3, 2)$. The focus is $(3, 4)$. The directrix is $y = 0$. The axis of symmetry is $x = 3$. The graph is a vertical shrink by a factor of $\frac{1}{2}$ followed by a translation 3 units right and 2 units up.
42. The vertex is $(-2, 1)$. The focus is $(-2, 0)$. The directrix is $y = 2$. The axis of symmetry is $x = -2$. The graph is a reflection in the x-axis and a translation 2 units left and 1 unit up.
43. The vertex is $(1, 3)$. The focus is $(5, 3)$. The directrix is $x = -3$. The axis of symmetry is $y = 3$. The graph is a horizontal shrink by a factor of $\frac{1}{4}$ followed by a translation 1 unit right and 3 units up.
44. The vertex is $(-3, -5)$. The focus is $(-3, -4.75)$. The directrix is $y = -5.25$. The axis of symmetry is $x = -3$. The graph is a vertical stretch by a factor of 4 followed by a translation 3 units left and 5 units down.

---

45. The vertex is $(2, -4)$. The focus is $\left(\frac{23}{12}, -4\right)$. The directrix is $x = \frac{25}{12}$. The axis of symmetry is $y = -4$. The graph is a horizontal stretch by a factor of 12 followed by a reflection in the y-axis and a translation 2 units right and 4 units down.

46. The vertex is $(-1, -5)$. The focus is $\left(-\frac{15}{16}, -5\right)$. The directrix is $x = -\frac{17}{16}$. The axis of symmetry is $y = -5$. The graph is a horizontal stretch by a factor of 16 followed by a translation 1 unit left and 5 units down.

47. $x = \frac{1}{5.2}y^2$; about 3.08 in.

ANSWERS

48. $y = \frac{1}{6.8}x^2$; The domain is $-2.9 \le x \le 2.9$ and the range is $0 \le y \le 1.7$; The domain represents the width of the trough, and the range represents the height of the trough.

49. As $|p|$ increases, the graph gets wider; As $|p|$ increases, the constant in the function gets smaller which results in a vertical shrink, making the graph wider.

50. a. B is the vertex, C is the focus, and A is a point on the directrix.
    b. The focus and directrix will both be shifted up 3 units.

51. $y = \frac{1}{4}x^2$

52. Sample answer: One equation is $y = \frac{1}{8}(x-a)^2 + (b-2)$ with a directrix of $y = b - 4$. Another equation is $y = -\frac{1}{8}(x-a)^2 + (b+2)$ with a directrix of $y = b + 4$.

53. $x = \frac{1}{4p}y^2$ 54. 8

55–58. See Additional Answers.

Mini-Assessment

1. Use the Distance Formula to write an equation of the parabola with focus $F(0, 8)$ and directrix $y = -8$. $y = \frac{1}{32}x^2$

2. Identify the focus, directrix, and axis of symmetry of $8y = x^2$. Graph the equation. The focus is $F(0, 2)$. The directrix is $y = -2$. The axis of symmetry is the y-axis, $x = 0$.

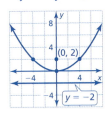

3. Write an equation of the parabola with vertex at $(0, 0)$ and directrix $y = 6$. $y = -\frac{1}{24}x^2$

4. Write an equation of the parabola with vertex $(2, 10)$ and focus $(2, 9.5)$. $y = -0.5(x-2)^2 + 10$

5. An arch is 10 feet across at ground level. Write an equation that represents the cross section of the arch with its vertex at $(0, 0)$ and its focus 0.5 foot below the vertex. What is the height of the arch? $y = -0.5x^2$; 12.5 feet

168   Chapter 3

---

48. **MODELING WITH MATHEMATICS** Solar energy can be concentrated using long troughs that have a parabolic cross section as shown in the figure. Write an equation to represent the cross section of the trough. What are the domain and range in this situation? What do they represent?

49. **ABSTRACT REASONING** As $|p|$ increases, how does the width of the graph of the equation $y = \frac{1}{4p}x^2$ change? Explain your reasoning.

50. **HOW DO YOU SEE IT?** The graph shows the path of a volleyball served from an initial height of 6 feet as it travels over a net.

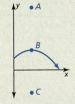

a. Label the vertex, focus, and a point on the directrix.
b. An underhand serve follows the same parabolic path but is hit from a height of 3 feet. How does this affect the focus? the directrix?

51. **CRITICAL THINKING** The distance from point P to the directrix is 2 units. Write an equation of the parabola.

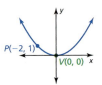

52. **THOUGHT PROVOKING** Two parabolas have the same focus (a, b) and focal length of 2 units. Write an equation of each parabola. Identify the directrix of each parabola.

53. **REPEATED REASONING** Use the Distance Formula to derive the equation of a parabola that opens to the right with vertex $(0, 0)$, focus $(p, 0)$, and directrix $x = -p$.

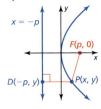

54. **PROBLEM SOLVING** The *latus rectum* of a parabola is the line segment that is parallel to the directrix, passes through the focus, and has endpoints that lie on the parabola. Find the length of the latus rectum of the parabola shown.

Maintaining Mathematical Proficiency — Reviewing what you learned in previous grades and lessons

Show that an exponential model fits the data. Then write a recursive rule that models the data. *(Section 1.6)*

55.
n	0	1	2	3	4	5
f(n)	0.25	0.5	1	2	4	8

56.
n	0	1	2	3	4	5
f(n)	2	6	18	54	162	486

57.
n	0	1	2	3	4	5
f(n)	144	72	36	18	9	4.5

58.
n	0	1	2	3	4	5
f(n)	1250	250	50	10	2	$\frac{2}{5}$

168   Chapter 3   Graphing Quadratic Functions

If students need help...	If students got it...
Resources by Chapter • Practice A and Practice B • Puzzle Time	**Resources by Chapter** • Enrichment and Extension • Cumulative Review
Student Journal • Practice	Start the *next* Section
Differentiating the Lesson Skills Review Handbook	

Laurie's Notes

Dynamic Teaching Tools
- Dynamic Assessment & Progress Monitoring Tool
- Lesson Planning Tool
- Interactive Whiteboard Lesson Library
- Dynamic Classroom with Dynamic Investigations

Overview of Section 3.7

Introduction

- The last lesson in this chapter is designed to help students take a look graphically and numerically at the three main functions they have studied so far.
- Given a set of data, students can graph the data to determine the type of function represented, provided sufficient behavior is evident from the graph. If the graph is incomplete—for instance, if only the increasing portion of a quadratic is displayed—it would be difficult to determine that the function is a quadratic.
- Examining the consecutive y-values in a data set is another technique for determining the type of function a data set represents. This is done by looking for patterns in the first and second differences of y-values, or by looking for a common ratio between consecutive y-values. This technique is used to help write a function for a data set.
- The lesson ends with a look at using average rates of change to compare different function types. The *Teaching Strategy* below helps explain this technique.

Teaching Strategy

- In the exploration, students compare the behavior of three functions over the domain $0 \le x \le 1$ and then for an expanded domain of $1 \le x \le 5$. The first function is linear ($y = x$), the second function is exponential ($y = 2^x - 1$), and the third is quadratic ($y = x^2$). These are represented in Y_1, Y_2, and Y_3, respectively, in Figure 1.
- Notice that over the domain $0 \le x \le 1$, for each value of x, the linear function has the greatest y-values, followed by the exponential, and then the quadratic. (See Figure 2 and Figure 3.)

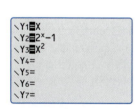

Figure 1

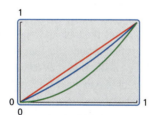

Figure 2

X	Y1	Y2	Y3
0	0	0	0
.1	.1	.07177	.01
.2	.2	.1487	.04
.3	.3	.23114	.09
.4	.4	.31951	.16
.5	.5	.41421	.25
.6	.6	.51572	.36
X=0			

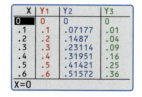

Figure 3

- Now change the domain to $1 \le x \le 4$. For each value of x, the quadratic function has the greatest y-values, followed by the exponential, and then the linear. (See Figure 4.) The final views (Figures 5 and 6) are for a domain $4 \le x \le 9$, which shows that the exponential function has y-values that will continue to grow at a much faster rate than the quadratic or linear functions. *Note:* In Exploration 2 on page 169, students will use a domain of $1 \le x \le 5$.

Figure 4

Figure 5

X	Y1	Y2	Y3
4	4	15	16
4.5	4.5	21.627	20.25
5	5	31	25
5.5	5.5	44.255	30.25
6	6	63	36
6.5	6.5	89.51	42.25
7	7	127	49
X=4			

Figure 6

Pacing Suggestion

- This is a long lesson with the explorations helping students to sense the behavior of linear, exponential, and quadratic functions. Refer to the explorations as you work through the formal lesson.

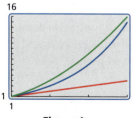

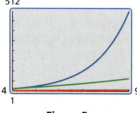

What Your Students Will Learn

- Choose from linear, exponential, or quadratic functions to model data.
- Write functions to model data and write a recursive rule for a quadratic function.
- Compare nonlinear functions using average rates of change over the same interval.

Laurie's Notes

Exploration

Motivate
- **Story Time:** Tell students that you test drove a sports car recently—a Lamborghini Reventón. Any student who follows cars will know that you are fibbing. The base price of a Reventón is approximately $1.6 million, a bit much for a teaching salary. So, instead of test driving, tell students you read about the car and the fact that it can go from 0 to 60 miles per hour in 3.3 seconds with a top speed of 211 miles per hour!
- A car that is a bit more affordable would be a Chevrolet Camaro ZL1, with a base price of about $53,000. It can go from 0 to 60 miles per hour in 3.9 seconds with a top speed of 184 miles per hour.
- Today, students will investigate some fast cars, but no speeding!

Exploration Note
- **Model with Mathematics:** Linear, exponential, and quadratic functions are common functions that can be used to model real-life applications. Mathematically proficient students can identify each of these functions from an equation, a complete graph, or a table of values. When examining a table of values, the behavior of the function becomes evident.
- In this exploration and lesson, students will compare linear, exponential, and quadratic functions.

Exploration 1
- This first exploration reacquaints students with the three types of functions they have studied so far: linear, exponential, and quadratic. Students will explore the growth rates of the three functions over a small, finite domain.
- Students work with partners to complete the table of values for each function.
- A common belief among students is that exponential functions grow most rapidly. What students discover quickly is that the functions all intersect at (0, 0) and (1, 1). It is the behavior of the functions between these two points that is of interest.
- ? "Were there any surprises?" Students should say yes.
- The car with a constant speed should be relatively easy to determine.
- Discuss the explanations offered for which car is accelerating the most.

Exploration 2
- Ask students to describe the three different speeds. Students can visually interpret the speeds in a graph, or they can calculate the speed of each car. Remember, speed is distance divided by time.
- If time permits, set up a table to show the change in average speed for each car. This connects to the answers in Questions 3 and 4.
- Ask students to interpret the table. For many students, the pattern may be hard to describe. The average speed of the car does not change for the linear function; the average speed "increases at an increasing rate" for the exponential function; and the average speed increases at a constant rate for the quadratic function.

Communicate Your Answer
- Students should discuss the questions with partners and then have a class discussion.

Connecting to Next Step
- These explorations help focus attention on the differences, numerically and graphically, between the three types of functions. In the formal lesson, students will do additional work with these three functions.

3.7 Comparing Linear, Exponential, and Quadratic Functions

Essential Question How can you compare the growth rates of linear, exponential, and quadratic functions?

EXPLORATION 1 Comparing Speeds

Work with a partner. Three cars start traveling at the same time. The distance traveled in t minutes is y miles. Complete each table and sketch all three graphs in the same coordinate plane. Compare the speeds of the three cars. Which car has a constant speed? Which car is accelerating the most? Explain your reasoning.

t	$y = t$
0	
0.2	
0.4	
0.6	
0.8	
1.0	

t	$y = 2^t - 1$
0	
0.2	
0.4	
0.6	
0.8	
1.0	

t	$y = t^2$
0	
0.2	
0.4	
0.6	
0.8	
1.0	

COMPARING PREDICTIONS

To be proficient in math, you need to visualize the results of varying assumptions, explore consequences, and compare predictions with data.

EXPLORATION 2 Comparing Speeds

Work with a partner. Analyze the speeds of the three cars over the given time periods. The distance traveled in t minutes is y miles. Which car eventually overtakes the others?

t	$y = t$
1.0	
1.5	
2.0	
2.5	
3.0	
3.5	
4.0	
4.5	
5.0	

t	$y = 2^t - 1$
1.0	
1.5	
2.0	
2.5	
3.0	
3.5	
4.0	
4.5	
5.0	

t	$y = t^2$
1.0	
1.5	
2.0	
2.5	
3.0	
3.5	
4.0	
4.5	
5.0	

Communicate Your Answer

3. How can you compare the growth rates of linear, exponential, and quadratic functions?

4. Which function has a growth rate that is eventually much greater than the growth rates of the other two functions? Explain your reasoning.

English Language Learners

Build on Past Knowledge
Remind students that they can identify the type of function by using both its graph and the form of its equation. The graph of a *linear* function is a *line*. The variable *x* in an *exponential* function equation is an *exponent*.

Extra Example 1

Plot the points. Tell whether the points appear to represent a *linear*, an *exponential*, or a *quadratic* function.

a. (8, 4), (4, 0), (0, −4), (−2, −6), (−4, −8)

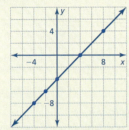

linear function

b. (−3, 6), (−2, 1), (0, −3), (2, 1), (3, 6)

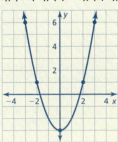

quadratic function

c. (−3, 9), (−2, 5), (−1, 3), (0, 2), (1, 1.5)

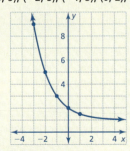

exponential function

MONITORING PROGRESS ANSWERS

1–3. See Additional Answers.

170   Chapter 3

3.7 Lesson

Core Vocabulary

Previous
average rate of change
slope

What You Will Learn

▸ Choose functions to model data.
▸ Write functions to model data.
▸ Compare functions using average rates of change.

Choosing Functions to Model Data

So far, you have studied linear functions, exponential functions, and quadratic functions. You can use these functions to model data.

Core Concept

Linear, Exponential, and Quadratic Functions

Linear Function	Exponential Function	Quadratic Function
$y = mx + b$	$y = ab^x$	$y = ax^2 + bx + c$

EXAMPLE 1 Using Graphs to Identify Functions

Plot the points. Tell whether the points appear to represent a *linear*, an *exponential*, or a *quadratic* function.

a. (4, 4), (2, 0), (0, 0), $\left(1, -\tfrac{1}{2}\right)$, (−2, 4)

b. (0, 1), (2, 4), (4, 7), (−2, −2), (−4, −5)

c. (0, 2), (2, 8), (1, 4), (−1, 1), $\left(-2, \tfrac{1}{2}\right)$

SOLUTION

a.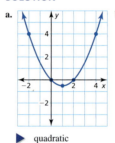
▸ quadratic

b.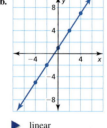
▸ linear

c.
▸ exponential

Monitoring Progress Help in English and Spanish at *BigIdeasMath.com*

Plot the points. Tell whether the points appear to represent a *linear*, an *exponential*, or a *quadratic* function.

1. (−1, 5), (2, −1), (0, −1), (3, 5), (1, −3)
2. (−1, 2), (−2, 8), (−3, 32), $\left(0, \tfrac{1}{2}\right)$, $\left(1, \tfrac{1}{8}\right)$
3. (−3, 5), (0, −1), (2, −5), (−4, 7), (1, −3)

170   Chapter 3   Graphing Quadratic Functions

Laurie's Notes Teacher Actions

- The *Core Concept* presents a summary of the three types of functions. It would be beneficial to briefly review what the parameters tell you about the graph of each function.
- **Thumbs Up:** Have students work independently on Example 1. Have students self-assess with a *Thumbs Up* signal.
- **Use Appropriate Tools Strategically:** If time permits, have students use a graphing calculator to enter the data and plot. If they know the related functions, enter them in the equation editor.

STUDY TIP
The first differences for exponential and quadratic functions are *not* constant.

Core Concept

Differences and Ratios of Functions
You can use patterns between consecutive data pairs to determine which type of function models the data. The differences of consecutive *y*-values are called *first differences*. The differences of consecutive first differences are called *second differences*.

- **Linear Function** The first differences are constant.
- **Exponential Function** Consecutive *y*-values have a common *ratio*.
- **Quadratic Function** The second differences are constant.

In all cases, the differences of consecutive *x*-values need to be constant.

EXAMPLE 2 Using Differences or Ratios to Identify Functions

Tell whether each table of values represents a *linear*, an *exponential*, or a *quadratic* function.

STUDY TIP
First determine that the differences of consecutive *x*-values are constant. Then check whether the first differences are constant or consecutive *y*-values have a common ratio. If neither of these is true, check whether the second differences are constant.

a.
x	−3	−2	−1	0	1
y	11	8	5	2	−1

b.
x	−2	−1	0	1	2
y	1	2	4	8	16

c.
x	−2	−1	0	1	2
y	−1	−2	−1	2	7

SOLUTION

a.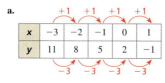

▶ The first differences are constant. So, the table represents a linear function.

b.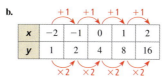

▶ Consecutive *y*-values have a common ratio. So, the table represents an exponential function.

c.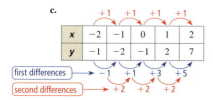

▶ The second differences are constant. So, the table represents a quadratic function.

Monitoring Progress Help in English and Spanish at *BigIdeasMath.com*

4. Tell whether the table of values represents a *linear*, an *exponential*, or a *quadratic* function.

x	−1	0	1	2	3
y	1	3	9	27	81

Section 3.7 Comparing Linear, Exponential, and Quadratic Functions 171

Differentiated Instruction

Inclusion
Some students may not understand the concept of "second difference." Give those students several tables of values with constant consecutive *x*-values. Have them find the *y*-value differences. Then have them find the differences of the *y*-value differences, or the "second differences."

Extra Example 2
Tell whether each table of values represents a *linear*, an *exponential*, or a *quadratic* function.

a.
x	−2	−1	0	1	2
y	11	5	3	5	11

quadratic function

b.
x	−2	−1	0	1	2
y	6	8	10	12	14

linear function

c.
x	0	1	2	3	4
y	0.5	3	18	108	648

exponential function

MONITORING PROGRESS ANSWER

4. exponential

Laurie's Notes Teacher Actions

- Equations and complete graphs can help determine what type of function you have. Differences and ratios can as well.
- Write the *Core Concept*. First differences and the ratio of consecutive *y*-values should make sense from the definitions of linear and exponential functions.
- In Example 2, note that consecutive *x*-values are used.
- Explain that often you can tell by inspection whether the first differences of the *y*-values are constant. Likewise, you might suspect a possible common ratio by inspection. When neither is true, try second differences.

Extra Example 3

Tell whether the table of values represents a *linear*, an *exponential*, or a *quadratic* function. Then write the function.

x	0	1	2	3	4
f(x)	−16	−6	0	2	0

quadratic function;
$f(x) = -2x^2 + 12x - 16$

Extra Example 4

Write a recursive rule for the quadratic function f in Extra Example 3.
$f(0) = -16, f(n) = f(n-1) - 4n + 14$

MONITORING PROGRESS ANSWERS

5. exponential; $y = 8\left(\frac{1}{2}\right)^x$
6. $f(0) = -8, f(n) = f(n-1) + 4n + 3$

Writing Functions to Model Data

EXAMPLE 3 Writing a Function to Model Data

x	0	1	2	3	4
f(x)	12	0	−4	0	12

Tell whether the table of values represents a *linear*, an *exponential*, or a *quadratic* function. Then write the function.

SOLUTION

Step 1 Determine which type of function the table of values represents.

The second differences are constant. So, the table represents a quadratic function.

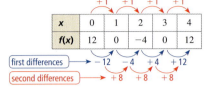

Step 2 Write an equation of the quadratic function. Using the table, notice that the x-intercepts are 1 and 3. So, use intercept form to write a function.

$f(x) = a(x - 1)(x - 3)$ Substitute for p and q in intercept form.

Use another point from the table, such as (0, 12), to find a.

$12 = a(0 - 1)(0 - 3)$ Substitute 0 for x and 12 for $f(x)$.

$4 = a$ Solve for a.

Use the value of a to write the function.

$f(x) = 4(x - 1)(x - 3)$ Substitute 4 for a.

$= 4x^2 - 16x + 12$ Use the FOIL Method and combine like terms.

▶ So, the quadratic function is $f(x) = 4x^2 - 16x + 12$.

STUDY TIP
To check your function in Example 3, substitute the other points from the table to verify that they satisfy the function.

Previously, you wrote recursive rules for linear and exponential functions. You can use the pattern in the first differences to write recursive rules for quadratic functions.

EXAMPLE 4 Writing a Recursive Rule

Write a recursive rule for the quadratic function f in Example 3.

SOLUTION

An expression for the nth term of the sequence of first differences is $-12 + 8(n-1)$, or $8n - 20$. Notice that $f(n) - f(n-1) = 8n - 20$.

▶ So, a recursive rule for the quadratic function is $f(0) = 12$,
$f(n) = f(n-1) + 8n - 20$.

Monitoring Progress   Help in English and Spanish at *BigIdeasMath.com*

5. Tell whether the table of values represents a *linear*, an *exponential*, or a *quadratic* function. Then write the function.

x	−1	0	1	2	3
y	16	8	4	2	1

6. Write a recursive rule for the function represented by the table of values.

x	0	1	2	3	4
f(x)	−8	−1	10	25	44

172   Chapter 3   Graphing Quadratic Functions

Laurie's Notes   Teacher Actions

- Present the data set in Example 3.
- **Turn and Talk:** "What type of function do the data represent?" Give time for partners to discuss. Once they decide it is quadratic, ask, "What can you do with data points (1, 0) and (3, 0)?" Identify 1 and 3 as the x-intercepts and substitute in the intercept form.
- "What do you notice about the sequence −12, −4, 4, 12, …?" It is an arithmetic sequence with a common difference of 8 and a beginning value of −12. "Can you write this sequence recursively?" $a_0 = -12$ and $d = 8$ so $f(n) - f(n-1) = -12 + 8(n-1)$.
- Continue to solve Example 4 as shown.

Comparing Functions Using Average Rates of Change

For nonlinear functions, the rate of change is not constant. You can compare two nonlinear functions over the same interval using their average rates of change. Recall that the average rate of change of a function $y = f(x)$ between $x = a$ and $x = b$ is the slope of the line through $(a, f(a))$ and $(b, f(b))$.

$$\text{average rate of change} = \frac{\text{change in } y}{\text{change in } x} = \frac{f(b) - f(a)}{b - a}$$

Exponential Function

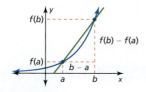

Quadratic Function

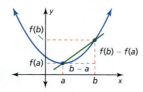

EXAMPLE 5 Using and Interpreting Average Rates of Change

The table and graph show the populations of two species of fish introduced into a lake. Compare the populations by calculating and interpreting the average rates of change from Year 0 to Year 4.

Species A

Year, x	Population, y
0	100
2	264
4	428
6	592
8	756
10	920

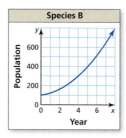

SOLUTION

Calculate the average rates of change by using the points whose x-coordinates are 0 and 4.

Species A: Use (0, 100) and (4, 428).

$$\text{average rate of change} = \frac{f(b) - f(a)}{b - a} = \frac{428 - 100}{4 - 0} = 82$$

Species B: Use the graph to estimate the points when $x = 0$ and $x = 4$. Use (0, 100) and (4, 300).

$$\text{average rate of change} = \frac{f(b) - f(a)}{b - a} \approx \frac{300 - 100}{4 - 0} = 50$$

▶ From Year 0 to Year 4, the population of Species A increases at an average rate of 82 fish per year, and the population of Species B increases at an average rate of about 50 fish per year. So, the population of Species A is growing faster.

Monitoring Progress Help in English and Spanish at BigIdeasMath.com

7. Compare the populations by calculating and interpreting the average rates of change from Year 4 to Year 8.

Extra Example 5

The table and graph show the numbers of members of two news websites after opening their websites to the public. Compare the websites by calculating and interpreting the average rates of change from Day 10 to Day 20.

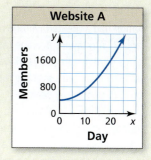

Website B	
Day, x	Members, y
0	500
5	850
10	1200
15	1550
20	1900
25	2250

From Day 10 to Day 20, Website A membership increases at an average rate of about 90 people per day, and Website B membership increases at an average rate of 70 people per day. So, Website A membership is growing faster.

MONITORING PROGRESS ANSWER

7. Species A: 82 fish per year; Species B: about 125 fish per year; The population of Species B is growing faster.

Laurie's Notes — Teacher Actions

- Explain how average rates of change are found for nonlinear functions when comparing them over the same interval.
- **Whiteboarding:** Pose Example 5 and have partners work to answer the question as you circulate. Students could use *whiteboards* to record their work. Do not rush in to rescue students. This is a problem where productive struggle will lead to a deeper understanding of average rate of change.
- When interpreting the results, make sure students understand that an average rate of increase of 82 fish per year means that there were years when there were fewer or more than 82 new fish. Connect these rates to the table of values and the graph.

Extra Example 6

Let x represent the number of years since 1900. The function $Y(x) = 4x^2 + 150$ represents the population of Yorktown. In 1900, Penn Hill had a population of 75 people. Penn Hill's population increased by 10% each year.

a. From 1900 to 1950, which town's population had a greater average rate of change? **Yorktown's average rate of change (200 people per year) was greater than Penn Hill's average rate of change (175 people per year).**

b. Which town will eventually have a greater population? Explain. **Because Penn Hill's population is given by an increasing exponential function and Yorktown's population is given by an increasing quadratic function, Penn Hill will eventually have a greater population. Penn Hill's population caught up to and exceeded Yorktown's population between $x = 52$ and $x = 53$, which corresponds to 1952.**

MONITORING PROGRESS ANSWER

8. Pine Valley

Core Concept

Comparing Functions Using Average Rates of Change

- As a and b increase, the average rate of change between $x = a$ and $x = b$ of an increasing exponential function $y = f(x)$ will eventually exceed the average rate of change between $x = a$ and $x = b$ of an increasing quadratic function $y = g(x)$ or an increasing linear function $y = h(x)$. So, as x increases, $f(x)$ will eventually exceed $g(x)$ or $h(x)$.
- As a and b increase, the average rate of change between $x = a$ and $x = b$ of an increasing quadratic function $y = g(x)$ will eventually exceed the average rate of change between $x = a$ and $x = b$ of an increasing linear function $y = h(x)$. So, as x increases, $g(x)$ will eventually exceed $h(x)$.

STUDY TIP
You can explore these concepts using a graphing calculator.

EXAMPLE 6 Comparing Different Function Types

Let x represent the number of years since 1900. The function $C(x) = 12x^2 + 12x + 100$ represents the population of Cedar Ridge. In 1900, Pine Valley had a population of 50 people. Pine Valley's population increased by 9% each year.

a. From 1900 to 1950, which town's population had a greater average rate of change?
b. Which town will eventually have a greater population? Explain.

SOLUTION

a. Write the function that represents the population of Cedar Ridge and a function to model the population of Pine Valley.

Cedar Ridge: $C(x) = 12x^2 + 12x + 100$ Quadratic function
Pine Valley: $P(x) = 50(1.09)^x$ Exponential function

Find the average rate of change from 1900 to 1950 for the population of each town.

Cedar Ridge
$$\frac{C(50) - C(0)}{50 - 0} = 612$$

Pine Valley
$$\frac{P(50) - P(0)}{50 - 0} \approx 73$$

▶ From 1900 to 1950, the average rate of change of Cedar Ridge's population was greater.

b. Because Pine Valley's population is given by an increasing exponential function and Cedar Ridge's population is given by an increasing quadratic function, Pine Valley will eventually have a greater population. Using a graphing calculator, you can see that Pine Valley's population caught up to and exceeded Cedar Ridge's population between $x = 87$ and $x = 88$, which corresponds to 1987.

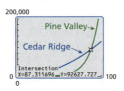

Monitoring Progress Help in English and Spanish at *BigIdeasMath.com*

8. From 1950 to 2000, which town's population had a greater average rate of change?

Laurie's Notes — Teacher Actions

- To help students understand the *Core Concept,* you might use a graphing calculator. See the *Teaching Strategy* on page T-168.
- **Make Sense of Problems and Persevere in Solving Them:** Read through the problem. Answer questions about language.
- **Popsicle Sticks:** Have partners work on part (a). Use *Popsicle Sticks* to solicit a solution.
- **Use Appropriate Tools Strategically:** Use technology to work through part (b) of the problem, as shown. Technology can help students explore the behavior of the two functions as x continues to increase. Use the table of values and the graph.

Closure

- Write a linear, an exponential, and a quadratic function that each has a y-intercept of 2.
Sample answer: $y = \frac{1}{2}x + 2$, $y = 2(3)^x$, $y = x^2 - 4x + 2$

3.7 Exercises

Dynamic Solutions available at *BigIdeasMath.com*

Vocabulary and Core Concept Check

1. **WRITING** Name three types of functions that you can use to model data. Describe the equation and graph of each type of function.

2. **WRITING** How can you decide whether to use a linear, an exponential, or a quadratic function to model a data set?

3. **VOCABULARY** Describe how to find the average rate of change of a function $y = f(x)$ between $x = a$ and $x = b$.

4. **WHICH ONE DOESN'T BELONG?** Which graph does *not* belong with the other three? Explain your reasoning.

Monitoring Progress and Modeling with Mathematics

In Exercises 5–8, tell whether the points appear to represent a *linear*, an *exponential*, or a *quadratic* function.

5.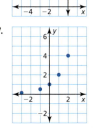

6.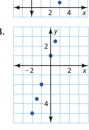

7.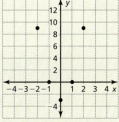

8.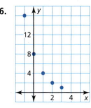

In Exercises 9–14, plot the points. Tell whether the points appear to represent a *linear*, an *exponential*, or a *quadratic* function. *(See Example 1.)*

9. $(-2, -1), (-1, 0), (1, 2), (2, 3), (0, 1)$

10. $\left(0, \frac{1}{4}\right), (1, 1), (2, 4), (3, 16), \left(-1, \frac{1}{16}\right)$

11. $(0, -3), (1, 0), (2, 9), (-2, 9), (-1, 0)$

12. $(-1, -3), (-3, 5), (0, -1), (1, 5), (2, 15)$

13. $(-4, -4), (-2, -3.4), (0, -3), (2, -2.6), (4, -2)$

14. $(0, 8), (-4, 0.25), (-3, 0.4), (-2, 1), (-1, 3)$

In Exercises 15–18, tell whether the table of values represents a *linear*, an *exponential*, or a *quadratic* function. *(See Example 2.)*

15.
x	−2	−1	0	1	2
y	0	0.5	1	1.5	2

16.
x	−1	0	1	2	3
y	0.2	1	5	25	125

17.
x	2	3	4	5	6
y	2	6	18	54	162

18.
x	−3	−2	−1	0	1
y	2	4.5	8	12.5	18

Section 3.7 Comparing Linear, Exponential, and Quadratic Functions 175

11.
quadratic

12.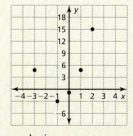
quadratic

Assignment Guide and Homework Check

ASSIGNMENT

Basic: 1–4, 5–31 odd, 35, 37, 38, 43, 46–53

Average: 1–4, 6–38 even, 41, 43, 46–53

Advanced: 1–4, 12–38 even, 39–53

HOMEWORK CHECK

Basic: 9, 15, 21, 27, 35, 37

Average: 12, 16, 22, 28, 34, 36

Advanced: 14, 18, 24, 28, 34, 36

ANSWERS

1. linear, exponential, quadratic; $y = mx + b$, $y = ab^x$, $y = ax^2 + bx + c$; a straight line, a continuously increasing or decreasing curve, a parabola

2. *Sample answer:* Plot the points from the data set and determine what shape they appear to take.

3. Find the slope of the line through $(a, f(a))$ and $(b, f(b))$.

4. *n*; *n* is an exponential function, but the rest are quadratic functions.

5. quadratic
6. exponential
7. exponential
8. linear
9.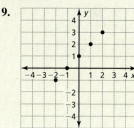
linear

10.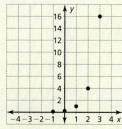
exponential

13–18. See Additional Answers.

Section 3.7   175

ANSWERS

19. linear; The first differences are constant.
20. quadratic; The second differences are constant.
21. quadratic; $y = 2x^2 - 2x - 4$
22. exponential; $y = \left(\frac{1}{2}\right)^x$
23. linear; $y = -3x - 2$
24. exponential; $y = 5(2)^x$
25. quadratic; $y = -3x^2$
26. linear; $y = -2x - 2$
27. $f(0) = 2, f(n) = f(n-1) + 2n - 3$
28. $f(0) = 3, f(n) = 2 \cdot f(n-1)$
29. Consecutive y-values have a constant ratio. They do not change by a constant amount; Consecutive y-values change by a constant ratio. So, the table represents an exponential function.
30. The factor for the first x-intercept should be $(x + 2)$, not $(x - 2)$; $f(x) = a(x + 2)(x - 1)$; $4 = a(-3 + 2)(-3 - 1)$; $1 = a$; $f(x) = 1(x + 2)(x - 1)$; $= x^2 + x - 2$; So, the function is $f(x) = x^2 + x - 2$.
31. a.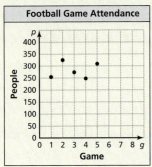

    b. no; The points do not appear to follow the shape of any of these types of functions.

---

19. MODELING WITH MATHEMATICS A student takes a subway to a public library. The table shows the distances d (in miles) the student travels in t minutes. Let the time t represent the independent variable. Tell whether the data can be modeled by a *linear*, an *exponential*, or a *quadratic* function. Explain.

Time, t	0.5	1	3	5
Distance, d	0.335	0.67	2.01	3.35

20. MODELING WITH MATHEMATICS A store sells custom circular rugs. The table shows the costs c (in dollars) of rugs that have diameters of d feet. Let the diameter d represent the independent variable. Tell whether the data can be modeled by a *linear*, an *exponential*, or a *quadratic* function. Explain.

Diameter, d	3	4	5	6
Cost, c	63.90	113.60	177.50	255.60

In Exercises 21–26, tell whether the data represent a *linear*, an *exponential*, or a *quadratic* function. Then write the function. *(See Example 3.)*

21. $(-2, 8), (-1, 0), (0, -4), (1, -4), (2, 0), (3, 8)$

22. $(-3, 8), (-2, 4), (-1, 2), (0, 1), (1, 0.5)$

23.
x	-2	-1	0	1	2
y	4	1	-2	-5	-8

24.
x	-1	0	1	2	3
y	2.5	5	10	20	40

25.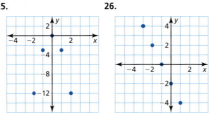

26.

In Exercises 27 and 28, write a recursive rule for the function represented by the table of values. *(See Example 4.)*

27.
x	0	1	2	3	4
$f(x)$	2	1	2	5	10

28.
x	0	1	2	3	4
$f(x)$	3	6	12	24	48

29. ERROR ANALYSIS Describe and correct the error in determining whether the table represents a linear, an exponential, or a quadratic function.

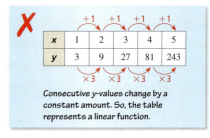

30. ERROR ANALYSIS Describe and correct the error in writing the function represented by the table.

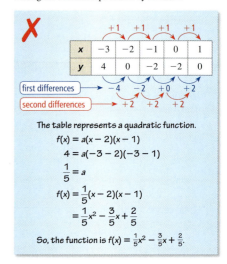

31. REASONING The table shows the numbers of people attending the first five football games at a high school.

Game, g	1	2	3	4	5
People, p	252	325	270	249	310

a. Plot the points. Let the game g represent the independent variable.

b. Can a linear, an exponential, or a quadratic function represent this situation? Explain.

32. MODELING WITH MATHEMATICS The table shows the breathing rates y (in liters of air per minute) of a cyclist traveling at different speeds x (in miles per hour).

Speed, x	20	21	22	23	24
Breathing rate, y	51.4	57.1	63.3	70.3	78.0

a. Plot the points. Let the speed x represent the independent variable. Then determine the type of function that best represents this situation.

b. Write a function that models the data.

c. Find the breathing rate of a cyclist traveling 18 miles per hour. Round your answer to the nearest tenth.

33. ANALYZING RATES OF CHANGE The function $f(t) = -16t^2 + 48t + 3$ represents the height (in feet) of a volleyball t seconds after it is hit into the air.

a. Copy and complete the table.

t	0	0.5	1	1.5	2	2.5	3
$f(t)$							

b. Plot the ordered pairs and draw a smooth curve through the points.

c. Describe where the function is increasing and decreasing.

d. Find the average rate of change for each 0.5-second interval in the table. What do you notice about the average rates of change when the function is increasing? decreasing?

34. ANALYZING RELATIONSHIPS The population of Species A in 2010 was 100. The population of Species A increased by 6% every year. Let x represent the number of years since 2010. The graph shows the population of Species B. Compare the populations of the species by calculating and interpreting the average rates of change from 2010 to 2016. *(See Example 5.)*

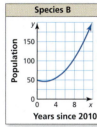

35. ANALYZING RELATIONSHIPS Three charities are holding telethons. Charity A begins with one pledge, and the number of pledges triples each hour. The table shows the numbers of pledges received by Charity B. The graph shows the numbers of pledges received by Charity C.

Time (hours), t	Number of pledges, y
0	12
1	24
2	36
3	48
4	60
5	72
6	84

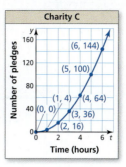

a. What type of function represents the numbers of pledges received by Charity A? B? C?

b. Find the average rates of change of each function for each 1-hour interval from $t = 0$ to $t = 6$.

c. For which function does the average rate of change increase most quickly? What does this tell you about the numbers of pledges received by the three charities?

36. COMPARING FUNCTIONS Let x represent the number of years since 1900. The function

$$H(x) = 10x^2 + 10x + 500$$

represents the population of Oak Hill. In 1900, Poplar Grove had a population of 200 people. Poplar Grove's population increased by 8% each year. *(See Example 6.)*

a. From 1900 to 1950, which town's population had a greater average rate of change?

b. Which town will eventually have a greater population? Explain.

37. COMPARING FUNCTIONS Let x represent the number of years since 2000. The function

$$R(x) = 0.01x^2 + 0.22x + 1.08$$

represents the revenue (in millions of dollars) of Company A. In 2000, Company B had a revenue of $2.12 million. Company B's revenue increased by $0.32 million each year.

a. From 2000 to 2015, which company's revenue had a greater average rate of change?

b. Which company will eventually have a greater revenue? Explain.

ANSWERS

32. a.

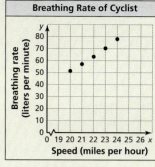

exponential function

b. $y = 51.4(1.11)^{x-20}$

c. about 41.7 L/min

33. a. 3, 23, 35, 39, 35, 23, 3

b.

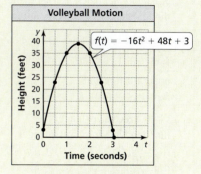

c. The function is increasing between 0 and 1.5 seconds, and the function is decreasing between 1.5 seconds and about 3 seconds.

d. 40, 24, 8, −8, −24, −40; The average rate of change decreases when the function is increasing, and the average rate of change increases in the negative direction when the function is decreasing.

34. Species A: about 7 per year; Species B: about 4 per year; The population of Species A is growing faster.

35. a. exponential; linear; quadratic

b. Charity A: 2, 6, 18, 54, 162, 486; Charity B: 12, 12, 12, 12, 12, 12; Charity C: 4, 12, 20, 28, 36, 44

c. Charity A; Charity A will have the most pledges followed by Charity C, then Charity B.

36. a. Oak Hill

b. Poplar Grove; Poplar Grove's population can be modeled by an increasing exponential function.

37. a. Company A

b. Company A; Company B's revenue can be modeled by an increasing linear function.

ANSWERS

38–53. See Additional Answers.

Mini-Assessment

1. Tell whether the table of values represents a *linear*, an *exponential*, or a *quadratic* function. Then write the function.

x	0	1	2	3	4
f(x)	8	3	0	−1	0

   quadratic function;
   $f(x) = x^2 - 6x + 8$

2. Write a recursive rule for the function f in Mini-Assessment 1.
   $f(0) = 8, f(n) = f(n-1) + 2n - 7$

3. Let x represent the number of years since 1900. The function $S(x) = x^2 + 2x + 2000$ represents the population of Smalltown. In 1900, Growton had a population of 500 people. Growton's population increased by 3% each year.

   a. From 1900 to 1950, which town's population had a greater average rate of change? Smalltown's average rate of change (52 people per year) was greater than Growton's average rate of change (34 people per year).

   b. Which town will eventually have a greater population? Explain. Because Growton's population is given by an increasing exponential function and Smalltown's population is given by an increasing quadratic function, Growton will eventually have a greater population. Growton's population catches up to and exceeds Smalltown's population between $x = 117$ and $x = 118$, which corresponds to 2017.

---

38. HOW DO YOU SEE IT? Match each graph with its function. Explain your reasoning.

a. b.

c. d.

A. $y = 2x^2 - 4$ B. $y = 2^x - 1$

C. $y = 2x - 1$ D. $y = -2x^2 + 4$

39. REASONING Explain why the average rate of change of a linear function is constant and the average rate of change of a quadratic or exponential function is not constant.

40. CRITICAL THINKING In the ordered pairs below, the y-values are given in terms of n. Tell whether the ordered pairs represent a *linear*, an *exponential*, or a *quadratic* function. Explain.

$(1, 3n - 1), (2, 10n + 2), (3, 26n),$
$(4, 51n - 7), (5, 85n - 19)$

41. USING STRUCTURE Write a function that has constant second differences of 3.

Maintaining Mathematical Proficiency   Reviewing what you learned in previous grades and lessons

Evaluate the expression. *(Section 1.5)*

46. $\sqrt{121}$
47. $\sqrt[3]{125}$
48. $\sqrt[3]{512}$
49. $\sqrt[5]{243}$

Find the product. *(Section 2.3)*

50. $(x + 6)(x - 6)$
51. $(2y + 5)(2y - 5)$
52. $(4c - 3d)(4c + 3d)$
53. $(-3s + 8t)(-3s - 8t)$

178   Chapter 3   Graphing Quadratic Functions

---

42. CRITICAL THINKING Is the graph of a set of points enough to determine whether the points represent a linear, an exponential, or a quadratic function? Justify your answer.

43. MAKING AN ARGUMENT Function p is an exponential function and function q is a quadratic function. Your friend says that after about $x = 3$, function q will always have a greater y-value than function p. Is your friend correct? Explain.

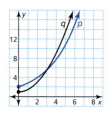

44. THOUGHT PROVOKING Find three different patterns in the figure. Determine whether each pattern represents a *linear*, an *exponential*, or a *quadratic* function. Write a model for each pattern.

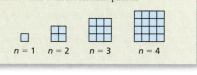

$n = 1$ $n = 2$ $n = 3$ $n = 4$

45. USING TOOLS The table shows the amount a (in billions of dollars) United States residents spent on pets or pet-related products and services each year for a 5-year period. Let the year x represent the independent variable. Using technology, find a function that models the data. How did you choose the model? Predict how much residents will spend on pets or pet-related products and services in Year 7.

Year, x	1	2	3	4	5
Amount, a	53.1	56.9	61.8	65.7	67.1

If students need help...	If students got it...
Resources by Chapter • Practice A and Practice B • Puzzle Time	**Resources by Chapter** • Enrichment and Extension • Cumulative Review
Student Journal • Practice	Start the *next* Section
Differentiating the Lesson Skills Review Handbook	

3.4–3.7 What Did You Learn?

Core Vocabulary

even function, *p. 146*
odd function, *p. 146*
vertex form (of a quadratic function), *p. 148*
intercept form, *p. 154*
focus, *p. 162*
directrix, *p. 162*

Core Concepts

Section 3.4
Even and Odd Functions, *p. 146*
Graphing $f(x) = a(x − h)^2$, *p. 147*
Graphing $f(x) = a(x − h)^2 + k$, *p. 148*
Writing Quadratic Functions of the Form $f(x) = a(x − h)^2 + k$, *p. 149*

Section 3.5
Graphing $f(x) = a(x − p)(x − q)$, *p. 154*
Factors and Zeros, *p. 156*
Using Characteristics to Graph and Write Quadratic Functions, *p. 156*

Section 3.6
Standard Equations of a Parabola with Vertex at the Origin, *p. 163*
Standard Equations of a Parabola with Vertex at (h, k), *p. 164*

Section 3.7
Linear, Exponential, and Quadratic Functions, *p. 170*
Differences and Ratios of Functions, *p. 171*
Writing Functions to Model Data, *p. 172*
Comparing Functions Using Average Rates of Change, *p. 173*

Mathematical Practices

1. How can you use technology to confirm your answer in Exercise 64 on page 152?
2. How did you use the structure of the equation in Exercise 65 on page 159 to solve the problem?
3. Describe why your answer makes sense considering the context of the data in Exercise 20 on page 176.

Performance Task:
Solar Energy

There are many factors that influence the design of a product, such as aesthetics, available space, and special properties. What are the special properties of a parabola that contribute to its use? How do parabolas help create power from solar energy?

To explore the answers to these questions and more, check out the Performance Task and Real-Life STEM video at *BigIdeasMath.com*.

ANSWERS

1.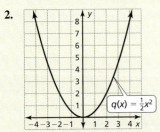
The graph of p is a vertical stretch by a factor of 7 of the graph of f.

2.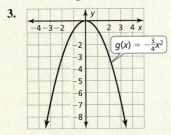
The graph of q is a vertical shrink by a factor of $\frac{1}{2}$ of the graph of f.

3.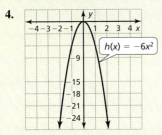
The graph of g is a vertical shrink by a factor of $\frac{3}{4}$ and a reflection in the x-axis of the graph of f.

4.
The graph of h is a vertical stretch by a factor of 6 and a reflection in the x-axis of the graph of f.

5. The vertex is $(1, -3)$. The axis of symmetry is $x = 1$. The domain is all real numbers. The range is $y \geq -3$. When $x < 1$, y increases as x decreases. When $x > 1$, y increases as x increases.

---

3 Chapter Review

Dynamic Solutions available at BigIdeasMath.com

3.1 Graphing $f(x) = ax^2$ (pp. 123–128)

Graph $g(x) = -4x^2$. Compare the graph to the graph of $f(x) = x^2$.

Step 1 Make a table of values.

x	−2	−1	0	1	2
g(x)	−16	−4	0	−4	−16

Step 2 Plot the ordered pairs.

Step 3 Draw a smooth curve through the points.

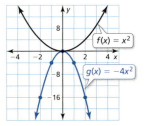

▶ The graphs have the same vertex, $(0, 0)$, and the same axis of symmetry, $x = 0$, but the graph of g opens down and is narrower than the graph of f. So, the graph of g is a vertical stretch by a factor of 4 and a reflection in the x-axis of the graph of f.

Graph the function. Compare the graph to the graph of $f(x) = x^2$.

1. $p(x) = 7x^2$
2. $q(x) = \frac{1}{2}x^2$
3. $g(x) = -\frac{3}{4}x^2$
4. $h(x) = -6x^2$

5. Identify characteristics of the quadratic function and its graph.

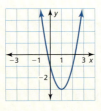

3.2 Graphing $f(x) = ax^2 + c$ (pp. 129–134)

Graph $g(x) = 2x^2 + 3$. Compare the graph to the graph of $f(x) = x^2$.

Step 1 Make a table of values.

x	−2	−1	0	1	2
g(x)	11	5	3	5	11

Step 2 Plot the ordered pairs.

Step 3 Draw a smooth curve through the points.

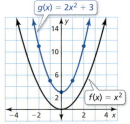

▶ Both graphs open up and have the same axis of symmetry, $x = 0$. The graph of g is narrower, and its vertex, $(0, 3)$, is above the vertex of the graph of f, $(0, 0)$. So, the graph of g is a vertical stretch by a factor of 2 and a vertical translation 3 units up of the graph of f.

Graph the function. Compare the graph to the graph of $f(x) = x^2$.

6. $g(x) = x^2 + 5$
7. $h(x) = -x^2 - 4$
8. $m(x) = -2x^2 + 6$
9. $n(x) = \frac{1}{3}x^2 - 5$

180 Chapter 3 Graphing Quadratic Functions

6.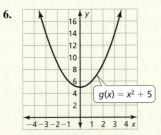
The graph of g is a vertical translation 5 units up of the graph of f.

7.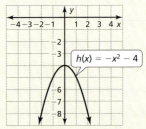
The graph of h is a reflection in the x-axis, and a vertical translation 4 units down of the graph of f.

8–9. See Additional Answers.

3.3 Graphing $f(x) = ax^2 + bx + c$ (pp. 135–142)

Graph $f(x) = 4x^2 + 8x - 1$. Describe the domain and range.

Step 1 Find and graph the axis of symmetry: $x = -\dfrac{b}{2a} = -\dfrac{8}{2(4)} = -1$.

Step 2 Find and plot the vertex. The axis of symmetry is $x = -1$. So, the x-coordinate of the vertex is -1. The y-coordinate of the vertex is $f(-1) = 4(-1)^2 + 8(-1) - 1 = -5$. So, the vertex is $(-1, -5)$.

Step 3 Use the y-intercept to find two more points on the graph. Because $c = -1$, the y-intercept is -1. So, $(0, -1)$ lies on the graph. Because the axis of symmetry is $x = -1$, the point $(-2, -1)$ also lies on the graph.

Step 4 Draw a smooth curve through the points.

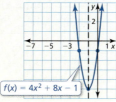

▶ The domain is all real numbers. The range is $y \geq -5$.

Graph the function. Describe the domain and range.

10. $y = x^2 - 2x + 7$ 11. $f(x) = -3x^2 + 3x - 4$ 12. $y = \tfrac{1}{2}x^2 - 6x + 10$

13. The function $f(t) = -16t^2 + 88t + 12$ represents the height (in feet) of a pumpkin t seconds after it is launched from a catapult. When does the pumpkin reach its maximum height? What is the maximum height of the pumpkin?

3.4 Graphing $f(x) = a(x - h)^2 + k$ (pp. 145–152)

Determine whether $f(x) = 2x^2 + 4$ is *even*, *odd*, or *neither*.

$f(x) = 2x^2 + 4$ Write the original function.
$f(-x) = 2(-x)^2 + 4$ Substitute $-x$ for x.
$\qquad = 2x^2 + 4$ Simplify.
$\qquad = f(x)$ Substitute $f(x)$ for $2x^2 + 4$.

▶ Because $f(-x) = f(x)$, the function is even.

Determine whether the function is *even*, *odd*, or *neither*.

14. $w(x) = 5^x$ 15. $r(x) = -8x$ 16. $h(x) = 3x^2 - 2x$

Graph the function. Compare the graph to the graph of $f(x) = x^2$.

17. $h(x) = 2(x - 4)^2$ 18. $g(x) = \tfrac{1}{2}(x - 1)^2 + 1$ 19. $q(x) = -(x + 4)^2 + 7$

20. Consider the function $g(x) = -3(x + 2)^2 - 4$. Graph $h(x) = g(x - 1)$.

21. Write a quadratic function whose graph has a vertex of $(3, 2)$ and passes through the point $(4, 7)$.

Chapter 3 Chapter Review 181

ANSWERS

10.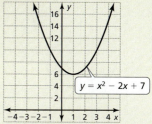
domain: all real numbers, range: $y \geq 6$

11.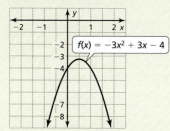
domain: all real numbers, range: $y \leq -\dfrac{13}{4}$

12.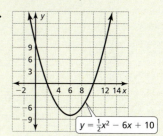
domain: all real numbers, range: $y \geq -8$

13. 2.75 sec; 133 ft

14. neither

15. odd   16. neither

17.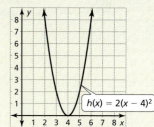
The graph of h is a vertical stretch by a factor of 2 and a horizontal translation 4 units right of the graph of f.

18.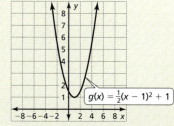
The graph of g is a vertical shrink by a factor of $\tfrac{1}{2}$, and a translation 1 unit right and 1 unit up of the graph of f.

19.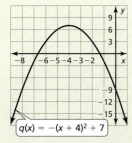
The graph of q is a reflection in the x-axis, and a translation 4 units left and 7 units up of the graph of f.

20.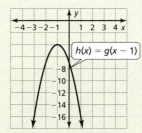

21. $f(x) = 5(x - 3)^2 + 2$

Chapter 3 181

3.5 Graphing $f(x) = a(x - p)(x - q)$ (pp. 153–160)

a. **Graph $f(x) = -(x + 4)(x - 2)$. Describe the domain and range.**

Step 1 Identify the x-intercepts. Because the x-intercepts are $p = -4$ and $q = 2$, plot $(-4, 0)$ and $(2, 0)$.

Step 2 Find and graph the axis of symmetry.

$$x = \frac{p + q}{2} = \frac{-4 + 2}{2} = -1$$

Step 3 Find and plot the vertex.

The x-coordinate of the vertex is -1. To find the y-coordinate of the vertex, substitute -1 for x and simplify.

$$f(-1) = -(-1 + 4)(-1 - 2) = 9$$

So, the vertex is $(-1, 9)$.

Step 4 Draw a parabola through the vertex and the points where the x-intercepts occur.

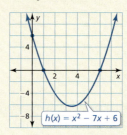

▶ The domain is all real numbers. The range is $y \leq 9$.

b. **Use zeros to graph $h(x) = x^2 - 7x + 6$.**

The function is in standard form. The parabola opens up ($a > 0$), and the y-intercept is 6. So, plot $(0, 6)$.

The polynomial that defines the function is factorable. So, write the function in intercept form and identify the zeros.

$h(x) = x^2 - 7x + 6$ Write the function.

$ = (x - 6)(x - 1)$ Factor the trinomial.

The zeros of the function are 1 and 6. So, plot $(1, 0)$ and $(6, 0)$. Draw a parabola through the points.

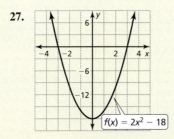

Graph the quadratic function. Label the vertex, axis of symmetry, and x-intercepts. Describe the domain and range of the function.

22. $y = (x - 4)(x + 2)$ 23. $f(x) = -3(x + 3)(x + 1)$ 24. $y = x^2 - 8x + 15$

Use zeros to graph the function.

25. $y = -2x^2 + 6x + 8$ 26. $f(x) = x^2 + x - 2$ 27. $f(x) = 2x^2 - 18$

28. Write a quadratic function in standard form whose graph passes through $(4, 0)$ and $(6, 0)$.

ANSWERS

22.
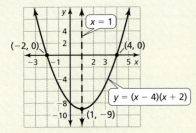

domain: all real numbers, range: $y \geq -9$

23.
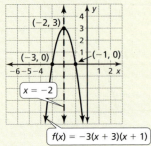

domain: all real numbers, range: $y \leq 3$

24.
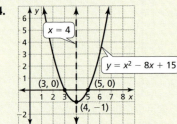

domain: all real numbers, range: $y \geq -1$

25.
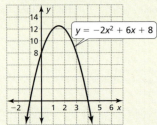

26.
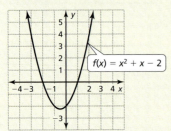

27.
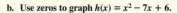

28. *Sample answer:* $x^2 - 10x + 24$

3.6 Focus of a Parabola (pp. 161–168)

a. Identify the focus, directrix, and axis of symmetry of $8x = y^2$. Graph the equation.

Step 1 Rewrite the equation in standard form.

$8x = y^2$ Write the original equation.

$x = \frac{1}{8}y^2$ Divide each side by 8.

Step 2 Identify the focus, directrix, and axis of symmetry. The equation has the form $x = \frac{1}{4p}y^2$, where $p = 2$. The focus is $(p, 0)$, or $(2, 0)$. The directrix is $x = -p$, or $x = -2$. Because y is squared, the axis of symmetry is the x-axis.

Step 3 Use a table of values to graph the equation. Notice that it is easier to substitute y-values and solve for x.

y	0	±2	±4	±6
x	0	0.5	2	4.5

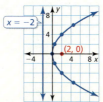

b. Write an equation of the parabola shown.

Because the vertex is not at the origin and the axis of symmetry is vertical, the equation has the form $y = \frac{1}{4p}(x - h)^2 + k$. The vertex (h, k) is $(2, 3)$ and the focus $(h, k + p)$ is $(2, 4)$, so $h = 2$, $k = 3$, and $p = 1$. Substitute these values to write an equation of the parabola.

$y = \frac{1}{4(1)}(x - 2)^2 + 3 = \frac{1}{4}(x - 2)^2 + 3$

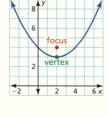

▶ An equation of the parabola is $y = \frac{1}{4}(x - 2)^2 + 3$.

29. You can make a solar hot-dog cooker by shaping foil-lined cardboard into a parabolic trough and passing a wire through the focus of each end piece. For the trough shown, how far from the bottom should the wire be placed?

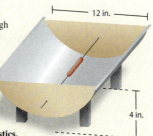

30. Graph the equation $36y = x^2$. Identify the focus, directrix, and axis of symmetry.

Write an equation of the parabola with the given characteristics.

31. vertex: $(0, 0)$
directrix: $x = 2$

32. focus: $(2, 2)$
vertex: $(2, 6)$

ANSWERS

29. 2.25 in.

30. The focus is $(0, 9)$, the directrix is $y = -9$, and the axis of symmetry is $x = 0$.

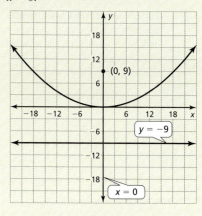

31. $x = -\frac{1}{8}y^2$

32. $y = -\frac{1}{16}(x - 2)^2 + 6$

ANSWERS

33. exponential; $y = 128\left(\frac{1}{4}\right)^x$

34. $f(0) = 5, f(n) = f(n-1) + 4n - 3$

35. From year 2 to year 7, your account increased at an average rate of about $29.55 per year and your friend's account increased at an average rate of $20 per year. So, your account is growing faster.

3.7 Comparing Linear, Exponential, and Quadratic Functions (pp. 169–178)

Tell whether the data represent a *linear*, an *exponential*, or a *quadratic* function.

a. $(-4, 1), (-3, -2), (-2, -3)$
$(-1, -2), (0, 1)$

b. $(-2, 16), (-1, 8), (0, 4)$
$(1, 2), (2, 1)$

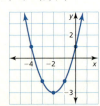

▶ The points appear to represent a quadratic function.

▶ The points appear to represent an exponential function.

c.

x	−1	0	1	2	3
y	15	8	1	−6	−13

d.

x	0	1	2	3	4
y	1	5	6	4	−1

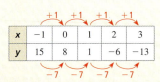

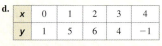
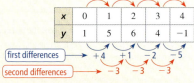

▶ The first differences are constant. So, the table represents a linear function.

▶ The second differences are constant. So, the table represents a quadratic function.

33. Tell whether the table of values represents a *linear*, an *exponential*, or a *quadratic* function. Then write the function.

x	−1	0	1	2	3
y	512	128	32	8	2

34. Write a recursive rule for the function represented by the table of values.

x	0	1	2	3	4
f(x)	5	6	11	20	33

35. The balance y (in dollars) of your savings account after t years is represented by $y = 200(1.1)^t$. The beginning balance of your friend's account is $250, and the balance increases by $20 each year. Compare the account balances by calculating and interpreting the average rates of change from $t = 2$ to $t = 7$.

3 Chapter Test

Graph the function. Compare the graph to the graph of $f(x) = x^2$.

1. $h(x) = 2x^2 - 3$
2. $g(x) = -\frac{1}{2}x^2$
3. $p(x) = \frac{1}{2}(x+1)^2 - 1$

4. Consider the graph of the function f.

   a. Find the domain, range, and zeros of the function.
   b. Write the function f in standard form.
   c. Compare the graph of f to the graph of $g(x) = x^2$.
   d. Graph $h(x) = f(x - 6)$.

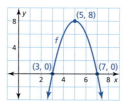

Use zeros to graph the function. Describe the domain and range of the function.

5. $f(x) = 2x^2 - 8x + 8$
6. $y = -(x+5)(x-1)$
7. $h(x) = 16x^2 - 4$

8. Identify an focus, directrix, and axis of symmetry of $x = 2y^2$. Graph the equation.

Write an equation of the parabola. Justify your answer.

9.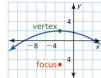
10.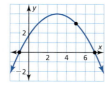
11.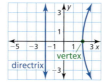

Tell whether the table of values represents a *linear*, an *exponential*, or a *quadratic* function. Explain your reasoning. Then write the function.

12.
x	−1	0	1	2	3
y	4	8	16	32	64

13.
x	−2	−1	0	1	2
y	−8	−2	0	−2	−8

14. You are playing tennis with a friend. The path of the tennis ball after you return a serve can be modeled by the function $y = -0.005x^2 + 0.17x + 3$, where x is the horizontal distance (in feet) from where you hit the ball and y is the height (in feet) of the ball.

   a. What is the maximum height of the tennis ball?
   b. You are standing 30 feet from the net, which is 3 feet high. Will the ball clear the net? Explain your reasoning.

15. Find values of a, b, and c so that the function $f(x) = ax^2 + bx + c$ is (a) even, (b) odd, and (c) neither even nor odd.

16. Consider the function $f(x) = x^2 + 4$. Find the average rate of change from $x = 0$ to $x = 1$, from $x = 1$ to $x = 2$, and from $x = 2$ to $x = 3$. What do you notice about the average rates of change when the function is increasing?

ANSWERS

1.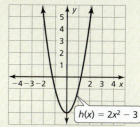

   The graph of h is a vertical stretch by a factor of 2 and a vertical translation 3 units down of the graph of f.

2.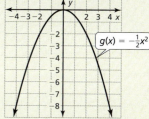

   The graph of g is a vertical shrink by a factor of $\frac{1}{2}$ and a reflection in the x-axis of the graph of f.

3.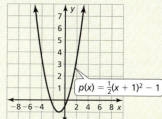

   The graph of p is a vertical shrink by a factor of $\frac{1}{2}$, and a translation 1 unit left and 1 unit down of the graph of f.

4. a. domain: all real numbers, range: $y \leq 8$; 3, 7
   b. $f(x) = -2x^2 + 20x - 42$
   c. The graph of f is a vertical stretch by a factor of 2, a reflection in the x-axis, and a translation 5 units right and 8 units up of the graph of g.
   d.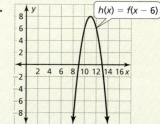

5–16. See Additional Answers.

If students need help...	If students got it...
Lesson Tutorials	Resources by Chapter • Enrichment and Extension • Cumulative Review
Skills Review Handbook	Performance Task
BigIdeasMath.com	Start the *next* Section

3 Cumulative Assessment

1. Which function is represented by the graph?

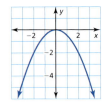

   Ⓐ $y = \frac{1}{2}x^2$
   Ⓑ $y = 2x^2$
   Ⓒ $y = -\frac{1}{2}x^2$
   Ⓓ $y = -2x^2$

2. A linear function f has the following values at $x = 2$ and $x = 5$.
   $$f(2) = 6, \ f(5) = 3$$
   What is the value of the inverse of f at $x = 4$?

3. The function $f(t) = -16t^2 + v_0 t + s_0$ represents the height (in feet) of a ball t seconds after it is thrown from an initial height s_0 (in feet) with an initial vertical velocity v_0 (in feet per second). The ball reaches its maximum height after $\frac{7}{8}$ second when it is thrown with an initial vertical velocity of _____ feet per second.

4. Order the following parabolas from widest to narrowest.

   A. focus: $(0, -3)$; directrix: $y = 3$
   B. $y = \frac{1}{16}x^2 + 4$
   C. $x = \frac{1}{8}y^2$
   D. $y = \frac{1}{4}(x - 2)^2 + 3$

5. Which polynomial represents the area (in square feet) of the shaded region of the figure?

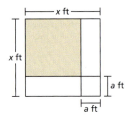

   Ⓐ $a^2 - x^2$
   Ⓑ $x^2 - a^2$
   Ⓒ $x^2 - 2ax + a^2$
   Ⓓ $x^2 + 2ax + a^2$

6. Consider the functions represented by the tables.

x	0	1	2	3
$p(x)$	−4	−16	−28	−40

x	1	2	3	4
$r(x)$	0	15	40	75

x	1	2	3	4
$s(x)$	72	36	18	9

x	1	3	5	7
$t(x)$	3	−5	−21	−45

a. Classify each function as *linear*, *exponential*, or *quadratic*.

b. Order the functions from least to greatest according to the average rates of change between $x = 1$ and $x = 3$.

7. Which expressions are equivalent to $14x - 15 + 8x^2$?

$(4x + 3)(2x - 5)$	$(4x - 3)(2x + 5)$
$(4x + 3)(2x + 5)$	$(4x - 3)(2x - 5)$
$8x^2 - 14x - 15$	$8x^2 + 14x - 15$

8. Complete each function using the symbols + or −, so that the graph of the quadratic function satisfies the given conditions.

a. $f(x) = 5(x \;\square\; 3)^2 \;\square\; 4$; vertex: $(-3, 4)$

b. $g(x) = -(x \;\square\; 2)(x \;\square\; 8)$; x-intercepts: -8 and 2

c. $h(x) = \;\square\; 3x^2 \;\square\; 6$; range: $y \geq -6$

d. $j(x) = \;\square\; 4(x \;\square\; 1)(x \;\square\; 1)$; range: $y \leq 4$

9. Which expressions are equivalent to $(b^{-5})^{-4}$?

b^{-20}	$b^{-6}b^{-14}$	$(b^{-4})^{-5}$	b^{-9}	$(b^{-2})^{-7}$
$b^{-5}b^{-4}$	$(b^{10})^2$	b^{20}	$(b^{-12})^3$	$b^{12}b^8$

ANSWERS

6. a. p is linear, r is quadratic, s is exponential, t is quadratic

b. s, p, t, r

7. $(4x - 3)(2x + 5)$, $8x^2 + 14x - 15$

8. a. +; +

b. −; +

c. +; −

d. −; +; − or −; −; +

9. $(b^{-4})^{-5}$, $(b^{10})^2$, b^{20}, $b^{12}b^8$

Chapter 4 Pacing Guide	
Chapter Opener/ Mathematical Practices	0.5 Day
Section 1	1.5 Days
Section 2	2 Days
Section 3	2 Days
Section 4	2 Days
Section 5	2 Days
Quiz	0.5 Day
Section 6	1.5 Days
Section 7	1 Day
Section 8	2 Days
Section 9	2 Days
Chapter Review/ Chapter Tests	2 Days
Total Chapter 4	19 Days
Year-to-Date	66 Days

4 Solving Quadratic Equations

- 4.1 Properties of Radicals
- 4.2 Solving Quadratic Equations by Graphing
- 4.3 Solving Quadratic Equations Using Square Roots
- 4.4 Solving Quadratic Equations by Completing the Square
- 4.5 Solving Quadratic Equations Using the Quadratic Formula
- 4.6 Complex Numbers
- 4.7 Solving Quadratic Equations with Complex Solutions
- 4.8 Solving Nonlinear Systems of Equations
- 4.9 Quadratic Inequalities

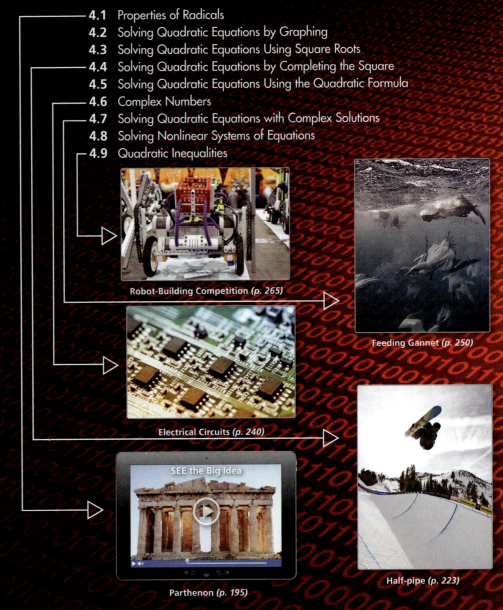

Robot-Building Competition (p. 265)

Electrical Circuits (p. 240)

SEE the Big Idea

Parthenon (p. 195)

Feeding Gannet (p. 250)

Half-pipe (p. 223)

Laurie's Notes

Chapter Summary

- This is a third chapter generally about polynomials and more specifically, about quadratics. In the previous chapters, students factored quadratics and graphed quadratic equations in different forms: standard, vertex, and intercept.
- This chapter is about solving quadratic equations, and depending upon the form in which the equations are written, different techniques are used.
- The chapter begins with a lesson on square roots and how to simplify them. These skills are needed for work later in the chapter.
- The next four lessons present different ways in which a quadratic can be solved: graphing, using square roots, completing the square, and using the Quadratic Formula. Students learn to distinguish between finding the zero of a function and finding the x-intercept of a graph. As the number of strategies increases in the chapter, students should be making informed choices as to which strategy to use given the equation.
- In the second half of the chapter, complex numbers are defined and operations on complex numbers presented. This naturally leads to solving quadratic equations with real and imaginary roots.
- The last two lessons extend work with solving quadratic equations to solving nonlinear systems and solving quadratic inequalities. Each of these topics requires recall of connected skills from work with linear equations. For example, nonlinear systems are solved by methods of graphing, substitution, and elimination.
- It is important throughout the chapter to be clear with students your expectation of the role of technology versus computational and analytical skills.

What Your Students Have Learned

Middle School
- Find square roots and evaluate expressions involving square roots.
- Solve linear equations algebraically and graphically.
- Add, subtract, and multiply rational numbers.
- Write, graph, and solve linear inequalities in one variable.

Math I
- Solve exponential functions algebraically and graphically.
- Solve systems of linear equations in two variables by graphing, substitution, and elimination.
- Graph linear inequalities in one variable and in two variables in the coordinate plane.

What Your Students Will Learn

Math II
- Simplify radical expressions and perform addition, subtraction, and multiplication with radicals.
- Solve quadratic equations with real and complex solutions by graphing, using square roots, by completing the square, and using the Quadratic Formula.
- Interpret the discriminant to determine the number of real solutions of quadratic equations.
- Choose efficient methods for solving quadratic equations.
- Add, subtract, and multiply complex numbers.
- Solve systems of nonlinear equations in two variables by graphing, substitution, and elimination.
- Graph quadratic inequalities in two variables in the coordinate plane.
- Solve quadratic inequalities in one variable algebraically and graphically.

Dynamic Teaching Tools
Dynamic Assessment & Progress Monitoring Tool
Lesson Planning Tool
Interactive Whiteboard Lesson Library
Dynamic Classroom with Dynamic Investigations
Real-Life STEM Videos

Scaffolding in the Classroom
Graphic Organizers: Y-Chart
A Y-Chart can be used to compare two topics. Students list differences between the two topics in the branches of the Y and similarities in the base of the Y. A Y-Chart serves as a good tool for assessing students' knowledge of a pair of topics that have subtle but important differences. You can include blank Y-Charts on tests or quizzes for this purpose.

Questioning in the Classroom

What I really want to know is …
Have each student write a "pertinent" question on a 3 × 5 card. Collect the cards (check the cards before using them) and randomly select a question to ask the class. After class discussion, the last person to participate randomly selects another question to ask the class.

Laurie's Notes

Maintaining Mathematical Proficiency

Factoring Perfect Square Trinomials

- Remind students that the sign of the middle term of a perfect square trinomial determines whether they should use addition or subtraction in the factors. For example, the second term, $14x$, in $x^2 + 14x + 49$ is positive, so addition is used in the factors: $(x + 7)(x + 7) = (x + 7)^2$.

  COMMON ERROR Students confuse perfect square trinomials with the difference of two squares and write two different factors. For example, students may write $(x + 10)(x - 10)$ instead of $(x - 10)^2$ as the factors of $x^2 - 20x + 100$ in Exercise 2.

Solving Systems of Linear Equations by Graphing

- Point out that the equations are in slope-intercept form. Remind students that they can plot the y-intercept of each equation and then use the slope to find a second point on each line.

  COMMON ERROR Students graph one or both of the equations incorrectly and do not check their solution in both equations.

Mathematical Practices (continued on page 190)

- The *Mathematical Practices* page focuses attention on how mathematics is learned—process versus content. This page demonstrates how a mathematically proficient student goes through the process of making adjustments in his or her thinking. Reasoning plays a role in these adjustments.
- Use the *Mathematical Practices* page to help students develop mathematical habits of mind—how mathematics can be explored and how mathematics is thought about.
- Students will be working with irrational numbers in this chapter. In the *Core Concept*, a decimal approximation of $\sqrt{2}$ is being found. Consider how students think about and understand irrational numbers. For instance, ask students the side length of a square with an area of 4 square units and they quickly reply 2 units. Repeat the same question for an area of 2 square units and common answers are 1 or 1 by 2. (They have forgotten it is a square.) By checking their answers students start to understand that they are looking for a number that when multiplied by itself has a product of 2. Their guess-and-check process will look much like the *Core Concept*.

If students need help...	If students got it...
Student Journal • Maintaining Mathematical Proficiency	Game Closet at *BigIdeasMath.com*
Lesson Tutorials	Start the *next* Section
Skills Review Handbook	

Maintaining Mathematical Proficiency

Factoring Perfect Square Trinomials

Example 1 Factor $x^2 + 14x + 49$.

$$x^2 + 14x + 49 = x^2 + 2(x)(7) + 7^2 \quad \text{Write as } a^2 + 2ab + b^2.$$
$$= (x + 7)^2 \quad \text{Perfect square trinomial pattern}$$

Factor the trinomial.

1. $x^2 + 10x + 25$
2. $x^2 - 20x + 100$
3. $x^2 + 12x + 36$
4. $x^2 - 18x + 81$
5. $x^2 + 16x + 64$
6. $x^2 - 30x + 225$

Solving Systems of Linear Equations by Graphing

Example 2 Solve the system of linear equations by graphing.

$y = 2x + 1$ Equation 1
$y = -\frac{1}{3}x + 8$ Equation 2

Step 1 Graph each equation.

Step 2 Estimate the point of intersection. The graphs appear to intersect at (3, 7).

Step 3 Check your point from Step 2.

Equation 1   Equation 2

$y = 2x + 1$ $y = -\frac{1}{3}x + 8$

$7 \stackrel{?}{=} 2(3) + 1$ $7 \stackrel{?}{=} -\frac{1}{3}(3) + 8$

$7 = 7$ ✓ $7 = 7$ ✓

▶ The solution is (3, 7).

Solve the system of linear equations by graphing.

7. $y = -5x + 3$
   $y = 2x - 4$
8. $y = \frac{3}{2}x - 2$
   $y = -\frac{1}{4}x + 5$
9. $y = \frac{1}{2}x + 4$
   $y = -3x - 3$

10. **ABSTRACT REASONING** What value of c makes $x^2 + bx + c$ a perfect square trinomial?

Dynamic Solutions available at *BigIdeasMath.com*

MONITORING PROGRESS ANSWERS

1. $x \approx -1.618$
2. $x \approx 1.303$, $x \approx -2.303$

Mathematical Practices

Mathematically proficient students monitor their work and change course as needed.

Problem-Solving Strategies

Core Concept

Guess, Check, and Revise

When solving a problem in mathematics, it is often helpful to estimate a solution and then observe how close that solution is to being correct. For instance, you can use the guess, check, and revise strategy to find a decimal approximation of the square root of 2.

	Guess	Check	How to revise
1.	1.4	$1.4^2 = 1.96$	Increase guess.
2.	1.41	$1.41^2 = 1.9881$	Increase guess.
3.	1.415	$1.415^2 = 2.002225$	Decrease guess.

By continuing this process, you can determine that the square root of 2 is approximately 1.4142.

EXAMPLE 1 Approximating a Solution of an Equation

The graph of $y = x^2 + x - 1$ is shown. Approximate the positive solution of the equation $x^2 + x - 1 = 0$ to the nearest thousandth.

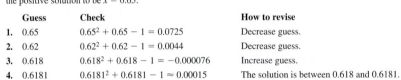

SOLUTION

Using the graph, you can make an initial estimate of the positive solution to be $x = 0.65$.

	Guess	Check	How to revise
1.	0.65	$0.65^2 + 0.65 - 1 = 0.0725$	Decrease guess.
2.	0.62	$0.62^2 + 0.62 - 1 = 0.0044$	Decrease guess.
3.	0.618	$0.618^2 + 0.618 - 1 = -0.000076$	Increase guess.
4.	0.6181	$0.6181^2 + 0.6181 - 1 \approx 0.00015$	The solution is between 0.618 and 0.6181.

▶ So, to the nearest thousandth, the positive solution of the equation is $x = 0.618$.

Monitoring Progress

1. Use the graph in Example 1 to approximate the negative solution of the equation $x^2 + x - 1 = 0$ to the nearest thousandth.

2. The graph of $y = x^2 + x - 3$ is shown. Approximate both solutions of the equation $x^2 + x - 3 = 0$ to the nearest thousandth.

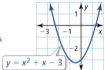

Laurie's Notes Mathematical Practices (continued from page T-189)

- Pose Example 1, and ask students to solve this without using the zero function on their calculators. Do they use correct reasoning in the process?
- Give time for students to work through the questions in the *Monitoring Progress*, and then discuss as a class. Have students make a list of their trials so that their reasoning and process is clear. It is not about random guessing!

Laurie's Notes

Overview of Section 4.1

Introduction
- Students have worked with radicals in an earlier lesson, Section 1.5. This included work with cube roots.
- This is a very long lesson that presents many skills and techniques. There are many connections to properties of exponents and real number properties.
- In the first part of the lesson, students will use the properties of radicals to simplify expressions. This includes work with both square roots and cube roots. This is followed by rationalizing the denominator by multiplying by the conjugate of the denominator. The last part of the lesson is on performing operations with radicals.
- The work in this lesson will help students when working with the Quadratic Formula later in the chapter.

Resources
- Example 7 uses the *golden rectangle*. In geometry, a golden rectangle is a rectangle whose side lengths are in the *golden ratio*, $1 : \frac{1 + \sqrt{5}}{2}$, which is 1: φ (the Greek letter phi), where φ is approximately 1.618. Artists and architects have been fascinated by the idea that the golden rectangle is considered aesthetically pleasing. The proportions of the golden rectangle have been observed in many famous buildings and works of art, such as the Parthenon, Egyptian pyramids, the Taj Mahal, Michelangelo's *David*, and Da Vinci's *Mona Lisa*.
- You might consider having students do a report or construction project related to the golden rectangle or golden ratio.

Formative Assessment Tips
- **Writing Prompt:** This technique asks students to give feedback at the end of the lesson, activity, or learning experience. The *Writing Prompt* allows you to collect feedback in a short period of time (two minutes) on student learning.
- Used at the end of the class, a *Writing Prompt* is similar to an *Exit Ticket*. The written responses give students time to reflect on their learning, how well they believe they understand the concept or skill, and where they are still uncertain.
- It is important on the following day to share responses with students and let them know how the lesson has been adjusted to reflect their responses. Students need to know that you value their responses and that instruction is modified accordingly. Students will take the writing prompts more seriously when they are valued and used.

Pacing Suggestion
- This is a very long lesson with many skills presented. Judge pacing and need for all of the *Monitoring Progress* questions based on frequent student self-assessment.

Dynamic Teaching Tools
- Dynamic Assessment & Progress Monitoring Tool
- Lesson Planning Tool
- Interactive Whiteboard Lesson Library
- Dynamic Classroom with Dynamic Investigations

What Your Students Will Learn

- Use properties of square roots and cube roots to simplify radical expressions.
- Simplify radical expressions by rationalizing the denominator and using conjugates when the denominator contains a sum or difference involving a square root.
- Perform addition, subtraction, and multiplication with like radicals.

Laurie's Notes

Exploration

Motivate

- As a warm-up, write several numeric sequences and ask students to *Think-Pair-Share* a description of the patterns. Examples:
  2, 5, 11, 23, ... double the previous term and add 1
  2, 3, 5, 8, 13, ... add the two previous terms
  2, 3, 5, 7, ... prime numbers; Students may ask for more information.
  1, 4, 9, 16, ... perfect squares
- Ask partners to share their descriptions.
- Be sure to include the last example. Perfect squares will be part of the investigation today.

Exploration 1

- Calculators are not necessary to perform these operations. The numbers involved are all perfect squares.
- Introduce the exploration by posing a question about a numeric expression that involves two operations, such as addition and multiplication.
- **?** "Does the order in which the operations are performed matter?" yes
- Explain that in each part of this exploration, you will determine whether order matters when combining one of the four basic operations with taking square roots.
- **Construct Viable Arguments and Critique the Reasoning of Others:** Encourage students to build strong arguments to support their conclusions. They should understand that one counterexample is sufficient to say that order does matter. However, they should try other examples to support their reasoning when they feel that order does not matter. Suggest that working with perfect squares is helpful. For instance, they may compare $\sqrt{16} \cdot \sqrt{25}$ to $\sqrt{16 \cdot 25}$, and so on.
- Ask students to summarize their findings and discuss their arguments. They are often surprised by the results. They try to make the connection to the Commutative and Associative Properties, where addition and multiplication work but subtraction and division do not.

Exploration 2

- Students should recognize that using perfect squares in their counterexamples will help facilitate the computation.
- **Whiteboarding:** Have students record their counterexamples on a whiteboard so that they can be shared quickly and efficiently.

Communicate Your Answer

- Students may not be able to phrase the statement efficiently, but listen for "The product of the square roots of two numbers is equal to the square root of the product of the two numbers." A similar statement can be made for quotients.

Connecting to Next Step

- Exploration 1 allows students to make sense of the first two properties stated in the formal lesson, providing a great introduction to their learning.

4.1 Properties of Radicals

Essential Question How can you multiply and divide square roots?

EXPLORATION 1   Operations with Square Roots

Work with a partner. For each operation with square roots, compare the results obtained using the two indicated orders of operations. What can you conclude?

a. Square Roots and Addition

Is $\sqrt{36} + \sqrt{64}$ equal to $\sqrt{36 + 64}$?

In general, is $\sqrt{a} + \sqrt{b}$ equal to $\sqrt{a + b}$? Explain your reasoning.

b. Square Roots and Multiplication

Is $\sqrt{4} \cdot \sqrt{9}$ equal to $\sqrt{4 \cdot 9}$?

In general, is $\sqrt{a} \cdot \sqrt{b}$ equal to $\sqrt{a \cdot b}$? Explain your reasoning.

c. Square Roots and Subtraction

Is $\sqrt{64} - \sqrt{36}$ equal to $\sqrt{64 - 36}$?

In general, is $\sqrt{a} - \sqrt{b}$ equal to $\sqrt{a - b}$? Explain your reasoning.

d. Square Roots and Division

Is $\dfrac{\sqrt{100}}{\sqrt{4}}$ equal to $\sqrt{\dfrac{100}{4}}$?

In general, is $\dfrac{\sqrt{a}}{\sqrt{b}}$ equal to $\sqrt{\dfrac{a}{b}}$? Explain your reasoning.

> **REASONING ABSTRACTLY**
> To be proficient in math, you need to recognize and use counterexamples.

EXPLORATION 2   Writing Counterexamples

Work with a partner. A **counterexample** is an example that proves that a general statement is *not* true. For each general statement in Exploration 1 that is not true, write a counterexample different from the example given.

Communicate Your Answer

3. How can you multiply and divide square roots?

4. Give an example of multiplying square roots and an example of dividing square roots that are different from the examples in Exploration 1.

5. Write an algebraic rule for each operation.

    a. the product of square roots

    b. the quotient of square roots

Dynamic Teaching Tools

Dynamic Assessment & Progress Monitoring Tool
Lesson Planning Tool
Interactive Whiteboard Lesson Library
Dynamic Classroom with Dynamic Investigations

ANSWERS

1. a. no; no; Because $\sqrt{36} + \sqrt{64} \neq \sqrt{36 + 64}$, the general statement cannot be true.

    b. yes; yes; By the Power of a Product Property, $a^{1/2} \cdot b^{1/2} = (a \cdot b)^{1/2}$.

    c. no; no; Because $\sqrt{64} - \sqrt{36} \neq \sqrt{64 - 36}$, the general statement cannot be true.

    d. yes; yes; By the Power of a Quotient Property, $\dfrac{a^{1/2}}{b^{1/2}} = \left(\dfrac{a}{b}\right)^{1/2}$.

2. Sample answer: $\sqrt{9} + \sqrt{16} \neq \sqrt{25}$, $\sqrt{16} - \sqrt{9} \neq \sqrt{7}$

3. Multiply or divide the numbers inside the square root symbols and take the square root of the product or quotient.

4. Sample answer:
   $\sqrt{9} \cdot \sqrt{16} = \sqrt{9 \cdot 16} = \sqrt{144} = 12$,
   $\dfrac{\sqrt{16}}{\sqrt{4}} = \sqrt{\dfrac{16}{4}} = \sqrt{4} = 2$

5. a. $\sqrt{a} \cdot \sqrt{b} = \sqrt{a \cdot b}$

   b. $\dfrac{\sqrt{a}}{\sqrt{b}} = \sqrt{\dfrac{a}{b}}$

Differentiated Instruction

Inclusion

Some students may have difficulty remembering frequently used square roots and cube roots such as $\sqrt{144} = 12$ and $\sqrt[3]{64} = 4$. Have these students make a list in their notebooks of the first eight perfect squares and their square roots and the first eight perfect cubes and their cube roots. Students can refer to the list as they simplify radicals.

Extra Example 1

Simplify each expression.

a. $\sqrt{147}$ $7\sqrt{3}$

b. $\sqrt{25x^5}$ $5x^2\sqrt{x}$

MONITORING PROGRESS ANSWERS

1. $2\sqrt{6}$
2. $-4\sqrt{5}$
3. $7x\sqrt{x}$
4. $5n^2\sqrt{3n}$

4.1 Lesson

What You Will Learn

- Use properties of radicals to simplify expressions.
- Simplify expressions by rationalizing the denominator.
- Perform operations with radicals.

Core Vocabulary

counterexample, *p. 191*
radical expression, *p. 192*
simplest form of a radical, *p. 192*
rationalizing the denominator, *p. 194*
conjugates, *p. 194*
like radicals, *p. 196*

Previous
radicand
perfect cube

Using Properties of Radicals

A **radical expression** is an expression that contains a radical. An expression involving a radical with index n is in **simplest form** when these three conditions are met.

- No radicands have perfect nth powers as factors other than 1.
- No radicands contain fractions.
- No radicals appear in the denominator of a fraction.

You can use the property below to simplify radical expressions involving square roots.

Core Concept

Product Property of Square Roots

Words   The square root of a product equals the product of the square roots of the factors.

Numbers $\sqrt{9 \cdot 5} = \sqrt{9} \cdot \sqrt{5} = 3\sqrt{5}$

Algebra $\sqrt{ab} = \sqrt{a} \cdot \sqrt{b}$, where $a, b \geq 0$

STUDY TIP

There can be more than one way to factor a radicand. An efficient method is to find the greatest perfect square factor.

EXAMPLE 1   Using the Product Property of Square Roots

a. $\sqrt{108} = \sqrt{36 \cdot 3}$ Factor using the greatest perfect square factor.

$\phantom{\sqrt{108}} = \sqrt{36} \cdot \sqrt{3}$ Product Property of Square Roots

$\phantom{\sqrt{108}} = 6\sqrt{3}$ Simplify.

b. $\sqrt{9x^3} = \sqrt{9 \cdot x^2 \cdot x}$ Factor using the greatest perfect square factor.

$\phantom{\sqrt{9x^3}} = \sqrt{9} \cdot \sqrt{x^2} \cdot \sqrt{x}$ Product Property of Square Roots

$\phantom{\sqrt{9x^3}} = 3x\sqrt{x}$ Simplify.

STUDY TIP

In this course, whenever a variable appears in the radicand, assume that it has only nonnegative values.

Monitoring Progress Help in English and Spanish at BigIdeasMath.com

Simplify the expression.

1. $\sqrt{24}$ 2. $-\sqrt{80}$ 3. $\sqrt{49x^3}$ 4. $\sqrt{75n^5}$

Core Concept

Quotient Property of Square Roots

Words   The square root of a quotient equals the quotient of the square roots of the numerator and denominator.

Numbers $\sqrt{\dfrac{3}{4}} = \dfrac{\sqrt{3}}{\sqrt{4}} = \dfrac{\sqrt{3}}{2}$ **Algebra** $\sqrt{\dfrac{a}{b}} = \dfrac{\sqrt{a}}{\sqrt{b}}$, where $a \geq 0$ and $b > 0$

192   Chapter 4   Solving Quadratic Equations

Laurie's Notes   Teacher Actions

- Define *radical expression* and what it means for a radical to be in simplest form.
- Write both Core Concepts (product and quotient) because students have just completed the exploration.
- Work through Example 1, and make note of the *Study Tips*.
- **? Probing Question:** In lieu of the *Monitoring Progress*, write $4\sqrt{8x^2}$, $2\sqrt{32x^2}$, $x\sqrt{128}$, and $2\sqrt{16x^2}$. Ask, "Which one doesn't belong? Explain." The expression $2\sqrt{16x^2}$ simplifies to $8x$ and doesn't belong. The other three expressions simplify to $8x\sqrt{2}$.

EXAMPLE 2 Using the Quotient Property of Square Roots

a. $\sqrt{\dfrac{15}{64}} = \dfrac{\sqrt{15}}{\sqrt{64}}$ Quotient Property of Square Roots

$= \dfrac{\sqrt{15}}{8}$ Simplify.

b. $\sqrt{\dfrac{81}{x^2}} = \dfrac{\sqrt{81}}{\sqrt{x^2}}$ Quotient Property of Square Roots

$= \dfrac{9}{x}$ Simplify.

You can extend the Product and Quotient Properties of Square Roots to other radicals, such as cube roots. When using these *properties of cube roots*, the radicands may contain negative numbers.

EXAMPLE 3 Using Properties of Cube Roots

STUDY TIP
To write a cube root in simplest form, find factors of the radicand that are perfect cubes.

a. $\sqrt[3]{-128} = \sqrt[3]{-64 \cdot 2}$ Factor using the greatest perfect cube factor.

$= \sqrt[3]{-64} \cdot \sqrt[3]{2}$ Product Property of Cube Roots

$= -4\sqrt[3]{2}$ Simplify.

b. $\sqrt[3]{125x^7} = \sqrt[3]{125 \cdot x^6 \cdot x}$ Factor using the greatest perfect cube factors.

$= \sqrt[3]{125} \cdot \sqrt[3]{x^6} \cdot \sqrt[3]{x}$ Product Property of Cube Roots

$= 5x^2\sqrt[3]{x}$ Simplify.

c. $\sqrt[3]{\dfrac{y}{216}} = \dfrac{\sqrt[3]{y}}{\sqrt[3]{216}}$ Quotient Property of Cube Roots

$= \dfrac{\sqrt[3]{y}}{6}$ Simplify.

d. $\sqrt[3]{\dfrac{8x^4}{27y^3}} = \dfrac{\sqrt[3]{8x^4}}{\sqrt[3]{27y^3}}$ Quotient Property of Cube Roots

$= \dfrac{\sqrt[3]{8 \cdot x^3 \cdot x}}{\sqrt[3]{27 \cdot y^3}}$ Factor using the greatest perfect cube factors.

$= \dfrac{\sqrt[3]{8} \cdot \sqrt[3]{x^3} \cdot \sqrt[3]{x}}{\sqrt[3]{27} \cdot \sqrt[3]{y^3}}$ Product Property of Cube Roots

$= \dfrac{2x\sqrt[3]{x}}{3y}$ Simplify.

Monitoring Progress Help in English and Spanish at *BigIdeasMath.com*

Simplify the expression.

5. $\sqrt{\dfrac{23}{9}}$ 6. $-\sqrt{\dfrac{17}{100}}$ 7. $\sqrt{\dfrac{36}{z^2}}$ 8. $\sqrt{\dfrac{4x^2}{64}}$

9. $\sqrt[3]{54}$ 10. $\sqrt[3]{16x^4}$ 11. $\sqrt[3]{\dfrac{a}{-27}}$ 12. $\sqrt[3]{\dfrac{25c^7d^3}{64}}$

Section 4.1 Properties of Radicals 193

English Language Learners

Notebook Development
Have students create a radicals page in their notebooks. They should include the vocabulary and properties used in this section. For each property, have students include an example. Be sure they include, when appropriate, a cube root example, as well as a square root example.

Extra Example 2
Simplify each expression.

a. $\sqrt{\dfrac{11}{36}}$ $\dfrac{\sqrt{11}}{6}$

b. $\sqrt{\dfrac{144}{n^2}}$ $\dfrac{12}{n}$

Extra Example 3
Simplify each expression.

a. $\sqrt[3]{135}$ $3\sqrt[3]{5}$

b. $\sqrt[3]{8x^4}$ $2x\sqrt[3]{x}$

c. $\sqrt[3]{\dfrac{n^2}{-1000}}$ $-\dfrac{\sqrt[3]{n^2}}{10}$

d. $\sqrt[3]{\dfrac{343a^8}{b^3}}$ $\dfrac{7a^2\sqrt[3]{a^2}}{b}$

MONITORING PROGRESS ANSWERS

5. $\dfrac{\sqrt{23}}{3}$

6. $-\dfrac{\sqrt{17}}{10}$

7. $\dfrac{6}{z}$

8. $\dfrac{x}{4}$

9. $3\sqrt[3]{2}$

10. $2x\sqrt[3]{2x}$

11. $-\dfrac{\sqrt[3]{a}}{3}$

12. $\dfrac{c^2d\sqrt[3]{25c}}{4}$

Laurie's Notes Teacher Actions

? Probing Question: Write $\sqrt{64}$, $\sqrt{-64}$, $\sqrt[3]{64}$, and $\sqrt[3]{-64}$. Ask, "Which one doesn't belong? Explain." All have real solutions except $\sqrt{-64}$. This question prompts memory of cube roots.

• Explain that the Product and Quotient Properties extend to cube roots. Students should be thinking perfect cubes: 1, 8, 27, 64, ….

• These expressions in Example 3 should be worked through slowly, and in some cases, showing alternate approaches. Explain why $\sqrt[3]{x^6} = x^2$.

• **Think-Pair-Share:** Give a *Three-Minute Pause* and question time before having students work independently on Questions 5–12.

Extra Example 4

Simplify each expression.

a. $\dfrac{\sqrt{7}}{\sqrt{2x}}$ $\dfrac{\sqrt{14x}}{2x}$

b. $\dfrac{5}{\sqrt[3]{4}}$ $\dfrac{5\sqrt[3]{2}}{2}$

Extra Example 5

Simplify $\dfrac{4}{\sqrt{3}+1}$. $2\sqrt{3} - 2$

MONITORING PROGRESS ANSWERS

13. $\dfrac{\sqrt{5}}{5}$

14. $\dfrac{\sqrt{30}}{3}$

15. $\dfrac{7\sqrt{2x}}{2x}$

16. $\dfrac{y\sqrt{6}}{3}$

17. $\dfrac{5\sqrt[3]{2}}{4}$

18. $-4 + 4\sqrt{3}$

19. $\sqrt{65} + 2\sqrt{13}$

20. $-\dfrac{12\sqrt{2} - 12\sqrt{7}}{5}$

Rationalizing the Denominator

When a radical is in the denominator of a fraction, you can multiply the fraction by an appropriate form of 1 to eliminate the radical from the denominator. This process is called **rationalizing the denominator**.

EXAMPLE 4 Rationalizing the Denominator

STUDY TIP
Rationalizing the denominator works because you multiply the numerator and denominator by the same nonzero number a, which is the same as multiplying by $\dfrac{a}{a}$, or 1.

a. $\dfrac{\sqrt{5}}{\sqrt{3n}} = \dfrac{\sqrt{5}}{\sqrt{3n}} \cdot \dfrac{\sqrt{3n}}{\sqrt{3n}}$ Multiply by $\dfrac{\sqrt{3n}}{\sqrt{3n}}$.

$= \dfrac{\sqrt{15n}}{\sqrt{9n^2}}$ Product Property of Square Roots

$= \dfrac{\sqrt{15n}}{\sqrt{9} \cdot \sqrt{n^2}}$ Product Property of Square Roots

$= \dfrac{\sqrt{15n}}{3n}$ Simplify.

b. $\dfrac{2}{\sqrt[3]{9}} = \dfrac{2}{\sqrt[3]{9}} \cdot \dfrac{\sqrt[3]{3}}{\sqrt[3]{3}}$ Multiply by $\dfrac{\sqrt[3]{3}}{\sqrt[3]{3}}$.

$= \dfrac{2\sqrt[3]{3}}{\sqrt[3]{27}}$ Product Property of Cube Roots

$= \dfrac{2\sqrt[3]{3}}{3}$ Simplify.

The binomials $a\sqrt{b} + c\sqrt{d}$ and $a\sqrt{b} - c\sqrt{d}$, where a, b, c, and d are rational numbers, are called **conjugates**. You can use conjugates to simplify radical expressions that contain a sum or difference involving square roots in the denominator.

EXAMPLE 5 Rationalizing the Denominator Using Conjugates

LOOKING FOR STRUCTURE
Notice that the product of two conjugates $a\sqrt{b} + c\sqrt{d}$ and $a\sqrt{b} - c\sqrt{d}$ does not contain a radical and is a *rational* number.
$(a\sqrt{b} + c\sqrt{d})(a\sqrt{b} - c\sqrt{d})$
$= (a\sqrt{b})^2 - (c\sqrt{d})^2$
$= a^2b - c^2d$

Simplify $\dfrac{7}{2 - \sqrt{3}}$.

SOLUTION

$\dfrac{7}{2 - \sqrt{3}} = \dfrac{7}{2 - \sqrt{3}} \cdot \dfrac{2 + \sqrt{3}}{2 + \sqrt{3}}$ The conjugate of $2 - \sqrt{3}$ is $2 + \sqrt{3}$.

$= \dfrac{7(2 + \sqrt{3})}{2^2 - (\sqrt{3})^2}$ Sum and difference pattern

$= \dfrac{14 + 7\sqrt{3}}{1}$ Simplify.

$= 14 + 7\sqrt{3}$ Simplify.

Monitoring Progress  Help in English and Spanish at *BigIdeasMath.com*

Simplify the expression.

13. $\dfrac{1}{\sqrt{5}}$ 14. $\dfrac{\sqrt{10}}{\sqrt{3}}$ 15. $\dfrac{7}{\sqrt{2x}}$ 16. $\sqrt{\dfrac{2y^2}{3}}$

17. $\dfrac{5}{\sqrt[3]{32}}$ 18. $\dfrac{8}{1 + \sqrt{3}}$ 19. $\dfrac{\sqrt{13}}{\sqrt{5} - 2}$ 20. $\dfrac{12}{\sqrt{2} + \sqrt{7}}$

194   Chapter 4   Solving Quadratic Equations

Laurie's Notes — Teacher Actions

- **Look For and Make Use of Structure:** Be sure students see the connection between rationalizing the denominator and writing equivalent fractions. See the *Study Tip*.
- **Teaching Tip:** Model Example 4(a) and then give partners sufficient time to work Example 4(b).
- **Look For and Make Use of Structure** and **Connection:** Multiplying by the conjugate is connected to the Sum and Difference Pattern: $(a + b)(a - b) = a^2 - b^2$.
- **Thumbs Up:** Work through Example 5 as shown. Ask students to self-assess with a *Thumbs Up*. Address questions and concerns.

EXAMPLE 6 Solving a Real-Life Problem

The distance d (in miles) that you can see to the horizon with your eye level h feet above the water is given by $d = \sqrt{\dfrac{3h}{2}}$. How far can you see when your eye level is 5 feet above the water?

SOLUTION

$$d = \sqrt{\dfrac{3(5)}{2}} \quad \text{Substitute 5 for } h.$$

$$= \dfrac{\sqrt{15}}{\sqrt{2}} \quad \text{Quotient Property of Square Roots}$$

$$= \dfrac{\sqrt{15}}{\sqrt{2}} \cdot \dfrac{\sqrt{2}}{\sqrt{2}} \quad \text{Multiply by } \dfrac{\sqrt{2}}{\sqrt{2}}.$$

$$= \dfrac{\sqrt{30}}{2} \quad \text{Simplify.}$$

▶ You can see $\dfrac{\sqrt{30}}{2}$, or about 2.74 miles.

EXAMPLE 7 Modeling with Mathematics

The ratio of the length to the width of a *golden rectangle* is $(1 + \sqrt{5}) : 2$. The dimensions of the face of the Parthenon in Greece form a golden rectangle. What is the height h of the Parthenon?

SOLUTION

1. **Understand the Problem** Think of the length and height of the Parthenon as the length and width of a golden rectangle. The length of the rectangular face is 31 meters. You know the ratio of the length to the height. Find the height h.

2. **Make a Plan** Use the ratio $(1 + \sqrt{5}) : 2$ to write a proportion and solve for h.

3. **Solve the Problem**

$$\dfrac{1 + \sqrt{5}}{2} = \dfrac{31}{h} \quad \text{Write a proportion.}$$

$$h(1 + \sqrt{5}) = 62 \quad \text{Cross Products Property}$$

$$h = \dfrac{62}{1 + \sqrt{5}} \quad \text{Divide each side by } 1 + \sqrt{5}.$$

$$h = \dfrac{62}{1 + \sqrt{5}} \cdot \dfrac{1 - \sqrt{5}}{1 - \sqrt{5}} \quad \text{Multiply the numerator and denominator by the conjugate.}$$

$$h = \dfrac{62 - 62\sqrt{5}}{-4} \quad \text{Simplify.}$$

$$h \approx 19.16 \quad \text{Use a calculator.}$$

▶ The height is about 19 meters.

4. **Look Back** $\dfrac{1 + \sqrt{5}}{2} \approx 1.62$ and $\dfrac{31}{19.16} \approx 1.62$. So, your answer is reasonable.

Monitoring Progress Help in English and Spanish at *BigIdeasMath.com*

21. **WHAT IF?** In Example 6, how far can you see when your eye level is 35 feet above the water?

22. The dimensions of a dance floor form a golden rectangle. The shorter side of the dance floor is 50 feet. What is the length of the longer side of the dance floor?

Section 4.1 Properties of Radicals 195

Extra Example 6

The distance d (in miles) that you can see to the horizon with your eye level h feet above the water is given by $d = \sqrt{\dfrac{3h}{2}}$. How far can you see when your eye level is 15 feet above the water?

You can see $\dfrac{3\sqrt{10}}{2}$, or about 4.74 miles.

Extra Example 7

The ratio of the length to the width of a *golden rectangle* is $(1 + \sqrt{5}) : 2$. The dimensions of the picture frame form a golden rectangle. What is the height h of the frame?

24 in.

The height is about 15 inches.

MONITORING PROGRESS ANSWERS

21. about 7.25 mi
22. about 81 ft

Laurie's Notes   Teacher Actions

- **Attend to Precision:** In Example 6, an exact answer such as $\dfrac{\sqrt{30}}{2}$ is not meaningful in answering the question. Use the approximate answer of 2.74 miles.
- Give a bit of background on the *golden rectangle*. (See the *Resources* on page T-190.)
- Although the terms *length* and *width* are interchangeable, in this example the length is the longer of the two sides. Students should be comfortable setting up a proportion.

COMMON ERROR When finding the cross product, h is multiplied by the quantity $(1 + \sqrt{5})$, resulting in $h(1 + \sqrt{5})$, not $h + \sqrt{5}$.

Extra Example 8

Simplify each expression.

a. $3\sqrt{15} + 2\sqrt{13} - 7\sqrt{15}$
   $-4\sqrt{15} + 2\sqrt{13}$

b. $5\sqrt{7} + \sqrt{63}$ $8\sqrt{7}$

c. $4\sqrt[3]{2x} + 5\sqrt[3]{2x}$ $9\sqrt[3]{2x}$

Extra Example 9

Simplify $\sqrt{7}(\sqrt{3} + \sqrt{48})$. $5\sqrt{21}$

MONITORING PROGRESS ANSWERS

23. $13\sqrt{2} - \sqrt{6}$
24. $-14\sqrt{7}$
25. $-7\sqrt[3]{5x}$
26. $36\sqrt{6}$
27. $36 - 16\sqrt{5}$
28. 2

Performing Operations with Radicals

Radicals with the same index and radicand are called **like radicals**. You can add and subtract like radicals the same way you combine like terms by using the Distributive Property.

STUDY TIP
Do not assume that radicals with different radicands cannot be added or subtracted. Always check to see whether you can simplify the radicals. In some cases, the radicals will become like radicals.

EXAMPLE 8 Adding and Subtracting Radicals

a. $5\sqrt{7} + \sqrt{11} - 8\sqrt{7} = 5\sqrt{7} - 8\sqrt{7} + \sqrt{11}$ Commutative Property of Addition
$\phantom{5\sqrt{7} + \sqrt{11} - 8\sqrt{7}} = (5-8)\sqrt{7} + \sqrt{11}$ Distributive Property
$\phantom{5\sqrt{7} + \sqrt{11} - 8\sqrt{7}} = -3\sqrt{7} + \sqrt{11}$ Subtract.

b. $10\sqrt{5} + \sqrt{20} = 10\sqrt{5} + \sqrt{4 \cdot 5}$ Factor using the greatest perfect square factor.
$\phantom{10\sqrt{5} + \sqrt{20}} = 10\sqrt{5} + \sqrt{4} \cdot \sqrt{5}$ Product Property of Square Roots
$\phantom{10\sqrt{5} + \sqrt{20}} = 10\sqrt{5} + 2\sqrt{5}$ Simplify.
$\phantom{10\sqrt{5} + \sqrt{20}} = (10 + 2)\sqrt{5}$ Distributive Property
$\phantom{10\sqrt{5} + \sqrt{20}} = 12\sqrt{5}$ Add.

c. $6\sqrt[3]{x} + 2\sqrt[3]{x} = (6+2)\sqrt[3]{x}$ Distributive Property
$\phantom{6\sqrt[3]{x} + 2\sqrt[3]{x}} = 8\sqrt[3]{x}$ Add.

EXAMPLE 9 Multiplying Radicals

Simplify $\sqrt{5}(\sqrt{3} - \sqrt{75})$.

SOLUTION

Method 1 $\sqrt{5}(\sqrt{3} - \sqrt{75}) = \sqrt{5} \cdot \sqrt{3} - \sqrt{5} \cdot \sqrt{75}$ Distributive Property
$\phantom{\sqrt{5}(\sqrt{3} - \sqrt{75})} = \sqrt{15} - \sqrt{375}$ Product Property of Square Roots
$\phantom{\sqrt{5}(\sqrt{3} - \sqrt{75})} = \sqrt{15} - 5\sqrt{15}$ Simplify.
$\phantom{\sqrt{5}(\sqrt{3} - \sqrt{75})} = (1 - 5)\sqrt{15}$ Distributive Property
$\phantom{\sqrt{5}(\sqrt{3} - \sqrt{75})} = -4\sqrt{15}$ Subtract.

Method 2 $\sqrt{5}(\sqrt{3} - \sqrt{75}) = \sqrt{5}(\sqrt{3} - 5\sqrt{3})$ Simplify $\sqrt{75}$.
$\phantom{\sqrt{5}(\sqrt{3} - \sqrt{75})} = \sqrt{5}[(1-5)\sqrt{3}]$ Distributive Property
$\phantom{\sqrt{5}(\sqrt{3} - \sqrt{75})} = \sqrt{5}(-4\sqrt{3})$ Subtract.
$\phantom{\sqrt{5}(\sqrt{3} - \sqrt{75})} = -4\sqrt{15}$ Product Property of Square Roots

Monitoring Progress Help in English and Spanish at *BigIdeasMath.com*

Simplify the expression.

23. $3\sqrt{2} - \sqrt{6} + 10\sqrt{2}$
24. $4\sqrt{7} - 6\sqrt{63}$
25. $4\sqrt[3]{5x} - 11\sqrt[3]{5x}$
26. $\sqrt{3}(8\sqrt{2} + 7\sqrt{32})$
27. $(2\sqrt{5} - 4)^2$
28. $\sqrt[3]{-4}(\sqrt[3]{2} - \sqrt[3]{16})$

Laurie's Notes Teacher Actions

- **Look For and Make Use of Structure:** Like radicals and like terms can be combined using the Distributive Property.
- **Teaching Tip:** Model Examples 8(a) and 8(b), and then give partners sufficient time to work Example 8(c). *Popsicle Stick* a response.
- Pose Example 9, and give independent time for students to work. If both approaches do not surface, share the alternate method.

Closure

- **Writing Prompt:** This was a very long lesson. Right now I am feeling …

4.1 Exercises

Dynamic Solutions available at *BigIdeasMath.com*

Vocabulary and Core Concept Check

1. **COMPLETE THE SENTENCE** The process of eliminating a radical from the denominator of a radical expression is called _____.

2. **VOCABULARY** What is the conjugate of the binomial $\sqrt{6} + 4$?

3. **WRITING** Are the expressions $\frac{1}{3}\sqrt{2x}$ and $\sqrt{\frac{2x}{9}}$ equivalent? Explain your reasoning.

4. **WHICH ONE DOESN'T BELONG?** Which expression does *not* belong with the other three? Explain your reasoning.

   $-\frac{1}{3}\sqrt{6}$ $6\sqrt{3}$ $\frac{1}{6}\sqrt{3}$ $-3\sqrt{3}$

Monitoring Progress and Modeling with Mathematics

In Exercises 5–12, determine whether the expression is in simplest form. If the expression is not in simplest form, explain why.

5. $\sqrt{19}$
6. $\sqrt{\frac{1}{7}}$
7. $\sqrt{48}$
8. $\sqrt{34}$
9. $\frac{5}{\sqrt{2}}$
10. $\frac{3\sqrt{10}}{4}$
11. $\frac{1}{2+\sqrt[3]{2}}$
12. $6 - \sqrt[3]{54}$

In Exercises 13–20, simplify the expression. (See Example 1.)

13. $\sqrt{20}$
14. $\sqrt{32}$
15. $\sqrt{128}$
16. $-\sqrt{72}$
17. $\sqrt{125b}$
18. $\sqrt{4x^2}$
19. $-\sqrt{81m^3}$
20. $\sqrt{48n^5}$

In Exercises 21–28, simplify the expression. (See Example 2.)

21. $\sqrt{\frac{4}{49}}$
22. $-\sqrt{\frac{7}{81}}$
23. $-\sqrt{\frac{23}{64}}$
24. $\sqrt{\frac{65}{121}}$
25. $\sqrt{\frac{a^3}{49}}$
26. $\sqrt{\frac{144}{k^2}}$
27. $\sqrt{\frac{100}{4x^2}}$
28. $\sqrt{\frac{25v^2}{36}}$

In Exercises 29–36, simplify the expression. (See Example 3.)

29. $\sqrt[3]{16}$
30. $\sqrt[3]{-108}$
31. $\sqrt[3]{-64x^5}$
32. $-\sqrt[3]{343n^2}$
33. $\sqrt[3]{\frac{6c}{-125}}$
34. $\sqrt[3]{\frac{8h^4}{27}}$
35. $-\sqrt[3]{\frac{81y^2}{1000x^3}}$
36. $\sqrt[3]{\frac{21}{-64a^3b^6}}$

ERROR ANALYSIS In Exercises 37 and 38, describe and correct the error in simplifying the expression.

37.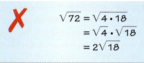

38.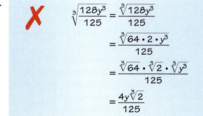

Section 4.1   Properties of Radicals   197

Dynamic Teaching Tools
- Dynamic Assessment & Progress Monitoring Tool
- Interactive Whiteboard Lesson Library
- Dynamic Classroom with Dynamic Investigations

ANSWERS

39. $\dfrac{\sqrt{6}}{\sqrt{6}}$
40. $\dfrac{\sqrt{13z}}{\sqrt{13z}}$
41. $\dfrac{\sqrt[3]{x}}{\sqrt[3]{x}}$
42. $\dfrac{\sqrt[3]{2}}{\sqrt[3]{2}}$
43. $\dfrac{\sqrt{5}+8}{\sqrt{5}+8}$
44. $\dfrac{\sqrt{3}-\sqrt{7}}{\sqrt{3}-\sqrt{7}}$
45. $\sqrt{2}$
46. $\dfrac{4\sqrt{3}}{3}$
47. $\dfrac{\sqrt{15}}{12}$
48. $\dfrac{\sqrt{13}}{13}$
49. $\dfrac{3\sqrt{a}}{a}$
50. $\dfrac{\sqrt{2x}}{2x}$
51. $\dfrac{d\sqrt{15}}{5}$
52. $\dfrac{2\sqrt{6n}}{3n^2}$
53. $\dfrac{4\sqrt[3]{5}}{5}$
54. $\dfrac{\sqrt[3]{2y}}{6y}$
55. $\dfrac{\sqrt{7}-1}{6}$
56. $\dfrac{5+\sqrt{3}}{11}$
57. $\dfrac{7\sqrt{10}+2\sqrt{5}}{47}$
58. $\dfrac{6\sqrt{5}-5}{31}$
59. $\sqrt{5}+\sqrt{2}$
60. $\dfrac{\sqrt{21}-3}{4}$
61. a. about 1.85 sec
    b. about 0.41 sec
62. a. $P = d\sqrt{d}$
    b. about 11.86 Earth years

In Exercises 39–44, write a factor that you can use to rationalize the denominator of the expression.

39. $\dfrac{4}{\sqrt{6}}$
40. $\dfrac{1}{\sqrt{13z}}$
41. $\dfrac{2}{\sqrt[3]{x^2}}$
42. $\dfrac{3m}{\sqrt[3]{4}}$
43. $\dfrac{\sqrt{2}}{\sqrt{5}-8}$
44. $\dfrac{5}{\sqrt{3}+\sqrt{7}}$

In Exercises 45–54, simplify the expression. (See Example 4.)

45. $\dfrac{2}{\sqrt{2}}$
46. $\dfrac{4}{\sqrt{3}}$
47. $\dfrac{\sqrt{5}}{\sqrt{48}}$
48. $\sqrt{\dfrac{4}{52}}$
49. $\dfrac{3}{\sqrt{a}}$
50. $\dfrac{1}{\sqrt{2x}}$
51. $\sqrt{\dfrac{3d^2}{5}}$
52. $\dfrac{\sqrt{8}}{\sqrt{3n^3}}$
53. $\dfrac{4}{\sqrt[3]{25}}$
54. $\sqrt[3]{\dfrac{1}{108y^2}}$

In Exercises 55–60, simplify the expression. (See Example 5.)

55. $\dfrac{1}{\sqrt{7}+1}$
56. $\dfrac{2}{5-\sqrt{3}}$
57. $\dfrac{\sqrt{10}}{7-\sqrt{2}}$
58. $\dfrac{\sqrt{5}}{6+\sqrt{5}}$
59. $\dfrac{3}{\sqrt{5}-\sqrt{2}}$
60. $\dfrac{\sqrt{3}}{\sqrt{7}+\sqrt{3}}$

61. **MODELING WITH MATHEMATICS** The time t (in seconds) it takes an object to hit the ground is given by $t = \sqrt{\dfrac{h}{16}}$, where h is the height (in feet) from which the object was dropped. (See Example 6.)

    a. How long does it take an earring to hit the ground when it falls from the roof of the building?

    b. How much sooner does the earring hit the ground when it is dropped from two stories (22 feet) below the roof? 55 ft

198   Chapter 4   Solving Quadratic Equations

62. **MODELING WITH MATHEMATICS** The orbital period of a planet is the time it takes the planet to travel around the Sun. You can find the orbital period P (in Earth years) using the formula $P = \sqrt{d^3}$, where d is the average distance (in astronomical units, abbreviated AU) of the planet from the Sun.

    a. Simplify the formula.
    b. What is Jupiter's orbital period?

63. **MODELING WITH MATHEMATICS** The electric current I (in amperes) an appliance uses is given by the formula $I = \sqrt{\dfrac{P}{R}}$, where P is the power (in watts) and R is the resistance (in ohms). Find the current an appliance uses when the power is 147 watts and the resistance is 5 ohms.

64. **MODELING WITH MATHEMATICS** You can find the average annual interest rate r (in decimal form) of a savings account using the formula $r = \sqrt{\dfrac{V_2}{V_0}} - 1$, where V_0 is the initial investment and V_2 is the balance of the account after 2 years. Use the formula to compare the savings accounts. In which account would you invest money? Explain.

Account	Initial investment	Balance after 2 years
1	$275	$293
2	$361	$382
3	$199	$214
4	$254	$272
5	$386	$406

63. about 5.42 amperes
64. Account 3; It has the greatest interest rate.

In Exercises 65–68, evaluate the function for the given value of x. Write your answer in simplest form and in decimal form rounded to the nearest hundredth.

65. $h(x) = \sqrt{5x}$; $x = 10$ **66.** $g(x) = \sqrt{3x}$; $x = 60$

67. $r(x) = \sqrt{\dfrac{3x}{3x^2 + 6}}$; $x = 4$

68. $p(x) = \sqrt{\dfrac{x-1}{5x}}$; $x = 8$

In Exercises 69–72, evaluate the expression when $a = -2$, $b = 8$, and $c = \frac{1}{2}$. Write your answer in simplest form and in decimal form rounded to the nearest hundredth.

69. $\sqrt{a^2 + bc}$ **70.** $-\sqrt{4c - 6ab}$

71. $-\sqrt{2a^2 + b^2}$ **72.** $\sqrt{b^2 - 4ac}$

73. MODELING WITH MATHEMATICS The text in the book shown forms a golden rectangle. What is the width w of the text? *(See Example 7.)*

74. MODELING WITH MATHEMATICS The flag of Togo is approximately the shape of a golden rectangle. What is the width w of the flag?

In Exercises 75–82, simplify the expression. *(See Example 8.)*

75. $\sqrt{3} - 2\sqrt{2} + 6\sqrt{2}$ **76.** $\sqrt{5} - 5\sqrt{13} - 8\sqrt{5}$

77. $2\sqrt{6} - 5\sqrt{54}$ **78.** $9\sqrt{32} + \sqrt{2}$

79. $\sqrt{12} + 6\sqrt{3} + 2\sqrt{6}$ **80.** $3\sqrt{7} - 5\sqrt{14} + 2\sqrt{28}$

81. $\sqrt[3]{-81} + 4\sqrt[3]{3}$ **82.** $6\sqrt[3]{128t} - 2\sqrt[3]{2t}$

In Exercises 83–90, simplify the expression. *(See Example 9.)*

83. $\sqrt{2}(\sqrt{45} + \sqrt{5})$ **84.** $\sqrt{3}(\sqrt{72} - 3\sqrt{2})$

85. $\sqrt{5}(2\sqrt{6x} - \sqrt{96x})$ **86.** $\sqrt{7y}(\sqrt{27y} + 5\sqrt{12y})$

87. $(4\sqrt{2} - \sqrt{98})^2$ **88.** $(\sqrt{3} + \sqrt{48})(\sqrt{20} - \sqrt{5})$

89. $\sqrt[3]{3}(\sqrt[3]{4} + \sqrt[3]{32})$ **90.** $\sqrt[3]{2}(\sqrt[3]{135} - 4\sqrt[3]{5})$

91. MODELING WITH MATHEMATICS The circumference C of the art room in a mansion is approximated by the formula $C \approx 2\pi\sqrt{\dfrac{a^2 + b^2}{2}}$. Approximate the circumference of the room.

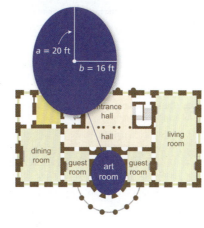

92. CRITICAL THINKING Determine whether each expression represents a *rational* or an *irrational* number. Justify your answer.

a. $4 + \sqrt{6}$ b. $\dfrac{\sqrt{48}}{\sqrt{3}}$

c. $\dfrac{8}{\sqrt{12}}$ d. $\sqrt{3} + \sqrt{7}$

e. $\dfrac{a}{\sqrt{10} - \sqrt{2}}$, where a is a positive integer

f. $\dfrac{2 + \sqrt{5}}{2b + \sqrt{5b^2}}$, where b is a positive integer

In Exercises 93–98, simplify the expression.

93. $\sqrt[5]{\dfrac{13}{5x^5}}$ **94.** $\sqrt[4]{\dfrac{10}{81}}$

95. $\sqrt[4]{256y}$ **96.** $\sqrt[5]{160x^6}$

97. $6\sqrt[4]{9} - \sqrt[5]{9} + 3\sqrt[4]{9}$ **98.** $\sqrt[5]{2}(\sqrt[4]{7} + \sqrt[5]{16})$

Section 4.1   Properties of Radicals   **199**

ANSWERS

65. $5\sqrt{2}$, about 7.07
66. $6\sqrt{5}$, about 13.42
67. $\dfrac{\sqrt{2}}{3}$, about 0.47
68. $\dfrac{\sqrt{70}}{20}$, about 0.42
69. $2\sqrt{2}$, about 2.83
70. $-7\sqrt{2}$, about -9.90
71. $-6\sqrt{2}$, about -8.49
72. $2\sqrt{17}$, about 8.25
73. about 3.71 in.
74. about 25.96 in.
75. $\sqrt{3} + 4\sqrt{2}$
76. $-7\sqrt{5} - 5\sqrt{13}$
77. $-13\sqrt{6}$
78. $37\sqrt{2}$
79. $8\sqrt{3} + 2\sqrt{6}$
80. $7\sqrt{7} - 5\sqrt{14}$
81. $\sqrt[3]{3}$
82. $22\sqrt[3]{2t}$
83. $4\sqrt{10}$
84. $3\sqrt{6}$
85. $-2\sqrt{30x}$
86. $13y\sqrt{21}$
87. 18
88. $5\sqrt{15}$
89. $3\sqrt[3]{12}$
90. $-\sqrt[3]{10}$
91. about 114 ft
92. a. irrational; Six is not a perfect square.
 b. rational; Four is a rational number.
 c. irrational; Twelve is not a perfect square.
 d. irrational; Three and seven are not perfect squares.
 e. irrational; Two and ten are not perfect squares.
 f. rational; $\dfrac{1}{b}$ is rational when b is a positive integer.

93. $\dfrac{\sqrt[5]{8125}}{5x}$
94. $\dfrac{\sqrt[4]{10}}{3}$
95. $4\sqrt[4]{y}$
96. $2x\sqrt[5]{5x}$
97. $9\sqrt[4]{9} - \sqrt[5]{9}$
98. $\sqrt[5]{2}\sqrt[4]{7} + 2$

ANSWERS

99. a. $4, 2\frac{1}{4}, 2, 2+\sqrt{3}, 2-\sqrt{3}, 2+\pi;$

$2\frac{1}{4}, \frac{1}{2}, \frac{1}{4}, \frac{1}{4}+\sqrt{3}, \frac{1}{4}-\sqrt{3}, \frac{1}{4}+\pi;$

$2, \frac{1}{4}, 0, \sqrt{3}, -\sqrt{3}, \pi;$

$2+\sqrt{3}, \frac{1}{4}+\sqrt{3}, \sqrt{3}, 2\sqrt{3}, 0,$
$\pi+\sqrt{3};$

$2-\sqrt{3}, \frac{1}{4}-\sqrt{3}, -\sqrt{3}, 0, -2\sqrt{3},$
$\pi-\sqrt{3};$

$2+\pi, \frac{1}{4}+\pi, \pi, \pi+\sqrt{3},$
$\pi-\sqrt{3}, 2\pi$

b. $4, \frac{1}{2}, 0, 2\sqrt{3}, -2\sqrt{3}, 2\pi;$

$\frac{1}{2}, \frac{1}{16}, 0, \frac{\sqrt{3}}{4}, -\frac{\sqrt{3}}{4}, \frac{\pi}{4};$

$0, 0, 0, 0, 0, 0;$

$2\sqrt{3}, \frac{\sqrt{3}}{4}, 0, 3, -3, \pi\sqrt{3};$

$-2\sqrt{3}, -\frac{\sqrt{3}}{4}, 0, -3, 3, -\pi\sqrt{3};$

$2\pi, \frac{\pi}{4}, 0, \pi\sqrt{3}, -\pi\sqrt{3}, \pi^2$

100. a. always; The sum of two fractions can be written as a fraction.

b. always; The sum cannot be written as the ratio of two integers.

c. sometimes; $\sqrt{3}+\sqrt{3}=2\sqrt{3}$, but $\sqrt{3}+(-\sqrt{3})=0.$

d. always; The product of two fractions can be written as a fraction.

e. always; The product cannot be written as the ratio of two integers.

f. sometimes; $\sqrt{3} \cdot \pi = \pi\sqrt{3}$, but $\sqrt{3} \cdot \sqrt{3} = 3.$

101. odd; even; When m is even, 2^m is a perfect square.

102. Sample answer: $\sqrt[3]{2}$

103. $a^2 < ab < b^2$ when $a < b.$

104. no; Applying the sum and difference pattern to the denominator results in $16 - \sqrt[3]{5^2}$, which still contains a radical.

105–115. See Additional Answers.

Mini-Assessment

Simplify the expression.

1. $\sqrt{144x^3}$ $12x\sqrt{x}$
2. $\sqrt[3]{-54x^3}$ $-3x\sqrt[3]{2}$
3. $\dfrac{7}{\sqrt{2}+3}$ $-\sqrt{2}+3$
4. $4\sqrt{5} - \sqrt{3} + 3\sqrt{5}$ $7\sqrt{5} - \sqrt{3}$
5. $\sqrt{3}(\sqrt{2} - \sqrt{8})$ $-\sqrt{6}$

REASONING In Exercises 99 and 100, use the table shown.

	2	$\frac{1}{4}$	0	$\sqrt{3}$	$-\sqrt{3}$	π
2						
$\frac{1}{4}$						
0						
$\sqrt{3}$						
$-\sqrt{3}$						
π						

99. Copy and complete the table by (a) finding each sum $(2 + 2, 2 + \frac{1}{4}, \text{etc.})$ and (b) finding each product $(2 \cdot 2, 2 \cdot \frac{1}{4}, \text{etc.}).$

100. Use your answers in Exercise 99 to determine whether each statement is *always*, *sometimes*, or *never* true. Justify your answer.

  a. The sum of a rational number and a rational number is rational.

  b. The sum of a rational number and an irrational number is irrational.

  c. The sum of an irrational number and an irrational number is irrational.

  d. The product of a rational number and a rational number is rational.

  e. The product of a nonzero rational number and an irrational number is irrational.

  f. The product of an irrational number and an irrational number is irrational.

101. REASONING Let m be a positive integer. For what values of m will the simplified form of the expression $\sqrt{2^m}$ contain a radical? For what values will it *not* contain a radical? Explain.

Maintaining Mathematical Proficiency Reviewing what you learned in previous grades and lessons

Graph the linear equation. Identify the x-intercept. *(Skills Review Handbook)*

108. $y = x - 4$ **109.** $y = -2x + 6$ **110.** $y = -\frac{1}{3}x - 1$ **111.** $y = \frac{3}{2}x + 6$

Find the product. *(Section 2.3)*

112. $(z + 3)^2$ **113.** $(3a - 5b)^2$ **114.** $(x + 8)(x - 8)$ **115.** $(4y + 2)(4y - 2)$

200   Chapter 4   Solving Quadratic Equations

---

102. HOW DO YOU SEE IT? The edge length s of a cube is an irrational number, the surface area is an irrational number, and the volume is a rational number. Give a possible value of s.

103. REASONING Let a and b be positive numbers. Explain why $\sqrt{ab}$ lies between a and b on a number line. (*Hint:* Let $a < b$ and multiply each side of $a < b$ by a. Then let $a < b$ and multiply each side by b.)

104. MAKING AN ARGUMENT Your friend says that you can rationalize the denominator of the expression $\dfrac{2}{4 + \sqrt[3]{5}}$ by multiplying the numerator and denominator by $4 - \sqrt[3]{5}$. Is your friend correct? Explain.

105. PROBLEM SOLVING The ratio of consecutive terms $\dfrac{a_n}{a_{n-1}}$ in the Fibonacci sequence gets closer and closer to the golden ratio $\dfrac{1 + \sqrt{5}}{2}$ as n increases. Find the term that precedes 610 in the sequence.

106. THOUGHT PROVOKING Use the golden ratio $\dfrac{1 + \sqrt{5}}{2}$ and the golden ratio conjugate $\dfrac{1 - \sqrt{5}}{2}$ for each of the following.

  a. Show that the golden ratio and golden ratio conjugate are both solutions of $x^2 - x - 1 = 0$.

  b. Construct a geometric diagram that has the golden ratio as the length of a part of the diagram.

107. CRITICAL THINKING Use the special product pattern $(a + b)(a^2 - ab + b^2) = a^3 + b^3$ to simplify the expression $\dfrac{2}{\sqrt[3]{x} + 1}$. Explain your reasoning.

If students need help...	If students got it...
Resources by Chapter • Practice A and Practice B • Puzzle Time	**Resources by Chapter** • Enrichment and Extension • Cumulative Review
Student Journal • Practice	Start the *next* Section
Differentiating the Lesson Skills Review Handbook	

Laurie's Notes

Overview of Section 4.2

Introduction
- In this lesson, students will use their prior knowledge of factoring quadratics and graphing quadratic equations in order to solve a quadratic equation in one variable and to find the zeros of a quadratic function.
- Equations and functions presented have 0, 1, or 2 solutions or zeros.
- Real-life problems do not always have whole number or rational solutions, so students will judge what degree of precision is required for the problem.

Formative Assessment Tips
- **Response Cards:** This technique is used as a quick check to see whether students' knowledge of a skill, technique, precept, or procedure is correct. Students are given cards at the beginning of class that are held up in front of them in response to a question. The cards are prepared in advance with particular responses (A, B, C, D, or 1, 2, 3, or True, False) or may be left blank for students to write their response on.
- This is the low-tech version of a clicker response system. Whiteboards could be used for students to write responses on. If cards are going to be used multiple times, and for several years, consider laminating the cards.
- This technique gives all students the opportunity to participate in the lesson. As a teacher, you are soliciting information from all students and not just those who raise their hands. Because the cards are held facing the teacher, when all students are also facing the teacher, it is a private way to gather quick information about students' understanding.
- The information learned from the responses should guide your next instructional move. If the responses are evenly distributed, you could gather like responses together and ask each group to present to the others the reason(s) for their response. If most of the responses are incorrect, there is reteaching to do. If only a few students have a wrong response, you might probe for the reason or be mindful of their response as the next question is posed that might clarify earlier confusion.
- In this lesson, response cards with 0, 1, and 2 are distributed, which will refer to the number of solutions for the problems presented.

Pacing Suggestion
- If you are confident in students' ability to graph quadratics, you might have students use a graphing calculator for the explorations. These explorations allow students to discover that quadratic equations in one variable can have 0, 1, or 2 solutions. Transition to the formal lesson after the explorations have been discussed.

Dynamic Teaching Tools
Dynamic Assessment & Progress Monitoring Tool
Lesson Planning Tool
Interactive Whiteboard Lesson Library
Dynamic Classroom with Dynamic Investigations

What Your Students Will Learn

- Solve quadratic equations by graphing and finding the *x*-intercepts.
- Use graphs to find and approximate the zeros of functions.
- Solve real-life problems using tables and graphs of quadratic functions.

Laurie's Notes

Exploration

Motivate
- Show a picture of a projectile being launched from a trebuchet or a catapult.
- Ask students what they would like to know about the projectile. Hopefully, a student will ask how long it takes for the projectile to land.
- Explain to students that in this section they will answer that type of question.

Discuss
- Connect the projectile motion to the *x*-intercept dialogue.
- Also, connect the previous two chapters to this section by reminding students that they solved quadratic equations by factoring and by graphing quadratic functions.

Exploration 1
- **?** "What do you know about the graph of $y = x^2 - 2x$?" opens up; *y*-intercept is 0; *x*-coordinate of the vertex is 1; axis of symmetry is $x = 1$
- Students should make an input-output table that includes *x*-values that show the key features of the graph.
- While students work, question them about the vertex and the minimum value.
- **Big Idea:** In part (d), students should discuss evaluating the left side of the equation for the *x*-values found in part (c), and they should also discuss the *x*-intercept. It is not enough to solve the equation graphically because computational errors can influence the graph. An algebraic check is also important.
- **Make Sense of Problems and Persevere in Solving Them:** This approach to solving a quadratic equation should make sense to students if they think back to solving systems of equations. They can think of solving $x^2 - 2x = 0$ as solving the system $y = x^2 - 2x$ and $y = 0$. The graph of $y = 0$ is the *x*-axis, which intersects the graph of $y = x^2 - 2x$ at its *x*-intercepts.

Exploration 2
- As they are solving each equation by graphing, students should be checking the reasonableness of their graphs.
- While students work, probe different pairs of students about the vertex, the minimum or maximum value, the *y*-intercept, and the general shape of the graph.
- The equations in parts (e) and (f) are different from the first four equations. It should occur to students to try factoring, which does not yield a solution. Next, graphing by finding the vertex and a few additional points should be possible. Students will recognize that the parabolas do not intersect the *x*-axis.
- **?** "In Exploration 1, the equation had two solutions. Does each equation in Exploration 2 have two solutions?" no; The equation in part (d) has only one solution, and the equations in parts (e) and (f) have no solution.
- **?** "Do you think it is possible to predict how many solutions a quadratic equation in one variable will have?" Answers will vary.

Communicate Your Answer
- **Make Sense of Problems and Persevere in Solving Them:** In Question 4, students are making sense of a problem in more than one way. It is important for students to realize that the *x*-intercepts correspond to the solutions of the equation, so substituting these values for *x* should result in a true equation.

Connecting to Next Step
- Exploration 1 is important in establishing what it means to solve a quadratic by graphing. This will be stated at the beginning of the formal lesson.

4.2 Solving Quadratic Equations by Graphing

Essential Question How can you use a graph to solve a quadratic equation in one variable?

Based on the definition of an x-intercept of a graph, it follows that the x-intercept of the graph of the linear equation

$$y = ax + b \quad \text{2 variables}$$

is the same value as the solution of

$$ax + b = 0. \quad \text{1 variable}$$

You can use similar reasoning to solve *quadratic equations*.

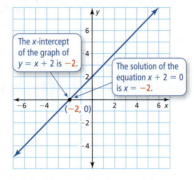

The x-intercept of the graph of $y = x + 2$ is -2.

The solution of the equation $x + 2 = 0$ is $x = -2$.

EXPLORATION 1 Solving a Quadratic Equation by Graphing

Work with a partner.

a. Sketch the graph of $y = x^2 - 2x$.

b. What is the definition of an x-intercept of a graph? How many x-intercepts does this graph have? What are they?

c. What is the definition of a solution of an equation in x? How many solutions does the equation $x^2 - 2x = 0$ have? What are they?

d. Explain how you can verify the solutions you found in part (c).

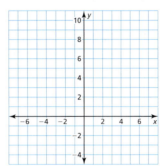

EXPLORATION 2 Solving Quadratic Equations by Graphing

Work with a partner. Solve each equation by graphing.

a. $x^2 - 4 = 0$
b. $x^2 + 3x = 0$
c. $-x^2 + 2x = 0$
d. $x^2 - 2x + 1 = 0$
e. $x^2 - 3x + 5 = 0$
f. $-x^2 + 3x - 6 = 0$

MAKING SENSE OF PROBLEMS

To be proficient in math, you need to check your answers to problems using a different method and continually ask yourself, "Does this make sense?"

Communicate Your Answer

3. How can you use a graph to solve a quadratic equation in one variable?

4. After you find a solution graphically, how can you check your result algebraically? Check your solutions for parts (a)–(d) in Exploration 2 algebraically.

5. How can you determine graphically that a quadratic equation has no solution?

Dynamic Teaching Tools

Dynamic Assessment & Progress Monitoring Tool
Lesson Planning Tool
Interactive Whiteboard Lesson Library
Dynamic Classroom with Dynamic Investigations

ANSWERS

1. a.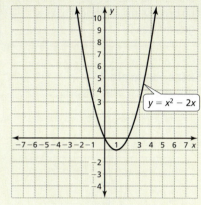

   b. the x-coordinate of a point where the graph crosses the x-axis; 2; 0, 2

   c. a value of x that makes the equation true; 2; 0, 2

   d. Substitute the solutions into the equation.

2. a. $x = 2$, $x = -2$
   b. $x = 0$, $x = -3$
   c. $x = 0$, $x = 2$
   d. $x = 1$
   e. no solution
   f. no solution

3. Write the equation in standard form $ax^2 + bx + c = 0$, graph the related function $y = ax^2 + bx + c$, and find the x-intercepts.

4. Substitute the solutions into the equation.

5. The related graph will have no x-intercepts.

Differentiated Instruction

Auditory

Have students verbally describe the steps they would use to graph a quadratic equation such as $y = x^2 - 3$. Ask them to explain how they can predict whether the graph will open up or down. Be sure students understand that this graph will open up because the coefficient of x^2 is greater than 0.

Extra Example 1

Solve $x^2 + 2x = 8$ by graphing.

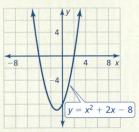

The solutions are $x = -4$ and $x = 2$.

MONITORING PROGRESS ANSWERS

1. $x = 2, x = -1$
2. $x = -5, x = -2$
3. $x = -4, x = 3$

4.2 Lesson

Core Vocabulary

quadratic equation, *p. 202*

Previous
x-intercept
root
zero of a function

What You Will Learn

- Solve quadratic equations by graphing.
- Use graphs to find and approximate the zeros of functions.
- Solve real-life problems using graphs of quadratic functions.

Solving Quadratic Equations by Graphing

A **quadratic equation** is a nonlinear equation that can be written in the standard form $ax^2 + bx + c = 0$, where $a \neq 0$.

In Chapter 2, you solved quadratic equations by factoring. You can also solve quadratic equations by graphing.

🄲 Core Concept

Solving Quadratic Equations by Graphing

Step 1 Write the equation in standard form, $ax^2 + bx + c = 0$.

Step 2 Graph the related function $y = ax^2 + bx + c$.

Step 3 Find the x-intercepts, if any.

The solutions, or *roots*, of $ax^2 + bx + c = 0$ are the x-intercepts of the graph.

EXAMPLE 1   Solving a Quadratic Equation: Two Real Solutions

Solve $x^2 + 2x = 3$ by graphing.

SOLUTION

Step 1 Write the equation in standard form.

$x^2 + 2x = 3$ Write original equation.

$x^2 + 2x - 3 = 0$ Subtract 3 from each side.

Step 2 Graph the related function $y = x^2 + 2x - 3$.

Step 3 Find the x-intercepts. The x-intercepts are -3 and 1.

▶ So, the solutions are $x = -3$ and $x = 1$.

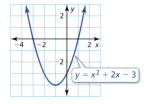

Check

$x^2 + 2x = 3$	Original equation	$x^2 + 2x = 3$
$(-3)^2 + 2(-3) \stackrel{?}{=} 3$	Substitute.	$1^2 + 2(1) \stackrel{?}{=} 3$
$3 = 3$ ✓	Simplify.	$3 = 3$ ✓

Monitoring Progress 🔊 Help in English and Spanish at *BigIdeasMath.com*

Solve the equation by graphing. Check your solutions.

1. $x^2 - x - 2 = 0$ 2. $x^2 + 7x = -10$ 3. $x^2 + x = 12$

202   Chapter 4   Solving Quadratic Equations

Laurie's Notes   Teacher Actions

- **Make Sense of Problems and Persevere in Solving Them:** Throughout this lesson, discuss with students the different graphing methods they can use to solve these types of equations, as well as the different checks they can use: factoring, using a graphing utility, or substituting solutions back into the original equation.

- ❓ "What is the first step in solving $x^2 + 2x = 3$ by graphing?" Write the equation in standard form. "What technique can be used to graph $y = x^2 + 2x - 3$?" Listen for valid methods, including factoring into $y = (x + 3)(x - 1)$. Discuss methods for checking the solutions.

ANOTHER WAY

You can also solve the equation in Example 2 by factoring.

$x^2 - 8x + 16 = 0$

$(x - 4)(x - 4) = 0$

So, $x = 4$.

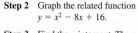 **Solving a Quadratic Equation: One Real Solution**

Solve $x^2 - 8x = -16$ by graphing.

SOLUTION

Step 1 Write the equation in standard form.

$x^2 - 8x = -16$ Write original equation.

$x^2 - 8x + 16 = 0$ Add 16 to each side.

Step 2 Graph the related function $y = x^2 - 8x + 16$.

Step 3 Find the x-intercept. The only x-intercept is at the vertex, $(4, 0)$.

▶ So, the solution is $x = 4$.

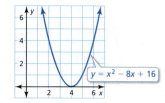

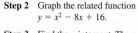 **Solving a Quadratic Equation: No Real Solutions**

Solve $-x^2 = 2x + 4$ by graphing.

SOLUTION

Method 1 Write the equation in standard form, $x^2 + 2x + 4 = 0$. Then graph the related function $y = x^2 + 2x + 4$, as shown at the left.

▶ There are no x-intercepts. So, $-x^2 = 2x + 4$ has no real solutions.

Method 2 Graph each side of the equation.

$y = -x^2$ Left side
$y = 2x + 4$ Right side

▶ The graphs do not intersect. So, $-x^2 = 2x + 4$ has no real solutions.

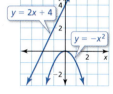

Monitoring Progress Help in English and Spanish at *BigIdeasMath.com*

Solve the equation by graphing.

4. $x^2 + 36 = 12x$ **5.** $x^2 + 4x = 0$ **6.** $x^2 + 10x = -25$

7. $x^2 = 3x - 3$ **8.** $x^2 + 7x = -6$ **9.** $2x + 5 = -x^2$

Concept Summary

Number of Solutions of a Quadratic Equation

A quadratic equation has:

- two real solutions when the graph of its related function has two x-intercepts.
- one real solution when the graph of its related function has one x-intercept.
- no real solutions when the graph of its related function has no x-intercepts.

Section 4.2 Solving Quadratic Equations by Graphing 203

Laurie's Notes — Teacher Actions

- Pose Example 2, and have students work independently to solve.
- **Response Cards:** Have students hold up the *Response Card* showing the number of solutions found. Solicit the solution, $x = 4$.
- Pose Example 3, and have students work independently to solve.
- **Response Cards:** Again use the *Response Cards*.
- Explain Method 2, as it is likely that students would not have used this approach. It is a technique they will use frequently in future mathematics classes.

English Language Learners

Pair Activity

Have students work in pairs to solve quadratic equations by graphing. Each student solves a different equation. When both students are done, they take turns explaining their solution while the other person follows along.

Extra Example 2

Solve $x^2 + 4x = -4$ by graphing.

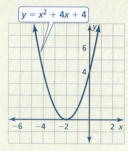

The solution is $x = -2$.

Extra Example 3

Solve $-x^2 = 4x + 8$ by graphing.

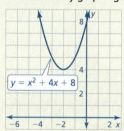

$-x^2 = 4x + 8$ has no real solutions.

MONITORING PROGRESS ANSWERS

4. $x = 6$
5. $x = 0, x = -4$
6. $x = -5$
7. no solution
8. $x = -6, x = -1$
9. no solution

Section 4.2 203

Extra Example 4

The graph of $f(x) = -x^2 + x + 6$ is shown. Find the zeros of f.

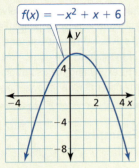

The zeros of f are -2 and 3.

Extra Example 5

The graph of $f(x) = x^2 - x - 5$ is shown. Approximate the zeros of f to the nearest tenth.

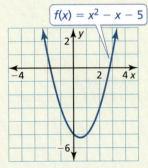

The zeros of f are about -1.8 and 2.8.

MONITORING PROGRESS ANSWERS

10.

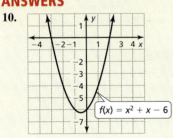

$-3, 2$

11.

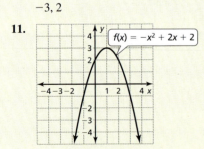

about -0.7, about 2.7

Finding Zeros of Functions

Recall that a zero of a function is an x-intercept of the graph of the function.

EXAMPLE 4 Finding the Zeros of a Function

The graph of $f(x) = -x^2 + 2x + 3$ is shown. Find the zeros of f.

SOLUTION

The x-intercepts are -1 and 3.

▶ So, the zeros of f are -1 and 3.

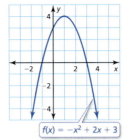

Check
$f(-1) = -(-1)^2 + 2(-1) + 3 = 0$ ✓
$f(3) = -3^2 + 2(3) + 3 = 0$ ✓

The zeros of a function are not necessarily integers. To approximate zeros, analyze the signs of function values. When two function values have different signs, a zero lies between the x-values that correspond to the function values.

EXAMPLE 5 Approximating the Zeros of a Function

The graph of $f(x) = x^2 + 4x + 1$ is shown. Approximate the zeros of f to the nearest tenth.

SOLUTION

There are two x-intercepts: one between -4 and -3, and another between -1 and 0.

Make tables using x-values between -4 and -3, and between -1 and 0. Use an increment of 0.1. Look for a change in the signs of the function values.

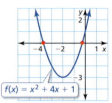

x	-3.9	-3.8	-3.7	-3.6	-3.5	-3.4	-3.3	-3.2	-3.1
$f(x)$	0.61	0.24	-0.11	-0.44	-0.75	-1.04	-1.31	-1.56	-1.79

change in signs

ANOTHER WAY

You could approximate one zero using a table and then use the axis of symmetry to find the other zero.

x	-0.9	-0.8	-0.7	-0.6	-0.5	-0.4	-0.3	-0.2	-0.1
$f(x)$	-1.79	-1.56	-1.31	-1.04	-0.75	-0.44	-0.11	0.24	0.61

The function values that are closest to 0 correspond to x-values that best approximate the zeros of the function.

change in signs

▶ In each table, the function value closest to 0 is -0.11. So, the zeros of f are about -3.7 and -0.3.

Monitoring Progress 🔊 Help in English and Spanish at BigIdeasMath.com

10. Graph $f(x) = x^2 + x - 6$. Find the zeros of f.

11. Graph $f(x) = -x^2 + 2x + 2$. Approximate the zeros of f to the nearest tenth.

Laurie's Notes Teacher Actions

- ❓ "What is another way to write $f(x) = -x^2 + 2x + 3$?" Answers will vary. One approach is to factor completely: $f(x) = -1(x + 1)(x - 3)$.
- ❓ **Assessing Question:** "Can you factor every quadratic expression?" Students will assume rational solutions and say no.
- **Use Appropriate Tools Strategically:** In approximating the zeros of the function, use the table feature on the graphing calculator.
- **Big Idea:** As students adjust the increment, they are focusing on the entries where the function value changes sign.

Solving Real-Life Problems

EXAMPLE 6 Real-Life Application

A football player kicks a football 2 feet above the ground with an initial vertical velocity of 75 feet per second. The function $h = -16t^2 + 75t + 2$ represents the height h (in feet) of the football after t seconds. (a) Find the height of the football each second after it is kicked. (b) Use the results of part (a) to estimate when the height of the football is 50 feet. (c) Using a graph, after how many seconds is the football 50 feet above the ground?

SOLUTION

a. Make a table of values starting with $t = 0$ seconds using an increment of 1. Continue the table until a function value is negative.

Seconds, t	Height, h
0	2
1	61
2	88
3	83
4	46
5	−23

▶ The height of the football is 61 feet after 1 second, 88 feet after 2 seconds, 83 feet after 3 seconds, and 46 feet after 4 seconds.

b. From part (a), you can estimate that the height of the football is 50 feet between 0 and 1 second and between 3 and 4 seconds.

▶ Based on the function values, it is reasonable to estimate that the height of the football is 50 feet slightly less than 1 second and slightly less than 4 seconds after it is kicked.

c. To determine when the football is 50 feet above the ground, find the t-values for which $h = 50$. So, solve the equation $-16t^2 + 75t + 2 = 50$ by graphing.

Step 1 Write the equation in standard form.

$$-16t^2 + 75t + 2 = 50 \quad \text{Write the equation.}$$
$$-16t^2 + 75t - 48 = 0 \quad \text{Subtract 50 from each side.}$$

Step 2 Use a graphing calculator to graph the related function $h = -16t^2 + 75t - 48$.

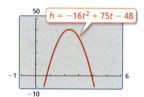

REMEMBER
Equations have *solutions*, or *roots*. Graphs have *x-intercepts*. Functions have *zeros*.

Step 3 Use the *zero* feature to find the zeros of the function.

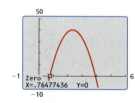
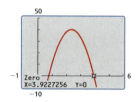

▶ The football is 50 feet above the ground after about 0.8 second and about 3.9 seconds, which supports the estimates in part (b).

Monitoring Progress Help in English and Spanish at BigIdeasMath.com

12. **WHAT IF?** After how many seconds is the football 65 feet above the ground?

Extra Example 6

A soccer player kicks a soccer ball 2 feet above the ground with an initial vertical velocity of 60 feet per second. The function $h = -16t^2 + 60t + 2$ represents the height h (in feet) of the soccer ball after t seconds.

a. Find the height of the soccer ball each second after it is kicked. The height of the soccer ball is 46 feet after 1 second, 58 feet after 2 seconds, and 38 feet after 3 seconds.

b. Use the results of part (a) to estimate when the height of the soccer ball is 40 feet. The height of the soccer ball is 40 feet slightly less than 1 second and slightly less than 3 seconds after it is kicked.

c. Using a graph, after how many seconds is the soccer ball 40 feet above the ground? The soccer ball is 40 feet above the ground after about 0.8 second and about 2.9 seconds.

MONITORING PROGRESS ANSWER

12. about 1.1 sec, about 3.6 sec

Laurie's Notes — Teacher Actions

- Remind students that the ordered pair (x, y) is replaced by (t, h) representing (time, height), with units associated with each variable.
- **Misconception:** The graph shows the height of the object over time, not the path of the object.
- **Connection:** This example could also be solved like Example 3, Method 2: Graph $y = -16t^2 + 75t + 2$ and $y = 50$.

Closure

- When a quadratic equation has two solutions, what do you know about the graph of its related function? *It has two x-intercepts.*

Assignment Guide and Homework Check

ASSIGNMENT

Basic: 1–4, 5–29 odd, 37–49 odd, 53, 58, 62, 66, 67

Average: 1–4, 6–58 even, 62, 66, 67

Advanced: 1–4, 8, 12, 20–32 even, 40–58 even, 59–67

HOMEWORK CHECK

Basic: 13, 19, 37, 43, 53

Average: 14, 20, 38, 44, 54

Advanced: 20, 24, 40, 44, 54

ANSWERS

1. an equation that can be written in the standard form $ax^2 + bx + c = 0$, where $a \neq 0$
2. $x^2 + x - 4 = 0$; It is the only equation written in standard form.
3. The number of x-intercepts is the number of solutions.
4. They are the same.
5. $x = 3, x = -1$
6. $x = 4, x = 2$
7. $x = -4$
8. no solution
9. $4x^2 - 12 = 0$ or $-4x^2 + 12 = 0$
10. $x^2 + 15 = 0$ or $-x^2 - 15 = 0$
11. $x^2 - 2x + 1 = 0$ or $-x^2 + 2x - 1 = 0$
12. $3x^2 - x - 5 = 0$ or $-3x^2 + x + 5 = 0$
13. $x = 0, x = 5$
14. $x = 2$
15. no solution
16. $x = 7, x = -1$
17. $x = 3$
18. no solution
19. $x = -1$
20. no solution
21. $x = -6, x = 2$
22. $x = 3, x = 2$
23. $x = -2, x = 1$
24. $x = -4$

4.2 Exercises

Dynamic Solutions available at BigIdeasMath.com

Vocabulary and Core Concept Check

1. **VOCABULARY** What is a quadratic equation?

2. **WHICH ONE DOESN'T BELONG?** Which equation does *not* belong with the other three? Explain your reasoning.

   $x^2 + 5x = 20$ $x^2 + x - 4 = 0$ $x^2 - 6 = 4x$ $7x + 12 = x^2$

3. **WRITING** How can you use a graph to find the number of solutions of a quadratic equation?

4. **WRITING** How are solutions, roots, x-intercepts, and zeros related?

Monitoring Progress and Modeling with Mathematics

In Exercises 5–8, use the graph to solve the equation.

5. $-x^2 + 2x + 3 = 0$
6. $x^2 - 6x + 8 = 0$

7. $x^2 + 8x + 16 = 0$
8. $-x^2 - 4x - 6 = 0$

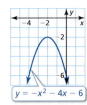

In Exercises 9–12, write the equation in standard form.

9. $4x^2 = 12$ 10. $-x^2 = 15$

11. $2x - x^2 = 1$ 12. $5 + x = 3x^2$

In Exercises 13–24, solve the equation by graphing. *(See Examples 1, 2, and 3.)*

13. $x^2 - 5x = 0$ 14. $x^2 - 4x + 4 = 0$
15. $x^2 - 2x + 5 = 0$ 16. $x^2 - 6x - 7 = 0$
17. $x^2 = 6x - 9$ 18. $-x^2 = 8x + 20$

19. $x^2 = -1 - 2x$ 20. $x^2 = -x - 3$
21. $4x - 12 = -x^2$ 22. $5x - 6 = x^2$
23. $x^2 - 2 = -x$ 24. $16 + x^2 = -8x$

25. **ERROR ANALYSIS** Describe and correct the error in solving $x^2 + 3x = 18$ by graphing.

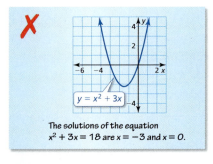

The solutions of the equation $x^2 + 3x = 18$ are $x = -3$ and $x = 0$.

26. **ERROR ANALYSIS** Describe and correct the error in solving $x^2 + 6x + 9 = 0$ by graphing.

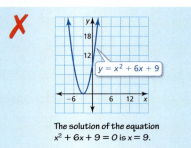

The solution of the equation $x^2 + 6x + 9 = 0$ is $x = 9$.

206   Chapter 4   Solving Quadratic Equations

---

25. The equation needs to be in standard form; $x^2 + 3x - 18 = 0$

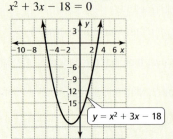

The solutions are $x = -6$ and $x = 3$.

26. The solution is the x-intercept, not the y-intercept; The solution is $x = -3$.

27. MODELING WITH MATHEMATICS The height y (in yards) of a flop shot in golf can be modeled by $y = -x^2 + 5x$, where x is the horizontal distance (in yards).

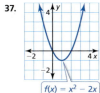

a. Interpret the x-intercepts of the graph of the equation.

b. How far away does the golf ball land?

28. MODELING WITH MATHEMATICS The height h (in feet) of an underhand volleyball serve can be modeled by $h = -16t^2 + 30t + 4$, where t is the time (in seconds).

a. Do both t-intercepts of the graph of the function have meaning in this situation? Explain.

b. No one receives the serve. After how many seconds does the volleyball hit the ground?

In Exercises 29–36, solve the equation by using Method 2 from Example 3.

29. $x^2 = 10 - 3x$
30. $2x - 3 = x^2$
31. $5x - 7 = x^2$
32. $x^2 = 6x - 5$
33. $x^2 + 12x = -20$
34. $x^2 + 8x = 9$
35. $-x^2 - 5 = -2x$
36. $-x^2 - 4 = -4x$

In Exercises 37–42, find the zero(s) of f. (See Example 4.)

37.
$f(x) = x^2 - 2x$

38.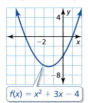
$f(x) = x^2 + 3x - 4$

39.
$f(x) = x^2 + 2x + 1$

40.
$f(x) = -x^2 + 6x - 9$

41.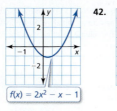
$f(x) = 2x^2 - x - 1$

42.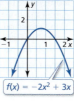
$f(x) = -2x^2 + 3x$

In Exercises 43–46, approximate the zeros of f to the nearest tenth. (See Example 5.)

43.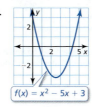
$f(x) = x^2 - 5x + 3$

44.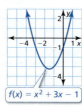
$f(x) = x^2 + 3x - 1$

45.
$f(x) = -x^2 + 2x + 1$

46.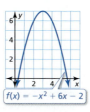
$f(x) = -x^2 + 6x - 2$

In Exercises 47–52, graph the function. Approximate the zeros of the function to the nearest tenth, if necessary.

47. $f(x) = x^2 + 6x + 1$
48. $f(x) = x^2 - 3x + 2$
49. $y = -x^2 + 4x - 2$
50. $y = -x^2 + 9x - 6$
51. $f(x) = \frac{1}{2}x^2 + 2x - 5$
52. $f(x) = -3x^2 + 4x + 3$

53. MODELING WITH MATHEMATICS At a Civil War reenactment, a cannonball is fired into the air with an initial vertical velocity of 128 feet per second. The release point is 6 feet above the ground. The function $h = -16t^2 + 128t + 6$ represents the height h (in feet) of the cannonball after t seconds. (See Example 6.)

a. Find the height of the cannonball each second after it is fired.

b. Use the results of part (a) to estimate when the height of the cannonball is 150 feet.

c. Using a graph, after how many seconds is the cannonball 150 feet above the ground?

Section 4.2   Solving Quadratic Equations by Graphing   207

49.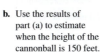
about 3.4, about 0.6

50.
about 8.3, about 0.7

51–53. See Additional Answers.

ANSWERS

54. a. 29 ft, 21 ft
    b. about 0.5 sec, about 2.3 sec
    c. about 0.3 sec, about 2.2 sec
55–67. See Additional Answers.

Mini-Assessment

1. Solve $x^2 + 4x = 5$ by graphing.

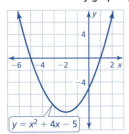

   $x = -5$ and $x = 1$

2. Use the graph to find the zeros of $f(x) = x^2 + 3x + 2$.

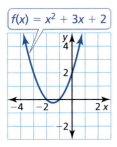

   The zeros are -2 and -1.

3. A tennis ball is thrown straight up into the air with an initial vertical velocity of 50 feet per second. The release point is 4 feet above the ground. The function $h = -16t^2 + 50t + 4$ represents the height h (in feet) of the tennis ball after t seconds.

   a. Find the height of the tennis ball each second after it is thrown.
      38 feet after 1 second, 40 feet after 2 seconds, 10 feet after 3 seconds

   b. Use the results of part (a) to estimate when the height of the tennis ball is 30 feet.
      slightly less than 1 second and slightly more than 2 seconds after it is thrown

   c. Using a graph, after how many seconds is the tennis ball 30 feet above the ground? after about 0.7 second and about 2.5 seconds

208   Chapter 4

---

54. **MODELING WITH MATHEMATICS** You throw a softball straight up into the air with an initial vertical velocity of 40 feet per second. The release point is 5 feet above the ground. The function $h = -16t^2 + 40t + 5$ represents the height h (in feet) of the softball after t seconds.

    a. Find the height of the softball each second after it is released.

    b. Use the results of part (a) to estimate when the height of the softball is 15 feet.

    c. Using a graph, after how many seconds is the softball 15 feet above the ground?

MATHEMATICAL CONNECTIONS In Exercises 55 and 56, use the given surface area S of the cylinder to find the radius r to the nearest tenth.

55. $S = 225$ ft²   56. $S = 750$ m²

57. **WRITING** Explain how to approximate zeros of a function when the zeros are not integers.

58. **HOW DO YOU SEE IT?** Consider the graph shown.

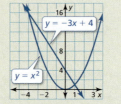

    a. How many solutions does the quadratic equation $x^2 = -3x + 4$ have? Explain.

    b. Without graphing, describe what you know about the graph of $y = x^2 + 3x - 4$.

59. **COMPARING METHODS** Example 3 shows two methods for solving a quadratic equation. Which method do you prefer? Explain your reasoning.

60. **THOUGHT PROVOKING** How many different parabolas have -2 and 2 as x-intercepts? Sketch examples of parabolas that have these two x-intercepts.

61. **MODELING WITH MATHEMATICS** To keep water off a road, the surface of the road is shaped like a parabola. A cross section of the road is shown in the diagram. The surface of the road can be modeled by $y = -0.0017x^2 + 0.041x$, where x and y are measured in feet. Find the width of the road to the nearest tenth of a foot.

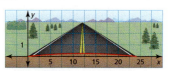

62. **MAKING AN ARGUMENT** A stream of water from a fire hose can be modeled by $y = -0.003x^2 + 0.58x + 3$, where x and y are measured in feet. A firefighter is standing 57 feet from a building and is holding the hose 3 feet above the ground. The bottom of a window of the building is 26 feet above the ground. Your friend claims the stream of water will pass through the window. Is your friend correct? Explain.

REASONING In Exercises 63–65, determine whether the statement is *always*, *sometimes*, or *never* true. Justify your answer.

63. The graph of $y = ax^2 + c$ has two x-intercepts when a is negative.

64. The graph of $y = ax^2 + c$ has no x-intercepts when a and c have the same sign.

65. The graph of $y = ax^2 + bx + c$ has more than two x-intercepts when $a \neq 0$.

Maintaining Mathematical Proficiency Reviewing what you learned in previous grades and lessons

Show that an exponential model fits the data. Then write a recursive rule that models the data. *(Section 1.6)*

66.
n	0	1	2	3
f(n)	18	3	$\frac{1}{2}$	$\frac{1}{12}$

67.
n	0	1	2	3
f(n)	2	8	32	128

208   Chapter 4   Solving Quadratic Equations

If students need help...	If students got it...
Resources by Chapter • Practice A and Practice B • Puzzle Time	Resources by Chapter • Enrichment and Extension • Cumulative Review
Student Journal • Practice	Start the *next* Section
Differentiating the Lesson Skills Review Handbook	

Laurie's Notes

Overview of Section 4.3

Introduction
- In this lesson, students will solve square root equations using the same techniques they have used in solving linear equations. The one difference is that there is one additional step of taking the square root of each side of the equation.
- Not all square root equations have solutions that are rational numbers, and therefore in this lesson, students will approximate solutions.

Teaching Strategy
- In Example 1, students are asked to solve $3x^2 - 27 = 0$. After adding 27 to each side of the equation, the equation becomes $3x^2 = 27$. This problem is like many that students will encounter in a mathematics class.
- A technique that can be used is to treat each side of the equation as a function. Graph each function and look for solutions to the system of equations, meaning where the two graphs intersect.
- In the graph shown, the original equation is shown, a parabola with a y-intercept at -27. The other two graphs represent the system with equations $y = 3x^2$ and $y = 27$.

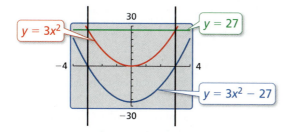

- The connection you want students to make is that the intersection points of the graphs have the same x-values as the x-intercepts of the original equation.

Pacing Suggestion
- Once students have worked the explorations, continue with the formal lesson.

What Your Students Will Learn

- Solve quadratic equations of the form $ax^2 + c = 0$ using square roots.
- Approximate the solutions of quadratic equations.

Laurie's Notes

Exploration

Motivate
- All of today's graphs will be symmetric about the *y*-axis.
- Draw two collections of shapes. In the first, each shape has one line of symmetry. In the second, each shape has no symmetry or more than one line of symmetry.

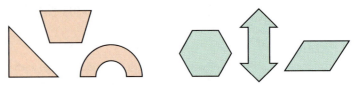

- Ask students to figure out how the groups have been sorted.
- ❓ "In which group would you place a parabola?" group on left

Discuss
- Today, you want students to think about the function $y = x^2$ (one *x*-intercept) and the effect of adding a constant *c* that shifts the graph up (no *x*-intercepts) or down (two *x*-intercepts).

Exploration 1
- Refer to the related function in part (c), $y = x^2$, as the parent function or basic quadratic function.
- **Construct Viable Arguments and Critique the Reasoning of Others:** Conjectures will likely relate the number of solutions to the number of *x*-intercepts of the graph.
- Students might go further and say that when *c* is positive, there are no solutions, and when *c* is negative, there are two solutions. Note, however, that this is not true when $a < 0$.

Exploration 2
- A calculator is a helpful tool for this exploration.
- ❓ "In part (a), what is happening to the *x*-values?" increasing by 0.01
- ❓ "How do the tables help you estimate the solutions of $x^2 - 5 = 0$?" The consecutive *x*-values at which the expression values change in sign from negative to positive indicate an approximate solution.
- **Reason Abstractly and Quantitatively:** In exploring the table of values, you are asking your students to reason quantitatively.

Exploration 3
- Students are using a calculator to confirm the estimates in Exploration 2.
- **Attend to Precision:** In part (b), students should have obtained approximate solutions such as $x \approx \pm 2.236067977$. In part (c), the exact solutions are $x = \pm\sqrt{5}$. Discuss the use of exact and approximate solutions. This draws attention to the concept of precision.

Communicate Your Answer
- **Reason Abstractly and Quantitatively:** Listen for correct reasoning as students share their answers.

Connecting to Next Step
- Students have now made the connection between the graph of the quadratic equation and the number of solutions the related quadratic equation has. This connection is stated in the formal lesson.

4.3 Solving Quadratic Equations Using Square Roots

Essential Question How can you determine the number of solutions of a quadratic equation of the form $ax^2 + c = 0$?

EXPLORATION 1 — The Number of Solutions of $ax^2 + c = 0$

Work with a partner. Solve each equation by graphing. Explain how the number of solutions of $ax^2 + c = 0$ relates to the graph of $y = ax^2 + c$.

a. $x^2 - 4 = 0$
b. $2x^2 + 5 = 0$
c. $x^2 = 0$
d. $x^2 - 5 = 0$

EXPLORATION 2 — Estimating Solutions

Work with a partner. Complete each table. Use the completed tables to estimate the solutions of $x^2 - 5 = 0$. Explain your reasoning.

a.
x	$x^2 - 5$
2.21	
2.22	
2.23	
2.24	
2.25	
2.26	

b.
x	$x^2 - 5$
−2.21	
−2.22	
−2.23	
−2.24	
−2.25	
−2.26	

ATTENDING TO PRECISION
To be proficient in math, you need to calculate accurately and express numerical answers with a level of precision appropriate for the problem's context.

EXPLORATION 3 — Using Technology to Estimate Solutions

Work with a partner. Two equations are equivalent when they have the same solutions.

a. Are the equations $x^2 - 5 = 0$ and $x^2 = 5$ equivalent? Explain your reasoning.

b. Use the square root key on a calculator to estimate the solutions of $x^2 - 5 = 0$. Describe the accuracy of your estimates in Exploration 2.

c. Write the exact solutions of $x^2 - 5 = 0$.

Communicate Your Answer

4. How can you determine the number of solutions of a quadratic equation of the form $ax^2 + c = 0$?

5. Write the exact solutions of each equation. Then use a calculator to estimate the solutions.

   a. $x^2 - 2 = 0$
   b. $3x^2 - 18 = 0$
   c. $x^2 = 8$

Section 4.3 Solving Quadratic Equations Using Square Roots 209

Extra Example 1

a. Solve $2x^2 - 32 = 0$ using square roots. The solutions are $x = 4$ and $x = -4$.

b. Solve $x^2 - 8 = -8$ using square roots. The only solution is $x = 0$.

c. Solve $-2x^2 + 3 = 27$ using square roots. The equation has no real solutions.

4.3 Lesson

What You Will Learn

▶ Solve quadratic equations using square roots.
▶ Approximate the solutions of quadratic equations.

Core Vocabulary
Previous
square root
zero of a function

Solving Quadratic Equations Using Square Roots

Earlier in this chapter, you studied properties of square roots. Now you will use square roots to solve quadratic equations of the form $ax^2 + c = 0$. First isolate x^2 on one side of the equation to obtain $x^2 = d$. Then solve by taking the square root of each side.

Core Concept

Solutions of $x^2 = d$
- When $d > 0$, $x^2 = d$ has two real solutions, $x = \pm\sqrt{d}$.
- When $d = 0$, $x^2 = d$ has one real solution, $x = 0$.
- When $d < 0$, $x^2 = d$ has no real solutions.

ANOTHER WAY
You can also solve $3x^2 - 27 = 0$ by factoring.
$3(x^2 - 9) = 0$
$3(x - 3)(x + 3) = 0$
$x = 3$ or $x = -3$

EXAMPLE 1 Solving Quadratic Equations Using Square Roots

a. Solve $3x^2 - 27 = 0$ using square roots.

$3x^2 - 27 = 0$	Write the equation.
$3x^2 = 27$	Add 27 to each side.
$x^2 = 9$	Divide each side by 3.
$x = \pm\sqrt{9}$	Take the square root of each side.
$x = \pm 3$	Simplify.

▶ The solutions are $x = 3$ and $x = -3$.

b. Solve $x^2 - 10 = -10$ using square roots.

$x^2 - 10 = -10$	Write the equation.
$x^2 = 0$	Add 10 to each side.
$x = 0$	Take the square root of each side.

▶ The only solution is $x = 0$.

c. Solve $-5x^2 + 11 = 16$ using square roots.

$-5x^2 + 11 = 16$	Write the equation.
$-5x^2 = 5$	Subtract 11 from each side.
$x^2 = -1$	Divide each side by -5.

▶ The square of a real number cannot be negative. So, the equation has no real solutions.

Laurie's Notes   Teacher Actions

- **Look For and Make Use of Structure:** Discuss similarities in solving $3x - 27 = 0$ and $3x^2 - 27 = 0$. The steps are the same, with one extra for $3x^2 - 27 = 0$.
- Note the side column suggestion for *Another Way* to solve. See the *Teaching Strategy* on page T-208 for a calculator technique.
- **Teaching Tip** and **Whiteboarding:** Model Example 1(a) and then give partners sufficient time for the remaining problems as you circulate. Use *whiteboards* for students to quickly share their work.

STUDY TIP
Each side of the equation $(x - 1)^2 = 25$ is a square. So, you can still solve by taking the square root of each side.

EXAMPLE 2 Solving a Quadratic Equation Using Square Roots

Solve $(x - 1)^2 = 25$ using square roots.

SOLUTION

$(x - 1)^2 = 25$	Write the equation.
$x - 1 = \pm 5$	Take the square root of each side.
$x = 1 \pm 5$	Add 1 to each side.

▶ So, the solutions are $x = 1 + 5 = 6$ and $x = 1 - 5 = -4$.

Check
Use a graphing calculator to check your answer. Rewrite the equation as $(x - 1)^2 - 25 = 0$. Graph the related function $f(x) = (x - 1)^2 - 25$ and find the zeros of the function. The zeros are -4 and 6.

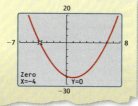

Monitoring Progress Help in English and Spanish at *BigIdeasMath.com*

Solve the equation using square roots.

1. $-3x^2 = -75$
2. $x^2 + 12 = 10$
3. $4x^2 - 15 = -15$
4. $(x + 7)^2 = 0$
5. $4(x - 3)^2 = 9$
6. $(2x + 1)^2 = 36$

Approximating Solutions of Quadratic Equations

EXAMPLE 3 Approximating Solutions of a Quadratic Equation

Solve $4x^2 - 13 = 15$ using square roots. Round to the nearest hundredth.

SOLUTION

$4x^2 - 13 = 15$	Write the equation.
$4x^2 = 28$	Add 13 to each side.
$x^2 = 7$	Divide each side by 4.
$x = \pm\sqrt{7}$	Take the square root of each side.
$x \approx \pm 2.65$	Use a calculator.

▶ The solutions are $x \approx -2.65$ and $x \approx 2.65$.

Check
Graph each side of the equation and find the points of intersection. The x-values of the points of intersection are about -2.65 and 2.65.

Monitoring Progress Help in English and Spanish at *BigIdeasMath.com*

Solve the equation using square roots. Round your solutions to the nearest hundredth.

7. $x^2 + 8 = 19$
8. $5x^2 - 2 = 0$
9. $3x^2 - 30 = 4$

Section 4.3 Solving Quadratic Equations Using Square Roots 211

English Language Learners

Pair Activity
Pair an English learner with an English speaker. One student solves a quadratic equation using square roots. The other student solves the same equation using a graphing calculator. Partners compare solutions and then take turns explaining their solutions to one another. Students should repeat the activity with a different quadratic equation and should swap solution methods.

Extra Example 2
Solve $(x - 3)^2 = 16$ using square roots. The solutions are $x = 7$ and $x = -1$.

Extra Example 3
Solve $2x^2 + 10 = 32$ using square roots. Round the solutions to the nearest hundredth. The solutions are $x \approx -3.32$ and $x \approx 3.32$.

MONITORING PROGRESS ANSWERS

1. $x = 5, x = -5$
2. no real solutions
3. $x = 0$
4. $x = -7$
5. $x = \frac{9}{2}, x = \frac{3}{2}$
6. $x = \frac{5}{2}, x = -\frac{7}{2}$
7. $x \approx 3.32, x \approx -3.32$
8. $x \approx 0.63, x \approx -0.63$
9. $x \approx 3.37, x \approx -3.37$

Laurie's Notes Teacher Actions

- **? Attend to Precision:** "How do you read $(x - 1)^2 = 25$?" Listen for something like "the quantity x minus 1 squared equals 25."
- **Connection:** Checking with a graphing calculator, the x-intercepts at -4 and 6 occur equidistant from the x-value of the vertex.
- **Think-Pair-Share:** Have students answer Questions 1–6, and then share and discuss as a class.
- **Turn and Talk:** Pose Example 3. "How do you solve $4x^2 - 13 = 15$, and how is this like solving $4x - 13 = 15$?"

Extra Example 4

A storage container has the shape of a rectangular prism. Its height is 6 feet. Its length is two times its width. The volume is 288 cubic feet. Find the length and width of the container.

width: $\sqrt{24} \approx 4.9$ feet;
length: $2\sqrt{24} \approx 9.8$ feet

Extra Example 5

The surface area S of a cube with side length a is given by the formula $S = 6a^2$. Solve the formula for a. Then approximate the side length of a cube with a surface area of 724 square inches.

$a = \sqrt{\dfrac{S}{6}}$; about 11 inches

MONITORING PROGRESS ANSWERS

10. width: about 5.9 ft, length: about 17.7 ft

11. $r = \sqrt{\dfrac{S}{4\pi}}$; about 8 in.

EXAMPLE 4 Solving a Real-Life Problem

A touch tank has a height of 3 feet. Its length is three times its width. The volume of the tank is 270 cubic feet. Find the length and width of the tank.

SOLUTION

The length ℓ is three times the width w, so $\ell = 3w$. Write an equation using the formula for the volume of a rectangular prism.

$V = \ell w h$	Write the formula.
$270 = 3w(w)(3)$	Substitute 270 for V, $3w$ for ℓ, and 3 for h.
$270 = 9w^2$	Multiply.
$30 = w^2$	Divide each side by 9.
$\pm\sqrt{30} = w$	Take the square root of each side.

INTERPRETING MATHEMATICAL RESULTS
Use the positive square root because negative solutions do not make sense in this context. Length and width cannot be negative.

The solutions are $\sqrt{30}$ and $-\sqrt{30}$. Use the positive solution.

▶ So, the width is $\sqrt{30} \approx 5.5$ feet and the length is $3\sqrt{30} \approx 16.4$ feet.

EXAMPLE 5 Rearranging and Evaluating a Formula

The area A of an equilateral triangle with side length s is given by the formula $A = \dfrac{\sqrt{3}}{4}s^2$. Solve the formula for s. Then approximate the side length of the traffic sign that has an area of 390 square inches.

SOLUTION

Step 1 Solve the formula for s.

$A = \dfrac{\sqrt{3}}{4}s^2$	Write the formula.
$\dfrac{4A}{\sqrt{3}} = s^2$	Multiply each side by $\dfrac{4}{\sqrt{3}}$.
$\sqrt{\dfrac{4A}{\sqrt{3}}} = s$	Take the positive square root of each side.

ANOTHER WAY
Notice that you can rewrite the formula as $s = \dfrac{2}{3^{1/4}}\sqrt{A}$, or $s \approx 1.52\sqrt{A}$. This can help you efficiently find the value of s for various values of A.

Step 2 Substitute 390 for A in the new formula and evaluate.

$$s = \sqrt{\dfrac{4A}{\sqrt{3}}} = \sqrt{\dfrac{4(390)}{\sqrt{3}}} = \sqrt{\dfrac{1560}{\sqrt{3}}} \approx 30 \qquad \text{Use a calculator.}$$

▶ The side length of the traffic sign is about 30 inches.

Monitoring Progress Help in English and Spanish at *BigIdeasMath.com*

10. **WHAT IF?** In Example 4, the volume of the tank is 315 cubic feet. Find the length and width of the tank.

11. The surface area S of a sphere with radius r is given by the formula $S = 4\pi r^2$. Solve the formula for r. Then find the radius of a globe with a surface area of 804 square inches.

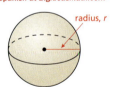

Laurie's Notes — Teacher Actions

? "What is a reasonable estimate for $\sqrt{30}$? Explain." About 5.5, because $\sqrt{25} = 5$ and $\sqrt{36} = 6$, and 30 is about halfway between 25 and 36.

- Solving for s in Example 5 can be challenging for students because of the fraction and the radical. Explain that you still want to multiply by the reciprocal of the coefficient. Having an expression with a radical as a radicand can confuse students.
- Discuss *Another Way*.

Closure

- **Exit Ticket:** State the number of solutions for each equation.
  a. $2x^2 + 8 = 40$ 2   **b.** $2x^2 - 8 = -40$ 0   **c.** $2x^2 = 0$ 1

4.3 Exercises

Vocabulary and Core Concept Check

1. **COMPLETE THE SENTENCE** The equation $x^2 = d$ has _____ real solutions when $d > 0$.

2. **DIFFERENT WORDS, SAME QUESTION** Which is different? Find "both" answers.

   | Solve $x^2 = 144$ using square roots. | Solve $x^2 - 144 = 0$ using square roots. |
   | Solve $x^2 + 146 = 2$ using square roots. | Solve $x^2 + 2 = 146$ using square roots. |

Monitoring Progress and Modeling with Mathematics

In Exercises 3–8, determine the number of real solutions of the equation. Then solve the equation using square roots.

3. $x^2 = 25$
4. $x^2 = -36$
5. $x^2 = -21$
6. $x^2 = 400$
7. $x^2 = 0$
8. $x^2 = 169$

In Exercises 9–18, solve the equation using square roots. *(See Example 1.)*

9. $x^2 - 16 = 0$
10. $x^2 + 6 = 0$
11. $3x^2 + 12 = 0$
12. $x^2 - 55 = 26$
13. $2x^2 - 98 = 0$
14. $-x^2 + 9 = 9$
15. $-3x^2 - 5 = -5$
16. $4x^2 - 371 = 29$
17. $4x^2 + 10 = 11$
18. $9x^2 - 35 = 14$

In Exercises 19–24, solve the equation using square roots. *(See Example 2.)*

19. $(x + 3)^2 = 0$
20. $(x - 1)^2 = 4$
21. $(2x - 1)^2 = 81$
22. $(4x + 5)^2 = 9$
23. $9(x + 1)^2 = 16$
24. $4(x - 2)^2 = 25$

In Exercises 25–30, solve the equation using square roots. Round your solutions to the nearest hundredth. *(See Example 3.)*

25. $x^2 + 6 = 13$
26. $x^2 + 11 = 24$
27. $2x^2 - 9 = 11$
28. $5x^2 + 2 = 6$
29. $-21 = 15 - 2x^2$
30. $2 = 4x^2 - 5$

31. **ERROR ANALYSIS** Describe and correct the error in solving the equation $2x^2 - 33 = 39$ using square roots.

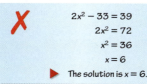

32. **MODELING WITH MATHEMATICS** An in-ground pond has the shape of a rectangular prism. The pond has a depth of 24 inches and a volume of 72,000 cubic inches. The length of the pond is two times its width. Find the length and width of the pond. *(See Example 4.)*

33. **MODELING WITH MATHEMATICS** A person sitting in the top row of the bleachers at a sporting event drops a pair of sunglasses from a height of 24 feet. The function $h = -16x^2 + 24$ represents the height h (in feet) of the sunglasses after x seconds. How long does it take the sunglasses to hit the ground?

Assignment Guide and Homework Check

ASSIGNMENT

Basic: 1, 2, 3–33 odd, 34, 36, 40, 45–50

Average: 1, 2–30 even, 31–36, 40, 45–50

Advanced: 1, 2, 6, 16–30 even, 31, 32–36 even, 37–50

HOMEWORK CHECK

Basic: 9, 19, 25, 33, 36

Average: 10, 20, 26, 32, 36

Advanced: 16, 22, 28, 32, 36

ANSWERS

1. two
2. Solve $x^2 + 146 = 2$ using square roots; no real solutions; $x = 12, x = -12$
3. 2; $x = 5, x = -5$
4. 0; no real solutions
5. 0; no real solutions
6. 2; $x = 20, x = -20$
7. 1; $x = 0$
8. 2; $x = 13, x = -13$
9. $x = 4, x = -4$
10. no real solutions
11. no real solutions
12. $x = 9, x = -9$
13. $x = 7, x = -7$
14. $x = 0$
15. $x = 0$
16. $x = 10, x = -10$
17. $x = \frac{1}{2}, x = -\frac{1}{2}$
18. $x = \frac{7}{3}, x = -\frac{7}{3}$
19. $x = -3$
20. $x = 3, x = -1$
21. $x = 5, x = -4$
22. $x = -\frac{1}{2}, x = -2$
23. $x = \frac{1}{3}, x = -\frac{7}{3}$
24. $x = \frac{9}{2}, x = -\frac{1}{2}$
25. $x \approx 2.65, x \approx -2.65$
26. $x \approx 3.61, x \approx -3.61$
27. $x \approx 3.16, x \approx -3.16$
28. $x \approx 0.89, x \approx -0.89$
29. $x \approx 4.24, x \approx -4.24$
30. $x \approx 1.32, x \approx -1.32$

31. The number 36 has both a positive and negative square root; $x = \pm 6$
32. length: about 77.5 in., width: about 38.7 in.
33. about 1.2 sec

Dynamic Teaching Tools
Dynamic Assessment & Progress Monitoring Tool
Interactive Whiteboard Lesson Library
Dynamic Classroom with Dynamic Investigations

ANSWERS
34. your cousin; Subtracting 4 from each side of the equation gives $x^2 = -4$, and a negative number has no real square roots.
35. 3 ft
36. a. $r = \sqrt{\dfrac{A}{\pi}}$
    b. about 6 ft; about 24 in., about 13 m
    c. The steps for solving only need to be completed once.
37. Sample answer: Use a calculator.
38. a. a and c have opposite signs.
    b. a is not 0, and c is 0.
    c. a and c have the same sign.
39. $(3, 9), (-3, 9)$; When $y = 9$, $x = \pm 3$.
40. 1; The graph has one x-intercept.
41. $x = 1.2$, $x = -1.2$; $1.2^2 = 1.44$
42–50. See Additional Answers.

Mini-Assessment
Solve the equation using square roots.
1. $3x^2 - 108 = 0$ $x = 6$ and $x = -6$
2. $(x - 5)^2 = 81$ $x = 14$ and $x = -4$
3. Solve $4x^2 - 3 = 17$ using square roots. Round the solutions to the nearest hundredth.
   $x \approx -2.24$ and $x \approx 2.24$
4. A cardboard box has the shape of a rectangular prism. Its height is 10 inches. Its length is three times its width. The volume is 540 cubic inches. Find the length and width of the box.
   width: $\sqrt{18} \approx 4.2$ inches;
   length: $3\sqrt{18} \approx 12.7$ inches
5. The volume V of a cylinder with radius r and height h is given by the formula $V = \pi r^2 h$. Solve the formula for r. Then find the radius of a cylinder with a height of 15 centimeters and a volume of 750 cubic centimeters.
   $r = \sqrt{\dfrac{V}{\pi h}}$; about 4 centimeters

---

34. **MAKING AN ARGUMENT** Your friend says that the solution of the equation $x^2 + 4 = 0$ is $x = 0$. Your cousin says that the equation has no real solutions. Who is correct? Explain your reasoning.

35. **MODELING WITH MATHEMATICS** The design of a square rug for your living room is shown. You want the area of the inner square to be 25% of the total area of the rug. Find the side length x of the inner square.

6 ft

36. **MATHEMATICAL CONNECTIONS** The area A of a circle with radius r is given by the formula $A = \pi r^2$. *(See Example 5.)*
    a. Solve the formula for r.
    b. Use the formula from part (a) to find the radius of each circle.

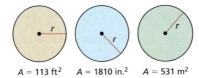

   $A = 113$ ft^2 $A = 1810$ in.^{2} $A = 531$ m^2

    c. Explain why it is beneficial to solve the formula for r before finding the radius.

37. **WRITING** How can you approximate the roots of a quadratic equation when the roots are not integers?

38. **WRITING** Given the equation $ax^2 + c = 0$, describe the values of a and c so the equation has the following number of solutions.
    a. two real solutions
    b. one real solution
    c. no real solutions

39. **REASONING** Without graphing, where do the graphs of $y = x^2$ and $y = 9$ intersect? Explain.

40. **HOW DO YOU SEE IT?** The graph represents the function $f(x) = (x - 1)^2$. How many solutions does the equation $(x - 1)^2 = 0$ have? Explain.

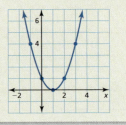

41. **REASONING** Solve $x^2 = 1.44$ without using a calculator. Explain your reasoning.

42. **THOUGHT PROVOKING** The quadratic equation
    $$ax^2 + bx + c = 0$$
    can be rewritten in the following form.
    $$\left(x + \frac{b}{2a}\right)^2 = \frac{b^2 - 4ac}{4a^2}$$
    Use this form to write the solutions of the equation.

43. **REASONING** An equation of the graph shown is $y = \frac{1}{2}(x - 2)^2 + 1$. Two points on the parabola have y-coordinates of 9. Find the x-coordinates of these points.

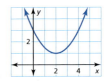

44. **CRITICAL THINKING** Solve each equation without graphing.
    a. $x^2 - 12x + 36 = 64$
    b. $x^2 + 14x + 49 = 16$

Maintaining Mathematical Proficiency   *Reviewing what you learned in previous grades and lessons*

Factor the polynomial. *(Section 2.7)*

45. $x^2 + 8x + 16$
46. $x^2 - 4x + 4$
47. $x^2 - 14x + 49$
48. $x^2 + 18x + 81$
49. $x^2 + 12x + 36$
50. $x^2 - 22x + 121$

214   Chapter 4   Solving Quadratic Equations

If students need help...	If students got it...
Resources by Chapter • Practice A and Practice B • Puzzle Time	**Resources by Chapter** • Enrichment and Extension • Cumulative Review
Student Journal • Practice	Start the *next* Section
Differentiating the Lesson Skills Review Handbook	

Laurie's Notes

Overview of Section 4.4

Introduction
- In this lesson, the *completing the square* technique is used to solve quadratic equations and to rewrite quadratic equations from standard form to vertex form. When the equation is in vertex form, the maximum or minimum value of the function can be found.
- Real-life problems are solved by using the *completing the square* technique.
- In this lesson, students are expected to interpret information about a quadratic function from parameters in the equation.

Formative Assessment Tips
- **Pass the Problem:** This technique provides students the opportunity to work with others on the solution of a problem or a proof that requires more than a few steps. Seeing how others approach and work the problem helps students to reflect on their own process.
- There are different ways, or configurations, to using this technique. Begin by posing a problem that individuals or pairs begin to work on. After a fixed amount of time, the problem is swapped with partners or another pair of students. The recipients finish solving the problem or make modifications or corrections to the problem. If changes are made, they must explain why there was an error or why the strategy that was begun is not going to be followed.
- This technique gives all students the opportunity to participate in the lesson. All students receive feedback on their work, not just one student that the teacher might have called on. If the work passed to another is not clear because steps are missing, students receive this feedback from peers, not just the teacher. If students or partners were unable to get started on a problem, they will hopefully be exchanging with others who can help them make sense of the problem.
- When students have finished the problem, the two or four people involved in swapping problems will confer with one another to discuss the problem and offer additional feedback. One thing you hope to hear is positive feedback on the clarity of thinking that was recorded, allowing the recipients to make sense of the work.

Pacing Suggestion
- The explorations help students visualize what completing the square means. The explorations should not take long. Transition to the formal lesson.

Dynamic Teaching Tools
Dynamic Assessment & Progress Monitoring Tool
Lesson Planning Tool
Interactive Whiteboard Lesson Library
Dynamic Classroom with Dynamic Investigations

What Your Students Will Learn

- Complete the square for expressions of the form $ax^2 + bx$ by adding a constant c to make a perfect square trinomial.
- Solve quadratic equations of the form $ax^2 + bx + c = 0$ by completing the square and then using square roots.
- Find and use maximum and minimum values of quadratic functions by writing the functions in vertex form.
- Solve real-life problems by modeling the problems with a quadratic equation and completing the square.

Laurie's Notes

Exploration

Motivate
- Write the sequence on the board: 1, 4, ___, 16, ___, 36, ...
- ❓ "What numbers are missing and what is the pattern? Explain." 9 and 25; The terms of the sequence are perfect squares.
- ❓ "What is a perfect square trinomial?" a trinomial that can be factored as $(a \pm b)^2$

Exploration 1
- Review the names and dimensions of the algebra tiles.
- Students should model the equation with the goal of adding 1-tiles to form a square array on the left side. To make this possible, start by arranging half of the x-tiles vertically and half horizontally.
- Add the number of 1-tiles needed to form a square on the left side. Add the same number of 1-tiles to the right side.
- Remind students how multiplication of binomials was modeled (rectangular array).
- Explain how to use the dimensions of the square formed on the left side of the equation to write it as the square of a binomial. The vertical and horizontal dimensions are each $x + 2$, so the left side of the equation represents $(x + 2)^2$.
- **Reason Abstractly and Quantitatively:** Students have obtained a concrete model of $(x + 2)^2 = 2$. They will now reason abstractly, using algebra to manipulate the equation.

Exploration 2
- Having been guided through the first exploration, students should be able to work with partners on this exploration.
- ❓ "What equation is modeled by the algebra tiles?" $x^2 + 6x = -5$
- ❓ "What equation is represented by the tiles after completing the square?" $x^2 + 6x + 9 = 4$
- **Construct Viable Arguments and Critique the Reasoning of Others:** Asking students to explain the thinking behind their solutions is helpful for them and for their peers.
- It is important to work through the checking process.

Communicate Your Answer
- **Construct Viable Arguments and Critique the Reasoning of Others:** In Question 3, answers should include an understanding of writing the quadratic equation in a form that allows you to solve it using square roots.

Connecting to Next Step
- Students have now seen a visual model of what it means to "complete the square," and they should be ready to work through the same process symbolically in the formal lesson.

4.4 Solving Quadratic Equations by Completing the Square

Essential Question How can you use "completing the square" to solve a quadratic equation?

EXPLORATION 1 Solving by Completing the Square

Work with a partner.

a. Write the equation modeled by the algebra tiles. This is the equation to be solved.

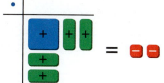

b. Four algebra tiles are added to the left side to "complete the square." Why are four algebra tiles also added to the right side?

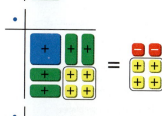

c. Use algebra tiles to label the dimensions of the square on the left side and simplify on the right side.

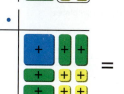

d. Write the equation modeled by the algebra tiles so that the left side is the square of a binomial. Solve the equation using square roots.

MAKING SENSE OF PROBLEMS
To be proficient in math, you need to explain to yourself the meaning of a problem. After that, you need to look for entry points to its solution.

EXPLORATION 2 Solving by Completing the Square

Work with a partner.

a. Write the equation modeled by the algebra tiles.

b. Use algebra tiles to "complete the square."

c. Write the solutions of the equation.

d. Check each solution in the original equation.

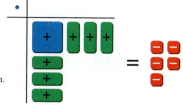

Communicate Your Answer

3. How can you use "completing the square" to solve a quadratic equation?

4. Solve each quadratic equation by completing the square.

   a. $x^2 - 2x = 1$ b. $x^2 - 4x = -1$ c. $x^2 + 4x = -3$

Section 4.4 Solving Quadratic Equations by Completing the Square 215

ANSWERS

1. a. $x^2 + 4x = -2$
   b. To be equivalent equations, the same number must be added to each side.
   c.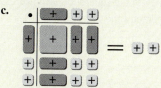
   d. $(x + 2)^2 = 2$; $x \approx -0.59$, $x \approx -3.41$

2. a. $x^2 + 6x = -5$
   b.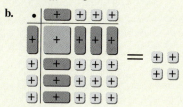
   c. $x = -1$, $x = -5$
   d. $(-1)^2 + 6(-1) = -5$; $(-5)^2 + 6(-5) = -5$

3. Write the equation in the form $x^2 + bx = d$. Add $\left(\dfrac{b}{2}\right)^2$ to each side of the equation. Factor the resulting expression on the left side as the square of a binomial. Solve the resulting equation using square roots.

4. a. $x \approx 2.41$, $x \approx -0.41$
   b. $x \approx 3.73$, $x \approx 0.27$
   c. $x = -1$, $x = -3$

English Language Learners

Organization
Students will benefit by writing down the steps for completing the square shown in the Core Concept box. Have students write the steps in their notebooks. A poster with the steps could be posted in the classroom.

Extra Example 1
Complete the square for each expression. Then factor the trinomial.

a. $x^2 + 18x$ $x^2 + 18x + 81; (x + 9)^2$
b. $x^2 - 17x$ $x^2 - 17x + \frac{289}{4}; \left(x - \frac{17}{2}\right)^2$

MONITORING PROGRESS ANSWERS
1. $x^2 + 10x + 25; (x + 5)^2$
2. $x^2 - 4x + 4; (x - 2)^2$
3. $x^2 + 7x + \frac{49}{4}; \left(x + \frac{7}{2}\right)^2$

4.4 Lesson

What You Will Learn
- Complete the square for expressions of the form $x^2 + bx$.
- Solve quadratic equations by completing the square.
- Find and use maximum and minimum values.
- Solve real-life problems by completing the square.

Core Vocabulary
completing the square, p. 216
Previous
perfect square trinomial
coefficient
maximum value
minimum value
vertex form of a quadratic function

Completing the Square

For an expression of the form $x^2 + bx$, you can add a constant c to the expression so that $x^2 + bx + c$ is a perfect square trinomial. This process is called **completing the square**.

Core Concept

Completing the Square

Words To complete the square for an expression of the form $x^2 + bx$, follow these steps.

    Step 1 Find one-half of b, the coefficient of x.
    Step 2 Square the result from Step 1.
    Step 3 Add the result from Step 2 to $x^2 + bx$.

Factor the resulting expression as the square of a binomial.

Algebra $x^2 + bx + \left(\frac{b}{2}\right)^2 = \left(x + \frac{b}{2}\right)^2$

JUSTIFYING STEPS
In each diagram below, the combined area of the shaded regions is $x^2 + bx$.
Adding $\left(\frac{b}{2}\right)^2$ completes the square in the second diagram.

EXAMPLE 1 Completing the Square

Complete the square for each expression. Then factor the trinomial.

a. $x^2 + 6x$ b. $x^2 - 9x$

SOLUTION

a. **Step 1** Find one-half of b. $\frac{b}{2} = \frac{6}{2} = 3$

    Step 2 Square the result from Step 1. $3^2 = 9$

    Step 3 Add the result from Step 2 to $x^2 + bx$. $x^2 + 6x + 9$

▶ $x^2 + 6x + 9 = (x + 3)^2$

b. **Step 1** Find one-half of b. $\frac{b}{2} = \frac{-9}{2}$

    Step 2 Square the result from Step 1. $\left(\frac{-9}{2}\right)^2 = \frac{81}{4}$

    Step 3 Add the result from Step 2 to $x^2 + bx$. $x^2 - 9x + \frac{81}{4}$

▶ $x^2 - 9x + \frac{81}{4} = \left(x - \frac{9}{2}\right)^2$

Monitoring Progress Help in English and Spanish at *BigIdeasMath.com*

Complete the square for the expression. Then factor the trinomial.

1. $x^2 + 10x$ 2. $x^2 - 4x$ 3. $x^2 + 7x$

Laurie's Notes Teacher Actions

- Write the *Core Concept*. The diagrams help to explain why $\left(\frac{b}{2}\right)^2$ is the amount added to complete the square.
- The examples help students become familiar with the technique of completing the square before they use it to solve an equation.
- It may be helpful to ask students what is missing that would make the expression a perfect square trinomial: $x^2 + 6x + \underline{?} = (x + \underline{?})^2$.
- **Think-Pair-Share:** Have students answer Questions 1–3, and then share and discuss as a class.

Solving Quadratic Equations by Completing the Square

The method of completing the square can be used to solve any quadratic equation. To solve a quadratic equation by completing the square, you must write the equation in the form $x^2 + bx = d$.

EXAMPLE 2 Solving a Quadratic Equation: $x^2 + bx = d$

Solve $x^2 - 16x = -15$ by completing the square.

SOLUTION

$x^2 - 16x = -15$	Write the equation.
$x^2 - 16x + (-8)^2 = -15 + (-8)^2$	Complete the square by adding $\left(\frac{-16}{2}\right)^2$, or $(-8)^2$, to each side.
$(x - 8)^2 = 49$	Write the left side as the square of a binomial.
$x - 8 = \pm 7$	Take the square root of each side.
$x = 8 \pm 7$	Add 8 to each side.

▶ The solutions are $x = 8 + 7 = 15$ and $x = 8 - 7 = 1$.

Check

$x^2 - 16x = -15$ Original equation $x^2 - 16x = -15$
$15^2 - 16(15) \stackrel{?}{=} -15$ Substitute. $1^2 - 16(1) \stackrel{?}{=} -15$
$-15 = -15$ ✓ Simplify. $-15 = -15$ ✓

COMMON ERROR
When completing the square to solve an equation, be sure to add $\left(\frac{b}{2}\right)^2$ to each side of the equation.

EXAMPLE 3 Solving a Quadratic Equation: $ax^2 + bx + c = 0$

Solve $2x^2 + 20x - 8 = 0$ by completing the square.

SOLUTION

$2x^2 + 20x - 8 = 0$	Write the equation.
$2x^2 + 20x = 8$	Add 8 to each side.
$x^2 + 10x = 4$	Divide each side by 2.
$x^2 + 10x + 5^2 = 4 + 5^2$	Complete the square by adding $\left(\frac{10}{2}\right)^2$, or 5^2, to each side.
$(x + 5)^2 = 29$	Write the left side as the square of a binomial.
$x + 5 = \pm\sqrt{29}$	Take the square root of each side.
$x = -5 \pm \sqrt{29}$	Subtract 5 from each side.

▶ The solutions are $x = -5 + \sqrt{29} \approx 0.39$ and $x = -5 - \sqrt{29} \approx -10.39$.

COMMON ERROR
Before you complete the square, be sure that the coefficient of the x^2-term is 1.

Monitoring Progress Help in English and Spanish at *BigIdeasMath.com*

Solve the equation by completing the square. Round your solutions to the nearest hundredth, if necessary.

4. $x^2 - 2x = 3$ 5. $m^2 + 12m = -8$ 6. $3g^2 - 24g + 27 = 0$

Differentiated Instruction

Inclusion
Have advanced students work in pairs to write and solve equations like those in Examples 2 and 3. Each student should write an equation, exchange equations with his or her partner, and solve the equation. After solving, students should exchange equations again and check the solutions.

Extra Example 2
Solve $x^2 - 18x = -17$ by completing the square. The solutions are $x = 9 + 8 = 17$ and $x = 9 - 8 = 1$.

Extra Example 3
Solve $2x^2 + 12x - 10 = 0$ by completing the square. The solutions are $x = -3 + \sqrt{14} \approx 0.74$ and $x = -3 - \sqrt{14} \approx -6.74$.

MONITORING PROGRESS ANSWERS

4. $x = 3, x = -1$
5. $m \approx -0.71, m \approx -11.29$
6. $g \approx 6.65, g \approx 1.35$

Laurie's Notes   Teacher Actions

- Write $x^2 - 16x + \underline{?} = -15 + \underline{?}$ on the board. This helps students remember to add $\left(\frac{b}{2}\right)^2$ to both sides of the equation.
- Continue to solve the equation as shown. Remind students that when square rooting each side of the equation, there are + and − cases.
- **Pass the Problem:** Explain this technique and then pose Example 3. Give partners 1 to 2 minutes to discuss and begin a solution. Call "swap" and have partners continue to solve. Both groups debrief. Deciding how to work with the coefficient of x^2 should be a rich conversation.

Extra Example 4
Find the minimum value of $y = x^2 + 8x + 5$. The function has a minimum value of -11.

Extra Example 5
Find the maximum value of $y = -x^2 + 4x + 2$. The function has a maximum value of 6.

MONITORING PROGRESS ANSWERS
7. maximum value; 8
8. minimum value; 4
9. minimum value; -3

Finding and Using Maximum and Minimum Values
One way to find the maximum or minimum value of a quadratic function is to write the function in vertex form by completing the square. Recall that the vertex form of a quadratic function is $y = a(x - h)^2 + k$, where $a \neq 0$. The vertex of the graph is (h, k).

EXAMPLE 4 Finding a Minimum Value

Find the minimum value of $y = x^2 + 4x - 1$.

SOLUTION
Write the function in vertex form.

$y = x^2 + 4x - 1$	Write the function.
$y + 1 = x^2 + 4x$	Add 1 to each side.
$y + 1 + 4 = x^2 + 4x + 4$	Complete the square for $x^2 + 4x$.
$y + 5 = x^2 + 4x + 4$	Simplify the left side.
$y + 5 = (x + 2)^2$	Write the right side as the square of a binomial.
$y = (x + 2)^2 - 5$	Write in vertex form.

The vertex is $(-2, -5)$. Because a is positive ($a = 1$), the parabola opens up and the y-coordinate of the vertex is the minimum value.

▶ So, the function has a minimum value of -5.

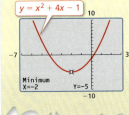

Check
$y = x^2 + 4x - 1$
Minimum
X=-2   Y=-5

EXAMPLE 5 Finding a Maximum Value

Find the maximum value of $y = -x^2 + 2x + 7$.

SOLUTION
Write the function in vertex form.

$y = -x^2 + 2x + 7$	Write the function.
$y - 7 = -x^2 + 2x$	Subtract 7 from each side.
$y - 7 = -(x^2 - 2x)$	Factor out -1.
$y - 7 - 1 = -(x^2 - 2x + 1)$	Complete the square for $x^2 - 2x$.
$y - 8 = -(x^2 - 2x + 1)$	Simplify the left side.
$y - 8 = -(x - 1)^2$	Write $x^2 - 2x + 1$ as the square of a binomial.
$y = -(x - 1)^2 + 8$	Write in vertex form.

The vertex is $(1, 8)$. Because a is negative ($a = -1$), the parabola opens down and the y-coordinate of the vertex is the maximum value.

▶ So, the function has a maximum value of 8.

STUDY TIP
Adding 1 inside the parentheses results in subtracting 1 from the right side of the equation.

Monitoring Progress Help in English and Spanish at BigIdeasMath.com

Determine whether the quadratic function has a maximum or minimum value. Then find the value.

7. $y = -x^2 - 4x + 4$ 8. $y = x^2 + 12x + 40$ 9. $y = x^2 - 2x - 2$

Laurie's Notes   Teacher Actions

- **Connection:** Completing the square is a technique that can be used to rewrite a quadratic from standard form to vertex form.
- **?** Pose Example 4. "Does this quadratic have a maximum or minimum value? Explain." *It has a minimum value because it opens upward.*
- **Make Sense of Problems and Persevere in Solving Them:** Have partners work together on Example 5. Circulate, but do not rush in to solve the example for them.
- **Probing Question:** Ask probing questions (see the *Study Tip*) and expect students to persevere.
- Graphing the original equation and the vertex form will tell students whether they are correct.

EXAMPLE 6 Interpreting Forms of Quadratic Functions

Which of the functions could be represented by the graph? Explain.

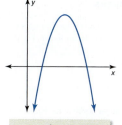

$f(x) = -\frac{1}{2}(x + 4)^2 + 8$

$g(x) = -(x - 5)^2 + 9$

$m(x) = (x - 3)(x - 12)$

$p(x) = -(x - 2)(x - 8)$

SOLUTION

You do not know the scale of either axis. To eliminate functions, consider the characteristics of the graph and information provided by the form of each function. The graph appears to be a parabola that opens down, which means the function has a maximum value. The vertex of the graph is in the first quadrant. Both x-intercepts are positive.

- The graph of f opens down because $a < 0$, which means f has a maximum value. However, the vertex $(-4, 8)$ of the graph of f is in the second quadrant. So, the graph does not represent f.

- The graph of g opens down because $a < 0$, which means g has a maximum value. The vertex $(5, 9)$ of the graph of g is in the first quadrant. By solving $0 = -(x - 5)^2 + 9$, you see that the x-intercepts of the graph of g are 2 and 8. So, the graph could represent g.

- The graph of m has two positive x-intercepts. However, its graph opens up because $a > 0$, which means m has a minimum value. So, the graph does not represent m.

- The graph of p has two positive x-intercepts, and its graph opens down because $a < 0$. This means that p has a maximum value and the vertex must be in the first quadrant. So, the graph could represent p.

▶ The graph could represent function g or function p.

EXAMPLE 7 Real-Life Application

The function $y = -16x^2 + 96x$ represents the height y (in feet) of a model rocket x seconds after it is launched. (a) Find the maximum height of the rocket. (b) Find and interpret the axis of symmetry.

STUDY TIP
Adding 9 inside the parentheses results in subtracting 144 from the right side of the equation.

SOLUTION

a. To find the maximum height, identify the maximum value of the function.

$y = -16x^2 + 96x$ Write the function.

$y = -16(x^2 - 6x)$ Factor out −16.

$y - 144 = -16(x^2 - 6x + 9)$ Complete the square for $x^2 - 6x$.

$y = -16(x - 3)^2 + 144$ Write in vertex form.

▶ Because the maximum value is 144, the model rocket reaches a maximum height of 144 feet.

b. The vertex is $(3, 144)$. So, the axis of symmetry is $x = 3$. On the left side of $x = 3$, the height increases as time increases. On the right side of $x = 3$, the height decreases as time increases.

Monitoring Progress Help in English and Spanish at *BigIdeasMath.com*

Determine whether the function could be represented by the graph in Example 6. Explain.

10. $h(x) = (x - 8)^2 + 10$ **11.** $n(x) = -2(x - 5)(x - 20)$

12. WHAT IF? Repeat Example 7 when the function is $y = -16x^2 + 128x$.

Section 4.4 Solving Quadratic Equations by Completing the Square

---

Extra Example 6

Which of the functions could be represented by the graph? Explain.

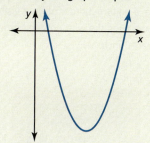

$f(x) = (x - 4)^2 + 9$
$g(x) = -(x - 3)^2 + 5$
$m(x) = -\frac{1}{4}(x + 2)(x + 6)$
$p(x) = (x - 1)(x - 7)$

The graph could represent function p.

The graph does not represent function f. The graph of f opens up because $a > 0$. However, the vertex $(4, 9)$ is in the first quadrant, so the graph of f does not cross the x-axis.

The graph does not represent functions g or m because $a < 0$ in both functions and these graphs open down.

The graph of p opens up because $a > 0$. It has two positive x-intercepts. So, p has a minimum value and could represent the graph shown.

Extra Example 7

The function $y = -16x^2 + 160x$ represents the height y (in feet) of a model rocket x seconds after it is launched.

a. Find the maximum height of the rocket. 400 feet

b. Find and interpret the axis of symmetry. The axis of symmetry is $x = 5$. On the left side of $x = 5$, the height increases as time increases. On the right side of $x = 5$, the height decreases as time increases.

MONITORING PROGRESS ANSWERS

10. no; The graph opens up.

11. yes; The graph has two positive x-intercepts and opens down.

12. a. 256 ft

    b. $x = 4$; On the left side of $x = 4$, the height increases as time increases. On the right side of $x = 4$, the height decreases as time increases.

Laurie's Notes — Teacher Actions

- **Reason Abstractly and Quantitatively** and **Construct Viable Arguments and Critique the Reasoning of Others:** Example 6 is done without calculators. Partners should discuss each function. In addition to selecting which function(s) could be represented by the graph, students should also explain why the other function(s) could not be represented by the graph.
- **? Popsicle Sticks:** "Could f(x) be the function for this graph? Explain." Listen for reasoning associated with information deduced from the function.
- **Look For and Make Use of Structure:** In Example 7, students should recognize that the Distributive Property is used in completing the square. Discuss the *Study Tip*.

Extra Example 8

You decide to put a rug on a floor. You want the rug to cover 100 square feet and to have a uniform border of floor visible, as shown. Find the width of the border to the nearest inch.

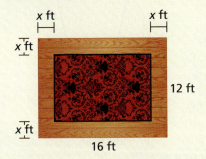

The width of the border should be about 23 inches.

MONITORING PROGRESS ANSWER

13. about 13 in.

Solving Real-Life Problems

EXAMPLE 8 Modeling with Mathematics

You decide to use chalkboard paint to create a chalkboard on a door. You want the chalkboard to cover 6 square feet and to have a uniform border, as shown. Find the width of the border to the nearest inch.

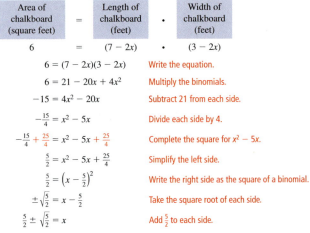

SOLUTION

1. **Understand the Problem** You know the dimensions (in feet) of the door from the diagram. You also know the area (in square feet) of the chalkboard and that it will have a uniform border. You are asked to find the width of the border to the nearest inch.

2. **Make a Plan** Use a verbal model to write an equation that represents the area of the chalkboard. Then solve the equation.

3. **Solve the Problem**

Let x be the width (in feet) of the border, as shown in the diagram.

Area of chalkboard (square feet)	=	Length of chalkboard (feet)	·	Width of chalkboard (feet)
6	=	$(7 - 2x)$	·	$(3 - 2x)$

$6 = (7 - 2x)(3 - 2x)$ Write the equation.

$6 = 21 - 20x + 4x^2$ Multiply the binomials.

$-15 = 4x^2 - 20x$ Subtract 21 from each side.

$-\frac{15}{4} = x^2 - 5x$ Divide each side by 4.

$-\frac{15}{4} + \frac{25}{4} = x^2 - 5x + \frac{25}{4}$ Complete the square for $x^2 - 5x$.

$\frac{5}{2} = x^2 - 5x + \frac{25}{4}$ Simplify the left side.

$\frac{5}{2} = \left(x - \frac{5}{2}\right)^2$ Write the right side as the square of a binomial.

$\pm\sqrt{\frac{5}{2}} = x - \frac{5}{2}$ Take the square root of each side.

$\frac{5}{2} \pm \sqrt{\frac{5}{2}} = x$ Add $\frac{5}{2}$ to each side.

The solutions of the equation are $x = \frac{5}{2} + \sqrt{\frac{5}{2}} \approx 4.08$ and $x = \frac{5}{2} - \sqrt{\frac{5}{2}} \approx 0.92$.

It is not possible for the width of the border to be 4.08 feet because the width of the door is 3 feet. So, the width of the border is about 0.92 foot.

$0.92 \text{ ft} \cdot \dfrac{12 \text{ in.}}{1 \text{ ft}} = 11.04 \text{ in.}$ Convert 0.92 foot to inches.

▶ The width of the border should be about 11 inches.

4. **Look Back** When the width of the border is slightly less than 1 foot, the length of the chalkboard is slightly more than 5 feet and the width of the chalkboard is slightly more than 1 foot. Multiplying these dimensions gives an area close to 6 square feet. So, an 11-inch border is reasonable.

Monitoring Progress Help in English and Spanish at *BigIdeasMath.com*

13. **WHAT IF?** You want the chalkboard to cover 4 square feet. Find the width of the border to the nearest inch.

Laurie's Notes Teacher Actions

- **Model with Mathematics** and **Thumbs Up:** Read through the problem. Students should self-assess with *Thumbs Up* to indicate that they understand the problem. Discuss a plan for solving the problem.
- **Pass the Problem:** Follow the protocol for this formative assessment technique. Circulate and listen to student dialogue. When students have finished, the groups should debrief one another's work.

Closure

- **Point of Most Significance:** Ask students to identify, aloud or written, the most significant point in the lesson that aided their learning.

4.4 Exercises

Vocabulary and Core Concept Check

1. **COMPLETE THE SENTENCE** The process of adding a constant c to the expression $x^2 + bx$ so that $x^2 + bx + c$ is a perfect square trinomial is called _____.

2. **VOCABULARY** Explain how to complete the square for an expression of the form $x^2 + bx$.

3. **WRITING** Is it more convenient to complete the square for $x^2 + bx$ when b is odd or when b is even? Explain.

4. **WRITING** Describe how you can use the process of completing the square to find the maximum or minimum value of a quadratic function.

Monitoring Progress and Modeling with Mathematics

In Exercises 5–10, find the value of c that completes the square.

5. $x^2 - 8x + c$
6. $x^2 - 2x + c$
7. $x^2 + 4x + c$
8. $x^2 + 12x + c$
9. $x^2 - 15x + c$
10. $x^2 + 9x + c$

In Exercises 11–16, complete the square for the expression. Then factor the trinomial. *(See Example 1.)*

11. $x^2 - 10x$
12. $x^2 - 40x$
13. $x^2 + 16x$
14. $x^2 + 22x$
15. $x^2 + 5x$
16. $x^2 - 3x$

In Exercises 17–22, solve the equation by completing the square. Round your solutions to the nearest hundredth, if necessary. *(See Example 2.)*

17. $x^2 + 14x = 15$
18. $x^2 - 6x = 16$
19. $x^2 - 4x = -2$
20. $x^2 + 2x = 5$
21. $x^2 - 5x = 8$
22. $x^2 + 11x = -10$

23. **MODELING WITH MATHEMATICS** The area of the patio is 216 square feet.

    a. Write an equation that represents the area of the patio.
    b. Find the dimensions of the patio by completing the square.

24. **MODELING WITH MATHEMATICS** Some sand art contains sand and water sealed in a glass case, similar to the one shown. When the art is turned upside down, the sand and water fall to create a new picture. The glass case has a depth of 1 centimeter and a volume of 768 cubic centimeters.

   a. Write an equation that represents the volume of the glass case.
   b. Find the dimensions of the glass case by completing the square.

In Exercises 25–32, solve the equation by completing the square. Round your solutions to the nearest hundredth, if necessary. *(See Example 3.)*

25. $x^2 - 8x + 15 = 0$
26. $x^2 + 4x - 21 = 0$
27. $2x^2 + 20x + 44 = 0$
28. $3x^2 - 18x + 12 = 0$
29. $-3x^2 - 24x + 17 = -40$
30. $-5x^2 - 20x + 35 = 30$
31. $2x^2 - 14x + 10 = 26$
32. $4x^2 + 12x - 15 = 5$

30. $x \approx 0.24, x \approx -4.24$
31. $x = 8, x = -1$
32. $x \approx 1.19, x \approx -4.19$

Assignment Guide and Homework Check

ASSIGNMENT
Basic: 1–4, 5–33 odd, 39–47 odd, 51–55 odd, 63, 66, 71, 75–80
Average: 1–4, 6–66 even, 71, 75–80
Advanced: 1–4, 10, 16, 22, 24, 30–38 even, 44–66 even, 67–80

HOMEWORK CHECK
Basic: 11, 17, 25, 41, 53
Average: 12, 20, 26, 42, 54
Advanced: 16, 22, 30, 44, 54

ANSWERS

1. completing the square
2. Add $\left(\dfrac{b}{2}\right)^2$ to $x^2 + bx$.
3. even; When b is even, $\dfrac{b}{2}$ is an integer.
4. Use completing the square to write the function in vertex form. The maximum or minimum is the y-coordinate of the vertex.
5. 16
6. 1
7. 4
8. 36
9. $\dfrac{225}{4}$
10. $\dfrac{81}{4}$
11. $x^2 - 10x + 25; (x - 5)^2$
12. $x^2 - 40x + 400; (x - 20)^2$
13. $x^2 + 16x + 64; (x + 8)^2$
14. $x^2 + 22x + 121; (x + 11)^2$
15. $x^2 + 5x + \dfrac{25}{4}; \left(x + \dfrac{5}{2}\right)^2$
16. $x^2 - 3x + \dfrac{9}{4}; \left(x - \dfrac{3}{2}\right)^2$
17. $x = 1, x = -15$
18. $x = 8, x = -2$
19. $x \approx 3.41, x \approx 0.59$
20. $x \approx 1.45, x \approx -3.45$
21. $x \approx 6.27, x \approx -1.27$
22. $x = -1, x = -10$
23. a. $x^2 + 6x = 216$
    b. width: 12 ft, length: 18 ft
24. a. $x^2 - 8x = 768$
    b. length: 32 cm, height: 24 cm
25. $x = 5, x = 3$
26. $x = 3, x = -7$
27. $x \approx -3.27, x \approx -6.73$
28. $x \approx 5.24, x \approx 0.76$
29. $x \approx 1.92, x \approx -9.92$

ANSWERS

33. The number 16 should be added to each side of the equation;
$x^2 + 8x + 16 = 10 + 16$;
$x = -4 \pm \sqrt{26}$

34. The leading coefficient should be 1 before completing the square;
$x^2 - x = 2$; $x^2 - x + \frac{1}{4} = 2 + \frac{1}{4}$

35. $b = 10$, $b = -10$; In a perfect square trinomial $c = \left(\frac{b}{2}\right)^2$, so $b = \pm 2\sqrt{c}$.

36. Divide each side of the equation by 3.

37. $y = (x + 3)^2 - 6$; D

38. $y = -(x - 4)^2 + 4$; A

39. $y = -(x + 2)^2 + 2$; B

40. $y = (x - 1)^2 + 3$; C

41. minimum value; -6

42. minimum value; 1

43. maximum value; -5

44. maximum value; 15

45. maximum value; -6

46. minimum value; -17

47. yes; The graph has two negative x-intercepts and it opens down.

48. no; The x-intercept is negative.

49. no; The x-intercepts are both positive.

50. yes; The graph has one positive x-intercept and one negative x-intercept, and it opens down.

51. f, m; The graph has two negative x-intercepts and it opens up.

52. r, n; The graph has one positive x-intercept and one negative x-intercept, and it opens down.

53. a. 36 ft

b. $x = \frac{3}{2}$; On the left side of $x = \frac{3}{2}$, the height increases as time increases. On the right side of $x = \frac{3}{2}$, the height decreases as time increases.

---

33. ERROR ANALYSIS Describe and correct the error in solving $x^2 + 8x = 10$ by completing the square.

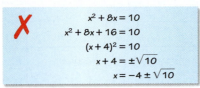

34. ERROR ANALYSIS Describe and correct the error in the first two steps of solving $2x^2 - 2x - 4 = 0$ by completing the square.

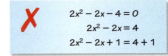

35. NUMBER SENSE Find all values of b for which $x^2 + bx + 25$ is a perfect square trinomial. Explain how you found your answer.

36. REASONING You are completing the square to solve $3x^2 + 6x = 12$. What is the first step?

In Exercises 37–40, write the function in vertex form by completing the square. Then match the function with its graph.

37. $y = x^2 + 6x + 3$ **38.** $y = -x^2 + 8x - 12$

39. $y = -x^2 - 4x - 2$ **40.** $y = x^2 - 2x + 4$

A. 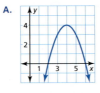   B.

C.   D.

In Exercises 41–46, determine whether the quadratic function has a maximum or minimum value. Then find the value. *(See Examples 4 and 5.)*

41. $y = x^2 - 4x - 2$ **42.** $y = x^2 + 6x + 10$

43. $y = -x^2 - 10x - 30$ **44.** $y = -x^2 + 14x - 34$

45. $f(x) = -3x^2 - 6x - 9$ **46.** $f(x) = 4x^2 - 28x + 32$

In Exercises 47–50, determine whether the graph could represent the function. Explain.

47. $y = -(x + 8)(x + 3)$ **48.** $y = (x - 5)^2$

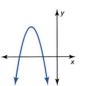

49. $y = \frac{1}{4}(x + 2)^2 - 4$ **50.** $y = -2(x - 1)(x + 2)$

In Exercises 51 and 52, determine which of the functions could be represented by the graph. Explain. *(See Example 6.)*

51.

$h(x) = (x + 2)^2 + 3$
$f(x) = 2(x + 3)^2 - 2$
$g(x) = -\frac{1}{2}(x - 8)(x - 4)$
$m(x) = (x + 2)(x + 4)$

52.

$r(x) = -\frac{1}{3}(x - 5)(x + 1)$
$p(x) = -2(x - 2)(x - 6)$
$q(x) = (x + 1)^2 + 4$
$n(x) = -(x - 2)^2 + 9$

53. MODELING WITH MATHEMATICS The function $h = -16t^2 + 48t$ represents the height h (in feet) of a kickball t seconds after it is kicked from the ground. *(See Example 7.)*

a. Find the maximum height of the kickball.

b. Find and interpret the axis of symmetry.

54. MODELING WITH MATHEMATICS
You throw a stone from a height of 16 feet with an initial vertical velocity of 32 feet per second. The function $h = -16t^2 + 32t + 16$ represents the height h (in feet) of the stone after t seconds.

a. Find the maximum height of the stone.

b. Find and interpret the axis of symmetry.

55. MODELING WITH MATHEMATICS You are building a rectangular brick patio surrounded by a crushed stone border with a uniform width, as shown. You purchase patio bricks to cover 140 square feet. Find the width of the border. *(See Example 8.)*

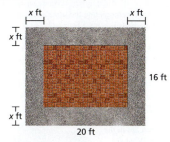

56. MODELING WITH MATHEMATICS
You are making a poster that will have a uniform border, as shown. The total area of the poster is 722 square inches. Find the width of the border to the nearest inch.

MATHEMATICAL CONNECTIONS In Exercises 57 and 58, find the value of x. Round your answer to the nearest hundredth, if necessary.

57. $A = 108$ m²   58. $A = 288$ in.²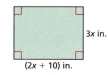

In Exercises 59–62, solve the equation by completing the square. Round your solutions to the nearest hundredth, if necessary.

59. $0.5x^2 + x - 2 = 0$

60. $0.75x^2 + 1.5x = 4$

61. $\frac{8}{3}x - \frac{2}{3}x^2 = -\frac{5}{6}$

62. $\frac{1}{4}x^2 + \frac{1}{2}x - \frac{5}{4} = 0$

63. PROBLEM SOLVING The distance d (in feet) that it takes a car to come to a complete stop can be modeled by $d = 0.05s^2 + 2.2s$, where s is the speed of the car (in miles per hour). A car has 168 feet to come to a complete stop. Find the maximum speed at which the car can travel.

64. PROBLEM SOLVING During a "big air" competition, snowboarders launch themselves from a half-pipe, perform tricks in the air, and land back in the half-pipe. The height h (in feet) of a snowboarder above the bottom of the half-pipe can be modeled by $h = -16t^2 + 24t + 16.4$, where t is the time (in seconds) after the snowboarder launches into the air. The snowboarder lands 3.2 feet lower than the height of the launch. How long is the snowboarder in the air? Round your answer to the nearest tenth of a second.

Cross section of a half-pipe

65. PROBLEM SOLVING You have 80 feet of fencing to make a rectangular horse pasture that covers 750 square feet. A barn will be used as one side of the pasture, as shown.

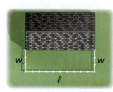

a. Write equations for the amount of fencing to be used and the area enclosed by the fencing.

b. Use substitution to solve the system of equations from part (a). What are the possible dimensions of the pasture?

ANSWERS

54. a. 32 ft
    b. $x = 1$; On the left side of $x = 1$, the height increases as time increases. On the right side of $x = 1$, the height decreases as time increases.
55. 3 ft
56. about 1 in.
57. 12
58. about 4.87
59. $x \approx 1.24, x \approx -3.24$
60. $x \approx 1.52, x \approx -3.52$
61. $x \approx 4.29, x \approx -0.29$
62. $x \approx 1.45, x \approx -3.45$
63. 40 mi/h
64. about 1.6 sec
65. a. $\ell + 2w = 80, \ell w = 750$
    b. length: 30 ft, width: 25 ft; length: 50 ft, width: 15 ft

ANSWERS

66. a. 3, 1
   b. Substitute 3 for y in the equation, then solve the equation by completing the square.

67. a. $x \approx 0.8, x \approx -12.8$
   b. $x \approx 0.78, x \approx -12.78$
   c. *Sample answer:* completing the square; The result is more accurate.

68. See Additional Answers.

69. $x(x+2) = 48$, 6 and 8

70. $x(x+2) = 195$; $x = -15$ and $x = -13$

71–80. See Additional Answers.

Mini-Assessment

1. Complete the square for the expression $x^2 + 24x$. Then factor the trinomial.
   $x^2 + 24x + 144$; $(x + 12)^2$

2. Solve $x^2 - 12x = 45$ by completing the square. $x = 15$ and $x = -3$

3. Find the minimum value of $y = x^2 - 20x + 75$. -25

4. The function $y = -16x^2 + 32x$ represents the height y (in feet) of a football x seconds after it is kicked off the ground.
   a. Find the maximum height of the football. **16 feet**
   b. Find and interpret the axis of symmetry. $x = 1$; The height increases when $x < 1$ and decreases when $x > 1$.

5. A rectangular city garden is surrounded by a sidewalk with a uniform width, as shown. The garden covers 340 square yards. Find the width of the sidewalk to the nearest foot.

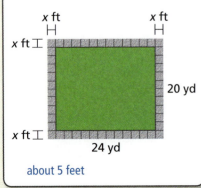

about 5 feet

---

66. HOW DO YOU SEE IT? The graph represents the quadratic function $y = x^2 - 4x + 6$.

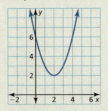

a. Use the graph to estimate the x-values for which $y = 3$.

b. Explain how you can use the method of completing the square to check your estimates in part (a).

67. COMPARING METHODS Consider the quadratic equation $x^2 + 12x + 2 = 12$.
   a. Solve the equation by graphing.
   b. Solve the equation by completing the square.
   c. Compare the two methods. Which do you prefer? Explain.

68. THOUGHT PROVOKING Sketch the graph of the equation $x^2 - 2xy + y^2 - x - y = 0$. Identify the graph.

69. REASONING The product of two consecutive even integers that are positive is 48. Write and solve an equation to find the integers.

70. REASONING The product of two consecutive odd integers that are negative is 195. Write and solve an equation to find the integers.

71. MAKING AN ARGUMENT You purchase stock for $16 per share. You sell the stock 30 days later for $23.50 per share. The price y (in dollars) of a share during the 30-day period can be modeled by $y = -0.025x^2 + x + 16$, where x is the number of days after the stock is purchased. Your friend says you could have sold the stock earlier for $23.50 per share. Is your friend correct? Explain.

72. REASONING You are solving the equation $x^2 + 9x = 18$. What are the advantages of solving the equation by completing the square instead of using other methods you have learned?

73. PROBLEM SOLVING You are knitting a rectangular scarf. The pattern results in a scarf that is 60 inches long and 4 inches wide. However, you have enough yarn to knit 396 square inches. You decide to increase the dimensions of the scarf so that you will use all your yarn. The increase in the length is three times the increase in the width. What are the dimensions of your scarf?

74. WRITING How many solutions does $x^2 + bx = c$ have when $c < -\left(\dfrac{b}{2}\right)^2$? Explain.

Maintaining Mathematical Proficiency *Reviewing what you learned in previous grades and lessons*

Write a piecewise function for the graph. *(Section 1.2)*

75.   76.   77.

Simplify the expression $\sqrt{b^2 - 4ac}$ for the given values. *(Section 4.1)*

78. $a = 3, b = -6, c = 2$ 79. $a = -2, b = 4, c = 7$ 80. $a = 1, b = 6, c = 4$

Laurie's Notes

Overview of Section 4.5

Introduction
- The Quadratic Formula is presented as the fifth method for solving quadratic equations. The formula is derived in the explorations and stated formally in the lesson. The formula is then used to solve quadratic equations.
- The discriminant is discussed and then interpreted to determine the number of zeros of an equation or the number of *x*-intercepts of a graph.
- The lesson ends with a summary of the five methods used to solve quadratics, and then students are asked to choose efficient methods for solving various quadratic equations.

Resources
- There are many online videos of students singing the Quadratic Formula to the tune of "Pop Goes the Weasel." Share one with students. You could return to the video at the end of the period when students may be ready to sing along.

Formative Assessment Tips
- **Paired Verbal Fluency (PVF):** This technique is used between two partners where each person takes a turn speaking, uninterrupted, for a specified period of time. The roles reverse and the listener then speaks, uninterrupted, for the same amount of time.
- The technique can be used at the beginning, middle, or end of instruction. Used at the beginning of instruction, students share their prior knowledge about a particular topic, skill, or concept. Used in the middle of instruction, students can share their understanding of a particular problem. Used at the end of instruction, students reflect on learning that occurred during the lesson or at the end of a connected group of lessons.
- Verbalizing their understanding and being an attentive listener will activate student thinking and should help identify areas of difficulty or uncertainty.
- In this lesson, students will be asked to reflect on their understanding of solving quadratics using a variety of techniques.

Pacing Suggestion
- The explorations provide an opportunity for students to derive the Quadratic Formula and then practice using the formula to solve a few simple quadratic equations. You may want to omit Example 1 and begin with the application problem in Example 2.

Dynamic Teaching Tools
Dynamic Assessment & Progress Monitoring Tool
Lesson Planning Tool
Interactive Whiteboard Lesson Library
Dynamic Classroom with Dynamic Investigations

What Your Students Will Learn

- Solve quadratic equations of the form $ax^2 + bx + c = 0$ using the Quadratic Formula.
- Interpret the discriminant to determine the number of real solutions of quadratic equations.
- Choose efficient methods for solving quadratic equations and explain the advantages of the methods.

Laurie's Notes

Exploration

Motivate
- Share with students some history about the Quadratic Formula.
  - Around 400 B.C., the Babylonians and Chinese use a method called "completing the square" to solve problems involving areas.
  - Around 300 B.C., the Greek mathematicians Pythagoras and Euclid use geometry to find a general procedure for solving a quadratic equation, but their methods are not considered useful.
  - Around A.D. 700, a Hindu mathematician named Brahmagupta finds the general solution for the quadratic equation. He uses irrational numbers and also recognizes the two roots in the solution.
  - Around A.D. 1100, another Hindu mathematician named Baskhara finds the complete solution. He recognizes that any positive number has two square roots.
  - In 1637, the French mathematician René Descartes publishes *La Géométrie*, which presents the Quadratic Formula in its present form.

Discuss
- Tell students that today they will see one way of deriving the Quadratic Formula algebraically. Then they will derive the Quadratic Formula on their own by completing the square. They will also learn about the discriminant and discover how it relates to the number of solutions.

Exploration 1
- A derivation of the Quadratic Formula is shown in this exploration. Students justify each step.
- If students get stuck on a step, ask them what is different from the last step. Mathematically, what has changed?
- When students are finished, discuss the justifications.
- Students sometimes lose sight of what they have accomplished. Starting with the quadratic equation in general form, they solve for *x* algebraically to produce a formula for finding the solutions of any quadratic equation.

Exploration 2
- In the last lesson, students learned how to solve a quadratic equation by completing the square. They are using that method to derive the Quadratic Formula and compare it to the method used in Exploration 1.
- **Make Sense of Problems and Persevere in Solving Them:** Do not be too quick to rescue your students. Give them time to wrestle with the derivation. If they get stuck, have them refer back to their notes from the last section. Believe that your students can persevere in deriving the formula. Unlike many problems, they know what the formula should look like when they finish because of Exploration 1.

Communicate Your Answer
- To solve each quadratic equation in Question 4, tell students they need to evaluate the Quadratic Formula for the values of *a*, *b*, and *c* from the equation.

Connecting to Next Step
- Students have now derived the Quadratic Formula. The formula will be stated at the beginning of the formal lesson.

4.5 Solving Quadratic Equations Using the Quadratic Formula

Essential Question How can you derive a formula that can be used to write the solutions of any quadratic equation in standard form?

EXPLORATION 1 Deriving the Quadratic Formula

Work with a partner. The following steps show a method of solving $ax^2 + bx + c = 0$. Explain what was done in each step.

$ax^2 + bx + c = 0$ ← 1. Write the equation.

$4a^2x^2 + 4abx + 4ac = 0$ ← 2. What was done?

$4a^2x^2 + 4abx + 4ac + b^2 = b^2$ ← 3. What was done?

$4a^2x^2 + 4abx + b^2 = b^2 - 4ac$ ← 4. What was done?

$(2ax + b)^2 = b^2 - 4ac$ ← 5. What was done?

$2ax + b = \pm\sqrt{b^2 - 4ac}$ ← 6. What was done?

$2ax = -b \pm \sqrt{b^2 - 4ac}$ ← 7. What was done?

Quadratic Formula: $x = \dfrac{-b \pm \sqrt{b^2 - 4ac}}{2a}$ ← 8. What was done?

EXPLORATION 2 Deriving the Quadratic Formula by Completing the Square

Work with a partner.

a. Solve $ax^2 + bx + c = 0$ by completing the square. (*Hint:* Subtract c from each side, divide each side by a, and then proceed by completing the square.)

b. Compare this method with the method in Exploration 1. Explain why you think $4a$ and b^2 were chosen in Steps 2 and 3 of Exploration 1.

USING TOOLS STRATEGICALLY
To be proficient in math, you need to identify relevant external mathematical resources.

Communicate Your Answer

3. How can you derive a formula that can be used to write the solutions of any quadratic equation in standard form?

4. Use the Quadratic Formula to solve each quadratic equation.

   a. $x^2 + 2x - 3 = 0$ b. $x^2 - 4x + 4 = 0$ c. $x^2 + 4x + 5 = 0$

5. Use the Internet to research *imaginary numbers*. How are they related to quadratic equations?

Dynamic Teaching Tools

Dynamic Assessment & Progress Monitoring Tool
Lesson Planning Tool
Interactive Whiteboard Lesson Library
Dynamic Classroom with Dynamic Investigations

ANSWERS

1. 2. Multiply each side by $4a$.
   3. Add b^2 to each side.
   4. Subtract $4ac$ from each side.
   5. Write the left side in factored form.
   6. Take the square root of each side.
   7. Subtract b from each side.
   8. Divide each side by $2a$.

2. a. $x = \dfrac{-b \pm \sqrt{b^2 - 4ac}}{2a}$

   b. to obtain a perfect square trinomial on the left side of the equation

3. Complete the square of the general form of the quadratic equation, $ax^2 + bx + c = 0$.

4. a. $x = 1, x = -3$
   b. $x = 2$
   c. no real solution

5. The imaginary number, i, is $\sqrt{-1}$. Quadratic equations with no real solution have complex solutions, which include an imaginary part.

Differentiated Instruction

Organization/Inclusion
Some students may view the Quadratic Formula as complicated and may not know what to do first when trying to use it. Suggest that they complete this table before they begin solving a quadratic equation in standard form.

Coefficient	Value
a	
b	
c	

Term	Value
$-b$	
b^2	
$4ac$	
$2a$	

After completing the table, students can substitute the appropriate values into the formula and simplify.

Extra Example 1

Solve $12x^2 - 4x - 5 = 0$ using the Quadratic Formula.
The solutions are $x = \dfrac{4 + 16}{24} = \dfrac{5}{6}$ and $x = \dfrac{4 - 16}{24} = -\dfrac{1}{2}$.

MONITORING PROGRESS ANSWERS

1. $x = 5, x = 1$
2. $x \approx 3.6, x \approx -5.6$
3. $x \approx -1.2, x \approx 1.9$
4. $x = \dfrac{1}{2}$

4.5 Lesson

What You Will Learn
- Solve quadratic equations using the Quadratic Formula.
- Interpret the discriminant.
- Choose efficient methods for solving quadratic equations.

Core Vocabulary
Quadratic Formula, p. 226
discriminant, p. 228

Using the Quadratic Formula

By completing the square for the quadratic equation $ax^2 + bx + c = 0$, you can develop a formula that gives the solutions of any quadratic equation in standard form. This formula is called the **Quadratic Formula**.

Core Concept

Quadratic Formula

The real solutions of the quadratic equation $ax^2 + bx + c = 0$ are

$$x = \dfrac{-b \pm \sqrt{b^2 - 4ac}}{2a} \qquad \text{Quadratic Formula}$$

where $a \neq 0$ and $b^2 - 4ac \geq 0$.

EXAMPLE 1 Using the Quadratic Formula

Solve $2x^2 - 5x + 3 = 0$ using the Quadratic Formula.

SOLUTION

$$x = \dfrac{-b \pm \sqrt{b^2 - 4ac}}{2a} \qquad \text{Quadratic Formula}$$

$$= \dfrac{-(-5) \pm \sqrt{(-5)^2 - 4(2)(3)}}{2(2)} \qquad \text{Substitute 2 for } a, -5 \text{ for } b, \text{ and } 3 \text{ for } c.$$

$$= \dfrac{5 \pm \sqrt{1}}{4} \qquad \text{Simplify.}$$

$$= \dfrac{5 \pm 1}{4} \qquad \text{Evaluate the square root.}$$

 So, the solutions are $x = \dfrac{5+1}{4} = \dfrac{3}{2}$ and $x = \dfrac{5-1}{4} = 1$.

STUDY TIP
You can use the roots of a quadratic equation to factor the related expression. In Example 1, you can use 1 and $\tfrac{3}{2}$ to factor $2x^2 - 5x + 3$ as $(x - 1)(2x - 3)$.

Check

$2x^2 - 5x + 3 = 0$	Original equation	$2x^2 - 5x + 3 = 0$
$2\left(\tfrac{3}{2}\right)^2 - 5\left(\tfrac{3}{2}\right) + 3 \stackrel{?}{=} 0$	Substitute.	$2(1)^2 - 5(1) + 3 \stackrel{?}{=} 0$
$\tfrac{9}{2} - \tfrac{15}{2} + 3 \stackrel{?}{=} 0$	Simplify.	$2 - 5 + 3 \stackrel{?}{=} 0$
$0 = 0$ ✓	Simplify.	$0 = 0$ ✓

Monitoring Progress Help in English and Spanish at BigIdeasMath.com

Solve the equation using the Quadratic Formula. Round your solutions to the nearest tenth, if necessary.

1. $x^2 - 6x + 5 = 0$
2. $\tfrac{1}{2}x^2 + x - 10 = 0$
3. $-3x^2 + 2x + 7 = 0$
4. $4x^2 - 4x = -1$

226   Chapter 4   Solving Quadratic Equations

Laurie's Notes   Teacher Actions

- **?** Write the quadratic equation. "Why must one side of the equation be equal to 0?" So you can determine the values of a, b, and c.
- **?** "Why is there a restriction that $a \neq 0$?" When $a = 0$, you do not have a quadratic equation. Also, the denominator would be 0 in the Quadratic Formula and you cannot divide by 0.
- **?** "Why is there a restriction that $b^2 - 4ac \geq 0$?" You cannot take the square root of a negative number.
- Work through Example 1. Review order of operations as you simplify. Explain how to interpret the $\pm$ symbol.

EXAMPLE 2 Modeling With Mathematics

The number y of Northern Rocky Mountain wolf breeding pairs x years since 1990 can be modeled by the function $y = 0.20x^2 + 1.8x - 3$. When were there about 35 breeding pairs?

Wolf Breeding Pairs

$y = 0.20x^2 + 1.8x - 3$

SOLUTION

1. **Understand the Problem** You are given a quadratic function that represents the number of wolf breeding pairs for years after 1990. You need to use the model to determine when there were 35 wolf breeding pairs.

2. **Make a Plan** To determine when there were 35 wolf breeding pairs, find the x-values for which $y = 35$. So, solve the equation $35 = 0.20x^2 + 1.8x - 3$.

3. **Solve the Problem**

$$35 = 0.20x^2 + 1.8x - 3 \quad \text{Write the equation.}$$
$$0 = 0.20x^2 + 1.8x - 38 \quad \text{Write in standard form.}$$
$$x = \frac{-b \pm \sqrt{b^2 - 4ac}}{2a} \quad \text{Quadratic Formula}$$
$$= \frac{-1.8 \pm \sqrt{1.8^2 - 4(0.2)(-38)}}{2(0.2)} \quad \text{Substitute 0.2 for } a, 1.8 \text{ for } b, \text{ and } -38 \text{ for } c.$$
$$= \frac{-1.8 \pm \sqrt{33.64}}{0.4} \quad \text{Simplify.}$$
$$= \frac{-1.8 \pm 5.8}{0.4} \quad \text{Simplify.}$$

The solutions are $x = \frac{-1.8 + 5.8}{0.4} = 10$ and $x = \frac{-1.8 - 5.8}{0.4} = -19$.

> **INTERPRETING MATHEMATICAL RESULTS**
> You can ignore the solution $x = -19$ because -19 represents the year 1971, which is not in the given time period.

Because x represents the number of years since 1990, x is greater than or equal to zero. So, there were about 35 breeding pairs 10 years after 1990, in 2000.

4. **Look Back** Use a graphing calculator to graph the equations $y = 0.20x^2 + 1.8x - 3$ and $y = 35$. Then use the *intersect* feature to find the point of intersection. The graphs intersect at (10, 35).

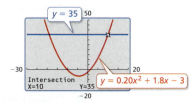

Monitoring Progress Help in English and Spanish at *BigIdeasMath.com*

5. **WHAT IF?** When were there about 60 wolf breeding pairs?

6. The number y of bald eagle nesting pairs in a state x years since 2000 can be modeled by the function $y = 0.34x^2 + 13.1x + 51$.

   a. When were there about 160 bald eagle nesting pairs?

   b. How many bald eagle nesting pairs were there in 2000?

English Language Learners

Forming Answers

Encourage English learners to form complete sentences in their responses to real-life problems. Demonstrate how to use the question to help students form the answer.

Question: When were there about 35 breeding pairs?

Answer: There were about 35 breeding pairs 10 years after 1990, in the year 2000.

Extra Example 2

The number y of Northern Rocky Mountain wolf breeding pairs x years since 1990 can be modeled by the function $y = 0.20x^2 + 1.8x - 3$. When were there about 77 breeding pairs of wolves? There were about 77 breeding pairs 16 years after 1990, in 2006.

MONITORING PROGRESS ANSWERS

5. 2004
6. a. 2007
   b. 51 bald eagle nesting pairs

Laurie's Notes Teacher Actions

- **Reason Abstractly and Quantitatively:** Students decontextualize the model to solve algebraically. Once the solutions are found, the context is considered again.
- Consider using technology to graph the function $y = 0.20x^2 + 1.8x - 38$. Use the graph to approximate the zeros. Then solve the corresponding equation by using the Quadratic Formula.
- **?** "Do both solutions make sense in the context of the problem? Explain." no; Only the positive solution makes sense.
- **?** "Can the related equation be factored using only integers? Explain." yes; Dividing the equation by 0.2 results in the equation $x^2 + 9x - 190 = 0$, which can be factored as $(x + 19)(x - 10) = 0$.

Extra Example 3

a. Determine the number of real solutions of $2x^2 - 3x - 1 = 0$.

The discriminant is greater than 0, so the equation has two real solutions.

b. Determine the number of real solutions of $x^2 + 4 = x$.

The discriminant is less than 0, so the equation has no real solutions.

MONITORING PROGRESS ANSWERS

7. one real solution
8. no real solutions
9. two real solutions

Interpreting the Discriminant

The expression $b^2 - 4ac$ in the Quadratic Formula is called the **discriminant**.

$$x = \frac{-b \pm \sqrt{b^2 - 4ac}}{2a} \leftarrow \text{discriminant}$$

Because the discriminant is under the radical symbol, you can use the value of the discriminant to determine the number of real solutions of a quadratic equation and the number of x-intercepts of the graph of the related function.

Core Concept

Interpreting the Discriminant

$b^2 - 4ac > 0$	$b^2 - 4ac = 0$	$b^2 - 4ac < 0$
		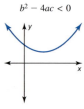
• two real solutions	• one real solution	• no real solutions
• two x-intercepts	• one x-intercept	• no x-intercepts

EXAMPLE 3 Determining the Number of Real Solutions

a. Determine the number of real solutions of $x^2 + 8x - 3 = 0$.

$b^2 - 4ac = 8^2 - 4(1)(-3)$ Substitute 1 for *a*, 8 for *b*, and -3 for *c*.

$= 64 + 12$ Simplify.

$= 76$ Add.

▶ The discriminant is greater than 0. So, the equation has two real solutions.

b. Determine the number of real solutions of $9x^2 + 1 = 6x$.

Write the equation in standard form: $9x^2 - 6x + 1 = 0$.

$b^2 - 4ac = (-6)^2 - 4(9)(1)$ Substitute 9 for *a*, -6 for *b*, and 1 for *c*.

$= 36 - 36$ Simplify.

$= 0$ Subtract.

▶ The discriminant is 0. So, the equation has one real solution.

Monitoring Progress Help in English and Spanish at *BigIdeasMath.com*

Determine the number of real solutions of the equation.

7. $-x^2 + 4x - 4 = 0$

8. $6x^2 + 2x = -1$

9. $\frac{1}{2}x^2 = 7x - 1$

Laurie's Notes — Teacher Actions

- **Connection:** Write the *Core Concept*. Discuss connections to previous lessons in the chapter and ways in which quadratics have been solved. Ask a volunteer to discuss the ways to recognize each number of solutions when solving by graphing, factoring, and the Quadratic Formula.
- ❓ After writing the equation in part (b) ask, "What are the values of *a*, *b*, and *c*?" Check that students subtract the 6x term first! $a = 9, b = -6, c = 1$
- **Think-Pair-Share:** Have students answer Questions 7–9, and then share and discuss as a class.

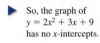 **Finding the Number of *x*-Intercepts of a Parabola**

Find the number of *x*-intercepts of the graph of $y = 2x^2 + 3x + 9$.

SOLUTION

Determine the number of real solutions of $0 = 2x^2 + 3x + 9$.

$b^2 - 4ac = 3^2 - 4(2)(9)$ Substitute 2 for *a*, 3 for *b*, and 9 for *c*.

$ = 9 - 72$ Simplify.

$ = -63$ Subtract.

Because the discriminant is less than 0, the equation has no real solutions.

▶ So, the graph of $y = 2x^2 + 3x + 9$ has no *x*-intercepts.

Check
Use a graphing calculator to check your answer. Notice that the graph of $y = 2x^2 + 3x + 9$ has no *x*-intercepts.

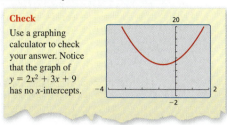

Monitoring Progress Help in English and Spanish at *BigIdeasMath.com*

Find the number of *x*-intercepts of the graph of the function.

10. $y = -x^2 + x - 6$ **11.** $y = x^2 - x$ **12.** $f(x) = x^2 + 12x + 36$

Choosing an Efficient Method

The table shows five methods for solving quadratic equations. For a given equation, it may be more efficient to use one method instead of another. Some advantages and disadvantages of each method are shown.

Methods for Solving Quadratic Equations

Method	Advantages	Disadvantages
Factoring (Lessons 2.5–2.8)	• Straightforward when the equation can be factored easily	• Some equations are not factorable.
Graphing (Lesson 4.2)	• Can easily see the number of solutions • Use when approximate solutions are sufficient. • Can use a graphing calculator	• May not give exact solutions
Using Square Roots (Lesson 4.3)	• Use to solve equations of the form $x^2 = d$.	• Can only be used for certain equations
Completing the Square (Lesson 4.4)	• Best used when $a = 1$ and *b* is even	• May involve difficult calculations
Quadratic Formula (Lesson 4.5)	• Can be used for any quadratic equation • Gives exact solutions	• Takes time to do calculations

---

Extra Example 4
Find the number of *x*-intercepts of the graph of $y = 2x^2 - 4x - 3$.
The graph of $y = 2x^2 - 4x - 3$ has two *x*-intercepts.

MONITORING PROGRESS ANSWERS

10. no *x*-intercepts

11. two *x*-intercepts

12. one *x*-intercept

Laurie's Notes — Teacher Actions

- **Connection:** In Example 3, students are finding the number of real solutions of an equation, and in Example 4, they are finding the number of *x*-intercepts of a graph. Discuss how the Quadratic Formula is used to solve both problems and how the problems are related.
- **Paired Verbal Fluency:** Have students pair up and follow the protocol described on page T-224. Ask students to describe the various methods they have learned to solve quadratic equations, noting the advantages and disadvantages of each method. Listen to students' dialogue.

Extra Example 5

Solve the equation using any method. Explain your choice of method.

a. $x^2 - 4x - 7 = 0$
$x = 2 + \sqrt{11} \approx 5.3$ and
$x = 2 - \sqrt{11} \approx -1.3$; The coefficient of the x^2-term is 1, and the coefficient of the x-term is an even number. So, solve by completing the square.

b. $(x - 2)^2 = 64$
$x = 10$ and $x = -6$; The equation has the form $x^2 = d$. So, solve by using square roots.

c. $10x^2 + 7x - 12 = 0$
$x = \dfrac{-7 + 23}{20} = \dfrac{4}{5}$ and
$x = \dfrac{-7 - 23}{20} = -\dfrac{3}{2}$; The equation is not easily factorable, and the numbers are somewhat large. So, solve using the Quadratic Formula.

MONITORING PROGRESS ANSWERS

13. $x = -12, x = 1$; *Sample answer:* The equation is easily factorable, so solve by factoring.

14. $x = 1, x = -1$; *Sample answer:* The equation can be written in the form $x^2 = d$, so solve using square roots.

15. $x \approx 0.56, x \approx -0.36$; *Sample answer:* The equation is not factorable and the coefficient of the x^2-term is not 1, so solve using the Quadratic Formula.

16. no real solutions; *Sample answer:* The coefficient of the x^2-term is 1 and b is even, so solve by completing the square.

EXAMPLE 5 Choosing a Method

Solve the equation using any method. Explain your choice of method.

a. $x^2 - 10x = 1$ b. $2x^2 - 13x - 24 = 0$ c. $x^2 + 8x + 12 = 0$

SOLUTION

a. The coefficient of the x^2-term is 1, and the coefficient of the x-term is an even number. So, solve by completing the square.

$x^2 - 10x = 1$	Write the equation.
$x^2 - 10x + 25 = 1 + 25$	Complete the square for $x^2 - 10x$.
$(x - 5)^2 = 26$	Write the left side as the square of a binomial.
$x - 5 = \pm\sqrt{26}$	Take the square root of each side.
$x = 5 \pm \sqrt{26}$	Add 5 to each side.

▶ So, the solutions are $x = 5 + \sqrt{26} \approx 10.1$ and $x = 5 - \sqrt{26} \approx -0.1$.

b. The equation is not easily factorable, and the numbers are somewhat large. So, solve using the Quadratic Formula.

$x = \dfrac{-b \pm \sqrt{b^2 - 4ac}}{2a}$	Quadratic Formula
$= \dfrac{-(-13) \pm \sqrt{(-13)^2 - 4(2)(-24)}}{2(2)}$	Substitute 2 for a, -13 for b, and -24 for c.
$= \dfrac{13 \pm \sqrt{361}}{4}$	Simplify.
$= \dfrac{13 \pm 19}{4}$	Evaluate the square root.

▶ So, the solutions are $x = \dfrac{13 + 19}{4} = 8$ and $x = \dfrac{13 - 19}{4} = -\dfrac{3}{2}$.

c. The equation is easily factorable. So, solve by factoring.

$x^2 + 8x + 12 = 0$	Write the equation.
$(x + 2)(x + 6) = 0$	Factor the polynomial.
$x + 2 = 0$ or $x + 6 = 0$	Zero-Product Property
$x = -2$ or $x = -6$	Solve for x.

▶ The solutions are $x = -2$ and $x = -6$.

Check

Graph the related function $f(x) = x^2 + 8x + 12$ and find the zeros. The zeros are -6 and -2.

Monitoring Progress

Solve the equation using any method. Explain your choice of method.

13. $x^2 + 11x - 12 = 0$
14. $9x^2 - 5 = 4$
15. $5x^2 - x - 1 = 0$
16. $x^2 = 2x - 5$

Laurie's Notes — Teacher Actions

- There is more than one method that could be used to solve the three quadratic equations. As students examine the equations, they should be considering why one method would be a wise choice versus another method.
- **Construct Viable Arguments and Critique the Reasoning of Others:** Have students critique each of the choices made by classmates. Allow personal preference. Listen for comments that are thoughtful and reasonable.

Closure

- Write a quadratic equation that you would *not* solve using (a) square roots, (b) factoring, and (c) completing the square. Check students' work. *Sample answers:* (a) $x^2 + 12x + 8 = 0$; (b) $x^2 = 211$; (c) $x^2 - 7x + 12 = 0$

4.5 Exercises

Dynamic Solutions available at *BigIdeasMath.com*

Vocabulary and Core Concept Check

1. **VOCABULARY** What formula can you use to solve any quadratic equation? Write the formula.
2. **VOCABULARY** In the Quadratic Formula, what is the discriminant? What does the value of the discriminant determine?

Monitoring Progress and Modeling with Mathematics

In Exercises 3–8, write the equation in standard form. Then identify the values of *a*, *b*, and *c* that you would use to solve the equation using the Quadratic Formula.

3. $x^2 = 7x$
4. $x^2 - 4x = -12$
5. $-2x^2 + 1 = 5x$
6. $3x + 2 = 4x^2$
7. $4 - 3x = -x^2 + 3x$
8. $-8x - 1 = 3x^2 + 2$

In Exercises 9–22, solve the equation using the Quadratic Formula. Round your solutions to the nearest tenth, if necessary. *(See Example 1.)*

9. $x^2 - 12x + 36 = 0$
10. $x^2 + 7x + 16 = 0$
11. $x^2 - 10x - 11 = 0$
12. $2x^2 - x - 1 = 0$
13. $2x^2 - 6x + 5 = 0$
14. $9x^2 - 6x + 1 = 0$
15. $6x^2 - 13x = -6$
16. $-3x^2 + 6x = 4$
17. $1 - 8x = -16x^2$
18. $x^2 - 5x + 3 = 0$
19. $x^2 + 2x = 9$
20. $5x^2 - 2 = 4x$
21. $2x^2 + 9x + 7 = 3$
22. $8x^2 + 8 = 6 - 9x$

23. **MODELING WITH MATHEMATICS** A dolphin jumps out of the water, as shown in the diagram. The function $h = -16t^2 + 26t$ models the height *h* (in feet) of the dolphin after *t* seconds. After how many seconds is the dolphin at a height of 5 feet? *(See Example 2.)*

24. **MODELING WITH MATHEMATICS** The amount of trout *y* (in tons) caught in a lake from 1995 to 2014 can be modeled by the equation $y = -0.08x^2 + 1.6x + 10$, where *x* is the number of years since 1995.

   a. When were about 15 tons of trout caught in the lake?
   b. Do you think this model can be used to determine the amounts of trout caught in future years? Explain your reasoning.

In Exercises 25–30, determine the number of real solutions of the equation. *(See Example 3.)*

25. $x^2 - 6x + 10 = 0$
26. $x^2 - 5x - 3 = 0$
27. $2x^2 - 12x = -18$
28. $4x^2 = 4x - 1$
29. $-\frac{1}{4}x^2 + 4x = -2$
30. $-5x^2 + 8x = 9$

In Exercises 31–36, find the number of *x*-intercepts of the graph of the function. *(See Example 4.)*

31. $y = x^2 + 5x - 1$
32. $y = 4x^2 + 4x + 1$
33. $y = -6x^2 + 3x - 4$
34. $y = -x^2 + 5x + 13$
35. $f(x) = 4x^2 + 3x - 6$
36. $f(x) = 2x^2 + 8x + 8$

In Exercises 37–44, solve the equation using any method. Explain your choice of method. *(See Example 5.)*

37. $-10x^2 + 13x = 4$
38. $x^2 - 3x - 40 = 0$
39. $x^2 + 6x = 5$
40. $-5x^2 = -25$
41. $x^2 + x - 12 = 0$
42. $x^2 - 4x + 1 = 0$
43. $4x^2 - x = 17$
44. $x^2 + 6x + 9 = 16$

Section 4.5 Solving Quadratic Equations Using the Quadratic Formula 231

Assignment Guide and Homework Check

ASSIGNMENT

Basic: 1, 2, 3–49 odd, 71, 74, 83–89
Average: 1, 2–48 even, 49, 50–62 even, 72, 74, 83–89
Advanced: 1, 2, 8, 20–30 even, 34–48 even, 49, 50–76 even, 77–89

HOMEWORK CHECK

Basic: 9, 23, 25, 31, 37
Average: 10, 24, 26, 32, 38
Advanced: 20, 24, 28, 34, 38

ANSWERS

1. the Quadratic Formula; $x = \dfrac{-b \pm \sqrt{b^2 - 4ac}}{2a}$
2. $b^2 - 4ac$; the number of real solutions of a quadratic equation
3. $x^2 - 7x = 0$; $a = 1, b = -7, c = 0$ or $-x^2 + 7x = 0$; $a = -1, b = 7, c = 0$
4. $x^2 - 4x + 12 = 0$; $a = 1, b = -4, c = 12$ or $-x^2 + 4x - 12 = 0$; $a = -1, b = 4, c = -12$
5. $-2x^2 - 5x + 1 = 0$; $a = -2, b = -5, c = 1$ or $2x^2 + 5x - 1 = 0$; $a = 2, b = 5, c = -1$
6. $-4x^2 + 3x + 2 = 0$; $a = -4, b = 3, c = 2$ or $4x^2 - 3x - 2 = 0$; $a = 4, b = -3, c = -2$
7. $x^2 - 6x + 4 = 0$; $a = 1, b = -6, c = 4$ or $-x^2 + 6x - 4 = 0$; $a = -1, b = 6, c = -4$
8. $-3x^2 - 8x - 3 = 0$; $a = -3, b = -8, c = -3$ or $3x^2 + 8x + 3 = 0$; $a = 3, b = 8, c = 3$
9. $x = 6$
10. no real solutions
11. $x = 11, x = -1$
12. $x = 1, x = -\frac{1}{2}$
13. no real solutions
14. $x = \frac{1}{3}$
15. $x = \frac{3}{2}, x = \frac{2}{3}$
16. no real solutions
17. $x = \frac{1}{4}$
18. $x \approx 4.3, x \approx 0.7$
19. $x \approx 2.2, x \approx -4.2$
20. $x \approx 1.1, x \approx -0.3$
21. $x = -\frac{1}{2}, x = -4$
22. $x \approx -0.3, x \approx -0.8$
23. about 0.2 sec, about 1.4 sec

24. a. 1999, 2011
    b. no; According to the model, the amount of trout caught after 2020 is negative.
25. no real solutions
26. two real solutions
27. one real solution
28. one real solution
29. two real solutions
30. no real solutions
31. two *x*-intercepts
32. one *x*-intercept
33. no *x*-intercepts
34. two *x*-intercepts
35. two *x*-intercepts
36. one *x*-intercept
37. $x = \frac{1}{2}, x = \frac{4}{5}$; *Sample answer:* The equation is not easily factorable and $a \ne 1$, so solve using the Quadratic Formula.
38. $x = -5, x = 8$; *Sample answer:* The equation is easily factorable, so solve by factoring.
39–44. See Additional Answers.

Section 4.5 231

Dynamic Teaching Tools

Dynamic Assessment & Progress Monitoring Tool

Interactive Whiteboard Lesson Library

Dynamic Classroom with Dynamic Investigations

ANSWERS

45. $-b$ should be $-(-7)$, not -7;

$x = \dfrac{-(-7) \pm \sqrt{(-7)^2 - 4(3)(-6)}}{2(3)}$;

$x = 3$ and $x = -\dfrac{2}{3}$

46. The equation needs to be in the form $ax^2 + bx + c = 0$, so $c = -4$ not 4;

$x = \dfrac{-9 \pm \sqrt{9^2 - 4(-2)(-4)}}{2(-2)}$;

$x = \dfrac{1}{2}$ and $x = 4$

47. yes; about 42 ft, about 158 ft

48. a. no
    b. yes; 2 days

49. no; The discriminant is -47, so the equation has no real solutions.

50. no; Substituting 4 for y in the model results in no real solutions.

51. 5; length: 13 m, width: 7 m

52. 4; length: 19 ft, width: 11 ft

53. a–c. $x = -2$

    Sample answer: factoring; The equation is easily factorable.

54. a–c. $x = -\dfrac{2}{3}, x = -3$

    Sample answer: Quadratic Formula; The equation is not easily factorable and one x-intercept is not an integer.

55. 2; When a and c have different signs, ac is negative, so the discriminant is positive.

56. rational; When the discriminant is a perfect square, the square root of the discriminant is an integer.

57. a. Sample answer: $\dfrac{1}{2}$
    b. 1
    c. Sample answer: 2

58. a. Sample answer: 1
    b. 16
    c. Sample answer: 17

59. a. Sample answer: 1
    b. 9
    c. Sample answer: 10

---

45. ERROR ANALYSIS Describe and correct the error in solving the equation $3x^2 - 7x - 6 = 0$ using the Quadratic Formula.

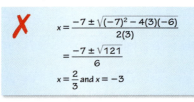

46. ERROR ANALYSIS Describe and correct the error in solving the equation $-2x^2 + 9x = 4$ using the Quadratic Formula.

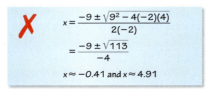

47. MODELING WITH MATHEMATICS A fountain shoots a water arc that can be modeled by the graph of the equation $y = -0.006x^2 + 1.2x + 10$, where x is the horizontal distance (in feet) from the river's north shore and y is the height (in feet) above the river. Does the water arc reach a height of 50 feet? If so, about how far from the north shore is the water arc 50 feet above the water?

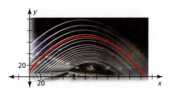

48. MODELING WITH MATHEMATICS Between the months of April and September, the number y of hours of daylight per day in Seattle, Washington, can be modeled by $y = -0.00046x^2 + 0.076x + 13$, where x is the number of days since April 1.

a. Do any of the days between April and September in Seattle have 17 hours of daylight? If so, how many?

b. Do any of the days between April and September in Seattle have 14 hours of daylight? If so, how many?

49. MAKING AN ARGUMENT Your friend uses the discriminant of the equation $2x^2 - 5x - 2 = -11$ and determines that the equation has two real solutions. Is your friend correct? Explain your reasoning.

232   Chapter 4   Solving Quadratic Equations

---

50. MODELING WITH MATHEMATICS The frame of the tent shown is defined by a rectangular base and two parabolic arches that connect the opposite corners of the base. The graph of $y = -0.18x^2 + 1.6x$ models the height y (in feet) of one of the arches x feet along the diagonal of the base. Can a child who is 4 feet tall walk under one of the arches without having to bend over? Explain.

MATHEMATICAL CONNECTIONS In Exercises 51 and 52, use the given area A of the rectangle to find the value of x. Then give the dimensions of the rectangle.

51. $A = 91$ m^2

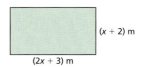

52. $A = 209$ ft^2

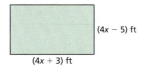

COMPARING METHODS In Exercises 53 and 54, solve the equation by (a) graphing, (b) factoring, and (c) using the Quadratic Formula. Which method do you prefer? Explain your reasoning.

53. $x^2 + 4x + 4 = 0$ 54. $3x^2 + 11x + 6 = 0$

55. REASONING How many solutions does the equation $ax^2 + bx + c = 0$ have when a and c have different signs? Explain your reasoning.

56. REASONING When the discriminant is a perfect square, are the solutions of $ax^2 + bx + c = 0$ rational or irrational? (Assume a, b, and c are integers.) Explain your reasoning.

REASONING In Exercises 57–59, give a value of c for which the equation has (a) two solutions, (b) one solution, and (c) no solutions.

57. $x^2 - 2x + c = 0$

58. $x^2 - 8x + c = 0$

59. $4x^2 + 12x + c = 0$

60. REPEATED REASONING You use the Quadratic Formula to solve an equation.

   a. You obtain solutions that are integers. Could you have used factoring to solve the equation? Explain your reasoning.

   b. You obtain solutions that are fractions. Could you have used factoring to solve the equation? Explain your reasoning.

   c. Make a generalization about quadratic equations with rational solutions.

61. MODELING WITH MATHEMATICS The fuel economy y (in miles per gallon) of a car can be modeled by the equation $y = -0.013x^2 + 1.25x + 5.6$, where $5 \leq x \leq 75$ and x is the speed (in miles per hour) of the car. Find the speed(s) at which you can travel and have a fuel economy of 32 miles per gallon.

62. MODELING WITH MATHEMATICS The depth d (in feet) of a river can be modeled by the equation $d = -0.25t^2 + 1.7t + 3.5$, where $0 \leq t \leq 7$ and t is the time (in hours) after a heavy rain begins. When is the river 6 feet deep?

ANALYZING EQUATIONS In Exercises 63–68, tell whether the vertex of the graph of the function lies above, below, or on the x-axis. Explain your reasoning without using a graph.

63. $y = x^2 - 3x + 2$ **64.** $y = 3x^2 - 6x + 3$

65. $y = 6x^2 - 2x + 4$ **66.** $y = -15x^2 + 10x - 25$

67. $f(x) = -3x^2 - 4x + 8$

68. $f(x) = 9x^2 - 24x + 16$

69. REASONING NASA creates a weightless environment by flying a plane in a series of parabolic paths. The height h (in feet) of a plane after t seconds in a parabolic flight path can be modeled by $h = -11t^2 + 700t + 21{,}000$. The passengers experience a weightless environment when the height of the plane is greater than or equal to 30,800 feet. For approximately how many seconds do passengers experience weightlessness on such a flight? Explain.

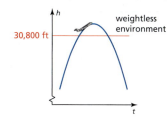

70. WRITING EQUATIONS Use the numbers to create a quadratic equation with the solutions $x = -1$ and $x = -\frac{1}{4}$.

___x^2 + ___x + ___ = 0

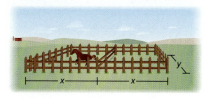

71. PROBLEM SOLVING A rancher constructs two rectangular horse pastures that share a side, as shown. The pastures are enclosed by 1050 feet of fencing. Each pasture has an area of 15,000 square feet.

   a. Show that $y = 350 - \frac{4}{3}x$.

   b. Find the possible lengths and widths of each pasture.

72. PROBLEM SOLVING A kicker punts a football from a height of 2.5 feet above the ground with an initial vertical velocity of 45 feet per second.

   a. Write an equation that models this situation using the function $h = -16t^2 + v_0t + s_0$, where h is the height (in feet) of the football, t is the time (in seconds) after the football is punted, v_0 is the initial vertical velocity (in feet per second), and s_0 is the initial height (in feet).

   b. The football is caught 5.5 feet above the ground, as shown in the diagram. Find the amount of time that the football is in the air.

73. CRITICAL THINKING The solutions of the quadratic equation $ax^2 + bx + c = 0$ are $x = \dfrac{-b + \sqrt{b^2 - 4ac}}{2a}$ and $x = \dfrac{-b - \sqrt{b^2 - 4ac}}{2a}$. Find the mean of the solutions. How is the mean of the solutions related to the graph of $y = ax^2 + bx + c$? Explain.

ANSWERS

74. a. C; The graph has two x-intercepts.
    b. A; The graph has one x-intercept.
    c. B; The graph has no x-intercepts.
75. about 24.7 ft/sec
76. $x = -\dfrac{b}{2a}$ is the axis of symmetry, and $\dfrac{\sqrt{b^2 - 4ac}}{2a}$ is the horizontal distance from the axis of symmetry to each x-intercept.
77. $\dfrac{-b}{a}$; $\dfrac{c}{a}$; Sample answer: $2x^2 - 4x + 1 = 0$
78. $x = \dfrac{-1 \pm \sqrt{1 - 4ac}}{2a}$; $x \approx -1.77$, $x \approx 2.27$
79. See Additional Answers.
80. a. $k < \dfrac{1}{24}$
    b. $k = \dfrac{1}{24}$
    c. $k > \dfrac{1}{24}$
81. a. $k < -3$ or $k > 3$
    b. $k = -3$ or $k = 3$
    c. $-3 < k < 3$
82–89. See Additional Answers.

Mini-Assessment

1. Solve $2x^2 - 5x + 3 = 0$ using the Quadratic Formula.
   $x = \dfrac{3}{2}$ and $x = 1$
2. Determine the number of real solutions of $-x^2 - 25 = -10x$.
   one real solution
3. Find the number of x-intercepts of the graph of $y = x^2 - 7x + 14$.
   no x-intercepts

Solve the equation using any method. Explain your choice of method.

4. $x^2 - 6x = 4$
   $x = 3 + \sqrt{13} \approx 6.6$ and $x = 3 - \sqrt{13} \approx -0.6$; The coefficient of the x^2-term is 1, and the coefficient of the x-term is an even number. So, solve by completing the square.
5. $x^2 - 2x = 15$
   $x = -3$ and $x = 5$; The equation is easily factorable to $(x - 5)(x + 3) = 0$. So, solve by factoring.

---

74. HOW DO YOU SEE IT? Match each graph with its discriminant. Explain your reasoning.

A.

B.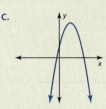

C.

a. $b^2 - 4ac > 0$
b. $b^2 - 4ac = 0$
c. $b^2 - 4ac < 0$

75. CRITICAL THINKING You are trying to hang a tire swing. To get the rope over a tree branch that is 15 feet high, you tie the rope to a weight and throw it over the branch. You release the weight at a height s_0 of 5.5 feet. What is the minimum initial vertical velocity v_0 needed to reach the branch? (*Hint*: Use the equation $h = -16t^2 + v_0 t + s_0$.)

Maintaining Mathematical Proficiency *Reviewing what you learned in previous grades and lessons*

Find the sum or difference. *(Section 2.1)*

83. $(x^2 + 2) + (2x^2 - x)$
84. $(x^3 + x^2 - 4) + (3x^2 + 10)$
85. $(-2x + 1) - (-3x^2 + x)$
86. $(-3x^3 + x^2 - 12x) - (-6x^2 + 3x - 9)$

Find the product. *(Section 2.2 and Section 2.3)*

87. $(x + 2)(x - 2)$
88. $2x(3 - x + 5x^2)$
89. $(7 - x)(x - 1)$

---

76. THOUGHT PROVOKING Consider the graph of the standard form of a quadratic function $y = ax^2 + bx + c$. Then consider the Quadratic Formula as given by

$$x = -\dfrac{b}{2a} \pm \dfrac{\sqrt{b^2 - 4ac}}{2a}.$$

Write a graphical interpretation of the two parts of this formula.

77. ANALYZING RELATIONSHIPS Find the sum and product of $\dfrac{-b + \sqrt{b^2 - 4ac}}{2a}$ and $\dfrac{-b - \sqrt{b^2 - 4ac}}{2a}$. Then write a quadratic equation whose solutions have a sum of 2 and a product of $\dfrac{1}{2}$.

78. WRITING A FORMULA Derive a formula that can be used to find solutions of equations that have the form $ax^2 + x + c = 0$. Use your formula to solve $-2x^2 + x + 8 = 0$.

79. MULTIPLE REPRESENTATIONS If p is a solution of a quadratic equation $ax^2 + bx + c = 0$, then $(x - p)$ is a factor of $ax^2 + bx + c$.

a. Copy and complete the table for each pair of solutions.

Solutions	Factors	Quadratic equation
3, 4	$(x - 3)$, $(x - 4)$	$x^2 - 7x + 12 = 0$
$-1, 6$		
0, 2		
$-\dfrac{1}{2}, 5$		

b. Graph the related function for each equation. Identify the zeros of the function.

CRITICAL THINKING In Exercises 80–82, find all values of k for which the equation has (a) two solutions, (b) one solution, and (c) no solutions.

80. $2x^2 + x + 3k = 0$
81. $x^2 - 4kx + 36 = 0$
82. $kx^2 + 5x - 16 = 0$

4.1–4.5 What Did You Learn?

Core Vocabulary

counterexample, *p. 191*
radical expression, *p. 192*
simplest form of a radical, *p. 192*
rationalizing the denominator, *p. 194*

conjugates, *p. 194*
like radicals, *p. 196*
quadratic equation, *p. 202*
completing the square, *p. 216*

Quadratic Formula, *p. 226*
discriminant, *p. 228*

Core Concepts

Section 4.1
Product Property of Square Roots, *p. 192*
Quotient Property of Square Roots, *p. 192*
Rationalizing the Denominator, *p. 194*
Performing Operations with Radicals, *p. 196*

Section 4.2
Solving Quadratic Equations by Graphing, *p. 202*
Number of Solutions of a Quadratic Equation, *p. 203*
Finding Zeros of Functions, *p. 204*

Section 4.3
Solutions of $x^2 = d$, *p. 210*
Approximating Solutions of Quadratic Equations, *p. 211*

Section 4.4
Completing the Square, *p. 216*
Finding and Using Maximum and Minimum Values, *p. 218*

Section 4.5
Quadratic Formula, *p. 226*
Interpreting the Discriminant, *p. 228*

Mathematical Practices

1. For each part of Exercise 100 on page 200 that is *sometimes* true, list all examples and counterexamples from the table that represent the sum or product being described.

2. Describe how solving a simpler equation can help you solve the equation in Exercise 41 on page 214.

Keeping a Positive Attitude

Do you ever feel frustrated or overwhelmed by math? You're not alone. Just take a deep breath and assess the situation. Try to find a productive study environment, review your notes and the examples in the textbook, and ask your teacher or friends for help.

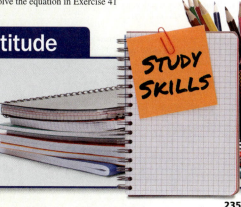

Dynamic Teaching Tools
Dynamic Assessment & Progress Monitoring Tool
Interactive Whiteboard Lesson Library
Dynamic Classroom with Dynamic Investigations

ANSWERS

1. part (c) examples:
   $\sqrt{3} + \sqrt{3} = 2\sqrt{3}$, $\sqrt{3} + \pi = \pi + \sqrt{3}$,
   $-\sqrt{3} + (-\sqrt{3}) = -2\sqrt{3}$,
   $-\sqrt{3} + \pi = \pi - \sqrt{3}$,
   $\pi + \pi = 2\pi$; part (c)
   counterexample: $\sqrt{3} + (-\sqrt{3}) = 0$,
   part (f) examples: $\sqrt{3} \cdot \pi = \pi\sqrt{3}$,
   $-\sqrt{3} \cdot \pi = -\pi\sqrt{3}$, $\pi \cdot \pi = \pi^2$;
   part (f) counterexamples:
   $\sqrt{3} \cdot \sqrt{3} = 3$, $\sqrt{3} \cdot (-\sqrt{3}) = -3$,
   $-\sqrt{3} \cdot (-\sqrt{3}) = 3$

2. Example 6 on page 205 or Example 5 on page 204

3. Solving the simpler equation $x^2 = 144$ helps because the solution can be found by moving the decimal point in the solution to the simpler equation.

ANSWERS

1. $4x\sqrt{7x}$
2. $\dfrac{\sqrt{2}}{3}$
3. $-5\sqrt[3]{5}$
4. $\dfrac{4\sqrt{11}}{11}$
5. $\dfrac{3x\sqrt[3]{2x}}{7y^2}$
6. $\dfrac{15 - 3\sqrt{3}}{11}$
7. $-4\sqrt{5} + 7\sqrt{10}$
8. $30\sqrt{2}$
9. $x = -1, x = 3$
10. no real solutions
11. $x = -5$
12. $x = 4, x = -4$
13. no real solutions
14. $x = 9, x = 7$
15. $x = 2, x = 4$
16. $x = -6 \pm 4\sqrt{2}$
17. no real solutions
18. $-7, -2$
19. $-4.2, 0.2$
20. $-2.8, 1.3$
21. *Sample answer:* factoring; The equation is easily factorable.
22. length: about 17.4 m, width: about 4.4 m
23. a. 0.5 sec, 1 sec
    b. about 1.65 sec

4.1–4.5 Quiz

Simplify the expression. *(Section 4.1)*

1. $\sqrt{112x^3}$
2. $\sqrt{\dfrac{18}{81}}$
3. $\sqrt[3]{-625}$
4. $\dfrac{4}{\sqrt{11}}$
5. $\sqrt[3]{\dfrac{54x^4}{343y^6}}$
6. $\dfrac{6}{5 + \sqrt{3}}$
7. $2\sqrt{5} + 7\sqrt{10} - 3\sqrt{20}$
8. $\sqrt{6}(7\sqrt{12} - 4\sqrt{3})$

Use the graph to solve the equation. *(Section 4.2)*

9. $x^2 - 2x - 3 = 0$

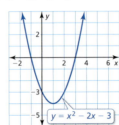

10. $x^2 - 2x + 3 = 0$

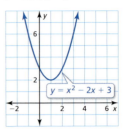

11. $x^2 + 10x + 25 = 0$

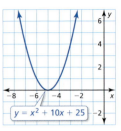

Solve the equation using square roots. *(Section 4.3)*

12. $4x^2 = 64$
13. $-3x^2 + 6 = 10$
14. $(x - 8)^2 = 1$

Solve the equation by completing the square. Round your solutions to the nearest hundredth, if necessary. *(Section 4.4)*

15. $x^2 - 6x + 8 = 0$
16. $x^2 + 12x + 4 = 0$
17. $4x(x + 6) = -40$

Solve the equation using the Quadratic Formula. Round your solutions to the nearest tenth, if necessary. *(Section 4.5)*

18. $x^2 + 9x + 14 = 0$
19. $x^2 + 4x = 1$
20. $-2x^2 + 7 = 3x$

21. Which method would you use to solve $2x^2 + 3x + 1 = 0$? Explain your reasoning. *(Section 4.5)*

22. The length of a rectangular prism is four times its width. The volume of the prism is 380 cubic meters. Find the length and width of the prism. *(Section 4.3)*

23. You cast a fishing lure into the water from a height of 4 feet above the water. The height h (in feet) of the fishing lure after t seconds can be modeled by the equation $h = -16t^2 + 24t + 4$. *(Section 4.5)*

    a. After how many seconds does the fishing lure reach a height of 12 feet?
    b. After how many seconds does the fishing lure hit the water?

Laurie's Notes

Overview of Section 4.6

Introduction
- This lesson introduces the *imaginary unit i* and operations with *imaginary numbers* and *complex numbers*.
- **FYI:** You might want to discuss the different ways to write complex numbers. For instance, in most precalculus texts, solutions $3i\sqrt{2}$ and $-3i\sqrt{2}$ would be written as $3\sqrt{2}i$ and $-3\sqrt{2}i$. Some students may be taking precalculus after Math III, so this might come up. There are some authors who prefer to write complex numbers as $a + ib$ (the i before the b). Also, in electrical engineering (and some other disciplines), j is used for the imaginary unit because i is used for electrical current.

Formative Assessment Tips
- **No-Hands Questioning:** Typically, when you ask a question, there are hands that immediately go up—often the same hands each time. Some students need a longer time to process a question and think through their response. This technique instructs students not to put their hands in the air when the question is posed. *Wait Time** is exercised.
- This technique encourages all students to be active and engaged in the lesson. Students who need additional think time are provided that opportunity. Teachers can then use *Popsicle Sticks* to call on students for a response, or they can purposely call on those students whose voices they do not hear enough.
- To be effective, the question(s) posed during *No-Hands Questioning* have to require more than a simple response.

*See Section 4.8 for a description of *Wait Time*.

Pacing Suggestion
- Complete the two explorations and then start the formal lesson by sharing information about the *imaginary unit i* and stating the *Core Concept*.

Dynamic Teaching Tools
Dynamic Assessment & Progress Monitoring Tool
Lesson Planning Tool
Interactive Whiteboard Lesson Library
Dynamic Classroom with Dynamic Investigations

What Your Students Will Learn

- Define and use the imaginary unit i and write complex numbers in standard form $a + bi$.
- Add and subtract complex numbers.
- Multiply complex numbers and complex conjugates.

Laurie's Notes

Exploration

Motivate
- Share a bit of the history of imaginary numbers with students.
- Complex numbers arose from the need to solve cubic equations, and not, as it is commonly believed, quadratic equations. The first to use imaginary numbers in the solution of a cubic equation was Italian mathematician Scipione del Ferro (1465–1526). The solution was published by Girolamo Cardano (1501–1576) in *Ars Magna* (1545).
- The term *imaginary* was used by René Descartes (1596–1650) in *La Géométrie* (1637) in describing numbers that many thought were useless.
- Leonhard Euler (1707–1783) introduced the notation $i = \sqrt{-1}$ and visualized complex numbers as points with rectangular coordinates.

Discuss
- This is students' first introduction to imaginary numbers.
- The flowchart identifies the sets of numbers. Be sure that students understand that in interpreting the flowchart, when a number is a member of one set, it is a member of all sets that connect to it from above.
- A Venn diagram is often used to represent the relationships between the different sets of numbers.

Exploration 1
- Check that students understand the directions. They are asked to determine to which *subsets* each number belongs. They are not naming the most specific set.
- Students may say that they do not know what $i = \sqrt{-1}$ means. Assure students that the definition of *imaginary numbers* will be made clear in the formal lesson.

Exploration 2
- Students will multiply complex numbers in this lesson. This exploration is very short, yet very important.
- Do not rush in and offer answers to students. Instead, reassure students that they have indeed solved problems of this type before. Trust that at least one student will think back to Chapter 1 when they worked with rational exponents and nth roots.

Communicate Your Answer
- Each of the three questions can develop into a rich discussion. Students should recognize the importance of knowing definitions when trying to justify answers.
- Question 5 may cause some cognitive dissonance for students. Have students refer back to Section 1.5. If i is defined to be $\sqrt{-1}$, then $i^2 = \left(\sqrt[2]{-1}\right)^2 = (-1)^{2/2} = -1$.

Connecting to Next Step
- The explorations help students recall previously learned subsets of numbers and introduce evaluating the power of a complex number.

4.6 Complex Numbers

Essential Question What are the subsets of the set of complex numbers?

In your study of mathematics, you have probably worked with only *real numbers*, which can be represented graphically on the real number line. In this lesson, the system of numbers is expanded to include *imaginary numbers*. The real numbers and imaginary numbers compose the set of *complex numbers*.

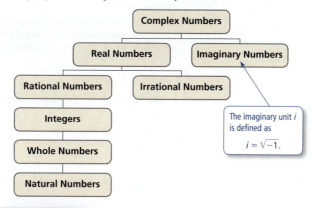

The imaginary unit i is defined as $i = \sqrt{-1}$.

EXPLORATION 1 Classifying Numbers

Work with a partner. Determine which subsets of the set of complex numbers contain each number.

a. $\sqrt{9}$ b. $\sqrt{0}$ c. $-\sqrt{4}$

d. $\sqrt{\dfrac{4}{9}}$ e. $\sqrt{2}$ f. $\sqrt{-1}$

ATTENDING TO PRECISION
To be proficient in math, you need to use clear definitions in your reasoning and discussions with others.

EXPLORATION 2 Simplifying i^2

Work with a partner. Justify each step in the simplification of i^2.

Algebraic Step	Justification
$i^2 = (\sqrt{-1})^2$	
$= -1$	

Communicate Your Answer

3. What are the subsets of the set of complex numbers? Give an example of a number in each subset.

4. Is it possible for a number to be both whole and natural? natural and rational? rational and irrational? real and imaginary? Explain your reasoning.

5. Your friend claims that the conclusion in Exploration 2 is incorrect because $i^2 = i \cdot i = \sqrt{-1} \cdot \sqrt{-1} = \sqrt{-1(-1)} = \sqrt{1} = 1$. Is your friend correct? Explain.

Section 4.6 Complex Numbers 237

Dynamic Teaching Tools

Dynamic Assessment & Progress Monitoring Tool
Lesson Planning Tool
Interactive Whiteboard Lesson Library
Dynamic Classroom with Dynamic Investigations

ANSWERS

1. a. natural numbers, whole numbers, integers, rational numbers, and real numbers
   b. whole numbers, integers, rational numbers, and real numbers
   c. integers, rational numbers, and real numbers
   d. rational numbers and real numbers
   e. irrational numbers and real numbers
   f. imaginary numbers

2. Definition of i; Evaluate the power using inverse operations.

3. natural numbers: *Sample answer:* 4; whole numbers: *Sample answer:* 0; integers: *Sample answer:* -3; rational numbers: *Sample answer:* $\frac{3}{5}$; irrational numbers: *Sample answer:* $\sqrt{5}$; real numbers: *Sample answer:* $\frac{1}{4} + \sqrt{10}$; imaginary numbers: *Sample answer:* $3i$

4. yes; yes; no; no; It is possible for a number to be whole and natural, or natural and rational because one is a subset of the other, but it is not possible for a number to be rational and irrational, or real and imaginary because they are separate categories.

5. no; *Sample answer:* $\sqrt{a}\sqrt{b} = \sqrt{ab}$ only for $a, b \geq 0$

Section 4.6 237

English Language Learners

Vocabulary
Have students draw a two-column, four-row table. In the rows of the first column, have students write *imaginary unit*, *complex number*, *imaginary number*, and *pure imaginary number*. In the second column, have students write an example that corresponds to the words in each row.

Extra Example 1

Find the square root of each number.

a. $\sqrt{-81}$ $9i$
b. $\sqrt{-56}$ $2i\sqrt{14}$
c. $-7\sqrt{-12}$ $-14i\sqrt{3}$

MONITORING PROGRESS ANSWERS

1. $2i$
2. $2i\sqrt{3}$
3. $-6i$
4. $6i\sqrt{6}$

4.6 Lesson

Core Vocabulary
imaginary unit i, p. 238
complex number, p. 238
imaginary number, p. 238
pure imaginary number, p. 238
complex conjugates, p. 241

What You Will Learn

▶ Define and use the imaginary unit i.
▶ Add and subtract complex numbers.
▶ Multiply complex numbers.

The Imaginary Unit i

Not all quadratic equations have real-number solutions. For example, $x^2 = -3$ has no real-number solutions because the square of any real number is never a negative number.

To overcome this problem, mathematicians created an expanded system of numbers using the **imaginary unit i**, defined as $i = \sqrt{-1}$. Note that $i^2 = -1$. The imaginary unit i can be used to write the square root of *any* negative number.

Core Concept

The Square Root of a Negative Number

Property	Example
1. If r is a positive real number, then $\sqrt{-r} = i\sqrt{r}$.	$\sqrt{-3} = i\sqrt{3}$
2. By the first property, it follows that $(i\sqrt{r})^2 = -r$.	$(i\sqrt{3})^2 = i^2 \cdot 3 = -3$

EXAMPLE 1 Finding Square Roots of Negative Numbers

Find the square root of each number.

a. $\sqrt{-25}$ b. $\sqrt{-72}$ c. $-5\sqrt{-9}$

SOLUTION

a. $\sqrt{-25} = \sqrt{25} \cdot \sqrt{-1} = 5i$
b. $\sqrt{-72} = \sqrt{72} \cdot \sqrt{-1} = \sqrt{36} \cdot \sqrt{2} \cdot i = 6\sqrt{2}\,i = 6i\sqrt{2}$
c. $-5\sqrt{-9} = -5\sqrt{9} \cdot \sqrt{-1} = -5 \cdot 3 \cdot i = -15i$

Monitoring Progress Help in English and Spanish at *BigIdeasMath.com*

Find the square root of the number.

1. $\sqrt{-4}$ 2. $\sqrt{-12}$ 3. $-\sqrt{-36}$ 4. $2\sqrt{-54}$

A **complex number** written in *standard form* is a number $a + bi$ where a and b are real numbers. The number a is the *real part*, and the number bi is the *imaginary part*.

$$a + bi$$

If $b \neq 0$, then $a + bi$ is an **imaginary number**. If $a = 0$ and $b \neq 0$, then $a + bi$ is a **pure imaginary number**. The diagram shows how different types of complex numbers are related.

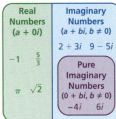

238   Chapter 4   Solving Quadratic Equations

Laurie's Notes — Teacher Actions

- Use the *Motivate* to give an introduction to *imaginary numbers*.
- Use properties of radicals to explain the *Core Concept*.
- **Attend to Precision:** Students should identify the property of radicals they are using as they find the square root of negative numbers in Example 1.
- ❓ Define *complex numbers*. Ask, "Are real numbers complex numbers? Explain." yes; Real numbers are always complex numbers, where $b = 0$. Relate this to integers, which are always rational numbers. Rational numbers can be written in the form $\frac{a}{b}$ and when $b = 1$, we call the number an integer.

Two complex numbers $a + bi$ and $c + di$ are equal if and only if $a = c$ and $b = d$.

EXAMPLE 2 Equality of Two Complex Numbers

Find the values of x and y that satisfy the equation $2x - 7i = 10 + yi$.

SOLUTION

Set the real parts equal to each other and the imaginary parts equal to each other.

| $2x = 10$ | Equate the real parts. | $-7i = yi$ | Equate the imaginary parts. |
| $x = 5$ | Solve for x. | $-7 = y$ | Solve for y. |

▶ So, $x = 5$ and $y = -7$.

Monitoring Progress  Help in English and Spanish at *BigIdeasMath.com*

Find the values of x and y that satisfy the equation.

5. $x + 3i = 9 - yi$
6. $9 + 4yi = -2x + 3i$

Adding and Subtracting Complex Numbers

Core Concept

Sums and Differences of Complex Numbers

To add (or subtract) two complex numbers, add (or subtract) their real parts and their imaginary parts separately.

Sum of complex numbers: $(a + bi) + (c + di) = (a + c) + (b + d)i$
Difference of complex numbers: $(a + bi) - (c + di) = (a - c) + (b - d)i$

EXAMPLE 3 Adding and Subtracting Complex Numbers

Add or subtract. Write the answer in standard form.

a. $(8 - i) + (5 + 4i)$
b. $(7 - 6i) - (3 - 6i)$
c. $13 - (2 + 7i) + 5i$

SOLUTION

a. $(8 - i) + (5 + 4i) = (8 + 5) + (-1 + 4)i$ — Definition of complex addition
    $= 13 + 3i$ — Write in standard form.
b. $(7 - 6i) - (3 - 6i) = (7 - 3) + (-6 + 6)i$ — Definition of complex subtraction
    $= 4 + 0i$ — Simplify.
    $= 4$ — Write in standard form.
c. $13 - (2 + 7i) + 5i = [(13 - 2) - 7i] + 5i$ — Definition of complex subtraction
    $= (11 - 7i) + 5i$ — Simplify.
    $= 11 + (-7 + 5)i$ — Definition of complex addition
    $= 11 - 2i$ — Write in standard form.

Differentiated Instruction

Visual

When students are adding or subtracting complex numbers, encourage them to highlight like terms. Students can highlight the real parts in one color and the imaginary parts in another color.

Extra Example 2

Find the values of x and y that satisfy the equation $5x - 20i = 35 + yi$.
$x = 7$ and $y = -20$

Extra Example 3

Add or subtract. Write the answer in standard form.
a. $(2 + 9i) + (11 - i)$ $13 + 8i$
b. $(13 + 4i) - (8 + 4i)$ 5
c. $16 - (13 + 6i) + 8i$ $3 + 2i$

MONITORING PROGRESS ANSWERS

5. $x = 9$ and $y = -3$
6. $x = -4.5$ and $y = 0.75$

Laurie's Notes — Teacher Actions

- Discuss how to determine whether two complex numbers are equal, and apply this in Example 2. Students often think the problem is too easy and therefore they must be doing something wrong. Assure them that it is indeed easy.
- **No-Hands Questioning:** Explain that you want to add two complex numbers. "What would make sense for $(6 + 3i) + (2 - 5i)$?" Wait an appropriate amount of time and ask for an answer. Add the real parts and add the imaginary parts.
- Work through Example 3.

Extra Example 4

Use the table and information in Example 4. A different series circuit is shown. Find the impedance of the circuit.

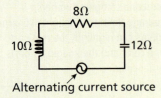

Alternating current source

$(8 - 2i)$ ohms

MONITORING PROGRESS ANSWERS

7. $3 + 6i$
8. $-5 + 9i$
9. $-10 - 10i$
10. $-4 + 9i$
11. $(5 - 4i)$ ohms

EXAMPLE 4 Solving a Real-Life Problem

Electrical circuit components, such as resistors, inductors, and capacitors, all oppose the flow of current. This opposition is called *resistance* for resistors and *reactance* for inductors and capacitors. Each of these quantities is measured in ohms. The symbol used for ohms is Ω, the uppercase Greek letter omega.

Component and symbol	Resistor —\/\/\/—	Inductor —⌇⌇⌇⌇—	Capacitor —⊢⊢—
Resistance or reactance (in ohms)	R	L	C
Impedance (in ohms)	R	Li	$-Ci$

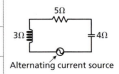

Alternating current source

The table shows the relationship between a component's resistance or reactance and its contribution to impedance. A *series circuit* is also shown with the resistance or reactance of each component labeled. The impedance for a series circuit is the sum of the impedances for the individual components. Find the impedance of the circuit.

SOLUTION

The resistor has a resistance of 5 ohms, so its impedance is 5 ohms. The inductor has a reactance of 3 ohms, so its impedance is $3i$ ohms. The capacitor has a reactance of 4 ohms, so its impedance is $-4i$ ohms.

Impedance of circuit $= 5 + 3i + (-4i) = 5 - i$

▶ The impedance of the circuit is $(5 - i)$ ohms.

Monitoring Progress Help in English and Spanish at *BigIdeasMath.com*

Add or subtract. Write the answer in standard form.

7. $(9 - i) + (-6 + 7i)$
8. $(3 + 7i) - (8 - 2i)$
9. $-4 - (1 + i) - (5 + 9i)$
10. $5 + (-9 + 3i) + 6i$

11. **WHAT IF?** In Example 4, what is the impedance of the circuit when the capacitor is replaced with one having a reactance of 7 ohms?

Multiplying Complex Numbers

Many properties of real numbers are also valid for complex numbers, such as those used below to multiply two complex numbers.

$(a + bi)(c + di) = a(c + di) + bi(c + di)$	Distributive Property
$= ac + (ad)i + (bc)i + (bd)i^2$	Distributive Property
$= ac + (ad)i + (bc)i + (bd)(-1)$	Use $i^2 = -1$.
$= ac - bd + (ad)i + (bc)i$	Commutative Property
$= (ac - bd) + (ad + bc)i$	Associative Property

This result shows a pattern that you could use to multiply two complex numbers. However, it is often more practical to use the Distributive Property or the FOIL Method, just as you do when multiplying real numbers or algebraic expressions.

Laurie's Notes Teacher Actions

- Have students read Example 4. The bottom row of the table represents the impedance. Note the symbols used at the top of each column. The same symbols are used in the circuit figure.
- **Think-Pair-Share:** Have students work independently on Questions 7–11 and then share with partners. Discuss as a class.
- **Look For and Make Use of Structure:** Help students see the connection between multiplying binomials and multiplying complex numbers.

STUDY TIP

When simplifying an expression that involves complex numbers, be sure to simplify i^2 as -1.

EXAMPLE 5 Multiplying Complex Numbers

Multiply. Write the answer in standard form.

a. $4i(-6 + i)$
b. $(9 - 2i)(-4 + 7i)$

SOLUTION

a. $4i(-6 + i) = -24i + 4i^2$ Distributive Property
$= -24i + 4(-1)$ Use $i^2 = -1$.
$= -4 - 24i$ Write in standard form.

b. $(9 - 2i)(-4 + 7i) = -36 + 63i + 8i - 14i^2$ Multiply using FOIL.
$= -36 + 71i - 14(-1)$ Simplify and use $i^2 = -1$.
$= -36 + 71i + 14$ Simplify.
$= -22 + 71i$ Write in standard form.

Monitoring Progress Help in English and Spanish at *BigIdeasMath.com*

Multiply. Write the answer in standard form.

12. $(-3i)(10i)$ **13.** $i(8 - i)$ **14.** $(3 + i)(5 - i)$

Pairs of complex numbers of the forms $a + bi$ and $a - bi$, where $b \neq 0$, are called **complex conjugates**. Consider the product of complex conjugates below.

$(a + bi)(a - bi) = a^2 - (ab)i + (ab)i - b^2i^2$ Multiply using FOIL.
$= a^2 - b^2(-1)$ Simplify and use $i^2 = -1$.
$= a^2 + b^2$ Simplify.

Because a and b are real numbers, $a^2 + b^2$ is a real number. So, the product of complex conjugates is a real number.

LOOKING FOR STRUCTURE

You can use the pattern $(a + bi)(a - bi) = a^2 + b^2$, where $a = 5$ and $b = 2$, to find the product in Example 6.

$(5 + 2i)(5 - 2i) = 5^2 + 2^2$
$= 25 + 4$
$= 29$

EXAMPLE 6 Multiplying Complex Conjugates

Multiply $5 + 2i$ by its complex conjugate.

SOLUTION

The complex conjugate of $5 + 2i$ is $5 - 2i$.

$(5 + 2i)(5 - 2i) = 25 - 10i + 10i - 4i^2$ Multiply using FOIL.
$= 25 - 4(-1)$ Simplify and use $i^2 = -1$.
$= 29$ Simplify.

Monitoring Progress Help in English and Spanish at *BigIdeasMath.com*

Multiply the complex number by its complex conjugate.

15. $1 + i$ **16.** $4 - 7i$ **17.** $-3 - 2i$

Extra Example 5

Multiply. Write the answer in standard form.

a. $15i(-1 + 2i)$ $-30 - 15i$
b. $(4 - 12i)(11 + 8i)$ $140 - 100i$

Extra Example 6

Multiply $3 - 4i$ by its complex conjugate. **25**

MONITORING PROGRESS ANSWERS

12. 30
13. $1 + 8i$
14. $16 + 2i$
15. 2
16. 65
17. 13

Laurie's Notes — Teacher Actions

- **? Turn and Talk:** Pose Example 5 and have students *Turn and Talk:* "Describe how these two multiplication problems can be solved." Circulate and listen for students mentioning the Distributive Property and multiplying two binomials.
- **Think-Pair-Share:** Have students work independently on Questions 12–14 and then share with partners. Discuss as a class.
- **?** "What is the product of $(a + b)(a - b)$?" $a^2 - b^2$
- Define complex conjugates and make the connection to Difference of Two Squares.

Closure

- **Writing Prompt:** To add or subtract complex numbers you… add (or subtract) their real parts and their imaginary parts separately.
- **Writing Prompt:** To multiply complex numbers you… use the Distributive Property or FOIL.

Assignment Guide and Homework Check

ASSIGNMENT

Basic: 1–4, 5–29 odd, 33–43 odd, 49–53 odd, 59, 62, 64, 75–80

Average: 1–4, 6–30 even, 34–44 even, 50–54 even, 58–64 even, 75–80

Advanced: 1–4, 12, 20, 30–72 even, 73–80

HOMEWORK CHECK

Basic: 7, 15, 21, 41, 51

Average: 12, 16, 26, 42, 52

Advanced: 12, 20, 30, 44, 52

ANSWERS

1. $i = \sqrt{-1}$ and is used to write the square root of any negative number.
2. $2i$; 5
3. Add the real parts and the imaginary parts separately.
4. $3 + 0i$; It is the only one that is not an imaginary number.
5. $6i$
6. $8i$
7. $3i\sqrt{2}$
8. $2i\sqrt{6}$
9. $8i$
10. $-21i$
11. $-16i\sqrt{2}$
12. $18i\sqrt{7}$
13. $x = 2$ and $y = 2$
14. $x = 9$ and $y = 6$
15. $x = -2$ and $y = 4$
16. $x = -4$ and $y = -3$
17. $x = 7$ and $y = -12$
18. $x = -5$ and $y = -13$
19. $x = 6$ and $y = 28$
20. $x = 30$ and $y = -\frac{2}{3}$
21. $13 + 2i$
22. $20 + 7i$
23. $9 + 11i$
24. $-2 - 20i$
25. 19
26. 14
27. $4 + 2i$
28. $14 + 2i$
29. $-4 - 14i$
30. $5 + 9i$
31. a. $-4 + 5i$
    b. $2\sqrt{2} + 10i$

4.6 Exercises

Dynamic Solutions available at BigIdeasMath.com

Vocabulary and Core Concept Check

1. **VOCABULARY** What is the imaginary unit i defined as and how can you use i?

2. **COMPLETE THE SENTENCE** For the complex number $5 + 2i$, the imaginary part is _____ and the real part is _____.

3. **WRITING** Describe how to add complex numbers.

4. **WHICH ONE DOESN'T BELONG?** Which number does *not* belong with the other three? Explain your reasoning.

   $3 + 0i$ $2 + 5i$ $\sqrt{3} + 6i$ $0 - 7i$

Monitoring Progress and Modeling with Mathematics

In Exercises 5–12, find the square root of the number. (*See Example 1.*)

5. $\sqrt{-36}$
6. $\sqrt{-64}$
7. $\sqrt{-18}$
8. $\sqrt{-24}$
9. $2\sqrt{-16}$
10. $-3\sqrt{-49}$
11. $-4\sqrt{-32}$
12. $6\sqrt{-63}$

In Exercises 13–20, find the values of x and y that satisfy the equation. (*See Example 2.*)

13. $4x + 2i = 8 + yi$
14. $3x + 6i = 27 + yi$
15. $-10x + 12i = 20 + 3yi$
16. $9x - 18i = -36 + 6yi$
17. $2x - yi = 14 + 12i$
18. $-12x + yi = 60 - 13i$
19. $54 - \frac{1}{7}yi = 9x - 4i$
20. $15 - 3yi = \frac{1}{2}x + 2i$

In Exercises 21–30, add or subtract. Write the answer in standard form. (*See Example 3.*)

21. $(6 - i) + (7 + 3i)$
22. $(9 + 5i) + (11 + 2i)$
23. $(12 + 4i) - (3 - 7i)$
24. $(2 - 15i) - (4 + 5i)$
25. $(12 - 3i) + (7 + 3i)$
26. $(16 - 9i) - (2 - 9i)$
27. $7 - (3 + 4i) + 6i$
28. $16 - (2 - 3i) - i$
29. $-10 + (6 - 5i) - 9i$
30. $-3 + (8 + 2i) + 7i$

31. **USING STRUCTURE** Write each expression as a complex number in standard form.
    a. $\sqrt{-9} + \sqrt{-4} - \sqrt{16}$
    b. $\sqrt{-16} + \sqrt{8} + \sqrt{-36}$

32. **REASONING** The additive inverse of a complex number z is a complex number z_a such that $z + z_a = 0$. Find the additive inverse of each complex number.
    a. $z = 1 + i$ b. $z = 3 - i$ c. $z = -2 + 8i$

In Exercises 33–36, find the impedance of the series circuit. (*See Example 4.*)

33.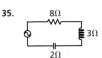
34.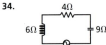
35.
36.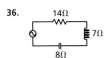

242   Chapter 4   Solving Quadratic Equations

32. a. $z_a = -1 - i$
    b. $z_a = -3 + i$
    c. $z_a = 2 - 8i$
33. $(12 + 2i)$ ohms
34. $(4 - 3i)$ ohms
35. $(8 + i)$ ohms
36. $(14 - i)$ ohms

In Exercises 37–44, multiply. Write the answer in standard form. *(See Example 5.)*

37. $3i(-5 + i)$
38. $2i(7 - i)$
39. $(3 - 2i)(4 + i)$
40. $(7 + 5i)(8 - 6i)$
41. $(5 - 2i)(-2 - 3i)$
42. $(-1 + 8i)(9 + 3i)$
43. $(3 - 6i)^2$
44. $(8 + 3i)^2$

JUSTIFYING STEPS In Exercises 45 and 46, justify each step in performing the operation.

45. $11 - (4 + 3i) + 5i$

$= [(11 - 4) - 3i] + 5i$

$= (7 - 3i) + 5i$

$= 7 + (-3 + 5)i$

$= 7 + 2i$

46. $(3 + 2i)(7 - 4i)$

$= 21 - 12i + 14i - 8i^2$

$= 21 + 2i - 8(-1)$

$= 21 + 2i + 8$

$= 29 + 2i$

REASONING In Exercises 47 and 48, place the tiles in the expression to make a true statement.

47. $(\_\_ - \_\_i) - (\_\_ - \_\_i) = 2 - 4i$

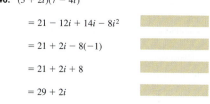

48. $\_\_i(\_\_ + \_\_i) = -18 - 10i$

In Exercises 49–54, multiply the complex number by its complex conjugate. *(See Example 6.)*

49. $1 - i$
50. $8 + i$
51. $4 + 2i$
52. $5 - 6i$
53. $-2 + 2i$
54. $-1 - 9i$

55. **OPEN-ENDED** Write a pair of imaginary numbers whose product is 80.

56. **NUMBER SENSE** Write the complex conjugate of $1 - \sqrt{-12}$. Then find the product of the complex conjugates.

57. **USING STRUCTURE** Expand $(a - bi)^2$ and write the result in standard form. Use your result to check your answer in Exercise 43.

58. **USING STRUCTURE** Expand $(a + bi)^2$ and write the result in standard form. Use your result to check your answer in Exercise 44.

ERROR ANALYSIS In Exercises 59 and 60, describe and correct the error in performing the operation and writing the answer in standard form.

59. ✗
$(3 + 2i)(5 - i) = 15 - 3i + 10i - 2i^2$
$= 15 + 7i - 2i^2$
$= -2i^2 + 7i + 15$

60. ✗
$(4 + 6i)^2 = (4)^2 + (6i)^2$
$= 16 + 36i^2$
$= 16 + (36)(-1)$
$= -20$

61. **NUMBER SENSE** Simplify each expression. Then classify your results in the table below.

a. $(-4 + 7i) + (-4 - 7i)$
b. $(2 - 6i) - (-10 + 4i)$
c. $(25 + 15i) - (25 - 6i)$
d. $(5 + i)(8 - i)$
e. $(17 - 3i) + (-17 - 6i)$
f. $(-1 + 2i)(11 - i)$
g. $(7 + 5i) + (7 - 5i)$
h. $(-3 + 6i) - (-3 - 8i)$

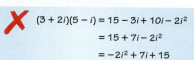

Real numbers	Imaginary numbers	Pure imaginary numbers

62. **MAKING AN ARGUMENT** The Product Property of Square Roots states $\sqrt{a} \cdot \sqrt{b} = \sqrt{ab}$. Your friend concludes $\sqrt{-4} \cdot \sqrt{-9} = \sqrt{36} = 6$. Is your friend correct? Explain.

Section 4.6 Complex Numbers 243

Dynamic Teaching Tools

Dynamic Assessment & Progress Monitoring Tool
Interactive Whiteboard Lesson Library
Dynamic Classroom with Dynamic Investigations

ANSWERS

37. $-3 - 15i$
38. $2 + 14i$
39. $14 - 5i$
40. $86 - 2i$
41. $-16 - 11i$
42. $-33 + 69i$
43. $-27 - 36i$
44. $55 + 48i$
45. Distributive Property; Simplify; Definition of complex addition; Write in standard form.
46. Multiply using FOIL; Simplify and use $i^2 = -1$; Simplify; Write in standard form.
47. $(6 - 7i) - (4 - 3i) = 2 - 4i$
48. $2i(-5 + 9i) = -18 - 10i$
49. 2
50. 65
51. 20
52. 61
53. 8
54. 82
55. Sample answer: $8 + 4i$, $8 - 4i$
56. $1 + \sqrt{-12}$; 13
57. $a^2 - 2abi - b^2$
58. $a^2 + 2abi - b^2$
59. i^2 can be simplified; $15 - 3i + 10i - 2i^2 = 15 + 7i + 2 = 17 + 7i$
60. Squaring a complex number requires FOIL; $(4 + 6i)(4 + 6i) = 16 + 24i + 24i + 36i^2 = 16 + 48i - 36 = -20 + 48i$
61. a. -8
    b. $12 - 10i$
    c. $21i$
    d. $41 + 3i$
    e. $-9i$
    f. $-9 + 23i$
    g. 14

61. h. $14i$

Real numbers	Imaginary numbers	Pure imaginary numbers
-8	$12 - 10i$	$21i$
14	$14 + 3i$	$-9i$
	$-9 + 23i$	$14i$

62. no, $\sqrt{-4} \cdot \sqrt{-9} = -6$; Simplifying results in $2i \cdot 3i = 6i^2$. Using $i^2 = -1$ results in -6.

ANSWERS

63. See Additional Answers.
64. a. D
    b. F
    c. B
    d. E
    e. A
    f. C
65. $-28 + 27i$
66. $-18 - 2i$
67. $-15 - 25i$
68. $-24 - 10i$
69. $9 + 5i$
70. $-2 - i$
71. $-2 - 7i, -2 + 7i$
72. Method 1 distributes $4i$ to each term, then simplifies. Method 2 factors $4i$ out of each term, combines like terms, and simplifies; *Sample answer:* Method 2; There are fewer computations.
73. a. false; *Sample answer:*
       $(3 - 5i) + (4 + 5i) = 7$
    b. true; *Sample answer:*
       $(3i)(2i) = 6i^2 = -6$
    c. true; *Sample answer:*
       $3i = 0 + 3i$
    d. false; *Sample answer:* $1 + 8i$
74. *Sample answer:*

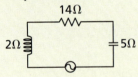

75. $y = (x - 1)^2 + 2$
76. $y = 2(x + 3)^2 - 3$
77. $y = -(x - 2)^2 - 1$
78. 16
79. 100
80. $\frac{49}{4}$

Mini-Assessment

1. Find the values of x and y that satisfy the equation
   $14x - 64i = 42 + yi$.
   $x = 3$ and $y = -64$

2. Simplify $24 - (9 + 21i) + 9i$. Write the answer in standard form.
   $15 - 12i$

3. Multiply $(9 - 3i)(-4 + 7i)$. Write the answer in standard form.
   $-15 + 75i$

4. Multiply $5 + 3i$ by its complex conjugate. 34

244   Chapter 4

---

63. FINDING A PATTERN Make a table that shows the powers of i from i^1 to i^8 in the first row and the simplified forms of these powers in the second row. Describe the pattern you observe in the table. Verify the pattern continues by evaluating the next four powers of i.

64. HOW DO YOU SEE IT? The coordinate system shown below is called the *complex plane*. In the complex plane, the point that corresponds to the complex number $a + bi$ is (a, b). Match each complex number with its corresponding point.

a. 2
b. $2i$
c. $4 - 2i$
d. $3 + 3i$
e. $-2 + 4i$
f. $-3 - 3i$

In Exercises 65–70, write the expression as a complex number in standard form.

65. $(3 + 4i) - (7 - 5i) + 2i(9 + 12i)$
66. $3i(2 + 5i) + (6 - 7i) - (9 + i)$
67. $(3 + 5i)(2 - 7i^4)$
68. $2i^3(5 - 12i)$
69. $(2 + 4i^5) + (1 - 9i^6) - (3 + i^7)$
70. $(8 - 2i^4) + (3 - 7i^8) - (4 + i^9)$

Maintaining Mathematical Proficiency
Reviewing what you learned in previous grades and lessons

Write a quadratic function in vertex form whose graph is shown. *(Section 3.4)*

75.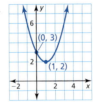
76.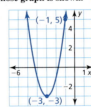
77.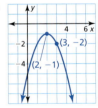

Find the value of c that completes the square. *(Section 4.4)*

78. $x^2 + 8x + c$
79. $x^2 - 20x + c$
80. $x^2 - 7x + c$

244   Chapter 4   Solving Quadratic Equations

---

71. NUMBER SENSE Write a pair of complex numbers whose sum is -4 and whose product is 53.

72. COMPARING METHODS Describe the two different methods shown for writing the complex expression in standard form. Which method do you prefer? Explain.

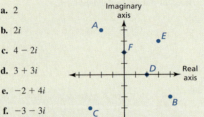

Method 1
$4i(2 - 3i) + 4i(1 - 2i) = 8i - 12i^2 + 4i - 8i^2$
$= 8i - 12(-1) + 4i - 8(-1)$
$= 20 + 12i$

Method 2
$4i(2 - 3i) + 4i(1 - 2i) = 4i[(2 - 3i) + (1 - 2i)]$
$= 4i[3 - 5i]$
$= 12i - 20i^2$
$= 12i - 20(-1)$
$= 20 + 12i$

73. CRITICAL THINKING Determine whether each statement is *true* or *false*. If it is true, give an example. If it is false, give a counterexample.

a. The sum of two imaginary numbers is an imaginary number.
b. The product of two pure imaginary numbers is a real number.
c. A pure imaginary number is an imaginary number.
d. A complex number is a real number.

74. THOUGHT PROVOKING Create a circuit that has an impedance of $14 - 3i$.

If students need help...	If students got it...
Resources by Chapter • Practice A and Practice B • Puzzle Time	Resources by Chapter • Enrichment and Extension • Cumulative Review
Student Journal • Practice	Start the *next* Section
Differentiating the Lesson Skills Review Handbook	

Laurie's Notes

Overview of Section 4.7

Introduction
- Earlier lessons were devoted to strategies for solving quadratic equations. Three of the strategies are reviewed in this lesson: using square roots, completing the square, and using the Quadratic Formula. These strategies are used to find all solutions, real and imaginary, of a quadratic equation.
- The lesson ends with a real-life application that uses the launched-object model.

Formative Assessment Tips
- **I Used to Think ... But Now I Know:** This technique asks students to consider how their thinking about a concept or skill has changed from the beginning of instruction to the end of instruction. This can be done orally or in writing. It is important for students to be able to self-assess and reflect on their own learning.
- Use this technique at the end of the formal lesson. If time permits, have students discuss with one another or the whole class how their understanding developed and/or changed. They may be able to identify the example or class discussion that helped to promote the learning.

Applications
- The real-life application, the launched-object model, is one that students should know and be comfortable with.

Another Way
- Discuss with students the simplification of $\sqrt{20}$. Some would argue that $\sqrt{20}$ is simpler than $2\sqrt{5}$.

Pacing Suggestion
- As students work through the explorations, listen and probe for recall of prior skills and then continue with the formal lesson.

Dynamic Teaching Tools
Dynamic Assessment & Progress Monitoring Tool
Lesson Planning Tool
Interactive Whiteboard Lesson Library
Dynamic Classroom with Dynamic Investigations

What Your Students Will Learn

- Solve quadratic equations and find zeros, both real and imaginary, of quadratic functions.
- Use the discriminant to model quadratic equations with two imaginary solutions.

Laurie's Notes

Exploration

Motivate
- **Whiteboarding:** Write the following quadratic functions on the board and ask students to describe what they know about the graphs of the functions.
  $f(x) = 3x^2 - 5x - 2$ The graph of f is a parabola that opens upward, is narrower than the graph of $f(x) = x^2$, and has a y-intercept of -2.
  $g(x) = -2(x - 3)(x + 1)$ The graph of g is a parabola that opens downward, is narrower than the graph of $f(x) = x^2$, and has x-intercepts of 3 and -1.
  $h(x) = \frac{1}{2}(x - 4)^2 + 1$ The graph of h is a parabola that opens upward, is wider than the graph of $f(x) = x^2$, and has its vertex at (4, 1).

Exploration 1
- The goal of the exploration is for students to use their knowledge of how to interpret the discriminant from Section 4.5 and the effect of the sign of the leading coefficient in order to match the equation and the graph.
- Students may suggest that they would do the matching by also using information about the y-intercept. They could make an accurate guess about the value of the real roots by looking at the graph.
- Equations (a), (b), and (c) open upward, as do graphs (C), (D), and (E). Equations (a), (b), and (c) have y-intercepts 0, 1, and 2, respectively, allowing for fairly quick matching.
- Equations (d), (e), and (f) open downward, as do graphs (A), (B), and (F). Equations (d), (e), and (f) have y-intercepts 0, -1, and -2, respectively, allowing for fairly quick matching.
- The number of x-intercepts is easy to determine by inspection, though students should be cautious when the vertex appears to be tangent to the x-axis, as is the case in graphs (D) and (F).
- **?** "Looking at graph (D), how do you know there is only one x-intercept?" Answers will vary.
- **?** "In what other form could you write the function $f(x) = -x^2 + 2x$?" factored or vertex form "Would this help you determine the number of x-intercepts? Explain." yes; If it is in vertex form and the y-coordinate is 0, then there must be only one x-intercept.

Exploration 2
- Students already know that when the discriminant is negative there are no real roots. This exploration is pushing students to think further about what it means when you have a quadratic with no real roots, such as $f(x) = x^2 + 1$.
- Students are expected to actually use the Quadratic Formula and what they know from the last lesson about complex numbers to conclude that when a quadratic has a discriminant that is negative, there will be two imaginary roots.

Communicate Your Answer
- Students should describe the relationship between the sign of the discriminant and the type of solutions.
- **Think-Pair-Share:** Have students solve Question 4, check with their partners, and then discuss as a whole class.

Connecting to Next Step
- The explorations should be a review of finding the discriminant of a quadratic equation and using the value of the discriminant to determine whether there are two real, two imaginary, or no solutions. Quickly transition to the formal lesson.

4.7 Solving Quadratic Equations with Complex Solutions

Essential Question How can you determine whether a quadratic equation has real solutions or imaginary solutions?

EXPLORATION 1 Using Graphs to Solve Quadratic Equations

Work with a partner. Use the discriminant of $f(x) = 0$ and the sign of the leading coefficient of $f(x)$ to match each quadratic function with its graph. Explain your reasoning. Then find the real solution(s) (if any) of each quadratic equation $f(x) = 0$.

a. $f(x) = x^2 - 2x$
b. $f(x) = x^2 - 2x + 1$
c. $f(x) = x^2 - 2x + 2$
d. $f(x) = -x^2 + 2x$
e. $f(x) = -x^2 + 2x - 1$
f. $f(x) = -x^2 + 2x - 2$

A.
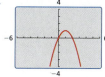

B.
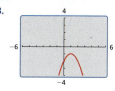

C.
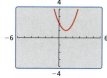

D.
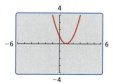

E.
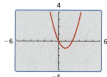

F.
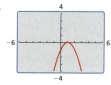

EXPLORATION 2 Finding Imaginary Solutions

MAKING SENSE OF PROBLEMS

To be proficient in math, you need to make conjectures about the form and meaning of solutions.

Work with a partner. What do you know about the discriminants of quadratic equations that have no real solutions? Use the Quadratic Formula and what you learned about the imaginary unit i to find the *imaginary* solutions of each equation in Exploration 1 that has no real solutions. Use substitution to check your answers.

Communicate Your Answer

3. How can you determine whether a quadratic equation has real solutions or imaginary solutions?

4. Describe the number and type of solutions of $x^2 + 2x + 3 = 0$. How do you know? What are the solutions?

Dynamic Teaching Tools

Dynamic Assessment & Progress Monitoring Tool
Lesson Planning Tool
Interactive Whiteboard Lesson Library
Dynamic Classroom with Dynamic Investigations

ANSWERS

1. a. E; $b^2 - 4ac > 0$ and $a > 0$, so the graph opens up and has two x-intercepts; $x = 0, x = 2$
   b. D; $b^2 - 4ac = 0$ and $a > 0$, so the graph opens up and has one x-intercept; $x = 1$
   c. C; $b^2 - 4ac < 0$ and $a > 0$, so the graph opens up and has no x-intercepts; no real solutions
   d. A; $b^2 - 4ac > 0$ and $a < 0$, so the graph opens down and has two x-intercepts; $x = 0, x = 2$
   e. F; $b^2 - 4ac = 0$ and $a < 0$, so the graph opens down and has one x-intercept; $x = 1$
   f. B; $b^2 - 4ac < 0$ and $a < 0$, so the graph opens down and has no x-intercepts; no real solutions

2. $b^2 - 4ac < 0$
   c. $1 - i, 1 + i$
   f. $1 - i, 1 + i$

3. When the discriminant is negative, the solutions are imaginary.

4. 2 imaginary solutions;
   $2^2 - 4(1)(3) < 0; -1 - \sqrt{2}i, -1 + \sqrt{2}i$

Extra Example 1

Solve each equation.

a. $-x^2 - 64 = 0$ $x = \pm 8i$
b. $x^2 + 6x + 25 = 0$ $x = -3 \pm 4i$
c. $9x^2 - 6x + 5 = 0$ $x = \dfrac{1 \pm 2i}{3}$

4.7 Lesson

What You Will Learn

▶ Solve quadratic equations and find zeros of quadratic functions.
▶ Use the discriminant.

Finding Solutions and Zeros

Previously, you learned that you can use the discriminant of a quadratic equation to determine whether the equation has two real solutions, one real solution, or no real solutions. When the discriminant is negative, you can use the imaginary unit i to write two *imaginary* solutions of the equation. So, all quadratic equations have complex number solutions.

You have solved quadratic equations with real solutions. Now you will solve quadratic equations with *imaginary* solutions.

> **STUDY TIP**
> In general, every polynomial equation has complex number solutions. This is implied by the *Fundamental Theorem of Algebra*. You will learn more about this theorem in a future course.

EXAMPLE 1 Solving Quadratic Equations

Solve each equation.

a. $x^2 + 9 = 0$ b. $x^2 + 4x + 5 = 0$ c. $5x^2 - 4x + 1 = 0$

SOLUTION

a. The equation does not have an x-term. So, solve using square roots.

$x^2 + 9 = 0$	Write the equation.
$x^2 = -9$	Subtract 9 from each side.
$x = \pm\sqrt{-9}$	Take the square root of each side.
$x = \pm 3i$	Write in terms of i.

> **LOOKING FOR STRUCTURE**
> You can use the pattern $(a + bi)(a - bi) = a^2 + b^2$ to rewrite $x^2 + 9 = 0$ as $(x + 3i)(x - 3i) = 0$.
> So, $x = \pm 3i$.

b. The coefficient of the x^2-term is 1, and the coefficient of the x-term is an even number. So, solve by completing the square.

$x^2 + 4x + 5 = 0$
$x^2 + 4x = -5$
$x^2 + 4x + 4 = -5 + 4$
$(x + 2)^2 = -1$
$x + 2 = \pm\sqrt{-1}$
$x = -2 \pm \sqrt{-1}$
$x = -2 \pm i$

> **Check** You can check imaginary solutions algebraically. The check for one of the imaginary solutions, $-2 + i$, is shown.
>
> $(-2 + i)^2 + 4(-2 + i) + 5 \stackrel{?}{=} 0$
> $3 - 4i - 8 + 4i + 5 \stackrel{?}{=} 0$
> $0 = 0$ ✓

c. The equation is not factorable, and completing the square would result in fractions. So, solve using the Quadratic Formula.

$x = \dfrac{-(-4) \pm \sqrt{(-4)^2 - 4(5)(1)}}{2(5)}$	Substitute 5 for a, -4 for b, and 1 for c.
$x = \dfrac{4 \pm \sqrt{-4}}{10}$	Simplify.
$x = \dfrac{4 \pm 2i}{10}$	Write in terms of i.
$x = \dfrac{2 \pm i}{5}$	Simplify.

Laurie's Notes — Teacher Actions

- Discuss the *Study Tip* to give students understanding to where this is content leads to in Math III.
- **❓ Turn and Talk:** "What methods have you used for solving quadratic equations?" Circulate and listen for graphing, using square roots, factoring, completing the square, and using the Quadratic Formula.
- **Teaching Tip** and **Whiteboarding:** Having reviewed the methods for solving quadratics, you might have partners or groups work through the three examples independently and share results using *whiteboards*. Alternately, you could ask students what method(s) they would use in solving each quadratic, why they chose the method(s), and then have them work with their partners or groups.

Check

$f(i\sqrt{5}) = 4(i\sqrt{5})^2 + 20$
$= 4(-5) + 20$
$= 0$ ✓

$f(-i\sqrt{5}) = 4(-i\sqrt{5})^2 + 20$
$= 4(-5) + 20$
$= 0$ ✓

EXAMPLE 2 Finding Zeros of a Quadratic Function

Find the zeros of $f(x) = 4x^2 + 20$.

SOLUTION

$4x^2 + 20 = 0$	Set $f(x)$ equal to 0.
$4x^2 = -20$	Subtract 20 from each side.
$x^2 = -5$	Divide each side by 4.
$x = \pm\sqrt{-5}$	Take the square root of each side.
$x = \pm i\sqrt{5}$	Write in terms of i.

▶ So, the zeros of f are $i\sqrt{5}$ and $-i\sqrt{5}$.

Monitoring Progress Help in English and Spanish at *BigIdeasMath.com*

Solve the equation using any method. Explain your choice of method.

1. $-x^2 - 25 = 0$
2. $x^2 - 4x + 8 = 0$
3. $8x^2 + 5 = 12x$

Find the zeros of the function.

4. $f(x) = -2x^2 - 18$
5. $f(x) = 9x^2 + 1$
6. $f(x) = x^2 - 6x + 10$

Using the Discriminant

EXAMPLE 3 Writing an Equation

Find a possible pair of integer values for a and c so that the equation $ax^2 - 4x + c = 0$ has two imaginary solutions. Then write the equation.

SOLUTION

For the equation to have two imaginary solutions, the discriminant must be less than zero.

$b^2 - 4ac < 0$	Write the discriminant.
$(-4)^2 - 4ac < 0$	Substitute -4 for b.
$16 - 4ac < 0$	Evaluate the power.
$-4ac < -16$	Subtract 16 from each side.
$ac > 4$	Divide each side by -4. Reverse inequality symbol.

ANOTHER WAY

Another possible equation in Example 3 is $3x^2 - 4x + 2 = 0$. You can obtain this equation by letting $a = 3$ and $c = 2$.

Because $ac > 4$, choose two integers whose product is greater than 4, such as $a = 2$ and $c = 3$.

▶ So, one possible equation is $2x^2 - 4x + 3 = 0$.

Check The graph of $y = 2x^2 - 4x + 3$ does not have any x-intercepts. ✓

Monitoring Progress Help in English and Spanish at *BigIdeasMath.com*

7. Find a possible pair of integer values for a and c so that the equation $ax^2 + 3x + c = 0$ has two imaginary solutions. Then write the equation.

Section 4.7 Solving Quadratic Equations with Complex Solutions 247

Extra Example 4
The juggler in Example 4 tosses the ball into the air again. This time it leaves the juggler's hand 3 feet above the ground and has an initial vertical velocity of 20 feet per second. Does the ball reach a height of 12 feet? 8 feet? **The ball reaches a height of 8 feet, but it does not reach a height of 12 feet.**

MONITORING PROGRESS ANSWER
8. no; yes; When $h = 30$, the equation has two imaginary solutions. When $h = 20$, $t = 0.5$ and $t = 2$.

The function $h = -16t^2 + s_0$ is used to model the height of a *dropped* object, where h is the height (in feet), t is the time in motion (in seconds), and s_0 is the initial height (in feet). For an object that is *launched* or *thrown*, an extra term $v_0 t$ must be added to the model to account for the object's initial vertical velocity v_0 (in feet per second).

$h = -16t^2 + s_0$ Object is dropped.
$h = -16t^2 + v_0 t + s_0$ Object is launched or thrown.

> **STUDY TIP**
> These models assume that the force of air resistance on the object is negligible. Also, these models apply only to objects on Earth. For planets with stronger or weaker gravitational forces, different models are used.

As shown below, the value of v_0 can be positive, negative, or zero depending on whether the object is launched upward, downward, or parallel to the ground.

$v_0 > 0$

$v_0 < 0$

$v_0 = 0$

EXAMPLE 4   Modeling a Launched Object

A juggler tosses a ball into the air. The ball leaves the juggler's hand 4 feet above the ground and has an initial vertical velocity of 30 feet per second. Does the ball reach a height of 25 feet? 10 feet? Explain your reasoning.

SOLUTION

Because the ball is *thrown*, use the model $h = -16t^2 + v_0 t + s_0$ to write a function that represents the height of the ball.

$h = -16t^2 + v_0 t + s_0$ Write the height model.
$h = -16t^2 + 30t + 4$ Substitute 30 for v_0 and 4 for s_0.

To determine whether the ball reaches each height, substitute each height for h to create two equations. Then solve each equation using the Quadratic Formula.

$25 = -16t^2 + 30t + 4$ $10 = -16t^2 + 30t + 4$
$0 = -16t^2 + 30t - 21$ $0 = -16t^2 + 30t - 6$

$t = \dfrac{-30 \pm \sqrt{30^2 - 4(-16)(-21)}}{2(-16)}$ $t = \dfrac{-30 \pm \sqrt{30^2 - 4(-16)(-6)}}{2(-16)}$

$t = \dfrac{-30 \pm \sqrt{-444}}{-32}$ $t = \dfrac{-30 \pm \sqrt{516}}{-32}$

Check The graph shows that the ball reaches a height of 10 feet but not 25 feet.

When $h = 25$, the equation has two imaginary solutions because the discriminant is negative. When $h = 10$, the equation has two real solutions, $t \approx 0.23$ and $t \approx 1.65$.

▶ So, the ball reaches a height of 10 feet, but it does not reach a height of 25 feet.

Monitoring Progress   Help in English and Spanish at *BigIdeasMath.com*

8. The ball leaves the juggler's hand with an initial vertical velocity of 40 feet per second. Does the ball reach a height of 30 feet? 20 feet? Explain.

248   Chapter 4   Solving Quadratic Equations

Laurie's Notes   Teacher Actions

- **Model with Mathematics:** Re-introduce the launched-object model and discuss the three difference cases of the initial velocity ($v_0 > 0$, $v_0 < 0$, $v_0 = 0$).
- ❓ "What is known in this problem?" **The initial height is 4 feet and the initial velocity is 30 feet per second.** "What are you trying to solve?" **whether the ball ever reaches a height of 25 feet or 10 feet**
- Give partners sufficient time to solve the problem.
- **Connection:** A graphing calculator can be used to verify result.

Closure
- **I Used to Think...But Now I Know:** Take time for students to reflect on their current understanding of solutions of quadratic equations.

4.7 Exercises

Dynamic Solutions available at *BigIdeasMath.com*

Vocabulary and Core Concept Check

1. **COMPLETE THE SENTENCE** When the graph of a quadratic function $y = f(x)$ has no x-intercepts, the equation $f(x) = 0$ has two _____ solutions.

2. **WRITING** Can a quadratic equation with real coefficients have one imaginary solution? Explain.

Monitoring Progress and Modeling with Mathematics

ANALYZING EQUATIONS In Exercises 3–6, use the discriminant to match the quadratic equation with the graph of the related function. Then describe the number and type of solutions of the equation.

3. $x^2 - 6x + 25 = 0$
4. $2x^2 - 20x + 50 = 0$
5. $3x^2 + 6x - 9 = 0$
6. $5x^2 - 10x - 35 = 0$

A.
B.
C.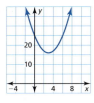
D.

In Exercises 7–20, solve the equation using any method. Explain your choice of method. *(See Example 1.)*

7. $x^2 + 49 = 0$
8. $2x^2 - 7 = -3$
9. $x^2 - 4x + 3 = 0$
10. $3x^2 + 6x + 3 = 0$
11. $x^2 + 6x + 15 = 0$
12. $6x^2 - 2x + 1 = 0$
13. $9x^2 + 17 = 24x$
14. $-3x = 2x^2 - 4$
15. $-10x = -25 - x^2$
16. $-2x^2 - 5 = -2x$
17. $-4x^2 + 3x = -5$
18. $3x^2 + 87 = 30x$
19. $-z^2 = -12z + 6$
20. $-7w + 6 = -4w^2$

21. **ERROR ANALYSIS** Describe and correct the error in solving the equation.

$x^2 + 10x + 74 = 0$

$x = \dfrac{-10 \pm \sqrt{10^2 - 4(1)(74)}}{2(1)}$

$= \dfrac{-10 \pm \sqrt{-196}}{2}$

$= \dfrac{-10 \pm 14}{2}$

$= -12 \text{ or } 2$

22. **REASONING** Write a quadratic equation in the form $ax^2 + bx + c = 0$ that has the solutions $x = 1 \pm i$.

In Exercises 23–28, find the zeros of the function. *(See Example 2.)*

23. $f(x) = 5x^2 + 35$
24. $g(x) = -3x^2 + 24$
25. $h(x) = x^2 + 8x - 13$
26. $r(x) = 8x^2 + 4x + 5$
27. $m(x) = -5x^2 + 50x - 135$
28. $r(x) = 4x^2 + 9x + 3$

OPEN-ENDED In Exercises 29–32, find a possible pair of integer values for a and c so that the quadratic equation has the given solution(s). Then write the equation. *(See Example 3.)*

29. $ax^2 + 4x + c = 0$; two imaginary solutions
30. $ax^2 - 8x + c = 0$; two real solutions
31. $ax^2 + 10x = c$; one real solution
32. $-4x + c = -ax^2$; two imaginary solutions

Section 4.7 Solving Quadratic Equations with Complex Solutions 249

Assignment Guide and Homework Check

ASSIGNMENT
Basic: 1–6, 7–37 odd, 38, 42–48
Average: 1–6, 8–20 even, 21, 22–36 even, 37, 38, 42–48
Advanced: 1–6, 18, 20–22, 26–32 even, 33–48

HOMEWORK CHECK
Basic: 3, 11, 25, 29
Average: 4, 12, 26, 30
Advanced: 4, 12, 26, 32

ANSWERS

1. imaginary
2. no; *Sample answer:* Using the Quadratic Formula, a quadratic equation has either two real solutions, one real solution, or two imaginary solutions.
3. C; two imaginary solutions
4. D; one real solution
5. A; two real solutions
6. B; two real solutions
7. $x = \pm 7i$; *Sample answer:* The equation does not have an x-term, so solve using square roots.
8. $x = \pm \sqrt{2}$; *Sample answer:* The equation does not have an x-term, so solve using square roots.
9. $x = 3$ and $x = 1$; *Sample answer:* The equation is factorable, so solve by factoring.
10. $x = -1$; *Sample answer:* The equation is factorable, so solve by factoring.
11. $x = -3 \pm i\sqrt{6}$; *Sample answer:* $a = 1$ and b is even, so solve by completing the square.
12. $x = \dfrac{1 \pm i\sqrt{5}}{6}$; *Sample answer:* The equation is not easily factorable and $a \neq 1$, so solve using the Quadratic Formula.
13. $x = \dfrac{4 \pm i}{3}$; *Sample answer:* The equation is not easily factorable and $a \neq 1$, so solve using the Quadratic Formula.
14. $x = \dfrac{-3 \pm \sqrt{41}}{4}$; *Sample answer:* The equation is not easily factorable and $a \neq 1$, so solve using the Quadratic Formula.
15. $x = 5$; *Sample answer:* The equation is factorable, so solve by factoring.

---

16. $x = \dfrac{1 \pm 3i}{2}$; *Sample answer:* The equation is not easily factorable and $a \neq 1$, so solve using the Quadratic Formula.

17. $x = \dfrac{3 \pm \sqrt{89}}{8}$; *Sample answer:* The equation is not easily factorable and $a \neq 1$, so solve using the Quadratic Formula.

18. $x = 5 \pm 2i$; *Sample answer:* $a = 1$ and b is even, so solve by completing the square.

19. $z = 6 \pm \sqrt{30}$; *Sample answer:* $a = 1$ and b is even, so solve by completing the square.

20. $w = \dfrac{7 \pm i\sqrt{47}}{8}$; *Sample answer:* The equation is not easily factorable and $a \neq 1$, so solve using the Quadratic Formula.

21. The i was left out after taking the square root;

$x = \dfrac{-10 \pm \sqrt{-196}}{2} = \dfrac{-10 \pm 14i}{2}$

$= -5 \pm 7i$

22–32. See Additional Answers.

Section 4.7 249

Dynamic Teaching Tools
- Dynamic Assessment & Progress Monitoring Tool
- Interactive Whiteboard Lesson Library
- Dynamic Classroom with Dynamic Investigations

ANSWERS

33. $h = -16t^2 - 88t + 100$
34. $h = -16t^2 + 5$
35. no; yes; *Sample answer:* When $h = 30$, the equation has two imaginary solutions. When $h = 26$, $t = 1$ and $t \approx 1.19$.
36. **a.** yes; no; *Sample answer:* When $h = 200$, the equation that models your friend's rocket has two imaginary solutions and the equation that models your rocket has two real solutions.
    b. your rocket; about 0.3 sec
37. *Sample answer:* Using the Quadratic Formula, the solutions are $x = \dfrac{-b \pm \sqrt{b^2 - 4ac}}{2a}$
$= \dfrac{-b}{2a} \pm \dfrac{\sqrt{b^2 - 4ac}}{2a}$, so when $b^2 - 4ac < 0$, the solutions are complex conjugates; Because $3i$ and $-2i$ are not complex conjugates, they cannot both be solutions of the same quadratic equation with real coefficients.
38–48. See Additional Answers.

Mini-Assessment

1. Solve $3x^2 + 324 = 0$.
   $6i\sqrt{3}$ and $-6i\sqrt{3}$

2. Find the zeros of $f(x) = 25x^2 + 175$.
   $i\sqrt{7}$ and $-i\sqrt{7}$

3. Find a possible pair of integer values for a and c so that the equation $ax^2 + 6x + c = 0$ has two imaginary solutions. Then write the equation.
   Sample answer: $a = 2$, $c = 5$; $2x^2 + 6x + 5 = 0$

4. A ballplayer tosses a ball into the air. The ball leaves the ballplayer's hand 5 feet above the ground and has an initial vertical velocity of 25 feet per second. Does the ball reach a height of 20 feet? 12 feet?
   no; yes

250 Chapter 4

MODELING WITH MATHEMATICS In Exercises 33 and 34, write a function that represents the situation.

33. A gannet is a bird that feeds on fish by diving into the water. A gannet spots a fish on the surface of the water and dives 100 feet to catch it. The bird plunges toward the water with an initial vertical velocity of -88 feet per second.

34. An archer is shooting at targets. The height of the arrow is 5 feet above the ground. Due to safety rules, the archer must aim the arrow parallel to the ground.

35. **MODELING WITH MATHEMATICS** A lacrosse player throws a ball in the air from an initial height of 7 feet. The ball has an initial vertical velocity of 35 feet per second. Does the ball reach a height of 30 feet? 26 feet? Explain your reasoning. *(See Example 4.)*

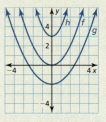

36. **PROBLEM SOLVING** A rocketry club is launching model rockets. The launching pad is 30 feet above the ground. Your model rocket has an initial vertical velocity of 105 feet per second. Your friend's model rocket has an initial vertical velocity of 100 feet per second.
    a. Does your rocket reach a height of 200 feet? Does your friend's rocket? Explain your reasoning.
    b. Which rocket is in the air longer? How much longer?

37. **CRITICAL THINKING** When a quadratic equation with real coefficients has imaginary solutions, why are the solutions complex conjugates? As part of your explanation, show that there is no such equation with solutions of $3i$ and $-2i$.

38. **HOW DO YOU SEE IT?** The graphs of three functions are shown. Which function(s) has real zeros? imaginary zeros? Explain your reasoning.

39. **USING STRUCTURE** Use the Quadratic Formula to write a quadratic equation that has the solutions $x = \dfrac{-8 \pm \sqrt{-176}}{-10}$.

40. **THOUGHT PROVOKING** Describe a real-life story that could be modeled by $h = -16t^2 + v_0 t + s_0$. Write the height model for your story and determine how long your object is in the air.

41. **MODELING WITH MATHEMATICS** The Stratosphere Tower in Las Vegas is 921 feet tall and has a "needle" at its top that extends even higher into the air. A thrill ride called Big Shot catapults riders 160 feet up the needle and then lets them fall back to the launching pad.
    a. The height h (in feet) of a rider on the Big Shot can be modeled by $h = -16t^2 + v_0 t + 921$, where t is the elapsed time (in seconds) after launch and v_0 is the initial vertical velocity (in feet per second). Find v_0 using the fact that the maximum value of h is $921 + 160 = 1081$ feet.
    b. A brochure for the Big Shot states that the ride up the needle takes 2 seconds. Compare this time to the time given by the model $h = -16t^2 + v_0 t + 921$, where v_0 is the value you found in part (a). Discuss the accuracy of the model.

Maintaining Mathematical Proficiency
Reviewing what you learned in previous grades and lessons

Solve the system of linear equations using any method. Explain why you chose the method. *(Skills Review Handbook)*

42. $y = -x + 4$
    $y = 2x - 8$

43. $x = 16 - 4y$
    $3x + 4y = 8$

44. $2x - y = 7$
    $2x + 7y = 31$

45. $3x - 2y = -20$
    $x + 1.2y = 6.4$

Find (a) the axis of symmetry and (b) the vertex of the graph of the function. *(Section 3.3)*

46. $y = -x^2 + 2x + 1$

47. $y = 2x^2 - x + 3$

48. $f(x) = 0.5x^2 + 2x + 5$

250 Chapter 4 Solving Quadratic Equations

If students need help...	If students got it...
Resources by Chapter • Practice A and Practice B • Puzzle Time	**Resources by Chapter** • Enrichment and Extension • Cumulative Review
Student Journal • Practice	Start the *next* Section
Differentiating the Lesson Skills Review Handbook	

Laurie's Notes

Overview of Section 4.8

Introduction
- This lesson revisits techniques studied in Math I for solving systems of linear equations: graphing, substitution, and elimination.
- The systems in this lesson contain at least one nonlinear equation. In some cases, the solutions are not rational and students will need to approximate the solution.

Formative Assessment Tips
- **Wait Time:** Wait time is the interval between a question being posed and a student (or the teacher) response. Silence can be uncomfortable in a classroom, but research has shown that increasing wait time increases class participation and answers are more detailed. For complex, higher-order thinking questions, increased wait time is necessary.
- **Reason Abstractly and Quantitatively** and **Construct Viable Arguments and Critique the Reasoning of Others:** If we want students to reason abstractly and quantitatively, and to construct viable arguments, we need to first pose complex questions. Students then need additional *Wait Time* to allow for thinking and formulation of responses. With increased participation, teachers then learn about students' progress and their learning.

Pacing Suggestion
- The explorations provide an opportunity for students to recall methods for solving a system of equations. Transition to the formal lesson as soon as students have discussed each exploration.

Dynamic Teaching Tools
Dynamic Assessment & Progress Monitoring Tool
Lesson Planning Tool
Interactive Whiteboard Lesson Library
Dynamic Classroom with Dynamic Investigations

What Your Students Will Learn

- Solve systems of nonlinear equations in two variables by graphing.
- Solve systems of nonlinear equations in two variables algebraically by substitution and by elimination.
- Approximate solutions of nonlinear systems and equations.

Laurie's Notes

Exploration

Motivate
- **Story Time:** Share a story about shooting clay pigeons launched into the air by a machine. In the last round, you hit the target on the way down. (Sketch the diagram shown below.)

- Connect this to today's exploration. Explain that you prefer to hit the target on the way up, because it is closer to you.

Discuss
- ❓ "What is a system of linear equations?" a set of two or more linear equations in the same variables
- ❓ "How do you solve a system of linear equations?" graphing, substitution, or elimination
- Explain that today's exploration is about systems of equations that include quadratic equations.

Exploration 1
- As students graph the two equations, you should hear comments about slope, y-intercepts, the parabola opening up, and so on.
- A careful graph of the two equations should allow the solutions to be read.
- **Attend to Precision:** Students should confirm the solutions by substituting them into each equation to verify that they are indeed solutions.

Exploration 2
- In Exploration 1, the focus was on the different solution methods. In Exploration 2, the focus is on the number of solutions.
- The graphs visually suggest three different cases for the number of solutions. These are similar to the three cases for the number of x-intercepts of the graph of a quadratic function.
- **Attend to Precision:** Students could approximate the solution(s) for each system. Instead, make sure students find the exact solution(s).
- ❓ "What happened when you solved the system in part (c) algebraically?" found that the system has no real solutions

Communicate Your Answer
- Question 4 takes time, and students are likely to begin by using trial and error.
- ❓ "In Question 4, which of the three cases was the most challenging and why?" Answers will vary. Many students will say that finding a system with one solution was the most challenging.

Connecting to Next Step
- The explorations should remind students of the various ways in which systems of equations are solved. In the formal lesson, all the methods will be reviewed.

4.8 Solving Nonlinear Systems of Equations

Essential Question How can you solve a system of two equations when one is linear and the other is quadratic?

EXPLORATION 1 Solving a System of Equations

Work with a partner. Solve the system of equations by graphing each equation and finding the points of intersection.

System of Equations

$y = x + 2$ Linear

$y = x^2 + 2x$ Quadratic

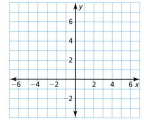

EXPLORATION 2 Analyzing Systems of Equations

Work with a partner. Match each system of equations with its graph. Then solve the system of equations.

a. $y = x^2 - 4$
   $y = -x - 2$

b. $y = x^2 - 2x + 2$
   $y = 2x - 2$

c. $y = x^2 + 1$
   $y = x - 1$

d. $y = x^2 - x - 6$
   $y = 2x - 2$

A.
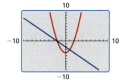

B.
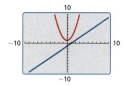

C.
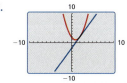

D.
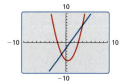

MAKING SENSE OF PROBLEMS

To be proficient in math, you need to analyze givens, relationships, and goals.

Communicate Your Answer

3. How can you solve a system of two equations when one is linear and the other is quadratic?

4. Write a system of equations (one linear and one quadratic) that has (a) no solutions, (b) one solution, and (c) two solutions. Your systems should be different from those in Explorations 1 and 2.

Section 4.8   Solving Nonlinear Systems of Equations   251

English Language Learners

Build on Past Knowledge

Remind students that a solution for a system of linear equations is an ordered pair that is a solution of each equation in the system. A linear system can have no solution, one solution, or infinitely many solutions. Explain that a nonlinear system that involves a linear equation and a quadratic equation may have no solution, one solution, or two solutions.

Extra Example 1

Solve the system by graphing.
$y = x^2 - 6x + 5$
$y = x - 1$

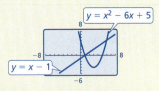

The solutions are (1, 0) and (6, 5).

MONITORING PROGRESS ANSWERS

1. $(-1, -7)$
2. no solutions
3. $(2, -9), (8, 9)$

4.8 Lesson

Core Vocabulary

system of nonlinear equations, *p. 252*

Previous
system of linear equations

STUDY TIP
In this section, *solutions* of a nonlinear system refer to the *real* solutions of the system.

What You Will Learn

- Solve systems of nonlinear equations by graphing.
- Solve systems of nonlinear equations algebraically.
- Approximate solutions of nonlinear systems and equations.

Solving Nonlinear Systems by Graphing

The methods for solving systems of linear equations can also be used to solve *systems of nonlinear equations*. A **system of nonlinear equations** is a system in which at least one of the equations is nonlinear.

When a nonlinear system consists of a linear equation and a quadratic equation, the graphs can intersect in zero, one, or two points. So, the system can have zero, one, or two solutions, as shown.

No solutions   One solution   Two solutions

EXAMPLE 1   Solving a Nonlinear System by Graphing

Solve the system by graphing.

$y = 2x^2 + 5x - 1$ **Equation 1**
$y = x - 3$ **Equation 2**

SOLUTION

Step 1 Graph each equation.

Step 2 Estimate the point of intersection. The graphs appear to intersect at $(-1, -4)$.

Step 3 Check the point from Step 2 by substituting the coordinates into each of the original equations.

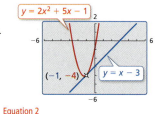

Equation 1   Equation 2

$y = 2x^2 + 5x - 1$ $y = x - 3$
$-4 \stackrel{?}{=} 2(-1)^2 + 5(-1) - 1$ $-4 \stackrel{?}{=} -1 - 3$
$-4 = -4$ ✓ $-4 = -4$ ✓

▶ The solution is $(-1, -4)$.

Monitoring Progress Help in English and Spanish at *BigIdeasMath.com*

Solve the system by graphing.

1. $y = x^2 + 4x - 4$
   $y = 2x - 5$

2. $y = -x + 6$
   $y = -2x^2 - x + 3$

3. $y = 3x - 15$
   $y = \frac{1}{2}x^2 - 2x - 7$

252   Chapter 4   Solving Quadratic Equations

Laurie's Notes   Teacher Actions

- From the explorations, students know what a *system of nonlinear equations* is and strategies for solving the system.
- ❓ Pose Example 1. "Can you determine the solution(s) in a standard viewing window?" **Answers will vary. Students may say no and that they need to zoom in.**
- Students can zoom in near the point of intersection and check the table of values. Students should be able to reason why there is only one point of intersection and not two—two that would be very close together.

> **REMEMBER**
> The algebraic procedures that you use to solve nonlinear systems are similar to the procedures that you used to solve linear systems.

Solving Nonlinear Systems Algebraically

EXAMPLE 2 Solving a Nonlinear System by Substitution

Solve the system by substitution.

$y = x^2 + x - 1$ Equation 1
$y = -2x + 3$ Equation 2

SOLUTION

Step 1 The equations are already solved for y.

Step 2 Substitute $-2x + 3$ for y in Equation 1 and solve for x.

$-2x + 3 = x^2 + x - 1$ Substitute $-2x + 3$ for y in Equation 1.
$3 = x^2 + 3x - 1$ Add $2x$ to each side.
$0 = x^2 + 3x - 4$ Subtract 3 from each side.
$0 = (x + 4)(x - 1)$ Factor the polynomial.
$x + 4 = 0$ or $x - 1 = 0$ Zero-Product Property
$x = -4$ or $x = 1$ Solve for x.

Step 3 Substitute -4 and 1 for x in Equation 2 and solve for y.

$y = -2(-4) + 3$ Substitute for x in Equation 2. $y = -2(1) + 3$
$= 11$ Simplify. $= 1$

▶ So, the solutions are $(-4, 11)$ and $(1, 1)$.

> **Check**
> Use a graphing calculator to check your answer. Notice that the graphs have two points of intersection at $(-4, 11)$ and $(1, 1)$.
>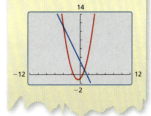

EXAMPLE 3 Solving a Nonlinear System by Elimination

Solve the system by elimination.

$y = x^2 - 3x - 2$ Equation 1
$y = -3x - 8$ Equation 2

SOLUTION

Step 1 Because the coefficients of the y-terms are the same, you do not need to multiply either equation by a constant.

Step 2 Subtract Equation 2 from Equation 1.

$y = x^2 - 3x - 2$ Equation 1
$\underline{y = -3x - 8}$ Equation 2
$0 = x^2 + 6$ Subtract the equations.

Step 3 Solve for x.

$0 = x^2 + 6$ Resulting equation from Step 2
$-6 = x^2$ Subtract 6 from each side.

▶ The square of a real number cannot be negative. So, the system has no real solutions.

> **Check**
> Use a graphing calculator to check your answer. The graphs do not intersect.

Section 4.8 Solving Nonlinear Systems of Equations 253

Extra Example 4

Approximate the solution(s) of the system to the nearest thousandth.

$y = x^2 - 2x - 5$
$y = 2^x + 1$

The solution of the system is about $(-1.703, 1.307)$.

MONITORING PROGRESS ANSWERS

4. $(0, 9)$
5. $(-3, 15), (1, -5)$
6. no solutions
7. about $(2.24, 7.24)$, about $(-2.24, 2.76)$
8. $\left(\frac{1}{3}, -\frac{7}{3}\right), \left(-\frac{2}{3}, -\frac{22}{3}\right)$
9. no solutions

Monitoring Progress Help in English and Spanish at *BigIdeasMath.com*

Solve the system by substitution.

4. $y = x^2 + 9$
   $y = 9$
5. $y = -5x$
   $y = x^2 - 3x - 3$
6. $y = -3x^2 + 2x + 1$
   $y = 5 - 3x$

Solve the system by elimination.

7. $y = x^2 + x$
   $y = x + 5$
8. $y = 9x^2 + 8x - 6$
   $y = 5x - 4$
9. $y = 2x + 5$
   $y = -3x^2 + x - 4$

Approximating Solutions

When you cannot find the exact solution(s) of a system of equations, you can analyze output values to approximate the solution(s).

EXAMPLE 4 Approximating Solutions of a Nonlinear System

Approximate the solution(s) of the system to the nearest thousandth.

$y = \frac{1}{2}x^2 + 3$ Equation 1
$y = 3^x$ Equation 2

SOLUTION

Sketch a graph of the system. You can see that the system has one solution between $x = 1$ and $x = 2$.

Substitute 3^x for y in Equation 1 and rewrite the equation.

$3^x = \frac{1}{2}x^2 + 3$ Substitute 3^x for y in Equation 1.

$3^x - \frac{1}{2}x^2 - 3 = 0$ Rewrite the equation.

Because you do not know how to solve this equation algebraically, let $f(x) = 3^x - \frac{1}{2}x^2 - 3$. Then evaluate the function for x-values between 1 and 2.

$f(1.1) \approx -0.26$
$f(1.2) \approx 0.02$ Because $f(1.1) < 0$ and $f(1.2) > 0$, the zero is between 1.1 and 1.2.

$f(1.2)$ is closer to 0 than $f(1.1)$, so decrease your guess and evaluate $f(1.19)$.

$f(1.19) \approx -0.012$ Because $f(1.19) < 0$ and $f(1.2) > 0$, the zero is between 1.19 and 1.2. So, increase guess.

$f(1.191) \approx -0.009$ Result is negative. Increase guess.
$f(1.192) \approx -0.006$ Result is negative. Increase guess.
$f(1.193) \approx -0.003$ Result is negative. Increase guess.
$f(1.194) \approx -0.0002$ Result is negative. Increase guess.
$f(1.195) \approx 0.003$ Result is positive.

Because $f(1.194)$ is closest to 0, $x \approx 1.194$.

Substitute $x = 1.194$ into one of the original equations and solve for y.

$y = \frac{1}{2}x^2 + 3 = \frac{1}{2}(1.194)^2 + 3 \approx 3.713$

▶ So, the solution of the system is about $(1.194, 3.713)$.

REMEMBER
The function values that are closest to 0 correspond to x-values that best approximate the zeros of the function.

Laurie's Notes — Teacher Actions

- **Think-Pair-Share:** Have students answer Questions 4–9, as needed, and then share and discuss as a class.
- **? Reason Abstractly and Quantitatively** and **Attend to Precision:** Pose Example 4. "Describe this system. What are the intersection possibilities?" *The system has a quadratic equation and exponential equation. There may be 0, 1, 2, or 3 points of intersection.*
- Have students use the table feature of their calculators to find an approximate solution. They should be able to explain their reasoning as they get closer and closer to an approximation of the solution.

REMEMBER

When entering the equations, be sure to use an appropriate viewing window that shows all the points of intersection. For this system, an appropriate viewing window is $-4 \le x \le 4$ and $-4 \le y \le 4$.

Recall that you can use systems of equations to solve equations with variables on both sides. To solve $f(x) = g(x)$, graph the system of equations $y = f(x)$ and $y = g(x)$. The x-value of each solution of the system is a solution of $f(x) = g(x)$.

EXAMPLE 5 Approximating Solutions of an Equation

Solve $-2(4)^x + 3 = 0.5x^2 - 2x$.

SOLUTION

You do not know how to solve this equation algebraically. So, use each side of the equation to write the system $y = -2(4)^x + 3$ and $y = 0.5x^2 - 2x$.

Method 1 Use a graphing calculator to graph the system. Then use the *intersect* feature to find the coordinates of each point of intersection.

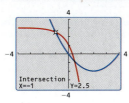
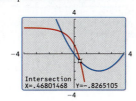

One point of intersection is $(-1, 2.5)$.

The other point of intersection is about $(0.47, -0.83)$.

▶ So, the solutions of the equation are $x = -1$ and $x \approx 0.47$.

Method 2 Use the *table* feature to create a table of values for the equations. Find the x-values for which the corresponding y-values are approximately equal.

STUDY TIP

You can use the differences between the corresponding y-values to determine the best approximation of a solution.

X	Y1	Y2
-1.03	2.5204	2.5905
-1.02	2.5137	2.5602
-1.01	2.5069	2.5301
-1	**2.5**	**2.5**
-.99	2.493	2.4701
-.98	2.4859	2.4402
-.97	2.4788	2.4105
X=-1		

X	Y1	Y2
.44	-.6808	-.7832
.45	-.7321	-.7988
.46	-.7842	-.8142
.47	**-.8371**	**-.8296**
.48	-.8906	-.8448
.49	-.9449	-.86
.50	-1	-.875
X=.47		

When $x = -1$, the corresponding y-values are 2.5.

When $x = 0.47$, the corresponding y-values are approximately -0.83.

▶ So, the solutions of the equation are $x = -1$ and $x \approx 0.47$.

Monitoring Progress Help in English and Spanish at BigIdeasMath.com

Use the method in Example 4 to approximate the solution(s) of the system to the nearest thousandth.

10. $y = 4^x$
    $y = x^2 + x + 3$

11. $y = 4x^2 - 1$
    $y = -2(3)^x + 4$

12. $y = x^2 + 3x$
    $y = -x^2 + x + 10$

Solve the equation. Round your solution(s) to the nearest hundredth.

13. $3^x - 1 = x^2 - 2x + 5$

14. $4x^2 + x = -2\left(\frac{1}{2}\right)^x + 5$

Extra Example 5

Solve $-(3^x) + 3 = x^2 - x$.
The solutions of the equation are $x \approx -1.23$ and $x = 1$.

MONITORING PROGRESS ANSWERS

10. about $(1.285, 5.938)$
11. about $(-1.045, 3.368)$, about $(0.565, 0.278)$
12. about $(1.791, 8.581)$, about $(-2.791, -0.583)$
13. $x \approx 1.51$
14. $x \approx 0.87, x \approx -0.77$

Laurie's Notes   Teacher Actions

- **Wait Time** and **Popsicle Sticks:** "What are the intersection possibilities for an exponential and a quadratic equation? Explain." Give *Wait Time* while students consider the question. It is likely they are considering the problem visually versus symbolically. *Popsicle Stick* responses.
- Pose Example 5 and give partners time to work the problem. Different strategies may be used, similar to the methods shown. Solicit responses.

Closure

- **Exit Ticket:** Solve the system using any method.
  $y = 2x + 2$
  $y = x^2 - x - 2$ $(-1, 0)$ and $(4, 10)$

Assignment Guide and Homework Check

ASSIGNMENT

Basic: 1, 2, 3–45 odd, 52, 56, 63–70

Average: 1, 2–56 even, 63–70

Advanced: 1, 2, 12, 18, 24–30 even, 38, 44–54 even, 55–70

HOMEWORK CHECK

Basic: 7, 13, 19, 33, 39

Average: 8, 14, 20, 34, 40

Advanced: 12, 18, 24, 38, 44

ANSWERS

1. Solve one of the equations for one of the variables and substitute into the other equation and solve.
2. *Sample answer:* Both can be solved by graphing; Some nonlinear systems cannot be solved algebraically.
3. B; (0, 1), (3, 4)
4. D; no solutions
5. A; (0, −1)
6. C; (−2, 5), (1, 2)
7. (2, 9), (−1, 6)
8. no solutions
9. (−1, 2)
10. (6, 4), (2, 0)
11. (3, 6), (−3, −6)
12. (−3, 14), (1, 2)
13. (0, −5), (−3, −8)
14. (−1, −3)
15. no solutions
16. (−3, −2), (1, 6)
17. (0, 5)
18. (2, 10), $\left(-\frac{3}{2}, -4\right)$
19. (4, −11), (−4, 29)
20. no solutions
21. (0, 2), (−6, −22)
22. no solutions
23. (1, 1)
24. no solutions
25. about (0.87, −1.74), about (−6.87, 13.74)
26. about (6.19, 5.38), about (0.81, −5.38)

4.8 Exercises

Dynamic Solutions available at *BigIdeasMath.com*

Vocabulary and Core Concept Check

1. **VOCABULARY** Describe how to use substitution to solve a system of nonlinear equations.
2. **WRITING** How is solving a system of nonlinear equations similar to solving a system of linear equations? How is it different?

Monitoring Progress and Modeling with Mathematics

In Exercises 3–6, match the system of equations with its graph. Then solve the system.

3. $y = x^2 - 2x + 1$
   $y = x + 1$

4. $y = x^2 + 3x + 2$
   $y = -x - 3$

5. $y = x - 1$
   $y = -x^2 + x - 1$

6. $y = -x + 3$
   $y = -x^2 - 2x + 5$

A.

B.

C.

D.

In Exercises 7–12, solve the system by graphing. *(See Example 1.)*

7. $y = 3x^2 - 2x + 1$
   $y = x + 7$

8. $y = x^2 + 2x + 5$
   $y = -2x - 5$

9. $y = -2x^2 - 4x$
   $y = 2$

10. $y = \frac{1}{2}x^2 - 3x + 4$
    $y = x - 2$

11. $y = \frac{1}{3}x^2 + 2x - 3$
    $y = 2x$

12. $y = 4x^2 + 5x - 7$
    $y = -3x + 5$

In Exercises 13–18, solve the system by substitution. *(See Example 2.)*

13. $y = x - 5$
    $y = x^2 + 4x - 5$

14. $y = -3x^2$
    $y = 6x + 3$

15. $y = -x + 7$
    $y = -x^2 - 2x - 1$

16. $y = -x^2 + 7$
    $y = 2x + 4$

17. $y - 5 = -x^2$
    $y = 5$

18. $y = 2x^2 + 3x - 4$
    $y - 4x = 2$

In Exercises 19–26, solve the system by elimination. *(See Example 3.)*

19. $y = x^2 - 5x - 7$
    $y = -5x + 9$

20. $y = -3x^2 + x + 2$
    $y = x + 4$

21. $y = -x^2 - 2x + 2$
    $y = 4x + 2$

22. $y = -2x^2 + x - 3$
    $y = 2x - 2$

23. $y = 2x - 1$
    $y = x^2$

24. $y = x^2 + x + 1$
    $y = -x - 2$

25. $y + 2x = 0$
    $y = x^2 + 4x - 6$

26. $y = 2x - 7$
    $y + 5x = x^2 - 2$

27. **ERROR ANALYSIS** Describe and correct the error in solving the system of equations by graphing.

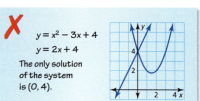

$y = x^2 - 3x + 4$
$y = 2x + 4$

The only solution of the system is (0, 4).

256   Chapter 4   Solving Quadratic Equations

27. The graph does not show both solutions.

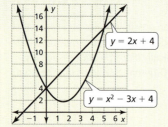

The solutions are (0, 4) and (5, 14).

28. **ERROR ANALYSIS** Describe and correct the error in solving for one of the variables in the system.

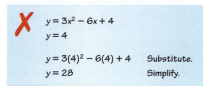

$y = 3x^2 - 6x + 4$
$y = 4$

$y = 3(4)^2 - 6(4) + 4$ Substitute.
$y = 28$ Simplify.

In Exercises 29–32, use the table to describe the locations of the zeros of the quadratic function f.

29.
x	−4	−3	−2	−1	0	1
f(x)	−2	2	4	4	2	−2

30.
x	−1	0	1	2	3	4
f(x)	11	5	1	−1	−1	1

31.
x	−4	−3	−2	−1	0	1
f(x)	3	−1	−1	3	11	23

32.
x	1	2	3	4	5	6
f(x)	−25	−9	1	5	3	−5

In Exercises 33–38, use the method in Example 4 to approximate the solution(s) of the system to the nearest thousandth. *(See Example 4.)*

33. $y = x^2 + 2x + 3$
    $y = 3^x$

34. $y = 2^x + 5$
    $y = x^2 - 3x + 1$

35. $y = 2(4)^x - 1$
    $y = 3x^2 + 8x$

36. $y = -x^2 - 4x - 4$
    $y = -5^x - 2$

37. $y = -x^2 - x + 5$
    $y = 2x^2 + 6x - 3$

38. $y = 2x^2 + x - 8$
    $y = x^2 - 5$

In Exercises 39–46, solve the equation. Round your solution(s) to the nearest hundredth. *(See Example 5.)*

39. $3x + 1 = x^2 + 7x - 1$

40. $-x^2 + 2x = -2x + 5$

41. $x^2 - 6x + 4 = -x^2 - 2x$

42. $2x^2 + 8x + 10 = -x^2 - 2x + 5$

43. $-4\left(\frac{1}{2}\right)^x = -x^2 - 5$

44. $1.5(2)^x - 3 = -x^2 + 4x$

45. $8^{x-2} + 3 = 2\left(\frac{3}{2}\right)^x$

46. $-0.5(4)^x = 5^x - 6$

47. **COMPARING METHODS** Solve the system in Exercise 37 using substitution. Compare the exact solutions to the approximated solutions.

48. **COMPARING METHODS** Solve the system in Exercise 38 using elimination. Compare the exact solutions to the approximated solutions.

49. **MODELING WITH MATHEMATICS** The attendances y for two movies can be modeled by the following equations, where x is the number of days since the movies opened.

    $y = -x^2 + 35x + 100$ Movie A
    $y = -5x + 275$ Movie B

    When is the attendance for each movie the same?

50. **MODELING WITH MATHEMATICS** You and a friend are driving boats on the same lake. Your path can be modeled by the equation $y = -x^2 - 4x - 1$, and your friend's path can be modeled by the equation $y = 2x + 8$. Do your paths cross each other? If so, what are the coordinates of the point(s) where the paths meet?

51. **MODELING WITH MATHEMATICS** The arch of a bridge can be modeled by $y = -0.002x^2 + 1.06x$, where x is the distance (in meters) from the left pylons and y is the height (in meters) of the arch above the water. The road can be modeled by the equation $y = 52$. To the nearest meter, how far from the left pylons are the two points where the road intersects the arch of the bridge?

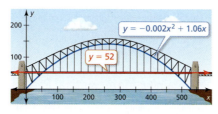

52. **MAKING AN ARGUMENT** Your friend says that a system of equations consisting of a linear equation and a quadratic equation can have zero, one, two, or infinitely many solutions. Is your friend correct? Explain.

ANSWERS

53. a–c. $(2, 11), (-2, -5)$
  Sample answer: elimination; The resulting equation can be written in the form $x^2 = d$.

54. a–c. $(-4, 11), (3, 4)$
  Sample answer: substitution; The resulting equation is easy to factor.

55. a. $y = 30x + 290$
  b. $(1, 320), (34, 1310)$

56. a. two solutions
  b. no solutions

57. See Additional Answers.

58. a. year 84
  b. yes; year 115

59. a. $(0, 4)$
  b. $(5, 9)$

60. yes; yes; Sample answer: When one equation is quadratic in x and the other is quadratic in y, there can be three or four solutions.

61. $(1, -6)$

62. $(4, -5), (-5, 4)$

63–70. See Additional Answers.

Mini-Assessment

1. Solve the system by graphing.
  $y = 2x^2 - 3x - 3$
  $y = x + 3$

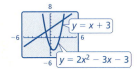

  $(-1, 2)$ and $(3, 6)$

2. Solve the system by substitution.
  $y = x^2 + 5x - 1$
  $y = x - 4$
  $(-3, -7)$ and $(-1, -5)$

3. Solve the system by elimination.
  $y = x^2 + 3x + 2$
  $y = 3x + 6$
  $(-2, 0)$ and $(2, 12)$

4. Approximate the solution(s) of the system to the nearest thousandth.
  $y = x^2 + x + 2$
  $y = 3^x - 3$
  about $(2.309, 9.642)$

5. Solve $2^x + 1 = x^2 + 3x - 1$.
  $x \approx -3.58$ and $x = 1$

258   Chapter 4

COMPARING METHODS In Exercises 53 and 54, solve the system of equations by (a) graphing, (b) substitution, and (c) elimination. Which method do you prefer? Explain your reasoning.

53. $y = 4x + 3$
  $y = x^2 + 4x - 1$

54. $y = x^2 - 5$
  $y = -x + 7$

55. MODELING WITH MATHEMATICS The function $y = -x^2 + 65x + 256$ models the number y of subscribers to a website, where x is the number of days since the website launched. The number of subscribers to a competitor's website can be modeled by a linear function. The websites have the same number of subscribers on Days 1 and 34.

  a. Write a linear function that models the number of subscribers to the competitor's website.
  b. Solve the system to verify the function from part (a).

56. HOW DO YOU SEE IT? The diagram shows the graphs of two equations in a system that has one solution.

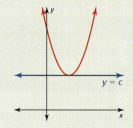

  a. How many solutions will the system have when you change the linear equation to $y = c + 2$?
  b. How many solutions will the system have when you change the linear equation to $y = c - 2$?

57. WRITING A system of equations consists of a quadratic equation whose graph opens up and a quadratic equation whose graph opens down. Describe the possible numbers of solutions of the system. Sketch examples to justify your answer.

Maintaining Mathematical Proficiency
Reviewing what you learned in previous grades and lessons

Graph the inequality in a coordinate plane. *(Skills Review Handbook)*

63. $y \leq 4x - 1$
64. $y > -\frac{1}{2}x + 3$
65. $4y - 12 \geq 8x$
66. $3y + 3 < x$

Graph the function. Describe the domain and range. *(Section 3.3)*

67. $y = 3x^2 + 2$
68. $y = -x^2 - 6x$
69. $y = -2x^2 + 12x - 7$
70. $y = 5x^2 + 10x - 3$

258   Chapter 4   Solving Quadratic Equations

---

58. PROBLEM SOLVING The population of a country is 2 million people and increases by 3% each year. The country's food supply is sufficient to feed 3 million people and increases at a constant rate that feeds 0.25 million additional people each year.

  a. When will the country first experience a food shortage?
  b. The country doubles the rate at which its food supply increases. Will food shortages still occur? If so, in what year?

59. ANALYZING GRAPHS Use the graphs of the linear and quadratic functions.

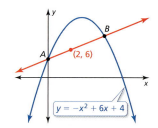

  a. Find the coordinates of point A.
  b. Find the coordinates of point B.

60. THOUGHT PROVOKING Is it possible for a system of two quadratic equations to have exactly three solutions? exactly four solutions? Explain your reasoning.

61. PROBLEM SOLVING Solve the system of three equations shown.
  $y = 2x - 8$
  $y = x^2 - 4x - 3$
  $y = -3(2)^x$

62. PROBLEM SOLVING Find the point(s) of intersection, if any, of the line $y = -x - 1$ and the circle $x^2 + y^2 = 41$.

If students need help...	If students got it...
Resources by Chapter • Practice A and Practice B • Puzzle Time	Resources by Chapter • Enrichment and Extension • Cumulative Review
Student Journal • Practice	Start the *next* Section
Differentiating the Lesson Skills Review Handbook	

Laurie's Notes

Overview of Section 4.9

Introduction
- **Connection:** In Math I, students solved linear inequalities in one and two variables. In one variable, a number line was used, and in two variables, the coordinate plane was used. The same types of problems and approaches are found with quadratic inequalities in this lesson.
- The lesson begins with solving a quadratic inequality graphically and then this is extended to a system of quadratic inequalities. The graphs are drawn, a test point is tried, and the appropriate region is shaded.
- In the second half of the lesson, an algebraic approach is used to solve quadratic inequalities in one variable. Although an algebraic approach is presented, a variety of strategies are possible.

Teaching Strategy
- **Selective Responses:** When students are working alone or with partners on a problem, teachers have the opportunity to circulate about the room to see different approaches. Students' work varies, revealing their thinking and conceptions about the mathematics. Discourse about the mathematics and student comprehension can both be improved if we are thoughtful in how responses (students' work) are displayed.
- I carry a clipboard with me as I circulate so that I can make notes about the order in which I want to call on students. If I leave it to chance when I call on students to share their solutions, then I lose the ability to control the sequence of responses. I do not want the first response to be the most polished, efficient, or eloquent. Knowing the outcome(s) I want from the activity or problem, I look for particular work that clearly demonstrates the outcome. If there is a particular mathematical practice that is evident in the students' work, I also want to highlight that.

Pacing Suggestion
- Once students have worked the explorations, continue with the formal lesson.

Dynamic Teaching Tools
Dynamic Assessment & Progress Monitoring Tool
Lesson Planning Tool
Interactive Whiteboard Lesson Library
Dynamic Classroom with Dynamic Investigations

What Your Students Will Learn

- Graph quadratic inequalities in two variables in the coordinate plane.
- Solve quadratic inequalities in one variable algebraically and graphically.

Laurie's Notes

Exploration

Motivate
- Ask whether any students enjoy rock climbing, and if so, what factors they think are important in making sure a rope will safely support their weight. Tell them that a quadratic inequality can be used to determine when a rope is strong enough to support them.

Discuss
- In Math I, students graphed linear inequalities in two variables. They used solid or dashed lines depending on whether the inequality included the equal to symbol. Using a test point, they shaded either above or below the boundary line. A similar process is used when graphing quadratic inequalities in two variables.

Exploration 1
- **Selective Responses:** As you circulate around you may hear different thinking about what it means for a quadratic expression to be less than or equal to 0, and for some the thinking may be incorrect.
- You may hear students discuss what $x^2 + 2x - 3 = 0$ means and then build on this understanding to think about what $x^2 + 2x - 3 \leq 0$ means.

COMMON ERROR Some students may think this means y-values less than 0, meaning Quadrants III and IV. Other students may think this means x-values less than 0, meaning Quadrants II and III.

- Students may enter the quadratic equation in their calculator and look at the table of values. They will observe that when $-3 \leq x \leq 1$, the value of the function is negative or 0.

Exploration 2
- **Reason Abstractly and Quantitatively:** Students have prior experience with matching graphs and equations. The purpose is not to use a calculator and quickly match. Instead you are listening for students to reason based on their knowledge of quadratic equations in standard form.
- You should expect students to make observations such as only one graph is symmetric about the y-axis, so it must match inequality (f), the only quadratic without a linear term. Two of the graphs have a positive y-intercept.
- Following the reasoning of Exploration 1, students should be able to solve the inequality once matched with the correct graph.

Communicate Your Answer
- **Turn and Talk:** Have partners discuss Question 4.

Connecting to Next Step
- The explorations help students develop an understanding of quadratic inequalities in one variable. In the formal lesson, students will solve quadratic inequalities in two variables.

4.9 Quadratic Inequalities

Essential Question How can you solve a quadratic inequality?

EXPLORATION 1 Solving a Quadratic Inequality

Work with a partner. The graphing calculator screen shows the graph of

$$f(x) = x^2 + 2x - 3.$$

Explain how you can use the graph to solve the inequality

$$x^2 + 2x - 3 \leq 0.$$

Then solve the inequality.

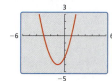

USING TOOLS STRATEGICALLY

To be proficient in math, you need to use technological tools to explore your understanding of concepts.

EXPLORATION 2 Solving Quadratic Inequalities

Work with a partner. Match each inequality with the graph of its related quadratic function. Then use the graph to solve the inequality.

a. $x^2 - 3x + 2 > 0$ b. $x^2 - 4x + 3 \leq 0$ c. $x^2 - 2x - 3 < 0$

d. $x^2 + x - 2 \geq 0$ e. $x^2 - x - 2 < 0$ f. $x^2 - 4 > 0$

A. 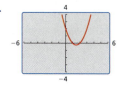   B.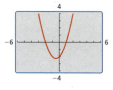

C. 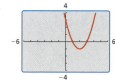   D.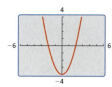

E. 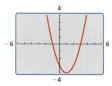   F.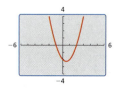

Communicate Your Answer

3. How can you solve a quadratic inequality?

4. Explain how you can use the graph in Exploration 1 to solve each inequality. Then solve each inequality.

   a. $x^2 + 2x - 3 > 0$ b. $x^2 + 2x - 3 < 0$ c. $x^2 + 2x - 3 \geq 0$

Dynamic Teaching Tools

Dynamic Assessment & Progress Monitoring Tool
Lesson Planning Tool
Interactive Whiteboard Lesson Library
Dynamic Classroom with Dynamic Investigations

ANSWERS

1. The x-values for which the graph touches or is below the x-axis are solutions; $-3 \leq x \leq 1$

2. a. A; $x < 1$ or $x > 2$
   b. C; $1 \leq x \leq 3$
   c. E; $-1 < x < 3$
   d. B; $x \leq -2$ or $x \geq 1$
   e. F; $-1 < x < 2$
   f. D; $x < -2$ or $x > 2$

3. Rewrite the inequality so that one side is 0. Graph the related quadratic function to see where the graph is above or below the x-axis.

4. Use the graph to determine for what x-values the function is positive and negative.
   a. $x < -3$ or $x > 1$
   b. $-3 < x < 1$
   c. $x \leq -3$ or $x \geq 1$

Differentiated Instruction

Visual

Have students visually compare the graphs of a quadratic equation and a quadratic inequality with the same terms. Have them identify the points that are solutions to each. Guide students to recognize that there are points that satisfy the inequality but not the equation. Explain that the dotted line in a graph is significant because it indicates that the points that satisfy the equation do not satisfy the inequality.

Extra Example 1

Graph $y > -x^2 - 2x + 15$.

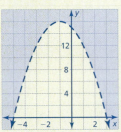

4.9 Lesson

What You Will Learn

- Graph quadratic inequalities in two variables.
- Solve quadratic inequalities in one variable.

Core Vocabulary

quadratic inequality in two variables, *p. 260*
quadratic inequality in one variable, *p. 262*

Previous
linear inequality in two variables

Graphing Quadratic Inequalities in Two Variables

A **quadratic inequality in two variables** can be written in one of the following forms, where a, b, and c are real numbers and $a \neq 0$.

$$y < ax^2 + bx + c \qquad y > ax^2 + bx + c$$
$$y \leq ax^2 + bx + c \qquad y \geq ax^2 + bx + c$$

The graph of any such inequality consists of all solutions (x, y) of the inequality.

Previously, you graphed linear inequalities in two variables. You can use a similar procedure to graph quadratic inequalities in two variables.

Core Concept

Graphing a Quadratic Inequality in Two Variables

To graph a quadratic inequality in one of the forms above, follow these steps.

Step 1 Graph the parabola with the equation $y = ax^2 + bx + c$. Make the parabola *dashed* for inequalities with $<$ or $>$ and *solid* for inequalities with $\leq$ or $\geq$.

Step 2 Test a point (x, y) inside the parabola to determine whether the point is a solution of the inequality.

Step 3 Shade the region inside the parabola if the point from Step 2 is a solution. Shade the region outside the parabola if it is not a solution.

EXAMPLE 1 Graphing a Quadratic Inequality in Two Variables

Graph $y < -x^2 - 2x - 1$.

SOLUTION

Step 1 Graph $y = -x^2 - 2x - 1$. Because the inequality symbol is $<$, make the parabola dashed.

Step 2 Test a point inside the parabola, such as $(0, -3)$.

$$y < -x^2 - 2x - 1$$
$$-3 \stackrel{?}{<} -0^2 - 2(0) - 1$$
$$-3 < -1 \checkmark$$

So, $(0, -3)$ is a solution of the inequality.

Step 3 Shade the region inside the parabola.

LOOKING FOR STRUCTURE

Notice that testing a point is less complicated when the *x*-value is 0 (the point is on the *y*-axis).

260   Chapter 4   Solving Quadratic Equations

Laurie's Notes   Teacher Actions

- **Connection:** A line divides the plane into two half-planes. A parabola divides the coordinate plane into two regions, the ordered pairs inside the parabola and the ordered pairs outside the parabola. Dashed and solid lines were used when working with linear inequalities in two variables. Dashed and solid parabolas will be used for quadratic inequalities in two variables.
- ❓ "What would be a good test point to use and why?" *a point on the y-axis where $x = 0$, so that the computation is easy*

EXAMPLE 2 Using a Quadratic Inequality in Real Life

A manila rope used for rappelling down a cliff can safely support a weight W (in pounds) provided

$$W \leq 1480d^2$$

where d is the diameter (in inches) of the rope. Graph the inequality and interpret the solution.

SOLUTION

Graph $W = 1480d^2$ for nonnegative values of d. Because the inequality symbol is $\leq$, make the parabola solid. Test a point inside the parabola, such as (1, 3000).

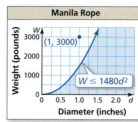

Manila Rope

$$W \leq 1480d^2$$
$$3000 \overset{?}{\leq} 1480(1)^2$$
$$3000 \not\leq 1480$$

▶ Because (1, 3000) is not a solution, shade the region outside the parabola. The shaded region represents weights that can be supported by ropes with various diameters.

Graphing a *system* of quadratic inequalities is similar to graphing a system of linear inequalities. First graph each inequality in the system. Then identify the region in the coordinate plane common to all of the graphs. This region is called the *graph of the system*.

EXAMPLE 3 Graphing a System of Quadratic Inequalities

Graph the system of quadratic inequalities.

$$y < -x^2 + 3 \quad \text{Inequality 1}$$
$$y \geq x^2 + 2x - 3 \quad \text{Inequality 2}$$

SOLUTION

Step 1 Graph $y < -x^2 + 3$. The graph is the red region inside (but not including) the parabola $y = -x^2 + 3$.

Step 2 Graph $y \geq x^2 + 2x - 3$. The graph is the blue region inside and including the parabola $y = x^2 + 2x - 3$.

Step 3 Identify the purple region where the two graphs overlap. This region is the graph of the system.

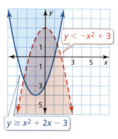

Check

Check that a point in the solution region, such as (0, 0), is a solution of the system.

$$y < -x^2 + 3$$
$$0 \overset{?}{<} -0^2 + 3$$
$$0 < 3 \checkmark$$

$$y \geq x^2 + 2x - 3$$
$$0 \overset{?}{\geq} 0^2 + 2(0) - 3$$
$$0 \geq -3 \checkmark$$

Monitoring Progress Help in English and Spanish at *BigIdeasMath.com*

Graph the inequality.

1. $y \geq x^2 + 2x - 8$
2. $y \leq 2x^2 - x - 1$
3. $y > -x^2 + 2x + 4$

4. Graph the system of inequalities consisting of $y \leq -x^2$ and $y > x^2 - 3$.

English Language Learners

Visual Aids

Encourage students to use a different color of shading or different pattern of shading for the graphed solution of each inequality in the system. Emphasize that the graph of the system is the overlapped area that contains both colors (or patterns) of shading.

Extra Example 2

The allowable bending load M (in pounds) of a certain size concrete beam can be expressed as $M \leq 540d^2$, where d is the diameter (in inches) of the steel reinforcing bar. Graph the inequality and interpret the solution.

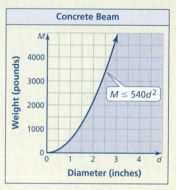

The shaded region represents the allowable bending loads for the concrete beam.

Extra Example 3

Graph the system of quadratic inequalities.
$$y > x^2 - 4$$
$$y \leq -x^2 + 3x + 4$$

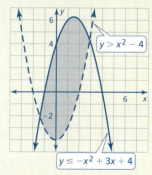

MONITORING PROGRESS ANSWERS

1–4. See Additional Answers.

Laurie's Notes — Teacher Actions

- **Reason Abstractly and Quantitatively** and **Use Appropriate Tools Strategically**: "What would be a good viewing window for $y \leq 1480x^2$?" Answers will vary. Students should mention the context and reasonable values for x and y. Only first quadrant values make sense in the context of the situation.
- ❓ "In Example 2, what would an ordered pair inside the parabola represent?" an unsafe weight for the diameter of the rope
- ❓ "In Example 3, is (0, 3) a solution of the system? Explain." no; (0, 3) is a solution of $y \geq x^2 + 2x - 3$ but is not a solution of $y < -x^2 + 3$.

Extra Example 4
Solve $x^2 + 4x - 32 > 0$ algebraically.
$x < -8$ or $x > 4$

Extra Example 5
Solve $2x^2 + 4x - 3 \leq 0$ by graphing.

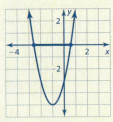

approximately $-2.58 \leq x \leq 0.58$

Solving Quadratic Inequalities in One Variable

A **quadratic inequality in one variable** can be written in one of the following forms, where a, b, and c are real numbers and $a \neq 0$.

$ax^2 + bx + c < 0 \qquad ax^2 + bx + c > 0 \qquad ax^2 + bx + c \leq 0 \qquad ax^2 + bx + c \geq 0$

You can solve quadratic inequalities using algebraic methods or graphs.

EXAMPLE 4 Solving a Quadratic Inequality Algebraically

Solve $x^2 - 3x - 4 < 0$ algebraically.

SOLUTION

First, write and solve the equation obtained by replacing $<$ with $=$.

$x^2 - 3x - 4 = 0$ Write the related equation.
$(x - 4)(x + 1) = 0$ Factor.
$x = 4 \quad \text{or} \quad x = -1$ Zero-Product Property

The numbers -1 and 4 are the *critical values* of the original inequality. Plot -1 and 4 on a number line, using open dots because the values do not satisfy the inequality. The critical x-values partition the number line into three intervals. Test an x-value in each interval to determine whether it satisfies the inequality.

$(-2)^2 - 3(-2) - 4 = 6 \not< 0 \qquad 0^2 - 3(0) - 4 = -4 < 0 \checkmark \qquad 5^2 - 3(5) - 4 = 6 \not< 0$

▶ So, the solution is $-1 < x < 4$.

Another way to solve $ax^2 + bx + c < 0$ is to first graph the related function $y = ax^2 + bx + c$. Then, because the inequality symbol is $<$, identify the x-values for which the graph lies *below* the x-axis. You can use a similar procedure to solve quadratic inequalities that involve $\leq$, $>$, or $\geq$.

EXAMPLE 5 Solving a Quadratic Inequality by Graphing

Solve $3x^2 - x - 5 \geq 0$ by graphing.

SOLUTION

The solution consists of the x-values for which the graph of $y = 3x^2 - x - 5$ lies on or above the x-axis. Find the x-intercepts of the graph by letting $y = 0$ and using the Quadratic Formula to solve $0 = 3x^2 - x - 5$ for x.

$x = \dfrac{-(-1) \pm \sqrt{(-1)^2 - 4(3)(-5)}}{2(3)}$ $a = 3, b = -1, c = -5$

$x = \dfrac{1 \pm \sqrt{61}}{6}$ Simplify.

The solutions are $x \approx -1.14$ and $x \approx 1.47$. Sketch a parabola that opens up and has -1.14 and 1.47 as x-intercepts. The graph lies on or above the x-axis to the left of (and including) $x = -1.14$ and to the right of (and including) $x = 1.47$.

▶ The solution of the inequality is approximately $x \leq -1.14$ or $x \geq 1.47$.

Laurie's Notes — Teacher Actions

- **Turn and Talk:** Pose Example 4 and give time for students to discuss strategies with their partners. Discuss as a class.
- **Selective Responses:** This is an example in which students may use different strategies. Allow ample work time before asking selected students to share their approach. Some may use a table of values, look at the graph, or factor and use a number line.
- Different approaches are possible for Example 5. Again, circulate to determine the order for soliciting answers.

EXAMPLE 6 Modeling with Mathematics

A rectangular parking lot must have a perimeter of 440 feet and an area of at least 8000 square feet. Describe the possible lengths of the parking lot.

SOLUTION

1. **Understand the Problem** You are given the perimeter and the minimum area of a parking lot. You are asked to determine the possible lengths of the parking lot.
2. **Make a Plan** Use the perimeter and area formulas to write a quadratic inequality describing the possible lengths of the parking lot. Then solve the inequality.
3. **Solve the Problem** Let ℓ represent the length (in feet) and let w represent the width (in feet) of the parking lot.

$$\text{Perimeter} = 440 \qquad \text{Area} \geq 8000$$
$$2\ell + 2w = 440 \qquad \ell w \geq 8000$$

Solve the perimeter equation for w to obtain $w = 220 - \ell$. Substitute this into the area inequality to obtain a quadratic inequality in one variable.

$$\ell w \geq 8000 \qquad \text{Write the area inequality.}$$
$$\ell(220 - \ell) \geq 8000 \qquad \text{Substitute } 220 - \ell \text{ for } w.$$
$$220\ell - \ell^2 \geq 8000 \qquad \text{Distributive Property}$$
$$-\ell^2 + 220\ell - 8000 \geq 0 \qquad \text{Write in standard form.}$$

Use a graphing calculator to find the ℓ-intercepts of $y = -\ell^2 + 220\ell - 8000$.

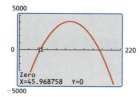 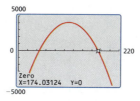

The ℓ-intercepts are $\ell \approx 45.97$ and $\ell \approx 174.03$. The solution consists of the ℓ-values for which the graph lies on or above the ℓ-axis. The graph lies on or above the ℓ-axis when $45.97 \leq \ell \leq 174.03$.

▶ So, the approximate length of the parking lot is at least 46 feet and at most 174 feet.

4. **Look Back** Choose a length in the solution region, such as $\ell = 100$, and find the width. Then check that the dimensions satisfy the original area inequality.

$$2\ell + 2w = 440 \qquad \ell w \geq 8000$$
$$2(100) + 2w = 440 \qquad 100(120) \stackrel{?}{\geq} 8000$$
$$w = 120 \qquad 12{,}000 \geq 8000 \checkmark$$

ANOTHER WAY
You can graph each side of $220\ell - \ell^2 = 8000$ and use the intersection points to determine when $220\ell - \ell^2$ is greater than or equal to 8000.

USING TECHNOLOGY
Variables displayed when using technology may not match the variables used in applications. In the graphs shown, the length ℓ corresponds to the independent variable x.

Monitoring Progress Help in English and Spanish at *BigIdeasMath.com*

Solve the inequality.

5. $2x^2 + 3x \leq 2$ 6. $-3x^2 - 4x + 1 < 0$ 7. $2x^2 + 2 > -5x$

8. **WHAT IF?** In Example 6, the area must be at least 8500 square feet. Describe the possible lengths of the parking lot.

Section 4.9   Quadratic Inequalities   263

Extra Example 6
A rectangular playground must have a perimeter of 320 meters and an area of at least 5000 square meters. Describe the possible lengths of the playground.
The approximate length is at least 43 meters and at most 117 meters.

MONITORING PROGRESS ANSWERS

5. $-2 \leq x \leq \frac{1}{2}$
6. about $x < -1.55$ or $x > 0.22$
7. $x < -2$ or $x > -\frac{1}{2}$
8. at least 50 ft and at most 170 ft

Laurie's Notes   **Teacher Actions**

- Have students write an equation for perimeter and an inequality for area and then discuss an approach with their partners. Some students may graph the two equations even though they have not worked with rational equations $\left(y = \dfrac{8000}{x}\right)$. Other students may use substitution, factor, and solve analytically. Take time to share a variety of approaches.
- Discuss the *Using Technology* note in the side column.

Closure

- **Exit Ticket:** Solve algebraically and graphically: $x^2 + 10x + 9 \geq 0$. $x \leq -9$ or $x \geq -1$

Section 4.9   263

Assignment Guide and Homework Check

ASSIGNMENT

Basic: 1, 2, 3–41 odd, 48, 50, 54–60

Average: 1, 2–44 even, 45, 48, 50, 54–60

Advanced: 1, 2, 14–18 even, 19, 20–42 even, 44–52, 54–60

HOMEWORK CHECK

Basic: 11, 19, 21, 29, 35

Average: 12, 20, 24, 32, 40

Advanced: 14, 19, 26, 34, 42

ANSWERS

1. The graph of a quadratic inequality in one variable consists of a number line, but the graph of a quadratic inequality in two variables consists of both the x- and y-axis.

2. To solve using algebraic methods, solve the equation $x^2 + 6x - 8 = 0$ to find the critical values of $x \approx 1.12$ and $x \approx -7.12$, and create three intervals. Choose a value within each interval to test the inequality. To solve graphically, graph the related quadratic function to determine for which intervals the graph is below the x-axis.

3. C; The x-intercepts are $x = -1$ and $x = -3$. The test point $(-2, 5)$ does not satisfy the inequality.

4. A; The x-intercepts are $x = 1$ and $x = 3$. The test point $(2, 5)$ satisfies the inequality.

5. B; The x-intercepts are $x = 1$ and $x = 3$. The test point $(2, 5)$ does not satisfy the inequality.

6. D; The x-intercepts are $x = -1$ and $x = -3$. The test point $(-2, 5)$ satisfies the inequality.

7.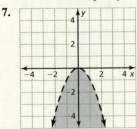

4.9 Exercises

Dynamic Solutions available at BigIdeasMath.com

Vocabulary and Core Concept Check

1. **WRITING** Compare the graph of a quadratic inequality in one variable to the graph of a quadratic inequality in two variables.

2. **WRITING** Explain how to solve $x^2 + 6x - 8 < 0$ using algebraic methods and using graphs.

Monitoring Progress and Modeling with Mathematics

In Exercises 3–6, match the inequality with its graph. Explain your reasoning.

3. $y \leq x^2 + 4x + 3$
4. $y > -x^2 + 4x - 3$
5. $y < x^2 - 4x + 3$
6. $y \geq x^2 + 4x + 3$

A.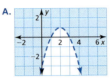
B.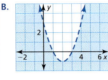
C.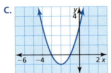
D.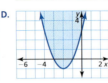

In Exercises 7–14, graph the inequality. (See Example 1.)

7. $y < -x^2$
8. $y \geq 4x^2$
9. $y > x^2 - 9$
10. $y < x^2 + 5$
11. $y \leq x^2 + 5x$
12. $y \geq -2x^2 + 9x - 4$
13. $y > 2(x + 3)^2 - 1$
14. $y \leq \left(x - \frac{1}{2}\right)^2 + \frac{5}{2}$

ANALYZING RELATIONSHIPS In Exercises 15 and 16, use the graph to write an inequality in terms of $f(x)$ so point P is a solution.

15.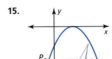
16.

ERROR ANALYSIS In Exercises 17 and 18, describe and correct the error in graphing $y \geq x^2 + 2$.

17.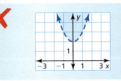

18.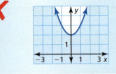

19. **MODELING WITH MATHEMATICS** A hardwood shelf in a wooden bookcase can safely support a weight W (in pounds) provided $W \leq 115x^2$, where x is the thickness (in inches) of the shelf. Graph the inequality and interpret the solution. (See Example 2.)

20. **MODELING WITH MATHEMATICS** A wire rope can safely support a weight W (in pounds) provided $W \leq 8000d^2$, where d is the diameter (in inches) of the rope. Graph the inequality and interpret the solution.

In Exercises 21–26, graph the system of quadratic inequalities. (See Example 3.)

21. $y \geq 2x^2$
    $y < -x^2 + 1$

22. $y > -5x^2$
    $y > 3x^2 - 2$

23. $y \leq -x^2 + 4x - 4$
    $y < x^2 + 2x - 8$

24. $y \geq x^2 - 4$
    $y \leq -2x^2 + 7x + 4$

25. $y \geq 2x^2 + x - 5$
    $y < -x^2 + 5x + 10$

26. $y \geq x^2 - 3x - 6$
    $y \geq x^2 + 7x + 6$

264 Chapter 4 Solving Quadratic Equations

8.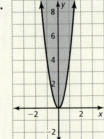

9.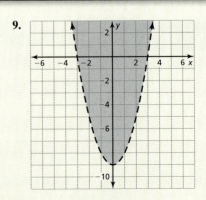

10–26. See Additional Answers.

264 Chapter 4

In Exercises 27–34, solve the inequality algebraically.
(See Example 4.)

27. $4x^2 < 25$
28. $x^2 + 10x + 9 < 0$
29. $x^2 - 11x \geq -28$
30. $3x^2 - 13x > -10$
31. $2x^2 - 5x - 3 \leq 0$
32. $4x^2 + 8x - 21 \geq 0$
33. $\frac{1}{2}x^2 - x > 4$
34. $-\frac{1}{2}x^2 + 4x \leq 1$

In Exercises 35–42, solve the inequality by graphing.
(See Example 5.)

35. $x^2 - 3x + 1 < 0$
36. $x^2 - 4x + 2 > 0$
37. $x^2 + 8x > -7$
38. $x^2 + 6x < -3$
39. $3x^2 - 8 \leq -2x$
40. $3x^2 + 5x - 3 < 1$
41. $\frac{1}{3}x^2 + 2x \geq 2$
42. $\frac{3}{4}x^2 + 4x \geq 3$

43. **DRAWING CONCLUSIONS** Consider the graph of the function $f(x) = ax^2 + bx + c$.

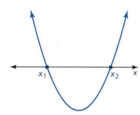

a. What are the solutions of $ax^2 + bx + c < 0$?
b. What are the solutions of $ax^2 + bx + c > 0$?
c. The graph of g represents a reflection in the x-axis of the graph of f. For which values of x is $g(x)$ positive?

44. **MODELING WITH MATHEMATICS** A rectangular fountain display has a perimeter of 400 feet and an area of at least 9100 feet. Describe the possible widths of the fountain. (See Example 6.)

45. **MODELING WITH MATHEMATICS** The arch of the Sydney Harbor Bridge in Sydney, Australia, can be modeled by $y = -0.00211x^2 + 1.06x$, where x is the distance (in meters) from the left pylons and y is the height (in meters) of the arch above the water. For what distances x is the arch above the road?

46. **PROBLEM SOLVING** The number T of teams that have participated in a robot-building competition for high-school students over a recent period of time x (in years) can be modeled by

$$T(x) = 17.155x^2 + 193.68x + 235.81,\ 0 \leq x \leq 6.$$

After how many years is the number of teams greater than 1000? Justify your answer.

47. **PROBLEM SOLVING** A study found that a driver's reaction time $A(x)$ to audio stimuli and his or her reaction time $V(x)$ to visual stimuli (both in milliseconds) can be modeled by

$$A(x) = 0.0051x^2 - 0.319x + 15,\ 16 \leq x \leq 70$$
$$V(x) = 0.005x^2 - 0.23x + 22,\ 16 \leq x \leq 70$$

where x is the age (in years) of the driver.

a. Write an inequality that you can use to find the x-values for which $A(x)$ is less than $V(x)$.
b. Use a graphing calculator to solve the inequality $A(x) < V(x)$. Describe how you used the domain $16 \leq x \leq 70$ to determine a reasonable solution.
c. Based on your results from parts (a) and (b), do you think a driver would react more quickly to a traffic light changing from green to yellow or to the siren of an approaching ambulance? Explain.

ANSWERS

27. $-\frac{5}{2} < x < \frac{5}{2}$
28. $-9 < x < -1$
29. $x \leq 4$ or $x \geq 7$
30. $x < 1$ or $x > \frac{10}{3}$
31. $-0.5 \leq x \leq 3$
32. $x \leq -3.5$ or $x \geq 1.5$
33. $x < -2$ or $x > 4$
34. about $x \leq 0.26$ or $x \geq 7.74$
35. about $0.38 < x < 2.62$
36. about $x < 0.59$ or $x > 3.41$
37. $x < -7$ or $x > -1$
38. about $-5.45 < x < 0.55$
39. $-2 \leq x \leq \frac{4}{3}$
40. about $-2.26 < x < 0.59$
41. about $x \leq -6.87$ or $x \geq 0.87$
42. $x \leq -6$ or $x \geq \frac{2}{3}$
43. a. $x_1 < x < x_2$
    b. $x < x_1$ or $x > x_2$
    c. $x_1 < x < x_2$
44. at least 70 ft and at most 130 ft
45. about 55 m from the left pylon to about 447 m from the left pylon
46. 4; $T(x) > 1000$ when $x > 3.1$.
47. a. $0.0051x^2 - 0.319x + 15 < 0.005x^2 - 0.23x + 22$, $16 \leq x \leq 70$
    b. $A(x) < V(x)$ for $16 \leq x \leq 70$; Graph the inequalities only on $16 \leq x \leq 70$. $A(x)$ is always less than $V(x)$.
    c. The driver would react more quickly to the siren of an approaching ambulance; The reaction time to audio stimuli is always less.

ANSWERS

48. **a.** Sample answer: (2, 0) and (3, 1)
  b. no; The lines are dashed.
  c. no; Because both points are points of intersection, they are either both solutions or both not solutions.

49–60. See Additional Answers.

Mini-Assessment

1. Graph $y < -x^2 - 4x + 12$.

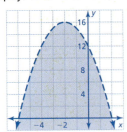

2. Graph the system of quadratic inequalities.
$y < -x^2 + 1$
$y > x^2 + 2x - 5$

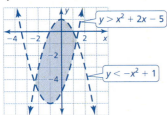

3. Solve $x^2 + 5x - 24 \leq 0$ algebraically. $-8 \leq x \leq 3$

4. Solve $4x^2 + 2x - 6 > 0$ by graphing.

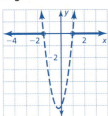

$x < -1.5$ or $x > 1$

5. A rectangular tree lot must have a perimeter of 100 yards and an area of at least 500 square yards. Describe the possible lengths of the tree lot. The approximate length is at least 14 yards and at most 36 yards.

266   Chapter 4

---

48. HOW DO YOU SEE IT? The graph shows a system of quadratic inequalities.

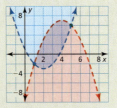

  a. Identify two solutions of the system.
  b. Are the points (1, −2) and (5, 6) solutions of the system? Explain.
  c. Is it possible to change the inequality symbol(s) so that one, but not both, of the points in part (b) is a solution of the system? Explain.

49. MODELING WITH MATHEMATICS The length L (in millimeters) of the larvae of the black porgy fish can be modeled by
$$L(x) = 0.00170x^2 + 0.145x + 2.35, \ 0 \leq x \leq 40$$
where x is the age (in days) of the larvae. Write and solve an inequality to find at what ages a larva's length tends to be greater than 10 millimeters. Explain how the given domain affects the solution.

50. MAKING AN ARGUMENT You claim the system of inequalities below, where a and b are real numbers, has no solution. Your friend claims the system will always have at least one solution. Who is correct? Explain.
$y < (x + a)^2$
$y < (x + b)^2$

51. MATHEMATICAL CONNECTIONS The area A of the region bounded by a parabola and a horizontal line can be modeled by $A = \frac{2}{3}bh$, where b and h are as defined in the diagram. Find the area of the region determined by each pair of inequalities.

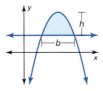

  a. $y \leq -x^2 + 4x$
     $y \geq 0$
  b. $y \geq x^2 - 4x - 5$
     $y \leq 7$

52. THOUGHT PROVOKING Draw a company logo that is created by the intersection of two quadratic inequalities. Justify your answer.

53. REASONING A truck that is 11 feet tall and 7 feet wide is traveling under an arch. The arch can be modeled by $y = -0.0625x^2 + 1.25x + 5.75$, where x and y are measured in feet.

  a. Will the truck fit under the arch? Explain.
  b. What is the maximum width that a truck 11 feet tall can have and still make it under the arch?
  c. What is the maximum height that a truck 7 feet wide can have and still make it under the arch?

Maintaining Mathematical Proficiency *Reviewing what you learned in previous grades and lessons*

Tell whether the function has a minimum value or maximum value. Then find the value. *(Section 3.3)*

54. $f(x) = -x^2 - 6x - 10$
55. $f(x) = -x^2 - 4x + 21$
56. $h(x) = \frac{1}{2}x^2 + 2x + 2$
57. $h(x) = x^2 + 3x - 18$

Find the zeros of the function. *(Section 3.5)*

58. $f(x) = (x + 7)(x - 9)$
59. $g(x) = x^2 - 4x$
60. $h(x) = -x^2 + 5x - 6$

266   Chapter 4   Solving Quadratic Equations

If students need help...	If students got it...
Resources by Chapter • Practice A and Practice B • Puzzle Time	**Resources by Chapter** • Enrichment and Extension • Cumulative Review
Student Journal • Practice	Start the *next* Section
Differentiating the Lesson Skills Review Handbook	

4.6–4.9 What Did You Learn?

Core Vocabulary

imaginary unit *i*, *p. 238*
complex number, *p. 238*
imaginary number, *p. 238*
pure imaginary number, *p. 238*
complex conjugates, *p. 241*
system of nonlinear equations, *p. 252*
quadratic inequality in two variables, *p. 260*
quadratic inequality in one variable, *p. 262*

Core Concepts

Section 4.6
The Square Root of a Negative Number, *p. 238*
Sums and Differences of Complex Numbers, *p. 239*
Multiplying Complex Numbers, *p. 240*

Section 4.7
Finding Solutions and Zeros, *p. 246*
Using the Discriminant, *p. 247*

Section 4.8
Solving Nonlinear Systems, *p. 252*
Approximating Solutions, *p. 254*

Section 4.9
Graphing a Quadratic Inequality in Two Variables, *p. 260*
Solving Quadratic Inequalities in One Variable, *p. 262*

Mathematical Practices

1. How can you use technology to determine whose rocket lands first in part (b) of Exercise 36 on page 250?
2. Compare the methods used to solve Exercise 53 on page 258. Discuss the similarities and differences among the methods.
3. Explain your plan to find the possible widths of the fountain in Exercise 44 on page 265.

Performance Task:

The Golden Ratio

The golden ratio is one of the most famous numbers in history. It can be seen in the simplicity of a seashell as well as in the grandeur of the Greek Parthenon. What is this number and how can it help you decorate your room?

To explore the answers to these questions and more, check out the Performance Task and Real-Life STEM video at **BigIdeasMath.com**.

Dynamic Teaching Tools

Dynamic Assessment & Progress Monitoring Tool
Interactive Whiteboard Lesson Library
Dynamic Classroom with Dynamic Investigations

ANSWERS

1. Use a graphing calculator to see where each parabola crosses the *x*-axis. The parabola that crosses the *x*-axis at the lesser *x*-value lands first.
2. *Sample answer:* All the methods had the same results. Substitution and elimination both resulted in equations that could easily be put in the form $x^2 = d$. Graphing did not require algebraic manipulation.
3. Set up an equation for the perimeter, $2\ell + 2w = 400$, and an inequality for the area $\ell w > 9100$. Solve the perimeter equation for ℓ and substitute the result into the inequality for the area. Solve the resulting inequality.

ANSWERS

1. $6p^3\sqrt{2p}$
2. $\dfrac{3\sqrt{35y}}{7y}$
3. $\dfrac{5x^3\sqrt[3]{2x^2}}{2}$
4. $4\sqrt{6} - 8$
5. $14\sqrt{3}$
6. $9\sqrt[3]{2}$
7. $88 + 30\sqrt{7}$
8. $10\sqrt{3}$
9. $x = 6, x = 3$
10. no solutions
11. $x = -4$
12. $-5, -2$
13.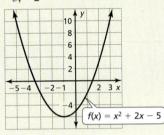

$-3.4, 1.4$

4 Chapter Review

Dynamic Solutions available at *BigIdeasMath.com*

4.1 Properties of Radicals (pp. 191–200)

a. Simplify $\sqrt[3]{27x^{10}}$.

$\sqrt[3]{27x^{10}} = \sqrt[3]{27 \cdot x^9 \cdot x}$ — Factor using the greatest perfect cube factors.

$= \sqrt[3]{27} \cdot \sqrt[3]{x^9} \cdot \sqrt[3]{x}$ — Product Property of Cube Roots

$= 3x^3\sqrt[3]{x}$ — Simplify.

b. Simplify $\dfrac{12}{3 + \sqrt{5}}$.

$\dfrac{12}{3+\sqrt{5}} = \dfrac{12}{3+\sqrt{5}} \cdot \dfrac{3-\sqrt{5}}{3-\sqrt{5}}$ — The conjugate of $3 + \sqrt{5}$ is $3 - \sqrt{5}$.

$= \dfrac{12(3-\sqrt{5})}{3^2 - (\sqrt{5})^2}$ — Sum and difference pattern

$= \dfrac{36 - 12\sqrt{5}}{4}$ — Simplify.

$= 9 - 3\sqrt{5}$ — Simplify.

Simplify the expression.

1. $\sqrt{72p^7}$
2. $\sqrt{\dfrac{45}{7y}}$
3. $\sqrt[3]{\dfrac{125x^{11}}{4}}$
4. $\dfrac{8}{\sqrt{6}+2}$
5. $4\sqrt{3} + 5\sqrt{12}$
6. $15\sqrt[3]{2} - 2\sqrt[3]{54}$
7. $(3\sqrt{7} + 5)^2$
8. $\sqrt{6}(\sqrt{18} + \sqrt{8})$

4.2 Solving Quadratic Equations by Graphing (pp. 201–208)

Solve $x^2 + 3x = 4$ by graphing.

Step 1 Write the equation in standard form.

$x^2 + 3x = 4$ — Write original equation.

$x^2 + 3x - 4 = 0$ — Subtract 4 from each side.

Step 2 Graph the related function $y = x^2 + 3x - 4$.

Step 3 Find the x-intercepts. The x-intercepts are -4 and 1.

▶ So, the solutions are $x = -4$ and $x = 1$.

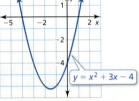

Solve the equation by graphing.

9. $x^2 - 9x + 18 = 0$
10. $x^2 - 2x = -4$
11. $-8x - 16 = x^2$
12. The graph of $f(x) = x^2 + 7x + 10$ is shown. Find the zeros of f.
13. Graph $f(x) = x^2 + 2x - 5$. Approximate the zeros of f to the nearest tenth.

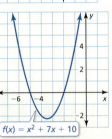

268 Chapter 4 Solving Quadratic Equations

4.3 Solving Quadratic Equations Using Square Roots (pp. 209–214)

A sprinkler sprays water that covers a circular region of 90π square feet. Find the diameter of the circle.

Write an equation using the formula for the area of a circle.

$A = \pi r^2$	Write the formula.
$90\pi = \pi r^2$	Substitute 90π for A.
$90 = r^2$	Divide each side by π.
$\pm\sqrt{90} = r$	Take the square root of each side.
$\pm 3\sqrt{10} = r$	Simplify.

A diameter cannot be negative, so use the positive square root. The diameter is twice the radius. So, the diameter is $6\sqrt{10}$.

▶ The diameter of the circle is $6\sqrt{10} \approx 19$ feet.

Solve the equation using square roots. Round your solutions to the nearest hundredth, if necessary.

14. $x^2 + 5 = 17$
15. $x^2 - 14 = -14$
16. $(x + 2)^2 = 64$
17. $4x^2 + 25 = 75$
18. $(x - 1)^2 = 0$
19. $19 = 30 - 5x^2$

4.4 Solving Quadratic Equations by Completing the Square (pp. 215–224)

Solve $x^2 - 6x + 4 = 11$ by completing the square.

$x^2 - 6x + 4 = 11$	Write the equation.
$x^2 - 6x = 7$	Subtract 4 from each side.
$x^2 - 6x + (-3)^2 = 7 + (-3)^2$	Complete the square by adding $\left(\frac{-6}{2}\right)^2$, or $(-3)^2$, to each side.
$(x - 3)^2 = 16$	Write the left side as the square of a binomial.
$x - 3 = \pm 4$	Take the square root of each side.
$x = 3 \pm 4$	Add 3 to each side.

▶ The solutions are $x = 3 + 4 = 7$ and $x = 3 - 4 = -1$.

Solve the equation by completing the square. Round your solutions to the nearest hundredth, if necessary.

20. $x^2 + 6x - 40 = 0$
21. $x^2 + 2x + 5 = 4$
22. $2x^2 - 4x = 10$

Determine whether the quadratic function has a maximum or minimum value. Then find the value.

23. $y = -x^2 + 6x - 1$
24. $f(x) = x^2 + 4x + 11$
25. $y = 3x^2 - 24x + 15$

26. The width w of a credit card is 3 centimeters shorter than the length ℓ. The area is 46.75 square centimeters. Find the perimeter.

ANSWERS

14. $x \approx 3.46, x \approx -3.46$
15. $x = 0$
16. $x = 6, x = -10$
17. $x \approx 3.54, x \approx -3.54$
18. $x = 1$
19. $x \approx 1.48, x \approx -1.48$
20. $x = 4, x = -10$
21. $x = -1$
22. $x \approx 3.45, x \approx -1.45$
23. maximum value; 8
24. minimum value; 7
25. minimum value; -33
26. 28 cm

ANSWERS

27. $x = 3, x = -5$
28. $x \approx 2.3, x \approx -1.8$
29. $x = 1$
30. one x-intercept
31. no x-intercepts
32. two x-intercepts
33. $x = 9$ and $y = -3$
34. $5 - 3i$
35. $11 + 10i$
36. $-62 + 11i$
37. 68

4.5 Solving Quadratic Equations Using the Quadratic Formula (pp. 225–234)

Solve $-3x^2 + x = -8$ using the Quadratic Formula.

$-3x^2 + x = -8$ Write the equation.

$-3x^2 + x + 8 = 0$ Write in standard form.

$x = \dfrac{-b \pm \sqrt{b^2 - 4ac}}{2a}$ Quadratic Formula

$x = \dfrac{-1 \pm \sqrt{1^2 - 4(-3)(8)}}{2(-3)}$ Substitute -3 for a, 1 for b, and 8 for c.

$x = \dfrac{-1 \pm \sqrt{97}}{-6}$ Simplify.

▶ So, the solutions are $x = \dfrac{-1 + \sqrt{97}}{-6} \approx -1.5$ and $x = \dfrac{-1 - \sqrt{97}}{-6} \approx 1.8$.

Solve the equation using the Quadratic Formula. Round your solutions to the nearest tenth, if necessary.

27. $x^2 + 2x - 15 = 0$
28. $2x^2 - x + 8 = 16$
29. $-5x^2 + 10x = 5$

Find the number of x-intercepts of the graph of the function.

30. $y = -x^2 + 6x - 9$
31. $y = 2x^2 + 4x + 8$
32. $y = -\frac{1}{2}x^2 + 2x$

4.6 Complex Numbers (pp. 237–244)

Perform each operation. Write the answer in standard form.

a. $(3 - 6i) - (7 + 2i) = (3 - 7) + (-6 - 2)i$ Definition of complex subtraction

                   $= -4 - 8i$ Write in standard form.

b. $5i(4 + 5i) = 20i + 25i^2$ Distributive Property

            $= 20i + 25(-1)$ Use $i^2 = -1$.

            $= -25 + 20i$ Write in standard form.

c. $(2 + 5i)(1 - 3i) = 2 - 6i + 5i - 15i^2$ Multiply using FOIL.

                   $= 2 - i - 15(-1)$ Simplify and use $i^2 = -1$.

                   $= 2 - i + 15$ Simplify.

                   $= 17 - i$ Write in standard form.

33. Find the values of x and y that satisfy the equation $36 - yi = 4x + 3i$.

Perform the operation. Write the answer in standard form.

34. $(-2 + 3i) + (7 - 6i)$
35. $(9 + 3i) - (-2 - 7i)$
36. $(5 + 6i)(-4 + 7i)$

37. Multiply $8 - 2i$ by its complex conjugate.

4.7 Solving Quadratic Equations with Complex Solutions (pp. 245–250)

Solve each equation.

a. $x^2 - 14x + 53 = 0$

The coefficient of the x^2-term is 1, and the coefficient of the x-term is an even number. So, solve by completing the square.

$x^2 - 14x + 53 = 0$	Write the equation.
$x^2 - 14x = -53$	Subtract 53 from each side.
$x^2 - 14x + 49 = -53 + 49$	Complete the square for $x^2 - 14x$.
$(x - 7)^2 = -4$	Write the left side as the square of a binomial.
$x - 7 = \pm\sqrt{-4}$	Take the square root of each side.
$x = 7 \pm \sqrt{-4}$	Add 7 to each side.
$x = 7 \pm 2i$	Simplify the radical.

▶ The solutions are $7 + 2i$ and $7 - 2i$.

b. $9x^2 - 6x + 5 = 0$

The equation is not factorable, and completing the square would result in fractions. So, solve using the Quadratic Formula.

$x = \dfrac{-(-6) \pm \sqrt{(-6)^2 - 4(9)(5)}}{2(9)}$	Substitute 9 for a, -6 for b, and 5 for c.
$x = \dfrac{6 \pm \sqrt{-144}}{18}$	Simplify.
$x = \dfrac{6 \pm 12i}{18}$	Write in terms of i.
$x = \dfrac{1 \pm 2i}{3}$	Simplify.

▶ The solutions are $\dfrac{1 + 2i}{3}$ and $\dfrac{1 - 2i}{3}$.

Solve the equation using any method. Explain your choice of method.

38. $-x^2 - 100 = 0$ **39.** $4x^2 + 53 = -11$ **40.** $x^2 + 16x - 17 = 0$

41. $-2x^2 + 12x = 36$ **42.** $2x^2 + 5x = 4$ **43.** $3x^2 - 7x + 13 = 0$

Find the zeros of the function.

44. $f(x) = x^2 + 81$ **45.** $f(x) = x^2 - 2x + 9$ **46.** $f(x) = -8x^2 + 4x - 1$

47. While marching, a drum major tosses a baton into the air from an initial height of 6 feet. The baton has an initial vertical velocity of 32 feet per second. Does the baton reach a height of 25 feet? 20 feet? Explain your reasoning.

Chapter 4 Chapter Review 271

ANSWERS

38. $x = \pm 10i$; *Sample answer:* The equation does not have an x-term, so solve using square roots.

39. $x = \pm 4i$; *Sample answer:* The equation does not have an x-term, so solve using square roots.

40. $x = -17, 1$; *Sample answer:* The equation is factorable, so solve by factoring.

41. $x = 3 \pm 3i$; *Sample answer:* The equation is not easily factorable and $a \neq 1$, so solve using the Quadratic Formula.

42. $x = \dfrac{-5 \pm \sqrt{57}}{4}$; *Sample answer:* The equation is not easily factorable and $a \neq 1$, so solve using the Quadratic Formula.

43. $x = \dfrac{7 \pm \sqrt{107}\,i}{6}$; *Sample answer:* The equation is not easily factorable and $a \neq 1$, so solve using the Quadratic Formula.

44. $\pm 9i$

45. $1 \pm 2\sqrt{2}\,i$

46. $\dfrac{1 \pm i}{4}$

47. no; yes; *Sample answer:* When $h = 25$, the equation has two imaginary solutions. When $h = 20$, $t \approx 0.65$ and $t \approx 1.35$.

Chapter 4   271

ANSWERS

48. $(1, -5)$

49. about $(4.87, 14.75)$, about $(-2.87, -0.75)$

50. about $(-1.88, 2.35)$, about $(2.48, -4.64)$

51.
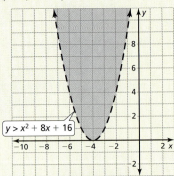

52.
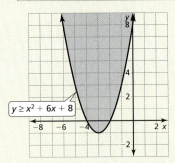

53.
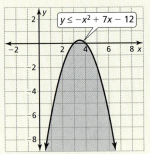

54.
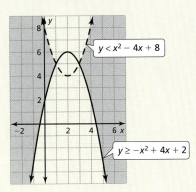

4.8 Solving Nonlinear Systems of Equations (pp. 251–258)

Solve the system by substitution. $y = x^2 - 5$ Equation 1
$y = -x + 1$ Equation 2

Step 1 The equations are already solved for y.

Step 2 Substitute $-x + 1$ for y in Equation 1 and solve for x.

$-x + 1 = x^2 - 5$	Substitute $-x + 1$ for y in Equation 1.
$1 = x^2 + x - 5$	Add x to each side.
$0 = x^2 + x - 6$	Subtract 1 from each side.
$0 = (x + 3)(x - 2)$	Factor the polynomial.
$x + 3 = 0$ or $x - 2 = 0$	Zero-Product Property
$x = -3$ or $x = 2$	Solve for x.

Step 3 Substitute -3 and 2 for x in Equation 2 and solve for y.

| $y = -(-3) + 1$ | Substitute for x in Equation 2. | $y = -2 + 1$ |
| $= 4$ | Simplify. | $= -1$ |

▶ So, the solutions are $(-3, 4)$ and $(2, -1)$.

Solve the system using any method.

48. $y = x^2 - 2x - 4$
$y = -5$

49. $y = x^2 - 9$
$y = 2x + 5$

50. $y = 2\left(\frac{1}{2}\right)^x - 5$
$y = -x^2 - x + 4$

4.9 Quadratic Inequalities (pp. 259–266)

Graph the system of quadratic inequalities.

$y > x^2 - 2$ Inequality 1
$y \leq -x^2 - 3x + 4$ Inequality 2

Step 1 Graph $y > x^2 - 2$. The graph is the red region inside (but not including) the parabola $y = x^2 - 2$.

Step 2 Graph $y \leq -x^2 - 3x + 4$. The graph is the blue region inside and including the parabola $y = -x^2 - 3x + 4$.

Step 3 Identify the purple region where the two graphs overlap. This region is the graph of the system.

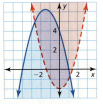

Graph the inequality.

51. $y > x^2 + 8x + 16$
52. $y \geq x^2 + 6x + 8$
53. $x^2 + y \leq 7x - 12$

Graph the system of quadratic inequalities.

54. $x^2 - 4x + 8 > y$
$-x^2 + 4x + 2 \leq y$

55. $2x^2 - x \geq y - 5$
$0.5x^2 > y - 2x - 1$

56. $-3x^2 - 2x \leq y + 1$
$-2x^2 + x - 5 > -y$

57. Solve (a) $3x^2 + 2 \leq 5x$ and (b) $-x^2 - 10x < 21$.

55.
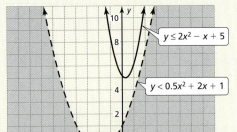

56. See Additional Answers.

57. a. $\frac{2}{3} \leq x \leq 1$
  b. $x < -7$ or $x > -3$

4 Chapter Test

Solve the equation using any method. Explain your choice of method.

1. $x^2 + 121 = 0$
2. $x^2 - 6x = -58$
3. $-2x^2 + 3x + 7 = 0$
4. $x^2 - 7x + 12 = 0$
5. $5x^2 + x + 4 = 0$
6. $(4x + 3)^2 = 16$

Perform the operation. Write the answer in standard form.

7. $(2 + 5i) + (-4 + 3i)$
8. $(3 + 9i) - (1 - 7i)$
9. $(2 + 4i)(-3 - 5i)$

Solve the system using any method.

10. $y = x^2 - 4x - 2$
    $y = -4x + 2$
11. $y = -5x^2 + x - 1$
    $y = -7$
12. $y = \frac{1}{2}(4)^x + 1$
    $y = x^2 - 2x + 4$

13. Graph the system of inequalities consisting of $y \geq 2x^2 - 3$ and $y < -x^2 + x + 4$.

14. Write an expression involving radicals in which a conjugate can be used to simplify the expression. Then simplify the expression.

15. Describe the value(s) of c for which $x^2 + 4x = c$ has (a) two real solutions, (b) one real solution, and (c) two imaginary solutions.

16. Simplify the expression $\sqrt{30x^7} \cdot \frac{36}{\sqrt{3}}$.

17. A skier leaves an 8-foot-tall ramp with an initial vertical velocity of 28 feet per second. The skier has a perfect landing. How many points does the skier earn?

Criteria	Scoring
Maximum height	1 point per foot
Time in air	5 points per second
Perfect landing	25 points

18. You are playing a game of horseshoes. One of your tosses is modeled in the diagram, where x is the horseshoe's horizontal position (in feet) and y is the corresponding height (in feet). Does the horseshoe reach a height of 8 feet? 4 feet? Explain your reasoning.

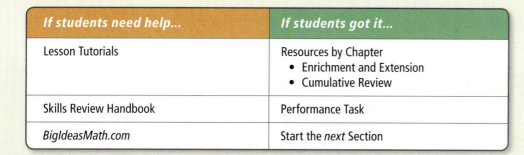

19. The numbers y of two types of bacteria after x hours are represented by the models below.

    $y = 3x^2 + 8x + 20$ Type A
    $y = 27x + 60$ Type B

    a. When are there 400 Type A bacteria?
    b. When are the number of Type A and Type B bacteria the same?
    c. When are there more Type A bacteria than Type B? When are there more Type B bacteria than Type A? Use a graph to support your answer.

ANSWERS

1. $x = \pm 11i$; *Sample answer:* The equation does not have an x-term, so solve using square roots.
2. $x = 3 \pm 7i$; *Sample answer:* $a = 1$ and b is even, so solve by completing the square.
3. $x \approx -1.27, x \approx 2.77$; *Sample answer:* The equation is not factorable and $a \neq 1$, so solve using the Quadratic Formula.
4. $x = 4, x = 3$; *Sample answer:* The equation is easily factorable, so solve by factoring.
5. $x = \dfrac{-1 \pm \sqrt{79}\,i}{10}$; *Sample answer:* The equation is not easily factorable and $a \neq 1$, so solve using the Quadratic Formula.
6. $x = \frac{1}{4}, x = -\frac{7}{4}$; *Sample answer:* The equation is in the form $x^2 = d$ where x is a binomial, so solve using square roots.
7. $-2 + 8i$
8. $2 + 16i$
9. $14 - 22i$
10. $(2, -6), (-2, 10)$
11. $(-1, -7), \left(\frac{6}{5}, -7\right)$
12. $(1, 3)$
13.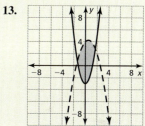
14. *Sample answer:* $\dfrac{1}{2 + \sqrt{3}}$; $2 - \sqrt{3}$
15. a. $c > -4$
    b. $c = -4$
    c. $c < -4$
16. $36x^3\sqrt{10x}$
17. 55 points
18. no; yes; *Sample answer:* When $y = 8$, the equation has two imaginary solutions. When $y = 4$, $x = 10$ and $x = 20$.
19. a. after 10 h
    b. after 8 h
    c. after 8 h; before 8 h

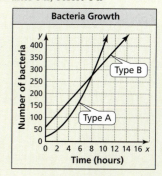

4 Cumulative Assessment

1. The graphs of four quadratic functions are shown. Determine whether the discriminants of the equations $f(x) = 0$, $g(x) = 0$, $h(x) = 0$, and $j(x) = 0$ are positive, negative, or zero.

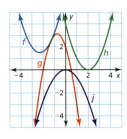

2. Your friend claims to be able to find the radius r of each figure, given the surface area S. Do you support your friend's claim? Justify your answer.

   a.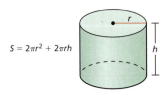

   b.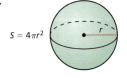

3. Choose values for the constants h and k in the equation $x = \frac{1}{4}(y - k)^2 + h$ so that each statement is true.

   a. The graph of $x = \frac{1}{4}\left(y - \boxed{}\right)^2 + \boxed{}$ is a parabola with its vertex in the second quadrant.

   b. The graph of $x = \frac{1}{4}\left(y - \boxed{}\right)^2 + \boxed{}$ is a parabola with its focus in the first quadrant.

   c. The graph of $x = \frac{1}{4}\left(y - \boxed{}\right)^2 + \boxed{}$ is a parabola with its focus in the third quadrant.

4. The graph of which inequality is shown?

   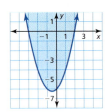

   Ⓐ $y > x^2 + x - 6$

   Ⓑ $y \geq x^2 + x - 6$

   Ⓒ $y > x^2 - x - 6$

   Ⓓ $y \geq x^2 - x - 6$

5. Use the graphs of the functions to answer each question.

   a. Are there any values of x greater than 0 where $f(x) > h(x)$? Explain.
   b. Are there any values of x greater than 1 where $g(x) > f(x)$? Explain.
   c. Are there any values of x greater than 0 where $g(x) > h(x)$? Explain.

6. Which statement best describes the solution(s) of the system of equations?

   $y = x^2 + 2x - 8$
   $y = 5x + 2$

   Ⓐ The graphs intersect at one point, $(-2, -8)$. So, there is one solution.

   Ⓑ The graphs intersect at two points, $(-2, -8)$ and $(5, 27)$. So, there are two solutions.

   Ⓒ The graphs do not intersect. So, there is no solution.

   Ⓓ The graph of $y = x^2 + 2x - 8$ has two x-intercepts. So, there are two solutions.

7. Which expressions are in simplest form?

8. You are making a tabletop with a tiled center and a uniform mosaic border.

   a. Write the polynomial in standard form that represents the perimeter of the tabletop.

   b. Write the polynomial in standard form that represents the area of the tabletop.

   c. The perimeter of the tabletop is less than 80 inches, and the area of tabletop is at least 252 square inches. Select all the possible values of x.

9. Let $f(x) = 2x - 5$. Solve $y = f(x)$ for x. Then find the input when the output is -4.

Chapter 5 Pacing Guide

Chapter Opener/Mathematical Practices	0.5 Day
Section 1	1.5 Days
Section 2	2 Days
Section 3	2 Days
Quiz	0.5 Day
Section 4	1.5 Days
Section 5	2 Days
Section 6	2 Days
Chapter Review/Chapter Tests	2 Days
Total Chapter 5	14 Days
Year-to-Date	80 Days

5 Probability

- **5.1** Sample Spaces and Probability
- **5.2** Independent and Dependent Events
- **5.3** Two-Way Tables and Probability
- **5.4** Probability of Disjoint and Overlapping Events
- **5.5** Permutations and Combinations
- **5.6** Binomial Distributions

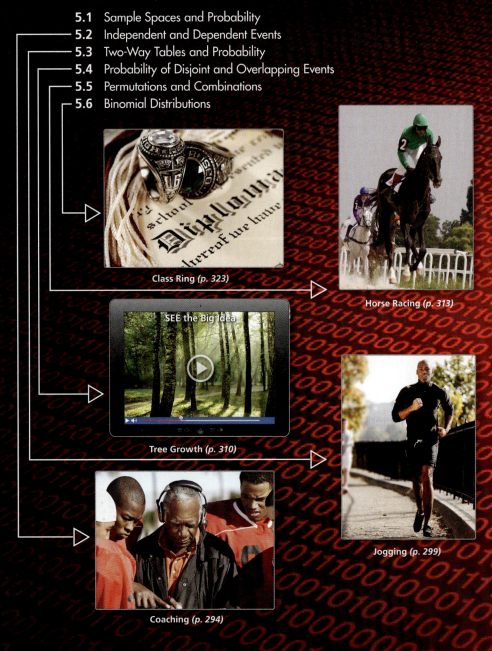

Class Ring (p. 323)

Horse Racing (p. 313)

SEE the Big Idea

Tree Growth (p. 310)

Jogging (p. 299)

Coaching (p. 294)

Laurie's Notes

Chapter Summary

- This chapter on probability presumes knowledge of basic probability concepts from middle school and some data analysis techniques from Math I.
- The chapter begins with finding sample spaces for a variety of classic contexts. The next three lessons focus on computing the probability of independent and dependent events, and compound events. Students construct and interpret two-way frequency tables of data and use the entries to compute joint and marginal relative frequencies.
- Permutations and combinations are introduced to allow students to compute probabilities of compound events and to solve real-life problems.
- The last lesson of the chapter introduces students to probability distributions. One type of probability distribution is a binomial distribution. Another type, a normal distribution, will be introduced in Math III.

What Your Students Have Learned

Middle School
- Identify favorable outcomes of events.
- Understand that probability is the likelihood of an event occurring, expressed as a number from 0 to 1.
- Develop probability models and use them to find probabilities.
- Construct and analyze two-way tables.

Math I
- Describe the shapes of data distributions, choose appropriate measures to describe them, and compare data distributions.
- Make two-way tables, find and interpret marginal frequencies, and find relative and conditional frequencies.
- Recognize associations in data from reading two-way tables.

What Your Students Will Learn

Math II
- Find theoretical and experimental probabilities.
- Find and compare probabilities of independent and dependents events.
- Find conditional probabilities when events are dependent.
- Find relative and conditional relative frequencies and use conditional relative frequencies to find conditional probabilities.
- Find probabilities of compound events.
- Use the formulas for the number of permutations and the number of combinations.
- Construct and interpret probability distributions and binomial distributions.

Dynamic Teaching Tools
Dynamic Assessment & Progress Monitoring Tool
Lesson Planning Tool
Interactive Whiteboard Lesson Library
Dynamic Classroom with Dynamic Investigations
Real-Life STEM Videos

Scaffolding in the Classroom
Begin with a worked-out problem and work backwards.
Ask students to explain what happened in step 5 to get to step 6. Then ask what happened in step 4 to get to step5, etc.

Questioning in the Classroom

Wait Time

Students need the opportunity to process the question before formulating an answer. Although the silence is difficult, make certain to give sufficient time before calling on someone or answering yourself.

Laurie's Notes

Maintaining Mathematical Proficiency

Finding a Percent
- Review with students what each part of the percent proportion means.
- Tell students that they can also find the percent using an equation.

COMMON ERROR Students may invert the values for a and w.

Making a Histogram
- Review the differences between bar graphs and histograms.
- Discuss with students how individual pieces of data are not shown in a histogram.

COMMON ERROR Students may not use equal intervals for the bars in their histograms.

Mathematical Practices (continued on page 278)

- The *Mathematical Practices* page focuses attention on how mathematics is learned—process versus content. This page demonstrates how mathematically proficient students apply the mathematics they know to solve real-life applications. Applying mathematics to solve everyday problems is an important skill.
- Use the *Mathematical Practices* page to help students develop mathematical habits of mind—how mathematics can be explored and how mathematics is thought about.

If students need help...	If students got it...
Student Journal • Maintaining Mathematical Proficiency	Game Closet at *BigIdeasMath.com*
Lesson Tutorials	Start the *next* Section
Skills Review Handbook	

Maintaining Mathematical Proficiency

Finding a Percent

Example 1 What percent of 12 is 9?

$$\frac{a}{w} = \frac{p}{100}$$ Write the percent proportion.

$$\frac{9}{12} = \frac{p}{100}$$ Substitute 9 for a and 12 for w.

$$100 \cdot \frac{9}{12} = 100 \cdot \frac{p}{100}$$ Multiplication Property of Equality.

$$75 = p$$ Simplify.

▶ So, 9 is 75% of 12.

Write and solve a proportion to answer the question.

1. What percent of 30 is 6?
2. What number is 68% of 25?
3. 34.4 is what percent of 86?

Making a Histogram

Example 2 The frequency table shows the ages of people at a gym. Display the data in a histogram.

Age	Frequency
10–19	7
20–29	12
30–39	6
40–49	4
50–59	0
60–69	3

Step 1 Draw and label the axes.

Step 2 Draw a bar to represent the frequency of each interval.

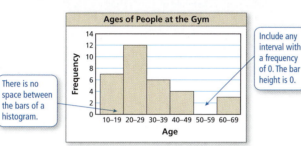

There is no space between the bars of a histogram.

Include any interval with a frequency of 0. The bar height is 0.

Display the data in a histogram.

4.

Movies Watched per Week			
Movies	0–1	2–3	4–5
Frequency	35	11	6

5. **ABSTRACT REASONING** You want to purchase either a sofa or an arm chair at a furniture store. Each item has the same retail price. The sofa is 20% off. The arm chair is 10% off, and you have a coupon to get an additional 10% off the discounted price of the chair. Are the items equally priced after the discounts are applied? Explain.

Dynamic Solutions available at *BigIdeasMath.com*

What Your Students Have Learned

- Use proportionality to solve percent problems.
- Make a histogram to display the frequency of data values.

ANSWERS

1. $\frac{6}{30} = \frac{p}{100}$, 20%

2. $\frac{a}{25} = \frac{68}{100}$, 17

3. $\frac{34.4}{86} = \frac{p}{100}$, 40%

4.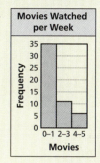

5. no; The sofa will cost 80% of the retail price and the arm chair will cost 81% of the retail price.

Vocabulary Review

Have students make a Notetaking Organizer for a histogram. Include the following words.
- Frequency
- Interval

Chapter 5

MONITORING PROGRESS ANSWERS

1. equally likely to happen or not happen; The probability of the event is 0.5.
2. unlikely; The probability of the event is $(0.5)(0.5) = 0.25$.
3. *Sample answer:* rolling a number less than 7 on a six-sided die

Mathematical Practices

Mathematically proficient students apply the mathematics they know to solve real-life problems.

Modeling with Mathematics

Core Concept

Likelihoods and Probabilities

The **probability of an event** is a measure of the likelihood that the event will occur. Probability is a number from 0 to 1, including 0 and 1. The diagram relates *likelihoods* (described in words) and probabilities.

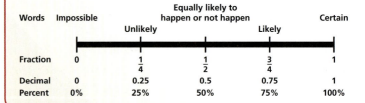

EXAMPLE 1 Describing Likelihoods

Describe the likelihood of each event.

Probability of an Asteroid or a Meteoroid Hitting Earth			
Name	Diameter	Probability of impact	Flyby date
a. Meteoroid	6 in.	0.75	Any day
b. Apophis	886 ft	0	2029
c. 2000 SG344	121 ft	$\frac{1}{435}$	2068–2110

SOLUTION

a. On any given day, it is *likely* that a meteoroid of this size will enter Earth's atmosphere. If you have ever seen a "shooting star," then you have seen a meteoroid.

b. A probability of 0 means this event is *impossible*.

c. With a probability of $\frac{1}{435} \approx 0.23\%$, this event is very *unlikely*. Of 435 identical asteroids, you would expect only one of them to hit Earth.

Monitoring Progress

In Exercises 1 and 2, describe the event as unlikely, equally likely to happen or not happen, or likely. Explain your reasoning.

1. The oldest child in a family is a girl.
2. The two oldest children in a family with three children are girls.
3. Give an example of an event that is certain to occur.

Laurie's Notes Mathematical Practices (continued from page T-277)

- Students will be working with probability in this chapter. In the *Core Concept*, the probability of an event is described. The likelihood that the event will occur is a number from 0 to 1, inclusive.
- **?** Pose Example 1. Ask, "What is the likelihood of an asteroid or meteoroid hitting Earth?" It depends on the size of the asteroid or meteoroid. Have students describe the likelihood of each event.
- Give time for students to work through the *Monitoring Progress* questions, and then discuss as a class.

Laurie's Notes

Overview of Section 5.1

Introduction
- Students should have some beginning knowledge of probability from earlier grades.
- This lesson begins with finding sample spaces for a variety of events, a needed skill in finding theoretical and experimental probabilities.

Formative Assessment Tips
- **Predict, Explain, Observe (P-E-O Probe):** This technique provides students the opportunity to make a prediction or to select a prediction from a set of options. Students explain or give reasons why their prediction makes sense, and then they test out their prediction. Finally, they analyze the results against their prediction and make adjustments to their thinking.
- This technique gives students the opportunity to reason about a problem, situation, or concept that is being probed, and students are asked to make a prediction related to the probe. Explaining the reason for their prediction is an important step. Curiosity piqued, students think of how to investigate their prediction. Students carry out their investigation and analyze the results. If their prediction is correct and it was based on valid reasoning (versus an educated guess), then their reasoning has been confirmed. If the results are not exactly what they had predicted, then the cognitive dissonance created should lead to more discussion, rethinking, and a new prediction.
- This technique is best used when you want to probe students' ideas. This technique does not need to be done only with individuals. Partners or small groups can make predictions, discussing their reasoning and rethinking their prediction after an investigation.

Pacing Suggestion
- Each one of the explorations presents a classic context for exploring the concept of probability. If you sense that your students have little familiarity with these contexts, take time for all of the explorations. Students should be comfortable with finding the sample space of an experiment before beginning the formal lesson.

Dynamic Teaching Tools
- Dynamic Assessment & Progress Monitoring Tool
- Lesson Planning Tool
- Interactive Whiteboard Lesson Library
- Dynamic Classroom with Dynamic Investigations

What Your Students Will Learn

- Find sample spaces (the sets of all possible outcomes) of probability experiments.
- Find theoretical probabilities that events will occur.
- Find experimental probabilities that events will occur based on repeated trials of probability experiments.
- Find odds in favor of events and odds against events.

Laurie's Notes

Exploration

Motivate
- ? Hold up a used lottery ticket and ask, "What do you think my chances are of holding a winning ticket?" Answers will vary.
- Most lottery or scratch-off tickets have a statement on them about the "odds of winning," which is quite close to the probability of winning.
- Explain to students that in this chapter they will learn more about probability and how the calculations are done.

Exploration Note
- The explorations on this page present four common contexts for introducing probability. It is important for students to work through each exploration so that they gain confidence and experience in determining possible outcomes in the sample space.

Exploration 1
- **Common Misconception:** Students may believe that HTT is the same outcome as THT because there is one head and two tails. One way to help students understand that these two outcomes are different is to use coins with three different dates. Using columns to list the outcomes of each coin will help as well.
- **Extension:** Look for patterns where you have 1 coin, 2 coins, 3 coins, 4 coins, and so on.

Exploration 2
- **Teaching Tip:** Use two colors of dice to help students distinguish 1-6 and 6-1.
- **Use Appropriate Tools Strategically:** A matrix is a common way to help organize the possible outcomes.

Exploration 3
- Students may believe that the location of the numbers influences the outcomes, meaning two 5's adjacent to one another is different from two 5's that are nonadjacent.
- ? "Are each of the outcomes equally likely to occur? Explain." No; There are four 5's and only two 4's.

Exploration 4
- The numbers on the marbles will help students recognize that there are multiple ways (18) to draw two red marbles.

Communicate Your Answer
- Have students work independently to answer Questions 5 and 6, and then discuss as a whole class.

Connecting to Next Step
- Listing the possible outcomes in a sample space is necessary in finding a theoretical probability, the focus of the formal lesson.

5.1 Sample Spaces and Probability

Essential Question How can you list the possible outcomes in the sample space of an experiment?

The **sample space** of an experiment is the set of all possible outcomes for that experiment.

EXPLORATION 1 Finding the Sample Space of an Experiment

Work with a partner. In an experiment, three coins are flipped. List the possible outcomes in the sample space of the experiment.

EXPLORATION 2 Finding the Sample Space of an Experiment

Work with a partner. List the possible outcomes in the sample space of the experiment.

a. One six-sided die is rolled.

b. Two six-sided dice are rolled.

EXPLORATION 3 Finding the Sample Space of an Experiment

Work with a partner. In an experiment, a spinner is spun.

a. How many ways can you spin a 1? 2? 3? 4? 5?

b. List the sample space.

c. What is the total number of outcomes?

EXPLORATION 4 Finding the Sample Space of an Experiment

Work with a partner. In an experiment, a bag contains 2 blue marbles and 5 red marbles. Two marbles are drawn from the bag.

a. How many ways can you choose two blue? a red then blue? a blue then red? two red?

b. List the sample space.

c. What is the total number of outcomes?

LOOKING FOR A PATTERN

To be proficient in math, you need to look closely to discern a pattern or structure.

Communicate Your Answer

5. How can you list the possible outcomes in the sample space of an experiment?

6. For Exploration 3, find the ratio of the number of each possible outcome to the total number of outcomes. Then find the sum of these ratios. Repeat for Exploration 4. What do you observe?

Section 5.1 Sample Spaces and Probability 279

Dynamic Teaching Tools
- Dynamic Assessment & Progress Monitoring Tool
- Lesson Planning Tool
- Interactive Whiteboard Lesson Library
- Dynamic Classroom with Dynamic Investigations

ANSWERS

1. HHH, HHT, HTH, THH, HTT, THT, TTH, TTT

2. a. 1, 2, 3, 4, 5, 6
   b. 1-1, 1-2, 1-3, 1-4, 1-5, 1-6, 2-1, 2-2, 2-3, 2-4, 2-5, 2-6, 3-1, 3-2, 3-3, 3-4, 3-5, 3-6, 4-1, 4-2, 4-3, 4-4, 4-5, 4-6, 5-1, 5-2, 5-3, 5-4, 5-5, 5-6, 6-1, 6-2, 6-3, 6-4, 6-5, 6-6

3. a. 1; 2; 3; 2; 4
   b. 1, 2, 2, 3, 3, 3, 4, 4, 5, 5, 5, 5
   c. 12

4. a. 2; 10; 10; 20
   b. B1-B2, B2-B1, B1-R1, B1-R2, B1-R3, B1-R4, B1-R5, B2-R1, B2-R2, B2-R3, B2-R4, B2-R5, R1-B1, R1-B2, R2-B1, R2-B2, R3-B1, R3-B2, R4-B1, R4-B2, R5-B1, R5-B2, R1-R2, R1-R3, R1-R4, R1-R5, R2-R1, R2-R3, R2-R4, R2-R5, R3-R1, R3-R2, R3-R4, R3-R5, R4-R1, R4-R2, R4-R3, R4-R5, R5-R1, R5-R2, R5-R3, R5-R4
   c. 42

5. *Sample answer:* Make a table or diagram to show all of the possible outcomes.

6. The sum of the ratios is 1.

Differentiated Instruction

Organization

Help students create a step-by-step process for making a tree diagram. Guide students to see that the first branch of the diagram is a list of all of the possible outcomes for the first event. Then, the next branch is built off of the first branch, listing all the possible outcomes for the second event.

Extra Example 1

You spin this spinner and flip a coin. How many possible outcomes are in the sample space? List the possible outcomes.

The sample space has 8 possible outcomes. They are 1H, 1T, 2H, 2T, 3H, 3T, 4H, and 4T.

MONITORING PROGRESS ANSWERS

1. 4; HH, HT, TH, TT
2. 24; HH1, HH2, HH3, HH4, HH5, HH6, HT1, HT2, HT3, HT4, HT5, HT6, TH1, TH2, TH3, TH4, TH5, TH6, TT1, TT2, TT3, TT4, TT5, TT6

5.1 Lesson

Core Vocabulary
probability experiment, *p. 280*
outcome, *p. 280*
event, *p. 280*
sample space, *p. 280*
probability of an event, *p. 280*
theoretical probability, *p. 281*
experimental probability, *p. 282*
odds, *p. 283*

Previous
tree diagram

What You Will Learn
▶ Find sample spaces.
▶ Find theoretical probabilities.
▶ Find experimental probabilities.
▶ Find odds.

Sample Spaces

A **probability experiment** is an action, or trial, that has varying results. The possible results of a probability experiment are **outcomes**. For instance, when you roll a six-sided die, there are 6 possible outcomes: 1, 2, 3, 4, 5, or 6. A collection of one or more outcomes is an **event**, such as rolling an odd number. The set of all possible outcomes is called a **sample space**.

EXAMPLE 1 Finding a Sample Space

You flip a coin and roll a six-sided die. How many possible outcomes are in the sample space? List the possible outcomes.

SOLUTION

Use a tree diagram to find the outcomes in the sample space.

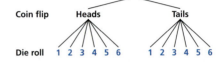

ANOTHER WAY

Using H for "heads" and T for "tails," you can list the outcomes as shown below.

H1 H2 H3 H4 H5 H6
T1 T2 T3 T4 T5 T6

▶ The sample space has 12 possible outcomes. They are listed below.

Heads, 1   Heads, 2   Heads, 3   Heads, 4   Heads, 5   Heads, 6
Tails, 1   Tails, 2   Tails, 3   Tails, 4   Tails, 5   Tails, 6

Monitoring Progress Help in English and Spanish at *BigIdeasMath.com*

Find the number of possible outcomes in the sample space. Then list the possible outcomes.

1. You flip two coins.
2. You flip two coins and roll a six-sided die.

Theoretical Probabilities

The **probability of an event** is a measure of the likelihood, or chance, that the event will occur. Probability is a number from 0 to 1, including 0 and 1, and can be expressed as a decimal, fraction, or percent.

280 Chapter 5 Probability

Laurie's Notes · Teacher Actions

- **Paired Verbal Fluency:** Have students pair up and follow the protocol described on page T-224. Ask students what they recall about probability from earlier courses. Expect to hear about theoretical versus experimental, sample spaces, likelihood, and outcomes.
- **Think-Pair-Share:** Have students work independently to solve Example 1. Share thinking with a neighbor.
- ? "How many outcomes were there?" 12
- ? "If it had been a ten-sided die, how many outcomes would you expect?" 20

ATTENDING TO PRECISION

Notice that the question uses the phrase "exactly two answers." This phrase is more precise than saying "two answers," which may be interpreted as "at least two" or as "exactly two."

The outcomes for a specified event are called *favorable outcomes*. When all outcomes are equally likely, the **theoretical probability** of the event can be found using the following.

$$\text{Theoretical probability} = \frac{\text{Number of favorable outcomes}}{\text{Total number of outcomes}}$$

The probability of event A is written as $P(A)$.

EXAMPLE 2 Finding a Theoretical Probability

A student taking a quiz randomly guesses the answers to four true-false questions. What is the probability of the student guessing exactly two correct answers?

SOLUTION

Step 1 Find the outcomes in the sample space. Let C represent a correct answer and I represent an incorrect answer. The possible outcomes are:

Number correct	Outcome
0	IIII
1	CIII ICII IICI IIIC
2	IICC ICIC ICCI CIIC CICI CCII
3	ICCC CICC CCIC CCCI
4	CCCC

exactly two correct

Step 2 Identify the number of favorable outcomes and the total number of outcomes. There are 6 favorable outcomes with exactly two correct answers and the total number of outcomes is 16.

Step 3 Find the probability of the student guessing exactly two correct answers. Because the student is randomly guessing, the outcomes should be equally likely. So, use the theoretical probability formula.

$$P(\text{exactly two correct answers}) = \frac{\text{Number of favorable outcomes}}{\text{Total number of outcomes}}$$
$$= \frac{6}{16}$$
$$= \frac{3}{8}$$

▶ The probability of the student guessing exactly two correct answers is $\frac{3}{8}$, or 37.5%.

The sum of the probabilities of all outcomes in a sample space is 1. So, when you know the probability of event A, you can find the probability of the *complement* of event A. The *complement* of event A consists of all outcomes that are not in A and is denoted by $\overline{A}$. The notation $\overline{A}$ is read as "A bar." You can use the following formula to find $P(\overline{A})$.

Core Concept

Probability of the Complement of an Event

The probability of the complement of event A is

$$P(\overline{A}) = 1 - P(A).$$

English Language Learners

Vocabulary
Ask students to state the definitions of *complement* with which they are familiar. As students state the definitions, write on the board the spelling of the word that aligns to their definition. Explain that words that are spelled differently but sound the same are called *homophones*. Guide students to create a definition of the word *complement* in their own words as it relates to probability.

Extra Example 2
A coin is flipped four times. What is the probability that it lands on heads exactly three times?
The probability that the coin lands on heads exactly three times is $\frac{1}{4}$, or 25%.

Laurie's Notes — Teacher Actions

- **Predict, Explain, Observe:** Group students and ask, "If you randomly guess on a four-question true-false quiz, what is the probability of answering exactly two questions correctly? Explain your reasoning." Give groups sufficient time to consider the question and to consider their reasoning. Often they predict a 50% chance of exactly two being correct. You want them to think about the sample space and the language of the question. The results are very different when asking about two or more questions being correct. Complements of events naturally arise in the discussion.

Extra Example 3

Find the probability of each event when two six-sided dice are rolled.

a. The sum is greater than 8.
 $P(\text{sum} > 8) = \frac{10}{36} \approx 0.278$

b. The sum is not 2.
 $P(\text{sum is not 2}) = 1 - \frac{1}{36} \approx 0.972$

Extra Example 4

A spinner has 5 sections, colored red, green, blue, yellow, and purple. Each section of the spinner has the same area. The table shows the results of 50 spins. For which color is the experimental probability of stopping on the color the same as the theoretical probability?

Spinner Results	
red	5
green	20
blue	3
yellow	10
purple	12

The experimental probability of stopping on yellow is the same as the theoretical probability.

MONITORING PROGRESS ANSWERS

3. $\frac{1}{12}$
4. 0.55
5. $\frac{3}{4}$
6. 0
7. 0.97
8. green

EXAMPLE 3 Finding Probabilities of Complements

You roll two six-sided dice. Find the probability of each event.

a. The sum is not 6.   b. The sum is less than or equal to 9.

SOLUTION

Find the outcomes in the sample space, as shown. There are 36 possible outcomes.

a. $P(\text{sum is not 6}) = 1 - P(\text{sum is 6}) = 1 - \frac{5}{36} = \frac{31}{36} \approx 0.861$

b. $P(\text{sum} \leq 9) = 1 - P(\text{sum} > 9) = 1 - \frac{6}{36} = \frac{30}{36} = \frac{5}{6} \approx 0.833$

Monitoring Progress Help in English and Spanish at *BigIdeasMath.com*

3. You flip a coin and roll a six-sided die. What is the probability that the coin shows tails and the die shows 4?

Find $P(\overline{A})$.

4. $P(A) = 0.45$ 5. $P(A) = \frac{1}{4}$ 6. $P(A) = 1$ 7. $P(A) = 0.03$

Experimental Probabilities

An **experimental probability** is based on repeated *trials* of a probability experiment. The number of trials is the number of times the probability experiment is performed. Each trial in which a favorable outcome occurs is called a *success*. The experimental probability can be found using the following.

$$\text{Experimental probability} = \frac{\text{Number of successes}}{\text{Number of trials}}$$

EXAMPLE 4 Finding an Experimental Probability

Spinner Results			
red	green	blue	yellow
5	9	3	3

Each section of the spinner shown has the same area. The spinner was spun 20 times. The table shows the results. For which color is the experimental probability of stopping on the color the same as the theoretical probability?

SOLUTION

The theoretical probability of stopping on each of the four colors is $\frac{1}{4}$. Use the outcomes in the table to find the experimental probabilities.

$P(\text{red}) = \frac{5}{20} = \frac{1}{4}$ $P(\text{green}) = \frac{9}{20}$ $P(\text{blue}) = \frac{3}{20}$ $P(\text{yellow}) = \frac{3}{20}$

▶ The experimental probability of stopping on red is the same as the theoretical probability.

Monitoring Progress Help in English and Spanish at *BigIdeasMath.com*

8. For which color is the experimental probability of stopping on the color greater than the theoretical probability?

282   Chapter 5   Probability

Laurie's Notes Teacher Actions

? Reason Abstractly and Quantitatively: "When is finding the probability of the complement of an event a helpful technique?" It is helpful when the probability of the complement can be found more easily than the probability of the event.

• Probe to ensure that students recognize the difference between experimental and theoretical probability. Example 4 assesses this.

EXAMPLE 5 Solving a Real-Life Problem

In the United States, a survey of 2184 adults ages 18 and over found that 1328 of them have at least one pet. The types of pets these adults have are shown in the figure. What is the probability that a pet-owning adult chosen at random has a dog?

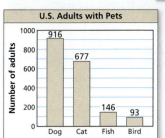
U.S. Adults with Pets

SOLUTION

The number of trials is the number of pet-owning adults, 1328. A success is a pet-owning adult who has a dog. From the graph, there are 916 adults who have a dog.

$$P(\text{pet-owning adult has a dog}) = \frac{916}{1328} = \frac{229}{332} \approx 0.690$$

▶ The probability that a pet-owning adult chosen at random has a dog is about 69%.

Monitoring Progress Help in English and Spanish at *BigIdeasMath.com*

9. What is the probability that a pet-owning adult chosen at random owns a fish?

Odds

The **odds** of an event compare the number of favorable and unfavorable outcomes when all outcomes are equally likely.

$$\text{Odds } in \text{ favor} = \frac{\text{Number of favorable outcomes}}{\text{Number of unfavorable outcomes}} \qquad \text{Odds } against = \frac{\text{Number of unfavorable outcomes}}{\text{Number of favorable outcomes}}$$

EXAMPLE 6 Finding Odds

In Example 4, find the odds in favor of stopping on yellow.

READING
Odds are read as the ratio of one number to another.

SOLUTION

The 4 possible outcomes are all equally likely. Yellow is the 1 favorable outcome and the other 3 colors are unfavorable outcomes.

$$\text{Odds in favor of yellow} = \frac{\text{Number of favorable outcomes}}{\text{Number of unfavorable outcomes}} = \frac{1}{3}, \text{ or } 1:3$$

EXAMPLE 7 Making a Decision

The probability of winning a balloon dart game is 0.2. The odds of winning a ring toss game are 1:5. Which game gives you a better chance of winning?

STUDY TIP
Using the given odds, there is 1 favorable outcome and 5 unfavorable outcomes. So, the total number of outcomes is 6.

SOLUTION

Find the probability of winning the ring toss game. Then compare the probabilities.

$$P(\text{winning ring toss game}) = \frac{\text{Number of favorable outcomes}}{\text{Total number of outcomes}} = \frac{1}{6} \approx 0.17$$

▶ Because $0.2 > 0.17$, the balloon dart game gives you a better chance of winning.

Monitoring Progress Help in English and Spanish at *BigIdeasMath.com*

10. In Example 4, find the odds against stopping on red.

11. **WHAT IF?** The odds of winning the ring toss game are 2:7. Does this change your answer in Example 7? Explain your reasoning.

Extra Example 5
Use the survey and data display in Example 5. What is the probability that a pet-owning adult chosen at random has a bird? The probability that a pet-owning adult chosen at random has a bird is ≈ 0.07003, or about 7%.

Extra Example 6
In Example 4, find the odds against stopping on blue. $\frac{3}{1}$ or $3:1$

Extra Example 7
The probability of winning a balloon dart game is 0.25. The odds of winning a ring toss game are 2:7. Which game gives you a better chance of winning? Because $0.25 > 0.22$, the balloon dart game gives you a better chance at winning.

MONITORING PROGRESS ANSWERS
9. about 11%
10. 3 : 1
11. yes; $\frac{2}{9} > 0.2$

Laurie's Notes   Teacher Actions

? Attend to Precision: In Example 5, pet owners are the number of trials versus the 2184 adults that were initially surveyed. Ask, "What is the probability that an adult who was surveyed has a dog?" $\frac{916}{2184} \approx 0.419$

COMMON ERROR Students and adults often use the words *probability* and *odds* interchangeably. To introduce the term and to distinguish the difference between the terms, use a simple example. The probability of rolling a 4 on a standard number cube is $\frac{1}{6}$. The odds of rolling a 4 are $\frac{1}{5}$.

• Work through Examples 6 and 7 as shown.

Closure
• **Exit Ticket:** You toss two dice. Find (a) $P(\text{sum is not } 7)$ and (b) $P(\text{sum is more than } 4)$.
(a) $\frac{5}{6} \approx 0.833$ (b) $\frac{5}{6} \approx 0.833$

Assignment Guide and Homework Check

ASSIGNMENT

Basic: 1, 2, 3–13 odd, 17–23 odd, 28, 30, 33–39

Average: 1, 2, 4–26 even, 27, 28, 33–39

Advanced: 1, 2, 4–28 even, 29–39

HOMEWORK CHECK

Basic: 3, 7, 9, 13, 19

Average: 4, 8, 10, 14, 20

Advanced: 6, 8, 10, 14, 20

ANSWERS

1. Theoretical probability is based on the number of outcomes and experimental probability is based on repeated trials.

2. *Sample answer:* The probability compares the number of favorable outcomes to the total number of outcomes while the odds compare the number of favorable outcomes to the number of unfavorable outcomes.

3. 48; 1HHH, 1HHT, 1HTH, 1THH, 1HTT, 1THT, 1TTH, 1TTT, 2HHH, 2HHT, 2HTH, 2THH, 2HTT, 2THT, 2TTH, 2TTT, 3HHH, 3HHT, 3HTH, 3THH, 3HTT, 3THT, 3TTH, 3TTT, 4HHH, 4HHT, 4HTH, 4THH, 4HTT, 4THT, 4TTH, 4TTT, 5HHH, 5HHT, 5HTH, 5THH, 5HTT, 5THT, 5TTH, 5TTT, 6HHH, 6HHT, 6HTH, 6THH, 6HTT, 6THT, 6TTH, 6TTT

4. 6; HP, HP, HW, TP, TP, TW

5. 12; R1, R2, R3, R4, W1, W2, W3, W4, B1, B2, B3, B4

6. 42; GG, GG, GG, GG, GG, GG, GB, GB, GB, GB, GB, GB, GB, GB, GB, GB, GB, GB, BG, BG, BG, BG, BG, BG, BG, BG, BG, BG, BG, BG, BB, BB, BB, BB, BB, BB, BB, BB, BB, BB, BB, BB

7. $\frac{5}{16}$, or about 31.25%

8. $\frac{1}{16}$, or 6.25%

9. a. $\frac{11}{12}$, or about 92%
   b. $\frac{13}{18}$, or about 72%

10. a. 80%
    b. 26%

11. There are 4 outcomes, not 3; The probability is $\frac{1}{4}$.

12. The event should be that the number is less than or equal to 4; $\frac{13}{15}$

5.1 Exercises

Dynamic Solutions available at *BigIdeasMath.com*

Vocabulary and Core Concept Check

1. **WRITING** Describe the difference between theoretical probability and experimental probability.

2. **WRITING** Explain how the probability of an event differs from the odds in favor of the event when all outcomes are equally likely.

Monitoring Progress and Modeling with Mathematics

In Exercises 3–6, find the number of possible outcomes in the sample space. Then list the possible outcomes. *(See Example 1.)*

3. You roll a die and flip three coins.

4. You flip a coin and draw a marble at random from a bag containing two purple marbles and one white marble.

5. A bag contains four red cards numbered 1 through 4, four white cards numbered 1 through 4, and four black cards numbered 1 through 4. You choose a card at random.

6. You draw two marbles without replacement from a bag containing three green marbles and four black marbles.

7. **PROBLEM SOLVING** A game show airs on television five days per week. Each day, a prize is randomly placed behind one of two doors. The contestant wins the prize by selecting the correct door. What is the probability that exactly two of the five contestants win a prize during a week? *(See Example 2.)*

8. **PROBLEM SOLVING** Your friend has two standard decks of 52 playing cards and asks you to randomly draw one card from each deck. What is the probability that you will draw two spades?

9. **PROBLEM SOLVING** You roll two six-sided dice. Find the probability that (a) the sum is not 4 and (b) the sum is greater than 5. *(See Example 3.)*

10. **PROBLEM SOLVING** The age distribution of a population is shown. Find the probability of each event.

    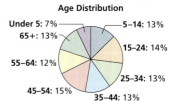
    Age Distribution
    Under 5: 7%
    5–14: 13%
    15–24: 14%
    25–34: 13%
    35–44: 13%
    45–54: 15%
    55–64: 12%
    65+: 13%

    a. A person chosen at random is at least 15 years old.
    b. A person chosen at random is from 25 to 44 years old.

11. **ERROR ANALYSIS** A student randomly guesses the answers to two true-false questions. Describe and correct the error in finding the probability of the student guessing both answers correctly.

     The student can either guess two incorrect answers, two correct answers, or one of each. So the probability of guessing both answers correctly is $\frac{1}{3}$.

12. **ERROR ANALYSIS** A student randomly draws a number between 1 and 30. Describe and correct the error in finding the probability that the number drawn is greater than 4.

     The probability that the number is less than 4 is $\frac{3}{30}$, or $\frac{1}{10}$. So, the probability that the number is greater than 4 is $1 - \frac{1}{10}$, or $\frac{9}{10}$.

284   Chapter 5   Probability

13. **DRAWING CONCLUSIONS** You roll a six-sided die 60 times. The table shows the results. For which number is the experimental probability of rolling the number the same as the theoretical probability? *(See Example 4.)*

Six-sided Die Results					
⚀	⚁	⚂	⚃	⚄	⚅
11	14	7	10	6	12

14. **DRAWING CONCLUSIONS** A bag contains 5 marbles that are each a different color. A marble is drawn, its color is recorded, and then the marble is placed back in the bag. This process is repeated until 30 marbles have been drawn. The table shows the results. For which marble is the experimental probability of drawing the marble the same as the theoretical probability?

Drawing Results				
white	black	red	green	blue
5	6	8	2	9

15. **REASONING** Refer to the spinner shown. The spinner is divided into sections with the same area.

a. What is the theoretical probability that the spinner stops on a multiple of 3?

b. You spin the spinner 30 times. It stops on a multiple of 3 twenty times. What is the experimental probability of stopping on a multiple of 3?

c. Explain why the probability you found in part (b) is different than the probability you found in part (a).

d. You spin the spinner 60 times. It stops on a multiple of 10 six times. How does the experimental probability of stopping on a multiple of 10 compare to the theoretical probability that the spinner stops on a multiple of 10?

16. **OPEN-ENDED** Describe a real-life event that has a probability of 0. Then describe a real-life event that has a probability of 1.

17. **DRAWING CONCLUSIONS** A survey of 2237 adults ages 18 and over asked which sport is their favorite. The results are shown in the figure. What is the probability that an adult chosen at random prefers auto racing? *(See Example 5.)*

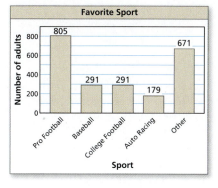

18. **DRAWING CONCLUSIONS** A survey of 2392 adults ages 18 and over asked what type of food they would be most likely to choose at a restaurant. The results are shown in the figure. What is the probability that an adult chosen at random prefers Italian food?

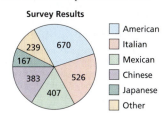

In Exercises 19–22, use the spinner in Exercise 15 to find the given odds. *(See Example 6.)*

19. in favor of stopping on a multiple of 6

20. in favor of stopping on a number greater than 10

21. against stopping on a number less than 9

22. against stopping on a number that is divisible by 7

23. **PROBLEM SOLVING** The odds of winning a lottery game are 3 : 5. The probability of winning a bingo game is 35%. Which game gives you a better chance of winning? Explain. *(See Example 7.)*

24. **PROBLEM SOLVING** The probability of winning a carnival game is $\frac{1}{4}$. The odds of winning a raffle are 1 : 3. For which do you have a better chance of winning? Explain.

ANSWERS

13. 4
14. black
15. a. $\frac{9}{10}$, or 90%
    b. $\frac{2}{3}$, or about 67%
    c. The probability in part (b) is based on trials, not possible outcomes.
    d. They are both $\frac{1}{10}$.
16. *Sample answer:* drawing an orange marble from a bag containing blue and green marbles; drawing a red marble from a bag containing red marbles
17. about 0.08, or about 8%
18. about 0.22, or about 22%
19. 1 : 1
20. 7 : 3
21. 4 : 1
22. 9 : 1
23. lottery game; $\frac{3}{8} > 0.35$
24. neither; Both probabilities are $\frac{1}{4}$.

ANSWERS

25. Sample answer: the odds of stopping on a number less than 16
26. a. B, A, C, D
    b. Friday: 1 : 19, Saturday: 3 : 7, Sunday: 4 : 1, Monday: 9 : 1
27. a. 2, 3, 4, 5, 6, 7, 8, 9, 10, 11, 12
    b. 2: $\frac{1}{36}$, 3: $\frac{1}{18}$, 4: $\frac{1}{12}$, 5: $\frac{1}{9}$, 6: $\frac{5}{36}$, 7: $\frac{1}{6}$, 8: $\frac{5}{36}$, 9: $\frac{1}{9}$, 10: $\frac{1}{12}$, 11: $\frac{1}{18}$, 12: $\frac{1}{36}$
    c. Sample answer: The probabilities are similar.
28. no; Your friend calculated the experimental probability. The theoretical probability of the coin landing heads up is $\frac{1}{2}$.
29. a. 50; 47 + 3 = 50
    b. 0.44; The probability of not drawing a white marble is $\frac{47}{50}$ = 0.94 and 0.94 − 0.5 = 0.44
30–39. See Additional Answers.

Mini-Assessment

A bag contains five cards numbered 1 through 5. You flip a coin and draw a card at random from the bag.

1. Find the number of possible outcomes in the sample space. Then list the outcomes. **10; H1, H2, H3, H4, H5, T1, T2, T3, T4, T5**

2. What is the probability that the coin lands on heads and you draw a card with an odd number on it? **$\frac{3}{10}$, or 30%**

3. A bag contains 5 red, 6 white, and 5 blue marbles. A marble is drawn from the bag and replaced 32 times. The table shows the results. For which color is the experimental probability of drawing the color the same as the theoretical probability?

red	white	blue
12	12	8

a white marble

4. Use the graph in Example 5. What is the probability that a pet-owning adult chosen at random has a cat? **≈0.5098, or about 51%**

5. Use the spinner in Exercise 15. Find the odds in favor of stopping on a number less than 10. **3 : 7**

286   Chapter 5

---

25. **REASONING** Use the spinner in Exercise 15 to give an example of an event whose odds are 1 : 1.

26. **ANALYZING RELATIONSHIPS** Refer to the chart below.

    Four-Day Forecast
Friday	Saturday	Sunday	Monday
Chance of Rain 5%	Chance of Rain 30%	Chance of Rain 80%	Chance of Rain 90%

    a. Order the following events from least likely to most likely.
       A. It rains on Sunday.
       B. It does not rain on Saturday.
       C. It rains on Monday.
       D. It does not rain on Friday.
    b. Find the odds in favor of rain for each day.

27. **USING TOOLS** Use the figure in Example 3 to answer each question.
    a. List the possible sums that result from rolling two six-sided dice.
    b. Find the theoretical probability of rolling each sum.
    c. The table below shows a simulation of rolling two six-sided dice three times. Use a random number generator to simulate rolling two six-sided dice 50 times. Compare the experimental probabilities of rolling each sum with the theoretical probabilities.

	A	B	C
1	First Die	Second Die	Sum
2	4	6	10
3	3	5	8
4	1	6	7
5			

28. **MAKING AN ARGUMENT** You flip a coin three times. It lands on heads twice and on tails once. Your friend concludes that the theoretical probability of the coin landing heads up is $P(\text{heads up}) = \frac{2}{3}$. Is your friend correct? Explain your reasoning.

29. **CRITICAL THINKING** You randomly draw a marble from a bag containing red, white, and blue marbles. The odds against drawing a white marble are 47 : 3.
    a. There are less than 100 marbles in the bag. How many marbles are in the bag? Justify your answer.
    b. The probability of drawing a red marble is 0.5. What is the probability of drawing a blue marble? Explain your reasoning.

30. **HOW DO YOU SEE IT?** Consider the graph of f shown. What is the probability that the graph of $y = f(x) + c$ intersects the x-axis when c is a randomly chosen integer from 1 to 6? Explain.

31. **DRAWING CONCLUSIONS** A manufacturer tests 1200 computers and finds that 9 of them have defects. Find the probability that a computer chosen at random has a defect. Predict the number of computers with defects in a shipment of 15,000 computers. Explain your reasoning.

32. **THOUGHT PROVOKING** The tree diagram shows a sample space. Write a probability problem that can be represented by the sample space. Then write the answer(s) to the problem.

Box A	Box B	Outcomes	Sum	Product
1	1	(1, 1)	2	1
	2	(1, 2)	3	2
2	1	(2, 1)	3	2
	2	(2, 2)	4	4
3	1	(3, 1)	4	3
	2	(3, 2)	5	6

Maintaining Mathematical Proficiency
Reviewing what you learned in previous grades and lessons

Factor the polynomial. *(Section 2.6)*

33. $3x^2 + 12x − 36$ 34. $2x^2 − 11x + 9$ 35. $4x^2 + 7x − 15$

Solve the equation using square roots. *(Section 4.3)*

36. $x^2 − 36 = 0$ 37. $5x^2 − 20 = 0$ 38. $(x + 4)^2 = 81$ 39. $25(x − 2)^2 = 9$

286   Chapter 5   Probability

If students need help...	If students got it...
Resources by Chapter • Practice A and Practice B • Puzzle Time	**Resources by Chapter** • Enrichment and Extension • Cumulative Review
Student Journal • Practice	Start the *next* Section
Differentiating the Lesson Skills Review Handbook	

Laurie's Notes

Overview of Section 5.2

Introduction
- In this lesson, the idea of independent and dependent events is introduced.
- Students will first prove whether events are independent or dependent by using the sample space. Students should use logical reasoning to confirm that their answer is correct.
- The probabilities of independent and dependent events are calculated, along with calculating conditional probabilities.

Teaching Strategy
- The graphing calculator can be used to simulate, and hence estimate, experimental probabilities. This strategy uses the random number generator to simulate events such as spinning a spinner.
- Example 3 in the lesson uses an eight-section spinner with the numbers 1–8. The problem asks students to calculate the probability of getting a 5 on the first spin and a number greater than 3 on the second spin.
- The steps below simulate 100 trials of each event, meaning spinning 100 times for the first spin, followed by another 100 spins for the second spin.
  - Generate two lists of 100 random integers from 1 to 8 by entering randInt(1,8,100) into lists L_1 and L_2. L_1 shows the outcomes of the first spin, and L_2 shows the outcomes of the second spin. (See Figure 1.)
  - To facilitate counting the number of outcomes with a 5 on the first spin, sort the two lists while keeping the paired data together by entering SortA(L_1,L_2) on the home screen. (See Figure 2.)
  - Scroll down L_1 to count the number of successful trials with 5 on the first spin and a number greater than 3 on the second spin. (See Figure 3.)

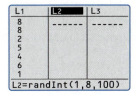 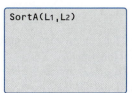 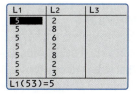

      Figure 1   Figure 2   Figure 3

- To calculate the experimental probability, divide the number of successes by 100.

Pacing Suggestion
- The explorations provide students the opportunity to explore the difference between independent and dependent events and to calculate probabilities of each.

Dynamic Teaching Tools
Dynamic Assessment & Progress Monitoring Tool
Lesson Planning Tool
Interactive Whiteboard Lesson Library
Dynamic Classroom with Dynamic Investigations

What Your Students Will Learn

- Determine whether events are independent events using probability.
- Find and compare probabilities of independent and dependents events.
- Find conditional probabilities when events are dependent.

Laurie's Notes

Exploration

Motivate
- Before class, cut 25 pieces of paper. Because this motivator is to model Example 4 in the lesson, you could cut the paper to the size of dollar bills. Write $1 on 20 of the pieces of paper and $100 on the other 5 pieces of paper.
- Display the paper money to the students.
- Select one student and say, "If you select two slips of paper and end up with $200, there will be no homework tonight! If you end up with $101, you will do two extra homework problems. If you end up with $2, you will do five extra homework problems."
- ? "What do you think your chances are of having no homework?" Answers will vary. Explain to students that in today's lesson they will calculate the probability of drawing $200.

Discuss
- Discuss independent and dependent events before students begin the explorations.

Exploration 1
- The two events may look very similar to students because they both involve the numbers 1 to 6.
- It is possible to end up with two 4s when rolling the dice. It is not possible when the pieces of paper are drawn from a bag without replacement.

Exploration 2
- Answers will vary in terms of the types of experiments students will design. If students are familiar with the random number generator on a graphing calculator, they may use it as part of the experiment.
- Have students share the description of their experiment and the results.

Exploration 3
- Before students find the two probabilities, ask them to make a guess as to which event has the greater probability, the dice or the paper.
- ? "What is the probability of rolling a sum of 7?" $\frac{6}{36} = \frac{1}{6}$
- ? "What is the probability of drawing two numbers that sum to 7?" $\frac{6}{30} = \frac{1}{5}$
- **Big Idea:** Students should recognize that the sample space is less for the dependent event than for the independent event.

Communicate Your Answer
- Have students work independently to answer Questions 4 and 5, and then discuss as a whole class.

Connecting to Next Step
- It should not take long for students to decide whether an event is independent or dependent. In the formal lesson, students will first prove whether an event is independent or dependent.

5.2 Independent and Dependent Events

Essential Question
How can you determine whether two events are independent or dependent?

Two events are **independent events** when the occurrence of one event does not affect the occurrence of the other event. Two events are **dependent events** when the occurrence of one *does* affect the occurrence of the other event.

EXPLORATION 1 — Identifying Independent and Dependent Events

Work with a partner. Determine whether the events are independent or dependent. Explain your reasoning.

a. Two six-sided dice are rolled.

b. Six pieces of paper, numbered 1 through 6, are in a bag. Two pieces of paper are selected one at a time without replacement.

> **REASONING ABSTRACTLY**
> To be proficient in math, you need to make sense of quantities and their relationships in problem situations.

EXPLORATION 2 — Finding Experimental Probabilities

Work with a partner.

a. In Exploration 1(a), experimentally estimate the probability that the sum of the two numbers rolled is 7. Describe your experiment.

b. In Exploration 1(b), experimentally estimate the probability that the sum of the two numbers selected is 7. Describe your experiment.

EXPLORATION 3 — Finding Theoretical Probabilities

Work with a partner.

a. In Exploration 1(a), find the theoretical probability that the sum of the two numbers rolled is 7. Then compare your answer with the experimental probability you found in Exploration 2(a).

b. In Exploration 1(b), find the theoretical probability that the sum of the two numbers selected is 7. Then compare your answer with the experimental probability you found in Exploration 2(b).

c. Compare the probabilities you obtained in parts (a) and (b).

Communicate Your Answer

4. How can you determine whether two events are independent or dependent?

5. Determine whether the events are independent or dependent. Explain your reasoning.

   a. You roll a 4 on a six-sided die and spin red on a spinner.

   b. Your teacher chooses a student to lead a group, chooses another student to lead a second group, and chooses a third student to lead a third group.

Dynamic Teaching Tools
- Dynamic Assessment & Progress Monitoring Tool
- Lesson Planning Tool
- Interactive Whiteboard Lesson Library
- Dynamic Classroom with Dynamic Investigations

ANSWERS

1. a. independent; The occurence of one event does not affect the occurence of the other event.
   b. dependent; The occurence of one event does affect the occurence of the other event.

2. a. *Sample answer:* 0.2; Two six-sided dice were rolled, and the sum of the numbers was recorded.
   b. *Sample answer:* 0.2; Two pieces of paper were selected one at a time from pieces numbered 1 through 6, and the sum of the numbers was recorded.

3. a. $\frac{6}{36} = \frac{1}{6} \approx 0.167$; *Sample answer:* This is less than the probability in Exploration 2(a).
   b. $\frac{6}{30} = \frac{1}{5} = 0.2$; *Sample answer:* This is the same as the probability in Exploration 2(b).
   c. The probability in part (a) is less than the probability in part (b).

4. Determine whether the occurence of one event affects the occurence of the other event.

5. a. independent; The occurence of one event does not affect the occurence of the other event.
   b. dependent; The occurence of one event does affect the occurence of the other event.

English Language Learners

Pair Activity
Pair an English learner with an English speaker. Have pairs create real-world examples of events that are considered to be independent events and those that are dependent events. Have pairs share their examples with the class.

Extra Example 1
A bag contains six pieces of paper, numbered 1 through 6. A student randomly selects a piece of paper, replaces it, and randomly selects another piece of paper. Use a sample space to determine whether randomly selecting a 5 first and randomly selecting an odd number second are independent events.
$P(5) = \frac{1}{6}$; $P(\text{odd number}) = \frac{1}{2}$;
$P(5, \text{then odd number}) = \frac{1}{12}$. Because $\frac{1}{6} \cdot \frac{1}{2} = \frac{1}{12}$, the events are independent.

Extra Example 2
A bag contains six pieces of paper, numbered 1 through 6. A student randomly selects a piece of paper, does not replace it, and randomly selects another piece of paper. Use a sample space to determine whether randomly selecting an even number first and randomly selecting a 4 second are independent events.
$P(\text{even number}) = \frac{1}{2}$; $P(4) = \frac{1}{6}$;
$P(\text{even number, then } 4) = \frac{1}{15}$.
Because $\frac{1}{2} \cdot \frac{1}{6} \neq \frac{1}{15}$, the events are not independent.

5.2 Lesson

Core Vocabulary
independent events, *p. 288*
dependent events, *p. 289*
conditional probability, *p. 289*
Previous
probability
sample space

What You Will Learn
- Determine whether events are independent events.
- Find probabilities of independent and dependent events.
- Find conditional probabilities.

Determining Whether Events Are Independent
Two events are **independent events** when the occurrence of one event does not affect the occurrence of the other event.

> **Core Concept**
>
> **Probability of Independent Events**
> **Words** Two events A and B are independent events if and only if the probability that both events occur is the product of the probabilities of the events.
> **Symbols** $P(A \text{ and } B) = P(A) \cdot P(B)$

EXAMPLE 1 Determining Whether Events Are Independent

A student taking a quiz randomly guesses the answers to four true-false questions. Use a sample space to determine whether guessing Question 1 correctly and guessing Question 2 correctly are independent events.

SOLUTION

Using the sample space in Example 2 on page 281:

$P(\text{correct on Question 1}) = \frac{8}{16} = \frac{1}{2}$ $P(\text{correct on Question 2}) = \frac{8}{16} = \frac{1}{2}$

$P(\text{correct on Question 1 and correct on Question 2}) = \frac{4}{16} = \frac{1}{4}$

▶ Because $\frac{1}{2} \cdot \frac{1}{2} = \frac{1}{4}$, the events are independent.

EXAMPLE 2 Determining Whether Events Are Independent

A group of four students includes one boy and three girls. The teacher randomly selects one of the students to be the speaker and a different student to be the recorder. Use a sample space to determine whether randomly selecting a girl first and randomly selecting a girl second are independent events.

SOLUTION

Let B represent the boy. Let G_1, G_2, and G_3 represent the three girls. Use a table to list the outcomes in the sample space.

Number of girls	Outcome	
1	G_1B	BG_1
1	G_2B	BG_2
1	G_3B	BG_3
2	G_1G_2	G_2G_1
2	G_1G_3	G_3G_1
2	G_2G_3	G_3G_2

Using the sample space:

$P(\text{girl first}) = \frac{9}{12} = \frac{3}{4}$ $P(\text{girl second}) = \frac{9}{12} = \frac{3}{4}$

$P(\text{girl first and girl second}) = \frac{6}{12} = \frac{1}{2}$

▶ Because $\frac{3}{4} \cdot \frac{3}{4} \neq \frac{1}{2}$, the events are not independent.

Laurie's Notes — Teacher Actions

? Attend to Precision: Write the *Core Concept* and ask, "What two statements result from the *if and only if* language?" If two events are independent, then $P(A \text{ and } B) = P(A) \cdot P(B)$, and if $P(A \text{ and } B) = P(A) \cdot P(B)$, then the events are independent.

- The probabilities can be used to test whether two events are independent or not. That is what Examples 1 and 2 do.
- **Use Appropriate Tools Strategically:** Have students write the sample space for Example 2. Coding the girls G_1, G_2, and G_3 is a helpful tool to organize and write results.

Monitoring Progress Help in English and Spanish at BigIdeasMath.com

1. In Example 1, determine whether guessing Question 1 incorrectly and guessing Question 2 correctly are independent events.
2. In Example 2, determine whether randomly selecting a girl first and randomly selecting a boy second are independent events.

Finding Probabilities of Events

In Example 1, it makes sense that the events are independent because the second guess should not be affected by the first guess. In Example 2, however, the selection of the second person *depends* on the selection of the first person because the same person cannot be selected twice. These events are *dependent*. Two events are **dependent events** when the occurrence of one event *does* affect the occurrence of the other event.

The probability that event B occurs given that event A has occurred is called the **conditional probability** of B given A and is written as $P(B|A)$.

MAKING SENSE OF PROBLEMS

One way that you can find $P(\text{girl second} | \text{girl first})$ is to list the 9 outcomes in which a girl is chosen first and then find the fraction of these outcomes in which a girl is chosen second:

$G_1B \quad G_2B \quad G_3B$
$G_1G_2 \quad G_2G_1 \quad G_3G_1$
$G_1G_3 \quad G_2G_3 \quad G_3G_2$

Core Concept

Probability of Dependent Events

Words If two events A and B are dependent events, then the probability that both events occur is the product of the probability of the first event and the conditional probability of the second event given the first event.

Symbols $P(A \text{ and } B) = P(A) \cdot P(B|A)$

Example Using the information in Example 2:

$P(\text{girl first and girl second}) = P(\text{girl first}) \cdot P(\text{girl second} | \text{girl first})$

$= \frac{9}{12} \cdot \frac{6}{9} = \frac{1}{2}$

EXAMPLE 3 Finding the Probability of Independent Events

As part of a board game, you need to spin the spinner, which is divided into equal parts. Find the probability that you get a 5 on your first spin and a number greater than 3 on your second spin.

SOLUTION

Let event A be "5 on first spin" and let event B be "greater than 3 on second spin."

The events are independent because the outcome of your second spin is not affected by the outcome of your first spin. Find the probability of each event and then multiply the probabilities.

$P(A) = \frac{1}{8}$ 1 of the 8 sections is a "5."

$P(B) = \frac{5}{8}$ 5 of the 8 sections (4, 5, 6, 7, 8) are greater than 3.

$P(A \text{ and } B) = P(A) \cdot P(B) = \frac{1}{8} \cdot \frac{5}{8} = \frac{5}{64} \approx 0.078$

▶ So, the probability that you get a 5 on your first spin and a number greater than 3 on your second spin is about 7.8%.

Extra Example 4
Nine women and six men are on a committee. Two people are randomly chosen from the committee members to serve as a chairperson and a treasurer. Find the probability that both events A and B will occur.
A: The chairperson is a man.
B: The treasurer is a woman.
The probability that the chairperson is a man and the treasurer is a woman is $\frac{9}{35}$, or about 25.7%.

Extra Example 5
A bag contains 10 red marbles and 5 blue marbles. You randomly select 3 marbles from the bag. What is the probability that all 3 marbles are blue when (a) you replace each marble before selecting the next one and (b) you do not replace each marble before selecting the next one? Compare the probabilities.

a. The probability that all three marbles are blue, with replacement, is $\frac{1}{27} \approx 0.037$.

b. The probability that all three marbles are blue, without replacement, is $\frac{2}{91} \approx 0.022$.

You are $\frac{1}{27} \div \frac{2}{91} \approx 1.7$ times more likely to select three blue marbles when you replace each marble before you select the next marble.

MONITORING PROGRESS ANSWERS
3. 25%
4. about 63.3%
5. a. about 42.2%
   b. about 41.4%
   You are about 1.02 times more likely to select 3 cards that are not hearts when you replace each card before you select the next card.

EXAMPLE 4 Finding the Probability of Dependent Events

A bag contains twenty $1 bills and five $100 bills. You randomly draw a bill from the bag, set it aside, and then randomly draw another bill from the bag. Find the probability that both events A and B will occur.

Event A: The first bill is $100.   Event B: The second bill is $100.

SOLUTION
The events are dependent because there is one less bill in the bag on your second draw than on your first draw. Find $P(A)$ and $P(B|A)$. Then multiply the probabilities.

$P(A) = \frac{5}{25}$ 5 of the 25 bills are $100 bills.

$P(B|A) = \frac{4}{24}$ 4 of the remaining 24 bills are $100 bills.

$P(A \text{ and } B) = P(A) \cdot P(B|A) = \frac{5}{25} \cdot \frac{4}{24} = \frac{1}{5} \cdot \frac{1}{6} = \frac{1}{30} \approx 0.033.$

▶ So, the probability that you draw two $100 bills is about 3.3%.

EXAMPLE 5 Comparing Independent and Dependent Events

You randomly select 3 cards from a standard deck of 52 playing cards. What is the probability that all 3 cards are hearts when (a) you replace each card before selecting the next card, and (b) you do not replace each card before selecting the next card? Compare the probabilities.

SOLUTION
Let event A be "first card is a heart," event B be "second card is a heart," and event C be "third card is a heart."

a. Because you replace each card before you select the next card, the events are independent. So, the probability is

$P(A \text{ and } B \text{ and } C) = P(A) \cdot P(B) \cdot P(C) = \frac{13}{52} \cdot \frac{13}{52} \cdot \frac{13}{52} = \frac{1}{64} \approx 0.016.$

b. Because you do not replace each card before you select the next card, the events are dependent. So, the probability is

$P(A \text{ and } B \text{ and } C) = P(A) \cdot P(B|A) \cdot P(C|A \text{ and } B)$

$= \frac{13}{52} \cdot \frac{12}{51} \cdot \frac{11}{50} = \frac{11}{850} \approx 0.013.$

▶ So, you are $\frac{1}{64} \div \frac{11}{850} \approx 1.2$ times more likely to select 3 hearts when you replace each card before you select the next card.

STUDY TIP
The formulas for finding probabilities of independent and dependent events can be extended to three or more events.

Monitoring Progress Help in English and Spanish at *BigIdeasMath.com*

3. In Example 3, what is the probability that you spin an even number and then an odd number?
4. In Example 4, what is the probability that both bills are $1 bills?
5. In Example 5, what is the probability that none of the cards drawn are hearts when (a) you replace each card, and (b) you do not replace each card? Compare the probabilities.

Laurie's Notes   **Teacher Actions**

- **Model with Mathematics:** Model Example 4 as suggested in the *Motivate* on page T-287.
- Students find it difficult to trust the formula for finding the probability of dependent events. You do not want to write the sample space for Example 4, so referring back to a simpler example is helpful.
- Work through Example 5. Students find it intuitive that when you replace the cards each time, the probability of a success is greater.

Finding Conditional Probabilities

EXAMPLE 6 Using a Table to Find Conditional Probabilities

	Pass	Fail
Defective	3	36
Non-defective	450	11

A quality-control inspector checks for defective parts. The table shows the results of the inspector's work. Find (a) the probability that a defective part "passes," and (b) the probability that a non-defective part "fails."

SOLUTION

a. $P(\text{pass} \mid \text{defective}) = \dfrac{\text{Number of defective parts "passed"}}{\text{Total number of defective parts}}$

$= \dfrac{3}{3+36} = \dfrac{3}{39} = \dfrac{1}{13} \approx 0.077$, or about 7.7%

b. $P(\text{fail} \mid \text{non-defective}) = \dfrac{\text{Number of non-defective parts "failed"}}{\text{Total number of non-defective parts}}$

$= \dfrac{11}{450+11} = \dfrac{11}{461} \approx 0.024$, or about 2.4%

STUDY TIP
Note that when A and B are independent, this rule still applies because $P(B) = P(B \mid A)$.

You can rewrite the formula for the probability of dependent events to write a rule for finding conditional probabilities.

$P(A) \cdot P(B \mid A) = P(A \text{ and } B)$ Write formula.

$P(B \mid A) = \dfrac{P(A \text{ and } B)}{P(A)}$ Divide each side by $P(A)$.

EXAMPLE 7 Finding a Conditional Probability

At a school, 60% of students buy a school lunch. Only 10% of students buy lunch and dessert. What is the probability that a student who buys lunch also buys dessert?

SOLUTION

Let event A be "buys lunch" and let event B be "buys dessert." You are given $P(A) = 0.6$ and $P(A \text{ and } B) = 0.1$. Use the formula to find $P(B \mid A)$.

$P(B \mid A) = \dfrac{P(A \text{ and } B)}{P(A)}$ Write formula for conditional probability.

$= \dfrac{0.1}{0.6}$ Substitute 0.1 for $P(A \text{ and } B)$ and 0.6 for $P(A)$.

$= \dfrac{1}{6} \approx 0.167$ Simplify.

▶ So, the probability that a student who buys lunch also buys dessert is about 16.7%.

Monitoring Progress Help in English and Spanish at *BigIdeasMath.com*

6. In Example 6, find (a) the probability that a non-defective part "passes," and (b) the probability that a defective part "fails."

7. At a coffee shop, 80% of customers order coffee. Only 15% of customers order coffee and a bagel. What is the probability that a customer who orders coffee also orders a bagel?

Assignment Guide and Homework Check

ASSIGNMENT

Basic: 1, 2, 3–15 odd, 19–23 odd, 26, 27, 31–33

Average: 1, 2–26 even, 27, 31–33

Advanced: 1, 2, 8–26 even, 27–33

HOMEWORK CHECK

Basic: 7, 11, 13, 19, 21

Average: 8, 12, 14, 20, 22

Advanced: 10, 12, 14, 22, 24

ANSWERS

1. When two events are dependent, the occurrence of one event affects the other. When two events are independent, the occurrence of one event does not affect the other. *Sample answer:* choosing two marbles from a bag without replacement; rolling two dice
2. conditional probability; $P(B \mid A)$
3. dependent; The occurrence of event A affects the occurrence of event B.
4. independent; The occurrence of event A does not affect the occurrence of event B.
5. dependent; The occurrence of event A affects the occurrence of event B.
6. independent; The occurrence of event A does not affect the occurrence of event B.
7. yes
8. no
9. yes
10. no
11. about 2.8%

5.2 Exercises

Dynamic Solutions available at *BigIdeasMath.com*

Vocabulary and Core Concept Check

1. **WRITING** Explain the difference between dependent events and independent events, and give an example of each.

2. **COMPLETE THE SENTENCE** The probability that event *B* will occur given that event *A* has occurred is called the _____ of *B* given *A* and is written as _____.

Monitoring Progress and Modeling with Mathematics

In Exercises 3–6, tell whether the events are independent or dependent. Explain your reasoning.

3. A box of granola bars contains an assortment of flavors. You randomly choose a granola bar and eat it. Then you randomly choose another bar.
   Event A: You choose a coconut almond bar first.
   Event B: You choose a cranberry almond bar second.

4. You roll a six-sided die and flip a coin.
   Event A: You get a 4 when rolling the die.
   Event B: You get tails when flipping the coin.

5. Your MP3 player contains hip-hop and rock songs. You randomly choose a song. Then you randomly choose another song without repeating song choices.
   Event A: You choose a hip-hop song first.
   Event B: You choose a rock song second.

6. There are 22 novels of various genres on a shelf. You randomly choose a novel and put it back. Then you randomly choose another novel.
   Event A: You choose a mystery novel.
   Event B: You choose a science fiction novel.

In Exercises 7–10, determine whether the events are independent. *(See Examples 1 and 2.)*

7. You play a game that involves spinning a wheel. Each section of the wheel shown has the same area. Use a sample space to determine whether randomly spinning blue and then green are independent events.

8. You have one red apple and three green apples in a bowl. You randomly select one apple to eat now and another apple for your lunch. Use a sample space to determine whether randomly selecting a green apple first and randomly selecting a green apple second are independent events.

9. A student is taking a multiple-choice test where each question has four choices. The student randomly guesses the answers to the five-question test. Use a sample space to determine whether guessing Question 1 correctly and Question 2 correctly are independent events.

10. A vase contains four white roses and one red rose. You randomly select two roses to take home. Use a sample space to determine whether randomly selecting a white rose first and randomly selecting a white rose second are independent events.

11. **PROBLEM SOLVING** You play a game that involves spinning the money wheel shown. You spin the wheel twice. Find the probability that you get more than $500 on your first spin and then go bankrupt on your second spin. *(See Example 3.)*

292   Chapter 5   Probability

12. **PROBLEM SOLVING** You play a game that involves drawing two numbers from a hat. There are 25 pieces of paper numbered from 1 to 25 in the hat. Each number is replaced after it is drawn. Find the probability that you will draw the 3 on your first draw and a number greater than 10 on your second draw.

13. **PROBLEM SOLVING** A drawer contains 12 white socks and 8 black socks. You randomly choose 1 sock and do not replace it. Then you randomly choose another sock. Find the probability that both events A and B will occur. *(See Example 4.)*

    Event A: The first sock is white.

    Event B: The second sock is white.

14. **PROBLEM SOLVING** A word game has 100 tiles, 98 of which are letters and 2 of which are blank. The numbers of tiles of each letter are shown. You randomly draw 1 tile, set it aside, and then randomly draw another tile. Find the probability that both events A and B will occur.

    Event A: The first tile is a consonant.

    Event B: The second tile is a vowel.

A – 9	H – 2	O – 8	V – 2
B – 2	I – 9	P – 2	W – 2
C – 2	J – 1	Q – 1	X – 1
D – 4	K – 1	R – 6	Y – 2
E – 12	L – 4	S – 4	Z – 1
F – 2	M – 2	T – 6	– 2
G – 3	N – 6	U – 4	Blank

15. **ERROR ANALYSIS** Events A and B are independent. Describe and correct the error in finding $P(A \text{ and } B)$.

    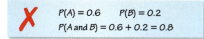

16. **ERROR ANALYSIS** A shelf contains 3 fashion magazines and 4 health magazines. You randomly choose one to read, set it aside, and randomly choose another for your friend to read. Describe and correct the error in finding the probability that both events A and B occur.

    Event A: The first magazine is fashion.

    Event B: The second magazine is health.

    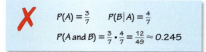

17. **NUMBER SENSE** Events A and B are independent. Suppose $P(B) = 0.4$ and $P(A \text{ and } B) = 0.13$. Find $P(A)$.

18. **NUMBER SENSE** Events A and B are dependent. Suppose $P(B|A) = 0.6$ and $P(A \text{ and } B) = 0.15$. Find $P(A)$.

19. **ANALYZING RELATIONSHIPS** You randomly select three cards from a standard deck of 52 playing cards. What is the probability that all three cards are face cards when (a) you replace each card before selecting the next card, and (b) you do not replace each card before selecting the next card? Compare the probabilities. *(See Example 5.)*

20. **ANALYZING RELATIONSHIPS** A bag contains 9 red marbles, 4 blue marbles, and 7 yellow marbles. You randomly select three marbles from the bag. What is the probability that all three marbles are red when (a) you replace each marble before selecting the next marble, and (b) you do not replace each marble before selecting the next marble? Compare the probabilities.

21. **ATTEND TO PRECISION** The table shows the number of species in the United States listed as endangered and threatened. Find (a) the probability that a randomly selected endangered species is a bird, and (b) the probability that a randomly selected mammal is endangered. *(See Example 6.)*

	Endangered	Threatened
Mammals	70	16
Birds	80	16
Other	318	142

22. **ATTEND TO PRECISION** The table shows the number of tropical cyclones that formed during the hurricane seasons over a 12-year period. Find (a) the probability to predict whether a future tropical cyclone in the Northern Hemisphere is a hurricane, and (b) the probability to predict whether a hurricane is in the Southern Hemisphere.

Type of Tropical Cyclone	Northern Hemisphere	Southern Hemisphere
tropical depression	100	107
tropical storm	342	487
hurricane	379	525

23. **PROBLEM SOLVING** At a school, 43% of students attend the homecoming football game. Only 23% of students go to the game and the homecoming dance. What is the probability that a student who attends the football game also attends the dance? *(See Example 7.)*

ANSWERS

12. 2.4%
13. about 34.7%
14. about 23.8%
15. The probabilities were added instead of multiplied; $P(A \text{ and } B) = (0.6)(0.2) = 0.12$
16. $P(B|A)$ is incorrect; $P(B|A) = \frac{4}{6}$; $P(A \text{ and } B) = \frac{2}{7} \approx 0.286$
17. 0.325
18. 0.25
19. a. about 1.2%
    b. about 1.0%
    You are about 1.2 times more likely to select 3 face cards when you replace each card before you select the next card.
20. a. about 9.1%
    b. about 7.4%
    You are about 1.23 times more likely to select 3 red marbles when you replace each marble before you select the next marble.
21. a. about 17.1%
    b. about 81.4%
22. a. about 46.2%
    b. about 58.1%
23. about 53.5%

ANSWERS
24. about 6.0%

25–33. See Additional Answers.

Mini-Assessment

A bin contains 6 blue socks and 4 black socks.

1. You randomly select a sock, put it aside, and then randomly select a second sock from the bin. Use a sample space to determine whether selecting a black sock first and a blue sock second are independent events.
$P(black) = \frac{2}{5}$; $P(blue) = \frac{3}{5}$; $P(black,$ then $blue) = \frac{4}{15}$; not independent: $\frac{2}{5} \cdot \frac{3}{5} \neq \frac{4}{15}$

2. You randomly select 3 socks from the bin. What is the probability that all three socks are black for each situation?

    a. You replace each sock before selecting the next one.
    $\frac{8}{125} = 0.064$

    b. You do not replace each sock before selecting the next one.
    $\frac{1}{30} \approx 0.033$

3. At a school, all 9th and 10th grade students are required to take either band or chorus. The table shows the enrollments. Find the probability that a 10th grade student is taking band.

	9th grade	10th grade
Band	57	48
Chorus	63	77

    $\frac{48}{125}$, or 38.4%

4. In a science class, 27% of the students complete an extra-credit project. 18% of the students complete the extra-credit project and get an A in the class. What is the probability that a student who completes an extra-credit project also gets an A?
$\frac{2}{3}$, or about 66.7%

---

24. PROBLEM SOLVING At a gas station, 84% of customers buy gasoline. Only 5% of customers buy gasoline and a beverage. What is the probability that a customer who buys gasoline also buys a beverage?

25. PROBLEM SOLVING You and 19 other students volunteer to present the "Best Teacher" award at a school banquet. One student volunteer will be chosen to present the award. Each student worked at least 1 hour in preparation for the banquet. You worked for 4 hours, and the group worked a combined total of 45 hours. For each situation, describe a process that gives you a "fair" chance to be chosen, and find the probability that you are chosen.

  a. "Fair" means equally likely.

  b. "Fair" means proportional to the number of hours each student worked in preparation.

26. HOW DO YOU SEE IT? A bag contains one red marble and one blue marble. The diagrams show the possible outcomes of randomly choosing two marbles using different methods. For each method, determine whether the marbles were selected with or without replacement.

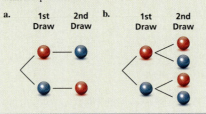

a. 1st Draw 2nd Draw   b. 1st Draw 2nd Draw

27. MAKING AN ARGUMENT A meteorologist claims that there is a 70% chance of rain. When it rains, there is a 75% chance that your softball game will be rescheduled. Your friend believes the game is more likely to be rescheduled than played. Is your friend correct? Explain your reasoning.

28. THOUGHT PROVOKING Two six-sided dice are rolled once. Events A and B are represented by the diagram. Describe each event. Are the two events dependent or independent? Justify your reasoning.

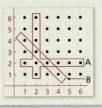

29. MODELING WITH MATHEMATICS A football team is losing by 14 points near the end of a game. The team scores two touchdowns (worth 6 points each) before the end of the game. After each touchdown, the coach must decide whether to go for 1 point with a kick (which is successful 99% of the time) or 2 points with a run or pass (which is successful 45% of the time).

  a. If the team goes for 1 point after each touchdown, what is the probability that the team wins? loses? ties?

  b. If the team goes for 2 points after each touchdown, what is the probability that the team wins? loses? ties?

  c. Can you develop a strategy so that the coach's team has a probability of winning the game that is greater than the probability of losing? If so, explain your strategy and calculate the probabilities of winning and losing the game.

30. ABSTRACT REASONING Assume that A and B are independent events.

  a. Explain why $P(B) = P(B|A)$ and $P(A) = P(A|B)$.

  b. Can $P(A \text{ and } B)$ also be defined as $P(B) \cdot P(A|B)$? Justify your reasoning.

Maintaining Mathematical Proficiency — Reviewing what you learned in previous grades and lessons

Solve the equation. Check your solution. *(Skills Review Handbook)*

31. $\frac{9}{10}x = 0.18$

32. $\frac{1}{4}x + 0.5x = 1.5$

33. $0.3x - \frac{3}{5}x + 1.6 = 1.555$

If students need help...	If students got it...
Resources by Chapter • Practice A and Practice B • Puzzle Time	Resources by Chapter • Enrichment and Extension • Cumulative Review
Student Journal • Practice	Start the *next* Section
Differentiating the Lesson Skills Review Handbook	

Laurie's Notes

Overview of Section 5.3

Introduction
- Students were introduced to two-way tables in Math I. In this lesson, the entries are converted to relative and conditional relative frequencies.
- This lesson connects to earlier work with conditional probabilities.

Formative Assessment Tips
- **Visitor Explanation:** This technique simulates what it would look like if a visitor were to enter your classroom during the middle of an exploration or activity. Could your students explain what they are doing and why they are doing it?
- This technique lets you know whether students understand the goal, or the essential question related to the exploration. Are students simply following directions or are they aware of the goal for today's exploration? Students are more engaged and their learning improves when the learning objective or purpose of the exploration is understood.
- This technique is best used when students are actively engaged in an exploration. This technique can be done as a *Think-Pair-Share* or as a *Writing Prompt*.
- The learning objective or essential question should have been made known to the students at the outset. This technique is not used to see whether students figured out the goal of the exploration or activity.

Pacing Suggestion
- The explorations are relatively short and serve to remind students of how two-way tables are constructed and analyzed. You may assign Exploration 3 as an out-of-class assignment.

Dynamic Teaching Tools
Dynamic Assessment & Progress Monitoring Tool
Lesson Planning Tool
Interactive Whiteboard Lesson Library
Dynamic Classroom with Dynamic Investigations

What Your Students Will Learn

- Make two-way tables to display data collected from one source that belongs to two different categories.
- Find relative and conditional relative frequencies of entries in a two-way table.
- Use conditional relative frequencies to find conditional probabilities.

Laurie's Notes

Exploration

Motivate
- Ask whether there are any students who keep track of their training or practice routines. If there are serious athletes in the class, then chances are they do track their efforts.
- Explain to students that in this lesson they will study a structure that is used to calculate and compare when goals were met for different exercise options.

For Your Information
- Two-way tables are explained in the first exploration, although students should be familiar with these tables from Math I.

Exploration 1
- ❓ "How many distinct regions are there in the Venn diagram? Name them." **four regions: play an instrument, speak a foreign language, do both, and do neither**
- **Model with Mathematics:** A Venn diagram is one way to represent the results of the survey. The two-way table is a second method.
- ❓ "What information is immediately known in the two-way table that is not displayed in the Venn diagram?" **the totals in each category**

Exploration 2
- Students should have little difficulty completing this exploration.
- Discuss the results as a class.

Exploration 3
- You might first ask students what categories they are curious to survey classmates about, though this can potentially lead to matters of privacy or discomfort for some students.
- Consider having safe yes-no questions ready to suggest: play video games, live more than x miles from school, and had a school-year job.

Communicate Your Answer
- Students should work independently to answer Questions 4 and 5. Discuss as a class.

Connecting to Next Step
- The explorations serve to quickly review how to construct and use a two-way table. Formal language describing the entries of the two-way table is presented in the formal lesson.

5.3 Two-Way Tables and Probability

Essential Question How can you construct and interpret a two-way table?

EXPLORATION 1 Completing and Using a Two-Way Table

Work with a partner. A *two-way table* displays the same information as a Venn diagram. In a two-way table, one category is represented by the rows and the other category is represented by the columns.

The Venn diagram shows the results of a survey in which 80 students were asked whether they play a musical instrument and whether they speak a foreign language. Use the Venn diagram to complete the two-way table. Then use the two-way table to answer each question.

Survey of 80 Students

Play an instrument: 25 | 16 | Speak a foreign language: 30
Outside: 9

	Play an Instrument	Do Not Play an Instrument	Total
Speak a Foreign Language			
Do Not Speak a Foreign Language			
Total			

a. How many students play an instrument?

b. How many students speak a foreign language?

c. How many students play an instrument and speak a foreign language?

d. How many students do not play an instrument and do not speak a foreign language?

e. How many students play an instrument and do not speak a foreign language?

EXPLORATION 2 Two-Way Tables and Probability

Work with a partner. In Exploration 1, one student is selected at random from the 80 students who took the survey. Find the probability that the student

a. plays an instrument.

b. speaks a foreign language.

c. plays an instrument and speaks a foreign language.

d. does not play an instrument and does not speak a foreign language.

e. plays an instrument and does not speak a foreign language.

EXPLORATION 3 Conducting a Survey

Work with your class. Conduct a survey of the students in your class. Choose two categories that are different from those given in Explorations 1 and 2. Then summarize the results in both a Venn diagram and a two-way table. Discuss the results.

> **MODELING WITH MATHEMATICS**
> To be proficient in math, you need to identify important quantities in a practical situation and map their relationships using such tools as diagrams and two-way tables.

Communicate Your Answer

4. How can you construct and interpret a two-way table?

5. How can you use a two-way table to determine probabilities?

ANSWERS

1. See Additional Answers.

2. a. $\frac{41}{80}$

   b. $\frac{23}{40}$

   c. $\frac{1}{5}$

   d. $\frac{9}{80}$

   e. $\frac{5}{16}$

3. See Additional Answers.

4. *Sample answer:* You can use a Venn diagram to construct the two-way table. Each entry represents the number of people in each category.

5. Divide each number in each category by the total surveyed.

Differentiated Instruction

Visual

Help students read the data in a two-way table. Direct them to use a piece of paper to cover all but one row and one column at a time. For each cell, ask students to describe what the number in the cell represents. Repeat with the other combinations of rows and columns.

Extra Example 1

There are 16 juniors and 24 seniors on a debate team. Of these, 7 juniors and 19 seniors qualify for a state debate competition. Organize this information in a two-way table. Then find and interpret the marginal frequencies.

In the table, Qual. means "qualified" and DNQ is "did not qualify."

	Qual.	DNQ	Total
Junior	7	9	16
Senior	19	5	24
Total	26	14	40

26 students qualified.
14 students did not qualify.
40 students are on the debate team.

MONITORING PROGRESS ANSWER

1.

		Response		
		In Favor	Against	Total
Gender	Boy	61	35	96
	Girl	71	17	88
	Total	132	52	184

184 students were surveyed, 132 students are in favor, 52 students are against, 96 boys were surveyed, 88 girls were surveyed.

5.3 Lesson

Core Vocabulary

two-way table, *p. 296*
joint frequency, *p. 296*
marginal frequency, *p. 296*
joint relative frequency, *p. 297*
marginal relative frequency, *p. 297*
conditional relative frequency, *p. 297*

Previous
conditional probability

READING

A two-way table is also called a *contingency table*, or a *two-way frequency table*.

What You Will Learn

- Make two-way tables.
- Find relative and conditional relative frequencies.
- Use conditional relative frequencies to find conditional probabilities.

Making Two-Way Tables

A **two-way table** is a frequency table that displays data collected from one source that belong to two different categories. One category of data is represented by rows and the other is represented by columns. Suppose you randomly survey freshmen and sophomores about whether they are attending a school concert. A two-way table is one way to organize your results.

Each entry in the table is called a **joint frequency**. The sums of the rows and columns are called **marginal frequencies**, which you will find in Example 1.

		Attendance	
		Attending	Not Attending
Class	Freshman	25	44
	Sophomore	80	32

joint frequency

EXAMPLE 1 Making a Two-Way Table

In another survey similar to the one above, 106 juniors and 114 seniors respond. Of those, 42 juniors and 77 seniors plan on attending. Organize these results in a two-way table. Then find and interpret the marginal frequencies.

SOLUTION

Step 1 Find the joint frequencies. Because 42 of the 106 juniors are attending, $106 - 42 = 64$ juniors are not attending. Because 77 of the 114 seniors are attending, $114 - 77 = 37$ seniors are not attending. Place each joint frequency in its corresponding cell.

Step 2 Find the marginal frequencies. Create a new column and row for the sums. Then add the entries and interpret the results.

		Attendance		
		Attending	Not Attending	Total
Class	Junior	42	64	106
	Senior	77	37	114
	Total	119	101	220

106 juniors responded.
114 seniors responded.
220 students were surveyed.
119 students are attending.
101 students are not attending.

Step 3 Find the sums of the marginal frequencies. Notice the sums $106 + 114 = 220$ and $119 + 101 = 220$ are equal. Place this value at the bottom right.

Monitoring Progress

Help in English and Spanish at *BigIdeasMath.com*

1. You randomly survey students about whether they are in favor of planting a community garden at school. Of 96 boys surveyed, 61 are in favor. Of 88 girls surveyed, 17 are against. Organize the results in a two-way table. Then find and interpret the marginal frequencies.

Laurie's Notes Teacher Actions

- Students can have difficulty discerning the categories when constructing a two-way table. For instance, in the introduction students may group freshmen and sophomores together as simply students who are attending a concert or not. In all problems, identify the categories.
- Define joint frequency and marginal frequency.
- **Turn and Talk:** "What is the fewest number of pieces of information needed to complete a two-category table?" Listen for student understanding of the relationships in the columns and rows. Four pieces of information are needed.

Finding Relative and Conditional Relative Frequencies

You can display values in a two-way table as frequency counts (as in Example 1) or as *relative frequencies*.

STUDY TIP
Two-way tables can display relative frequencies based on the total number of observations, the row totals, or the column totals.

Core Concept

Relative and Conditional Relative Frequencies

A **joint relative frequency** is the ratio of a frequency that is not in the total row or the total column to the total number of values or observations.

A **marginal relative frequency** is the sum of the joint relative frequencies in a row or a column.

A **conditional relative frequency** is the ratio of a joint relative frequency to the marginal relative frequency. You can find a conditional relative frequency using a row total or a column total of a two-way table.

EXAMPLE 2 Finding Joint and Marginal Relative Frequencies

Use the survey results in Example 1 to make a two-way table that shows the joint and marginal relative frequencies.

SOLUTION

To find the joint relative frequencies, divide each frequency by the total number of students in the survey. Then find the sum of each row and each column to find the marginal relative frequencies.

INTERPRETING MATHEMATICAL RESULTS
Relative frequencies can be interpreted as probabilities. The probability that a randomly selected student is a junior and is *not* attending the concert is 29.1%.

		Attendance		
		Attending	Not Attending	Total
Class	Junior	$\frac{42}{220} \approx 0.191$	$\frac{64}{220} \approx 0.291$	0.482
	Senior	$\frac{77}{220} = 0.35$	$\frac{37}{220} \approx 0.168$	0.518
	Total	0.541	0.459	1

About 29.1% of the students in the survey are juniors and are *not* attending the concert.

About 51.8% of the students in the survey are seniors.

EXAMPLE 3 Finding Conditional Relative Frequencies

Use the survey results in Example 1 to make a two-way table that shows the conditional relative frequencies based on the row totals.

SOLUTION

Use the marginal relative frequency of each *row* to calculate the conditional relative frequencies.

		Attendance	
		Attending	Not Attending
Class	Junior	$\frac{0.191}{0.482} \approx 0.396$	$\frac{0.291}{0.482} \approx 0.604$
	Senior	$\frac{0.35}{0.518} \approx 0.676$	$\frac{0.168}{0.518} \approx 0.324$

Given that a student is a senior, the conditional relative frequency that he or she is *not* attending the concert is about 32.4%.

English Language Learners

Analyzing Word Problems
Give students a copy of the problems in this section on paper with extra spacing and wide margins. Demonstrate how to underline key words, phrases, and numbers to find the important information in the problem. When students organize the given information in a table, have them include extra space for the probability calculations and space for additional rows and columns in the table if needed.

Extra Example 2

Use the debate competition information in Extra Example 1 to make a two-way table that shows the joint and marginal relative frequencies.

Junior: $\frac{7}{40} = 0.175$, $\frac{9}{40} = 0.225$

Senior: $\frac{19}{40} = 0.475$, $\frac{5}{40} = 0.125$

	Qual.	DNQ	Total
Junior	0.175	0.225	0.4
Senior	0.475	0.125	0.6
Total	0.65	0.35	1

Extra Example 3

Use the debate competition information in Extra Example 1 to make a two-way table that shows the conditional relative frequencies based on the row totals.

Junior: $\frac{0.175}{0.4} \approx 0.438$, $\frac{0.225}{0.4} \approx 0.563$

Senior: $\frac{0.475}{0.6} \approx 0.792$, $\frac{0.125}{0.6} \approx 0.208$

	Qual.	DNQ
Junior	0.438	0.563
Senior	0.792	0.208

Laurie's Notes — Teacher Actions

- Discuss the difference between relative counts and relative frequencies. Write the *Core Concept*.
- **?** "In Example 1, there were 220 students surveyed. How do you calculate the joint relative frequencies?" Divide each entry by 220.
- **?** "What does the 0.396 entry in the two-way table mean in Example 3?" Given that the student is a junior, the conditional relative frequency that he or she is attending the concert is 39.6%.

Extra Example 4

A store surveys customers of different ages. The survey asks whether they would like to see the store expand its toy department. The results, given as joint relative frequencies, are shown in the two-way table.

	Age (in years)		
	< 10	10–20	> 20
Yes	0.27	0.06	0.23
No	0.09	0.17	0.18

a. What is the probability that a randomly selected customer whose age is between 10 and 20 would not like to see the toy department expanded?
$\frac{0.17}{0.17 + 0.06}$, or about 73.9%

b. What is the probability that a randomly selected customer who would like to see the toy department expanded is younger than 10?
$\frac{0.27}{0.27 + 0.06 + 0.23}$, or about 48.2%

c. Determine whether replying "Yes" and being younger than 10 are independent events.
$P(\text{younger than 10}) = 0.36$;
$P(\text{younger than 10} \mid \text{Yes}) \approx 0.48$;
Because $0.36 \neq 0.48$, the two events are not independent.

MONITORING PROGRESS ANSWERS

2.

		Response		
		In Favor	Against	Total
Gender	Boy	0.332	0.190	0.522
	Girl	0.386	0.092	0.478
	Total	0.718	0.282	1

3–4. See Additional Answers.

Monitoring Progress  Help in English and Spanish at *BigIdeasMath.com*

2. Use the survey results in Monitoring Progress Question 1 to make a two-way table that shows the joint and marginal relative frequencies.

3. Use the survey results in Example 1 to make a two-way table that shows the conditional relative frequencies based on the column totals. Interpret the conditional relative frequencies in the context of the problem.

4. Use the survey results in Monitoring Progress Question 1 to make a two-way table that shows the conditional relative frequencies based on the row totals. Interpret the conditional relative frequencies in the context of the problem.

Finding Conditional Probabilities

You can use conditional relative frequencies to find conditional probabilities.

EXAMPLE 4 Finding Conditional Probabilities

A satellite TV provider surveys customers in three cities. The survey asks whether they would recommend the TV provider to a friend. The results, given as joint relative frequencies, are shown in the two-way table.

		Location		
		Glendale	Santa Monica	Long Beach
Response	Yes	0.29	0.27	0.32
	No	0.05	0.03	0.04

a. What is the probability that a randomly selected customer who is located in Glendale will recommend the provider?

b. What is the probability that a randomly selected customer who will not recommend the provider is located in Long Beach?

c. Determine whether recommending the provider to a friend and living in Long Beach are independent events.

SOLUTION

INTERPRETING MATHEMATICAL RESULTS
The probability 0.853 is a conditional relative frequency based on a column total. The condition is that the customer lives in Glendale.

a. $P(\text{yes} \mid \text{Glendale}) = \frac{P(\text{Glendale and yes})}{P(\text{Glendale})} = \frac{0.29}{0.29 + 0.05} \approx 0.853$

▶ So, the probability that a customer who is located in Glendale will recommend the provider is about 85.3%.

b. $P(\text{Long Beach} \mid \text{no}) = \frac{P(\text{no and Long Beach})}{P(\text{no})} = \frac{0.04}{0.05 + 0.03 + 0.04} \approx 0.333$

▶ So, the probability that a customer who will not recommend the provider is located in Long Beach is about 33.3%.

c. Use the formula $P(B) = P(B \mid A)$ and compare $P(\text{Long Beach})$ and $P(\text{Long Beach} \mid \text{yes})$.

$P(\text{Long Beach}) = 0.32 + 0.04 = 0.36$

$P(\text{Long Beach} \mid \text{yes}) = \frac{P(\text{Yes and Long Beach})}{P(\text{yes})} = \frac{0.32}{0.29 + 0.27 + 0.32} \approx 0.36$

▶ Because $P(\text{Long Beach}) \approx P(\text{Long Beach} \mid \text{yes})$, the two events are independent.

Laurie's Notes   Teacher Actions

- **Think-Pair-Share:** Have students answer Questions 2–4, and then share and discuss as a class.
- Work through Example 4, finding conditional probabilities from the table of joint relative frequencies.
- **Visitor Explanation:** Ask, "If a visitor entered the room right now, how would you explain what you are doing and why are you doing it?" Students should practice explanations with partners. Solicit oral responses or do a one-minute write.

Monitoring Progress Help in English and Spanish at *BigIdeasMath.com*

5. In Example 4, what is the probability that a randomly selected customer who is located in Santa Monica will not recommend the provider to a friend?

6. In Example 4, determine whether recommending the provider to a friend and living in Santa Monica are independent events. Explain your reasoning.

EXAMPLE 5 Comparing Conditional Probabilities

A jogger wants to burn a certain number of calories during his workout. He maps out three possible jogging routes. Before each workout, he randomly selects a route, and then determines the number of calories he burns and whether he reaches his goal. The table shows his findings. Which route should he use?

	Reaches Goal	Does Not Reach Goal
Route A	|||| ||||		||||	
Route B	|||| ||||		||||
Route C	|||| |||| ||	||||	

SOLUTION

Step 1 Use the findings to make a two-way table that shows the joint and marginal relative frequencies. There are a total of 50 observations in the table.

		Result		
		Reaches Goal	Does Not Reach Goal	Total
Route	A	0.22	0.12	0.34
	B	0.22	0.08	0.30
	C	0.24	0.12	0.36
	Total	0.68	0.32	1

Step 2 Find the conditional probabilities by dividing each joint relative frequency in the "Reaches Goal" column by the marginal relative frequency in its corresponding row.

$P(\text{reaches goal} \mid \text{Route A}) = \dfrac{P(\text{Route A and reaches goal})}{P(\text{Route A})} = \dfrac{0.22}{0.34} \approx 0.647$

$P(\text{reaches goal} \mid \text{Route B}) = \dfrac{P(\text{Route B and reaches goal})}{P(\text{Route B})} = \dfrac{0.22}{0.30} \approx 0.733$

$P(\text{reaches goal} \mid \text{Route C}) = \dfrac{P(\text{Route C and reaches goal})}{P(\text{Route C})} = \dfrac{0.24}{0.36} \approx 0.667$

▶ Based on the sample, the probability that he reaches his goal is greatest when he uses Route B. So, he should use Route B.

Monitoring Progress Help in English and Spanish at *BigIdeasMath.com*

7. A manager is assessing three employees in order to offer one of them a promotion. Over a period of time, the manager records whether the employees meet or exceed expectations on their assigned tasks. The table shows the manager's results. Which employee should be offered the promotion? Explain.

	Exceed Expectations	Meet Expectations
Joy	|||| ||||	||||	
Elena	|||| |||| ||	|||| |||
Sam	|||| ||||		|||| ||

Section 5.3 Two-Way Tables and Probability 299

Extra Example 5

An airline company strives not to lose luggage for passengers. A manager at the company randomly selects three flights to check on lost luggage. At the end of each day, the manager determines whether or not there was luggage lost on the flight that day. The table shows the findings. Which flight does the best at not losing luggage?

	Lost Luggage	No Lost Luggage
Flight 1	|||| ||||	|||
Flight 2	|||| ||||	||||
Flight 3	|||| |||| ||		

Based on the sample, the probability that a flight will not lose luggage is greatest for Flight 2. So, Flight 2 does the best at not losing luggage.

MONITORING PROGRESS ANSWERS

5. 0.1

6. The events are independent; $P(\text{Santa Monica}) \approx P(\text{Santa Monica} \mid \text{yes})$

7. Sam; Sam has the greatest probability of exceeding expectations.

Laurie's Notes — Teacher Actions

- **Look For and Make Use of Structure:** Example 5 helps students recognize that the number of rows and columns in the table can be increased as needed for the context.
- Have students use the tally information to construct a two-way table of joint and marginal relative frequencies.
- **Turn and Talk:** "How do you now find the conditional probabilities of reaching his goal given a particular route?" Listen for correct reasoning.

Closure

- **Point of Most Significance:** Ask students to identify, aloud or on paper, the most significant point (or part) in the lesson that aided their learning.

Assignment Guide and Homework Check

ASSIGNMENT

Basic: 1, 2, 3–17 odd, 20, 21, 27–29

Average: 1, 2, 4–20 even, 21–23, 27–29

Advanced: 1, 2, 6–20 even, 21–29

HOMEWORK CHECK

Basic: 5, 9, 11, 13, 17

Average: 6, 10, 12, 14, 18

Advanced: 6, 10, 12, 14, 18

ANSWERS

1. two-way table

2. A joint relative frequency is the ratio of an entry in the table (that is not a total) to the survey total. A marginal relative frequency is the sum of the joint relative frequencies in any row or column. A conditional relative frequency is the ratio of the joint relative frequency to the marginal relative frequency.

3. 34; 40; 4; 6; 12

4. 42; 98; 3; 59; 108

5.

Response	Gender		
	Male	Female	Total
Yes	132	151	283
No	39	29	68
Total	171	180	351

351 people were surveyed, 171 males were surveyed, 180 females were surveyed, 283 people said yes, 68 people said no.

6.

Response	Role		
	Teachers	Parents	Total
Yes	49	18	67
No	11	30	41
Total	60	48	108

108 people were surveyed, 60 teachers were surveyed, 48 parents were surveyed, 67 people said yes, 41 people said no.

5.3 Exercises

Dynamic Solutions available at BigIdeasMath.com

Vocabulary and Core Concept Check

1. **COMPLETE THE SENTENCE** A(n) _____ displays data collected from the same source that belongs to two different categories.

2. **WRITING** Compare the definitions of joint relative frequency, marginal relative frequency, and conditional relative frequency.

Monitoring Progress and Modeling with Mathematics

In Exercises 3 and 4, complete the two-way table.

3.

Grade	Preparation		
	Studied	Did Not Study	Total
Pass		6	
Fail			10
Total	38		50

4.

Role	Response		
	Yes	No	Total
Student	56		
Teacher		7	10
Total		49	

5. **MODELING WITH MATHEMATICS** You survey 171 males and 180 females at Grand Central Station in New York City. Of those, 132 males and 151 females wash their hands after using the public rest rooms. Organize these results in a two-way table. Then find and interpret the marginal frequencies. *(See Example 1.)*

6. **MODELING WITH MATHEMATICS** A survey asks 60 teachers and 48 parents whether school uniforms reduce distractions in school. Of those, 49 teachers and 18 parents say uniforms reduce distractions in school. Organize these results in a two-way table. Then find and interpret the marginal frequencies.

USING STRUCTURE In Exercises 7 and 8, use the two-way table to create a two-way table that shows the joint and marginal relative frequencies.

7.

Gender	Dominant Hand		
	Left	Right	Total
Female	11	104	115
Male	24	92	116
Total	35	196	231

8.

Experience	Gender		
	Male	Female	Total
Expert	62	6	68
Average	275	24	299
Novice	40	3	43
Total	377	33	410

9. **MODELING WITH MATHEMATICS** Use the survey results from Exercise 5 to make a two-way table that shows the joint and marginal relative frequencies. *(See Example 2.)*

10. **MODELING WITH MATHEMATICS** In a survey, 49 people received a flu vaccine before the flu season and 63 people did not receive the vaccine. Of those who receive the flu vaccine, 16 people got the flu. Of those who did not receive the vaccine, 17 got the flu. Make a two-way table that shows the joint and marginal relative frequencies.

300   Chapter 5   Probability

7.

Gender	Dominant Hand		
	Left	Right	Total
Female	0.048	0.450	0.498
Male	0.104	0.398	0.502
Total	0.152	0.848	1

8.

Experience	Gender		
	Male	Female	Total
Expert	0.151	0.015	0.166
Average	0.670	0.059	0.729
Novice	0.098	0.007	0.105
Total	0.919	0.081	1

9–10. See Additional Answers.

11. **MODELING WITH MATHEMATICS** A survey finds that 110 people ate breakfast and 30 people skipped breakfast. Of those who ate breakfast, 10 people felt tired. Of those who skipped breakfast, 10 people felt tired. Make a two-way table that shows the conditional relative frequencies based on the breakfast totals. *(See Example 3.)*

12. **MODELING WITH MATHEMATICS** Use the survey results from Exercise 10 to make a two-way table that shows the conditional relative frequencies based on the flu vaccine totals.

13. **PROBLEM SOLVING** Three different local hospitals in New York surveyed their patients. The survey asked whether the patient's physician communicated efficiently. The results, given as joint relative frequencies, are shown in the two-way table. *(See Example 4.)*

		Location		
		Glens Falls	Saratoga	Albany
Response	Yes	0.123	0.288	0.338
	No	0.042	0.077	0.131

   a. What is the probability that a randomly selected patient located in Saratoga was satisfied with the communication of the physician?

   b. What is the probability that a randomly selected patient who was not satisfied with the physician's communication is located in Glens Falls?

   c. Determine whether being satisfied with the communication of the physician and living in Saratoga are independent events.

14. **PROBLEM SOLVING** A researcher surveys a random sample of high school students in seven states. The survey asks whether students plan to stay in their home state after graduation. The results, given as joint relative frequencies, are shown in the two-way table.

		Location		
		Nebraska	North Carolina	Other States
Response	Yes	0.044	0.051	0.056
	No	0.400	0.193	0.256

   a. What is the probability that a randomly selected student who lives in Nebraska plans to stay in his or her home state after graduation?

   b. What is the probability that a randomly selected student who does not plan to stay in his or her home state after graduation lives in North Carolina?

   c. Determine whether planning to stay in their home state and living in Nebraska are independent events.

ERROR ANALYSIS In Exercises 15 and 16, describe and correct the error in finding the given conditional probability.

		City			
		Tokyo	London	Washington, D.C.	Total
Response	Yes	0.049	0.136	0.171	0.356
	No	0.341	0.112	0.191	0.644
	Total	0.39	0.248	0.362	1

15. $P(\text{yes} \mid \text{Tokyo})$

   ✗ $P(\text{yes} \mid \text{Tokyo}) = \dfrac{P(\text{Tokyo and yes})}{P(\text{Tokyo})}$
   $= \dfrac{0.049}{0.356} \approx 0.138$

16. $P(\text{London} \mid \text{no})$

   ✗ $P(\text{London} \mid \text{no}) = \dfrac{P(\text{no and London})}{P(\text{London})}$
   $= \dfrac{0.112}{0.248} \approx 0.452$

17. **PROBLEM SOLVING** You want to find the quickest route to school. You map out three routes. Before school, you randomly select a route and record whether you are late or on time. The table shows your findings. Assuming you leave at the same time each morning, which route should you use? Explain. *(See Example 5.)*

	On Time	Late
Route A	ⅢⅡ ⅠⅠ	ⅠⅠⅠⅠ
Route B	ⅢⅡ ⅢⅡ Ⅰ	ⅠⅠⅠ
Route C	ⅢⅡ ⅢⅡ ⅠⅠ	ⅠⅠⅠⅠ

18. **PROBLEM SOLVING** A teacher is assessing three groups of students in order to offer one group a prize. Over a period of time, the teacher records whether the groups meet or exceed expectations on their assigned tasks. The table shows the teacher's results. Which group should be awarded the prize? Explain.

	Exceed Expectations	Meet Expectations
Group 1	ⅢⅡ ⅢⅡ ⅠⅠ	ⅠⅠⅠⅠ
Group 2	ⅢⅡ ⅠⅠⅠ	ⅢⅡ
Group 3	ⅢⅡ ⅠⅠⅠⅠ	ⅢⅡ Ⅰ

Section 5.3   Two-Way Tables and Probability   301

Dynamic Teaching Tools
Dynamic Assessment & Progress Monitoring Tool
Interactive Whiteboard Lesson Library
Dynamic Classroom with Dynamic Investigations

ANSWERS

11.

		Breakfast	
		Ate	Did Not Eat
Feeling	Tired	0.091	0.333
	Not Tired	0.909	0.667

12.

		Vaccination	
		Received	Not Received
Health	Flu	0.327	0.270
	No Flu	0.673	0.730

13. a. about 0.789
    b. 0.168
    c. The events are independent.

14. a. about 0.10
    b. about 0.227
    c. The events are not independent.

15. The value for $P(\text{yes})$ was used in the denominator instead of the value for $P(\text{Tokyo})$; $\dfrac{0.049}{0.39} \approx 0.126$

16. The denominator should have been $P(\text{no})$; $\dfrac{0.112}{0.644} \approx 0.174$

17. Route B; It has the best probability of getting to school on time.

18. Group 1; It has the greatest probability of exceeding expectations.

ANSWERS

19–29. See Additional Answers.

Mini-Assessment

A radio station surveys listeners in three cities: Austin, Houston, and San Antonio. The survey asks whether they have installed the station's app on their mobile devices. The results are shown in the two-way table.

	AUS	HOU	SA
Yes	152	232	168
No	64	96	88

1. Use the survey results to make a two-way table that shows the joint and marginal relative frequencies.

	AUS	HOU	SA	Total
Yes	0.19	0.29	0.21	0.69
No	0.08	0.12	0.11	0.31
Total	0.27	0.41	0.32	1

2. Given that a listener downloaded the app, what is the conditional relative frequency that he or she lives in Austin? $\frac{0.19}{0.69}$, or about 27.5%

3. What is the probability that a randomly selected customer who is not located in Houston will not download the app? $\frac{0.19}{0.59}$, or about 32.2%

4. For a science project, a student plants seeds in 3 different soil mixes and observes whether the seeds sprout within 10 days. The table shows the results. Which soil mix has the highest probability of the seeds sprouting within 10 days?

	Sprouted	Did Not Sprout
Soil Mix 1	𝍩𝍩 𝍩𝍩 |	𝍩𝍩 ||||
Soil Mix 2	𝍩𝍩 𝍩𝍩 |||	𝍩𝍩 𝍩𝍩 |
Soil Mix 3	𝍩𝍩 𝍩𝍩	𝍩𝍩 |

Soil Mix 3

---

19. **OPEN-ENDED** Create and conduct a survey in your class. Organize the results in a two-way table. Then create a two-way table that shows the joint and marginal frequencies.

20. **HOW DO YOU SEE IT?** A research group surveys parents and coaches of high school students about whether competitive sports are important in school. The two-way table shows the results of the survey.

		Role		
		Parent	Coach	Total
Important	Yes	880	456	1336
	No	120	45	165
	Total	1000	501	1501

  a. What does 120 represent?
  b. What does 1336 represent?
  c. What does 1501 represent?

21. **MAKING AN ARGUMENT** Your friend uses the table below to determine which workout routine is the best. Your friend decides that Routine B is the best option because it has the fewest tally marks in the "Does Not Reach Goal" column. Is your friend correct? Explain your reasoning.

	Reached Goal	Does Not Reach Goal
Routine A	𝍩𝍩	|||
Routine B	||||	||
Routine C	𝍩𝍩 ||	||||

22. **MODELING WITH MATHEMATICS** A survey asks students whether they prefer math class or science class. Of the 150 male students surveyed, 62% prefer math class over science class. Of the female students surveyed, 74% prefer math. Construct a two-way table to show the number of students in each category if 350 students were surveyed.

23. **MULTIPLE REPRESENTATIONS** Use the Venn diagram to construct a two-way table. Then use your table to answer the questions.

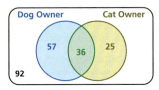

  a. What is the probability that a randomly selected person does not own either pet?
  b. What is the probability that a randomly selected person who owns a dog also owns a cat?

24. **WRITING** Compare two-way tables and Venn diagrams. Then describe the advantages and disadvantages of each.

25. **PROBLEM SOLVING** A company creates a new snack, N, and tests it against its current leader, L. The table shows the results.

	Prefer L	Prefer N
Current L Consumer	72	46
Not Current L Consumer	52	114

The company is deciding whether it should try to improve the snack before marketing it, and to whom the snack should be marketed. Use probability to explain the decisions the company should make when the total size of the snack's market is expected to (a) change very little, and (b) expand very rapidly.

26. **THOUGHT PROVOKING** Bayes' Theorem is given by
$$P(A|B) = \frac{P(B|A) \cdot P(A)}{P(B)}.$$
Use a two-way table to write an example of Bayes' Theorem.

Maintaining Mathematical Proficiency Reviewing what you learned in previous grades and lessons

Draw a Venn diagram of the sets described. *(Skills Review Handbook)*

27. Of the positive integers less than 15, set A consists of the factors of 15 and set B consists of all odd numbers.

28. Of the positive integers less than 14, set A consists of all prime numbers and set B consists of all even numbers.

29. Of the positive integers less than 24, set A consists of the multiples of 2 and set B consists of all the multiples of 3.

If students need help...	If students got it...
Resources by Chapter • Practice A and Practice B • Puzzle Time	Resources by Chapter • Enrichment and Extension • Cumulative Review
Student Journal • Practice	Start the *next* Section
Differentiating the Lesson Skills Review Handbook	

5.1–5.3 What Did You Learn?

Core Vocabulary

probability experiment, *p. 280*
outcome, *p. 280*
event, *p. 280*
sample space, *p. 280*
probability of an event, *p. 280*
theoretical probability, *p. 281*

experimental probability, *p. 282*
odds, *p. 283*
independent events, *p. 288*
dependent events, *p. 289*
conditional probability, *p. 289*
two-way table, *p. 296*

joint frequency, *p. 296*
marginal frequency, *p. 296*
joint relative frequency, *p. 297*
marginal relative frequency, *p. 297*
conditional relative frequency, *p. 297*

Core Concepts

Section 5.1
Theoretical Probabilities, *p. 280*
Probability of the Complement of an Event, *p. 281*
Experimental Probabilities, *p. 282*
Odds, *p. 283*

Section 5.2
Probability of Independent Events, *p. 288*
Probability of Dependent Events, *p. 289*
Finding Conditional Probabilities, *p. 291*

Section 5.3
Making Two-Way Tables, *p. 296*
Relative and Conditional Relative Frequencies, *p. 297*

Mathematical Practices

1. How can you use a number line to analyze the error in Exercise 12 on page 284?
2. Explain how you used probability to correct the flawed logic of your friend in Exercise 21 on page 302.

Making a Mental Cheat Sheet

- Write down important information on note cards.
- Memorize the information on the note cards, placing the ones containing information you know in one stack and the ones containing information you do not know in another stack. Keep working on the information you do not know.

Study Skills

Dynamic Teaching Tools
Dynamic Assessment & Progress Monitoring Tool
Interactive Whiteboard Lesson Library
Dynamic Classroom with Dynamic Investigations

ANSWERS
1. *Sample answer:* Draw a number line and then count the possibilities.
2. Create a table showing the conditional probability of each routine, and choose the routine with the highest probability of reaching the goal.

5.1–5.3 Quiz

1. You randomly draw a marble out of a bag containing 8 green marbles, 4 blue marbles, 12 yellow marbles, and 10 red marbles. Find the probability of drawing a marble that is not yellow. *(Section 5.1)*

Find $P(\overline{A})$. *(Section 5.1)*

2. $P(A) = 0.32$
3. $P(A) = \frac{8}{9}$
4. $P(A) = 0.01$

5. You roll a six-sided die 30 times. A 5 is rolled 8 times. What is the theoretical probability of rolling a 5? What is the experimental probability of rolling a 5? *(Section 5.1)*

6. You randomly draw one card from a standard deck of 52 playing cards. *(Section 5.1)*
   a. What are the odds that it is a diamond?
   b. What are the odds that it is a 7?

7. Events *A* and *B* are independent. Find the missing probability. *(Section 5.2)*
   $P(A) = 0.25$
   $P(B) = \underline{\hspace{1cm}}$
   $P(A \text{ and } B) = 0.05$

8. Events *A* and *B* are dependent. Find the missing probability. *(Section 5.2)*
   $P(A) = 0.6$
   $P(B|A) = 0.2$
   $P(A \text{ and } B) = \underline{\hspace{1cm}}$

You roll a red six-sided die and a blue six-sided die. Find the probability. *(Section 5.2)*

9. $P(\text{red odd and blue odd})$
10. $P(\text{red and blue both less than 3})$
11. $P(\text{red 2 and blue greater than 1})$

12. A survey asks 13-year-old and 15-year-old students about their eating habits. Four hundred students are surveyed, 100 male students and 100 female students from each age group. The bar graph shows the number of students who said they eat fruit every day. *(Section 5.2)*
    a. Find the probability that a female student, chosen at random from the students surveyed, eats fruit every day.
    b. Find the probability that a 15-year-old student, chosen at random from the students surveyed, eats fruit every day.

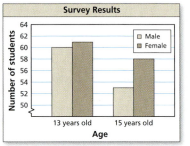

13. At school, 55% of students use public transportation. Twenty-five percent of students use public transportation and participate in an extracurricular activity after school. What is the probability that a student who uses public transportation also participates in an extracurricular activity after school? *(Section 5.2)*

14. There are 14 boys and 18 girls in a class. The teacher allows the students to vote whether they want to take a test on Friday or on Monday. A total of 6 boys and 10 girls vote to take the test on Friday. Organize the information in a two-way table. Then find and interpret the marginal frequencies. *(Section 5.3)*

15. Three schools compete in a cross country invitational. Of the 15 athletes on your team, 9 achieve their goal times. Of the 20 athletes on the home team, 6 achieve their goal times. On your rival's team, 8 of the 13 athletes achieve their goal times. Organize the information in a two-way table. Then determine the probability that a randomly selected runner who achieves his or her goal time is from your school. *(Section 5.3)*

Laurie's Notes

Overview of Section 5.4

Introduction
- Venn diagrams are used to introduce compound events. The events may be disjoint, or mutually exclusive. The events can also be overlapping.
- Probability rules are developed for both types of compound events, and students must first be able to distinguish between disjoint and overlapping events.
- Students will recognize that earlier skills such as determining the sample space of an event or finding the probability of a complement of an event are necessary skills in this lesson.
- In solving real-life applications, more than one type of probability rule may be needed.

Formative Assessment Tips
- **Partner Speaks:** This strategy provides a student the opportunity to share his or her thinking about a problem or concept with a partner. The partner then shares this thinking with the whole class.
- Thinking through a problem with a partner is less intimidating for many students than sharing it with an entire class. For the listener, he or she needs to pay attention to the thoughts of the speaker and set aside his or her own thinking about the problem. As students are engaged in this dialogue, circulate to hear the discussion and to gain an understanding of how students are thinking about the problem or concept. You will be able to judge appropriate next steps.
- The opportunity to speak to a partner and then have him or her share the student's thinking should not be used for simple, less challenging problems. It should be used as a way to pair students differently from the usual pairing of a student having a conversation with the person to his or her right.

Pacing Suggestion
- Have students work through all of the explorations before transitioning to the formal lesson.

Dynamic Teaching Tools
Dynamic Assessment & Progress Monitoring Tool
Lesson Planning Tool
Interactive Whiteboard Lesson Library
Dynamic Classroom with Dynamic Investigations

What Your Students Will Learn

- Find probabilities of compound events, disjoint and overlapping.
- Use more than one probability rule to solve real-life problems.

Laurie's Notes

Exploration

Motivate
- Have students divide into two groups, girls and boys. Draw a Venn diagram of this relationship.
- Now ask students to divide into two more groups, those with female cousins and those without.
- ❓ "What would the Venn diagram look like for this situation?" two intersecting circles where the overlapping region represents girls or boys in the class who have female cousins Draw the result.
- Explain to students that the two Venn diagrams represent the types of events they will study in this lesson.

Exploration Note
- Students should read the introduction, which describes new vocabulary about events: disjoint, mutually exclusive, and overlapping.

Exploration 1
- This exploration should not take students long to complete. Solicit a volunteer to draw the two Venn diagrams.
- ❓ "Are the events in part (a) disjoint or overlapping?" overlapping
- ❓ "Are the events in part (b) disjoint or overlapping?" disjoint

Exploration 2
- Students use the possible outcomes shown in the Venn diagrams in Exploration 1 to find probabilities. The same four questions are answered for parts (a) and (b).
- Students will use the results of this exploration to make a conjecture about a possible rule in the next exploration.
- Solicit responses for each part of the exploration.

Exploration 3
- **Construct Viable Arguments and Critique the Reasoning of Others:** Students may not yet have precise language to describe the rules, but listen for correct reasoning. The Venn diagram should be helpful in explaining why you need to subtract the probability of both events occurring.
- **Teaching Tip:** You could use the graphing calculator technique described in the *Teaching Strategy* on page T-286 to simulate the rolling of a six-sided die 50 times.

Communicate Your Answer
- Students should be able to describe disjoint events and overlapping events, along with how to find the probabilities of each type of event. Listen for reasoning discussed in Exploration 3.

Connecting to Next Step
- The explorations help students discover how to find the probability of compound events. The formula is stated in the formal lesson.

5.4 Probability of Disjoint and Overlapping Events

Essential Question How can you find probabilities of disjoint and overlapping events?

Two events are **disjoint**, or **mutually exclusive**, when they have no outcomes in common. Two events are **overlapping** when they have one or more outcomes in common.

EXPLORATION 1 Disjoint Events and Overlapping Events

Work with a partner. A six-sided die is rolled. Draw a Venn diagram that relates the two events. Then decide whether the events are disjoint or overlapping.

a. Event *A*: The result is an even number.
Event *B*: The result is a prime number.

b. Event *A*: The result is 2 or 4.
Event *B*: The result is an odd number.

EXPLORATION 2 Finding the Probability that Two Events Occur

Work with a partner. A six-sided die is rolled. For each pair of events, find (a) $P(A)$, (b) $P(B)$, (c) $P(A \text{ and } B)$, and (d) $P(A \text{ or } B)$.

a. Event *A*: The result is an even number.
Event *B*: The result is a prime number.

b. Event *A*: The result is 2 or 4.
Event *B*: The result is an odd number.

EXPLORATION 3 Discovering Probability Formulas

Work with a partner.

a. In general, if event *A* and event *B* are disjoint, then what is the probability that event *A* or event *B* will occur? Use a Venn diagram to justify your conclusion.

b. In general, if event *A* and event *B* are overlapping, then what is the probability that event *A* or event *B* will occur? Use a Venn diagram to justify your conclusion.

c. Conduct an experiment using a six-sided die. Roll the die 50 times and record the results. Then use the results to find the probabilities described in Exploration 2. How closely do your experimental probabilities compare to the theoretical probabilities you found in Exploration 2?

Communicate Your Answer

4. How can you find probabilities of disjoint and overlapping events?

5. Give examples of disjoint events and overlapping events that do not involve dice.

MODELING WITH MATHEMATICS
To be proficient in math, you need to map the relationships between important quantities in a practical situation using such tools as diagrams.

ANSWERS

1. a.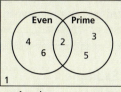
overlapping

b.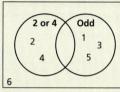
disjoint

2. a. $\frac{1}{2}, \frac{1}{2}, \frac{1}{6}, \frac{5}{6}$
b. $\frac{1}{3}, \frac{1}{2}, 0, \frac{5}{6}$

3. a. $P(A) + P(B)$; Sample answer:
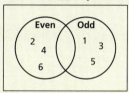

b. $P(A) + P(B) - P(A \text{ and } B)$; Sample answer:
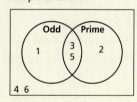

c. Sample answer:
$P(\text{even}) = \frac{25}{50} = \frac{1}{2}$, $P(\text{prime}) = \frac{19}{50}$,
$P(\text{even and prime}) = \frac{5}{50} = \frac{1}{10}$,
$P(\text{even or prime}) = \frac{39}{50}$,
$P(\text{odd}) = \frac{25}{50} = \frac{1}{2}$,
$P(2 \text{ or } 4) = \frac{16}{50} = \frac{8}{25}$,
$P(\text{odd and } 2 \text{ or } 4) = 0$,
$P(\text{odd or } 2 \text{ or } 4) = \frac{41}{50}$;
The probabilities are similar.

4. For disjoint events, add the probabilities of each event. For overlapping events, add the probabilities of each event, and subtract the probability that both events occur.

5. *Sample answer:* disjoint events: picking an ace and picking a queen; overlapping events: picking an ace and picking a heart

English Language Learners

Comprehension

Some students may assume that the use of the word "and" indicates that a probability will increase, and that conversely the use of the word "or" indicates a limited, or decreased, probability. Emphasize that using the word "or" means including more outcomes or individuals, not fewer.

Extra Example 1

Two six-sided dice are rolled. What is the probability that the sum of the numbers rolled is a multiple of 4 *or* is 5?
$\frac{9}{36} + \frac{4}{36} = \frac{13}{36} \approx 0.361$

5.4 Lesson

Core Vocabulary

compound event, *p. 306*
overlapping events, *p. 306*
disjoint or mutually exclusive events, *p. 306*

Previous
Venn diagram

What You Will Learn

- Find probabilities of compound events.
- Use more than one probability rule to solve real-life problems.

Compound Events

When you consider all the outcomes for either of two events A and B, you form the *union* of A and B, as shown in the first diagram. When you consider only the outcomes shared by both A and B, you form the *intersection* of A and B, as shown in the second diagram. The union or intersection of two events is called a **compound event**.

Union of A and B Intersection of A and B Intersection of A and B is empty.

To find $P(A \text{ or } B)$ you must consider what outcomes, if any, are in the intersection of A and B. Two events are **overlapping** when they have one or more outcomes in common, as shown in the first two diagrams. Two events are **disjoint**, or **mutually exclusive**, when they have no outcomes in common, as shown in the third diagram.

STUDY TIP

If two events A and B are overlapping, then the outcomes in the intersection of A and B are counted *twice* when $P(A)$ and $P(B)$ are added. So, $P(A \text{ and } B)$ must be subtracted from the sum.

Core Concept

Probability of Compound Events

If A and B are any two events, then the probability of A or B is
$$P(A \text{ or } B) = P(A) + P(B) - P(A \text{ and } B).$$

If A and B are disjoint events, then the probability of A or B is
$$P(A \text{ or } B) = P(A) + P(B).$$

EXAMPLE 1   Finding the Probability of Disjoint Events

A card is randomly selected from a standard deck of 52 playing cards. What is the probability that it is a 10 *or* a face card?

SOLUTION

Let event A be selecting a 10 and event B be selecting a face card. From the diagram, A has 4 outcomes and B has 12 outcomes. Because A and B are disjoint, the probability is

$P(A \text{ or } B) = P(A) + P(B)$ Write disjoint probability formula.

$= \frac{4}{52} + \frac{12}{52}$ Substitute known probabilities.

$= \frac{16}{52}$ Add.

$= \frac{4}{13}$ Simplify.

$\approx 0.308.$ Use a calculator.

Laurie's Notes   Teacher Actions

- **Model with Mathematics:** Review union and intersection of sets by drawing Venn diagrams and relating them to disjoint and overlapping events.
- Write the *Core Concept*. Both formulas are used to find the probability of event A or event B occurring.
- ? "How are the formulas different?" *If the events are not disjoint, you need to subtract the probability of both events occurring.*
- ? "In Example 1, why are the two events disjoint?" *A card cannot be both a 10 and a face card.*

COMMON ERROR
When two events A and B overlap, as in Example 2, $P(A$ or $B)$ does not equal $P(A) + P(B)$.

EXAMPLE 2 Finding the Probability of Overlapping Events

A card is randomly selected from a standard deck of 52 playing cards. What is the probability that it is a face card *or* a spade?

SOLUTION

Let event A be selecting a face card and event B be selecting a spade. From the diagram, A has 12 outcomes and B has 13 outcomes. Of these, 3 outcomes are common to A and B. So, the probability of selecting a face card or a spade is

$P(A$ or $B) = P(A) + P(B) - P(A$ and $B)$ Write general formula.

$= \dfrac{12}{52} + \dfrac{13}{52} - \dfrac{3}{52}$ Substitute known probabilities.

$= \dfrac{22}{52}$ Add.

$= \dfrac{11}{26}$ Simplify.

$\approx 0.423.$ Use a calculator.

EXAMPLE 3 Using a Formula to Find $P(A$ and $B)$

Out of 200 students in a senior class, 113 students are either varsity athletes or on the honor roll. There are 74 seniors who are varsity athletes and 51 seniors who are on the honor roll. What is the probability that a randomly selected senior is both a varsity athlete *and* on the honor roll?

SOLUTION

Let event A be selecting a senior who is a varsity athlete and event B be selecting a senior on the honor roll. From the given information, you know that $P(A) = \dfrac{74}{200}$, $P(B) = \dfrac{51}{200}$, and $P(A$ or $B) = \dfrac{113}{200}$. The probability that a randomly selected senior is both a varsity athlete *and* on the honor roll is $P(A$ and $B)$.

$P(A$ or $B) = P(A) + P(B) - P(A$ and $B)$ Write general formula.

$\dfrac{113}{200} = \dfrac{74}{200} + \dfrac{51}{200} - P(A$ and $B)$ Substitute known probabilities.

$P(A$ and $B) = \dfrac{74}{200} + \dfrac{51}{200} - \dfrac{113}{200}$ Solve for $P(A$ and $B)$.

$P(A$ and $B) = \dfrac{12}{200}$ Simplify.

$P(A$ and $B) = \dfrac{3}{50},$ or 0.06 Simplify.

Monitoring Progress Help in English and Spanish at BigIdeasMath.com

A card is randomly selected from a standard deck of 52 playing cards. Find the probability of the event.

1. selecting an ace *or* an 8
2. selecting a 10 *or* a diamond
3. **WHAT IF?** In Example 3, suppose 32 seniors are in the band and 64 seniors are in the band or on the honor roll. What is the probability that a randomly selected senior is both in the band and on the honor roll?

Section 5.4   Probability of Disjoint and Overlapping Events   307

Extra Example 2
A bag contains cards numbered 1 through 20. One card is randomly selected. What is the probability that the number on the card is a multiple of 3 *or* a multiple of 4?
$\dfrac{6}{20} + \dfrac{5}{20} - \dfrac{1}{20} = \dfrac{10}{20} = 0.5$

Extra Example 3
Out of 45 customers at a breakfast café, 42 customers bought either coffee or orange juice. There were 30 customers who bought orange juice and 40 customers who bought coffee. What is the probability that a randomly selected customer bought both coffee and orange juice?
$\dfrac{30}{45} + \dfrac{40}{45} - \dfrac{42}{45} = \dfrac{28}{45} \approx 0.62$

MONITORING PROGRESS ANSWERS

1. $\dfrac{2}{13}$
2. $\dfrac{4}{13}$
3. $\dfrac{19}{200},$ or 0.095

Laurie's Notes   Teacher Actions

- **Popsicle Sticks:** Pose Example 2, and have students work with their partners. Use *Popsicle Sticks* to solicit a solution.
- Have partners work independently on Example 3.
- Circulate and observe solution strategies. Students may draw a Venn diagram, or they may use mental math to determine the number that are athletes and on the honor roll. Others may use the formula for the probability of compound events.
- **Construct Viable Arguments and Critique the Reasoning of Others:** Have students share their approaches to increase the number of strategies students will be familiar with.

Extra Example 4

A medical association estimates that 10.9% of the people in the United States have a thyroid disorder. Suppose a medical lab has developed a simple diagnostic test that is 96% accurate for people who have the disorder and 99% accurate for people who do not have it. The medical lab gives the test to a randomly selected person. What is the probability that the diagnosis is correct? The probability that the diagnosis is correct is about 0.987, or 98.7%.

MONITORING PROGRESS ANSWERS

4. about 0.048, or 4.8%
5. 0.52, or 52%

Using More Than One Probability Rule

In the first four sections of this chapter, you have learned several probability rules. The solution to some real-life problems may require the use of two or more of these probability rules, as shown in the next example.

EXAMPLE 4 Solving a Real-Life Problem

The American Diabetes Association estimates that 8.3% of people in the United States have diabetes. Suppose that a medical lab has developed a simple diagnostic test for diabetes that is 98% accurate for people who have the disease and 95% accurate for people who do not have it. The medical lab gives the test to a randomly selected person. What is the probability that the diagnosis is correct?

SOLUTION

Let event A be "person has diabetes" and event B be "correct diagnosis." Notice that the probability of B depends on the occurrence of A, so the events are dependent. When A occurs, $P(B) = 0.98$. When A does not occur, $P(B) = 0.95$.

A probability tree diagram, where the probabilities are given along the branches, can help you see the different ways to obtain a correct diagnosis. Use the complements of events A and B to complete the diagram, where $\overline{A}$ is "person does not have diabetes" and $\overline{B}$ is "incorrect diagnosis." Notice that the probabilities for all branches from the same point must sum to 1.

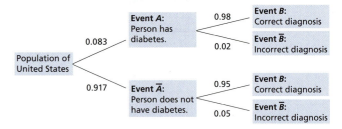

To find the probability that the diagnosis is correct, follow the branches leading to event B.

$P(B) = P(A \text{ and } B) + P(\overline{A} \text{ and } B)$ Use tree diagram.

$\qquad = P(A) \cdot P(B|A) + P(\overline{A}) \cdot P(B|\overline{A})$ Probability of dependent events

$\qquad = (0.083)(0.98) + (0.917)(0.95)$ Substitute.

$\qquad \approx 0.952$ Use a calculator.

▶ The probability that the diagnosis is correct is about 0.952, or 95.2%.

Monitoring Progress  Help in English and Spanish at *BigIdeasMath.com*

4. In Example 4, what is the probability that the diagnosis is *incorrect*?
5. A high school basketball team leads at halftime in 60% of the games in a season. The team wins 80% of the time when they have the halftime lead, but only 10% of the time when they do not. What is the probability that the team wins a particular game during the season?

Laurie's Notes — Teacher Actions

- Example 4 connects skills from earlier lessons.
- Pose the problem and say, "Work independently to answer the problem. Then you will have time to share your thinking with partners."
- **Partner Speaks:** After students have had sufficient time, pair students. Only one person speaks. The listener then asks for clarification or gives feedback. Have the listener describe the solution method or thinking used by his or her partner.

Closure

- **Exit Ticket:** A card is drawn from a standard deck of 52 playing cards. What is the probability of drawing a 4 *or* a red card? $\frac{28}{52} = \frac{7}{13} \approx 0.538$

5.4 Exercises

Dynamic Solutions available at *BigIdeasMath.com*

Vocabulary and Core Concept Check

1. **WRITING** Are the events A and $\overline{A}$ disjoint? Explain. Then give an example of a real-life event and its complement.

2. **DIFFERENT WORDS, SAME QUESTION** Which is different? Find "both" answers.

   How many outcomes are in the intersection of A and B?

   How many outcomes are shared by both A and B?

   How many outcomes are in the union of A and B?

   How many outcomes in B are also in A?

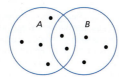

Monitoring Progress and Modeling with Mathematics

In Exercises 3–6, events A and B are disjoint. Find $P(A \text{ or } B)$.

3. $P(A) = 0.3$, $P(B) = 0.1$
4. $P(A) = 0.55$, $P(B) = 0.2$
5. $P(A) = \frac{1}{3}$, $P(B) = \frac{1}{4}$
6. $P(A) = \frac{2}{3}$, $P(B) = \frac{1}{5}$

7. **PROBLEM SOLVING** Your dart is equally likely to hit any point inside the board shown. You throw a dart and pop a balloon. What is the probability that the balloon is red or blue? *(See Example 1.)*

8. **PROBLEM SOLVING** You and your friend are among several candidates running for class president. You estimate that there is a 45% chance you will win and a 25% chance your friend will win. What is the probability that you or your friend win the election?

9. **PROBLEM SOLVING** You are performing an experiment to determine how well plants grow under different light sources. Of the 30 plants in the experiment, 12 receive visible light, 15 receive ultraviolet light, and 6 receive both visible and ultraviolet light. What is the probability that a plant in the experiment receives visible or ultraviolet light? *(See Example 2.)*

10. **PROBLEM SOLVING** Of 162 students honored at an academic awards banquet, 48 won awards for mathematics and 78 won awards for English. There are 14 students who won awards for both mathematics and English. A newspaper chooses a student at random for an interview. What is the probability that the student interviewed won an award for English or mathematics?

ERROR ANALYSIS In Exercises 11 and 12, describe and correct the error in finding the probability of randomly drawing the given card from a standard deck of 52 playing cards.

11. ✗
$$P(\text{heart or face card})$$
$$= P(\text{heart}) + P(\text{face card})$$
$$= \frac{13}{52} + \frac{12}{52} = \frac{25}{52}$$

12. ✗
$$P(\text{club or 9})$$
$$= P(\text{club}) + P(9) + P(\text{club and 9})$$
$$= \frac{13}{52} + \frac{4}{52} + \frac{1}{52} = \frac{9}{26}$$

In Exercises 13 and 14, you roll a six-sided die. Find $P(A \text{ or } B)$.

13. Event A: Roll a 6.
    Event B: Roll a prime number.

14. Event A: Roll an odd number.
    Event B: Roll a number less than 5.

Section 5.4   Probability of Disjoint and Overlapping Events   309

Assignment Guide and Homework Check

ASSIGNMENT
Basic: 1, 2, 3–17 odd, 20, 23–26
Average: 1, 2–20 even, 23–26
Advanced: 1, 2, 8–20 even, 21–26

HOMEWORK CHECK
Basic: 3, 7, 9, 15, 17
Average: 8, 10, 14, 16, 18
Advanced: 8, 10, 16, 18, 23

ANSWERS

1. yes; $\overline{A}$ is everything not in A; Sample answer: event A: you win the game, event $\overline{A}$: you do not win the game
2. How many outcomes are in the union of A and B?; 9; 2
3. 0.4
4. 0.75
5. $\frac{7}{12}$, or about 0.58
6. $\frac{13}{15}$, or about 0.87
7. $\frac{9}{20}$, or 0.45
8. 70%
9. $\frac{7}{10}$, or 0.7
10. $\frac{56}{81}$, or about 0.69
11. forgot to subtract $P(\text{heart and face card})$; $P(\text{heart}) + P(\text{face card}) - P(\text{heart and face card}) = \frac{11}{26}$
12. added instead of subtracted $P(\text{club and 9})$; $P(\text{club}) + P(9) - P(\text{club and 9}) = \frac{4}{13}$
13. $\frac{2}{3}$
14. $\frac{5}{6}$

Dynamic Teaching Tools

- Dynamic Assessment & Progress Monitoring Tool
- Interactive Whiteboard Lesson Library
- Dynamic Classroom with Dynamic Investigations

ANSWERS

15. 10%
16. $\frac{1}{6}$
17. 0.4742, or 47.42%
18. a. 0.09
    b. 0.12
    c. The coach should leave the goalie in the game.
19. $\frac{13}{18}$
20. no; The intersection of A and B is not empty.
21. $\frac{3}{20}$
22. See Additional Answers.
23. no; Until all cards, numbers, and colors are known, the conclusion cannot be made.
24-26. See Additional Answers.

Mini-Assessment

One card is randomly selected from a standard deck of 52 playing cards.

1. What is the probability that the card is a 4 *or* a queen? $\frac{2}{13} \approx 0.154$
2. What is the probability that the card is a diamond *or* a jack? $\frac{4}{13} \approx 0.308$
3. There are 63 cars in a used car lot, 60 of which have air conditioning or are silver. There are 18 silver cars and 57 cars with air conditioning. What is the probability that a car selected at random from the lot is silver *and* has air conditioning? $\frac{5}{21} \approx 0.238$
4. An airline estimates that 90% of its eastbound flights arrive on time and 72% of its westbound flights arrive on time. This morning, 42% of the airline's schedule is westbound flights. If a flight is chosen at random, what is the probability that the flight does not arrive on time? 0.1756, or about 17.6%

310   Chapter 5

---

15. **DRAWING CONCLUSIONS** A group of 40 trees in a forest are not growing properly. A botanist determines that 34 of the trees have a disease or are being damaged by insects, with 18 trees having a disease and 20 being damaged by insects. What is the probability that a randomly selected tree has both a disease and is being damaged by insects? *(See Example 3.)*

16. **DRAWING CONCLUSIONS** A company paid overtime wages or hired temporary help during 9 months of the year. Overtime wages were paid during 7 months, and temporary help was hired during 4 months. At the end of the year, an auditor examines the accounting records and randomly selects one month to check the payroll. What is the probability that the auditor will select a month in which the company paid overtime wages and hired temporary help?

17. **DRAWING CONCLUSIONS** A company is focus testing a new type of fruit drink. The focus group is 47% male. Of the responses, 40% of the males and 54% of the females said they would buy the fruit drink. What is the probability that a randomly selected person would buy the fruit drink? *(See Example 4.)*

18. **DRAWING CONCLUSIONS** The Redbirds trail the Bluebirds by one goal with 1 minute left in the hockey game. The Redbirds' coach must decide whether to remove the goalie and add a frontline player. The probabilities of each team scoring are shown in the table.

	Goalie	No goalie
Redbirds score	0.1	0.3
Bluebirds score	0.1	0.6

a. Find the probability that the Redbirds score and the Bluebirds do not score when the coach leaves the goalie in.

b. Find the probability that the Redbirds score and the Bluebirds do not score when the coach takes the goalie out.

c. Based on parts (a) and (b), what should the coach do?

Maintaining Mathematical Proficiency
Reviewing what you learned in previous grades and lessons

Graph the function. Compare the graph to the graph of $f(x) = x^2$. *(Section 3.1)*

24. $r(x) = 3x^2$
25. $g(x) = \frac{3}{4}x^2$
26. $h(x) = -5x^2$

310   Chapter 5   Probability

---

19. **PROBLEM SOLVING** You can win concert tickets from a radio station if you are the first person to call when the song of the day is played, or if you are the first person to correctly answer the trivia question. The song of the day is announced at a random time between 7:00 and 7:30 A.M. The trivia question is asked at a random time between 7:15 and 7:45 A.M. You begin listening to the radio station at 7:20. Find the probability that you miss the announcement of the song of the day or the trivia question.

20. **HOW DO YOU SEE IT?** Are events A and B disjoint events? Explain your reasoning.

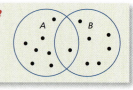

21. **PROBLEM SOLVING** You take a bus from your neighborhood to your school. The express bus arrives at your neighborhood at a random time between 7:30 and 7:36 A.M. The local bus arrives at your neighborhood at a random time between 7:30 and 7:40 A.M. You arrive at the bus stop at 7:33 A.M. Find the probability that you missed both the express bus and the local bus.

22. **THOUGHT PROVOKING** Write a general rule for finding $P(A \text{ or } B \text{ or } C)$ for (a) disjoint and (b) overlapping events A, B, and C.

23. **MAKING AN ARGUMENT** A bag contains 40 cards numbered 1 through 40 that are either red or blue. A card is drawn at random and placed back in the bag. This is done four times. Two red cards are drawn, numbered 31 and 19, and two blue cards are drawn, numbered 22 and 7. Your friend concludes that red cards and even numbers must be mutually exclusive. Is your friend correct? Explain.

If students need help...	If students got it...
Resources by Chapter • Practice A and Practice B • Puzzle Time	Resources by Chapter • Enrichment and Extension • Cumulative Review
Student Journal • Practice	Start the *next* Section
Differentiating the Lesson Skills Review Handbook	

Laurie's Notes

Overview of Section 5.5

Introduction
- The Fundamental Counting Principle was introduced in grade 7 and is connected to the counting techniques students will learn in this lesson.
- Permutations (order matters) are introduced in contextual problems, and a formula is developed for calculating the number of ways n objects can be selected r at a time.
- Combinations (order does not matter) are introduced in contextual problems, and a formula is developed for calculating the number of ways n objects can be selected r at a time.

Teaching Strategy
- Formulas for permutations and combinations of n objects taken r at a time are presented in this lesson. Each of them involves finding factorials ($n!$) of a number.
- Most graphing calculators have built-in factorial, permutation, and combination functions.
- When you first introduce $n!$, students naturally want to know the largest factorial the calculator can compute before it reaches an overflow error. This exploration will not take students long. Students should also recognize that they can perform calculations with factorials.

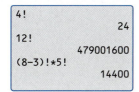

Factorials

- Permutations can be computed as well as combinations.

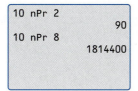

Permutations   Combinations

- If time permits, you might have students look at patterns with combinations, such as $_nC_r = {_nC_{n-r}}$.

Pacing Suggestion
- The explorations are a quick and fun way to help students recall how to determine the number of possible outcomes for a compound event. They provide a good introduction to permutations in the formal lesson.

Dynamic Teaching Tools
Dynamic Assessment & Progress Monitoring Tool
Lesson Planning Tool
Interactive Whiteboard Lesson Library
Dynamic Classroom with Dynamic Investigations

What Your Students Will Learn

- Use the formula for the number of permutations to find the number of ways objects can be chosen when the order matters.
- Use the formula for the number of combinations to find the number of ways objects can be chosen when the order does not matter.

Laurie's Notes

Exploration

Motivate
- **Story Time:** Tell students that your friend could not resist when the jackpot got really high, so she bought a lottery ticket. Your friend had to pick 6 numbers correctly out of the numbers from 1 to 40.
- ❓ "What is the probability my friend will win the jackpot?" Students may be willing to take a guess. Many will say the probability is very small!
- Tell students that in today's lesson they will learn how to calculate the probability of winning the jackpot.

For Your Information
- Tree diagrams and the Fundamental Counting Principle were introduced in grade 7.

Exploration 1
- Students should be familiar with how to construct and read a tree diagram.
- ❓ "How many events are there in this compound event?" three; flip a coin, flip a coin, and spin a spinner
- Students should be efficient with listing the outcomes, such as HH1 versus heads-heads-one.

Exploration 2
- Knowing how to read the tree diagram means it is not necessary to know what the actual events are in order to know the number of possible outcomes of the event.
- Discuss student responses and clarify any differences of opinion.

Exploration 3
- **Construct Viable Arguments and Critique the Reasoning of Others:** Students should recall what the Fundamental Counting Principle says, even if they have forgotten its name.
- A tree diagram is a helpful model in explaining why the number of possible outcomes of each event are multiplied together.

Communicate Your Answer
- Students should work independently to answer Questions 4 and 5 and then compare answers and language with their neighbor.

Connecting to Next Step
- The explorations will help students recall the use of tree diagrams to count outcomes of a compound event. Knowledge of the Fundamental Counting Principle is assumed in the formal lesson.

5.5 Permutations and Combinations

Essential Question How can a tree diagram help you visualize the number of ways in which two or more events can occur?

EXPLORATION 1 Reading a Tree Diagram

Work with a partner. Two coins are flipped and the spinner is spun. The tree diagram shows the possible outcomes.

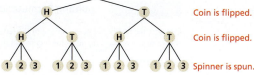

a. How many outcomes are possible?

b. List the possible outcomes.

EXPLORATION 2 Reading a Tree Diagram

Work with a partner. Consider the tree diagram below.

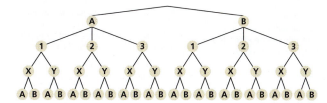

a. How many events are shown?
b. What outcomes are possible for each event?
c. How many outcomes are possible?
d. List the possible outcomes.

EXPLORATION 3 Writing a Conjecture

Work with a partner.

a. Consider the following general problem: Event 1 can occur in m ways and event 2 can occur in n ways. Write a conjecture about the number of ways the two events can occur. Explain your reasoning.

b. Use the conjecture you wrote in part (a) to write a conjecture about the number of ways *more than* two events can occur. Explain your reasoning.

c. Use the results of Explorations 1(a) and 2(c) to verify your conjectures.

CONSTRUCTING VIABLE ARGUMENTS

To be proficient in math, you need to make conjectures and build a logical progression of statements to explore the truth of your conjectures.

Communicate Your Answer

4. How can a tree diagram help you visualize the number of ways in which two or more events can occur?

5. In Exploration 1, the spinner is spun a second time. How many outcomes are possible?

Dynamic Teaching Tools
- Dynamic Assessment & Progress Monitoring Tool
- Lesson Planning Tool
- Interactive Whiteboard Lesson Library
- Dynamic Classroom with Dynamic Investigations

ANSWERS

1. a. 12
   b. HH1, HH2, HH3, HT1, HT2, HT3, TH1, TH2, TH3, TT1, TT2, TT3

2. a. 4
   b. event 1: A, B; event 2: 1, 2, 3; event 3: X, Y; event 4: A, B
   c. 24
   d. A1XA, A1XB, A1YA, A1YB, A2XA, A2XB, A2YA, A2YB, A3XA, A3XB, A3YA, A3YB, B1XA, B1XB, B1YA, B1YB, B2XA, B2XB, B2YA, B2YB, B3XA, B3XB, B3YA, B3YB

3. a. *Sample answer:* The two events can occur in mn ways; This is the Fundamental Counting Principle.
   b. *Sample answer:* Find the product of the number of ways each event can occur; The Fundamental Counting Principle works for more than two events.
   c. 1(a): (2)(2)(3) = 12; 2(c): (2)(3)(2)(2) = 24

4. *Sample answer:* Tree diagrams show all possibilities.

5. 36

Differentiated Instruction

Kinesthetic
Give students four index cards and direct them to write one letter of the word LAKE on each card. Have students work together in small groups to rearrange the letters to find the number of permutations they can make. Designate one member of each group to record all the rearrangement possibilities.

Extra Example 1

Consider the permutations of the letters in the word PENCILS.

a. In how many ways can you arrange all the letters? *There are 5040 ways that you can arrange all the letters in the word PENCILS.*

b. In how many ways can you arrange 3 of the letters? *There are 210 ways you can arrange 3 of the letters in the word PENCILS.*

MONITORING PROGRESS ANSWERS

1. 120
2. 60

5.5 Lesson

Core Vocabulary
permutation, p. 312
n factorial, p. 312
combination, p. 314
Previous
Fundamental Counting Principle

REMEMBER
Fundamental Counting Principle: If one event can occur in m ways and another event can occur in n ways, then the number of ways that both events can occur is $m \cdot n$. The Fundamental Counting Principle can be extended to three or more events.

What You Will Learn

▶ Use the formula for the number of permutations.
▶ Use the formula for the number of combinations.

Permutations

A **permutation** is an arrangement of objects in which order is important. For instance, the 6 possible permutations of the letters A, B, and C are shown.

$$ABC \quad ACB \quad BAC \quad BCA \quad CAB \quad CBA$$

EXAMPLE 1 Counting Permutations

Consider the number of permutations of the letters in the word JULY. In how many ways can you arrange (a) all of the letters and (b) 2 of the letters?

SOLUTION

a. Use the Fundamental Counting Principle to find the number of permutations of the letters in the word JULY.

$$\begin{aligned} \text{Number of permutations} &= \binom{\text{Choices for}}{\text{1st letter}} \binom{\text{Choices for}}{\text{2nd letter}} \binom{\text{Choices for}}{\text{3rd letter}} \binom{\text{Choices for}}{\text{4th letter}} \\ &= 4 \cdot 3 \cdot 2 \cdot 1 \\ &= 24 \end{aligned}$$

▶ There are 24 ways you can arrange all of the letters in the word JULY.

b. When arranging 2 letters of the word JULY, you have 4 choices for the first letter and 3 choices for the second letter.

$$\begin{aligned} \text{Number of permutations} &= \binom{\text{Choices for}}{\text{1st letter}} \binom{\text{Choices for}}{\text{2nd letter}} \\ &= 4 \cdot 3 \\ &= 12 \end{aligned}$$

▶ There are 12 ways you can arrange 2 of the letters in the word JULY.

Monitoring Progress Help in English and Spanish at *BigIdeasMath.com*

1. In how many ways can you arrange the letters in the word HOUSE?
2. In how many ways can you arrange 3 of the letters in the word MARCH?

In Example 1(a), you evaluated the expression $4 \cdot 3 \cdot 2 \cdot 1$. This expression can be written as 4! and is read "4 *factorial*." For any positive integer n, the product of the integers from 1 to n is called *n* **factorial** and is written as

$$n! = n \cdot (n-1) \cdot (n-2) \cdot \cdots \cdot 3 \cdot 2 \cdot 1.$$

As a special case, the value of 0! is defined to be 1.

In Example 1(b), you found the permutations of 4 objects taken 2 at a time. You can find the number of permutations using the formulas on the next page.

312 Chapter 5 Probability

Laurie's Notes — Teacher Actions

❓ "How can you tell that order is important for this example?" *The word* arrange *means to put in order, so order is important.*

❓ "Why are there 3 choices for the second letter?" *There are 4 letters in the word JULY. Once one has been chosen for the first letter, there are 3 letters left.*

• Introduce *n* factorial and the notation. See the *Teaching Strategy* on page T-310. Say, "When you have four objects taken four at a time and order matters, this is a permutation and can be found by multiplying $4 \cdot 3 \cdot 2 \cdot 1 = 4!$"

USING A GRAPHING CALCULATOR

Most graphing calculators can calculate permutations.

```
4 nPr 4
              24
4 nPr 2
              12
```

Core Concept

Permutations

Formulas

The number of permutations of n objects is given by
$$_nP_n = n!.$$

The number of permutations of n objects taken r at a time, where $r \leq n$, is given by
$$_nP_r = \frac{n!}{(n-r)!}.$$

Examples

The number of permutations of 4 objects is
$$_4P_4 = 4! = 4 \cdot 3 \cdot 2 \cdot 1 = 24.$$

The number of permutations of 4 objects taken 2 at a time is
$$_4P_2 = \frac{4!}{(4-2)!} = \frac{4 \cdot 3 \cdot 2!}{2!} = 12.$$

EXAMPLE 2 Using a Permutations Formula

Ten horses are running in a race. In how many different ways can the horses finish first, second, and third? (Assume there are no ties.)

SOLUTION

To find the number of permutations of 3 horses chosen from 10, find $_{10}P_3$.

$_{10}P_3 = \dfrac{10!}{(10-3)!}$ Permutations formula

$= \dfrac{10!}{7!}$ Subtract.

$= \dfrac{10 \cdot 9 \cdot 8 \cdot 7!}{7!}$ Expand factorial. Divide out common factor, 7!.

$= 720$ Simplify.

▶ There are 720 ways for the horses to finish first, second, and third.

STUDY TIP
When you divide out common factors, remember that 7! is a factor of 10!.

EXAMPLE 3 Finding a Probability Using Permutations

For a town parade, you will ride on a float with your soccer team. There are 12 floats in the parade, and their order is chosen at random. Find the probability that your float is first and the float with the school chorus is second.

SOLUTION

Step 1 Write the number of possible outcomes as the number of permutations of the 12 floats in the parade. This is $_{12}P_{12} = 12!$.

Step 2 Write the number of favorable outcomes as the number of permutations of the other floats, given that the soccer team is first and the chorus is second. This is $_{10}P_{10} = 10!$.

Step 3 Find the probability.

$P(\text{soccer team is 1st, chorus is 2nd}) = \dfrac{10!}{12!}$ Form a ratio of favorable to possible outcomes.

$= \dfrac{10!}{12 \cdot 11 \cdot 10!}$ Expand factorial. Divide out common factor, 10!.

$= \dfrac{1}{132}$ Simplify.

Section 5.5 Permutations and Combinations 313

Extra Example 2

Eight people serve on a committee. In how many different ways can a chairperson, a recorder, and a treasurer be chosen from the committee members?

There are 336 ways for a chairperson, a recorder, and a treasurer to be chosen from the committee members.

Extra Example 3

Drea and Bryan are auditioning for a part in the school play. There are 15 people auditioning, and the order of their auditions is chosen at random. Find the probability that Drea auditions last and Bryan auditions second to last.

$P(\text{Drea is last, Bryan is 2nd to last}) = \dfrac{13!}{15!} = \dfrac{1}{210}$

Laurie's Notes — Teacher Actions

? "How many ways can you arrange 10 objects taken 2 at a time?" $10 \cdot 9 = 90$

- Write the *Core Concept*. Explore the role of the denominator.

$$_{10}P_2 = \frac{10!}{(10-2)!} = \frac{10 \cdot 9 \cdot 8!}{8!} = 10 \cdot 9 = 90$$

Partially expanding the numerator allows common factors to divide out.

- **Another Way:** In Example 3, you could find that $_{12}P_2 = 132$, the number of permutations for the first two positions in the parade. Of those, only one has your team first followed by the school chorus, and therefore the probability is $\frac{1}{132}$.

English Language Learners

Vocabulary
In this lesson, students learn to distinguish between *permutations* and *combinations*. Help students interpret the definitions by comparing the two formulas in symbols and with numbers. Remind students that if a problem describes items that must be put in order, then the permutations formula is used, and if the order is not important, then the combinations formula is used.

Extra Example 4

Count the possible combinations of 4 letters chosen from the list P, Q, R, S, T, U.
There are 15 possible combinations.

MONITORING PROGRESS ANSWERS
3. 336
4. $\frac{1}{182}$
5. 10

Monitoring Progress  Help in English and Spanish at BigIdeasMath.com

3. **WHAT IF?** In Example 2, suppose there are 8 horses in the race. In how many different ways can the horses finish first, second, and third? (Assume there are no ties.)

4. **WHAT IF?** In Example 3, suppose there are 14 floats in the parade. Find the probability that the soccer team is first and the chorus is second.

Combinations

A **combination** is a selection of objects in which order is *not* important. For instance, in a drawing for 3 identical prizes, you would use combinations, because the order of the winners would not matter. If the prizes were different, then you would use permutations, because the order would matter.

EXAMPLE 4   Counting Combinations

Count the possible combinations of 2 letters chosen from the list A, B, C, D.

SOLUTION

List all of the permutations of 2 letters from the list A, B, C, D. Because order is not important in a combination, cross out any duplicate pairs.

    AB   AC   AD   B̶A̶   BC   BD
    C̶A̶   C̶B̶   CD   D̶A̶   D̶B̶   D̶C̶

BD and DB are the same pair.

▶ There are 6 possible combinations of 2 letters from the list A, B, C, D.

Monitoring Progress  Help in English and Spanish at BigIdeasMath.com

5. Count the possible combinations of 3 letters chosen from the list A, B, C, D, E.

In Example 4, you found the number of combinations of objects by making an organized list. You can also find the number of combinations using the following formula.

USING A GRAPHING CALCULATOR

Most graphing calculators can calculate combinations.

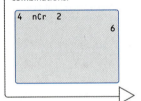

Core Concept

Combinations

Formula   The number of combinations of n objects taken r at a time, where $r \le n$, is given by
$$_nC_r = \frac{n!}{(n-r)! \cdot r!}.$$

Example   The number of combinations of 4 objects taken 2 at a time is
$$_4C_2 = \frac{4!}{(4-2)! \cdot 2!} = \frac{4 \cdot 3 \cdot 2!}{2! \cdot (2 \cdot 1)} = 6.$$

Laurie's Notes   Teacher Actions

? "Two players from a 12-person team are picked as co-captains. Does the order matter in terms of which player is picked first?" no

- Introduce combinations.
- **Predict, Explain, Observe:** Group students and ask, "If order does not matter in a combination, how will the number of outcomes compare to the permutation using the same numbers? Explain your reasoning." Give groups sufficient time to consider the question and their reasoning. Students often say that you need to subtract the arrangements of the lesser number. (For "12 take 2," students may say the number of combinations is the number of permutations minus the arrangements of 2 people.)

EXAMPLE 5 **Using the Combinations Formula**

You order a sandwich at a restaurant. You can choose 2 side dishes from a list of 8. How many combinations of side dishes are possible?

SOLUTION

The order in which you choose the side dishes is not important. So, to find the number of combinations of 8 side dishes taken 2 at a time, find $_8C_2$.

Check

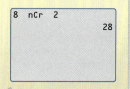

$$_8C_2 = \frac{8!}{(8-2)! \cdot 2!}$$ Combinations formula

$$= \frac{8!}{6! \cdot 2!}$$ Subtract.

$$= \frac{8 \cdot 7 \cdot \cancel{6!}}{\cancel{6!} \cdot (2 \cdot 1)}$$ Expand factorials. Divide out common factor, 6!.

$$= 28$$ Multiply.

▶ There are 28 different combinations of side dishes you can order.

EXAMPLE 6 **Finding a Probability Using Combinations**

A yearbook editor has selected 14 photos, including one of you and one of your friend, to use in a collage for the yearbook. The photos are placed at random. There is room for 2 photos at the top of the page. What is the probability that your photo and your friend's photo are the 2 placed at the top of the page?

SOLUTION

Step 1 Write the number of possible outcomes as the number of combinations of 14 photos taken 2 at a time, or $_{14}C_2$, because the order in which the photos are chosen is not important.

$$_{14}C_2 = \frac{14!}{(14-2)! \cdot 2!}$$ Combinations formula

$$= \frac{14!}{12! \cdot 2!}$$ Subtract.

$$= \frac{14 \cdot 13 \cdot \cancel{12!}}{\cancel{12!} \cdot (2 \cdot 1)}$$ Expand factorials. Divide out common factor, 12!.

$$= 91$$ Multiply.

Step 2 Find the number of favorable outcomes. Only one of the possible combinations includes your photo and your friend's photo.

Step 3 Find the probability.

$$P(\text{your photo and your friend's photos are chosen}) = \frac{1}{91}$$

Monitoring Progress 🔊 Help in English and Spanish at *BigIdeasMath.com*

6. **WHAT IF?** In Example 5, suppose you can choose 3 side dishes out of the list of 8 side dishes. How many combinations are possible?

7. **WHAT IF?** In Example 6, suppose there are 20 photos in the collage. Find the probability that your photo and your friend's photo are the 2 placed at the top of the page.

Extra Example 5
You are listening to music. You have time to hear only 3 songs from your playlist of 16 songs. How many combinations of 3 songs are possible?
560 combinations

Extra Example 6
The art teacher has selected 13 projects, including one of yours and one of your friend's, to use in a display case in the hallway. The projects are placed at random. There is room for 2 projects in the middle row of the case. What is the probability that your project and your friend's project are the two placed in the middle row? $\frac{1}{78}$

MONITORING PROGRESS ANSWERS
6. 56
7. $\frac{1}{190}$

Laurie's Notes   Teacher Actions

- **? Reason Abstractly and Quantitatively** and **Wait Time:** "Is $_nP_r$ always, sometimes, or never greater than $_nC_r$? Explain." Give *Wait Time* before soliciting a response. *Always true unless $r = 1$. If $r = 1$, then $_nP_r = _nC_r$.*
- **Thumbs Up:** Work through Example 5 as shown. Have students self-assess with *Thumbs Up*.
- **?** "How would you list the 28 combinations?" *Answers will vary.*
- **?** "How can you tell order is not important for Example 6?" *The problem does not specify that either picture must be first.*
- **Predict, Explain, Observe:** "Which combination increases more, when you increase n by 1 or r by 1?" For example, for $_{14}C_2$, which combination increases more, $_{15}C_2$ or $_{14}C_3$?

Closure
- **Exit Ticket:** Calculate the probability of picking all 6 numbers out of 40 in a lottery. $\frac{1}{3,838,380}$

Assignment Guide and Homework Check

ASSIGNMENT

Basic: 1, 2, 3–35 odd, 45, 52, 59, 60

Average: 1, 2–42 even, 46, 49, 52, 59, 60

Advanced: 1, 2, 8, 14–26 even, 34–46 even, 47–49, 52–60

HOMEWORK CHECK

Basic: 5, 13, 17, 27, 33

Average: 8, 18, 20, 34, 40

Advanced: 18, 20, 34, 46, 49

ANSWERS

1. permutation
2. $\dfrac{7!}{(7-2)!}$; It is the only expression that does not equal 21.
3. a. 2
   b. 2
4. a. 6
   b. 6
5. a. 24
   b. 12
6. a. 120
   b. 20
7. a. 720
   b. 30
8. a. 5040
   b. 42
9. 20
10. 210
11. 9
12. 720
13. 20,160
14. 1
15. 870
16. 6,375,600
17. 990
18. 720
19. $\dfrac{1}{56}$
20. $\dfrac{1}{720}$
21. 4
22. 6
23. 20
24. 5
25. 5
26. 56
27. 1

28. 28
29. 220
30. 330
31. 6435
32. 15,504
33. 635,376

316   Chapter 5

5.5 Exercises

Dynamic Solutions available at *BigIdeasMath.com*

Vocabulary and Core Concept Check

1. **COMPLETE THE SENTENCE** An arrangement of objects in which order is important is called a(n) _____.

2. **WHICH ONE DOESN'T BELONG?** Which expression does *not* belong with the other three? Explain your reasoning.

   $\dfrac{7!}{2! \cdot 5!}$ $_7C_5$ $_7C_2$ $\dfrac{7!}{(7-2)!}$

Monitoring Progress and Modeling with Mathematics

In Exercises 3–8, find the number of ways you can arrange (a) all of the letters and (b) 2 of the letters in the given word. *(See Example 1.)*

3. AT
4. TRY
5. ROCK
6. WATER
7. FAMILY
8. FLOWERS

In Exercises 9–16, evaluate the expression.

9. $_5P_2$
10. $_7P_3$
11. $_9P_1$
12. $_6P_5$
13. $_8P_6$
14. $_{12}P_0$
15. $_{30}P_2$
16. $_{25}P_5$

17. **PROBLEM SOLVING** Eleven students are competing in an art contest. In how many different ways can the students finish first, second, and third? *(See Example 2.)*

18. **PROBLEM SOLVING** Six friends go to a movie theater. In how many different ways can they sit together in a row of 6 empty seats?

19. **PROBLEM SOLVING** You and your friend are 2 of 8 servers working a shift in a restaurant. At the beginning of the shift, the manager randomly assigns one section to each server. Find the probability that you are assigned Section 1 and your friend is assigned Section 2. *(See Example 3.)*

20. **PROBLEM SOLVING** You make 6 posters to hold up at a basketball game. Each poster has a letter of the word TIGERS. You and 5 friends sit next to each other in a row. The posters are distributed at random. Find the probability that TIGERS is spelled correctly when you hold up the posters.

In Exercises 21–24, count the possible combinations of *r* letters chosen from the given list. *(See Example 4.)*

21. A, B, C, D; *r* = 3
22. L, M, N, O; *r* = 2
23. U, V, W, X, Y, Z; *r* = 3
24. D, E, F, G, H; *r* = 4

In Exercises 25–32, evaluate the expression.

25. $_5C_1$
26. $_8C_5$
27. $_9C_9$
28. $_8C_6$
29. $_{12}C_3$
30. $_{11}C_4$
31. $_{15}C_8$
32. $_{20}C_5$

33. **PROBLEM SOLVING** Each year, 64 golfers participate in a golf tournament. The golfers play in groups of 4. How many groups of 4 golfers are possible? *(See Example 5.)*

316   Chapter 5   Probability

34. **PROBLEM SOLVING** You want to purchase vegetable dip for a party. A grocery store sells 7 different flavors of vegetable dip. You have enough money to purchase 2 flavors. How many combinations of 2 flavors of vegetable dip are possible?

ERROR ANALYSIS In Exercises 35 and 36, describe and correct the error in evaluating the expression.

35.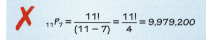
$$_{11}P_7 = \frac{11!}{(11-7)} = \frac{11!}{4} = 9{,}979{,}200$$

36.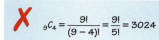
$$_9C_4 = \frac{9!}{(9-4)!} = \frac{9!}{5!} = 3024$$

REASONING In Exercises 37–40, tell whether the question can be answered using *permutations* or *combinations*. Explain your reasoning. Then answer the question.

37. To complete an exam, you must answer 8 questions from a list of 10 questions. In how many ways can you complete the exam?

38. Ten students are auditioning for 3 different roles in a play. In how many ways can the 3 roles be filled?

39. Fifty-two athletes are competing in a bicycle race. In how many orders can the bicyclists finish first, second, and third? (Assume there are no ties.)

40. An employee at a pet store needs to catch 5 tetras in an aquarium containing 27 tetras. In how many groupings can the employee capture 5 tetras?

41. **CRITICAL THINKING** Compare the quantities $_{50}C_9$ and $_{50}C_{41}$ without performing any calculations. Explain your reasoning.

42. **CRITICAL THINKING** Show that each identity is true for any whole numbers r and n, where $0 \le r \le n$.
    a. $_nC_n = 1$
    b. $_nC_r = {_nC_{n-r}}$
    c. $_{n+1}C_r = {_nC_r} + {_nC_{r-1}}$

43. **REASONING** Complete the table for each given value of r. Then write an inequality relating $_nP_r$ and $_nC_r$. Explain your reasoning.

	$r=0$	$r=1$	$r=2$	$r=3$
$_3P_r$				
$_3C_r$				

44. **REASONING** Write an equation that relates $_nP_r$ and $_nC_r$. Then use your equation to find and interpret the value of $\dfrac{_{182}P_4}{_{182}C_4}$.

45. **PROBLEM SOLVING** You and your friend are in the studio audience on a television game show. From an audience of 300 people, 2 people are randomly selected as contestants. What is the probability that you and your friend are chosen? *(See Example 6.)*

46. **PROBLEM SOLVING** You work 5 evenings each week at a bookstore. Your supervisor assigns you 5 evenings at random from the 7 possibilities. What is the probability that your schedule does not include working on the weekend?

REASONING In Exercises 47 and 48, find the probability of winning a lottery using the given rules. Assume that lottery numbers are selected at random.

47. You must correctly select 6 numbers, each an integer from 0 to 49. The order is not important.

48. You must correctly select 4 numbers, each an integer from 0 to 9. The order is important.

49. **MATHEMATICAL CONNECTIONS** A polygon is convex when no line that contains a side of the polygon contains a point in the interior of the polygon. Consider a convex polygon with n sides.

    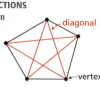

    a. Use the combinations formula to write an expression for the number of diagonals in an n-sided polygon.
    b. Use your result from part (a) to write a formula for the number of diagonals of an n-sided convex polygon.

50. **PROBLEM SOLVING** You are ordering a burrito with 2 main ingredients and 3 toppings. The menu below shows the possible choices. How many different burritos are possible?

ANSWERS

51. 30
52. a. 1; Each outcome has the same three marbles.
    b. 6; Each outcome has a different permutation.
53. a. $\frac{1}{2}$
    b. $\frac{1}{2}$; The probabilities are the same.
54. $\frac{1061}{1250}$
55. a. $\frac{1}{90}$
    b. $\frac{9}{10}$
56. 376
57. $\frac{1}{406}$; There are $_{30}C_5$ possible groups. The number of groups that will have you and your two best friends is $_{27}C_2$.
58. a. about 0.04; about 0.12
    b. $1 - \frac{_{365}P_n}{365^n}$
    c. 23 people
59. $\frac{1}{5}$
60. TH

Mini-Assessment

1. Consider the letters in the word KINDLY.
   a. In how many ways can you arrange all the letters?
      720 ways
   b. In how many ways can you arrange 3 of the letters?
      120 ways

2. Nine people, including Becky and Samir, are being interviewed for a scholarship. If the order is chosen at random, what is the probability that Becky will be interviewed first and Samir will be interviewed second? **$\frac{1}{72}$**

3. You are preparing for a trip. You have a list of 7 books to read, but you can bring only 3 with you on the trip. How many combinations of books are possible?
   35 different combinations

4. A restaurant is giving free pizzas to 2 customers selected at random from a group of 30. You and a friend enter the drawing. What is the probability that both of you win a free pizza? **$\frac{1}{435}$**

---

51. **PROBLEM SOLVING** You want to purchase 2 different types of contemporary music CDs and 1 classical music CD from the music collection shown. How many different sets of music types can you choose for your purchase?

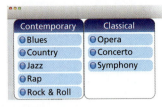

52. **HOW DO YOU SEE IT?** A bag contains one green marble, one red marble, and one blue marble. The diagram shows the possible outcomes of randomly drawing three marbles from the bag without replacement.

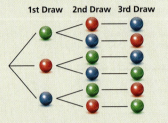

   a. How many combinations of three marbles can be drawn from the bag? Explain.
   b. How many permutations of three marbles can be drawn from the bag? Explain.

53. **PROBLEM SOLVING** Every student in your history class is required to present a project in front of the class. Each day, 4 students make their presentations in an order chosen at random by the teacher. You make your presentation on the first day.
   a. What is the probability that you are chosen to be the first or second presenter on the first day?
   b. What is the probability that you are chosen to be the second or third presenter on the first day? Compare your answer with that in part (a).

54. **PROBLEM SOLVING** The organizer of a cast party for a drama club asks each of the 6 cast members to bring 1 food item from a list of 10 items. Assuming each member randomly chooses a food item to bring, what is the probability that at least 2 of the 6 cast members bring the same item?

55. **PROBLEM SOLVING** You are one of 10 students performing in a school talent show. The order of the performances is determined at random. The first 5 performers go on stage before the intermission.
   a. What is the probability that you are the last performer before the intermission and your rival performs immediately before you?
   b. What is the probability that you are *not* the first performer?

56. **THOUGHT PROVOKING** How many integers, greater than 999 but not greater than 4000, can be formed with the digits 0, 1, 2, 3, and 4? Repetition of digits is allowed.

57. **PROBLEM SOLVING** There are 30 students in your class. Your science teacher chooses 5 students at random to complete a group project. Find the probability that you and your 2 best friends in the science class are chosen to work in the group. Explain how you found your answer.

58. **PROBLEM SOLVING** Follow the steps below to explore a famous probability problem called the *birthday problem*. (Assume there are 365 equally likely birthdays possible.)
   a. What is the probability that at least 2 people share the same birthday in a group of 6 randomly chosen people? in a group of 10 randomly chosen people?
   b. Generalize the results from part (a) by writing a formula for the probability $P(n)$ that at least 2 people in a group of n people share the same birthday. (*Hint*: Use $_nP_r$ notation in your formula.)
   c. Enter the formula from part (b) into a graphing calculator. Use the *table* feature to make a table of values. For what group size does the probability that at least 2 people share the same birthday first exceed 50%?

Maintaining Mathematical Proficiency — Reviewing what you learned in previous grades and lessons

59. A bag contains 12 white marbles and 3 black marbles. You pick 1 marble at random. What is the probability that you pick a black marble? *(Section 5.1)*

60. The table shows the results of flipping two coins 12 times. For what outcome is the experimental probability the same as the theoretical probability? *(Section 5.1)*

HH	HT	TH	TT
2	6	3	1

If students need help...	If students got it...
Resources by Chapter • Practice A and Practice B • Puzzle Time	Resources by Chapter • Enrichment and Extension • Cumulative Review
Student Journal • Practice	Start the *next* Section
Differentiating the Lesson Skills Review Handbook	

Laurie's Notes

Overview of Section 5.6

Introduction
- Students have worked with probability throughout this chapter. This last lesson introduces the probability distribution function that displays the probability of each possible value of a random variable.
- Students construct and interpret probability distributions.
- A special case of a probability distribution is a binomial distribution where the independent trials have only two possible outcomes: success and failure.
- Students construct and interpret binomial distributions.

Teaching Strategy
- The values of a binomial distribution can be calculated using a function (such as *binompdf*) found on most graphing calculators. This function is especially useful when there are numerous probabilities to calculate.
- Laurie's Notes on page 321 present a simple example for introducing the probability of a binomial experiment. The steps below correspond to this example.
    - Press 2ND VARS for the [DISTR] menu. Scroll down to *A:binompdf*. (See Figure 1.) Press ENTER.
    - The syntax for the binomial probability function will ask you to enter three values:
      3   the number of trials *n*,
      0.8   the probability of a success *p*, and
      2   the number of successes *k* (often *x* on the calculator).
      (See Figure 2.)
    - Paste the function to the home screen and press ENTER to calculate the probability of 2 successes. (See Figure 3.)

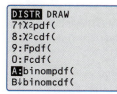 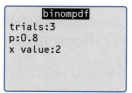 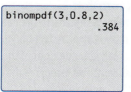

      Figure 1   Figure 2   Figure 3

- Probability distribution histograms can also be constructed using a graphing calculator.

Pacing Suggestion
- The explorations give students a chance to review skills from this chapter along with working with binomial distributions. Gaining a sense of what binomial distributions are will be helpful in the formal lesson. Begin the formal lesson with the definition.

Dynamic Teaching Tools
Dynamic Assessment & Progress Monitoring Tool
Lesson Planning Tool
Interactive Whiteboard Lesson Library
Dynamic Classroom with Dynamic Investigations

What Your Students Will Learn

- Construct and interpret probability distributions of the outcomes of an experiment.
- Construct and interpret binomial distributions of the outcomes of a binomial experiment.

Laurie's Notes

Exploration

Motivate
- ❓ "What would be a reasonable free throw percentage for a professional basketball player?" Answers will vary.
- The NBA player with the highest career free throw percentage is Steve Nash, who over 18 years in the NBA has averaged 0.9043 at the free throw line.
- "If Nash goes to the free throw line 8 times in one game, what is the probability that he will make 6 of the 8 shots?" Students will have guesses about how to answer this question. Explain to students that at the end of this lesson they will be able to answer this question and others similar to it.

Exploration Note
- If you have two-color counters, students can use them to model the heads and tails that result when there are n coins flipped.

Exploration 1
- Give students time to study the histograms. They will be able to figure out what information is being conveyed without being told.
- Students will notice the symmetry of the histograms but may not recognize any patterns for the heights of each interval. When students get to part (b), they will find it a bit more challenging. Allow for productive struggle. You can also return to this problem after Exploration 2 when a numeric pattern may become evident.
- ❓ "What would the histograms look like for 'Number of Tails' when n coins are flipped?" The same as the histograms for number of heads when n coins are flipped.
- ❓ "How many different outcomes are there when n coins are flipped?" 2^n

Exploration 2
- Students should use the data from the histograms in Exploration 1 to complete this table.
- ❓ "If two coins are flipped, how many occurrences are there of 2 heads?" 1 This result could be added to the front of the table.
- Students will recognize the pattern of add 2, add 3, add 4, add 5, and so on.
- ❓ "Can you think of a way to generalize this in terms of n?" Answers will vary.
- Some students may recognize these numbers from working with combinations in the last lesson, or they may know that this set of numbers is known as the triangular numbers.

Communicate Your Answer
- Listen for valid reasoning as students discuss their answers.

Connecting to Next Step
- The explorations have students working with binomial distributions without naming or defining them. This will be done in the formal lesson.

5.6 Binomial Distributions

Essential Question How can you determine the frequency of each outcome of an event?

EXPLORATION 1 Analyzing Histograms

Work with a partner. The histograms show the results when n coins are flipped.

STUDY TIP

When 4 coins are flipped ($n = 4$), the possible outcomes are

TTTT TTTH TTHT TTHH
THTT THTH THHT THHH
HTTT HTTH HTHT HTHH
HHTT HHTH HHHT HHHH.

The histogram shows the numbers of outcomes having 0, 1, 2, 3, and 4 heads.

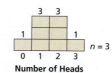
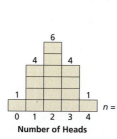
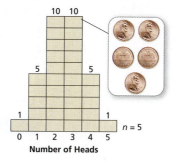
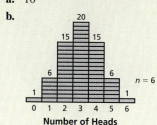

a. In how many ways can 3 heads occur when 5 coins are flipped?

b. Draw a histogram that shows the numbers of heads that can occur when 6 coins are flipped.

c. In how many ways can 3 heads occur when 6 coins are flipped?

EXPLORATION 2 Determining the Number of Occurrences

Work with a partner.

LOOKING FOR A PATTERN

To be proficient in math, you need to look closely to discern a pattern or structure.

a. Complete the table showing the numbers of ways in which 2 heads can occur when n coins are flipped.

n	3	4	5	6	7
Occurrences of 2 heads					

b. Determine the pattern shown in the table. Use your result to find the number of ways in which 2 heads can occur when 8 coins are flipped.

Communicate Your Answer

3. How can you determine the frequency of each outcome of an event?

4. How can you use a histogram to find the probability of an event?

Section 5.6 Binomial Distributions 319

Extra Example 1

The spinner is divided into equal three equal parts. Let x be a random variable that represents the sum when the spinner is spun twice. Make a table and draw a histogram showing the probability distribution for x.

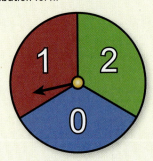

x (sum)	0	1	2	3	4
Outcomes	1	2	3	2	1
$P(x)$	$\frac{1}{9}$	$\frac{2}{9}$	$\frac{1}{3}$	$\frac{2}{9}$	$\frac{1}{9}$

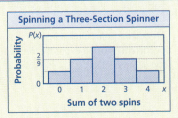

5.6 Lesson

Core Vocabulary
random variable, p. 320
probability distribution, p. 320
binomial distribution, p. 321
binomial experiment, p. 321
Previous
histogram

What You Will Learn

▶ Construct and interpret probability distributions.
▶ Construct and interpret binomial distributions.

Probability Distributions

A **random variable** is a variable whose value is determined by the outcomes of a probability experiment. For example, when you roll a six-sided die, you can define a random variable x that represents the number showing on the die. So, the possible values of x are 1, 2, 3, 4, 5, and 6. For every random variable, a *probability distribution* can be defined.

Core Concept

Probability Distributions

A **probability distribution** is a function that gives the probability of each possible value of a random variable. The sum of all the probabilities in a probability distribution must equal 1.

Probability Distribution for Rolling a Six-Sided Die

x	1	2	3	4	5	6
$P(x)$	$\frac{1}{6}$	$\frac{1}{6}$	$\frac{1}{6}$	$\frac{1}{6}$	$\frac{1}{6}$	$\frac{1}{6}$

EXAMPLE 1 Constructing a Probability Distribution

Let x be a random variable that represents the sum when two six-sided dice are rolled. Make a table and draw a histogram showing the probability distribution for x.

SOLUTION

Step 1 Make a table. The possible values of x are the integers from 2 to 12. The table shows how many outcomes of rolling two dice produce each value of x. Divide the number of outcomes for x by 36 to find $P(x)$.

x (sum)	2	3	4	5	6	7	8	9	10	11	12
Outcomes	1	2	3	4	5	6	5	4	3	2	1
$P(x)$	$\frac{1}{36}$	$\frac{1}{18}$	$\frac{1}{12}$	$\frac{1}{9}$	$\frac{5}{36}$	$\frac{1}{6}$	$\frac{5}{36}$	$\frac{1}{9}$	$\frac{1}{12}$	$\frac{1}{18}$	$\frac{1}{36}$

Step 2 Draw a histogram where the intervals are given by x and the frequencies are given by $P(x)$.

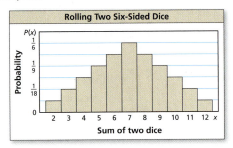

STUDY TIP
Recall that there are 36 possible outcomes when rolling two six-sided dice. These are listed in Example 3 on page 282.

Laurie's Notes — Teacher Actions

- Use a context, such as the outcomes of a six-sided die, to discuss random variables. A probability distribution is a function.
- ❓ "What are the possible sums when two six-sided dice are rolled?" 2, 3, 4, 5, 6, 7, 8, 9, 10, 11, 12
- There are many ways to simulate the sum of two dice. See the *Teaching Strategy* on page T-286, or search the Internet.
- The experimental probability should approximate the theoretical probability for large trials.
- **Connection:** The probability distribution shown has the same shape as the frequency distribution for the same event.

EXAMPLE 2 Interpreting a Probability Distribution

Use the probability distribution in Example 1 to answer each question.

a. What is the most likely sum when rolling two six-sided dice?
b. What is the probability that the sum of the two dice is at least 10?

SOLUTION

a. The most likely sum when rolling two six-sided dice is the value of *x* for which $P(x)$ is greatest. This probability is greatest for $x = 7$. So, when rolling the two dice, the most likely sum is 7.

b. The probability that the sum of the two dice is at least 10 is

$$P(x \geq 10) = P(x = 10) + P(x = 11) + P(x = 12)$$
$$= \tfrac{3}{36} + \tfrac{2}{36} + \tfrac{1}{36}$$
$$= \tfrac{6}{36}$$
$$= \tfrac{1}{6}$$
$$\approx 0.167.$$

▶ The probability is about 16.7%.

Monitoring Progress  Help in English and Spanish at *BigIdeasMath.com*

An octahedral die has eight sides numbered 1 through 8. Let *x* be a random variable that represents the sum when two such dice are rolled.

1. Make a table and draw a histogram showing the probability distribution for *x*.
2. What is the most likely sum when rolling the two dice?
3. What is the probability that the sum of the two dice is at most 3?

Binomial Distributions

One type of probability distribution is a **binomial distribution**. A binomial distribution shows the probabilities of the outcomes of a *binomial experiment*.

Core Concept

Binomial Experiments

A **binomial experiment** meets the following conditions.
- There are *n* independent trials.
- Each trial has only two possible outcomes: success and failure.
- The probability of success is the same for each trial. This probability is denoted by *p*. The probability of failure is $1 - p$.

For a binomial experiment, the probability of exactly *k* successes in *n* trials is

$$P(k \text{ successes}) = {}_nC_k\, p^k (1-p)^{n-k}.$$

Section 5.6 Binomial Distributions 321

Differentiated Instruction

Inclusion
Students who struggle with the formula for the probability of *k* successes in *n* trials of a binomial experiment may benefit by having access to graphing calculators to compute the probabilities. Students can enter the number of trials and the probability of success and view a list of the coefficients given by the formula.

Extra Example 2
Use the probability distribution in Extra Example 1 to answer each question.

a. What is the most likely sum when spinning a 3-section spinner twice? The most likely sum is 2.
b. What is the probability that the sum of the two spins is odd? The probability is $\tfrac{4}{9}$, or about 44.4%.

MONITORING PROGRESS ANSWERS

1.

x (sum)	2	3	4	5	6	7	8	9
Outcomes	1	2	3	4	5	6	7	8
P(x)	$\tfrac{1}{64}$	$\tfrac{1}{32}$	$\tfrac{3}{64}$	$\tfrac{1}{16}$	$\tfrac{5}{64}$	$\tfrac{3}{32}$	$\tfrac{7}{64}$	$\tfrac{1}{8}$

x (sum)	10	11	12	13	14	15	16
Outcomes	7	6	5	4	3	2	1
P(x)	$\tfrac{7}{64}$	$\tfrac{3}{32}$	$\tfrac{5}{64}$	$\tfrac{1}{16}$	$\tfrac{3}{64}$	$\tfrac{1}{32}$	$\tfrac{1}{64}$

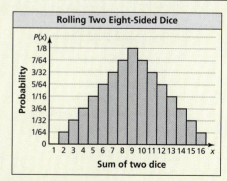

2. 9
3. $\tfrac{3}{64}$

Laurie's Notes — Teacher Actions

- **Thumbs Up:** Have students work independently to answer Example 2. Students should self-assess with *Thumbs Up*.
- **Binomial Experiment:** Define a binomial experiment such as tossing a coin. Discuss the three parts of the formula $P(k \text{ successes}) = {}_nC_k\, p^k (1-p)^{n-k}$. If the probability of a success is 0.8, then the probability of failure is 0.2. The event is repeated 3 times. Then $P(2 \text{ successes}) = (0.8)(0.8)(0.2) + (0.8)(0.2)(0.8) + (0.2)(0.8)(0.8)$. There are ${}_3C_2 = 3$ combinations of 2 successes and 1 failure. See the *Teaching Strategy* on page T-318.

Extra Example 3

According to a survey, about 62% of adults have visited a dentist in the past year. You ask 5 randomly selected adults whether they have had a dentist visit in the past year. Draw a histogram of the binomial distribution for your survey.

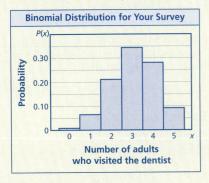

Extra Example 4

Use the binomial distribution in Extra Example 3 to answer each question.

a. What is the most likely outcome of the survey? **The most likely outcome is that 3 of the 5 adults surveyed visited a dentist in the past year.**

b. What is the probability that more than 3 people surveyed have visited a dentist in the past year? **about 37.3%**

MONITORING PROGRESS ANSWERS

4.
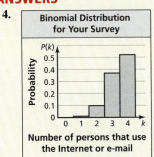

5. The most likely outcome is that 4 of the 4 people use the Internet or e-mail.

6. about 0.11

EXAMPLE 3 Constructing a Binomial Distribution

According to a survey, about 33% of people ages 16 and older in the U.S. own an electronic book reading device, or e-reader. You ask 6 randomly chosen people (ages 16 and older) whether they own an e-reader. Draw a histogram of the binomial distribution for your survey.

SOLUTION

The probability that a randomly selected person has an e-reader is $p = 0.33$. Because you survey 6 people, $n = 6$.

$P(k = 0) = {}_6C_0(0.33)^0(0.67)^6 \approx 0.090$

$P(k = 1) = {}_6C_1(0.33)^1(0.67)^5 \approx 0.267$

$P(k = 2) = {}_6C_2(0.33)^2(0.67)^4 \approx 0.329$

$P(k = 3) = {}_6C_3(0.33)^3(0.67)^3 \approx 0.216$

$P(k = 4) = {}_6C_4(0.33)^4(0.67)^2 \approx 0.080$

$P(k = 5) = {}_6C_5(0.33)^5(0.67)^1 \approx 0.016$

$P(k = 6) = {}_6C_6(0.33)^6(0.67)^0 \approx 0.001$

ATTENDING TO PRECISION
When probabilities are rounded, the sum of the probabilities may differ slightly from 1.

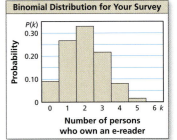

A histogram of the distribution is shown.

EXAMPLE 4 Interpreting a Binomial Distribution

Use the binomial distribution in Example 3 to answer each question.

a. What is the most likely outcome of the survey?

b. What is the probability that at most 2 people have an e-reader?

COMMON ERROR
Because a person may not have an e-reader, be sure you include $P(k = 0)$ when finding the probability that at most 2 people have an e-reader.

SOLUTION

a. The most likely outcome of the survey is the value of k for which $P(k)$ is greatest. This probability is greatest for $k = 2$. The most likely outcome is that 2 of the 6 people own an e-reader.

b. The probability that at most 2 people have an e-reader is

$P(k \leq 2) = P(k = 0) + P(k = 1) + P(k = 2)$

$\approx 0.090 + 0.267 + 0.329$

$\approx 0.686.$

▶ The probability is about 68.6%.

Monitoring Progress Help in English and Spanish at *BigIdeasMath.com*

According to a survey, about 85% of people ages 18 and older in the U.S. use the Internet or e-mail. You ask 4 randomly chosen people (ages 18 and older) whether they use the Internet or e-mail.

4. Draw a histogram of the binomial distribution for your survey.

5. What is the most likely outcome of your survey?

6. What is the probability that at most 2 people you survey use the Internet or e-mail?

Laurie's Notes — Teacher Actions

- **Use Appropriate Tools Strategically:** Students should use calculators to generate the probability values for each possible outcome in Example 3. Students should be efficient in changing the parameters when editing the previous command.
- **Think-Pair-Share:** Have students answer Questions 4–6, and then share and discuss as a class.

Closure

- **Exit Ticket:** Have students answer the question posed at the end of the *Motivate* on page T-319. **about 14%**

5.6 Exercises

Dynamic Solutions available at BigIdeasMath.com

Vocabulary and Core Concept Check

1. **VOCABULARY** What is a random variable?
2. **WRITING** Give an example of a binomial experiment and describe how it meets the conditions of a binomial experiment.

Monitoring Progress and Modeling with Mathematics

In Exercises 3–6, make a table and draw a histogram showing the probability distribution for the random variable. *(See Example 1.)*

3. x = the number on a table tennis ball randomly chosen from a bag that contains 5 balls labeled "1," 3 balls labeled "2," and 2 balls labeled "3."

4. $c = 1$ when a randomly chosen card out of a standard deck of 52 playing cards is a heart and $c = 2$ otherwise.

5. $w = 1$ when a randomly chosen letter from the English alphabet is a vowel and $w = 2$ otherwise.

6. n = the number of digits in a random integer from 0 through 999.

In Exercises 7 and 8, use the probability distribution to determine (a) the number that is most likely to be spun on a spinner, and (b) the probability of spinning an even number. *(See Example 2.)*

7.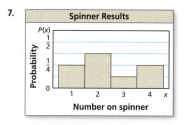

8.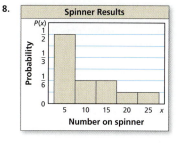

USING EQUATIONS In Exercises 9–12, calculate the probability of flipping a coin 20 times and getting the given number of heads.

9. 1
10. 4
11. 18
12. 20

13. **MODELING WITH MATHEMATICS** According to a survey, 27% of high school students in the United States buy a class ring. You ask 6 randomly chosen high school students whether they own a class ring. *(See Examples 3 and 4.)*

   a. Draw a histogram of the binomial distribution for your survey.
   b. What is the most likely outcome of your survey?
   c. What is the probability that at most 2 people have a class ring?

14. **MODELING WITH MATHEMATICS** According to a survey, 48% of adults in the United States believe that Unidentified Flying Objects (UFOs) are observing our planet. You ask 8 randomly chosen adults whether they believe UFOs are watching Earth.

   a. Draw a histogram of the binomial distribution for your survey.
   b. What is the most likely outcome of your survey?
   c. What is the probability that at most 3 people believe UFOs are watching Earth?

Section 5.6   Binomial Distributions   323

Assignment Guide and Homework Check

ASSIGNMENT
Basic: 1, 2, 3–15 odd, 18, 19, 23, 24
Average: 1, 2–16 even, 17–19, 23, 24
Advanced: 1, 2–16 even, 17–24

HOMEWORK CHECK
Basic: 3, 5, 7, 13, 18
Average: 4, 8, 12, 14, 18
Advanced: 6, 8, 14, 18, 19

ANSWERS

1. a variable whose value is determined by the outcomes of a probability experiment

2. *Sample answer:* According to a survey, about 30% of households in America own at least one cat. You ask 5 randomly chosen people (who live in separate households) whether they own a cat. There are 5 independent trials, and each trial has only two possible outcomes.

3.
x (value)	1	2	3
Outcomes	5	3	2
$P(x)$	$\frac{1}{2}$	$\frac{3}{10}$	$\frac{1}{5}$

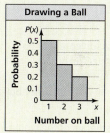

4.
c (value)	1	2
Outcomes	13	39
$P(c)$	$\frac{1}{4}$	$\frac{3}{4}$

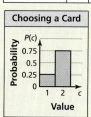

5–6. See Additional Answers.

7. a. 2
   b. $\frac{5}{8}$

8. a. 5
   b. $\frac{1}{4}$

9. about 0.00002
10. about 0.0046
11. about 0.00018
12. about 0.000001
13. a.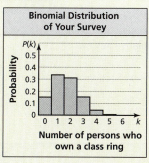

   b. The most likely outcome is that 1 of the 6 students owns a ring.
   c. about 0.798

14. a. See Additional Answers.
   b. The most likely outcome is that 4 of the 8 adults believe UFOs are watching Earth.
   c. about 0.407

Section 5.6   323

Dynamic Teaching Tools

- Dynamic Assessment & Progress Monitoring Tool
- Interactive Whiteboard Lesson Library
- Dynamic Classroom with Dynamic Investigations

ANSWERS

15. The exponents are switched;
$P(k = 3) = {}_5C_3\left(\frac{1}{6}\right)^3\left(\frac{5}{6}\right)^{5-3} \approx 0.032$

16. The combination part of the formula is missing;
$P(k = 3) = {}_5C_3\left(\frac{1}{6}\right)^3\left(\frac{5}{6}\right)^{5-3} \approx 0.032$

17–24. See Additional Answers.

Mini-Assessment

1. Use the probability distribution to find the probability that the spinner will land on a vowel.

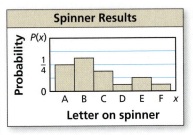

$\frac{3}{8} = 37.5\%$

According to a survey, about 65% of the students at a school buy their lunch on campus. You ask 4 randomly selected students whether they buy lunch on campus.

2. Draw a histogram of the binomial distribution for your survey.

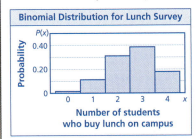

3. What is the most likely outcome of your survey? **3 of the 4 students surveyed buy their lunch on campus.**

4. What is the probability that fewer than 2 of the students you survey buy lunch on campus? **about 12.6%**

324   Chapter 5

ERROR ANALYSIS In Exercises 15 and 16, describe and correct the error in calculating the probability of rolling a 1 exactly 3 times in 5 rolls of a six-sided die.

15.

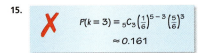

$P(k = 3) = {}_5C_3\left(\frac{1}{6}\right)^{5-3}\left(\frac{5}{6}\right)^3$

≈ 0.161

16.

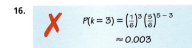

$P(k = 3) = \left(\frac{1}{6}\right)^3\left(\frac{5}{6}\right)^{5-3}$

≈ 0.003

17. **MATHEMATICAL CONNECTIONS** At most 7 gopher holes appear each week on the farm shown. Let x represent how many of the gopher holes appear in the carrot patch. Assume that a gopher hole has an equal chance of appearing at any point on the farm.

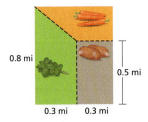

a. Find $P(x)$ for $x = 0, 1, 2, \ldots, 7$. (*Note*: The probability of a gopher hole appearing in the carrot patch is equal to the ratio of the area of the carrot patch to the area of the entire farm.)

b. Make a table showing the probability distribution for x.

c. Make a histogram showing the probability distribution for x.

18. **HOW DO YOU SEE IT?** Complete the probability distribution for the random variable x. What is the probability the value of x is greater than 2?

x	1	2	3	4
$P(x)$	0.1	0.3	0.4	

Maintaining Mathematical Proficiency Reviewing what you learned in previous grades and lessons

List the possible outcomes for the situation. *(Section 5.1)*

23. guessing the gender of three children

24. picking one of two doors and one of three curtains

324   Chapter 5   Probability

---

19. **MAKING AN ARGUMENT** The binomial distribution shows the results of a binomial experiment. Your friend claims that the probability p of a success must be greater than the probability $1 - p$ of a failure. Is your friend correct? Explain your reasoning.

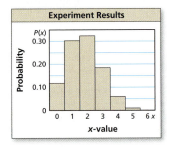

20. **THOUGHT PROVOKING** There are 100 coins in a bag. Only one of them has a date of 2010. You choose a coin at random, check the date, and then put the coin back in the bag. You repeat this 100 times. Are you certain of choosing the 2010 coin at least once? Explain your reasoning.

21. **MODELING WITH MATHEMATICS** Assume that having a male and having a female child are independent events, and that the probability of each is 0.5.

a. A couple has 4 male children. Evaluate the validity of this statement: "The first 4 kids were all boys, so the next one will probably be a girl."

b. What is the probability of having 4 male children and then a female child?

c. Let x be a random variable that represents the number of children a couple already has when they have their first female child. Draw a histogram of the distribution of $P(x)$ for $0 \leq x \leq 10$. Describe the shape of the histogram.

22. **CRITICAL THINKING** An entertainment system has n speakers. Each speaker will function properly with probability p, independent of whether the other speakers are functioning. The system will operate effectively when at least 50% of its speakers are functioning. For what values of p is a 5-speaker system more likely to operate than a 3-speaker system?

If students need help...	If students got it...
Resources by Chapter • Practice A and Practice B • Puzzle Time	**Resources by Chapter** • Enrichment and Extension • Cumulative Review
Student Journal • Practice	Start the *next* Section
Differentiating the Lesson Skills Review Handbook	

5.4–5.6 What Did You Learn?

Core Vocabulary

compound event, *p. 306*
overlapping events, *p. 306*
disjoint events, *p. 306*
mutually exclusive events, *p. 306*
permutation, *p. 312*
n factorial, *p. 312*
combination, *p. 314*
random variable, *p. 320*
probability distribution, *p. 320*
binomial distribution, *p. 321*
binomial experiment, *p. 321*

Core Concepts

Section 5.4
Probability of Compound Events, *p. 306*

Section 5.5
Permutations, *p. 313*
Combinations, *p. 314*

Section 5.6
Probability Distributions, *p. 320*
Binomial Experiments, *p. 321*

Mathematical Practices

1. How can you use diagrams to understand the situation in Exercise 22 on page 310?
2. Describe a relationship between the results in part (a) and part (b) in Exercise 52 on page 318.
3. Explain how you were able to break the situation into cases to evaluate the validity of the statement in part (a) of Exercise 21 on page 324.

Performance Task:

Risk Analysis

An arborist performs an extensive risk analysis to help identify and treat tree diseases. How can this process be applied to other things?

To explore the answer to this question and more, check out the Performance Task and Real-Life STEM video at *BigIdeasMath.com*.

Dynamic Teaching Tools

Dynamic Assessment & Progress Monitoring Tool
Interactive Whiteboard Lesson Library
Dynamic Classroom with Dynamic Investigations

ANSWERS

1. Use a Venn diagram to show that when *A*, *B*, and *C* are disjoint, you must add the probability of each, and when *A*, *B*, and *C* are overlapping, you must subtract parts so you do not count anything more than once.
2. *Sample answer:* The same marbles are drawn in both parts.
3. The probability of having four boys and then a girl is equal to the probability of having four boys and then another boy, so the statement is invalid.

ANSWERS
1. 6; FY, FN, FM, MY, MN, MM
2. $\frac{2}{9}, \frac{7}{9}$
3. 1 : 4; 4 : 1
4. a. 0.15625
   b. about 0.1667
   You are about 1.07 times more likely to pick a red then a green if you do not replace the first marble.
5. a. about 0.0586
   b. 0.0625
   You are about 1.07 times more likely to pick a blue then a red if you do not replace the first marble.
6. a. 0.25
   b. about 0.2333
   You are about 1.07 times more likely to pick a green and then another green if you replace the first marble.

5 Chapter Review

Dynamic Solutions available at *BigIdeasMath.com*

5.1 Sample Spaces and Probability (pp. 279–286)

Each section of the spinner shown has the same area. The spinner was spun 30 times. The table shows the results. For which color is the experimental probability of stopping on the color the same as the theoretical probability?

The theoretical probability of stopping on each of the five colors is $\frac{1}{5}$. Use the outcomes in the table to find the experimental probabilities.

Spinner Results	
green	4
orange	6
red	9
blue	8
yellow	3

$P(\text{green}) = \frac{4}{30} = \frac{2}{15}$ $P(\text{orange}) = \frac{6}{30} = \frac{1}{5}$ $P(\text{red}) = \frac{9}{30} = \frac{3}{10}$ $P(\text{blue}) = \frac{8}{30} = \frac{4}{15}$ $P(\text{yellow}) = \frac{3}{30} = \frac{1}{10}$

▶ The experimental probability of stopping on orange is the same as the theoretical probability.

1. You record a person's gender and whether they respond *yes*, *no*, or *maybe* to a survey question. How many possible outcomes are in the sample space? List the possible outcomes.
2. A bag contains 9 tiles, one for each letter in the word HAPPINESS. You choose a tile at random. What is the probability that you choose a tile with the letter S? What is the probability that you choose a tile with a letter other than P?
3. Using the spinner above, what are the odds in favor of stopping on yellow? What are the odds against stopping on blue?

5.2 Independent and Dependent Events (pp. 287–294)

You randomly select 2 cards from a standard deck of 52 playing cards. What is the probability that both cards are jacks when (a) you replace the first card before selecting the second, and (b) you do not replace the first card. Compare the probabilities.

Let event *A* be "first card is a jack" and event *B* be "second card is a jack."

a. Because you replace the first card before you select the second card, the events are independent. So, the probability is

$$P(A \text{ and } B) = P(A) \cdot P(B) = \frac{4}{52} \cdot \frac{4}{52} = \frac{16}{2704} = \frac{1}{169} \approx 0.006.$$

b. Because you do not replace the first card before you select the second card, the events are dependent. So, the probability is

$$P(A \text{ and } B) = P(A) \cdot P(B|A) = \frac{4}{52} \cdot \frac{3}{51} = \frac{12}{2652} = \frac{1}{221} \approx 0.005.$$

▶ So, you are $\frac{1}{169} \div \frac{1}{221} \approx 1.3$ times more likely to select 2 jacks when you replace the first card before you select the second card.

Find the probability of randomly selecting the given marbles from a bag of 5 red, 8 green, and 3 blue marbles when (a) you replace the first marble before drawing the second, and (b) you do not replace the first marble. Compare the probabilities.

4. red, then green
5. blue, then red
6. green, then green

326   Chapter 5   Probability

5.3 Two-Way Tables and Probability (pp. 295–302)

A survey asks residents of the east and west sides of a city whether they support the construction of a bridge. The results, given as joint relative frequencies, are shown in the two-way table. What is the probability that a randomly selected resident from the east side will support the project?

		Location	
		East Side	West Side
Response	Yes	0.47	0.36
	No	0.08	0.09

Find the joint and marginal relative frequencies. Then use these values to find the conditional probability.

$$P(\text{yes} \mid \text{east side}) = \frac{P(\text{east side and yes})}{P(\text{east side})} = \frac{0.47}{0.47 + 0.08} \approx 0.855$$

▶ So, the probability that a resident of the east side of the city will support the project is about 85.5%.

7. What is the probability that a randomly selected resident who does not support the project in the example above is from the west side?

8. After a conference, 220 men and 270 women respond to a survey. Of those, 200 men and 230 women say the conference was impactful. Organize these results in a two-way table. Then find and interpret the marginal frequencies.

5.4 Probability of Disjoint and Overlapping Events (pp. 305–310)

Let A and B be events such that $P(A) = \frac{2}{3}$, $P(B) = \frac{1}{2}$, and $P(A \text{ and } B) = \frac{1}{3}$. Find $P(A \text{ or } B)$.

$P(A \text{ or } B) = P(A) + P(B) - P(A \text{ and } B)$ Write general formula.

$\qquad = \frac{2}{3} + \frac{1}{2} - \frac{1}{3}$ Substitute known probabilities.

$\qquad = \frac{5}{6}$ Simplify.

$\qquad \approx 0.833$ Use a calculator.

9. Let A and B be events such that $P(A) = 0.32$, $P(B) = 0.48$, and $P(A \text{ and } B) = 0.12$. Find $P(A \text{ or } B)$.

10. Out of 100 employees at a company, 92 employees either work part time or work 5 days each week. There are 14 employees who work part time and 80 employees who work 5 days each week. What is the probability that a randomly selected employee works both part time and 5 days each week?

ANSWERS

7. about 0.529

8.

		Gender		
		Men	Women	Total
Response	Yes	200	230	430
	No	20	40	60
	Total	220	270	490

About 44.9% of responders were men, about 55.1% of responders were women, about 87.8% of responders thought it was impactful, about 12.2% of responders thought it was not impactful.

9. 0.68

10. 0.02

ANSWERS

11. 5040
12. 1,037,836,800
13. 15
14. 70
15. a. 120
    b. 60
16. $\frac{1}{84}$
17. about 0.12
18.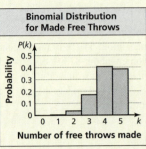

The most likely outcome is that 4 of the 5 free throw shots will be made.

5.5 Permutations and Combinations (pp. 311–318)

A 5-digit code consists of 5 different integers from 0 to 9. How many different codes are possible?

To find the number of permutations of 5 integers chosen from 10, find $_{10}P_5$.

$$_{10}P_5 = \frac{10!}{(10-5)!}$$ Permutations formula

$$= \frac{10!}{5!}$$ Subtract.

$$= \frac{10 \cdot 9 \cdot 8 \cdot 7 \cdot 6 \cdot 5!}{5!}$$ Expand factorials. Divide out common factor, 5!.

$$= 30{,}240$$ Simplify.

▶ There are 30,240 possible codes.

Evaluate the expression.

11. $_7P_6$ 12. $_{13}P_{10}$ 13. $_6C_2$ 14. $_8C_4$

15. In how many ways can you arrange (a) all of the letters and (b) 3 of the letters in the word UNCLE?

16. A random drawing will determine which 3 people in a group of 9 will win concert tickets. What is the probability that you and your 2 friends will win the tickets?

5.6 Binomial Distributions (pp. 319–324)

According to a survey, about 21% of adults in the U.S. visited an art museum last year. You ask 4 randomly chosen adults whether they visited an art museum last year. Draw a histogram of the binomial distribution for your survey.

The probability that a randomly selected person visited an art museum is $p = 0.21$. Because you survey 4 people, $n = 4$.

$P(k=0) = {_4C_0}(0.21)^0(0.79)^4 \approx 0.390$

$P(k=1) = {_4C_1}(0.21)^1(0.79)^3 \approx 0.414$

$P(k=2) = {_4C_2}(0.21)^2(0.79)^2 \approx 0.165$

$P(k=3) = {_4C_3}(0.21)^3(0.79)^1 \approx 0.029$

$P(k=4) = {_4C_4}(0.21)^4(0.79)^0 \approx 0.002$

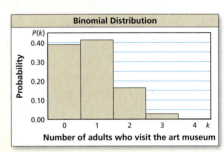

17. Find the probability of flipping a coin 12 times and getting exactly 4 heads.

18. A basketball player makes a free throw 82.6% of the time. The player attempts 5 free throws. Draw a histogram of the binomial distribution of the number of successful free throws. What is the most likely outcome?

5 Chapter Test

You roll a six-sided die. Find the probability of the event described. Explain your reasoning.

1. You roll a number less than 5.
2. You roll a multiple of 3.

Evaluate the expression.

3. $_7P_2$
4. $_8P_3$
5. $_6C_3$
6. $_{12}C_7$

7. The odds in favor of an event are 2 : 9. What are the odds against the event?

8. You find the probability $P(A \text{ or } B)$ by using the equation $P(A \text{ or } B) = P(A) + P(B) - P(A \text{ and } B)$. Describe why it is necessary to subtract $P(A \text{ and } B)$ when the events A and B are overlapping. Then describe why it is *not* necessary to subtract $P(A \text{ and } B)$ when the events A and B are disjoint.

9. Is it possible to use the formula $P(A \text{ and } B) = P(A) \cdot P(B|A)$ when events A and B are independent? Explain your reasoning.

10. According to a survey, about 58% of families sit down for a family dinner at least four times per week. You ask 5 randomly chosen families whether they have a family dinner at least four times per week.
    a. Draw a histogram of the binomial distribution for the survey.
    b. What is the most likely outcome of the survey?
    c. What is the probability that at least 3 families have a family dinner four times per week?

11. You are choosing a cell phone company to sign with for the next 2 years. The three plans you consider are equally priced. You ask several of your neighbors whether they are satisfied with their current cell phone company. The table shows the results. According to this survey, which company should you choose?

	Satisfied	Not Satisfied
Company A	IIII	II
Company B	IIII	III
Company C	HHT I	HHT

12. Out of 10,500 people that attend a math conference, 7315 people are either college professors or female. There are 4680 college professors and 5140 females. What is the probability that a randomly selected person is both a college professor and female?

13. Consider a bag that contains all the chess pieces in a set, as shown in the diagram.

	King	Queen	Bishop	Rook	Knight	Pawn
Black	1	1	2	2	2	8
White	1	1	2	2	2	8

    a. You choose one piece at random. Find the probability that you choose a black piece or a queen.
    b. You choose one piece at random, do not replace it, then choose a second piece at random. Find the probability that you choose a king, then a pawn.

14. Three volunteers are chosen at random from a group of 12 to help at a summer camp.
    a. What is the probability that you, your brother, and your friend are chosen?
    b. The first person chosen will be a counselor, the second will be a lifeguard, and the third will be a cook. What is the probability that you are the cook, your brother is the lifeguard, and your friend is the counselor?

ANSWERS

1. $\frac{2}{3}$
2. $\frac{1}{3}$
3. 42
4. 336
5. 20
6. 792
7. 9 : 2
8. $P(A \text{ and } B)$ is counted twice when adding $P(A)$ and $P(B)$; When events A and B are disjoint, $P(A \text{ and } B) = 0$.
9. yes; If events A and B are independent, then $P(B \mid A) = P(B)$.
10. a.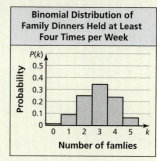

    b. The most likely outcome is that 3 of the 5 families will sit down for a family dinner at least 4 times per week.
    c. about 0.6474
11. Company A
12. about 23.86%
13. a. $\frac{17}{32}$
    b. $\frac{1}{31}$
14. a. $\frac{1}{220}$
    b. $\frac{1}{1320}$

If students need help...	If students got it...
Lesson Tutorials	Resources by Chapter • Enrichment and Extension • Cumulative Review
Skills Review Handbook	Performance Task
BigIdeasMath.com	Start the *next* Section

5 Cumulative Assessment

1. According to a survey, 63% of Americans consider themselves sports fans. You randomly select 14 Americans to survey.

    a. Draw a histogram of the binomial distribution of your survey.

    b. What is the most likely number of Americans who consider themselves sports fans?

    c. What is the probability at least 7 Americans consider themselves sports fans?

2. Place each function into one of the three categories.

No zeros	One zero	Two zeros

    $f(x) = 3x^2 + 4x + 2$ $f(x) = 4x^2 - 8x + 4$ $f(x) = 7x^2$
    $f(x) = -x^2 + 2x$ $f(x) = x^2 - 3x - 21$ $f(x) = -6x^2 - 5$

3. You order a fruit smoothie made with 2 liquid ingredients and 3 fruit ingredients from the menu shown. How many different fruit smoothies can you order?

    Liquids
    - Water
    - Tea
    - Coconut Water
    - Almond Milk
    - Apple Juice
    - Orange Juice

    Fruits
    - Orange
    - Banana
    - Pineapple
    - Cantaloupe
    - Strawberry
    - Watermelon
    - Kiwi
    - Peach
    - Blueberry
    - Pomegranate

4. Select all the numbers that are in the range of the function shown.

    $$y = \begin{cases} x^2 + 4x + 7, & \text{if } x \leq -1 \\ \frac{1}{2}x + 2, & \text{if } x > -1 \end{cases}$$

    0 $\frac{1}{2}$ 1 $1\frac{1}{2}$ 2 $2\frac{1}{2}$ 3 $3\frac{1}{2}$ 4

5. Complete the equation so that the solutions of the system of equations are $(-2, 4)$ and $(1, -5)$.

    $y = \Box\, x + \Box$
    $y = 2x^2 - x - 6$

ANSWERS

1. a.

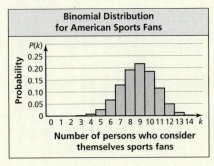

   b. 9
   c. about 0.8988
2. See Additional Answers.
3. 1800
4. $2, 2\frac{1}{2}, 3, 3\frac{1}{2}, 4$
5. $-3; -2$

6. Use the diagram to explain why the equation is true.

   $P(A) + P(B) = P(A \text{ or } B) + P(A \text{ and } B)$

7. Which polynomial represents the product of $2x - 4$ and $x^2 + 6x - 2$?

   Ⓐ $2x^3 + 8x^2 - 4x + 8$

   Ⓑ $2x^3 + 8x^2 - 28x + 8$

   Ⓒ $2x^3 + 8$

   Ⓓ $2x^3 - 24x - 2$

8. A survey asked male and female students about whether they prefer to take gym class or choir. The table shows the results of the survey.

		Class		
		Gym	Choir	Total
Gender	Male			50
	Female	23		
	Total		49	106

   a. Complete the two-way table.

   b. What is the probability that a randomly selected student is female and prefers choir?

   c. What is the probability that a randomly selected male student prefers gym class?

9. The owner of a lawn-mowing business has three mowers. As long as one of the mowers is working, the owner can stay productive. One of the mowers is unusable 10% of the time, one is unusable 8% of the time, and one is unusable 18% of the time.

   a. Find the probability that all three mowers are unusable on a given day.

   b. Find the probability that at least one of the mowers is unusable on a given day.

   c. Suppose the least-reliable mower stops working completely. How does this affect the probability that the lawn-mowing business can be productive on a given day?

10. Write a system of quadratic inequalities whose solution is represented in the graph.

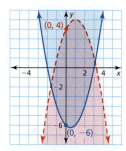

Chapter 6 Pacing Guide

Chapter Opener/Mathematical Practices	0.5 Day
Section 1	1.5 Days
Section 2	2 Days
Section 3	2 Days
Section 4	2 Days
Quiz	0.5 Day
Section 5	1.5 Days
Section 6	2 Days
Section 7	1 Day
Chapter Review/Chapter Tests	2 Days
Total Chapter 6	15 Days
Year-to-Date	95 Days

6 Relationships Within Triangles

- 6.1 Proving Geometric Relationships
- 6.2 Perpendicular and Angle Bisectors
- 6.3 Bisectors of Triangles
- 6.4 Medians and Altitudes of Triangles
- 6.5 The Triangle Midsegment Theorem
- 6.6 Indirect Proof and Inequalities in One Triangle
- 6.7 Inequalities in Two Triangles

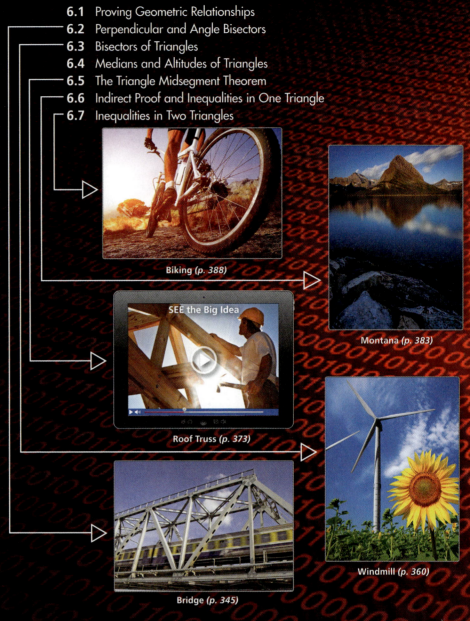

Biking (p. 388)

Montana (p. 383)

SEE the Big Idea

Roof Truss (p. 373)

Windmill (p. 360)

Bridge (p. 345)

Laurie's Notes

Chapter Summary

- This chapter uses the deductive skills developed in Math I to explore special segments in a triangle. These segments include perpendicular bisectors, angle bisectors, medians, altitudes, and midsegments.
- The first lesson serves as an introduction to proofs for students who did not cover this topic in Math I, and as a review for students who did.
- Students are able to discover special properties of these segments by using dynamic geometry software. To prove these relationships, a variety of proof formats and approaches are used: transformational, synthetic, analytic, and paragraph.
- The last two lessons in the chapter are about inequalities within one triangle and in two triangles. The indirect proof is introduced and used to prove several of the theorems in these lessons.
- Triangles have been the geometric structure used to help students develop their deductive reasoning skills. In the next chapter, quadrilaterals and other polygons are studied.

What Your Students Have Learned

Middle School
- Draw geometric shapes with given conditions, focusing on triangles and quadrilaterals.
- Draw polygons in the coordinate plane given coordinates for the vertices, and find lengths of sides.
- Identify congruent figures.

Math I
- Construct congruent segments and angles, and bisect segments and angles.
- Write two-column, paragraph, flowchart, and coordinate proofs to prove geometric relationships.
- Construct parallel lines, perpendicular lines, and perpendicular bisectors.
- Use SAS, SSS, HL, ASA, and AAS Congruence Theorems to prove two triangles are congruent.
- Use congruent triangles to solve problems and write proofs.

What Your Students Will Learn

Math II
- Write two-column, paragraph, flowchart, and coordinate proofs to prove geometric relationships.
- Use perpendicular bisectors and angle bisectors to find measures.
- Use and find the circumcenters, incenters, centroids, and orthocenters of triangles.
- Use the Triangle Midsegment Theorem to find distances.
- Prove geometric relationships using indirect proofs.
- Relate sides and angles of a triangle and use the Triangle Inequality Theorem to find possible side lengths.
- Use the Hinge Theorem to compare angle measures and side lengths between two triangles.

Dynamic Teaching Tools
Dynamic Assessment & Progress Monitoring Tool
Lesson Planning Tool
Interactive Whiteboard Lesson Library
Dynamic Classroom with Dynamic Investigations
Real-Life STEM Videos

Scaffolding in the Classroom
Graphic Organizers: Word Magnet
A Word Magnet can be used to organize information associated with a vocabulary word or term. Students write the word or term inside the magnet. Students write associated information on the blank lines that "radiate" from the magnet. Associated information can include, but is not limited to: other vocabulary words or terms, definitions, formulas, procedures, examples, and visuals. This type of organizer serves as a good summary tool because any information related to a topic can be included.

Questioning in the Classroom

What do you think?
Allow students time to write their answers on individual whiteboards. After the allotted time, they can hold up their boards. The teacher can see what students are thinking and follow up with discussion.

Laurie's Notes

Maintaining Mathematical Proficiency

Writing an Equation of a Perpendicular Line

- An equation written in slope-intercept form, $y = mx + b$, has a slope m and a y-intercept b.
- Students should remember that in a coordinate plane, two nonvertical lines are perpendicular if and only if the product of their slopes is -1.

COMMON ERROR Students substitute incorrectly when finding the value of b. Have them write the slope-intercept form of an equation of a line, $y = mx + b$. Then remind students to substitute the first coordinate of the given point for x and the second coordinate for y.

Writing Compound Inequalities

- Be sure students understand the implications of the words *and* and *or*.
- Remind students that when a compound inequality uses the word *and*, the variable must satisfy both inequalities. When a compound inequality uses the word *or*, the variable needs to satisfy only one of the inequalities.

COMMON ERROR Students sometimes switch *and* and *or* when writing inequalities, which leads to confusion when they have to solve a compound inequality.

Mathematical Practices (continued on page 334)

- The *Mathematical Practices* page focuses attention on how mathematics is learned—process versus content. This page demonstrates that by using technological tools, students can explore and discover geometric relationships. Although these relationships can be explored in other manners, dynamic geometry software can be very effective and efficient.
- Use the *Mathematical Practices* page to help students develop mathematical habits of mind—how mathematics can be explored and how mathematics is thought about.

If students need help...	If students got it...
Student Journal • Maintaining Mathematical Proficiency	Game Closet at *BigIdeasMath.com*
Lesson Tutorials	Start the *next* Section
Skills Review Handbook	

Maintaining Mathematical Proficiency

Writing an Equation of a Perpendicular Line (Math I)

Example 1 Write the equation of a line passing through the point $(-2, 0)$ that is perpendicular to the line $y = 2x + 8$.

Step 1 Find the slope m of the perpendicular line. The line $y = 2x + 8$ has a slope of 2. Use the Slopes of Perpendicular Lines Theorem.

$2 \cdot m = -1$ The product of the slopes of $\perp$ lines is -1.

$m = -\frac{1}{2}$ Divide each side by 2.

Step 2 Find the y-intercept b by using $m = -\frac{1}{2}$ and $(x, y) = (-2, 0)$.

$y = mx + b$ Use the slope-intercept form.

$0 = -\frac{1}{2}(-2) + b$ Substitute for m, x, and y.

$-1 = b$ Solve for b.

▶ Because $m = -\frac{1}{2}$ and $b = -1$, an equation of the line is $y = -\frac{1}{2}x - 1$.

Write an equation of the line passing through point P that is perpendicular to the given line.

1. $P(3, 1), y = \frac{1}{3}x - 5$
2. $P(4, -3), y = -x - 5$
3. $P(-1, -2), y = -4x + 13$

Writing Compound Inequalities (Math I)

Example 2 Write each sentence as an inequality.

a. A number x is greater than or equal to -1 and less than 6.

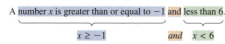

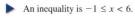   and $x < 6$

▶ An inequality is $-1 \leq x < 6$.

b. A number y is at most 4 or at least 9.

$y \leq 4$ or $y \geq 9$

▶ An inequality is $y \leq 4 \text{ or } y \geq 9$.

Write the sentence as an inequality.

4. A number w is at least -3 and no more than 8.
5. A number m is more than 0 and less than 11.
6. A number s is less than or equal to 5 or greater than 2.
7. A number d is fewer than 12 or no less than -7.

8. **ABSTRACT REASONING** Is it possible for the solution of a compound inequality to be all real numbers? Explain your reasoning.

Dynamic Solutions available at *BigIdeasMath.com*

Vocabulary Review

Have students make a Summary Triangle to explain how to write an equation of a perpendicular line. Include the following terms.
- Slope
- Slope-intercept form
- y-intercept

MONITORING PROGRESS ANSWERS

1. A perpendicular bisector is perpendicular to a side of the triangle at its midpoint.
2. An angle bisector divides an angle of the triangle into two congruent adjacent angles.
3. A median of a triangle is a segment from a vertex to the midpoint of the opposite side.
4. An altitude of a triangle is the perpendicular segment from a vertex to the opposite side or to the line that contains the opposite side.
5. A midsegment of a triangle is a segment that connects the midpoints of two sides of the triangle.

Mathematical Practices

Mathematically proficient students use technological tools to explore concepts.

Lines, Rays, and Segments in Triangles

Core Concept

Lines, Rays, and Segments in Triangles

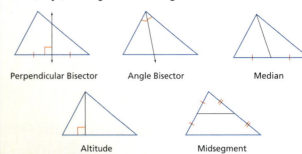

Perpendicular Bisector   Angle Bisector   Median

Altitude   Midsegment

EXAMPLE 1 Drawing a Perpendicular Bisector

Use dynamic geometry software to construct the perpendicular bisector of one of the sides of the triangle with vertices $A(-1, 2)$, $B(5, 4)$, and $C(4, -1)$. Find the lengths of the two segments of the bisected side.

SOLUTION

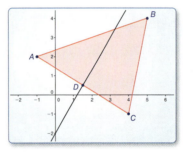

Sample
Points
$A(-1, 2)$
$B(5, 4)$
$C(4, -1)$
Line
$-5x + 3y = -6$
Segments
$AD = 2.92$
$CD = 2.92$

▶ The two segments of the bisected side have the same length, $AD = CD = 2.92$ units.

Monitoring Progress

Refer to the figures at the top of the page to describe each type of line, ray, or segment in a triangle.

1. perpendicular bisector
2. angle bisector
3. median
4. altitude
5. midsegment

Laurie's Notes   Mathematical Practices *(continued from page T-333)*

- There are many definitions that students will encounter in their study of geometry. Exploring with technology can help students discover the meaning of a word.
- Give time for students to work through the questions in the *Monitoring Progress*, and then discuss as a class.

Laurie's Notes

Overview of Section 6.1

Introduction
- All of the content in this lesson was studied in Math I. This lesson serves to review the styles of proof: two-column, flowchart, paragraph, and coordinate presented in Math I as well as some of the geometric content.
- Students who studied in the Math I book should find the material familiar. Students who did not study previously from the Math I book may need additional support.

Formative Assessment Tips
- **Learning Goals Inventory:** At the beginning of a chapter or unit of study, it is important for students to identify their level of understanding or knowledge of the learning intentions (goals) for the chapter. Taking an inventory of the explicit goals for the chapter informs students what you want them to learn—they know the target. On one side of a sheet of paper, I type the assignment guide for the chapter. On the other side, I list the learning intentions for the chapter and ask students to do a self-assessment for each learning goal.
    - 0—no knowledge, understanding, or ability
    - 1—beginning to get it
    - 2—feeling okay
    - 3—very confident
- Students return to the *Learning Goals Inventory* midway through the chapter and after the chapter assessment. Students recognize their own growth in completing the inventory multiple times.

Pacing Suggestion
- The formal lesson is fairly long; however, experience with the explorations helps students recall the manner in which proofs can be written in a two-column format. Have students complete the two explorations, and then begin with Example 1 in the formal lesson.

Dynamic Teaching Tools
Dynamic Assessment & Progress Monitoring Tool
Lesson Planning Tool
Interactive Whiteboard Lesson Library
Dynamic Classroom with Dynamic Investigations

What Your Students Will Learn

- Write two-column proofs to prove geometric relationships.
- Write paragraph proofs to prove geometric relationships.
- Write flowchart proofs to prove geometric relationships.
- Write coordinate proofs to prove geometric relationships.

Laurie's Notes

Exploration

Motivate
- ❓ "What does it mean to prove something?" Answers will vary.
- ❓ "Can you think of examples from the field of science where a theory was proven?" Answers will vary.
- ❓ "What styles of proof do you recall from Math I?" Answers will vary and should include two-column, flowchart, paragraph, and coordinate proofs.
- Explain that in geometry they will have opportunities to prove something, a statement that they believe is true. Just as a scientist has protocols that they follow to prove a theory, we will have protocols that we will follow to prove statements.

For Your Information
- If students are new to your program and did not take Math I, this lesson is designed to introduce what it means to prove a statement. It will also serve to review this content for students who did study Math I. Regardless of what math program students were in during prior years, it is hoped that there would have been some focus on reasoning and proof.

Exploration 1
- Explain that the template, meaning the T-diagram and the numbering of statements and reasons, is used to write down the steps in the proof, just as statements and reasons are written when solving an algebraic problem. We want to be able to justify the statement we are making as we write a logical sequence of steps.
- ❓ "What does AD mean?" The length of segment AD.
- ❓ "What does $AB + AC$ mean?" The length of segment AB added to the length of segment AC.
- Translate and say, "The length of segment AD is equal to the length of segment AB added to the length of segment AC."
- ❓ "What is the justification for saying that $AC + CD = AD$?" Segment Addition Postulate
- ❓ "How does the third statement follow from the first two statements?" Substitution; $AB + AC$ is substituted for AD.
- ❓ "How does the fourth statement follow from the third statement?" The length of segment AC is subtracted from each side of the equation.
- **Teaching Tip:** You may want to circle, highlight, or underline the AD in steps 1 and 2 and then draw an arrow to show the substitution that is being done.

Exploration 2
- Students should be able to read through the proof with their partners and make sense of the reasoning.
- **Wait Time:** Give partners time to write the statements for this example. Do not "rush in" to answer questions.

Communicate Your Answer
- Listen for students to describe the process of writing logical statements and being able to justify each one.

Connecting to Next Step
- Discuss the process of writing statements and a justification for each step. This will lead to a discussion of proof in the formal lesson.

6.1 Proving Geometric Relationships

Essential Question How can you prove a mathematical statement?

A **proof** is a logical argument that uses deductive reasoning to show that a statement is true.

EXPLORATION 1 Writing Reasons in a Proof

Work with a partner. Four steps of a proof are shown. Write the reasons for each statement.

Given $AD = AB + AC$

Prove $CD = AB$

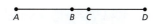

> **REASONING ABSTRACTLY**
> To be proficient in math, you need to know and be able to use algebraic properties.

STATEMENTS	REASONS
1. $AD = AB + AC$	1. Given
2. $AC + CD = AD$	2.
3. $AC + CD = AB + AC$	3.
4. $CD = AB$	4.

EXPLORATION 2 Writing Steps in a Proof

Work with a partner. Five steps of a proof are shown. Complete the statements that correspond to each reason.

Given $m\angle ABD = m\angle CBE$

Prove $m\angle 1 = m\angle 3$

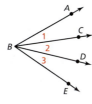

STATEMENTS	REASONS
1. $m\angle ABD = m\angle 1 + m\angle 2$	1. Angle Addition Postulate
2. $m\angle CBE =$	2. Angle Addition Postulate
3.	3. Given
4. $m\angle 1 + m\angle 2 = m\angle 2 + m\angle 3$	4. Substitution Property of Equality
5.	5. Subtraction Property of Equality

Communicate Your Answer

3. How can you prove a mathematical statement?

4. In Exploration 2, can you prove that $m\angle 1 = m\angle 2$? Explain your reasoning.

Differentiated Instruction

Inclusion
Some students may wonder why the properties of congruence related to segments and angles are theorems, and the properties of equality are not. Remind them that the properties of equality are true for all real numbers. Segment lengths and angle measures are real numbers, so these properties of equality could be accepted as true. The properties of congruence were proven using facts previously accepted to be true.

Extra Example 1

Write a two-column proof.
Given ∠1 is complementary to ∠3.
∠2 is complementary to ∠3.
Prove ∠1 ≅ ∠2

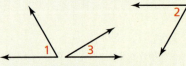

Statements (Reasons)
1. ∠1 is complementary to ∠3. (Given)
2. $m\angle 1 + m\angle 3 = 90°$ (Def. comp. ∠s)
3. ∠2 is complementary to ∠3. (Given)
4. $m\angle 2 + m\angle 3 = 90°$ (Def. comp. ∠s)
5. $m\angle 1 + m\angle 3 = m\angle 2 + m\angle 3$ (Subst. Prop. =)
6. $m\angle 1 = m\angle 2$ (Subtr. Prop. =)
7. ∠1 ≅ ∠2 (Def. congruent. ∠s)

MONITORING PROGRESS ANSWER

1. Given; Substitution Property of Equality; $2x = 14$; Division Property of Equality

---

6.1 Lesson

What You Will Learn

▶ Write two-column proofs to prove geometric relationships.
▶ Write paragraph proofs to prove geometric relationships.
▶ Write flowchart proofs to prove geometric relationships.
▶ Write coordinate proofs to prove geometric relationships.

Core Vocabulary

proof, p. 336
two-column proof, p. 336
paragraph proof, p. 337
flowchart proof, or flow proof, p. 338
coordinate proof, p. 339

Writing Two-Column Proofs

A **proof** is a logical argument that uses deductive reasoning to show that a statement is true. There are several formats for proofs. A **two-column proof** has numbered statements and corresponding reasons that show an argument in a logical order.

In a two-column proof, each statement in the left-hand column is either given information or the result of applying a known property or fact to statements already made. Each reason in the right-hand column is the explanation for the corresponding statement.

EXAMPLE 1 Writing a Two-Column Proof

Write a two-column proof of the Vertical Angles Congruence Theorem.

Given ∠1 and ∠3 are vertical angles.
Prove ∠1 ≅ ∠3

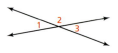

STATEMENTS	REASONS
1. ∠1 and ∠3 are vertical angles.	1. Given
2. ∠1 and ∠2 are a linear pair. ∠2 and ∠3 are a linear pair.	2. Definition of linear pair, as shown in the diagram
3. ∠1 and ∠2 are supplementary. ∠2 and ∠3 are supplementary.	3. Linear Pair Postulate
4. ∠1 ≅ ∠3	4. Congruent Supplements Theorem

Monitoring Progress   Help in English and Spanish at BigIdeasMath.com

1. Six steps of a two-column proof are shown. Copy and complete the proof.

Given B is the midpoint of $\overline{AC}$.
Prove $x = 7$

A ——— 6x ——— B ——— 4x + 14 ——— C

STATEMENTS	REASONS
1. B is the midpoint of $\overline{AC}$.	1. _____
2. $\overline{AB} \cong \overline{BC}$	2. Definition of midpoint
3. $AB = BC$	3. Definition of congruent segments
4. $6x = 4x + 14$	4. _____
5. _____	5. Subtraction Property of Equality
6. $x = 7$	6. _____

336   Chapter 6   Relationships Within Triangles

Laurie's Notes — Teacher Actions

- Begin with a discussion of *proof* and a reminder of what deductive reasoning is. Explain that statements are either given information or they result from applying a known property or fact to statements already made. Reasons are the explanations for the corresponding statements.
- ❓ "What are vertical angles?" *The nonadjacent angles formed when two lines intersect.* "What do you recall about vertical angles?" *Vertical angles are congruent.*
- This proof was completed in Math I, so spend whatever time is necessary in reviewing the reasoning, vocabulary, and strategy for completing this proof.

Writing Paragraph Proofs

Another proof format is a **paragraph proof**, which presents the statements and reasons of a proof as sentences in a paragraph. It uses words to explain the logical flow of the argument.

EXAMPLE 2 Writing a Paragraph Proof

Use the given paragraph proof to write a two-column proof of the Corresponding Angles Theorem.

Given $\overrightarrow{AB} \parallel \overrightarrow{CD}$

Prove $\angle 1 \cong \angle 2$

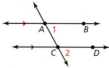

Paragraph Proof

Because $\overrightarrow{AB} \parallel \overrightarrow{CD}$ and translations map lines to parallel lines, a translation along $\overrightarrow{AC}$ maps $\overrightarrow{AB}$ to $\overrightarrow{CD}$. Because translations are rigid motions, angle measures are preserved, which means the angles formed by $\overrightarrow{AB}$ and $\overrightarrow{AC}$ are congruent to the corresponding angles formed by $\overrightarrow{CD}$ and $\overrightarrow{AC}$. So, $\angle 1 \cong \angle 2$.

Two-Column Proof

STATEMENTS	REASONS
1. $\overrightarrow{AB} \parallel \overrightarrow{CD}$	1. Given
2. A translation along $\overrightarrow{AC}$ maps $\overrightarrow{AB}$ to $\overrightarrow{CD}$.	2. Translations map lines to parallel lines.
3. Angles formed by $\overrightarrow{AB}$ and $\overrightarrow{AC}$ are congruent to the corresponding angles formed by $\overrightarrow{CD}$ and $\overrightarrow{AC}$.	3. Translations are rigid motions.
4. $\angle 1 \cong \angle 2$	4. $\angle 1$ and $\angle 2$ are corresponding angles.

Monitoring Progress Help in English and Spanish at *BigIdeasMath.com*

2. Copy and complete the paragraph proof of the Alternate Interior Angles Theorem. Then write a two-column proof.

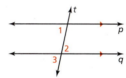

Given $p \parallel q$

Prove $\angle 1 \cong \angle 2$

$\angle 1 \cong \angle 3$ by the _____. $\angle 3 \cong \angle 2$ by the Vertical Angles Congruence Theorem. $\angle 1 \cong \angle 2$ by the _____.

Section 6.1   Proving Geometric Relationships   337

English Language Learners

Notebook Development

Have students record the postulates and theorems from this chapter in their notebooks. For each, include a labeled figure and an example.

Extra Example 2

Write a paragraph proof for Extra Example 1.
Because $\angle 1$ and $\angle 3$ are complementary, the sum of $m\angle 1$ and $m\angle 3$ is 90° by the definition of complementary angles. Because $\angle 2$ and $\angle 3$ are complementary, the sum of $m\angle 2$ and $m\angle 3$ is 90° by the definition of complementary angles. By the Substitution Property of Equality, the sum of $m\angle 1$ and $m\angle 3$ equals the sum of $m\angle 2$ and $m\angle 3$. Using the Subtraction Property of Equality, $m\angle 3$ can be subtracted from each side so that $m\angle 1$ equals $m\angle 2$. So, $\angle 1$ is congruent to $\angle 2$ by the definition of congruent angles.

MONITORING PROGRESS ANSWER

2. Corresponding Angles Theorem; Transitive Property of Congruence

STATEMENTS	REASONS
1. $p \parallel q$	1. Given
2. $\angle 1 \cong \angle 3$	2. Corresponding Angles Theorem
3. $\angle 3 \cong \angle 2$	3. Vertical Angles Congruence Theorem
4. $\angle 1 \cong \angle 2$	4. Transitive Property of Congruence

---

Laurie's Notes   Teacher Actions

- Example 2 reviews a second proof format, the paragraph proof. The example also reviews a theorem about the corresponding angles formed when parallel lines are intersected by a transversal.
- After completing the paragraph proof and two-column proof ask, "What other angle pair relationships do you recall when parallel lines are intersected by a transversal?" Listen for congruency relationships (alternate interior angles and alternate exterior angles) and supplementary relationships (interior angles on the same side of the transversal and exterior angles on the same side of the transversal).
- **Think-Pair-Share:** Have students complete the proof in the *Monitoring Progress* and compare with their neighbors.

Section 6.1   337

Extra Example 3
Write a flowchart proof for Extra Example 1.

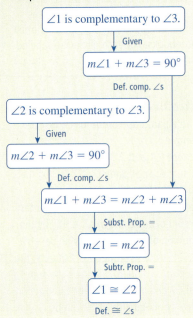

MONITORING PROGRESS ANSWER

3. $\angle CAD \cong \angle BAD$; SAS Congruence Theorem

STATEMENTS	REASONS
1. Draw $\overline{AD}$, the angle bisector of $\angle CAB$.	1. Construction of angle bisector
2. $\angle CAD \cong \angle BAD$	2. Definition of angle bisector
3. $\overline{AB} \cong \overline{AC}$	3. Given
4. $\overline{DA} \cong \overline{DA}$	4. Reflexive Property of Congruence
5. $\triangle ADB \cong \triangle ADC$	5. SAS Congruence Theorem
6. $\angle B \cong \angle C$	6. Corresponding parts of congruent triangles are congruent.

Writing Flowchart Proofs

Another proof format is a **flowchart proof**, or **flow proof**, which uses boxes and arrows to show the flow of a logical argument. Each reason is below the statement it justifies.

EXAMPLE 3 Writing a Flowchart Proof

Use the given flowchart proof to write a two-column proof of the Triangle Sum Theorem.

Given $\triangle ABC$

Prove $m\angle 1 + m\angle 2 + m\angle 3 = 180°$

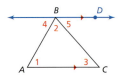

Flowchart Proof

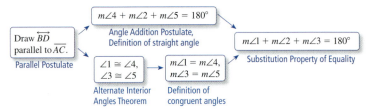

Two-Column Proof

STATEMENTS	REASONS
1. Draw $\overleftrightarrow{BD}$ parallel to $\overline{AC}$.	1. Parallel Postulate
2. $m\angle 4 + m\angle 2 + m\angle 5 = 180°$	2. Angle Addition Postulate and definition of straight angle
3. $\angle 1 \cong \angle 4, \angle 3 \cong \angle 5$	3. Alternate Interior Angles Theorem
4. $m\angle 1 = m\angle 4, m\angle 3 = m\angle 5$	4. Definition of congruent angles
5. $m\angle 1 + m\angle 2 + m\angle 3 = 180°$	5. Substitution Property of Equality

Monitoring Progress Help in English and Spanish at *BigIdeasMath.com*

3. Copy and complete the flowchart proof of the Base Angles Theorem. Then write a two-column proof.

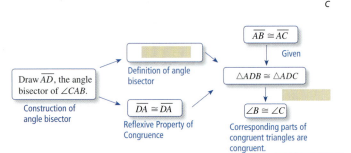

338 Chapter 6 Relationships Within Triangles

Laurie's Notes — Teacher Actions

- Students will be familiar with the flowchart proof format and the theorem about the sum of the angles of a triangle.
- **Whiteboarding:** Sketch the steps in the proof and ask partners to provide reasons for each statement on their *whiteboards*. This can be done for the flowchart proof or the two-column proof.
- ? "What does the Base Angles Theorem state?" If two sides of a triangle are congruent, then the angles opposite them are congruent. "What name do you give this triangle?" isosceles triangle
- ? **Reason Abstractly and Quantitatively:** "An equilateral triangle is *always, sometimes,* or *never* an isosceles triangle?" always

Writing Coordinate Proofs

A **coordinate proof** involves placing geometric figures in a coordinate plane. When you use variables to represent the coordinates of a figure in a coordinate proof, the results are true for all figures of that type.

EXAMPLE 4 Writing a Coordinate Proof

Write a coordinate proof.

Given $P(0, k)$, $Q(h, 0)$, $R(-h, 0)$

Prove $\triangle PQR$ is isosceles.

By the Distance Formula, $PR = \sqrt{h^2 + k^2}$ and $PQ = \sqrt{h^2 + k^2}$. So, $\triangle PQR$ is isosceles by definition.

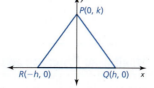

Monitoring Progress Help in English and Spanish at BigIdeasMath.com

4. Write a coordinate proof.

   Given $T(m, 0)$, $U(0, -m)$, $V(-m, 0)$, $W(0, m)$

   Prove $\triangle TUV \cong \triangle VWT$

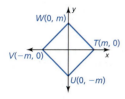

Concept Summary

Types of Proofs

Symmetric Property of Angle Congruence

   Given $\angle 1 \cong \angle 2$

   Prove $\angle 2 \cong \angle 1$

Two-Column Proof

STATEMENTS	REASONS
1. $\angle 1 \cong \angle 2$	1. Given
2. $m\angle 1 = m\angle 2$	2. Definition of congruent angles
3. $m\angle 2 = m\angle 1$	3. Symmetric Property of Equality
4. $\angle 2 \cong \angle 1$	4. Definition of congruent angles

Flowchart Proof

| $\angle 1 \cong \angle 2$ | $m\angle 1 = m\angle 2$ | $m\angle 2 = m\angle 1$ | $\angle 2 \cong \angle 1$ |
| Given | Definition of congruent angles | Symmetric Property of Equality | Definition of congruent angles |

Paragraph Proof

$\angle 1$ is congruent to $\angle 2$. By the definition of congruent angles, the measure of $\angle 1$ is equal to the measure of $\angle 2$. The measure of $\angle 2$ is equal to the measure of $\angle 1$ by the Symmetric Property of Equality. Then by the definition of congruent angles, $\angle 2$ is congruent to $\angle 1$.

Extra Example 4

Write a coordinate proof.

Given $D(0, 0)$, $E(a, b)$, $F(2a, 0)$

Prove $\triangle DEF$ is isosceles.

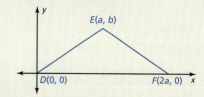

By the Distance Formula, $DE = \sqrt{a^2 + b^2}$ and $EF = \sqrt{a^2 + b^2}$. So, $\triangle DEF$ is isosceles by definition.

MONITORING PROGRESS ANSWER

4. By the Distance Formula, $TU = m\sqrt{2}$, $UV = m\sqrt{2}$, $VW = m\sqrt{2}$, and $WT = m\sqrt{2}$. So, $\overline{TU} \cong \overline{VW}$ and $\overline{UV} \cong \overline{WT}$. By the Reflexive Property of Congruence, $\overline{TV} \cong \overline{VT}$. So, $\triangle TUV \cong \triangle VWT$ by the SSS Congruence Theorem.

Laurie's Notes — Teacher Actions

- Example 4 serves to remind students of how to write a coordinate proof. It also involves a review of the Distance Formula.
- **Three-Minute Pause:** Pause 3 minutes for students to read through the *Concept Summary* and confer with partners or small groups. Have them discuss the styles of the proofs and the theorems and properties reviewed. What content are they feeling unsure of? Are all styles of proof feeling familiar at this point?

Closure

- **Writing Prompt:** Ask students to do a free write about how they are feeling about the styles of proofs and the geometric content reviewed.

Assignment Guide and Homework Check

ASSIGNMENT

Basic: 1, 2, 3–13 odd, 14, 15, 17–19
Average: 1, 2–12 even, 13–15, 17–19
Advanced: 1, 2, 6–12 even, 13–19

HOMEWORK CHECK

Basic: 5, 7, 9, 11
Average: 6, 8, 10, 12
Advanced: 6, 8, 10, 12

ANSWERS

1. Paragraph proofs present reasons as sentences. Flow chart proofs present reasons below the statements they justify.
2. statements; reasons
3. Given; $m\angle 1 = m\angle 2$
4. $\angle 1 \cong \angle 3$; Transitive Property of Congruence
5. Vertical Angles Congruence Theorem; $\angle 2 \cong \angle 3$
6. $\angle 1 \cong \angle 3$; Vertical Angles Congruence Theorem

6.1 Exercises

Dynamic Solutions available at BigIdeasMath.com

Vocabulary and Core Concept Check

1. **WRITING** Describe how paragraph proofs and flowchart proofs differ in how they present the reason for each step.
2. **COMPLETE THE SENTENCE** In a two-column proof, the _____ are on the left and the _____ are on the right.

Monitoring Progress and Modeling with Mathematics

PROOF In Exercises 3–6, copy and complete the proof. *(See Example 1.)*

3. Right Angles Congruence Theorem
   Given $\angle 1$ and $\angle 2$ are right angles.
   Prove $\angle 1 \cong \angle 2$

STATEMENTS	REASONS
1. $\angle 1$ and $\angle 2$ are right angles.	1. _____
2. $m\angle 1 = 90°$, $m\angle 2 = 90°$	2. Definition of right angle
3. _____	3. Transitive Property of Equality
4. $\angle 1 \cong \angle 2$	4. Definition of congruent angles

4. Statement about congruent angles
   Given $\angle 1 \cong \angle 4$
   Prove $\angle 2 \cong \angle 3$

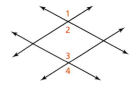

STATEMENTS	REASONS
1. $\angle 1 \cong \angle 4$	1. Given
2. $\angle 1 \cong \angle 2$, $\angle 3 \cong \angle 4$	2. Vertical Angles Congruence Theorem
3. _____	3. Transitive Property of Congruence
4. $\angle 2 \cong \angle 3$	4. _____

5. Statement about congruent angles
   Given $\angle 1 \cong \angle 3$
   Prove $\angle 2 \cong \angle 4$

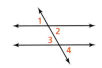

STATEMENTS	REASONS
1. $\angle 1 \cong \angle 3$	1. Given
2. $\angle 1 \cong \angle 2$, $\angle 3 \cong \angle 4$	2. _____
3. _____	3. Transitive Property of Congruence
4. $\angle 2 \cong \angle 4$	4. Transitive Property of Congruence

6. Alternate Exterior Angles Theorem
   Given $p \parallel q$
   Prove $\angle 1 \cong \angle 2$

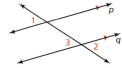

STATEMENTS	REASONS
1. $p \parallel q$	1. Given
2. _____	2. Corresponding Angles Theorem
3. $\angle 3 \cong \angle 2$	3. _____
4. $\angle 1 \cong \angle 2$	4. Transitive Property of Congruence

PROVING A THEOREM In Exercises 7 and 8, copy and complete the paragraph proof. Then write a two-column proof. *(See Example 2.)*

7. **Consecutive Interior Angles Theorem**
   Given $n \parallel p$
   Prove $\angle 1$ and $\angle 2$ are supplementary.

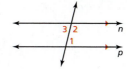

Line n and line p are parallel. By the _____, $\angle 1 \cong \angle 3$. By the definition of congruent angles, $m\angle 1 = m\angle 3$. By the Linear Pair Postulate, _____. By the definition of supplementary angles, _____ = 180°. By substitution, $m\angle 2 + m\angle 1 = 180°$. So, $\angle 1$ and $\angle 2$ are supplementary by the definition of supplementary angles.

8. **Perpendicular Transversal Theorem**
   Given $h \parallel k, j \perp h$
   Prove $j \perp k$

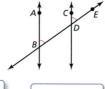

Line h and line k are parallel, and line j and line h are perpendicular. By the definition of perpendicular lines, $m\angle 2 =$ ____. By the _____, $\angle 2 \cong \angle 6$. By the definition of congruent angles, _____. By the Transitive Property of Equality, _____ = 90°. By the _____, $j \perp k$.

PROVING A THEOREM In Exercises 9 and 10, copy and complete the flowchart proof. Then write a two-column proof. *(See Example 3.)*

9. **Corresponding Angles Converse**
   Given $\angle ABD \cong \angle CDE$
   Prove $\overleftrightarrow{AB} \parallel \overleftrightarrow{CD}$

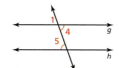

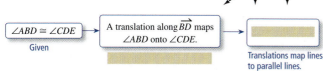

10. **Alternate Interior Angles Converse**
    Given $\angle 4 \cong \angle 5$
    Prove $g \parallel h$

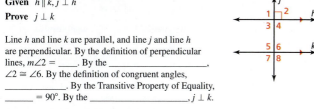

Section 6.1 Proving Geometric Relationships 341

9. Translations are rigid motions; $\overleftrightarrow{AB} \parallel \overleftrightarrow{CD}$

STATEMENTS	REASONS
1. $\angle ABD \cong \angle CDE$	1. Given
2. A translation along $\overrightarrow{BD}$ maps $\angle ABD$ onto $\angle CDE$	2. Translations are rigid motions
3. $\overleftrightarrow{AB} \parallel \overleftrightarrow{CD}$	3. Translations map lines to parallel lines

10. Vertical Angles Congruence Theorem; $\angle 1 \cong \angle 5$

STATEMENTS	REASONS
1. $\angle 4 \cong \angle 5$	1. Given
2. $\angle 1 \cong \angle 4$	2. Vertical Angles Congruence Theorem
3. $\angle 1 \cong \angle 5$	3. Transitive Property of Congruence
4. $g \parallel h$	4. Corresponding Angles Converse

ANSWERS

11–15. See Additional Answers.

16. *Sample answer:* Place the figure with a vertex on the origin and one or two sides on the axes.

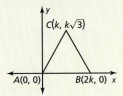

17. $x = 7$
18. $46°$
19. 142

Mini-Assessment

Prove the Exterior Angle Theorem.
Given $\triangle ABC$, exterior $\angle BCD$
Prove $m\angle A + m\angle B = m\angle BCD$

1. Write a two-column proof.

Statements (Reasons)

1. $\triangle ABC$, exterior $\angle BCD$ (Given)
2. $m\angle A + m\angle B + m\angle BCA = 180°$ (Triangle Sum Thm.)
3. $\angle BCA$ and $\angle BCD$ form a linear pair. (Def. of linear pair)
4. $m\angle BCA + m\angle BCD = 180°$ (Linear Pair Post.)
5. $m\angle A + m\angle B + m\angle BCA = m\angle BCA + m\angle BCD$ (Trans. Prop. =)
6. $m\angle A + m\angle B = m\angle BCD$ (Subtr. Prop. =)

2. Write a paragraph proof.

Given $\triangle ABC$, the sum of $m\angle A$, $m\angle B$, and $m\angle BCA$ is $180°$ by the Triangle Sum Theorem. Given exterior $\angle BCD$, $\angle BCA$ and $\angle BCD$ form a linear pair by the definition of linear pair. Using the Linear Pair Postulate, the sum of $m\angle BCA$ and $m\angle BCD$ is $180°$. Then, $m\angle A + m\angle B + m\angle BCA = m\angle BCA + m\angle BCD$ by the Transitive Property of Equality. So, $m\angle A + m\angle B = m\angle BCD$ by the Subtraction Property of Equality.

3. Write a flow chart proof.
See Additional Answers.

PROOF In Exercises 11 and 12, write a coordinate proof. (See Example 4.)

11. **Given** Coordinates of vertices of $\triangle OBC$ and $\triangle ODC$
Prove $\triangle OBC$ and $\triangle ODC$ are isosceles triangles.

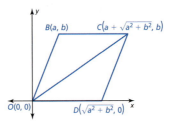

12. **Given** X is the midpoint of $\overline{YW}$.
Prove $\triangle XYZ \cong \triangle XWO$

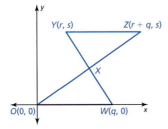

13. ERROR ANALYSIS In the diagram, $\overline{AE} \cong \overline{DE}$ and $\overline{AE} \cong \overline{BE}$. Describe and correct the error in the reasoning.

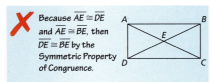

Maintaining Mathematical Proficiency *Reviewing what you learned in previous grades and lessons*

$\overrightarrow{BD}$ bisects $\angle ABC$ such that $m\angle ABD = (4x - 5)°$ and $m\angle DBC = (3x + 2)°$. *(Skills Review Handbook)*

17. Find the value of x.   **18.** Find $m\angle ABC$.

19. Point M is the midpoint of $\overline{AB}$. Find the length of $\overline{AB}$. *(Skills Review Handbook)*

$A \underset{6x+5}{\longmapsto} M \underset{7x-6}{\longmapsto} B$

342 Chapter 6   Relationships Within Triangles

---

14. HOW DO YOU SEE IT? Use the figure to write each proof.

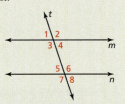

a. Alternate Exterior Angles Converse
   Given $\angle 2 \cong \angle 7$
   Prove $m \parallel n$

b. Consecutive Interior Angles Converse
   Given $\angle 3$ and $\angle 5$ are supplementary.
   Prove $m \parallel n$

15. MAKING AN ARGUMENT Your friend says that there is enough information to prove that $\angle ABC$ and $\angle DBE$ are vertical angles. Is your friend correct? Explain your reasoning.

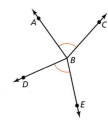

16. THOUGHT PROVOKING Describe convenient ways in which you can place a figure in a coordinate plane to complete a coordinate proof. Then place an equilateral triangle in a coordinate plane and label the vertices.

If students need help...	If students got it...
Resources by Chapter • Practice A and Practice B • Puzzle Time	**Resources by Chapter** • Enrichment and Extension • Cumulative Review
Student Journal • Practice	Start the *next* Section
Differentiating the Lesson Skills Review Handbook	

Laurie's Notes

Overview of Section 6.2

Introduction
- Segment bisectors, perpendicular bisectors, and angle bisectors were defined in an earlier course.
- In this lesson, properties of perpendicular bisectors and angle bisectors are presented.
- The Perpendicular Bisector Theorem and its converse and the Angle Bisector Theorem and its converse are stated. One of the four theorems is proven in the lesson, with the other three proofs left for the exercises.
- The lesson ends with a real-life application of the Angle Bisector Theorem and a problem integrating previously learned algebra skills.

Resources
- Tracing paper is helpful when the explorations are done by paper folding.

Teaching Strategy
- If you do not have access to dynamic geometry software, an alternate way to introduce this lesson would be through a paper-folding activity or a compass construction activity. It is powerful for students to explore and discover the properties of perpendicular bisectors and angle bisectors.
- Give directions similar to what is written in the explorations for students to paper fold (or construct) the perpendicular bisector (Exploration 1) and the angle bisector (Exploration 2). Distances can be measured using a compass or ruler. Although precision may be a problem, the focus is on the discovery. A deductive proof follows in the lesson.

Pacing Suggestion
- The explorations allow students to explore and discover two relationships that will be stated as theorems in the formal lesson. The constructions and investigation are not long and should be followed with the statement of the Perpendicular Bisector Theorem in the lesson.

Dynamic Teaching Tools
Dynamic Assessment & Progress Monitoring Tool
Lesson Planning Tool
Interactive Whiteboard Lesson Library
Dynamic Classroom with Dynamic Investigations

What Your Students Will Learn

- Use perpendicular bisectors to find measures in planar figures.
- Use angle bisectors to find measures and distance relationships in planar figures.
- Write equations for perpendicular bisectors in the coordinate plane.

Laurie's Notes

Exploration

Motivate
- You will need four volunteers and some yarn or string.
- Have two students (A and B) position themselves on opposite sides of the classroom holding a piece of string between them. They represent the endpoints of a segment.
- Have the other two students (C and D) position themselves anywhere in the classroom as long as they are the same distance from A and B. If you wish, you could have additional students (E, F, G, and H) join in and use the same condition—they must be the same distance from A and B.
- **?** "What do you notice about everyone except A and B?" *Students will describe them as being in a line.* "Do you notice anything special about the line?" *It is the perpendicular bisector of $\overline{AB}$.*
- Explain to students that they have just modeled one of the theorems they will work with in this lesson, the Converse of the Perpendicular Bisector Theorem.

Exploration 1
- This is a quick construction using dynamic geometric software. It is not necessary for the coordinate axes to be visible.
- Students should click and drag on point C, moving it to both sides of $\overline{AB}$.
- **Use Appropriate Tools Strategically:** Students can quickly explore other segments by clicking and dragging on point A or point B.
- **? Construct Viable Arguments and Critique the Reasoning of Others:** "What conjecture(s) can you make?" *$\overline{CA}$ and $\overline{CB}$ are always the same length.*
- Students may make other observations about what a perpendicular bisector is and may also mention that the base angles, $\angle A$ and $\angle B$, are congruent.

Exploration 2
- When constructing the perpendiculars from point D to the sides of the angle, lines will be drawn, not segments. You may want to show students how to construct the perpendiculars, draw the segments, and then hide the lines.
- Students should click and drag on point D, moving it closer and further away from the vertex, point A.
- **Use Appropriate Tools Strategically:** Students can quickly explore other angles by clicking and dragging on point B or point C.
- **? Construct Viable Arguments and Critique the Reasoning of Others:** "What conjecture(s) can you make?" *$\overline{DE}$ and $\overline{DF}$ are always the same length.*
- Students may make other observations about what an angle bisector is. They may also mention that the right triangles, $\triangle AED$ and $\triangle AFD$, are congruent.

Communicate Your Answer
- **Construct Viable Arguments and Critique the Reasoning of Others:** Students should state conjectures that are complete sentences and complete thoughts. Instead of saying, "They are congruent," students should say, "If point C is a point on the perpendicular bisector of $\overline{AB}$, then $\overline{CA}$ and $\overline{CB}$ are always congruent."

Connecting to Next Step
- The explorations are related to the theorems and converses presented in the lesson.

6.2 Perpendicular and Angle Bisectors

Essential Question What conjectures can you make about a point on the perpendicular bisector of a segment and a point on the bisector of an angle?

EXPLORATION 1 Points on a Perpendicular Bisector

Work with a partner. Use dynamic geometry software.

a. Draw any segment and label it $\overline{AB}$. Construct the perpendicular bisector of $\overline{AB}$.

b. Label a point C that is on the perpendicular bisector of $\overline{AB}$ but is not on $\overline{AB}$.

c. Draw $\overline{CA}$ and $\overline{CB}$ and find their lengths. Then move point C to other locations on the perpendicular bisector and note the lengths of $\overline{CA}$ and $\overline{CB}$.

d. Repeat parts (a)–(c) with other segments. Describe any relationship(s) you notice.

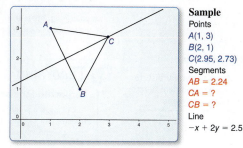

Sample
Points
$A(1, 3)$
$B(2, 1)$
$C(2.95, 2.73)$
Segments
$AB = 2.24$
$CA = ?$
$CB = ?$
Line
$-x + 2y = 2.5$

EXPLORATION 2 Points on an Angle Bisector

USING TOOLS STRATEGICALLY
To be proficient in math, you need to visualize the results of varying assumptions, explore consequences, and compare predictions with data.

Work with a partner. Use dynamic geometry software.

a. Draw two rays $\overrightarrow{AB}$ and $\overrightarrow{AC}$ to form $\angle BAC$. Construct the bisector of $\angle BAC$.

b. Label a point D on the bisector of $\angle BAC$.

c. Construct and find the lengths of the perpendicular segments from D to the sides of $\angle BAC$. Move point D along the angle bisector and note how the lengths change.

d. Repeat parts (a)–(c) with other angles. Describe any relationship(s) you notice.

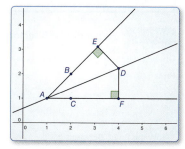

Sample
Points
$A(1, 1)$
$B(2, 2)$
$C(2, 1)$
$D(4, 2.24)$
Rays
$AB = -x + y = 0$
$AC = y = 1$
Line
$-0.38x + 0.92y = 0.54$

Communicate Your Answer

3. What conjectures can you make about a point on the perpendicular bisector of a segment and a point on the bisector of an angle?

4. In Exploration 2, what is the distance from point D to $\overrightarrow{AB}$ when the distance from D to $\overrightarrow{AC}$ is 5 units? Justify your answer.

Section 6.2   Perpendicular and Angle Bisectors   343

Dynamic Teaching Tools
Dynamic Assessment & Progress Monitoring Tool
Lesson Planning Tool
Interactive Whiteboard Lesson Library
Dynamic Classroom with Dynamic Investigations

ANSWERS

1. a. Check students' work.
   b. Check students' work.
   c. Check students' work (for sample in text, $CA \approx 1.97$, $CB \approx 1.97$); For all locations of C, $\overline{CA}$ and $\overline{CB}$ have the same measure.
   d. Every point on the perpendicular bisector of a segment is equidistant from the endpoints of the segment.

2. a. Check students' work.
   b. Check students' work.
   c. Check students' work (for sample in text, $DE \approx 1.24$, $DF \approx 1.24$); For all locations of D on the angle bisector, $\overline{ED}$ and $\overline{FD}$ have the same measure.
   d. Every point on an angle bisector is equidistant from both sides of the angle.

3. If a point is on the perpendicular bisector of a segment, then it is equidistant from the endpoints of the segment. Every point on the bisector of an angle is equidistant from the sides of the angle.

4. 5 units; Point D is on the angle bisector, so it is equidistant from either side of the angle.

> **English Language Learners**
>
> **Vocabulary**
> For English learners, relate the term *equidistant* with the phrase *equal distances*. Likewise, equilateral figures have sides with equal lengths, and equiangular figures have interior angles with equal measures.

6.2 Lesson

Core Vocabulary

equidistant, *p. 344*
Previous
perpendicular bisector
angle bisector

What You Will Learn

▸ Use perpendicular bisectors to find measures.
▸ Use angle bisectors to find measures and distance relationships.
▸ Write equations for perpendicular bisectors.

Using Perpendicular Bisectors

Previously, you learned that a *perpendicular bisector* of a line segment is the line that is perpendicular to the segment at its midpoint.

A point is **equidistant** from two figures when the point is the *same distance* from each figure.

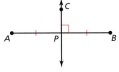

$\overleftrightarrow{CP}$ is a ⊥ bisector of $\overline{AB}$.

> **STUDY TIP**
> A perpendicular bisector can be a segment, a ray, a line, or a plane.

Theorems

Perpendicular Bisector Theorem

In a plane, if a point lies on the perpendicular bisector of a segment, then it is equidistant from the endpoints of the segment.

If $\overleftrightarrow{CP}$ is the ⊥ bisector of $\overline{AB}$, then $CA = CB$.

Proof p. 344

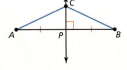

Converse of the Perpendicular Bisector Theorem

In a plane, if a point is equidistant from the endpoints of a segment, then it lies on the perpendicular bisector of the segment.

If $DA = DB$, then point D lies on the ⊥ bisector of $\overline{AB}$.

Proof Ex. 32, p. 350

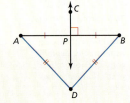

PROOF **Perpendicular Bisector Theorem**

Given $\overleftrightarrow{CP}$ is the perpendicular bisector of $\overline{AB}$.

Prove $CA = CB$

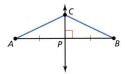

Paragraph Proof Because $\overleftrightarrow{CP}$ is the perpendicular bisector of $\overline{AB}$, $\overleftrightarrow{CP}$ is perpendicular to $\overline{AB}$ and point P is the midpoint of $\overline{AB}$. By the definition of midpoint, $AP = BP$, and by the definition of perpendicular lines, $m\angle CPA = m\angle CPB = 90°$. Then by the definition of segment congruence, $\overline{AP} \cong \overline{BP}$, and by the definition of angle congruence, $\angle CPA \cong \angle CPB$. By the Reflexive Property of Congruence, $\overline{CP} \cong \overline{CP}$. So, $\triangle CPA \cong \triangle CPB$ by the SAS Congruence Theorem, and $\overline{CA} \cong \overline{CB}$ because corresponding parts of congruent triangles are congruent. So, $CA = CB$ by the definition of segment congruence.

344 Chapter 6 Relationships Within Triangles

Laurie's Notes | Teacher Actions

- Ask what the word *equidistant* means. State the Perpendicular Bisector Theorem, which students discovered in the exploration.
- **?** "Is the converse of a statement *always, sometimes,* or *never* true?" sometimes true
- **Turn and Talk:** "What would the converse of the Perpendicular Bisector Theorem say, and is it true?" Use *Popsicle Sticks* to solicit students' thoughts about the converse. Write the Converse of the Perpendicular Bisector Theorem.
- Have partners discuss and outline a proof for the Perpendicular Bisector Theorem. Compare with the paragraph proof.

EXAMPLE 1 Using the Perpendicular Bisector Theorems

Find each measure.

a. *RS*

From the figure, $\overleftrightarrow{SQ}$ is the perpendicular bisector of $\overline{PR}$. By the Perpendicular Bisector Theorem, $PS = RS$.

▶ So, $RS = PS = 6.8$.

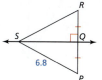

b. *EG*

Because $EH = GH$ and $\overleftrightarrow{HF} \perp \overline{EG}$, $\overleftrightarrow{HF}$ is the perpendicular bisector of $\overline{EG}$ by the Converse of the Perpendicular Bisector Theorem. By the definition of segment bisector, $EG = 2GF$.

▶ So, $EG = 2(9.5) = 19$.

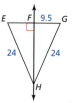

c. *AD*

From the figure, $\overleftrightarrow{BD}$ is the perpendicular bisector of $\overline{AC}$.

$AD = CD$ Perpendicular Bisector Theorem
$5x = 3x + 14$ Substitute.
$x = 7$ Solve for x.

▶ So, $AD = 5x = 5(7) = 35$.

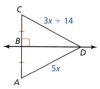

EXAMPLE 2 Solving a Real-Life Problem

Is there enough information in the diagram to conclude that point N lies on the perpendicular bisector of $\overline{KM}$?

SOLUTION

It is given that $\overline{KL} \cong \overline{ML}$. So, $\overline{LN}$ is a segment bisector of $\overline{KM}$. You do not know whether $\overline{LN}$ is perpendicular to $\overline{KM}$ because it is not indicated in the diagram.

▶ So, you cannot conclude that point N lies on the perpendicular bisector of $\overline{KM}$.

Monitoring Progress 🔊 Help in English and Spanish at BigIdeasMath.com

Use the diagram and the given information to find the indicated measure.

1. $\overleftrightarrow{ZX}$ is the perpendicular bisector of $\overline{WY}$, and $YZ = 13.75$. Find WZ.

2. $\overleftrightarrow{ZX}$ is the perpendicular bisector of $\overline{WY}$, $WZ = 4n - 13$, and $YZ = n + 17$. Find YZ.

3. Find WX when $WZ = 20.5$, $WY = 14.8$, and $YZ = 20.5$.

Differentiated Instruction

Kinesthetic

Discuss the difference between a *perpendicular bisector* and an *angle bisector*. Have students draw a triangle in which the perpendicular bisector of each side is also an angle bisector. Have students draw a triangle in which no perpendicular bisector is also an angle bisector.

Extra Example 1

Find each measure.

a. *CD*

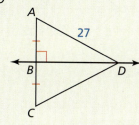

$CD = 27$

b. *PR*

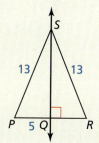

$PR = 10$

c. *GH*

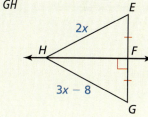

$GH = 16$

Extra Example 2

Is there enough information given in the diagram in Example 2 to conclude that point L lies on the perpendicular bisector of $\overline{KM}$? Yes. Because $LK = LM$, point L must lie on the perpendicular bisector of $\overline{KM}$ by the Converse of the Perpendicular Bisector Theorem.

MONITORING PROGRESS ANSWERS

1–3. See Additional Answers.

Laurie's Notes   **Teacher Actions**

- Example 1 problems are applications of the two previous theorems. Pose the problems and have students work independently to solve.
- ❓ "In Example 2, what information is needed to know that $\overline{LN}$ is perpendicular to $\overline{KM}$?" Either the perpendicular symbol is present or $\overline{NK} \cong \overline{NM}$.
- **COMMON ERROR** Example 2 highlights the common error of assuming that because a segment looks perpendicular, it must be.
- **Think-Pair-Share:** Have students answer Questions 1–3, and then share and discuss as a class.

Extra Example 3

Find each measure.

a. $m\angle ABC$

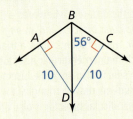

$m\angle ABC = 112°$

b. JM

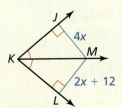

$JM = 24$

MONITORING PROGRESS ANSWERS

4. 6.9
5. 19
6. 78°

Using Angle Bisectors

Previously, you learned that an *angle bisector* is a ray that divides an angle into two congruent adjacent angles. You also know that the *distance from a point to a line* is the length of the perpendicular segment from the point to the line. So, in the figure, $\overrightarrow{AD}$ is the bisector of $\angle BAC$, and the distance from point D to $\overrightarrow{AB}$ is DB, where $\overline{DB} \perp \overrightarrow{AB}$.

Theorems

Angle Bisector Theorem

If a point lies on the bisector of an angle, then it is equidistant from the two sides of the angle.

If $\overrightarrow{AD}$ bisects $\angle BAC$ and $\overline{DB} \perp \overrightarrow{AB}$ and $\overline{DC} \perp \overrightarrow{AC}$, then $DB = DC$.

Proof Ex. 33(a), p. 350

Converse of the Angle Bisector Theorem

If a point is in the interior of an angle and is equidistant from the two sides of the angle, then it lies on the bisector of the angle.

If $\overline{DB} \perp \overrightarrow{AB}$ and $\overline{DC} \perp \overrightarrow{AC}$ and $DB = DC$, then $\overrightarrow{AD}$ bisects $\angle BAC$.

Proof Ex. 33(b), p. 350

EXAMPLE 3 Using the Angle Bisector Theorems

Find each measure.

a. $m\angle GFJ$

Because $\overline{JG} \perp \overrightarrow{FG}$ and $\overline{JH} \perp \overrightarrow{FH}$ and $JG = JH = 7$, $\overrightarrow{FJ}$ bisects $\angle GFH$ by the Converse of the Angle Bisector Theorem.

▶ So, $m\angle GFJ = m\angle HFJ = 42°$.

b. RS

$PS = RS$	Angle Bisector Theorem
$5x = 6x - 5$	Substitute.
$5 = x$	Solve for x.

▶ So, $RS = 6x - 5 = 6(5) - 5 = 25$.

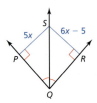

Monitoring Progress Help in English and Spanish at *BigIdeasMath.com*

Use the diagram and the given information to find the indicated measure.

4. $\overrightarrow{BD}$ bisects $\angle ABC$, and $DC = 6.9$. Find DA.
5. $\overrightarrow{BD}$ bisects $\angle ABC$, $AD = 3z + 7$, and $CD = 2z + 11$. Find CD.
6. Find $m\angle ABC$ when $AD = 3.2$, $CD = 3.2$, and $m\angle DBC = 39°$.

346 Chapter 6 Relationships Within Triangles

Laurie's Notes — Teacher Actions

- Ask how the altitude of an airplane is measured. It is a perpendicular distance from the plane to the ground. When students find the distance from a point on the angle bisector to the side of the angle, remind them that it needs to be a perpendicular segment.
- State the Angle Bisector Theorem and its converse. Take time to solicit ideas about an outline for the proof of each theorem.
- Example 3 problems are applications of the two previous theorems. Pose the problems and have students work independently to solve.

> **EXAMPLE 4** Solving a Real-Life Problem

A soccer goalie's position relative to the ball and goalposts forms congruent angles, as shown. Will the goalie have to move farther to block a shot toward the right goalpost R or the left goalpost L?

SOLUTION

The congruent angles tell you that the goalie is on the bisector of $\angle LBR$. By the Angle Bisector Theorem, the goalie is equidistant from $\overrightarrow{BR}$ and $\overrightarrow{BL}$.

▶ So, the goalie must move the same distance to block either shot.

Writing Equations for Perpendicular Bisectors

> **EXAMPLE 5** Writing an Equation for a Bisector

Write an equation of the perpendicular bisector of the segment with endpoints $P(-2, 3)$ and $Q(4, 1)$.

SOLUTION

Step 1 Graph $\overline{PQ}$. By definition, the perpendicular bisector of $\overline{PQ}$ is perpendicular to $\overline{PQ}$ at its midpoint.

Step 2 Find the midpoint M of $\overline{PQ}$.

$$M\left(\frac{-2+4}{2}, \frac{3+1}{2}\right) = M\left(\frac{2}{2}, \frac{4}{2}\right) = M(1, 2)$$

Step 3 Find the slope of the perpendicular bisector.

$$\text{slope of } \overline{PQ} = \frac{1-3}{4-(-2)} = \frac{-2}{6} = -\frac{1}{3}$$

Because the slopes of perpendicular lines are negative reciprocals, the slope of the perpendicular bisector is 3.

Step 4 Write an equation. The bisector of $\overline{PQ}$ has slope 3 and passes through $(1, 2)$.

$y = mx + b$ Use slope-intercept form.
$2 = 3(1) + b$ Substitute for m, x, and y.
$-1 = b$ Solve for b.

▶ So, an equation of the perpendicular bisector of $\overline{PQ}$ is $y = 3x - 1$.

Monitoring Progress 🔊 Help in English and Spanish at *BigIdeasMath.com*

7. Do you have enough information to conclude that $\overrightarrow{QS}$ bisects $\angle PQR$? Explain.

8. Write an equation of the perpendicular bisector of the segment with endpoints $(-1, -5)$ and $(3, -1)$.

Section 6.2 Perpendicular and Angle Bisectors 347

Extra Example 4

A parachutist lands in a triangular field at point P, and then walks to a road. Will he have to walk further to Wells Road or to Turner Road?

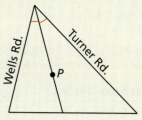

Neither. The distances to both roads are the same.

Extra Example 5

Write an equation of the perpendicular bisector of the segment with endpoints $D(5, -1)$ and $E(-11, 3)$. $y = 4x + 13$

MONITORING PROGRESS ANSWERS

7. no; In order to use the Converse of the Angle Bisector Theorem, $\overrightarrow{PS}$ would have to be perpendicular to $\overrightarrow{QP}$, and $\overrightarrow{RS}$ would have to be perpendicular to $\overrightarrow{QR}$.

8. $y = -x - 2$

Laurie's Notes Teacher Actions

- Ask whether any student plays the goalie position in any sport. If so, he or she can explain why goalies position themselves on the angle bisector formed by the opponent and the goalposts. This can also be modeled using dynamic geometry software.
- Pose Example 5. Use whiteboards and have partners work to write the equation of the perpendicular bisector.

Closure

- **Writing Prompt:** The *Big Ideas* presented in today's lesson …

Assignment Guide and Homework Check

ASSIGNMENT

Basic: 1, 2, 3–25 odd, 31, 34, 39–44

Average: 1, 2, 6–30 even, 31, 32, 34, 39–44

Advanced: 1, 2, 6–26 even, 29–32, 34–44

HOMEWORK CHECK

Basic: 3, 7, 13, 19, 25

Average: 6, 14, 22, 26, 30

Advanced: 6, 14, 22, 26, 30

ANSWERS

1. bisector

2. Is point *B* collinear with *X* and *Z*?; no; yes

3. 4.6; Because $GK = KJ$ and $\overleftrightarrow{HK} \perp \overleftrightarrow{GJ}$, point *H* is on the perpendicular bisector of $\overline{GJ}$. So, by the Perpendicular Bisector Theorem, $GH = HJ = 4.6$.

4. 1.3; Because point *T* is equidistant from *Q* and *S*, point *T* is on the perpendicular bisector of $\overline{QS}$ by the Converse of the Perpendicular Bisector Theorem. So, by definition of segment bisector, $QR = RS = 1.3$.

5. 15; Because $\overleftrightarrow{DB} \perp \overleftrightarrow{AC}$ and point *D* is equidistant from *A* and *C*, point *D* is on the perpendicular bisector of $\overline{AC}$ by the Converse of the Perpendicular Bisector Theorem. By definition of segment bisector, $AB = BC$. So, $5x = 4x + 3$, and the solution is $x = 3$. So, $AB = 5x = 5(3) = 15$.

6. 55; Because $\overline{VD} \cong \overline{WD}$ and $\overleftrightarrow{UX} \perp \overleftrightarrow{VW}$, point *U* is on the perpendicular bisector of $\overline{VW}$. So, by the Perpendicular Bisector Theorem, $VU = WU$. So, $9x + 1 = 7x + 13$, and the solution is $x = 6$, which means that $UW = 7x + 13 = 7(6) + 13 = 55$.

7. yes; Because point *N* is equidistant from *L* and *M*, point *N* is on the perpendicular bisector of $\overline{LM}$ by the Converse of the Perpendicular Bisector Theorem. Because only one line can be perpendicular to $\overline{LM}$ at point *K*, $\overline{NK}$ must be the perpendicular bisector of $\overline{LM}$, and *P* is on $\overrightarrow{NK}$.

8. no; You would need to know that either $LN = MN$ or $LP = MP$.

6.2 Exercises

Dynamic Solutions available at BigIdeasMath.com

Vocabulary and Core Concept Check

1. **COMPLETE THE SENTENCE** Point *C* is in the interior of ∠*DEF*. If ∠*DEC* and ∠*CEF* are congruent, then $\overrightarrow{EC}$ is the _____ of ∠*DEF*.

2. **DIFFERENT WORDS, SAME QUESTION** Which is different? Find "both" answers.

   Is point *B* the same distance from both *X* and *Z*?

   Is point *B* equidistant from *X* and *Z*?

   Is point *B* collinear with *X* and *Z*?

   Is point *B* on the perpendicular bisector of $\overline{XZ}$?

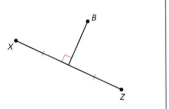

Monitoring Progress and Modeling with Mathematics

In Exercises 3–6, find the indicated measure. Explain your reasoning. *(See Example 1.)*

3. *GH*

4. *QR*

5. *AB*

6. *UW*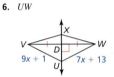

In Exercises 7–10, tell whether the information in the diagram allows you to conclude that point *P* lies on the perpendicular bisector of $\overline{LM}$. Explain your reasoning. *(See Example 2.)*

7.

8.

9.

10.

In Exercises 11–14, find the indicated measure. Explain your reasoning. *(See Example 3.)*

11. $m\angle ABD$

12. *PS*

13. $m\angle KJL$

14. *FG*

348 Chapter 6 Relationships Within Triangles

---

9. no; You would need to know that $\overleftrightarrow{PN} \perp \overleftrightarrow{ML}$.

10. yes; Because point *P* is equidistant from *L* and *M*, point *P* is on the perpendicular bisector of $\overline{LM}$ by the Converse of the Perpendicular Bisector Theorem. Also, $\overline{LN} \cong \overline{MN}$, so $\overleftrightarrow{PN}$ is a bisector of $\overline{LM}$. Because *P* can only be on one of the bisectors, $\overleftrightarrow{PN}$ is the perpendicular bisector of $\overline{LM}$.

11. 20°; Because *D* is equidistant from $\overrightarrow{BC}$ and $\overrightarrow{BA}$, $\overrightarrow{BD}$ bisects ∠*ABC* by the Converse of the Angle Bisector Theorem. So, $m\angle ABD = m\angle CBD = 20°$.

12. 12; $\overrightarrow{QS}$ is an angle bisector of ∠*PQR*, $\overline{PS} \perp \overline{QP}$, and $\overline{SR} \perp \overline{QR}$. So, by the Angle Bisector Theorem, $PS = RS = 12$.

13. 28°; Because *L* is equidistant from $\overrightarrow{JK}$ and $\overrightarrow{JM}$, $\overrightarrow{JL}$ bisects ∠*KJM* by the Angle Bisector Theorem. This means that $7x = 3x + 16$, and the solution is $x = 4$. So, $m\angle KJL = 7x = 7(4) = 28°$.

14. See Additional Answers.

In Exercises 15 and 16, tell whether the information in the diagram allows you to conclude that $\overrightarrow{EH}$ bisects $\angle FEG$. Explain your reasoning. (See Example 4.)

15.   16.

In Exercises 17 and 18, tell whether the information in the diagram allows you to conclude that $DB = DC$. Explain your reasoning.

17.   18.

In Exercises 19–22, write an equation of the perpendicular bisector of the segment with the given endpoints. (See Example 5.)

19. $M(1, 5), N(7, -1)$ 20. $Q(-2, 0), R(6, 12)$

21. $U(-3, 4), V(9, 8)$ 22. $Y(10, -7), Z(-4, 1)$

ERROR ANALYSIS In Exercises 23 and 24, describe and correct the error in the student's reasoning.

23.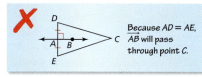

24.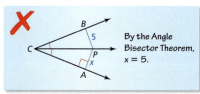

25. **MODELING MATHEMATICS** In the photo, the road is perpendicular to the support beam and $\overline{AB} \cong \overline{CB}$. Which theorem allows you to conclude that $\overline{AD} \cong \overline{CD}$?

26. **MODELING WITH MATHEMATICS** The diagram shows the position of the goalie and the puck during a hockey game. The goalie is at point G, and the puck is at point P.

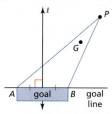

a. What should be the relationship between $\overrightarrow{PG}$ and $\angle APB$ to give the goalie equal distances to travel on each side of $\overrightarrow{PG}$?

b. How does $m\angle APB$ change as the puck gets closer to the goal? Does this change make it easier or more difficult for the goalie to defend the goal? Explain your reasoning.

27. **CONSTRUCTION** Use a compass and straightedge to construct a copy of $\overline{XY}$. Construct a perpendicular bisector and plot a point Z on the bisector so that the distance between point Z and $\overline{XY}$ is 3 centimeters. Measure $\overline{XZ}$ and $\overline{YZ}$. Which theorem does this construction demonstrate?

28. **WRITING** Explain how the Converse of the Perpendicular Bisector Theorem is related to the construction of a perpendicular bisector.

29. **REASONING** What is the value of x in the diagram?

Ⓐ 13
Ⓑ 18
Ⓒ 33
Ⓓ not enough information

 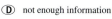

30. **REASONING** Which point lies on the perpendicular bisector of the segment with endpoints $M(7, 5)$ and $N(-1, 5)$?

Ⓐ $(2, 0)$ Ⓑ $(3, 9)$
Ⓒ $(4, 1)$ Ⓓ $(1, 3)$

31. **MAKING AN ARGUMENT** Your friend says it is impossible for an angle bisector of a triangle to be the same line as the perpendicular bisector of the opposite side. Is your friend correct? Explain your reasoning.

ANSWERS

32.

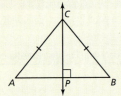

Given isosceles △ACB, construct $\overleftrightarrow{CP}$ such that point P is on $\overline{AB}$ and $\overleftrightarrow{CP} \perp \overline{AB}$. So, ∠CPB and ∠CPA are right angles by the definition of perpendicular lines, and △CPB and △CPA are right triangles. Also, because $\overline{AC} \cong \overline{BC}$ and $\overline{CP} \cong \overline{CP}$ by the Reflexive Property of Congruence, △CPB ≅ △CPA by the HL Congruence Theorem. So, $\overline{AP} \cong \overline{BP}$ because corresponding parts of congruent triangles are congruent, which means that point P is the midpoint of $\overline{AB}$, and $\overleftrightarrow{CP}$ is the perpendicular bisector of $\overline{AB}$.

33–44. See Additional Answers.

Mini-Assessment

1. Write an equation of the perpendicular bisector of the segment with endpoints A(–1, –1) and B(5, 3). $y = -1.5x + 4$

2. Find AD.

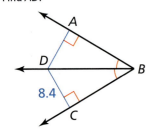

AD = 8.4

3. Find YZ.

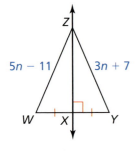

YZ = 34

350   Chapter 6

---

32. PROVING A THEOREM Prove the Converse of the Perpendicular Bisector Theorem. (*Hint:* Construct a line through point C perpendicular to $\overline{AB}$ at point P.)

Given CA = CB

Prove Point C lies on the perpendicular bisector of $\overline{AB}$.

33. PROVING A THEOREM Use a congruence theorem to prove each theorem.

a. Angle Bisector Theorem
b. Converse of the Angle Bisector Theorem

34. HOW DO YOU SEE IT? The figure shows a map of a city. The city is arranged so each block north to south is the same length and each block east to west is the same length.

a. Which school is approximately equidistant from both hospitals? Explain your reasoning.
b. Is the museum approximately equidistant from Wilson School and Roosevelt School? Explain your reasoning.

Maintaining Mathematical Proficiency Reviewing what you learned in previous grades and lessons

Classify the triangle by its sides. (*Skills Review Handbook*)

39.   40.   41.

Classify the triangle by its angles. (*Skills Review Handbook*)

42.   43.   44.

350   Chapter 6   Relationships Within Triangles

---

35. MATHEMATICAL CONNECTIONS Write an equation whose graph consists of all the points in the given quadrants that are equidistant from the x- and y-axes.

a. I and III   b. II and IV   c. I and II

36. THOUGHT PROVOKING The postulates and theorems you have learned represent Euclidean geometry. In spherical geometry, all points are on the surface of a sphere. A line is a circle on the sphere whose diameter is equal to the diameter of the sphere. In spherical geometry, is it possible for two lines to be perpendicular but not bisect each other? Explain your reasoning.

37. PROOF Use the information in the diagram to prove that $\overline{AB} \cong \overline{CB}$ if and only if points D, E, and B are collinear.

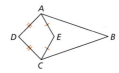

38. PROOF Prove the statements in parts (a)–(c).

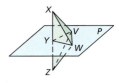

Given Plane P is a perpendicular bisector of $\overline{XZ}$ at point Y.

Prove   a. $\overline{XW} \cong \overline{ZW}$
b. $\overline{XV} \cong \overline{ZV}$
c. ∠VXW ≅ ∠VZW

If students need help...	If students got it...
Resources by Chapter • Practice A and Practice B • Puzzle Time	**Resources by Chapter** • Enrichment and Extension • Cumulative Review
Student Journal • Practice	Start the *next* Section
Differentiating the Lesson Skills Review Handbook	

Laurie's Notes

Overview of Section 6.3

Introduction
- In the last lesson, perpendicular bisectors and angle bisectors were reviewed and their properties explored.
- In this lesson, the three perpendicular bisectors in a triangle are constructed and the circumcenter, the *point of concurrency*, is found. The *circumcenter* is the point that is equidistant from all three vertices of the triangle.
- In the second part of the lesson, the three angle bisectors in a triangle are constructed and the incenter, the *point of concurrency*, is found. The *incenter* is the point that is equidistant from all three sides of the triangle.

Resources
- Tracing paper is helpful when the explorations are done by paper folding.

Teaching Strategy
- For either of the constructions in this lesson (circumcenter and incenter), consider using a local map that your students would be familiar with. Take a screen shot of the image and import it to your dynamic geometry software.
- In the example shown, I currently live at point *A*, I lived for many years at point *B*, and I work at point *C*.
- Think of the type of applications for the circumcenter and incenter, and select an image that will make sense.

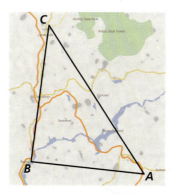

Pacing Suggestion
- The explorations allow students to explore and discover two relationships that will be stated as theorems in the formal lesson. The constructions and investigations are not long and should be followed with the statement of the Circumcenter Theorem in the lesson. You may choose to omit the two compass constructions in the lesson.

Dynamic Teaching Tools
Dynamic Assessment & Progress Monitoring Tool
Lesson Planning Tool
Interactive Whiteboard Lesson Library
Dynamic Classroom with Dynamic Investigations

What Your Students Will Learn

- Use and find the circumcenters of triangles by finding the intersection of the perpendicular bisectors of the sides of the triangle.
- Use and find the incenters of triangles by finding the intersection of the angle bisectors of the angles of the triangle.

Laurie's Notes

Exploration

Motivate
- There are three high schools located in different parts of the city. You want to open an after-school tutoring service for students from all of the schools. Where would you locate your service so that any student has the same distance to drive?
- Let students discuss the scenario.
- Explain to students that after this lesson they should be able to give an expert opinion about the location of the tutoring service.

Exploration 1
- This is a quick construction using dynamic geometric software. It is not necessary for the coordinate axes to be visible.
- Students observe that the three perpendicular bisectors appear to intersect at one point.
- **Turn and Talk:** "How could you show that the lines intersect at just one point and not at points that are simply very close together?" One way is to label the points of intersection of each pair of lines D, E, and F. If coordinates appear, students will note that D, E, and F are the same point. If the construction was not done in a coordinate plane, measure $\overline{DE}$, $\overline{DF}$, and $\overline{FE}$. All segments will have a length of 0, meaning D, E, and F are the same point.
- **? Construct Viable Arguments and Critique the Reasoning of Others:** Draw the circle. "What conjecture(s) can you make?" The circle goes through the three vertices of the triangle.
- **Use Appropriate Tools Strategically:** Students can quickly explore other triangles by clicking and dragging on any vertex of the triangle.
- Students may make other observations, for instance that point D is the same distance from all three vertices because D is the center of the circle.

Exploration 2
- This also is a quick construction.
- Students observe that the three angle bisectors appear to intersect at one point. Have students confirm that there is but one point and not three points that are simply close together.
- **? Construct Viable Arguments and Critique the Reasoning of Others:** Draw the circle. "What conjecture(s) can you make?" The circle touches each side of the triangle.
- **Use Appropriate Tools Strategically:** Students can quickly explore other triangles by clicking and dragging on any vertex of the triangle.
- Students may make other observations—for instance, that the circle stays inside the sides of the triangle.

Communicate Your Answer
- Students should state that the three perpendicular bisectors of the sides of a triangle intersect at one point, and the three angle bisectors of a triangle intersect at one point.

Connecting to Next Step
- The explorations are related to the theorems presented in the lesson, helping students make sense of the theorems before proving them to be true.

6.3 Bisectors of Triangles

Essential Question What conjectures can you make about the perpendicular bisectors and the angle bisectors of a triangle?

> **EXPLORATION 1** **Properties of the Perpendicular Bisectors of a Triangle**

Work with a partner. Use dynamic geometry software. Draw any △ABC.

a. Construct the perpendicular bisectors of all three sides of △ABC. Then drag the vertices to change △ABC. What do you notice about the perpendicular bisectors?

b. Label a point D at the intersection of the perpendicular bisectors.

c. Draw the circle with center D through vertex A of △ABC. Then drag the vertices to change △ABC. What do you notice?

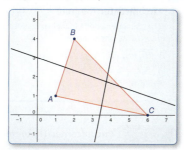

Sample
Points
$A(1, 1)$
$B(2, 4)$
$C(6, 0)$
Segments
$BC = 5.66$
$AC = 5.10$
$AB = 3.16$
Lines
$x + 3y = 9$
$-5x + y = -17$

LOOKING FOR STRUCTURE

To be proficient in math, you need to see complicated things as single objects or as being composed of several objects.

> **EXPLORATION 2** **Properties of the Angle Bisectors of a Triangle**

Work with a partner. Use dynamic geometry software. Draw any △ABC.

a. Construct the angle bisectors of all three angles of △ABC. Then drag the vertices to change △ABC. What do you notice about the angle bisectors?

b. Label a point D at the intersection of the angle bisectors.

c. Find the distance between D and $\overline{AB}$. Draw the circle with center D and this distance as a radius. Then drag the vertices to change △ABC. What do you notice?

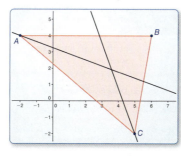

Sample
Points
$A(-2, 4)$
$B(6, 4)$
$C(5, -2)$
Segments
$BC = 6.08$
$AC = 9.22$
$AB = 8$
Lines
$0.35x + 0.94y = 3.06$
$-0.94x - 0.34y = -4.02$

Communicate Your Answer

3. What conjectures can you make about the perpendicular bisectors and the angle bisectors of a triangle?

Dynamic Teaching Tools

Dynamic Assessment & Progress Monitoring Tool
Lesson Planning Tool
Interactive Whiteboard Lesson Library
Dynamic Classroom with Dynamic Investigations

ANSWERS

1. **a–c.** *Sample answer:*

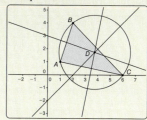

a. The perpendicular bisectors all intersect at one point.

c. The circle passes through all three vertices of △ABC.

2. **a–c.** *Sample answer:*

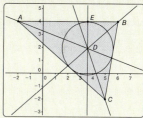

a. The angle bisectors all intersect at one point.

c. distance ≈ 2.06; The circle passes through exactly one point of each side of △ABC.

3. The perpendicular bisectors of the sides of a triangle meet at a point that is the same distance from each vertex of the triangle. The angle bisectors of a triangle meet at a point that is the same distance from each side of the triangle.

English Language Learners

Notebook Development

Have students record the Circumcenter Theorem and the Incenter Theorem from this section in their notebooks. For each theorem, include a sketch of the point of concurrency and an example of how to use the theorem to find unknown measures.

6.3 Lesson

Core Vocabulary

concurrent, p. 352
point of concurrency, p. 352
circumcenter, p. 352
incenter, p. 355

Previous
perpendicular bisector
angle bisector

What You Will Learn

- Use and find the circumcenter of a triangle.
- Use and find the incenter of a triangle.

Using the Circumcenter of a Triangle

When three or more lines, rays, or segments intersect in the same point, they are called **concurrent** lines, rays, or segments. The point of intersection of the lines, rays, or segments is called the **point of concurrency**.

In a triangle, the three perpendicular bisectors are concurrent. The point of concurrency is the **circumcenter** of the triangle.

Theorems

Circumcenter Theorem

The circumcenter of a triangle is equidistant from the vertices of the triangle.

If $\overline{PD}, \overline{PE},$ and $\overline{PF}$ are perpendicular bisectors, then $PA = PB = PC$.

Proof p. 352

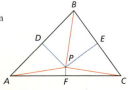

PROOF — Circumcenter Theorem

Given △ABC; the perpendicular bisectors of $\overline{AB}, \overline{BC},$ and $\overline{AC}$

Prove The perpendicular bisectors intersect in a point; that point is equidistant from A, B, and C.

Plan for Proof Show that P, the point of intersection of the perpendicular bisectors of $\overline{AB}$ and $\overline{BC}$, also lies on the perpendicular bisector of $\overline{AC}$. Then show that point P is equidistant from the vertices of the triangle.

STUDY TIP
Use diagrams like the one below to help visualize your proof.

Plan in Action

STATEMENTS	REASONS
1. △ABC; the perpendicular bisectors of $\overline{AB}, \overline{BC},$ and $\overline{AC}$	1. Given
2. The perpendicular bisectors of $\overline{AB}$ and $\overline{BC}$ intersect at some point P.	2. Because the sides of a triangle cannot be parallel, these perpendicular bisectors must intersect in some point. Call it P.
3. Draw $\overline{PA}, \overline{PB},$ and $\overline{PC}$.	3. Two Point Postulate
4. $PA = PB, PB = PC$	4. Perpendicular Bisector Theorem
5. $PA = PC$	5. Transitive Property of Equality
6. P is on the perpendicular bisector of $\overline{AC}$.	6. Converse of the Perpendicular Bisector Theorem
7. $PA = PB = PC$. So, P is equidistant from the vertices of the triangle.	7. From the results of Steps 4 and 5 and the definition of equidistant

352 Chapter 6 Relationships Within Triangles

Laurie's Notes — Teacher Actions

- Ask what the word *concurrent* means. Relate *point of concurrency* with the two explorations.
- ? Draw △ABC and sketch one perpendicular bisector. "What do you know about any point on this perpendicular bisector?" *It is equidistant from the two vertices of the bisected side.*
- ? Repeat for a second perpendicular bisector. Where the two bisectors intersect ask, "What is true about this point?" *It is equidistant from all three vertices of the triangle.*
- State the Circumcenter Theorem and discuss the proof.

EXAMPLE 1 Solving a Real-Life Problem

Three snack carts sell frozen yogurt from points *A*, *B*, and *C* outside a city. Each of the three carts is the same distance from the frozen yogurt distributor.

Find the location of the distributor.

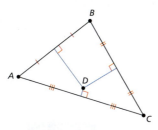

SOLUTION

The distributor is equidistant from the three snack carts. The Circumcenter Theorem shows that you can find a point equidistant from three points by using the perpendicular bisectors of the triangle formed by those points.

Copy the positions of points *A*, *B*, and *C* and connect the points to draw △*ABC*. Then use a ruler and protractor to draw the three perpendicular bisectors of △*ABC*. The circumcenter *D* is the location of the distributor.

READING
The prefix *circum-* means "around" or "about," as in *circumference* (distance around a circle).

Monitoring Progress Help in English and Spanish at BigIdeasMath.com

1. Three snack carts sell hot pretzels from points *A*, *B*, and *E*. What is the location of the pretzel distributor if it is equidistant from the three carts? Sketch the triangle and show the location.

The circumcenter *P* is equidistant from the three vertices, so *P* is the center of a circle that passes through all three vertices. As shown below, the location of *P* depends on the type of triangle. The circle with center *P* is said to be *circumscribed* about the triangle.

Acute triangle
P is inside triangle.

Right triangle
P is on triangle.

Obtuse triangle
P is outside triangle.

Section 6.3 Bisectors of Triangles 353

Extra Example 1
A carnival operator wants to locate a food stand so that it is the same distance from the carousel (*C*), the Ferris wheel (*F*), and the bumper cars (*B*). Find the location of the food stand (*S*).

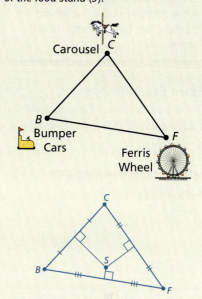

The circumcenter *S* of △*BCF* is the location of the food stand.

MONITORING PROGRESS ANSWER

1. The pretzel distributor is located at point *F*, which is the circumcenter of △*ABE*.

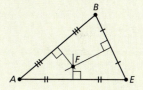

Laurie's Notes Teacher Actions

- **Extension:** To make Example 1 more interesting, print a street map of your area. Locate three points that are known to students and describe a plausible context. Locate the circumcenter for the triangle formed by the three fixed points.
- **? Wait Time:** "The circumcenter could be located on the side of a triangle. What type of triangle would it be?" *If students explored the circumcenter with dynamic geometry software, they likely know the triangle would be a right triangle.* Note the summary at the bottom of the page.

Section 6.3 353

> **Differentiated Instruction**
>
> **Auditory**
> Have students work in pairs. One student should read aloud the steps in circumscribing a circle about a triangle, while the other student carries out the construction. Then students should reverse roles.

Extra Example 2
Find the coordinates of the circumcenter of △DEF with vertices D(6, 4), E(−2, 4), and F(−2, −2). (2, 1)

MONITORING PROGRESS ANSWERS
2. (−4, 2)
3. (0, −1)

CONSTRUCTION Circumscribing a Circle About a Triangle

Use a compass and straightedge to construct a circle that is circumscribed about △ABC.

SOLUTION

Step 1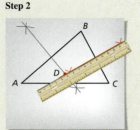
Draw a bisector Draw the perpendicular bisector of $\overline{AB}$.

Step 2
Draw a bisector Draw the perpendicular bisector of $\overline{BC}$. Label the intersection of the bisectors D. This is the circumcenter.

Step 3
Draw a circle Place the compass at D. Set the width by using any vertex of the triangle. This is the radius of the *circumcircle*. Draw the circle. It should pass through all three vertices A, B, and C.

> **STUDY TIP**
> Note that you only need to find the equations for *two* perpendicular bisectors. You can use the perpendicular bisector of the third side to verify your result.

EXAMPLE 2 Finding the Circumcenter of a Triangle

Find the coordinates of the circumcenter of △ABC with vertices A(0, 3), B(0, −1), and C(6, −1).

SOLUTION

Step 1 Graph △ABC.

Step 2 Find equations for two perpendicular bisectors. Use the Slopes of Perpendicular Lines Theorem, which states that horizontal lines are perpendicular to vertical lines.

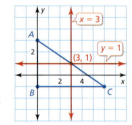

The midpoint of $\overline{AB}$ is (0, 1). The line through (0, 1) that is perpendicular to $\overline{AB}$ is $y = 1$.

The midpoint of $\overline{BC}$ is (3, −1). The line through (3, −1) that is perpendicular to $\overline{BC}$ is $x = 3$.

> **MAKING SENSE OF PROBLEMS**
> Because △ABC is a right triangle, the circumcenter lies on the triangle.

Step 3 Find the point where $x = 3$ and $y = 1$ intersect. They intersect at (3, 1).

▶ So, the coordinates of the circumcenter are (3, 1).

Monitoring Progress Help in English and Spanish at BigIdeasMath.com

Find the coordinates of the circumcenter of the triangle with the given vertices.

2. R(−2, 5), S(−6, 5), T(−2, −1)
3. W(−1, 4), X(1, 4), Y(1, −6)

354 Chapter 6 Relationships Within Triangles

Laurie's Notes Teacher Actions

- Construct the circumcenter. Vary the type of triangle (right, obtuse, acute) used by students.
- You might choose to modify Example 2 so that the perpendicular bisectors are not vertical or horizontal. Students should have the prerequisite skills to find the point of intersection for a more challenging problem. Alternately, consider using dynamic geometry software with a map that has been imported. See the *Teaching Strategy* on page T-350.

Using the Incenter of a Triangle

Just as a triangle has three perpendicular bisectors, it also has three angle bisectors. The angle bisectors of a triangle are also concurrent. This point of concurrency is the **incenter** of the triangle. For any triangle, the incenter always lies inside the triangle.

Theorem

Incenter Theorem

The incenter of a triangle is equidistant from the sides of the triangle.

If $\overline{AP}, \overline{BP},$ and $\overline{CP}$ are angle bisectors of $\triangle ABC$, then $PD = PE = PF$.

Proof Ex. 38, p. 359

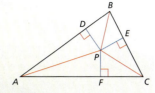

EXAMPLE 3 **Using the Incenter of a Triangle**

In the figure shown, $ND = 5x - 1$ and $NE = 2x + 11$.

a. Find NF.

b. Can NG be equal to 18? Explain your reasoning.

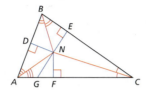

SOLUTION

a. N is the incenter of $\triangle ABC$ because it is the point of concurrency of the three angle bisectors. So, by the Incenter Theorem, $ND = NE = NF$.

Step 1 Solve for x.

$ND = NE$ Incenter Theorem

$5x - 1 = 2x + 11$ Substitute.

$x = 4$ Solve for x.

Step 2 Find ND (or NE).

$ND = 5x - 1 = 5(4) - 1 = 19$

▶ So, because $ND = NF$, $NF = 19$.

b. Recall that the shortest distance between a point and a line is a perpendicular segment. In this case, the perpendicular segment is $\overline{NF}$, which has a length of 19. Because $18 < 19$, NG cannot be equal to 18.

Monitoring Progress 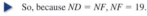 Help in English and Spanish at BigIdeasMath.com

4. In the figure shown, $QM = 3x + 8$ and $QN = 7x + 2$. Find QP.

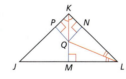

Extra Example 3

In the figure shown, $NE = 6x + 1$ and $NF = 4x + 15$.

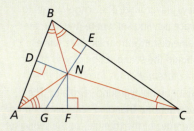

a. Find ND.   43

b. Can $NB = 40$? Explain your reasoning.
No. The shortest distance from a point to a line is the length of the perpendicular segment from the point to the line. In this case, the perpendicular segment is $\overline{ND}$, which has a length of 43. Because $40 < 43$, NB cannot equal 40.

MONITORING PROGRESS ANSWER

4. 12.5

Laurie's Notes   Teacher Actions

? Construct Viable Arguments and Critique the Reasoning of Others and Assessing Question: "The incenter of a triangle is always in the interior of the triangle. Explain why." Listen for correct reasoning involving the three angle bisectors intersecting at a point in the interior of all three angles.

- **Extension:** To make Example 3 more interesting, print a street map of your area. Locate three points that are known to students and describe a plausible context. Locate the incenter for the triangle formed by the three fixed points.

Extra Example 4
A school has fenced in an area in the shape of a scalene triangle to use for a new playground. The school wants to place a swing set where it will be the same distance from all three fences. Should the swing set be placed at the *circumcenter* or the *incenter* of the triangular playground? Explain. The incenter of a triangle is equidistant from the sides of the triangle, so the swing set should be at the incenter.

MONITORING PROGRESS ANSWER
5.
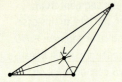

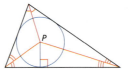

Because the incenter P is equidistant from the three sides of the triangle, a circle drawn using P as the center and the distance to one side of the triangle as the radius will just touch the other two sides of the triangle. The circle is said to be *inscribed* within the triangle.

CONSTRUCTION Inscribing a Circle Within a Triangle

Use a compass and straightedge to construct a circle that is inscribed within $\triangle ABC$.

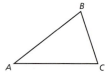

SOLUTION

Step 1
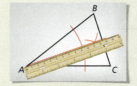
Draw a bisector Draw the angle bisector of $\angle A$.

Step 2
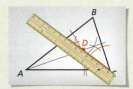
Draw a bisector Draw the angle bisector of $\angle C$. Label the intersection of the bisectors D. This is the incenter.

Step 3
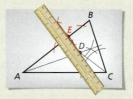
Draw a perpendicular line Draw the perpendicular line from D to $\overline{AB}$. Label the point where it intersects $\overline{AB}$ as E.

Step 4
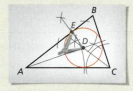
Draw a circle Place the compass at D. Set the width to E. This is the radius of the *incircle*. Draw the circle. It should touch each side of the triangle.

EXAMPLE 4 Solving a Real-Life Problem

A city wants to place a lamppost on the boulevard shown so that the lamppost is the same distance from all three streets. Should the location of the lamppost be at the *circumcenter* or *incenter* of the triangular boulevard? Explain.

ATTENDING TO PRECISION
Pay close attention to how a problem is stated. The city wants the lamppost to be the *same distance* from the three streets, not from where the streets intersect.

SOLUTION
Because the shape of the boulevard is an obtuse triangle, its circumcenter lies outside the triangle. So, the location of the lamppost cannot be at the circumcenter. The city wants the lamppost to be the same distance from all three streets. By the Incenter Theorem, the incenter of a triangle is equidistant from the sides of a triangle.

▶ So, the location of the lamppost should be at the incenter of the boulevard.

Monitoring Progress Help in English and Spanish at BigIdeasMath.com

5. Draw a sketch to show the location L of the lamppost in Example 4.

Laurie's Notes   **Teacher Actions**

- Construct the incenter. Vary the type of triangle (right, obtuse, acute) used by students.
- **Attend to Precision:** Note that Example 4 states that the lamppost is to be the same distance from the three streets, not from where they intersect.
- ❓ "Can you think of other contexts where knowing the incenter would be necessary?" Answers will vary.

Closure

- **Exit Ticket:** Explain what the circumcenter and incenter of a triangle are and how they can be found. The circumcenter of a triangle is equidistant from the vertices of the triangle. To find the circumcenter, find the intersection of the perpendicular bisectors of the sides of the triangle. The incenter of a triangle is equidistant from the sides of the triangle. To find the incenter, find the intersection of the angle bisectors of the interior angles of the triangle.

6.3 Exercises

Dynamic Solutions available at BigIdeasMath.com

Vocabulary and Core Concept Check

1. **VOCABULARY** When three or more lines, rays, or segments intersect in the same point, they are called _____ lines, rays, or segments.

2. **WHICH ONE DOESN'T BELONG?** Which triangle does *not* belong with the other three? Explain your reasoning.

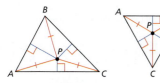 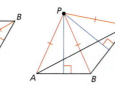 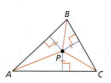

Monitoring Progress and Modeling with Mathematics

In Exercises 3 and 4, the perpendicular bisectors of △ABC intersect at point G and are shown in blue. Find the indicated measure.

3. Find BG.   4. Find GA.

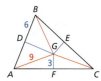 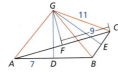

In Exercises 5 and 6, the angle bisectors of △XYZ intersect at point P and are shown in red. Find the indicated measure.

5. Find PB.   6. Find HP.

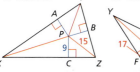 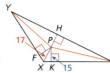

In Exercises 7–10, find the coordinates of the circumcenter of the triangle with the given vertices. *(See Example 2.)*

7. $A(2, 6), B(8, 6), C(8, 10)$

8. $D(-7, -1), E(-1, -1), F(-7, -9)$

9. $H(-10, 7), J(-6, 3), K(-2, 3)$

10. $L(3, -6), M(5, -3), N(8, -6)$

In Exercises 11–14, N is the incenter of △ABC. Use the given information to find the indicated measure. *(See Example 3.)*

11. $ND = 6x - 2$
$NE = 3x + 7$
Find NF.

12. $NG = x + 3$
$NH = 2x - 3$
Find NJ.

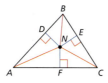

13. $NK = 2x - 2$
$NL = -x + 10$
Find NM.

14. $NQ = 2x$
$NR = 3x - 2$
Find NS.

 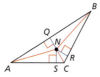

15. P is the circumcenter of △XYZ. Use the given information to find PZ.

$PX = 3x + 2$
$PY = 4x - 8$

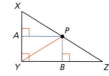

Section 6.3   Bisectors of Triangles   357

Assignment Guide and Homework Check

ASSIGNMENT
Basic: 1, 2, 3–27 odd, 28, 45, 46, 52–59
Average: 1, 2–28 even, 41–45 odd, 46, 52–59
Advanced: 1, 2–8 even, 14–42 even, 43–59

HOMEWORK CHECK
Basic: 3, 5, 7, 13, 28
Average: 4, 6, 8, 14, 18
Advanced: 6, 14, 18, 34, 43

ANSWERS

1. concurrent
2. the fourth triangle; This triangle shows the incenter of the triangle, but the other three show the circumcenter.
3. 9
4. 11
5. 9
6. 15
7. (5, 8)
8. (−4, −5)
9. (−4, 9)
10. (5.5, −5.5)
11. 16
12. 9
13. 6
14. 4
15. 32

Dynamic Teaching Tools

Dynamic Assessment & Progress Monitoring Tool
Interactive Whiteboard Lesson Library
Dynamic Classroom with Dynamic Investigations

ANSWERS

16. 31

17. Sample answer:

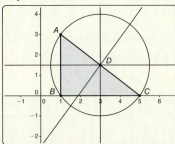

18. Sample answer:

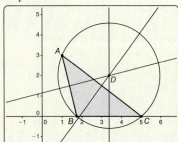

19. Sample answer:

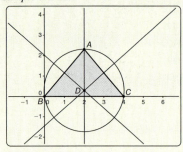

20. Sample answer:

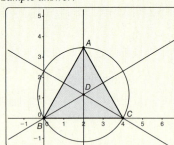

---

16. P is the circumcenter of $\triangle XYZ$. Use the given information to find PY.

    $PX = 4x + 3$
    $PZ = 6x - 11$

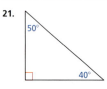

CONSTRUCTION In Exercises 17–20, draw a triangle of the given type. Find the circumcenter. Then construct the circumscribed circle.

17. right
18. obtuse
19. acute isosceles
20. equilateral

CONSTRUCTION In Exercises 21–24, copy the triangle with the given angle measures. Find the incenter. Then construct the inscribed circle.

21.

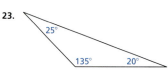

22.

23.

24.

ERROR ANALYSIS In Exercises 25 and 26, describe and correct the error in identifying equal distances inside the triangle.

25.

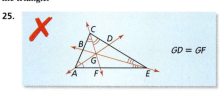

26.

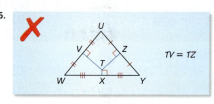

27. **MODELING WITH MATHEMATICS** You and two friends plan to meet to walk your dogs together. You want the meeting place to be the same distance from each person's house. Explain how you can use the diagram to locate the meeting place. *(See Example 1.)*

28. **MODELING WITH MATHEMATICS** You are placing a fountain in a triangular koi pond. You want the fountain to be the same distance from each edge of the pond. Where should you place the fountain? Explain your reasoning. Use a sketch to support your answer. *(See Example 4.)*

CRITICAL THINKING In Exercises 29–32, complete the statement with *always*, *sometimes*, or *never*. Explain your reasoning.

29. The circumcenter of a scalene triangle is _____ inside the triangle.

30. If the perpendicular bisector of one side of a triangle intersects the opposite vertex, then the triangle is _____ isosceles.

31. The perpendicular bisectors of a triangle intersect at a point that is _____ equidistant from the midpoints of the sides of the triangle.

32. The angle bisectors of a triangle intersect at a point that is _____ equidistant from the sides of the triangle.

358   Chapter 6   Relationships Within Triangles

---

21. Sample answer:

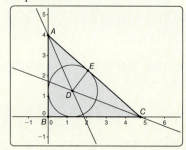

22. Sample answer:

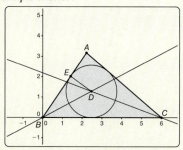

23–32. See Additional Answers.

CRITICAL THINKING In Exercises 33 and 34, find the coordinates of the circumcenter of the triangle with the given vertices.

33. $A(2, 5)$, $B(6, 6)$, $C(12, 3)$

34. $D(-9, -5)$, $E(-5, -9)$, $F(-2, -2)$

MATHEMATICAL CONNECTIONS In Exercises 35 and 36, find the value of x that makes N the incenter of the triangle.

35.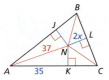

36.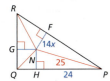

37. **PROOF** Where is the circumcenter located in any right triangle? Write a coordinate proof of this result.

38. **PROVING A THEOREM** Write a proof of the Incenter Theorem.

    Given $\triangle ABC$, $\overline{AD}$ bisects $\angle CAB$, $\overline{BD}$ bisects $\angle CBA$, $\overline{DE} \perp \overline{AB}$, $\overline{DF} \perp \overline{BC}$, and $\overline{DG} \perp \overline{CA}$

    Prove The angle bisectors intersect at D, which is equidistant from $\overline{AB}$, $\overline{BC}$, and $\overline{CA}$.

    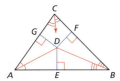

39. **WRITING** Explain the difference between the circumcenter and the incenter of a triangle.

40. **REASONING** Is the incenter of a triangle ever located outside the triangle? Explain your reasoning.

41. **MODELING WITH MATHEMATICS** You are installing a circular pool in the triangular courtyard shown. You want to have the largest pool possible on the site without extending into the walkway.

    a. Copy the triangle and show how to install the pool so that it just touches each edge. Then explain how you can be sure that you could not fit a larger pool on the site.

    b. You want to have the largest pool possible while leaving at least 1 foot of space around the pool. Would the center of the pool be in the same position as in part (a)? Justify your answer.

42. **MODELING WITH MATHEMATICS** Archaeologists find three stones. They believe that the stones were once part of a circle of stones with a community fire pit at its center. They mark the locations of stones A, B, and C on a graph, where distances are measured in feet.

    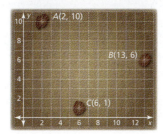

    a. Explain how archaeologists can use a sketch to estimate the center of the circle of stones.

    b. Copy the diagram and find the approximate coordinates of the point at which the archaeologists should look for the fire pit.

43. **REASONING** Point P is inside $\triangle ABC$ and is equidistant from points A and B. On which of the following segments must P be located?

    Ⓐ $\overline{AB}$
    Ⓑ the perpendicular bisector of $\overline{AB}$
    Ⓒ $\overline{AC}$
    Ⓓ the perpendicular bisector of $\overline{AC}$

ANSWERS

33. $\left(\frac{35}{6}, -\frac{11}{6}\right)$

34. $(-4.9, -4.9)$

35. $x = 6$

36. $x = \frac{1}{2}$

37. The circumcenter of any right triangle is located at the midpoint of the hypotenuse of the triangle. Let $A(0, 2b)$, $B(0, 0)$, and $C(2a, 0)$ represent the vertices of a right triangle where $\angle B$ is the right angle. The midpoint of $\overline{AB}$ is $M_{\overline{AB}}(0, b)$. The midpoint of $\overline{BC}$ is $M_{\overline{BC}}(a, 0)$. The midpoint of $\overline{AC}$ is $M_{\overline{AC}}(a, b)$. Because $\overline{AB}$ is vertical, its perpendicular bisector is horizontal. So, the equation of the horizontal line passing through $M_{\overline{AB}}(0, b)$ is $y = b$. Because $\overline{BC}$ is horizontal, its perpendicular bisector is vertical. So, the equation of the vertical line passing through $M_{\overline{BC}}(a, 0)$ is $x = a$. The circumcenter of $\triangle ABC$ is the intersection of perpendicular bisectors, $y = b$ and $x = a$, which is (a, b). This point is also the midpoint of $\overline{AC}$.

38. Because $\overline{DE} \perp \overline{AB}$, $\overline{DF} \perp \overline{BC}$, and $\overline{DG} \perp \overline{CA}$, $\angle DFB$, $\angle DEB$, $\angle DEA$, and $\angle DGA$ are congruent right angles. Also, by definition of angle bisector, $\angle DBF \cong \angle DBE$ and $\angle DAE \cong \angle DAG$. In addition, $\overline{DB} \cong \overline{DB}$ and $\overline{DA} \cong \overline{DA}$ by the Reflexive Property of Congruence. So, $\triangle DFB \cong \triangle DEB$ and $\triangle DEA \cong \triangle DGA$ by the AAS Congruence Theorem. Next, because corresponding parts of congruent triangles are congruent, $\overline{DF} \cong \overline{DE}$ and $\overline{DG} \cong \overline{DE}$. By the Transitive Property of Congruence, $\overline{DF} \cong \overline{DE} \cong \overline{DG}$. So, point D is equidistant from $\overline{AB}$, $\overline{BC}$, and $\overline{CA}$. Because D is equidistant from $\overline{CA}$ and $\overline{CB}$, by the Converse of the Angle Bisector Theorem, point D is on the angle bisector of $\angle ACB$. So, the angle bisectors intersect at point D.

39. The circumcenter is the point of intersection of the perpendicular bisectors of the sides of a triangle, and it is equidistant from the vertices of the triangle. In contrast, the incenter is the point of intersection of the angle bisectors of a triangle, and it is equidistant from the sides of the triangle.

40. no; Because the incenter is the center of an inscribed circle, it must be inside the triangle.

41. a.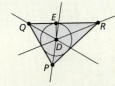

    Because this circle is inscribed in the triangle, it is the largest circle that fits inside the triangle without extending into the boundaries.

    b. yes; You would keep the center of the pool as the incenter of the triangle, but you would make the radius of the pool at least 1 foot shorter.

42. a. The archaeologists need to locate the circumcenter of the three stones, because that will be the center of the circle that contains all three stones. In order to locate the circumcenter, the archaeologists need to find the point of concurrency of the perpendicular bisectors of the sides of the triangle formed by the three stones.

    42b–43. See Additional Answers.

ANSWERS

44. no; When you find the circumcenter of three of the points and draw the circle that circumscribes those three points, it does not pass through the fourth point. An example of one circle is shown.

45. yes; In an equilateral triangle, each perpendicular bisector passes through the opposite vertex and divides the triangle into two congruent triangles. So, it is also an angle bisector.

46. incenter

47. a. equilateral; 3; In an equilateral triangle, each perpendicular bisector also bisects the opposite angle.

    b. scalene; 6; In a scalene triangle, none of the perpendicular bisectors will also bisect an angle.

48–59. See Additional Answers.

Mini-Assessment

1. Find the coordinates of the circumcenter of △ABC with vertices $A(0, 5)$, $B(4, 5)$, and $C(4, -1)$. **(2, 2)**

2. In the figure shown, $QP = 2x + 9$ and $QM = 5x - 3$. Find QN.

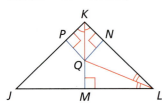

$QN = 17$

3. In the figure shown, P is the circumcenter of △XYZ, $PY = 5x - 4$, and $PZ = 4x + 11$. Find PX.

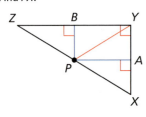

$PX = 71$

360   Chapter 6

---

44. CRITICAL THINKING A high school is being built for the four towns shown on the map. Each town agrees that the school should be an equal distance from each of the four towns. Is there a single point where they could agree to build the school? If so, find it. If not, explain why not. Justify your answer with a diagram.

45. MAKING AN ARGUMENT Your friend says that the circumcenter of an equilateral triangle is also the incenter of the triangle. Is your friend correct? Explain your reasoning.

46. HOW DO YOU SEE IT? The arms of the windmill are the angle bisectors of the red triangle. What point of concurrency is the point that connects the three arms?

47. ABSTRACT REASONING You are asked to draw a triangle and all its perpendicular bisectors and angle bisectors.

a. For which type of triangle would you need the fewest segments? What is the minimum number of segments you would need? Explain.

b. For which type of triangle would you need the most segments? What is the maximum number of segments you would need? Explain.

48. THOUGHT PROVOKING The diagram shows an official hockey rink used by the National Hockey League. Create a triangle using hockey players as vertices in which the center circle is inscribed in the triangle. The center dot should be the incenter of your triangle. Sketch a drawing of the locations of your hockey players. Then label the actual lengths of the sides and the angle measures in your triangle.

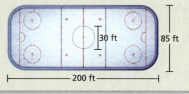

COMPARING METHODS In Exercises 49 and 50, state whether you would use *perpendicular bisectors* or *angle bisectors*. Then solve the problem.

49. You need to cut the largest circle possible from an isosceles triangle made of paper whose sides are 8 inches, 12 inches, and 12 inches. Find the radius of the circle.

50. On a map of a camp, you need to create a circular walking path that connects the pool at (10, 20), the nature center at (16, 2), and the tennis court at (2, 4). Find the coordinates of the center of the circle and the radius of the circle.

51. **CRITICAL THINKING** Point D is the incenter of △ABC. Write an expression for the length x in terms of the three side lengths AB, AC, and BC.

Maintaining Mathematical Proficiency
Reviewing what you learned in previous grades and lessons

The endpoints of $\overline{AB}$ are given. Find the coordinates of the midpoint M. Then find AB. *(Skills Review Handbook)*

52. $A(-3, 5), B(3, 5)$

53. $A(2, -1), B(10, 7)$

54. $A(-5, 1), B(4, -5)$

55. $A(-7, 5), B(5, 9)$

Write an equation of the line passing through point P that is perpendicular to the given line. Graph the equations of the lines to check that they are perpendicular. *(Skills Review Handbook)*

56. $P(2, 8), y = 2x + 1$

57. $P(6, -3), y = -5$

58. $P(-8, -6), 2x + 3y = 18$

59. $P(-4, 1), y + 3 = -4(x + 3)$

If students need help...
Resources by Chapter
- Practice A and Practice B
- Puzzle Time

Student Journal
- Practice

Differentiating the Lesson
Skills Review Handbook

If students got it...
Resources by Chapter
- Enrichment and Extension
- Cumulative Review

Start the *next* Section

Laurie's Notes

Overview of Section 6.4

Introduction
- In this lesson, the *medians of a triangle* are constructed and the centroid, the point of concurrency, is found. The *centroid* is the point that is $\frac{2}{3}$ the distance from the vertex to the midpoint of the opposite side.
- In the second part of the lesson, the three *altitudes of a triangle* are constructed and the *orthocenter*, the point of concurrency, is found.
- The lesson also integrates algebraic skills as students find the coordinates of the centroid and orthocenter.

Resources
- Tracing paper is helpful when the explorations are done by paper folding.

Teaching Strategy
- Like the last lesson, the points of concurrency in this lesson are quickly constructed using dynamic geometry software as shown in the explorations.
- If you do not have access to software, an alternate way to introduce this lesson would be through a paper-folding activity or a compass construction activity. It is powerful for students to explore and discover the properties of medians and altitudes.
- Give directions similar to what is written in the explorations for students to paper fold (or construct) the medians (Exploration 1) and the altitudes (Exploration 2). Distances can be measured using a compass or ruler. Although precision may be a problem, the focus is on the discovery.

Extensions
- There is a famous construction called the Nine-Point Circle for which you can find references in many places. When your students have finished this lesson, they will have the vocabulary necessary to perform this construction. It can be done with a compass and a straightedge or with dynamic geometry software.

Pacing Suggestion
- The explorations allow students to explore and discover two relationships that will be stated formally in the lesson. The constructions and investigations are not long and should be followed with the statement of the Centroid Theorem in the lesson. You may choose to omit the compass construction in the lesson.

Dynamic Teaching Tools
Dynamic Assessment & Progress Monitoring Tool
Lesson Planning Tool
Interactive Whiteboard Lesson Library
Dynamic Classroom with Dynamic Investigations

What Your Students Will Learn

- Use and find the centroids of triangles by finding the intersection of the medians of the triangle.
- Use and find the orthocenters of triangles by finding the intersection of the altitudes of the triangle.

Laurie's Notes

Exploration

Motivate
- Before students arrive, cut a triangle out of heavier weight paper so that it remains rigid when held. You will also need a sharp pencil.
- When students arrive, ask them whether they think you could balance the triangle on the tip of your pencil.
- Explain to students that in this lesson they will learn about a special point in a triangle that is called the *centroid*, which is the balancing point of the triangle. (See Exercise 50.)

Exploration 1
- This is a quick and straightforward construction using dynamic geometric software. Students should be comfortable at this point knowing that the triangle will be manipulated so that various cases can be explored.
- Students observe that the three medians appear to intersect at one point.
- **Turn and Talk:** "How could you show that the segments intersect at just one point and not at points that are simply very close together?" One way is to label the points of intersection of each pair of segments G, H, and I. If coordinates appear, students will note that G, H, and I are the same point. If the construction was not done in a coordinate plane, measure $\overline{GH}$, $\overline{HI}$, and $\overline{IG}$. All segments will have a length of 0, meaning G, H, and I are the same point.
- **?** **Construct Viable Arguments and Critique the Reasoning of Others:** Measure the segments identified and compute the ratios. "What conjecture(s) can you make?" *Sample answer*: The ratio of the length of the longer segment to the length of the median is 2 : 3.
- **Extension:** If time permits, have students construct △*DEF*, where *D*, *E*, and *F* are the midpoints of the three sides of △*ABC*. Ask students to make observations about the central triangle, all four small triangles, and the original △*ABC*.

Exploration 2
- When constructing the perpendicular line from a vertex to the opposite side of the triangle, a line will be drawn, not a segment. You may want to show students how to construct the perpendicular, draw the segment, and then hide the line.
- **Construct Viable Arguments and Critique the Reasoning of Others:** Students will observe that the lines containing the three altitudes are concurrent. They should also notice when they click and drag on the vertices that the point of concurrency is not always in the interior of the triangle.

Communicate Your Answer
- Students should state that the three medians of a triangle intersect at one point, and the lines containing the three altitudes of a triangle intersect at one point.

Connecting to Next Step
- The explorations are related to the theorem and properties presented in the lesson, helping students make sense of these before proving them to be true.

6.4 Medians and Altitudes of Triangles

Essential Question What conjectures can you make about the medians and altitudes of a triangle?

EXPLORATION 1
Finding Properties of the Medians of a Triangle

Work with a partner. Use dynamic geometry software. Draw any △ABC.

a. Plot the midpoint of $\overline{BC}$ and label it D. Draw $\overline{AD}$, which is a *median* of △ABC. Construct the medians to the other two sides of △ABC.

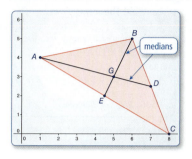

Sample
Points
A(1, 4)
B(6, 5)
C(8, 0)
D(7, 2.5)
E(4.5, 2)
G(5, 3)

b. What do you notice about the medians? Drag the vertices to change △ABC. Use your observations to write a conjecture about the medians of a triangle.

c. In the figure above, point G divides each median into a shorter segment and a longer segment. Find the ratio of the length of each longer segment to the length of the whole median. Is this ratio always the same? Justify your answer.

EXPLORATION 2
Finding Properties of the Altitudes of a Triangle

Work with a partner. Use dynamic geometry software. Draw any △ABC.

a. Construct the perpendicular segment from vertex A to $\overline{BC}$. Label the endpoint D. $\overline{AD}$ is an *altitude* of △ABC.

b. Construct the altitudes to the other two sides of △ABC. What do you notice?

c. Write a conjecture about the altitudes of a triangle. Test your conjecture by dragging the vertices to change △ABC.

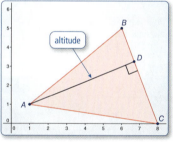

LOOKING FOR STRUCTURE

To be proficient in math, you need to look closely to discern a pattern or structure.

Communicate Your Answer

3. What conjectures can you make about the medians and altitudes of a triangle?

4. The length of median $\overline{RU}$ in △RST is 3 inches. The point of concurrency of the three medians of △RST divides $\overline{RU}$ into two segments. What are the lengths of these two segments?

Dynamic Teaching Tools
Dynamic Assessment & Progress Monitoring Tool
Lesson Planning Tool
Interactive Whiteboard Lesson Library
Dynamic Classroom with Dynamic Investigations

ANSWERS

1. a. *Sample answer:*

   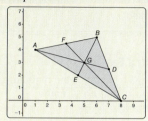

   b. The medians of a triangle are concurrent at a point inside the triangle.

   c. 2 : 3; yes; The ratio is the same for each median and does not change when you change the triangle.

2. a. Check students' work.

   b. *Sample answer:*

   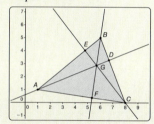

   They meet at the same point.

   c. The altitudes of a triangle meet at a point that may be inside, on, or outside of the triangle.

3. The medians meet at a point inside the triangle that divides each median into two segments whose lengths have the ratio 1 : 2. The altitudes meet at a point inside, on, or outside the triangle depending on whether the triangle is acute, right, or obtuse.

4. 1 in. and 2 in.

Differentiated Instruction

Kinesthetic
The centroid of a triangle is its center of gravity. Distribute cardboard to the class. Have each student draw a large triangle, construct its centroid, cut out the triangle, and try to balance it on a pencil point. The tip of the pencil will coincide with the centroid of the triangle.

Extra Example 1
In $\triangle RST$, point Q is the centroid, and $VQ = 5$. Find RQ and RV.

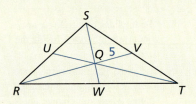

$RQ = 10$, $RV = 15$

6.4 Lesson

What You Will Learn
- Use medians and find the centroids of triangles.
- Use altitudes and find the orthocenters of triangles.

Core Vocabulary
median of a triangle, p. 362
centroid, p. 362
altitude of a triangle, p. 363
orthocenter, p. 363

Previous
midpoint
concurrent
point of concurrency

Using the Median of a Triangle

A **median of a triangle** is a segment from a vertex to the midpoint of the opposite side. The three medians of a triangle are concurrent. The point of concurrency, called the **centroid**, is inside the triangle.

Theorem

Centroid Theorem

The centroid of a triangle is two-thirds of the distance from each vertex to the midpoint of the opposite side.

The medians of $\triangle ABC$ meet at point P, and $AP = \frac{2}{3}AE$, $BP = \frac{2}{3}BF$, and $CP = \frac{2}{3}CD$.

Proof BigIdeasMath.com

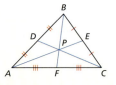

CONSTRUCTION   Finding the Centroid of a Triangle

Use a compass and straightedge to construct the medians of $\triangle ABC$.

SOLUTION

Step 1
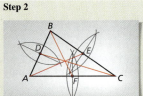

Find midpoints Draw $\triangle ABC$. Find the midpoints of $\overline{AB}, \overline{BC}$, and $\overline{AC}$. Label the midpoints of the sides D, E, and F, respectively.

Step 2

Draw medians Draw $\overline{AE}, \overline{BF}$, and $\overline{CD}$. These are the three medians of $\triangle ABC$.

Step 3
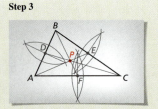

Label a point Label the point where $\overline{AE}, \overline{BF}$, and $\overline{CD}$ intersect as P. This is the centroid.

EXAMPLE 1   Using the Centroid of a Triangle

In $\triangle RST$, point Q is the centroid, and $SQ = 8$. Find QW and SW.

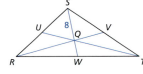

SOLUTION

$SQ = \frac{2}{3}SW$ Centroid Theorem

$8 = \frac{2}{3}SW$ Substitute 8 for SQ.

$12 = SW$ Multiply each side by the reciprocal, $\frac{3}{2}$.

Then $QW = SW - SQ = 12 - 8 = 4$.

▶ So, $QW = 4$ and $SW = 12$.

362   Chapter 6   Relationships Within Triangles

Laurie's Notes   Teacher Actions

- The Centroid Theorem follows from the first exploration. The centroid divides the median into two pieces that are in the ratio of 1:2. This also means that the ratio of the longer segment to the median is 2:3.
- **Extension:** You can find lots of information about centroids, balancing points, and centers of gravity. The U.S. Census Bureau also defines a population centroid. Search the Internet for "population centroid."

FINDING AN ENTRY POINT

The median $\overline{SV}$ is chosen in Example 2 because it is easier to find a distance on a vertical segment.

EXAMPLE 2 Finding the Centroid of a Triangle

Find the coordinates of the centroid of △RST with vertices R(2, 1), S(5, 8), and T(8, 3).

SOLUTION

Step 1 Graph △RST.

Step 2 Use the Midpoint Formula to find the midpoint V of $\overline{RT}$ and sketch median $\overline{SV}$.

$$V\left(\frac{2+8}{2}, \frac{1+3}{2}\right) = (5, 2)$$

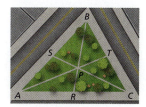

JUSTIFYING CONCLUSIONS

You can check your result by using a different median to find the centroid.

Step 3 Find the centroid. It is two-thirds of the distance from each vertex to the midpoint of the opposite side.

The distance from vertex S(5, 8) to V(5, 2) is 8 − 2 = 6 units. So, the centroid is $\frac{2}{3}(6) = 4$ units down from vertex S on $\overline{SV}$.

▶ So, the coordinates of the centroid P are (5, 8 − 4), or (5, 4).

Monitoring Progress   Help in English and Spanish at *BigIdeasMath.com*

There are three paths through a triangular park. Each path goes from the midpoint of one edge to the opposite corner. The paths meet at point P.

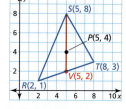

1. Find PS and PC when SC = 2100 feet.
2. Find TC and BC when BT = 1000 feet.
3. Find PA and TA when PT = 800 feet.

Find the coordinates of the centroid of the triangle with the given vertices.

4. F(2, 5), G(4, 9), H(6, 1)   5. X(−3, 3), Y(1, 5), Z(−1, −2)

READING

In the area formula for a triangle, $A = \frac{1}{2}bh$, you can use the length of any side for the base b. The height h is the length of the altitude to that side from the opposite vertex.

Using the Altitude of a Triangle

An **altitude of a triangle** is the perpendicular segment from a vertex to the opposite side or to the line that contains the opposite side.

Core Concept

Orthocenter

The lines containing the altitudes of a triangle are concurrent. This point of concurrency is the **orthocenter** of the triangle.

The lines containing $\overline{AF}$, $\overline{BD}$, and $\overline{CE}$ meet at the orthocenter G of △ABC.

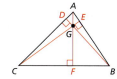

Section 6.4 Medians and Altitudes of Triangles 363

Extra Example 2

Find the coordinates of the centroid of △ABC with vertices A(0, 4), B(−4, −2), and C(7, 1). (1, 1)

MONITORING PROGRESS ANSWERS

1. 700 ft, 1400 ft
2. 1000 ft, 2000 ft
3. 1600 ft, 2400 ft
4. (4, 5)
5. (−1, 2)

Laurie's Notes — Teacher Actions

- You might consider having different groups of students use different medians in Example 2 when finding the coordinates of the centroid.
- **Think-Pair-Share:** Have students answer Questions 1–5, and then share and discuss as a class.
- **Teaching Tip:** To sketch the three altitudes, students find it helpful to rotate their paper so that they are drawing a vertical line from the vertex, perpendicular to the opposite side. They rotate the triangle for each altitude sketched.

English Language Learners

Vocabulary
Sections 6.3 and 6.4 contain many unfamiliar terms. Make sure students copy the chart on page 365 into their notebooks, adding any notes that will help them remember the definitions of the terms *circumcenter*, *incenter*, *centroid*, and *orthocenter*.

Extra Example 3
Find the coordinates of the orthocenter of △DEF with vertices D(0, 6), E(−4, −2), and F(4, 6). (−4, 10)

MONITORING PROGRESS ANSWERS
6. outside; (−1, −3)
7. on; (−3, 4)

As shown below, the location of the orthocenter P of a triangle depends on the type of triangle.

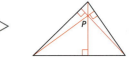

Acute triangle — P is inside triangle.   Right triangle — P is on triangle.   Obtuse triangle — P is outside triangle.

READING
The altitudes are shown in red. Notice that in the right triangle, the legs are also altitudes. The altitudes of the obtuse triangle are extended to find the orthocenter.

EXAMPLE 3   Finding the Orthocenter of a Triangle

Find the coordinates of the orthocenter of △XYZ with vertices $X(-5, -1)$, $Y(-2, 4)$, and $Z(3, -1)$.

SOLUTION

Step 1 Graph △XYZ.

Step 2 Find an equation of the line that contains the altitude from Y to $\overline{XZ}$. Because $\overline{XZ}$ is horizontal, the altitude is vertical. The line that contains the altitude passes through $Y(-2, 4)$. So, the equation of the line is $x = -2$.

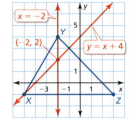

Step 3 Find an equation of the line that contains the altitude from X to $\overline{YZ}$.

$$\text{slope of } \overleftrightarrow{YZ} = \frac{-1 - 4}{3 - (-2)} = -1$$

Because the product of the slopes of two perpendicular lines is −1, the slope of a line perpendicular to $\overleftrightarrow{YZ}$ is 1. The line passes through $X(-5, -1)$.

$y = mx + b$ Use slope-intercept form.
$-1 = 1(-5) + b$ Substitute −1 for *y*, 1 for *m*, and −5 for *x*.
$4 = b$ Solve for *b*.

So, the equation of the line is $y = x + 4$.

Step 4 Find the point of intersection of the graphs of the equations $x = -2$ and $y = x + 4$.

Substitute −2 for *x* in the equation $y = x + 4$. Then solve for *y*.

$y = x + 4$ Write equation.
$y = -2 + 4$ Substitute −2 for *x*.
$y = 2$ Solve for *y*.

So, the coordinates of the orthocenter are $(-2, 2)$.

Monitoring Progress   Help in English and Spanish at *BigIdeasMath.com*

Tell whether the orthocenter of the triangle with the given vertices is *inside*, *on*, or *outside* the triangle. Then find the coordinates of the orthocenter.

6. $A(0, 3)$, $B(0, -2)$, $C(6, -3)$ 7. $J(-3, -4)$, $K(-3, 4)$, $L(5, 4)$

364   Chapter 6   Relationships Within Triangles

Laurie's Notes   Teacher Actions

? "Is the orthocenter always located in the interior of the triangle?" no; It is in the interior for an acute triangle, at the vertex of the right angle for a right triangle, and in the exterior for an obtuse triangle.

- **Thumbs Up:** You might consider having different groups of students use different pairs of altitudes in Example 3 when finding the coordinates of the orthocenter. Ask students to give a *Thumbs Up* assessment of the process before they begin.

In an isosceles triangle, the perpendicular bisector, angle bisector, median, and altitude from the vertex angle to the base are all the same segment. In an equilateral triangle, this is true for any vertex.

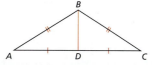 **Proving a Property of Isosceles Triangles**

Prove that the median from the vertex angle to the base of an isosceles triangle is an altitude.

SOLUTION

Given △ABC is isosceles, with base $\overline{AC}$. $\overline{BD}$ is the median to base $\overline{AC}$.

Prove $\overline{BD}$ is an altitude of △ABC.

Paragraph Proof Legs $\overline{AB}$ and $\overline{BC}$ of isosceles △ABC are congruent. $\overline{CD} \cong \overline{AD}$ because $\overline{BD}$ is the median to $\overline{AC}$. Also, $\overline{BD} \cong \overline{BD}$ by the Reflexive Property of Congruence. So, △ABD ≅ △CBD by the SSS Congruence Theorem. ∠ADB ≅ ∠CDB because corresponding parts of congruent triangles are congruent. Also, ∠ADB and ∠CDB are a linear pair. $\overline{BD}$ and $\overline{AC}$ intersect to form a linear pair of congruent angles, so $\overline{BD} \perp \overline{AC}$ and $\overline{BD}$ is an altitude of △ABC.

Monitoring Progress Help in English and Spanish at *BigIdeasMath.com*

8. **WHAT IF?** In Example 4, you want to show that median $\overline{BD}$ is also an angle bisector. How would your proof be different?

Concept Summary

Segments, Lines, Rays, and Points in Triangles

	Example	Point of Concurrency	Property	Example
perpendicular bisector		circumcenter	The circumcenter *P* of a triangle is equidistant from the vertices of the triangle.	
angle bisector		incenter	The incenter *I* of a triangle is equidistant from the sides of the triangle.	
median		centroid	The centroid *R* of a triangle is two thirds of the distance from each vertex to the midpoint of the opposite side.	
altitude		orthocenter	The lines containing the altitudes of a triangle are concurrent at the orthocenter *O*.	

Section 6.4 Medians and Altitudes of Triangles 365

Extra Example 4
Prove that the bisector of the vertex angle of an isosceles triangle is an altitude.

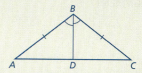

Given △ABC is isosceles, $\overline{BD}$ bisects ∠ABC.

Prove $\overline{BD}$ is an altitude of △ABC.

Paragraph Proof $\overline{AB} \cong \overline{BC}$ by definition of isosceles triangle. ∠ABD ≅ ∠CBD by definition of angle bisector. $\overline{BD} \cong \overline{BD}$ by the Reflex. Prop. ≅. So △ABD ≅ △CBD by SAS ≅ Thm. ∠ADB ≅ ∠CDB by CPCTC. Also, they are a linear pair. $\overline{BD}$ and $\overline{AC}$ intersect to form a linear pair of ≅ ∠s, so $\overline{BD} \perp \overline{AC}$, and $\overline{BD}$ is an altitude of △ABC.

MONITORING PROGRESS ANSWER

8. Proving △ABD ≅ △CBD by the SSS Congruence Theorem at the beginning of the proof would be the same. But then you would state that ∠ABD ≅ ∠CBD because corresponding parts of congruent triangles are congruent. This means that $\overline{BD}$ is also an angle bisector by definition.

Laurie's Notes Teacher Actions

- **Turn and Talk:** "Are the four points of concurrency distinct points in every triangle?" Listen for valid reasoning.
- **Whiteboarding:** Pose Example 4. Use whiteboards and have partners work on the proof. Compare and contrast proofs from different pairs of students.
- Students should find the *Concept Summary* helpful.

Closure

- **Exit Ticket:** Draw a right scalene triangle. Sketch the three medians. Draw an obtuse isosceles triangle. Sketch the three altitudes. *Check students' work.*

Assignment Guide and Homework Check

ASSIGNMENT

Basic: 1, 2, 3–29 odd, 41, 48, 49, 55–58

Average: 1, 2, 6–24 even, 28–42 even, 48, 49, 55–58

Advanced: 1, 2, 14–34 even, 39–49, 52, 55–58

HOMEWORK CHECK

Basic: 3, 7, 11, 15, 29

Average: 14, 18, 20, 30, 42

Advanced: 14, 18, 20, 30, 46

ANSWERS

1. circumcenter, incenter, centroid, orthocenter; Perpendicular bisectors form the circumcenter, angle bisectors form the incenter, medians form the centroid, and altitudes form the orthocenter.
2. $\frac{2}{3}$
3. 6, 3
4. 14, 7
5. 20, 10
6. 28, 14
7. 10, 15
8. 22, 33
9. 18, 27
10. 30, 45
11. 12
12. 9
13. 10
14. 5
15. $\left(5, \frac{11}{3}\right)$
16. $\left(-\frac{7}{3}, 5\right)$
17. $(5, 1)$
18. $\left(\frac{10}{3}, 3\right)$
19. outside; $(0, -5)$
20. on; $(-3, 2)$
21. inside; $(-1, 2)$
22. inside; $\left(0, \frac{7}{3}\right)$
23.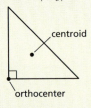

6.4 Exercises

Dynamic Solutions available at BigIdeasMath.com

Vocabulary and Core Concept Check

1. **VOCABULARY** Name the four types of points of concurrency. Which lines intersect to form each of the points?
2. **COMPLETE THE SENTENCE** The length of a segment from a vertex to the centroid is _____ the length of the median from that vertex.

Monitoring Progress and Modeling with Mathematics

In Exercises 3–6, point P is the centroid of $\triangle LMN$. Find PN and QP. (See Example 1.)

3. $QN = 9$ 4. $QN = 21$

5. $QN = 30$ 6. $QN = 42$

In Exercises 7–10, point D is the centroid of $\triangle ABC$. Find CD and CE.

7. $DE = 5$ 8. $DE = 11$

9. $DE = 9$ 10. $DE = 15$

 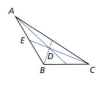

In Exercises 11–14, point G is the centroid of $\triangle ABC$. $BG = 6$, $AF = 12$, and $AE = 15$. Find the length of the segment.

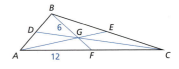

11. $\overline{FC}$ 12. $\overline{BF}$

13. $\overline{AG}$ 14. $\overline{GE}$

In Exercises 15–18, find the coordinates of the centroid of the triangle with the given vertices. (See Example 2.)

15. $A(2, 3)$, $B(8, 1)$, $C(5, 7)$
16. $F(1, 5)$, $G(-2, 7)$, $H(-6, 3)$
17. $S(5, 5)$, $T(11, -3)$, $U(-1, 1)$
18. $X(1, 4)$, $Y(7, 2)$, $Z(2, 3)$

In Exercises 19–22, tell whether the orthocenter is *inside*, *on*, or *outside* the triangle. Then find the coordinates of the orthocenter. (See Example 3.)

19. $L(0, 5)$, $M(3, 1)$, $N(8, 1)$
20. $X(-3, 2)$, $Y(5, 2)$, $Z(-3, 6)$
21. $A(-4, 0)$, $B(1, 0)$, $C(-1, 3)$
22. $T(-2, 1)$, $U(2, 1)$, $V(0, 4)$

CONSTRUCTION In Exercises 23–26, draw the indicated triangle and find its centroid and orthocenter.

23. isosceles right triangle   24. obtuse scalene triangle

25. right scalene triangle   26. acute isosceles triangle

366   Chapter 6   Relationships Within Triangles

24. Sample answer:

25. Sample answer:

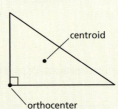

26. Sample answer:

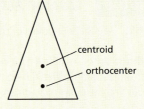

ERROR ANALYSIS In Exercises 27 and 28, describe and correct the error in finding DE. Point D is the centroid of $\triangle ABC$.

27.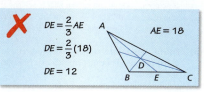

28.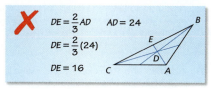

PROOF In Exercises 29 and 30, write a proof of the statement. *(See Example 4.)*

29. The angle bisector from the vertex angle to the base of an isosceles triangle is also a median.

30. The altitude from the vertex angle to the base of an isosceles triangle is also a perpendicular bisector.

CRITICAL THINKING In Exercises 31–36, complete the statement with *always*, *sometimes*, or *never*. Explain your reasoning.

31. The centroid is _____ on the triangle.

32. The orthocenter is _____ outside the triangle.

33. A median is _____ the same line segment as a perpendicular bisector.

34. An altitude is _____ the same line segment as an angle bisector.

35. The centroid and orthocenter are _____ the same point.

36. The centroid is _____ formed by the intersection of the three medians.

37. **WRITING** Compare an altitude of a triangle with a perpendicular bisector of a triangle.

38. **WRITING** Compare a median, an altitude, and an angle bisector of a triangle.

39. **MODELING WITH MATHEMATICS** Find the area of the triangular part of the paper airplane wing that is outlined in red. Which special segment of the triangle did you use?

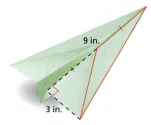

40. **ANALYZING RELATIONSHIPS** Copy and complete the statement for $\triangle DEF$ with centroid K and medians $\overline{DH}, \overline{EJ},$ and $\overline{FG}$.

   a. $EJ = $ _____ KJ b. $DK = $ _____ KH

   c. $FG = $ _____ KF d. $KG = $ _____ FG

MATHEMATICAL CONNECTIONS In Exercises 41–44, point D is the centroid of $\triangle ABC$. Use the given information to find the value of x.

41. $BD = 4x + 5$ and $BF = 9x$

42. $GD = 2x - 8$ and $GC = 3x + 3$

43. $AD = 5x$ and $DE = 3x - 2$

44. $DF = 4x - 1$ and $BD = 6x + 4$

45. **MATHEMATICAL CONNECTIONS** Graph the lines on the same coordinate plane. Find the centroid of the triangle formed by their intersections.

   $y_1 = 3x - 4$

   $y_2 = \frac{3}{4}x + 5$

   $y_3 = -\frac{3}{2}x - 4$

46. **CRITICAL THINKING** In what type(s) of triangles can a vertex be one of the points of concurrency of the triangle? Explain your reasoning.

Section 6.4 Medians and Altitudes of Triangles 367

ANSWERS

27. The length of $\overline{DE}$ should be $\frac{1}{3}$ of the length of $\overline{AE}$ because it is the shorter segment from the centroid to the side;
   $DE = \frac{1}{3}AE$
   $DE = \frac{1}{3}(18)$
   $DE = 6$

28. The length of $\overline{DE}$ is $\frac{1}{2}$ of the length of $\overline{AD}$ because $DE = \frac{1}{3}AE$ and $AD = \frac{2}{3}AE$;
   $DE = \frac{1}{2}AD$
   $DE = \frac{1}{2}(24)$
   $DE = 12$

29.

Legs $\overline{AB}$ and $\overline{BC}$ of isosceles $\triangle ABC$ are congruent. $\angle ABD \cong \angle CBD$ because $\overline{BD}$ is an angle bisector of vertex angle $\angle ABC$. Also, $\overline{BD} \cong \overline{BD}$ by the Reflexive Property of Congruence. So, $\triangle ABD \cong \triangle CBD$ by the SAS Congruence Theorem. $\overline{AD} \cong \overline{CD}$ because corresponding parts of congruent triangles are congruent. So, $\overline{BD}$ is a median.

30.

Legs $\overline{AB}$ and $\overline{BC}$ of isosceles $\triangle ABC$ are congruent. $\angle ADB$ and $\angle CDB$ are right angles because $\overline{BD}$ is an altitude from vertex angle $\angle ABC$ to base $\overline{AC}$ of $\triangle ABC$. So, $\triangle ABD$ and $\triangle CBD$ are right triangles. Also, $\overline{BD} \cong \overline{BD}$ by the Reflexive Property of Congruence. So, $\triangle ABD \cong \triangle CBD$ by the HL Congruence Theorem. $\overline{AD} \cong \overline{CD}$ because corresponding parts of congruent triangles are congruent. So, $\overline{BD}$ is a perpendicular bisector.

---

31. never; Because medians are always inside a triangle, and the centroid is the point of concurrency of the medians, it will always be inside the triangle.

32. sometimes; An orthocenter can be inside, on, or outside the triangle depending on whether the triangle is acute, right, or obtuse.

33. sometimes; A median is the same line segment as the perpendicular bisector if the triangle is equilateral or if the segment is connecting the vertex angle to the base of an isosceles triangle. Otherwise, the median and the perpendicular bisectors are not the same segment.

34. sometimes; An altitude is the same line segment as the angle bisector if the triangle is equilateral or if the segment is connecting the vertex angle to the base of an isosceles triangle. Otherwise, the altitude and the angle bisector are not the same segment.

35. sometimes; The centroid and the orthocenter are not the same point unless the triangle is equilateral.

36–46. See Additional Answers.

ANSWERS

47. $PE = \frac{1}{3}AE$, $PE = \frac{1}{2}AP$, $PE = AE - AP$

48. a. median; centroid
    b. altitude; orthocenter
    c. △JKM and △KLM both have an area of 4.5h; yes; The triangles formed by the median will always have the same area, because they will have the same base length and height.

49. yes; If the triangle is equilateral, then the perpendicular bisectors, angle bisectors, medians, and altitudes will all be the same three segments.

50. centroid; Because the triangles formed by the median of any triangle will always be congruent, the mass of the triangle on either side of the median is the same. So, the centroid is the point that has an equal distribution of mass on all sides.

51–54. See Additional Answers.

55. yes
56. no
57. no
58. yes

Mini-Assessment

1. Find the coordinates of the centroid of △ABC with vertices A(−1, 5), B(−3, −2), and C(1, 0). (−1, 1)

2. Point P is the centroid of △LMN, and QN = 12.6. Find PN and QP.

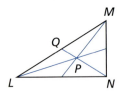

   PN = 8.4, QP = 4.2

3. Find the coordinates of the orthocenter of △DEF with vertices D(−1, −3), E(−1, 5), and F(5, −1). (1, −1)

368 Chapter 6

---

47. WRITING EQUATIONS Use the numbers and symbols to write three different equations for PE.

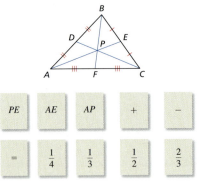

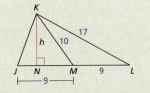

48. HOW DO YOU SEE IT? Use the figure.

a. What type of segment is $\overline{KM}$? Which point of concurrency lies on $\overline{KM}$?

b. What type of segment is $\overline{KN}$? Which point of concurrency lies on $\overline{KN}$?

c. Compare the areas of △JKM and △KLM. Do you think the areas of the triangles formed by the median of any triangle will always compare this way? Explain your reasoning.

49. MAKING AN ARGUMENT Your friend claims that it is possible for the circumcenter, incenter, centroid, and orthocenter to all be the same point. Do you agree? Explain your reasoning.

Maintaining Mathematical Proficiency
Reviewing what you learned in previous grades and lessons

Determine whether $\overline{AB}$ is parallel to $\overline{CD}$. *(Skills Review Handbook)*

55. A(5, 6), B(−1, 3), C(−4, 9), D(−16, 3)
56. A(−3, 6), B(5, 4), C(−14, −10), D(−2, −7)
57. A(6, −3), B(5, 2), C(−4, −4), D(−5, 2)
58. A(−5, 6), B(−7, 2), C(7, 1), D(4, −5)

368 Chapter 6 Relationships Within Triangles

---

50. DRAWING CONCLUSIONS The center of gravity of a triangle, the point where a triangle can balance on the tip of a pencil, is one of the four points of concurrency. Draw and cut out a large scalene triangle on a piece of cardboard. Which of the four points of concurrency is the center of gravity? Explain.

51. PROOF Prove that a median of an equilateral triangle is also an angle bisector, perpendicular bisector, and altitude.

52. THOUGHT PROVOKING Construct an acute scalene triangle. Find the orthocenter, centroid, and circumcenter. What can you conclude about the three points of concurrency?

53. CONSTRUCTION Follow the steps to construct a nine-point circle. Why is it called a nine-point circle?

Step 1 Construct a large acute scalene triangle.
Step 2 Find the orthocenter and circumcenter of the triangle.
Step 3 Find the midpoint between the orthocenter and circumcenter.
Step 4 Find the midpoint between each vertex and the orthocenter.
Step 5 Construct a circle. Use the midpoint in Step 3 as the center of the circle, and the distance from the center to the midpoint of a side of the triangle as the radius.

54. PROOF Prove the statements in parts (a)−(c).

Given $\overline{LP}$ and $\overline{MQ}$ are medians of scalene △LMN. Point R is on $\overrightarrow{LP}$ such that $\overline{LP} \cong \overline{PR}$. Point S is on $\overrightarrow{MQ}$ such that $\overline{MQ} \cong \overline{QS}$.

Prove a. $\overline{NS} \cong \overline{NR}$
       b. $\overline{NS}$ and $\overline{NR}$ are both parallel to $\overline{LM}$.
       c. R, N, and S are collinear.

If students need help...	If students got it...
Resources by Chapter • Practice A and Practice B • Puzzle Time	**Resources by Chapter** • Enrichment and Extension • Cumulative Review
Student Journal • Practice	Start the *next* Section
Differentiating the Lesson Skills Review Handbook	

6.1–6.4 What Did You Learn?

Core Vocabulary

proof, *p. 336*
two-column proof, *p. 336*
paragraph proof, *p. 337*
flowchart proof, or flow proof, *p. 338*
coordinate proof, *p. 339*
equidistant, *p. 344*
concurrent, *p. 352*
point of concurrency, *p. 352*
circumcenter, *p. 352*
incenter, *p. 355*
median of a triangle, *p. 362*
centroid, *p. 362*
altitude of a triangle, *p. 363*
orthocenter, *p. 363*

Core Concepts

Section 6.1
Writing Two-Column Proofs, *p. 336*
Writing Paragraph Proofs, *p. 337*
Writing Flowchart Proofs, *p. 338*
Writing Coordinate Proofs, *p. 339*

Section 6.2
Perpendicular Bisector Theorem, *p. 344*
Converse of the Perpendicular Bisector Theorem, *p. 344*
Angle Bisector Theorem, *p. 346*
Converse of the Angle Bisector Theorem, *p. 346*

Section 6.3
Circumcenter Theorem, *p. 352*
Incenter Theorem, *p. 355*

Section 6.4
Centroid Theorem, *p. 362*
Orthocenter, *p. 363*
Segments, Lines, Rays, and Points in Triangles, *p. 365*

Mathematical Practices

1. Did you make a plan before completing your proof in Exercise 37 on page 350? Describe your thought process.

2. What tools did you use to complete Exercises 17–20 on page 358? Describe how you could use technological tools to complete these exercises.

Rework Your Notes

A good way to reinforce concepts and put them into your long-term memory is to rework your notes. When you take notes, leave extra space on the pages. You can go back after class and fill in

- important definitions and rules,
- additional examples, and
- questions you have about the material.

6.1–6.4 Quiz

Find the indicated measure. Explain your reasoning. *(Section 6.2)*

1. UV
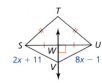

2. QP
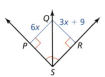

3. $m\angle GJK$

Find the coordinates of the circumcenter of the triangle with the given vertices. *(Section 6.3)*

4. $A(-4, 2)$, $B(-4, -4)$, $C(0, -4)$

5. $D(3, 5)$, $E(7, 9)$, $F(11, 5)$

The incenter of $\triangle ABC$ is point N. Use the given information to find the indicated measure. *(Section 6.3)*

6. $NQ = 2x + 1$, $NR = 4x - 9$
   Find NS.
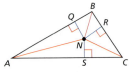

7. $NU = -3x + 6$, $NV = -5x$
   Find NT.
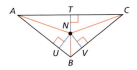

8. $NZ = 4x - 10$, $NY = 3x - 1$
   Find NW.
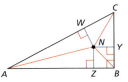

Find the coordinates of the centroid of the triangle with the given vertices. *(Section 6.4)*

9. $J(-1, 2)$, $K(5, 6)$, $L(5, -2)$

10. $M(-8, -6)$, $N(-4, -2)$, $P(0, -4)$

Tell whether the orthocenter is *inside*, *on*, or *outside* the triangle. Then find its coordinates. *(Section 6.4)*

11. $T(-2, 5)$, $U(0, 1)$, $V(2, 5)$

12. $X(-1, -4)$, $Y(7, -4)$, $Z(7, 4)$

13. A woodworker is cutting the largest wheel possible from a triangular scrap of wood. The wheel just touches each side of the triangle, as shown. *(Section 6.1 and Section 6.3)*

    a. Which point of concurrency is the center of the circle? What type of segments are $\overline{BG}$, $\overline{CG}$, and $\overline{AG}$?

    b. Prove that $\triangle BGF \cong \triangle BGE$.

    c. Find the radius of the wheel to the nearest tenth of a centimeter. Justify your answer.

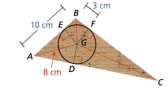

14. The Deer County Parks Committee plans to build a park at point P, equidistant from the three largest cities labeled X, Y, and Z. The map shown was created by the committee. *(Section 6.3 and Section 6.4)*

    a. Which point of concurrency did the committee use as the location of the park?

    b. Did the committee use the best point of concurrency for the location of the park? If not, which point would be better to use? Explain.

Laurie's Notes

Overview of Section 6.5

Introduction
- This lesson presents the last special segment in a triangle, the midsegment. The *midsegment* connects the midpoints of two sides of a triangle. It is parallel to the third side and half its length.
- The Triangle Midsegment Theorem is proven with a coordinate proof, integrating algebraic skills.

Resources
- Tracing paper is helpful when the explorations are done by paper folding.

Teaching Strategy
- If you do not have access to software, an alternate way to introduce this lesson would be through a paper-folding activity or a compass construction activity. It is powerful for students to explore and discover the properties of a midsegment.
- Give directions similar to what is written in the explorations for students to paper fold (or construct) the midsegments. Once all three midsegments are folded (or constructed), the construction can be used for Exploration 2.
- **Extension:** Paper fold the three midsegments. Cut out the original triangle. The three outer triangles will fold up to make a pyramid.

Extensions
- Have students investigate the Sierpinski triangle. The images are related to the midsegments of a triangle. (See Exercise 25.)

Pacing Suggestion
- The explorations allow students to explore and discover two properties of a midsegment that will be stated as a theorem in the formal lesson. When students have finished the explorations, transition to the formal lesson.

Dynamic Teaching Tools
Dynamic Assessment & Progress Monitoring Tool
Lesson Planning Tool
Interactive Whiteboard Lesson Library
Dynamic Classroom with Dynamic Investigations

What Your Students Will Learn

- Use midsegments of triangles in the coordinate plane.
- Use the Triangle Midsegment Theorem to find distances in planar figures.

Laurie's Notes

Exploration

Motivate

- Show a photo of a roof truss that contains a midsegment. More than likely, it will have two midsegments, as shown at the right in the Howe roof truss design.
- Trace lines over the midsegments and ask students whether they have any observations about the segments (not the lines).
- Explain to students that in this lesson they will explore properties of the midsegment.

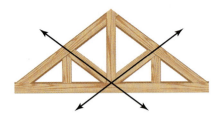

Exploration 1

- This is a straightforward construction using dynamic geometric software. Students should be comfortable at this point knowing that the triangle will be manipulated so that various cases can be explored.
- **?** "What conjecture can you make about midsegment $\overline{DE}$?" Students should mention that it is parallel to side $\overline{AC}$ and also half the length of side $\overline{AC}$.
- **Extension:** If time permits, have students construct the midsegment of △DBE. Ask students whether their conjecture for midsegment $\overline{DE}$ is also valid for the new midsegment.

Exploration 2

- This exploration is an extension of the previous exploration.
- **Construct Viable Arguments and Critique the Reasoning of Others:** There are a number of conjectures that students will make about the triangles formed when all three midsegments are constructed. The conjectures are likely connected to the conjectures made for a single midsegment. In each case, ask students to consider how they might prove their conjectures.
- Students may state that the four smaller triangles formed are congruent, that the four triangles have the same areas, and they have the same perimeters. If students compare the small triangles to △ABC, they may also say that the perimeter of each small triangle is $\frac{1}{2}$ the perimeter of △ABC and the area of each triangle is $\frac{1}{4}$ the area of △ABC.
- Be sure that students click and drag on different vertices to see whether their conjectures hold true for various triangles.

Communicate Your Answer

- Students should state that the midsegment of a triangle is parallel to the third side of the triangle and half the length of the third side.

Connecting to Next Step

- The explorations are related to the theorem presented in the lesson, helping students make sense of the theorem before proving it to be true.

6.5 The Triangle Midsegment Theorem

Essential Question How are the midsegments of a triangle related to the sides of the triangle?

EXPLORATION 1   Midsegments of a Triangle

Work with a partner. Use dynamic geometry software. Draw any △ABC.

a. Plot midpoint D of $\overline{AB}$ and midpoint E of $\overline{BC}$. Draw $\overline{DE}$, which is a midsegment of △ABC.

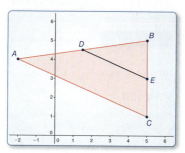

Sample
Points
A(−2, 4)
B(5, 5)
C(5, 1)
D(1.5, 4.5)
E(5, 3)
Segments
BC = 4
AC = 7.62
AB = 7.07
DE = ?

b. Compare the slope and length of $\overline{DE}$ with the slope and length of $\overline{AC}$.

c. Write a conjecture about the relationships between the midsegments and sides of a triangle. Test your conjecture by drawing the other midsegments of △ABC, dragging vertices to change △ABC, and noting whether the relationships hold.

CONSTRUCTING VIABLE ARGUMENTS
To be proficient in math, you need to make conjectures and build a logical progression of statements to explore the truth of your conjectures.

EXPLORATION 2   Midsegments of a Triangle

Work with a partner. Use dynamic geometry software. Draw any △ABC.

a. Draw all three midsegments of △ABC.

b. Use the drawing to write a conjecture about the triangle formed by the midsegments of the original triangle.

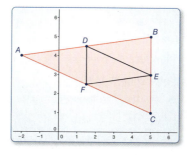

Sample
Points
A(−2, 4)
B(5, 5)
C(5, 1)
D(1.5, 4.5)
E(5, 3)

Segments
BC = 4
AC = 7.62
AB = 7.07
DE = ?
DF = ?
EF = ?

Communicate Your Answer

3. How are the midsegments of a triangle related to the sides of the triangle?

4. In △RST, $\overline{UV}$ is the midsegment connecting the midpoints of $\overline{RS}$ and $\overline{ST}$. Given UV = 12, find RT.

Section 6.5   The Triangle Midsegment Theorem   371

Dynamic Teaching Tools
Dynamic Assessment & Progress Monitoring Tool
Lesson Planning Tool
Interactive Whiteboard Lesson Library
Dynamic Classroom with Dynamic Investigations

ANSWERS

1. a. Check students' work.
   b. *Sample answer:* The slopes are the same (−0.43). Because 3.81 = $\frac{1}{2}$(7.62), DE = $\frac{1}{2}$AC.
   c. The segment connecting the midpoints of two sides of a triangle is parallel to the third side and is half as long as that side.

2. a. Check students' work.
   b. The triangle formed by the midsegments of a triangle, △EFD, is similar to the original triangle, △ABC, and has a scale factor of $\frac{1}{2}$.

3. Each midsegment is parallel to a side of the triangle and is half as long as that side.

4. 24

Differentiated Instruction

Kinesthetic
Some students might benefit by having access to a ruler, compass, and protractor. Students can use these tools to construct the midsegment of a triangle and then verify the Triangle Midsegment Theorem.

Extra Example 1

In △RST, show that midsegment $\overline{MN}$ is parallel to $\overline{RS}$ and that $MN = \frac{1}{2}RS$.

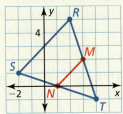

The vertices of the triangle are $R(2, 5)$, $S(-2, 1)$, and $T(4, -1)$.

Find the coordinates of midpoints M and N. $M = \left(\frac{2+4}{2}, \frac{5+(-1)}{2}\right) = (3, 2)$

and $N = \left(\frac{-2+4}{2}, \frac{1+(-1)}{2}\right) = (1, 0)$.

Compare the slopes of $\overline{MN}$ and $\overline{RS}$.

The slope of $\overline{MN} = \frac{2-0}{3-1} = 1$.

The slope of $\overline{RS} = \frac{5-1}{2-(-2)} = 1$. Because the slopes are the same, $\overline{MN} \parallel \overline{RS}$.

Compare the lengths of $\overline{MN}$ and $\overline{RS}$.

$MN = \sqrt{(3-1)^2 + (2-0)^2} = \sqrt{8} = 2\sqrt{2}$

and $RS = \sqrt{(2-(-2))^2 + (5-1)^2}$
$= \sqrt{32} = 4\sqrt{2}$.

Because $2\sqrt{2} = \frac{1}{2}(4\sqrt{2})$, $MN = \frac{1}{2}RS$.

MONITORING PROGRESS ANSWERS

1. Because the slopes of $\overline{DE}$ and $\overline{AC}$ are the same $\left(\frac{3}{2}\right)$, $\overline{DE} \parallel \overline{AC}$.
$DE = \sqrt{13}$ and $AC = 2\sqrt{13}$. Because $\sqrt{13} = \frac{1}{2}(2\sqrt{13})$, $DE = \frac{1}{2}AC$.

2. $E(2, 2)$, $F(3, -3)$; Because the slopes of $\overline{EF}$ and $\overline{AB}$ are the same (-5), $\overline{EF} \parallel \overline{AB}$. $EF = \sqrt{26}$ and $AB = 2\sqrt{26}$. Because $\sqrt{26} = \frac{1}{2}(2\sqrt{26})$, $EF = \frac{1}{2}AB$.

6.5 Lesson

What You Will Learn

- Use midsegments of triangles in the coordinate plane.
- Use the Triangle Midsegment Theorem to find distances.

Core Vocabulary

midsegment of a triangle, p. 372

Previous
midpoint
parallel
slope
coordinate proof

Using the Midsegment of a Triangle

A **midsegment of a triangle** is a segment that connects the midpoints of two sides of the triangle. Every triangle has three midsegments, which form the *midsegment triangle*.

The midsegments of △ABC at the right are $\overline{MP}$, $\overline{MN}$, and $\overline{NP}$. The *midsegment triangle* is △MNP.

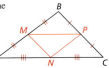

EXAMPLE 1 Using Midsegments in the Coordinate Plane

In △JKL, show that midsegment $\overline{MN}$ is parallel to $\overline{JL}$ and that $MN = \frac{1}{2}JL$.

SOLUTION

Step 1 Find the coordinates of M and N by finding the midpoints of $\overline{JK}$ and $\overline{KL}$.

$M\left(\frac{-6+(-2)}{2}, \frac{1+5}{2}\right) = M\left(\frac{-8}{2}, \frac{6}{2}\right) = M(-4, 3)$

$N\left(\frac{-2+2}{2}, \frac{5+(-1)}{2}\right) = N\left(\frac{0}{2}, \frac{4}{2}\right) = N(0, 2)$

Step 2 Find and compare the slopes of $\overline{MN}$ and $\overline{JL}$.

slope of $\overline{MN} = \frac{2-3}{0-(-4)} = -\frac{1}{4}$ slope of $\overline{JL} = \frac{-1-1}{2-(-6)} = -\frac{2}{8} = -\frac{1}{4}$

▶ Because the slopes are the same, $\overline{MN}$ is parallel to $\overline{JL}$.

Step 3 Find and compare the lengths of $\overline{MN}$ and $\overline{JL}$.

$MN = \sqrt{[0-(-4)]^2 + (2-3)^2} = \sqrt{16+1} = \sqrt{17}$

$JL = \sqrt{[2-(-6)]^2 + (-1-1)^2} = \sqrt{64+4} = \sqrt{68} = 2\sqrt{17}$

▶ Because $\sqrt{17} = \frac{1}{2}(2\sqrt{17})$, $MN = \frac{1}{2}JL$.

READING
In the figure for Example 1, midsegment $\overline{MN}$ can be called "the midsegment opposite $\overline{JL}$."

Monitoring Progress Help in English and Spanish at *BigIdeasMath.com*

Use the graph of △ABC.

1. In △ABC, show that midsegment $\overline{DE}$ is parallel to $\overline{AC}$ and that $DE = \frac{1}{2}AC$.

2. Find the coordinates of the endpoints of midsegment $\overline{EF}$, which is opposite $\overline{AB}$. Show that $\overline{EF}$ is parallel to $\overline{AB}$ and that $EF = \frac{1}{2}AB$.

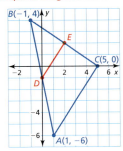

372   Chapter 6   Relationships Within Triangles

Laurie's Notes — Teacher Actions

- Students will be familiar with midsegments and the midsegment triangle from the explorations.
- **Thumbs Up:** Example 1 tests the conjectures made about a midsegment. Verify with *Thumbs Up* that students understand the problem. Give partners time to complete the example.
- Students may not recognize that $MN = \frac{1}{2}JL$ unless they simplify the radical $\sqrt{68}$. Review this skill if needed.

Using the Triangle Midsegment Theorem

Theorem

Triangle Midsegment Theorem

The segment connecting the midpoints of two sides of a triangle is parallel to the third side and is half as long as that side.

$\overline{DE}$ is a midsegment of $\triangle ABC$, $\overline{DE} \parallel \overline{AC}$, and $DE = \frac{1}{2}AC$.

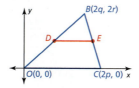

Proof Example 2, p. 373; Monitoring Progress Question 3, p. 373; Ex. 22, p. 376

EXAMPLE 2 Proving the Triangle Midsegment Theorem

Write a coordinate proof of the Triangle Midsegment Theorem for one midsegment.

Given $\overline{DE}$ is a midsegment of $\triangle OBC$.

Prove $\overline{DE} \parallel \overline{OC}$ and $DE = \frac{1}{2}OC$

STUDY TIP
When assigning coordinates, try to choose coordinates that make some of the computations easier. In Example 2, you can avoid fractions by using $2p$, $2q$, and $2r$.

SOLUTION

Step 1 Place $\triangle OBC$ in a coordinate plane and assign coordinates. Because you are finding midpoints, use $2p$, $2q$, and $2r$. Then find the coordinates of D and E.

$D\left(\dfrac{2q + 0}{2}, \dfrac{2r + 0}{2}\right) = D(q, r)$ $E\left(\dfrac{2q + 2p}{2}, \dfrac{2r + 0}{2}\right) = E(q + p, r)$

Step 2 Prove $\overline{DE} \parallel \overline{OC}$. The y-coordinates of D and E are the same, so $\overline{DE}$ has a slope of 0. $\overline{OC}$ is on the x-axis, so its slope is 0.

▶ Because their slopes are the same, $\overline{DE} \parallel \overline{OC}$.

Step 3 Prove $DE = \frac{1}{2}OC$. Use the Ruler Postulate to find DE and OC.

$DE = |(q + p) - q| = p$ $OC = |2p - 0| = 2p$

▶ Because $p = \frac{1}{2}(2p)$, $DE = \frac{1}{2}OC$.

Monitoring Progress Help in English and Spanish at *BigIdeasMath.com*

3. In Example 2, find the coordinates of F, the midpoint of $\overline{OC}$. Show that $\overline{FE} \parallel \overline{OB}$ and $FE = \frac{1}{2}OB$.

EXAMPLE 3 Using the Triangle Midsegment Theorem

Triangles are used for strength in roof trusses. In the diagram, $\overline{UV}$ and $\overline{VW}$ are midsegments of $\triangle RST$. Find UV and RS.

SOLUTION

$UV = \frac{1}{2} \cdot RT = \frac{1}{2}(90 \text{ in.}) = 45 \text{ in.}$

$RS = 2 \cdot VW = 2(57 \text{ in.}) = 114 \text{ in.}$

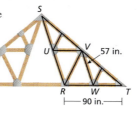

Extra Example 2
Use the diagram in Example 2. Find the coordinates of F, the midpoint of $\overline{OC}$. Show that $\overline{FD} \parallel \overline{BC}$ and $FD = \frac{1}{2}BC$.
The coordinates of F are $(p, 0)$. The slope of $\overline{FD} = \dfrac{r - 0}{q - p} = \dfrac{r}{q - p}$. The slope of $\overline{BC} = \dfrac{2r - 0}{2q - 2p} = \dfrac{2r}{2(q - p)} = \dfrac{r}{q - p}$.
Because their slopes are =, $\overline{FD} \parallel \overline{BC}$.
By the Distance Formula,
$FD = \sqrt{(q - p)^2 + r^2}$ and
$BC = \sqrt{(2q - 2p)^2 + (2r)^2}$
$= \sqrt{[2(q - p)]^2 + 4r^2}$
$= \sqrt{2^2(q - p)^2 + 4r^2}$
$= \sqrt{4}\sqrt{(q - p)^2 + r^2}$
$= 2\sqrt{(q - p)^2 + r^2}$.
Therefore, $FD = \frac{1}{2}BC$.

Extra Example 3
$\overline{DE}$ is a midsegment of $\triangle ABC$. Find AC.

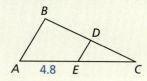

$AC = 9.6$

MONITORING PROGRESS ANSWER

3. $F(p, 0)$; Because the slopes of $\overline{FE}$ and $\overline{OB}$ are the same $\left(\dfrac{r}{q}\right)$, $\overline{FE} \parallel \overline{OB}$. Using the Distance Formula,
$FE = \sqrt{q^2 + r^2}$
and $OB = 2\sqrt{q^2 + r^2}$. Because $\sqrt{q^2 + r^2} = \frac{1}{2}\left(2\sqrt{q^2 + r^2}\right)$,
$FE = \frac{1}{2}OB$.

Laurie's Notes   Teacher Actions

- State the Triangle Midsegment Theorem. *Note:* This theorem is sometimes presented later in the course after similar triangles have been introduced. Logically, it fits with the other special segments in a triangle.
- Explain the *Study Tip* to students.
- ? "How can you find the coordinates of points D and E?" *Use the Midpoint Formula.* "How will you show that the midsegment is parallel to $\overline{OC}$?" *Find the slope of each segment.* "How will you show $DE = \frac{1}{2}OC$?" *Use the Distance Formula.* Give partners time to complete the proof.

Extra Example 4
In the figure in Example 4, $CF = FB$ and $CD = DA$. Which segments must be parallel? $\overline{DF}$ and $\overline{AB}$

Extra Example 5
A walking path $\overline{BD}$ in a park intersects the sides of the park at their midpoints. You walk from park corner A to walking path $\overline{BD}$, over the path to the other side of the park, up to corner E, and then back down to your starting point. How many yards do you walk?

402 yards

MONITORING PROGRESS ANSWERS

4.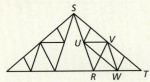

   midsegment: $\overline{UW}$; 81 in.

5. $\overline{DF}$ is a midsegment, $\overline{DF} \parallel \overline{AB}$, $\overline{DF} = \frac{1}{2}AB$

6. yes; The distance along the new route is 6.775 miles.

EXAMPLE 4 Using the Triangle Midsegment Theorem

In the kaleidoscope image, $\overline{AE} \cong \overline{BE}$ and $\overline{AD} \cong \overline{CD}$. Show that $\overline{CB} \parallel \overline{DE}$.

SOLUTION

Because $\overline{AE} \cong \overline{BE}$ and $\overline{AD} \cong \overline{CD}$, E is the midpoint of $\overline{AB}$ and D is the midpoint of $\overline{AC}$ by definition. Then $\overline{DE}$ is a midsegment of $\triangle ABC$ by definition and $\overline{CB} \parallel \overline{DE}$ by the Triangle Midsegment Theorem.

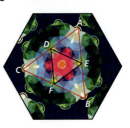

EXAMPLE 5 Modeling with Mathematics

Pear Street intersects Cherry Street and Peach Street at their midpoints. Your home is at point P. You leave your home and jog down Cherry Street to Plum Street, over Plum Street to Peach Street, up Peach Street to Pear Street, over Pear Street to Cherry Street, and then back home up Cherry Street. About how many miles do you jog?

SOLUTION

1. **Understand the Problem** You know the distances from your home to Plum Street along Peach Street, from Peach Street to Cherry Street along Plum Street, and from Pear Street to your home along Cherry Street. You need to find the other distances on your route, then find the total number of miles you jog.

2. **Make a Plan** By definition, you know that Pear Street is a midsegment of the triangle formed by the other three streets. Use the Triangle Midsegment Theorem to find the length of Pear Street and the definition of midsegment to find the length of Cherry Street. Then add the distances along your route.

3. **Solve the Problem**

   length of Pear Street $= \frac{1}{2} \cdot$ (length of Plum St.) $= \frac{1}{2}(1.4\ \text{mi}) = 0.7\ \text{mi}$

   length of Cherry Street $= 2 \cdot$ (length from P to Pear St.) $= 2(1.3\ \text{mi}) = 2.6\ \text{mi}$

   distance along your route: $2.6 + 1.4 + \frac{1}{2}(2.25) + 0.7 + 1.3 = 7.125$

   ▶ So, you jog about 7 miles.

4. **Look Back** Use compatible numbers to check that your answer is reasonable.

   total distance:

   $2.6 + 1.4 + \frac{1}{2}(2.25) + 0.7 + 1.3 \approx 2.5 + 1.5 + 1 + 0.5 + 1.5 = 7$

Monitoring Progress Help in English and Spanish at BigIdeasMath.com

4. Copy the diagram in Example 3. Draw and name the third midsegment. Then find the length of $\overline{VS}$ when the length of the third midsegment is 81 inches.

5. In Example 4, if F is the midpoint of $\overline{CB}$, what do you know about $\overline{DF}$?

6. **WHAT IF?** In Example 5, you jog down Peach Street to Plum Street, over Plum Street to Cherry Street, up Cherry Street to Pear Street, over Pear Street to Peach Street, and then back home up Peach Street. Do you jog more miles in Example 5? Explain.

374 Chapter 6 Relationships Within Triangles

Laurie's Notes — Teacher Actions

- If you have a kaleidoscope, bring it to class for students to view. Traditional kaleidoscopes will have several midsegments visible.
- **Think-Pair-Share:** Have students work independently to answer Example 5. When students have finished, discuss the answers.
- **? Probing Question:** "If you found the segment midway between Pear Street and Plum Street and parallel to both, what would its length be?" 1.05 miles

Closure

- A midsegment triangle has side lengths of 3 inches, 4 inches, and 5 inches. Sketch the original triangle. What are the lengths of its sides? Check students' work. The triangle will have side lengths of 6 inches, 8 inches, and 10 inches.

6.5 Exercises

Dynamic Solutions available at BigIdeasMath.com

Vocabulary and Core Concept Check

1. **VOCABULARY** The _____ of a triangle is a segment that connects the midpoints of two sides of the triangle.

2. **COMPLETE THE SENTENCE** If $\overline{DE}$ is the midsegment opposite $\overline{AC}$ in $\triangle ABC$, then $\overline{DE} \parallel \overline{AC}$ and $DE = \underline{\quad} AC$ by the Triangle Midsegment Theorem.

Monitoring Progress and Modeling with Mathematics

In Exercises 3–6, use the graph of $\triangle ABC$ with midsegments $\overline{DE}, \overline{EF},$ and $\overline{DF}$. (See Example 1.)

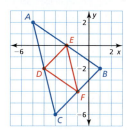

3. Find the coordinates of points D, E, and F.

4. Show that $\overline{DE}$ is parallel to $\overline{CB}$ and that $DE = \frac{1}{2}CB$.

5. Show that $\overline{EF}$ is parallel to $\overline{AC}$ and that $EF = \frac{1}{2}AC$.

6. Show that $\overline{DF}$ is parallel to $\overline{AB}$ and that $DF = \frac{1}{2}AB$.

In Exercises 7–10, $\overline{DE}$ is a midsegment of $\triangle ABC$. Find the value of x. (See Example 3.)

7. 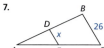   8.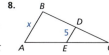

9. 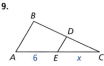   10.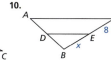

In Exercises 11–16, $\overline{XJ} \cong \overline{JY}, \overline{YL} \cong \overline{LZ},$ and $\overline{XK} \cong \overline{KZ}$. Copy and complete the statement. (See Example 4.)

11. $\overline{JK} \parallel$ ___   12. $\overline{JL} \parallel$ ___

13. $\overline{XY} \parallel$ ___   14. $\overline{JY} \cong$ ___ $\cong$ ___

15. $\overline{JL} \cong$ ___ $\cong$ ___   16. $\overline{JK} \cong$ ___ $\cong$ ___

MATHEMATICAL CONNECTIONS In Exercises 17–19, use $\triangle GHJ$, where A, B, and C are midpoints of the sides.

17. When $AB = 3x + 8$ and $GJ = 2x + 24$, what is AB?

18. When $AC = 3y - 5$ and $HJ = 4y + 2$, what is HB?

19. When $GH = 7z - 1$ and $CB = 4z - 3$, what is GA?

20. **ERROR ANALYSIS** Describe and correct the error.

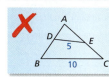

Section 6.5   The Triangle Midsegment Theorem   375

Assignment Guide and Homework Check

ASSIGNMENT
Basic: 1, 2, 3–19 odd, 20, 21, 26, 28, 29
Average: 1–3, 6–22 even, 26–29
Advanced: 1–3, 10–22 even, 23–29

HOMEWORK CHECK
Basic: 3, 5, 9, 15, 17
Average: 3, 6, 8, 18, 22
Advanced: 10, 18, 20, 22, 25

ANSWERS

1. midsegment
2. $\frac{1}{2}$
3. $D(-4, -2), E(-2, 0), F(-1, -4)$
4. Because the slopes of $\overline{DE}$ and $\overline{CB}$ are the same (1), $\overline{DE} \parallel \overline{CB}$. $DE = 2\sqrt{2}$ and $CB = 4\sqrt{2}$. Because $2\sqrt{2} = \frac{1}{2}(4\sqrt{2})$, $DE = \frac{1}{2}CB$.
5. Because the slopes of $\overline{EF}$ and $\overline{AC}$ are the same (-4), $\overline{EF} \parallel \overline{AC}$. $EF = \sqrt{17}$ and $AC = 2\sqrt{17}$. Because $\sqrt{17} = \frac{1}{2}(2\sqrt{17})$, $EF = \frac{1}{2}AC$.
6. Because the slopes of $\overline{DF}$ and $\overline{AB}$ are the same $\left(-\frac{2}{3}\right)$, $\overline{DF} \parallel \overline{AB}$. $DF = \sqrt{13}$ and $AB = 2\sqrt{13}$. Because $\sqrt{13} = \frac{1}{2}(2\sqrt{13})$, $DF = \frac{1}{2}AB$.
7. $x = 13$
8. $x = 10$
9. $x = 6$
10. $x = 8$
11. $\overline{JK} \parallel \overline{YZ}$
12. $\overline{JL} \parallel \overline{XZ}$
13. $\overline{XY} \parallel \overline{KL}$
14. $\overline{JY} \cong \overline{JX} \cong \overline{KL}$
15. $\overline{JL} \cong \overline{XK} \cong \overline{KZ}$
16. $\overline{JK} \cong \overline{YL} \cong \overline{LZ}$
17. 14
18. 13
19. 17
20. $\overline{DE}$ is not parallel to $\overline{BC}$. So, $\overline{DE}$ is not a midsegment. So, according to the contrapositive of the Triangle Midsegment Theorem, $\overline{DE}$ does not connect the midpoints of $\overline{AC}$ and $\overline{AB}$.

Dynamic Teaching Tools
- Dynamic Assessment & Progress Monitoring Tool
- Interactive Whiteboard Lesson Library
- Dynamic Classroom with Dynamic Investigations

ANSWERS
21. 45 ft

22–29. See Additional Answers.

Mini-Assessment

1. A, B, and C are the midpoints of the sides of △GHJ. AB = 3.4, BC = 4.2, and AC = 3.8. Find the perimeter of △GHJ.

22.8

2. In △RST, show that midsegment $\overline{MN}$ is parallel to $\overline{ST}$.

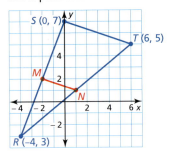

The coordinates of midpoints M and N are M(−2, 2) and N(1, 1). The slopes of both $\overline{MN}$ and $\overline{ST} = -\frac{1}{3}$. Because the slopes are the same, $\overline{MN}$ and $\overline{ST}$ are parallel.

3. $\overline{DE}$ is a midsegment of △ABC. $\overline{FG}$ is a midsegment of △EDC. AE = 22. Find FC.

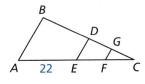

FC = 11

376   Chapter 6

---

21. **MODELING WITH MATHEMATICS** The distance between consecutive bases on a baseball field is 90 feet. A second baseman stands halfway between first base and second base, a shortstop stands halfway between second base and third base, and a pitcher stands halfway between first base and third base. Find the distance between the shortstop and the pitcher. *(See Example 5.)*

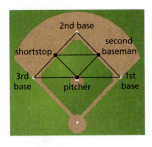

22. **PROVING A THEOREM** Use the figure from Example 2 to prove the Triangle Midsegment Theorem for midsegment $\overline{DF}$, where F is the midpoint of $\overline{OC}$. *(See Example 2.)*

23. **CRITICAL THINKING** $\overline{XY}$ is a midsegment of △LMN. Suppose $\overline{DE}$ is called a "quarter segment" of △LMN. What do you think an "eighth segment" would be? Make conjectures about the properties of a quarter segment and an eighth segment. Use variable coordinates to verify your conjectures.

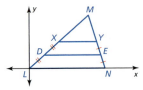

24. **THOUGHT PROVOKING** Find a real-life object that uses midsegments as part of its structure. Print a photograph of the object and identify the midsegments of one of the triangles in the structure.

Maintaining Mathematical Proficiency Reviewing what you learned in previous grades and lessons

Find a counterexample to show that the conjecture is false. *(Skills Review Handbook)*

28. The difference of two numbers is always less than the greater number.

29. An isosceles triangle is always equilateral.

376   Chapter 6   Relationships Within Triangles

---

25. **ABSTRACT REASONING** To create the design shown, shade the triangle formed by the three midsegments of the triangle. Then repeat the process for each unshaded triangle.

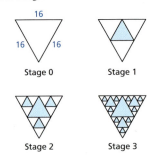

a. What is the perimeter of the shaded triangle in Stage 1?

b. What is the total perimeter of all the shaded triangles in Stage 2?

c. What is the total perimeter of all the shaded triangles in Stage 3?

26. **HOW DO YOU SEE IT?** Explain how you know that the yellow triangle is the midsegment triangle of the red triangle in the pattern of floor tiles shown.

27. **ATTENDING TO PRECISION** The points P(2, 1), Q(4, 5), and R(7, 4) are the midpoints of the sides of a triangle. Graph the three midsegments. Then show how to use your graph and the properties of midsegments to draw the original triangle. Give the coordinates of each vertex.

If students need help...	If students got it...
Resources by Chapter • Practice A and Practice B • Puzzle Time	**Resources by Chapter** • Enrichment and Extension • Cumulative Review
Student Journal • Practice	Start the *next* Section
Differentiating the Lesson Skills Review Handbook	

Laurie's Notes

Overview of Section 6.6

Introduction
- This lesson presents two theorems that relate the side lengths and angle measures of a triangle. This relationship is first discovered in the explorations and then proven in the lesson using an indirect proof.
- The indirect proof method can be challenging for students to understand and grasp. It feels unnatural to assume something is true that is the negation of what you were first trying to prove true.
- The Triangle Inequality Theorem is introduced, and students use the theorem to find possible lengths for the third side of a triangle when two side lengths are known.

Formative Assessment Tips
- **Muddiest Point:** This technique is the opposite of *Point of Most Significance* (Section 2.3). Students are asked to reflect on the most difficult or confusing point in the lesson. Their written comments are collected at the end of the lesson so that the following day's instruction can address the confusion.
- This technique can be used at any time during instruction.
- It is important for teachers to know whether there was a point in the lesson that was confusing for students so that the lesson can be modified. Share with students what you learned from their reflections. Students will take the reflections more seriously when they are valued and used.

Applications
- When discussing the Triangle Inequality Theorem, ask students how many have cut across lawns or parking lots instead of following sidewalks. It is likely that all students have. Long before giving a name to the theorem, students know that cutting across a diagonal is faster (shorter distance) than sticking to the sidewalks.

Pacing Suggestion
- The explorations allow students to explore and discover relationships between angle measures and side lengths of a triangle that will be stated as theorems in the formal lesson. When students have finished the explorations, transition to the formal lesson.

Dynamic Teaching Tools
Dynamic Assessment & Progress Monitoring Tool
Lesson Planning Tool
Interactive Whiteboard Lesson Library
Dynamic Classroom with Dynamic Investigations

What Your Students Will Learn

- Prove geometric relationship using indirect proofs.
- List sides and angles of a triangle in order by size and determine relationships between the sides and angles.
- Use the Triangle Inequality Theorem to find possible side lengths of triangles and write inequalities comparing side lengths.

Laurie's Notes

Exploration

Motivate
- Give each student three straws of various lengths, some of which form triangles and some of which do not. Have students investigate when three straws can form a triangle and when they cannot. Tell students that they will describe and examine this property in the lesson.

Exploration 1
- Students will be able to quickly measure the sides and angles of scalene $\triangle ABC$.
- Students may not observe a connection between the angle measures and side lengths. When students order the side lengths, have them write the name of the side, not simply the measure. When they order the angle measures, have them write the name of the angle, not simply the measure.
- If students do not observe any relationship, suggest that they complete part (c).
- **? Probing Question:** "If $\angle A$ has the greatest measure, what side would you expect to be the longest?" This may result in students observing the relationship between angle measures and side lengths in a triangle.

Exploration 2
- **Attend to Precision:** When asked to compare each side length with the sum of the lengths of the other two sides, students should be thinking about less than (<) and greater than (>). This is not always obvious to students who try to make more out of "compare" than it is.
- When you click and drag a vertex toward the opposite side so that the triangle begins to collapse, the sum of the two side lengths gets very close to the length of the third side.

Communicate Your Answer
- Students should state that the longest side of a triangle is opposite the largest angle and the shortest side is opposite the smallest angle.

Connecting to Next Step
- The explorations are related to the theorems presented in the lesson, one of which will be proven using an indirect proof.

6.6 Indirect Proof and Inequalities in One Triangle

Essential Question How are the sides related to the angles of a triangle? How are any two sides of a triangle related to the third side?

EXPLORATION 1 Comparing Angle Measures and Side Lengths

Work with a partner. Use dynamic geometry software. Draw any scalene △ABC.

a. Find the side lengths and angle measures of the triangle.

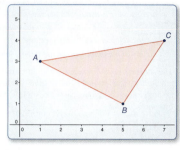
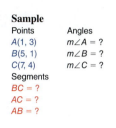

Sample
Points
A(1, 3)
B(5, 1)
C(7, 4)
Segments
BC = ?
AC = ?
AB = ?

Angles
m∠A = ?
m∠B = ?
m∠C = ?

ATTENDING TO PRECISION
To be proficient in math, you need to express numerical answers with a degree of precision appropriate for the content.

b. Order the side lengths. Order the angle measures. What do you observe?

c. Drag the vertices of △ABC to form new triangles. Record the side lengths and angle measures in a table. Write a conjecture about your findings.

EXPLORATION 2 A Relationship of the Side Lengths of a Triangle

Work with a partner. Use dynamic geometry software. Draw any △ABC.

a. Find the side lengths of the triangle.

b. Compare each side length with the sum of the other two side lengths.

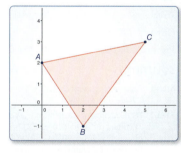

Sample
Points
A(0, 2)
B(2, −1)
C(5, 3)
Segments
BC = ?
AC = ?
AB = ?

c. Drag the vertices of △ABC to form new triangles and repeat parts (a) and (b). Organize your results in a table. Write a conjecture about your findings.

Communicate Your Answer

3. How are the sides related to the angles of a triangle? How are any two sides of a triangle related to the third side?

4. Is it possible for a triangle to have side lengths of 3, 4, and 10? Explain.

Section 6.6 Indirect Proof and Inequalities in One Triangle 377

ANSWERS

1. a. Check students' work. (For sample in text, $AB \approx 4.47$, $AC \approx 6.08$, $BC \approx 3.61$, $m\angle A \approx 36.03°$, $m\angle B \approx 97.13°$, and $m\angle C \approx 46.85°$.)

   b. Check students' work. (For sample in text, $BC < AB < AC$ and $m\angle A < m\angle C < m\angle B$.); The shortest side is across from the smallest angle, and the longest side is across from the largest angle.

   c. See Additional Answers.

2. a. Check students' work. (For sample in text, $AB \approx 3.61$, $AC \approx 5.10$, and $BC = 5$.)

   b. For the sample in the text, $BC = 5 < 8.71 = AC + AB$, $AC = 5.10 < 8.61 = BC + AB$, and $AB = 3.61 < 10.10 = BC + AC$.

   c. See Additional Answers.

3. The largest angle is opposite the longest side, and the smallest angle is opposite the shortest side; The sum of any two side lengths is greater than the third side length.

4. no; The sum $3 + 4$ is not greater than 10, and it is not possible to form a triangle when the sum of the lengths of the two sides is less than the length of the third side.

English Language Learners

Graphic Organizer
Have students create a Process Diagram to organize the steps for writing an indirect proof.

Extra Example 1

Write an indirect proof.
Given Line ℓ is not parallel to line k.
Prove $\angle 3$ and $\angle 5$ are not supplementary.

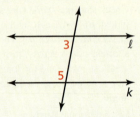

Assume temporarily that $\angle 3$ and $\angle 5$ are supplementary. By the Converse of the Consecutive Interior Angles Theorem, line ℓ is parallel to line k. This contradicts the given information. So, the assumption must be false, which proves that $\angle 3$ and $\angle 5$ are not supplementary.

MONITORING PROGRESS ANSWER

1. Assume temporarily that $\triangle ABC$ is a scalene triangle with $\angle A \cong \angle B$. By the Converse of Base Angles Theorem, if $\angle A \cong \angle B$, then the opposite sides are congruent: $\overline{BC} \cong \overline{AC}$. A scalene triangle cannot have two congruent sides. So, this contradicts the given information. So, the assumption that $\triangle ABC$ is a scalene triangle with two congruent angles must be false, which proves that a scalene triangle cannot have two congruent angles.

6.6 Lesson

Core Vocabulary
indirect proof, p. 378
Previous
proof
inequality

What You Will Learn

- Write indirect proofs.
- List sides and angles of a triangle in order by size.
- Use the Triangle Inequality Theorem to find possible side lengths of triangles.

Writing an Indirect Proof

Suppose a student looks around the cafeteria, concludes that hamburgers are not being served, and explains as follows.

At first, I assumed that we are having hamburgers because today is Tuesday, and Tuesday is usually hamburger day.

There is always ketchup on the table when we have hamburgers, so I looked for the ketchup, but I didn't see any.

So, my assumption that we are having hamburgers must be false.

The student uses *indirect* reasoning. In an **indirect proof**, you start by making the temporary assumption that the desired conclusion is false. By then showing that this assumption leads to a logical impossibility, you prove the original statement true *by contradiction*.

Core Concept

How to Write an Indirect Proof (Proof by Contradiction)

Step 1 Identify the statement you want to prove. Assume temporarily that this statement is false by assuming that its opposite is true.

Step 2 Reason logically until you reach a contradiction.

Step 3 Point out that the desired conclusion must be true because the contradiction proves the temporary assumption false.

EXAMPLE 1 Writing an Indirect Proof

Write an indirect proof that in a given triangle, there can be at most one right angle.

Given $\triangle ABC$

Prove $\triangle ABC$ can have at most one right angle.

SOLUTION

Step 1 Assume temporarily that $\triangle ABC$ has two right angles. Then assume $\angle A$ and $\angle B$ are right angles.

Step 2 By the definition of right angle, $m\angle A = m\angle B = 90°$. By the Triangle Sum Theorem, $m\angle A + m\angle B + m\angle C = 180°$. Using the Substitution Property of Equality, $90° + 90° + m\angle C = 180°$. So, $m\angle C = 0°$ by the Subtraction Property of Equality. A triangle cannot have an angle measure of $0°$. So, this contradicts the given information.

Step 3 So, the assumption that $\triangle ABC$ has two right angles must be false, which proves that $\triangle ABC$ can have at most one right angle.

READING
You have reached a *contradiction* when you have two statements that cannot both be true at the same time.

Monitoring Progress Help in English and Spanish at *BigIdeasMath.com*

1. Write an indirect proof that a scalene triangle cannot have two congruent angles.

378   Chapter 6   Relationships Within Triangles

Laurie's Notes | Teacher Actions

- A scenario is suggested as a way of introducing indirect reasoning. You may have your own favorite example. It is helpful to have a contextual, non-geometric example to discuss before explaining the steps of an indirect proof.
- **Reason Abstractly and Quantitatively:** Say, "Either a triangle has at most one right angle or more than one right angle. Both of these statements cannot be true. By showing that the latter leads to a contradiction, the former must be true." Work through Example 1, identifying the steps of an indirect proof.

Relating Sides and Angles of a Triangle

EXAMPLE 2 Relating Side Length and Angle Measure

Draw an obtuse scalene triangle. Find the largest angle and longest side and mark them in red. Find the smallest angle and shortest side and mark them in blue. What do you notice?

SOLUTION

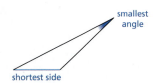

The longest side and largest angle are opposite each other.

The shortest side and smallest angle are opposite each other.

COMMON ERROR
Be careful not to confuse the symbol ∠ meaning *angle* with the symbol < meaning *is less than*. Notice that the bottom edge of the angle symbol is horizontal.

The relationships in Example 2 are true for all triangles, as stated in the two theorems below. These relationships can help you decide whether a particular arrangement of side lengths and angle measures in a triangle may be possible.

Theorems

Triangle Longer Side Theorem

If one side of a triangle is longer than another side, then the angle opposite the longer side is larger than the angle opposite the shorter side.

Proof Ex. 43, p. 384

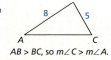

$AB > BC$, so $m\angle C > m\angle A$.

Triangle Larger Angle Theorem

If one angle of a triangle is larger than another angle, then the side opposite the larger angle is longer than the side opposite the smaller angle.

Proof p. 379

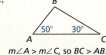

$m\angle A > m\angle C$, so $BC > AB$.

PROOF Triangle Larger Angle Theorem

Given $m\angle A > m\angle C$

Prove $BC > AB$

Indirect Proof

Step 1 Assume temporarily that $BC \not> AB$. Then it follows that either $BC < AB$ or $BC = AB$.

Step 2 If $BC < AB$, then $m\angle A < m\angle C$ by the Triangle Longer Side Theorem. If $BC = AB$, then $m\angle A = m\angle C$ by the Base Angles Theorem.

Step 3 Both conclusions contradict the given statement that $m\angle A > m\angle C$. So, the temporary assumption that $BC \not> AB$ cannot be true. This proves that $BC > AB$.

COMMON ERROR
Be sure to consider all cases when assuming the opposite is true.

Section 6.6 Indirect Proof and Inequalities in One Triangle 379

Differentiated Instruction

Inclusion
Have students describe how they use indirect reasoning in everyday life.

Extra Example 2
Draw an obtuse isosceles triangle. Find the largest angle and the longest side and mark them in red. Find the smallest angle and shortest side and mark them in blue. What do you notice?
Check students' drawings. Sample drawing of an obtuse isosceles triangle:

The longest side is opposite the largest angle, which is the obtuse angle. There are two congruent smallest angles. They are the base angles of the isosceles triangle, because the two sides are congruent. The two congruent sides are the smallest sides.

Laurie's Notes — Teacher Actions

- If students have worked through the explorations, the example and theorems will make sense.
- **?** "What would it sound like if you combined both theorems into one statement?" Answers will vary. Listen for something like "when you put the sides in order from longest to shortest, the angles opposite them will be in order from largest to smallest."

COMMON ERROR "If $BC \not> AB$, what do you know?" Either $BC = AB$ or $BC < AB$. Students often forget the case $BC = AB$.

Section 6.6 379

Extra Example 3

List the angles of △ABC in order from smallest to largest.

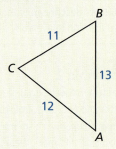

∠A, ∠B, ∠C

Extra Example 4

List the sides of △ABC in order from shortest to longest.

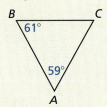

$\overline{BC}, \overline{AB}, \overline{AC}$

MONITORING PROGRESS ANSWERS

2. ∠Q, ∠P, ∠R
3. $\overline{ST}, \overline{RS}, \overline{RT}$

EXAMPLE 3 Ordering Angle Measures of a Triangle

You are constructing a stage prop that shows a large triangular mountain. The bottom edge of the mountain is about 32 feet long, the left slope is about 24 feet long, and the right slope is about 26 feet long. List the angles of △JKL in order from smallest to largest.

SOLUTION

Draw the triangle that represents the mountain. Label the side lengths.

The sides from shortest to longest are $\overline{JK}$, $\overline{KL}$, and $\overline{JL}$. The angles opposite these sides are ∠L, ∠J, and ∠K, respectively.

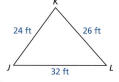

▶ So, by the Triangle Longer Side Theorem, the angles from smallest to largest are ∠L, ∠J, and ∠K.

EXAMPLE 4 Ordering Side Lengths of a Triangle

List the sides of △DEF in order from shortest to longest.

SOLUTION

First, find $m\angle F$ using the Triangle Sum Theorem.

$m\angle D + m\angle E + m\angle F = 180°$

$51° + 47° + m\angle F = 180°$

$m\angle F = 82°$

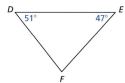

The angles from smallest to largest are ∠E, ∠D, and ∠F. The sides opposite these angles are $\overline{DF}$, $\overline{EF}$, and $\overline{DE}$, respectively.

▶ So, by the Triangle Larger Angle Theorem, the sides from shortest to longest are $\overline{DF}, \overline{EF},$ and $\overline{DE}$.

Monitoring Progress Help in English and Spanish at *BigIdeasMath.com*

2. List the angles of △PQR in order from smallest to largest.

3. List the sides of △RST in order from shortest to longest.

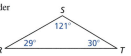

380 Chapter 6 Relationships Within Triangles

Laurie's Notes — Teacher Actions

- **Popsicle Sticks:** Pose Example 3 and have students work with their partners to solve. Use *Popsicle Sticks* for students to share their results.
- **? Reason Abstractly and Quantitatively** and **Construct Viable Arguments and Critique the Reasoning of Others:** "In Example 4, the diagram may not be drawn to scale. How do you know ∠F is the largest angle without computing its measure?" All angles are 60° in an equiangular triangle. Because ∠D and ∠E are each less than 60°, ∠F has to be greater than 60°.
- **Think-Pair-Share:** Have students answer Questions 2 and 3, and then share and discuss as a class.

Using the Triangle Inequality Theorem

Not every group of three segments can be used to form a triangle. The lengths of the segments must fit a certain relationship. For example, three attempted triangle constructions using segments with given lengths are shown below. Only the first group of segments forms a triangle.

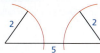
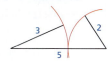

When you start with the longest side and attach the other two sides at its endpoints, you can see that the other two sides are not long enough to form a triangle in the second and third figures. This leads to the *Triangle Inequality Theorem*.

Theorem

Triangle Inequality Theorem

The sum of the lengths of any two sides of a triangle is greater than the length of the third side.

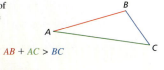

$AB + BC > AC$ $AC + BC > AB$ $AB + AC > BC$

Proof Ex. 47, p. 384

EXAMPLE 5   Finding Possible Side Lengths

A triangle has one side of length 14 and another side of length 9. Describe the possible lengths of the third side.

SOLUTION

Let x represent the length of the third side. Draw diagrams to help visualize the small and large values of x. Then use the Triangle Inequality Theorem to write and solve inequalities.

Small values of x

$x + 9 > 14$
$x > 5$

Large values of x

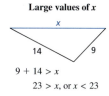

$9 + 14 > x$
$23 > x$, or $x < 23$

 The length of the third side must be greater than 5 and less than 23.

CONNECTIONS TO ALGEBRA

You can combine the two inequalities, $x > 5$ and $x < 23$, to write the compound inequality $5 < x < 23$. This can be read as x is between 5 and 23.

Monitoring Progress Help in English and Spanish at BigIdeasMath.com

4. A triangle has one side of length 12 inches and another side of length 20 inches. Describe the possible lengths of the third side.

Decide whether it is possible to construct a triangle with the given side lengths. Explain your reasoning.

5. 4 ft, 9 ft, 10 ft   6. 8 m, 9 m, 18 m   7. 5 cm, 7 cm, 12 cm

Assignment Guide and Homework Check

ASSIGNMENT

Basic: 1, 2, 3–29 odd, 35, 42, 45, 50–53

Average: 1, 2, 8–30 even, 36–42 even, 45, 50–53

Advanced: 1, 2, 8, 16–32 even, 37–40, 42, 45–48, 50–53

HOMEWORK CHECK

Basic: 3, 7, 11, 13, 17
Average: 8, 12, 14, 22, 40
Advanced: 8, 22, 37, 40, 47

ANSWERS

1. In an indirect proof, rather than proving a statement directly, you show that when the statement is false, it leads to a contradiction.

2. The longest side is opposite the largest angle, and the shortest side is opposite the smallest angle.

3. Assume temporarily that $WV = 7$ inches.

4. Assume temporarily that xy is even.

5. Assume temporarily that $\angle B$ is a right angle.

6. Assume temporarily that $\overline{JM}$ is not a median.

7. A and C; The angles of an equilateral triangle are always 60°. So, an equilateral triangle cannot have a 90° angle, and cannot be a right triangle.

8. B and C; If both $\angle X$ and $\angle Y$ have measures less than 30°, then their total is less than 60°. So, the sum of their measures cannot be 62°.

9.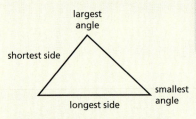

The longest side is across from the largest angle, and the shortest side is across from the smallest angle.

10.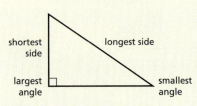

6.6 Exercises

Dynamic Solutions available at BigIdeasMath.com

Vocabulary and Core Concept Check

1. **VOCABULARY** Why is an indirect proof also called a *proof by contradiction*?

2. **WRITING** How can you tell which side of a triangle is the longest from the angle measures of the triangle? How can you tell which side is the shortest?

Monitoring Progress and Modeling with Mathematics

In Exercises 3–6, write the first step in an indirect proof of the statement. *(See Example 1.)*

3. If $WV + VU \neq 12$ inches and $VU = 5$ inches, then $WV \neq 7$ inches.

4. If x and y are odd integers, then xy is odd.

5. In $\triangle ABC$, if $m\angle A = 100°$, then $\angle B$ is not a right angle.

6. In $\triangle JKL$, if M is the midpoint of $\overline{KL}$, then $\overline{JM}$ is a median.

In Exercises 7 and 8, determine which two statements contradict each other. Explain your reasoning.

7. Ⓐ $\triangle LMN$ is a right triangle.
   Ⓑ $\angle L \cong \angle N$
   Ⓒ $\triangle LMN$ is equilateral.

8. Ⓐ Both $\angle X$ and $\angle Y$ have measures greater than 20°.
   Ⓑ Both $\angle X$ and $\angle Y$ have measures less than 30°.
   Ⓒ $m\angle X + m\angle Y = 62°$

In Exercises 9 and 10, use a ruler and protractor to draw the given type of triangle. Mark the largest angle and longest side in red and the smallest angle and shortest side in blue. What do you notice? *(See Example 2.)*

9. acute scalene
10. right scalene

In Exercises 11 and 12, list the angles of the given triangle from smallest to largest. *(See Example 3.)*

11.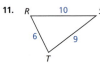
12. (triangle K, J, L with sides 28, 13, 25)

382 Chapter 6 Relationships Within Triangles

In Exercises 13–16, list the sides of the given triangle from shortest to longest. *(See Example 4.)*

13. (triangle ABC: ∠B = 67°, ∠A = 62°, ∠C = 51°)
14.
15. (triangle MNP: ∠N = 127°, ∠P = 29°)
16. (triangle FGD: right angle at F, ∠G = 33°)

In Exercises 17–20, describe the possible lengths of the third side of the triangle given the lengths of the other two sides. *(See Example 5.)*

17. 5 inches, 12 inches
18. 12 feet, 18 feet
19. 2 feet, 40 inches
20. 25 meters, 25 meters

In Exercises 21–24, is it possible to construct a triangle with the given side lengths? If not, explain why not.

21. 6, 7, 11
22. 3, 6, 9
23. 28, 17, 46
24. 35, 120, 125

25. **ERROR ANALYSIS** Describe and correct the error in writing the first step of an indirect proof.

> ✗ Show that $\angle A$ is obtuse.
> **Step 1** Assume temporarily that $\angle A$ is acute.

The longest side is the hypotenuse, and it is across from the right angle, which is the largest angle. The shortest side is across from the smallest angle.

11. $\angle S, \angle R, \angle T$
12. $\angle J, \angle K, \angle L$
13. $\overline{AB}, \overline{BC}, \overline{AC}$
14. $\overline{XY}, \overline{YZ}, \overline{XZ}$
15. $\overline{NP}, \overline{MN}, \overline{MP}$
16. $\overline{FD}, \overline{FG}, \overline{DG}$
17. 7 in. $< x <$ 17 in.
18. 6 ft $< x <$ 30 ft
19. 16 in. $< x <$ 64 in.
20. 0 m $< x <$ 50 m
21. yes
22. no; $3 + 6 \not> 9$
23. no; $28 + 17 \not> 46$
24. yes
25. An angle that is not obtuse could be acute or right; Assume temporarily that $\angle A$ is not obtuse.

26. **ERROR ANALYSIS** Describe and correct the error in labeling the side lengths 1, 2, and √3 on the triangle.

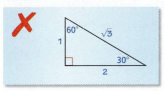

27. **REASONING** You are a lawyer representing a client who has been accused of a crime. The crime took place in Los Angeles, California. Security footage shows your client in New York at the time of the crime. Explain how to use indirect reasoning to prove your client is innocent.

28. **REASONING** Your class has fewer than 30 students. The teacher divides your class into two groups. The first group has 15 students. Use indirect reasoning to show that the second group must have fewer than 15 students.

29. **PROBLEM SOLVING** Which statement about △TUV is false?

    Ⓐ $UV > TU$
    Ⓑ $UV + TV > TU$
    Ⓒ $UV < TV$
    Ⓓ △TUV is isosceles.

30. **PROBLEM SOLVING** In △RST, which is a possible side length for ST? Select all that apply.

    Ⓐ 7
    Ⓑ 8
    Ⓒ 9
    Ⓓ 10

31. **PROOF** Write an indirect proof that an odd number is not divisible by 4.

32. **PROOF** Write an indirect proof of the statement "In △QRS, if $m\angle Q + m\angle R = 90°$, then $m\angle S = 90°$."

33. **WRITING** Explain why the hypotenuse of a right triangle must always be longer than either leg.

34. **CRITICAL THINKING** Is it possible to decide if three side lengths form a triangle without checking all three inequalities shown in the Triangle Inequality Theorem? Explain your reasoning.

35. **MODELING WITH MATHEMATICS** You can estimate the width of the river from point A to the tree at point B by measuring the angle to the tree at several locations along the riverbank. The diagram shows the results for locations C and D.

    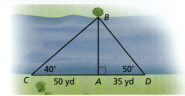

    a. Using △BCA and △BDA, determine the possible widths of the river. Explain your reasoning.
    b. What could you do if you wanted a closer estimate?

36. **MODELING WITH MATHEMATICS** You travel from Fort Peck Lake to Glacier National Park and from Glacier National Park to Granite Peak.

    a. Write two inequalities to represent the possible distances from Granite Peak back to Fort Peck Lake.
    b. How is your answer to part (a) affected if you know that $m\angle 2 < m\angle 1$ and $m\angle 2 < m\angle 3$?

37. **REASONING** In the figure, $\overline{XY}$ bisects ∠WYZ. List all six angles of △XYZ and △WXY in order from smallest to largest. Explain your reasoning.

38. **MATHEMATICAL CONNECTIONS** In △DEF, $m\angle D = (x + 25)°$, $m\angle E = (2x - 4)°$, and $m\angle F = 63°$. List the side lengths and angle measures of the triangle in order from least to greatest.

Section 6.6   Indirect Proof and Inequalities in One Triangle   383

ANSWERS

39. By the Exterior Angle Theorem, $m\angle 1 = m\angle A + m\angle B$. Then by the Subtraction Property of Equality, $m\angle 1 - m\angle B = m\angle A$. If you assume temporarily that $m\angle 1 \leq m\angle B$, then $m\angle A \leq 0$. Because the measure of any angle in a triangle must be a positive number, the assumption must be false. So, $m\angle 1 > m\angle B$. Similarly, by the Subtraction Property of Equality, $m\angle 1 - m\angle A = m\angle B$. If you assume temporarily that $m\angle 1 \leq m\angle A$, then $m\angle B \leq 0$. Because the measure of any angle in a triangle must be a positive number, the assumption must be false. So, $m\angle 1 > m\angle A$.

40. $2 < x < 15$

41. $2\frac{1}{7} < x < 13$

42–53. See Additional Answers.

Mini-Assessment

1. A triangle has one side of length 46 and another side of length 49. Describe the possible lengths of the third side. *The length of the third side must be greater than 3 and less than 95.*

2. Write the first step in an indirect proof of the following statement: If $AM > MB$, then M is not the midpoint of $\overline{AB}$.
   Assume temporarily that M is the midpoint of $\overline{AB}$.

3. List the sides of $\triangle DFG$ in order from shortest to longest.

   $\overline{DF}, \overline{FG}, \overline{DG}$

4. List the angles of the triangle in order from largest to smallest.

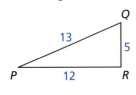

   $\angle R, \angle Q, \angle P$

384   Chapter 6

---

39. ANALYZING RELATIONSHIPS Another triangle inequality relationship is given by the Exterior Angle Inequality Theorem. It states:

The measure of an exterior angle of a triangle is greater than the measure of either of the nonadjacent interior angles.

Explain how you know that $m\angle 1 > m\angle A$ and $m\angle 1 > m\angle B$ in $\triangle ABC$ with exterior angle $\angle 1$.

MATHEMATICAL CONNECTIONS In Exercises 40 and 41, describe the possible values of x.

40.

41.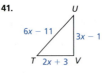

42. HOW DO YOU SEE IT? Your house is on the corner of Hill Street and Eighth Street. The library is on the corner of View Street and Seventh Street. What is the shortest route to get from your house to the library? Explain your reasoning.

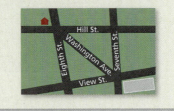

43. PROVING A THEOREM Use the diagram to prove the Triangle Longer Side Theorem.

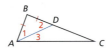

Given $BC > AB, BD = BA$
Prove $m\angle BAC > m\angle C$

Maintaining Mathematical Proficiency
Reviewing what you learned in previous grades and lessons

Name the included angle between the pair of sides given. *(Skills Review Handbook)*

50. $\overline{AE}$ and $\overline{BE}$ **51.** $\overline{AC}$ and $\overline{DC}$

52. $\overline{AD}$ and $\overline{DC}$ **53.** $\overline{CE}$ and $\overline{BE}$

384   Chapter 6   Relationships Within Triangles

---

44. USING STRUCTURE The length of the base of an isosceles triangle is ℓ. Describe the possible lengths for each leg. Explain your reasoning.

45. MAKING AN ARGUMENT Your classmate claims to have drawn a triangle with one side length of 13 inches and a perimeter of 2 feet. Is this possible? Explain your reasoning.

46. THOUGHT PROVOKING Cut two pieces of string that are each 24 centimeters long. Construct an isosceles triangle out of one string and a scalene triangle out of the other. Measure and record the side lengths. Then classify each triangle by its angles.

47. PROVING A THEOREM Prove the Triangle Inequality Theorem.
   Given $\triangle ABC$
   Prove $AB + BC > AC, AC + BC > AB,$ and $AB + AC > BC$

48. ATTENDING TO PRECISION The perimeter of $\triangle HGF$ must be between what two integers? Explain your reasoning.

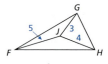

49. PROOF Write an indirect proof that a perpendicular segment is the shortest segment from a point to a plane.

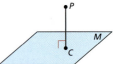

Given $\overline{PC} \perp$ plane M
Prove $\overline{PC}$ is the shortest segment from P to plane M.

If students need help...	If students got it...
Resources by Chapter • Practice A and Practice B • Puzzle Time	Resources by Chapter • Enrichment and Extension • Cumulative Review
Student Journal • Practice	Start the *next* Section
Differentiating the Lesson Skills Review Handbook	

Laurie's Notes

Overview of Section 6.7

Introduction
- The last lesson looked at inequalities within one triangle. How are sides and angles related when the side lengths are not the same measure?
- In this lesson, inequalities in two triangles are explored. When two sides of one triangle are congruent to two sides in another triangle, what is the relationship between the third sides and the angles opposite them?
- Indirect proof and paragraph proof are each used in this lesson.

Formative Assessment Tips
- **Look Back:** This technique is used when you want students to look back over a period of instruction and summarize their learning. What do they understand now that they did not before? It is helpful for students to articulate what helped them to learn a concept or skill.
- You may want students to focus on just the last two lessons, inequalities in one or two triangles. How are the side and angle relationships the same? How were indirect proofs used in the lessons? Having students summarize what they understand about indirect proofs and what helped them learn it will enable you to make instruction more effective for your students in the future.
- If this is the first time you have used this technique, allow 6–8 minutes of class time to model what it looks like. Say, "Over the last two lessons, we have studied inequalities in one or two triangles and indirect proof. Take a piece of paper and draw a vertical line down the middle. I'm going to give you 5 minutes. On the left side, make a list of the things you learned that you did not know or understand before we began. On the right side, describe how you learned these things. What technique(s) were effective in helping you learn?"
- Evaluate what worked for your students: particular examples, talking with partners, making tables, using a model, exploring relationships with dynamic geometry software, discussion of a particular problem, and so on.

Pacing Suggestion
- The exploration will enable students to make sense of the Hinge Theorem and its converse, which are presented in the formal lesson. When students have finished the exploration, transition to the formal lesson.

Dynamic Teaching Tools

Dynamic Assessment & Progress Monitoring Tool
Lesson Planning Tool
Interactive Whiteboard Lesson Library
Dynamic Classroom with Dynamic Investigations

What Your Students Will Learn

- Use the Hinge Theorem and its converse to compare angle measures and side lengths of two triangles.
- Solve real-life problems using the Hinge Theorem.

Laurie's Notes

Exploration

Motivate

- If you have two door hinges or similar objects, bring them to class. Students know how a hinge operates. I tie a rubber band on the outer edges of each hinge to simulate the third side of each triangle.
- **?** Have two students hold the hinges so that the angle of the opened hinges is the same. "What do you observe about the third side (rubber bands)?" The third sides are the same length.
- **?** Have one student open his or her hinge wider. "What is true about the angle of the hinges now?" One hinge angle is greater than the other hinge angle. "What do you observe about the third side of the hinge that is opened wider?" The third side is longer.
- Explain to students that in the exploration and in the lesson they will make conjectures about sides and angles of two triangles.

Exploration 1

- It will be helpful if students use the same lettering as the diagram so that when they collect data in the table, the column headings will not need to be changed.
- **?** Once the basic construction is completed ask, "What is true about the sides of $\triangle ACB$ and $\triangle DCB$?" The two triangles have two pairs of sides that are congruent. $\overline{AC} \cong \overline{DC}$ because they are radii of the same circle, and $\overline{BC} \cong \overline{BC}$ because of the Reflexive Property.
- **?** "Could the third sides of the two triangles be congruent?" yes "Do they have to be?" no
- Have students gather data by dragging point D to different positions.
- **Construct Viable Arguments and Critique the Reasoning of Others:** Ask students to make a conjecture after looking for patterns in the table. The organization of the columns in the table should help students observe patterns.
- In addition to recognizing the pattern known as the Hinge Theorem, students might also observe other relationships, such as when side $\overline{AB}$ increases, $\angle ACB$ also increases.

Communicate Your Answer

- You should hear a number of observations from students about the third sides of the two triangles: they may be congruent or not; when they are congruent, the angle opposite each is congruent; when they are not congruent, the angles opposite are not congruent and the largest angle will be opposite the longest side.

Connecting to Next Step

- The exploration is related to the theorem and its converse presented in the lesson, helping students make sense of the theorems before proving them to be true.

6.7 Inequalities in Two Triangles

Essential Question If two sides of one triangle are congruent to two sides of another triangle, what can you say about the third sides of the triangles?

EXPLORATION 1 Comparing Measures in Triangles

Work with a partner. Use dynamic geometry software.

a. Draw △ABC, as shown below.
b. Draw the circle with center C(3, 3) through the point A(1, 3).
c. Draw △DBC so that D is a point on the circle.

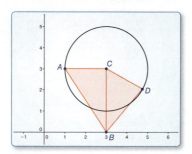

Sample
Points
A(1, 3)
B(3, 0)
C(3, 3)
D(4.75, 2.03)
Segments
BC = 3
AC = 2
DC = 2
AB = 3.61
DB = 2.68

d. Which two sides of △ABC are congruent to two sides of △DBC? Justify your answer.
e. Compare the lengths of $\overline{AB}$ and $\overline{DB}$. Then compare the measures of ∠ACB and ∠DCB. Are the results what you expected? Explain.
f. Drag point D to several locations on the circle. At each location, repeat part (e). Copy and record your results in the table below.

	D	AC	BC	AB	BD	m∠ACB	m∠BCD
1.	(4.75, 2.03)	2	3				
2.		2	3				
3.		2	3				
4.		2	3				
5.		2	3				

CONSTRUCTING VIABLE ARGUMENTS
To be proficient in math, you need to make conjectures and build a logical progression of statements to explore the truth of your conjectures.

g. Look for a pattern of the measures in your table. Then write a conjecture that summarizes your observations.

Communicate Your Answer

2. If two sides of one triangle are congruent to two sides of another triangle, what can you say about the third sides of the triangles?
3. Explain how you can use the hinge shown at the left to model the concept described in Question 2.

Section 6.7 Inequalities in Two Triangles 385

ANSWERS

1. a. Check students' work.
   b. Check students' work.
   c. Check students' work.
   d. $\overline{AC} \cong \overline{DC}$, because all points on a circle are equidistant from the center; $\overline{BC} \cong \overline{BC}$ by the Reflexive Property of Congruence.
   e. AB > DB; m∠ACB > m∠DCB; yes; The triangle with the longer third side has the larger angle opposite the third side.
   f. See Additional Answers.
   g. If two sides of one triangle are congruent to two sides of another triangle, and the included angle of the first is larger than the included angle of the second, then the third side of the first is longer than the third side of the second.

2. If the included angle of one is larger than the included angle of the other, then the third side of the first is longer than the third side of the second. If the included angles are congruent, then you already know that the triangles are congruent by the SAS Congruence Theorem. Therefore the third sides are congruent because corresponding parts of congruent triangles are congruent.

3. Because the sides of the hinge do not change in length, the angle of the hinge can model the included angle and the distance between the opposite ends of the hinge can model the third side. When the hinge is open wider, the angle is larger and the ends of the hinge are farther apart. If the hinge is open less, the ends are closer together.

Differentiated Instruction

Visual

Ask students to use their thumb and forefinger as the two sides of a triangle. Ask students how the distance between the tips of their fingers changes as they increase or decrease the angle between their thumb and forefinger. Explain that because the lengths of their thumb and forefinger do not change, they are demonstrating the Hinge Theorem. You can also use a chalkboard compass or a large pair of scissors to demonstrate the Hinge Theorem to the class.

Extra Example 1

Given that $\overline{AB} \cong \overline{DE}$ and $\overline{BC} \cong \overline{EF}$, how does $m\angle B$ compare to $m\angle E$?

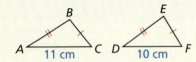

$m\angle B > m\angle E$

6.7 Lesson

Core Vocabulary

Previous
indirect proof
inequality

What You Will Learn

▶ Compare measures in triangles.
▶ Solve real-life problems using the Hinge Theorem.

Comparing Measures in Triangles

Imagine a gate between fence posts A and B that has hinges at A and swings open at B.

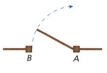

As the gate swings open, you can think of $\triangle ABC$, with side $\overline{AC}$ formed by the gate itself, side $\overline{AB}$ representing the distance between the fence posts, and side $\overline{BC}$ representing the opening between post B and the outer edge of the gate.

Notice that as the gate opens wider, both the measure of $\angle A$ and the distance BC increase. This suggests the *Hinge Theorem*.

Theorems

Hinge Theorem

If two sides of one triangle are congruent to two sides of another triangle, and the included angle of the first is larger than the included angle of the second, then the third side of the first is longer than the third side of the second.

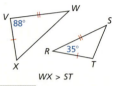

Proof BigIdeasMath.com

$WX > ST$

Converse of the Hinge Theorem

If two sides of one triangle are congruent to two sides of another triangle, and the third side of the first is longer than the third side of the second, then the included angle of the first is larger than the included angle of the second.

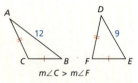

Proof Example 3, p. 387

$m\angle C > m\angle F$

EXAMPLE 1   Using the Converse of the Hinge Theorem

Given that $\overline{ST} \cong \overline{PR}$, how does $m\angle PST$ compare to $m\angle SPR$?

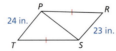

SOLUTION

You are given that $\overline{ST} \cong \overline{PR}$, and you know that $\overline{PS} \cong \overline{PS}$ by the Reflexive Property of Congruence. Because 24 inches > 23 inches, $PT > SR$. So, two sides of $\triangle STP$ are congruent to two sides of $\triangle PRS$ and the third side of $\triangle STP$ is longer.

▶ By the Converse of the Hinge Theorem, $m\angle PST > m\angle SPR$.

386   Chapter 6   Relationships Within Triangles

Laurie's Notes — Teacher Actions

- If you have not done the *Motivate* on page T-385, use it to introduce the formal lesson. You can use the classroom door as a hinge.
- Write the Hinge Theorem and its converse. Both theorems are wordy but readable. Have students annotate the theorems with marked diagrams that convey the meaning of the theorems.
- **Turn and Talk:** Pose Example 1. Give students time to talk with their partners.
- **Construct Viable Arguments and Critique the Reasoning of Others:** Solicit an answer and explanation.

EXAMPLE 2 Using the Hinge Theorem

Given that $\overline{JK} \cong \overline{LK}$, how does JM compare to LM?

SOLUTION

You are given that $\overline{JK} \cong \overline{LK}$, and you know that $\overline{KM} \cong \overline{KM}$ by the Reflexive Property of Congruence. Because $64° > 61°$, $m\angle JKM > m\angle LKM$. So, two sides of $\triangle JKM$ are congruent to two sides of $\triangle LKM$, and the included angle in $\triangle JKM$ is larger.

▶ By the Hinge Theorem, $JM > LM$.

Monitoring Progress Help in English and Spanish at BigIdeasMath.com

Use the diagram.

1. If $PR = PS$ and $m\angle QPR > m\angle QPS$, which is longer, $\overline{SQ}$ or $\overline{RQ}$?
2. If $PR = PS$ and $RQ < SQ$, which is larger, $\angle RPQ$ or $\angle SPQ$?

EXAMPLE 3 Proving the Converse of the Hinge Theorem

Write an indirect proof of the Converse of the Hinge Theorem.

Given $\overline{AB} \cong \overline{DE}, \overline{BC} \cong \overline{EF}, AC > DF$

Prove $m\angle B > m\angle E$

Indirect Proof

Step 1 Assume temporarily that $m\angle B \not> m\angle E$. Then it follows that either $m\angle B < m\angle E$ or $m\angle B = m\angle E$.

Step 2 If $m\angle B < m\angle E$, then $AC < DF$ by the Hinge Theorem.
If $m\angle B = m\angle E$, then $\angle B \cong \angle E$. So, $\triangle ABC \cong \triangle DEF$ by the SAS Congruence Theorem and $AC = DF$.

Step 3 Both conclusions contradict the given statement that $AC > DF$. So, the temporary assumption that $m\angle B \not> m\angle E$ cannot be true. This proves that $m\angle B > m\angle E$.

EXAMPLE 4 Proving Triangle Relationships

Write a paragraph proof.

Given $\angle XWY \cong \angle XYW, WZ > YZ$

Prove $m\angle WXZ > m\angle YXZ$

Paragraph Proof Because $\angle XWY \cong \angle XYW$, $\overline{XY} \cong \overline{XW}$ by the Converse of the Base Angles Theorem. By the Reflexive Property of Congruence, $\overline{XZ} \cong \overline{XZ}$. Because $WZ > YZ$, $m\angle WXZ > m\angle YXZ$ by the Converse of the Hinge Theorem.

Monitoring Progress Help in English and Spanish at BigIdeasMath.com

3. Write a temporary assumption you can make to prove the Hinge Theorem indirectly. What two cases does that assumption lead to?

Extra Example 2

Given that $\overline{AB} \cong \overline{DE}$ and $\overline{BC} \cong \overline{EC}$, how does AC compare to DC?

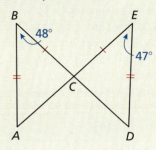

$AC > DC$

Extra Example 3

What can you conclude about the measures of $\angle A$ and $\angle Q$ in this figure? Explain.

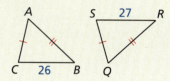

$m\angle A < m\angle Q$, by the Converse of the Hinge Theorem because $BC < RS$.

Extra Example 4

Write a paragraph proof.
Given $\overline{AB} \cong \overline{BC}, AD > CD$
Prove $m\angle ABD > m\angle CBD$

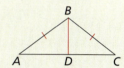

It is given that $\overline{AB} \cong \overline{BC}$. By the Reflexive Property of Congruence, $\overline{BD} \cong \overline{BD}$. Because $AD > CD$, $m\angle ABD > m\angle CBD$ by the Converse of the Hinge Theorem.

MONITORING PROGRESS ANSWERS

1. $\overline{RQ}$
2. $\angle SPQ$
3. Assume temporarily that the third side of the first triangle is not longer than the third side of the second triangle. The third side of the first triangle is either equal to the third side of the second triangle or less than the third side of the second triangle.

Laurie's Notes — Teacher Actions

- In proving the Converse of the Hinge Theorem (Example 3), state the given information. The congruent sides are marked on the diagram, but the given information that $AC > DF$ is not. It is important that this is written. The contradiction relates back to this piece of given information.
- **Construct Viable Arguments and Critique the Reasoning of Others** and **Attend to Precision:** Have partners work collaboratively on Example 4. Have several pairs share their proofs. Critique the reasoning used in each proof and the clarity and precision of the language used.

Extra Example 5

In Example 5, Group D leaves camp and travels 2 miles due south, then turns 25° toward east and travels 1.2 miles. Is Group D farther from camp than Group A, Group B, both groups, or neither group? Explain your reasoning. **Both groups. The included angle for Group D is 155°, which is greater than the included angle for Group A and for Group B. So, Group D is farthest from camp.**

MONITORING PROGRESS ANSWER

4. Group B is the farthest from camp, Group C is next, and Group A is the closest to camp.

Solving Real-Life Problems

EXAMPLE 5 Solving a Real-Life Problem

Two groups of bikers leave the same camp heading in opposite directions. Each group travels 2 miles, then changes direction and travels 1.2 miles. Group A starts due east and then turns 45° toward north. Group B starts due west and then turns 30° toward south. Which group is farther from camp? Explain your reasoning.

SOLUTION

1. **Understand the Problem** You know the distances and directions that the groups of bikers travel. You need to determine which group is farther from camp. You can interpret a turn of 45° toward north, as shown.

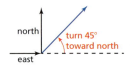

2. **Make a Plan** Draw a diagram that represents the situation and mark the given measures. The distances that the groups bike and the distances back to camp form two triangles. The triangles have two congruent side lengths of 2 miles and 1.2 miles. Include the third side of each triangle in the diagram.

3. **Solve the Problem** Use linear pairs to find the included angles for the paths that the groups take.

   Group A: $180° − 45° = 135°$ **Group B:** $180° − 30° = 150°$

   The included angles are 135° and 150°.

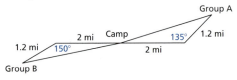

Because 150° > 135°, the distance Group B is from camp is greater than the distance Group A is from camp by the Hinge Theorem.

▶ So, Group B is farther from camp.

4. **Look Back** Because the included angle for Group A is 15° less than the included angle for Group B, you can reason that Group A would be closer to camp than Group B. So, Group B is farther from camp.

Monitoring Progress Help in English and Spanish at *BigIdeasMath.com*

4. **WHAT IF?** In Example 5, Group C leaves camp and travels 2 miles due north, then turns 40° toward east and travels 1.2 miles. Compare the distances from camp for all three groups.

388   Chapter 6   Relationships Within Triangles

Laurie's Notes — Teacher Actions

- **Model with Mathematics:** Read Example 5 and ask students to make a diagram of the problem. Have students compare their diagrams and modify as needed. Once a diagram has been drawn, ask partners to write an explanation of which group is farther from the camp.
- **Think-Pair-Share:** Have students answer Question 4, and then share and discuss as a class.

Closure

- **Look Back:** Have students *Look Back* over Sections 6.6 and 6.7. Have them summarize what they understand about inequalities in one triangle and inequalities in two triangles, as well as what helped them learn it.

6.7 Exercises

Dynamic Solutions available at BigIdeasMath.com

Vocabulary and Core Concept Check

1. **WRITING** Explain why the Hinge Theorem has that name.
2. **COMPLETE THE SENTENCE** In △ABC and △DEF, $\overline{AB} \cong \overline{DE}$, $\overline{BC} \cong \overline{EF}$, and $AC < DF$. So m∠____ > m∠____ by the Converse of the Hinge Theorem.

Monitoring Progress and Modeling with Mathematics

In Exercises 3–6, copy and complete the statement with <, >, or =. Explain your reasoning. (See Example 1.)

3. m∠1 ____ m∠2
4. m∠1 ____ m∠2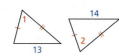
5. m∠1 ____ m∠2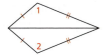
6. m∠1 ____ m∠2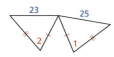

In Exercises 7–10, copy and complete the statement with <, >, or =. Explain your reasoning. (See Example 2.)

7. AD ____ CD
8. MN ____ LK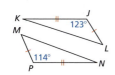
9. TR ____ UR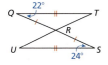
10. AC ____ DC

PROOF In Exercises 11 and 12, write a proof. (See Example 4.)

11. Given $\overline{XY} \cong \overline{YZ}$, m∠WYZ > m∠WYX
    Prove WZ > WX

12. Given $\overline{BC} \cong \overline{DA}$, DC < AB
    Prove m∠BCA > m∠DAC

In Exercises 13 and 14, you and your friend leave on different flights from the same airport. Determine which flight is farther from the airport. Explain your reasoning. (See Example 5.)

13. Your flight: Flies 100 miles due west, then turns 20° toward north and flies 50 miles.
    Friend's flight: Flies 100 miles due north, then turns 30° toward east and flies 50 miles.

14. Your flight: Flies 210 miles due south, then turns 70° toward west and flies 80 miles.
    Friend's flight: Flies 80 miles due north, then turns 50° toward east and flies 210 miles.

Section 6.7 Inequalities in Two Triangles 389

---

10. AC > DC; By the Hinge Theorem, because $\overline{AC}$ is the third side of the triangle with the larger included angle, it is longer than $\overline{DC}$.

11. $\overline{XY} \cong \overline{YZ}$ and m∠WYZ > m∠WYX are given. By the Reflexive Property of Congruence, $\overline{WY} \cong \overline{WY}$. So, by the Hinge Theorem, WZ > WX.

12. $\overline{BC} \cong \overline{DA}$ and DC < AB are given. By the Reflexive Property of Congruence, $\overline{AC} \cong \overline{AC}$. So, by the Converse of the Hinge Theorem, m∠BCA > m∠DAC.

13. your flight; Because 160° > 150°, the distance you flew is a greater distance than the distance your friend flew by the Hinge Theorem.

14. friend's flight; Because 130° > 110°, the distance your friend flew is a greater distance than the distance you flew by the Hinge Theorem.

Assignment Guide and Homework Check

ASSIGNMENT
Basic: 1, 2, 3–17 odd, 22, 25–28
Average: 1, 2, 6–14 even, 15–19 odd, 22, 25–28
Advanced: 1, 2, 6–14 even, 15–28

HOMEWORK CHECK
Basic: 3, 7, 11, 13, 22
Average: 6, 8, 12, 14, 19
Advanced: 6, 8, 12, 21, 24

ANSWERS

1. The Hinge Theorem refers to two angles with two pairs of sides that have the same measure, just like two hinges whose sides are the same length. Then, the angle whose measure is greater is opposite a longer side, just like the ends of a hinge are farther apart when the hinge is open wider.

2. m∠E > m∠B

3. m∠1 > m∠2; By the Converse of the Hinge Theorem, because ∠1 is the included angle in the triangle with the longer third side, its measure is greater than that of ∠2.

4. m∠1 < m∠2; By the Converse of the Hinge Theorem, because ∠1 is the included angle in the triangle with the shorter third side, its measure is less than that of ∠2.

5. m∠1 = m∠2; The triangles are congruent by the SSS Congruence Theorem. So, ∠1 ≅ ∠2 because corresponding parts of congruent triangles are congruent.

6. m∠1 > m∠2; By the Converse of the Hinge Theorem, because ∠1 is the included angle in the triangle with the longer third side, its measure is greater than that of ∠2.

7. AD > CD; By the Hinge Theorem, because $\overline{AD}$ is the third side of the triangle with the larger included angle, it is longer than $\overline{CD}$.

8. MN < LK; By the Hinge Theorem, because $\overline{MN}$ is the third side of the triangle with the smaller included angle, it is shorter than $\overline{LK}$.

9. TR < UR; By the Hinge Theorem, because $\overline{TR}$ is the third side of the triangle with the smaller included angle, it is shorter than $\overline{UR}$.

Section 6.7 389

Dynamic Teaching Tools
- Dynamic Assessment & Progress Monitoring Tool
- Interactive Whiteboard Lesson Library
- Dynamic Classroom with Dynamic Investigations

ANSWERS

15. The measure of the included angle in $\triangle PSQ$ is greater than the measure of the included angle in $\triangle SQR$; By the Hinge Theorem, $PQ > SR$.

16. A, B

17. $m\angle EGF > m\angle DGE$ by the Hinge Theorem.

18–24. See Additional Answers.

25. $x = 38$
26. $x = 72$
27. $x = 60$
28. $x = 108$

Mini-Assessment

Use the figure.

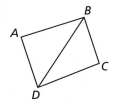

1. Given $AB = DC$, $m\angle ABD = 35°$, and $m\angle BDC = 25°$. How does AD compare to BC? $AD > BC$

2. Given $AB = DC$, $AD = 14$, and $BC = 15$. How does $m\angle ABD$ compare to $m\angle BDC$? $m\angle ABD < m\angle BDC$

3. Complete the statement with $<$, $>$, or $=$.
   $m\angle P$ ___ $m\angle R$

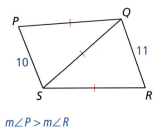

$m\angle P > m\angle R$

390   Chapter 6

---

15. ERROR ANALYSIS Describe and correct the error in using the Hinge Theorem.

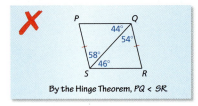

By the Hinge Theorem, $PQ < SR$.

16. REPEATED REASONING Which is a possible measure for $\angle JKM$? Select all that apply.

(A) 15°   (B) 22°   (C) 25°   (D) 35°

17. DRAWING CONCLUSIONS The path from E to F is longer than the path from E to D. The path from G to D is the same length as the path from G to F. What can you conclude about the angles of the paths? Explain your reasoning.

18. ABSTRACT REASONING In $\triangle EFG$, the bisector of $\angle F$ intersects the bisector of $\angle G$ at point H. Explain why $\overline{FG}$ must be longer than $\overline{FH}$ or $\overline{HG}$.

19. ABSTRACT REASONING $\overline{NR}$ is a median of $\triangle NPQ$, and $NQ > NP$. Explain why $\angle NRQ$ is obtuse.

Maintaining Mathematical Proficiency Reviewing what you learned in previous grades and lessons

Find the value of x. *(Skills Review Handbook)*

25.   26.   27.   28.

390   Chapter 6   Relationships Within Triangles

MATHEMATICAL CONNECTIONS In Exercises 20 and 21, write and solve an inequality for the possible values of x.

20. 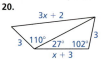   21.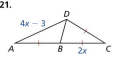

22. HOW DO YOU SEE IT? In the diagram, triangles are formed by the locations of the players on the basketball court. The dashed lines represent the possible paths of the basketball as the players pass. How does $m\angle ACB$ compare with $m\angle ACD$?

23. CRITICAL THINKING In $\triangle ABC$, the altitudes from B and C meet at point D, and $m\angle BAC > m\angle BDC$. What is true about $\triangle ABC$? Justify your answer.

24. THOUGHT PROVOKING The postulates and theorems you have learned represent Euclidean geometry. In spherical geometry, all points are on the surface of a sphere. A line is a circle on the sphere whose diameter is equal to the diameter of the sphere. In spherical geometry, state an inequality involving the sum of the angles of a triangle. Find a formula for the area of a triangle in spherical geometry.

If students need help...	If students got it...
Resources by Chapter • Practice A and Practice B • Puzzle Time	**Resources by Chapter** • Enrichment and Extension • Cumulative Review
Student Journal • Practice	Start the *next* Section
Differentiating the Lesson Skills Review Handbook	

6.5–6.7 What Did You Learn?

Core Vocabulary

midsegment of a triangle, *p. 372*
indirect proof, *p. 378*

Core Concepts

Section 6.5
Using the Midsegment of a Triangle, *p. 372*
Triangle Midsegment Theorem, *p. 373*

Section 6.6
How to Write an Indirect Proof (Proof by Contradiction), *p. 378*
Triangle Longer Side Theorem, *p. 379*
Triangle Larger Angle Theorem, *p. 379*
Triangle Inequality Theorem, *p. 381*

Section 6.7
Hinge Theorem, *p. 386*
Converse of the Hinge Theorem, *p. 386*

Mathematical Practices

1. In Exercise 25 on page 376, analyze the relationship between the stage and the total perimeter of all the shaded triangles at that stage. Then predict the total perimeter of all the shaded triangles in Stage 4.
2. In Exercise 17 on page 382, write all three inequalities using the Triangle Inequality Theorem. Determine the reasonableness of each one. Why do you only need to use two of the three inequalities?
3. In Exercise 23 on page 390, try all three cases of triangles (acute, right, obtuse) to gain insight into the solution.

Performance Task:

Building a Roof Truss

The simple roof truss is also called a planar truss because all its components lie in a two-dimensional plane. How can this structure be extended to three-dimensional space? What applications would this type of structure be used for?

To explore the answers to these questions and more, check out the Performance Task and Real-Life STEM video at *BigIdeasMath.com*.

Dynamic Teaching Tools

Dynamic Assessment & Progress Monitoring Tool
Interactive Whiteboard Lesson Library
Dynamic Classroom with Dynamic Investigations

ANSWERS

1. Let n be the stage, then the side length of the new triangles in each stage is 2^{4-n}. So, the perimeter of each new triangle is $(3 \cdot 2^{4-n})$. The number of new triangles is given by (3^{n-1}). So, to find the perimeter of all the shaded triangles in each stage, start with the total from the previous stage and add $(3 \cdot 2^{4-n})(3^{n-1})$. The perimeter of the new triangles in stage 4 will be $(3 \cdot 2^{4-4})(3^{4-1}) = 81$. The total perimeter of the new triangles and old triangles is $81 + 114 = 195$ units.

2. $x + 5 > 12, x + 12 > 5, 5 + 12 > x$; Because the length of the third side has to be a positive value, the inequality $x + 12 > 5$ will always be true. So, you do not have to consider this inequality in determining the possible values of x. Solve the other two inequalities to find that the length of the third side must be greater than 7 and less than 19.

3. If $m\angle BAC < m\angle BDC$, then the orthocenter D is inside the triangle, and $\triangle ABC$ is an acute triangle. If $m\angle BAC = m\angle BDC$, then the orthocenter D is on vertex A, and $\triangle ABC$ is a right triangle, with $\angle A$ being the right angle. If $m\angle BAC > m\angle BDC$, then the orthocenter D is outside the triangle, and $\triangle ABC$ is an obtuse triangle, with $\angle A$ being the obtuse angle.

6 Chapter Review

Dynamic Solutions available at *BigIdeasMath.com*

6.1 Proving Geometric Relationships (pp. 335–342)

a. Rewrite the two-column proof into a paragraph proof.

Given $\angle 6 \cong \angle 7$
Prove $\angle 7 \cong \angle 2$

Two-Column Proof

STATEMENTS	REASONS
1. $\angle 6 \cong \angle 7$	1. Given
2. $\angle 6 \cong \angle 2$	2. Vertical Angles Congruence Theorem
3. $\angle 7 \cong \angle 2$	3. Transitive Property of Congruence

Paragraph Proof

$\angle 6$ and $\angle 7$ are congruent. By the Vertical Angles Congruence Theorem, $\angle 6 \cong \angle 2$. So, by the Transitive Property of Congruence, $\angle 7 \cong \angle 2$.

b. Write a coordinate proof.

Given Coordinates of vertices of $\triangle OAC$ and $\triangle BAC$
Prove $\triangle OAC \cong \triangle BAC$

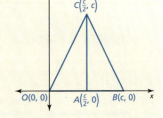

Segments $\overline{OA}$ and $\overline{BA}$ have the same length, so $\overline{OA} \cong \overline{BA}$.

$OA = \left|\dfrac{c}{2} - 0\right| = \dfrac{c}{2}$

$BA = \left|c - \dfrac{c}{2}\right| = \dfrac{c}{2}$

Segments $\overline{OC}$ and $\overline{BC}$ have the same length, so $\overline{OC} \cong \overline{BC}$.

$OC = \sqrt{\left(0 - \dfrac{c}{2}\right)^2 + (0-c)^2} = \sqrt{\left(-\dfrac{c}{2}\right)^2 + (-c)^2} = \sqrt{\dfrac{5c^2}{4}} = \dfrac{1}{2}c\sqrt{5}$

$BC = \sqrt{\left(c - \dfrac{c}{2}\right)^2 + (0-c)^2} = \sqrt{\left(\dfrac{c}{2}\right)^2 + (-c)^2} = \sqrt{\dfrac{5c^2}{4}} = \dfrac{1}{2}c\sqrt{5}$

By the Reflexive Property of Congruence, $\overline{AC} \cong \overline{AC}$.

▶ So, you can apply the SSS Congruence Theorem to conclude that $\triangle OAC \cong \triangle BAC$.

1. Use the figure in part (a). Write a proof using any format.

Given $\angle 5$ and $\angle 6$ are complementary.
$m\angle 5 + m\angle 7 = 90°$

Prove $\angle 6 \cong \angle 7$

2. Use the figure at the right. Write a coordinate proof.

Given Coordinates of vertices of $\triangle OBC$
Prove $\triangle OBC$ is isosceles.

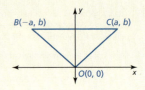

ANSWERS

1.

STATEMENTS	REASONS
1. $\angle 5$ and $\angle 6$ are complementary, $m\angle 5 + m\angle 7 = 90°$	1. Given
2. $\angle 5$ and $\angle 7$ are complementary	2. Definition of complementary angles
3. $\angle 6 \cong \angle 7$	3. Congruent Complements Theorem

2. By the Distance Formula, $OB = \sqrt{a^2 + b^2}$ and $OC = \sqrt{a^2 + b^2}$. So, $\triangle OBC$ is isosceles by the definition of isosceles triangle.

6.2 Perpendicular and Angle Bisectors (pp. 343–350)

Find each measure.

a. AD

From the figure, $\overrightarrow{AC}$ is the perpendicular bisector of $\overline{BD}$.

$AB = AD$	Perpendicular Bisector Theorem
$4x + 3 = 6x - 9$	Substitute.
$x = 6$	Solve for x.

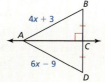

▶ So, $AD = 6(6) - 9 = 27$.

b. FG

Because $EH = GH$ and $\overrightarrow{HF} \perp \overline{EG}$, $\overrightarrow{HF}$ is the perpendicular bisector of $\overline{EG}$ by the Converse of the Perpendicular Bisector Theorem. By the definition of segment bisector, $EF = FG$.

▶ So, $FG = EF = 6$.

c. LM

$JM = LM$	Angle Bisector Theorem
$8x - 15 = 3x$	Substitute.
$x = 3$	Solve for x.

▶ So, $LM = 3x = 3(3) = 9$.

d. $m\angle XWZ$

Because $\overline{ZX} \perp \overrightarrow{WX}$ and $\overline{ZY} \perp \overrightarrow{WY}$ and $ZX = ZY = 8$, $\overrightarrow{WZ}$ bisects $\angle XWY$ by the Converse of the Angle Bisector Theorem.

▶ So, $m\angle XWZ = m\angle YWZ = 46°$.

Find the indicated measure. Explain your reasoning.

3. DC

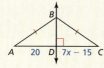

4. RS

5. $m\angle JFH$

ANSWERS

6. $(-3, -3)$
7. $(4, 3)$
8. $x = 5$

6.3 Bisectors of Triangles (pp. 351–360)

a. Find the coordinates of the circumcenter of $\triangle QRS$ with vertices $Q(3, 3)$, $R(5, 7)$, and $S(9, 3)$.

Step 1 Graph $\triangle QRS$.

Step 2 Find equations for two perpendicular bisectors.

The midpoint of $\overline{QS}$ is $(6, 3)$. The line through $(6, 3)$ that is perpendicular to $\overline{QS}$ is $x = 6$.

The midpoint of $\overline{QR}$ is $(4, 5)$. The line through $(4, 5)$ that is perpendicular to $\overline{QR}$ is $y = -\frac{1}{2}x + 7$.

Step 3 Find the point where $x = 6$ and $y = -\frac{1}{2}x + 7$ intersect. They intersect at $(6, 4)$.

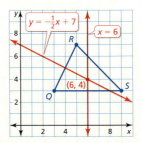

▶ So, the coordinates of the circumcenter are $(6, 4)$.

b. Point N is the incenter of $\triangle ABC$. In the figure shown, $ND = 4x + 7$ and $NE = 3x + 9$. Find NF.

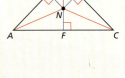

Step 1 Solve for x.

$ND = NE$ Incenter Theorem
$4x + 7 = 3x + 9$ Substitute.
$x = 2$ Solve for x.

Step 2 Find ND (or NE).

$ND = 4x + 7 = 4(2) + 7 = 15$

▶ So, because $ND = NF$, $NF = 15$.

Find the coordinates of the circumcenter of the triangle with the given vertices.

6. $T(-6, -5)$, $U(0, -1)$, $V(0, -5)$

7. $X(-2, 1)$, $Y(2, -3)$, $Z(6, -3)$

8. Point D is the incenter of $\triangle LMN$. Find the value of x.

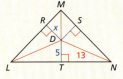

6.4 Medians and Altitudes of Triangles (pp. 361–368)

Find the coordinates of the centroid of △TUV with vertices T(1, −8), U(4, −1), and V(7, −6).

Step 1 Graph △TUV.

Step 2 Use the Midpoint Formula to find the midpoint W of $\overline{TV}$. Sketch median $\overline{UW}$.

$$W\left(\frac{1+7}{2}, \frac{-8+(-6)}{2}\right) = (4, -7)$$

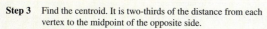

Step 3 Find the centroid. It is two-thirds of the distance from each vertex to the midpoint of the opposite side.

The distance from vertex U(4, −1) to W(4, −7) is −1 − (−7) = 6 units.
So, the centroid is $\frac{2}{3}$(6) = 4 units down from vertex U on $\overline{UW}$.

▶ So, the coordinates of the centroid P are (4, −1 − 4), or (4, −5).

Find the coordinates of the centroid of the triangle with the given vertices.

9. A(−10, 3), B(−4, 5), C(−4, 1)
10. D(2, −8), E(2, −2), F(8, −2)

Tell whether the orthocenter of the triangle with the given vertices is *inside*, *on*, or *outside* the triangle. Then find the coordinates of the orthocenter.

11. G(1, 6), H(5, 6), J(3, 1)
12. K(−8, 5), L(−6, 3), M(0, 5)

6.5 The Triangle Midsegment Theorem (pp. 371–376)

In △JKL, show that midsegment $\overline{MN}$ is parallel to $\overline{JL}$ and that $MN = \frac{1}{2}JL$.

Step 1 Find the coordinates of M and N by finding the midpoints of $\overline{JK}$ and $\overline{KL}$.

$$M\left(\frac{-8+(-4)}{2}, \frac{1+7}{2}\right) = M\left(\frac{-12}{2}, \frac{8}{2}\right) = M(-6, 4)$$

$$N\left(\frac{-4+(-2)}{2}, \frac{7+3}{2}\right) = N\left(\frac{-6}{2}, \frac{10}{2}\right) = N(-3, 5)$$

Step 2 Find and compare the slopes of $\overline{MN}$ and $\overline{JL}$.

slope of $\overline{MN} = \frac{5-4}{-3-(-6)} = \frac{1}{3}$

slope of $\overline{JL} = \frac{3-1}{-2-(-8)} = \frac{2}{6} = \frac{1}{3}$

▶ Because the slopes are the same, $\overline{MN}$ is parallel to $\overline{JL}$.

Step 3 Find and compare the lengths of $\overline{MN}$ and $\overline{JL}$.

$MN = \sqrt{[-3-(-6)]^2 + (5-4)^2} = \sqrt{9+1} = \sqrt{10}$

$JL = \sqrt{[-2-(-8)]^2 + (3-1)^2} = \sqrt{36+4} = \sqrt{40} = 2\sqrt{10}$

▶ Because $\sqrt{10} = \frac{1}{2}(2\sqrt{10})$, $MN = \frac{1}{2}JL$.

Find the coordinates of the vertices of the midsegment triangle for the triangle with the given vertices.

13. A(−6, 8), B(−6, 4), C(0, 4)
14. D(−3, 1), E(3, 5), F(1, −5)

ANSWERS
9. (−6, 3)
10. (4, −4)
11. inside; (3, 5.2)
12. outside; (−6, −1)
13. (−6, 6), (−3, 6), (−3, 4)
14. (0, 3), (2, 0), (−1, −2)

ANSWERS

15. 4 in. < x < 12 in.
16. 3 m < x < 15 m
17. 7 ft < x < 29 ft
18. Assume temporarily that $YZ \not> 4$. Then, it follows that either $YZ < 4$ or $YZ = 4$. If $YZ < 4$, then $XY + YZ < XZ$ because $4 + YZ < 8$ when $YZ < 4$. If $YZ = 4$, then $XY + YZ = XZ$ because $4 + 4 = 8$. Both conclusions contradict the Triangle Inequality Theorem, which says that $XY + YZ > XZ$. So, the temporary assumption that $YZ \not> 4$ cannot be true. This proves that in $\triangle XYZ$, if $XY = 4$ and $XZ = 8$, then $YZ > 4$.
19. $QT > ST$
20. $m\angle QRT > m\angle SRT$

6.6 Indirect Proof and Inequalities in One Triangle (pp. 377–384)

a. List the sides of $\triangle ABC$ in order from shortest to longest.

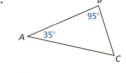

First, find $m\angle C$ using the Triangle Sum Theorem.

$$m\angle A + m\angle B + m\angle C = 180°$$
$$35° + 95° + m\angle C = 180°$$
$$m\angle C = 50°$$

The angles from smallest to largest are $\angle A$, $\angle C$, and $\angle B$. The sides opposite these angles are $\overline{BC}$, $\overline{AB}$, and $\overline{AC}$, respectively.

▶ So, by the Triangle Larger Angle Theorem, the sides from shortest to longest are $\overline{BC}$, $\overline{AB}$, and $\overline{AC}$.

b. List the angles of $\triangle DEF$ in order from smallest to largest.

The sides from shortest to longest are $\overline{DF}$, $\overline{EF}$, and $\overline{DE}$. The angles opposite these sides are $\angle E$, $\angle D$, and $\angle F$, respectively.

▶ So, by the Triangle Longer Side Theorem, the angles from smallest to largest are $\angle E$, $\angle D$, and $\angle F$.

Describe the possible lengths of the third side of the triangle given the lengths of the other two sides.

15. 4 inches, 8 inches
16. 6 meters, 9 meters
17. 11 feet, 18 feet
18. Write an indirect proof of the statement "In $\triangle XYZ$, if $XY = 4$ and $XZ = 8$, then $YZ > 4$."

6.7 Inequalities in Two Triangles (pp. 385–390)

Given that $\overline{WZ} \cong \overline{YZ}$, how does XY compare to XW?

You are given that $\overline{WZ} \cong \overline{YZ}$, and you know that $\overline{XZ} \cong \overline{XZ}$ by the Reflexive Property of Congruence.

Because $90° > 80°$, $m\angle XZY > m\angle XZW$. So, two sides of $\triangle XZY$ are congruent to two sides of $\triangle XZW$ and the included angle in $\triangle XZY$ is larger.

▶ By the Hinge Theorem, $XY > XW$.

Use the diagram.

19. If $RQ = RS$ and $m\angle QRT > m\angle SRT$, then how does $\overline{QT}$ compare to $\overline{ST}$?
20. If $RQ = RS$ and $QT > ST$, then how does $\angle QRT$ compare to $\angle SRT$?

6 Chapter Test

In Exercises 1 and 2, $\overline{MN}$ is a midsegment of $\triangle JKL$. Find the value of x.

1.
2.

Find the indicated measure. Identify the theorem you use.

3. ST
4. WY
5. BW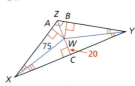

Copy and complete the statement with <, >, or =.

6. AB ___ CB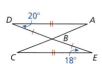
7. $m\angle 1$ ___ $m\angle 2$
8. $m\angle MNP$ ___ $m\angle NPM$

9. Find the coordinates of the circumcenter, orthocenter, and centroid of the triangle with vertices $A(0, -2)$, $B(4, -2)$, and $C(0, 6)$.

10. Write an indirect proof of the Corollary to the Base Angles Theorem:
    If $\triangle PQR$ is equilateral, then it is equiangular.

11. $\triangle DEF$ is a right triangle with area A. Use the area for $\triangle DEF$ to write an expression for the area of $\triangle GEH$. Justify your answer.

12. Prove the Corollary to the Triangle Sum Theorem using any format.
    Given $\triangle ABC$ is a right triangle.
    Prove $\angle A$ and $\angle B$ are complementary.

In Exercises 13–15, use the map.

13. Describe the possible lengths of Pine Avenue.

14. You ride your bike along a trail that represents the shortest distance from the beach to Main Street. You end up exactly halfway between your house and the movie theatre. How long is Pine Avenue? Explain.

15. A market is the same distance from your house, the movie theater, and the beach. Copy the map and locate the market.

ANSWERS

1. $x = 6$
2. $x = 9$
3. $ST = 17$; Perpendicular Bisector Theorem
4. $WY = 32$; Angle Bisector Theorem
5. $BW = 20$; Incenter Theorem
6. $AB > CB$
7. $m\angle 1 < m\angle 2$
8. $m\angle MNP < m\angle NPM$
9. $(2, 2); (0, -2); \left(\frac{4}{3}, \frac{2}{3}\right)$
10. Assume temporarily that $\triangle PQR$ is equilateral and equiangular. Then it follows that $m\angle P \neq m\angle Q$, $m\angle Q \neq m\angle R$, or $m\angle P \neq m\angle R$. By the contrapositive of the Base Angles Theorem, if $m\angle P \neq m\angle Q$, then $PR \neq QR$, if $m\angle Q \neq m\angle R$, then $QP \neq RP$, and if $m\angle P \neq m\angle R$, then $PQ \neq RQ$. All three conclusions contradict the fact that $\triangle PQR$ is equilateral. So, the temporary conclusion must be false. This proves that if $\triangle PQR$ is equilateral, it must also be equiangular.
11. area of $\triangle GEH = \frac{1}{4}A$; By the Triangle Midsegment Theorem, $GH = \frac{1}{2}FD$. By the markings $EG = GD$. By the Segment Addition Postulate, $EG + GD = ED$. So, when you substitute EG for GD, you get $EG + EG = ED$, or $2(EG) = ED$, which means that $EG = \frac{1}{2}ED$. So, the area of $\triangle GEH = \frac{1}{2}bh$
    $= \frac{1}{2}(EG)(GH)$
    $= \frac{1}{2}\left(\frac{1}{2}ED\right)\left(\frac{1}{2}FD\right)$
    $= \frac{1}{8}(ED)(FD)$.
    Note that the area of $\triangle DEF = \frac{1}{2}bh = \frac{1}{2}(ED)(FD)$.
    So, the area of $\triangle GEH = \frac{1}{8}(ED)(FD) = \frac{1}{4}\left[\frac{1}{2}(ED)(FD)\right] = \frac{1}{4}A$.
12. See Additional Answers.
13. Pine Avenue must be longer than 2 miles and shorter than 16 miles.
14. 9 mi; Because the path represents the shortest distance from the beach to Main Street, it must be perpendicular to Main Street, and you ended up at the midpoint between your house and the movie theater. So, the trail must be the perpendicular bisector of the portion of Main Street between your house and the movie theater. By the Perpendicular Bisector Theorem, the beach must be the same distance from your house and the movie theater. So, Pine Avenue is the same length as the 9-mile portion of Hill Street between your house and the beach.
15. See Additional Answers.

If students need help...	If students got it...
Lesson Tutorials	Resources by Chapter • Enrichment and Extension • Cumulative Review
Skills Review Handbook	Performance Task
BigIdeasMath.com	Start the *next* Section

6 Cumulative Assessment

1. Which definition(s) and/or theorem(s) do you need to use to prove the Converse of the Perpendicular Bisector Theorem? Select all that apply.

   Given $CA = CB$
   Prove Point C lies on the perpendicular bisector of $\overline{AB}$.

   - definition of perpendicular bisector
   - definition of angle bisector
   - definition of segment congruence
   - definition of angle congruence
   - Base Angles Theorem
   - Converse of the Base Angles Theorem
   - ASA Congruence Theorem
   - AAS Congruence Theorem

2. Which of the following values are x-coordinates of the solutions of the system?

   $y = x^2 - 3x + 3$
   $y = 3x - 5$

   | -7 | -4 | -3 | -2 | -1 |
   | 1 | 2 | 3 | 4 | 7 |

3. What are the coordinates of the centroid of $\triangle LMN$?

   Ⓐ (2, 5)
   Ⓑ (3, 5)
   Ⓒ (4, 5)
   Ⓓ (5, 5)

   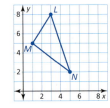

4. Use the steps in the construction to explain how you know that the circle is circumscribed about $\triangle ABC$.

   Step 1

   Step 2

   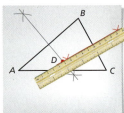

   Step 3

   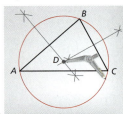

398   Chapter 6   Relationships Within Triangles

5. According to a survey, 58% of voting-age citizens in a town are planning to vote in an upcoming election. You randomly select 10 citizens to survey.

   a. What is the probability that less than 4 citizens are planning to vote?
   b. What is the probability that exactly 4 citizens are planning to vote?
   c. What is the probability that more than 4 citizens are planning to vote?

6. What are the solutions of $2x^2 + 18 = -6$?

   Ⓐ $i\sqrt{6}$ and $-i\sqrt{6}$

   Ⓑ $2i\sqrt{3}$ and $-2i\sqrt{3}$

   Ⓒ $i\sqrt{15}$ and $-i\sqrt{15}$

   Ⓓ $-i\sqrt{21}$ and $i\sqrt{21}$

7. Use the graph of $\triangle QRS$.

   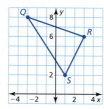

   a. Find the coordinates of the vertices of the midsegment triangle. Label the vertices T, U, and V.
   b. Show that each midsegment joining the midpoints of two sides is parallel to the third side and is equal to half the length of the third side.

8. The graph of which inequality is shown?

   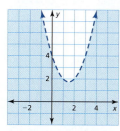

   Ⓐ $y > x^2 - 3x + 4$

   Ⓑ $y \geq x^2 - 3x + 4$

   Ⓒ $y < x^2 - 3x + 4$

   Ⓓ $y \leq x^2 - 3x + 4$

ANSWERS

5. a. about 7.12%
   b. about 13.04%
   c. about 79.84%
6. B
7. a. $T(0, 7)$, $U(2, 4)$, $V(-1, 5)$

   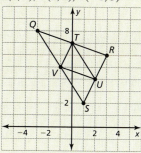

   b. slope of $\overline{TV} = 2$, slope of $\overline{SR} = 2$,
   Because the slopes are the same, $\overline{TV} \parallel \overline{RS}$.
   slope of $\overline{TU} = -\frac{3}{2}$, slope of $\overline{QS} = -\frac{3}{2}$,
   Because the slopes are the same, $\overline{TU} \parallel \overline{QS}$.
   slope of $\overline{VU} = -\frac{1}{3}$, slope of $\overline{QR} = -\frac{1}{3}$
   Because the slopes are the same, $\overline{VU} \parallel \overline{QR}$.
   $TV = \sqrt{5}$, $SR = 2\sqrt{5}$,
   Because $\sqrt{5} = \frac{1}{2}(2\sqrt{5})$,
   $TV = \frac{1}{2}SR$.
   $TU = \sqrt{13}$, $QS = 2\sqrt{13}$,
   Because $\sqrt{13} = \frac{1}{2}(2\sqrt{13})$,
   $TU = \frac{1}{2}QS$.
   $VU = \sqrt{10}$, $QR = 2\sqrt{10}$,
   Because $\sqrt{10} = \frac{1}{2}(2\sqrt{10})$,
   $VU = \frac{1}{2}QR$.
8. C

Chapter 7 Pacing Guide

Chapter Opener/Mathematical Practices	0.5 Day
Section 1	1.5 Days
Section 2	2 Days
Section 3	2 Days
Quiz	0.5 Day
Section 4	1.5 Days
Section 5	2 Days
Chapter Review/Chapter Tests	2 Days
Total Chapter 7	12 Days
Year-to-Date	107 Days

7 Quadrilaterals and Other Polygons

- 7.1 Angles of Polygons
- 7.2 Properties of Parallelograms
- 7.3 Proving That a Quadrilateral Is a Parallelogram
- 7.4 Properties of Special Parallelograms
- 7.5 Properties of Trapezoids and Kites

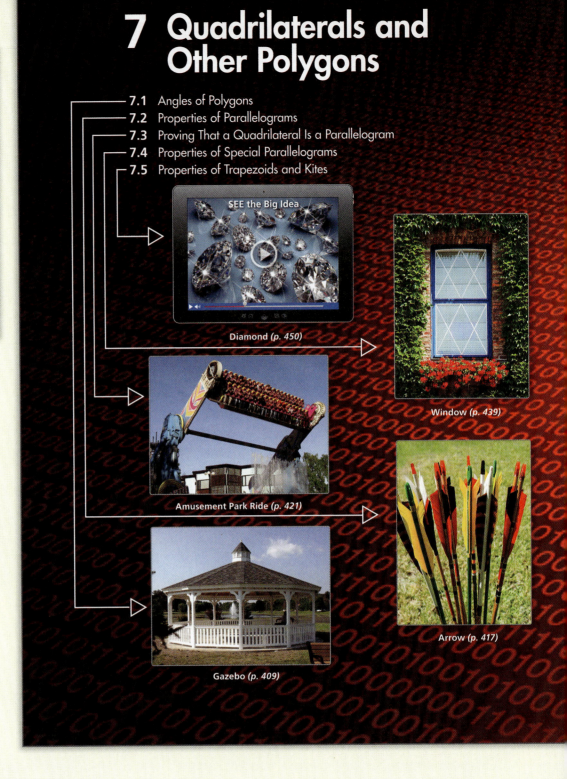

Diamond (p. 450)

Window (p. 439)

Amusement Park Ride (p. 421)

Arrow (p. 417)

Gazebo (p. 409)

Laurie's Notes

Chapter Summary

- Students have studied the special quadrilaterals in middle school and should be familiar with their definitions and some of their properties. Special quadrilaterals include parallelograms, rectangles, rhombuses, squares, trapezoids, isosceles trapezoids, and kites. Students would have investigated the properties of these special quadrilaterals using inductive reasoning.
- In this chapter, students are able to investigate the properties of special quadrilaterals by using dynamic geometry software. The properties are then proven using a variety of proof formats: transformational, synthetic, analytic, and paragraph.
- The first lesson in the chapter is about the angle measures in a polygon. First, students derive a formula for the sum of the interior angles and then a formula for the sum of the exterior angles.
- The last four lessons are about the special quadrilaterals.
- In Math I, students proved triangles congruent given different hypotheses. The reasoning used in that chapter is applied to proofs involving quadrilaterals in this chapter.

What Your Students Have Learned

Middle School
- Draw points, lines, line segments, rays, angles, parallel lines, and perpendicular lines.
- Draw geometric shapes with given conditions, focusing on triangles and quadrilaterals.
- Find interior and exterior angle measures of polygons.
- Draw polygons in the coordinate plane and find the distance between points with the same x- or y-coordinate.

Math I
- Classify polygons by the number of sides.
- Find perimeters and areas of polygons in the coordinate plane.
- Identify and use corresponding parts of congruent figures.
- Use theorems to find angle measures.
- Use SAS, SSS, HL, ASA, and AAS Congruence Theorems to prove two triangles are congruent.
- Use congruent triangles to solve problems and write proofs.

What Your Students Will Learn

Math II
- Find and use the interior and exterior angle measures of polygons.
- Use properties of parallelograms, rhombuses, rectangles, squares, trapezoids, and kites.
- Prove that quadrilaterals with certain properties are parallelograms.
- Use coordinate geometry to identify special types of parallelograms.
- Identify quadrilaterals using the most specific name based on given information.

Dynamic Teaching Tools
Dynamic Assessment & Progress Monitoring Tool
Lesson Planning Tool
Interactive Whiteboard Lesson Library
Dynamic Classroom with Dynamic Investigations
Real-Life STEM Videos

Scaffolding in the Classroom

Graphic Organizers: Summary Triangle
A Summary Triangle can be used to explain a concept. Typically, the Summary Triangle is divided into 3 or 4 parts. In the top part, students write the concept being explained. In the middle part(s), students write any procedure, explanation, description, definition, theorem, and/or formula(s). In the bottom part, students write an example to illustrate the concept. A Summary Triangle can be used as an assessment tool, in which blanks are left for students to complete. Also, students can place their Summary Triangles on note cards to use as quick study references.

Questioning in the Classroom

How will this work?
Look for examples, exercises, problems, and questions that do not have a specific solution or answer. Ask students to explain how to solve the problem.

Laurie's Notes

Maintaining Mathematical Proficiency

Using Structure to Solve a Multi-Step Equation

- Some students may find it easier to substitute z for the expression in parentheses, solve the equation for z, and then use the value of z to find the value of x.
- **COMMON ERROR** Students may have trouble understanding the meaning of the direction line and use the Distributive Property as the first step. Explain that the method shown does *not* involve distributing the factor 3 but instead performs inverse operations on the expression $(2 + x)$.

Identifying Parallel and Perpendicular Lines

- Review the relationships between slopes of parallel lines and slopes of perpendicular lines.
- In a coordinate plane, two nonvertical lines are parallel if and only if they have the same slope.
- In a coordinate plane, two nonvertical lines are perpendicular if and only if the product of their slopes is -1.
- **COMMON ERROR** Some students may forget the slope formula. Remind students that they can also use the expression $\frac{\text{rise}}{\text{run}}$ to find the slope.

Mathematical Practices (continued on page 402)

- The *Mathematical Practices* page focuses attention on how mathematics is learned—process versus content. This page explains how statements can be proven true using a method called *mathematical induction*. Students have previously used deductive methods in two-column, flowchart, and paragraph formats, along with transformational proofs. Mathematical induction is a technique that is used to prove statements involving natural numbers.
- Use the *Mathematical Practices* page to help students develop mathematical habits of mind—how mathematics can be explored and how mathematics is thought about.

If students need help...	If students got it...
Student Journal • Maintaining Mathematical Proficiency	Game Closet at *BigIdeasMath.com*
Lesson Tutorials	Start the *next* Section
Skills Review Handbook	

Maintaining Mathematical Proficiency

Using Structure to Solve a Multi-Step Equation

Example 1 Solve $3(2 + x) = -9$ by interpreting the expression $2 + x$ as a single quantity.

$$3(2 + x) = -9 \quad \text{Write the equation.}$$
$$\frac{3(2 + x)}{3} = \frac{-9}{3} \quad \text{Divide each side by 3.}$$
$$2 + x = -3 \quad \text{Simplify.}$$
$$\underline{-2} \quad \underline{-2} \quad \text{Subtract 2 from each side.}$$
$$x = -5 \quad \text{Simplify.}$$

Solve the equation by interpreting the expression in parentheses as a single quantity.

1. $4(7 - x) = 16$
2. $7(1 - x) + 2 = -19$
3. $3(x - 5) + 8(x - 5) = 22$

Identifying Parallel and Perpendicular Lines

Example 2 Determine which of the lines are parallel and which are perpendicular.

Find the slope of each line.

Line a: $m = \dfrac{3 - (-3)}{-4 - (-2)} = -3$

Line b: $m = \dfrac{-1 - (-4)}{1 - 2} = -3$

Line c: $m = \dfrac{2 - (-2)}{3 - 4} = -4$

Line d: $m = \dfrac{2 - 0}{2 - (-4)} = \dfrac{1}{3}$

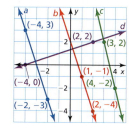

Because lines a and b have the same slope, lines a and b are parallel. Because $\frac{1}{3}(-3) = -1$, lines a and d are perpendicular and lines b and d are perpendicular.

Determine which lines are parallel and which are perpendicular.

4.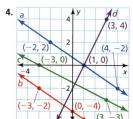
5.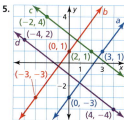
6.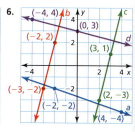

7. **ABSTRACT REASONING** Explain why interpreting an expression as a single quantity does not contradict the order of operations.

Dynamic Solutions available at *BigIdeasMath.com*

MONITORING PROGRESS ANSWERS

1. When $n = 1$,

   $\dfrac{1(1+1)(2(1)+1)}{6} = 1 = 1^2$. So, the statement is true for $n = 1$. Assume that $1^2 + 2^2 + 3^2 + \cdots + k^2$

   $= \dfrac{k(k+1)(2k+1)}{6}$ for a natural number $k \geq 1$. Then

   $1^2 + 2^2 + 3^2 + \ldots + k^2 + (k+1)^2$

   $= (1^2 + 2^2 + 3^2 + \ldots + k^2) + (k+1)^2$

   $= \dfrac{k(k+1)(2k+1)}{6} + (k+1)^2$

   $= \dfrac{k(k+1)(2k+1) + 6(k+1)^2}{6}$

   $= \dfrac{(k+1)[k(2k+1) + 6(k+1)]}{6}$

   $= \dfrac{(k+1)(k+2)(2k+3)}{6}$.

   So, the statement is true for all natural numbers $n \geq 1$.

2. When $n = 1$, $\dfrac{1^2(1+1)^2}{4} = 1 = 1^3$.

   So, the statement is true for $n = 1$. Assume that

   $1^3 + 2^3 + 3^3 + \cdots + k^3 = \dfrac{k^2(k+1)^2}{4}$

   for a natural number $k \geq 1$. Then

   $1^3 + 2^3 + 3^3 + \cdots + k^3 + (k+1)^3$

   $= (1^3 + 2^3 + 3^3 + \ldots + k^3) + (k+1)^3$

   $= \dfrac{k^2(k+1)^2}{4} + (k+1)^3$

   $= \dfrac{k^2(k+1)^2 + 4(k+1)^3}{4}$

   $= \dfrac{(k+1)^2[k^2 + 4(k+1)]}{4}$

   $= \dfrac{(k+1)^2(k+2)^2}{4}$.

   So, the statement is true for all natural numbers $n \geq 1$.

Mathematical Practices

Mathematically proficient students understand and use previously established results.

Proving by Mathematical Induction

Core Concept

Mathematical Induction

Mathematical induction is a technique that you can use to prove statements involving natural numbers. A proof by mathematical induction has two steps.

Base Case: Show that the statement P_n is true for the first case, usually $n = 1$.

Inductive Step: Assume that P_n is true for a natural number $n = k$. This assumption is called the *inductive hypothesis*. Use the inductive hypothesis to prove that P_n is true for the next natural number, $n = k + 1$.

EXAMPLE 1 Proving by Mathematical Induction

Prove that $1 + 2 + 3 + \cdots + n = \dfrac{n(n+1)}{2}$ for all natural numbers $n \geq 1$.

SOLUTION

Base Case: Let $n = 1$. Show that the statement is true.

$1 \stackrel{?}{=} \dfrac{1(1+1)}{2}$

$1 = 1$ ✓

So, the statement is true for $n = 1$.

Inductive Step: Assume that $1 + 2 + 3 + \cdots + k = \dfrac{k(k+1)}{2}$ for a natural number $k \geq 1$.

Prove that $1 + 2 + 3 + \cdots + k + (k+1) = \dfrac{(k+1)[(k+1)+1]}{2}$, or $\dfrac{(k+1)(k+2)}{2}$.

$1 + 2 + 3 + \cdots + k + (k+1) = (1 + 2 + 3 + \cdots + k) + (k+1)$ Assoc. Prop. of Addition

$= \dfrac{k(k+1)}{2} + (k+1)$ Inductive hypothesis

$= \dfrac{k(k+1)}{2} + \dfrac{2(k+1)}{2}$ Rewrite using LCD.

$= \dfrac{k(k+1) + 2(k+1)}{2}$ Add.

$= \dfrac{(k+1)(k+2)}{2}$ Factor.

So, if the statement is true for a natural number $k \geq 1$, then it is true for $k + 1$.

▶ So, you can conclude that the statement is true for all natural numbers $n \geq 1$.

Monitoring Progress

Use mathematical induction to prove the statement.

1. $1^2 + 2^2 + 3^2 + \cdots + n^2 = \dfrac{n(n+1)(2n+1)}{6}$

2. $1^3 + 2^3 + 3^3 + \cdots + n^3 = \dfrac{n^2(n+1)^2}{4}$

Laurie's Notes — Mathematical Practices (continued from page T-401)

- Explain the two steps (base case and inductive step) involved in the technique of mathematical induction.
- Write the statement you are trying to prove: $1 + 2 + 3 + \cdots + n = \dfrac{n(n+1)}{2}$.

  It is likely that students have seen this formula before, and understand that they are summing the numbers from 1 to some ending number.
- Work through Example 1 together, highlighting the two steps of a proof by mathematical induction.
- The two questions in *Monitoring Progress* are proven in a similar fashion. The symbolic manipulation involved should be within the ability of your students. You may wish to assign the problems as extra credit.

Laurie's Notes

Overview of Section 7.1

Introduction
- Students should be familiar with the content in this lesson from middle school mathematics.
- The lesson presents theorems for the sum of the interior angles and the sum of the exterior angles of a polygon. Interior and exterior angles of regular polygons are also discussed.

Teaching Strategy
- The explorations focus on discovering the relationship between the number of sides in a polygon and the sum of the interior angles.
- There are several ways in which the sum of the exterior angles can be explored.
- The low-tech version is to draw a polygon and then extend each side to form an exterior angle at each vertex. Be sure to extend in the same direction (clockwise or counterclockwise) at each vertex. Place a pencil or pointer along one side of the polygon. Simulate what happens when the pencil travels along each side of the polygon without being lifted. The pencil turns at the vertices, rotating through each exterior angle. When the pencil returns to the starting position, it has rotated through a total of 360°.
- A second way to investigate exterior angles is to use a slider and create dilations of the polygon. Place the point of dilation at the center of the polygon and draw smaller and smaller images. The polygon collapses to the point of dilation. The result is a very visual demonstration that the sum of the exterior angles is 360°.

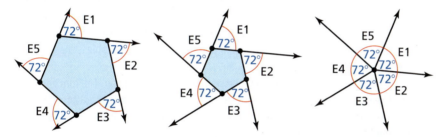

- You can find simulations of these methods online by searching for "simulation sum of exterior angles of a polygon."

Pacing Suggestion
- Exploration 1 is long when done in its entirety. You could assign Exploration 2 for homework. If students work through Exploration 1 and they have familiarity with the content from middle school, then you might be selective with the number of examples modeled.

Dynamic Teaching Tools
Dynamic Assessment & Progress Monitoring Tool
Lesson Planning Tool
Interactive Whiteboard Lesson Library
Dynamic Classroom with Dynamic Investigations

What Your Students Will Learn

- Find and use the interior and exterior angle measures to determine characteristics of polygons.

Laurie's Notes

Exploration

Motivate
- Capture and display images from the Internet of common objects that are shaped like various polygons.
- *Examples:* pentagon: top of fire hydrant; hexagon: honeycomb; octagon: stop sign or tiles; decagon: concave five-pointed star
- Explain to students that in this lesson they will be investigating properties of polygons.

Exploration Note
- Depending on the software and preference settings, you may have reflex angles (angles with measures greater than 180°) that result when students click and drag on a vertex. Students should include observations about concave polygons.
- Depending on the software, spreadsheet functionality may be built in. If not, students should use a spreadsheet to work with the data analytically.

Exploration 1
- The summing process should be set up dynamically so that students can click and drag on a vertex and observe that, although the angle measures change, the sum remains a constant. This is true even for concave polygons.
- This is a longer exploration, but it is important for students to experience the dynamic features of this investigation. For instance, what changes and what stays the same when you drag on one vertex of a polygon?
- **?** "What is the independent variable, and what is the dependent variable?" The independent variable is the number of sides of the polygon, and the dependent variable is the sum of the angle measures.
- Students should not need to perform a regression to write the function. The function they write will likely be $S = 180n - 360$. The focus is on the sum increasing by a constant rate (slope) of 180 each time another side is added to the polygon. So it makes sense that their equation would be in slope-intercept form.
- If this exploration had used the approach of drawing in all the diagonals from one vertex to form triangular regions inside the polygon, the equation the students write would have been $S = 180(n - 2)$. The $(n - 2)$ represents the number of triangles formed inside an n-gon when all the diagonals from one vertex are drawn.
- Make sure students see the connection between the two forms of the equation.

Exploration 2
- **?** "What is a regular polygon?" a polygon whose sides are all congruent and whose angles are all congruent
- If the data from the first exploration have been entered into a spreadsheet, students can add an additional column that divides the sum by the number of the sides, which is also the number of angles.

Communicate Your Answer
- If students have written an equation for the sum of the interior angles of a polygon, they only need to evaluate the equation for $n = 12$.

Connecting to Next Step
- If students have worked through the first exploration completely, they could do the first four examples in the lesson independently and quickly.

7.1 Angles of Polygons

Essential Question What is the sum of the measures of the interior angles of a polygon?

EXPLORATION 1 — The Sum of the Angle Measures of a Polygon

Work with a partner. Use dynamic geometry software.

a. Draw a quadrilateral and a pentagon. Find the sum of the measures of the interior angles of each polygon.

Sample

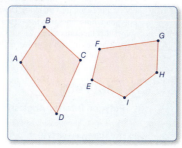

b. Draw other polygons and find the sums of the measures of their interior angles. Record your results in the table below.

CONSTRUCTING VIABLE ARGUMENTS
To be proficient in math, you need to reason inductively about data.

Number of sides, n	3	4	5	6	7	8	9
Sum of angle measures, S							

c. Plot the data from your table in a coordinate plane.

d. Write a function that fits the data. Explain what the function represents.

EXPLORATION 2 — Measure of One Angle in a Regular Polygon

Work with a partner.

a. Use the function you found in Exploration 1 to write a new function that gives the measure of one interior angle in a regular polygon with n sides.

b. Use the function in part (a) to find the measure of one interior angle of a regular pentagon. Use dynamic geometry software to check your result by constructing a regular pentagon and finding the measure of one of its interior angles.

c. Copy your table from Exploration 1 and add a row for the measure of one interior angle in a regular polygon with n sides. Complete the table. Use dynamic geometry software to check your results.

Communicate Your Answer

3. What is the sum of the measures of the interior angles of a polygon?

4. Find the measure of one interior angle in a regular dodecagon (a polygon with 12 sides).

Section 7.1 Angles of Polygons 403

Dynamic Teaching Tools

Dynamic Assessment & Progress Monitoring Tool
Lesson Planning Tool
Interactive Whiteboard Lesson Library
Dynamic Classroom with Dynamic Investigations

ANSWERS

1. a. 360°, 540°

   b. 180°, 360°, 540°, 720°, 900°, 1080°, 1260°

   c.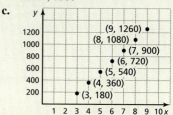

   d. $S = (n - 2) \cdot 180$; Let $n = $ the number of sides of the polygon. If you subtract 2 and multiply the difference by 180°, then you get the sum of the measures of the interior angles of a polygon.

2. a. $S = \dfrac{(n - 2) \cdot 180}{n}$

   b. 108°

   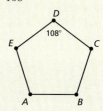

   c.

Number of Sides, n	3	4	5	6
Sum of Angle Measures, S	180°	360°	540°	720°
Interior Angle	60°	90°	108°	120°

Number of Sides, n	7	8	9
Sum of Angle Measures, S	900°	1080°	1260°
Interior Angle	128.57°	135°	140°

3. The sum S of the measures of the interior angles of a polygon with n sides is given by $S = (n - 2)180$.

4. 150°

English Language Learners

Visual
Students may mistakenly use the expression $n \cdot 180°$ instead of $(n-2) \cdot 180°$ to find the sum of the measures of the interior angles in a convex polygon. Remind students that a set of diagonals drawn from a single vertex in a convex polygon divides it into $(n-2)$ triangles. Have students illustrate this by drawing several polygons and the diagonals from one vertex to each of the other vertices in the polygon. Then count the number of triangles formed.

Extra Example 1

Find the sum of the measures of the interior angles of the figure.

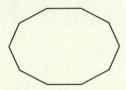

1440°

MONITORING PROGRESS ANSWER

1. 1620°

7.1 Lesson

What You Will Learn

▶ Use the interior angle measures of polygons.
▶ Use the exterior angle measures of polygons.

Core Vocabulary

diagonal, *p. 404*
equilateral polygon, *p. 405*
equiangular polygon, *p. 405*
regular polygon, *p. 405*

Previous
polygon
convex
interior angles
exterior angles

Using Interior Angle Measures of Polygons

In a polygon, two vertices that are endpoints of the same side are called *consecutive vertices*. A **diagonal** of a polygon is a segment that joins two nonconsecutive vertices.

As you can see, the diagonals from one vertex divide a polygon into triangles. Dividing a polygon with n sides into $(n-2)$ triangles shows that the sum of the measures of the interior angles of a polygon is a multiple of 180°.

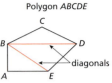

Polygon ABCDE
A and B are consecutive vertices.
Vertex B has two diagonals, $\overline{BD}$ and $\overline{BE}$.

> **REMEMBER**
> A polygon is *convex* when no line that contains a side of the polygon contains a point in the interior of the polygon.

Theorem

Polygon Interior Angles Theorem

The sum of the measures of the interior angles of a convex n-gon is $(n-2) \cdot 180°$.

$$m\angle 1 + m\angle 2 + \cdots + m\angle n = (n-2) \cdot 180°$$

Proof Ex. 42, p. 409

$n = 6$

EXAMPLE 1 Finding the Sum of Angle Measures in a Polygon

Find the sum of the measures of the interior angles of the figure.

SOLUTION

The figure is a convex octagon. It has 8 sides. Use the Polygon Interior Angles Theorem.

$(n-2) \cdot 180° = (8-2) \cdot 180°$ Substitute 8 for n.
$ = 6 \cdot 180°$ Subtract.
$ = 1080°$ Multiply.

▶ The sum of the measures of the interior angles of the figure is 1080°.

Monitoring Progress   Help in English and Spanish at *BigIdeasMath.com*

1. The coin shown is in the shape of an 11-gon. Find the sum of the measures of the interior angles.

404 Chapter 7 Quadrilaterals and Other Polygons

Laurie's Notes   Teacher Actions

- The Polygon Interior Angles Theorem is a statement of the results found by students in the first exploration. Note that it is written in the factored form. Connect this to the number of triangles formed when the diagonals from one vertex are drawn.
- Note that the theorem is stated for convex n-gons. While the formula is valid for concave polygons, reflex angles have not been defined.
- **Think-Pair-Share:** Have students work independently on Example 1 and then compare with partners.

EXAMPLE 2 **Finding the Number of Sides of a Polygon**

The sum of the measures of the interior angles of a convex polygon is 900°. Classify the polygon by the number of sides.

SOLUTION

Use the Polygon Interior Angles Theorem to write an equation involving the number of sides n. Then solve the equation to find the number of sides.

$(n-2) \cdot 180° = 900°$	Polygon Interior Angles Theorem
$n - 2 = 5$	Divide each side by 180°.
$n = 7$	Add 2 to each side.

▶ The polygon has 7 sides. It is a heptagon.

Corollary

Corollary to the Polygon Interior Angles Theorem
The sum of the measures of the interior angles of a quadrilateral is 360°.

Proof Ex. 43, p. 410

EXAMPLE 3 **Finding an Unknown Interior Angle Measure**

Find the value of x in the diagram.

SOLUTION

The polygon is a quadrilateral. Use the Corollary to the Polygon Interior Angles Theorem to write an equation involving x. Then solve the equation.

$x° + 108° + 121° + 59° = 360°$	Corollary to the Polygon Interior Angles Theorem
$x + 288 = 360$	Combine like terms.
$x = 72$	Subtract 288 from each side.

▶ The value of x is 72.

Monitoring Progress 🔊 Help in English and Spanish at *BigIdeasMath.com*

2. The sum of the measures of the interior angles of a convex polygon is 1440°. Classify the polygon by the number of sides.

3. The measures of the interior angles of a quadrilateral are $x°$, $3x°$, $5x°$, and $7x°$. Find the measures of all the interior angles.

In an **equilateral polygon**, all sides are congruent.

In an **equiangular polygon**, all angles in the interior of the polygon are congruent.

A **regular polygon** is a convex polygon that is both equilateral and equiangular.

Section 7.1 Angles of Polygons 405

Extra Example 2
The sum of the measures of the interior angles of a convex polygon is 1800°. Classify the polygon by the number of sides. **12 sides; It is a dodecagon.**

Extra Example 3
Find the value of x in the diagram.

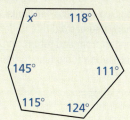

$x = 107$

MONITORING PROGRESS ANSWERS
2. decagon
3. 22.5°, 67.5°, 112.5°, 157.5°

Laurie's Notes — Teacher Actions

- Students become adept at remembering the list of polygon sums and may recall the number of sides in the polygon without solving.
- ❓ "What is a corollary?" It is a statement that follows directly from a theorem. Because quadrilaterals are studied in the remainder of the chapter, this corollary is stated for use in future problems.
- **Turn and Talk:** "Is it possible for a polygon to be equilateral and not equiangular? Equiangular and not equilateral? Both? Neither?"

Differentiated Instruction

Kinesthetic

To demonstrate the Polygon Exterior Angles Theorem, draw a large convex polygon on the floor and label one vertex A. Have a student start at vertex A and walk along the sides of the polygon, returning to the starting position. Ask the class "Through how many degrees did this student turn to start from and return to A?" The discussion should lead to the answer 360°.

Extra Example 4

A polygon is shown.

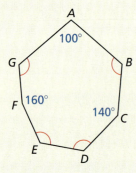

a. Is the polygon regular? Explain your reasoning.
   The polygon is not equiangular, so it is not regular.

b. Find the measures of ∠B, ∠D, ∠E, and ∠G.
   $m\angle B = m\angle D = m\angle E = m\angle G = 125°$

MONITORING PROGRESS ANSWERS

4. $m\angle S = m\angle T = 103°$

5.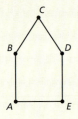

EXAMPLE 4 Finding Angle Measures in Polygons

A home plate for a baseball field is shown.

a. Is the polygon regular? Explain your reasoning.
b. Find the measures of ∠C and ∠E.

SOLUTION

a. The polygon is not equilateral or equiangular. So, the polygon is not regular.

b. Find the sum of the measures of the interior angles.

$(n - 2) \cdot 180° = (5 - 2) \cdot 180° = 540°$ Polygon Interior Angles Theorem

Then write an equation involving x and solve the equation.

$x° + x° + 90° + 90° + 90° = 540°$ Write an equation.
$2x + 270 = 540$ Combine like terms.
$x = 135$ Solve for x.

▶ So, $m\angle C = m\angle E = 135°$.

Monitoring Progress   Help in English and Spanish at *BigIdeasMath.com*

4. Find $m\angle S$ and $m\angle T$ in the diagram.

5. Sketch a pentagon that is equilateral but not equiangular.

Using Exterior Angle Measures of Polygons

Unlike the sum of the interior angle measures of a convex polygon, the sum of the exterior angle measures does *not* depend on the number of sides of the polygon. The diagrams suggest that the sum of the measures of the exterior angles, one angle at each vertex, of a pentagon is 360°. In general, this sum is 360° for any convex polygon.

JUSTIFYING STEPS

To help justify this conclusion, you can visualize a circle containing two straight angles. So, there are 180° + 180°, or 360°, in a circle.

Step 1 Shade one exterior angle at each vertex.

Step 2 Cut out the exterior angles.

Step 3 Arrange the exterior angles to form 360°.

Theorem

Polygon Exterior Angles Theorem

The sum of the measures of the exterior angles of a convex polygon, one angle at each vertex, is 360°.

$m\angle 1 + m\angle 2 + \cdots + m\angle n = 360°$

$n = 5$

Proof Ex. 51, p. 410

406   Chapter 7   Quadrilaterals and Other Polygons

Laurie's Notes   Teacher Actions

? Construct Viable Arguments and Critique the Reasoning of Others: After Example 4 ask, "Is it possible for a pentagon to have four right angles? Explain." Answers will vary. Listen for knowledge that the answer is no. Students may reference that the last angle would be 180° or that it is also impossible to construct.

• Students can discover that the measures of the exterior angles of a polygon sum to 360° using the method shown. Alternate approaches are suggested in the *Teaching Strategy* on page T-402. Write the Polygon Exterior Angles Theorem.

EXAMPLE 5 Finding an Unknown Exterior Angle Measure

Find the value of x in the diagram.

SOLUTION

Use the Polygon Exterior Angles Theorem to write and solve an equation.

$x° + 2x° + 89° + 67° = 360°$ Polygon Exterior Angles Theorem
$3x + 156 = 360$ Combine like terms.
$x = 68$ Solve for x.

▶ The value of x is 68.

> **REMEMBER**
> A *dodecagon* is a polygon with 12 sides and 12 vertices.

EXAMPLE 6 Finding Angle Measures in Regular Polygons

The trampoline shown is shaped like a regular dodecagon.

a. Find the measure of each interior angle.
b. Find the measure of each exterior angle.

SOLUTION

a. Use the Polygon Interior Angles Theorem to find the sum of the measures of the interior angles.

$(n - 2) \cdot 180° = (12 - 2) \cdot 180°$
$= 1800°$

Then find the measure of one interior angle. A regular dodecagon has 12 congruent interior angles. Divide 1800° by 12.

$\dfrac{1800°}{12} = 150°$

▶ The measure of each interior angle in the dodecagon is 150°.

b. By the Polygon Exterior Angles Theorem, the sum of the measures of the exterior angles, one angle at each vertex, is 360°. Divide 360° by 12 to find the measure of one of the 12 congruent exterior angles.

$\dfrac{360°}{12} = 30°$

▶ The measure of each exterior angle in the dodecagon is 30°.

Monitoring Progress Help in English and Spanish at *BigIdeasMath.com*

6. A convex hexagon has exterior angles with measures 34°, 49°, 58°, 67°, and 75°. What is the measure of an exterior angle at the sixth vertex?

7. An interior angle and an adjacent exterior angle of a polygon form a linear pair. How can you use this fact as another method to find the measure of each exterior angle in Example 6?

Section 7.1   Angles of Polygons   407

Extra Example 5
Find the value of x in the diagram.

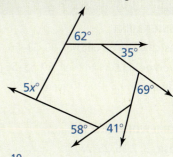

$x = 19$

Extra Example 6
Each face of the dodecahedron is shaped like a regular pentagon.

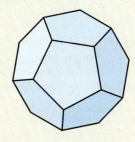

a. Find the measure of each interior angle of a regular pentagon. 108°
b. Find the measure of each exterior angle of a regular pentagon. 72°

MONITORING PROGRESS ANSWERS

6. 77°
7. You can find the measure of each exterior angle by subtracting the measure of the interior angle from 180°. In Example 6, the measure of each exterior angle is $180° - 150° = 30°$.

Laurie's Notes — Teacher Actions

- **Think-Pair-Share:** Have students work independently on Example 5 and then compare with partners.
- **Turn and Talk:** "How can you find the measure of each interior angle of a regular polygon? Each exterior angle?"
- **? Probing Question:** "If you know that the measure of an interior angle of a regular decagon is 144°, how can you find the exterior angle measure?" The interior and exterior angles are supplementary, so the exterior angle would be 36°.

Closure

- **Writing Prompt:** To find the sum of the measures of the interior angles of an n-gon …

Assignment Guide and Homework Check

ASSIGNMENT

Basic: 1, 2, 3–33 odd, 44, 50, 53–56

Average: 1, 2, 6–44 even, 50, 53–56

Advanced: 1, 2, 10–40 even, 41–50, 53–56

HOMEWORK CHECK

Basic: 3, 7, 15, 25, 27

Average: 6, 10, 14, 26, 36

Advanced: 18, 28, 36, 45, 50

ANSWERS

1. A segment connecting consecutive vertices is a side of the polygon, not a diagonal.
2. the sum of the measures of the interior angles of a pentagon; This sum is 540°, but the other sums are 360°.
3. 1260°
4. 2160°
5. 2520°
6. 3240°
7. hexagon
8. octagon
9. 16-gon
10. 20-gon
11. $x = 64$
12. $x = 66$
13. $x = 89$
14. $x = 99$
15. $x = 70$
16. $x = 117$
17. $x = 150$
18. $x = 88\frac{1}{3}$
19. $m\angle X = m\angle Y = 92°$
20. $m\angle X = m\angle Y = 142°$
21. $m\angle X = m\angle Y = 100.5°$
22. $m\angle X = m\angle Y = 135°$

7.1 Exercises

Dynamic Solutions available at *BigIdeasMath.com*

Vocabulary and Core Concept Check

1. **VOCABULARY** Why do vertices connected by a diagonal of a polygon have to be nonconsecutive?

2. **WHICH ONE DOESN'T BELONG?** Which sum does *not* belong with the other three? Explain your reasoning.

   the sum of the measures of the interior angles of a quadrilateral

   the sum of the measures of the exterior angles of a quadrilateral

   the sum of the measures of the interior angles of a pentagon

   the sum of the measures of the exterior angles of a pentagon

Monitoring Progress and Modeling with Mathematics

In Exercises 3–6, find the sum of the measures of the interior angles of the indicated convex polygon. (See Example 1.)

3. nonagon
4. 14-gon
5. 16-gon
6. 20-gon

In Exercises 7–10, the sum of the measures of the interior angles of a convex polygon is given. Classify the polygon by the number of sides. (See Example 2.)

7. 720°
8. 1080°
9. 2520°
10. 3240°

In Exercises 11–14, find the value of *x*. (See Example 3.)

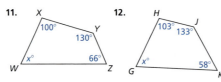

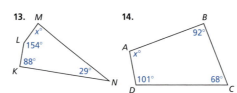

In Exercises 15–18, find the value of *x*.

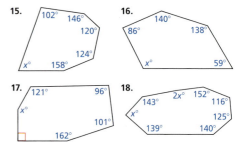

In Exercises 19–22, find the measures of ∠*X* and ∠*Y*. (See Example 4.)

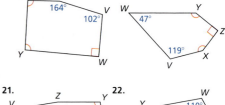

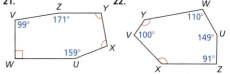

In Exercises 23–26, find the value of *x*. (See Example 5.)

23.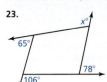

24.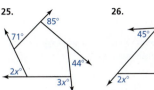

25.

26.

In Exercises 27–30, find the measure of each interior angle and each exterior angle of the indicated regular polygon. (See Example 6.)

27. pentagon
28. 18-gon
29. 45-gon
30. 90-gon

ERROR ANALYSIS In Exercises 31 and 32, describe and correct the error in finding the measure of one exterior angle of a regular pentagon.

31. ✗
$(n - 2) \cdot 180° = (5 - 2) \cdot 180°$
$= 3 \cdot 180°$
$= 540°$
The sum of the measures of the angles is 540°. There are five angles, so the measure of one exterior angle is $\frac{540°}{5} = 108°$.

32. ✗
There are a total of 10 exterior angles, two at each vertex, so the measure of one exterior angle is $\frac{360°}{10} = 36°$.

33. **MODELING WITH MATHEMATICS** The base of a jewelry box is shaped like a regular hexagon. What is the measure of each interior angle of the jewelry box base?

34. **MODELING WITH MATHEMATICS** The floor of the gazebo shown is shaped like a regular decagon. Find the measure of each interior angle of the regular decagon. Then find the measure of each exterior angle.

35. **WRITING A FORMULA** Write a formula to find the number of sides *n* in a regular polygon given that the measure of one interior angle is *x*°.

36. **WRITING A FORMULA** Write a formula to find the number of sides *n* in a regular polygon given that the measure of one exterior angle is *x*°.

REASONING In Exercises 37–40, find the number of sides for the regular polygon described.

37. Each interior angle has a measure of 156°.
38. Each interior angle has a measure of 165°.
39. Each exterior angle has a measure of 9°.
40. Each exterior angle has a measure of 6°.

41. **DRAWING CONCLUSIONS** Which of the following angle measures are possible interior angle measures of a regular polygon? Explain your reasoning. Select all that apply.

   Ⓐ 162°   Ⓑ 171°   Ⓒ 75°   Ⓓ 40°

42. **PROVING A THEOREM** The Polygon Interior Angles Theorem states that the sum of the measures of the interior angles of a convex *n*-gon is $(n - 2) \cdot 180°$.

   a. Write a paragraph proof of this theorem for the case when $n = 5$.

   b. You proved statements using mathematical induction on page 402. Prove this theorem for $n \geq 3$ using mathematical induction and the figure shown.

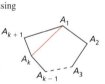

Dynamic Teaching Tools

Dynamic Assessment & Progress Monitoring Tool
Interactive Whiteboard Lesson Library
Dynamic Classroom with Dynamic Investigations

ANSWERS

23. $x = 111$
24. $x = 53$
25. $x = 32$
26. $x = 66$
27. 108°, 72°
28. 160°, 20°
29. 172°, 8°
30. 176°, 4°
31. The measure of one interior angle of a regular pentagon was found, but the exterior angle should be found by dividing 360° by the number of angles; $\frac{360°}{5} = 72°$
32. There is one exterior angle at each vertex, so the measure of one exterior angle is found by dividing 360° by the number of vertices; $\frac{360°}{5} = 72°$
33. 120°
34. 144°; 36°
35. $n = \frac{360}{180 - x}$
36. $n = \frac{360}{x}$
37. 15
38. 24
39. 40
40. 60
41. A, B; Solving the equation found in Exercise 35 for *n* yields a positive integer greater than or equal to 3 for A and B, but not for C and D.

42. a. In a pentagon, when all the diagonals from one vertex are drawn, the polygon is divided into three triangles. Because the sum of the measures of the interior angles of each triangle is 180°, the sum of the measures of the interior angles of the pentagon is $(5 - 2) \cdot 180° = 3 \cdot 180° = 540°$.

b. When $n = 3$, $(3 - 2) \cdot 180° = (1)180° = 180°$. The sum of the angles of a triangle is 180°, so the statement is true for $n = 3$. Assume that $(k - 2) \cdot 180°$ is true for a natural number $k \geq 3$. The vertices $A_1, A_2, A_3, \ldots A_k$ form a convex *k*-gon. So, the sum of the angles is $(k - 2) \cdot 180°$. Because the vertices A_1, A_k, A_{k+1} form a triangle, the sum of the angles is 180°. So, $(k - 2) \cdot 180° + 180° = (k - 1) \cdot 180°$. So, the Polygon Interior Angles Theorem is true for all natural numbers $n \geq 3$.

ANSWERS

43. In a quadrilateral, when all the diagonals from one vertex are drawn, the polygon is divided into two triangles. Because the sum of the measures of the interior angles of each triangle is 180°, the sum of the measures of the interior angles of the quadrilateral is 2 • 180° = 360°.

44. yes; Because an interior angle and an adjacent exterior angle of a polygon form a linear pair, you can use the Polygon Exterior Angles Theorem to find the measure of the exterior angles, and then you can subtract this value from 180° to find the interior angle measures of a regular polygon.

45. 21°, 21°, 21°, 21°, 138°, 138°

46. yes; The measure of the angle where the polygon caves in is greater than 180° but less than 360°.

47–56. See Additional Answers.

Mini-Assessment

1. A bumper pool table is shaped like a regular octagon.

   a. Find the measure of each interior angle. 135°
   b. Find the measure of each exterior angle. 45°

2. The sum of the measures of the interior angles of a convex polygon is 2700°. Classify the polygon by the number of sides. The polygon has 17 sides, so it is a 17-gon.

3. The measures of three of the exterior angles of a convex quadrilateral are 90°, 76°, and 110°. What is the measure of the exterior angle at the fourth vertex? 84°

---

43. **PROVING A COROLLARY** Write a paragraph proof of the Corollary to the Polygon Interior Angles Theorem.

44. **MAKING AN ARGUMENT** Your friend claims that to find the interior angle measures of a regular polygon, you do not have to use the Polygon Interior Angles Theorem. You instead can use the Polygon Exterior Angles Theorem and then the Linear Pair Postulate. Is your friend correct? Explain your reasoning.

45. **MATHEMATICAL CONNECTIONS** In an equilateral hexagon, four of the exterior angles each have a measure of $x°$. The other two exterior angles each have a measure of twice the sum of x and 48. Find the measure of each exterior angle.

46. **THOUGHT PROVOKING** For a concave polygon, is it true that at least one of the interior angle measures must be greater than 180°? If not, give an example. If so, explain your reasoning.

47. **WRITING EXPRESSIONS** Write an expression to find the sum of the measures of the interior angles for a concave polygon. Explain your reasoning.

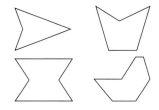

48. **ANALYZING RELATIONSHIPS** Polygon $ABCDEFGH$ is a regular octagon. Suppose sides $\overline{AB}$ and $\overline{CD}$ are extended to meet at a point P. Find $m\angle BPC$. Explain your reasoning. Include a diagram with your answer.

49. **MULTIPLE REPRESENTATIONS** The formula for the measure of each interior angle in a regular polygon can be written in function notation.

   a. Write a function $h(n)$, where n is the number of sides in a regular polygon and $h(n)$ is the measure of any interior angle in the regular polygon.
   b. Use the function to find $h(9)$.
   c. Use the function to find n when $h(n) = 150°$.
   d. Plot the points for $n = 3, 4, 5, 6, 7,$ and 8. What happens to the value of $h(n)$ as n gets larger?

50. **HOW DO YOU SEE IT?** Is the hexagon a regular hexagon? Explain your reasoning.

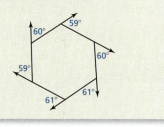

51. **PROVING A THEOREM** Write a paragraph proof of the Polygon Exterior Angles Theorem. (Hint: In a convex n-gon, the sum of the measures of an interior angle and an adjacent exterior angle at any vertex is 180°.)

52. **ABSTRACT REASONING** You are given a convex polygon. You are asked to draw a new polygon by increasing the sum of the interior angle measures by 540°. How many more sides does your new polygon have? Explain your reasoning.

Maintaining Mathematical Proficiency *Reviewing what you learned in previous grades and lessons*

Find the value of x. *(Skills Review Handbook)*

53. $x°$, $79°$

54. $113°$, $x°$

55. $(8x − 16)°$, $(3x + 20)°$

56.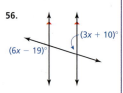
$(3x + 10)°$, $(6x − 19)°$

If students need help...

Resources by Chapter
• Practice A and Practice B
• Puzzle Time

Student Journal
• Practice

Differentiating the Lesson
Skills Review Handbook

If students got it...

Resources by Chapter
• Enrichment and Extension
• Cumulative Review

Start the *next* Section

Laurie's Notes

Overview of Section 7.2

Introduction
- This lesson is about the properties of a parallelogram. The properties are investigated in the explorations.
- Thinking about the hierarchy of quadrilaterals, the parallelogram is just one type of quadrilateral. Other quadrilaterals that are not parallelograms are presented in Section 7.5.

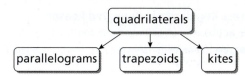

- Synthetic, transformational, and analytic proofs could be used for different theorems in this lesson.

Teaching Strategy
- **Transformational Proof:** Here is a proof using transformations of the Parallelogram Opposite Sides Theorem, which states that the opposite sides of a parallelogram are congruent.

  Given $ABCD$ is a parallelogram.
  Prove $\overline{AB} \cong \overline{DC}$, $\overline{BC} \cong \overline{AD}$
  - Draw diagonal $\overline{BD}$ and let P be the midpoint of $\overline{BD}$.
  - Rotate the figure 180° about point P.
  - $\overline{BD}$ rotates to itself.
  - Because P is the midpoint of $\overline{BD}$, $\overline{PB} \cong \overline{PD}$ and B and D rotate to each other.
  - By definition of a parallelogram, $\overline{AB} \parallel \overline{DC}$ and $\overline{BC} \parallel \overline{AD}$, so $\angle ABD \cong \angle CDB$ and $\angle ADB \cong \angle CBD$ by the Alternate Interior Angles Theorem. Therefore, the two pairs of angles, $\angle ABD$ and $\angle CDB$, and $\angle ADB$ and $\angle CBD$, rotate to each other.
  - Because $\angle ABD$ and $\angle CDB$ coincide, $\overrightarrow{BA}$ and $\overrightarrow{DC}$ coincide. Because $\angle ADB$ and $\angle CBD$ coincide, $\overrightarrow{DA}$ and $\overrightarrow{BC}$ coincide.
  - Because two lines intersect in only one point, the intersection of $\overrightarrow{BA}$ and $\overrightarrow{DA}$, point A, rotates to the intersection of $\overrightarrow{DC}$ and $\overrightarrow{BC}$, point C, and vice versa.
  - Therefore, the image of parallelogram $ABCD$ is parallelogram $CDAB$.
  - Based on what coincides, $\overline{AB} \cong \overline{DC}$ and $\overline{BC} \cong \overline{AD}$.

Pacing Suggestion
- The explorations provide an opportunity for students to discover the properties of a parallelogram. Transition to the formal lesson as soon as students have discussed each exploration.

Dynamic Teaching Tools
Dynamic Assessment & Progress Monitoring Tool
Lesson Planning Tool
Interactive Whiteboard Lesson Library
Dynamic Classroom with Dynamic Investigations

Section 7.2

What Your Students Will Learn

- Use properties to find side lengths and angles of parallelograms.
- Use properties of parallelograms to identify coordinates in the coordinate plane.

Laurie's Notes

Exploration

Motivate
- Capture and display images from the Internet of parallelograms used in structural and artistic designs. Refer back to these images at the end of the lesson, and identify the properties of parallelograms.

Exploration Note
- **Look For and Express Regularity in Repeated Reasoning:** Mathematically proficient students do not base a conjecture on one construction. The construction must be repeated several times, each time with the hypothesis being satisfied. Each time, look for patterns.

Exploration 1
- There is a difference between constructing a quadrilateral that looks like a parallelogram and constructing a parallelogram. When it only looks like a parallelogram, it will certainly change when you drag on a vertex!
- The construction will show lines versus segments. Students do not need to hide the rays that extend beyond the vertices of the parallelogram. Time is not well spent doing this.
- ? "What observations do you have about the angles of a parallelogram?" Opposite angles are congruent. Adjacent angles are supplementary.
- ? "What observations do you have about the sides of a parallelogram?" Opposite sides are congruent.

Exploration 2
- In this exploration, you could have students construct the diagonals and make observations versus telling students to measure particular segments.
- ? "What observations do you have about the diagonals of a parallelogram?" The diagonals bisect each other.
- ? "Are the diagonals perpendicular?" sometimes "Are the diagonals congruent?" sometimes

Communicate Your Answer
- There are many properties of parallelograms that students may mention. Make a list of the properties suggested by students.

Connecting to Next Step
- The properties discovered by students will be stated and proven as theorems in the formal lesson.

7.2 Properties of Parallelograms

Essential Question What are the properties of parallelograms?

EXPLORATION 1 Discovering Properties of Parallelograms

Work with a partner. Use dynamic geometry software.

a. Construct any parallelogram and label it *ABCD*. Explain your process.

Sample

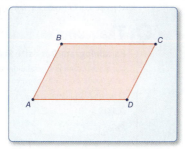

b. Find the angle measures of the parallelogram. What do you observe?
c. Find the side lengths of the parallelogram. What do you observe?
d. Repeat parts (a)–(c) for several other parallelograms. Use your results to write conjectures about the angle measures and side lengths of a parallelogram.

EXPLORATION 2 Discovering a Property of Parallelograms

Work with a partner. Use dynamic geometry software.

a. Construct any parallelogram and label it *ABCD*.
b. Draw the two diagonals of the parallelogram. Label the point of intersection *E*.

Sample

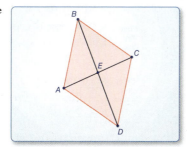

MAKING SENSE OF PROBLEMS

To be proficient in math, you need to analyze givens, constraints, relationships, and goals.

c. Find the segment lengths *AE*, *BE*, *CE*, and *DE*. What do you observe?
d. Repeat parts (a)–(c) for several other parallelograms. Use your results to write a conjecture about the diagonals of a parallelogram.

Communicate Your Answer

3. What are the properties of parallelograms?

ANSWERS

1. a. Check students' work; Construct $\overleftrightarrow{AB}$ and a line parallel to $\overleftrightarrow{AB}$ through point *C*. Construct $\overleftrightarrow{BC}$ and a line parallel to $\overleftrightarrow{BC}$ through point *A*. Construct a point *D* at the intersection of the line drawn parallel to $\overleftrightarrow{AB}$ and the line drawn parallel to $\overleftrightarrow{BC}$. Finally, construct $\overline{AB}$, $\overline{BC}$, $\overline{CD}$, and $\overline{DA}$ by removing the rest of the parallel lines drawn.

   b. Check students' work. (For sample in text, $m\angle A = m\angle C = 63.43°$ and $m\angle B = m\angle D = 116.57°$.); Opposite angles are congruent, and consecutive angles are supplementary.

   c. Check students' work. (For sample in text, $AB = CD = 2.24$ and $BC = AD = 4$.); Opposite sides are congruent.

   d. Check students' work; Opposite angles of a parallelogram are congruent. Consecutive angles of a parallelogram are supplementary. Opposite sides of a parallelogram are congruent.

2. a. Check students' work.
   b. Check students' work.
   c. Check student's work. (For sample in text, $AE = CE = 1.58$ and $BE = DE = 2.55$.); Point *E* bisects $\overline{AC}$ and $\overline{BD}$.
   d. The diagonals of a parallelogram bisect each other.

3. A parallelogram is a quadrilateral where both pairs of opposite sides are congruent and parallel, opposite angles are congruent, consecutive angles are supplementary, and the diagonals bisect each other.

English Language Learners

Notebook Development
Have students record the theorems on pages 412 and 413 in their notebooks. For each theorem, sketch a parallelogram and explain what the theorem says about the figure.

7.2 Lesson

What You Will Learn
- Use properties to find side lengths and angles of parallelograms.
- Use parallelograms in the coordinate plane.

Core Vocabulary
parallelogram, p. 412
Previous
quadrilateral
diagonal
interior angles
segment bisector

Using Properties of Parallelograms

A **parallelogram** is a quadrilateral with both pairs of opposite sides parallel. In $\square PQRS$, $\overline{PQ} \parallel \overline{RS}$ and $\overline{QR} \parallel \overline{PS}$ by definition. The theorems below describe other properties of parallelograms.

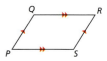

Theorems

Parallelogram Opposite Sides Theorem

If a quadrilateral is a parallelogram, then its opposite sides are congruent.

If $PQRS$ is a parallelogram, then $\overline{PQ} \cong \overline{RS}$ and $\overline{QR} \cong \overline{SP}$.

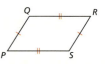

Proof p. 412

Parallelogram Opposite Angles Theorem

If a quadrilateral is a parallelogram, then its opposite angles are congruent.

If $PQRS$ is a parallelogram, then $\angle P \cong \angle R$ and $\angle Q \cong \angle S$.

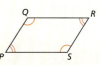

Proof Ex. 37, p. 417

PROOF Parallelogram Opposite Sides Theorem

Given $PQRS$ is a parallelogram.
Prove $\overline{PQ} \cong \overline{RS}, \overline{QR} \cong \overline{SP}$

Plan for Proof
a. Draw diagonal $\overline{QS}$ to form $\triangle PQS$ and $\triangle RSQ$.
b. Use the ASA Congruence Theorem to show that $\triangle PQS \cong \triangle RSQ$.
c. Use congruent triangles to show that $\overline{PQ} \cong \overline{RS}$ and $\overline{QR} \cong \overline{SP}$.

Plan in Action

	STATEMENTS	REASONS
	1. $PQRS$ is a parallelogram.	1. Given
a.	2. Draw $\overline{QS}$.	2. Through any two points, there exists exactly one line.
	3. $\overline{PQ} \parallel \overline{RS}, \overline{QR} \parallel \overline{PS}$	3. Definition of parallelogram
b.	4. $\angle PQS \cong \angle RSQ$, $\angle PSQ \cong \angle RQS$	4. Alternate Interior Angles Theorem
	5. $\overline{QS} \cong \overline{SQ}$	5. Reflexive Property of Congruence
	6. $\triangle PQS \cong \triangle RSQ$	6. ASA Congruence Theorem
c.	7. $\overline{PQ} \cong \overline{RS}, \overline{QR} \cong \overline{SP}$	7. Corresponding parts of congruent triangles are congruent.

412   Chapter 7   Quadrilaterals and Other Polygons

Laurie's Notes — Teacher Actions

- **Attend to Precision:** Urge students to be careful with the definition of *parallelogram*. "A quadrilateral with parallel sides" is not the same as saying "a quadrilateral with both pairs of opposite sides parallel." The first phrase could mean it is sufficient for only one pair of sides to be parallel.
- **Whiteboarding:** Discuss the first theorem. Have partners write a proof.
- **Construct Viable Arguments and Critique the Reasoning of Others:** Compare and critique the proofs offered by several volunteers.
- **Extension:** You might also write a transformational proof of the Parallelogram Opposite Sides Theorem. See the *Teaching Strategy* on page T-410 for a sample proof.

EXAMPLE 1 Using Properties of Parallelograms

Find the values of x and y.

SOLUTION

$ABCD$ is a parallelogram by the definition of a parallelogram. Use the Parallelogram Opposite Sides Theorem to find the value of x.

$AB = CD$ Opposite sides of a parallelogram are congruent.
$x + 4 = 12$ Substitute $x + 4$ for AB and 12 for CD.
$x = 8$ Subtract 4 from each side.

By the Parallelogram Opposite Angles Theorem, $\angle A \cong \angle C$, or $m\angle A = m\angle C$. So, $y° = 65°$.

▶ In $\square ABCD$, $x = 8$ and $y = 65$.

Monitoring Progress Help in English and Spanish at *BigIdeasMath.com*

1. Find FG and $m\angle G$.

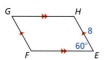

2. Find the values of x and y.

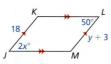

The Consecutive Interior Angles Theorem states that if two parallel lines are cut by a transversal, then the pairs of consecutive interior angles formed are supplementary.

A pair of consecutive angles in a parallelogram is like a pair of consecutive interior angles between parallel lines. This similarity suggests the Parallelogram Consecutive Angles Theorem.

Theorems

Parallelogram Consecutive Angles Theorem

If a quadrilateral is a parallelogram, then its consecutive angles are supplementary.

If $PQRS$ is a parallelogram, then $x° + y° = 180°$.

Proof Ex. 38, p. 417

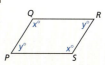

Parallelogram Diagonals Theorem

If a quadrilateral is a parallelogram, then its diagonals bisect each other.

If $PQRS$ is a parallelogram, then $\overline{QM} \cong \overline{SM}$ and $\overline{PM} \cong \overline{RM}$.

Proof p. 414

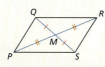

Section 7.2 Properties of Parallelograms 413

Differentiated Instruction

Visual

Ask students whether or not the diagonals of a parallelogram always bisect the opposite angles of the parallelogram. If not, have them identify a counterexample (non-square rectangle). Ask students to name the types of parallelogram for which this statement is always true (rhombus, square).

Extra Example 1

Find the values of x and y.

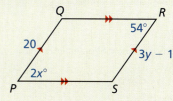

$x = 27, y = 7$

MONITORING PROGRESS ANSWERS

1. $FG = 8, m\angle G = 60°$
2. $x = 25, y = 15$

Laurie's Notes — Teacher Actions

- **Think-Pair-Share:** Have students work independently on Example 1 and then compare with partners.
- ❓ "If you know the measure of one angle of a parallelogram, how can you find the measure of the two adjacent angles?" *The adjacent angles will be supplementary to the known angle, because the consecutive interior angles formed by parallel lines cut by a transversal are supplementary.*
- ❓ "What do you know about the diagonals of a parallelogram?" *They bisect each other.*

Extra Example 2

In parallelogram PQRS, m∠P is four times m∠Q. Find m∠P. 144°

Extra Example 3

Use the figure in Example 3. Write a two-column proof.
Given ABCD and GDEF are parallelograms.
Prove ∠C ≅ ∠G
Statements (Reasons)
1. ABCD and GDEF are parallelograms. (Given)
2. ∠C and ∠CDA are supplementary, ∠G and ∠EDG are supplementary. (Parallelogram Consec. ∠s Thm.)
3. ∠CDA ≅ ∠EDG (Vert. ∠s ≅ Thm.)
4. ∠C ≅ ∠G (≅ Supplements Thm.)

MONITORING PROGRESS ANSWERS

3. 60°
4. See Additional Answers.

PROOF Parallelogram Diagonals Theorem

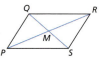

Given PQRS is a parallelogram. Diagonals $\overline{PR}$ and $\overline{QS}$ intersect at point M.
Prove M bisects $\overline{QS}$ and $\overline{PR}$.

STATEMENTS	REASONS
1. PQRS is a parallelogram.	1. Given
2. $\overline{PQ} \parallel \overline{RS}$	2. Definition of a parallelogram
3. ∠QPR ≅ ∠SRP, ∠PQS ≅ ∠RSQ	3. Alternate Interior Angles Theorem
4. $\overline{PQ} \cong \overline{RS}$	4. Parallelogram Opposite Sides Theorem
5. △PMQ ≅ △RMS	5. ASA Congruence Theorem
6. $\overline{QM} \cong \overline{SM}$, $\overline{PM} \cong \overline{RM}$	6. Corresponding parts of congruent triangles are congruent.
7. M bisects $\overline{QS}$ and $\overline{PR}$.	7. Definition of segment bisector

EXAMPLE 2 Using Properties of a Parallelogram

As shown, part of the extending arm of a desk lamp is a parallelogram. The angles of the parallelogram change as the lamp is raised and lowered. Find m∠BCD when m∠ADC = 110°.

SOLUTION

By the Parallelogram Consecutive Angles Theorem, the consecutive angle pairs in ▱ABCD are supplementary. So, m∠ADC + m∠BCD = 180°. Because m∠ADC = 110°, m∠BCD = 180° − 110° = 70°.

EXAMPLE 3 Writing a Two-Column Proof

Write a two-column proof.
Given ABCD and GDEF are parallelograms.
Prove ∠B ≅ ∠F

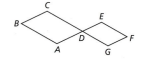

STATEMENTS	REASONS
1. ABCD and GDEF are parallelograms.	1. Given
2. ∠CDA ≅ ∠B, ∠EDG ≅ ∠F	2. If a quadrilateral is a parallelogram, then its opposite angles are congruent.
3. ∠CDA ≅ ∠EDG	3. Vertical Angles Congruence Theorem
4. ∠B ≅ ∠F	4. Transitive Property of Congruence

Monitoring Progress Help in English and Spanish at *BigIdeasMath.com*

3. **WHAT IF?** In Example 2, find m∠BCD when m∠ADC is twice the measure of ∠BCD.
4. Using the figure and the given statement in Example 3, prove that ∠C and ∠F are supplementary angles.

414 Chapter 7 Quadrilaterals and Other Polygons

Laurie's Notes Teacher Actions

❓ "What is the given information in the proof of the Parallelogram Diagonals Theorem and what do we need to prove?" We have a parallelogram and we want to prove that the diagonals bisect each other.
• There are four pairs of non-overlapping triangles that could be proven congruent. Have students discuss what pairs might be helpful, meaning corresponding parts that yield $\overline{QM} \cong \overline{SM}$ and $\overline{PM} \cong \overline{RM}$. Have partners work on a proof of one pair of triangles.
❓ In Example 2 ask, "Does the measure of ∠ADC always equal 110°?" No, it changes when the lamp is repositioned.

Using Parallelograms in the Coordinate Plane

EXAMPLE 4 Using Parallelograms in the Coordinate Plane

Find the coordinates of the intersection of the diagonals of ▱LMNO with vertices L(1, 4), M(7, 4), N(6, 0), and O(0, 0).

SOLUTION

By the Parallelogram Diagonals Theorem, the diagonals of a parallelogram bisect each other. So, the coordinates of the intersection are the midpoints of diagonals $\overline{LN}$ and $\overline{OM}$.

coordinates of midpoint of $\overline{OM} = \left(\dfrac{7+0}{2}, \dfrac{4+0}{2}\right) = \left(\dfrac{7}{2}, 2\right)$ Midpoint Formula

▶ The coordinates of the intersection of the diagonals are $\left(\dfrac{7}{2}, 2\right)$. You can check your answer by graphing ▱LMNO and drawing the diagonals. The point of intersection appears to be correct.

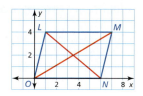

JUSTIFYING STEPS
In Example 4, you can use either diagonal to find the coordinates of the intersection. Using diagonal $\overline{OM}$ helps simplify the calculation because one endpoint is (0, 0).

EXAMPLE 5 Using Parallelograms in the Coordinate Plane

Three vertices of ▱WXYZ are W(−1, −3), X(−3, 2), and Z(4, −4). Find the coordinates of vertex Y.

SOLUTION

Step 1 Graph the vertices W, X, and Z.

Step 2 Find the slope of $\overline{WX}$.

slope of $\overline{WX} = \dfrac{2-(-3)}{-3-(-1)} = \dfrac{5}{-2} = -\dfrac{5}{2}$

Step 3 Start at Z(4, −4). Use the rise and run from Step 2 to find vertex Y.

A rise of 5 represents a change of 5 units up. A run of −2 represents a change of 2 units left.

So, plot the point that is 5 units up and 2 units left from Z(4, −4). The point is (2, 1). Label it as vertex Y.

Step 4 Find the slopes of $\overline{XY}$ and $\overline{WZ}$ to verify that they are parallel.

slope of $\overline{XY} = \dfrac{1-2}{2-(-3)} = \dfrac{-1}{5} = -\dfrac{1}{5}$ slope of $\overline{WZ} = \dfrac{-4-(-3)}{4-(-1)} = \dfrac{-1}{5} = -\dfrac{1}{5}$

▶ So, the coordinates of vertex Y are (2, 1).

REMEMBER
When graphing a polygon in the coordinate plane, the name of the polygon gives the order of the vertices.

Monitoring Progress Help in English and Spanish at BigIdeasMath.com

5. Find the coordinates of the intersection of the diagonals of ▱STUV with vertices S(−2, 3), T(1, 5), U(6, 3), and V(3, 1).

6. Three vertices of ▱ABCD are A(2, 4), B(5, 2), and C(3, −1). Find the coordinates of vertex D.

Assignment Guide and Homework Check

ASSIGNMENT

Basic: 1, 2, 3–29 odd, 33, 42, 48–50

Average: 1, 2, 6–32 even, 33, 36, 39–42, 45, 48–50

Advanced: 1, 2, 6–26 even, 30–34, 40–50

HOMEWORK CHECK

Basic: 5, 9, 19, 25, 27

Average: 6, 10, 20, 24, 26

Advanced: 10, 20, 26, 45, 47

ANSWERS

1. In order to be a quadrilateral, a polygon must have 4 sides, and parallelograms always have 4 sides. In order to be a parallelogram, a polygon must have 4 sides with opposite sides parallel. Quadrilaterals always have 4 sides, but do not always have opposite sides parallel.

2. The two angles that are consecutive to the given angle are supplementary to it. So, you can find each of their measures by subtracting the measure of the given angle from 180°. The angle opposite the given angle is congruent and therefore has the same measure.

3. $x = 9, y = 15$
4. $m = 5, n = 12$
5. $d = 126, z = 28$
6. $g = 61, h = 9$
7. 129°
8. 85°
9. 13; By the Parallelogram Opposite Sides Theorem, $LM = QN$.
10. 7; By the Parallelogram Diagonals Theorem, $LP = PN$.
11. 8; By the Parallelogram Opposite Sides Theorem, $LQ = MN$.
12. 16.4; By the Parallelogram Diagonals Theorem, $MP = PQ$. So, $MQ = 2 \cdot 8.2$.
13. 80°; By the Parallelogram Consecutive Angles Theorem, $\angle QLM$ and $\angle LMN$ are supplementary. So, $m\angle LMN = 180° - 100°$.
14. 80°; By the Parallelogram Consecutive Angles Theorem, $\angle QLM$ and $\angle NQL$ are supplementary. So, $m\angle NQL = 180° - 100°$.

7.2 Exercises

Dynamic Solutions available at *BigIdeasMath.com*

Vocabulary and Core Concept Check

1. **VOCABULARY** Why is a parallelogram always a quadrilateral, but a quadrilateral is only sometimes a parallelogram?

2. **WRITING** You are given one angle measure of a parallelogram. Explain how you can find the other angle measures of the parallelogram.

Monitoring Progress and Modeling with Mathematics

In Exercises 3–6, find the value of each variable in the parallelogram. *(See Example 1.)*

3.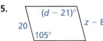

4.

5.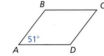

6.

In Exercises 7 and 8, find the measure of the indicated angle in the parallelogram. *(See Example 2.)*

7. Find $m\angle B$.

8. Find $m\angle N$.

In Exercises 9–16, find the indicated measure in $\square LMNQ$. Explain your reasoning.

9. LM
10. LP
11. LQ
12. MQ
13. $m\angle LMN$
14. $m\angle NQL$
15. $m\angle MNQ$
16. $m\angle LMQ$

In Exercises 17–20, find the value of each variable in the parallelogram.

17.

18.

19.

20.

ERROR ANALYSIS In Exercises 21 and 22, describe and correct the error in using properties of parallelograms.

21.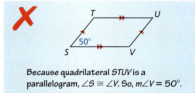

Because quadrilateral STUV is a parallelogram, $\angle S \cong \angle V$. So, $m\angle V = 50°$.

22.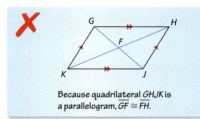

Because quadrilateral GHJK is a parallelogram, $\overline{GF} \cong \overline{FH}$.

416   Chapter 7   Quadrilaterals and Other Polygons

---

15. 100°; By the Parallelogram Opposite Angles Theorem, $m\angle QLM = m\angle MNQ$.
16. 29°; By the Alternate Interior Angles Theorem, $m\angle LMQ = m\angle MQN$.
17. $m = 35, n = 110$
18. $b = 90, c = 80, d = 100$
19. $k = 7, m = 8$
20. $u = 4, v = 18$
21. In a parallelogram, consecutive angles are supplementary; Because quadrilateral STUV is a parallelogram, $\angle S$ and $\angle V$ are supplementary. So, $m\angle V = 180° - 50° = 130°$.
22. In a parallelogram, the diagonals bisect each other. So the two parts of $\overline{GJ}$ are congruent to each other; Because quadrilateral GHJK is a parallelogram, $\overline{GF} \cong \overline{FJ}$.

PROOF In Exercises 23 and 24, write a two-column proof. *(See Example 3.)*

23. Given *ABCD* and *CEFD* are parallelograms.

    Prove $\overline{AB} \cong \overline{FE}$

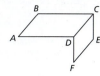

24. Given *ABCD*, *EBGF*, and *HJKD* are parallelograms.

    Prove $\angle 2 \cong \angle 3$

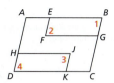

In Exercises 25 and 26, find the coordinates of the intersection of the diagonals of the parallelogram with the given vertices. *(See Example 4.)*

25. $W(-2, 5)$, $X(2, 5)$, $Y(4, 0)$, $Z(0, 0)$

26. $Q(-1, 3)$, $R(5, 2)$, $S(1, -2)$, $T(-5, -1)$

In Exercises 27–30, three vertices of ▱*DEFG* are given. Find the coordinates of the remaining vertex. *(See Example 5.)*

27. $D(0, 2)$, $E(-1, 5)$, $G(4, 0)$

28. $D(-2, -4)$, $F(0, 7)$, $G(1, 0)$

29. $D(-4, -2)$, $E(-3, 1)$, $F(3, 3)$

30. $E(-1, 4)$, $F(5, 6)$, $G(8, 0)$

MATHEMATICAL CONNECTIONS In Exercises 31 and 32, find the measure of each angle.

31. The measure of one interior angle of a parallelogram is 0.25 times the measure of another angle.

32. The measure of one interior angle of a parallelogram is 50 degrees more than 4 times the measure of another angle.

33. **MAKING AN ARGUMENT** In quadrilateral *ABCD*, $m\angle B = 124°$, $m\angle A = 56°$, and $m\angle C = 124°$. Your friend claims quadrilateral *ABCD* could be a parallelogram. Is your friend correct? Explain your reasoning.

34. **ATTENDING TO PRECISION** $\angle J$ and $\angle K$ are consecutive angles in a parallelogram, $m\angle J = (3x + 7)°$, and $m\angle K = (5x - 11)°$. Find the measure of each angle.

35. **CONSTRUCTION** Construct any parallelogram and label it *ABCD*. Draw diagonals $\overline{AC}$ and $\overline{BD}$. Explain how to use paper folding to verify the Parallelogram Diagonals Theorem for ▱*ABCD*.

36. **MODELING WITH MATHEMATICS** The feathers on an arrow form two congruent parallelograms. The parallelograms are reflections of each other over the line that contains their shared side. Show that $m\angle 2 = 2m\angle 1$.

37. **PROVING A THEOREM** Use the diagram to write a two-column proof of the Parallelogram Opposite Angles Theorem.

Given *ABCD* is a parallelogram.

Prove $\angle A \cong \angle C$, $\angle B \cong \angle D$

38. **PROVING A THEOREM** Use the diagram to write a two-column proof of the Parallelogram Consecutive Angles Theorem.

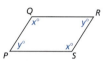

Given *PQRS* is a parallelogram.

Prove $x° + y° = 180°$

39. **PROBLEM SOLVING** The sides of ▱*MNPQ* are represented by the expressions below. Sketch ▱*MNPQ* and find its perimeter.

    $MQ = -2x + 37$ $QP = y + 14$
    $NP = x - 5$ $MN = 4y + 5$

40. **PROBLEM SOLVING** In ▱*LMNP*, the ratio of *LM* to *MN* is 4 : 3. Find *LM* when the perimeter of ▱*LMNP* is 28.

ANSWERS

41. no; Two parallelograms with congruent corresponding sides may or may not have congruent corresponding angles.

42. a. decreases; Because $\angle P$ and $\angle Q$ are supplementary, as $m\angle Q$ increases, $m\angle P$ must decrease so that their total is still 180°.

   b. increases; As $m\angle Q$ decreases, the parallelogram gets skinnier, which means that Q and S get farther apart and QS increases.

   c. The mirror gets closer to the wall; As $m\angle Q$ decreases, the parallelograms get skinnier, which means that P, R, and the other corresponding vertices all get closer together. So, the distance between the mirror and the wall gets smaller.

43. 16°

44–50. See Additional Answers.

Mini-Assessment

1. Find the coordinates of the point of intersection of the diagonals of $\square ABCD$ with vertices $A(-1, 3)$, $B(3, 3)$, $C(5, -1)$, and $D(1, -1)$. **(2, 1)**

2. Three vertices of $\square PQRS$ are $P(-3, 2)$, $Q(2, 6)$, and $R(3, 3)$. Find the coordinates of vertex S. **$(-2, -1)$**

Use parallelogram STUV.

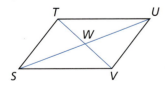

3. $ST = 29$ and $VU = 2x + 5$. Find the value of x. **12**

4. $m\angle STU = 112°$ and $m\angle UVS = (5y - 3)°$. Find the value of y. **23**

5. $SW = 10$ and $TW = 7$. Find SU. **20**

418   Chapter 7

---

41. ABSTRACT REASONING Can you prove that two parallelograms are congruent by proving that all their corresponding sides are congruent? Explain your reasoning.

42. HOW DO YOU SEE IT? The mirror shown is attached to the wall by an arm that can extend away from the wall. In the figure, points P, Q, R, and S are the vertices of a parallelogram. This parallelogram is one of several that change shape as the mirror is extended.

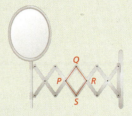

   a. What happens to $m\angle P$ as $m\angle Q$ increases? Explain.
   b. What happens to QS as $m\angle Q$ decreases? Explain.
   c. What happens to the overall distance between the mirror and the wall when $m\angle Q$ decreases? Explain.

43. MATHEMATICAL CONNECTIONS In $\square STUV$, $m\angle TSU = 32°$, $m\angle USV = (x^2)°$, $m\angle TUV = 12x°$, and $\angle TUV$ is an acute angle. Find $m\angle USV$.

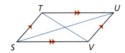

Maintaining Mathematical Proficiency
Reviewing what you learned in previous grades and lessons

Determine whether lines ℓ and m are parallel. Explain your reasoning. *(Skills Review Handbook)*

48.   49.    50.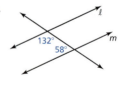

418   Chapter 7   Quadrilaterals and Other Polygons

---

44. THOUGHT PROVOKING Is it possible that any triangle can be partitioned into four congruent triangles that can be rearranged to form a parallelogram? Explain your reasoning.

45. CRITICAL THINKING Points $W(1, 2)$, $X(3, 6)$, and $Y(6, 4)$ are three vertices of a parallelogram. How many parallelograms can be created using these three vertices? Find the coordinates of each point that could be the fourth vertex.

46. PROOF In the diagram, $\overline{EK}$ bisects $\angle FEH$, and $\overline{FJ}$ bisects $\angle EFG$. Prove that $\overline{EK} \perp \overline{FJ}$. (*Hint*: Write equations using the angle measures of the triangles and quadrilaterals formed.)

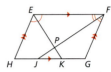

47. PROOF Prove the *Congruent Parts of Parallel Lines Corollary*: If three or more parallel lines cut off congruent segments on one transversal, then they cut off congruent segments on every transversal.

Given $\overleftrightarrow{GH} \parallel \overleftrightarrow{JK} \parallel \overleftrightarrow{LM}$, $\overline{GJ} \cong \overline{JL}$

Prove $\overline{HK} \cong \overline{KM}$

(*Hint*: Draw $\overline{KP}$ and $\overline{MQ}$ such that quadrilateral $GPKJ$ and quadrilateral $JQML$ are parallelograms.)

If students need help...	If students got it...
Resources by Chapter • Practice A and Practice B • Puzzle Time	**Resources by Chapter** • Enrichment and Extension • Cumulative Review
Student Journal • Practice	Start the *next* Section
Differentiating the Lesson Skills Review Handbook	

Laurie's Notes

Overview of Section 7.3

Introduction
- There are four theorems presented in this lesson, three of which are converses of theorems in the last lesson. Students should recall that although a statement is true, its converse may not be.
- Each of the four theorems can be used to show that a quadrilateral is a parallelogram.

Teaching Strategy
- All the theorems in this lesson can be introduced in a very low-tech fashion that will give students a chance to visualize and think about the given conditions (the hypotheses) that can be used to show that a quadrilateral is a parallelogram.
- Make a collection of pieces that can be used to make a quadrilateral. The pieces are used to make the sides, angles, and diagonals of the quadrilateral.
- My collection is made on transparencies using a permanent marker. Make pairs of segments in various lengths (2, 3, 4, 5, and 6 inches) and pairs of angles of various measures (30°, 60°, 120°, and 150°).
- Give students the parts (i.e., opposite sides congruent), and have them place the pieces on the overhead to make a quadrilateral. Does the quadrilateral appear to be a parallelogram? How might you prove it?
- This same approach can be done using an interactive whiteboard where the pieces can be manipulated.

Pacing Suggestion
- The explorations provide an opportunity for students to consider how they can prove they have a parallelogram. Transition to the formal lesson as soon as students have discussed each exploration.

Dynamic Teaching Tools
Dynamic Assessment & Progress Monitoring Tool
Lesson Planning Tool
Interactive Whiteboard Lesson Library
Dynamic Classroom with Dynamic Investigations

What Your Students Will Learn

- Prove that quadrilaterals with certain properties are parallelograms.
- Show that a quadrilateral is a parallelogram in the coordinate plane.

Laurie's Notes

Exploration

Motivate
- Ask for a volunteer. Hand the student two 5-inch straws and two 3-inch straws. Ask him or her to arrange the straws to make a quadrilateral.
- There are three possibilities. The student will make a quadrilateral that appears to be parallelogram, a convex kite, or a concave kite.
- Focus on the quadrilateral with opposite sides congruent, and ask students whether it could be proven that this is a parallelogram. Give time for discussion.
- Explain to students that in the lesson they will complete the proof!

Exploration Note
- **Look For and Express Regularity in Repeated Reasoning:** Mathematically proficient students do not base a conjecture on one construction. The construction must be repeated several times, each time with the hypothesis being satisfied. Each time, look for patterns.

Exploration 1
- There are different ways in which students could construct the quadrilateral and satisfy the hypothesis. Knowing the features of the software that your students work with is important. An alternate approach with the focus on angles is explained in Exploration 2.
- The software I use allows me to create four consecutive segments using five points, with $AB = CD$ and $BC = DE$. I then wrap the segment around until points E and A coincide.
- **? Construct Viable Arguments and Critique the Reasoning of Others:** "How did you decide whether the quadrilateral was a parallelogram?" Listen for valid reasoning. Most students measure the angles and determine that opposite sides are parallel because interior angles on the same side of the transversal are supplementary. They also could find the slopes of opposite sides.
- Suggest to students that they begin with a quadrilateral having sides parallel to the x- or y-axis.
- Students should conclude that when a quadrilateral has opposite sides congruent, the quadrilateral is a parallelogram.

Exploration 2
- The second construction can be done by rotating a triangle 180° about the midpoint of one side.
- **? Construct Viable Arguments and Critique the Reasoning of Others:** "How did you decide whether the quadrilateral was a parallelogram?" Listen for valid reasoning similar to that in Exploration 1.
- Students should conclude that when the opposite angles of a quadrilateral are congruent, the quadrilateral is a parallelogram.

Communicate Your Answer
- Listen for valid reasoning from students. They may already sense that this lesson is very much connected to the properties studied in the last lesson.

Connecting to Next Step
- The conjectures made by students will be stated and proven as theorems in the formal lesson.

7.3 Proving That a Quadrilateral Is a Parallelogram

Essential Question How can you prove that a quadrilateral is a parallelogram?

EXPLORATION 1 Proving That a Quadrilateral Is a Parallelogram

Work with a partner. Use dynamic geometry software.

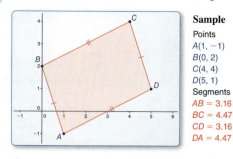

Sample
Points
$A(1, -1)$
$B(0, 2)$
$C(4, 4)$
$D(5, 1)$
Segments
$AB = 3.16$
$BC = 4.47$
$CD = 3.16$
$DA = 4.47$

REASONING ABSTRACTLY
To be proficient in math, you need to know and flexibly use different properties of objects.

a. Construct any quadrilateral $ABCD$ whose opposite sides are congruent.
b. Is the quadrilateral a parallelogram? Justify your answer.
c. Repeat parts (a) and (b) for several other quadrilaterals. Then write a conjecture based on your results.
d. Write the converse of your conjecture. Is the converse true? Explain.

EXPLORATION 2 Proving That a Quadrilateral Is a Parallelogram

Work with a partner. Use dynamic geometry software.

a. Construct any quadrilateral $ABCD$ whose opposite angles are congruent.
b. Is the quadrilateral a parallelogram? Justify your answer.

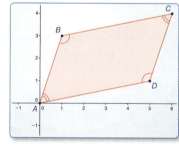

Sample
Points
$A(0, 0)$
$B(1, 3)$
$C(6, 4)$
$D(5, 1)$
Angles
$\angle A = 60.26°$
$\angle B = 119.74°$
$\angle C = 60.26°$
$\angle D = 119.74°$

c. Repeat parts (a) and (b) for several other quadrilaterals. Then write a conjecture based on your results.
d. Write the converse of your conjecture. Is the converse true? Explain.

Communicate Your Answer

3. How can you prove that a quadrilateral is a parallelogram?
4. Is the quadrilateral at the left a parallelogram? Explain your reasoning.

Section 7.3 Proving That a Quadrilateral Is a Parallelogram 419

Dynamic Teaching Tools
Dynamic Assessment & Progress Monitoring Tool
Lesson Planning Tool
Interactive Whiteboard Lesson Library
Dynamic Classroom with Dynamic Investigations

ANSWERS

1. a. Check students' work.
   b. yes; Because they have the same slope, opposite sides are parallel.
   c. Check students' work. If the opposite sides of a quadrilateral are congruent, then the quadrilateral is a parallelogram.
   d. If a quadrilateral is a parallelogram, then its opposite sides are congruent; yes; This is the Parallelogram Opposite Sides Theorem.

2. a. Check students' work.
   b. yes; Because they have the same slope, opposite sides are parallel.
   c. Check students' work. If the opposite angles of a quadrilateral are congruent, then the quadrilateral is a parallelogram.
   d. If a quadrilateral is a parallelogram, then its opposite angles are congruent; yes; This is the Parallelogram Opposite Angles Theorem.

3. Show that the opposite sides or opposite angles are congruent.

4. yes; The opposite angles are congruent.

English Language Learners

Build on Past Knowledge

Before beginning Section 7.3, write the four theorems (Parallelogram Opposite Sides Theorem, Parallelogram Opposite Angles Theorem, Parallelogram Consecutive Angles Theorem, and Parallelogram Diagonals Theorem) from the previous lesson on the board. Ask students to state the converse of each of these theorems and write these on the board. Then ask students whether they think each converse is true. At the end of the lesson, go back and review the truth of these converse statements.

7.3 Lesson

Core Vocabulary

Previous
diagonal
parallelogram

What You Will Learn

▶ Identify and verify parallelograms.
▶ Show that a quadrilateral is a parallelogram in the coordinate plane.

Identifying and Verifying Parallelograms

Given a parallelogram, you can use the Parallelogram Opposite Sides Theorem and the Parallelogram Opposite Angles Theorem to prove statements about the sides and angles of the parallelogram. The converses of the theorems are stated below. You can use these and other theorems in this lesson to prove that a quadrilateral with certain properties is a parallelogram.

🔄 Theorems

Parallelogram Opposite Sides Converse

If both pairs of opposite sides of a quadrilateral are congruent, then the quadrilateral is a parallelogram.

If $\overline{AB} \cong \overline{CD}$ and $\overline{BC} \cong \overline{DA}$, then $ABCD$ is a parallelogram.

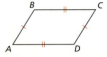

Parallelogram Opposite Angles Converse

If both pairs of opposite angles of a quadrilateral are congruent, then the quadrilateral is a parallelogram.

If $\angle A \cong \angle C$ and $\angle B \cong \angle D$, then $ABCD$ is a parallelogram.

Proof Ex. 39, p. 427

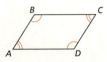

PROOF Parallelogram Opposite Sides Converse

Given $\overline{AB} \cong \overline{CD}, \overline{BC} \cong \overline{DA}$

Prove $ABCD$ is a parallelogram.

Plan for Proof
a. Draw diagonal $\overline{AC}$ to form $\triangle ABC$ and $\triangle CDA$.
b. Use the SSS Congruence Theorem to show that $\triangle ABC \cong \triangle CDA$.
c. Use the Alternate Interior Angles Converse to show that opposite sides are parallel.

Plan in Action

	STATEMENTS	REASONS
a.	1. $\overline{AB} \cong \overline{CD}, \overline{BC} \cong \overline{DA}$	1. Given
	2. Draw $\overline{AC}$.	2. Through any two points, there exists exactly one line.
	3. $\overline{AC} \cong \overline{CA}$	3. Reflexive Property of Congruence
b.	4. $\triangle ABC \cong \triangle CDA$	4. SSS Congruence Theorem
c.	5. $\angle BAC \cong \angle DCA$, $\angle BCA \cong \angle DAC$	5. Corresponding parts of congruent triangles are congruent.
	6. $\overline{AB} \parallel \overline{CD}, \overline{BC} \parallel \overline{DA}$	6. Alternate Interior Angles Converse
	7. $ABCD$ is a parallelogram.	7. Definition of parallelogram

---

Laurie's Notes Teacher Actions

? Sketch quadrilateral $ABCD$ that appears to be a parallelogram. "What information would you need in order to know that $ABCD$ is a parallelogram?" *The opposite sides are parallel.* If students start to list additional properties, remind them of the definition.

- Explain to students that in this lesson they will look back at the theorems of the last lesson to see whether the converse of each is true.
- **Whiteboarding:** Discuss the Parallelogram Opposite Sides Converse. Have partners write a proof.
- **Construct Viable Arguments and Critique the Reasoning of Others:** Compare and critique the proofs offered by several volunteers.

> **EXAMPLE 1** Identifying a Parallelogram

An amusement park ride has a moving platform attached to four swinging arms. The platform swings back and forth, higher and higher, until it goes over the top and around in a circular motion. In the diagram below, $\overline{AD}$ and $\overline{BC}$ represent two of the swinging arms, and $\overline{DC}$ is parallel to the ground (line ℓ). Explain why the moving platform $\overline{AB}$ is always parallel to the ground.

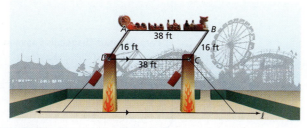

SOLUTION

The shape of quadrilateral $ABCD$ changes as the moving platform swings around, but its side lengths do not change. Both pairs of opposite sides are congruent, so $ABCD$ is a parallelogram by the Parallelogram Opposite Sides Converse.

By the definition of a parallelogram, $\overline{AB} \parallel \overline{DC}$. Because $\overline{DC}$ is parallel to line ℓ, $\overline{AB}$ is also parallel to line ℓ by the Transitive Property of Parallel Lines. So, the moving platform is parallel to the ground.

Monitoring Progress Help in English and Spanish at *BigIdeasMath.com*

1. In quadrilateral $WXYZ$, $m\angle W = 42°$, $m\angle X = 138°$, and $m\angle Y = 42°$. Find $m\angle Z$. Is $WXYZ$ a parallelogram? Explain your reasoning.

> **EXAMPLE 2** Finding Side Lengths of a Parallelogram

For what values of x and y is quadrilateral $PQRS$ a parallelogram?

SOLUTION

By the Parallelogram Opposite Sides Converse, if both pairs of opposite sides of a quadrilateral are congruent, then the quadrilateral is a parallelogram. Find x so that $\overline{PQ} \cong \overline{SR}$.

$PQ = SR$	Set the segment lengths equal.
$x + 9 = 2x - 1$	Substitute $x + 9$ for PQ and $2x - 1$ for SR.
$10 = x$	Solve for x.

When $x = 10$, $PQ = 10 + 9 = 19$ and $SR = 2(10) - 1 = 19$. Find y so that $\overline{PS} \cong \overline{QR}$.

$PS = QR$	Set the segment lengths equal.
$y = x + 7$	Substitute y for PS and $x + 7$ for QR.
$y = 10 + 7$	Substitute 10 for x.
$y = 17$	Add.

When $x = 10$ and $y = 17$, $PS = 17$ and $QR = 10 + 7 = 17$.

▶ Quadrilateral $PQRS$ is a parallelogram when $x = 10$ and $y = 17$.

Extra Example 1

In quadrilateral $ABCD$, $AB = BC$ and $CD = AD$. Is $ABCD$ a parallelogram? Explain your reasoning. *You cannot tell. Two pairs of opposite sides must be congruent, not two pairs of adjacent sides.*

Extra Example 2

For what values of x and y is quadrilateral $STUV$ a parallelogram?

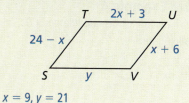

$x = 9, y = 21$

MONITORING PROGRESS ANSWER

1. 138°; yes; Because opposite angles are congruent, quadrilateral $WXYZ$ is a parallelogram.

Laurie's Notes Teacher Actions

- Discuss the amusement park ride to be sure students know how it operates. Listen for student justification of why $\overline{AB}$ is parallel to the ground.
- **Extension:** There may be students who want to construct a model of this ride using dynamic geometry software.

COMMON ERROR In Example 2, students may try to work with all four expressions simultaneously. They need to work with the pair involving only x first, and then move to the other pair of expressions.

Extra Example 3
Use the photograph in Example 3. Explain how you know that ∠S ≅ ∠U. By the Opposite Sides Parallel and Congruent Theorem, quadrilateral STUV is a parallelogram. By the Parallelogram Opposite Angles Theorem, the opposite angles of a parallelogram are congruent. So, ∠S ≅ ∠U.

Extra Example 4
For what value of x is quadrilateral CDEF a parallelogram?

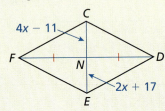

x = 14

Theorems

Opposite Sides Parallel and Congruent Theorem
If one pair of opposite sides of a quadrilateral are congruent and parallel, then the quadrilateral is a parallelogram.

If $\overline{BC} \parallel \overline{AD}$ and $\overline{BC} \cong \overline{AD}$, then ABCD is a parallelogram.

Proof Ex. 40, p. 427

Parallelogram Diagonals Converse
If the diagonals of a quadrilateral bisect each other, then the quadrilateral is a parallelogram.

If $\overline{BD}$ and $\overline{AC}$ bisect each other, then ABCD is a parallelogram.

Proof Ex. 41, p. 427

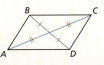

EXAMPLE 3 Identifying a Parallelogram

The doorway shown is part of a building in England. Over time, the building has leaned sideways. Explain how you know that SV = TU.

SOLUTION
In the photograph, $\overline{ST} \parallel \overline{UV}$ and $\overline{ST} \cong \overline{UV}$. By the Opposite Sides Parallel and Congruent Theorem, quadrilateral STUV is a parallelogram. By the Parallelogram Opposite Sides Theorem, you know that opposite sides of a parallelogram are congruent. So, SV = TU.

EXAMPLE 4 Finding Diagonal Lengths of a Parallelogram

For what value of x is quadrilateral CDEF a parallelogram?

SOLUTION
By the Parallelogram Diagonals Converse, if the diagonals of CDEF bisect each other, then it is a parallelogram. You are given that $\overline{CN} \cong \overline{EN}$. Find x so that $\overline{FN} \cong \overline{DN}$.

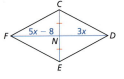

FN = DN	Set the segment lengths equal.
5x − 8 = 3x	Substitute 5x − 8 for FN and 3x for DN.
2x − 8 = 0	Subtract 3x from each side.
2x = 8	Add 8 to each side.
x = 4	Divide each side by 2.

When x = 4, FN = 5(4) − 8 = 12 and DN = 3(4) = 12.

▶ Quadrilateral CDEF is a parallelogram when x = 4.

Laurie's Notes — Teacher Actions

- The Opposite Sides Parallel and Congruent Theorem is not a converse of an earlier theorem. This hypothesis can easily be demonstrated using dynamic geometry software. As long as two segments remain congruent and parallel, the quadrilateral formed when joining adjacent endpoints will be a parallelogram.

COMMON ERROR Students sometimes use faulty reasoning in a problem like Example 3. They say STUV is a parallelogram because both pairs of opposite sides are parallel, and they say SV = TU because opposite sides of a parallelogram are congruent.

Monitoring Progress

2. For what values of *x* and *y* is quadrilateral *ABCD* a parallelogram? Explain your reasoning.

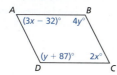

State the theorem you can use to show that the quadrilateral is a parallelogram.

3. 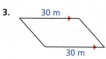   4. 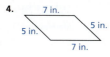   5.

6. For what value of *x* is quadrilateral *MNPQ* a parallelogram? Explain your reasoning.

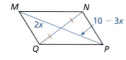

Concept Summary

Ways to Prove a Quadrilateral Is a Parallelogram

1. Show that both pairs of opposite sides are parallel. *(Definition)*

2. Show that both pairs of opposite sides are congruent.
   (Parallelogram Opposite Sides Converse)

3. Show that both pairs of opposite angles are congruent.
   (Parallelogram Opposite Angles Converse)

4. Show that one pair of opposite sides are congruent and parallel.
   (Opposite Sides Parallel and Congruent Theorem)

5. Show that the diagonals bisect each other.
   (Parallelogram Diagonals Converse)

Differentiated Instruction

Organization
Have students copy the Ways to Prove a Quadrilateral Is a Parallelogram table into their notebooks. Ask students to explain which methods might be best suited and least suited to use in a coordinate proof.

MONITORING PROGRESS ANSWERS

2. $x = 32$, $y = 29$; By the Parallelogram Opposite Angles Converse, if both pairs of opposite angles of a quadrilateral are congruent, then the quadrilateral is a parallelogram. So, solve $3x - 32 = 2x$ for *x*, and $4y = y + 87$ for *y*.

3. Opposite Sides Parallel and Congruent Theorem

4. Parallelogram Opposite Sides Converse

5. Parallelogram Opposite Angles Converse

6. $x = 2$; By the Parallelogram Diagonals Converse, if the diagonals of a quadrilateral bisect each other, then the quadrilateral is a parallelogram. So, solve $2x = 10 - 3x$ for *x*.

Laurie's Notes — Teacher Actions

- **Think-Pair-Share:** Have students answer Questions 2–6, and then share and discuss as a class.
- Have students list a summary of different ways to prove that a quadrilateral is a parallelogram.
- ❓ "How might you know that the opposite sides are parallel?" Listen for suggestions such as knowing something about their slopes or knowing the measures of interior angles.

Extra Example 5

Show that quadrilateral ABCD is a parallelogram.

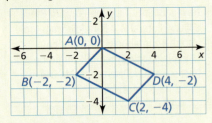

Sample answer: Show that the diagonals $\overline{AC}$ and $\overline{BD}$ bisect each other. Midpoint of $\overline{AC}$ is $\left(\dfrac{0+2}{2}, \dfrac{0+(-4)}{2}\right) = (1, -2)$. Midpoint of $\overline{BD}$ is $\left(\dfrac{-2+4}{2}, \dfrac{-2+(-2)}{2}\right)$ $= (1, -2)$. The diagonals have the same midpoint, so they bisect each other. So, by the Parallelogram Diagonals Converse, quadrilateral ABCD is a parallelogram.

MONITORING PROGRESS ANSWERS

7. Sample answer (using Method 1): Because $JK = LM = \sqrt{37}$, $\overline{JK} \cong \overline{LM}$. Because the slopes of $\overline{JK}$ and $\overline{LM}$ are both -6, they are parallel. So, $\overline{JK}$ and $\overline{LM}$ are congruent and parallel, which means that JKLM is a parallelogram by the Opposite Sides Parallel and Congruent Theorem.

8. Sample answer: Find the slopes of all four sides and show that opposite sides are parallel. Another way is to find the point of intersection of the diagonals and show that the diagonals bisect each other.

Using Coordinate Geometry

EXAMPLE 5 Identifying a Parallelogram in the Coordinate Plane

Show that quadrilateral ABCD is a parallelogram.

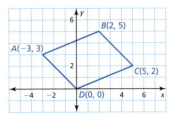

SOLUTION

Method 1 Show that a pair of sides are congruent and parallel. Then apply the Opposite Sides Parallel and Congruent Theorem.

First, use the Distance Formula to show that $\overline{AB}$ and $\overline{CD}$ are congruent.

$AB = \sqrt{[2-(-3)]^2 + (5-3)^2} = \sqrt{29}$

$CD = \sqrt{(5-0)^2 + (2-0)^2} = \sqrt{29}$

Because $AB = CD = \sqrt{29}$, $\overline{AB} \cong \overline{CD}$.

Then, use the slope formula to show that $\overline{AB} \parallel \overline{CD}$.

slope of $\overline{AB} = \dfrac{5-3}{2-(-3)} = \dfrac{2}{5}$

slope of $\overline{CD} = \dfrac{2-0}{5-0} = \dfrac{2}{5}$

Because $\overline{AB}$ and $\overline{CD}$ have the same slope, they are parallel.

 $\overline{AB}$ and $\overline{CD}$ are congruent and parallel. So, ABCD is a parallelogram by the Opposite Sides Parallel and Congruent Theorem.

Method 2 Show that opposite sides are congruent. Then apply the Parallelogram Opposite Sides Converse. In Method 1, you already have shown that because $AB = CD = \sqrt{29}$, $\overline{AB} \cong \overline{CD}$. Now find AD and BC.

$AD = \sqrt{(-3-0)^2 + (3-0)^2} = 3\sqrt{2}$

$BC = \sqrt{(2-5)^2 + (5-2)^2} = 3\sqrt{2}$

Because $AD = BC = 3\sqrt{2}$, $\overline{AD} \cong \overline{BC}$.

 $\overline{AB} \cong \overline{CD}$ and $\overline{AD} \cong \overline{BC}$. So, ABCD is a parallelogram by the Parallelogram Opposite Sides Converse.

Monitoring Progress Help in English and Spanish at *BigIdeasMath.com*

7. Show that quadrilateral JKLM is a parallelogram.

8. Refer to the Concept Summary on page 423. Explain two other methods you can use to show that quadrilateral ABCD in Example 5 is a parallelogram.

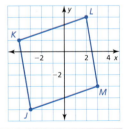

Laurie's Notes — Teacher Actions

- Pose Example 5, and then give partners time to work on the problem.
- **Make Sense of Problems and Persevere in Solving Them:** Do not rush in to solve this for them. Students know how to determine distances, find a slope, and compute the midpoint.
- **Selective Responses:** This is a problem in which students may use different strategies to prove ABCD is a parallelogram. Allow ample work time before asking selected students to share.

Closure

- **Point of Most Significance:** Ask students to identify, aloud or on a paper to be collected, the most significant point (or part) in the lesson that aided their learning.

7.3 Exercises

Dynamic Solutions available at BigIdeasMath.com

Vocabulary and Core Concept Check

1. **WRITING** A quadrilateral has four congruent sides. Is the quadrilateral a parallelogram? Justify your answer.

2. **DIFFERENT WORDS, SAME QUESTION** Which is different? Find "both" answers.

Construct a quadrilateral with opposite sides congruent.	Construct a quadrilateral with one pair of parallel sides.
Construct a quadrilateral with opposite angles congruent.	Construct a quadrilateral with one pair of opposite sides congruent and parallel.

Monitoring Progress and Modeling with Mathematics

In Exercises 3–8, state which theorem you can use to show that the quadrilateral is a parallelogram. *(See Examples 1 and 3.)*

3.
4.
5.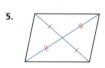
6.
7.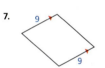
8.

In Exercises 9–12, find the values of x and y that make the quadrilateral a parallelogram. *(See Example 2.)*

9.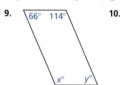
10.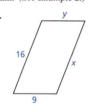
11.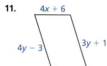
12. (4x + 13)° (4y + 7)° (3x − 8)° (5x − 12)°

In Exercises 13–16, find the value of x that makes the quadrilateral a parallelogram. *(See Example 4.)*

13.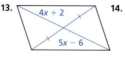
14.
15.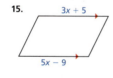
16.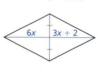

In Exercises 17–20, graph the quadrilateral with the given vertices in a coordinate plane. Then show that the quadrilateral is a parallelogram. *(See Example 5.)*

17. $A(0, 1)$, $B(4, 4)$, $C(12, 4)$, $D(8, 1)$
18. $E(-3, 0)$, $F(-3, 4)$, $G(3, -1)$, $H(3, -5)$
19. $J(-2, 3)$, $K(-5, 7)$, $L(3, 6)$, $M(6, 2)$
20. $N(-5, 0)$, $P(0, 4)$, $Q(3, 0)$, $R(-2, -4)$

Section 7.3 Proving That a Quadrilateral Is a Parallelogram 425

17.

Because $BC = AD = 8$, $\overline{BC} \cong \overline{AD}$. Because both $\overline{BC}$ and $\overline{AD}$ are horizontal lines, their slope is 0, and they are parallel. $\overline{BC}$ and $\overline{AD}$ are opposite sides that are both congruent and parallel. So, $ABCD$ is a parallelogram by the Opposite Sides Parallel and Congruent Theorem.

18–20. See Additional Answers.

Assignment Guide and Homework Check

ASSIGNMENT
Basic: 1, 2, 3–21 odd, 24, 28–34 even, 44, 51–54
Average: 1, 2, 8–38 even, 43, 44, 51–54
Advanced: 1, 2, 8–42 even, 43–47, 49, 51–54

HOMEWORK CHECK
Basic: 3, 7, 11, 13, 17
Average: 8, 12, 18, 32, 43
Advanced: 12, 20, 36, 40, 47

ANSWERS

1. yes; If all four sides are congruent, then both pairs of opposite sides are congruent. So, the quadrilateral is a parallelogram by the Parallelogram Opposite Sides Converse.

2. Construct a quadrilateral with one pair of parallel sides;

3. Parallelogram Opposite Angles Converse
4. Parallelogram Opposite Sides Converse
5. Parallelogram Diagonals Converse
6. Parallelogram Opposite Angles Converse
7. Opposite Sides Parallel and Congruent Theorem
8. Parallelogram Diagonals Converse
9. $x = 114$, $y = 66$
10. $x = 16$, $y = 9$
11. $x = 3$, $y = 4$
12. $x = 25$, $y = 15$
13. $x = 8$
14. $x = 4$
15. $x = 7$
16. $x = \frac{2}{3}$

Section 7.3 425

Dynamic Teaching Tools

Dynamic Assessment & Progress Monitoring Tool
Interactive Whiteboard Lesson Library
Dynamic Classroom with Dynamic Investigations

ANSWERS

21. In order to be a parallelogram, the quadrilateral must have two pairs of opposite sides that are congruent, not consecutive sides; *DEFG* is not a parallelogram.

22. In order to determine that *JKLM* is a parallelogram, using the Opposite Sides Parallel and Congruent Theorem, you would need to know that $\overline{JM} \parallel \overline{KL}$; There is not enough information provided to determine whether *JKLM* is a parallelogram.

23. $x = 5$; The diagonals must bisect each other so you could solve for x using either $2x + 1 = x + 6$ or $4x - 2 = 3x + 3$. Also, the opposite sides must be congruent, so you could solve for x using either $3x + 1 = 4x - 4$ or $3x + 10 = 5x$.

24. yes; By the Consecutive Interior Angles Converse, $\overline{WX} \parallel \overline{ZY}$. Because $\overline{WX}$ and $\overline{ZY}$ are also congruent, *WXYZ* is a parallelogram by the Opposite Sides Parallel and Congruent Theorem.

25. A quadrilateral is a parallelogram if and only if both pairs of opposite sides are congruent.

26. A quadrilateral is a parallelogram if and only if both pairs of opposite angles are congruent.

27. A quadrilateral is a parallelogram if and only if the diagonals bisect each other.

28. *Sample answer:* Draw two horizontal segments that are the same length and connect the endpoints.

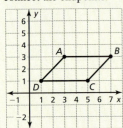

29. Check students' work; Because the diagonals bisect each other, this quadrilateral is a parallelogram by the Parallelogram Diagonals Converse.

ERROR ANALYSIS In Exercises 21 and 22, describe and correct the error in identifying a parallelogram.

21.

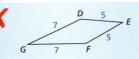

DEFG is a parallelogram by the Parallelogram Opposite Sides Converse.

22.

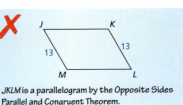

JKLM is a parallelogram by the Opposite Sides Parallel and Congruent Theorem.

23. **MATHEMATICAL CONNECTIONS** What value of x makes the quadrilateral a parallelogram? Explain how you found your answer.

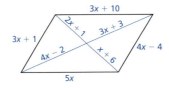

24. **MAKING AN ARGUMENT** Your friend says you can show that quadrilateral *WXYZ* is a parallelogram by using the Consecutive Interior Angles Converse and the Opposite Sides Parallel and Congruent Theorem. Is your friend correct? Explain your reasoning.

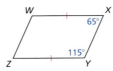

ANALYZING RELATIONSHIPS In Exercises 25–27, write the indicated theorems as a biconditional statement.

25. Parallelogram Opposite Sides Theorem and Parallelogram Opposite Sides Converse

26. Parallelogram Opposite Angles Theorem and Parallelogram Opposite Angles Converse

27. Parallelogram Diagonals Theorem and Parallelogram Diagonals Converse

426   Chapter 7   Quadrilaterals and Other Polygons

28. **CONSTRUCTION** Describe a method that uses the Opposite Sides Parallel and Congruent Theorem to construct a parallelogram. Then construct a parallelogram using your method.

29. **REASONING** Follow the steps below to construct a parallelogram. Explain why this method works. State a theorem to support your answer.

Step 1 Use a ruler to draw two segments that intersect at their midpoints.

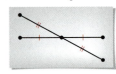

Step 2 Connect the endpoints of the segments to form a parallelogram.

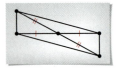

30. **MAKING AN ARGUMENT** Your brother says to show that quadrilateral *QRST* is a parallelogram, you must show that $\overline{QR} \parallel \overline{TS}$ and $\overline{QT} \parallel \overline{RS}$. Your sister says that you must show that $\overline{QR} \cong \overline{TS}$ and $\overline{QT} \cong \overline{RS}$. Who is correct? Explain your reasoning.

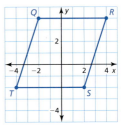

REASONING In Exercises 31 and 32, your classmate incorrectly claims that the marked information can be used to show that the figure is a parallelogram. Draw a quadrilateral with the same marked properties that is clearly *not* a parallelogram.

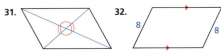

30. both; If you show that $\overline{QR} \parallel \overline{TS}$ and $\overline{QT} \parallel \overline{RS}$, then *QRST* is a parallelogram by definition. If you show that $QR \cong TS$ and $QT \cong RS$, then *QRST* is a parallelogram by the Parallelogram Opposite Sides Converse.

31. *Sample answer:*

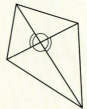

32. *Sample answer:*

426   Chapter 7

33. **MODELING WITH MATHEMATICS** You shoot a pool ball, and it rolls back to where it started, as shown in the diagram. The ball bounces off each wall at the same angle at which it hits the wall.

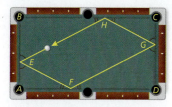

a. The ball hits the first wall at an angle of 63°. So $m\angle AEF = m\angle BEH = 63°$. What is $m\angle AFE$? Explain your reasoning.

b. Explain why $m\angle FGD = 63°$.

c. What is $m\angle GHC$? $m\angle EHB$?

d. Is quadrilateral $EFGH$ a parallelogram? Explain your reasoning.

34. **MODELING WITH MATHEMATICS** In the diagram of the parking lot shown, $m\angle JKL = 60°$, $JK = LM = 21$ feet, and $KL = JM = 9$ feet.

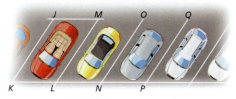

a. Explain how to show that parking space $JKLM$ is a parallelogram.

b. Find $m\angle JML$, $m\angle KJM$, and $m\angle KLM$.

c. $\overline{LM} \parallel \overline{NO}$ and $\overline{NO} \parallel \overline{PQ}$. Which theorem could you use to show that $\overline{JK} \parallel \overline{PQ}$?

REASONING In Exercises 35–37, describe how to prove that $ABCD$ is a parallelogram.

35.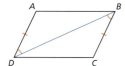

36.

37.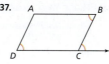

38. **REASONING** Quadrilateral $JKLM$ is a parallelogram. Describe how to prove that $\triangle MGJ \cong \triangle KHL$.

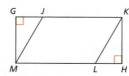

39. **PROVING A THEOREM** Prove the Parallelogram Opposite Angles Converse. (*Hint:* Let $x°$ represent $m\angle A$ and $m\angle C$. Let $y°$ represent $m\angle B$ and $m\angle D$. Write and simplify an equation involving x and y.)

Given $\angle A \cong \angle C, \angle B \cong \angle D$

Prove $ABCD$ is a parallelogram.

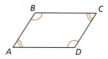

40. **PROVING A THEOREM** Use the diagram of $PQRS$ with the auxiliary line segment drawn to prove the Opposite Sides Parallel and Congruent Theorem.

Given $\overline{QR} \parallel \overline{PS}, \overline{QR} \cong \overline{PS}$

Prove $PQRS$ is a parallelogram.

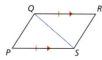

41. **PROVING A THEOREM** Prove the Parallelogram Diagonals Converse.

Given Diagonals $\overline{JL}$ and $\overline{KM}$ bisect each other.

Prove $JKLM$ is a parallelogram.

ANSWERS

33. a. 27°; Because $\angle EAF$ is a right angle, the other two angles of $\triangle EAF$ must be complementary. So, $m\angle AFE = 90° - 63° = 27°$.

b. Because $\angle GDF$ is a right angle, the other two angles of $\triangle GDF$ must be complementary. So, $m\angle FGD = 90° - 27° = 63°$.

c. 27°; 27°

d. yes; $\angle HEF \cong \angle HGF$ because they both are adjacent to two congruent angles that together add up to 180°, and $\angle EHG \cong \angle GFE$ for the same reason. So, $EFGH$ is a parallelogram by the Parallelogram Opposite Angles Converse.

34. a. Because $\overline{JK} \cong \overline{LM}$ and $\overline{KL} \cong \overline{JM}$, $JKLM$ is a parallelogram by the Parallelogram Opposite Sides Converse.

b. 60°, 120°, 120°

c. Transitive Property of Parallel Lines

35. You can use the Alternate Interior Angles Converse to show that $\overline{AD} \parallel \overline{BC}$. Then, $\overline{AD}$ and $\overline{BC}$ are both congruent and parallel. So, $ABCD$ is a parallelogram by the Opposite Sides Parallel and Congruent Theorem.

36. You can use the Alternate Interior Angles Converse to show that $\overline{AB} \parallel \overline{DC}$ and $\overline{AD} \parallel \overline{BC}$. Because both pairs of opposite sides are parallel, $ABCD$ is a parallelogram by definition.

37. First, you can use the Linear Pair Postulate and the Congruent Supplements Theorem to show that $\angle ABC$ and $\angle DCB$ are supplementary. Then, you can use the Consecutive Interior Angles Converse to show that $\overline{AB} \parallel \overline{DC}$ and $\overline{AD} \parallel \overline{BC}$. So, $ABCD$ is a parallelogram by definition.

38. By the Parallelogram Opposite Sides Theorem, $\overline{JM} \cong \overline{LK}$. Also, you can use the Linear Pair Postulate and the Congruent Supplements Theorem to show that $\angle GJM \cong \angle HLK$. Because $\angle JGM$ and $\angle LHK$ are congruent right angles, you can now state that $\triangle MGJ \cong \triangle KHL$ by the AAS Congruence Theorem.

39–41. See Additional Answers.

ANSWERS

42. See Additional Answers.
43. no; The fourth angle will be 113° because of the Corollary to the Polygon Interior Angles Theorem, but these could also be the angle measures of an isosceles trapezoid with base angles that are each 67°.
44. $\overline{AE} \parallel \overline{DF}, \overline{EB} \parallel \overline{FC}, \overline{AD} \parallel \overline{EF} \parallel \overline{BC}$
45–54. See Additional Answers.

Mini-Assessment

1. Show that quadrilateral *ABCD* is a parallelogram.

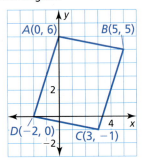

Sample answer: Use slope to show that both pairs of opposite sides are parallel.
Slope of $\overline{AB}$ = slope of $\overline{DC}$ = −0.2
Slope of $\overline{BC}$ = slope of $\overline{AD}$ = 3
Both pairs of opposite sides of *ABCD* are parallel. So, *ABCD* is a parallelogram by definition.

Is the given information sufficient to prove that quadrilateral *JKLM* is a parallelogram? Explain your reasoning.

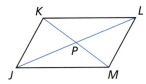

2. *JK* = *LM* and *KL* = *JM* yes; Parallelogram Opposite Sides Converse
3. *JK* = *LM* and $\overline{KL} \parallel \overline{JM}$ no; by Opp. Sides $\parallel$ and $\cong$ Thm, the congruent sides must also be parallel.
4. *JP* = *LP* and *KP* = *MP* yes; Parallelogram Diagonals Converse

428   Chapter 7

---

42. PROOF Write a proof.

Given *DEBF* is a parallelogram.
AE = *CF*

Prove *ABCD* is a parallelogram.

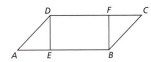

43. REASONING Three interior angle measures of a quadrilateral are 67°, 67°, and 113°. Is this enough information to conclude that the quadrilateral is a parallelogram? Explain your reasoning.

44. HOW DO YOU SEE IT? A music stand can be folded up, as shown. In the diagrams, *AEFD* and *EBCF* are parallelograms. Which labeled segments remain parallel as the stand is folded?

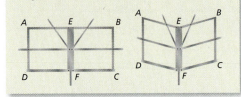

45. CRITICAL THINKING In the diagram, *ABCD* is a parallelogram, *BF* = *DE* = 12, and *CF* = 8. Find *AE*. Explain your reasoning.

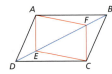

46. THOUGHT PROVOKING Create a regular hexagon using congruent parallelograms.

Maintaining Mathematical Proficiency Reviewing what you learned in previous grades and lessons

Classify the quadrilateral. (*Skills Review Handbook*)

51. 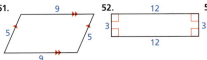 52. 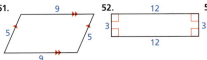 53.   54.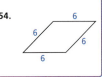

428   Chapter 7   Quadrilaterals and Other Polygons

---

47. WRITING The Parallelogram Consecutive Angles Theorem says that if a quadrilateral is a parallelogram, then its consecutive angles are supplementary. Write the converse of this theorem. Then write a plan for proving the converse. Include a diagram.

48. PROOF Write a proof.

Given *ABCD* is a parallelogram.
∠*A* is a right angle.

Prove ∠*B*, ∠*C*, and ∠*D* are right angles.

49. ABSTRACT REASONING The midpoints of the sides of a quadrilateral have been joined to form what looks like a parallelogram. Show that a quadrilateral formed by connecting the midpoints of the sides of any quadrilateral is always a parallelogram. (*Hint*: Draw a diagram. Include a diagonal of the larger quadrilateral. Show how two sides of the smaller quadrilateral relate to the diagonal.)

50. CRITICAL THINKING Show that if *ABCD* is a parallelogram with its diagonals intersecting at *E*, then you can connect the midpoints *F*, *G*, *H*, and *J* of $\overline{AE}, \overline{BE}, \overline{CE},$ and $\overline{DE}$, respectively, to form another parallelogram, *FGHJ*.

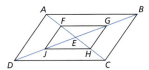

If students need help...	If students got it...
Resources by Chapter • Practice A and Practice B • Puzzle Time	**Resources by Chapter** • Enrichment and Extension • Cumulative Review
Student Journal • Practice	Start the *next* Section
Differentiating the Lesson Skills Review Handbook	

7.1–7.3 What Did You Learn?

Core Vocabulary
diagonal, *p. 404*
equilateral polygon, *p. 405*
equiangular polygon, *p. 405*
regular polygon, *p. 405*
parallelogram, *p. 412*

Core Concepts

Section 7.1
Polygon Interior Angles Theorem, *p. 404*
Corollary to the Polygon Interior Angles Theorem, *p. 405*
Polygon Exterior Angles Theorem, *p. 406*

Section 7.2
Parallelogram Opposite Sides Theorem, *p. 412*
Parallelogram Opposite Angles Theorem, *p. 412*
Parallelogram Consecutive Angles Theorem, *p. 413*
Parallelogram Diagonals Theorem, *p. 413*
Using Parallelograms in the Coordinate Plane, *p. 415*

Section 7.3
Parallelogram Opposite Sides Converse, *p. 420*
Parallelogram Opposite Angles Converse, *p. 420*
Opposite Sides Parallel and Congruent Theorem, *p. 422*
Parallelogram Diagonals Converse, *p. 422*
Ways to Prove a Quadrilateral is a Parallelogram, *p. 423*
Showing That a Quadrilateral Is a Parallelogram in the Coordinate Plane, *p. 424*

Mathematical Practices

1. In Exercise 52 on page 410, what is the relationship between the 540° increase and the answer?
2. Explain why the process you used works every time in Exercise 25 on page 417. Is there another way to do it?
3. In Exercise 23 on page 426, explain how you started the problem. Why did you start that way? Could you have started another way? Explain.

Study Skills: Keeping Your Mind Focused during Class

- When you sit down at your desk, get all other issues out of your mind by reviewing your notes from the last class and focusing on just math.
- Repeat in your mind what you are writing in your notes.
- When the math is particularly difficult, ask your teacher for another example.

Dynamic Teaching Tools
Dynamic Assessment & Progress Monitoring Tool
Interactive Whiteboard Lesson Library
Dynamic Classroom with Dynamic Investigations

ANSWERS

1. $\dfrac{540°}{180°} = 3$

2. By the Parallelogram Diagonals Theorem, the diagonals of a parallelogram bisect each other. So, the diagonals will have the same midpoint, and it will also be the point where the diagonals intersect. Therefore, with any parallelogram, you can find the midpoint of either diagonal, and it will be the coordinates of the intersection of the diagonals; Instead of this method, you could also find the equations of the lines that define each diagonal, set them equal to each other and solve for the values of the coordinates where the lines intersect.

3. *Sample answer:* Solve $5x = 3x + 10$; opposite sides of a parallelogram are congruent; yes; You could start by setting the two parts of either diagonal equal to each other by the Parallelogram Diagonals Theorem.

ANSWERS

1. $x = 80$
2. $x = 135$
3. $x = 97$
4. 144°, 36°
5. 156°, 24°
6. 165°, 15°
7. 174°, 6°
8. 16; By the Parallelogram Opposite Sides Theorem, $AB = CD$.
9. 7; By the Parallelogram Opposite Sides Theorem, $AD = BC$.
10. 7; By the Parallelogram Diagonals Theorem, $AE = EC$.
11. 20.4; By the Parallelogram Diagonals Theorem, $BE = ED$. So, $BD = 2 \cdot 10.2$.
12. 120°; By the Parallelogram Opposite Angles Theorem, $m\angle DAB = m\angle BCD$.
13. 60°; By the Parallelogram Consecutive Angles Theorem, $\angle DAB$ and $\angle ABC$ are supplementary. So, $m\angle ABC = 180° - 120°$.
14. 60°; By the Parallelogram Consecutive Angles Theorem, $\angle DAB$ and $\angle ADC$ are supplementary. So, $m\angle ADC = 180° - 120°$.
15. 43°; By the Alternate Interior Angles Theorem, $m\angle DBC = m\angle ADB$.
16. Opposite Sides Parallel and Congruent Theorem
17. Parallelogram Diagonals Converse
18. Parallelogram Opposite Angles Converse
19.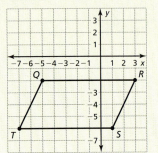

Because $QR = ST = 8, \overline{QR} \cong \overline{ST}$. Because both $\overline{QR}$ and $\overline{ST}$ are horizontal lines, their slope is 0, and they are parallel. $\overline{QR}$ and $\overline{ST}$ are opposite sides that are both congruent and parallel. So, $QRST$ is a parallelogram by the Opposite Sides Parallel and Congruent Theorem.

7.1–7.3 Quiz

Find the value of x. *(Section 7.1)*

1.
2.
3.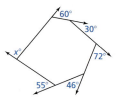

Find the measure of each interior angle and each exterior angle of the indicated regular polygon. *(Section 7.1)*

4. decagon
5. 15-gon
6. 24-gon
7. 60-gon

Find the indicated measure in $\square ABCD$. Explain your reasoning. *(Section 7.2)*

8. CD
9. AD
10. AE
11. BD
12. $m\angle BCD$
13. $m\angle ABC$
14. $m\angle ADC$
15. $m\angle DBC$

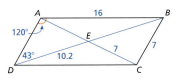

State which theorem you can use to show that the quadrilateral is a parallelogram. *(Section 7.3)*

16.
17.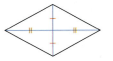
18.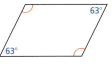

Graph the quadrilateral with the given vertices in a coordinate plane. Then show that the quadrilateral is a parallelogram. *(Section 7.3)*

19. $Q(-5, -2), R(3, -2), S(1, -6), T(-7, -6)$
20. $W(-3, 7), X(3, 3), Y(1, -3), Z(-5, 1)$

21. A stop sign is a regular polygon. *(Section 7.1)*
    a. Classify the stop sign by its number of sides.
    b. Find the measure of each interior angle and each exterior angle of the stop sign.

22. In the diagram of the staircase shown, $JKLM$ is a parallelogram, $\overline{QT} \parallel \overline{RS}$, $QT = RS = 9$ feet, $QR = 3$ feet, and $m\angle QRS = 123°$. *(Section 7.2 and Section 7.3)*
    a. List all congruent sides and angles in $\square JKLM$. Explain your reasoning.
    b. Which theorem could you use to show that $QRST$ is a parallelogram?
    c. Find ST, $m\angle QTS$, $m\angle TQR$, and $m\angle TSR$. Explain your reasoning.

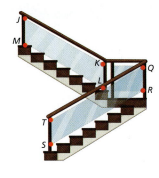

430 Chapter 7 Quadrilaterals and Other Polygons

20.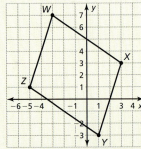

Because the slopes of $\overline{WX}$ and $\overline{YZ}$ are both $-\frac{2}{3}$, they are parallel. Because the slopes of $\overline{XY}$ and $\overline{WZ}$ are both 3, they are parallel. Because both pairs of opposite sides are parallel, $WXYZ$ is a parallelogram by definition.

21. a. octagon
    b. 135°, 45°
22. See Additional Answers.

Laurie's Notes

Overview of Section 7.4

Introduction
- This lesson is about special parallelograms: rectangles, rhombuses, and squares.
- Properties about the diagonals and opposite angles are presented as biconditional statements.

Resources
- Use bendable straws to make models of a rectangle and a rhombus. Use elastic thread (available where sewing notions are sold) tied at the vertices to represent the diagonals.
- By flexing the straws, it is possible to show how the diagonals of a parallelogram become congruent as the parallelogram is straightened to become a rectangle.
- For the model of the rhombus, the diagonals become perpendicular as the parallelogram is resized to become a rhombus.

Teaching Strategy
- **Draw a Diagram:** One way to show the relationships between quadrilaterals is to draw a Venn diagram. A flowchart is another strategy to show the hierarchical relationships of these quadrilaterals.
- Because rectangles, rhombuses, and squares are special parallelograms, they have all of the properties of parallelograms.
- The definition of each of the new quadrilaterals begins with, "A ___ is a parallelogram with …." Without having to repeat the definition or list any properties, these attributes are already known.
- You can discuss the properties of the diagonals of each quadrilateral in the flowchart.

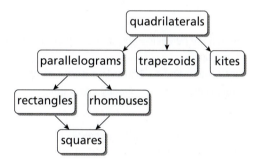

Pacing Suggestion
- The explorations provide an opportunity for students to consider how they can prove they have a rectangle or a rhombus. Transition to the formal lesson as soon as students have discussed each exploration.

What Your Students Will Learn

- Use properties of special parallelograms to identify rhombuses, rectangles, and squares.
- Use properties of diagonals of special parallelograms to identify rhombuses and rectangles.
- Use coordinate geometry to identify special types of parallelograms.

Laurie's Notes

Exploration

Motivate
- Draw and cut out a large copy of a rhombus. It should not be a square.
- **Attend to Precision:** Ask students to describe the symmetry of the rhombus, both rotational and line.
- The rhombus has 180° rotational symmetry and line symmetries through the opposite vertices.
- Fold and crease the rhombus on both lines of symmetry. If possible, display it under a document camera.
- **Turn and Talk:** "What does the symmetry tell you about the diagonals of a rhombus? About the opposite angles of a rhombus?"

Exploration Note
- **Look For and Express Regularity in Repeated Reasoning:** Mathematically proficient students do not base a conjecture on one construction. The construction must be repeated several times, each time with the hypothesis being satisfied. Each time, look for patterns.

Exploration 1
- Students are familiar with the definitions of *rhombus*, *rectangle*, and *square* from middle school.
- **?** It is important for students to recognize what the hypothesis is. "What is true about the quadrilateral you are investigating in this exploration?" The diagonals of the quadrilateral are congruent.
- **? Construct Viable Arguments and Critique the Reasoning of Others:** "How do you know the quadrilateral is a parallelogram?" The diagonals are bisected, so the quadrilateral must be a parallelogram.
- Students should observe that the parallelograms in this exploration are always rectangles, and in some cases they are squares.
- **? Construct Viable Arguments and Critique the Reasoning of Others:** "How did you determine that the parallelogram was a rectangle or a square?" Answers will vary.

Exploration 2
- **?** It is important for students to recognize what the hypothesis is. "What is true about the quadrilateral you are investigating in this exploration?" The diagonals of the quadrilateral are perpendicular bisectors of each other.
- **? Construct Viable Arguments and Critique the Reasoning of Others:** "How do you know the quadrilateral is a parallelogram?" The diagonals are bisected, so the quadrilateral must be a parallelogram.
- Students should observe that the parallelograms in this investigation are always rhombuses, and in some cases they are squares.
- **? Construct Viable Arguments and Critique the Reasoning of Others:** "How did you determine that the parallelogram was a rhombus or a square?" Answers will vary.

Communicate Your Answer
- Listen carefully as students distinguish between the various parallelograms.
  - If the diagonals of a parallelogram are congruent, then it is a rectangle.
  - If the diagonals of a parallelogram are perpendicular, then it is a rhombus.
  - If the diagonals of a parallelogram are congruent and perpendicular, then it is a square.

Connecting to Next Step
- The conjectures made by students will be stated and proven as theorems in the formal lesson.

7.4 Properties of Special Parallelograms

Essential Question What are the properties of the diagonals of rectangles, rhombuses, and squares?

Recall the three types of parallelograms shown below.

Rhombus   Rectangle   Square

EXPLORATION 1   Identifying Special Quadrilaterals

Work with a partner. Use dynamic geometry software.

a. Draw a circle with center *A*.

b. Draw two diameters of the circle. Label the endpoints *B*, *C*, *D*, and *E*.

c. Draw quadrilateral *BDCE*.

d. Is *BDCE* a parallelogram? rectangle? rhombus? square? Explain your reasoning.

e. Repeat parts (a)–(d) for several other circles. Write a conjecture based on your results.

Sample

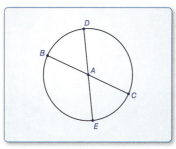

EXPLORATION 2   Identifying Special Quadrilaterals

Work with a partner. Use dynamic geometry software.

a. Construct two segments that are perpendicular bisectors of each other. Label the endpoints *A*, *B*, *D*, and *E*. Label the intersection *C*.

b. Draw quadrilateral *AEBD*.

c. Is *AEBD* a parallelogram? rectangle? rhombus? square? Explain your reasoning.

d. Repeat parts (a)–(c) for several other segments. Write a conjecture based on your results.

Sample

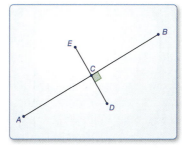

CONSTRUCTING VIABLE ARGUMENTS
To be proficient in math, you need to make conjectures and build a logical progression of statements to explore the truth of your conjectures.

Communicate Your Answer

3. What are the properties of the diagonals of rectangles, rhombuses, and squares?

4. Is *RSTU* a parallelogram? rectangle? rhombus? square? Explain your reasoning.

5. What type of quadrilateral has congruent diagonals that bisect each other?

ANSWERS

1. a. Check students' work.
   b. Check students' work.
   c.

   d. yes; yes; no; no; Because all points on a circle are the same distance from the center, $\overline{AB} \cong \overline{AE} \cong \overline{AC} \cong \overline{AD}$. So, the diagonals of quadrilateral *BDCE* bisect each other, which means it is a parallelogram by the Parallelogram Diagonals Converse. Because all 4 angles of *BDCE* are right angles, it is a rectangle. *BDCE* is neither a rhombus nor a square because $\overline{BD}$ and $\overline{EC}$ are not necessarily the same length as $\overline{BE}$ and $\overline{DC}$.

   e. Check students' work; The quadrilateral formed by the endpoints of two diameters is a rectangle (and a parallelogram). In other words, a quadrilateral is a rectangle if and only if its diagonals are congruent and bisect each other.

2. a. Check students' work.
   b.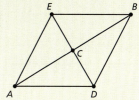

   c. yes; no; yes; no; Because the diagonals bisect each other, *AEBD* is a parallelogram by the Parallelogram Diagonals Converse. Because $EB = BD = AD = AE$, *AEBD* is a rhombus. *AEBD* is neither a rectangle nor a square because its angles are not necessarily right angles.

2. d. Check students' work; A quadrilateral is a rhombus if and only if the diagonals are perpendicular bisectors of each other.

3. Because rectangles, rhombuses, and squares are all parallelograms, their diagonals bisect each other by the Parallelogram Diagonals Theorem. The diagonals of a rectangle are congruent. The diagonals of a rhombus are perpendicular. The diagonals of a square are congruent and perpendicular.

4. yes; no; yes; no; *RSTU* is a parallelogram because the diagonals bisect each other. *RSTU* is not a rectangle because the diagonals are not congruent. *RSTU* is a rhombus because the diagonals are perpendicular. *RSTU* is not a square because the diagonals are not congruent.

5. rectangle

Differentiated Instruction

Inclusion

Point out to advanced learners that some textbooks define a rectangle as a parallelogram with one right angle. Have students prove that this condition of one right angle in a parallelogram is sufficient to prove that the parallelogram is a rectangle.

7.4 Lesson

What You Will Learn

▶ Use properties of special parallelograms.
▶ Use properties of diagonals of special parallelograms.
▶ Use coordinate geometry to identify special types of parallelograms.

Core Vocabulary

rhombus, *p. 432*
rectangle, *p. 432*
square, *p. 432*

Previous
quadrilateral
parallelogram
diagonal

Using Properties of Special Parallelograms

In this lesson, you will learn about three special types of parallelograms: *rhombuses*, *rectangles*, and *squares*.

Core Concept

Rhombuses, Rectangles, and Squares

A **rhombus** is a parallelogram with four congruent sides.

A **rectangle** is a parallelogram with four right angles.

A **square** is a parallelogram with four congruent sides and four right angles.

You can use the corollaries below to prove that a quadrilateral is a rhombus, rectangle, or square, without first proving that the quadrilateral is a parallelogram.

Corollaries

Rhombus Corollary

A quadrilateral is a rhombus if and only if it has four congruent sides.

$ABCD$ is a rhombus if and only if $\overline{AB} \cong \overline{BC} \cong \overline{CD} \cong \overline{AD}$.

Proof Ex. 81, p. 440

Rectangle Corollary

A quadrilateral is a rectangle if and only if it has four right angles.

$ABCD$ is a rectangle if and only if $\angle A$, $\angle B$, $\angle C$, and $\angle D$ are right angles.

Proof Ex. 82, p. 440

Square Corollary

A quadrilateral is a square if and only if it is a rhombus and a rectangle.

$ABCD$ is a square if and only if $\overline{AB} \cong \overline{BC} \cong \overline{CD} \cong \overline{AD}$ and $\angle A$, $\angle B$, $\angle C$, and $\angle D$ are right angles.

Proof Ex. 83, p. 440

432 Chapter 7 Quadrilaterals and Other Polygons

Laurie's Notes — Teacher Actions

- **Common Misconception:** Students often think that they should include everything they know about a polygon in its definition. While definitions may differ in language, it is generally agreed that good definitions are precise and without extraneous information.
- **Reason Abstractly and Quantitatively:** Be sure students understand how corollaries will be used.
- **Turn and Talk:** Have partners select a corollary and discuss how to prove it.
- **Selective Responses:** Students may use different strategies. Allow ample work time before asking selected students to share their proofs.

The Venn diagram below illustrates some important relationships among parallelograms, rhombuses, rectangles, and squares. For example, you can see that a square is a rhombus because it is a parallelogram with four congruent sides. Because it has four right angles, a square is also a rectangle.

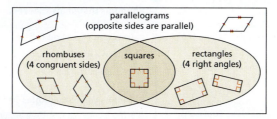

EXAMPLE 1 Using Properties of Special Quadrilaterals

For any rhombus *QRST*, decide whether the statement is *always* or *sometimes* true. Draw a diagram and explain your reasoning.

a. ∠Q ≅ ∠S   **b.** ∠Q ≅ ∠R

SOLUTION

a. By definition, a rhombus is a parallelogram with four congruent sides. By the Parallelogram Opposite Angles Theorem, opposite angles of a parallelogram are congruent. So, ∠Q ≅ ∠S. The statement is *always* true.

b. If rhombus *QRST* is a square, then all four angles are congruent right angles. So, ∠Q ≅ ∠R when *QRST* is a square. Because not all rhombuses are also squares, the statement is *sometimes* true.

EXAMPLE 2 Classifying Special Quadrilaterals

Classify the special quadrilateral. Explain your reasoning.

SOLUTION

The quadrilateral has four congruent sides. By the Rhombus Corollary, the quadrilateral is a rhombus. Because one of the angles is not a right angle, the rhombus cannot be a square.

Monitoring Progress Help in English and Spanish at *BigIdeasMath.com*

1. For any square *JKLM*, is it *always* or *sometimes* true that $\overline{JK} \perp \overline{KL}$? Explain your reasoning.
2. For any rectangle *EFGH*, is it *always* or *sometimes* true that $\overline{FG} \cong \overline{GH}$? Explain your reasoning.
3. A quadrilateral has four congruent sides and four congruent angles. Sketch the quadrilateral and classify it.

Section 7.4 Properties of Special Parallelograms 433

English Language Learners

Vocabulary
Some students may think that a square is *not* a rectangle. Have them prove that if quadrilateral *ABCD* is a square, then it is a rectangle.

Extra Example 1
For any rectangle *ABCD*, decide whether the statement is *always* or *sometimes* true. Explain your reasoning.

a. AB ≅ BC

If rectangle *ABCD* is a square, then all four sides are congruent. So, AB = BC when *ABCD* is a square. Because not all rectangles are also squares, the statement is *sometimes* true.

b. AB ≅ CD

By definition, a rectangle is a parallelogram. By the Parallelogram Opposite Sides Theorem, opposite sides of a parallelogram are congruent. So, AB = CD is *always* true.

Extra Example 2
Classify the special quadrilateral. Explain your reasoning.

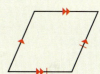

The figure is a rhombus. Two pairs of opposite sides are parallel. So, by definition, the figure is a parallelogram. In a parallelogram, opposite sides are congruent. The figure has two adjacent sides that are congruent. So, by the Transitive Property of Congruence, all four sides are congruent.

MONITORING PROGRESS ANSWERS

1. always; By definition, a square has four right angles.
2. sometimes; Some rectangles are squares.
3.

   square

Laurie's Notes Teacher Actions

- Relationships between various quadrilaterals can be shown in a Venn diagram or a flowchart. See the *Teaching Strategy* on page T-430.
- Example 1 can be difficult for students, as they often answer with their eyes versus reasoning with the given information.
- **? Assessing Questions:** Ask students whether the following statements are *always* true, *sometimes* true, or *never* true. In each case, students should provide their reasoning. "A rectangle is a rhombus." sometimes true "A square is a rectangle." always true "A trapezoid is a parallelogram." never true
- **Think-Pair-Share:** Have students answer Questions 1–3, and then share and discuss as a class.

Section 7.4 433

Extra Example 3

Use the diagram in Example 3. Find $m\angle ABC$ and $m\angle ACB$ in rhombus ABCD. $m\angle ABC = 58°$, $m\angle ACB = 61°$

Using Properties of Diagonals

Theorems

Rhombus Diagonals Theorem

A parallelogram is a rhombus if and only if its diagonals are perpendicular.

□ABCD is a rhombus if and only if $\overline{AC} \perp \overline{BD}$.

Proof p. 434; Ex. 72, p. 439

READING
Recall that biconditionals, such as the Rhombus Diagonals Theorem, can be rewritten as two parts. To prove a biconditional, you must prove both parts.

Rhombus Opposite Angles Theorem

A parallelogram is a rhombus if and only if each diagonal bisects a pair of opposite angles.

□ABCD is a rhombus if and only if $\overline{AC}$ bisects ∠BCD and ∠BAD, and $\overline{BD}$ bisects ∠ABC and ∠ADC.

Proof Exs. 73 and 74, p. 439

PROOF Part of Rhombus Diagonals Theorem

Given ABCD is a rhombus.

Prove $\overline{AC} \perp \overline{BD}$

ABCD is a rhombus. By the definition of a rhombus, $\overline{AB} \cong \overline{BC}$. Because a rhombus is a parallelogram and the diagonals of a parallelogram bisect each other, $\overline{BD}$ bisects $\overline{AC}$ at E. So, $\overline{AE} \cong \overline{EC}$. $\overline{BE} \cong \overline{BE}$ by the Reflexive Property of Congruence. So, △AEB ≅ △CEB by the SSS Congruence Theorem. ∠AEB ≅ ∠CEB because corresponding parts of congruent triangles are congruent. Then by the Linear Pair Postulate, ∠AEB and ∠CEB are supplementary. Two congruent angles that form a linear pair are right angles, so $m\angle AEB = m\angle CEB = 90°$ by the definition of a right angle. So, $\overline{AC} \perp \overline{BD}$ by the definition of perpendicular lines.

EXAMPLE 3 Finding Angle Measures in a Rhombus

Find the measures of the numbered angles in rhombus ABCD.

SOLUTION

Use the Rhombus Diagonals Theorem and the Rhombus Opposite Angles Theorem to find the angle measures.

$m\angle 1 = 90°$	The diagonals of a rhombus are perpendicular.
$m\angle 2 = 61°$	Alternate Interior Angles Theorem
$m\angle 3 = 61°$	Each diagonal of a rhombus bisects a pair of opposite angles, and $m\angle 2 = 61°$.
$m\angle 1 + m\angle 3 + m\angle 4 = 180°$	Triangle Sum Theorem
$90° + 61° + m\angle 4 = 180°$	Substitute 90° for $m\angle 1$ and 61° for $m\angle 3$.
$m\angle 4 = 29°$	Solve for $m\angle 4$.

▶ So, $m\angle 1 = 90°$, $m\angle 2 = 61°$, $m\angle 3 = 61°$, and $m\angle 4 = 29°$.

Laurie's Notes Teacher Actions

? "What do you know about the diagonals of a rhombus from the exploration?" When diagonals are perpendicular bisectors of each other, the quadrilateral is a rhombus.
- Review the language of a biconditional statement. To prove a biconditional, both parts must be proven.
- **Paired Verbal Fluency:** Have students pair up and follow the protocol described on page T-224. Have partner A describe how they would solve the first part of the Rhombus Diagonals Theorem. After one minute partner B offers feedback. Now reverse roles and have partner B describe how they would solve the second part of the theorem. After one minute partner A offers feedback.

Monitoring Progress Help in English and Spanish at *BigIdeasMath.com*

4. In Example 3, what is $m\angle ADC$ and $m\angle BCD$?

5. Find the measures of the numbered angles in rhombus *DEFG*.

Theorem

Rectangle Diagonals Theorem

A parallelogram is a rectangle if and only if its diagonals are congruent.

▱*ABCD* is a rectangle if and only if $\overline{AC} \cong \overline{BD}$.

Proof Exs. 87 and 88, p. 440

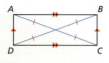

EXAMPLE 4 Identifying a Rectangle

You are building a frame for a window. The window will be installed in the opening shown in the diagram.

a. The opening must be a rectangle. Given the measurements in the diagram, can you assume that it is? Explain.

b. You measure the diagonals of the opening. The diagonals are 54.8 inches and 55.3 inches. What can you conclude about the shape of the opening?

SOLUTION

a. No, you cannot. The boards on opposite sides are the same length, so they form a parallelogram. But you do not know whether the angles are right angles.

b. By the Rectangle Diagonals Theorem, the diagonals of a rectangle are congruent. The diagonals of the quadrilateral formed by the boards are not congruent, so the boards do not form a rectangle.

EXAMPLE 5 Finding Diagonal Lengths in a Rectangle

In rectangle *QRST*, $QS = 5x - 31$ and $RT = 2x + 11$. Find the lengths of the diagonals of *QRST*.

SOLUTION

By the Rectangle Diagonals Theorem, the diagonals of a rectangle are congruent. Find x so that $\overline{QS} \cong \overline{RT}$.

$QS = RT$	Set the diagonal lengths equal.
$5x - 31 = 2x + 11$	Substitute $5x - 31$ for *QS* and $2x + 11$ for *RT*.
$3x - 31 = 11$	Subtract $2x$ from each side.
$3x = 42$	Add 31 to each side.
$x = 14$	Divide each side by 3.

When $x = 14$, $QS = 5(14) - 31 = 39$ and $RT = 2(14) + 11 = 39$.

▶ Each diagonal has a length of 39 units.

Section 7.4 Properties of Special Parallelograms 435

Extra Example 4

Use the diagram in Example 4. Suppose you measure one angle of the window opening and its measure is 90°. Can you conclude that the shape of the opening is a rectangle? Explain.

Yes. Both pairs of opposite sides are congruent, so the quadrilateral is a parallelogram by the Parallelogram Opposite Sides Converse. Opposite angles of a parallelogram are congruent by the Parallelogram Opposite Angles Theorem, so the parallelogram has two right angles. Consecutive angles of a parallelogram are supplementary by the Parallelogram Consecutive Angles Theorem, so the other two angles of the parallelogram are also right angles. A parallelogram with four right angles is a rectangle by definition.

Extra Example 5

In rectangle *ABCD*, $AC = 7x - 15$ and $BD = 2x + 25$. Find the lengths of the diagonals of *ABCD*.

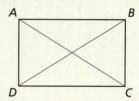

41 units

MONITORING PROGRESS ANSWERS

4. 58°; 122°
5. $m\angle 1 = 31°$; $m\angle 2 = 31°$; $m\angle 3 = 31°$; $m\angle 4 = 31°$

Laurie's Notes Teacher Actions

? "Are the diagonals of a rectangle (that is not a square) perpendicular?" no "Do they bisect one another?" yes

• Write the Rectangle Diagonals Theorem, and have partners discuss an outline of the proof.

• **Common Misconception:** In Example 4, the opposite sides are congruent and the shape appears to be a rectangle. Discuss that this is not a rigid shape and can sway to one side.

? "If the diagonals of another window frame were congruent and you didn't measure the sides of the frame would you be able to say the frame is a rectangle? Explain your reasoning." no; The Rectangle Diagonals Theorem requires that the quadrilateral be a parallelogram and you do not know if opposite sides are congruent.

Section 7.4 435

Extra Example 6
Decide whether ▱ABCD with vertices A(−2, 3), B(2, 2), C(1, −2), and D(−3, −1) is a *rectangle*, a *rhombus*, or a *square*. Give all names that apply. **a rectangle, a rhombus, and a square**

MONITORING PROGRESS ANSWERS
6. no; The quadrilateral might not be a parallelogram.
7. 77
8. rectangle, rhombus, square

Monitoring Progress Help in English and Spanish at BigIdeasMath.com

6. Suppose you measure only the diagonals of the window opening in Example 4 and they have the same measure. Can you conclude that the opening is a rectangle? Explain.

7. **WHAT IF?** In Example 5, $QS = 4x - 15$ and $RT = 3x + 8$. Find the lengths of the diagonals of QRST.

Using Coordinate Geometry

EXAMPLE 6 Identifying a Parallelogram in the Coordinate Plane

Decide whether ▱ABCD with vertices A(−2, 6), B(6, 8), C(4, 0), and D(−4, −2) is a *rectangle*, a *rhombus*, or a *square*. Give all names that apply.

SOLUTION

1. **Understand the Problem** You know the vertices of ▱ABCD. You need to identify the type of parallelogram.

2. **Make a Plan** Begin by graphing the vertices. From the graph, it appears that all four sides are congruent and there are no right angles.

   Check the lengths and slopes of the diagonals of ▱ABCD. If the diagonals are congruent, then ▱ABCD is a rectangle. If the diagonals are perpendicular, then ▱ABCD is a rhombus. If they are both congruent and perpendicular, then ▱ABCD is a rectangle, a rhombus, and a square.

3. **Solve the Problem** Use the Distance Formula to find AC and BD.

   $AC = \sqrt{(-2-4)^2 + (6-0)^2} = \sqrt{72} = 6\sqrt{2}$

   $BD = \sqrt{[6-(-4)]^2 + [8-(-2)]^2} = \sqrt{200} = 10\sqrt{2}$

   Because $6\sqrt{2} \neq 10\sqrt{2}$, the diagonals are not congruent. So, ▱ABCD is not a rectangle. Because it is not a rectangle, it also cannot be a square.

   Use the slope formula to find the slopes of the diagonals $\overline{AC}$ and $\overline{BD}$.

   slope of $\overline{AC} = \dfrac{6-0}{-2-4} = \dfrac{6}{-6} = -1$ slope of $\overline{BD} = \dfrac{8-(-2)}{6-(-4)} = \dfrac{10}{10} = 1$

   Because the product of the slopes of the diagonals is −1, the diagonals are perpendicular.

   ▶ So, ▱ABCD is a rhombus.

4. **Look Back** Check the side lengths of ▱ABCD. Each side has a length of $2\sqrt{17}$ units, so ▱ABCD is a rhombus. Check the slopes of two consecutive sides.

   slope of $\overline{AB} = \dfrac{8-6}{6-(-2)} = \dfrac{2}{8} = \dfrac{1}{4}$ slope of $\overline{BC} = \dfrac{8-0}{6-4} = \dfrac{8}{2} = 4$

   Because the product of these slopes is not −1, $\overline{AB}$ is not perpendicular to $\overline{BC}$.

   So, ∠ABC is not a right angle, and ▱ABCD cannot be a rectangle or a square. ✓

Monitoring Progress Help in English and Spanish at BigIdeasMath.com

8. Decide whether ▱PQRS with vertices P(−5, 2), Q(0, 4), R(2, −1), and S(−3, −3) is a *rectangle*, a *rhombus*, or a *square*. Give all names that apply.

Laurie's Notes — Teacher Actions

- **Make Sense of Problems and Persevere in Solving Them:** Give partners sufficient *Wait Time* to work on the example. When students do a quick plot, they may assume it is a rhombus. Let students decide what they need to calculate to determine that it is a rhombus.
- **Selective Responses:** This is a problem in which students may use different strategies. Allow ample work time before asking selected students to share.

Closure
- **Exit Ticket:** List different ways in which you can prove a quadrilateral is a rectangle, a rhombus, and a square.

7.4 Exercises

Vocabulary and Core Concept Check

1. **VOCABULARY** What is another name for an equilateral rectangle?
2. **WRITING** What should you look for in a parallelogram to know if the parallelogram is also a rhombus?

Monitoring Progress and Modeling with Mathematics

In Exercises 3–8, for any rhombus *JKLM*, decide whether the statement is *always* or *sometimes* true. Draw a diagram and explain your reasoning. (*See Example 1.*)

3. $\angle L \cong \angle M$
4. $\angle K \cong \angle M$
5. $\overline{JM} \cong \overline{KL}$
6. $\overline{JK} \cong \overline{KL}$
7. $\overline{JL} \cong \overline{KM}$
8. $\angle JKM \cong \angle LKM$

In Exercises 9–12, classify the quadrilateral. Explain your reasoning. (*See Example 2.*)

9.
10.
11.
12.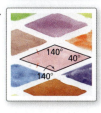

In Exercises 13–16, find the measures of the numbered angles in rhombus *DEFG*. (*See Example 3.*)

13.
14.
15.
16.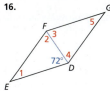

In Exercises 17–22, for any rectangle *WXYZ*, decide whether the statement is *always* or *sometimes* true. Draw a diagram and explain your reasoning.

17. $\angle W \cong \angle X$
18. $\overline{WX} \cong \overline{YZ}$
19. $\overline{WX} \cong \overline{XY}$
20. $\overline{WY} \cong \overline{XZ}$
21. $\overline{WY} \perp \overline{XZ}$
22. $\angle WXZ \cong \angle YXZ$

In Exercises 23 and 24, determine whether the quadrilateral is a rectangle. (*See Example 4.*)

23.

24.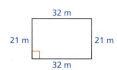

In Exercises 25–28, find the lengths of the diagonals of rectangle *WXYZ*. (*See Example 5.*)

25. $WY = 6x - 7$
    $XZ = 3x + 2$
26. $WY = 14x + 10$
    $XZ = 11x + 22$
27. $WY = 24x - 8$
    $XZ = -18x + 13$
28. $WY = 16x + 2$
    $XZ = 36x - 6$

Assignment Guide and Homework Check

ASSIGNMENT

Basic: 1, 2, 3–55 odd, 63–69 odd, 78, 84, 89–91

Average: 1, 2, 6–58 even, 64–78 even, 84, 89–91

Advanced: 1, 2, 8–70 even, 74–84, 88–91

HOMEWORK CHECK

Basic: 3, 11, 15, 25, 55

Average: 10, 16, 28, 50, 56

Advanced: 36, 60, 62, 78, 80

ANSWERS

1. square
2. two consecutive sides that are congruent
3. sometimes; Some rhombuses are squares.

4. always; A rhombus is a parallelogram, and the opposite angles of a parallelogram are congruent.

5. always; By definition, a rhombus is a parallelogram, and opposite sides of a parallelogram are congruent.

6. always; By definition, a rhombus is a parallelogram with four congruent sides.

7. sometimes; Some rhombuses are squares.

---

8. always; Each diagonal of a rhombus bisects a pair of opposite angles.

9. square; All of the sides are congruent, and all of the angles are congruent.
10. rectangle; Opposite sides are congruent and the angles are 90°.
11. rectangle; Opposite sides are parallel and the angles are 90°.
12. rhombus; Opposite angles are congruent and adjacent sides are congruent.
13. $m\angle 1 = m\angle 2 = m\angle 4 = 27°$, $m\angle 3 = 90°$; $m\angle 5 = m\angle 6 = 63°$
14. $m\angle 1 = 90°$; $m\angle 2 = m\angle 3 = 42°$; $m\angle 4 = m\angle 5 = 48°$
15. $m\angle 1 = m\angle 2 = m\angle 3 = m\angle 4 = 37°$; $m\angle 5 = 106°$
16. $m\angle 1 = m\angle 5 = 36°$; $m\angle 2 = m\angle 3 = m\angle 4 = 72°$
17–28. See Additional Answers.

Dynamic Teaching Tools

Dynamic Assessment & Progress Monitoring Tool
Interactive Whiteboard Lesson Library
Dynamic Classroom with Dynamic Investigations

ANSWERS

29. rectangle, square
30. square
31. rhombus, square
32. parallelogram, rectangle, rhombus, square
33. parallelogram, rectangle, rhombus, square
34. rhombus, square
35. Diagonals do not necessarily bisect opposite angles of a rectangle;
    $m\angle QSR = 90° - m\angle QSP$
    $x = 32$
36. $\angle QSP$ and $\angle RQS$ should be complementary because they are the two acute angles of a right triangle;
    $m\angle QSP = 90° - m\angle RQS$
    $x = 53$
37. 53°
38. 90°
39. 74°
40. 16
41. 6
42. 12
43. 56°
44. 34°
45. 56°
46. 5
47. 10
48. 5
49. 90°
50. 45°
51. 45°
52. 1
53. 2
54. 2
55. rectangle, rhombus, square; The diagonals are congruent and perpendicular.
56. rhombus; The diagonals are perpendicular and not congruent.
57. rectangle; The sides are perpendicular and not congruent.
58. rectangle; The diagonals are congruent and not perpendicular.
59. rhombus; The diagonals are perpendicular and not congruent.
60. rectangle, rhombus, square; The diagonals are perpendicular and congruent.

In Exercises 29–34, name each quadrilateral—*parallelogram*, *rectangle*, *rhombus*, or *square*—for which the statement is always true.

29. It is equiangular.
30. It is equiangular and equilateral.
31. The diagonals are perpendicular.
32. Opposite sides are congruent.
33. The diagonals bisect each other.
34. The diagonals bisect opposite angles.

35. **ERROR ANALYSIS** Quadrilateral *PQRS* is a rectangle. Describe and correct the error in finding the value of *x*.

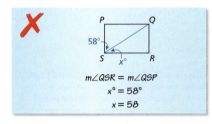

36. **ERROR ANALYSIS** Quadrilateral *PQRS* is a rhombus. Describe and correct the error in finding the value of *x*.

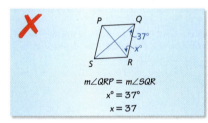

In Exercises 37–42, the diagonals of rhombus *ABCD* intersect at *E*. Given that $m\angle BAC = 53°$, $DE = 8$, and $EC = 6$, find the indicated measure.

37. $m\angle DAC$
38. $m\angle AED$
39. $m\angle ADC$
40. DB
41. AE
42. AC

438   Chapter 7   Quadrilaterals and Other Polygons

In Exercises 43–48, the diagonals of rectangle *QRST* intersect at *P*. Given that $m\angle PTS = 34°$ and $QS = 10$, find the indicated measure.

43. $m\angle QTR$
44. $m\angle QRT$
45. $m\angle SRT$
46. QP
47. RT
48. RP

In Exercises 49–54, the diagonals of square *LMNP* intersect at *K*. Given that $LK = 1$, find the indicated measure.

49. $m\angle MKN$
50. $m\angle LMK$
51. $m\angle LPK$
52. KN
53. LN
54. MP

In Exercises 55–60, decide whether ▱*JKLM* is a rectangle, a rhombus, or a square. Give all names that apply. Explain your reasoning. (See Example 6.)

55. $J(-4, 2)$, $K(0, 3)$, $L(1, -1)$, $M(-3, -2)$
56. $J(-2, 7)$, $K(7, 2)$, $L(-2, -3)$, $M(-11, 2)$
57. $J(3, 1)$, $K(3, -3)$, $L(-2, -3)$, $M(-2, 1)$
58. $J(-1, 4)$, $K(-3, 2)$, $L(2, -3)$, $M(4, -1)$
59. $J(5, 2)$, $K(1, 9)$, $L(-3, 2)$, $M(1, -5)$
60. $J(5, 2)$, $K(2, 5)$, $L(-1, 2)$, $M(2, -1)$

MATHEMATICAL CONNECTIONS In Exercises 61 and 62, classify the quadrilateral. Explain your reasoning. Then find the values of *x* and *y*.

61.

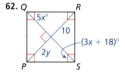

62.

---

61. rhombus; The sides are congruent; $x = 76$; $y = 4$
62. square; Because all four angles are 90° and the diagonals are perpendicular; $x = 9$; $y = 5$

63. **DRAWING CONCLUSIONS** In the window, $\overline{BD} \cong \overline{DF} \cong \overline{BH} \cong \overline{HF}$. Also, ∠HAB, ∠BCD, ∠DEF, and ∠FGH are right angles.

   a. Classify HBDF and ACEG. Explain your reasoning.
   b. What can you conclude about the lengths of the diagonals $\overline{AE}$ and $\overline{GC}$? Given that these diagonals intersect at J, what can you conclude about the lengths of $\overline{AJ}, \overline{JE}, \overline{CJ},$ and $\overline{JG}$? Explain.

64. **ABSTRACT REASONING** Order the terms in a diagram so that each term builds off the previous term(s). Explain why each figure is in the location you chose.

quadrilateral	square
rectangle	rhombus
parallelogram	

CRITICAL THINKING In Exercises 65–70, complete each statement with *always*, *sometimes*, or *never*. Explain your reasoning.

65. A square is _____ a rhombus.
66. A rectangle is _____ a square.
67. A rectangle _____ has congruent diagonals.
68. The diagonals of a square _____ bisect its angles.
69. A rhombus _____ has four congruent angles.
70. A rectangle _____ has perpendicular diagonals.

71. **USING TOOLS** You want to mark off a square region for a garden at school. You use a tape measure to mark off a quadrilateral on the ground. Each side of the quadrilateral is 2.5 meters long. Explain how you can use the tape measure to make sure that the quadrilateral is a square.

72. **PROVING A THEOREM** Use the plan for proof below to write a paragraph proof for one part of the Rhombus Diagonals Theorem.

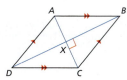

   Given ABCD is a parallelogram.
   $\overline{AC} \perp \overline{BD}$
   Prove ABCD is a rhombus.

   Plan for Proof Because ABCD is a parallelogram, its diagonals bisect each other at X. Use $\overline{AC} \perp \overline{BD}$ to show that △BXC ≅ △DXC. Then show that $\overline{BC} \cong \overline{DC}$. Use the properties of a parallelogram to show that ABCD is a rhombus.

PROVING A THEOREM In Exercises 73 and 74, write a proof for part of the Rhombus Opposite Angles Theorem.

73. **Given** PQRS is a parallelogram.
    $\overline{PR}$ bisects ∠SPQ and ∠QRS.
    $\overline{SQ}$ bisects ∠PSR and ∠RQP.
    Prove PQRS is a rhombus.

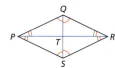

74. **Given** WXYZ is a rhombus.
    Prove $\overline{WY}$ bisects ∠ZWX and ∠XYZ.
    $\overline{ZX}$ bisects ∠WZY and ∠YXW.

ANSWERS

63. a. rhombus; rectangle; HBDF has four congruent sides; ACEG has four right angles.
    b. AE = GC; AJ = JE = CJ = JG; The diagonals of a rectangle are congruent and bisect each other.

64.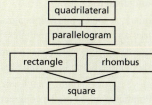

   All of the shapes have 4 sides. So, quadrilateral is at the top of the diagram. Because the rest all have two pairs of parallel sides, they are all parallelograms. Then, parallelograms with four right angles make one category (rectangle), while those with four congruent sides make another (rhombus), and if a parallelogram is both a rhombus and a rectangle, then it is a square.

65. always; By the Square Corollary, a square is a rhombus.
66. sometimes; Some rectangles are squares.
67. always; The diagonals of a rectangle are congruent by the Rectangle Diagonals Theorem.
68. always; A square is a rhombus.
69. sometimes; Some rhombuses are squares.
70. sometimes; Some rectangles are rhombuses.
71. Measure the diagonals to see if they are congruent.
72. Because ABCD is a parallelogram, its diagonals bisect each other by the Parallelogram Diagonals Theorem. So, $\overline{BX} \cong \overline{DX}$ by the definition of segment bisector. Because $\overline{AC} \perp \overline{BD}$, ∠DXC ≅ ∠BXC. By the Reflexive Property of Congruence, $\overline{XC} \cong \overline{XC}$. So, △BXC ≅ △DXC by the SAS Congruence Theorem. So, $\overline{BC} \cong \overline{DC}$ because corresponding parts of congruent triangles are congruent. Also, $\overline{AD} \cong \overline{BC}$ and $\overline{DC} \cong \overline{AB}$ because opposite sides of a parallelogram are congruent. So, by the Transitive Property of Congruence, $\overline{AB} \cong \overline{BC} \cong \overline{DC} \cong \overline{AD}$, which means that by the Rhombus Corollary, ABCD is a rhombus.

73–74. See Additional Answers.

ANSWERS

75. no; The diagonals of a square always create two right triangles.
76. yes; If the angles of a rhombus are 60°, 120°, 60°, and 120°, the diagonal that bisects the opposite 120° angles will divide the rhombus into two equilateral triangles.
77. square; A square has four congruent sides and four congruent angles.
78. *Sample answer:* You need to know whether the figure is a parallelogram.
79. no; yes; Corresponding angles of two rhombuses might not be congruent; Corresponding angles of two squares are congruent.
80. Because the line connecting a point with its preimage in a reflection is always perpendicular to the line of reflection, when a diagonal connecting two vertices is perpendicular to the other diagonal, both can be a line of symmetry.
81–83. See Additional Answers.
84. no; If a rhombus is a square, then it is also a rectangle.
85–88. See Additional Answers.
89. $x = 10$, $y = 8$
90. $x = 14$, $y = 6$
91. $x = 9$, $y = 26$

Mini-Assessment

1. Decide whether ▱ABCD with vertices $A(-3, 4)$, $B(3, 2)$, $C(2, -1)$, and $D(-4, 1)$ is a *rectangle*, a *rhombus*, or a *square*. Give all names that apply.
   a rectangle

   WXYZ is a rhombus.

   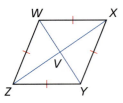

2. Find $m\angle WVX$. 90°
3. $m\angle ZYX = 128°$. Find $m\angle XYW$ and $m\angle ZXY$. $m\angle XYW = 64°$, $m\angle ZXY = 26°$
4. What can you conclude about a rhombus with congruent diagonals? It is also a rectangle and a square.

440   Chapter 7

---

75. **ABSTRACT REASONING** Will a diagonal of a square ever divide the square into two equilateral triangles? Explain your reasoning.

76. **ABSTRACT REASONING** Will a diagonal of a rhombus ever divide the rhombus into two equilateral triangles? Explain your reasoning.

77. **CRITICAL THINKING** Which quadrilateral could be called a regular quadrilateral? Explain your reasoning.

78. **HOW DO YOU SEE IT?** What other information do you need to determine whether the figure is a rectangle?

79. **REASONING** Are all rhombuses similar? Are all squares similar? Explain your reasoning.

80. **THOUGHT PROVOKING** Use the Rhombus Diagonals Theorem to explain why every rhombus has at least two lines of symmetry.

PROVING A COROLLARY In Exercises 81–83, write the corollary as a conditional statement and its converse. Then explain why each statement is true.

81. Rhombus Corollary
82. Rectangle Corollary
83. Square Corollary

84. **MAKING AN ARGUMENT** Your friend claims a rhombus will never have congruent diagonals because it would have to be a rectangle. Is your friend correct? Explain your reasoning.

85. **PROOF** Write a proof in the style of your choice.
    Given $\triangle XYZ \cong \triangle XWZ$, $\angle XYW \cong \angle ZWY$
    Prove WXYZ is a rhombus.

86. **PROOF** Write a proof in the style of your choice.
    Given $\overline{BC} \cong \overline{AD}, \overline{BC} \perp \overline{DC}, \overline{AD} \perp \overline{DC}$
    Prove ABCD is a rectangle.

PROVING A THEOREM In Exercises 87 and 88, write a proof for part of the Rectangle Diagonals Theorem.

87. Given PQRS is a rectangle.
    Prove $\overline{PR} \cong \overline{SQ}$

88. Given PQRS is a parallelogram.
    $\overline{PR} \cong \overline{SQ}$
    Prove PQRS is a rectangle.

Maintaining Mathematical Proficiency Reviewing what you learned in previous grades and lessons

$\overline{DE}$ is a midsegment of $\triangle ABC$. Find the values of x and y. *(Section 6.5)*

89.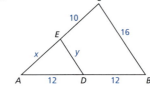

90.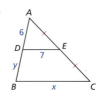

91.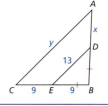

440   Chapter 7   Quadrilaterals and Other Polygons

If students need help...	If students got it...
Resources by Chapter • Practice A and Practice B • Puzzle Time	Resources by Chapter • Enrichment and Extension • Cumulative Review
Student Journal • Practice	Start the *next* Section
Differentiating the Lesson Skills Review Handbook	

Laurie's Notes

Overview of Section 7.5

Introduction
- This lesson is about two additional special quadrilaterals: trapezoids and kites.
- Properties about the diagonals and angles of these special quadrilaterals are presented, with some proven in the lesson and the remainder in the exercises.
- The midsegment of a trapezoid is defined. Students will see the connection to the midsegment of a triangle.
- The lesson ends with a flowchart summary of all of the special quadrilaterals presented in this chapter.

Teaching Strategy
- There are common ways in which a trapezoid is often drawn, leading students to believe that all trapezoids look like the one shown on the left. Be sure to draw trapezoids in different orientations and not all isosceles.

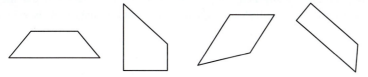

- A kite is defined in this lesson as a quadrilateral with two pairs of consecutive congruent sides, but opposite sides are not congruent. Here are two alternate definitions. Have students discuss each and decide what information is known about a kite from its definition.
  - A kite is a quadrilateral with two distinct pairs of adjacent sides that are congruent.
  - A kite is quadrilateral with perpendicular diagonals and exactly one line of symmetry.

Pacing Suggestion
- The explorations provide an opportunity for students to consider how they can prove they have a trapezoid or a kite. Transition to the formal lesson as soon as students have discussed each exploration.

Dynamic Teaching Tools
Dynamic Assessment & Progress Monitoring Tool
Lesson Planning Tool
Interactive Whiteboard Lesson Library
Dynamic Classroom with Dynamic Investigations

What Your Students Will Learn

- Use properties of trapezoids and kites to identify quadrilaterals.
- Use the Trapezoid Midsegment Theorem to find distances.
- Identify quadrilaterals using the most specific name based on given information.

Laurie's Notes

Exploration

Motivate
- Ask students to sketch quadrilaterals with different numbers of lines of symmetry: 0, 1, 2, 3, and 4.
- **Popsicle Sticks:** Use *Popsicle Sticks* to solicit responses for each. A square is the only quadrilateral with 4 lines of symmetry. No quadrilateral has 3 lines of symmetry. Rectangles and rhombuses have 2 lines of symmetry. Isosceles trapezoids and kites have 1 line of symmetry. Parallelograms and non-isosceles trapezoids have 0 lines of symmetry.
- Explain that today's lesson is about trapezoids and kites.

Exploration 1
- There are many different ways in which students could construct a trapezoid with base angles congruent.
- One method is to draw a line, $\overleftrightarrow{AB}$, and construct a line parallel to this line through a point C not on the line. Construct the perpendicular bisector of $\overline{AB}$ and reflect point C in the perpendicular bisector. Quadrilateral $ABC'C$ has congruent base angles and is an isosceles trapezoid.

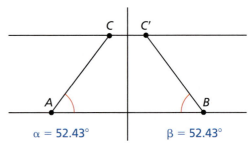

- **Construct Viable Arguments and Critique the Reasoning of Others:** Ask students to share methods used to construct a trapezoid with congruent base angles.
- Measuring the legs reveals that the trapezoid is isosceles.

Exploration 2
- Students should not have difficulty constructing a kite. Kites have exactly one line of symmetry. Begin with a scalene triangle and reflect the triangle in one side.

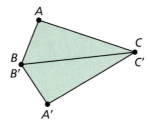

- If students use a reflection to construct the kite, they will easily recognize that one pair of opposite angles are congruent and the other pair of opposite angles are bisected by a diagonal.

Communicate Your Answer
- Listen for students to share the properties of trapezoids, isosceles trapezoids, and kites. Although the explorations did not explore diagonals, students may have conjectures.

Connecting to Next Step
- The conjectures made by students will be stated and proven as theorems in the formal lesson.

7.5 Properties of Trapezoids and Kites

Essential Question What are some properties of trapezoids and kites?

Recall the types of quadrilaterals shown below.

Trapezoid

Isosceles Trapezoid

Kite

PERSEVERE IN SOLVING PROBLEMS
To be proficient in math, you need to draw diagrams of important features and relationships, and search for regularity or trends.

EXPLORATION 1 Making a Conjecture about Trapezoids

Work with a partner. Use dynamic geometry software.

a. Construct a trapezoid whose base angles are congruent. Explain your process.

b. Is the trapezoid isosceles? Justify your answer.

c. Repeat parts (a) and (b) for several other trapezoids. Write a conjecture based on your results.

Sample

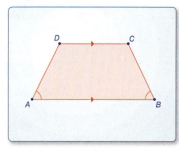

EXPLORATION 2 Discovering a Property of Kites

Work with a partner. Use dynamic geometry software.

a. Construct a kite. Explain your process.

b. Measure the angles of the kite. What do you observe?

c. Repeat parts (a) and (b) for several other kites. Write a conjecture based on your results.

Sample

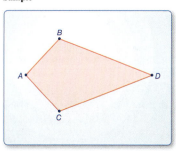

Communicate Your Answer

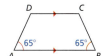

3. What are some properties of trapezoids and kites?

4. Is the trapezoid at the left isosceles? Explain.

5. A quadrilateral has angle measures of 70°, 70°, 110°, and 110°. Is the quadrilateral a kite? Explain.

English Language Learners

Build on Past Knowledge
Remind students that they have studied isosceles triangles. Have students compare the properties of isosceles triangles and isosceles trapezoids. *Sample answer:* Two sides of each figure are called legs and those sides are congruent. Both figures have bases and base angles. The base angles are congruent.

Extra Example 1
Show that *ABCD* is a trapezoid and decide whether it is isosceles.

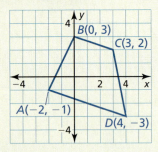

The slope of $\overline{BC}$ = slope of $\overline{AD} = -\frac{1}{3}$, so $\overline{BC} \parallel \overline{AD}$. The slope of $\overline{AB} = 2$ and the slope of $\overline{CD} = -5$, so $\overline{AB}$ is not parallel to $\overline{CD}$. With exactly one pair of parallel sides, *ABCD* is a trapezoid. The length of leg $\overline{AB}$ is $\sqrt{20}$, and the length of leg $\overline{CD}$ is $\sqrt{26}$. So, the legs are not congruent, and *ABCD* is not an isosceles trapezoid.

MONITORING PROGRESS ANSWER
1. slope of $\overline{AB}$ = slope of $\overline{DC}$ and slope of $\overline{AD} \neq$ slope of $\overline{BC}$. $AD = BC$, so *ABCD* is an isosceles trapezoid.

7.5 Lesson

What You Will Learn
- Use properties of trapezoids.
- Use the Trapezoid Midsegment Theorem to find distances.
- Use properties of kites.
- Identify quadrilaterals.

Core Vocabulary
trapezoid, p. 442
bases, p. 442
base angles, p. 442
legs, p. 442
isosceles trapezoid, p. 442
midsegment of a trapezoid, p. 444
kite, p. 445

Previous
diagonal
parallelogram

Using Properties of Trapezoids

A **trapezoid** is a quadrilateral with exactly one pair of parallel sides. The parallel sides are the **bases**.

Base angles of a trapezoid are two consecutive angles whose common side is a base. A trapezoid has two pairs of base angles. For example, in trapezoid *ABCD*, ∠*A* and ∠*D* are one pair of base angles, and ∠*B* and ∠*C* are the second pair. The nonparallel sides are the **legs** of the trapezoid.

If the legs of a trapezoid are congruent, then the trapezoid is an **isosceles trapezoid**.

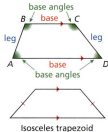

EXAMPLE 1 Identifying a Trapezoid in the Coordinate Plane

Show that *ORST* is a trapezoid. Then decide whether it is isosceles.

SOLUTION

Step 1 Compare the slopes of opposite sides.

slope of $\overline{RS} = \frac{4-3}{2-0} = \frac{1}{2}$

slope of $\overline{OT} = \frac{2-0}{4-0} = \frac{2}{4} = \frac{1}{2}$

The slopes of $\overline{RS}$ and $\overline{OT}$ are the same, so $\overline{RS} \parallel \overline{OT}$.

slope of $\overline{ST} = \frac{2-4}{4-2} = \frac{-2}{2} = -1$ slope of $\overline{RO} = \frac{3-0}{0-0} = \frac{3}{0}$ Undefined

The slopes of $\overline{ST}$ and $\overline{RO}$ are not the same, so $\overline{ST}$ is not parallel to $\overline{OR}$.

▶ Because *ORST* has exactly one pair of parallel sides, it is a trapezoid.

Step 2 Compare the lengths of legs $\overline{RO}$ and $\overline{ST}$.

$RO = |3 - 0| = 3$ $ST = \sqrt{(2-4)^2 + (4-2)^2} = \sqrt{8} = 2\sqrt{2}$

Because $RO \neq ST$, legs $\overline{RO}$ and $\overline{ST}$ are *not* congruent.

▶ So, *ORST* is not an isosceles trapezoid.

Monitoring Progress Help in English and Spanish at BigIdeasMath.com

1. The points *A*(−5, 6), *B*(4, 9), *C*(4, 4), and *D*(−2, 2) form the vertices of a quadrilateral. Show that *ABCD* is a trapezoid. Then decide whether it is isosceles.

442   Chapter 7   Quadrilaterals and Other Polygons

Laurie's Notes — Teacher Actions

- **Turn and Talk:** "What is a trapezoid and what properties does a trapezoid have?" From this assessing question, you'll find that students have prior knowledge about trapezoids. The theorems about trapezoids will make sense to students.

  COMMON ERROR In Example 1, students will find that the slopes of $\overline{RS}$ and $\overline{OT}$ are equal and conclude that *ORST* is a trapezoid. Remind students that they must also show that the other pair of opposite sides, $\overline{ST}$ and $\overline{OR}$, are not parallel.

Theorems

Isosceles Trapezoid Base Angles Theorem

If a trapezoid is isosceles, then each pair of base angles is congruent.

If trapezoid *ABCD* is isosceles, then ∠A ≅ ∠D and ∠B ≅ ∠C.

Proof Ex. 39, p. 449

Isosceles Trapezoid Base Angles Converse

If a trapezoid has a pair of congruent base angles, then it is an isosceles trapezoid.

If ∠A ≅ ∠D (or if ∠B ≅ ∠C), then trapezoid *ABCD* is isosceles.

Proof Ex. 40, p. 449

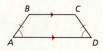

Isosceles Trapezoid Diagonals Theorem

A trapezoid is isosceles if and only if its diagonals are congruent.

Trapezoid *ABCD* is isosceles if and only if $\overline{AC} \cong \overline{BD}$.

Proof Ex. 51, p. 450

EXAMPLE 2 Using Properties of Isosceles Trapezoids

The stone above the arch in the diagram is an isosceles trapezoid. Find m∠K, m∠M, and m∠J.

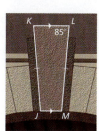

SOLUTION

Step 1 Find m∠K. *JKLM* is an isosceles trapezoid. So, ∠K and ∠L are congruent base angles, and m∠K = m∠L = 85°.

Step 2 Find m∠M. Because ∠L and ∠M are consecutive interior angles formed by $\overline{LM}$ intersecting two parallel lines, they are supplementary. So, m∠M = 180° − 85° = 95°.

Step 3 Find m∠J. Because ∠J and ∠M are a pair of base angles, they are congruent, and m∠J = m∠M = 95°.

▶ So, m∠K = 85°, m∠M = 95°, and m∠J = 95°.

Monitoring Progress Help in English and Spanish at BigIdeasMath.com

In Exercises 2 and 3, use trapezoid *EFGH*.

2. If *EG* = *FH*, is trapezoid *EFGH* isosceles? Explain.

3. If m∠*HEF* = 70° and m∠*FGH* = 110°, is trapezoid *EFGH* isosceles? Explain.

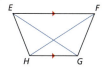

Section 7.5 Properties of Trapezoids and Kites 443

Extra Example 2

ABCD is an isosceles trapezoid, and m∠A = 42°. Find m∠B, m∠C, and m∠D.

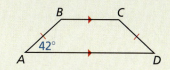

m∠B = 138°, m∠C = 138°, m∠D = 42°

MONITORING PROGRESS ANSWERS

2. yes, by the Isosceles Trapezoid Diagonals Theorem

3. yes; m∠GFE = 70°, ∠HEF ≅ ∠GFE

Laurie's Notes   Teacher Actions

- **Reason Abstractly and Quantitatively:** Discuss each theorem and have students label the diagrams.
- **Turn and Talk** and **Selective Responses:** Have partners select a theorem and discuss how to prove it. Students may use different strategies. Allow ample work time before asking selected students to share their proofs.
- ❓ "What is known in this example?" You have an isosceles trapezoid and the measure of one base angle.
- **Reason Abstractly and Quantitatively:** For each angle, students should give reasoning for how they know the angle measure. Can they use the language of the theorems correctly?

Extra Example 3

In the diagram, $\overline{MN}$ is the midsegment of trapezoid PQRS. Find MN.

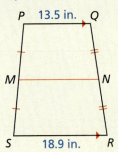

16.2 inches

Extra Example 4

Find the length of midsegment $\overline{YZ}$ in trapezoid PQRS.

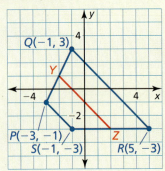

$4\sqrt{2}$ units

MONITORING PROGRESS ANSWERS

4.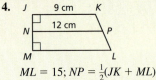

   $ML = 15$; $NP = \frac{1}{2}(JK + ML)$

5. Sample answer: Find the coordinates of Y and Z and calculate the distance between the points.

Using the Trapezoid Midsegment Theorem

Recall that a midsegment of a triangle is a segment that connects the midpoints of two sides of the triangle. The **midsegment of a trapezoid** is the segment that connects the midpoints of its legs. The theorem below is similar to the Triangle Midsegment Theorem.

> **READING**
> The midsegment of a trapezoid is sometimes called the *median* of the trapezoid.

Theorem

Trapezoid Midsegment Theorem

The midsegment of a trapezoid is parallel to each base, and its length is one-half the sum of the lengths of the bases.

If $\overline{MN}$ is the midsegment of trapezoid ABCD, then $\overline{MN} \parallel \overline{AB}$, $\overline{MN} \parallel \overline{DC}$, and $MN = \frac{1}{2}(AB + CD)$.

Proof Ex. 49, p. 450

EXAMPLE 3 Using the Midsegment of a Trapezoid

In the diagram, $\overline{MN}$ is the midsegment of trapezoid PQRS. Find MN.

SOLUTION

$MN = \frac{1}{2}(PQ + SR)$ Trapezoid Midsegment Theorem

$ = \frac{1}{2}(12 + 28)$ Substitute 12 for PQ and 28 for SR.

$ = 20$ Simplify.

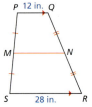

▶ The length of $\overline{MN}$ is 20 inches.

EXAMPLE 4 Using a Midsegment in the Coordinate Plane

Find the length of midsegment $\overline{YZ}$ in trapezoid STUV.

SOLUTION

Step 1 Find the lengths of $\overline{SV}$ and $\overline{TU}$.

$SV = \sqrt{(0-2)^2 + (6-2)^2} = \sqrt{20} = 2\sqrt{5}$

$TU = \sqrt{(8-12)^2 + (10-2)^2} = \sqrt{80} = 4\sqrt{5}$

Step 2 Multiply the sum of SV and TU by $\frac{1}{2}$.

$YZ = \frac{1}{2}(2\sqrt{5} + 4\sqrt{5}) = \frac{1}{2}(6\sqrt{5}) = 3\sqrt{5}$

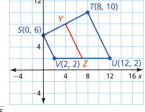

▶ So, the length of $\overline{YZ}$ is $3\sqrt{5}$ units.

Monitoring Progress   Help in English and Spanish at BigIdeasMath.com

4. In trapezoid JKLM, $\angle J$ and $\angle M$ are right angles, and $JK = 9$ centimeters. The length of midsegment $\overline{NP}$ of trapezoid JKLM is 12 centimeters. Sketch trapezoid JKLM and its midsegment. Find ML. Explain your reasoning.

5. Explain another method you can use to find the length of $\overline{YZ}$ in Example 4.

Laurie's Notes   Teacher Actions

- Use dynamic geometry software to investigate the midsegment of a trapezoid. If you click and drag on a vertex to decrease the length of a parallel base, the trapezoid begins to look like a triangle. Connect this to the Triangle Midsegment Theorem.

COMMON ERROR If students correctly find the length of each base in Example 4 and do not simplify the radicals, they may incorrectly say that the length of the midsegment is $\sqrt{50}$, reasoning that it is halfway between $\sqrt{20}$ and $\sqrt{80}$.

Using Properties of Kites

A **kite** is a quadrilateral that has two pairs of consecutive congruent sides, but opposite sides are not congruent.

Theorems

Kite Diagonals Theorem

If a quadrilateral is a kite, then its diagonals are perpendicular.

If quadrilateral $ABCD$ is a kite, then $\overline{AC} \perp \overline{BD}$.

Proof p. 445

STUDY TIP
The congruent angles of a kite are formed by the noncongruent adjacent sides.

Kite Opposite Angles Theorem

If a quadrilateral is a kite, then exactly one pair of opposite angles are congruent.

If quadrilateral $ABCD$ is a kite and $\overline{BC} \cong \overline{BA}$, then $\angle A \cong \angle C$ and $\angle B \not\cong \angle D$.

Proof Ex. 47, p. 450

PROOF Kite Diagonals Theorem

Given $ABCD$ is a kite, $\overline{BC} \cong \overline{BA}$, and $\overline{DC} \cong \overline{DA}$.
Prove $\overline{AC} \perp \overline{BD}$

STATEMENTS	REASONS
1. $ABCD$ is a kite with $\overline{BC} \cong \overline{BA}$ and $\overline{DC} \cong \overline{DA}$.	1. Given
2. B and D lie on the $\perp$ bisector of $\overline{AC}$.	2. Converse of the $\perp$ Bisector Theorem
3. $\overline{BD}$ is the $\perp$ bisector of $\overline{AC}$.	3. Through any two points, there exists exactly one line.
4. $\overline{AC} \perp \overline{BD}$	4. Definition of $\perp$ bisector

EXAMPLE 5 Finding Angle Measures in a Kite

Find $m\angle D$ in the kite shown.

SOLUTION

By the Kite Opposite Angles Theorem, $DEFG$ has exactly one pair of congruent opposite angles. Because $\angle E \not\cong \angle G$, $\angle D$ and $\angle F$ must be congruent. So, $m\angle D = m\angle F$. Write and solve an equation to find $m\angle D$.

$m\angle D + m\angle F + 115° + 73° = 360°$ — Corollary to the Polygon Interior Angles Theorem

$m\angle D + m\angle D + 115° + 73° = 360°$ — Substitute $m\angle D$ for $m\angle F$.

$2m\angle D + 188° = 360°$ — Combine like terms.

$m\angle D = 86°$ — Solve for $m\angle D$.

Differentiated Instruction

Kinesthetic
Have students use a compass and straightedge (or dynamic software application) and draw two intersecting circles of unequal radii. Draw the radii (sides) and the diagonals as shown. Measure the angles at the intersection of the diagonals. Ask students to write a conjecture based on their result.

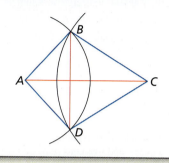

Extra Example 5

Find $m\angle C$ in the kite shown.

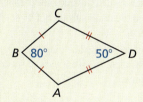

$m\angle C = 115°$

Laurie's Notes — Teacher Actions

- Kites may be convex or concave. The theorems hold true for each type. Only convex kites are considered in this lesson.
- **Turn and Talk:** "How could you prove that the diagonals of a kite are perpendicular?" Students often prove pairs of triangles congruent, and it is a much longer proof than the one shown. If time permits, have students share alternate methods.
- **Think-Pair-Share:** Have students work on Example 5, and then share and discuss as a class.

Extra Example 6

What is the most specific name for quadrilateral *JKLM*?

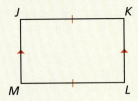

$\overline{JM} \parallel \overline{KL}$ and $JK = ML$. You do not know that the parallel sides are congruent or that the congruent sides are parallel. So, the most specific name for *JKLM* is a quadrilateral.

MONITORING PROGRESS ANSWERS

6. $x = 25$; 75°

7. parallelogram, rectangle, rhombus, square, and isosceles trapezoid

8. kite; Two pairs of consecutive sides are congruent and opposite sides are not congruent.

9. trapezoid; Two sides are parallel and the diagonals do not bisect each other.

10. quadrilateral; The markings are not sufficient to give it a more specific name.

Monitoring Progress  Help in English and Spanish at *BigIdeasMath.com*

6. In a kite, the measures of the angles are $3x°$, 75°, 90°, and 120°. Find the value of *x*. What are the measures of the angles that are congruent?

Identifying Special Quadrilaterals

The diagram shows relationships among the special quadrilaterals you have studied in this chapter. Each shape in the diagram has the properties of the shapes linked above it. For example, a rhombus has the properties of a parallelogram and a quadrilateral.

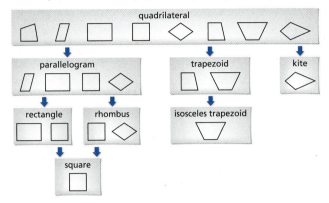

READING DIAGRAMS
In Example 6, *ABCD* looks like a square. But you must rely only on marked information when you interpret a diagram.

EXAMPLE 6 Identifying a Quadrilateral

What is the most specific name for quadrilateral *ABCD*?

SOLUTION

The diagram shows $\overline{AE} \cong \overline{CE}$ and $\overline{BE} \cong \overline{DE}$. So, the diagonals bisect each other. By the Parallelogram Diagonals Converse, *ABCD* is a parallelogram.

Rectangles, rhombuses, and squares are also parallelograms. However, there is no information given about the side lengths or angle measures of *ABCD*. So, you cannot determine whether it is a rectangle, a rhombus, or a square.

▶ So, the most specific name for *ABCD* is a parallelogram.

Monitoring Progress Help in English and Spanish at *BigIdeasMath.com*

7. Quadrilateral *DEFG* has at least one pair of opposite sides congruent. What types of quadrilaterals meet this condition?

Give the most specific name for the quadrilateral. Explain your reasoning.

8. [rhombus RSTU with sides 50, 50, 51, 51]
9. [quadrilateral VWXY with diagonals 64, 62, 75, 80]
10. [quadrilateral DEFC/G with sides marked congruent and diagonal 9]

446   Chapter 7   Quadrilaterals and Other Polygons

Laurie's Notes — Teacher Actions

- The flowchart shows how the special quadrilaterals are related. Any quadrilateral takes the properties and attributes of the quadrilateral above it and adds an additional property. It also shows that special quadrilaterals are parallelograms, trapezoids, or kites.
- Discuss the note in the side column. Students often assume from eyesight versus from reasoning.

Closure
- **Writing Prompt:** The big ideas presented in this lesson are . . .

446   Chapter 7

7.5 Exercises

Dynamic Solutions available at *BigIdeasMath.com*

Vocabulary and Core Concept Check

1. **WRITING** Describe the differences between a trapezoid and a kite.

2. **DIFFERENT WORDS, SAME QUESTION** Which is different? Find "both" answers.

   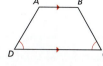

   Is there enough information to prove that trapezoid *ABCD* is isosceles?

   Is there enough information to prove that $\overline{AB} \cong \overline{DC}$?

   Is there enough information to prove that the non-parallel sides of trapezoid *ABCD* are congruent?

   Is there enough information to prove that the legs of trapezoid *ABCD* are congruent?

Monitoring Progress and Modeling with Mathematics

In Exercises 3–6, show that the quadrilateral with the given vertices is a trapezoid. Then decide whether it is isosceles. (*See Example 1.*)

3. $W(1, 4), X(1, 8), Y(-3, 9), Z(-3, 3)$

4. $D(-3, 3), E(-1, 1), F(1, -4), G(-3, 0)$

5. $M(-2, 0), N(0, 4), P(5, 4), Q(8, 0)$

6. $H(1, 9), J(4, 2), K(5, 2), L(8, 9)$

In Exercises 7 and 8, find the measure of each angle in the isosceles trapezoid. (*See Example 2.*)

7.   8.

In Exercises 9 and 10, find the length of the midsegment of the trapezoid. (*See Example 3.*)

9.   10.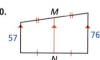

In Exercises 11 and 12, find *AB*.

11.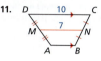

12. *D* 11.5 *C*, *M* 18.7 *N*, *A* *B*

In Exercises 13 and 14, find the length of the midsegment of the trapezoid with the given vertices. (*See Example 4.*)

13. $A(2, 0), B(8, -4), C(12, 2), D(0, 10)$

14. $S(-2, 4), T(-2, -4), U(3, -2), V(13, 10)$

In Exercises 15–18, find $m\angle G$. (*See Example 5.*)

15.   16. *F* 150° *G*, *E*, *H*

17.   18.

Section 7.5   Properties of Trapezoids and Kites   447

Dynamic Teaching Tools

Dynamic Assessment & Progress Monitoring Tool
Interactive Whiteboard Lesson Library
Dynamic Classroom with Dynamic Investigations

ANSWERS

19. Because $MN = \frac{1}{2}(AB + DC)$, when you solve for DC, you should get $DC = 2(MN) - AB$; $DC = 2(8) - 14 = 2$.

20. In the kite shown, $\angle B \cong \angle D$. Find $m\angle A$ by subtracting the measures of the other three angles from 360°, $m\angle A = 360° - 50° - 2(120°) = 70°$.

21. rectangle; $JKLM$ is a quadrilateral with 4 right angles.

22. trapezoid; $\overline{PS} \parallel \overline{QR}$ and $\angle QPS$ and $\angle PSR$ are not supplementary.

23. square; All four sides are congruent and the angles are 90°.

24. kite; $WXYZ$ has two pairs of consecutive congruent sides and opposite sides are not congruent.

25. no; It could be a kite.

26. no; It could be a rectangle.

27. 3

28. 6

29. 26 in.

30. 18 in., 29 in.; Consecutive sides are congruent.

31. $\angle A \cong \angle D$, or $\angle B \cong \angle C$; $\overline{AB} \parallel \overline{CD}$, so base angles need to be congruent.

32. Sample answer: $\overline{BC} \cong \overline{DC}$; Then $\triangle ABC \cong \triangle ADC$ and $ABCD$ has two pairs of consecutive congruent sides.

33. Sample answer: $\overline{BE} \cong \overline{DE}$; Then the diagonals bisect each other.

34. Sample answer: $\overline{AB} \cong \overline{BC}$; A rectangle with a pair of congruent adjacent sides is a square.

---

19. **ERROR ANALYSIS** Describe and correct the error in finding DC.

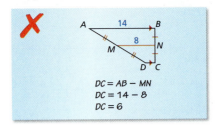

20. **ERROR ANALYSIS** Describe and correct the error in finding $m\angle A$.

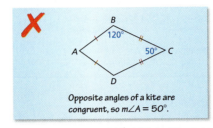

In Exercises 21–24, give the most specific name for the quadrilateral. Explain your reasoning. (See Example 6.)

21.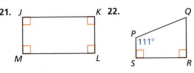

22.

23.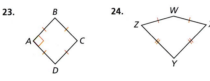

24.

REASONING In Exercises 25 and 26, tell whether enough information is given in the diagram to classify the quadrilateral by the indicated name. Explain.

25. rhombus

26. square

MATHEMATICAL CONNECTIONS In Exercises 27 and 28, find the value of x.

27.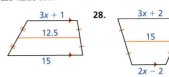

28.

29. **MODELING WITH MATHEMATICS** In the diagram, $NP = 8$ inches, and $LR = 20$ inches. What is the diameter of the bottom layer of the cake?

30. **PROBLEM SOLVING** You and a friend are building a kite. You need a stick to place from X to W and a stick to place from W to Z to finish constructing the frame. You want the kite to have the geometric shape of a kite. How long does each stick need to be? Explain your reasoning.

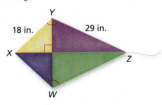

REASONING In Exercises 31–34, determine which pairs of segments or angles must be congruent so that you can prove that $ABCD$ is the indicated quadrilateral. Explain your reasoning. (There may be more than one right answer.)

31. isosceles trapezoid   32. kite

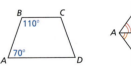

33. parallelogram   34. square

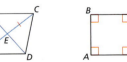

448   Chapter 7   Quadrilaterals and Other Polygons

35. PROOF Write a proof.

Given $\overline{JL} \cong \overline{LN}$, $\overline{KM}$ is a midsegment of $\triangle JLN$.

Prove Quadrilateral $JKMN$ is an isosceles trapezoid.

36. PROOF Write a proof.

Given $ABCD$ is a kite.
$\overline{AB} \cong \overline{CB}$, $\overline{AD} \cong \overline{CD}$

Prove $\overline{CE} \cong \overline{AE}$

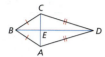

37. ABSTRACT REASONING Point U lies on the perpendicular bisector of $\overline{RT}$. Describe the set of points S for which $RSTU$ is a kite.

38. REASONING Determine whether the points $A(4, 5)$, $B(-3, 3)$, $C(-6, -13)$, and $D(6, -2)$ are the vertices of a kite. Explain your reasoning.

PROVING A THEOREM In Exercises 39 and 40, use the diagram to prove the given theorem. In the diagram, $\overline{EC}$ is drawn parallel to $\overline{AB}$.

39. Isosceles Trapezoid Base Angles Theorem

Given $ABCD$ is an isosceles trapezoid.
$\overline{BC} \parallel \overline{AD}$

Prove $\angle A \cong \angle D$, $\angle B \cong \angle BCD$

40. Isosceles Trapezoid Base Angles Converse

Given $ABCD$ is a trapezoid.
$\angle A \cong \angle D$, $\overline{BC} \parallel \overline{AD}$

Prove $ABCD$ is an isosceles trapezoid.

41. MAKING AN ARGUMENT Your cousin claims there is enough information to prove that $JKLM$ is an isosceles trapezoid. Is your cousin correct? Explain.

42. MATHEMATICAL CONNECTIONS The bases of a trapezoid lie on the lines $y = 2x + 7$ and $y = 2x - 5$. Write the equation of the line that contains the midsegment of the trapezoid.

43. CONSTRUCTION $\overline{AC}$ and $\overline{BD}$ bisect each other.

a. Construct quadrilateral $ABCD$ so that $\overline{AC}$ and $\overline{BD}$ are congruent, but not perpendicular. Classify the quadrilateral. Justify your answer.

b. Construct quadrilateral $ABCD$ so that $\overline{AC}$ and $\overline{BD}$ are perpendicular, but not congruent. Classify the quadrilateral. Justify your answer.

44. PROOF Write a proof.

Given $QRST$ is an isosceles trapezoid.

Prove $\angle TQS \cong \angle SRT$

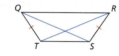

45. MODELING WITH MATHEMATICS A plastic spiderweb is made in the shape of a regular dodecagon (12-sided polygon). $\overline{AB} \parallel \overline{PQ}$, and X is equidistant from the vertices of the dodecagon.

a. Are you given enough information to prove that $ABPQ$ is an isosceles trapezoid?

b. What is the measure of each interior angle of $ABPQ$?

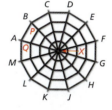

46. ATTENDING TO PRECISION In trapezoid $PQRS$, $\overline{PQ} \parallel \overline{RS}$ and $\overline{MN}$ is the midsegment of $PQRS$. If $RS = 5 \cdot PQ$, what is the ratio of MN to RS?

Ⓐ 3 : 5   Ⓑ 5 : 3

Ⓒ 1 : 2   Ⓓ 3 : 1

ANSWERS

35–36. See Additional Answers.

37. any point on $\overleftrightarrow{UV}$ such that $UV \neq SV$

38. yes; $AB = AD = \sqrt{53}$ and $BC = DC = \sqrt{265}$

39. Given isosceles trapezoid $ABCD$ with $\overline{BC} \parallel \overline{AD}$, construct $\overline{CE}$ parallel to $\overline{BA}$. Then, $ABCE$ is a parallelogram by definition, so $\overline{AB} \cong \overline{EC}$. Because $\overline{AB} \cong \overline{CD}$ by the definition of an isosceles trapezoid, $\overline{CE} \cong \overline{CD}$ by the Transitive Property of Congruence. So, $\angle CED \cong \angle D$ by the Base Angles Theorem and $\angle A \cong \angle CED$ by the Corresponding Angles Theorem. So, $\angle A \cong \angle D$ by the Transitive Property of Congruence. Next, by the Consecutive Interior Angles Theorem, $\angle B$ and $\angle A$ are supplementary and so are $\angle BCD$ and $\angle D$. So, $\angle B \cong \angle BCD$ by the Congruent Supplements Theorem.

40. Given trapezoid $ABCD$ with $\angle A \cong \angle D$ and $\overline{BC} \parallel \overline{AD}$, construct $\overline{CE}$ parallel to $\overline{BA}$. Then, $ABCE$ is a parallelogram by definition, so $\overline{AB} \cong \overline{EC}$. $\angle A \cong \angle CED$ by the Corresponding Angles Theorem, so $\angle CED \cong \angle D$ by the Transitive Property of Congruence. Then by the Converse of the Base Angles Theorem, $\overline{EC} \cong \overline{DC}$. So, $\overline{AB} \cong \overline{DC}$ by the Transitive Property of Congruence, and trapezoid $ABCD$ is isosceles.

41. no; It could be a square.

42. $y = 2x + 1$

43. a.

rectangle; The diagonals are congruent, but not perpendicular.

b.

rhombus; The diagonals are perpendicular, but not congruent.

44. See Additional Answers.

45. a. yes

b. 75°, 75°, 105°, 105°

46. A

ANSWERS

47. See Additional Answers.
48. a. trapezoid
    b. isosceles trapezoid
49. By the Triangle Midsegment Theorem, $\overline{BG} \parallel \overline{CD}$, $BG = \frac{1}{2}CD$, $\overline{GE} \parallel \overline{AF}$, and $GE = \frac{1}{2}AF$. By the Transitive Property of Parallel Lines, $\overline{CD} \parallel \overline{BE} \parallel \overline{AF}$. Also, by the Segment Addition Postulate, $BE = BG + GE$. So, by the Substitution Property of Equality, $BE = \frac{1}{2}CD + \frac{1}{2}AF = \frac{1}{2}(CD + AF)$.

50–58. See Additional Answers.

Mini-Assessment

1. What is the most specific name for quadrilateral ABCD? Explain your reasoning.

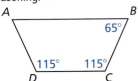

Because ∠B and ∠C are supplementary, but ∠C and ∠D are not supplementary, only $\overline{AB} \parallel \overline{CD}$. So, by definition, ABCD is a trapezoid. Because ∠C and ∠D are a pair of congruent base angles, ABCD is an isosceles trapezoid.

2. $\overline{MN}$ is the midsegment of trapezoid PQRS. Find PQ.

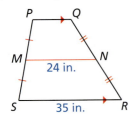

13 inches

3. Find m∠E.

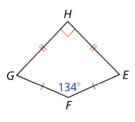

68°

450 Chapter 7

---

47. **PROVING A THEOREM** Use the plan for proof below to write a paragraph proof of the Kite Opposite Angles Theorem.

   Given EFGH is a kite.
   $\overline{EF} \cong \overline{FG}, \overline{EH} \cong \overline{GH}$

   Prove ∠E ≅ ∠G, ∠F ≇ ∠H

   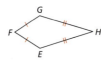

   Plan for Proof First show that ∠E ≅ ∠G. Then use an indirect argument to show that ∠F ≇ ∠H.

48. **HOW DO YOU SEE IT?** One of the earliest shapes used for cut diamonds is called the *table cut*, as shown in the figure. Each face of a cut gem is called a *facet*.

   a. $\overline{BC} \parallel \overline{AD}$, and $\overline{AB}$ and $\overline{DC}$ are not parallel. What shape is the facet labeled ABCD?

   b. $\overline{DE} \parallel \overline{GF}$, and $\overline{DG}$ and $\overline{EF}$ are congruent but not parallel. What shape is the facet labeled DEFG?

49. **PROVING A THEOREM** In the diagram below, $\overline{BG}$ is the midsegment of △ACD, and $\overline{GE}$ is the midsegment of △ADF. Use the diagram to prove the Trapezoid Midsegment Theorem.

   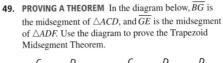

Maintaining Mathematical Proficiency
Reviewing what you learned in previous grades and lessons

Graph △PQR with vertices P(−3, 2), Q(2, 3), and R(4, −2) and its image after the translation. *(Skills Review Handbook)*

53. $(x, y) \rightarrow (x + 5, y + 8)$ 54. $(x, y) \rightarrow (x + 6, y - 3)$ 55. $(x, y) \rightarrow (x - 4, y - 7)$

Tell whether the two figures are similar. Explain your reasoning. *(Skills Review Handbook)*

56.   57. 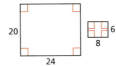   58.

450 Chapter 7 Quadrilaterals and Other Polygons

---

50. **THOUGHT PROVOKING** Is SSASS a valid congruence theorem for kites? Justify your answer.

51. **PROVING A THEOREM** To prove the biconditional statement in the Isosceles Trapezoid Diagonals Theorem, you must prove both parts separately.

   a. Prove part of the Isosceles Trapezoid Diagonals Theorem.

      Given JKLM is an isosceles trapezoid.
      $\overline{KL} \parallel \overline{JM}, \overline{JK} \cong \overline{LM}$

      Prove $\overline{JL} \cong \overline{KM}$

      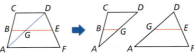

   b. Write the other part of the Isosceles Trapezoid Diagonals Theorem as a conditional. Then prove the statement is true.

52. **PROOF** What special type of quadrilateral is EFGH? Write a proof to show that your answer is correct.

   Given In the three-dimensional figure, $\overline{JK} \cong \overline{LM}$. E, F, G, and H are the midpoints of $\overline{JL}$, $\overline{KL}$, $\overline{KM}$, and $\overline{JM}$, respectively.

   Prove EFGH is a _____.

If students need help...	If students got it...
Resources by Chapter • Practice A and Practice B • Puzzle Time	**Resources by Chapter** • Enrichment and Extension • Cumulative Review
Student Journal • Practice	Start the *next* Section
Differentiating the Lesson Skills Review Handbook	

7.4–7.5 What Did You Learn?

Core Vocabulary

rhombus, p. 432
rectangle, p. 432
square, p. 432
trapezoid, p. 442
bases (of a trapezoid), p. 442
base angles (of a trapezoid), p. 442
legs (of a trapezoid), p. 442
isosceles trapezoid, p. 442
midsegment of a trapezoid, p. 444
kite, p. 445

Core Concepts

Section 7.4
Rhombus Corollary, p. 432
Rectangle Corollary, p. 432
Square Corollary, p. 432
Relationships between Special Parallelograms, p. 433
Rhombus Diagonals Theorem, p. 434
Rhombus Opposite Angles Theorem, p. 434
Rectangle Diagonals Theorem, p. 435
Identifying Special Parallelograms in the Coordinate Plane, p. 436

Section 7.5
Showing That a Quadrilateral Is a Trapezoid in the Coordinate Plane, p. 442
Isosceles Trapezoid Base Angles Theorem, p. 443
Isosceles Trapezoid Base Angles Converse, p. 443
Isosceles Trapezoid Diagonals Theorem, p. 443
Trapezoid Midsegment Theorem, p. 444
Kite Diagonals Theorem, p. 445
Kite Opposite Angles Theorem, p. 445
Identifying Special Quadrilaterals, p. 446

Mathematical Practices

1. In Exercise 14 on page 437, one reason $m\angle 4$, $m\angle 5$, and $m\angle DFE$ are all 48° is because diagonals of a rhombus bisect each other. What is another reason they are equal?

2. Explain how the diagram you created in Exercise 64 on page 439 can help you answer questions like Exercises 65–70.

3. In Exercise 29 on page 448, describe a pattern you can use to find the measure of a base of a trapezoid when given the length of the midsegment and the other base.

Performance Task:

Diamonds

Have you ever heard someone say, "Diamonds are a girl's best friend"? What are the properties of diamonds that make them shine? In what ways does mathematics contribute to the brilliance of a diamond?

To explore the answers to these questions and more, check out the Performance Task and Real-Life STEM video at *BigIdeasMath.com*.

451

ANSWERS

1. 5040°; 168°; 12°
2. 133
3. 82
4. 15
5. $a = 79, b = 101$
6. $a = 28, b = 87$
7. $c = 6, d = 10$
8. $(-2, -1)$
9. $M(2, -2)$

7 Chapter Review

Dynamic Solutions available at BigIdeasMath.com

7.1 Angles of Polygons (pp. 403–410)

Find the sum of the measures of the interior angles of the figure.

The figure is a convex hexagon. It has 6 sides. Use the Polygon Interior Angles Theorem.

$(n - 2) \cdot 180° = (6 - 2) \cdot 180°$ Substitute 6 for n.

$\qquad\qquad\qquad = 4 \cdot 180°$ Subtract.

$\qquad\qquad\qquad = 720°$ Multiply.

▶ The sum of the measures of the interior angles of the figure is 720°.

1. Find the sum of the measures of the interior angles of a regular 30-gon. Then find the measure of each interior angle and each exterior angle.

Find the value of x.

2.
3.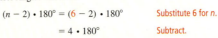
4.

7.2 Properties of Parallelograms (pp. 411–418)

Find the values of x and y.

ABCD is a parallelogram by the definition of a parallelogram. Use the Parallelogram Opposite Sides Theorem to find the value of x.

$AD = BC$ Opposite sides of a parallelogram are congruent.

$x + 6 = 9$ Substitute $x + 6$ for AD and 9 for BC.

$x = 3$ Subtract 6 from each side.

By the Parallelogram Opposite Angles Theorem, $\angle D \cong \angle B$, or $m\angle D = m\angle B$. So, $y° = 66°$.

▶ In □ABCD, $x = 3$ and $y = 66$.

Find the value of each variable in the parallelogram.

5.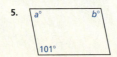
6.
7.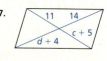

8. Find the coordinates of the intersection of the diagonals of □QRST with vertices $Q(-8, 1), R(2, 1), S(4, -3),$ and $T(-6, -3)$.

9. Three vertices of □JKLM are $J(1, 4), K(5, 3),$ and $L(6, -3)$. Find the coordinates of vertex M.

452 Chapter 7 Quadrilaterals and Other Polygons

7.3 Proving That a Quadrilateral Is a Parallelogram (pp. 419–428)

For what value of x is quadrilateral DEFG a parallelogram?

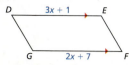

By the Opposite Sides Parallel and Congruent Theorem, if one pair of opposite sides are congruent and parallel, then DEFG is a parallelogram. You are given that $\overline{DE} \parallel \overline{FG}$. Find x so that $\overline{DE} \cong \overline{FG}$.

$DE = FG$	Set the segment lengths equal.
$3x + 1 = 2x + 7$	Substitute $3x + 1$ for DE and $2x + 7$ for FG.
$x + 1 = 7$	Subtract 2x from each side.
$x = 6$	Subtract 1 from each side.

When $x = 6$, $DE = 3(6) + 1 = 19$ and $FG = 2(6) + 7 = 19$.

▶ Quadrilateral DEFG is a parallelogram when $x = 6$.

State which theorem you can use to show that the quadrilateral is a parallelogram.

10.   11.   12.

13. Find the values of x and y that make the quadrilateral a parallelogram.

   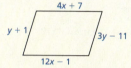

14. Find the value of x that makes the quadrilateral a parallelogram.

15. Show that quadrilateral WXYZ with vertices $W(-1, 6)$, $X(2, 8)$, $Y(1, 0)$, and $Z(-2, -2)$ is a parallelogram.

7.4 Properties of Special Parallelograms (pp. 431–440)

Classify the special quadrilateral. Explain your reasoning.

The quadrilateral has four right angles. By the Rectangle Corollary, the quadrilateral is a rectangle. Because the four sides are not marked as congruent, you cannot conclude that the rectangle is a square.

ANSWERS

16. rhombus; There are four congruent sides.
17. parallelogram; There are two pairs of parallel sides.
18. square; There are four congruent sides and the angles are 90°.
19. 10
20. rectangle, rhombus, square; The diagonals are congruent and perpendicular.
21. $m\angle Z = m\angle Y = 58°$, $m\angle W = m\angle X = 122°$
22. 26
23. $3\sqrt{5}$
24. $x = 15$; 105°
25. yes; Use the Isosceles Trapezoid Base Angles Converse.
26. trapezoid; There is one pair of parallel sides.
27. rhombus; There are four congruent sides.
28. rectangle; There are four right angles.

Classify the special quadrilateral. Explain your reasoning.

16.
17.
18.

19. Find the lengths of the diagonals of rectangle WXYZ where $WY = -2x + 34$ and $XZ = 3x - 26$.

20. Decide whether ▱JKLM with vertices J(5, 8), K(9, 6), L(7, 2), and M(3, 4) is a rectangle, a rhombus, or a square. Give all names that apply. Explain.

7.5 Properties of Trapezoids and Kites (pp. 441–450)

Find the length of midsegment $\overline{EF}$ in trapezoid ABCD.

Step 1 Find the lengths of $\overline{AD}$ and $\overline{BC}$.
$$AD = \sqrt{[1-(-5)]^2 + (-4-2)^2}$$
$$= \sqrt{72} = 6\sqrt{2}$$
$$BC = \sqrt{[1-(-1)]^2 + (0-2)^2}$$
$$= \sqrt{8} = 2\sqrt{2}$$

Step 2 Multiply the sum of AD and BC by $\frac{1}{2}$.
$$EF = \tfrac{1}{2}(6\sqrt{2} + 2\sqrt{2}) = \tfrac{1}{2}(8\sqrt{2}) = 4\sqrt{2}$$

▶ So, the length of $\overline{EF}$ is $4\sqrt{2}$ units.

21. Find the measure of each angle in the isosceles trapezoid WXYZ.

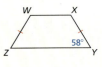

22. Find the length of the midsegment of trapezoid ABCD.

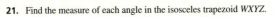

23. Find the length of the midsegment of trapezoid JKLM with vertices J(6, 10), K(10, 6), L(8, 2), and M(2, 2).

24. A kite has angle measures of $7x°$, 65°, 85°, and 105°. Find the value of x. What are the measures of the angles that are congruent?

25. Quadrilateral WXYZ is a trapezoid with one pair of congruent base angles. Is WXYZ an isosceles trapezoid? Explain your reasoning.

Give the most specific name for the quadrilateral. Explain your reasoning.

26.
27.
28.

7 Chapter Test

Find the value of each variable in the parallelogram.

1.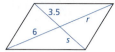
2.
3.

Give the most specific name for the quadrilateral. Explain your reasoning.

4.
5.
6.

7. In a convex octagon, three of the exterior angles each have a measure of $x°$. The other five exterior angles each have a measure of $(2x + 7)°$. Find the measure of each exterior angle.

8. Quadrilateral *PQRS* has vertices $P(5, 1)$, $Q(9, 6)$, $R(5, 11)$, and $S(1, 6)$. Classify quadrilateral *PQRS* using the most specific name.

Determine whether enough information is given to show that the quadrilateral is a parallelogram. Explain your reasoning.

9.
10.
11.

12. Explain why a parallelogram with one right angle must be a rectangle.

13. Summarize the ways you can prove that a quadrilateral is a square.

14. Three vertices of ▱*JKLM* are $J(-2, -1)$, $K(0, 2)$, and $L(4, 3)$.
    a. Find the coordinates of vertex *M*.
    b. Find the coordinates of the intersection of the diagonals of ▱*JKLM*.

15. You are building a plant stand with three equally-spaced circular shelves. The diagram shows a vertical cross section of the plant stand. What is the diameter of the middle shelf?

16. The Pentagon in Washington, D.C., is shaped like a regular pentagon. Find the measure of each interior angle.

17. You are designing a binocular mount. If $\overline{BC}$ is always vertical, the binoculars will point in the same direction while they are raised and lowered for different viewers. How can you design the mount so $\overline{BC}$ is always vertical? Justify your answer.

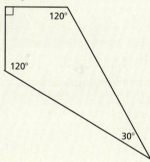

18. The measure of one angle of a kite is 90°. The measure of another angle in the kite is 30°. Sketch a kite that matches this description.

ANSWERS

1. $r = 6$, $s = 3.5$
2. $a = 79$, $b = 101$
3. $p = 5$, $q = 9$
4. trapezoid; There is one pair of parallel sides.
5. kite; There are two pairs of consecutive congruent sides but opposite sides are not congruent.
6. isosceles trapezoid; Base angles are congruent
7. 25°, 25°, 25°, 57°, 57°, 57°, 57°, 57°
8. rhombus
9. yes; The diagonals bisect each other.
10. no; $\overline{JK}$ and $\overline{ML}$ might not be parallel.
11. yes; Opposite angles are congruent.
12. Consecutive angles are supplementary.
13. Show that a quadrilateral is a parallelogram with four congruent sides and four right angles, or show that a quadrilateral is both a rectangle and a rhombus.
14. a. $M(2, 0)$
    b. $(1, 1)$
15. 10.5 in.
16. 108°
17. Design $AB = DC$ and $AD = BC$, then *ABCD* is a parallelogram.
18. Sample answer:

7 Cumulative Assessment

ANSWERS

1. Definition of parallelogram; Alternate Interior Angles Theorem; Reflexive Property of Congruence; Definition of congruent angles; Angle Addition Postulate; Transitive Property of Equality; Definition of congruent angles; ASA Congruence Theorem; Corresponding parts of congruent triangles are congruent

2. B

3. no; Account A can be modeled by an increasing linear function and Account B can be modeled by an increasing exponential function, so the balance of Account B will eventually exceed the balance of Account A.

4. no; You cannot use a theorem to prove itself.

5. $\frac{5}{6}$

---

1. Copy and complete the flowchart proof of the Parallelogram Opposite Angles Theorem.

   Given $ABCD$ is a parallelogram.
   Prove $\angle A \cong \angle C$, $\angle B \cong \angle D$

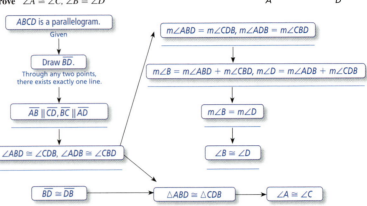

2. Which function is represented by the graph?

   Ⓐ $y = -\frac{1}{4}x^2$

   Ⓑ $y = \frac{1}{4}x^2$

   Ⓒ $y = -4x^2$

   Ⓓ $y = 4x^2$

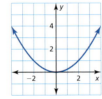

3. Account A has an initial balance of $1000 and increases by $50 each year. Account B has an initial balance of $500 and increases by 10% each year. Will Account A always have a greater balance than Account B? Explain.

4. Your friend claims that he can prove the Parallelogram Opposite Sides Theorem using the SSS Congruence Theorem and the Parallelogram Opposite Sides Theorem. Is your friend correct? Explain your reasoning.

5. You randomly select a vertex of polygon $QRSTUV$. What is the probability that the x-coordinate is greater than -3 or the y-coordinate is less than 4?

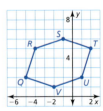

6. Choose the correct symbols to complete the proof of the Converse of the Hinge Theorem.

   Given $\overline{AB} \cong \overline{DE}, \overline{BC} \cong \overline{EF}, AC > DF$
   Prove $m\angle B > m\angle E$

   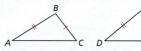

   Indirect Proof

   Step 1 Assume temporarily that $m\angle B \not> m\angle E$. Then it follows that either $m\angle B$ ___ $m\angle E$ or $m\angle B$ ___ $m\angle E$.

   Step 2 If $m\angle B$ ___ $m\angle E$, then AC ___ DF by the Hinge Theorem.
   If $m\angle B$ ___ $m\angle E$, then $\angle B$ ___ $\angle E$. So, $\triangle ABC$ ___ $\triangle DEF$ by the SAS Congruence Theorem and AC ___ DF.

   Step 3 Both conclusions contradict the given statement that AC ___ DF. So, the temporary assumption that $m\angle B \not> m\angle E$ cannot be true. This proves that $m\angle B$ ___ $m\angle E$.

7. Use the Isosceles Trapezoid Base Angles Converse to prove that $ABCD$ is an isosceles trapezoid.

   Given $\overline{BC} \parallel \overline{AD}, \angle EBC \cong \angle ECB, \angle ABE \cong \angle DCE$
   Prove $ABCD$ is an isosceles trapezoid.

   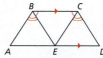

8. One part of the Rectangle Diagonals Theorem says, "If the diagonals of a parallelogram are congruent, then it is a rectangle." Using the reasons given, there are multiple ways to prove this part of the theorem. Provide a statement for each reason to form one possible proof of this part of the theorem.

   Given $QRST$ is a parallelogram.
   $\overline{QS} \cong \overline{RT}$

   Prove $QRST$ is a rectangle.

   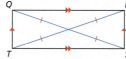

STATEMENTS	REASONS
1. $\overline{QS} \cong \overline{RT}$	1. Given
2. _____	2. Parallelogram Opposite Sides Theorem
3. _____	3. SSS Congruence Theorem
4. _____	4. Corresponding parts of congruent triangles are congruent.
5. _____	5. Parallelogram Consecutive Angles Theorem
6. _____	6. Congruent supplementary angles have the same measure.
7. _____	7. Parallelogram Consecutive Angles Theorem
8. _____	8. Subtraction Property of Equality
9. _____	9. Definition of a right angle
10. _____	10. Definition of a rectangle

ANSWERS
6. $<$; $=$; $<$; $<$; $=$; $\cong$; $\cong$; $=$; $>$; $>$
7. See Additional Answers.
8. Sample answer:
   2. $\overline{QT} \cong \overline{RS}, \overline{QR} \cong \overline{TS}$
   3. $\triangle QRS \cong \triangle RQT$
   4. $\angle QRS \cong \angle RQT$
   5. $m\angle QRS + m\angle RQT = 180°$
   6. $m\angle QRS = m\angle RQT = 90°$
   7. $m\angle TSR + 90° = 180°$, $m\angle STQ + 90° = 180°$
   8. $m\angle TSR = 90°, m\angle STQ = 90°$
   9. $\angle RQT, \angle QRS, \angle TSR,$ and $\angle STQ$ are right angles.
   10. $QRST$ is a rectangle.

Chapter 8 Pacing Guide	
Chapter Opener/ Mathematical Practices	0.5 Day
Section 1	1.5 Days
Section 2	1 Day
Section 3	2 Days
Section 4	1 Day
Quiz	0.5 Day
Section 5	1.5 Days
Section 6	1 Day
Chapter Review/ Chapter Tests	2 Days
Total Chapter 8	11 Days
Year-to-Date	118 Days

8 Similarity

8.1 Dilations
8.2 Similarity and Transformations
8.3 Similar Polygons
8.4 Proving Triangle Similarity by AA
8.5 Proving Triangle Similarity by SSS and SAS
8.6 Proportionality Theorems

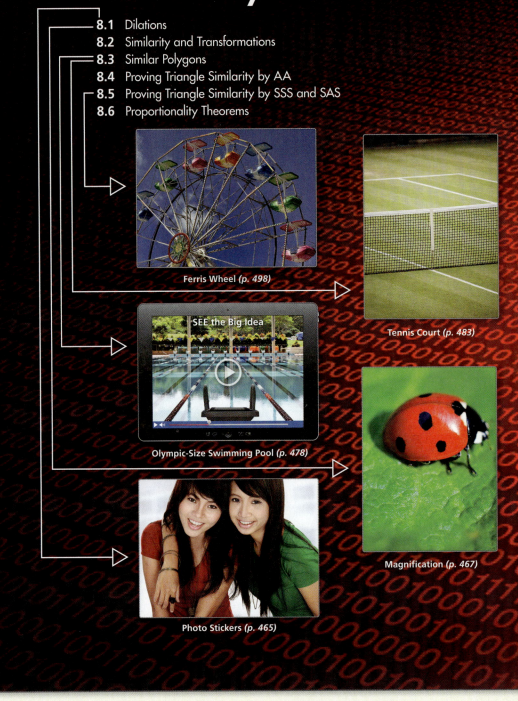

Ferris Wheel (p. 498)

Tennis Court (p. 483)

SEE the Big Idea

Olympic-Size Swimming Pool (p. 478)

Magnification (p. 467)

Photo Stickers (p. 465)

Laurie's Notes

Chapter Summary

- In Math I, students studied the rigid transformations of translations, reflections, and rotations. This chapter begins with a look at non-rigid transformations called dilations. In a rigid transformation the image and preimage are congruent. In a non-rigid transformation the image and preimage are similar.
- The next lesson introduces what it means for two polygons to be similar: corresponding sides are proportional and corresponding angles are congruent.
- The following two lessons present methods for proving two triangles are similar. One of the methods involves only angles (AA), with the other two methods involving only sides (SSS) or sides and the included angle (SAS). Unlike congruency methods, the corresponding sides must be proportional versus congruent.
- The last lesson presents several proportionality theorems, mostly connected to triangles.
- **Connection:** Properties of similar triangles will be needed in the next chapter when trigonometric ratios are defined.

What Your Students Have Learned

Middle School
- Compare ratios using proportions and solve proportions using a variety of strategies.
- Determine whether two figures are similar using proportions.
- Find ratios of perimeters and areas of similar figures.
- Draw dilations in the coordinate plane.

Math I
- Perform translations, reflections, rotations, and compositions of transformations.
- Describe and perform congruence transformations.

What Your Students Will Learn

Math II
- Identify and perform dilations.
- Describe and perform similarity transformations.
- Find perimeters and areas of similar polygons using proportions.
- Use the AA, SSS, and SAS Similarity Theorems to prove two triangles are similar.
- Use proportionality theorems to find lengths of segments.

Dynamic Teaching Tools
Dynamic Assessment & Progress Monitoring Tool
Lesson Planning Tool
Interactive Whiteboard Lesson Library
Dynamic Classroom with Dynamic Investigations
Real-Life STEM Videos

Scaffolding in the Classroom

Graphic Organizers: Example and Non-Example Chart
An Example and Non-Example Chart can be used to list examples and non-examples of a vocabulary word or term. Students write examples of the word or term in the left column and non-examples in the right column. This type of organizer serves as a good tool for assessing students' knowledge of pairs of topics that have subtle but important differences, such as complementary and supplementary angles. Blank Example and Non-Example Charts can be included on tests or quizzes for this purpose.

Questioning in the Classroom
How would you begin?
Before beginning the homework, select one challenging problem from the exercise set. Have students work in pairs or small groups to discuss how they would begin the problem. Students share their ideas before actually solving the problems, discussing the pros and cons of different methods.

Laurie's Notes

Maintaining Mathematical Proficiency

Determining Whether Ratios Form a Proportion
- Another way to check whether ratios form a proportion is to check their cross products: $\frac{a}{b} = \frac{c}{d}$ ($b \neq 0$, $d \neq 0$) if and only if $ad = bc$. Demonstrate this with Example 1.

COMMON ERROR Students often confuse *ratio* with *rate*. Remind them that a ratio is the relationship between two numbers, often written without units, such as 3 to 5. A rate is a ratio that compares two quantities with different units, such as 35 miles per hour.

Identifying Similar Figures
- Remind students that two figures are similar if and only if corresponding angles are congruent and the lengths of corresponding sides are proportional.

COMMON ERROR Students may forget that congruent figures are also similar figures. Remind students that the ratio of the lengths of corresponding sides in congruent figures is 1:1.

Mathematical Practices (continued on page 460)
- The *Mathematical Practices* page focuses attention on how mathematics is learned—process versus content. This page demonstrates how a mathematically proficient student uses structure to recognize a pattern in data. When perimeters and areas of similar polygons are explored, a pattern can be discerned.
- Use the *Mathematical Practices* page to help students develop mathematical habits of mind—how mathematics can be explored and how mathematics is thought about.

If students need help...	If students got it...
Student Journal • Maintaining Mathematical Proficiency	Game Closet at *BigIdeasMath.com*
Lesson Tutorials	Start the *next* Section
Skills Review Handbook	

Maintaining Mathematical Proficiency

Determining Whether Ratios Form a Proportion

Example 1 Tell whether $\frac{2}{8}$ and $\frac{3}{12}$ form a proportion.

Compare the ratios in simplest form.

$$\frac{2}{8} = \frac{2 \div 2}{8 \div 2} = \frac{1}{4} \qquad \frac{3}{12} = \frac{3 \div 3}{12 \div 3} = \frac{1}{4}$$

The ratios are equivalent.

▶ So, $\frac{2}{8}$ and $\frac{3}{12}$ form a proportion.

Tell whether the ratios form a proportion.

1. $\frac{5}{3}, \frac{35}{21}$
2. $\frac{9}{24}, \frac{24}{64}$
3. $\frac{8}{56}, \frac{6}{28}$
4. $\frac{18}{4}, \frac{27}{9}$
5. $\frac{15}{21}, \frac{55}{77}$
6. $\frac{26}{8}, \frac{39}{12}$

Identifying Similar Figures

Example 2 Which rectangle is similar to Rectangle A?

Rectangle A — 8 by 4

Rectangle B — 4 by 1

Rectangle C — 6 by 3

Each figure is a rectangle, so corresponding angles are congruent. Check to see whether corresponding side lengths are proportional.

Rectangle A and Rectangle B

$\frac{\text{Length of A}}{\text{Length of B}} = \frac{8}{4} = 2 \qquad \frac{\text{Width of A}}{\text{Width of B}} = \frac{4}{1} = 4$

not proportional

Rectangle A and Rectangle C

$\frac{\text{Length of A}}{\text{Length of C}} = \frac{8}{6} = \frac{4}{3} \qquad \frac{\text{Width of A}}{\text{Width of C}} = \frac{4}{3}$

proportional

▶ So, Rectangle C is similar to Rectangle A.

Tell whether the two figures are similar. Explain your reasoning.

7.

8.

9.

10. **ABSTRACT REASONING** Can you draw two squares that are not similar? Explain your reasoning.

Dynamic Solutions available at *BigIdeasMath.com*

459

What Your Students Have Learned

- Determine whether ratios form a proportion by simplifying the ratios.
- Determine whether two figures are similar using proportions.

ANSWERS

1. yes
2. yes
3. no
4. no
5. yes
6. yes
7. no; $\frac{12}{14} = \frac{6}{7} \neq \frac{5}{7}$, The sides are not proportional.
8. yes; The corresponding angles are congruent and the corresponding side lengths are proportional.
9. yes; The corresponding angles are congruent and the corresponding side lengths are proportional.
10. no; Squares have four right angles, so the corresponding angles are always congruent. Because all four sides are congruent, the corresponding sides will always be proportional.

Vocabulary Review

Have students make an Idea and Examples Chart for each of the following terms.

- Proportion
- Similar figures

Chapter 8

MONITORING PROGRESS ANSWERS

1. a. 32 cm, 48 cm²
   b. 48 cm, 108 cm²
   c. 64 cm, 192 cm²
2. a. 28 ft, 32 ft²
   b. 42 ft, 72 ft²
   c. 7 ft, 2 ft²
3. a. 376 in.², 480 in.³
   b. 846 in.², 1620 in.³
   c. 1504 in.², 3840 in.³

Mathematical Practices

Mathematically proficient students look for and make use of a pattern or structure.

Discerning a Pattern or Structure

Core Concept

Dilations, Perimeter, Area, and Volume

Consider a figure that is dilated by a scale factor of k.

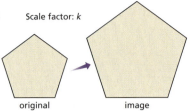

Scale factor: k

1. The perimeter of the image is k times the perimeter of the original figure.
2. The area of the image is k^2 times the area of the original figure.
3. If the original figure is three dimensional, then the volume of the image is k^3 times the volume of the original figure.

original   image

EXAMPLE 1   Finding Perimeter and Area after a Dilation

The triangle shown has side lengths of 3 inches, 4 inches, and 5 inches. Find the perimeter and area of the image when the triangle is dilated by a scale factor of (a) 2, (b) 3, and (c) 4.

SOLUTION

Perimeter: $P = 5 + 3 + 4 = 12$ in.   Area: $A = \frac{1}{2}(4)(3) = 6$ in.²

	Scale factor: k	Perimeter: kP	Area: k^2A
a.	2	$2(12) = 24$ in.	$(2^2)(6) = 24$ in.²
b.	3	$3(12) = 36$ in.	$(3^2)(6) = 54$ in.²
c.	4	$4(12) = 48$ in.	$(4^2)(6) = 96$ in.²

Monitoring Progress

1. Find the perimeter and area of the image when the trapezoid is dilated by a scale factor of (a) 2, (b) 3, and (c) 4.

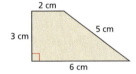

2. Find the perimeter and area of the image when the parallelogram is dilated by a scale factor of (a) 2, (b) 3, and (c) $\frac{1}{2}$.

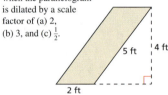

3. A rectangular prism is 3 inches wide, 4 inches long, and 5 inches tall. Find the surface area and volume of the image of the prism when it is dilated by a scale factor of (a) 2, (b) 3, and (c) 4.

Laurie's Notes   Mathematical Practices (continued from page T-459)

- The *Core Concept* states three relationships that students should be familiar with from middle school.
- Pose Example 1. Students may be tempted to write the dimensions of each similar triangle. They should not need to do this in order to find the related perimeters and areas.
- **Extension:** Use a scale factor that results in a reduction, $0 < k < 1$. Have students compute several perimeters and areas.
- Give time for students to work through the *Monitoring Progress* questions, and then discuss as a class.

Laurie's Notes

Overview of Section 8.1

Introduction
- Dilations were studied in middle school where students were expected to describe the effect of dilations on two-dimensional figures using coordinates. New in this lesson is that the centers of dilation do not need to be at the origin, and students are constructing dilations with a straightedge and a compass.
- The explorations provide an opportunity for students to explore different scale factors and centers of dilation. Encourage students to measure lengths and angles when they are using the dynamic geometry software.
- The scale factor of a dilation is the ratio of corresponding sides of the image and preimage. This ratio is used to solve real-life applications.

Resources
- A flashlight can be used to demonstrate the center of dilation. Darken the room. Use the flashlight to project the image of a figure onto a wall.

Formative Assessment Tips
- **Think-Alouds:** This technique is used when you want to hear how well partners comprehend a process involved with solving a problem. It is important to model the process first so that students have a sense of what is expected in *Think-Alouds*.
- *Think-Alouds* give students the opportunity to hear the metacognitive processes used by someone who is a proficient problem solver.
- **Make Sense of Problems and Persevere in Solving Them** and **Attend to Precision:** Hearing someone else describe a process using mathematical language will improve all students' problem-solving abilities; they can now apply the process with their partners to the problem being solved.
- Use this technique with a multi-step problem. Model using a starter sentence such as: "The problem is asking … ," "I can use the strategy of … ," "The steps I will use in solving this problem are … ," "This problem is similar to … ," and "I can check my answer by … ."
- Use this technique for a variety of problem types. Listen for comprehension of skills, concepts, and procedures as well as precision of language.

Applications
- Print and frame sizes for photographs are explored in Exercise 37 on page 467. Standard photographic sizes are: 2.5" $\times$ 3.5", 3.5" $\times$ 5", 4" $\times$ 6", 5" $\times$ 7", 8" $\times$ 10", 11" $\times$ 14", and 16" $\times$ 20". Ask students whether any of these could be a dilation of another.

Pacing Suggestion
- The formal lesson is long; however, experience with the explorations helps students develop essential understanding of dilations. You might consider having half the class do Exploration 1(a) and the other half do Exploration 1(b).

Dynamic Teaching Tools
Dynamic Assessment & Progress Monitoring Tool
Lesson Planning Tool
Interactive Whiteboard Lesson Library
Dynamic Classroom with Dynamic Investigations

What Your Students Will Learn

- Identify and perform dilations in which a figure is enlarged or reduced with respect to a fixed point C and a scale factor k.
- Solve real-life problems involving scale factors and dilations.

Laurie's Notes

Exploration

Motivate
- Cut a rectangle out of heavier card stock. Use a flashlight to cast a shadow of the rectangle onto the wall.
- **?** "Do the angles still appear to be right angles?" yes; They should be unless the rectangle is not parallel to the surface it is reflected onto.
- Vary the distance between the bulb of the flashlight and the rectangle. Discuss how this changes the shadow.
- **?** "Is the shadow always similar to the original figure?" It should be when the figure is held parallel to the wall and the flashlight is perpendicular to the wall.

Exploration 1
- This first exploration serves to familiarize students with the *dilate* command in the dynamic geometry software. Students should also be comfortable with the exploration posed verbally so that they are not looking at the result of the dilation in the book.
- Students should be comfortable clicking and dragging to change the shape and location of △ABC.
- **? Popsicle Sticks:** "What do you observe about the side lengths?" The sides of △A'B'C' are twice the length of the sides of △ABC. "What do you observe about the angle measures?" The measures of corresponding angles are equal.
- In part (b), students should again try different locations and shapes for △ABC and make observations about the coordinates of △A'B'C'.
- **Turn and Talk:** Discuss observations with partners when the center is (0, 0) and the scale factor is $\frac{1}{2}$.
- **? Extension:** "When the scale factor is 1, what will the image look like?" It will be congruent to the preimage.
- **Extension:** Usually there is at least one student who will enter a negative scale factor such as $-\frac{1}{2}$. If so, then the student should share his or her observations with the class. If time permits, have all students try a negative scale factor. Explain the effect of the negative sign.

Exploration 2
- You could pose this exploration to students verbally so that they are not looking at the result of the dilation in the book.
- Ask students to make a conjecture about what they think is going to happen when they dilate a line through the origin and a line not through the origin.
- When students dilate the line through the origin, they may say, "Nothing happened!" Do not give it away. Have them dilate a line not through the origin, which they will be able to see.
- **Big Idea:** A line through the origin has an equation $y = kx$ (direct variation). Multiplying (x, y) by a scale factor of 3 will result in ordered pairs that are on the same line. ($3y = 3kx$ simplifies to $y = kx$.) A line that does not pass through the origin has an equation $y = mx + b$, where $b \neq 0$. Multiplying (x, y) by a scale factor of 3 will result in ordered pairs that are on a line parallel to the original line.

Communicate Your Answer
- Ask different students to share their understanding of what it means to dilate a figure.

Connecting to Next Step
- Students have now been introduced to dilations. In the formal lesson, students will construct dilations with a compass and a straightedge and will construct dilations in the coordinate plane.

8.1 Dilations

Essential Question What does it mean to dilate a figure?

EXPLORATION 1 — Dilating a Triangle in a Coordinate Plane

Work with a partner. Use dynamic geometry software to draw any triangle and label it △ABC.

a. *Dilate* △ABC using a *scale factor* of 2 and a *center of dilation* at the origin to form △A'B'C'. Compare the coordinates, side lengths, and angle measures of △ABC and △A'B'C'.

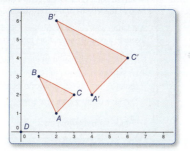

Sample
Points
A(2, 1)
B(1, 3)
C(3, 2)
Segments
AB = 2.24
BC = 2.24
AC = 1.41
Angles
m∠A = 71.57°
m∠B = 36.87°
m∠C = 71.57°

LOOKING FOR STRUCTURE
To be proficient in math, you need to look closely to discern a pattern or structure.

b. Repeat part (a) using a *scale factor* of $\frac{1}{2}$.

c. What do the results of parts (a) and (b) suggest about the coordinates, side lengths, and angle measures of the image of △ABC after a dilation with a scale factor of k?

EXPLORATION 2 — Dilating Lines in a Coordinate Plane

Work with a partner. Use dynamic geometry software to draw $\overleftrightarrow{AB}$ that passes through the origin and $\overleftrightarrow{AC}$ that does not pass through the origin.

a. *Dilate* $\overleftrightarrow{AB}$ using a *scale factor* of 3 and a *center of dilation* at the origin. Describe the image.

b. *Dilate* $\overleftrightarrow{AC}$ using a *scale factor* of 3 and a *center of dilation* at the origin. Describe the image.

c. Repeat parts (a) and (b) using a scale factor of $\frac{1}{4}$.

d. What do you notice about dilations of lines passing through the center of dilation and dilations of lines not passing through the center of dilation?

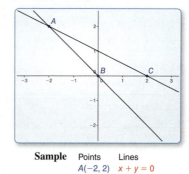

Sample
Points   Lines
A(−2, 2)   x + y = 0
B(0, 0)   x + 2y = 2
C(2, 0)

Communicate Your Answer

3. What does it mean to dilate a figure?

4. Repeat Exploration 1 using a center of dilation at a point other than the origin.

Section 8.1   Dilations   461

ANSWERS

1. a. Check students' work. The x-value of each vertex of △A'B'C' is twice the x-value of its corresponding vertex of △ABC, and the y-value of each vertex of △A'B'C' is twice the y-value of its corresponding vertex of △ABC. Each side of △A'B'C' is twice as long as its corresponding side of △ABC. Each angle of △A'B'C' is congruent to its corresponding angle of △ABC.

   b. Sample answer:

   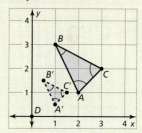

   The x-value of each vertex of △A'B'C' is half of the x-value of its corresponding vertex of △ABC, and the y-value of each vertex of △A'B'C' is half of the y-value of its corresponding vertex of △ABC. Each side of △A'B'C' is half as long as its corresponding side of △ABC. Each angle of △A'B'C' is congruent to its corresponding angle of △ABC.

   c. The x-value of each vertex of △A'B'C' is k times the x-value of its corresponding vertex of △ABC, and the y-value of each vertex of △A'B'C' is k times the y-value of its corresponding vertex of △ABC. Each side of △A'B'C' is k times as long as its corresponding side of △ABC. Each angle of △A'B'C' is congruent to its corresponding angle of △ABC.

2. a. The image is a line that coincides with $\overleftrightarrow{AB}$.

   b. The image is a line that is parallel to $\overleftrightarrow{AC}$. The x- and y-intercepts of the image are each three times the x- and y-intercepts of $\overleftrightarrow{AC}$.

   c. The image of $\overleftrightarrow{AB}$ is a line that coincides with $\overleftrightarrow{AB}$. The image of $\overleftrightarrow{AC}$ is a line that is parallel to $\overleftrightarrow{AC}$. The x- and y-intercepts of the image are each one-fourth of the x- and y-intercepts of $\overleftrightarrow{AC}$.

   d. When you dilate an image that passes through the center of dilation, the image coincides with the preimage. When you dilate a line that does not pass through the center of dilation, the image is parallel to the preimage, and the image has intercepts that can be found by multiplying the intercepts of the preimage by the constant of dilation.

3. to reduce or enlarge a figure so that the image is proportional to the preimage

4. See Additional Answers.

Differentiated Instruction
Auditory Some students may have difficulty distinguishing dilations from rigid transformations. Ask students for examples of dilations in the real world, and have them explain why these are dilations. Answers may include enlarging a photo, changing the size of a font on a computer, and zooming on a digital image. Ensure student explanations mention that the size has changed, but relative positions in a figure have not.

Extra Example 1
Find the scale factor of the dilation. Then tell whether the dilation is a *reduction* or an *enlargement*.

a.
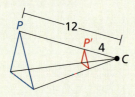
$k = \frac{1}{3}$; reduction

b.
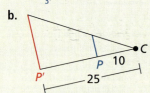
$k = \frac{5}{2}$; enlargement

MONITORING PROGRESS ANSWER
1. $k = \frac{1}{4}$; reduction

8.1 Lesson

What You Will Learn
- Identify and perform dilations.
- Solve real-life problems involving scale factors and dilations.

Core Vocabulary
dilation, *p. 462*
center of dilation, *p. 462*
scale factor, *p. 462*
enlargement, *p. 462*
reduction, *p. 462*

Identifying and Performing Dilations

Core Concept

Dilations
A **dilation** is a transformation in which a figure is enlarged or reduced with respect to a fixed point C called the **center of dilation** and a **scale factor** k, which is the ratio of the lengths of the corresponding sides of the image and the preimage.

A dilation with center of dilation C and scale factor k maps every point P in a figure to a point P' so that the following are true.

- If P is the center point C, then $P = P'$.
- If P is not the center point C, then the image point P' lies on $\overrightarrow{CP}$. The scale factor k is a positive number such that $k = \frac{CP'}{CP}$.
- Angle measures are preserved.

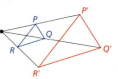

A dilation does not change any line that passes through the center of dilation. A dilation maps a line that does not pass through the center of dilation to a parallel line. In the figure above, $\overleftrightarrow{PR} \parallel \overleftrightarrow{P'R'}$, $\overleftrightarrow{PQ} \parallel \overleftrightarrow{P'Q'}$, and $\overleftrightarrow{QR} \parallel \overleftrightarrow{Q'R'}$.

When the scale factor $k > 1$, a dilation is an **enlargement**. When $0 < k < 1$, a dilation is a **reduction**.

EXAMPLE 1 Identifying Dilations

Find the scale factor of the dilation. Then tell whether the dilation is a *reduction* or an *enlargement*.

a. 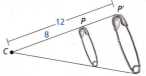   b.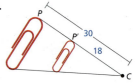

READING
The scale factor of a dilation can be written as a fraction, decimal, or percent.

SOLUTION
a. Because $\frac{CP'}{CP} = \frac{12}{8}$, the scale factor is $k = \frac{3}{2}$. So, the dilation is an enlargement.

b. Because $\frac{CP'}{CP} = \frac{18}{30}$, the scale factor is $k = \frac{3}{5}$. So, the dilation is a reduction.

Monitoring Progress Help in English and Spanish at *BigIdeasMath.com*

1. In a dilation, $CP' = 3$ and $CP = 12$. Find the scale factor. Then tell whether the dilation is a *reduction* or an *enlargement*.

462   Chapter 8   Similarity

Laurie's Notes   Teacher Actions

- Take time to fully discuss the definition of *dilation*. Connect to the explorations completed by students.
- ❓ "What will the *scale factor k* tell you about the dilation?" If students have done the explorations, they will know that when $k > 1$, the dilation will be an enlargement and when $0 < k < 1$, the dilation will be a reduction.
- Students may have tried a scale factor that was negative with the dynamic geometry software. This is discussed on page 464.
- **Turn and Talk:** Have students discuss how to find the scale factor for each dilation.

READING DIAGRAMS
In this chapter, for all of the dilations in the coordinate plane, the center of dilation is the origin unless otherwise noted.

Core Concept

Coordinate Rule for Dilations

If $P(x, y)$ is the preimage of a point, then its image after a dilation centered at the origin $(0, 0)$ with scale factor k is the point $P'(kx, ky)$.

$(x, y) \rightarrow (kx, ky)$

EXAMPLE 2 Dilating a Figure in the Coordinate Plane

Graph $\triangle ABC$ with vertices $A(2, 1)$, $B(4, 1)$, and $C(4, -1)$ and its image after a dilation with a scale factor of 2.

SOLUTION

Use the coordinate rule for a dilation with $k = 2$ to find the coordinates of the vertices of the image. Then graph $\triangle ABC$ and its image.

$(x, y) \rightarrow (2x, 2y)$
$A(2, 1) \rightarrow A'(4, 2)$
$B(4, 1) \rightarrow B'(8, 2)$
$C(4, -1) \rightarrow C'(8, -2)$

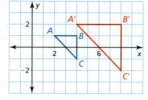

Notice the relationships between the lengths and slopes of the sides of the triangles in Example 2. Each side length of $\triangle A'B'C'$ is longer than its corresponding side by the scale factor. The corresponding sides are parallel because their slopes are the same.

EXAMPLE 3 Dilating a Figure in the Coordinate Plane

Graph quadrilateral $KLMN$ with vertices $K(-3, 6)$, $L(0, 6)$, $M(3, 3)$, and $N(-3, -3)$ and its image after a dilation with a scale factor of $\frac{1}{3}$.

SOLUTION

Use the coordinate rule for a dilation with $k = \frac{1}{3}$ to find the coordinates of the vertices of the image. Then graph quadrilateral $KLMN$ and its image.

$(x, y) \rightarrow \left(\frac{1}{3}x, \frac{1}{3}y\right)$
$K(-3, 6) \rightarrow K'(-1, 2)$
$L(0, 6) \rightarrow L'(0, 2)$
$M(3, 3) \rightarrow M'(1, 1)$
$N(-3, -3) \rightarrow N'(-1, -1)$

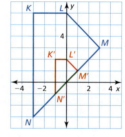

Monitoring Progress Help in English and Spanish at *BigIdeasMath.com*

Graph $\triangle PQR$ and its image after a dilation with scale factor k.

2. $P(-2, -1)$, $Q(-1, 0)$, $R(0, -1)$; $k = 4$
3. $P(5, -5)$, $Q(10, -5)$, $R(10, 5)$; $k = 0.4$

English Language Learners

Comprehension
Make sure students understand the relationship between the *scale factor* and the effects of a dilation. Use examples to guide students through a discussion developing the inequalities $0 < k < 1$ and $k < 0$ as guides for identifying whether a dilation with a scale factor k is a *reduction* or an *enlargement*.

Extra Example 2
Graph $\triangle PQR$ with vertices $P(0, 2)$, $Q(1, 0)$, and $R(2, 2)$ and its image after a dilation with scale factor 3.

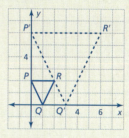

Extra Example 3
Graph $\triangle PQR$ with vertices $P(4, 6)$, $Q(-4, 2)$, and $R(2, -6)$ and its image after a dilation with scale factor 0.5.

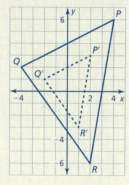

MONITORING PROGRESS ANSWERS

2.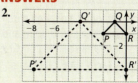

3. See Additional Answers.

Laurie's Notes — Teacher Actions

- Write the *Core Concept*.
- **Assessing Question:** "If the preimage is in Quadrant III, then the image will be in Quadrant III. Explain why." The point P' is on $\overrightarrow{CP}$, where C is the center of dilation, so P' will still be in Quadrant III.
- **Think-Alouds:** Pose Example 2 and say, "To solve this example, I need to … ." Ask partner A to think aloud for partner B to hear the problem-solving process. When students have finished the example, *Popsicle Sticks* a response.
- **Think-Alouds:** Pose Example 3 and say, "In this example, I predict … ." Ask partner B to think aloud for partner A to hear the prediction. When students have finished the example, *Popsicle Sticks* a response.

Extra Example 4
Graph △FGH with vertices F(3, 6), G(3, −3), and H(6, 6) and its image after a dilation with a scale factor of $-\frac{1}{3}$.

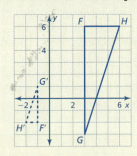

MONITORING PROGRESS ANSWERS

4.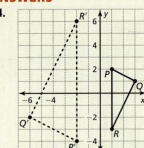

5. According to the Coordinate Rule for Dilations, if the origin P(0, 0) is the preimage of a point, then its image after a dilation centered at the origin with a scale factor k is the point P′(k · 0, k · 0), which is also the origin, or (0, 0).

CONSTRUCTION Constructing a Dilation

Use a compass and straightedge to construct a dilation of △PQR with a scale factor of 2. Use a point C outside the triangle as the center of dilation.

SOLUTION

Step 1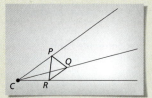

Step 2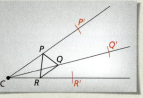

Step 3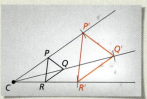

Draw a triangle Draw △PQR and choose the center of the dilation C outside the triangle. Draw rays from C through the vertices of the triangle.

Use a compass Use a compass to locate P′ on $\overrightarrow{CP}$ so that CP′ = 2(CP). Locate Q′ and R′ using the same method.

Connect points Connect points P′, Q′, and R′ to form △P′Q′R′.

In the coordinate plane, you can have scale factors that are negative numbers. When this occurs, the figure rotates 180°. So, when k > 0, a dilation with a scale factor of −k is the same as the composition of a dilation with a scale factor of k followed by a rotation of 180° about the center of dilation. Using the coordinate rules for a dilation and a rotation of 180°, you can think of the notation as

$(x, y) \rightarrow (kx, ky) \rightarrow (-kx, -ky)$.

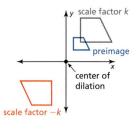

EXAMPLE 4 Using a Negative Scale Factor

Graph △FGH with vertices F(−4, −2), G(−2, 4), and H(−2, −2) and its image after a dilation with a scale factor of $-\frac{1}{2}$.

SOLUTION

Use the coordinate rule for a dilation with $k = -\frac{1}{2}$ to find the coordinates of the vertices of the image. Then graph △FGH and its image.

$(x, y) \rightarrow \left(-\frac{1}{2}x, -\frac{1}{2}y\right)$

$F(-4, -2) \rightarrow F'(2, 1)$
$G(-2, 4) \rightarrow G'(1, -2)$
$H(-2, -2) \rightarrow H'(1, 1)$

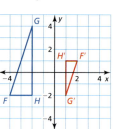

Monitoring Progress Help in English and Spanish at BigIdeasMath.com

4. Graph △PQR with vertices P(1, 2), Q(3, 1), and R(1, −3) and its image after a dilation with a scale factor of −2.

5. Suppose a figure containing the origin is dilated. Explain why the corresponding point in the image of the figure is also the origin.

Laurie's Notes **Teacher Actions**

- The construction of a dilation with scale factor of 2 is shown. You could differentiate at this point and hand slips of paper to different groups stating the scale factor. Groups of students ready for a more demanding task could use scale factors of $\frac{5}{2}$, 3, $\frac{1}{2}$, or $\frac{2}{3}$.
- **Use Appropriate Tools Strategically:** Use dynamic geometry software to demonstrate negative scale factors, or search the Internet ("GCSE transformations videos") for a video. Using a scalene figure will help students see the rotation of 180°.
- **Connection:** Rotation of 180° about the origin maps (x, y) → (−x, −y). Dilation by scale factor of −k centered at the origin maps (x, y) → (−kx, −ky).
- **? Advancing Question:** "A dilation of −k results in an image in Quadrant III." sometimes true; When the preimage is in Quadrant I, the image will be in Quadrant III; otherwise it will be in a different quadrant.
- Circulate as students try Example 4 with their partners.

Solving Real-Life Problems

EXAMPLE 5 Finding a Scale Factor

You are making your own photo stickers. Your photo is 4 inches by 4 inches. The image on the stickers is 1.1 inches by 1.1 inches. What is the scale factor of this dilation?

SOLUTION

The scale factor is the ratio of a side length of the sticker image to a side length of the original photo, or $\frac{1.1 \text{ in.}}{4 \text{ in.}}$.

▶ So, in simplest form, the scale factor is $\frac{11}{40}$.

EXAMPLE 6 Finding the Length of an Image

You are using a magnifying glass that shows the image of an object that is six times the object's actual size. Determine the length of the image of the spider seen through the magnifying glass.

1.5 cm

SOLUTION

$$\frac{\text{image length}}{\text{actual length}} = k$$

$$\frac{x}{1.5} = 6$$

$$x = 9$$

▶ So, the image length through the magnifying glass is 9 centimeters.

Monitoring Progress Help in English and Spanish at BigIdeasMath.com

12.6 cm

6. An optometrist dilates the pupils of a patient's eyes to get a better look at the back of the eyes. A pupil dilates from 4.5 millimeters to 8 millimeters. What is the scale factor of this dilation?

7. The image of a spider seen through the magnifying glass in Example 6 is shown at the left. Find the actual length of the spider.

When a transformation, such as a dilation, changes the shape or size of a figure, the transformation is *nonrigid*. In addition to dilations, there are many possible nonrigid transformations. Two examples are shown below. It is important to pay close attention to whether a nonrigid transformation preserves lengths and angle measures.

Horizontal Stretch   **Vertical Stretch**

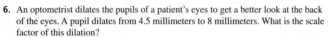

Extra Example 5
You are using word processing software to type the online school newsletter. You change the size of the text in one headline from 0.5 inch tall to 1.25 inches tall. What is the scale factor of this dilation? **2.5**

Extra Example 6
You are using a magnifying glass that shows the image of an object that is six times the object's actual size. The image of a spider seen through the magnifying glass is 13.5 centimeters long. Find the actual length of the spider. **2.25 centimeters**

MONITORING PROGRESS ANSWERS

6. $\frac{16}{9}$
7. 2.1 cm

READING
Scale factors are written so that the units in the numerator and denominator divide out.

Laurie's Notes Teacher Actions

- Help students understand that the original photo is the preimage. The image is smaller, meaning this is a reduction.
- Students may be confused because there is no "center" specified. The scale factor is the ratio of corresponding sides comparing preimage side length to image side length.
- **Look Back:** Give students time to reflect on this lesson. Previous transformations in Math I were rigid and dilations are not. The result of a dilation is an image that is the same shape but not necessarily the same size. A transformation might distort the shape as shown at the bottom of the page. A stretch horizontally or vertically does not preserve length or angle measure.

Closure
- **Writing Prompt:** How are dilations alike/different from rigid transformations? *Sample answer:* They are alike because they both preserve angle measure. They are different because the lengths of the sides are congruent for rigid transformations and proportional for dilations.

Assignment Guide and Homework Check

ASSIGNMENT

Basic: 1, 2, 3–35 odd, 38, 48, 52–57

Average: 1, 2, 4, 8, 14, 19, 24–34 even, 38–42, 48, 52–57

Advanced: 1, 2, 10, 16, 22–36 even, 38–49, 52–57

HOMEWORK CHECK

Basic: 5, 7, 21, 25, 31

Average: 14, 19, 24, 26, 39

Advanced: 22, 30, 39, 46, 47

ANSWERS

1. $P'(kx, ky)$
2. 60%; Because $0.6 < 1$, 60% is a scale factor for a reduction. The other three are scale factors for enlargements.
3. $\frac{3}{7}$; reduction
4. $\frac{8}{3}$; enlargement
5. $\frac{3}{5}$; reduction
6. $\frac{7}{2}$; enlargement
7–14. See Additional Answers.
15.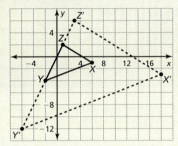
16.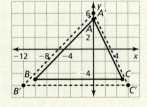

8.1 Exercises

Dynamic Solutions available at BigIdeasMath.com

Vocabulary and Core Concept Check

1. **COMPLETE THE SENTENCE** If $P(x, y)$ is the preimage of a point, then its image after a dilation centered at the origin $(0, 0)$ with scale factor k is the point _____.

2. **WHICH ONE DOESN'T BELONG?** Which scale factor does *not* belong with the other three? Explain your reasoning.

   $\frac{5}{4}$ 60%   115%   2

Monitoring Progress and Modeling with Mathematics

In Exercises 3–6, find the scale factor of the dilation. Then tell whether the dilation is a *reduction* or an *enlargement*. (See Example 1.)

3.
4.
5.
6.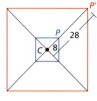

CONSTRUCTION In Exercises 7–10, copy the diagram. Then use a compass and straightedge to construct a dilation of $\triangle LMN$ with the given center and scale factor k.

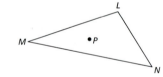

7. Center C, $k = 2$
8. Center P, $k = 3$
9. Center M, $k = \frac{1}{2}$
10. Center C, $k = 25\%$

CONSTRUCTION In Exercises 11–14, copy the diagram. Then use a compass and straightedge to construct a dilation of quadrilateral $RSTU$ with the given center and scale factor k.

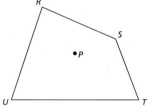

11. Center C, $k = 3$
12. Center P, $k = \frac{1}{3}$
13. Center R, $k = 0.25$
14. Center C, $k = 75\%$

In Exercises 15–18, graph the polygon and its image after a dilation with scale factor k. (See Examples 2 and 3.)

15. $X(6, -1)$, $Y(-2, -4)$, $Z(1, 2)$; $k = 3$
16. $A(0, 5)$, $B(-10, -5)$, $C(5, -5)$; $k = 120\%$
17. $T(9, -3)$, $U(6, 0)$, $V(3, 9)$, $W(0, 0)$; $k = \frac{2}{3}$
18. $J(4, 0)$, $K(-8, 4)$, $L(0, -4)$, $M(12, -8)$; $k = 0.25$

In Exercises 19–22, graph the polygon and its image after a dilation with scale factor k. (See Example 4.)

19. $B(-5, -10)$, $C(-10, 15)$, $D(0, 5)$; $k = -\frac{1}{5}$
20. $L(0, 0)$, $M(-4, 1)$, $N(-3, -6)$; $k = -3$
21. $R(-7, -1)$, $S(2, 5)$, $T(-2, -3)$, $U(-3, -3)$; $k = -4$
22. $W(8, -2)$, $X(6, 0)$, $Y(-6, 4)$, $Z(-2, 2)$; $k = -0.5$

466   Chapter 8   Similarity

17.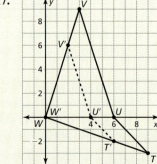
18.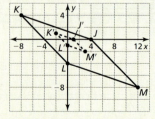

19–22. See Additional Answers.

ERROR ANALYSIS In Exercises 23 and 24, describe and correct the error in finding the scale factor of the dilation.

23.
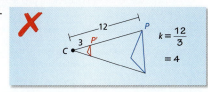

24.
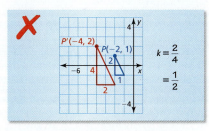

In Exercises 25–28, the red figure is the image of the blue figure after a dilation with center C. Find the scale factor of the dilation. Then find the value of the variable.

25.
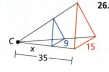

26.
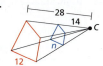

27.

28.

29. **FINDING A SCALE FACTOR** You receive wallet-sized photos of your school picture. The photo is 2.5 inches by 3.5 inches. You decide to dilate the photo to 5 inches by 7 inches at the store. What is the scale factor of this dilation? *(See Example 5.)*

30. **FINDING A SCALE FACTOR** Your visually impaired friend asked you to enlarge your notes from class so he can study. You took notes on 8.5-inch by 11-inch paper. The enlarged copy has a smaller side with a length of 10 inches. What is the scale factor of this dilation?

In Exercises 31–34, you are using a magnifying glass. Use the length of the insect and the magnification level to determine the length of the image seen through the magnifying glass. *(See Example 6.)*

31. emperor moth
Magnification: 5×

60 mm

32. ladybug
Magnification: 10×

4.5 mm

33. dragonfly
Magnification: 20×

47 mm

34. carpenter ant
Magnification: 15×

12 mm

35. **ANALYZING RELATIONSHIPS** Use the given actual and magnified lengths to determine which of the following insects were looked at using the same magnifying glass. Explain your reasoning.

grasshopper
Actual: 2 in.
Magnified: 15 in.

black beetle
Actual: 0.6 in.
Magnified: 4.2 in.

honeybee
Actual: $\frac{5}{8}$ in.
Magnified: $\frac{75}{16}$ in.

monarch butterfly
Actual: 3.9 in.
Magnified: 29.25 in.

36. **THOUGHT PROVOKING** Draw $\triangle ABC$ and $\triangle A'B'C'$ so that $\triangle A'B'C'$ is a dilation of $\triangle ABC$. Find the center of dilation and explain how you found it.

37. **REASONING** Your friend prints a 4-inch by 6-inch photo for you from the school dance. All you have is an 8-inch by 10-inch frame. Can you dilate the photo to fit the frame? Explain your reasoning.

Section 8.1   Dilations   467

ANSWERS

38. larger star; smaller star; Because the scale factor is between 0 and 1, the dilation is a reduction.

39. $x = 5, y = 25$

40. A figure that is 200% larger than the preimage will be twice as large.

41. original

42. dilated

43. original

44. dilated

45–49. See Additional Answers.

50. The center of dilation must be on that page. So, this point will be in the same place for both the original figure and the dilated figure.

51. $A'(4, 4), B'(4, 12), C'(10, 4)$

52. $A'(2, -5), B'(0, 0), C'(-3, 1)$

53. $A'(1, 2), B'(-1, 7), C'(-4, 8)$

54. $A'(5, -2), B'(3, 3), C'(0, 4)$

55. $A'(0, -1), B'(-2, 4), C'(-5, 5)$

56. $A'(3, -3), B'(1, 2), C'(-2, 3)$

57. $A'(-1, 0), B'(-3, 5), C'(-6, 6)$

Mini-Assessment

1. Find the scale factor of the dilation. Then tell whether it is a *reduction* or an *enlargement*.

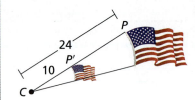

$k = \frac{5}{12}$, reduction

2. Graph $\triangle DEF$ with vertices $D(2, 6), E(2, 2),$ and $F(4, 2)$ and its image after a dilation with scale factor $-\frac{1}{2}$.

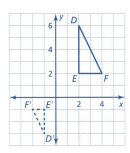

3. A photographer enlarges a 4 inch × 5 inch photo to an 8 inch × 10 inch photo. What is the scale factor of the dilation? 2

---

38. HOW DO YOU SEE IT? Point C is the center of dilation of the images. The scale factor is $\frac{1}{3}$. Which figure is the original figure? Which figure is the dilated figure? Explain your reasoning.

39. MATHEMATICAL CONNECTIONS The larger triangle is a dilation of the smaller triangle. Find the values of x and y.

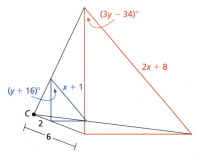

40. WRITING Explain why a scale factor of 2 is the same as 200%.

In Exercises 41–44, determine whether the dilated figure or the original figure is closer to the center of dilation. Use the given location of the center of dilation and scale factor k.

41. Center of dilation: inside the figure; $k = 3$

42. Center of dilation: inside the figure; $k = \frac{1}{2}$

43. Center of dilation: outside the figure; $k = 120\%$

44. Center of dilation: outside the figure; $k = 0.1$

Maintaining Mathematical Proficiency
Reviewing what you learned in previous grades and lessons

The vertices of $\triangle ABC$ are $A(2, -1), B(0, 4),$ and $C(-3, 5)$. Find the coordinates of the vertices of the image after the translation. *(Skills Review Handbook)*

52. $(x, y) \rightarrow (x, y - 4)$

53. $(x, y) \rightarrow (x - 1, y + 3)$

54. $(x, y) \rightarrow (x + 3, y - 1)$

55. $(x, y) \rightarrow (x - 2, y)$

56. $(x, y) \rightarrow (x + 1, y - 2)$

57. $(x, y) \rightarrow (x - 3, y + 1)$

---

45. ANALYZING RELATIONSHIPS Dilate the line through $O(0, 0)$ and $A(1, 2)$ using a scale factor of 2.

  a. What do you notice about the lengths of $\overline{O'A'}$ and $\overline{OA}$?

  b. What do you notice about $\overleftrightarrow{O'A'}$ and $\overleftrightarrow{OA}$?

46. ANALYZING RELATIONSHIPS Dilate the line through $A(0, 1)$ and $B(1, 2)$ using a scale factor of $\frac{1}{2}$.

  a. What do you notice about the lengths of $\overline{A'B'}$ and $\overline{AB}$?

  b. What do you notice about $\overleftrightarrow{A'B'}$ and $\overleftrightarrow{AB}$?

47. ATTENDING TO PRECISION You are making a blueprint of your house. You measure the lengths of the walls of your room to be 11 feet by 12 feet. When you draw your room on the blueprint, the lengths of the walls are 8.25 inches by 9 inches. What scale factor dilates your room to the blueprint?

48. MAKING AN ARGUMENT Your friend claims that dilating a figure by 1 is the same as dilating a figure by −1 because the original figure will not be enlarged or reduced. Is your friend correct? Explain your reasoning.

49. USING STRUCTURE Rectangle $WXYZ$ has vertices $W(-3, -1), X(-3, 3), Y(5, 3),$ and $Z(5, -1)$.

  a. Find the perimeter and area of the rectangle.

  b. Dilate the rectangle using a scale factor of 3. Find the perimeter and area of the dilated rectangle. Compare with the original rectangle. What do you notice?

  c. Repeat part (b) using a scale factor of $\frac{1}{4}$.

  d. Make a conjecture for how the perimeter and area change when a figure is dilated.

50. REASONING You put a reduction of a page on the original page. Explain why there is a point that is in the same place on both pages.

51. REASONING $\triangle ABC$ has vertices $A(4, 2), B(4, 6),$ and $C(7, 2)$. Find the coordinates of the vertices of the image after a dilation with center $(4, 0)$ and a scale factor of 2.

If students need help...

Resources by Chapter
- Practice A and Practice B
- Puzzle Time

Student Journal
- Practice

Differentiating the Lesson
Skills Review Handbook

If students got it...

Resources by Chapter
- Enrichment and Extension
- Cumulative Review

Start the *next* Section

Laurie's Notes

Overview of Section 8.2

Introduction
- In grade 8, similar figures were defined as figures that have the same shape but not necessarily the same size. Two figures are similar when corresponding side lengths are proportional and corresponding angles are congruent. Finally, perimeters and areas of similar figures and surface areas and volumes of similar solids were studied.
- In this section, similarity is defined in terms of similarity transformations. Students need to understand that a similarity transformation is a dilation or a composition of rigid motions and dilations. To decide (prove) whether two figures are similar, you need to map one figure onto the other by a single dilation or a composition of rigid motions and dilations. The composition of rigid motions and dilations may not be unique.

Resources
- Use dynamic geometry software to demonstrate, or perform, a similarity transformation. Use the software to determine whether there is a similarity transformation that maps one figure onto another.

Formative Assessment Tips
- **3-2-1:** This formative assessment strategy is useful in giving students a structured way in which to reflect on their learning, particularly at the conclusion of a unit of study. Students are asked to respond to three writing prompts, giving three responses to the first prompt, two responses to the second prompt, and one response to the third prompt. All six responses relate to what students have learned during the unit of study—in this case, a chapter on similarity.
- Distribute a printed *3-2-1* reflection sheet and give time for students to reflect on their learning and to write.
  - 3 new things (concepts, skills, procedures, …) I learned in this chapter
  - 2 things (concepts, skills, procedures, …) I am still struggling with
  - 1 thing that will help me tomorrow
- Collect and review student reflections and plan instruction for tomorrow accordingly. For students, the reflection allows them to see what learning has occurred and where they should focus their attention.

Pacing Suggestion
- Once students have worked the explorations, continue with the formal lesson. Alternatively, simply use the software to perform the similarity transformations in the formal lesson.

Dynamic Teaching Tools
Dynamic Assessment & Progress Monitoring Tool
Lesson Planning Tool
Interactive Whiteboard Lesson Library
Dynamic Classroom with Dynamic Investigations

What Your Students Will Learn

- Describe and perform similarity transformations that preserve shapes and angle measures, but not size.
- Prove that two figures are similar.

Laurie's Notes

Exploration

Motivate

- Draw a simple stick figure or other image on a stretchable surface, such as a balloon, physical therapy elastic, or play putty.
- ? Ask students what they think will happen to the figure when you pull the picture to the right. The image will be distorted. Pull one of the sides of the picture to confirm.
- Pull the top of the picture so that students see this result as the same.
- ? Ask students what they think will happen when the stretchable surface is pulled in both directions (right and up). The image will enlarge proportionally.
- **Alternate Approach:** If you can display a computer image to the class, you can drag a side to distort the image, or drag a corner to change the size of the image proportionally.

Discuss

- Similar figures can be congruent. Two congruent figures meet the definition of similarity: corresponding side lengths are proportional (1 : 1) and corresponding angles are congruent.
- Note that the definition of a dilation excludes a scale factor of 1. For that reason, the dilation is referred to as a nonrigid motion because the size changes.

Exploration 1

- Note that the center of dilation does not need to be the origin.
- ? **Construct Viable Arguments and Critique the Reasoning of Others:** "Are the triangles similar? How do you know?" Listen for the ratio of corresponding sides all equal 3, and the corresponding angles are congruent.
- **Extension:** Have students repeat the exploration (a) using one of the vertices as the center of dilation and (b) using a point inside the original triangle (but not the origin) as the center of dilation.
  - ? "When the center of dilation is a vertex of the original triangle, will the triangles be similar? Explain." yes; The ratio of corresponding sides all equal 3, and the corresponding angles are congruent.
  - ? "When the center of dilation is inside the original triangle, will the triangles be similar? Explain." yes; The ratio of corresponding sides all equal 3, and the corresponding angles are congruent.

Exploration 2

- In each construction, ask students how they know the preimage and the image are similar.
- **Reason Abstractly and Quantitatively** and **Construct Viable Arguments and Critique the Reasoning of Others:** What you hope is that students will say that translations, reflections, and rotations are isometries, congruence-preserving transformations. If two figures are congruent, then they are similar. They should not need to measure!
- ? **Probing Question:** "If two figures are congruent, then they are similar." always true
- ? **Probing Question:** "If two figures are similar, then they are congruent." sometimes true

Communicate Your Answer

- Listen for student understanding of rigid transformations producing congruent and, hence, similar images. Dilations are nonrigid, and the image is similar to the original figure.
- Question 4 prepares students for content in the formal lesson.

Connecting to Next Step

- The explorations should be a review of rigid and nonrigid transformations as well as the definition of similar figures. Quickly transition to the formal lesson.

8.2 Similarity and Transformations

Essential Question When a figure is translated, reflected, rotated, or dilated in the plane, is the image always similar to the original figure?

Two figures are *similar figures* when they have the same shape but not necessarily the same size.

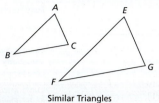

Similar Triangles

ATTENDING TO PRECISION
To be proficient in math, you need to use clear definitions in discussions with others and in your own reasoning.

EXPLORATION 1 Dilations and Similarity

Work with a partner.

a. Use dynamic geometry software to draw any triangle and label it △ABC.

b. Dilate the triangle using a scale factor of 3. Is the image similar to the original triangle? Justify your answer.

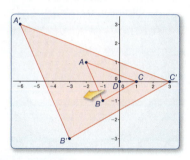

Sample
Points
A(−2, 1)
B(−1, −1)
C(1, 0)
D(0, 0)
Segments
AB = 2.24
BC = 2.24
AC = 3.16
Angles
m∠A = 45°
m∠B = 90°
m∠C = 45°

EXPLORATION 2 Rigid Motions and Similarity

Work with a partner.

a. Use dynamic geometry software to draw any triangle.

b. Copy the triangle and translate it 3 units left and 4 units up. Is the image similar to the original triangle? Justify your answer.

c. Reflect the triangle in the y-axis. Is the image similar to the original triangle? Justify your answer.

d. Rotate the original triangle 90° counterclockwise about the origin. Is the image similar to the original triangle? Justify your answer.

Communicate Your Answer

3. When a figure is translated, reflected, rotated, or dilated in the plane, is the image always similar to the original figure? Explain your reasoning.

4. A figure undergoes a composition of transformations, which includes translations, reflections, rotations, and dilations. Is the image similar to the original figure? Explain your reasoning.

Dynamic Teaching Tools
Dynamic Assessment & Progress Monitoring Tool
Lesson Planning Tool
Interactive Whiteboard Lesson Library
Dynamic Classroom with Dynamic Investigations

ANSWERS

1. a. Check students' work.
   b. Check students' work; yes; Each side of △A′B′C′ is three times as long as its corresponding side of △ABC. The corresponding angles are congruent. Because the corresponding sides are proportional and the corresponding angles are congruent, the image is similar to the original triangle.

2. a. Sample answer:

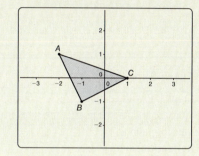

   b. Sample answer:

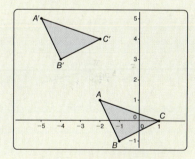

   yes; Because the corresponding sides are congruent and the corresponding angles are congruent, the image is similar to the original triangle.

2c–d. See Additional Answers.

3. yes; The corresponding sides are always congruent or proportional, and the corresponding angles are always congruent.

4. yes; According to Composition Theorem, the composition of two or more rigid motions is a rigid motion. Also, a dilation preserves angle measures and results in an image with lengths proportional to the preimage lengths. So, a composition of rigid motions or dilations will result in an image that has angle measures congruent to the corresponding angle measures of the original figure, and sides that are either congruent or proportional to the corresponding sides of the original figure.

Differentiated Instruction

Organization

Have students create a chart to compare the properties of congruence transformations and similarity transformations.

Transformation	Congruence	Similarity ($k \neq 1, -1$)
Preserves Angle Congruence		
Preserves Length		
Preserves Shape		
Preserves Size		

Extra Example 1

Graph $\overline{AB}$ with endpoints $A(12, -6)$ and $B(0, -3)$ and its image after the similarity transformation.
Reflection: in the y-axis
Dilation: $(x, y) \rightarrow \left(\frac{1}{3}x, \frac{1}{3}y\right)$

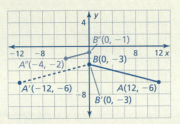

MONITORING PROGRESS ANSWERS

1.

2.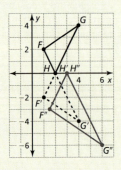

8.2 Lesson

Core Vocabulary

similarity transformation, *p. 470*
similar figures, *p. 470*

What You Will Learn

▶ Perform similarity transformations.
▶ Describe similarity transformations.
▶ Prove that figures are similar.

Performing Similarity Transformations

A dilation is a transformation that preserves shape but not size. So, a dilation is a nonrigid motion. A **similarity transformation** is a dilation or a composition of rigid motions and dilations. Two geometric figures are **similar figures** if and only if there is a similarity transformation that maps one of the figures onto the other. Similar figures have the same shape but not necessarily the same size.

Congruence transformations preserve length and angle measure. When the scale factor of the dilation(s) is not equal to 1 or -1, similarity transformations preserve angle measure only.

EXAMPLE 1 Performing a Similarity Transformation

Graph $\triangle ABC$ with vertices $A(-4, 1)$, $B(-2, 2)$, and $C(-2, 1)$ and its image after the similarity transformation.

Translation: $(x, y) \rightarrow (x + 5, y + 1)$
Dilation: $(x, y) \rightarrow (2x, 2y)$

SOLUTION

Step 1 Graph $\triangle ABC$.

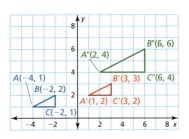

Step 2 Translate $\triangle ABC$ 5 units right and 1 unit up. $\triangle A'B'C'$ has vertices $A'(1, 2)$, $B'(3, 3)$, and $C'(3, 2)$.

Step 3 Dilate $\triangle A'B'C'$ using a scale factor of 2. $\triangle A''B''C''$ has vertices $A''(2, 4)$, $B''(6, 6)$, and $C''(6, 4)$.

Monitoring Progress Help in English and Spanish at *BigIdeasMath.com*

1. Graph $\overline{CD}$ with endpoints $C(-2, 2)$ and $D(2, 2)$ and its image after the similarity transformation.
   Rotation: 90° about the origin
   Dilation: $(x, y) \rightarrow \left(\frac{1}{2}x, \frac{1}{2}y\right)$

2. Graph $\triangle FGH$ with vertices $F(1, 2)$, $G(4, 4)$, and $H(2, 0)$ and its image after the similarity transformation.
   Reflection: in the x-axis
   Dilation: $(x, y) \rightarrow (1.5x, 1.5y)$

STUDY TIP
In this chapter, all rotations are counterclockwise unless otherwise noted.

Laurie's Notes — Teacher Actions

- Discuss dilations—they are nonrigid, meaning that the size will change and the scale factor k does not equal 1. Similar figures on the other hand can be congruent.
- Note that *similarity transformation* is defined and then used to determine whether two figures are similar. Again note the "if and only if" portion of the statement. To help demonstrate this, draw two similar trapezoids, rotating one 90°, on the board. The figures are similar *if* there are similarity transformations mapping one to the other, *and if* the figures are similar, then there are similarity transformations that would map one to the other.
- Pose Example 1 and ask for a *Thumbs Up* indication when partners are ready to begin the example.
- **Monitoring Progress:** Using whiteboards, have half the class do Question 1 and the other half do Question 2.

Describing Similarity Transformations

EXAMPLE 2 **Describing a Similarity Transformation**

Describe a similarity transformation that maps trapezoid *PQRS* to trapezoid *WXYZ*.

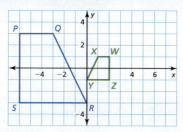

SOLUTION

$\overline{QR}$ falls from left to right, and $\overline{XY}$ rises from left to right. If you reflect trapezoid *PQRS* in the *y*-axis as shown, then the image, trapezoid $P'Q'R'S'$, will have the same orientation as trapezoid *WXYZ*.

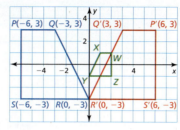

Trapezoid *WXYZ* appears to be about one-third as large as trapezoid $P'Q'R'S'$. Dilate trapezoid $P'Q'R'S'$ using a scale factor of $\frac{1}{3}$.

$$(x, y) \to \left(\tfrac{1}{3}x, \tfrac{1}{3}y\right)$$

$P'(6, 3) \to P''(2, 1)$

$Q'(3, 3) \to Q''(1, 1)$

$R'(0, -3) \to R''(0, -1)$

$S'(6, -3) \to S''(2, -1)$

The vertices of trapezoid $P''Q''R''S''$ match the vertices of trapezoid *WXYZ*.

 So, a similarity transformation that maps trapezoid *PQRS* to trapezoid *WXYZ* is a reflection in the *y*-axis followed by a dilation with a scale factor of $\frac{1}{3}$.

Monitoring Progress 🔊 Help in English and Spanish at *BigIdeasMath.com*

3. In Example 2, describe another similarity transformation that maps trapezoid *PQRS* to trapezoid *WXYZ*.

4. Describe a similarity transformation that maps quadrilateral *DEFG* to quadrilateral *STUV*.

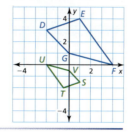

Extra Example 2
Describe a similarity transformation that maps trapezoid *WXYZ* to trapezoid *PQRS*.

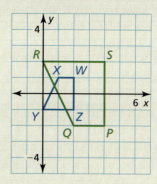

Sample answer: A reflection in the *x*-axis followed by a dilation with a scale factor of 2.

MONITORING PROGRESS ANSWERS

3. *Sample answer:* reflection in the *x*-axis followed by a dilation with a scale factor of $-\frac{1}{3}$

4. *Sample answer:* dilation with a scale factor of $\frac{1}{2}$ followed by a 180° rotation about the origin

---

Laurie's Notes — Teacher Actions

- Describing a similarity transformation is more challenging than performing a similarity transformation. Students are often uncertain where to begin.
- **Teaching Tip:** Have students focus on a pair of corresponding sides. What dilation about the origin will make them the same size? Does the orientation need to change (reflection)? Does the location need to change (translation)? Does the position need to change (rotation)?
- **Alternate Approach:** Have students use dynamic geometry software to answer the question.
- **Monitoring Progress:** Question 3 suggests that the solution (similarity transformation) is not unique! This is a *Big Idea* for students to explore.
- **Use Appropriate Tools Strategically:** If students are using dynamic geometry software, take time for them to generate a sample of different similarity transformations for Example 2.

Extra Example 3

Prove that square *ABCD* is similar to square *EFGH*.

Given Square *ABCD* with side length *s*, square *EFGH* with side length 2*s*, $\overline{AD} \parallel \overline{EH}$

Prove Square *ABCD* is similar to square *EFGH*.

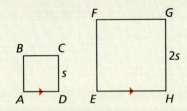

See Additional Answers.

MONITORING PROGRESS ANSWER

5. See Additional Answers.

Proving Figures Are Similar

To prove that two figures are similar, you must prove that a similarity transformation maps one of the figures onto the other.

EXAMPLE 3 Proving That Two Squares Are Similar

Prove that square *ABCD* is similar to square *EFGH*.

Given Square *ABCD* with side length *r*, square *EFGH* with side length *s*, $\overline{AD} \parallel \overline{EH}$

Prove Square *ABCD* is similar to square *EFGH*.

SOLUTION

Translate square *ABCD* so that point *A* maps to point *E*. Because translations map segments to parallel segments and $\overline{AD} \parallel \overline{EH}$, the image of $\overline{AD}$ lies on $\overline{EH}$.

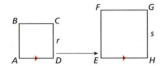

Because translations preserve length and angle measure, the image of *ABCD*, *EB'C'D'*, is a square with side length *r*. Because all the interior angles of a square are right angles, $\angle B'ED' \cong \angle FEH$. When $\overrightarrow{ED'}$ coincides with $\overrightarrow{EH}$, $\overrightarrow{EB'}$ coincides with $\overrightarrow{EF}$. So, $\overline{EB'}$ lies on $\overline{EF}$. Next, dilate square *EB'C'D'* using center of dilation *E*. Choose the scale factor to be the ratio of the side lengths of *EFGH* and *EB'C'D'*, which is $\frac{s}{r}$.

This dilation maps $\overline{ED'}$ to $\overline{EH}$ and $\overline{EB'}$ to $\overline{EF}$ because the images of $\overline{ED'}$ and $\overline{EB'}$ have side length $\frac{s}{r}(r) = s$ and the segments $\overline{ED'}$ and $\overline{EB'}$ lie on lines passing through the center of dilation. So, the dilation maps *B'* to *F* and *D'* to *H*. The image of *C'* lies $\frac{s}{r}(r) = s$ units to the right of the image of *B'* and $\frac{s}{r}(r) = s$ units above the image of *D'*. So, the image of *C'* is *G*.

▶ A similarity transformation maps square *ABCD* to square *EFGH*. So, square *ABCD* is similar to square *EFGH*.

Monitoring Progress Help in English and Spanish at BigIdeasMath.com

5. Prove that △*JKL* is similar to △*MNP*.

   Given Right isosceles △*JKL* with leg length *t*, right isosceles △*MNP* with leg length *v*, $\overline{LJ} \parallel \overline{PM}$

   Prove △*JKL* is similar to △*MNP*.

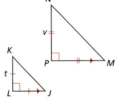

Laurie's Notes — Teacher Actions

- Remind students of the definition of a square.
- **"How can you prove that two squares are similar?"** You have to show that there is a similarity transformation that maps one square onto the other square.
- The proof may seem unnecessarily long and wordy to students, but it models what it means to prove that two figures are similar. We need to show that there is a sequence of similarity transformations that map square *ABCD* onto square *EFGH*. Note that if $\overline{AD}$ and $\overline{EH}$ were not given to be parallel, then a rotation would have been necessary as well.
- Work through the proof as shown. Ask, "Is there another series of similarity transformations that map square *ABCD* onto square *EFGH*? Explain." yes; One possibility would be to dilate first and then translate.

Closure

- **3-2-1:** Hand out a *3-2-1* reflection sheet as described on page T-468.

8.2 Exercises

Dynamic Solutions available at *BigIdeasMath.com*

Vocabulary and Core Concept Check

1. **VOCABULARY** What is the difference between *similar figures* and *congruent figures*?

2. **COMPLETE THE SENTENCE** A transformation that produces a similar figure, such as a dilation, is called a _____.

Monitoring Progress and Modeling with Mathematics

In Exercises 3–6, graph △FGH with vertices $F(-2, 2)$, $G(-2, -4)$, and $H(-4, -4)$ and its image after the similarity transformation. *(See Example 1.)*

3. Translation: $(x, y) \to (x + 3, y + 1)$
   Dilation: $(x, y) \to (2x, 2y)$

4. Dilation: $(x, y) \to \left(\frac{1}{2}x, \frac{1}{2}y\right)$
   Reflection: in the *y*-axis

5. Rotation: 90° about the origin
   Dilation: $(x, y) \to (3x, 3y)$

6. Dilation: $(x, y) \to \left(\frac{3}{4}x, \frac{3}{4}y\right)$
   Reflection: in the *x*-axis

In Exercises 7 and 8, describe a similarity transformation that maps the blue preimage to the green image. *(See Example 2.)*

7.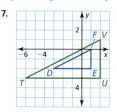

8.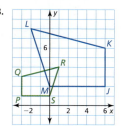

In Exercises 9–12, determine whether the polygons with the given vertices are similar. Use transformations to explain your reasoning.

9. $A(6, 0)$, $B(9, 6)$, $C(12, 6)$ and $D(0, 3)$, $E(1, 5)$, $F(2, 5)$

10. $Q(-1, 0)$, $R(-2, 2)$, $S(1, 3)$, $T(2, 1)$ and $W(0, 2)$, $X(4, 4)$, $Y(6, -2)$, $Z(2, -4)$

11. $G(-2, 3)$, $H(4, 3)$, $I(4, 0)$ and $J(1, 0)$, $K(6, -2)$, $L(1, -2)$

12. $D(-4, 3)$, $E(-2, 3)$, $F(-1, 1)$, $G(-4, 1)$ and $L(1, -1)$, $M(3, -1)$, $N(6, -3)$, $P(1, -3)$

In Exercises 13 and 14, prove that the figures are similar. *(See Example 3.)*

13. **Given** Right isosceles △ABC with leg length *j*, right isosceles △RST with leg length *k*, $\overline{CA} \parallel \overline{RT}$
    Prove △ABC is similar to △RST.

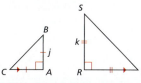

14. **Given** Rectangle JKLM with side lengths *x* and *y*, rectangle QRST with side lengths 2*x* and 2*y*
    Prove Rectangle JKLM is similar to rectangle QRST.

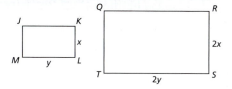

Section 8.2 Similarity and Transformations 473

Assignment Guide and Homework Check

ASSIGNMENT
Basic: 1, 2, 3–15 odd, 16–18, 23–26
Average: 1, 2, 4–14 even, 16–18, 21–26
Advanced: 1, 2, 8, 12, 14–18, 20–26

HOMEWORK CHECK
Basic: 5, 7, 9, 18
Average: 8, 10, 14, 18
Advanced: 8, 14, 18, 20

ANSWERS

1. Congruent figures have the same size and shape. Similar figures have the same shape, but not necessarily the same size.

2. similarity transformation

3.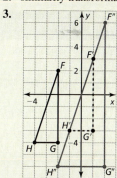

4.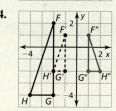

5.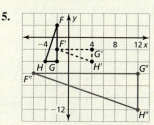

6.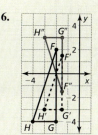

7. Sample answer: translation 1 unit down and 1 unit right followed by a dilation with center at $E(2, -3)$ and a scale factor of 2

8. Sample answer: dilation with center at the origin and a scale factor of $\frac{1}{2}$ followed by a reflection in the *y*-axis

9. yes; △ABC can be mapped to △DEF by a dilation with center at the origin and a scale factor of $\frac{1}{3}$ followed by a translation of 2 units left and 3 units up.

10. yes; □QRST can be mapped to □WXYZ by a 270° rotation about the origin followed by a dilation with center at the origin and a scale factor of 2.

11. no; The scale factor from $\overline{HI}$ to $\overline{JL}$ is $\frac{2}{3}$, but the scale factor from $\overline{GH}$ to $\overline{KL}$ is $\frac{5}{6}$.

12. no; The scale factor from $\overline{DG}$ to $\overline{LP}$ is 1, but the scale factor from $\overline{FG}$ to $\overline{NP}$ is $\frac{5}{3}$.

13–14. See Additional Answers.

Section 8.2 473

Dynamic Teaching Tools

Dynamic Assessment & Progress Monitoring Tool
Interactive Whiteboard Lesson Library
Dynamic Classroom with Dynamic Investigations

ANSWERS

15. yes; The stop sign sticker can be mapped to the regular-sized stop sign by translating the sticker to the left until the centers match, and then dilating the sticker with a scale factor of 3.15. Because there is a similarity transformation that maps one stop sign to the other, the sticker is similar to the regular-sized stop sign.

16. See Additional Answers.

17. no; The scale factor is 6 for both dimensions. So, the enlarged banner is proportional to the smaller one.

18–26. See Additional Answers.

Mini-Assessment

1. $\overline{AB}$ has endpoints $A(-8, 6)$ and $B(6, 0)$. Find the endpoints of its image after the similarity transformation.
   Translation: $(x, y) \rightarrow (x, y - 6)$
   Dilation: $(x, y) \rightarrow \left(\frac{1}{2}x, \frac{1}{2}y\right)$

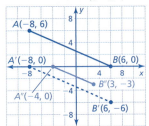

2. Describe a similarity transformation that maps quadrilateral STUV to quadrilateral DEFG.

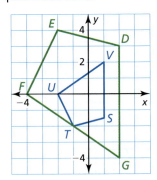

Sample answer: A reflection in the x-axis followed by a dilation with a scale factor of 2

---

15. **MODELING WITH MATHEMATICS** Determine whether the regular-sized stop sign and the stop sign sticker are similar. Use transformations to explain your reasoning.

16. **ERROR ANALYSIS** Describe and correct the error in comparing the figures.

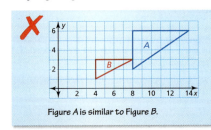

Figure A is similar to Figure B.

17. **MAKING AN ARGUMENT** A member of the homecoming decorating committee gives a printing company a banner that is 3 inches by 14 inches to enlarge. The committee member claims the banner she receives is distorted. Do you think the printing company distorted the image she gave it? Explain.

18. **HOW DO YOU SEE IT?** Determine whether each pair of figures is similar. Explain your reasoning.

19. **ANALYZING RELATIONSHIPS** Graph a polygon in a coordinate plane. Use a similarity transformation involving a dilation (where k is a whole number) and a translation to graph a second polygon. Then describe a similarity transformation that maps the second polygon onto the first.

20. **THOUGHT PROVOKING** Is the composition of a rotation and a dilation commutative? (In other words, do you obtain the same image regardless of the order in which you perform the transformations?) Justify your answer.

21. **MATHEMATICAL CONNECTIONS** Quadrilateral JKLM is mapped to quadrilateral $J'K'L'M'$ using the dilation $(x, y) \rightarrow \left(\frac{3}{2}x, \frac{3}{2}y\right)$. Then quadrilateral $J'K'L'M'$ is mapped to quadrilateral $J''K''L''M''$ using the translation $(x, y) \rightarrow (x + 3, y - 4)$. The vertices of quadrilateral $J'K'L'M'$ are $J'(-12, 0)$, $K'(-12, 18)$, $L'(-6, 18)$, and $M'(-6, 0)$. Find the coordinates of the vertices of quadrilateral JKLM and quadrilateral $J''K''L''M''$. Are quadrilateral JKLM and quadrilateral $J''K''L''M''$ similar? Explain.

22. **REPEATED REASONING** Use the diagram.

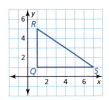

a. Connect the midpoints of the sides of $\triangle QRS$ to make another triangle. Is this triangle similar to $\triangle QRS$? Use transformations to support your answer.

b. Repeat part (a) for two other triangles. What conjecture can you make?

Maintaining Mathematical Proficiency — Reviewing what you learned in previous grades and lessons

Classify the angle as *acute*, *obtuse*, *right*, or *straight*. *(Skills Review Handbook)*

23. 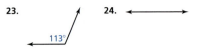 24. 25.   26.

474   Chapter 8   Similarity

If students need help...	If students got it...
Resources by Chapter • Practice A and Practice B • Puzzle Time	Resources by Chapter • Enrichment and Extension • Cumulative Review
Student Journal • Practice	Start the *next* Section
Differentiating the Lesson Skills Review Handbook	

Laurie's Notes

Overview of Section 8.3

Introduction
- Students should have recall of relationships between similar polygons from middle school.
- In this lesson, corresponding parts of similar triangles are defined by a similarity transformation. Corresponding side lengths are proportional, and corresponding angles are congruent.
- Perimeters and areas of similar polygons are calculated and used to solve real-life problems.

Resources
- Dynamic geometry software is helpful in demonstrating the connections between dilations and similar polygons. It is also helpful in developing spatial skills.

Teaching Strategy
- In middle school, students studied the relationship between the perimeters of similar figures and the areas of similar figures. These concepts are difficult for students to remember and perhaps some of the reason is that different units are involved (linear and square units).
- In the explorations, students will investigate properties of similar figures. Students may need to have a tactile experience as well.
- When the dimensions of a polygon are increased by a scale factor of k, it makes sense to students that the perimeter increases by a scale factor of k. It does not always make sense that the area increases by a scale factor of k^2.
- Square tiles can be used to investigate area of similar figures. You can also place a 3" × 5" index card under a document camera or on an overhead projector. Measure the dimensions of the projected image. Ask students whether the image and preimage are similar. Yes, the corresponding sides are proportional.
- I find that when students measure something so simple as a rectangle, and in this case the scale factor is rarely an integer, performing the computation of perimeter and area helps them better understand the difference between what happens to the perimeter and what happens to the area when a figure is enlarged (or reduced) proportionally.

Pacing Suggestion
- As students work through the explorations, listen and probe recall of prior knowledge, and then continue with the formal lesson. Be selective in the examples done based on students' recall of similar polygons.

Dynamic Teaching Tools
Dynamic Assessment & Progress Monitoring Tool
Lesson Planning Tool
Interactive Whiteboard Lesson Library
Dynamic Classroom with Dynamic Investigations

What Your Students Will Learn

- Use similarity statements to find the scale factor from one polygon to another polygon.
- Find corresponding lengths in similar polygons using proportions.
- Find perimeters and areas of similar polygons using proportions.
- Decide whether polygons are similar using ratios and rigid transformations.

Laurie's Notes

Exploration

Motivate
- Display an image of a Major League baseball diamond and a Little League baseball diamond.
- ? "Are these polygons similar? Explain." Students should be comfortable in knowing that all squares are similar.
- ? "How are the ratios of the perimeter of the base path to the distance from home plate to second base related?" The ratios are the same.
- ? "How is the the ratio of infield areas related to the ratio of the perimeters?" The ratio of the infield areas is the square of the ratio of the perimeters.
- Explain to students that in this lesson they will be working with relationships found in similar polygons.

For Your Information
- To save time, you might combine both explorations into one because the direction lines are the same.

Exploration 1
- Students should be very familiar with the functionality of dynamic geometry software. Note that the direction states that any scale factor and any center of dilation may be used.
- ? "What do you observe about the corresponding angles of similar triangles?" They are congruent.
- ? "What do you observe about the ratios of corresponding sides of similar triangles?" The ratios are all equal and the same as the scale factor.
- While the observations should not be new to students, this exploration helps students to review relationships found in similar triangles.

Exploration 2
- This exploration reviews the relationships about the perimeters and areas of similar triangles.
- **Extension:** You could have students begin with a quadrilateral versus a triangle.

Communicate Your Answer
- **Attend to Precision:** There are many relationships students should mention about similar polygons. Students should not say, "Similar polygons are the same shape but not necessarily the same size."

Connecting to Next Step
- The explorations should be a review of relationships found in similar polygons. Quickly transition to the formal lesson.

8.3 Similar Polygons

Essential Question How are similar polygons related?

EXPLORATION 1 Comparing Triangles after a Dilation

Work with a partner. Use dynamic geometry software to draw any △ABC. Dilate △ABC to form a similar △A'B'C' using any scale factor k and any center of dilation.

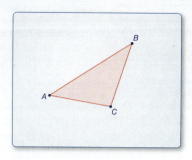

a. Compare the corresponding angles of △A'B'C' and △ABC.

b. Find the ratios of the lengths of the sides of △A'B'C' to the lengths of the corresponding sides of △ABC. What do you observe?

c. Repeat parts (a) and (b) for several other triangles, scale factors, and centers of dilation. Do you obtain similar results?

EXPLORATION 2 Comparing Triangles after a Dilation

Work with a partner. Use dynamic geometry software to draw any △ABC. Dilate △ABC to form a similar △A'B'C' using any scale factor k and any center of dilation.

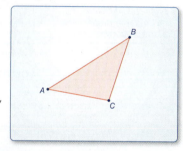

LOOKING FOR STRUCTURE

To be proficient in math, you need to look closely to discern a pattern or structure.

a. Compare the perimeters of △A'B'C' and △ABC. What do you observe?

b. Compare the areas of △A'B'C' and △ABC. What do you observe?

c. Repeat parts (a) and (b) for several other triangles, scale factors, and centers of dilation. Do you obtain similar results?

Communicate Your Answer

3. How are similar polygons related?

4. A △RST is dilated by a scale factor of 3 to form △R'S'T'. The area of △RST is 1 square inch. What is the area of △R'S'T'?

Section 8.3 Similar Polygons 475

Dynamic Teaching Tools

Dynamic Assessment & Progress Monitoring Tool
Lesson Planning Tool
Interactive Whiteboard Lesson Library
Dynamic Classroom with Dynamic Investigations

ANSWERS

1. a. The corresponding angles are congruent.
   b. The ratios are equal to the scale factor.
   c. yes
2. a. The perimeter of △A'B'C' is equal to k times the perimeter of △ABC.
   b. The area of △A'B'C' is equal to k^2 times the area of △ABC.
   c. yes
3. Corresponding angles are congruent, and corresponding side lengths are proportional.
4. 9 in.^{2}

English Language Learners

Vocabulary
Students may be familiar with the word *similar* as it is used in everyday language, meaning "alike" or "comparable." Review with students the fact that in mathematics, some words, like *similar*, have meanings different from their meanings in everyday language. Discuss the mathematical meaning of *similar*.

Extra Example 1
In the diagram, $\triangle ABC \sim \triangle JKL$.

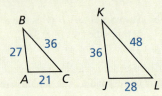

a. Find the scale factor from $\triangle ABC$ to $\triangle JKL$. $\frac{4}{3}$
b. List all pairs of congruent angles.
$\angle A \cong \angle J, \angle B \cong \angle K, \angle C \cong \angle L$
c. Write the ratios of the corresponding side lengths in a *statement of proportionality*.
Sample answer: $\frac{JK}{AB} = \frac{KL}{BC} = \frac{LJ}{CA}$

MONITORING PROGRESS ANSWER

1. $\frac{3}{2}$; $\angle J \cong \angle P$, $\angle K \cong \angle Q$,
$\angle L \cong \angle R$; $\frac{PQ}{JK} = \frac{QR}{KL} = \frac{PR}{JL}$

8.3 Lesson

What You Will Learn
- Use similarity statements.
- Find corresponding lengths in similar polygons.
- Find perimeters and areas of similar polygons.
- Decide whether polygons are similar.

Core Vocabulary
Previous
similar figures
similarity transformation
corresponding parts

Using Similarity Statements

Core Concept
Corresponding Parts of Similar Polygons
In the diagram below, $\triangle ABC$ is similar to $\triangle DEF$. You can write "$\triangle ABC$ is similar to $\triangle DEF$" as $\triangle ABC \sim \triangle DEF$. A similarity transformation preserves angle measure. So, corresponding angles are congruent. A similarity transformation also enlarges or reduces side lengths by a scale factor k. So, corresponding side lengths are proportional.

Corresponding angles
$\angle A \cong \angle D, \angle B \cong \angle E, \angle C \cong \angle F$

Ratios of corresponding side lengths
$\frac{DE}{AB} = \frac{EF}{BC} = \frac{FD}{CA} = k$

LOOKING FOR STRUCTURE
Notice that any two congruent figures are also similar. In $\triangle LMN$ and $\triangle WXY$ below, the scale factor is $\frac{5}{5} = \frac{6}{6} = \frac{7}{7} = 1$. So, you can write $\triangle LMN \sim \triangle WXY$ and $\triangle LMN \cong \triangle WXY$.

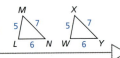

READING
In a *statement of proportionality*, any pair of ratios forms a true proportion.

EXAMPLE 1 Using Similarity Statements

In the diagram, $\triangle RST \sim \triangle XYZ$.

a. Find the scale factor from $\triangle RST$ to $\triangle XYZ$.
b. List all pairs of congruent angles.
c. Write the ratios of the corresponding side lengths in a *statement of proportionality*.

SOLUTION

a. $\frac{XY}{RS} = \frac{12}{20} = \frac{3}{5}$ $\frac{YZ}{ST} = \frac{18}{30} = \frac{3}{5}$ $\frac{ZX}{TR} = \frac{15}{25} = \frac{3}{5}$

So, the scale factor is $\frac{3}{5}$.

b. $\angle R \cong \angle X$, $\angle S \cong \angle Y$, and $\angle T \cong \angle Z$.

c. Because the ratios in part (a) are equal, $\frac{XY}{RS} = \frac{YZ}{ST} = \frac{ZX}{TR}$.

Monitoring Progress 🔊 Help in English and Spanish at *BigIdeasMath.com*

1. In the diagram, $\triangle JKL \sim \triangle PQR$. Find the scale factor from $\triangle JKL$ to $\triangle PQR$. Then list all pairs of congruent angles and write the ratios of the corresponding side lengths in a statement of proportionality.

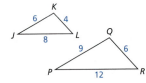

476   Chapter 8   Similarity

Laurie's Notes | Teacher Actions

- Write the *Core Concept*, noting the language.
- **Turn and Talk:** "Compare a congruency transformation (isometry) and a *similarity transformation*." After comparisons are discussed, ask, "Are congruent figures similar? Explain." Listen for correct reasoning.
- **Whiteboarding:** Have partners work independently on Example 1. Have partners share solutions. Circulate and listen to students' conversations.
- The *statement of proportionality* may be unfamiliar language.

Finding Corresponding Lengths in Similar Polygons

> **READING**
> Corresponding lengths in similar triangles include side lengths, altitudes, medians, and midsegments.

Core Concept

Corresponding Lengths in Similar Polygons

If two polygons are similar, then the ratio of any two corresponding lengths in the polygons is equal to the scale factor of the similar polygons.

EXAMPLE 2 Finding a Corresponding Length

In the diagram, $\triangle DEF \sim \triangle MNP$. Find the value of x.

SOLUTION

The triangles are similar, so the corresponding side lengths are proportional.

$\dfrac{MN}{DE} = \dfrac{NP}{EF}$ Write proportion.

$\dfrac{18}{15} = \dfrac{30}{x}$ Substitute.

$18x = 450$ Cross Products Property

$x = 25$ Solve for x.

▶ The value of x is 25.

> **FINDING AN ENTRY POINT**
> There are several ways to write the proportion. For example, you could write $\dfrac{DF}{MP} = \dfrac{EF}{NP}$.

EXAMPLE 3 Finding a Corresponding Length

In the diagram, $\triangle TPR \sim \triangle XPZ$. Find the length of the altitude $\overline{PS}$.

SOLUTION

First, find the scale factor from $\triangle XPZ$ to $\triangle TPR$.

$\dfrac{TR}{XZ} = \dfrac{6+6}{8+8} = \dfrac{12}{16} = \dfrac{3}{4}$

Because the ratio of the lengths of the altitudes in similar triangles is equal to the scale factor, you can write the following proportion.

$\dfrac{PS}{PY} = \dfrac{3}{4}$ Write proportion.

$\dfrac{PS}{20} = \dfrac{3}{4}$ Substitute 20 for PY.

$PS = 15$ Multiply each side by 20 and simplify.

▶ The length of the altitude $\overline{PS}$ is 15.

Monitoring Progress 🔊 Help in English and Spanish at BigIdeasMath.com

2. Find the value of x.

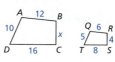

$ABCD \sim QRST$

3. Find KM.

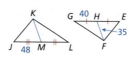

$\triangle JKL \sim \triangle EFG$

Section 8.3 Similar Polygons 477

Differentiated Instruction

Inclusion
In Example 2, point out that there are many ways to write a proportion correctly. Ask students to identify other correct proportions in $\triangle DEF$ and $\triangle MNP$. Have them write these proportions on the board.

Extra Example 2
In the diagram, $\triangle GHM \sim \triangle HKL$. Find the value of x.

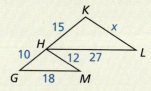

The value of x is 18.

Extra Example 3
In the diagram, $\triangle UVW \sim \triangle QRS$. Find the length of the median $\overline{ST}$.

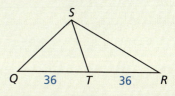

The length of the median $\overline{ST}$ is 27.

MONITORING PROGRESS ANSWERS
2. $x = 8$
3. $KM = 42$

Laurie's Notes   Teacher Actions

- **Common Misconception:** The *Core Concept* refers to corresponding lengths in similar polygons. Lengths are not restricted to only side lengths. Corresponding lengths also refer to special segments, such as altitudes, medians, and so on, as well as corresponding diagonals.
- Have partners work independently on Examples 2 and 3 and then compare how the proportions were set up.
- **Construct Viable Arguments and Critique the Reasoning of Others:** As students share their results, they should justify why they wrote their proportions as they did.

Extra Example 4

Your neighbor has decided to enlarge his garden. The garden is rectangular with width 6 feet and length 15 feet. The new garden will be similar to the original one, but will have a length of 35 feet. Find the perimeter of the original garden and the enlarged garden. *The perimeter of the original garden is 42 feet. The perimeter of the enlarged garden is 98 feet.*

MONITORING PROGRESS ANSWER

4. 46 m

Finding Perimeters and Areas of Similar Polygons

Theorem

Perimeters of Similar Polygons

If two polygons are similar, then the ratio of their perimeters is equal to the ratios of their corresponding side lengths.

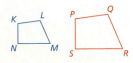

If $KLMN \sim PQRS$, then $\dfrac{PQ + QR + RS + SP}{KL + LM + MN + NK} = \dfrac{PQ}{KL} = \dfrac{QR}{LM} = \dfrac{RS}{MN} = \dfrac{SP}{NK}$.

Proof Ex. 52, p. 484; *BigIdeasMath.com*

ANALYZING RELATIONSHIPS

When two similar polygons have a scale factor of k, the ratio of their perimeters is equal to k.

EXAMPLE 4 Modeling with Mathematics

A town plans to build a new swimming pool. An Olympic pool is rectangular with a length of 50 meters and a width of 25 meters. The new pool will be similar in shape to an Olympic pool but will have a length of 40 meters. Find the perimeters of an Olympic pool and the new pool.

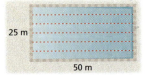

SOLUTION

1. **Understand the Problem** You are given the length and width of a rectangle and the length of a similar rectangle. You need to find the perimeters of both rectangles.

2. **Make a Plan** Find the scale factor of the similar rectangles and find the perimeter of an Olympic pool. Then use the Perimeters of Similar Polygons Theorem to write and solve a proportion to find the perimeter of the new pool.

3. **Solve the Problem** Because the new pool will be similar to an Olympic pool, the scale factor is the ratio of the lengths, $\frac{40}{50} = \frac{4}{5}$. The perimeter of an Olympic pool is $2(50) + 2(25) = 150$ meters. Write and solve a proportion to find the perimeter x of the new pool.

   $\dfrac{x}{150} = \dfrac{4}{5}$ Perimeters of Similar Polygons Theorem

   $x = 120$ Multiply each side by 150 and simplify.

   So, the perimeter of an Olympic pool is 150 meters, and the perimeter of the new pool is 120 meters.

4. **Look Back** Check that the ratio of the perimeters is equal to the scale factor.

   $\dfrac{120}{150} = \dfrac{4}{5}$ ✓

STUDY TIP

You can also write the scale factor as a decimal. In Example 4, you can write the scale factor as 0.8 and multiply by 150 to get $x = 0.8(150) = 120$.

Monitoring Progress Help in English and Spanish at *BigIdeasMath.com*

4. The two gazebos shown are similar pentagons. Find the perimeter of Gazebo A.

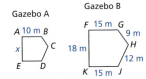

Laurie's Notes — Teacher Actions

- See the *Teaching Strategy* on page T-474.
- **?** Write the Perimeters of Similar Polygons Theorem. "What other name could you give for the ratio of corresponding side lengths?" scale factor
- **?** "Do you need to find the width of the new pool in order to find the perimeter? Explain." no; Because you can determine the perimeter of the Olympic-sized pool and the scale factor of the similar rectangles, you can find the perimeter of the new pool.
- Note that the *Study Tip* in the side margin suggests another way to solve the problem.

ANALYZING RELATIONSHIPS

When two similar polygons have a scale factor of k, the ratio of their areas is equal to k^2.

Theorem

Areas of Similar Polygons

If two polygons are similar, then the ratio of their areas is equal to the squares of the ratios of their corresponding side lengths.

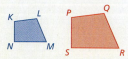

If $KLMN \sim PQRS$, then $\dfrac{\text{Area of } PQRS}{\text{Area of } KLMN} = \left(\dfrac{PQ}{KL}\right)^2 = \left(\dfrac{QR}{LM}\right)^2 = \left(\dfrac{RS}{MN}\right)^2 = \left(\dfrac{SP}{NK}\right)^2$.

Proof Ex. 53, p. 484; BigIdeasMath.com

EXAMPLE 5 Finding Areas of Similar Polygons

In the diagram, $\triangle ABC \sim \triangle DEF$. Find the area of $\triangle DEF$.

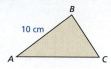

Area of $\triangle ABC = 36$ cm²

SOLUTION

Because the triangles are similar, the ratio of the area of $\triangle ABC$ to the area of $\triangle DEF$ is equal to the square of the ratio of AB to DE. Write and solve a proportion to find the area of $\triangle DEF$. Let A represent the area of $\triangle DEF$.

$\dfrac{\text{Area of } \triangle ABC}{\text{Area of } \triangle DEF} = \left(\dfrac{AB}{DE}\right)^2$ Areas of Similar Polygons Theorem

$\dfrac{36}{A} = \left(\dfrac{10}{5}\right)^2$ Substitute.

$\dfrac{36}{A} = \dfrac{100}{25}$ Square the right side of the equation.

$36 \cdot 25 = 100 \cdot A$ Cross Products Property

$900 = 100A$ Simplify.

$9 = A$ Solve for A.

▶ The area of $\triangle DEF$ is 9 square centimeters.

Monitoring Progress Help in English and Spanish at BigIdeasMath.com

5. In the diagram, $GHJK \sim LMNP$. Find the area of $LMNP$.

 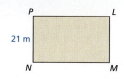

Area of $GHJK = 84$ m²

Section 8.3 Similar Polygons 479

Extra Example 5

In the diagram, $\triangle PQT \sim \triangle RST$, and the area of $\triangle RST$ is 75 square meters. Find the area of $\triangle PQT$.

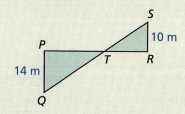

The area of $\triangle PQT$ is 147 square meters.

MONITORING PROGRESS ANSWER

5. 756 m²

Laurie's Notes   Teacher Actions

- See the *Teaching Strategy* on page T-474.
- Although the relationship between the areas of similar polygons was explored in middle school, students find it challenging to apply, particularly for non-integer scale factors.
- **Common Misconception:** Students are uncertain whether they can simplify a ratio written in a proportion. In Example 5, the ratio $\frac{10}{5}$ can be simplified to $\frac{2}{1}$. Have students work the example with and without simplifying the ratio to clarify the misconception.

Section 8.3   479

Extra Example 6
Decide whether *GNMH* and *MLKH* are similar. Explain your reasoning.

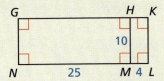

yes; Both quadrilaterals *GNMH* and *MLKH* are rectangles because each has four right angles. So, the corresponding angles are congruent. Sides $GN = HM = KL = 10$ because opposite sides in a rectangle are congruent. In rectangle *GNMH*, $\frac{GN}{NM} = \frac{10}{25} = \frac{2}{5}$, and in rectangle *MLKH*, $\frac{ML}{LK} = \frac{4}{10} = \frac{2}{5}$.

The lengths of the corresponding sides are in proportion. So, *GNMH* and *MLKH* are similar.

MONITORING PROGRESS ANSWERS

6. no; The blue hexagon is not regular, so corresponding side lengths are not proportional.

7. yes; Both hexagons are regular, so corresponding angles are congruent and corresponding side lengths are proportional.

Deciding Whether Polygons Are Similar

EXAMPLE 6 Deciding Whether Polygons Are Similar

Decide whether *ABCDE* and *KLQRP* are similar. Explain your reasoning.

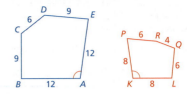

SOLUTION

Corresponding sides of the pentagons are proportional with a scale factor of $\frac{2}{3}$. However, this does not necessarily mean the pentagons are similar. A dilation with center *A* and scale factor $\frac{2}{3}$ moves *ABCDE* onto *AFGHJ*. Then a reflection moves *AFGHJ* onto *KLMNP*.

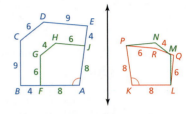

KLMNP does not exactly coincide with *KLQRP*, because not all the corresponding angles are congruent. (Only ∠*A* and ∠*K* are congruent.)

▶ Because angle measure is not preserved, the two pentagons are not similar.

Monitoring Progress Help in English and Spanish at *BigIdeasMath.com*

Refer to the floor tile designs below. In each design, the red shape is a regular hexagon.

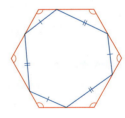 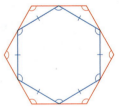

Tile Design 1   Tile Design 2

6. Decide whether the hexagons in Tile Design 1 are similar. Explain.
7. Decide whether the hexagons in Tile Design 2 are similar. Explain.

Laurie's Notes | Teacher Actions

? Agree-Disagree Statement: "The corresponding sides of two polygons are proportional, so the polygons are similar. Agree/Disagree. Explain." Students should discuss their understanding of similar figures. What students often forget is that corresponding angles must also be congruent.

- Now pose Example 6 for students to consider.

Closure

- **Writing Prompt:** If two polygons are similar, then …

8.3 Exercises

Dynamic Solutions available at BigIdeasMath.com

Vocabulary and Core Concept Check

1. **COMPLETE THE SENTENCE** For two figures to be similar, the corresponding angles must be _____, and the corresponding side lengths must be _____.

2. **DIFFERENT WORDS, SAME QUESTION** Which is different? Find "both" answers.

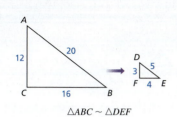

- What is the scale factor?
- What is the ratio of their areas?
- What is the ratio of their corresponding side lengths?
- What is the ratio of their perimeters?

△ABC ~ △DEF

Monitoring Progress and Modeling with Mathematics

In Exercises 3 and 4, find the scale factor. Then list all pairs of congruent angles and write the ratios of the corresponding side lengths in a statement of proportionality. *(See Example 1.)*

3. △ABC ~ △LMN

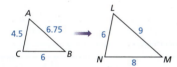

4. DEFG ~ PQRS

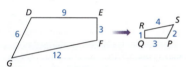

In Exercises 5–8, the polygons are similar. Find the value of x. *(See Example 2.)*

5.

6.

7.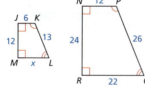

8.

Section 8.3 Similar Polygons 481

Assignment Guide and Homework Check

ASSIGNMENT

Basic: 1, 2, 3–25 odd, 29, 49, 50, 57–60

Average: 1, 2–38 even, 44, 49, 50, 57–60

Advanced: 1, 2, 4, 8, 10, 24–34 even, 38, 42, 48–60

HOMEWORK CHECK

Basic: 5, 9, 17, 21, 50

Average: 6, 10, 18, 26, 50

Advanced: 8, 10, 26, 38, 50

ANSWERS

1. congruent; proportional
2. What is the ratio of their areas?; $\frac{1}{16}$; $\frac{1}{4}$
3. $\frac{4}{3}$; $\angle A \cong \angle L$, $\angle B \cong \angle M$, $\angle C \cong \angle N$; $\frac{LM}{AB} = \frac{MN}{BC} = \frac{NL}{CA}$
4. $\frac{1}{3}$; $\angle D \cong \angle P$, $\angle E \cong \angle Q$, $\angle F \cong \angle R$, $\angle G \cong \angle S$; $\frac{PQ}{DE} = \frac{QR}{EF} = \frac{RS}{FG} = \frac{SP}{GD}$
5. $x = 30$
6. $x = 24$
7. $x = 11$
8. $x = 12$

Dynamic Teaching Tools
- Dynamic Assessment & Progress Monitoring Tool
- Interactive Whiteboard Lesson Library
- Dynamic Classroom with Dynamic Investigations

ANSWERS
9. altitude; 24
10. median; 9
11. 2 : 3
12. 4 : 3
13. 72 cm
14. 88 ft
15. 20 yd
16. 34 m
17. 288 ft, 259.2 ft
18. 130 ft, 52 ft
19. 108 ft^2
20. 90 cm^2
21. 4 in.^{2}
22. 6 cm^2
23. Because the first ratio has a side length of B over a side length of A, the second ratio should have the perimeter of B over the perimeter of A;
$$\frac{5}{10} = \frac{x}{28}$$
$$x = 14$$
24. The square of the ratio of their corresponding side lengths should be set equal to the ratio of their areas;
$$\left(\frac{6}{18}\right)^2 = \frac{24}{x}$$
$$x = 216$$

In Exercises 9 and 10, the black triangles are similar. Identify the type of segment shown in blue and find the value of the variable. *(See Example 3.)*

9.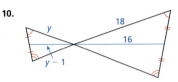

10.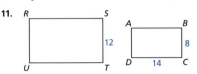

In Exercises 11 and 12, *RSTU ~ ABCD*. Find the ratio of their perimeters.

11.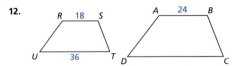

12.

In Exercises 13–16, two polygons are similar. The perimeter of one polygon and the ratio of the corresponding side lengths are given. Find the perimeter of the other polygon.

13. perimeter of smaller polygon: 48 cm; ratio: $\frac{2}{3}$
14. perimeter of smaller polygon: 66 ft; ratio: $\frac{3}{4}$
15. perimeter of larger polygon: 120 yd; ratio: $\frac{1}{6}$
16. perimeter of larger polygon: 85 m; ratio: $\frac{2}{5}$

17. **MODELING WITH MATHEMATICS** A school gymnasium is being remodeled. The basketball court will be similar to an NCAA basketball court, which has a length of 94 feet and a width of 50 feet. The school plans to make the width of the new court 45 feet. Find the perimeters of an NCAA court and of the new court in the school. *(See Example 4.)*

482   Chapter 8   Similarity

---

18. **MODELING WITH MATHEMATICS** Your family has decided to put a rectangular patio in your backyard, similar to the shape of your backyard. Your backyard has a length of 45 feet and a width of 20 feet. The length of your new patio is 18 feet. Find the perimeters of your backyard and of the patio.

In Exercises 19–22, the polygons are similar. The area of one polygon is given. Find the area of the other polygon. *(See Example 5.)*

19.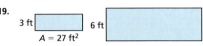

20.

21.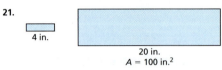

22.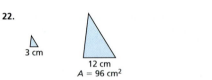

23. **ERROR ANALYSIS** Describe and correct the error in finding the perimeter of triangle B. The triangles are similar.

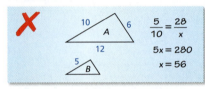

24. **ERROR ANALYSIS** Describe and correct the error in finding the area of rectangle B. The rectangles are similar.

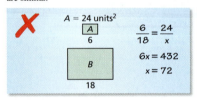

In Exercises 25 and 26, decide whether the red and blue polygons are similar. *(See Example 6.)*

25.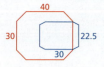

26.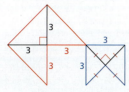

27. **REASONING** Triangles *ABC* and *DEF* are similar. Which statement is correct? Select all that apply.

    Ⓐ $\dfrac{BC}{EF} = \dfrac{AC}{DF}$ Ⓑ $\dfrac{AB}{DE} = \dfrac{CA}{FE}$

    Ⓒ $\dfrac{AB}{EF} = \dfrac{BC}{DE}$ Ⓓ $\dfrac{CA}{FD} = \dfrac{BC}{EF}$

ANALYZING RELATIONSHIPS In Exercises 28–34, *JKLM* ~ *EFGH*.

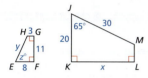

28. Find the scale factor of *JKLM* to *EFGH*.

29. Find the scale factor of *EFGH* to *JKLM*.

30. Find the values of *x*, *y*, and *z*.

31. Find the perimeter of each polygon.

32. Find the ratio of the perimeters of *JKLM* to *EFGH*.

33. Find the area of each polygon.

34. Find the ratio of the areas of *JKLM* to *EFGH*.

35. **USING STRUCTURE** Rectangle A is similar to rectangle B. Rectangle A has side lengths of 6 and 12. Rectangle B has a side length of 18. What are the possible values for the length of the other side of rectangle B? Select all that apply.

    Ⓐ 6   Ⓑ 9   Ⓒ 24   Ⓓ 36

36. **DRAWING CONCLUSIONS** In table tennis, the table is a rectangle 9 feet long and 5 feet wide. A tennis court is a rectangle 78 feet long and 36 feet wide. Are the two surfaces similar? Explain. If so, find the scale factor of the tennis court to the table.

MATHEMATICAL CONNECTIONS In Exercises 37 and 38, the two polygons are similar. Find the values of *x* and *y*.

37.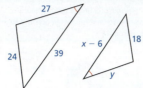

38.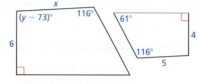

ATTENDING TO PRECISION In Exercises 39–42, the figures are similar. Find the missing corresponding side length.

39. Figure A has a perimeter of 72 meters and one of the side lengths is 18 meters. Figure B has a perimeter of 120 meters.

40. Figure A has a perimeter of 24 inches. Figure B has a perimeter of 36 inches and one of the side lengths is 12 inches.

41. Figure A has an area of 48 square feet and one of the side lengths is 6 feet. Figure B has an area of 75 square feet.

42. Figure A has an area of 18 square feet. Figure B has an area of 98 square feet and one of the side lengths is 14 feet.

ANSWERS

25. no; Corresponding angles are not congruent.
26. yes; Corresponding side lengths are proportional and corresponding angles are congruent.
27. A, D
28. $\dfrac{2}{5}$
29. $\dfrac{5}{2}$
30. $x = 27.5$, $y = 12$, $z = 65$
31. 34, 85
32. 5 : 2
33. 60.5, 378.125
34. 25 : 4
35. B, D
36. no; Corresponding side lengths are not proportional.
37. $x = 35.25$, $y = 20.25$
38. $x = 7.5$, $y = 166$
39. 30 m
40. 8 in.
41. 7.5 ft
42. 6 ft

ANSWERS
43. sometimes
44. sometimes
45. sometimes
46–60. See Additional Answers.

Mini-Assessment

1. In the diagram, △ABE ~ △DBC. Find the value of x.

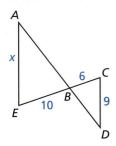

   x = 15

2. In the diagram, △NOW ~ △PQR. Identify the type of segment shown in blue and find the value of x.

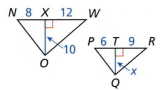

   $\overline{OX}$ and $\overline{QT}$ are altitudes; x = 7.5

3. The two rectangles are similar. The area of the smaller rectangle is 7.5 square inches. Find the area of the larger rectangle.

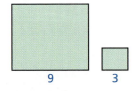

   67.5 square inches

4. True or false? All squares are similar. Explain your reasoning.
   True. A square has four right angles. So, the corresponding angles in all squares are congruent. A square has four congruent sides. If each side of one square has length s, and each side of a second square has length ks, then k is the scale factor, and the ratio between any two pairs of corresponding sides is $\frac{ks}{s} = k$. The squares are similar by definition.

484   Chapter 8

CRITICAL THINKING In Exercises 43–48, tell whether the polygons are *always*, *sometimes*, or *never* similar.

43. two isosceles triangles   44. two isosceles trapezoids

45. two rhombuses   46. two squares

47. two regular polygons

48. a right triangle and an equilateral triangle

49. **MAKING AN ARGUMENT** Your sister claims that when the side lengths of two rectangles are proportional, the two rectangles must be similar. Is she correct? Explain your reasoning.

50. **HOW DO YOU SEE IT?** You shine a flashlight directly on an object to project its image onto a parallel screen. Will the object and the image be similar? Explain your reasoning.

51. **MODELING WITH MATHEMATICS** During a total eclipse of the Sun, the moon is directly in line with the Sun and blocks the Sun's rays. The distance DA between Earth and the Sun is 93,000,000 miles, the distance DE between Earth and the moon is 240,000 miles, and the radius AB of the Sun is 432,500 miles. Use the diagram and the given measurements to estimate the radius EC of the moon.

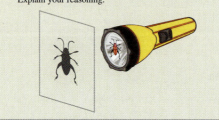

△DEC ~ △DAB   Not drawn to scale

52. **PROVING A THEOREM** Prove the Perimeters of Similar Polygons Theorem for similar rectangles. Include a diagram in your proof.

53. **PROVING A THEOREM** Prove the Areas of Similar Polygons Theorem for similar rectangles. Include a diagram in your proof.

54. **THOUGHT PROVOKING** The postulates and theorems you have learned represent Euclidean geometry. In spherical geometry, all points are points on the surface of a sphere. A line is a circle on the sphere whose diameter is equal to the diameter of the sphere. A plane is the surface of the sphere. In spherical geometry, is it possible that two triangles are similar but not congruent? Explain your reasoning.

55. **CRITICAL THINKING** In the diagram, PQRS is a square, and PLMS ~ LMRQ. Find the exact value of x. This value is called the *golden ratio*. Golden rectangles have their length and width in this ratio. Show that the similar rectangles in the diagram are golden rectangles.

56. **MATHEMATICAL CONNECTIONS** The equations of the lines shown are $y = \frac{4}{3}x + 4$ and $y = \frac{4}{3}x - 8$. Show that △AOB ~ △COD.

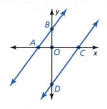

Maintaining Mathematical Proficiency Reviewing what you learned in previous grades and lessons

Find the value of x. *(Section 7.1)*

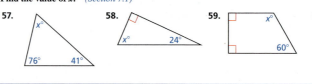

484   Chapter 8   Similarity

If students need help...	If students got it...
Resources by Chapter • Practice A and Practice B • Puzzle Time	Resources by Chapter • Enrichment and Extension • Cumulative Review
Student Journal • Practice	Start the *next* Section
Differentiating the Lesson Skills Review Handbook	

Laurie's Notes

Overview of Section 8.4

Introduction
- In Math I, students learned several ways to prove that two triangles are congruent. If two triangles are congruent, then they are similar. The converse is not true.
- This is a short lesson that introduces the first method for proving two triangles similar: Angle-Angle Similarity Theorem.
- Students will use AA to prove that triangles are similar. They will also apply the theorem in a real-life application to find the height of an object using indirect measure.

Resources
- Students will need a small mirror or a sunny day in order to find the height of an object using the strategy described below. A tape measure is needed for either method.

Teaching Strategy
- Students will enjoy being outside to find the height of an object using indirect measure. There are two common methods that use the Angle-Angle Similarity Theorem introduced in this lesson.
- On a sunny day, measure the shadows cast by the object and the student. Measure the student's height.
- **?** "Why are the two triangles similar?" Each triangle has a right angle, and the angle of elevation to the sun is the same for both triangles.

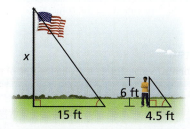

- A second method is to place a mirror on the ground between the person and the object. The student should position himself or herself so that he or she can see the top of the object in the middle of the mirror.
- **?** "Why are the two triangles similar?" Each triangle has a right angle, and the angle of incidence and the angle of reflection are the same for both triangles.

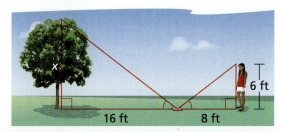

Pacing Suggestion
- Once students have worked the exploration, continue with the formal lesson.

Dynamic Teaching Tools
Dynamic Assessment & Progress Monitoring Tool
Lesson Planning Tool
Interactive Whiteboard Lesson Library
Dynamic Classroom with Dynamic Investigations

What Your Students Will Learn

- Use the Angle-Angle Similarity Theorem to determine whether two triangles are similar.
- Solve real-life problems using indirect measurement.

Laurie's Notes

Exploration

Motivate
- Refer to an object outside your school building, such as a flagpole, tree, or building.
- ❓ "How tall do you think the object is?" Answers will vary.
- Explain to students that in today's lesson they will learn a technique for measuring the height of an object using a method that uses *indirect measure.*

Exploration Note
- In Math I, students proved that if two angles of one triangle are congruent to two angles of a second triangle, then the third angles are congruent. In this exploration, they take it one step further and investigate the side lengths of the two triangles.

Exploration 1
- Constructing a pair of triangles with the specified angle measures should not take students very long.
- If the labels for the vertices of their triangles are not the same as what is shown in the table, either rename the labels on their construction or rename angles and segments in the table.
- Be sure that students do not omit part (e) of the exploration.
- When students have had sufficient time to gather data, solicit conjectures and justifications.

Communicate Your Answer
- Expect students to offer conclusions similar to the conjectures made at the end of the exploration.

Connecting to Next Step
- The exploration allows students to investigate the side lengths of two triangles that share two pairs of congruent angles. The Angle-Angle Similarity Theorem is stated in the formal lesson.

8.4 Proving Triangle Similarity by AA

Essential Question What can you conclude about two triangles when you know that two pairs of corresponding angles are congruent?

EXPLORATION 1   Comparing Triangles

Work with a partner. Use dynamic geometry software.

a. Construct △ABC and △DEF so that $m\angle A = m\angle D = 106°$, $m\angle B = m\angle E = 31°$, and △DEF is not congruent to △ABC.

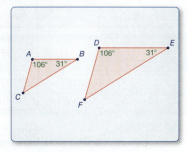

b. Find the third angle measure and the side lengths of each triangle. Copy the table below and record your results in column 1.

	1.	2.	3.	4.	5.	6.
m∠A, m∠D	106°	88°	40°			
m∠B, m∠E	31°	42°	65°			
m∠C						
m∠F						
AB						
DE						
BC						
EF						
AC						
DF						

CONSTRUCTING VIABLE ARGUMENTS

To be proficient in math, you need to understand and use stated assumptions, definitions, and previously established results in constructing arguments.

c. Are the two triangles similar? Explain.

d. Repeat parts (a)–(c) to complete columns 2 and 3 of the table for the given angle measures.

e. Complete each remaining column of the table using your own choice of two pairs of equal corresponding angle measures. Can you construct two triangles in this way that are *not* similar?

f. Make a conjecture about any two triangles with two pairs of congruent corresponding angles.

Communicate Your Answer

2. What can you conclude about two triangles when you know that two pairs of corresponding angles are congruent?

3. Find *RS* in the figure at the left.

Section 8.4   Proving Triangle Similarity by AA   **485**

ANSWERS

1. See Additional Answers.
2. They are similar.
3. $RS = 4$

Differentiated Instruction

Kinesthetic

Draw or photocopy and enlarge several pairs of similar triangles. Have students cut out the smaller triangle in each pair of similar triangles and align each of its angles in turn with each corresponding angle of the larger triangle. Encourage students to trace or copy diagrams of similar triangles given in the exercises and align their corresponding pairs of congruent angles.

8.4 Lesson

Core Vocabulary

Previous
similar figures
similarity transformation

What You Will Learn

- Use the Angle-Angle Similarity Theorem.
- Solve real-life problems.

Using the Angle-Angle Similarity Theorem

Theorem

Angle-Angle (AA) Similarity Theorem

If two angles of one triangle are congruent to two angles of another triangle, then the two triangles are similar.

If $\angle A \cong \angle D$ and $\angle B \cong \angle E$, then $\triangle ABC \sim \triangle DEF$.

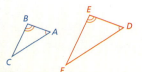

Proof p. 486

PROOF Angle-Angle (AA) Similarity Theorem

Given $\angle A \cong \angle D$, $\angle B \cong \angle E$

Prove $\triangle ABC \sim \triangle DEF$

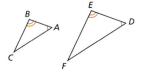

Dilate $\triangle ABC$ using a scale factor of $k = \dfrac{DE}{AB}$ and center A. The image of $\triangle ABC$ is $\triangle AB'C'$.

Because a dilation is a similarity transformation, $\triangle ABC \sim \triangle AB'C'$. Because the ratio of corresponding lengths of similar polygons equals the scale factor, $\dfrac{AB'}{AB} = \dfrac{DE}{AB}$. Multiplying each side by AB yields $AB' = DE$. By the definition of congruent segments, $\overline{AB'} \cong \overline{DE}$.

By the Reflexive Property of Congruence, $\angle A \cong \angle A$. Because corresponding angles of similar polygons are congruent, $\angle B' \cong \angle B$. Because $\angle B' \cong \angle B$ and $\angle B \cong \angle E$, $\angle B' \cong \angle E$ by the Transitive Property of Congruence.

Because $\angle A \cong \angle D$, $\angle B' \cong \angle E$, and $\overline{AB'} \cong \overline{DE}$, $\triangle AB'C' \cong \triangle DEF$ by the ASA Congruence Theorem. So, a composition of rigid motions maps $\triangle AB'C'$ to $\triangle DEF$.

Because a dilation followed by a composition of rigid motions maps $\triangle ABC$ to $\triangle DEF$, $\triangle ABC \sim \triangle DEF$.

Laurie's Notes — Teacher Actions

- State the Angle-Angle Similarity Theorem.
- **Turn and Talk:** "How can you prove this theorem?" Students will know that the third angles are congruent, and hopefully they will think about transformations as a way to prove the theorem.
- **Construct Viable Arguments and Critique the Reasoning of Others:** Give time for partners to sketch out a proof. Have a class discussion, and listen for correct reasoning.
- This is the first method for proving two triangles similar that are not necessarily congruent.

EXAMPLE 1 Using the AA Similarity Theorem

Determine whether the triangles are similar. If they are, write a similarity statement. Explain your reasoning.

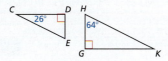

SOLUTION

Because they are both right angles, ∠D and ∠G are congruent.

By the Triangle Sum Theorem, 26° + 90° + m∠E = 180°, so m∠E = 64°. So, ∠E and ∠H are congruent.

▶ So, △CDE ~ △KGH by the AA Similarity Theorem.

VISUAL REASONING

Use colored pencils to show congruent angles. This will help you write similarity statements.

EXAMPLE 2 Using the AA Similarity Theorem

Show that the two triangles are similar.

a. △ABE ~ △ACD   b. △SVR ~ △UVT

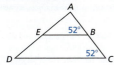 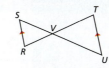

SOLUTION

a. Because m∠ABE and m∠C both equal 52°, ∠ABE ≅ ∠C. By the Reflexive Property of Congruence, ∠A ≅ ∠A.

▶ So, △ABE ~ △ACD by the AA Similarity Theorem.

b. You know ∠SVR ≅ ∠UVT by the Vertical Angles Congruence Theorem. The diagram shows $\overline{RS} \parallel \overline{UT}$, so ∠S ≅ ∠U by the Alternate Interior Angles Theorem.

▶ So, △SVR ~ △UVT by the AA Similarity Theorem.

VISUAL REASONING

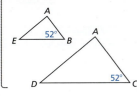

You may find it helpful to redraw the triangles separately.

Monitoring Progress Help in English and Spanish at BigIdeasMath.com

Show that the triangles are similar. Write a similarity statement.

1. △FGH and △RQS   2. △CDF and △DEF

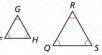 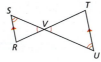

3. **WHAT IF?** Suppose that $\overline{SR} \not\parallel \overline{TU}$ in Example 2 part (b). Could the triangles still be similar? Explain.

Section 8.4 Proving Triangle Similarity by AA 487

Extra Example 1

Determine whether the triangles are similar. If they are, write a similarity statement. Explain your reasoning.

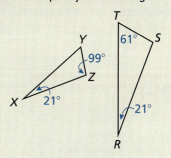

Sample answer: Given that m∠X = m∠R = 21°, so ∠X ≅ ∠R. By the Triangle Sum Theorem, m∠Y + 21° + 99° = 180° and m∠S + 21° + 61° = 180°. So, m∠Y = 60° and m∠S = 98°. That means only one pair of angles in the two triangles is congruent. By the definition of similar triangles, △XYZ and △RTS are not similar.

Extra Example 2

Show that △QPR ~ △QTP.

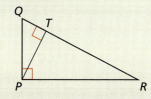

Sample answer: ∠Q ≅ ∠Q by the Reflexive Prop. of Angle Congruence. ∠QPR ≅ ∠QTP by the Right Angles Congruence Theorem. So, △QPR ~ △QTP by the AA Similarity Theorem.

MONITORING PROGRESS ANSWERS

1. ∠F ≅ ∠R and ∠G ≅ ∠Q, so △FGH ~ △RQS.
2. m∠CDF = 58° and m∠DFE = 90°, so ∠CDF ≅ ∠DEF and ∠CFD ≅ ∠DFE. So, △CDF ~ △DEF.
3. yes; If ∠S ≅ ∠T, then △SVR ~ △TVU.

Laurie's Notes Teacher Actions

- Pose Example 1, and have partners solve independently.
- ❓ "Will all similarity statements be the same? Explain." no; There are different but equivalent ways in which the similarity statements can be written.
- **Think-Alouds:** Model Example 2(a), and then do a *Think-Aloud* as partner A shows partner B how the two triangles are similar. "I know the triangles are similar because ...." Partners now reverse roles as partner B explains when the triangles in part (b) are similar.

Extra Example 3
A school flagpole casts a shadow that is 45 feet long. At the same time, a boy who is five feet eight inches tall casts a shadow that is 51 inches long. How tall is the flagpole to the nearest foot? **The flagpole is 60 feet tall.**

MONITORING PROGRESS ANSWERS

4. $36\frac{1}{4}$ in.

5. $\dfrac{\text{your height}}{\text{length of your shadow}} = \dfrac{x}{\text{length of tree's shadow}}$

Solving Real-Life Problems

Previously, you learned a way to use congruent triangles to find measurements indirectly. Another useful way to find measurements indirectly is by using similar triangles.

EXAMPLE 3   Modeling with Mathematics

A flagpole casts a shadow that is 50 feet long. At the same time, a woman standing nearby who is 5 feet 4 inches tall casts a shadow that is 40 inches long. How tall is the flagpole to the nearest foot?

Not drawn to scale

SOLUTION

1. **Understand the Problem** You are given the length of a flagpole's shadow, the height of a woman, and the length of the woman's shadow. You need to find the height of the flagpole.

2. **Make a Plan** Use similar triangles to write a proportion and solve for the height of the flagpole.

3. **Solve the Problem** The flagpole and the woman form sides of two right triangles with the ground. The Sun's rays hit the flagpole and the woman at the same angle. You have two pairs of congruent angles, so the triangles are similar by the AA Similarity Theorem.

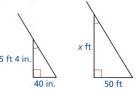

You can use a proportion to find the height x. Write 5 feet 4 inches as 64 inches so that you can form two ratios of feet to inches.

$\dfrac{x \text{ ft}}{64 \text{ in.}} = \dfrac{50 \text{ ft}}{40 \text{ in.}}$ Write proportion of side lengths.

$40x = 3200$ Cross Products Property

$x = 80$ Solve for x.

 The flagpole is 80 feet tall.

4. **Look Back** Attend to precision by checking that your answer has the correct units. The problem asks for the height of the flagpole to the nearest *foot*. Because your answer is 80 feet, the units match.

Also, check that your answer is reasonable in the context of the problem. A height of 80 feet makes sense for a flagpole. You can estimate that an eight-story building would be about 8(10 feet) = 80 feet, so it is reasonable that a flagpole could be that tall.

Monitoring Progress   Help in English and Spanish at BigIdeasMath.com

4. **WHAT IF?** A child who is 58 inches tall is standing next to the woman in Example 3. How long is the child's shadow?

5. You are standing outside, and you measure the lengths of the shadows cast by both you and a tree. Write a proportion showing how you could find the height of the tree.

488   Chapter 8   Similarity

Laurie's Notes   Teacher Actions

- Say, "Recall that we used indirect measure in working with congruent triangles. We were able to find distances indirectly because we had congruent triangles. A similar approach can be used with similar triangles." Work through the example as shown.
- **Model with Mathematics:** This example can be modeled on a sunny day or by using a mirror. See the *Teaching Strategy* on page T-484.

Closure
- **Exit Ticket:** One triangle has angles of 18° and 96°, and a second triangle has angles of 56° and 96°. Are the triangles similar? Explain.   no; The first triangle has angles of 18°, 66°, and 96°. The second triangle has angles of 28°, 56°, and 96°. The triangles have only one pair of angles that are congruent.

8.4 Exercises

Dynamic Solutions available at *BigIdeasMath.com*

Vocabulary and Core Concept Check

1. **COMPLETE THE SENTENCE** If two angles of one triangle are congruent to two angles of another triangle, then the triangles are _____.

2. **WRITING** Can you assume that corresponding sides and corresponding angles of any two similar triangles are congruent? Explain.

Monitoring Progress and Modeling with Mathematics

In Exercises 3–6, determine whether the triangles are similar. If they are, write a similarity statement. Explain your reasoning. *(See Example 1.)*

3. 4.

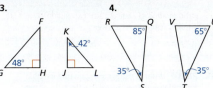

5. 6.

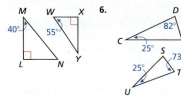

In Exercises 7–10, show that the two triangles are similar. *(See Example 2.)*

7. 8.

9. 10.

In Exercises 11–18, use the diagram to copy and complete the statement.

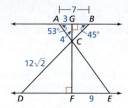

11. △CAG ~
12. △DCF ~ ▢
13. △ACB ~ ▢
14. m∠ECF = ▢
15. m∠ECD = ▢
16. CF = ▢
17. BC = ▢
18. DE = ▢

19. **ERROR ANALYSIS** Describe and correct the error in using the AA Similarity Theorem.

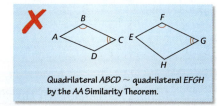

Quadrilateral ABCD ~ quadrilateral EFGH by the AA Similarity Theorem.

20. **ERROR ANALYSIS** Describe and correct the error in finding the value of x.

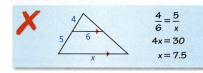

$\frac{4}{6} = \frac{5}{x}$

$4x = 30$

$x = 7.5$

Section 8.4 Proving Triangle Similarity by AA 489

Dynamic Teaching Tools

Dynamic Assessment & Progress Monitoring Tool
Interactive Whiteboard Lesson Library
Dynamic Classroom with Dynamic Investigations

ANSWERS

21. 78 m; Corresponding angles are congruent, so the triangles are similar.

22–36. See Additional Answers.

Mini-Assessment

1. Determine whether the triangles are similar. If they are, write a similarity statement. Explain your reasoning.

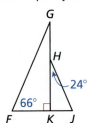

Sample answer: ∠GKF ≅ ∠HKJ by the Right ∠s ≅ Thm.
By the Triangle Sum Thm.
66° + 90° + m∠G = 180°, so m∠G = 24°. So, ∠G ≅ ∠JHK.
By the AA ~ Thm.
△FGK ~ △JHK.

2. Show that △LMN ~ △QPN without using vertical angles.

Sample answer: $\overline{LM} \parallel \overline{QP}$, so ∠L ≅ ∠Q and ∠M ≅ ∠P by the Alt. Int. ∠s Thm. So, by the AA ~ Thm. △LMN ~ △QPN.

3. The diagram shows a triangular ramp leading up to the entrance of a building. How long in feet is the vertical support shown in blue?

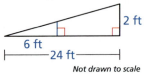

The ramp support is 0.5 foot.

490   Chapter 8

---

21. **MODELING WITH MATHEMATICS** You can measure the width of the lake using a surveying technique, as shown in the diagram. Find the width of the lake, WX. Justify your answer. *(See Example 3.)*

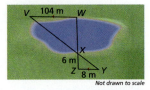

22. **MAKING AN ARGUMENT** You and your cousin are trying to determine the height of a telephone pole. Your cousin tells you to stand in the pole's shadow so that the tip of your shadow coincides with the tip of the pole's shadow. Your cousin claims to be able to use the distance between the tips of the shadows and you, the distance between you and the pole, and your height to estimate the height of the telephone pole. Is this possible? Explain. Include a diagram in your answer.

REASONING In Exercises 23–26, is it possible for △JKL and △XYZ to be similar? Explain your reasoning.

23. m∠J = 71°, m∠K = 52°, m∠X = 71°, and m∠Z = 57°

24. △JKL is a right triangle and m∠X + m∠Y = 150°.

25. m∠L = 87° and m∠Y = 94°

26. m∠J + m∠K = 85° and m∠Y + m∠Z = 80°

27. **MATHEMATICAL CONNECTIONS** Explain how you can use similar triangles to show that any two points on a line can be used to find its slope.

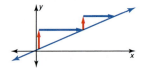

28. **HOW DO YOU SEE IT?** In the diagram, which triangles would you use to find the distance x between the shoreline and the buoy? Explain your reasoning.

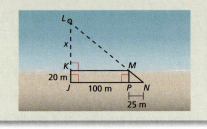

29. **WRITING** Explain why all equilateral triangles are similar.

30. **THOUGHT PROVOKING** Decide whether each is a valid method of showing that two quadrilaterals are similar. Justify your answer.

    a. AAA   b. AAAA

31. **PROOF** Without using corresponding lengths in similar polygons, prove that the ratio of two corresponding angle bisectors in similar triangles is equal to the scale factor.

32. **PROOF** Prove that if the lengths of two sides of a triangle are a and b, respectively, then the lengths of the corresponding altitudes to those sides are in the ratio $\frac{b}{a}$.

33. **MODELING WITH MATHEMATICS** A portion of an amusement park ride is shown. Find EF. Justify your answer.

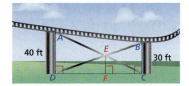

Maintaining Mathematical Proficiency
Reviewing what you learned in previous grades and lessons

Determine whether there is enough information to prove that the triangles are congruent. Explain your reasoning. *(Skills Review Handbook)*

34. 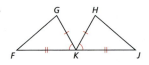   35.   36.

490   Chapter 8   Similarity

If students need help...	If students got it...
Resources by Chapter • Practice A and Practice B • Puzzle Time	Resources by Chapter • Enrichment and Extension • Cumulative Review
Student Journal • Practice	Start the *next* Section
Differentiating the Lesson Skills Review Handbook	

8.1–8.4 What Did You Learn?

Core Vocabulary

dilation, *p. 462*
center of dilation, *p. 462*
scale factor, *p. 462*
enlargement, *p. 462*
reduction, *p. 462*
similarity transformation, *p. 470*
similar figures, *p. 470*

Core Concepts

Section 8.1
Dilations and Scale Factor, *p. 462*
Coordinate Rule for Dilations, *p. 463*
Negative Scale Factors, *p. 464*

Section 8.2
Similarity Transformations, *p. 470*

Section 8.3
Corresponding Parts of Similar Polygons, *p. 476*
Corresponding Lengths in Similar Polygons, *p. 477*
Perimeters of Similar Polygons, *p. 478*
Areas of Similar Polygons, *p. 479*

Section 8.4
Angle-Angle (AA) Similarity Theorem, *p. 486*

Mathematical Practices

1. In Exercise 35 on page 483, why is there more than one correct answer for the length of the other side?

2. In Exercise 50 on page 484, how could you find the scale factor of the similar figures? Describe any tools that might be helpful.

3. In Exercise 21 on page 490, explain why the surveyor needs V, X, and Y to be collinear and Z, X, and W to be collinear.

Analyzing Your Errors

Misreading Directions

- **What Happens:** You incorrectly read or do not understand directions.
- **How to Avoid This Error:** Read the instructions for exercises at least twice and make sure you understand what they mean. Make this a habit and use it when taking tests.

Dynamic Teaching Tools

Dynamic Assessment & Progress Monitoring Tool
Interactive Whiteboard Lesson Library
Dynamic Classroom with Dynamic Investigations

ANSWERS

1. The side with length 18 could correspond to either side of Rectangle A.
2. Measure the object and the image with a tape measure.
3. so that vertical angles are formed at X

ANSWERS

1.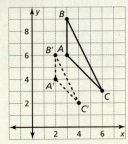

2.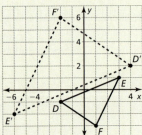

3.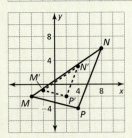

4.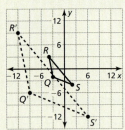

5. *Sample answer:* dilation centered at the origin with a scale factor of 2 followed by a translation 2 units right and 2 units up

6. *Sample answer:* dilation centered at the origin with a scale factor of $\frac{1}{2}$ followed by a reflection in the y-axis

7. $\angle B \cong \angle M, \angle D \cong \angle P,$
$\angle G \cong \angle Q; \dfrac{BD}{MP} = \dfrac{DG}{PQ} = \dfrac{GB}{QM}$

8.1–8.4 Quiz

Graph the triangle and its image after a dilation with scale factor k. *(Section 8.1)*

1. $A(3, 6), B(3, 9), C(6, 3); k = \frac{2}{3}$
2. $D(-2, -1), E(3, 1), F(1, -3); k = -2$
3. $M(-4, -2), N(8, 6), P(4, -4); k = 0.5$
4. $Q(-3, -2), R(-4, 3), S(2, -4); k = 300\%$

Describe a similarity transformation that maps △JKL to △MNP. *(Section 8.2)*

5. $J(2, 1), K(-3, -2), L(1, -3)$ and $M(6, 4), N(-4, -2), P(4, -4)$
6. $J(6, 2), K(-4, 2), L(-6, 4)$ and $M(-3, 1), N(2, 1), P(3, 2)$

List all pairs of congruent angles. Then write the ratios of the corresponding side lengths in a statement of proportionality. *(Section 8.3)*

7. △BDG ~ △MPQ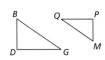
8. DEFG ~ HJKL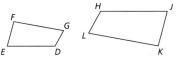

The polygons are similar. Find the value of x. *(Section 8.3)*

9.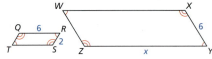
10.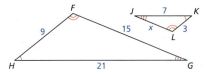

Show that the two triangles are similar. *(Section 8.4)*

11.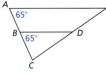
12.
13.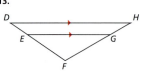

14. The dimensions of an official hockey rink used by the National Hockey League (NHL) are 200 feet by 85 feet. The dimensions of an air hockey table are 96 inches by 40.8 inches. Assume corresponding angles are congruent. *(Section 8.3)*
    a. Determine whether the two surfaces are similar.
    b. If the surfaces are similar, find the ratio of their perimeters and the ratio of their areas. If not, find the dimensions of an air hockey table that are similar to an NHL hockey rink.

15. You and a friend buy camping tents made by the same company but in different sizes and colors. Use the information given in the diagram to decide whether the triangular faces of the tents are similar. Explain your reasoning. *(Section 8.4)*

492   Chapter 8   Similarity

8. $\angle D \cong \angle H, \angle E \cong \angle J, \angle F \cong \angle K,$
$\angle G \cong \angle L;$
$\dfrac{DE}{HJ} = \dfrac{EF}{JK} = \dfrac{FG}{KL} = \dfrac{GD}{LH}$

9. $x = 18$
10. $x = 5$
11. $\angle CBD \cong \angle A$ and $\angle C \cong \angle C,$ so △ACE ~ △BCD.

12. $\angle G \cong \angle K$ and $\angle FHG \cong \angle JHK,$ so △FGH ~ △JKH.
13. $\angle GEF \cong \angle HDF$ and $\angle F \cong \angle F,$ so △DHF ~ △EGF.
14. a. yes
    b. $\dfrac{25}{1}; \dfrac{625}{1}$
15. yes; Corresponding angles are congruent.

Laurie's Notes

Overview of Section 8.5

Introduction
- Students have proven triangles similar by using the definition of similarity and by using the Angle-Angle Similarity Theorem.
- In this lesson, two additional methods are explored and then stated as valid methods: the Side-Side-Side Similarity Theorem and the Side-Angle-Side Similarity Theorem.

Formative Assessment Tips
- **Fact-First Questioning:** This is a higher-order questioning technique that goes beyond asking strait recall questions. Instead, this strategy allows you to assess students' growing understanding of a concept or skill.
- First, state a fact. Then ask students why, how, or to explain. Student thinking is activated and you gain insight into the depth of students' conceptual understanding.
- *Example:* Make the statement that all squares are rectangles. Then ask, "Why is this true?"

Pacing Suggestion
- In the explorations, students will investigate two new ways to establish that triangles are similar. After students have discussed the explorations, transition to the formal lesson.

Dynamic Teaching Tools
Dynamic Assessment & Progress Monitoring Tool
Lesson Planning Tool
Interactive Whiteboard Lesson Library
Dynamic Classroom with Dynamic Investigations

What Your Students Will Learn

- Use the Side-Side-Side Similarity Theorem to determine whether two triangles are similar.
- Use the Side-Angle-Side Similarity Theorem to determine whether two triangles are similar.

Laurie's Notes

Exploration

Motivate
- Gather pictures of quilts that contain what appear to be similar triangles and likely congruent triangles as well. Make a slide show of the quilts.
- Have students identify triangles that appear to be similar.
- Explain to students that in this lesson they will learn additional ways to prove triangles are similar besides the Angle-Angle Similarity Theorem.

Exploration Note
- Students may need guidance in constructing the triangles in both of the explorations, depending on the software being used and what functionality the software has.
- Before students begin, you could discuss techniques with students.

Exploration 1
- You may consider having different groups of students construct a subset of the problems and then share results with the whole class. For each pair of triangles, students not only construct the triangles from the given side lengths, they must also measure all six angles.
- ? "What do you observe about the corresponding angles when the corresponding sides are proportional?" They are congruent. "Are the triangles similar? Explain." Yes; by the definition of similar triangles, they are similar.
- ? "What do you observe about the corresponding angles when the corresponding sides are not proportional?" The angles are not congruent.

Exploration 2
- In constructing the triangles, students should be careful to make sure that the given angle is between the two given sides.
- Students should try values of k that result in enlargements and reductions.
- In deciding whether the triangles are similar, students could use the Angle-Angle Similarity Theorem from the last lesson.
- Students should conclude that the triangles are similar.

Communicate Your Answer
- **Construct Viable Arguments and Critique the Reasoning of Others:** In describing the results of the explorations, be sure that students do not confuse SSS and SAS congruency theorems with proportional side relationships in these explorations.

Connecting to Next Step
- The explorations help students develop an understanding of the theorems that will be stated in the formal lesson.

8.5 Proving Triangle Similarity by SSS and SAS

Essential Question What are two ways to use corresponding sides of two triangles to determine that the triangles are similar?

EXPLORATION 1   Deciding Whether Triangles Are Similar

Work with a partner. Use dynamic geometry software.

a. Construct △ABC and △DEF with the side lengths given in column 1 of the table below.

	1.	2.	3.	4.	5.	6.	7.
AB	5	5	6	15	9	24	
BC	8	8	8	20	12	18	
AC	10	10	10	10	8	16	
DE	10	15	9	12	12	8	
EF	16	24	12	16	15	6	
DF	20	30	15	8	10	8	
m∠A							
m∠B							
m∠C							
m∠D							
m∠E							
m∠F							

CONSTRUCTING VIABLE ARGUMENTS
To be proficient in math, you need to analyze situations by breaking them into cases and recognize and use counterexamples.

b. Copy the table and complete column 1.

c. Are the triangles similar? Explain your reasoning.

d. Repeat parts (a)–(c) for columns 2–6 in the table.

e. How are the corresponding side lengths related in each pair of triangles that are similar? Is this true for each pair of triangles that are not similar?

f. Make a conjecture about the similarity of two triangles based on their corresponding side lengths.

g. Use your conjecture to write another set of side lengths of two similar triangles. Use the side lengths to complete column 7 of the table.

EXPLORATION 2   Deciding Whether Triangles Are Similar

Work with a partner. Use dynamic geometry software. Construct any △ABC.

a. Find AB, AC, and m∠A. Choose any positive rational number k and construct △DEF so that DE = k • AB, DF = k • AC, and m∠D = m∠A.

b. Is △DEF similar to △ABC? Explain your reasoning.

c. Repeat parts (a) and (b) several times by changing △ABC and k. Describe your results.

Communicate Your Answer

3. What are two ways to use corresponding sides of two triangles to determine that the triangles are similar?

Section 8.5   Proving Triangle Similarity by SSS and SAS   493

Dynamic Teaching Tools
- Dynamic Assessment & Progress Monitoring Tool
- Lesson Planning Tool
- Interactive Whiteboard Lesson Library
- Dynamic Classroom with Dynamic Investigations

ANSWERS

1. a.

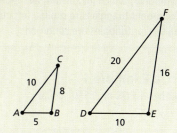

b. See Additional Answers.

c. yes; Corresponding angles are congruent.

d. See Additional Answers for table. The triangles are similar in columns 2–4 because corresponding angles are congruent. The triangles are not similar in columns 5 and 6 because corresponding angles are not congruent.

e. They are proportional; no

f. If the corresponding side lengths of two triangles are proportional, then the triangles are similar.

g. See Additional Answers.

2. a. *Sample answer:*

 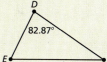

b. yes; Corresponding angles are congruent.

c. The triangles are similar in each case.

3. If all three pairs of corresponding side lengths of two triangles are proportional, then the triangles are similar. If an angle of one triangle is congruent to an angle of a second triangle and the lengths of the sides including these angles are proportional, then the triangles are similar.

Section 8.5   493

English Language Learners

Build on Past Knowledge
Remind students that they have learned different ways to prove triangles are congruent. Now they will learn new ways to prove two triangles are similar. Point out that the similarity symbol ~ is part of the congruence symbol ≅, which means that all congruent triangles are similar, but not all similar triangles are congruent.

Extra Example 1
Is either △PQR or △STU similar to △VWX?

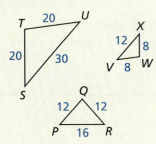

△STU ~ △VWX
△PQR is not similar to △VWX.

8.5 Lesson

Core Vocabulary
Previous
similar figures
corresponding parts
parallel lines

What You Will Learn
▶ Use the Side-Side-Side Similarity Theorem.
▶ Use the Side-Angle-Side Similarity Theorem.

Using the Side-Side-Side Similarity Theorem
In addition to using congruent corresponding angles to show that two triangles are similar, you can use proportional corresponding side lengths.

Theorem

Side-Side-Side (SSS) Similarity Theorem
If the corresponding side lengths of two triangles are proportional, then the triangles are similar.

If $\dfrac{AB}{RS} = \dfrac{BC}{ST} = \dfrac{CA}{TR}$, then △ABC ~ △RST.

Proof p. 495

EXAMPLE 1 Using the SSS Similarity Theorem

Is either △DEF or △GHJ similar to △ABC?

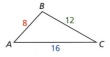 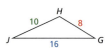

FINDING AN ENTRY POINT
When using the SSS Similarity Theorem, compare the shortest sides, the longest sides, and then the remaining sides.

SOLUTION
Compare △ABC and △DEF by finding ratios of corresponding side lengths.

Shortest sides	Longest sides	Remaining sides
$\dfrac{AB}{DE} = \dfrac{8}{6}$	$\dfrac{CA}{FD} = \dfrac{16}{12}$	$\dfrac{BC}{EF} = \dfrac{12}{9}$
$= \dfrac{4}{3}$	$= \dfrac{4}{3}$	$= \dfrac{4}{3}$

▶ All the ratios are equal, so △ABC ~ △DEF.

Compare △ABC and △GHJ by finding ratios of corresponding side lengths.

Shortest sides	Longest sides	Remaining sides
$\dfrac{AB}{GH} = \dfrac{8}{8}$	$\dfrac{CA}{JG} = \dfrac{16}{16}$	$\dfrac{BC}{HJ} = \dfrac{12}{10}$
$= 1$	$= 1$	$= \dfrac{6}{5}$

▶ The ratios are not all equal, so △ABC and △GHJ are not similar.

Laurie's Notes — Teacher Actions

- State the SSS Similarity Theorem. Distinguish the differences from the SSS Congruence Theorem.
- **Teaching Tip:** To help pair corresponding sides, use colored pencils and language such as "shortest side to shortest side," and so on.
- **? Fact-First Questioning** and **Construct Viable Arguments and Critique the Reasoning of Others:** "There is no SSSS Similarity Theorem for quadrilaterals. Why?" Give *Wait Time* before soliciting responses. Listen for an explanation, such as "A square and rhombus could have corresponding sides proportional but angles that are not congruent."

> **PROOF** SSS Similarity Theorem

Given $\dfrac{RS}{JK} = \dfrac{ST}{KL} = \dfrac{TR}{LJ}$

Prove $\triangle RST \sim \triangle JKL$

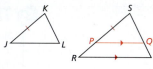

JUSTIFYING STEPS

The Parallel Postulate allows you to draw an auxiliary line $\overrightarrow{PQ}$ in $\triangle RST$. There is only one line through point P parallel to $\overline{RT}$, so you are able to draw it.

Locate P on $\overline{RS}$ so that $PS = JK$. Draw $\overline{PQ}$ so that $\overline{PQ} \parallel \overline{RT}$. Then $\triangle RST \sim \triangle PSQ$ by the AA Similarity Theorem, and $\dfrac{RS}{PS} = \dfrac{ST}{SQ} = \dfrac{TR}{QP}$. You can use the given proportion and the fact that $PS = JK$ to deduce that $SQ = KL$ and $QP = LJ$. By the SSS Congruence Theorem, it follows that $\triangle PSQ \cong \triangle JKL$. Finally, use the definition of congruent triangles and the AA Similarity Theorem to conclude that $\triangle RST \sim \triangle JKL$.

> **EXAMPLE 2** Using the SSS Similarity Theorem

Find the value of x that makes $\triangle ABC \sim \triangle DEF$.

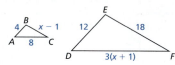

FINDING AN ENTRY POINT

You can use either $\dfrac{AB}{DE} = \dfrac{BC}{EF}$ or $\dfrac{AB}{DE} = \dfrac{AC}{DF}$ in Step 1.

SOLUTION

Step 1 Find the value of x that makes corresponding side lengths proportional.

$\dfrac{AB}{DE} = \dfrac{BC}{EF}$ Write proportion.

$\dfrac{4}{12} = \dfrac{x-1}{18}$ Substitute.

$4 \cdot 18 = 12(x-1)$ Cross Products Property

$72 = 12x - 12$ Simplify.

$7 = x$ Solve for x.

Step 2 Check that the side lengths are proportional when $x = 7$.

$BC = x - 1 = 6$ $DF = 3(x+1) = 24$

$\dfrac{AB}{DE} \stackrel{?}{=} \dfrac{BC}{EF} \Rightarrow \dfrac{4}{12} = \dfrac{6}{18}$ ✓ $\dfrac{AB}{DE} \stackrel{?}{=} \dfrac{AC}{DF} \Rightarrow \dfrac{4}{12} = \dfrac{8}{24}$ ✓

▶ When $x = 7$, the triangles are similar by the SSS Similarity Theorem.

Monitoring Progress Help in English and Spanish at *BigIdeasMath.com*

Use the diagram.

1. Which of the three triangles are similar? Write a similarity statement.
2. The shortest side of a triangle similar to $\triangle RST$ is 12 units long. Find the other side lengths of the triangle.

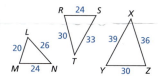

Section 8.5   Proving Triangle Similarity by SSS and SAS   495

Differentiated Instruction

Organization

Encourage students to list the different ways to prove that two triangles are similar and include a sketch for each method.

Extra Example 2

Find the value of x that makes $\triangle XYZ \sim \triangle HJK$.

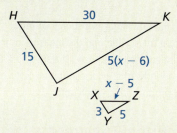

When $x = 11$, the triangles are similar by the SSS Similarity Theorem.

MONITORING PROGRESS ANSWERS

1. $\triangle LMN \sim \triangle YZX$
2. 15, 16.5

Laurie's Notes   **Teacher Actions**

- **Turn and Talk:** Have partners read the proof of the SSS Similarity Theorem. "What are the key steps in the proof narrative presented?" Listen for: drawing the auxiliary line so that the overlapping triangles are similar, showing smaller triangles are congruent, and using corresponding parts and the AA Similarity Theorem to show given triangles are similar.
- In Example 2, students may use the orientation of the triangles to set up the proportions. They should confirm with the triangle similarity statement.

Extra Example 3

The diagram is a scale drawing of a triangular roof truss. The lengths of the two upper sides of the actual truss are 18 feet and 40 feet. The actual truss and the scale drawing both have an included angle of 110°. Is the scale drawing of the truss similar to the actual truss? Explain.

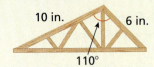

No, the drawing is not similar to the actual truss. Although the measure of the included angles between the sides is 110°, the lengths of the corresponding sides are not proportional: $\frac{10}{40} = \frac{1}{4}$ and $\frac{6}{18} = \frac{1}{3}$.

MONITORING PROGRESS ANSWERS

3. $\angle R \cong \angle N$ and $\frac{SR}{PN} = \frac{RT}{NQ} = \frac{4}{3}$, so $\triangle RST \sim \triangle NPQ$.

4. *Sample answer:* Corresponding side lengths are proportional, so $\triangle XYZ \sim \triangle WXZ$.

Using the Side-Angle-Side Similarity Theorem

Theorem

Side-Angle-Side (SAS) Similarity Theorem

If an angle of one triangle is congruent to an angle of a second triangle and the lengths of the sides including these angles are proportional, then the triangles are similar.

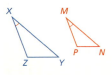

If $\angle X \cong \angle M$ and $\frac{ZX}{PM} = \frac{XY}{MN}$, then $\triangle XYZ \sim \triangle MNP$.

Proof Ex. 19, p. 498

EXAMPLE 3 Using the SAS Similarity Theorem

You are building a lean-to shelter starting from a tree branch, as shown. Can you construct the right end so it is similar to the left end using the angle measure and lengths shown?

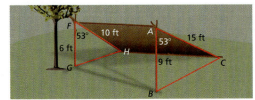

SOLUTION

Both $m\angle A$ and $m\angle F$ equal 53°, so $\angle A \cong \angle F$. Next, compare the ratios of the lengths of the sides that include $\angle A$ and $\angle F$.

Shorter sides

$$\frac{AB}{FG} = \frac{9}{6}$$
$$= \frac{3}{2}$$

Longer sides

$$\frac{AC}{FH} = \frac{15}{10}$$
$$= \frac{3}{2}$$

The lengths of the sides that include $\angle A$ and $\angle F$ are proportional. So, by the SAS Similarity Theorem, $\triangle ABC \sim \triangle FGH$.

▶ Yes, you can make the right end similar to the left end of the shelter.

Monitoring Progress Help in English and Spanish at *BigIdeasMath.com*

Explain how to show that the indicated triangles are similar.

3. $\triangle SRT \sim \triangle PNQ$

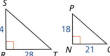

4. $\triangle XZW \sim \triangle YZX$

496   Chapter 8   Similarity

Laurie's Notes — Teacher Actions

- **State the SAS Similarity Theorem.** Distinguish the differences from the SAS Congruence Theorem.
- **Whiteboarding:** "Sketch a picture to show why there is no SAS Similarity Theorem for quadrilaterals." Use whiteboards to share sketches.
- **Popsicle Sticks:** Pose Example 3, and have partners independently solve the problem. Use *Popsicle Sticks* for steps in the solution.
- **Think-Pair-Share:** Have students answer Questions 3 and 4, and then share and discuss as a class.

Closure

- **Exit Ticket:** Draw a triangle with sides of 7 centimeters and 10 centimeters and an included angle of 42°. Draw and label a similar triangle with a side of 15 centimeters. Answers will vary. Check students' work.

8.5 Exercises

Dynamic Solutions available at *BigIdeasMath.com*

Vocabulary and Core Concept Check

1. **COMPLETE THE SENTENCE** You plan to show that △QRS is similar to △XYZ by the SSS Similarity Theorem. Copy and complete the proportion that you will use: $\frac{QR}{\square} = \frac{\square}{YZ} = \frac{QS}{\square}$.

2. **WHICH ONE DOESN'T BELONG?** Which triangle does *not* belong with the other three? Explain your reasoning.

  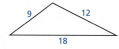

Monitoring Progress and Modeling with Mathematics

In Exercises 3 and 4, determine whether △JKL or △RST is similar to △ABC. *(See Example 1.)*

3.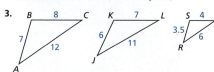

4.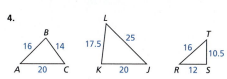

In Exercises 5 and 6, find the value of x that makes △DEF ~ △XYZ. *(See Example 2.)*

5.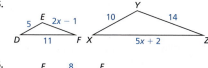

6.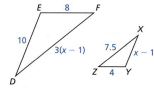

In Exercises 7 and 8, verify that △ABC ~ △DEF. Find the scale factor of △ABC to △DEF.

7. △ABC: BC = 18, AB = 15, AC = 12
   △DEF: EF = 12, DE = 10, DF = 8

8. △ABC: AB = 10, BC = 16, CA = 20
   △DEF: DE = 25, EF = 40, FD = 50

In Exercises 9 and 10, determine whether the two triangles are similar. If they are similar, write a similarity statement and find the scale factor of triangle B to triangle A. *(See Example 3.)*

9.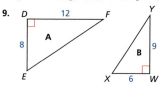

10.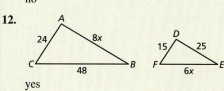

In Exercises 11 and 12, sketch the triangles using the given description. Then determine whether the two triangles can be similar.

11. In △RST, RS = 20, ST = 32, and m∠S = 16°. In △FGH, GH = 30, HF = 48, and m∠H = 24°.

12. The side lengths of △ABC are 24, 8x, and 48, and the side lengths of △DEF are 15, 25, and 6x.

Section 8.5   Proving Triangle Similarity by SSS and SAS   497

Assignment Guide and Homework Check

ASSIGNMENT

Basic: 1, 2, 3–15 odd, 20, 25–27

Average: 1, 2–12 even, 13, 15, 16–20 even, 25–27

Advanced: 1, 2–12 even, 13–15, 17, 20, 22, 25–27

HOMEWORK CHECK

Basic: 3, 5, 9, 20

Average: 4, 6, 10, 20

Advanced: 6, 12, 14, 20

ANSWERS

1. $\frac{QR}{XY} = \frac{RS}{YZ} = \frac{QS}{XZ}$

2. The triangle with side lengths 4, 6, and 8 does not belong because the other three are similar by the SSS Similarity Theorem.

3. △RST

4. △JKL

5. x = 4

6. x = 6

7. $\frac{12}{18} = \frac{10}{15} = \frac{8}{12} = \frac{2}{3}$

8. $\frac{25}{10} = \frac{40}{16} = \frac{50}{20} = \frac{5}{2}$

9. similar; △DEF ~ △WXY; $\frac{4}{3}$

10. not similar

11. *(sketch: △RST with 16°, RS = 20, ST = 32; △FGH with 30, 24°, 48)*
    no

12. *(sketch: △ABC with 24, 8x, 48; △DEF with 15, 25, 6x)*
    yes

13. **ERROR ANALYSIS** Describe and correct the error in writing a similarity statement.

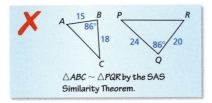

△ABC ~ △PQR by the SAS Similarity Theorem.

14. **MATHEMATICAL CONNECTIONS** Find the value of n that makes △DEF ~ △XYZ when $DE = 4$, $EF = 5$, $XY = 4(n + 1)$, $YZ = 7n - 1$, and $\angle E \cong \angle Y$. Include a sketch.

15. **MAKING AN ARGUMENT** Your friend claims that △JKL ~ △MNO by the SAS Similarity Theorem when $JK = 18$, $m\angle K = 130°$, $KL = 16$, $MN = 9$, $m\angle N = 65°$, and $NO = 8$. Do you support your friend's claim? Explain your reasoning.

16. **ANALYZING RELATIONSHIPS** Certain sections of stained glass are sold in triangular, beveled pieces. Which of the three beveled pieces, if any, are similar?

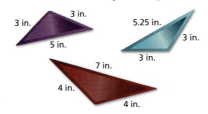

17. **ATTENDING TO PRECISION** In the diagram, $\dfrac{MN}{MR} = \dfrac{MP}{MQ}$. Which of the statements must be true? Select all that apply. Explain your reasoning.

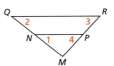

Ⓐ $\angle 1 \cong \angle 2$ Ⓑ $\overline{QR} \parallel \overline{NP}$
Ⓒ $\angle 1 \cong \angle 4$ Ⓓ △MNP ~ △MRQ

18. **WRITING** Are any two right triangles similar? Explain.

19. **PROVING A THEOREM** Write a two-column proof of the SAS Similarity Theorem.

Given $\angle A \cong \angle D$, $\dfrac{AB}{DE} = \dfrac{AC}{DF}$

Prove △ABC ~ △DEF

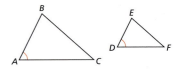

20. **HOW DO YOU SEE IT?** Which theorem could you use to show that △OPQ ~ △OMN in the portion of the Ferris wheel shown when $PM = QN = 5$ feet and $MO = NO = 10$ feet?

21. **DRAWING CONCLUSIONS** Explain why it is not necessary to have an Angle-Side-Angle Similarity Theorem.

22. **THOUGHT PROVOKING** Decide whether each is a valid method of showing that two quadrilaterals are similar. Justify your answer.

   a. SASA   b. SASAS   c. SSSS   d. SASSS

23. **WRITING** Can two triangles have all three ratios of corresponding angle measures equal to a value greater than 1? less than 1? Explain.

24. **MULTIPLE REPRESENTATIONS** Use a diagram to show why there is no Side-Side-Angle Similarity Theorem.

Maintaining Mathematical Proficiency Reviewing what you learned in previous grades and lessons

Find the slope of the line that passes through the given points. *(Skills Review Handbook)*

25. $(-3, 6), (2, 1)$ 26. $(-3, -5), (9, -1)$ 27. $(1, -2), (8, 12)$

Laurie's Notes

Overview of Section 8.6

Introduction
- Students will use their knowledge of solving proportions in this lesson.
- Four theorems are presented in this lesson, all involving proportions. The Triangle Proportionality Theorem and its converse involve a segment parallel to a side of a triangle. As noted in the *Teaching Strategy* below, there are different but equivalent ways in which the proportions can be written.
- Two additional proportionality theorems are also presented and used to solve problems.

Teaching Strategy
- **Proportional Triangles:** Given a proportion, students should be familiar with different but equivalent ways in which the proportion can be written.
- Use the sketch from the Triangle Proportionality Theorem. The conclusion of the theorem states that $\frac{RT}{TQ} = \frac{RU}{US}$.

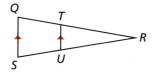

- Discuss with students additional and equivalent proportions.
$$\frac{RT}{TQ} = \frac{RU}{US} \rightarrow \frac{RT}{RU} = \frac{TQ}{US} \rightarrow \frac{RT}{RQ} = \frac{RU}{RS}$$
- Being comfortable with multiple ways in which a proportion can be correctly written to represent a situation will help students get started in solving the problem.

Pacing Suggestion
- The explorations provide an opportunity for students to investigate and make sense of two theorems in the formal lesson. Transition to the formal lesson as soon as students have discussed the explorations.

Dynamic Teaching Tools
Dynamic Assessment & Progress Monitoring Tool
Lesson Planning Tool
Interactive Whiteboard Lesson Library
Dynamic Classroom with Dynamic Investigations

What Your Students Will Learn

- Use the Triangle Proportionality Theorem and its converse to find lengths of segments.
- Use the Three Parallel Lines Theorem and the Triangle Angle Bisector Theorem to find lengths of segments.
- Partition directed line segments using slope and a ratio.

Laurie's Notes

Exploration

Motivate
- Display a street map, as shown, with parallel city streets that are intersected by nonparallel streets.

- Have students describe the orientations of the various streets. Explain to students that in this lesson they will look at proportional segments found in triangles.

Exploration 1
- This exploration can be constructed quickly using dynamic geometry software. It can then be explored dynamically, providing numerous cases to compare.
- Be sure that students explore multiple locations of the parallel segment within one triangle, followed by trials with different triangles. Students should not shy away from special cases such as right, equilateral, and obtuse triangles.
- **?** "What do you observe when you construct a segment parallel to one side of a triangle that intersects the other two sides?" The segment divides the other two sides proportionally.

Exploration 2
- Again, the results of this construction can be explored on multiple triangles utilizing the dynamic nature of the software. Be sure that students try special triangles such as right, equilateral, and obtuse triangles.
- **?** "What do you observe when you construct the angle bisector of one angle of a triangle and it intersects the opposite side of the triangle?" The angle bisector divides the opposite side into two segments whose lengths are proportional to the lengths of the other two sides.

Communicate Your Answer
- **Turn and Talk:** Have partners discuss the questions together before sharing as a class.

Connecting to Next Step
- Two of the four theorems in this lesson are investigated in these explorations. Refer to the explorations when stating the theorems.

8.6 Proportionality Theorems

Essential Question What proportionality relationships exist in a triangle intersected by an angle bisector or by a line parallel to one of the sides?

EXPLORATION 1   Discovering a Proportionality Relationship

Work with a partner. Use dynamic geometry software to draw any △ABC.

a. Construct $\overline{DE}$ parallel to $\overline{BC}$ with endpoints on $\overline{AB}$ and $\overline{AC}$, respectively.

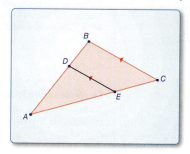

LOOKING FOR STRUCTURE

To be proficient in math, you need to look closely to discern a pattern or structure.

b. Compare the ratios of AD to BD and AE to CE.

c. Move $\overline{DE}$ to other locations parallel to $\overline{BC}$ with endpoints on $\overline{AB}$ and $\overline{AC}$, and repeat part (b).

d. Change △ABC and repeat parts (a)–(c) several times. Write a conjecture that summarizes your results.

EXPLORATION 2   Discovering a Proportionality Relationship

Work with a partner. Use dynamic geometry software to draw any △ABC.

a. Bisect ∠B and plot point D at the intersection of the angle bisector and $\overline{AC}$.

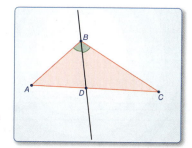

b. Compare the ratios of AD to DC and BA to BC.

c. Change △ABC and repeat parts (a) and (b) several times. Write a conjecture that summarizes your results.

Communicate Your Answer

3. What proportionality relationships exist in a triangle intersected by an angle bisector or by a line parallel to one of the sides?

4. Use the figure at the right to write a proportion.

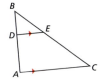

Section 8.6   Proportionality Theorems   499

ANSWERS

1. a. Check students' work.

    b. $\dfrac{AD}{BD} = \dfrac{AE}{CE}$

    c. Check students' work; $\dfrac{AD}{BD} = \dfrac{AE}{CE}$

    d. Check students' work; If $\overline{DE} \parallel \overline{AC}$ in △ABC, then $\dfrac{AD}{BD} = \dfrac{AE}{CE}$.

2. a. Check students' work.

    b. $\dfrac{AD}{DC} = \dfrac{BA}{BC}$

    c. Check students' work; If $\overline{BD}$ bisects ∠B in △ABC, $\dfrac{AD}{DC} = \dfrac{BA}{BC}$.

3. If a ray bisects an angle of a triangle, then the opposite segments are proportional to the lengths of the other two sides. If a line parallel to one side of a triangle intersects the other two sides, then it divides the two sides proportionally.

4. $\dfrac{BD}{DA} = \dfrac{BE}{EC}$

English Language Learners

Graphic Organizer
Remind students from their study of proportions that the terms of a proportion can be placed in different positions within the proportion as long as the relationships are the same. Students may recall that the numerical proportion $\frac{3}{6} = \frac{1}{2}$ can also be written as $\frac{3}{1} = \frac{6}{2}, \frac{1}{3} = \frac{2}{6},$ or $\frac{6}{3} = \frac{2}{1}$. Likewise, when the terms are side lengths, a proportion showing the relationship between the lengths can be written in different, but equivalent, ways.

Extra Example 1
In the diagram, $\overline{WZ} \parallel \overline{XY}$, $WX = 12$, $VZ = 10$, and $ZY = 8$. What is the length of $\overline{VW}$?

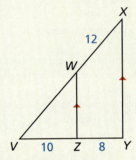

The length of $\overline{VW}$ is 15 units.

MONITORING PROGRESS ANSWER
1. $\frac{315}{11}$

8.6 Lesson

What You Will Learn
- Use proportionality theorems.
- Partition directed line segments.

Core Vocabulary
directed line segment, *p. 502*
Previous
corresponding angles
ratio
proportion

Using Proportionality Theorems

Theorems

Triangle Proportionality Theorem
If a line parallel to one side of a triangle intersects the other two sides, then it divides the two sides proportionally.

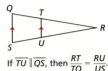

If $\overline{TU} \parallel \overline{QS}$, then $\frac{RT}{TQ} = \frac{RU}{US}$.

Proof Ex. 27, p. 505

Converse of the Triangle Proportionality Theorem
If a line divides two sides of a triangle proportionally, then it is parallel to the third side.

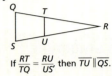

If $\frac{RT}{TQ} = \frac{RU}{US}$, then $\overline{TU} \parallel \overline{QS}$.

Proof Ex. 28, p. 505

EXAMPLE 1 Finding the Length of a Segment

In the diagram, $\overline{QS} \parallel \overline{UT}$, $RS = 4$, $ST = 6$, and $QU = 9$. What is the length of $\overline{RQ}$?

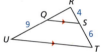

SOLUTION

$\frac{RQ}{QU} = \frac{RS}{ST}$ Triangle Proportionality Theorem

$\frac{RQ}{9} = \frac{4}{6}$ Substitute.

$RQ = 6$ Multiply each side by 9 and simplify.

▶ The length of $\overline{RQ}$ is 6 units.

Monitoring Progress   Help in English and Spanish at *BigIdeasMath.com*

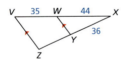

1. Find the length of $\overline{YZ}$.

The theorems above also imply the following:

Contrapositive of the Triangle Proportionality Theorem	**Inverse of the Triangle Proportionality Theorem**
If $\frac{RT}{TQ} \neq \frac{RU}{US}$, then $\overline{TU} \not\parallel \overline{QS}$.	If $\overline{TU} \not\parallel \overline{QS}$, then $\frac{RT}{TQ} \neq \frac{RU}{US}$.

500   Chapter 8   Similarity

Laurie's Notes   Teacher Actions

- State the Triangle Proportionality Theorem and its converse.
- **Turn and Talk:** "How would you prove these theorems?" Give *Wait Time* and then solicit thoughts from students. They should be gaining confidence in their ability to prove statements.
- Pose Example 1. Have partners work to solve. Share not only the solution, but also the original proportion.
- **Construct Viable Arguments and Critique the Reasoning of Others:** Discuss the equivalent proportions that students likely used. See the *Teaching Strategy* on page T-498.
- State the contrapositive of each theorem, and have students discuss with their partners whether these statements are valid.

EXAMPLE 2 Solving a Real-Life Problem

On the shoe rack shown, $BA = 33$ centimeters, $CB = 27$ centimeters, $CD = 44$ centimeters, and $DE = 25$ centimeters. Explain why the shelf is not parallel to the floor.

SOLUTION

Find and simplify the ratios of the lengths.

$$\frac{CD}{DE} = \frac{44}{25} \qquad \frac{CB}{BA} = \frac{27}{33} = \frac{9}{11}$$

▶ Because $\frac{44}{25} \neq \frac{9}{11}$, $\overline{BD}$ is not parallel to $\overline{AE}$. So, the shelf is not parallel to the floor.

Monitoring Progress Help in English and Spanish at *BigIdeasMath.com*

2. Determine whether $\overline{PS} \parallel \overline{QR}$.

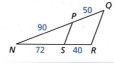

Theorem

Three Parallel Lines Theorem

If three parallel lines intersect two transversals, then they divide the transversals proportionally.

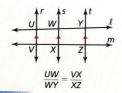

Proof Ex. 32, p. 505

EXAMPLE 3 Using the Three Parallel Lines Theorem

In the diagram, $\angle 1$, $\angle 2$, and $\angle 3$ are all congruent, $GF = 120$ yards, $DE = 150$ yards, and $CD = 300$ yards. Find the distance HF between Main Street and South Main Street.

SOLUTION

Corresponding angles are congruent, so $\overleftrightarrow{FE}$, $\overleftrightarrow{GD}$, and $\overleftrightarrow{HC}$ are parallel.

Use the Three Parallel Lines Theorem to set up a proportion.

$\dfrac{HG}{GF} = \dfrac{CD}{DE}$ Three Parallel Lines Theorem

$\dfrac{HG}{120} = \dfrac{300}{150}$ Substitute.

$HG = 240$ Multiply each side by 120 and simplify.

By the Segment Addition Postulate, $HF = HG + GF = 240 + 120 = 360$.

▶ The distance between Main Street and South Main Street is 360 yards.

Differentiated Instruction

Visual

Have students color-code their diagrams when solving problems involving proportions. In Example 3, for instance, students can color the line segments and use the same colors when setting up a proportion involving the lengths of the colored segments.

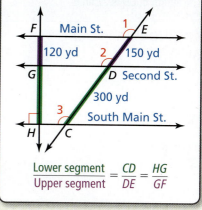

$\dfrac{\text{Lower segment}}{\text{Upper segment}} = \dfrac{CD}{DE} = \dfrac{HG}{GF}$

Extra Example 2

Use the diagram in Example 2. $BA = 35$ centimeters, $CB = 25$ centimeters, $CD = 20$ centimeters, and $DE = 28$ centimeters. Explain why the shelf is parallel to the floor.

The ratios of the corresponding lengths are $\dfrac{CD}{DE} = \dfrac{20}{28} = \dfrac{5}{7}$ and $\dfrac{CB}{BA} = \dfrac{25}{35} = \dfrac{5}{7}$. Because $\dfrac{5}{7} = \dfrac{5}{7}$, $\overline{BD} \parallel \overline{AE}$ by the Converse of the Triangle Proportionality Theorem. So, the shelf is parallel to the floor.

Extra Example 3

In the diagram, $\angle ADE$, $\angle BEF$, and $\angle CFG$ are all congruent. $AB = 30$, $BC = 12$, and $DE = 35$. Find DF.

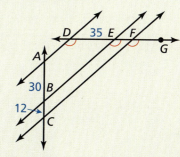

$DF = 49$ units

MONITORING PROGRESS ANSWER

2. yes

Laurie's Notes — Teacher Actions

- **Common Misconception:** Note that the ratio $\dfrac{27}{33}$ is simplified before being used in the proportion, which is mathematically correct.
- **Teaching Tip:** Using lined paper, have students trace any three of the lines on the paper. Use a straightedge to draw two lines that intersect the three parallel lines. Use this model to discuss the Three Parallel Lines Theorem.
- **Whiteboarding:** Pose Example 3. Use whiteboards and give partners time to solve the problem. There are different ways to write a proportion to find HG. Circulate and observe methods of solution. Solicit solutions from multiple students, assuring that more than one method will be demonstrated. Discuss.

Extra Example 4

In the diagram, $\angle BAC \cong \angle CAD$. Use the given lengths to find the length of $\overline{CD}$.

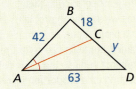

The length of $\overline{CD}$ is 27 units.

MONITORING PROGRESS ANSWERS

3. 12
4. 19.2
5. 28
6. $4\sqrt{2}$

Theorem

Triangle Angle Bisector Theorem

If a ray bisects an angle of a triangle, then it divides the opposite side into segments whose lengths are proportional to the lengths of the other two sides.

Proof Ex. 35, p. 506

EXAMPLE 4 Using the Triangle Angle Bisector Theorem

In the diagram, $\angle QPR \cong \angle RPS$. Use the given side lengths to find the length of $\overline{RS}$.

SOLUTION

Because $\overrightarrow{PR}$ is an angle bisector of $\angle QPS$, you can apply the Triangle Angle Bisector Theorem. Let $RS = x$. Then $RQ = 15 - x$.

$\dfrac{RQ}{RS} = \dfrac{PQ}{PS}$	Triangle Angle Bisector Theorem
$\dfrac{15-x}{x} = \dfrac{7}{13}$	Substitute.
$195 - 13x = 7x$	Cross Products Property
$9.75 = x$	Solve for x.

▶ The length of $\overline{RS}$ is 9.75 units.

Monitoring Progress Help in English and Spanish at *BigIdeasMath.com*

Find the length of the given line segment.

3. $\overline{BD}$ 4. $\overline{JM}$

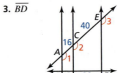 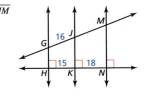

Find the value of the variable.

5.   6.

Partitioning a Directed Line Segment

A **directed line segment** AB is a segment that represents moving from point A to point B. The following example shows how to use slope to find a point on a directed line segment that partitions the segment in a given ratio.

502   Chapter 8   Similarity

Laurie's Notes — Teacher Actions

- Refer to Exploration 2 on page 499, and then state the Triangle Angle Bisector Theorem.
- ❓ "What is an equivalent proportion that results from this construction?" $\dfrac{AD}{CA} = \dfrac{DB}{CB}$
- ❓ "In Example 4, how can you represent the length of $\overline{RQ}$?" $15 - x$ Continue to work through the example as shown.
- ❓ Define a *directed line segment*.

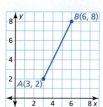

EXAMPLE 5 **Partitioning a Directed Line Segment**

Find the coordinates of point P along the directed line segment AB so that the ratio of AP to PB is 3 to 2.

SOLUTION

In order to divide the segment in the ratio 3 to 2, think of dividing, or *partitioning*, the segment into $3 + 2$, or 5 congruent pieces. Point P is the point that is $\frac{3}{5}$ of the way from point A to point B.

Find the rise and run from point A to point B. Leave the slope in terms of rise and run and do not simplify.

slope of $\overline{AB}$: $m = \dfrac{8-2}{6-3} = \dfrac{6}{3} = \dfrac{\text{rise}}{\text{run}}$

To find the coordinates of point P, add $\frac{3}{5}$ of the run to the x-coordinate of A, and add $\frac{3}{5}$ of the rise to the y-coordinate of A.

run: $\frac{3}{5}$ of $3 = \frac{3}{5} \cdot 3 = 1.8$ rise: $\frac{3}{5}$ of $6 = \frac{3}{5} \cdot 6 = 3.6$

▶ So, the coordinates of P are $(3 + 1.8, 2 + 3.6) = (4.8, 5.6)$. The ratio of AP to PB is 3 to 2.

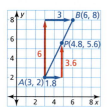

Monitoring Progress Help in English and Spanish at *BigIdeasMath.com*

Find the coordinates of point P along the directed line segment AB so that AP to PB is the given ratio.

7. $A(1, 3), B(8, 4)$; 4 to 1   **8.** $A(-2, 1), B(4, 5)$; 3 to 7

You can apply the Triangle Proportionality Theorem to construct a point along a directed line segment that partitions the segment in a given ratio.

CONSTRUCTION **Constructing a Point along a Directed Line Segment**

Construct the point L on $\overline{AB}$ so that the ratio of AL to LB is 3 to 1.

SOLUTION

Step 1

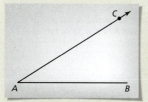

Draw a segment and a ray
Draw $\overline{AB}$ of any length. Choose any point C not on $\overline{AB}$. Draw $\overrightarrow{AC}$.

Step 2

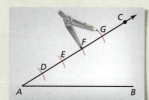

Draw arcs Place the point of a compass at A and make an arc of any radius intersecting $\overrightarrow{AC}$. Label the point of intersection D. Using the same compass setting, make three more arcs on $\overrightarrow{AC}$, as shown. Label the points of intersection E, F, and G and note that $AD = DE = EF = FG$.

Step 3

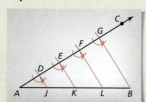

Draw a segment Draw $\overline{GB}$. Copy $\angle AGB$ and construct congruent angles at D, E, and F with sides that intersect $\overline{AB}$ at J, K, and L. Sides $\overline{DJ}, \overline{EK},$ and $\overline{FL}$ are all parallel, and they divide $\overline{AB}$ equally. So, $AJ = JK = KL = LB$. Point L divides directed line segment AB in the ratio 3 to 1.

Section 8.6 Proportionality Theorems **503**

Extra Example 5
Find the coordinates of point F along the directed line segment CD so that the ratio of CF to FD is 3 to 5.

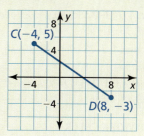

$F(0.5, 2)$

MONITORING PROGRESS ANSWERS

7. $P(6.6, 3.8)$
8. $P(-0.2, 2.2)$

Laurie's Notes — Teacher Actions

- **Turn and Talk:** Pose Example 5. Give time for partners to discuss strategies. If students are getting nowhere, ask a *Probing Question*: "How would you proceed if the segment were horizontal or vertical?" Walk away. Do not solve it for students at this point.
- **Make Sense of Problems and Persevere in Solving Them:** Trust that students have the tools to solve the problem when they persevere.
- **Monitoring Progress:** You may want to have some students do Question 7 and others Question 8.

? Have students perform the construction as shown. "Why are $\overline{DJ}, \overline{EK}$, and $\overline{FL}$ parallel?" Corresponding angles are congruent, so the lines are parallel.

? "Are there other methods you could use to construct a series of parallel segments?" Answers will vary.

Closure

- **3-2-1:** Hand out a *3-2-1* reflection sheet as described on page T-468.

Section 8.6 **503**

Assignment Guide and Homework Check

ASSIGNMENT

Basic: 1, 2, 3–9 odd, 10, 11–23 odd, 29, 34, 38, 41–45

Average: 1, 2–28 even, 29, 30, 34, 37, 38, 41–45

Advanced: 1, 2, 8, 12–30 even, 31, 33–38, 41–45

HOMEWORK CHECK

Basic: 3, 5, 11, 13, 17

Average: 4, 8, 12, 16, 18

Advanced: 8, 12, 16, 20, 30

ANSWERS

1. parallel, Converse of the Triangle Proportionality Theorem
2. $\dfrac{BR}{RC} = \dfrac{AB}{AC}$
3. 9
4. 21
5. yes
6. yes
7. no
8. no
9. CE
10. EG
11. 6
12. 15
13. 12
14. 1
15. 27
16. 9
17. $P(7, -0.4)$
18. $P(2.8, -1)$
19. $P(-1.5, -1.5)$
20. $P(-1, 0.5)$

8.6 Exercises

Dynamic Solutions available at *BigIdeasMath.com*

Vocabulary and Core Concept Check

1. **COMPLETE THE STATEMENT** If a line divides two sides of a triangle proportionally, then it is _____ to the third side. This theorem is known as the _____.

2. **VOCABULARY** In △ABC, point R lies on $\overline{BC}$ and $\overrightarrow{AR}$ bisects ∠CAB. Write the proportionality statement for the triangle that is based on the Triangle Angle Bisector Theorem.

Monitoring Progress and Modeling with Mathematics

In Exercises 3 and 4, find the length of $\overline{AB}$. (See Example 1.)

3. 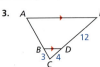   4.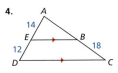

In Exercises 5–8, determine whether $\overline{KM} \parallel \overline{JN}$. (See Example 2.)

5.   6.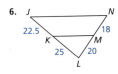

7. 8.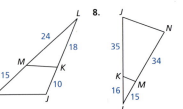

In Exercises 9 and 10, use the diagram to complete the proportion.

9. $\dfrac{BD}{BF} = \dfrac{\square}{CG}$ 10. $\dfrac{CG}{\square} = \dfrac{BF}{DF}$

In Exercises 11 and 12, find the length of the indicated line segment. (See Example 3.)

11. $\overline{VX}$ 12. $\overline{SU}$

In Exercises 13–16, find the value of the variable. (See Example 4.)

13.   14.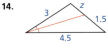

15. 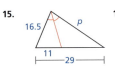   16.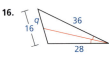

In Exercises 17–20, find the coordinates of point *P* along the directed line segment *AB* so that *AP* to *PB* is the given ratio. (See Example 5.)

17. $A(8, 0), B(3, -2)$; 1 to 4

18. $A(-2, -4), B(6, 1)$; 3 to 2

19. $A(1, 6), B(-2, -3)$; 5 to 1

20. $A(-3, 2), B(5, -4)$; 2 to 6

CONSTRUCTION In Exercises 21 and 22, draw a segment with the given length. Construct the point that divides the segment in the given ratio.

21. 3 in.; 1 to 4   22. 2 in.; 2 to 3

21.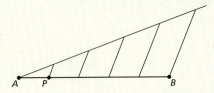

22.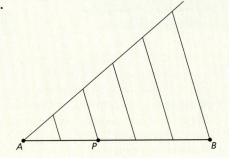

23. **ERROR ANALYSIS** Describe and correct the error in solving for x.

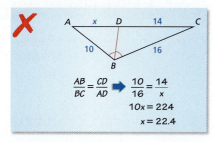

24. **ERROR ANALYSIS** Describe and correct the error in the student's reasoning.

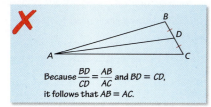

MATHEMATICAL CONNECTIONS In Exercises 25 and 26, find the value of x for which $\overline{PQ} \parallel \overline{RS}$.

25.   26.

27. **PROVING A THEOREM** Prove the Triangle Proportionality Theorem.

    Given $\overline{QS} \parallel \overline{TU}$

    Prove $\dfrac{QT}{TR} = \dfrac{SU}{UR}$

    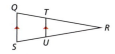

28. **PROVING A THEOREM** Prove the Converse of the Triangle Proportionality Theorem.

    Given $\dfrac{ZY}{YW} = \dfrac{ZX}{XV}$

    Prove $\overline{YX} \parallel \overline{WV}$

29. **MODELING WITH MATHEMATICS** The real estate term *lake frontage* refers to the distance along the edge of a piece of property that touches a lake.

    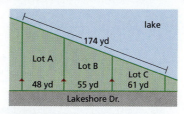

    a. Find the lake frontage (to the nearest tenth) of each lot shown.

    b. In general, the more lake frontage a lot has, the higher its selling price. Which lot(s) should be listed for the highest price?

    c. Suppose that lot prices are in the same ratio as lake frontages. If the least expensive lot is $250,000, what are the prices of the other lots? Explain your reasoning.

30. **MODELING WITH MATHEMATICS** Your school lies directly between your house and the movie theater. The distance from your house to the school is one-fourth of the distance from the school to the movie theater. What point on the graph represents your school?

    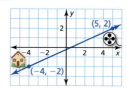

31. **REASONING** In the construction on page 503, explain why you can apply the Triangle Proportionality Theorem in Step 3.

32. **PROVING A THEOREM** Use the diagram with the auxiliary line drawn to write a paragraph proof of the Three Parallel Lines Theorem.

    Given $k_1 \parallel k_2 \parallel k_3$

    Prove $\dfrac{CB}{BA} = \dfrac{DE}{EF}$

    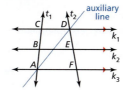

Section 8.6   Proportionality Theorems   505

ANSWERS

23. The proportion should show that AD corresponds with DC and BA corresponds with BC;

    $\dfrac{AD}{DC} = \dfrac{BA}{BC}$

    $\dfrac{x}{14} = \dfrac{10}{16}$

    $x = 8.75$

24. $\overline{AD}$ is a median, not an angle bisector.

25. $x = 3$

26. $x = 5$

27–28. See Additional Answers.

29. a. about 50.9 yd, about 58.4 yd, about 64.7 yd

    b. Lot C

    c. about $287,000, about $318,000; $\dfrac{50.9}{250,000} \approx \dfrac{58.4}{287,000}$ and $\dfrac{50.9}{250,000} \approx \dfrac{64.7}{318,000}$

30. $\left(-\dfrac{11}{5}, -\dfrac{6}{5}\right)$

31. Because $\overline{DJ}, \overline{EK}, \overline{FL},$ and $\overline{GB}$ are cut by a transversal $\overleftrightarrow{AC}$, and $\angle ADJ \cong \angle DEK \cong \angle EFL \cong \angle FGB$ by construction, $\overline{DJ} \parallel \overline{EK} \parallel \overline{FL} \parallel \overline{GB}$ by the Corresponding Angles Converse.

32. Let G be the point of intersection of $\overleftrightarrow{AD}$ and k_2. Because $k_2 \parallel k_3$, and $k_2 \parallel k_1$, $\dfrac{DG}{GA} = \dfrac{DE}{EF}$ and $\dfrac{CB}{BA} = \dfrac{DG}{GA}$ by the Triangle Proportionality Theorem. So, by the Transitive Property of Equality, $\dfrac{CB}{BA} = \dfrac{DE}{EF}$.

ANSWERS

33. isosceles; By the Triangle Angle Bisector Theorem (Thm. 8.9), the ratio of the lengths of the segments of $\overline{LN}$ equals the ratio of the other two side lengths. Because $\overline{LN}$ is bisected, the ratio is 1, and $ML = MN$.

34. Player 1; The angle is bisected, so the lengths are proportional.

35–45. See Additional Answers.

Mini-Assessment

1. In the diagram, $\overline{QT} \parallel \overline{RS}$, $PQ = 6$, $QR = 1.5$, and $PT = 12$. Find ST.

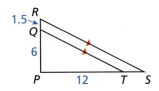

$ST = 3$ units

2. Find AC.

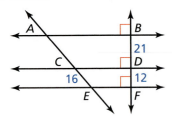

$AC = 28$ units

3. In the diagram, $\angle XWY \cong \angle ZWY$. Find XY.

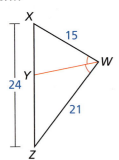

$XY = 10$ units

4. $\overline{QR}$ has endpoints $Q(8, 4)$ and $R(-2, -1)$. Find the coordinates of point P along the directed line segment QR so that the ratio of QP to PR is 1 to 4. $P(6, 3)$

---

33. **CRITICAL THINKING** In △LMN, the angle bisector of ∠M also bisects $\overline{LN}$. Classify △LMN as specifically as possible. Justify your answer.

34. **HOW DO YOU SEE IT?** During a football game, the quarterback throws the ball to the receiver. The receiver is between two defensive players, as shown. If Player 1 is closer to the quarterback when the ball is thrown and both defensive players move at the same speed, which player will reach the receiver first? Explain your reasoning.

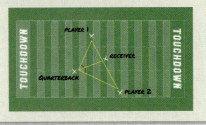

35. **PROVING A THEOREM** Use the diagram with the auxiliary lines drawn to write a paragraph proof of the Triangle Angle Bisector Theorem.

Given ∠YXW ≅ ∠WXZ

Prove $\dfrac{YW}{WZ} = \dfrac{XY}{XZ}$

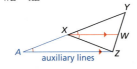

36. **THOUGHT PROVOKING** Write the converse of the Triangle Angle Bisector Theorem. Is the converse true? Justify your answer.

37. **REASONING** How is the Triangle Midsegment Theorem related to the Triangle Proportionality Theorem? Explain your reasoning.

Maintaining Mathematical Proficiency
Reviewing what you learned in previous grades and lessons

Use the triangle. *(Skills Review Handbook)*

41. Which sides are the legs?

42. Which side is the hypotenuse?

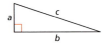

Solve the equation. *(Section 4.3)*

43. $x^2 = 121$
44. $x^2 + 16 = 25$
45. $36 + x^2 = 85$

---

38. **MAKING AN ARGUMENT** Two people leave points A and B at the same time. They intend to meet at point C at the same time. The person who leaves point A walks at a speed of 3 miles per hour. You and a friend are trying to determine how fast the person who leaves point B must walk. Your friend claims you need to know the length of $\overline{AC}$. Is your friend correct? Explain your reasoning.

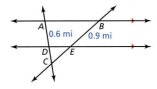

39. **CONSTRUCTION** Given segments with lengths r, s, and t, construct a segment of length x, such that $\dfrac{r}{s} = \dfrac{t}{x}$.

40. **PROOF** Prove *Ceva's Theorem*: If P is any point inside △ABC, then $\dfrac{AY}{YC} \cdot \dfrac{CX}{XB} \cdot \dfrac{BZ}{ZA} = 1$.

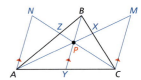

(*Hint*: Draw segments parallel to $\overline{BY}$ through A and C, as shown. Apply the Triangle Proportionality Theorem to △ACM. Show that △APN ~ △MPC, △CXM ~ △BXP, and △BZP ~ △AZN.)

8.5–8.6 What Did You Learn?

Core Vocabulary
directed line segment, *p. 502*

Core Concepts

Section 8.5
Side-Side-Side (SSS) Similarity Theorem, *p. 494*
Side-Angle-Side (SAS) Similarity Theorem, *p. 496*

Section 8.6
Triangle Proportionality Theorem, *p. 500*
Converse of the Triangle Proportionality Theorem, *p. 500*
Three Parallel Lines Theorem, *p. 501*
Triangle Angle Bisector Theorem, *p. 502*
Partitioning a Directed Line Segment, *p. 503*

Mathematical Practices

1. In Exercise 16 on page 498, why do you only need to find two ratios instead of three to compare the side lengths of a pair of triangles?

2. In Exercise 4 on page 504, is it better to use $\frac{7}{6}$ or 1.17 as your ratio of the lengths when finding the length of $\overline{AB}$? Explain your reasoning.

3. In Exercise 30 on page 505, a classmate tells you that your answer is incorrect because you should have divided the segment into four congruent pieces. Respond to your classmate's argument by justifying your original answer.

Performance Task:

Pool Maintenance

Your pool service company has been very successful at maintaining rectangular pools designed for regulation competitions. Now you want to expand your services to owners of private, recreational pools. How will your experience transfer to pools of various shapes and sizes?

To explore the answer to this question and more, check out the Performance Task and Real-Life STEM video at *BigIdeasMath.com*.

ANSWERS

1.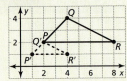

2.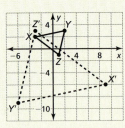

3. 1.9 cm
4. *Sample answer:* reflection in the line $x = -1$ followed by a dilation with center $(-3, 0)$ and $k = 3$
5. *Sample answer:* dilation with center at the origin and $k = \frac{1}{2}$, followed by a reflection in the line $y = x$
6. *Sample answer:* 270° rotation about the origin followed by a dilation with center at the origin and $k = 2$

8 Chapter Review
Dynamic Solutions available at *BigIdeasMath.com*

8.1 Dilations (pp. 461–468)

Graph trapezoid *ABCD* with vertices *A*(1, 1), *B*(1, 3), *C*(3, 2), and *D*(3, 1) and its image after a dilation with a scale factor of 2.

Use the coordinate rule for a dilation with $k = 2$ to find the coordinates of the vertices of the image. Then graph trapezoid *ABCD* and its image.

$(x, y) \rightarrow (2x, 2y)$

$A(1, 1) \rightarrow A'(2, 2)$
$B(1, 3) \rightarrow B'(2, 6)$
$C(3, 2) \rightarrow C'(6, 4)$
$D(3, 1) \rightarrow D'(6, 2)$

Graph the triangle and its image after a dilation with scale factor *k*.

1. $P(2, 2), Q(4, 4), R(8, 2); k = \frac{1}{2}$
2. $X(-3, 2), Y(2, 3), Z(1, -1); k = -3$
3. You are using a magnifying glass that shows the image of an object that is eight times the object's actual size. The image length is 15.2 centimeters. Find the actual length of the object.

8.2 Similarity and Transformations (pp. 469–474)

Describe a similarity transformation that maps △*FGH* to △*LMN*, as shown at the right.

$\overline{FG}$ is horizontal, and $\overline{LM}$ is vertical. If you rotate △*FGH* 90° about the origin as shown at the bottom right, then the image, △*F'G'H'*, will have the same orientation as △*LMN*. △*LMN* appears to be half as large as △*F'G'H'*. Dilate △*F'G'H'* using a scale factor of $\frac{1}{2}$.

$(x, y) \rightarrow \left(\frac{1}{2}x, \frac{1}{2}y\right)$

$F'(-2, 2) \rightarrow F''(-1, 1)$
$G'(-2, 6) \rightarrow G''(-1, 3)$
$H'(-6, 4) \rightarrow H''(-3, 2)$

The vertices of △*F"G"H"* match the vertices of △*LMN*.

▶ So, a similarity transformation that maps △*FGH* to △*LMN* is a rotation of 90° about the origin followed by a dilation with a scale factor of $\frac{1}{2}$.

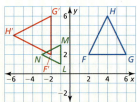

Describe a similarity transformation that maps △*ABC* to △*RST*.

4. $A(1, 0), B(-2, -1), C(-1, -2)$ and $R(-3, 0), S(6, -3), T(3, -6)$
5. $A(6, 4), B(-2, 0), C(-4, 2)$ and $R(2, 3), S(0, -1), T(1, -2)$
6. $A(3, -2), B(0, 4), C(-1, -3)$ and $R(-4, -6), S(8, 0), T(-6, 2)$

8.3 Similar Polygons (pp. 475–484)

In the diagram, EHGF ~ KLMN. Find the scale factor from EHGF to KLMN. Then list all pairs of congruent angles and write the ratios of the corresponding side lengths in a statement of proportionality.

From the diagram, you can see that $\overline{EH}$ and $\overline{KL}$ are corresponding sides. So, the scale factor of EHGF to KLMN is $\dfrac{KL}{EH} = \dfrac{18}{12} = \dfrac{3}{2}$.

$\angle E \cong \angle K$, $\angle H \cong \angle L$, $\angle G \cong \angle M$, and $\angle F \cong \angle N$.

$\dfrac{KL}{EH} = \dfrac{LM}{HG} = \dfrac{MN}{GF} = \dfrac{NK}{FE}$

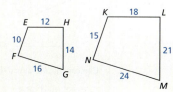

Find the scale factor. Then list all pairs of congruent angles and write the ratios of the corresponding side lengths in a statement of proportionality.

7. ABCD ~ EFGH

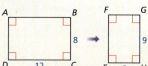

8. △XYZ ~ △RPQ

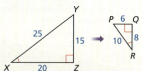

9. Two similar triangles have a scale factor of 3 : 5. The altitude of the larger triangle is 24 inches. What is the altitude of the smaller triangle?

10. Two similar triangles have a pair of corresponding sides of length 12 meters and 8 meters. The larger triangle has a perimeter of 48 meters and an area of 180 square meters. Find the perimeter and area of the smaller triangle.

8.4 Proving Triangle Similarity by AA (pp. 485–490)

Determine whether the triangles are similar. If they are, write a similarity statement. Explain your reasoning.

Because they are both right angles, $\angle F$ and $\angle B$ are congruent. By the Triangle Sum Theorem, $61° + 90° + m\angle E = 180°$, so $m\angle E = 29°$. So, $\angle E$ and $\angle A$ are congruent. So, △DFE ~ △CBA by the AA Similarity Theorem.

Show that the triangles are similar. Write a similarity statement.

11.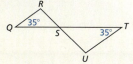

12.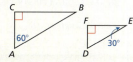

13. A cellular telephone tower casts a shadow that is 72 feet long, while a nearby tree that is 27 feet tall casts a shadow that is 6 feet long. How tall is the tower?

ANSWERS

7. $\dfrac{3}{4}$; $\angle A \cong \angle E$, $\angle B \cong \angle F$, $\angle C \cong \angle G$, $\angle D \cong \angle H$;

$\dfrac{EF}{AB} = \dfrac{FG}{BC} = \dfrac{GH}{CD} = \dfrac{EH}{AD}$

8. $\dfrac{2}{5}$; $\angle X \cong \angle R$, $\angle Y \cong \angle P$, $\angle Z \cong \angle Q$; $\dfrac{RP}{XY} = \dfrac{PQ}{YZ} = \dfrac{RQ}{XZ}$

9. 14.4 in.

10. $P = 32$ m; $A = 80$ m^2

11. $\angle Q \cong \angle T$ and $\angle RSQ \cong \angle UST$, so △RSQ ~ △UST.

12. $\angle C \cong \angle F$ and $\angle B \cong \angle E$, so △ABC ~ △DEF.

13. 324 ft

ANSWERS

14. $\angle C \cong \angle C$ and $\dfrac{CD}{CE} = \dfrac{CB}{CA}$, so $\triangle CBD \sim \triangle CAE$.

15. $\dfrac{QU}{QT} = \dfrac{QR}{QS} = \dfrac{UR}{TS}$, so $\triangle QUR \sim \triangle QTS$.

16. 10.5
17. 7.2

8.5 Proving Triangle Similarity by SSS and SAS (pp. 493–498)

Show that the triangles are similar.

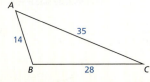

Compare $\triangle ABC$ and $\triangle DEF$ by finding ratios of corresponding side lengths.

Shortest sides	Longest sides	Remaining sides
$\dfrac{AB}{DE} = \dfrac{14}{6} = \dfrac{7}{3}$	$\dfrac{AC}{DF} = \dfrac{35}{15} = \dfrac{7}{3}$	$\dfrac{BC}{EF} = \dfrac{28}{12} = \dfrac{7}{3}$

▶ All the ratios are equal, so $\triangle ABC \sim \triangle DEF$ by the SSS Similarity Theorem.

Use the SSS Similarity Theorem or the SAS Similarity Theorem to show that the triangles are similar.

14. 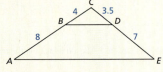 15.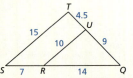

8.6 Proportionality Theorems (pp. 499–506)

In the diagram, $\overline{AD}$ bisects $\angle CAB$. Find the length of $\overline{DB}$.

Because $\overline{AD}$ is an angle bisector of $\angle CAB$, you can apply the Triangle Angle Bisector Theorem.

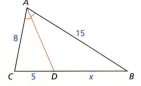

$\dfrac{DB}{DC} = \dfrac{AB}{AC}$ Triangle Angle Bisector Theorem

$\dfrac{x}{5} = \dfrac{15}{8}$ Substitute.

$8x = 75$ Cross Products Property

$9.375 = x$ Solve for x.

▶ The length of $\overline{DB}$ is 9.375 units.

Find the length of $\overline{AB}$.

16. 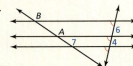 17.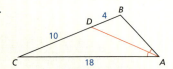

8 Chapter Test

Determine whether the triangles are similar. If they are, write a similarity statement. Explain your reasoning.

1.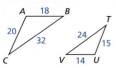
2.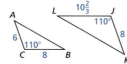
3.

Find the value of the variable.

4.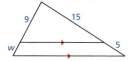
5.
6.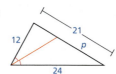

7. Given △QRS ~ △MNP, list all pairs of congruent angles. Then write the ratios of the corresponding side lengths in a statement of proportionality.

8. There is one slice of a large pizza and one slice of a small pizza in the box.
   a. Describe a similarity transformation that maps pizza slice ABC to pizza slice DEF.
   b. What is one possible scale factor for a medium slice of pizza? Explain your reasoning. (Use a dilation on the large slice of pizza.)

9. The original photograph shown is 4 inches by 6 inches.
   a. What transfomations can you use to produce the new photograph?
   b. You dilate the original photograph by a scale factor of $\frac{1}{2}$. What are the dimensions of the new photograph?
   c. You have a frame that holds photos that are 8.5 inches by 11 inches. Can you dilate the original photograph to fit the frame? Explain your reasoning.

10. In a *perspective drawing*, lines that are parallel in real life must meet at a vanishing point on the horizon. To make the train cars in the drawing appear equal in length, they are drawn so that the lines connecting the opposite corners of each car are parallel. Use the dimensions given and the yellow parallel lines to find the length of the bottom edge of the drawing of Car 2.

ANSWERS

1. no; Corresponding side lengths are not proportional.
2. yes; △ABC ~ △KLJ; $\frac{AC}{KJ} = \frac{BC}{LJ}$ and ∠C ≅ ∠J.
3. yes; △WXP ~ △WYZ; ∠X ≅ ∠Y and ∠W ≅ ∠W
4. w = 3
5. q = 27.5
6. p = 14
7. ∠Q ≅ ∠M, ∠R ≅ ∠N, ∠S ≅ ∠P; $\frac{QR}{MN} = \frac{RS}{NP} = \frac{QS}{MP}$
8. a. Sample answer: 270° rotation about the origin followed by a dilation with center at the origin and $k = \frac{1}{2}$, followed by the translation $(x, y) \rightarrow (x + 1, y - 1)$
   b. Sample answer: $k = \frac{3}{4}$; A medium slice would be between a small and a large, and $\frac{1}{2} < \frac{3}{4} < 1$.
9. a. reflection, reduction (dilation), and translation
   b. 2 in. by 3 in.
   c. no; The scale factor for the shorter sides is $\frac{17}{8}$, but the scale factor for the longer sides is $\frac{11}{6}$. So, the photo would have to be cropped or distorted in order to fit the frame.
10. about 4.3 cm

If students need help...	If students got it...
Lesson Tutorials	Resources by Chapter • Enrichment and Extension • Cumulative Review
Skills Review Handbook	Performance Task
BigIdeasMath.com	Start the *next* Section

8 Cumulative Assessment

1. Use the graph of quadrilaterals *ABCD* and *QRST*.

   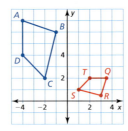

   a. Write a composition of transformations that maps quadrilateral *ABCD* to quadrilateral *QRST*.

   b. Are the quadrilaterals similar? Explain your reasoning.

2. In the diagram, *ABCD* is a parallelogram. Which congruence theorem(s) could you use to show that $\triangle AED \cong \triangle CEB$? Select all that apply.

   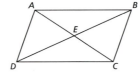

   SAS Congruence Theorem

   SSS Congruence Theorem

   HL Congruence Theorem

   ASA Congruence Theorem

   AAS Congruence Theorem

3. Which expression is *not* equivalent to $20x^2 + 27x - 14$?

   Ⓐ $27x + 20x^2 - 14$

   Ⓑ $(5x - 2)(4x + 7)$

   Ⓒ $(5x + 2)(4x - 7)$

   Ⓓ $(4x + 7)(5x - 2)$

4. A bag contains six tiles with a letter on each tile. The letters are A, B, C, D, E, and F. You randomly draw two tiles from the bag. Which expression represents the probability that you draw the A tile and the B tile?

   $\dfrac{2!}{6!}$ $\dfrac{1}{{}_6P_2}$ ${}_6P_2$ $\dfrac{1}{{}_6C_2}$ ${}_6C_2$ $\dfrac{{}_6C_2}{{}_6C_4}$

512   Chapter 8   Similarity

5. Enter a statement or reason in each blank to complete the two-column proof.

   Given $\dfrac{KJ}{KL} = \dfrac{KH}{KM}$

   Prove $\angle LMN \cong \angle JHG$

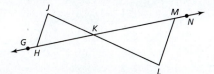

STATEMENTS	REASONS
1. $\dfrac{KJ}{KL} = \dfrac{KH}{KM}$	1. Given
2. $\angle JKH \cong \angle LKM$	2. _____
3. $\triangle JKH \sim \triangle LKM$	3. _____
4. $\angle KHJ \cong \angle KML$	4. _____
5. _____	5. Definition of congruent angles
6. $m\angle KHJ + m\angle JHG = 180°$	6. Linear Pair Postulate
7. $m\angle JHG = 180° - m\angle KHJ$	7. _____
8. $m\angle KML + m\angle LMN = 180°$	8. _____
9. _____	9. Subtraction Property of Equality
10. $m\angle LMN = 180° - m\angle KHJ$	10. _____
11. _____	11. Transitive Property of Equality
12. $\angle LMN \cong \angle JHG$	12. _____

6. The coordinates of the vertices of $\triangle DEF$ are $D(-8, 5)$, $E(-5, 8)$, and $F(-1, 4)$. The coordinates of the vertices of $\triangle JKL$ are $J(16, -10)$, $K(10, -16)$, and $L(2, -8)$. $\angle D \cong \angle J$. Can you show that $\triangle DEF \sim \triangle JKL$ by using the AA Similarity Theorem? If so, do so by listing the congruent corresponding angles and writing a similarity transformation that maps $\triangle DEF$ to $\triangle JKL$. If not, explain why not.

7. Classify the quadrilateral using the most specific name.

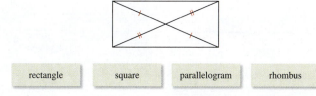

8. Your friend makes the statement "Quadrilateral $PQRS$ is similar to quadrilateral $WXYZ$." Describe the relationships between corresponding angles and between corresponding sides that make this statement true.

Chapter 9 Pacing Guide	
Chapter Opener/ Mathematical Practices	0.5 Day
Section 1	1.5 Days
Section 2	1 Day
Section 3	2 Days
Quiz	0.5 Day
Section 4	1.5 Days
Section 5	2 Days
Section 6	2 Days
Chapter Review/ Chapter Tests	2 Days
Total Chapter 9	13 Days
Year-to-Date	131 Days

9 Right Triangles and Trigonometry

9.1 The Pythagorean Theorem
9.2 Special Right Triangles
9.3 Similar Right Triangles
9.4 The Tangent Ratio
9.5 The Sine and Cosine Ratios
9.6 Solving Right Triangles

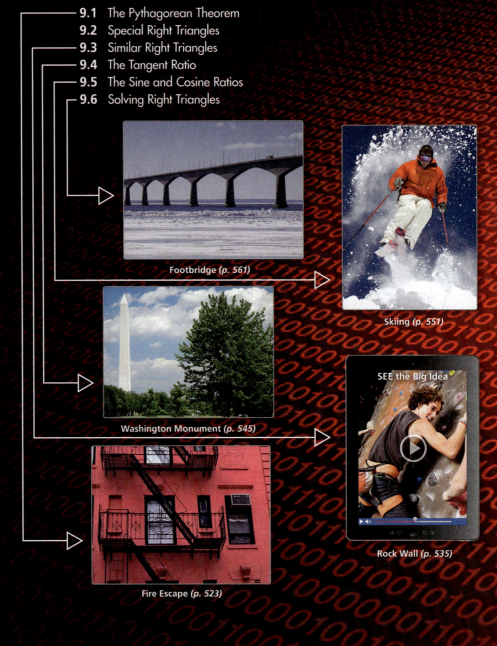

Footbridge (p. 561)

Skiing (p. 551)

Washington Monument (p. 545)

SEE the Big Idea

Rock Wall (p. 535)

Fire Escape (p. 523)

Laurie's Notes

Chapter Summary

- This is a fairly long chapter that introduces students to right triangle trigonometry. Students will encounter a more in-depth study of trigonometry in Math III.
- The first lesson on the Pythagorean Theorem will not be completely new to students who will have familiarity with this theorem from middle school.
- The next two lessons use knowledge of similar triangles to investigate relationships in special right triangles (30°-60°-90° and 45°-45°-90°) as well as similar triangles that are formed when the altitude to the hypotenuse is drawn in a right triangle. Being familiar with these relationships and solving for segment lengths in triangles will be helpful in subsequent lessons.
- The next three lessons present the tangent, sine, and cosine ratios. The focus of these lessons is to solve for parts of a right triangle. Many real-life applications are presented.
- Lesson 9.5 also introduces students to trigonometric identities.

What Your Students Have Learned

Middle School
- Use the Pythagorean Theorem to find distances between points in the coordinate plane.
- Compare ratios using proportions and solve proportions using a variety of strategies.
- Determine whether two figures are similar using proportions.

Math I
- Use properties of equality to justify steps in solving problems involving segment lengths and angle measures.
- Solve multi-step linear equations in one variable that have variables on both sides using inverse operations.

What Your Students Will Learn

Math II
- Use the Pythagorean Theorem and the converse of the Pythagorean Theorem.
- Find side lengths in 45°-45°-90° and 30°-60°-90° triangles.
- Use geometric means to write proportions of similar right triangles.
- Use tangent, sine, and cosine ratios to find the length of a leg or a hypotenuse of a right triangle.
- Find the sine and cosine of angle measures in 45°-45°-90° and 30°-60°-90° triangles.
- Use trigonometric identities involving sine and cosine to find trigonometric values.
- Use inverse trigonometric ratios to find angle measures of right triangles.
- Solve right triangles using the Pythagorean Theorem and inverse trigonometric ratios.

Dynamic Teaching Tools
Dynamic Assessment & Progress Monitoring Tool
Lesson Planning Tool
Interactive Whiteboard Lesson Library
Dynamic Classroom with Dynamic Investigations
Real-Life STEM Videos

Scaffolding in the Classroom
Build on prior knowledge.
Instead of reteaching a topic, use review. Begin with the statement, "Remember when we did _____?" This lets the students know that you know they have previously learned the concept or process. Use a problem to review. Then say, "Today we will use this idea to _____."

Questioning in the Classroom
What was that answer?
Once a student has answered a question, ask another student to paraphrase the answer. Ask yet another student to illustrate or explain what was said.

Laurie's Notes

Maintaining Mathematical Proficiency

Using Properties of Radicals

- Review the Product Property of Square Roots: For all real numbers $a \geq 0$ and $b \geq 0$, $\sqrt{ab} = \sqrt{a}\sqrt{b}$.
- Remind students that they need to find all perfect-square factors of the radicand. Then they can rewrite the radicand as a product of perfect-square factors and the remaining factors.
- For Example 2, point out that when students multiply by $\frac{\sqrt{5}}{\sqrt{5}}$, they are multiplying by 1, which does not change the value of the expression.

COMMON ERROR When rationalizing the denominator of a radical expression, students may forget to multiply *both* the numerator and denominator by the same number.

Solving Proportions

- Review the Cross Products Property: If $\frac{a}{b} = \frac{c}{d}$, then $ad = bc$ ($b \neq 0$, $d \neq 0$).
- Ask students how to solve the proportion in Example 3 by using the Multiplication Property of Equality instead of the Cross Products Property.

COMMON ERROR Some students may try to solve a proportion by multiplying the numerators together and the denominators together, and then setting these products equal. Encourage these students to use the Multiplication Property of Equality, instead of the Cross Products Property, to solve a proportion.

Mathematical Practices (continued on page 516)

- The *Mathematical Practices* page focuses attention on how mathematics is learned—process versus content. This page demonstrates that mathematically proficient students understand the importance of precise definitions. They also understand the difference between precise and approximate measures, and they know from the context of a problem what type of measurement is called for.
- Use the *Mathematical Practices* page to help students develop mathematical habits of mind—how mathematics can be explored and how mathematics is thought about.
- Students will be working with unit circle trigonometry on this page and therefore need to be familiar with the definition of a right triangle in *standard position*. In the Core Concept, the three necessary conditions for the standard position of a right triangle are described.

If students need help...	*If students got it...*
Student Journal • Maintaining Mathematical Proficiency	Game Closet at *BigIdeasMath.com*
Lesson Tutorials	Start the *next* Section
Skills Review Handbook	

Maintaining Mathematical Proficiency

Using Properties of Radicals

Example 1 Simplify $\sqrt{128}$.

$\sqrt{128} = \sqrt{64 \cdot 2}$ Factor using the greatest perfect square factor.

$= \sqrt{64} \cdot \sqrt{2}$ Product Property of Square Roots

$= 8\sqrt{2}$ Simplify.

Example 2 Simplify $\dfrac{4}{\sqrt{5}}$.

$\dfrac{4}{\sqrt{5}} = \dfrac{4}{\sqrt{5}} \cdot \dfrac{\sqrt{5}}{\sqrt{5}}$ Multiply by $\dfrac{\sqrt{5}}{\sqrt{5}}$.

$= \dfrac{4\sqrt{5}}{\sqrt{25}}$ Product Property of Square Roots

$= \dfrac{4\sqrt{5}}{5}$ Simplify.

Simplify the expression.

1. $\sqrt{75}$
2. $\sqrt{270}$
3. $\sqrt{135}$
4. $\dfrac{2}{\sqrt{7}}$
5. $\dfrac{5}{\sqrt{2}}$
6. $\dfrac{12}{\sqrt{6}}$

Solving Proportions

Example 3 Solve $\dfrac{x}{10} = \dfrac{3}{2}$.

$\dfrac{x}{10} = \dfrac{3}{2}$ Write the proportion.

$x \cdot 2 = 10 \cdot 3$ Cross Products Property

$2x = 30$ Multiply.

$\dfrac{2x}{2} = \dfrac{30}{2}$ Divide each side by 2.

$x = 15$ Simplify.

Solve the proportion.

7. $\dfrac{x}{12} = \dfrac{3}{4}$
8. $\dfrac{x}{3} = \dfrac{5}{2}$
9. $\dfrac{4}{x} = \dfrac{7}{56}$
10. $\dfrac{10}{23} = \dfrac{4}{x}$
11. $\dfrac{x+1}{2} = \dfrac{21}{14}$
12. $\dfrac{9}{3x-15} = \dfrac{3}{12}$

13. **ABSTRACT REASONING** The Product Property of Square Roots allows you to simplify the square root of a product. Are you able to simplify the square root of a sum? of a difference? Explain.

Dynamic Solutions available at *BigIdeasMath.com*

What Your Students Have Learned

- Use properties of square roots to simplify radical expressions.
- Simplify radical expressions by rationalizing the denominator.
- Solve proportions using the Cross Products Property.

ANSWERS

1. $5\sqrt{3}$
2. $3\sqrt{30}$
3. $3\sqrt{15}$
4. $\dfrac{2\sqrt{7}}{7}$
5. $\dfrac{5\sqrt{2}}{2}$
6. $2\sqrt{6}$
7. $x = 9$
8. $x = 7.5$
9. $x = 32$
10. $x = 9.2$
11. $x = 2$
12. $x = 17$
13. no; no; Because square roots have to do with factors, the rule allows you to simplify with products, not sums and differences.

Vocabulary Review

Have students make an Information Wheel for each of the following properties.

- Product Property of Square Roots
- Cross Products Property

MONITORING PROGRESS ANSWERS

1. Sample answer:

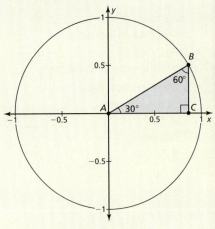

$A(0, 0)$, $B\left(\frac{\sqrt{3}}{2}, \frac{1}{2}\right)$, $C\left(\frac{\sqrt{3}}{2}, 0\right)$

2. Sample answer:

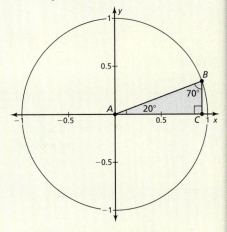

$A(0, 0)$, $B(0.94, 0.34)$, $C(0.94, 0)$

Mathematical Practices

Mathematically proficient students express numerical answers precisely.

Attending to Precision

Core Concept

Standard Position for a Right Triangle

In *unit circle trigonometry*, a right triangle is in **standard position** when:

1. The hypotenuse is a radius of the circle of radius 1 with center at the origin.
2. One leg of the right triangle lies on the *x*-axis.
3. The other leg of the right triangle is perpendicular to the *x*-axis.

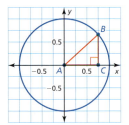

EXAMPLE 1 Drawing an Isosceles Right Triangle in Standard Position

Use dynamic geometry software to construct an isosceles right triangle in standard position. What are the exact coordinates of its vertices?

SOLUTION

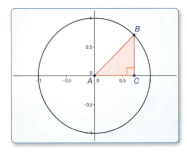

Sample
Points
$A(0, 0)$
$B(0.71, 0.71)$
$C(0.71, 0)$
Segments
$AB = 1$
$BC = 0.71$
$AC = 0.71$
Angle
$m\angle A = 45°$

To determine the exact coordinates of the vertices, label the length of each leg x. By the Pythagorean Theorem, which you will study in Section 9.1, $x^2 + x^2 = 1$. Solving this equation yields

$$x = \frac{1}{\sqrt{2}}, \text{ or } \frac{\sqrt{2}}{2}.$$

▶ So, the exact coordinates of the vertices are $A(0, 0)$, $B\left(\frac{\sqrt{2}}{2}, \frac{\sqrt{2}}{2}\right)$, and $C\left(\frac{\sqrt{2}}{2}, 0\right)$.

Monitoring Progress

1. Use dynamic geometry software to construct a right triangle with acute angle measures of 30° and 60° in standard position. What are the exact coordinates of its vertices?
2. Use dynamic geometry software to construct a right triangle with acute angle measures of 20° and 70° in standard position. What are the approximate coordinates of its vertices?

Laurie's Notes — Mathematical Practices (continued from page T-515)

- Example 1 shows the construction of a right isosceles triangle in standard position. The approximate coordinates can be found using a function of the software. To find the exact coordinate, the Pythagorean Theorem is needed. This theorem is formally presented in the first lesson of this chapter. However, students should be familiar with the theorem from middle school.
- Work through the steps in solving for the exact coordinates.
- Give time for students to work through the *Monitoring Progress* questions, and then discuss as a class.

Laurie's Notes

Dynamic Teaching Tools
Dynamic Assessment & Progress Monitoring Tool
Lesson Planning Tool
Interactive Whiteboard Lesson Library
Dynamic Classroom with Dynamic Investigations

Overview of Section 9.1

Introduction
- Students should certainly be familiar with the Pythagorean Theorem from middle school. In this lesson, the theorem and its converse are presented. Two proofs of the theorem are investigated in the explorations.
- Also in this lesson, students will use the Pythagorean Inequalities Theorem to classify triangles as right, acute, or obtuse when the side lengths are given. This theorem can easily be explored using dynamic geometry software.

Resources
- There are many online resources related to the Pythagorean Theorem. Some resources are applets that present a dynamic proof of the theorem. Others allow the user to input three side lengths, and the output is the type of triangle formed, classified by its angles.

Teaching Strategy
- Students should be familiar with the Pythagorean Theorem from middle school. They will likely recall the 3-4-5 right triangle and may recall other Pythagorean triples.
- In middle school, students also study similar polygons. In this lesson, the first few multiples of common Pythagorean triples are stated in the *Core Concept*. Students are generally comfortable when the Pythagorean triples are multiplied by factors of 2, 3, and x.
- Students are less certain if three numbers form a Pythagorean triple when the 3-4-5 triple is transformed in a different manner. I like to pose the following examples:

    Which of the following are Pythagorean triples? Explain.

    a. 0.3, 0.4, 0.5   Pythagorean triple

    b. $\frac{1}{3}, \frac{1}{4}, \frac{1}{5}$ not a Pythagorean triple

    c. $\sqrt{3}, \sqrt{4}, \sqrt{5}$ not a Pythagorean triple

    d. $\frac{3}{2}, \frac{4}{2}, \frac{5}{2}$ Pythagorean triple

    e. 3.2, 4.2, 5.2   not a Pythagorean triple

- The underlying concept of similarity should be evident in student responses.

Pacing Suggestion
- Students should be familiar with at least one proof of the Pythagorean Theorem. The explorations provide two proofs within the reach of students. Complete the explorations and transition to the formal lesson.

What Your Students Will Learn

- Use the Pythagorean Theorem to determine whether the side lengths of a right triangle form a Pythagorean triple.
- Use the converse of the Pythagorean Theorem to determine whether a triangle is a right triangle.
- Classify triangles as acute triangles or obtuse triangles using the Pythagorean Inequality Theorem.

Laurie's Notes

Exploration

Motivate
- Share information about Pythagoras, who was born in Greece in 569 B.C.
  - He is known as the Father of Numbers.
  - He traveled extensively in Egypt, learning math, astronomy, and music.
  - Pythagoras undertook a reform of the cultural life of Cretona, urging the citizens to follow his religious, political, and philosophical goals.
  - He created a school where his followers, known as Pythagoreans, lived and worked. They observed a rule of silence called echemythia, the breaking of which was punishable by death. One had to remain silent for five years before being allowed to contribute to the group.
  - Over the years, many mathematicians and non-mathematicians have given various proofs of the Pythagorean Theorem. One of our former presidents, President James Garfield, is credited with a proof.

Exploration Note
- The explorations on this page are two of the many ways in which the Pythagorean Theorem can be proven to be true. Consider having some students work on Exploration 1 while others work on Exploration 2. Students then present their results to the other group.
- To help facilitate the first exploration, make a template for students to work with for part of the exploration. Draw the large square with sides subdivided into segments a and b, as shown. From this template students can quickly create the four triangles and the small square.

- I have found that students are not always accurate enough in making copies of the right triangle to have this exploration work well, hence the initial template.

Exploration 1
- Distribute the template, a straightedge, and a pair of scissors to students so that the exploration can be completed efficiently.
- Students should be able to write equations for the areas of each of the polygons.
- ❓ "What is the area of the large square in terms of the area of the four triangles and the small square?" Answers will vary but should be equivalent to $4\left(\frac{1}{2}ab\right) + c^2 = 2ab + c^2$.
- ❓ "What is the area of the large square in terms of the area of the two rectangles and two smaller squares?" Answers will vary but should be equivalent to $2ab + a^2 + b^2$.
- **Construct Viable Arguments and Critique the Reasoning of Others:** Ask a volunteer to explain how this work is related to the Pythagorean Theorem.

Exploration 2
- Students must first determine why the three triangles are similar. If they need a hint, refer them to Section 8.4.
- Students should begin by writing proportional relationships that they are confident of. Substitute side names with a, b, and c whenever possible.

Communicate Your Answer
- Question 4 could be used as an extra credit problem.

Connecting to Next Step
- Students are familiar with the Pythagorean Theorem from middle school, but they would not have proven the theorem. Students should now be ready to work with the theorem in the formal lesson.

9.1 The Pythagorean Theorem

Essential Question How can you prove the Pythagorean Theorem?

EXPLORATION 1 Proving the Pythagorean Theorem without Words

Work with a partner.

a. Draw and cut out a right triangle with legs a and b, and hypotenuse c.

b. Make three copies of your right triangle. Arrange all four triangles to form a large square, as shown.

c. Find the area of the large square in terms of a, b, and c by summing the areas of the triangles and the small square.

d. Copy the large square. Divide it into two smaller squares and two equally-sized rectangles, as shown.

e. Find the area of the large square in terms of a and b by summing the areas of the rectangles and the smaller squares.

f. Compare your answers to parts (c) and (e). Explain how this proves the Pythagorean Theorem.

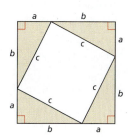

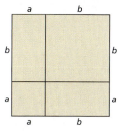

EXPLORATION 2 Proving the Pythagorean Theorem

Work with a partner.

a. Draw a right triangle with legs a and b, and hypotenuse c, as shown. Draw the altitude from C to $\overline{AB}$. Label the lengths, as shown.

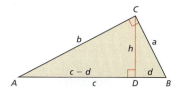

REASONING ABSTRACTLY

To be proficient in math, you need to know and flexibly use different properties of operations and objects.

b. Explain why $\triangle ABC$, $\triangle ACD$, and $\triangle CBD$ are similar.

c. Write a two-column proof using the similar triangles in part (b) to prove that $a^2 + b^2 = c^2$.

Communicate Your Answer

3. How can you prove the Pythagorean Theorem?

4. Use the Internet or some other resource to find a way to prove the Pythagorean Theorem that is different from Explorations 1 and 2.

Section 9.1   The Pythagorean Theorem   517

Dynamic Teaching Tools

Dynamic Assessment & Progress Monitoring Tool
Lesson Planning Tool
Interactive Whiteboard Lesson Library
Dynamic Classroom with Dynamic Investigations

ANSWERS

1. a. Check students' work.
   b. Check students' work.
   c. $\frac{1}{2}ab + \frac{1}{2}ab + \frac{1}{2}ab + \frac{1}{2}ab + c^2 = 2ab + c^2$
   d. Check students' work.
   e. $ab + ab + a^2 + b^2 = 2ab + a^2 + b^2$
   f. The equations from parts (c) and (e) both describe the area of the same square. So, they are equal, $2ab + c^2 = 2ab + a^2 + b^2$, and by the Subtraction Property of Equality, $c^2 = a^2 + b^2$.

2. a. Check students' work.
   b. $\triangle CBD \sim \triangle ABC$ by the AA Similarity Theorem because both triangles have a right angle, and both triangles include $\angle B$. $\triangle ACD \sim \triangle ABC$ by the AA Similarity Theorem because both triangles have a right angle, and both triangles include $\angle A$. $\triangle ACD \sim \triangle CBD$ by the AA Similarity Theorem because both triangles have a right angle, and both $\angle B$ and $\angle ACD$ are complementary to $\angle BCD$, so $\angle B \cong \angle ACD$.
   c. See Additional Answers.

3. You can create a physical model, a drawing, or a formal proof.

---

4. *Sample answer:* Arrange the same triangles from method 1 as shown at the right. The area of the large square is c^2. The area of the small square is $(b - a)^2 = b^2 - 2ab + a^2$, or it can be written as the area of the large square minus the area of the triangles: $c^2 - 4\left(\frac{1}{2}ab\right) = c^2 - 2ab$. Set these two expressions equal to each other to get $b^2 - 2ab + a^2 = c^2 - 2ab$, which becomes $a^2 + b^2 = c^2$, when $2ab$ is added to each side.

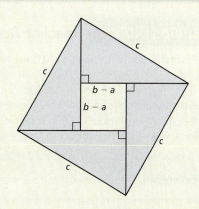

English Language Learners

Vocabulary
Some students may confuse the legs of an isosceles triangle with the legs of a right triangle. Have students practice classifying triangles and identifying the legs, base, and hypotenuse when present. Discuss the properties of each type of *leg* and ask students if it is possible for a single triangle to contain both types of legs.

Extra Example 1

Find the value of *x*. Then tell whether the side lengths form a Pythagorean triple.

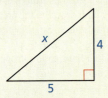

$x = \sqrt{41}$; $\sqrt{41}$ is not an integer, so the side lengths do not form a Pythagorean triple.

9.1 Lesson

What You Will Learn
- Use the Pythagorean Theorem.
- Use the Converse of the Pythagorean Theorem.
- Classify triangles.

Core Vocabulary
Pythagorean triple, *p. 518*
Previous
right triangle
legs of a right triangle
hypotenuse

Using the Pythagorean Theorem

One of the most famous theorems in mathematics is the Pythagorean Theorem, named for the ancient Greek mathematician Pythagoras. This theorem describes the relationship between the side lengths of a right triangle.

Theorem

Pythagorean Theorem

In a right triangle, the square of the length of the hypotenuse is equal to the sum of the squares of the lengths of the legs.

Proof Explorations 1 and 2, p. 517; Ex. 39, p. 538

$c^2 = a^2 + b^2$

A **Pythagorean triple** is a set of three positive integers *a*, *b*, and *c* that satisfy the equation $c^2 = a^2 + b^2$.

STUDY TIP
You may find it helpful to memorize the basic Pythagorean triples, shown in **bold**, for standardized tests.

Core Concept

Common Pythagorean Triples and Some of Their Multiples

3, 4, 5	**5, 12, 13**	**8, 15, 17**	**7, 24, 25**
6, 8, 10	10, 24, 26	16, 30, 34	14, 48, 50
9, 12, 15	15, 36, 39	24, 45, 51	21, 72, 75
$3x, 4x, 5x$	$5x, 12x, 13x$	$8x, 15x, 17x$	$7x, 24x, 25x$

The most common Pythagorean triples are in bold. The other triples are the result of multiplying each integer in a bold-faced triple by the same factor.

EXAMPLE 1 Using the Pythagorean Theorem

Find the value of *x*. Then tell whether the side lengths form a Pythagorean triple.

SOLUTION

$c^2 = a^2 + b^2$	Pythagorean Theorem
$x^2 = 5^2 + 12^2$	Substitute.
$x^2 = 25 + 144$	Multiply.
$x^2 = 169$	Add.
$x = 13$	Find the positive square root.

▶ The value of *x* is 13. Because the side lengths 5, 12, and 13 are integers that satisfy the equation $c^2 = a^2 + b^2$, they form a Pythagorean triple.

518   Chapter 9   Right Triangles and Trigonometry

Laurie's Notes   Teacher Actions

- **Paired Verbal Fluency** and **Attend to Precision:** Have students pair up and follow the protocol described on page T-224. Ask students to share what they recall about the Pythagorean Theorem from middle school. Students should be precise in their summaries.
- *Pythagorean triples* and their multiples are a handy list to remember. Suggest to students that they memorize the four key triples in bold.
- **? Fact-First Questioning:** Say, "If *p*, *q*, and *r* satisfy the Pythagorean Theorem, then $\frac{p}{2}$, $\frac{q}{2}$, and $\frac{r}{2}$ will also. Explain." Listen for valid reasoning.
- **Extension:** See the *Teaching Strategy* on page T-516 for an extension idea.

CONNECTIONS TO ALGEBRA

The Product Property of Square Roots on page 192 states that the square root of a product equals the product of the square roots of the factors.

EXAMPLE 2 Using the Pythagorean Theorem

Find the value of x. Then tell whether the side lengths form a Pythagorean triple.

SOLUTION

$c^2 = a^2 + b^2$	Pythagorean Theorem
$14^2 = 7^2 + x^2$	Substitute.
$196 = 49 + x^2$	Multiply.
$147 = x^2$	Subtract 49 from each side.
$\sqrt{147} = x$	Find the positive square root.
$\sqrt{49} \cdot \sqrt{3} = x$	Product Property of Square Roots
$7\sqrt{3} = x$	Simplify.

▶ The value of x is $7\sqrt{3}$. Because $7\sqrt{3}$ is not an integer, the side lengths do not form a Pythagorean triple.

EXAMPLE 3 Solving a Real-Life Problem

The skyscrapers shown are connected by a skywalk with support beams. Use the Pythagorean Theorem to approximate the length of each support beam.

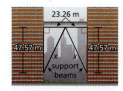

SOLUTION

Each support beam forms the hypotenuse of a right triangle. The right triangles are congruent, so the support beams are the same length.

$x^2 = (23.26)^2 + (47.57)^2$	Pythagorean Theorem
$x = \sqrt{(23.26)^2 + (47.57)^2}$	Find the positive square root.
$x \approx 52.95$	Use a calculator to approximate.

▶ The length of each support beam is about 52.95 meters.

Monitoring Progress Help in English and Spanish at *BigIdeasMath.com*

Find the value of x. Then tell whether the side lengths form a Pythagorean triple.

1.
2.
3. An anemometer is a device used to measure wind speed. The anemometer shown is attached to the top of a pole. Support wires are attached to the pole 5 feet above the ground. Each support wire is 6 feet long. How far from the base of the pole is each wire attached to the ground?

Differentiated Instruction

Inclusion
Have advanced students use the Pythagorean Theorem to develop a formula for the length of a diagonal of a cube with side length s.
diagonal length $= s\sqrt{3}$

Extra Example 2
Find the value of x. Then tell whether the side lengths form a Pythagorean triple.

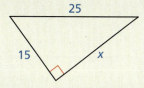

$x = 20$; Because the side lengths 15, 20, and 25 are integers that satisfy the equation $a^2 + b^2 = c^2$, they form a Pythagorean triple.

Extra Example 3
The flagpole shown is supported by two wires. Use the Pythagorean Theorem to approximate the length of each wire.

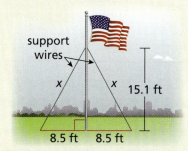

The length of each wire is about 17.3 feet.

MONITORING PROGRESS ANSWERS

1. $x = 2\sqrt{13} \approx 7.2$; no
2. $x = 4$; yes
3. about 3.32 ft

Laurie's Notes — Teacher Actions

- **Think-Pair-Share:** Pose Example 2, and have students work independently to solve.
- **Use Appropriate Tools Strategically:** Calculators are useful tools for this chapter.
- **Attend to Precision:** In Example 3, students are asked to approximate the length of each support beam. While a calculator could give a decimal answer with many place values, students should understand why a rounded answer is sufficient for the context.
- **Think-Pair-Share:** Have students answer Questions 1–3, and then share and discuss as a class.

Extra Example 4
Tell whether each triangle is a right triangle.
a.
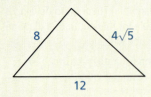
yes

b.
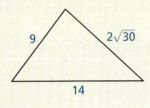
no

MONITORING PROGRESS ANSWERS
4. yes
5. no

Using the Converse of the Pythagorean Theorem
The converse of the Pythagorean Theorem is also true. You can use it to determine whether a triangle with given side lengths is a right triangle.

Theorem

Converse of the Pythagorean Theorem
If the square of the length of the longest side of a triangle is equal to the sum of the squares of the lengths of the other two sides, then the triangle is a right triangle.

If $c^2 = a^2 + b^2$, then $\triangle ABC$ is a right triangle.

Proof Ex. 39, p. 524

EXAMPLE 4 Verifying Right Triangles

Tell whether each triangle is a right triangle.

a.   b.

USING TOOLS STRATEGICALLY
Use a calculator to determine that $\sqrt{113} \approx 10.630$ is the length of the longest side in part (a).

SOLUTION
Let c represent the length of the longest side of the triangle. Check to see whether the side lengths satisfy the equation $c^2 = a^2 + b^2$.

a. $(\sqrt{113})^2 \stackrel{?}{=} 7^2 + 8^2$

   $113 \stackrel{?}{=} 49 + 64$

   $113 = 113$ ✓

▶ The triangle is a right triangle.

b. $(4\sqrt{95})^2 \stackrel{?}{=} 15^2 + 36^2$

   $4^2 \cdot (\sqrt{95})^2 \stackrel{?}{=} 15^2 + 36^2$

   $16 \cdot 95 \stackrel{?}{=} 225 + 1296$

   $1520 \neq 1521$ ✗

▶ The triangle is *not* a right triangle.

Monitoring Progress Help in English and Spanish at *BigIdeasMath.com*

Tell whether the triangle is a right triangle.

4.   5.

Laurie's Notes — Teacher Actions

? "Is the converse of every true statement also true?" **no**
? "What is the converse of the Pythagorean Theorem?" **If the sum of the squares of the lengths of the two shorter sides of a triangle equals the square of the length of the longest side, then the triangle is a right triangle.** Say, "The converse is true."

- **Think-Alouds:** In Example 4, have partner A *Think-Aloud* as he or she decides whether the triangle in part (a) is a right triangle. "If the triangle is a right triangle, then …." Partners now reverse roles as partner B decides whether the triangle in part (b) is a right triangle.

Classifying Triangles

The Converse of the Pythagorean Theorem is used to determine whether a triangle is a right triangle. You can use the theorem below to determine whether a triangle is acute or obtuse.

Theorem

Pythagorean Inequalities Theorem

For any $\triangle ABC$, where c is the length of the longest side, the following statements are true.

If $c^2 < a^2 + b^2$, then $\triangle ABC$ is acute.   If $c^2 > a^2 + b^2$, then $\triangle ABC$ is obtuse.

$c^2 < a^2 + b^2$

$c^2 > a^2 + b^2$

Proof Exs. 42 and 43, p. 524

REMEMBER

The Triangle Inequality Theorem on page 381 states that the sum of the lengths of any two sides of a triangle is greater than the length of the third side.

EXAMPLE 5 Classifying Triangles

Verify that segments with lengths of 4.3 feet, 5.2 feet, and 6.1 feet form a triangle. Is the triangle *acute*, *right*, or *obtuse*?

SOLUTION

Step 1 Use the Triangle Inequality Theorem to verify that the segments form a triangle.

$4.3 + 5.2 \stackrel{?}{>} 6.1$ $4.3 + 6.1 \stackrel{?}{>} 5.2$ $5.2 + 6.1 \stackrel{?}{>} 4.3$
$9.5 > 6.1$ ✓ $10.4 > 5.2$ ✓ $11.3 > 4.3$ ✓

▶ The segments with lengths of 4.3 feet, 5.2 feet, and 6.1 feet form a triangle.

Step 2 Classify the triangle by comparing the square of the length of the longest side with the sum of the squares of the lengths of the other two sides.

c^2		$a^2 + b^2$	Compare c^2 with $a^2 + b^2$.
6.1^2		$4.3^2 + 5.2^2$	Substitute.
37.21		$18.49 + 27.04$	Simplify.
37.21	$<$	45.53	c^2 is less than $a^2 + b^2$.

▶ The segments with lengths of 4.3 feet, 5.2 feet, and 6.1 feet form an acute triangle.

Monitoring Progress   Help in English and Spanish at BigIdeasMath.com

6. Verify that segments with lengths of 3, 4, and 6 form a triangle. Is the triangle *acute*, *right*, or *obtuse*?

7. Verify that segments with lengths of 2.1, 2.8, and 3.5 form a triangle. Is the triangle *acute*, *right*, or *obtuse*?

Extra Example 5

Verify that segments with lengths of 14 meters, 15 meters, and 11 meters form a triangle. Is the triangle *acute*, *right*, or *obtuse*?
The segments form a triangle, because $14 + 15 > 11$, $15 + 11 > 14$, and $14 + 11 > 15$. Because $15^2 < 14^2 + 11^2$, the triangle is acute.

MONITORING PROGRESS ANSWERS

6. yes; obtuse
7. yes; right

Laurie's Notes   Teacher Actions

- **Use Appropriate Tools Strategically:** Use dynamic geometry software to demonstrate the relationships stated in the Pythagorean Inequalities Theorem.
- In working independently on Example 5, students must reason about how to write the initial statement (equality or inequality). The sum of the squares of the lengths of the two shorter sides is compared to the square of the length of the longest side.

Closure

- **Exit Ticket:** Classify the triangles as acute, right, or obtuse. (a) sides 7, 9, and 11; (b) sides 6, 8, and 10 (a) acute triangle; (b) right triangle

Assignment Guide and Homework Check

ASSIGNMENT

Basic: 1, 2, 3–29 odd, 36, 41, 44–47

Average: 1, 2, 6–32 even, 36–38, 41, 44–47

Advanced: 1, 2, 9, 10–40 even, 41, 42, 44–47

HOMEWORK CHECK

Basic: 7, 11, 13, 15, 25

Average: 6, 10, 18, 28, 32

Advanced: 9, 18, 28, 34, 40

ANSWERS

1. A Pythagorean triple is a set of three positive integers a, b, and c that satisfy the equation $c^2 = a^2 + b^2$.
2. Find the length of the longest leg; 4; 5
3. $x = \sqrt{170} \approx 13.0$; no
4. $x = 34$; yes
5. $x = 41$; yes
6. $x = 2\sqrt{13} \approx 7.2$; no
7. $x = 15$; yes
8. $x = 3\sqrt{55} \approx 22.2$; no
9. $x = 14$; yes
10. $x = 4\sqrt{2} \approx 5.7$; no
11. Exponents cannot be distributed as shown in the third line; $c^2 = a^2 + b^2$; $x^2 = 7^2 + 24^2$; $x^2 = 49 + 576$; $x^2 = 625$; $x = 25$
12. Because 26 is the length of the hypotenuse, it should be substituted for c; $c^2 = a^2 + b^2$; $26^2 = 10^2 + x^2$; $676 = 100 + x^2$; $576 = x^2$; $24 = x$

9.1 Exercises

Dynamic Solutions available at *BigIdeasMath.com*

Vocabulary and Core Concept Check

1. **VOCABULARY** What is a Pythagorean triple?

2. **DIFFERENT WORDS, SAME QUESTION** Which is different? Find "both" answers.

   Find the length of the longest side.

   Find the length of the hypotenuse.

   Find the length of the longest leg.

   Find the length of the side opposite the right angle.

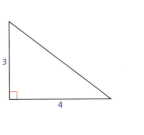

Monitoring Progress and Modeling with Mathematics

In Exercises 3–6, find the value of x. Then tell whether the side lengths form a Pythagorean triple. *(See Example 1.)*

3.

4.

5.

6.

In Exercises 7–10, find the value of x. Then tell whether the side lengths form a Pythagorean triple. *(See Example 2.)*

7.

8.

9.
10.

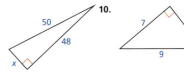

ERROR ANALYSIS In Exercises 11 and 12, describe and correct the error in using the Pythagorean Theorem.

11.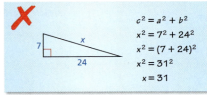

12.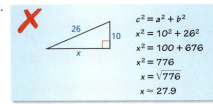

522 Chapter 9 Right Triangles and Trigonometry

13. MODELING WITH MATHEMATICS The fire escape forms a right triangle, as shown. Use the Pythagorean Theorem to approximate the distance between the two platforms. *(See Example 3.)*

14. MODELING WITH MATHEMATICS The backboard of the basketball hoop forms a right triangle with the supporting rods, as shown. Use the Pythagorean Theorem to approximate the distance between the rods where they meet the backboard.

In Exercises 15–20, tell whether the triangle is a right triangle. *(See Example 4.)*

15.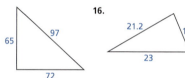

16. (21.2, 11.4, 23)

17.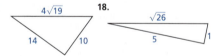

18. ($\sqrt{26}$, 5, 1)

19.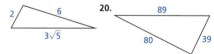

20. (89, 80, 39)

In Exercises 21–28, verify that the segment lengths form a triangle. Is the triangle *acute*, *right*, or *obtuse*? *(See Example 5.)*

21. 10, 11, and 14

22. 6, 8, and 10

23. 12, 16, and 20

24. 15, 20, and 36

25. 5.3, 6.7, and 7.8

26. 4.1, 8.2, and 12.2

27. 24, 30, and $6\sqrt{43}$

28. 10, 15, and $5\sqrt{13}$

29. MODELING WITH MATHEMATICS In baseball, the lengths of the paths between consecutive bases are 90 feet, and the paths form right angles. The player on first base tries to steal second base. How far does the ball need to travel from home plate to second base to get the player out?

30. REASONING You are making a canvas frame for a painting using stretcher bars. The rectangular painting will be 10 inches long and 8 inches wide. Using a ruler, how can you be certain that the corners of the frame are 90°?

In Exercises 31–34, find the area of the isosceles triangle.

31.

32.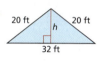

33.

34.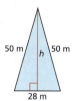

Dynamic Teaching Tools
Dynamic Assessment & Progress Monitoring Tool
Interactive Whiteboard Lesson Library
Dynamic Classroom with Dynamic Investigations

ANSWERS
13. about 14.1 ft
14. about 9.14 in.
15. yes
16. no
17. no
18. yes
19. no
20. yes
21. yes; acute
22. yes; right
23. yes; right
24. no
25. yes; acute
26. yes; obtuse
27. yes; obtuse
28. yes; right
29. about 127.3 ft
30. Each diagonal should form a right triangle with $c^2 = 8^2 + 10^2$, so each diagonal should be about 12.8 inches long.
31. 120 m²
32. 192 ft²
33. 48 cm²
34. 672 m²

ANSWERS

35. The horizontal distance between any two points is given by $(x_2 - x_1)$, and the vertical distance is given by $(y_2 - y_1)$. The horizontal and vertical segments that represent these distances form a right angle, with the segment between the two points being the hypotenuse. So, you can use the Pythagorean Theorem to say $d^2 = (x_2 - x_1)^2 + (y_2 - y_1)^2$, and when you solve for d, you get the distance formula:
$d = \sqrt{(x_2 - x_1)^2 + (y_2 - y_1)^2}$.

36. Because $10^2 = 6^2 + 8^2$, by the Converse of the Pythagorean Theorem, $\triangle ABC$ is a right triangle.

37. 2 packages

38–47. See Additional Answers.

Mini-Assessment

1. Tell whether a triangle with side lengths 8, $2\sqrt{65}$, and 14 is a right triangle. **yes**

2. Do segments with lengths 11 feet, 12 feet, and 20 feet form a triangle? If so, classify the triangle as *acute*, *right*, or *obtuse*. **Yes, the segments form a triangle because $11 + 12 > 20$, $11 + 20 > 12$, and $12 + 20 > 11$. Because $20^2 > 11^2 + 12^2$, the triangle is obtuse.**

3. A boat travels 1 mile east and then turns and travels 2.4 miles north. How far is the boat from its starting point? **2.6 miles**

4. The sides of a triangle have lengths 5, 7 and x.
   a. For what values of x is the triangle a right triangle? **$2\sqrt{6}$ and $\sqrt{74}$**
   b. Tell whether the side lengths form a Pythagorean triple. **no; The values of x are not integers.**

524   Chapter 9

---

35. ANALYZING RELATIONSHIPS Justify the Distance Formula using the Pythagorean Theorem.

36. HOW DO YOU SEE IT? How do you know $\angle C$ is a right angle without using the Pythagorean Theorem?

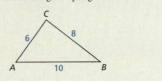

37. PROBLEM SOLVING You are making a kite and need to figure out how much binding to buy. You need the binding for the perimeter of the kite. The binding comes in packages of two yards. How many packages should you buy?

38. PROVING A THEOREM Use the Pythagorean Theorem to prove the Hypotenuse-Leg (HL) Congruence Theorem.

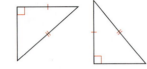

39. PROVING A THEOREM Prove the Converse of the Pythagorean Theorem. (*Hint:* Draw $\triangle ABC$ with side lengths a, b, and c, where c is the length of the longest side. Then draw a right triangle with side lengths a, b, and x, where x is the length of the hypotenuse. Compare lengths c and x.)

40. THOUGHT PROVOKING Consider two integers m and n, where $m > n$. Do the following expressions produce a Pythagorean triple? If yes, prove your answer. If no, give a counterexample.
$2mn$, $m^2 - n^2$, $m^2 + n^2$

41. MAKING AN ARGUMENT Your friend claims 72 and 75 cannot be part of a Pythagorean triple because $72^2 + 75^2$ does not equal a positive integer squared. Is your friend correct? Explain your reasoning.

42. PROVING A THEOREM Copy and complete the proof of the Pythagorean Inequalities Theorem when $c^2 < a^2 + b^2$.

Given In $\triangle ABC$, $c^2 < a^2 + b^2$, where c is the length of the longest side.
$\triangle PQR$ has side lengths a, b, and x, where x is the length of the hypotenuse, and $\angle R$ is a right angle.

Prove $\triangle ABC$ is an acute triangle.

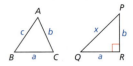

STATEMENTS	REASONS
1. In $\triangle ABC$, $c^2 < a^2 + b^2$, where c is the length of the longest side. $\triangle PQR$ has side lengths a, b, and x, where x is the length of the hypotenuse, and $\angle R$ is a right angle.	1. _____
2. $a^2 + b^2 = x^2$	2. _____
3. $c^2 < x^2$	3. _____
4. $c < x$	4. Take the positive square root of each side.
5. $m\angle R = 90°$	5. _____
6. $m\angle C < m\angle R$	6. Converse of the Hinge Theorem
7. $m\angle C < 90°$	7. _____
8. $\angle C$ is an acute angle.	8. _____
9. $\triangle ABC$ is an acute triangle.	9. _____

43. PROVING A THEOREM Prove the Pythagorean Inequalities Theorem when $c^2 > a^2 + b^2$. (*Hint:* Look back at Exercise 42.)

Maintaining Mathematical Proficiency — Reviewing what you learned in previous grades and lessons

Simplify the expression by rationalizing the denominator. (*Section 4.1*)

44. $\dfrac{7}{\sqrt{2}}$ **45.** $\dfrac{14}{\sqrt{3}}$ **46.** $\dfrac{8}{\sqrt{2}}$ **47.** $\dfrac{12}{\sqrt{3}}$

If students need help...	If students got it...
Resources by Chapter • Practice A and Practice B • Puzzle Time	**Resources by Chapter** • Enrichment and Extension • Cumulative Review
Student Journal • Practice	Start the *next* Section
Differentiating the Lesson Skills Review Handbook	

Laurie's Notes

Overview of Section 9.2

Introduction
- In this lesson, students will use their knowledge of similar triangles and the Pythagorean Theorem to deduce relationships about special right triangles.
- The two special triangles are the 45°-45°-90° and 30°-60°-90° triangles. The ratio of any two sides of a right isosceles triangle (i.e., hypotenuse to leg) will be the same regardless of the size of the right isosceles triangle. These relationships are stated as theorems.
- The lesson ends with using the side length relationships of special right triangles to solve real-life problems.

Teaching Strategy
- I have found that it is important for students to do some sort of investigation to appreciate and make sense of the relationship between the legs and the hypotenuse in a right isosceles triangle. A technological approach is presented in the first exploration. Alternately, the investigation can be done with paper folding.
- Begin with a 6-inch square piece of paper. Fold the square on its diagonal to form two isosceles right triangles. Measure and record the lengths of a leg and the hypotenuse in a table.
- Continue folding the paper into smaller isosceles right triangles and filling in the table.

Number of Folds	Leg Length	Hypotenuse	$\dfrac{\text{Hypotenuse}}{\text{Leg Length}}$
1			
2			
3			
4			
5			

- Use a calculator to find the ratios of hypotenuse to leg length. Depending upon the accuracy of students' measurements, the ratios should always be close to 1.41, the approximation of $\sqrt{2}$.
- Write $\dfrac{\text{hypotenuse}}{\text{leg}} = \sqrt{2}$. Solving for the length of the hypotenuse, hypotenuse = leg $\cdot \sqrt{2}$.

Pacing Suggestion
- Take time for students to explore and discover the relationships between the side lengths in special right triangles, and then transition to the formal lesson.

Dynamic Teaching Tools
Dynamic Assessment & Progress Monitoring Tool
Lesson Planning Tool
Interactive Whiteboard Lesson Library
Dynamic Classroom with Dynamic Investigations

What Your Students Will Learn

- Find side lengths in 45°-45°-90° and 30°-60°-90° triangles.
- Solve real-life problems involving special right triangles.

Laurie's Notes

Exploration

Motivate
- Pose a baseball problem for students. If you are located near a major league team, you could display the team's logo!
- "The distance between bases is 90 feet. What is the *exact* distance from home plate to second base?" Students should note the use of the word *exact*, and they should also realize that the problem involves finding the length of a diagonal of a square where the side lengths are known.
- This problem allows you to review simplifying expressions involving radicals, a technique that is needed in this lesson.
- The exact distance is $90\sqrt{2}$ feet.

Exploration Note
- There are different techniques students can use to construct the special triangles in the two explorations.
- One way is certainly to look at the coordinates in the diagram. Students should be able to reason why (0, 0), (4, 0), and (0, 4) are the coordinates of a right isosceles triangle.
- This technique is not possible for the 30°-60°-90° triangle because the y-coordinate of vertex A is not obvious. One way my students have efficiently constructed a 30°-60°-90° triangle is to construct an equilateral triangle and then construct one of the angle bisectors.
- Whatever method(s) students use, they should note that the construction and subsequent measurements are going to be repeated several times. Students might want to consider how to perform the construction so that the construction and measurements are dynamic.

Exploration 1
- Circulate as students construct the right isosceles triangle so that you know the various methods used by your students.
- The lengths of the sides are measured, and three ratios are calculated.
- ❓ "What did you observe about the computed ratios? Explain why this occurred." *The triangles are all similar by AA. The ratio of any two sides is going to stay the same as the triangle is enlarged or reduced proportionally.*
- **Attend to Precision:** Ask students to explain how to find the exact ratio versus an approximation. This will take some students longer to reason through. Give *Wait Time* for partners to discuss this. The Pythagorean Theorem should come to mind! Using one of the triangles where the legs are integer values will allow students to find the hypotenuse exactly.

Exploration 2
- Have students work through this exploration in a manner similar to the first. The difference being that to find the exact values, it will be the shorter leg and the hypotenuse that will be integer values.

Communicate Your Answer
- Listen for students to describe the relationships found while working through the explorations.

Connecting to Next Step
- Students have discovered the special relationships between the side lengths in 45°-45°-90° and 30°-60°-90° triangles. These will be stated as theorems in the formal lesson.

9.2 Special Right Triangles

Essential Question What is the relationship among the side lengths of 45°-45°-90° triangles? 30°-60°-90° triangles?

EXPLORATION 1 Side Ratios of an Isosceles Right Triangle

Work with a partner.

a. Use dynamic geometry software to construct an isosceles right triangle with a leg length of 4 units.

b. Find the acute angle measures. Explain why this triangle is called a 45°-45°-90° triangle.

ATTENDING TO PRECISION
To be proficient in math, you need to express numerical answers with a degree of precision appropriate for the problem context.

c. Find the exact ratios of the side lengths (using square roots).

$\dfrac{AB}{AC} =$

$\dfrac{AB}{BC} =$

$\dfrac{AC}{BC} =$

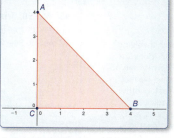

Sample
Points
A(0, 4)
B(4, 0)
C(0, 0)
Segments
AB = 5.66
BC = 4
AC = 4
Angles
m∠A = 45°
m∠B = 45°

d. Repeat parts (a) and (c) for several other isosceles right triangles. Use your results to write a conjecture about the ratios of the side lengths of an isosceles right triangle.

EXPLORATION 2 Side Ratios of a 30°-60°-90° Triangle

Work with a partner.

a. Use dynamic geometry software to construct a right triangle with acute angle measures of 30° and 60° (a 30°-60°-90° triangle), where the shorter leg length is 3 units.

b. Find the exact ratios of the side lengths (using square roots).

$\dfrac{AB}{AC} =$

$\dfrac{AB}{BC} =$

$\dfrac{AC}{BC} =$

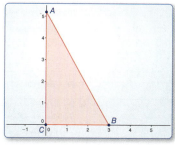

Sample
Points
A(0, 5.20)
B(3, 0)
C(0, 0)
Segments
AB = 6
BC = 3
AC = 5.20
Angles
m∠A = 30°
m∠B = 60°

c. Repeat parts (a) and (b) for several other 30°-60°-90° triangles. Use your results to write a conjecture about the ratios of the side lengths of a 30°-60°-90° triangle.

Communicate Your Answer

3. What is the relationship among the side lengths of 45°-45°-90° triangles? 30°-60°-90° triangles?

Section 9.2 Special Right Triangles 525

ANSWERS

1. a. Check students' work.
   b. 45° and 45°; 45°-45°-90° describes the angle measures.
   c. $\sqrt{2}$; $\sqrt{2}$; 1
   d. Check students' work. The ratio of the length of each leg to the length of the hypotenuse is $\sqrt{2}$. The ratio of the length of one leg to the other is 1.

2. a. Check students' work.
   b. $\dfrac{2}{\sqrt{3}}$; 2; $\sqrt{3}$
   c. Check students' work. The ratio of the length of the hypotenuse to the length of the longer leg is $\dfrac{2}{\sqrt{3}}$. The ratio of the length of the hypotenuse to the length of the shorter leg is 2. The ratio of the length of the longer leg to the length of the shorter leg is $\sqrt{3}$.

3. In a 45°-45°-90° triangle, the legs are the same length, and the length of the hypotenuse is $\sqrt{2}$ times as long as each leg. In a 30°-60°-90° triangle, the hypotenuse is twice as long as the shorter leg, and the longer leg is $\sqrt{3}$ times as long as the shorter leg.

English Language Learners

Building on Past Knowledge
Remind students that the properties of 45°-45°-90° and 30°-60°-90° triangles can be derived by applying the Pythagorean Theorem to the triangles formed by a diagonal in a square and the height of an equilateral triangle.

Extra Example 1
Find the value of *x*. Write your answer in simplest form.

a.
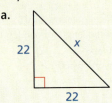

b.
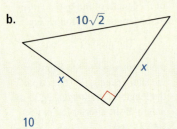

9.2 Lesson

Core Vocabulary
Previous
isosceles triangle

REMEMBER
An expression involving a radical with index 2 is in simplest form when no radicands have perfect squares as factors other than 1, no radicands contain fractions, and no radicals appear in the denominator of a fraction.

What You Will Learn

▶ Find side lengths in special right triangles.
▶ Solve real-life problems involving special right triangles.

Finding Side Lengths in Special Right Triangles

A 45°-45°-90° triangle is an *isosceles right triangle* that can be formed by cutting a square in half diagonally.

Theorem

45°-45°-90° Triangle Theorem
In a 45°-45°-90° triangle, the hypotenuse is $\sqrt{2}$ times as long as each leg.

hypotenuse = leg · $\sqrt{2}$

Proof Ex. 19, p. 530

EXAMPLE 1 Finding Side Lengths in 45°-45°-90° Triangles

Find the value of *x*. Write your answer in simplest form.

a.

b.

SOLUTION

a. By the Triangle Sum Theorem, the measure of the third angle must be 45°, so the triangle is a 45°-45°-90° triangle.

hypotenuse = leg · $\sqrt{2}$	45°-45°-90° Triangle Theorem
$x = 8 \cdot \sqrt{2}$	Substitute.
$x = 8\sqrt{2}$	Simplify.

▶ The value of *x* is $8\sqrt{2}$.

b. By the Base Angles Theorem and the Corollary to the Triangle Sum Theorem, the triangle is a 45°-45°-90° triangle.

hypotenuse = leg · $\sqrt{2}$	45°-45°-90° Triangle Theorem
$5\sqrt{2} = x \cdot \sqrt{2}$	Substitute.
$\dfrac{5\sqrt{2}}{\sqrt{2}} = \dfrac{x\sqrt{2}}{\sqrt{2}}$	Divide each side by $\sqrt{2}$.
$5 = x$	Simplify.

▶ The value of *x* is 5.

526 Chapter 9 Right Triangles and Trigonometry

Laurie's Notes — Teacher Actions

- State the theorem and connect to Exploration 1 done by students. See the *Teaching Strategy* on page T-524.
- In answering problems, such as Example 1, students will often start to use the Pythagorean Theorem to solve for the missing side length versus applying the 45°-45°-90° Triangle Theorem.
- **Reason Abstractly and Quantitatively:** Mathematically proficient students understand that all right isosceles triangles possess the relationship stated in the theorem. They use the theorem versus performing a computation each time.

Theorem

30°-60°-90° Triangle Theorem

In a 30°-60°-90° triangle, the hypotenuse is twice as long as the shorter leg, and the longer leg is $\sqrt{3}$ times as long as the shorter leg.

hypotenuse = shorter leg · 2
longer leg = shorter leg · $\sqrt{3}$

Proof Ex. 21, p. 530

REMEMBER
Because the angle opposite 9 is larger than the angle opposite x, the leg with length 9 is longer than the leg with length x by the Triangle Larger Angle Theorem.

EXAMPLE 2 Finding Side Lengths in a 30°-60°-90° Triangle

Find the values of x and y. Write your answer in simplest form.

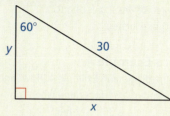

SOLUTION

Step 1 Find the value of x.

longer leg = shorter leg · $\sqrt{3}$	30°-60°-90° Triangle Theorem
$9 = x \cdot \sqrt{3}$	Substitute.
$\dfrac{9}{\sqrt{3}} = x$	Divide each side by $\sqrt{3}$.
$\dfrac{9}{\sqrt{3}} \cdot \dfrac{\sqrt{3}}{\sqrt{3}} = x$	Multiply by $\dfrac{\sqrt{3}}{\sqrt{3}}$.
$\dfrac{9\sqrt{3}}{3} = x$	Multiply fractions.
$3\sqrt{3} = x$	Simplify.

▶ The value of x is $3\sqrt{3}$.

Step 2 Find the value of y.

hypotenuse = shorter leg · 2	30°-60°-90° Triangle Theorem
$y = 3\sqrt{3} \cdot 2$	Substitute.
$y = 6\sqrt{3}$	Simplify.

▶ The value of y is $6\sqrt{3}$.

Monitoring Progress 🔊 Help in English and Spanish at *BigIdeasMath.com*

Find the value of the variable. Write your answer in simplest form.

1.

2.

3.

4.

Differentiated Instruction

Inclusion
Work with advanced students to develop the formula for the area A of a 30°-60°-90° triangle with hypotenuse of length s. Check to see whether they wrote $A = \dfrac{s^2\sqrt{3}}{8}$.

Extra Example 2

Find the values of x and y. Write your answer in simplest form.

$x = 15\sqrt{3}, y = 15$

MONITORING PROGRESS ANSWERS

1. $x = 2$
2. $y = 2$
3. $x = 3$
4. $h = 2\sqrt{3}$

Laurie's Notes Teacher Actions

- State the theorem and connect to Exploration 2 done by students. Stress that there are two relationships stated in this theorem.
- Discuss with students that the shorter leg is also referred to as the side opposite 30° and the longer leg is the side opposite 60°.
- ❓ "How do you simplify an expression that has a radical in the denominator?" Multiply by a factor of 1 — the radical divided by itself.
- **Think-Pair-Share:** Have students answer Questions 1–4, and then share and discuss as a class.

Extra Example 3
A warning sticker is shaped like an equilateral triangle with side length of 4 inches. Estimate the area of the sticker by finding the area of the equilateral triangle to the nearest tenth of an inch.

The area of the sign is about 6.9 square inches.

Extra Example 4
Use the diagram in Example 4. How high is the end of a 54-foot ramp when the tipping angle is 30°? 27 feet

MONITORING PROGRESS ANSWERS
5. about 15.59 cm^2
6. about 12 ft

Solving Real-Life Problems

EXAMPLE 3 Modeling with Mathematics

The road sign is shaped like an equilateral triangle. Estimate the area of the sign by finding the area of the equilateral triangle.

SOLUTION
First find the height h of the triangle by dividing it into two 30°-60°-90° triangles. The length of the longer leg of one of these triangles is h. The length of the shorter leg is 18 inches.

$h = 18 \cdot \sqrt{3} = 18\sqrt{3}$ 30°-60°-90° Triangle Theorem

 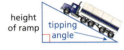

Use $h = 18\sqrt{3}$ to find the area of the equilateral triangle.

Area $= \frac{1}{2}bh = \frac{1}{2}(36)(18\sqrt{3}) \approx 561.18$

▶ The area of the sign is about 561 square inches.

EXAMPLE 4 Finding the Height of a Ramp

A tipping platform is a ramp used to unload trucks. How high is the end of an 80-foot ramp when the tipping angle is 30°? 45°?

SOLUTION
When the tipping angle is 30°, the height h of the ramp is the length of the shorter leg of a 30°-60°-90° triangle. The length of the hypotenuse is 80 feet.

$80 = 2h$ 30°-60°-90° Triangle Theorem

$40 = h$ Divide each side by 2.

When the tipping angle is 45°, the height h of the ramp is the length of a leg of a 45°-45°-90° triangle. The length of the hypotenuse is 80 feet.

$80 = h \cdot \sqrt{2}$ 45°-45°-90° Triangle Theorem

$\dfrac{80}{\sqrt{2}} = h$ Divide each side by $\sqrt{2}$.

$56.6 \approx h$ Use a calculator.

▶ When the tipping angle is 30°, the ramp height is 40 feet. When the tipping angle is 45°, the ramp height is about 56 feet 7 inches.

Monitoring Progress   Help in English and Spanish at BigIdeasMath.com

5. The logo on a recycling bin resembles an equilateral triangle with side lengths of 6 centimeters. Approximate the area of the logo.

6. The body of a dump truck is raised to empty a load of sand. How high is the 14-foot-long body from the frame when it is tipped upward by a 60° angle?

Laurie's Notes — Teacher Actions

? "In Example 3, why is the answer not left in exact form?" An estimate was asked for, and in using the answer for a contextual purpose, an approximation may be more helpful.

? "In Example 4, as the tipping angle increases, what happens to the height of the front of the truck?" It increases.

Closure
- **Exit Ticket:** (a) The hypotenuse of a right isosceles triangle is 12. Find the length of the legs. $6\sqrt{2}$ (b) The hypotenuse of a 30°-60°-90° triangle is 12. Find the lengths of the legs. 6 and $6\sqrt{3}$

9.2 Exercises

Dynamic Solutions available at *BigIdeasMath.com*

Vocabulary and Core Concept Check

1. **VOCABULARY** Name two special right triangles by their angle measures.
2. **WRITING** Explain why the acute angles in an isosceles right triangle always measure 45°.

Monitoring Progress and Modeling with Mathematics

In Exercises 3–6, find the value of x. Write your answer in simplest form. *(See Example 1.)*

3.
4.
5.
6.

In Exercises 7–10, find the values of x and y. Write your answers in simplest form. *(See Example 2.)*

7.
8.
9.
10.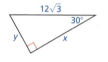

ERROR ANALYSIS In Exercises 11 and 12, describe and correct the error in finding the length of the hypotenuse.

11.

12.

In Exercises 13 and 14, sketch the figure that is described. Find the indicated length. Round decimal answers to the nearest tenth.

13. The side length of an equilateral triangle is 5 centimeters. Find the length of an altitude.
14. The perimeter of a square is 36 inches. Find the length of a diagonal.

In Exercises 15 and 16, find the area of the figure. Round decimal answers to the nearest tenth. *(See Example 3.)*

15.
16.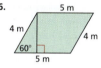

17. **PROBLEM SOLVING** Each half of the drawbridge is about 284 feet long. How high does the drawbridge rise when x is 30°? 45°? 60°? *(See Example 4.)*

Section 9.2 Special Right Triangles 529

Assignment Guide and Homework Check

ASSIGNMENT
Basic: 1, 2, 3–17 odd, 20, 24, 26, 27
Average: 1, 2, 6–20 even, 21, 24, 26, 27
Advanced: 1, 2, 6–18 even, 19–27

HOMEWORK CHECK
Basic: 5, 9, 13, 15, 20
Average: 6, 10, 14, 16, 21
Advanced: 6, 10, 14, 16, 22

ANSWERS

1. 45°-45°-90°, 30°-60°-90°
2. Because the acute angles of a right isosceles triangle must be congruent by the Base Angles Theorem, and complementary, the measures must be $\frac{90°}{2} = 45°$.
3. $x = 7\sqrt{2}$
4. $x = 10$
5. $x = 3$
6. $x = \dfrac{9\sqrt{2}}{2}$
7. $x = 9\sqrt{3}, y = 18$
8. $x = 3, y = 6$
9. $x = 12\sqrt{3}, y = 12$
10. $x = 18; y = 6\sqrt{3}$
11. The hypotenuse of a 30°-60°-90° triangle is equal to the shorter leg times 2; hypotenuse = shorter leg • 2 = 7 • 2 = 14; So, the length of the hypotenuse is 14 units.
12. The hypotenuse of a 45°-45°-90° triangle is equal to a leg times $\sqrt{2}$; hypotenuse = leg • $\sqrt{2}$ = $\sqrt{5}$ • $\sqrt{2}$ = $\sqrt{10}$; So, the length of the hypotenuse is $\sqrt{10}$ units.
13.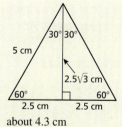
    about 4.3 cm

14.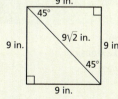
    about 12.7 in.
15. 32 ft²
16. about 17.3 m²
17. 142 ft; about 200.82 ft; about 245.95 ft

Section 9.2 529

Dynamic Teaching Tools

Dynamic Assessment & Progress Monitoring Tool
Interactive Whiteboard Lesson Library
Dynamic Classroom with Dynamic Investigations

ANSWERS

18. $x = \sqrt{3}$ cm
19. Because $\triangle DEF$ is a 45°-45°-90° triangle, by the Converse of the Base Angles Theorem, $\overline{DF} \cong \overline{FE}$. So, let $x = DF = FE$. By the Pythagorean Theorem, $x^2 + x^2 = c^2$, where c is the length of the hypotenuse. So, $2x^2 = c^2$ by the Distributive Property. Take the positive square root of each side to get $x\sqrt{2} = c$. So, the hypotenuse is $\sqrt{2}$ times as long as each leg.
20–25. See Additional Answers.
26. $x = 18$
27. $x = 2$

Mini-Assessment

1. This quilt is made up of cloth pieces shaped like isosceles right triangles. The hypotenuse of each triangle is 16 centimeters long. Find the area of one cloth triangle.

64 square centimeters

2. Use the diagram in Example 4. How high is the end of a 38-foot ramp when the tipping angle is 30°?

19 feet

3. Find the values of x and y.

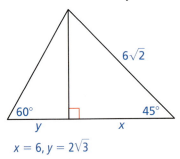

$x = 6, y = 2\sqrt{3}$

530   Chapter 9

---

18. **MODELING WITH MATHEMATICS** A nut is shaped like a regular hexagon with side lengths of 1 centimeter. Find the value of x. (*Hint*: A regular hexagon can be divided into six congruent triangles.)

19. **PROVING A THEOREM** Write a paragraph proof of the 45°-45°-90° Triangle Theorem.

    Given $\triangle DEF$ is a 45°-45°-90° triangle.
    Prove   The hypotenuse is $\sqrt{2}$ times as long as each leg.

20. **HOW DO YOU SEE IT?** The diagram shows part of the *Wheel of Theodorus*.

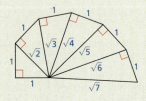

    a. Which triangles, if any, are 45°-45°-90° triangles?
    b. Which triangles, if any, are 30°-60°-90° triangles?

21. **PROVING A THEOREM** Write a paragraph proof of the 30°-60°-90° Triangle Theorem. (*Hint*: Construct $\triangle JML$ congruent to $\triangle JKL$.)

    Given $\triangle JKL$ is a 30°-60°-90° triangle.
    Prove   The hypotenuse is twice as long as the shorter leg, and the longer leg is $\sqrt{3}$ times as long as the shorter leg.

22. **THOUGHT PROVOKING** A special right triangle is a right triangle that has rational angle measures and each side length contains at most one square root. There are only three special right triangles. The diagram below is called the *Ailles rectangle*. Label the sides and angles in the diagram. Describe all three special right triangles.

23. **WRITING** Describe two ways to show that all isosceles right triangles are similar to each other.

24. **MAKING AN ARGUMENT** Each triangle in the diagram is a 45°-45°-90° triangle. At Stage 0, the legs of the triangle are each 1 unit long. Your brother claims the lengths of the legs of the triangles added are halved at each stage. So, the length of a leg of a triangle added in Stage 8 will be $\frac{1}{256}$ unit. Is your brother correct? Explain your reasoning.

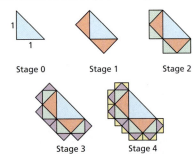

25. **USING STRUCTURE** $\triangle TUV$ is a 30°-60°-90° triangle, where two vertices are $U(3, -1)$ and $V(-3, -1)$, $\overline{UV}$ is the hypotenuse, and point T is in Quadrant I. Find the coordinates of T.

Maintaining Mathematical Proficiency
Reviewing what you learned in previous grades and lessons

Find the value of x. *(Section 8.3)*

26. $\triangle DEF \sim \triangle LMN$

27. $\triangle ABC \sim \triangle QRS$

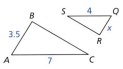

530   Chapter 9   Right Triangles and Trigonometry

If students need help...	If students got it...
Resources by Chapter • Practice A and Practice B • Puzzle Time	**Resources by Chapter** • Enrichment and Extension • Cumulative Review
Student Journal • Practice	Start the *next* Section
Differentiating the Lesson Skills Review Handbook	

Laurie's Notes

Overview of Section 9.3

Introduction
- This lesson applies students' understanding of similar triangles to the triangles formed when the altitude to the hypotenuse in a right triangle is drawn.
- The theorems that follow reference certain segments as being the *geometric mean* of two other segments.
- The geometric mean is used to solve real-life problems.

Teaching Strategy
- The challenge many students have with the Right Triangle Similarity Theorem is in correctly naming the pairs of triangles that are similar. One reason for the difficulty is that the triangles are in different orientations.
- It is helpful to actually have physical copies of the three triangles or to create them using dynamic geometry software or an interactive whiteboard.
- To make physical copies to manipulate, cut two congruent right scalene triangles. Draw the altitude to the hypotenuse in one of the right triangles. Cut along the altitude, making two smaller triangles.
- You can overlay the triangles to see the congruent angles. In doing so, it becomes obvious that the smaller triangles must be reflected in order to overlay the original triangle. It is this spatial manipulation that makes it difficult for students to write the similarity statements for these triangles.

Pacing Suggestion
- The two explorations provide an introduction to geometric means and the similar triangles found when the altitude to the hypotenuse is drawn in a right triangle. Transition to the formal lesson after the second exploration has been discussed.

Dynamic Teaching Tools
Dynamic Assessment & Progress Monitoring Tool
Lesson Planning Tool
Interactive Whiteboard Lesson Library
Dynamic Classroom with Dynamic Investigations

What Your Students Will Learn

- Identify similar right triangles formed by drawing the altitude perpendicular to the hypotenuse of a right triangle.
- Solve real-life problems involving similar right triangles.
- Use geometric means to write proportions of similar right triangles.

Laurie's Notes

Exploration

Motivate
- Pose the following problem: You invest $1000. The first year your money earns 10%, and the second year your money earns 20%. What is your average rate of return over the two years?
- Students may incorrectly say 15%. Perform the computation: $1000 \times 1.1 = \$1100$ at the end of year one. $\$1100 \times 1.2 = \$1320$ at the end of year two.
- The geometric mean of 1.1 (110%) and 1.2 (120%) is approximately 1.1489, which is the average rate of return over two years.
- Today's lesson involves finding geometric means.

Exploration 1
- The construction of the right triangle and altitude are straightforward. Students will need to read and understand the definition of a *geometric mean*, which is not difficult.
- A similarity statement is given for the two smaller triangles: $\triangle CBD \sim \triangle ACD$. Ask students how they could prove that the triangles are similar.
- In part (c), finding *CD* is done by setting up the proportion. The measure can be verified using the software.
- Some students may want to repeat the construction for several right triangles to ensure that they understand the relationship between the altitude and the segments formed on the hypotenuse. The generalization follows from the smaller triangles being similar.

Exploration 2
- This exploration can be done using a spreadsheet or lists in the graphing calculator.
- **Extension:** Students should be aware that there is a third type of (Pythagorean) mean—the harmonic mean. The harmonic mean for two positive numbers a and b is $\dfrac{2}{\frac{1}{a} + \frac{1}{b}}$. When compared to the arithmetic mean and the geometric mean, given two positive numbers, the following relationship is true: arithmetic mean $\geq$ geometric mean $\geq$ harmonic mean.

Communicate Your Answer
- Have students share their thinking about the relationship between altitudes and geometric means of right triangles.

Connecting to Next Step
- Students have now explored the altitude to the hypotenuse in a right triangle. The relationship between the altitude and the segments formed on the hypotenuse will be stated at the beginning of the formal lesson.

9.3 Similar Right Triangles

Essential Question How are altitudes and geometric means of right triangles related?

EXPLORATION 1 Writing a Conjecture

Work with a partner.

a. Use dynamic geometry software to construct right △ABC, as shown. Draw $\overline{CD}$ so that it is an altitude from the right angle to the hypotenuse of △ABC.

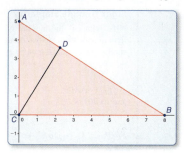

Points
A(0, 5)
B(8, 0)
C(0, 0)
D(2.25, 3.6)

Segments
AB = 9.43
BC = 8
AC = 5

b. The **geometric mean** of two positive numbers a and b is the positive number x that satisfies

$$\frac{a}{x} = \frac{x}{b}.$$ x is the geometric mean of a and b.

Write a proportion involving the side lengths of △CBD and △ACD so that CD is the geometric mean of two of the other side lengths. Use similar triangles to justify your steps.

c. Use the proportion you wrote in part (b) to find CD.

d. Generalize the proportion you wrote in part (b). Then write a conjecture about how the geometric mean is related to the altitude from the right angle to the hypotenuse of a right triangle.

CONSTRUCTING VIABLE ARGUMENTS
To be proficient in math, you need to understand and use stated assumptions, definitions, and previously established results in constructing arguments.

EXPLORATION 2 Comparing Geometric and Arithmetic Means

Work with a partner. Use a spreadsheet to find the arithmetic mean and the geometric mean of several pairs of positive numbers. Compare the two means. What do you notice?

	A	B	C	D
1	a	b	Arithmetic Mean	Geometric Mean
2	3	4	3.5	3.464
3	4	5		
4	6	7		
5	0.5	0.5		
6	0.4	0.8		
7	2	5		
8	1	4		
9	9	16		
10	10	100		
11				

Communicate Your Answer

3. How are altitudes and geometric means of right triangles related?

ANSWERS

1. a. Check students' work.

   b. $\dfrac{AD}{CD} = \dfrac{CD}{BD}$; By the Corollary to the Triangle Sum Theorem, two pairs of angles are complementary, ∠A and ∠ACD as well as ∠B and ∠BCD. Also, because adjacent angles ∠ACD and ∠BCD form a right angle, they are complementary. Then, by the Congruent Complements Theorem, ∠A ≅ ∠BCD and ∠B ≅ ∠ACD. So, △ACD ~ △BCD by the AA Similarity Theorem. Then, because corresponding sides of similar figures are proportional, $\dfrac{AD}{CD} = \dfrac{CD}{BD}$.

   c. about 4.24 units

   d. In a right triangle, △ABC, where $\overline{AB}$ is the hypotenuse, the altitude, $\overline{CD}$, from the right angle to the hypotenuse, divides the hypotenuse into two segments, $\overline{AD}$ and $\overline{BD}$. The length of the altitude is the geometric mean of the lengths of the two segments of the hypotenuse: $\dfrac{AD}{CD} = \dfrac{CD}{BD}$.

2.
	A	B	C	D
			Arithmetic	Geometric
1	a	b	Mean	Mean
2	3	4	3.5	3.464
3	4	5	4.5	4.472
4	6	7	6.5	6.481
5	0.5	0.5	0.5	0.5
6	0.4	0.8	0.6	0.566
7	2	5	3.5	3.162
8	1	4	2.5	2
9	9	16	12.5	12
10	10	100	55	31.623

Sample answer: The geometric mean is always less than or equal to the arithmetic mean. If the pair of positive numbers are closer together, so are the two means. If the pair of positive numbers are equal, they are also equal to each mean.

3. In a right triangle, the altitude from the right angle to the hypotenuse divides the hypotenuse into two segments. The length of the altitude is the geometric mean of the lengths of the two segments of the hypotenuse.

Differentiated Instruction

Auditory
Draw a version of this triangle on the board.

Have students work in pairs to verbally describe to each other how to find and list all the pairs of congruent angles in the figure. Have them use that information to identify three similar triangles, listing the corresponding vertices in the correct order.

Extra Example 1
Identify the similar triangles in the diagram.

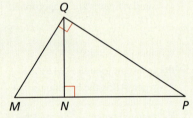

△MQP ~ △QNP ~ △MNQ

MONITORING PROGRESS ANSWERS
1. △QST ~ △SRT ~ △QRS
2. △GFH ~ △EGH ~ △EFG

9.3 Lesson

Core Vocabulary
geometric mean, *p. 534*
Previous
altitude of a triangle
similar figures

What You Will Learn
- Identify similar triangles.
- Solve real-life problems involving similar triangles.
- Use geometric means.

Identifying Similar Triangles
When the altitude is drawn to the hypotenuse of a right triangle, the two smaller triangles are similar to the original triangle and to each other.

Theorem

Right Triangle Similarity Theorem
If the altitude is drawn to the hypotenuse of a right triangle, then the two triangles formed are similar to the original triangle and to each other.

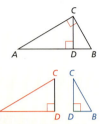

△CBD ~ △ABC, △ACD ~ △ABC, and △CBD ~ △ACD.

Proof Ex. 45, p. 538

EXAMPLE 1 Identifying Similar Triangles

Identify the similar triangles in the diagram.

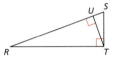

SOLUTION
Sketch the three similar right triangles so that the corresponding angles and sides have the same orientation.

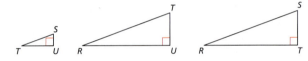

▶ △TSU ~ △RTU ~ △RST

Monitoring Progress Help in English and Spanish at *BigIdeasMath.com*

Identify the similar triangles.

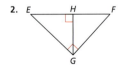

532 Chapter 9 Right Triangles and Trigonometry

Laurie's Notes — Teacher Actions

- The Right Triangle Similarity Theorem is easy to state. Writing the triangle similarity statements (Example 1) is challenging for students. See the *Teaching Strategy* on page T-530.
- **Teaching Tip:** Use a colored pencil to shade the angles of the triangles.
- Drawing the three triangles as non-overlapping, and in the same orientation, is helpful for students.

Solving Real-Life Problems

EXAMPLE 2 **Modeling with Mathematics**

A roof has a cross section that is a right triangle. The diagram shows the approximate dimensions of this cross section. Find the height h of the roof.

SOLUTION

1. **Understand the Problem** You are given the side lengths of a right triangle. You need to find the height of the roof, which is the altitude drawn to the hypotenuse.

2. **Make a Plan** Identify any similar triangles. Then use the similar triangles to write a proportion involving the height and solve for h.

3. **Solve the Problem** Identify the similar triangles and sketch them.

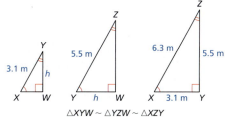

$\triangle XYW \sim \triangle YZW \sim \triangle XZY$

COMMON ERROR
Notice that if you tried to write a proportion using $\triangle XYW$ and $\triangle YZW$, then there would be two unknowns, so you would not be able to solve for h.

Because $\triangle XYW \sim \triangle XZY$, you can write a proportion.

$\dfrac{YW}{ZY} = \dfrac{XY}{XZ}$ Corresponding side lengths of similar triangles are proportional.

$\dfrac{h}{5.5} = \dfrac{3.1}{6.3}$ Substitute.

$h \approx 2.7$ Multiply each side by 5.5.

▶ The height of the roof is about 2.7 meters.

4. **Look Back** Because the height of the roof is a leg of right $\triangle YZW$ and right $\triangle XYW$, it should be shorter than each of their hypotenuses. The lengths of the two hypotenuses are $YZ = 5.5$ and $XY = 3.1$. Because $2.7 < 3.1$, the answer seems reasonable.

Monitoring Progress 🔊 Help in English and Spanish at BigIdeasMath.com

Find the value of x.

3.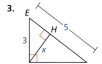

4.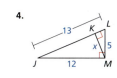

Section 9.3 Similar Right Triangles 533

Extra Example 2

A roof has a cross section that is a right triangle. The diagram shows the approximate dimensions of this cross section. Find the height h of the roof.

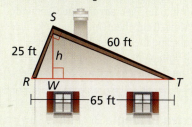

The height is about 23.1 feet.

MONITORING PROGRESS ANSWERS

3. $x = \dfrac{12}{5} = 2.4$

4. $x = \dfrac{60}{13} \approx 4.62$

Laurie's Notes — Teacher Actions

- Pose Example 2. Have students work with their partners to draw and label the three similar triangles.
- An alternative to writing three separate similarity statements for each pair of triangles is to write one statement showing the relationship between all three. In other words, instead of writing $\triangle 1 \sim \triangle 2$, $\triangle 2 \sim \triangle 3$, and $\triangle 1 \sim \triangle 3$, simply write $\triangle 1 \sim \triangle 2 \sim \triangle 3$.
- Labeling the known dimensions on each triangle will help students see how to use corresponding sides to write a solvable proportion.

English Language Learners

Graphic Organizer
Have students make a Summary Triangle for the Geometric Mean (Altitude) Theorem and the Geometric Mean (Leg) Theorem. Include examples of using the theorems to find the length of an altitude and finding the length of a leg.

Extra Example 3
Find the geometric mean of 8 and 10.
$4\sqrt{5} \approx 8.9$

Using a Geometric Mean

 **Core Concept**

Geometric Mean

The **geometric mean** of two positive numbers a and b is the positive number x that satisfies $\dfrac{a}{x} = \dfrac{x}{b}$. So, $x^2 = ab$ and $x = \sqrt{ab}$.

EXAMPLE 3 Finding a Geometric Mean

Find the geometric mean of 24 and 48.

SOLUTION

$x^2 = ab$	Definition of geometric mean
$x^2 = 24 \cdot 48$	Substitute 24 for a and 48 for b.
$x = \sqrt{24 \cdot 48}$	Take the positive square root of each side.
$x = \sqrt{24 \cdot 24 \cdot 2}$	Factor.
$x = 24\sqrt{2}$	Simplify.

▶ The geometric mean of 24 and 48 is $24\sqrt{2} \approx 33.9$.

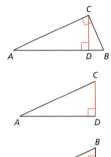

In right △ABC, altitude $\overline{CD}$ is drawn to the hypotenuse, forming two smaller right triangles that are similar to △ABC. From the Right Triangle Similarity Theorem, you know that △CBD ~ △ACD ~ △ABC. Because the triangles are similar, you can write and simplify the following proportions involving geometric means.

$$\dfrac{CD}{AD} = \dfrac{BD}{CD} \qquad \dfrac{CB}{DB} = \dfrac{AB}{CB} \qquad \dfrac{AC}{AD} = \dfrac{AB}{AC}$$

$$CD^2 = AD \cdot BD \qquad CB^2 = DB \cdot AB \qquad AC^2 = AD \cdot AB$$

Theorems

Geometric Mean (Altitude) Theorem

In a right triangle, the altitude from the right angle to the hypotenuse divides the hypotenuse into two segments.

The length of the altitude is the geometric mean of the lengths of the two segments of the hypotenuse.

Proof Ex. 41, p. 538

$CD^2 = AD \cdot BD$

Geometric Mean (Leg) Theorem

In a right triangle, the altitude from the right angle to the hypotenuse divides the hypotenuse into two segments.

The length of each leg of the right triangle is the geometric mean of the lengths of the hypotenuse and the segment of the hypotenuse that is adjacent to the leg.

Proof Ex. 42, p. 538

$CB^2 = DB \cdot AB$
$AC^2 = AD \cdot AB$

Laurie's Notes — Teacher Actions

- **Attend to Precision:** Write the *Core Concept*. Students should note that the *geometric mean* is the positive square root of the product ab.
- **Teaching Tip:** Students can get lost in the notation and words of these theorems. In the Geometric Mean (Leg) Theorem, help students see that the shortest side of the original right triangle is also the hypotenuse of the smallest right triangle, so that side is written twice in the proportion. Similarly, the altitude of the original triangle is the longer leg of the smallest triangle, which leads to the relationship given in the Geometric Mean (Altitude) Theorem.

EXAMPLE 4 Using a Geometric Mean

Find the value of each variable.

a.

b.

SOLUTION

COMMON ERROR
In Example 4(b), the Geometric Mean (Leg) Theorem gives $y^2 = 2 \cdot (5 + 2)$, not $y^2 = 5 \cdot (5 + 2)$, because the side with length y is adjacent to the segment with length 2.

a. Apply the Geometric Mean (Altitude) Theorem.

$x^2 = 6 \cdot 3$
$x^2 = 18$
$x = \sqrt{18}$
$x = \sqrt{9} \cdot \sqrt{2}$
$x = 3\sqrt{2}$

▶ The value of x is $3\sqrt{2}$.

b. Apply the Geometric Mean (Leg) Theorem.

$y^2 = 2 \cdot (5 + 2)$
$y^2 = 2 \cdot 7$
$y^2 = 14$
$y = \sqrt{14}$

▶ The value of y is $\sqrt{14}$.

EXAMPLE 5 Using Indirect Measurement

To find the cost of installing a rock wall in your school gymnasium, you need to find the height of the gym wall. You use a cardboard square to line up the top and bottom of the gym wall. Your friend measures the vertical distance from the ground to your eye and the horizontal distance from you to the gym wall. Approximate the height of the gym wall.

SOLUTION

By the Geometric Mean (Altitude) Theorem, you know that 8.5 is the geometric mean of w and 5.

$8.5^2 = w \cdot 5$ Geometric Mean (Altitude) Theorem
$72.25 = 5w$ Square 8.5.
$14.45 = w$ Divide each side by 5.

▶ The height of the wall is $5 + w = 5 + 14.45 = 19.45$ feet.

Monitoring Progress 🔊 Help in English and Spanish at BigIdeasMath.com

Find the geometric mean of the two numbers.

5. 12 and 27
6. 18 and 54
7. 16 and 18

8. Find the value of x in the triangle at the left.

9. **WHAT IF?** In Example 5, the vertical distance from the ground to your eye is 5.5 feet and the distance from you to the gym wall is 9 feet. Approximate the height of the gym wall.

Section 9.3 Similar Right Triangles 535

Extra Example 4
Find the value of each variable.

a.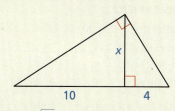

$x = 2\sqrt{10}$

b.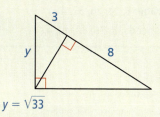

$y = \sqrt{33}$

Extra Example 5
Use the diagram and information in Example 5. The vertical distance from the ground to your eye is 5.4 feet and the distance from you to the gym wall is 8.1 feet. Approximate the height of the gym wall. 17.55 feet

MONITORING PROGRESS ANSWERS

5. 18
6. $18\sqrt{3} \approx 31.2$
7. $12\sqrt{2} \approx 17.0$
8. $x = 6$
9. about 20.2 ft

Laurie's Notes Teacher Actions

- **Think-Alouds:** In Example 4, have partner A *Think Aloud* as he or she finds the value of the variable. "Because x is the altitude of the right triangle, then ... ." Partners now reverse roles as partner B finds the value of y in part (b). Have students self-assess with *Thumbs Up*.
- **Attend to Precision:** Model Example 5 in the school gym or another similar area in the building.

Closure
- **Point of Most Significance:** Ask students to identify, aloud or on a paper, the most significant point (or part) in the lesson that aided their learning.

Section 9.3   535

Assignment Guide and Homework Check

ASSIGNMENT

Basic: 1, 2, 3–29 odd, 38, 40, 46–49

Average: 1, 2, 4–32 even, 38–41, 46–49

Advanced: 1, 2, 10–44 even, 45–49

HOMEWORK CHECK

Basic: 3, 5, 11, 19, 29

Average: 8, 18, 26, 32, 41

Advanced: 18, 26, 32, 40, 42

ANSWERS

1. each other
2. *Sample answer:* Geometric mean is a measure of center between two values. The ratio of the smaller of two values to the geometric mean is equal to the ratio of the geometric mean to the larger of two values.
3. △HFE ~ △GHE ~ △GFH
4. △MKN ~ △LMN ~ △LKM
5. $x = \frac{168}{25} = 6.72$
6. $x = \frac{48}{5} = 9.6$
7. $x = \frac{180}{13} \approx 13.8$
8. $x = \frac{240}{17} \approx 14.1$
9. about 11.2 ft
10. about 2.8 ft
11. 16
12. 12
13. $2\sqrt{70} \approx 16.7$
14. $5\sqrt{35} \approx 29.6$
15. 20
16. $4\sqrt{14} \approx 15.0$
17. $6\sqrt{17} \approx 24.7$
18. $6\sqrt{30} \approx 32.9$
19. $x = 8$
20. $y = 2\sqrt{10} \approx 6.3$
21. $y = 27$
22. $x = 4$
23. $x = 3\sqrt{5} \approx 6.7$
24. $b = 4\sqrt{22} \approx 18.8$
25. $z = \frac{729}{16} \approx 45.6$
26. $x = 4\sqrt{3} \approx 6.9$

9.3 Exercises

Dynamic Solutions available at *BigIdeasMath.com*

Vocabulary and Core Concept Check

1. **COMPLETE THE SENTENCE** If the altitude is drawn to the hypotenuse of a right triangle, then the two triangles formed are similar to the original triangle and _____.

2. **WRITING** In your own words, explain *geometric mean*.

Monitoring Progress and Modeling with Mathematics

In Exercises 3 and 4, identify the similar triangles. *(See Example 1.)*

3.

4.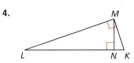

In Exercises 5–10, find the value of *x*. *(See Example 2.)*

5.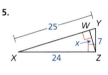

6.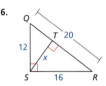

7. 8.

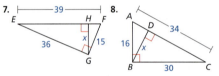

9. 10.

In Exercises 11–18, find the geometric mean of the two numbers. *(See Example 3.)*

11. 8 and 32
12. 9 and 16
13. 14 and 20
14. 25 and 35
15. 16 and 25
16. 8 and 28
17. 17 and 36
18. 24 and 45

In Exercises 19–26, find the value of the variable. *(See Example 4.)*

19.  20.

21.

22.

23. 24.

25.   26.

536   Chapter 9   Right Triangles and Trigonometry

ERROR ANALYSIS In Exercises 27 and 28, describe and correct the error in writing an equation for the given diagram.

27.
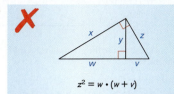
$z^2 = w \cdot (w + v)$

28.
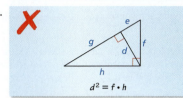
$d^2 = f \cdot h$

MODELING WITH MATHEMATICS In Exercises 29 and 30, use the diagram. *(See Example 5.)*

29. You want to determine the height of a monument at a local park. You use a cardboard square to line up the top and bottom of the monument, as shown at the above left. Your friend measures the vertical distance from the ground to your eye and the horizontal distance from you to the monument. Approximate the height of the monument.

30. Your classmate is standing on the other side of the monument. She has a piece of rope staked at the base of the monument. She extends the rope to the cardboard square she is holding lined up to the top and bottom of the monument. Use the information in the diagram above to approximate the height of the monument. Do you get the same answer as in Exercise 29? Explain your reasoning.

MATHEMATICAL CONNECTIONS In Exercises 31–34, find the value(s) of the variable(s).

31.   32.

33. 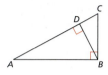   34.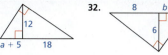

35. **REASONING** Use the diagram. Decide which proportions are true. Select all that apply.

Ⓐ $\dfrac{DB}{DC} = \dfrac{DA}{DB}$ Ⓑ $\dfrac{BA}{CB} = \dfrac{CB}{BD}$

Ⓒ $\dfrac{CA}{BA} = \dfrac{BA}{CA}$ Ⓓ $\dfrac{DB}{BC} = \dfrac{DA}{BA}$

36. **ANALYZING RELATIONSHIPS** You are designing a diamond-shaped kite. You know that $AD = 44.8$ centimeters, $DC = 72$ centimeters, and $AC = 84.8$ centimeters. You want to use a straight crossbar $\overline{BD}$. About how long should it be? Explain your reasoning.

37. **ANALYZING RELATIONSHIPS** Use both of the Geometric Mean Theorems to find AC and BD.

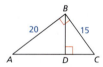

Section 9.3   Similar Right Triangles   537

ANSWERS

38. B

39. given; Geometric Mean (Leg) Theorem; a^2; Substitution Property of Equality; Distributive Property; c; Substitution Property of Equality

40. no; The geometric mean of 4 and 9 is 6, but the labels are incorrect on the triangle. The altitude could be 6, and would be the geometric mean of the two segments that make up the hypotenuse. Or, a leg could be 6, and would be the geometric mean of the length of the hypotenuse and the segment of the hypotenuse that is adjacent to the leg.

41–43. See Additional Answers.

44. $\dfrac{x+y}{2} \geq \sqrt{xy}$; They are only equal if $x = y$. Otherwise, the arithmetic mean is always greater than or equal to the geometric mean.

45–49. See Additional Answers.

Mini-Assessment

1. Find the values of x, y, and z.

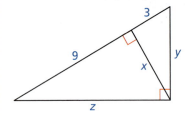

$x = 3\sqrt{3}$, $y = 6$, $z = 6\sqrt{3}$

2. Use the diagram and information in Example 5. The vertical distance from the ground to your eye is 5 feet and the distance from you to the gym wall is 9.5 feet. Approximate the height of the gym wall. **23.05 feet**

3. Identify the similar triangles.

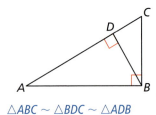

$\triangle ABC \sim \triangle BDC \sim \triangle ADB$

4. Find the geometric mean of $6\sqrt{3}$ and $2\sqrt{3}$. **6**

538   Chapter 9

---

38. HOW DO YOU SEE IT? In which of the following triangles does the Geometric Mean (Altitude) Theorem apply?

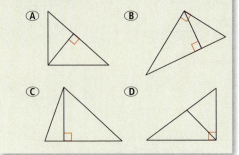

39. PROVING A THEOREM Use the diagram of $\triangle ABC$. Copy and complete the proof of the Pythagorean Theorem.

Given In $\triangle ABC$, $\angle BCA$ is a right angle.

Prove $c^2 = a^2 + b^2$

STATEMENTS	REASONS
1. In $\triangle ABC$, $\angle BCA$ is a right angle.	1. _____
2. Draw a perpendicular segment (altitude) from C to $\overline{AB}$.	2. Perpendicular Postulate
3. $ce = a^2$ and $cf = b^2$	3. _____
4. $ce + b^2 =$ ___ $+ b^2$	4. Addition Property of Equality
5. $ce + cf = a^2 + b^2$	5. _____
6. $c(e + f) = a^2 + b^2$	6. _____
7. $e + f =$ ___	7. Segment Addition Postulate
8. $c \cdot c = a^2 + b^2$	8. _____
9. $c^2 = a^2 + b^2$	9. Simplify.

Maintaining Mathematical Proficiency
Reviewing what you learned in previous grades and lessons

Solve the equation for x. *(Skills Review Handbook)*

46. $13 = \dfrac{x}{5}$ **47.** $29 = \dfrac{x}{4}$ **48.** $9 = \dfrac{78}{x}$ **49.** $30 = \dfrac{115}{x}$

538   Chapter 9   Right Triangles and Trigonometry

---

40. MAKING AN ARGUMENT Your friend claims the geometric mean of 4 and 9 is 6, and then labels the triangle, as shown. Is your friend correct? Explain your reasoning.

In Exercises 41 and 42, use the given statements to prove the theorem.

Given $\triangle ABC$ is a right triangle.
       Altitude $\overline{CD}$ is drawn to hypotenuse $\overline{AB}$.

41. PROVING A THEOREM Prove the Geometric Mean (Altitude) Theorem by showing that $CD^2 = AD \cdot BD$.

42. PROVING A THEOREM Prove the Geometric Mean (Leg) Theorem by showing that $CB^2 = DB \cdot AB$ and $AC^2 = AD \cdot AB$.

43. CRITICAL THINKING Draw a right isosceles triangle and label the two leg lengths x. Then draw the altitude to the hypotenuse and label its length y. Now, use the Right Triangle Similarity Theorem to draw the three similar triangles from the image and label any side length that is equal to either x or y. What can you conclude about the relationship between the two smaller triangles? Explain your reasoning.

44. THOUGHT PROVOKING The arithmetic mean and geometric mean of two nonnegative numbers x and y are shown.

arithmetic mean $= \dfrac{x+y}{2}$

geometric mean $= \sqrt{xy}$

Write an inequality that relates these two means. Justify your answer.

45. PROVING A THEOREM Prove the Right Triangle Similarity Theorem by proving three similarity statements.

Given $\triangle ABC$ is a right triangle.
       Altitude $\overline{CD}$ is drawn to hypotenuse $\overline{AB}$.

Prove $\triangle CBD \sim \triangle ABC$, $\triangle ACD \sim \triangle ABC$, $\triangle CBD \sim \triangle ACD$

If students need help...	If students got it...
Resources by Chapter • Practice A and Practice B • Puzzle Time	**Resources by Chapter** • Enrichment and Extension • Cumulative Review
Student Journal • Practice	Start the *next* Section
Differentiating the Lesson Skills Review Handbook	

9.1–9.3 What Did You Learn?

Core Vocabulary

Pythagorean triple, *p. 518* geometric mean, *p. 534*

Core Concepts

Section 9.1
Pythagorean Theorem, *p. 518*
Common Pythagorean Triples and Some of Their Multiples, *p. 518*
Converse of the Pythagorean Theorem, *p. 520*
Pythagorean Inequalities Theorem, *p. 521*

Section 9.2
45°-45°-90° Triangle Theorem, *p. 526*
30°-60°-90° Triangle Theorem, *p. 527*

Section 9.3
Right Triangle Similarity Theorem, *p. 532*
Geometric Mean (Altitude) Theorem, *p. 534*
Geometric Mean (Leg) Theorem, *p. 534*

Mathematical Practices

1. In Exercise 31 on page 523, describe the steps you took to find the area of the triangle.
2. In Exercise 23 on page 530, can one of the ways be used to show that all 30°-60°-90° triangles are similar? Explain.
3. Explain why the Geometric Mean (Altitude) Theorem does not apply to three of the triangles in Exercise 38 on page 538.

Form a Weekly Study Group, Set Up Rules

Consider using the following rules.
- Members must attend regularly, be on time, and participate.
- The sessions will focus on the key math concepts, not on the needs of one student.
- Students who skip classes will not be allowed to participate in the study group.
- Students who keep the group from being productive will be asked to leave the group.

Dynamic Teaching Tools

Dynamic Assessment & Progress Monitoring Tool
Interactive Whiteboard Lesson Library
Dynamic Classroom with Dynamic Investigations

ANSWERS

1. The altitude, h, divides the original triangle into two congruent smaller triangles by the HL Congruence Theorem. So, each of these smaller triangles has a side length of $\frac{16}{2} = 8$ meters. Use the Pythagorean Theorem, $h^2 + 8^2 = 17^2$, to solve for the height, $h = 15$ meters. The area of the original triangle is $A = \frac{1}{2}bh = \frac{1}{2}(16)(15) = 120$ square meters.

2. yes; You can also use the AA Similarity Theorem to conclude that all 30°-60°-90° triangles are similar because they have congruent corresponding angles.

3. The Geometric Mean (Altitude) Theorem applies when the triangle is a right triangle, and the altitude from the right angle to the hypotenuse divides the hypotenuse into two segments. Triangles A and C have altitudes drawn, but the original triangles are not right triangles. Triangle D is a right triangle, but the segment drawn from the right angle is not an altitude.

9.1–9.3 Quiz

Find the value of *x*. Tell whether the side lengths form a Pythagorean triple. *(Section 9.1)*

1.
2.
3.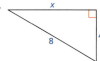

Verify that the segment lengths form a triangle. Is the triangle *acute*, *right*, or *obtuse*? *(Section 9.1)*

4. 24, 32, and 40
5. 7, 9, and 13
6. 12, 15, and $10\sqrt{3}$

Find the values of *x* and *y*. Write your answers in simplest form. *(Section 9.2)*

7.
8.
9.

Find the geometric mean of the two numbers. *(Section 9.3)*

10. 6 and 12
11. 15 and 20
12. 18 and 26

Identify the similar right triangles. Then find the value of the variable. *(Section 9.3)*

13.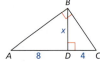
14.
15.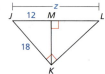

16. Television sizes are measured by the length of their diagonal. You want to purchase a television that is at least 40 inches. Should you purchase the television shown? Explain your reasoning. *(Section 9.1)*

17. Each triangle shown below is a right triangle. *(Sections 9.1–9.3)*
    a. Are any of the triangles special right triangles? Explain your reasoning.
    b. List all similar triangles, if any.
    c. Find the lengths of the altitudes of triangles *B* and *C*.

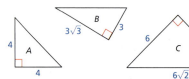 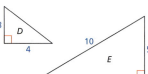

Laurie's Notes

Overview of Section 9.4

Introduction
- The tangent of an angle, the first of three trigonometric ratios studied, is defined and used in this lesson to solve application problems.

Resources
- Make clinometers that students can use to measure the angle of elevation (inclination) of an object so they can use a tangent ratio to determine its height.

Common Misconceptions
- Students can become sloppy and simply write $\tan = \dfrac{\text{opposite side}}{\text{adjacent side}}$, forgetting to specify the angle. Tangent is the name of the function, and the angle measure is the independent variable. Forgetting to write the angle measure is like writing $\sqrt[\square]{\square} = 4$, which does not give enough information to find either the radicand or the index of the radical.

Formative Assessment Tips
- **Which One Doesn't Belong?:** This technique is one that you should be quite familiar with because it is often used in the *Vocabulary and Core Concept Check* at the beginning of exercise sets! Students are presented with four expressions, quantities, images, or words and asked which one does not belong with the other three. They are also expected to give a reason for their choice.
- This technique gives students the opportunity to analyze and compare items in a set and determine what is alike in three of the four cases. This technique challenges students' reasoning and understanding of some aspect of the lesson they have just learned or what knowledge and conceptions they have about content to be learned.
- Used at the end of instruction, this technique informs you as to how students have conceptualized and made connections in their learning. The reasoning or justification for their choice can be quite informative. Used at the beginning of a lesson, this technique can inform you about what knowledge students already have about the topic.
- Select four items where it is not immediately obvious which one does not belong. You want to encourage deeper thinking.

Pacing Suggestion
- Take time for students to explore and discover the tangent ratio for acute angles, and then transition to the formal lesson.

Dynamic Teaching Tools
Dynamic Assessment & Progress Monitoring Tool
Lesson Planning Tool
Interactive Whiteboard Lesson Library
Dynamic Classroom with Dynamic Investigations

What Your Students Will Learn

- Use the tangent ratio to find the length of a leg of a right triangle.
- Solve real-life problems involving the tangent ratio.

Laurie's Notes

Exploration

Motivate
- Draw a right triangle and label the sides *a*, *b*, and *c*.
- ❓ "How many different ratios can be written using the sides *a*, *b*, and *c*? Explain." six: $a:b$, $a:c$, $b:a$, $b:c$, $c:a$, $c:b$
- Explain to students that today they will learn about the tangent of an angle, one of the six trigonometric ratios. Share a bit of the history of trigonometry if time permits.

Exploration 1
- The first exploration will take a little time for students to set up. To save a bit of time, you could create the template so that students spend their time with the measurements. *Note:* The template could also be used in the next lesson.
- Alternately, different groups of students could work with different angle measures for ∠*BAC*, noting that Exploration 2 would then need to be adjusted.
- This is a powerful exploration for students to experience. It is essential for students to recognize the similar triangles that have been constructed.
- ❓ "How can you show △*AQJ* ~ △*API*?" Use AA. Each triangle contains ∠*A* and a right angle.
- The second relationship for students to understand is that instead of comparing corresponding sides in similar triangles, the tangent ratios compare two sides in one triangle, the legs opposite and adjacent to a given angle.

Exploration 2
- **FYI:** Students' calculators need to be in degree mode, not radian mode.
- If other angle measures have been used in Exploration 1, make the appropriate adjustment.

Communicate Your Answer
- **Think-Pair-Share:** Have students work with their partners to answer Questions 3 and 4. Discuss as a class.

Connecting to Next Step
- The tangent ratio in a right triangle is defined in the formal lesson.

9.4 The Tangent Ratio

Essential Question How is a right triangle used to find the tangent of an acute angle? Is there a unique right triangle that must be used?

Let $\triangle ABC$ be a right triangle with acute $\angle A$. The *tangent* of $\angle A$ (written as tan A) is defined as follows.

$$\tan A = \frac{\text{length of leg opposite } \angle A}{\text{length of leg adjacent to } \angle A} = \frac{BC}{AC}$$

EXPLORATION 1 — Calculating a Tangent Ratio

Work with a partner. Use dynamic geometry software.

a. Construct $\triangle ABC$, as shown. Construct segments perpendicular to $\overline{AC}$ to form right triangles that share vertex A and are similar to $\triangle ABC$ with vertices, as shown.

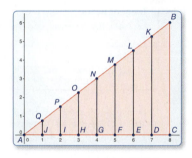

Sample
Points
$A(0, 0)$
$B(8, 6)$
$C(8, 0)$
Angle
$m\angle BAC = 36.87°$

b. Calculate each given ratio to complete the table for the decimal value of tan A for each right triangle. What can you conclude?

Ratio	$\frac{BC}{AC}$	$\frac{KD}{AD}$	$\frac{LE}{AE}$	$\frac{MF}{AF}$	$\frac{NG}{AG}$	$\frac{OH}{AH}$	$\frac{PI}{AI}$	$\frac{QJ}{AJ}$
tan A								

EXPLORATION 2 — Using a Calculator

Work with a partner. Use a calculator that has a tangent key to calculate the tangent of 36.87°. Do you get the same result as in Exploration 1? Explain.

Communicate Your Answer

3. Repeat Exploration 1 for $\triangle ABC$ with vertices $A(0, 0)$, $B(8, 5)$, and $C(8, 0)$. Construct the seven perpendicular segments so that not all of them intersect $\overline{AC}$ at integer values of x. Discuss your results.

4. How is a right triangle used to find the tangent of an acute angle? Is there a unique right triangle that must be used?

ATTENDING TO PRECISION
To be proficient in math, you need to express numerical answers with a degree of precision appropriate for the problem context.

English Language Learners

Class Activity
Draw the triangle on the board. Ask students to write and solve two different equations to find the value of x rounded to the nearest hundredth. Ask them if they see any advantages to using one method over the other.

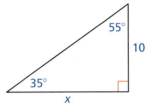

Sample answer: $\tan 35° = \dfrac{10}{x}$, $\tan 55° = \dfrac{x}{10}$; $x \approx 14.28$; Explanations will vary.

Extra Example 1
Find tan A and tan B. Write each answer as a fraction and as a decimal rounded to four places.

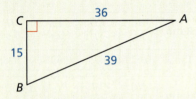

$\tan A = \dfrac{15}{36} = \dfrac{5}{12} \approx 0.4167$,

$\tan B = \dfrac{36}{15} = \dfrac{12}{5} = 2.4000$

MONITORING PROGRESS ANSWERS
1. $\dfrac{3}{4} = 0.7500$, $\dfrac{4}{3} \approx 1.3333$
2. $\dfrac{8}{15} \approx 0.5333$, $\dfrac{15}{8} = 1.8750$

9.4 Lesson

Core Vocabulary
trigonometric ratio, p. 542
tangent, p. 542
angle of elevation, p. 544

READING
Remember the following abbreviations.
tangent → tan
opposite → opp.
adjacent → adj.

What You Will Learn
▶ Use the tangent ratio.
▶ Solve real-life problems involving the tangent ratio.

Using the Tangent Ratio

A **trigonometric ratio** is a ratio of the lengths of two sides in a right triangle. All right triangles with a given acute angle are similar by the AA Similarity Theorem. So, △JKL ~ △XYZ, and you can write $\dfrac{KL}{YZ} = \dfrac{JL}{XZ}$. This can be rewritten as $\dfrac{KL}{JL} = \dfrac{YZ}{XZ}$, which is a trigonometric ratio. So, trigonometric ratios are constant for a given angle measure.

The **tangent** ratio is a trigonometric ratio for acute angles that involves the lengths of the legs of a right triangle.

🎯 Core Concept

Tangent Ratio

Let △ABC be a right triangle with acute ∠A.

The tangent of ∠A (written as tan A) is defined as follows.

$$\tan A = \dfrac{\text{length of leg opposite } \angle A}{\text{length of leg adjacent to } \angle A} = \dfrac{BC}{AC}$$

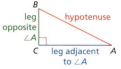

In the right triangle above, ∠A and ∠B are complementary. So, ∠B is acute. You can use the same diagram to find the tangent of ∠B. Notice that the leg adjacent to ∠A is the leg *opposite* ∠B and the leg opposite ∠A is the leg *adjacent* to ∠B.

ATTENDING TO PRECISION
Unless told otherwise, you should round the values of trigonometric ratios to four decimal places and round lengths to the nearest tenth.

EXAMPLE 1 Finding Tangent Ratios

Find tan S and tan R. Write each answer as a fraction and as a decimal rounded to four places.

SOLUTION

$\tan S = \dfrac{\text{opp. } \angle S}{\text{adj. to } \angle S} = \dfrac{RT}{ST} = \dfrac{80}{18} = \dfrac{40}{9} \approx 4.4444$

$\tan R = \dfrac{\text{opp. } \angle R}{\text{adj. to } \angle R} = \dfrac{ST}{RT} = \dfrac{18}{80} = \dfrac{9}{40} = 0.2250$

Monitoring Progress 🔊 Help in English and Spanish at *BigIdeasMath.com*

Find tan J and tan K. Write each answer as a fraction and as a decimal rounded to four places.

1.
2.

Laurie's Notes — Teacher Actions

- Discuss the introduction to the *tangent ratio*. All 10°-80°-90° triangles are similar, so the ratio of the sides opposite and adjacent to the 10° angle will be constant.
- Introduce the abbreviations for opposite, adjacent, and hypotenuse.
- **Common Misconception:** While the hypotenuse is adjacent to each acute angle, it is only referred to as the hypotenuse.
- **Attend to Precision:** Reference the *Attending to Precision* note so that students will know how precise their measures need to be.

EXAMPLE 2 Finding a Leg Length

Find the value of x. Round your answer to the nearest tenth.

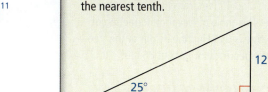

SOLUTION

Use the tangent of an acute angle to find a leg length.

$\tan 32° = \dfrac{\text{opp.}}{\text{adj.}}$ Write ratio for tangent of 32°.

$\tan 32° = \dfrac{11}{x}$ Substitute.

$x \cdot \tan 32° = 11$ Multiply each side by x.

$x = \dfrac{11}{\tan 32°}$ Divide each side by tan 32°.

$x \approx 17.6$ Use a calculator.

USING TOOLS STRATEGICALLY
You can also use the Table of Trigonometric Ratios available at *BigIdeasMath.com* to find the decimal approximations of trigonometric ratios.

▶ The value of x is about 17.6.

You can find the tangent of an acute angle measuring 30°, 45°, or 60° by applying what you know about special right triangles.

STUDY TIP
The tangents of all 60° angles are the same constant ratio. Any right triangle with a 60° angle can be used to determine this value.

EXAMPLE 3 Using a Special Right Triangle to Find a Tangent

Use a special right triangle to find the tangent of a 60° angle.

SOLUTION

Step 1 Because all 30°-60°-90° triangles are similar, you can simplify your calculations by choosing 1 as the length of the shorter leg. Use the 30°-60°-90° Triangle Theorem to find the length of the longer leg.

longer leg = shorter leg $\cdot \sqrt{3}$ 30°-60°-90° Triangle Theorem

$= 1 \cdot \sqrt{3}$ Substitute.

$= \sqrt{3}$ Simplify.

Step 2 Find tan 60°.

$\tan 60° = \dfrac{\text{opp.}}{\text{adj.}}$ Write ratio for tangent of 60°.

$\tan 60° = \dfrac{\sqrt{3}}{1}$ Substitute.

$\tan 60° = \sqrt{3}$ Simplify.

▶ The tangent of any 60° angle is $\sqrt{3} \approx 1.7321$.

Monitoring Progress Help in English and Spanish at *BigIdeasMath.com*

Find the value of x. Round your answer to the nearest tenth.

3.

4.

5. **WHAT IF?** In Example 3, the length of the shorter leg is 5 instead of 1. Show that the tangent of 60° is still equal to $\sqrt{3}$.

Extra Example 2

Find the value of x. Round your answer to the nearest tenth.

$x \approx 25.7$

Extra Example 3

Use a special right triangle to find the tangent of a 30° angle.
The tangent of any 30° angle is
$\dfrac{1}{\sqrt{3}} = \dfrac{\sqrt{3}}{3} \approx 0.5773$.

MONITORING PROGRESS ANSWERS

3. $x \approx 12.2$
4. $x \approx 19.3$
5. longer leg = shorter leg $\cdot \sqrt{3} = 5\sqrt{3}$,
 $\tan 60° = \dfrac{5\sqrt{3}}{5} = \dfrac{\sqrt{3}}{1} = \sqrt{3}$

Laurie's Notes   Teacher Actions

COMMON ERROR Solving the ratio $\tan 32° = \dfrac{11}{x}$, students incorrectly divide both sides by 11. Explain that this would still leave $11 \tan 32° = \dfrac{1}{x}$. This is the same error students will make in solving Question 3.

? Which One Doesn't Belong: Draw a 30°-60°-90° triangle and correctly label the sides 2, $2\sqrt{3}$, and 4. Write: $\tan 30° = \dfrac{\sqrt{3}}{3}$, $2\sqrt{3} \cdot \tan 60° = 2$, $2 \cdot \tan 60° = 2\sqrt{3}$, and $\sqrt{3} \cdot \tan 30° = 1$. Ask, "Which one doesn't belong? Explain." $2\sqrt{3} \cdot \tan 60° = 2$ is not a correct ratio. The rest are correct.

Extra Example 4

You are measuring the height of a tree. You stand 40 feet from the base of the tree. The angle of elevation to the top of the tree is 65°. Find the height of the tree to the nearest foot. about 86 feet

MONITORING PROGRESS ANSWER

6. about 110 in.

Solving Real-Life Problems

The angle that an upward line of sight makes with a horizontal line is called the **angle of elevation**.

EXAMPLE 4 Modeling with Mathematics

You are measuring the height of a spruce tree. You stand 45 feet from the base of the tree. You measure the angle of elevation from the ground to the top of the tree to be 59°. Find the height h of the tree to the nearest foot.

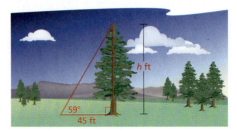

SOLUTION

1. **Understand the Problem** You are given the angle of elevation and the distance from the tree. You need to find the height of the tree to the nearest foot.

2. **Make a Plan** Write a trigonometric ratio for the tangent of the angle of elevation involving the height h. Then solve for h.

3. **Solve the Problem**

$\tan 59° = \dfrac{\text{opp.}}{\text{adj.}}$	Write ratio for tangent of 59°.
$\tan 59° = \dfrac{h}{45}$	Substitute.
$45 \cdot \tan 59° = h$	Multiply each side by 45.
$74.9 \approx h$	Use a calculator.

   ▶ The tree is about 75 feet tall.

4. **Look Back** Check your answer. Because 59° is close to 60°, the value of h should be close to the length of the longer leg of a 30°-60°-90° triangle, where the length of the shorter leg is 45 feet.

longer leg = shorter leg $\cdot \sqrt{3}$	30°-60°-90° Triangle Theorem
$= 45 \cdot \sqrt{3}$	Substitute.
≈ 77.9	Use a calculator.

   The value of 77.9 feet is close to the value of h. ✓

Monitoring Progress 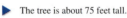 Help in English and Spanish at *BigIdeasMath.com*

6. You are measuring the height of a lamppost. You stand 40 inches from the base of the lamppost. You measure the angle of elevation from the ground to the top of the lamppost to be 70°. Find the height h of the lamppost to the nearest inch.

544   Chapter 9   Right Triangles and Trigonometry

Laurie's Notes — Teacher Actions

- Explain angle of elevation.
- **Wait Time:** Have partners work together to solve Example 4. Give sufficient *Wait Time*, and do not rush in to solve for students.
- **Attend to Precision:** Be sure that students label the answer with the correct units.
- **Connection:** Have students compare this method to what was done in the previous chapter in finding the height of a flagpole. If time permits, use this method outside to determine the height of an object.

Closure

- **Writing Prompt:** The tangent ratio is constant for an acute ∠A because …

9.4 Exercises

Dynamic Solutions available at *BigIdeasMath.com*

Vocabulary and Core Concept Check

1. **COMPLETE THE SENTENCE** The tangent ratio compares the length of _____ to the length of _____.

2. **WRITING** Explain how you know the tangent ratio is constant for a given angle measure.

Monitoring Progress and Modeling with Mathematics

In Exercises 3–6, find the tangents of the acute angles in the right triangle. Write each answer as a fraction and as a decimal rounded to four decimal places. *(See Example 1.)*

3.
4.
5.
6.

In Exercises 7–10, find the value of x. Round your answer to the nearest tenth. *(See Example 2.)*

7.
8.
9.
10.

ERROR ANALYSIS In Exercises 11 and 12, describe the error in the statement of the tangent ratio. Correct the error if possible. Otherwise, write not possible.

11.

12.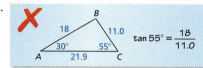

In Exercises 13 and 14, use a special right triangle to find the tangent of the given angle measure. *(See Example 3.)*

13. 45°
14. 30°

15. **MODELING WITH MATHEMATICS** A surveyor is standing 118 feet from the base of the Washington Monument. The surveyor measures the angle of elevation from the ground to the top of the monument to be 78°. Find the height h of the Washington Monument to the nearest foot. *(See Example 4.)*

16. **MODELING WITH MATHEMATICS** Scientists can measure the depths of craters on the moon by looking at photos of shadows. The length of the shadow cast by the edge of a crater is 500 meters. The angle of elevation of the rays of the Sun is 55°. Estimate the depth d of the crater.

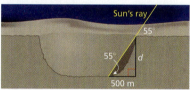

17. **USING STRUCTURE** Find the tangent of the smaller acute angle in a right triangle with side lengths 5, 12, and 13.

Section 9.4 The Tangent Ratio 545

Assignment Guide and Homework Check

ASSIGNMENT

Basic: 1, 2, 3–15 odd, 21, 22, 27–29
Average: 1, 2, 6–20 even, 21–23, 27–29
Advanced: 1, 2, 6–20 even, 21–29

HOMEWORK CHECK

Basic: 3, 9, 11, 15, 22
Average: 6, 10, 12, 16, 20
Advanced: 6, 10, 16, 20, 24

ANSWERS

1. the opposite leg, the adjacent leg
2. All right triangles with the given angle measure will be similar by the AA Similarity Theorem. So, the corresponding sides lengths will be proportional and the ratio will be constant.
3. $\tan R = \frac{45}{28} \approx 1.6071$, $\tan S = \frac{28}{45} \approx 0.6222$
4. $\tan D = \frac{7}{24} \approx 0.2917$, $\tan F = \frac{24}{7} \approx 3.4286$
5. $\tan G = \frac{2}{1} = 2.0000$, $\tan H = \frac{1}{2} = 0.5000$
6. $\tan J = \frac{5}{3} \approx 1.6667$, $\tan K = \frac{3}{5} = 0.6000$
7. $x \approx 13.8$
8. $x \approx 7.6$
9. $x \approx 13.7$
10. $x \approx 8.0$
11. The tangent ratio should be the length of the leg opposite ∠D to the length of the leg adjacent to ∠D, not the length of the hypotenuse; $\tan D = \frac{35}{12}$
12. Because △ABC is not a right triangle, the tangent ratio cannot be used; not possible
13. 1
14. $\frac{\sqrt{3}}{3} \approx 0.5774$
15. about 555 ft
16. about 714.1 m
17. $\frac{5}{12} \approx 0.4167$

Dynamic Teaching Tools

- Dynamic Assessment & Progress Monitoring Tool
- Interactive Whiteboard Lesson Library
- Dynamic Classroom with Dynamic Investigations

ANSWERS

18. $\frac{4}{3} \approx 1.3333$
19. it increases; The opposite side gets longer.
20–25. See Additional Answers.
26. about 128.0 units
27. $x = 2\sqrt{3} \approx 3.5$
28. $x = \frac{7\sqrt{3}}{3} \approx 4.0$
29. $x = 5\sqrt{2} \approx 7.1$

Mini-Assessment

1. You are measuring the height of a tree. You stand 50 feet from the base of the tree. The angle of elevation to the top of the tree is 47°. Find the height of the tree to the nearest foot. **about 54 feet**

2. Find tan D and tan E. Write each answer as a fraction and as a decimal rounded to four places.

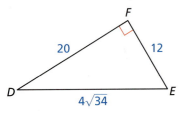

$\tan D = \frac{3}{5} = 0.6000$,

$\tan E = \frac{5}{3} \approx 1.6667$

3. Find the value of x. Round your answer to the nearest tenth.

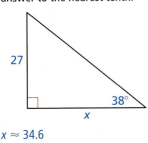

$x \approx 34.6$

---

18. **USING STRUCTURE** Find the tangent of the larger acute angle in a right triangle with side lengths 3, 4, and 5.

19. **REASONING** How does the tangent of an acute angle in a right triangle change as the angle measure increases? Justify your answer.

20. **CRITICAL THINKING** For what angle measure(s) is the tangent of an acute angle in a right triangle equal to 1? greater than 1? less than 1? Justify your answer.

21. **MAKING AN ARGUMENT** Your family room has a sliding-glass door. You want to buy an awning for the door that will be just long enough to keep the Sun out when it is at its highest point in the sky. The angle of elevation of the rays of the Sun at this point is 70°, and the height of the door is 8 feet. Your sister claims you can determine how far the overhang should extend by multiplying 8 by tan 70°. Is your sister correct? Explain.

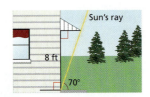

22. **HOW DO YOU SEE IT?** Write expressions for the tangent of each acute angle in the right triangle. Explain how the tangent of one acute angle is related to the tangent of the other acute angle. What kind of angle pair is ∠A and ∠B?

23. **REASONING** Explain why it is not possible to find the tangent of a right angle or an obtuse angle.

24. **THOUGHT PROVOKING** To create the diagram below, you begin with an isosceles right triangle with legs 1 unit long. Then the hypotenuse of the first triangle becomes the leg of a second triangle, whose remaining leg is 1 unit long. Continue the diagram until you have constructed an angle whose tangent is $\frac{1}{\sqrt{6}}$. Approximate the measure of this angle.

25. **PROBLEM SOLVING** Your class is having a class picture taken on the lawn. The photographer is positioned 14 feet away from the center of the class. The photographer turns 50° to look at either end of the class.

a. What is the distance between the ends of the class?

b. The photographer turns another 10° either way to see the end of the camera range. If each student needs 2 feet of space, about how many more students can fit at the end of each row? Explain.

26. **PROBLEM SOLVING** Find the perimeter of the figure, where $AC = 26$, $AD = BF$, and D is the midpoint of $\overline{AC}$.

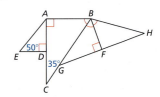

Maintaining Mathematical Proficiency
Reviewing what you learned in previous grades and lessons

Find the value of x. *(Section 9.2)*

27. (triangle with angle 30°, side $\sqrt{3}$, side x)

28. (right triangle with side 7, angle 60°, side x)

29. (triangle with side 5, angle 45°, side x)

546 Chapter 9   Right Triangles and Trigonometry

If students need help...	If students got it...
Resources by Chapter • Practice A and Practice B • Puzzle Time	**Resources by Chapter** • Enrichment and Extension • Cumulative Review
Student Journal • Practice	Start the *next* Section
Differentiating the Lesson Skills Review Handbook	

Laurie's Notes

Overview of Section 9.5

Introduction
- The sine and cosine ratios are defined and used in this lesson to determine the lengths of missing sides of a right triangle.
- Connections are made to a previous lesson as students find the sine and cosine of angle measures in special right triangles.
- The lesson continues with students using the ratios to solve a real-life application.
- Another connection to previous learning is made when the Pythagorean Theorem is used to write a Pythagorean identity, $\cos^2\theta + \sin^2\theta = 1$.

Formative Assessment Tips
- **Give Me Five:** This technique is public. Students are given a prompt and asked to think and write privately for a couple of minutes. You then ask five (or fewer) students to share aloud their reflection. In doing so, you are (a) continuing to encourage students to be reflective about their learning and (b) letting students know that you value their thoughts and insights.
- The feedback you receive will help guide subsequent instruction. If used at the end of the period, you will have insightful knowledge of students' learning that day and what subsequent instruction may be needed tomorrow. If used during the middle of the class after a critical point, be prepared to make course corrections to address what students have shared.
- As students share, you might have other students with similar thoughts indicate so by a show of hands. If many of the reflections are quite similar, it will not be necessary to have five students share.
- Sample prompts:
  - How did today's lesson help you better understand finding a trigonometric function of an acute angle?
  - How well do you feel you understand how to find the side lengths of a right triangle?
  - As we have worked on trigonometric functions and special right triangles this week, what has helped you the most in making sense of this topic?
  - How are similar triangles related to trigonometric functions?

Pacing Suggestion
- Take time for students to explore and discover the sine and cosine ratios for acute angles, and then transition to the formal lesson.

Dynamic Teaching Tools
Dynamic Assessment & Progress Monitoring Tool
Lesson Planning Tool
Interactive Whiteboard Lesson Library
Dynamic Classroom with Dynamic Investigations

What Your Students Will Learn

- Use the sine and cosine ratios to find the length of a leg or a hypotenuse of a right triangle.
- Find the sine and cosine of angle measures in 45°-45°-90° and 30°-60°-90° triangles.
- Solve real-life problems involving sine and cosine ratios.
- Use trigonometric identities involving sine and cosine to find trigonometric values.

Laurie's Notes

Exploration

Motivate
- **Story Time:** Share information with your students about a spring ski trip you (or your friends) are planning to Tuckerman Ravine, on the east side of Mt. Washington in the White Mountain National Forest of New Hampshire. Tuckerman Ravine is famous for its spectacular scenery, deep snow, and challenging terrain. Thousands of motivated skiers make the six-mile round trip to the floor of Tuckerman Ravine every year.
- There are no ski lifts—you hike in and hike out, thus you need to be motivated! To decide whether the hike is worth it, skiers want to know the length of the trail.
- Explain to students that in this lesson they will find a way to determine the length of a ski trail.

Exploration 1
- This exploration has the same setup as the exploration in the previous lesson, so students should be familiar with the setup and directions.
- If different groups of students worked with a different angle measure for ∠BAC in the last lesson, you might consider having groups of students do the same for this lesson.
- Like the last lesson, this is an important exploration for students to investigate. The ratio of the length of the opposite side to the length of the hypotenuse is the same for ∠A in every triangle in this exploration.

Communicate Your Answer
- **Think-Pair-Share:** Have students work with their partners to answer Questions 2 and 3. Discuss as a class.

Connecting to Next Step
- The sine and cosine ratios in a right triangle are defined in the formal lesson.

9.5 The Sine and Cosine Ratios

Essential Question How is a right triangle used to find the sine and cosine of an acute angle? Is there a unique right triangle that must be used?

Let $\triangle ABC$ be a right triangle with acute $\angle A$. The *sine* of $\angle A$ and *cosine* of $\angle A$ (written as sin A and cos A, respectively) are defined as follows.

$$\sin A = \frac{\text{length of leg opposite } \angle A}{\text{length of hypotenuse}} = \frac{BC}{AB}$$

$$\cos A = \frac{\text{length of leg adjacent to } \angle A}{\text{length of hypotenuse}} = \frac{AC}{AB}$$

EXPLORATION 1 Calculating Sine and Cosine Ratios

Work with a partner. Use dynamic geometry software.

a. Construct $\triangle ABC$, as shown. Construct segments perpendicular to $\overline{AC}$ to form right triangles that share vertex A and are similar to $\triangle ABC$ with vertices, as shown.

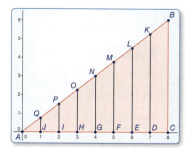

Sample
Points
$A(0, 0)$
$B(8, 6)$
$C(8, 0)$
Angle
$m\angle BAC = 36.87°$

b. Calculate each given ratio to complete the table for the decimal values of sin A and cos A for each right triangle. What can you conclude?

Sine ratio	$\frac{BC}{AB}$	$\frac{KD}{AK}$	$\frac{LE}{AL}$	$\frac{MF}{AM}$	$\frac{NG}{AN}$	$\frac{OH}{AO}$	$\frac{PI}{AP}$	$\frac{QJ}{AQ}$
sin A								
Cosine ratio	$\frac{AC}{AB}$	$\frac{AD}{AK}$	$\frac{AE}{AL}$	$\frac{AF}{AM}$	$\frac{AG}{AN}$	$\frac{AH}{AO}$	$\frac{AI}{AP}$	$\frac{AJ}{AQ}$
cos A								

LOOKING FOR STRUCTURE

To be proficient in math, you need to look closely to discern a pattern or structure.

Communicate Your Answer

2. How is a right triangle used to find the sine and cosine of an acute angle? Is there a unique right triangle that must be used?

3. In Exploration 1, what is the relationship between $\angle A$ and $\angle B$ in terms of their measures? Find sin B and cos B. How are these two values related to sin A and cos A? Explain why these relationships exist.

Section 9.5   The Sine and Cosine Ratios   547

Differentiated Instruction

Auditory
The mnemonic device SOH-CAH-TOA is a helpful way for students to remember the trigonometric ratios. "Sine is Opposite over Hypotenuse" (SOH), "Cosine is Adjacent over Hypotenuse" (CAH), and "Tangent is Opposite over Adjacent" (TOA). Write SOC-CAH-TOA on the board, explain how to use it, and then have the class state the term represented by each letter.

Extra Example 1
Find sin A, sin B, cos A, and cos B. Write each answer as a fraction and as a decimal rounded to four places.

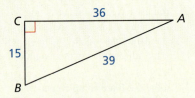

$\sin A = \cos B = \frac{15}{39} = \frac{5}{13} \approx 0.3846$,

$\sin B = \cos A = \frac{36}{39} = \frac{12}{13} \approx 0.9231$

9.5 Lesson

What You Will Learn
- Use the sine and cosine ratios.
- Find the sine and cosine of angle measures in special right triangles.
- Solve real-life problems involving sine and cosine ratios.
- Use a trigonometric identity.

Core Vocabulary
sine, *p. 548*
cosine, *p. 548*
angle of depression, *p. 551*

Using the Sine and Cosine Ratios

The **sine** and **cosine** ratios are trigonometric ratios for acute angles that involve the lengths of a leg and the hypotenuse of a right triangle.

READING
Remember the following abbreviations.
sine → sin
cosine → cos
hypotenuse → hyp.

Core Concept
Sine and Cosine Ratios

Let $\triangle ABC$ be a right triangle with acute $\angle A$. The sine of $\angle A$ and cosine of $\angle A$ (written as sin A and cos A) are defined as follows.

$\sin A = \dfrac{\text{length of leg opposite } \angle A}{\text{length of hypotenuse}} = \dfrac{BC}{AB}$

$\cos A = \dfrac{\text{length of leg adjacent to } \angle A}{\text{length of hypotenuse}} = \dfrac{AC}{AB}$

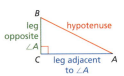

EXAMPLE 1 Finding Sine and Cosine Ratios

Find sin S, sin R, cos S, and cos R. Write each answer as a fraction and as a decimal rounded to four places.

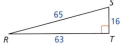

SOLUTION

$\sin S = \dfrac{\text{opp. } \angle S}{\text{hyp.}} = \dfrac{RT}{SR} = \dfrac{63}{65} \approx 0.9692 \qquad \sin R = \dfrac{\text{opp. } \angle R}{\text{hyp.}} = \dfrac{ST}{SR} = \dfrac{16}{65} \approx 0.2462$

$\cos S = \dfrac{\text{adj. to } \angle S}{\text{hyp.}} = \dfrac{ST}{SR} = \dfrac{16}{65} \approx 0.2462 \qquad \cos R = \dfrac{\text{adj. to } \angle R}{\text{hyp.}} = \dfrac{RT}{SR} = \dfrac{63}{65} \approx 0.9692$

In Example 1, notice that sin $S = \cos R$ and sin $R = \cos S$. This is true because the side opposite $\angle S$ is adjacent to $\angle R$ and the side opposite $\angle R$ is adjacent to $\angle S$. The relationship between the sine and cosine of $\angle S$ and $\angle R$ is true for all complementary angles.

Core Concept
Sine and Cosine of Complementary Angles

The sine of an acute angle is equal to the cosine of its complement. The cosine of an acute angle is equal to the sine of its complement.

Let A and B be complementary angles. Then the following statements are true.

$\sin A = \cos(90° - A) = \cos B \qquad \sin B = \cos(90° - B) = \cos A$

$\cos A = \sin(90° - A) = \sin B \qquad \cos B = \sin(90° - B) = \sin A$

Laurie's Notes — Teacher Actions

- Write the *Core Concept*, being sure to note the abbreviations and the need to specify the angle.
- ? "What is the same about the sine and cosine ratios?" **They have the same denominator, the length of the hypotenuse.**
- ? "If the sine and cosine ratios compare the length of a leg to the length of the hypotenuse, then what do you know about the ratios for acute angles?" **The ratios are less than 1.**
- **Reason Abstractly and Quantitatively** and **Turn and Talk:** "If A and B are complementary angles, then why does sin $A = \cos B$?" Listen for correct reasoning.

EXAMPLE 2 Rewriting Trigonometric Expressions

Write sin 56° in terms of cosine.

SOLUTION

Use the fact that the sine of an acute angle is equal to the cosine of its complement.

$$\sin 56° = \cos(90° - 56°) = \cos 34°$$

▶ The sine of 56° is the same as the cosine of 34°.

You can use the sine and cosine ratios to find unknown measures in right triangles.

EXAMPLE 3 Finding Leg Lengths

Find the values of x and y using sine and cosine. Round your answers to the nearest tenth.

SOLUTION

Step 1 Use a sine ratio to find the value of x.

$\sin 26° = \dfrac{\text{opp.}}{\text{hyp.}}$	Write ratio for sine of 26°.
$\sin 26° = \dfrac{x}{14}$	Substitute.
$14 \cdot \sin 26° = x$	Multiply each side by 14.
$6.1 \approx x$	Use a calculator.

▶ The value of x is about 6.1.

Step 2 Use a cosine ratio to find the value of y.

$\cos 26° = \dfrac{\text{adj.}}{\text{hyp.}}$	Write ratio for cosine of 26°.
$\cos 26° = \dfrac{y}{14}$	Substitute.
$14 \cdot \cos 26° = y$	Multiply each side by 14.
$12.6 \approx y$	Use a calculator.

▶ The value of y is about 12.6.

Monitoring Progress Help in English and Spanish at *BigIdeasMath.com*

1. Find sin D, sin F, cos D, and cos F. Write each answer as a fraction and as a decimal rounded to four places.

2. Write cos 23° in terms of sine.

3. Find the values of u and t using sine and cosine. Round your answers to the nearest tenth.

English Language Learners

Words and Abbreviations

This lesson uses many abbreviations for trigonometric ratios and parts of triangles. Encourage English learners to make a table in their notebooks showing the full word and the abbreviation.

Word	Abbreviation
sine	sin
cosine	cos
tangent	tan
adjacent	adj.
opposite	opp.
hypotenuse	hyp.

Extra Example 2
Write cos 69° in terms of sine.
sin 21°

Extra Example 3
Find the values of x and y using sine and cosine. Round your answers to the nearest tenth.

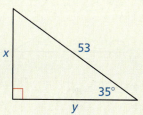

$x \approx 30.4, y \approx 43.4$

MONITORING PROGRESS ANSWERS

1. $\sin D = \dfrac{7}{25} = 0.2800$,
   $\sin F = \dfrac{24}{25} = 0.9600$,
   $\cos D = \dfrac{24}{25} = 0.9600$,
   $\cos F = \dfrac{7}{25} = 0.2800$

2. sin 67°

3. $u \approx 3.4, t \approx 7.3$

Laurie's Notes   Teacher Actions

❓ "In Example 3, what are the angle measures of the triangle?" 26°, 64°, and 90°

❓ "What trigonometric equation can you write to solve for x?" $\sin 26° = \dfrac{x}{14}$ or $\cos 64° = \dfrac{x}{14}$
  Ask a similar question in solving for y.

❓ **Extension:** After solving for x and y, ask, "How can you check to see whether your answers are reasonable?" Use the Pythagorean Theorem: $x^2 + y^2 = 14^2$. Accounting for rounding, the values for x and y should work in this equation.

Extra Example 4

Which ratios are equal to $\frac{\sqrt{2}}{2}$? Select all that apply.
- sin A
- cos A
- tan A
- sin B
- cos B
- tan B

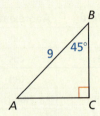

sin A, cos A, sin B, cos B

Extra Example 5

Which ratios are equal to $\frac{\sqrt{3}}{2}$? Select all that apply.
- sin M
- sin P
- cos M
- cos P

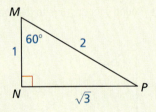

sin M, cos P

MONITORING PROGRESS ANSWER

4. $\sin 60° = \frac{\sqrt{3}}{2} \approx 0.8660,$

   $\cos 60° = \frac{1}{2} = 0.5000$

Finding Sine and Cosine in Special Right Triangles

EXAMPLE 4 Finding the Sine and Cosine of 45°

Find the sine and cosine of a 45° angle.

SOLUTION

Begin by sketching a 45°-45°-90° triangle. Because all such triangles are similar, you can simplify your calculations by choosing 1 as the length of each leg. Using the 45°-45°-90° Triangle Theorem, the length of the hypotenuse is $\sqrt{2}$.

STUDY TIP

Notice that
$\sin 45° = \cos(90 - 45)°$
$= \cos 45°.$

$\sin 45° = \frac{\text{opp.}}{\text{hyp.}}$ $\qquad \cos 45° = \frac{\text{adj.}}{\text{hyp.}}$

$\quad = \frac{1}{\sqrt{2}}$ $\qquad\qquad\qquad = \frac{1}{\sqrt{2}}$

$\quad = \frac{\sqrt{2}}{2}$ $\qquad\qquad\qquad = \frac{\sqrt{2}}{2}$

$\quad \approx 0.7071$ $\qquad\qquad\quad \approx 0.7071$

EXAMPLE 5 Finding the Sine and Cosine of 30°

Find the sine and cosine of a 30° angle.

SOLUTION

Begin by sketching a 30°-60°-90° triangle. Because all such triangles are similar, you can simplify your calculations by choosing 1 as the length of the shorter leg. Using the 30°-60°-90° Triangle Theorem, the length of the longer leg is $\sqrt{3}$ and the length of the hypotenuse is 2.

$\sin 30° = \frac{\text{opp.}}{\text{hyp.}}$ $\qquad \cos 30° = \frac{\text{adj.}}{\text{hyp.}}$

$\quad = \frac{1}{2}$ $\qquad\qquad\qquad = \frac{\sqrt{3}}{2}$

$\quad = 0.5000$ $\qquad\qquad\quad \approx 0.8660$

Monitoring Progress Help in English and Spanish at *BigIdeasMath.com*

4. Find the sine and cosine of a 60° angle.

Laurie's Notes — Teacher Actions

- Have students work in groups of two to four students.
- Say, "You have learned about special right triangles in an earlier lesson. Today, you have learned about the sine and cosine ratios. Work with your partners to find the exact and approximate values of the sine and cosine of 30°, 45°, and 60°."
- Let students work with one another as you circulate. Do not rush in to rescue! Trust that students have the necessary knowledge to answer the question. Solicit answers and summarize in a table. Look for patterns (i.e., $\sin \theta$ is an increasing function).

Solving Real-Life Problems

Recall from the previous lesson that the angle an upward line of sight makes with a horizontal line is called the *angle of elevation*. The angle that a downward line of sight makes with a horizontal line is called the **angle of depression**.

EXAMPLE 6   Modeling with Mathematics

You are skiing on a mountain with an altitude of 1200 feet. The angle of depression is 21°. Find the distance x you ski down the mountain to the nearest foot.

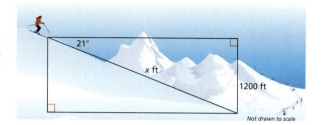

SOLUTION

1. **Understand the Problem** You are given the angle of depression and the altitude of the mountain. You need to find the distance that you ski down the mountain.

2. **Make a Plan** Write a trigonometric ratio for the sine of the angle of depression involving the distance x. Then solve for x.

3. **Solve the Problem**

   $\sin 21° = \dfrac{\text{opp.}}{\text{hyp.}}$ Write ratio for sine of 21°.

   $\sin 21° = \dfrac{1200}{x}$ Substitute.

   $x \cdot \sin 21° = 1200$ Multiply each side by x.

   $x = \dfrac{1200}{\sin 21°}$ Divide each side by sin 21°.

   $x \approx 3348.5$ Use a calculator.

   ▶ You ski about 3349 feet down the mountain.

4. **Look Back** Check your answer. The value of sin 21° is about 0.3584. Substitute for x in the sine ratio and compare the values.

   $\dfrac{1200}{x} \approx \dfrac{1200}{3348.5}$

   ≈ 0.3584

   This value is approximately the same as the value of sin 21°. ✓

Monitoring Progress   Help in English and Spanish at *BigIdeasMath.com*

5. **WHAT IF?** In Example 6, the angle of depression is 28°. Find the distance x you ski down the mountain to the nearest foot.

Extra Example 6

You are skiing down a hill with an altitude of 800 feet. The angle of depression is 15°. Find the distance x you ski down the hill to the nearest foot. **about 3091 feet**

MONITORING PROGRESS ANSWER

5. about 2556 ft

Laurie's Notes   Teacher Actions

- In the last lesson on the tangent ratio, only the leg lengths were used. Discuss with students that now problems also involve the hypotenuse.
- **Turn and Talk:** Pose Example 6. "How can you find the distance the skier skis?"
- **Make Sense of Problems and Persevere in Solving Them** and **Model with Mathematics:** Drawing a sketch is helpful in making sense of the problem. Note that students could use sine or cosine to solve.

Extra Example 7

Given that $\sin \theta = \frac{1}{3}$, find $\cos \theta$ and $\tan \theta$.

$\cos \theta = \frac{2\sqrt{2}}{3}$, $\tan \theta = \frac{\sqrt{2}}{4}$

MONITORING PROGRESS ANSWER

6. $\cos \theta = \frac{12}{13}$; $\tan \theta = \frac{5}{12}$

Using a Trigonometric Identity

In the figure, the point (x, y) is on a circle of radius 1 with center at the origin. Consider an angle θ (the Greek letter *theta*) with its vertex at the origin. You know that $x = \cos \theta$ and $y = \sin \theta$. By the Pythagorean Theorem,

$$x^2 + y^2 = 1.$$

So,

$$\cos^2 \theta + \sin^2 \theta = 1.$$

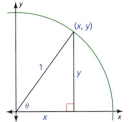

STUDY TIP
Note that $\sin^2 \theta$ represents $(\sin \theta)^2$ and $\cos^2 \theta$ represents $(\cos \theta)^2$.

The equation $\cos^2 \theta + \sin^2 \theta = 1$ is true for any value of θ. In this section, θ will always be an acute angle. A trigonometric equation that is true for all values of the variable for which both sides of the equation are defined is called a *trigonometric identity*. Because of its derivation above, $\cos^2 \theta + \sin^2 \theta = 1$ is also referred to as a *Pythagorean identity*.

EXAMPLE 7 Finding Trigonometric Values

Given that $\cos \theta = \frac{3}{5}$, find $\sin \theta$ and $\tan \theta$.

SOLUTION

Step 1 Find $\sin \theta$.

$\sin^2 \theta + \cos^2 \theta = 1$	Write Pythagorean identity.
$\sin^2 \theta + \left(\frac{3}{5}\right)^2 = 1$	Substitute $\frac{3}{5}$ for $\cos \theta$.
$\sin^2 \theta + \frac{9}{25} = 1$	Evaluate power.
$\sin^2 \theta = 1 - \frac{9}{25}$	Subtract $\frac{9}{25}$ from each side.
$\sin^2 \theta = \frac{16}{25}$	Subtract.
$\sin \theta = \frac{4}{5}$	Take the positive square root of each side.

STUDY TIP
Using the diagram above, you can write
$\tan \theta = \frac{y}{x} = \frac{\sin \theta}{\cos \theta}.$

Step 2 Find $\tan \theta$ using $\sin \theta$ and $\cos \theta$.

$$\tan \theta = \frac{\sin \theta}{\cos \theta} = \frac{\frac{4}{5}}{\frac{3}{5}} = \frac{4}{3}$$

Monitoring Progress Help in English and Spanish at *BigIdeasMath.com*

6. Given that $\sin \theta = \frac{5}{13}$, find $\cos \theta$ and $\tan \theta$.

Laurie's Notes — Teacher Actions

- **Connection:** Remind students that we refer to the equation $3(x − 2) = 3x − 6$ as an identity because it is true for all values of *x*. Trigonometric identities are true for all values of the variable for which both sides of the equation are defined.
- Derive the Pythagorean identity, $\cos^2 x + \sin^2 x = 1$ using the diagram shown.
- Pose Example 7 and set up the problem. Give partners time to work the problem themselves as you circulate. Solicit volunteers to share their work with the class.

Closure

- **Give Me Five:** Select one or more of the prompts on page T-546 for students to reflect on.

9.5 Exercises

Dynamic Solutions available at BigIdeasMath.com

Vocabulary and Core Concept Check

1. **VOCABULARY** The sine ratio compares the length of _____ to the length of _____.

2. **VOCABULARY** The _____ ratio compares the length of the adjacent leg to the length of the hypotenuse.

3. **WRITING** How are angles of elevation and angles of depression similar? How are they different?

4. **WHICH ONE DOESN'T BELONG?** Which ratio does *not* belong with the other three? Explain your reasoning.

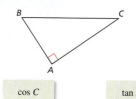

| sin B | cos C | tan B | $\frac{AC}{BC}$ |

Monitoring Progress and Modeling with Mathematics

In Exercises 5–10, find sin D, sin E, cos D, and cos E. Write each answer as a fraction and as a decimal rounded to four places. *(See Example 1.)*

5.
6.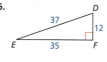
7.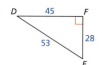
8.
9.
10.

In Exercises 11–14, write the expression in terms of cosine. *(See Example 2.)*

11. sin 37°
12. sin 81°
13. sin 29°
14. sin 64°

In Exercises 15–18, write the expression in terms of sine.

15. cos 59°
16. cos 42°
17. cos 73°
18. cos 18°

In Exercises 19–24, find the value of each variable using sine and cosine. Round your answers to the nearest tenth. *(See Example 3.)*

19.
20.
21.
22.
23.
24.

Section 9.5 The Sine and Cosine Ratios 553

Assignment Guide and Homework Check

ASSIGNMENT
Basic: 1–4, 5–27 odd, 30, 31–45 odd, 48, 54, 59–63

Average: 1–4, 10–46 even, 47–49, 54, 59–63

Advanced: 1–4, 10–28 even, 29, 30–46 even, 47–49, 52–58 even, 59-63

HOMEWORK CHECK
Basic: 9, 17, 23, 31, 37
Average: 10, 18, 24, 32, 38
Advanced: 10, 18, 24, 32, 38

ANSWERS

1. the opposite leg, the hypotenuse

2. cosine

3. *Sample answer:* Both angles are formed by a horizontal line and a line of sight. An angle of elevation is formed by an upward line of sight and an angle of depression is formed by a downward line of sight.

4. tan B; The other three are all equal to $\frac{AC}{BC}$, but $\tan B = \frac{AC}{AB}$.

5. $\sin D = \frac{4}{5} = 0.8000$,
   $\sin E = \frac{3}{5} = 0.6000$,
   $\cos D = \frac{3}{5} = 0.6000$,
   $\cos E = \frac{4}{5} = 0.8000$

6. $\sin D = \frac{35}{37} \approx 0.9459$,
   $\sin E = \frac{12}{37} \approx 0.3243$,
   $\cos D = \frac{12}{37} \approx 0.3243$,
   $\cos E = \frac{35}{37} \approx 0.9459$

7. $\sin D = \frac{28}{53} \approx 0.5283$,
   $\sin E = \frac{45}{53} \approx 0.8491$,
   $\cos D = \frac{45}{53} \approx 0.8491$,
   $\cos E = \frac{28}{53} \approx 0.5283$

8. $\sin D = \frac{4}{5} = 0.8000$,
   $\sin E = \frac{3}{5} = 0.6000$,
   $\cos D = \frac{3}{5} = 0.6000$,
   $\cos E = \frac{4}{5} = 0.8000$

9. $\sin D = \frac{\sqrt{3}}{2} \approx 0.8660$,
   $\sin E = \frac{1}{2} = 0.5000$,
   $\cos D = \frac{1}{2} = 0.5000$,
   $\cos E = \frac{\sqrt{3}}{2} \approx 0.8660$

10. $\sin D = \frac{8}{17} \approx 0.4706$, $\sin E = \frac{15}{17} \approx 0.8824$,
    $\cos D = \frac{15}{17} \approx 0.8824$, $\cos E = \frac{8}{17} \approx 0.4706$

11. cos 53°
12. cos 9°
13. cos 61°
14. cos 26°
15. sin 31°
16. sin 48°
17. sin 17°
18. sin 72°
19. $x \approx 9.5, y \approx 15.3$
20. $p \approx 30.6, q \approx 14.9$
21. $v \approx 4.7, w \approx 1.6$
22. $r \approx 19.0, s \approx 17.7$
23. $a \approx 14.9, b \approx 11.1$
24. $m \approx 6.7, n \approx 10.4$

Dynamic Teaching Tools

Dynamic Assessment & Progress Monitoring Tool
Interactive Whiteboard Lesson Library
Dynamic Classroom with Dynamic Investigations

ANSWERS

25. $\sin X = \cos X = \sin Z = \cos Z$
26. $\sin L$, $\cos J$
27. $\sin M$, $\cos N$
28. $\sin Q$, $\cos S$
29. Any given acute angle in a right triangle is formed by one leg and the hypotenuse. The hypotenuse is across from the right angle, and the leg is called the adjacent leg.
30. The sine of $\angle A$ should be equal to the ratio of the length of the leg opposite the angle, to the length of the hypotenuse; $\sin A = \frac{12}{13}$
31. about 15 ft
32. about 19.6 ft
33. $\sin \theta = \frac{\sqrt{21}}{5}$; $\tan \theta = \frac{\sqrt{21}}{2}$
34. $\sin \theta = \frac{2\sqrt{10}}{7}$; $\tan \theta = \frac{2\sqrt{10}}{3}$
35. $\sin \theta = \frac{2\sqrt{14}}{9}$; $\tan \theta = \frac{2\sqrt{14}}{5}$
36. $\sin \theta = \frac{\sqrt{57}}{11}$; $\tan \theta = \frac{\sqrt{57}}{8}$
37. $\sin \theta = \frac{\sqrt{15}}{8}$; $\tan \theta = \frac{\sqrt{15}}{7}$
38. $\sin \theta = \frac{\sqrt{19}}{10}$; $\tan \theta = \frac{\sqrt{19}}{9}$
39. $\cos \theta = \frac{\sqrt{3}}{2}$; $\tan \theta = \frac{\sqrt{3}}{3}$
40. $\cos \theta = \frac{\sqrt{5}}{3}$; $\tan \theta = \frac{2\sqrt{5}}{5}$
41. $\cos \theta = \frac{\sqrt{39}}{8}$; $\tan \theta = \frac{5\sqrt{39}}{39}$
42. $\cos \theta = \frac{\sqrt{51}}{10}$; $\tan \theta = \frac{7\sqrt{51}}{51}$
43. $\cos \theta = \frac{\sqrt{77}}{9}$; $\tan \theta = \frac{2\sqrt{77}}{77}$
44. $\cos \theta = \frac{\sqrt{13}}{7}$; $\tan \theta = \frac{6\sqrt{13}}{13}$

---

25. REASONING Which ratios are equal? Select all that apply. *(See Example 4.)*

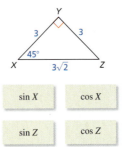

sin X   cos X

sin Z   cos Z

26. REASONING Which ratios are equal to $\frac{1}{2}$? Select all that apply. *(See Example 5.)*

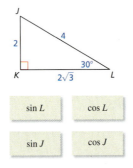

sin L   cos L

sin J   cos J

27. REASONING Write the trigonometric ratios that have a value of $\frac{\sqrt{3}}{2}$.

28. REASONING Write the trigonometric ratios that have a value of $\frac{1}{2}$.

29. WRITING Explain how to tell which side of a right triangle is adjacent to an angle and which side is the hypotenuse.

554   Chapter 9   Right Triangles and Trigonometry

---

30. ERROR ANALYSIS Describe and correct the error in finding sin A.

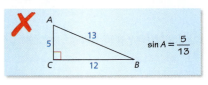

31. MODELING WITH MATHEMATICS The top of the slide is 12 feet from the ground and has an angle of depression of 53°. What is the length of the slide? *(See Example 6.)*

32. MODELING WITH MATHEMATICS Find the horizontal distance x the escalator covers.

In Exercises 33–38, find sin θ and tan θ. *(See Example 7.)*

33. $\cos \theta = \frac{2}{5}$ 34. $\cos \theta = \frac{3}{7}$

35. $\cos \theta = \frac{5}{9}$ 36. $\cos \theta = \frac{8}{11}$

37. $\cos \theta = \frac{7}{8}$ 38. $\cos \theta = \frac{9}{10}$

In Exercises 39–44, find cos θ and tan θ.

39. $\sin \theta = \frac{1}{2}$ 40. $\sin \theta = \frac{2}{3}$

41. $\sin \theta = \frac{5}{8}$ 42. $\sin \theta = \frac{7}{10}$

43. $\sin \theta = \frac{2}{9}$ 44. $\sin \theta = \frac{6}{7}$

45. ERROR ANALYSIS Describe and correct the error in finding $\cos \theta$ when $\sin \theta = \frac{8}{9}$.

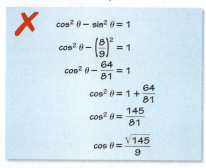

46. ERROR ANALYSIS Describe and correct the error in finding $\sin \theta$ when $\cos \theta = \frac{18}{25}$.

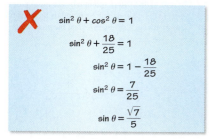

47. PROBLEM SOLVING You are flying a kite with 20 feet of string extended. The angle of elevation from the spool of string to the kite is 67°.

  a. Draw and label a diagram that represents the situation.

  b. How far off the ground is the kite if you hold the spool 5 feet off the ground? Describe how the height where you hold the spool affects the height of the kite.

48. MAKING AN ARGUMENT Your friend uses the equation $\sin 49° = \frac{x}{16}$ to find BC. Your cousin uses the equation $\cos 41° = \frac{x}{16}$ to find BC. Who is correct? Explain your reasoning.

49. MODELING WITH MATHEMATICS Planes that fly at high speeds and low elevations have radar systems that can determine the range of an obstacle and the angle of elevation to the top of the obstacle. The radar of a plane flying at an altitude of 20,000 feet detects a tower that is 25,000 feet away, with an angle of elevation of 1°.

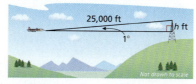

  a. How many feet must the plane rise to pass over the tower?

  b. Planes cannot come closer than 1000 feet vertically to any object. At what altitude must the plane fly in order to pass over the tower?

50. WRITING Describe what you must know about a triangle in order to use the sine ratio and what you must know about a triangle in order to use the cosine ratio.

51. MATHEMATICAL CONNECTIONS If $\triangle EQU$ is equilateral and $\triangle RGT$ is a right triangle with $RG = 2$, $RT = 1$, and $m\angle T = 90°$, show that $\sin E = \cos G$.

52. MODELING WITH MATHEMATICS Submarines use sonar systems, which are similar to radar systems, to detect obstacles. Sonar systems use sound to detect objects under water.

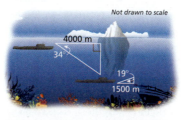

  a. You are traveling underwater in a submarine. The sonar system detects an iceberg 4000 meters ahead, with an angle of depression of 34° to the bottom of the iceberg. How many meters must the submarine lower to pass under the iceberg?

  b. The sonar system then detects a sunken ship 1500 meters ahead, with an angle of elevation of 19° to the highest part of the sunken ship. How many meters must the submarine rise to pass over the sunken ship?

Section 9.5 The Sine and Cosine Ratios 555

ANSWERS

45. The identity is $\cos^2 \theta + \sin^2 \theta = 1$, not $\cos^2 \theta - \sin^2 \theta = 1$; $\cos \theta = \frac{\sqrt{17}}{9}$

46. $\frac{18}{25}$ should be squared; $\sin \theta = \frac{\sqrt{301}}{25}$

47. a.

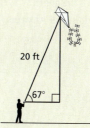

  b. about 23.4 ft; The higher you hold the spool, the farther the kite is from the ground.

48. both; The sine of an acute angle is equal to the cosine of its complement, so these two equations are equivalent.

49. a. at least 437 ft

  b. at least 21,437 ft

50. The sine function involves the measure of an acute angle, the length of the leg opposite that angle, and the length of the hypotenuse. You must know two out of three of those values to solve for the third (although you have not yet learned how to solve for the measure of the angle). The cosine function is the same, except it involves the measure of the leg adjacent to the acute angle.

51.

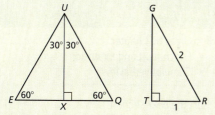

Because $\triangle EQU$ is an equilateral triangle, all three angles have a measure of 60°. When an altitude, $\overline{UX}$, is drawn from U to $\overline{EQ}$ as shown, two congruent 30°-60°-90° triangles are formed, where $m\angle E = 60°$.

So, $\sin E = \sin 60° = \frac{\sqrt{3}}{2}$. Also, in $\triangle RGT$, because the hypotenuse is twice as long as one of the legs, it is also a 30°-60°-90° triangle. Because $\angle G$ is across from the shorter leg, it must have a measure of 30°, which means that $\cos G = \cos 30° = \frac{\sqrt{3}}{2}$. So, $\sin E = \cos G$.

52. a. at least 2699 m

  b. at least 517 m

Section 9.5 555

ANSWERS

53–56. See Additional Answers.

57. If you knew how to take the inverse of the trigonometric ratios, you could first find the respective ratio of sides and then take the inverse of the trigonometric ratio to find the measure of the angle.

58–63. See Additional Answers.

Mini-Assessment

1. Find the values of x and y using sine and cosine. Round your answers to the nearest tenth.

   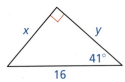

   $x \approx 10.5, y \approx 12.1$

2. Write cos 81° in terms of sine. $\sin 9°$

3. Is cos 45° greater than or less than cos 60°? Explain why. greater than; $\cos 45° = \frac{\sqrt{2}}{2} \approx 0.7071$, $\cos 60° = \frac{1}{2} = 0.5000$

4. The angle of elevation of a ladder leaning against a wall is 70°. The base of the ladder is 1.4 meters from the base of the wall. Find the length of the ladder to the nearest tenth of a meter.

   about 4.1 meters

5. Given that $\cos \theta = \frac{8}{17}$, find $\sin \theta$ and $\tan \theta$. $\sin \theta = \frac{15}{17}$, $\tan \theta = \frac{15}{8}$

---

53. **REASONING** Use $\sin A = \frac{\text{opp.}}{\text{hyp.}}$ and $\cos A = \frac{\text{adj.}}{\text{hyp.}}$ to show that $\tan A = \frac{\sin A}{\cos A}$.

54. **HOW DO YOU SEE IT?** Using only the given information, would you use a sine ratio or a cosine ratio to find the length of the hypotenuse? Explain your reasoning.

    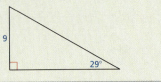

55. **MULTIPLE REPRESENTATIONS** You are standing on a cliff above an ocean. You see a sailboat from your vantage point 30 feet above the ocean.
    a. Draw and label a diagram of the situation.
    b. Make a table showing the angle of depression and the length of your line of sight. Use the angles 40°, 50°, 60°, 70°, and 80°.
    c. Graph the values you found in part (b), with the angle measures on the x-axis.
    d. Predict the length of the line of sight when the angle of depression is 30°.

56. **THOUGHT PROVOKING** One of the following infinite series represents sin x and the other one represents cos x (where x is measured in radians). Which is which? Justify your answer. Then use each series to approximate the sine and cosine of $\frac{\pi}{6}$.
    (Hints: $\pi = 180°$; $5! = 5 \cdot 4 \cdot 3 \cdot 2 \cdot 1$; Find the values that the sine and cosine ratios approach as the angle measure approaches zero.)

    a. $x - \frac{x^3}{3!} + \frac{x^5}{5!} - \frac{x^7}{7!} + \cdots$

    b. $1 - \frac{x^2}{2!} + \frac{x^4}{4!} - \frac{x^6}{6!} + \cdots$

57. **ABSTRACT REASONING** Make a conjecture about how you could use trigonometric ratios to find angle measures in a triangle.

58. **CRITICAL THINKING** Explain why the area of $\triangle ABC$ in the diagram can be found using the formula Area $= \frac{1}{2}ab \sin C$. Then calculate the area when $a = 4$, $b = 7$, and $m\angle C = 40°$.

    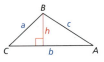

Maintaining Mathematical Proficiency
Reviewing what you learned in previous grades and lessons

Find the value of x. Tell whether the side lengths form a Pythagorean triple. *(Section 9.1)*

59.

60.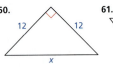

61. (triangle with 27, 36, x)

62.

63. Use the two-way table to create a two-way table that shows the joint and marginal relative frequencies. *(Section 5.3)*

		Gender		
		Male	Female	Total
Class	Sophomore	26	54	80
	Junior	32	48	80
	Senior	50	55	105
	Total	108	157	265

Laurie's Notes

Overview of Section 9.6

Introduction
- Students are familiar with sine, cosine, and tangent ratios and have used these ratios to solve for missing sides of a right triangle.
- In this lesson, the inverse trigonometric ratios are presented, allowing students to now solve for the missing angles of a right triangle.
- Given sufficient starting information, students solve a right triangle completely—meaning they are able to find all three angle measures and all three side lengths.

Teaching Strategy
- Students enjoy the opportunity to apply their learning to a familiar context. There are many ways in which students can practice using trigonometric ratios and inverse trigonometric ratios around school.
- Any context that can be represented as a right triangle can be investigated. In my own setting, I have students investigate the angle of the main staircase. My school is located in an 1860 Victorian house, built long before building codes were in effect. Students know the stairs are steeper than those in their own homes, and they investigate the angle of the stairway. Ramps for people with disabilities represent another opportunity for investigation.
- Measuring inaccessible heights (in the gymnasium, of a flagpole or a roofline) is possible if you use a clinometer (mentioned on page T-540).
- Consider asking students to look around school for scenarios where a right triangle could represent the situation. What information is known, and what information can be solved for?

Pacing Suggestion
- The two explorations provide an introduction to inverse trigonometric ratios. Transition to the formal lesson after discussion of Questions 3 and 4.

Dynamic Teaching Tools
Dynamic Assessment & Progress Monitoring Tool
Lesson Planning Tool
Interactive Whiteboard Lesson Library
Dynamic Classroom with Dynamic Investigations

What Your Students Will Learn

- Use inverse trigonometric ratios to find angle measures of right triangles.
- Solve right triangles using the Pythagorean Theorem and inverse trigonometric ratios.

Laurie's Notes

Exploration

Motivate
- Show the Table of Trigonometric Ratios available at *BigIdeasMath.com*. Explain that before scientific calculators were readily available, tables such as these were included in textbooks. Also printed in textbooks were tables of square roots and cube roots!

Exploration 1
- In this first exploration, students are constructing triangles using information from the diagrams.
- In part (a), the coordinates of the triangle can be determined: (0, 0), (4, 0), and (4, 4). Students should recognize this as a right isosceles triangle.
- In part (b), the first step is to construct a circle of radius 4 centered at (0, 2). The coordinates of point *B* in the diagram correspond to the intersection of the circle and the *x*-axis.
- Measuring the lengths of the sides, the sine and cosine ratios can be found. The values should sound familiar to students, who can then verify the angles by measuring.
- **Extension:** Construct another triangle, *EDC*, similar to triangle *ABC* in part (b). Students should recognize that the sine and cosine ratios will be the same. Moreover, the measures of the corresponding angles will be congruent.

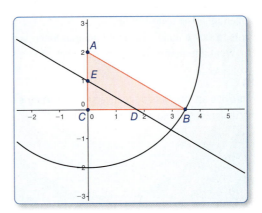

Exploration 2
- The triangles in this exploration have leg lengths that make finding the tangent ratio very easy.
- Students can compute the length of the hypotenuse using the software.
- Be sure that students have their calculators set for degree mode rather than radian mode.
- **Attend to Precision:** Tell students that $\sin^{-1}$ is read "inverse sine" and not "sine to the negative one."

Communicate Your Answer
- There are many online applets that simulate the leaning ladder problem. Although this problem specifies one exact location, it is an interesting problem for students to consider.

Connecting to Next Step
- Students have now been introduced to the inverse trigonometric ratios. In the formal lesson, these ratios will be defined formally and used to solve problems.

9.6 Solving Right Triangles

Essential Question When you know the lengths of the sides of a right triangle, how can you find the measures of the two acute angles?

EXPLORATION 1 Solving Special Right Triangles

Work with a partner. Use the figures to find the values of the sine and cosine of ∠A and ∠B. Use these values to find the measures of ∠A and ∠B. Use dynamic geometry software to verify your answers.

a.
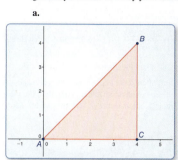

b.
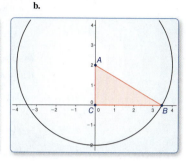

ATTENDING TO PRECISION
To be proficient in math, you need to calculate accurately and efficiently, expressing numerical answers with a degree of precision appropriate for the problem context.

EXPLORATION 2 Solving Right Triangles

Work with a partner. You can use a calculator to find the measure of an angle when you know the value of the sine, cosine, or tangent of the angle. Use the inverse sine, inverse cosine, or inverse tangent feature of your calculator to approximate the measures of ∠A and ∠B to the nearest tenth of a degree. Then use dynamic geometry software to verify your answers.

a.
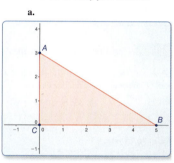

b.
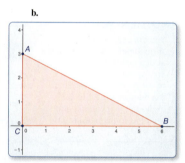

Communicate Your Answer

3. When you know the lengths of the sides of a right triangle, how can you find the measures of the two acute angles?

4. A ladder leaning against a building forms a right triangle with the building and the ground. The legs of the right triangle (in meters) form a 5-12-13 Pythagorean triple. Find the measures of the two acute angles to the nearest tenth of a degree.

Dynamic Teaching Tools

Dynamic Assessment & Progress Monitoring Tool
Lesson Planning Tool
Interactive Whiteboard Lesson Library
Dynamic Classroom with Dynamic Investigations

ANSWERS

1. a. $\sin A = \cos A = \sin B = \cos B = \frac{\sqrt{2}}{2}$; $m\angle A = m\angle B = 45°$

   b. $\sin A = \cos B = \frac{\sqrt{3}}{2}$, $\cos A = \sin B = \frac{1}{2}$; $m\angle A = 60°$, $m\angle B = 30°$

2. a. $m\angle A \approx 59.0°$, $m\angle B \approx 31.0°$
   b. $m\angle A \approx 63.4°$, $m\angle B \approx 26.6°$

3. You can find the ratio of two side lengths that gives the value of the sine, cosine, or tangent of an angle. If you recognize the ratio from a special right triangle, then you can find the measure of the angle that way. Otherwise, you can use the inverse sine, cosine, or tangent feature of your calculator to approximate the measure of the angle.

4. about 67.4°, about 22.6°

Extra Example 1
Determine which of the two acute angles has a sine of 0.4.

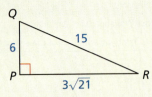

The acute angle that has a sine of 0.4 is $\angle R$.

Extra Example 2
Let $\angle A$, $\angle B$, and $\angle C$ be acute angles. Use a calculator to approximate the measures of $\angle A$, $\angle B$, and $\angle C$ to the nearest tenth of a degree.

a. $\tan A = 3.29$ $m\angle A \approx 73.1°$
b. $\sin B = 0.55$ $m\angle B \approx 33.4°$
c. $\cos C = 0.87$ $m\angle C \approx 29.5°$

MONITORING PROGRESS ANSWERS
1. $\angle E$
2. $\angle F$

9.6 Lesson

Core Vocabulary
inverse tangent, p. 558
inverse sine, p. 558
inverse cosine, p. 558
solve a right triangle, p. 559

What You Will Learn
- Use inverse trigonometric ratios.
- Solve right triangles.

Using Inverse Trigonometric Ratios

EXAMPLE 1 Identifying Angles from Trigonometric Ratios

Determine which of the two acute angles has a cosine of 0.5.

SOLUTION
Find the cosine of each acute angle.

$\cos A = \dfrac{\text{adj. to } \angle A}{\text{hyp.}} = \dfrac{\sqrt{3}}{2} \approx 0.8660$ $\cos B = \dfrac{\text{adj. to } \angle B}{\text{hyp.}} = \dfrac{1}{2} = 0.5$

▶ The acute angle that has a cosine of 0.5 is $\angle B$.

If the measure of an acute angle is 60°, then its cosine is 0.5. The converse is also true. If the cosine of an acute angle is 0.5, then the measure of the angle is 60°. So, in Example 1, the measure of $\angle B$ must be 60° because its cosine is 0.5.

Core Concept

Inverse Trigonometric Ratios
Let $\angle A$ be an acute angle.

READING
The expression "$\tan^{-1} x$" is read as "the inverse tangent of x."

Inverse Tangent If $\tan A = x$, then $\tan^{-1} x = m\angle A$. $\tan^{-1} \dfrac{BC}{AC} = m\angle A$

Inverse Sine If $\sin A = y$, then $\sin^{-1} y = m\angle A$. $\sin^{-1} \dfrac{BC}{AB} = m\angle A$

Inverse Cosine If $\cos A = z$, then $\cos^{-1} z = m\angle A$. $\cos^{-1} \dfrac{AC}{AB} = m\angle A$

ANOTHER WAY
You can use the Table of Trigonometric Ratios available at BigIdeasMath.com to approximate $\tan^{-1} 0.75$ to the nearest degree. Find the number closest to 0.75 in the tangent column and read the angle measure at the left.

EXAMPLE 2 Finding Angle Measures

Let $\angle A$, $\angle B$, and $\angle C$ be acute angles. Use a calculator to approximate the measures of $\angle A$, $\angle B$, and $\angle C$ to the nearest tenth of a degree.

a. $\tan A = 0.75$ b. $\sin B = 0.87$ c. $\cos C = 0.15$

SOLUTION
a. $m\angle A = \tan^{-1} 0.75 \approx 36.9°$
b. $m\angle B = \sin^{-1} 0.87 \approx 60.5°$
c. $m\angle C = \cos^{-1} 0.15 \approx 81.4°$

Monitoring Progress Help in English and Spanish at BigIdeasMath.com

Determine which of the two acute angles has the given trigonometric ratio.

1. The sine of the angle is $\dfrac{12}{13}$.
2. The tangent of the angle is $\dfrac{5}{12}$.

558 Chapter 9 Right Triangles and Trigonometry

Laurie's Notes | Teacher Actions

- Sketch the triangle shown in Example 1. Ask students to find the sine and cosine of angles A and B as well as the measure of each angle.
- State the *Core Concept*. Students find it helpful when you say, "$\tan^{-1}(1) = x$ means you want to find the angle whose tangent is 1." Another way of asking this question would be to write $\tan x = 1$.
- **Thumbs Up:** Have students work with partners to answer all parts of Example 2. Use *Thumbs Up* to assess students' understanding.

Monitoring Progress   Help in English and Spanish at *BigIdeasMath.com*

Let $\angle G$, $\angle H$, and $\angle K$ be acute angles. Use a calculator to approximate the measures of $\angle G$, $\angle H$, and $\angle K$ to the nearest tenth of a degree.

3. $\tan G = 0.43$ **4.** $\sin H = 0.68$ **5.** $\cos K = 0.94$

Solving Right Triangles

 Core Concept

Solving a Right Triangle

To **solve a right triangle** means to find all unknown side lengths and angle measures. You can solve a right triangle when you know either of the following.
- two side lengths
- one side length and the measure of one acute angle

EXAMPLE 3   Solving a Right Triangle

Solve the right triangle. Round decimal answers to the nearest tenth.

SOLUTION

Step 1 Use the Pythagorean Theorem to find the length of the hypotenuse.

$c^2 = a^2 + b^2$ Pythagorean Theorem
$c^2 = 3^2 + 2^2$ Substitute.
$c^2 = 13$ Simplify.
$c = \sqrt{13}$ Find the positive square root.
$c \approx 3.6$ Use a calculator.

ANOTHER WAY

You could also have found $m\angle A$ first by finding $\tan^{-1} \frac{3}{2} \approx 56.3°$.

Step 2 Find $m\angle B$.

$m\angle B = \tan^{-1} \frac{2}{3} \approx 33.7°$ Use a calculator.

Step 3 Find $m\angle A$.

Because $\angle A$ and $\angle B$ are complements, you can write

$m\angle A = 90° - m\angle B$
$\approx 90° - 33.7°$
$= 56.3°$.

▶ In $\triangle ABC$, $c \approx 3.6$, $m\angle B \approx 33.7°$, and $m\angle A \approx 56.3°$.

Monitoring Progress   Help in English and Spanish at *BigIdeasMath.com*

Solve the right triangle. Round decimal answers to the nearest tenth.

6.   **7.**

Section 9.6   Solving Right Triangles   559

Extra Example 4

Solve the right triangle. Round decimal answers to the nearest tenth.

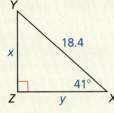

$x \approx 12.1$, $y \approx 13.9$, $m\angle Y = 49°$

Extra Example 5

Use the information in Example 5. Another raked stage is 25 feet long from front to back with a total rise of 1.5 feet. You want the rake to be 5° or less. Is the raked stage within your desired range? Explain. yes; The rake is about 3.4°, so it is within the desired range.

MONITORING PROGRESS ANSWERS

8. $XY \approx 13.8$, $YZ \approx 10.9$, $m\angle Y = 38°$
9. no

EXAMPLE 4 Solving a Right Triangle

Solve the right triangle. Round decimal answers to the nearest tenth.

SOLUTION

Use trigonometric ratios to find the values of g and h.

$$\sin H = \frac{\text{opp.}}{\text{hyp.}} \qquad \cos H = \frac{\text{adj.}}{\text{hyp.}}$$

$$\sin 25° = \frac{h}{13} \qquad \cos 25° = \frac{g}{13}$$

$$13 \cdot \sin 25° = h \qquad 13 \cdot \cos 25° = g$$

$$5.5 \approx h \qquad 11.8 \approx g$$

Because $\angle H$ and $\angle G$ are complements, you can write

$$m\angle G = 90° - m\angle H = 90° - 25° = 65°.$$

▶ In $\triangle GHJ$, $h \approx 5.5$, $g \approx 11.8$, and $m\angle G = 65°$.

READING
A *raked stage* slants upward from front to back to give the audience a better view.

EXAMPLE 5 Solving a Real-Life Problem

Your school is building a *raked stage*. The stage will be 30 feet long from front to back, with a total rise of 2 feet. You want the rake (angle of elevation) to be 5° or less for safety. Is the raked stage within your desired range?

SOLUTION

Use the inverse sine ratio to find the degree measure x of the rake.

$$x \approx \sin^{-1} \frac{2}{30} \approx 3.8$$

▶ The rake is about 3.8°, so it is within your desired range of 5° or less.

Monitoring Progress Help in English and Spanish at *BigIdeasMath.com*

8. Solve the right triangle. Round decimal answers to the nearest tenth.

9. **WHAT IF?** In Example 5, suppose another raked stage is 20 feet long from front to back with a total rise of 2 feet. Is the raked stage within your desired range?

560   Chapter 9   Right Triangles and Trigonometry

Laurie's Notes Teacher Actions

? "After solving a right triangle completely, how can you check your answers?" Answers will vary. Students may mention using the Pythagorean Theorem, or that the angles should still sum to 180°, or that the shortest side is opposite the least angle measure.

- **Model with Mathematics:** After working Example 5, consider any ramps on school property. See the *Teaching Strategy* on page T-556.

Closure

- **Exit Ticket:** A right triangle has legs of 8 centimeters and 13 centimeters. Solve the triangle completely. hypotenuse = $\sqrt{233}$ cm or ≈ 15.3 cm; angle measures $\approx 31.6°$, $\approx 58.4°$, 90°

9.6 Exercises

Dynamic Solutions available at *BigIdeasMath.com*

Vocabulary and Core Concept Check

1. **COMPLETE THE SENTENCE** To solve a right triangle means to find the measures of all its _____ and _____.

2. **WRITING** Explain when you can use a trigonometric ratio to find a side length of a right triangle and when you can use the Pythagorean Theorem.

Monitoring Progress and Modeling with Mathematics

In Exercises 3–6, determine which of the two acute angles has the given trigonometric ratio. *(See Example 1.)*

3. The cosine of the angle is $\frac{4}{5}$.

4. The sine of the angle is $\frac{5}{11}$.

5. The sine of the angle is 0.96.

6. The tangent of the angle is 1.5.

In Exercises 7–12, let $\angle D$ be an acute angle. Use a calculator to approximate the measure of $\angle D$ to the nearest tenth of a degree. *(See Example 2.)*

7. $\sin D = 0.75$
8. $\sin D = 0.19$
9. $\cos D = 0.33$
10. $\cos D = 0.64$
11. $\tan D = 0.28$
12. $\tan D = 0.72$

In Exercises 13–18, solve the right triangle. Round decimal answers to the nearest tenth. *(See Examples 3 and 4.)*

13.
14.
15.
16.
17.
18.

19. **ERROR ANALYSIS** Describe and correct the error in using an inverse trigonometric ratio.

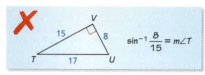

20. **PROBLEM SOLVING** In order to unload clay easily, the body of a dump truck must be elevated to at least 45°. The body of a dump truck that is 14 feet long has been raised 8 feet. Will the clay pour out easily? Explain your reasoning. *(See Example 5.)*

21. **PROBLEM SOLVING** You are standing on a footbridge that is 12 feet above a lake. You look down and see a duck in the water. The duck is 7 feet away from the footbridge. What is the angle of elevation from the duck to you?

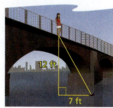

Assignment Guide and Homework Check

ASSIGNMENT
Basic: 1, 2, 3–21 odd, 22, 26, 33–36
Average: 1, 2, 6–18 even, 19–23, 26–28, 33–36
Advanced: 1, 2, 6, 10, 16, 18–24, 26–31, 33–36

HOMEWORK CHECK
Basic: 5, 13, 15, 21, 26
Average: 6, 18, 20, 26, 28
Advanced: 6, 16, 20, 26, 30

ANSWERS

1. sides, angles
2. You can use a trigonometric ratio to find a side length of a right triangle when you know one of the acute angle measures and one of the other side lengths. You can use the Pythagorean Theorem when you know the other two side lengths.
3. $\angle C$
4. $\angle A$
5. $\angle A$
6. $\angle C$
7. about 48.6°
8. about 11.0°
9. about 70.7°
10. about 50.2°
11. about 15.6°
12. about 35.8°
13. $AB = 15$, $m\angle A \approx 53.1°$, $m\angle B \approx 36.9°$
14. $DE \approx 16.1$, $m\angle E \approx 60.3°$, $m\angle D \approx 29.7°$
15. $YZ \approx 8.5$, $m\angle X \approx 70.5°$, $m\angle Z \approx 19.5°$
16. $HJ \approx 7.7$, $m\angle G \approx 29.0°$, $m\angle J \approx 61.0°$
17. $KL \approx 5.1$, $ML \approx 6.1$, $m\angle K = 50°$
18. $RS \approx 9.7$, $RT \approx 17.9$, $m\angle T = 33°$
19. The sine ratio should be the length of the opposite side to the length of the hypotenuse, not the adjacent side; $\sin^{-1} \frac{8}{17} = m\angle T$
20. no; The body of the dump truck has only been elevated to about 34.8°, which is less than the recommended minimum angle of 45°.
21. about 59.7°

22. **HOW DO YOU SEE IT?** Write three expressions that can be used to approximate the measure of ∠A. Which expression would you choose? Explain your choice.

23. **MODELING WITH MATHEMATICS** The Uniform Federal Accessibility Standards specify that a wheelchair ramp may not have an incline greater than 4.76°. You want to build a ramp with a vertical rise of 8 inches. You want to minimize the horizontal distance taken up by the ramp. Draw a diagram showing the approximate dimensions of your ramp.

24. **MODELING WITH MATHEMATICS** The horizontal part of a step is called the *tread*. The vertical part is called the *riser*. The recommended riser-to-tread ratio is 7 inches : 11 inches.

    a. Find the value of x for stairs built using the recommended riser-to-tread ratio.

    b. You want to build stairs that are less steep than the stairs in part (a). Give an example of a riser-to-tread ratio that you could use. Find the value of x for your stairs.

25. **USING TOOLS** Find the measure of ∠R without using a protractor. Justify your technique.

    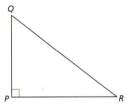

26. **MAKING AN ARGUMENT** Your friend claims that $\tan^{-1} x = \dfrac{1}{\tan x}$. Is your friend correct? Explain your reasoning.

USING STRUCTURE In Exercises 27 and 28, solve each triangle.

27. △JKM and △LKM

    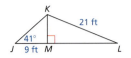

28. △TUS and △VTW

29. **MATHEMATICAL CONNECTIONS** Write an expression that can be used to find the measure of the acute angle formed by each line and the x-axis. Then approximate the angle measure to the nearest tenth of a degree.

    a. $y = 3x$

    b. $y = \frac{4}{3}x + 4$

30. **THOUGHT PROVOKING** Simplify each expression. Justify your answer.

    a. $\sin^{-1}(\sin x)$

    b. $\tan(\tan^{-1} y)$

    c. $\cos(\cos^{-1} z)$

31. **REASONING** Explain why the expression $\sin^{-1}(1.2)$ does not make sense.

32. **USING STRUCTURE** The perimeter of rectangle ABCD is 16 centimeters, and the ratio of its width to its length is 1 : 3. Segment BD divides the rectangle into two congruent triangles. Find the side lengths and angle measures of these two triangles.

Maintaining Mathematical Proficiency *Reviewing what you learned in previous grades and lessons*

Solve the equation. *(Skills Review Handbook)*

33. $\dfrac{12}{x} = \dfrac{3}{2}$

34. $\dfrac{13}{9} = \dfrac{x}{18}$

35. $\dfrac{x}{2.1} = \dfrac{4.1}{3.5}$

36. $\dfrac{5.6}{12.7} = \dfrac{4.9}{x}$

562   Chapter 9   Right Triangles and Trigonometry

9.4–9.6 What Did You Learn?

Core Vocabulary

trigonometric ratio, *p. 542*
tangent, *p. 542*
angle of elevation, *p. 544*
sine, *p. 548*
cosine, *p. 548*
angle of depression, *p. 551*
inverse tangent, *p. 558*
inverse sine, *p. 558*
inverse cosine, *p. 558*
solve a right triangle, *p. 559*

Core Concepts

Section 9.4
Tangent Ratio, *p. 542*

Section 9.5
Sine and Cosine Ratios, *p. 548*
Sine and Cosine of Complementary Angles, *p. 548*
Using a Trigonometric Identity, *p. 552*

Section 9.6
Inverse Trigonometric Ratios, *p. 558*
Solving a Right Triangle, *p. 559*

Mathematical Practices

1. In Exercise 21 on page 546, your brother claims that you could determine how far the overhang should extend by dividing 8 by tan 70°. Justify his conclusion and explain why it works.

2. In Exercise 47 on page 555, explain the flaw in the argument that the kite is 18.4 feet high.

Performance Task:

Challenging the Rock Wall

There are multiple ways to use indirect measurement when you cannot measure directly. When you look at a rock wall, do you wonder about its height before you attempt to climb it? What other situations require you to calculate a height using trigonometry and indirect measurement?

To explore the answers to these questions and more, check out the Performance Task and Real-Life STEM video at *BigIdeasMath.com*.

ANSWERS

1. $x = 2\sqrt{34} \approx 11.7$; no
2. $x = 12$; yes
3. $x = 2\sqrt{30} \approx 11.0$; no
4. yes; acute
5. yes; right
6. yes; obtuse
7. $x = 6\sqrt{2}$
8. $x = 7$
9. $x = 16\sqrt{3}$

9 Chapter Review

Dynamic Solutions available at *BigIdeasMath.com*

9.1 The Pythagorean Theorem (pp. 517–524)

Find the value of *x*. Then tell whether the side lengths form a Pythagorean triple.

$c^2 = a^2 + b^2$ Pythagorean Theorem
$x^2 = 15^2 + 20^2$ Substitute.
$x^2 = 225 + 400$ Multiply.
$x^2 = 625$ Add.
$x = 25$ Find the positive square root.

▶ The value of *x* is 25. Because the side lengths 15, 20, and 25 are integers that satisfy the equation $c^2 = a^2 + b^2$, they form a Pythagorean triple.

Find the value of *x*. Then tell whether the side lengths form a Pythagorean triple.

1.
2.
3.

Verify that the segment lengths form a triangle. Is the triangle *acute*, *right*, or *obtuse*?

4. 6, 8, and 9
5. 10, $2\sqrt{2}$, and $6\sqrt{3}$
6. 13, 18, and $3\sqrt{55}$

9.2 Special Right Triangles (pp. 525–530)

Find the value of *x*. Write your answer in simplest form.

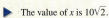

By the Triangle Sum Theorem, the measure of the third angle must be 45°, so the triangle is a 45°-45°-90° triangle.

hypotenuse = leg · $\sqrt{2}$ 45°-45°-90° Triangle Theorem
$x = 10 \cdot \sqrt{2}$ Substitute.
$x = 10\sqrt{2}$ Simplify.

▶ The value of *x* is $10\sqrt{2}$.

Find the value of *x*. Write your answer in simplest form.

7.
8.
9.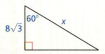

9.3 Similar Right Triangles (pp. 531–538)

Identify the similar triangles. Then find the value of *x*.

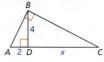

Sketch the three similar right triangles so that the corresponding angles and sides have the same orientation.

 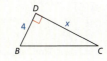 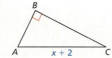

▶ △DBA ~ △DCB ~ △BCA

By the Geometric Mean (Altitude) Theorem, you know that 4 is the geometric mean of 2 and *x*.

$4^2 = 2 \cdot x$ Geometric Mean (Altitude) Theorem
$16 = 2x$ Square 4.
$8 = x$ Divide each side by 2.

▶ The value of *x* is 8.

Identify the similar triangles. Then find the value of *x*.

10.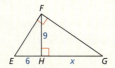
11.
12.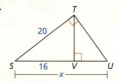

9.4 The Tangent Ratio (pp. 541–546)

Find tan *M* and tan *N*. Write each answer as a fraction and as a decimal rounded to four places.

$\tan M = \dfrac{\text{opp. } \angle M}{\text{adj. to } \angle M} = \dfrac{LN}{LM} = \dfrac{6}{8} = \dfrac{3}{4} = 0.7500$

$\tan N = \dfrac{\text{opp. } \angle N}{\text{adj. to } \angle N} = \dfrac{LM}{LN} = \dfrac{8}{6} = \dfrac{4}{3} \approx 1.3333$

13. Find tan *J* and tan *L*. Write each answer as a fraction and as a decimal rounded to four decimal places.

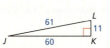

14. The angle between the bottom of a fence and the top of a tree is 75°. The tree is 4 feet from the fence. How tall is the tree? Round your answer to the nearest foot.

ANSWERS

10. △GFH ~ △FEH ~ △GEF; $x = 13.5$
11. △KLM ~ △JKM ~ △JLK; $x = 2\sqrt{6} \approx 4.9$
12. △TUV ~ △STV ~ △SUT; $x = 25$
13. $\tan J = \dfrac{11}{60} \approx 0.1833$, $\tan L = \dfrac{60}{11} \approx 5.4545$
14. about 15 ft

ANSWERS

15. $\sin X = \dfrac{7\sqrt{149}}{149} \approx 0.5735$,

$\sin Z = \dfrac{10\sqrt{149}}{149} \approx 0.8192$,

$\cos X = \dfrac{10\sqrt{149}}{149} \approx 0.8192$,

$\cos Z = \dfrac{7\sqrt{149}}{149} \approx 0.5735$

16. $s \approx 31.3$, $t \approx 13.3$

17. $r \approx 4.0$, $s \approx 2.9$

18. $v \approx 9.4$, $w \approx 3.4$

19. $\cos 18°$

20. $\cos \theta = \dfrac{7}{25}$, $\tan \theta = \dfrac{24}{7}$

21. $m\angle A \approx 48.2°$, $m\angle B \approx 41.8°$, $BC \approx 11.2$

22. $m\angle L = 53°$, $ML \approx 4.5$, $NL \approx 7.5$

23. $m\angle X \approx 46.1°$, $m\angle Z \approx 43.9°$, $XY \approx 17.3$

9.5 The Sine and Cosine Ratios (pp. 547–556)

Find sin A, sin B, cos A, and cos B. Write each answer as a fraction and as a decimal rounded to four places.

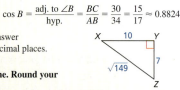

$\sin A = \dfrac{\text{opp. } \angle A}{\text{hyp.}} = \dfrac{BC}{AB} = \dfrac{30}{34} = \dfrac{15}{17} \approx 0.8824$

$\sin B = \dfrac{\text{opp. } \angle B}{\text{hyp.}} = \dfrac{AC}{AB} = \dfrac{16}{34} = \dfrac{8}{17} \approx 0.4706$

$\cos A = \dfrac{\text{adj. to } \angle A}{\text{hyp.}} = \dfrac{AC}{AB} = \dfrac{16}{34} = \dfrac{8}{17} \approx 0.4706$

$\cos B = \dfrac{\text{adj. to } \angle B}{\text{hyp.}} = \dfrac{BC}{AB} = \dfrac{30}{34} = \dfrac{15}{17} \approx 0.8824$

15. Find sin X, sin Z, cos X, and cos Z. Write each answer as a fraction and as a decimal rounded to four decimal places.

Find the value of each variable using sine and cosine. Round your answers to the nearest tenth.

16.   **17.**   **18.**

19. Write sin 72° in terms of cosine.   **20.** Given that $\sin \theta = \dfrac{24}{25}$, find $\cos \theta$ and $\tan \theta$.

9.6 Solving Right Triangles (pp. 557–562)

Solve the right triangle. Round decimal answers to the nearest tenth.

Step 1 Use the Pythagorean Theorem to find the length of the hypotenuse.

$c^2 = a^2 + b^2$ Pythagorean Theorem
$c^2 = 19^2 + 12^2$ Substitute.
$c^2 = 505$ Simplify.
$c = \sqrt{505}$ Find the positive square root.
$c \approx 22.5$ Use a calculator.

Step 2 Find $m\angle B$.

$m\angle B = \tan^{-1}\dfrac{12}{19} \approx 32.3°$ Use a calculator.

Step 3 Find $m\angle A$. Because $\angle A$ and $\angle B$ are complements, you can write $m\angle A = 90° - m\angle B \approx 90° - 32.3° = 57.7°$.

▶ In $\triangle ABC$, $c \approx 22.5$, $m\angle B \approx 32.3°$, and $m\angle A \approx 57.7°$.

Solve the right triangle. Round decimal answers to the nearest tenth.

21.   **22.**   **23.**

9 Chapter Test

Find the value of each variable. Round your answers to the nearest tenth.

1.
2.
3.

Verify that the segment lengths form a triangle. Is the triangle *acute*, *right*, or *obtuse*?

4. 16, 30, and 34
5. 4, $\sqrt{67}$, and 9
6. $\sqrt{5}$, 5, and 5.5

Solve △ABC. Round decimal answers to the nearest tenth.

7.
8.
9.

10. Find the geometric mean of 9 and 25.
11. Write cos 53° in terms of sine.

Find the value of each variable. Write your answers in simplest form.

12.
13.
14.

15. Given that $\cos \theta = \frac{5}{7}$, find $\sin \theta$ and $\tan \theta$.

16. You are given the measures of both acute angles of a right triangle. Can you determine the side lengths? Explain.

17. You are at a parade looking up at a large balloon floating directly above the street. You are 60 feet from a point on the street directly beneath the balloon. To see the top of the balloon, you look up at an angle of 53°. To see the bottom of the balloon, you look up at an angle of 29°. Estimate the height *h* of the balloon.

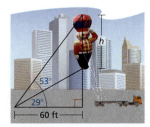

viewing angle

18. You want to take a picture of a statue on Easter Island, called a *moai*. The moai is about 13 feet tall. Your camera is on a tripod that is 5 feet tall. The vertical viewing angle of your camera is set at 90°. How far from the moai should you stand so that the entire height of the moai is perfectly framed in the photo?

ANSWERS

1. $s \approx 16.3$, $t \approx 7.6$
2. $x \approx 16.0$, $y \approx 14.9$
3. $j \approx 13.1$, $k \approx 8.4$
4. yes; right
5. yes; acute
6. yes; obtuse
7. $m\angle A \approx 24.4°$, $m\angle C \approx 65.6°$, $b \approx 12.1$
8. $m\angle B \approx 37.7°$, $m\angle C \approx 52.3°$, $c \approx 14.2$
9. $m\angle A \approx 29.6°$, $m\angle B \approx 60.4°$, $a \approx 4.5$
10. 15
11. sin 37°
12. $q = 16\sqrt{2}$, $r = 16$
13. $c = 12\frac{1}{4}$, $d = 3\frac{3}{4}$, $e = \frac{7\sqrt{15}}{44}$
14. $f = \frac{8\sqrt{3}}{3}$, $h = \frac{16\sqrt{3}}{3}$
15. $\sin \theta = \frac{2\sqrt{6}}{7}$; $\tan \theta = \frac{2\sqrt{6}}{5}$
16. no; There are infinitely many right triangles that can be created using the same acute angle measures. They are all similar by the AA Similarity Theorem, and their corresponding sides are proportional, but not congruent.
17. about 46.4 ft
18. about 6.3 ft

9 Cumulative Assessment

1. The size of a laptop screen is measured by the length of its diagonal. You want to purchase a laptop with the largest screen possible. Which laptop should you buy?

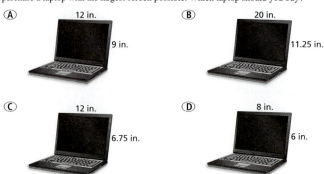

Ⓐ 12 in. / 9 in.
Ⓑ 20 in. / 11.25 in.
Ⓒ 12 in. / 6.75 in.
Ⓓ 8 in. / 6 in.

2. In △PQR and △SQT, S is between P and Q, T is between R and Q, and $\frac{QS}{SP} = \frac{QT}{TR}$. What must be true about $\overline{ST}$ and $\overline{PR}$? Select all that apply.

   $\overline{ST} \perp \overline{PR}$ $\overline{ST} \parallel \overline{PR}$ $ST = PR$ $ST = \frac{1}{2}PR$

3. In the diagram, △JKL ~ △QRS. Choose the symbol that makes each statement true.

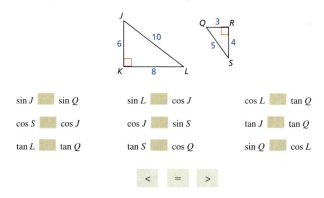

   sin J ☐ sin Q   sin L ☐ cos J   cos L ☐ tan Q
   cos S ☐ cos J   cos J ☐ sin S   tan J ☐ tan Q
   tan L ☐ tan Q   tan S ☐ cos Q   sin Q ☐ cos L

   <   =   >

4. Out of 60 students, 46 like fiction novels, 30 like nonfiction novels, and 22 like both. What is the probability that a randomly selected student likes fiction or nonfiction novels?

568   Chapter 9   Right Triangles and Trigonometry

5. Create as many true equations as possible.

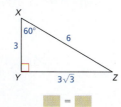

 =

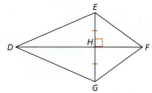

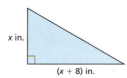

6. Prove that quadrilateral *DEFG* is a kite.

    Given $\overline{HE} \cong \overline{HG}, \overline{EG} \perp \overline{DF}$
    Prove $\overline{FE} \cong \overline{FG}, \overline{DE} \cong \overline{DG}$

7. The area of the triangle is 42 square inches. Find the value of *x*.

8. Classify the solution(s) of each equation as real numbers, imaginary numbers, or pure imaginary numbers. Justify your answers.

    a. $x + \sqrt{-16} = 0$
    b. $(11 - 2i) - (-3i + 6) = 8 + x$
    c. $3x^2 - 14 = -20$
    d. $x^2 + 2x = -3$
    e. $x^2 = 16$
    f. $x^2 - 5x - 8 = 0$

9. The Red Pyramid in Egypt has a square base. Each side of the base measures 722 feet. The height of the pyramid is 343 feet.

    a. Use the side length of the base, the height of the pyramid, and the Pythagorean Theorem to find the *slant height*, *AB*, of the pyramid.
    b. Find *AC*.
    c. Name three possible ways of finding $m\angle 1$. Then, find $m\angle 1$.

ANSWERS

5. $\sin X = \dfrac{YZ}{XZ}, \cos X = \dfrac{XY}{XZ},$
   $\tan X = \dfrac{YZ}{XY}, \sin Z = \dfrac{XY}{XZ},$
   $\cos Z = \dfrac{YZ}{XZ}, \tan Z = \dfrac{XY}{YZ},$
   $\sin X = \cos Z, \cos X = \sin Z$

6. See Additional Answers.
7. $x = 6$
8. a. pure imaginary; $x = -4i$
   b. imaginary; $x = i - 3$
   c. pure imaginary; $x = \pm\sqrt{2}i$
   d. imaginary; $x = -1 \pm \sqrt{2}i$
   e. real; $x = \pm 4$
   f. real; $x = \dfrac{5}{2} \pm \dfrac{\sqrt{57}}{2}$

9. a. about 498.0 ft
   b. about 615.1 ft
   c. Use any of the three trigonometric functions, sine, cosine, or tangent; about 35.9°

Chapter 10 Pacing Guide	
Chapter Opener/Mathematical Practices	0.5 Day
Section 1	1.5 Days
Section 2	1 Day
Section 3	1 Day
Quiz	0.5 Day
Section 4	1.5 Days
Section 5	2 Days
Section 6	1 Day
Section 7	1 Day
Chapter Review/Chapter Tests	2 Days
Total Chapter 10	12 Days
Year-to-Date	143 Days

10 Circles

10.1 Lines and Segments That Intersect Circles
10.2 Finding Arc Measures
10.3 Using Chords
10.4 Inscribed Angles and Polygons
10.5 Angle Relationships in Circles
10.6 Segment Relationships in Circles
10.7 Circles in the Coordinate Plane

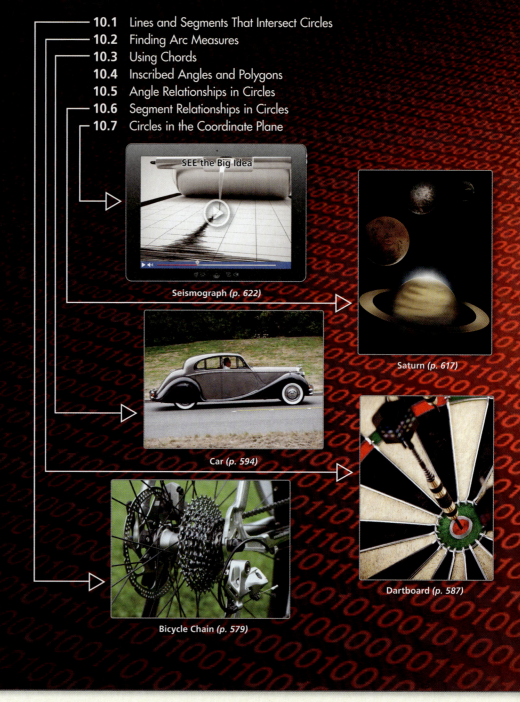

SEE the Big Idea

Seismograph (p. 622)

Saturn (p. 617)

Car (p. 594)

Dartboard (p. 587)

Bicycle Chain (p. 579)

Laurie's Notes

Chapter Summary

- This chapter is all about circles. Many of the theorems will be investigated using dynamic geometry software.
- The first two lessons introduce the vocabulary and symbols related to circles.
- They are followed by a lesson looking at circular arcs that are intercepted by chords.
- The next two lessons introduce all of the angle relationships that occur when two chords, secants, or tangents intersect a circle.
- An investigation of segment relationships that occur when two chords, secants, or tangents intersect a circle is the focus of the next lesson.
- In the last lesson, the circle is presented in the coordinate plane where the standard form of the equation is derived, and systems of equations that include circles are solved.
- Many lessons have been devoted to lines, triangles, and polygons. Circles are presented here so that measured attributes (arc length, area of a sector, circumference of a circle, and area of a circle) can be studied in the next chapter.

What Your Students Have Learned

Middle School
- Find circumferences and areas of circles.
- Identify congruent segments and congruent angles.
- Find interior and exterior angle measures of polygons.
- Solve systems of linear equations in two variables by graphing, substitution, and elimination.
- Solve systems of linear equations in two variables with no solution or infinitely many solutions.

Math I
- Solve multi-step linear equations in one variable that have variables on both sides using inverse operations.
- Solve systems of linear equations in two variables.
- Identify supplementary and vertical angles.
- Name points, lines, planes, segments, rays, and angles.
- Inscribe regular polygons in circles.
- Use properties of equality to justify steps in solving problems involving segment lengths and angle measures.

What Your Students Will Learn

Math II
- Identify radii, chords, diameters, secants, and tangents that intersect circles.
- Use chords of circles to find lengths and arc measures in circles.
- Use inscribed angles and circumscribed angles to find angle and arc measures in circles.
- Use inscribed polygons to find angle measures in polygons.
- Use segments of chords, tangents, and secants of circles to find lengths of line segments.
- Write and graph equations of circles in the coordinate plane.
- Solve nonlinear systems involving equations of circles and lines.

Dynamic Teaching Tools
Dynamic Assessment & Progress Monitoring Tool
Lesson Planning Tool
Interactive Whiteboard Lesson Library
Dynamic Classroom with Dynamic Investigations
Real-Life STEM Videos

Scaffolding in the Classroom

Graphic Organizers: Four Square
A Four Square can be used to organize information about a topic. Students write the topic in the "bubble" in the middle of the Four Square. Then students write concepts related to the topic in the four squares surrounding the bubble. Any concept related to the topic can be used. Encourage students to include concepts that will help them learn the topic. Students can place their Four Squares on note cards to use as a quick study reference.

Questioning in the Classroom
But why?
Ask questions that require critical thinking so that follow-up questions may be asked. Avoid questions having *yes* or *no* answers. If this cannot be avoided, always follow with *why*?

Laurie's Notes

Maintaining Mathematical Proficiency

Multiplying Binomials
- Remind students that when multiplying binomials, each term in the first binomial is distributed over the terms in the second binomial, resulting in the sum of four products.
- Explain that after finding the four products, students can combine like terms to simplify the expression.

COMMON ERROR Students may forget to include negative signs when distributing or students may combine like terms incorrectly.

Solving Quadratic Equations by Completing the Square
- Remind students that the goal in completing the square is to rewrite the equation in the form $(x + a)^2 = c$.
- Work through the example with students, emphasizing that the constant added to each side of the equation is the square of half of the coefficient of the linear term.
- Remind students to include both the positive and negative square roots when they take the square root of each side of the equation.

COMMON ERROR Students may add a constant to only one side of the equation or students may find half of the linear term's coefficient but forget to square it.

Mathematical Practices (continued on page 572)

- The *Mathematical Practices* page focuses attention on how mathematics is learned—process versus content. This page demonstrates that mathematically proficient students understand the importance of precise definitions. They also are able to use definitions to make sense of a stated problem in order to seek a solution.
- Use the *Mathematical Practices* page to help students develop mathematical habits of mind—how mathematics can be explored and how mathematics is thought about.
- Students will be working with circles throughout the chapter, learning theorems related to circles, and applying previously learned concepts to circles. In the *Core Concept*, definitions of a circle and tangent circles are given. Students should note the references to "points in a plane" or "coplanar circles" in each definition. This will be a consistent reference throughout the chapter.

If students need help...	If students got it...
Student Journal • Maintaining Mathematical Proficiency	Game Closet at *BigIdeasMath.com*
Lesson Tutorials	Start the *next* Section
Skills Review Handbook	

Maintaining Mathematical Proficiency

Multiplying Binomials

Example 1 Find the product $(x + 3)(2x - 1)$.

$$ \text{First} \quad \text{Outer} \quad \text{Inner} \quad \text{Last}$$

$(x + 3)(2x - 1) = x(2x) + x(-1) + 3(2x) + (3)(-1)$ FOIL Method

$ = 2x^2 + (-x) + 6x + (-3)$ Multiply.

$ = 2x^2 + 5x - 3$ Simplify.

▶ The product is $2x^2 + 5x - 3$.

Find the product.

1. $(x + 7)(x + 4)$
2. $(a + 1)(a - 5)$
3. $(q - 9)(3q - 4)$
4. $(2v - 7)(5v + 1)$
5. $(4h + 3)(2 + h)$
6. $(8 - 6b)(5 - 3b)$

Solving Quadratic Equations by Completing the Square

Example 2 Solve $x^2 + 8x - 3 = 0$ by completing the square.

$x^2 + 8x - 3 = 0$ Write original equation.

$x^2 + 8x = 3$ Add 3 to each side.

$x^2 + 8x + 4^2 = 3 + 4^2$ Complete the square by adding $\left(\frac{8}{2}\right)^2$, or 4^2, to each side.

$(x + 4)^2 = 19$ Write the left side as a square of a binomial.

$x + 4 = \pm\sqrt{19}$ Take the square root of each side.

$x = -4 \pm \sqrt{19}$ Subtract 4 from each side.

▶ The solutions are $x = -4 + \sqrt{19} \approx 0.36$ and $x = -4 - \sqrt{19} \approx -8.36$.

Solve the equation by completing the square. Round your answer to the nearest hundredth, if necessary.

7. $x^2 - 2x = 5$
8. $r^2 + 10r = -7$
9. $w^2 - 8w = 9$
10. $p^2 + 10p - 4 = 0$
11. $k^2 - 4k - 7 = 0$
12. $-z^2 + 2z = 1$

13. **ABSTRACT REASONING** Write an expression that represents the product of two consecutive positive odd integers. Explain your reasoning.

Dynamic Solutions available at *BigIdeasMath.com*

What Your Students Have Learned

- Multiply two binomials using the FOIL Method.
- Solve quadratic equations of the form $ax^2 + bx + c = 0$ by completing the square and then using square roots.

ANSWERS

1. $x^2 + 11x + 28$
2. $a^2 - 4a - 5$
3. $3q^2 - 31q + 36$
4. $10v^2 - 33v - 7$
5. $4h^2 + 11h + 6$
6. $18b^2 - 54b + 40$
7. $x \approx -1.45; x \approx 3.45$
8. $r \approx -9.24; r \approx -0.76$
9. $w = -1, w = 9$
10. $p \approx -10.39; p \approx 0.39$
11. $k \approx -1.32; k \approx 5.32$
12. $z = 1$
13. Sample answer: $(2n + 1)(2n + 3)$; $2n + 1$ is positive and odd when n is a nonnegative integer. The next positive, odd integer is $2n + 3$.

Vocabulary Review

Have students make Process Diagrams for the following procedures.

- Multiplying binomials
- Completing the square

Chapter 10

MONITORING PROGRESS ANSWERS

1.
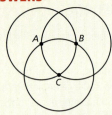

C must be 3 units from A and B, so C must lie on an intersection of circles A and B.

2.
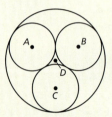

greater than; The radius of ⊙D is greater than the diameter of ⊙A.

Mathematical Practices

Mathematically proficient students make sense of problems and do not give up when faced with challenges.

Analyzing Relationships of Circles

Core Concept

Circles and Tangent Circles

A **circle** is the set of all points in a plane that are equidistant from a given point called the **center** of the circle. A circle with center D is called "circle D" and can be written as ⊙D.

circle D, or ⊙D

Coplanar circles that intersect in one point are called **tangent circles**.

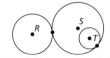

⊙R and ⊙S are tangent circles.
⊙S and ⊙T are tangent circles.

EXAMPLE 1 Relationships of Circles and Tangent Circles

a. Each circle at the right consists of points that are 3 units from the center. What is the greatest distance from any point on ⊙A to any point on ⊙B?

b. Three circles, ⊙C, ⊙D, and ⊙E, consist of points that are 3 units from their centers. The centers C, D, and E of the circles are collinear, ⊙C is tangent to ⊙D, and ⊙D is tangent to ⊙E. What is the distance from ⊙C to ⊙E?

SOLUTION

a. Because the points on each circle are 3 units from the center, the greatest distance from any point on ⊙A to any point on ⊙B is 3 + 3 + 3 = 9 units.

b. Because C, D, and E are collinear, ⊙C is tangent to ⊙D, and ⊙D is tangent to ⊙E, the circles are as shown. So, the distance from ⊙C to ⊙E is 3 + 3 = 6 units.

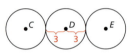

Monitoring Progress

Let ⊙A, ⊙B, and ⊙C consist of points that are 3 units from the centers.

1. Draw ⊙C so that it passes through points A and B in the figure at the right. Explain your reasoning.

2. Draw ⊙A, ⊙B, and ⊙C so that each is tangent to the other two. Draw a larger circle, ⊙D, that is tangent to each of the other three circles. Is the distance from point D to a point on ⊙D *less than*, *greater than*, or *equal to* 6? Explain.

572   Chapter 10   Circles

Laurie's Notes — Mathematical Practices (continued from page T-571)

- Example 1 applies the definitions of a circle and tangent circles in solving each problem. In solving the problems, it is helpful to be able to visualize the problem and draw a sketch.
- Work through each problem as shown.
- Give time for students to work through the *Monitoring Progress* questions. Students may find that a compass is a helpful tool. Discuss as a class.

Laurie's Notes

Overview of Section 10.1

Introduction
- This introductory lesson on circles presents the line and segment vocabulary associated with circles. Some of the vocabulary should be quite familiar to students.
- Internal and external lines and segments are drawn and identified.
- The second part of the lesson presents the properties of tangents, stated as theorems. The problems that follow integrate prior skills and content such as the Pythagorean Theorem and symbolic manipulation.

Teaching Strategy
- This lesson contains two theorems relating to tangents of a circle. Each theorem can be investigated using dynamic geometry software. This approach allows student discovery and extensions to problems.
- The External Tangent Congruence Theorem is easily constructed using most dynamic software. Clicking and dragging on the external point effectively demonstrates that the external tangent segments are always congruent regardless of the location of the external point.

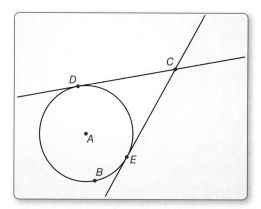

- Students in my class were curious about having two noncongruent tangent circles. Students wanted to select a point in the exterior of both circles and measure the tangent circles. In order to construct the tangent circles, they needed to use the first theorem in the lesson!

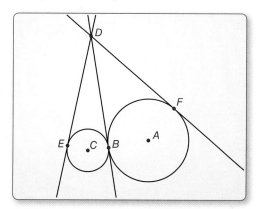

- Whenever possible, have students use dynamic geometry software to investigate and discover theorems.

Pacing Suggestion
- The explorations are a quick and engaging way for students to work with the vocabulary of a circle. Students should be ready to begin with Example 1 in the formal lesson.

Dynamic Teaching Tools
Dynamic Assessment & Progress Monitoring Tool
Lesson Planning Tool
Interactive Whiteboard Lesson Library
Dynamic Classroom with Dynamic Investigations

What Your Students Will Learn

- Identify special segments and lines that intersect circles.
- Draw coplanar circles, and draw and identify common tangents.
- Use properties of tangents of circles to find measures of segments.

Laurie's Notes

Exploration

Motivate
- The circle is a common shape and hence a very common image when you look around. I like to begin this chapter by asking students to make a list of the places where a circular shape is seen.
- Have each student share one or two examples from his or her list.
- Share a short photo album of images such as bicycle gears, a round stained-glass window, and manhole covers. These are images you can refer back to throughout the chapter.

Exploration 1
- **Attend to Precision:** This exploration gives students the opportunity to practice writing a definition based upon an observation (i.e., looking at a sketch). Remind students that a definition needs to be precise so that it leads to only one possible object.
- **Construct Viable Arguments and Critique the Reasoning of Others:** Have students compare and contrast definitions. In critiquing the definitions of others, students learn to be analytic and to apply reasoning skills.

Exploration 2
- This is a quick, kinesthetic exploration that reinforces the definition of a circle.
- **Teaching Tip:** Another way to draw an accurate circle is to use a paper clip. Anchor the paper clip at one end with a pencil. Use a second pencil at the other end of the paper clip to draw the circle. Although different lengths of paper clips are available, the radius is still rather limited.

Communicate Your Answer
- **Reason Abstractly and Quantitatively:** Question 4 helps focus attention on the relationships between the various segments and lines associated with a circle.

Connecting to Next Step
- If accurate definitions result from the first exploration, begin the formal lesson with Example 1.

10.1 Lines and Segments That Intersect Circles

Essential Question What are the definitions of the lines and segments that intersect a circle?

EXPLORATION 1 Lines and Line Segments That Intersect Circles

Work with a partner. The drawing at the right shows five lines or segments that intersect a circle. Use the relationships shown to write a definition for each type of line or segment. Then use the Internet or some other resource to verify your definitions.

Chord: _____

Secant: _____

Tangent: _____

Radius: _____

Diameter: _____

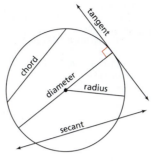

EXPLORATION 2 Using String to Draw a Circle

Work with a partner. Use two pencils, a piece of string, and a piece of paper.

a. Tie the two ends of the piece of string loosely around the two pencils.

b. Anchor one pencil on the paper at the center of the circle. Use the other pencil to draw a circle around the anchor point while using slight pressure to keep the string taut. Do not let the string wind around either pencil.

> **REASONING ABSTRACTLY**
> To be proficient in math, you need to know and flexibly use different properties of operations and objects.

c. Explain how the distance between the two pencil points as you draw the circle is related to two of the lines or line segments you defined in Exploration 1.

Communicate Your Answer

3. What are the definitions of the lines and segments that intersect a circle?

4. Of the five types of lines and segments in Exploration 1, which one is a subset of another? Explain.

5. Explain how to draw a circle with a diameter of 8 inches.

Section 10.1 Lines and Segments That Intersect Circles 573

Differentiated Instruction
Kinesthetic Draw a large circle on the floor or have students stand in a circle. Mark the center of the circle. Have students walk paths to represent diameters, radii, and chords of the circle. For example, to represent a radius, a student walks from the center to any spot on the circle.

Extra Example 1
Tell whether the line, ray, or segment is best described as a *radius, chord, diameter, secant,* or *tangent* of ⊙O.

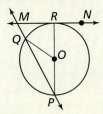

a. $\overline{PR}$ diameter
b. $\overleftrightarrow{MN}$ tangent
c. $\overleftrightarrow{PQ}$ secant
d. $\overline{QO}$ radius

MONITORING PROGRESS ANSWERS
1. chord; radius
2. Sample answer: $\overleftrightarrow{DE}, \overline{DB}$

10.1 Lesson

Core Vocabulary
circle, p. 574
center, p. 574
radius, p. 574
chord, p. 574
diameter, p. 574
secant, p. 574
tangent, p. 574
point of tangency, p. 574
tangent circles, p. 575
concentric circles, p. 575
common tangent, p. 575

READING
The words "radius" and "diameter" refer to lengths as well as segments. For a given circle, think of *a radius* and *a diameter* as segments and *the radius* and *the diameter* as lengths.

What You Will Learn
▶ Identify special segments and lines.
▶ Draw and identify common tangents.
▶ Use properties of tangents.

Identifying Special Segments and Lines

A **circle** is the set of all points in a plane that are equidistant from a given point called the **center** of the circle. A circle with center P is called "circle P" and can be written as ⊙P.

circle P, or ⊙P

🔑 Core Concept

Lines and Segments That Intersect Circles

A segment whose endpoints are the center and any point on a circle is a **radius**.

A **chord** is a segment whose endpoints are on a circle. A **diameter** is a chord that contains the center of the circle.

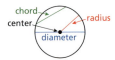

A **secant** is a line that intersects a circle in two points.

A **tangent** is a line in the plane of a circle that intersects the circle in exactly one point, the **point of tangency**. The *tangent ray* $\overrightarrow{AB}$ and the *tangent segment* $\overline{AB}$ are also called tangents.

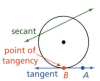

EXAMPLE 1 Identifying Special Segments and Lines

Tell whether the line, ray, or segment is best described as a *radius, chord, diameter, secant,* or *tangent* of ⊙C.

a. $\overline{AC}$
b. $\overline{AB}$
c. $\overrightarrow{DE}$
d. $\overleftrightarrow{AE}$

STUDY TIP
In this book, assume that all segments, rays, or lines that appear to be tangent to a circle are tangents.

SOLUTION

a. $\overline{AC}$ is a radius because C is the center and A is a point on the circle.
b. $\overline{AB}$ is a diameter because it is a chord that contains the center C.
c. $\overrightarrow{DE}$ is a tangent ray because it is contained in a line that intersects the circle in exactly one point.
d. $\overleftrightarrow{AE}$ is a secant because it is a line that intersects the circle in two points.

Monitoring Progress Help in English and Spanish at *BigIdeasMath.com*

1. In Example 1, what word best describes $\overline{AG}$? $\overline{CB}$?
2. In Example 1, name a tangent and a tangent segment.

574 Chapter 10 Circles

Laurie's Notes   Teacher Actions

- **Turn and Talk:** "What vocabulary of circles are you familiar with and what does it mean?" Circulate and listen. Solicit answers.
- ❓ "Are all chords diameters? Are all diameters chords?" *All diameters are chords, but not all chords are diameters.*
- ❓ **Reason Abstractly and Quantitatively:** "Why is it necessary to state that a tangent is in the same plane as a circle?" *Answers will vary.* Model what it would look like if that language were removed from the definition.

574 Chapter 10

Drawing and Identifying Common Tangents

Coplanar Circles and Common Tangents

In a plane, two circles can intersect in two points, one point, or no points. Coplanar circles that intersect in one point are called **tangent circles**. Coplanar circles that have a common center are called **concentric circles**.

A line or segment that is tangent to two coplanar circles is called a **common tangent**. A *common internal tangent* intersects the segment that joins the centers of the two circles. A *common external tangent* does not intersect the segment that joins the centers of the two circles.

EXAMPLE 2 Drawing and Identifying Common Tangents

Tell how many common tangents the circles have and draw them. Use blue to indicate common external tangents and red to indicate common internal tangents.

a.   b.   c.

SOLUTION

Draw the segment that joins the centers of the two circles. Then draw the common tangents. Use blue to indicate lines that do not intersect the segment joining the centers and red to indicate lines that intersect the segment joining the centers.

a. 4 common tangents   b. 3 common tangents   c. 2 common tangents

Monitoring Progress 🔊 *Help in English and Spanish at BigIdeasMath.com*

Tell how many common tangents the circles have and draw them. State whether the tangents are external tangents or internal tangents.

3.   4.   5.

Section 10.1 Lines and Segments That Intersect Circles 575

Extra Example 2
Tell how many common tangents the circles have and draw them.

a.

0 common tangents

b.

1 common external tangent

c.

4 common tangents: 2 internal, 2 external

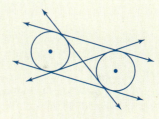

MONITORING PROGRESS ANSWERS

3. 2 internal, 2 external

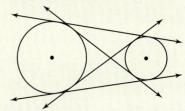

4. 1 external

5. See Additional Answers.

Laurie's Notes — Teacher Actions

- **? Whiteboarding:** "What are the intersection possibilities for two circles—meaning what could they look like?" Use whiteboards for partners to consider the question. Follow their responses with the vocabulary in the *Core Concept*. Again, note the reference to coplanar circles.
- **Thumbs Up:** Have students use whiteboards to sketch their solutions for Example 2. Ask students to self-assess with *Thumbs Up*.
- Omit *Monitoring Progress* questions if you sense students are secure in their understanding of internal and external tangents.

Section 10.1 575

Extra Example 3

Is $\overline{ST}$ tangent to $\odot P$?

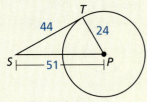

No, $\overline{ST}$ is not tangent to $\odot P$.

Extra Example 4

In the diagram, point P is a point of tangency. Find the radius r of $\odot O$.

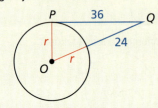

$r = 15$

Using Properties of Tangents

Theorems

Tangent Line to Circle Theorem

In a plane, a line is tangent to a circle if and only if the line is perpendicular to a radius of the circle at its endpoint on the circle.

Proof Ex. 47, p. 580

Line m is tangent to $\odot Q$ if and only if $m \perp \overline{QP}$.

External Tangent Congruence Theorem

Tangent segments from a common external point are congruent.

Proof Ex. 46, p. 580

If $\overline{SR}$ and $\overline{ST}$ are tangent segments, then $\overline{SR} \cong \overline{ST}$.

EXAMPLE 3 Verifying a Tangent to a Circle

Is $\overline{ST}$ tangent to $\odot P$?

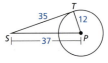

SOLUTION

Use the Converse of the Pythagorean Theorem. Because $12^2 + 35^2 = 37^2$, $\triangle PTS$ is a right triangle and $\overline{ST} \perp \overline{PT}$. So, $\overline{ST}$ is perpendicular to a radius of $\odot P$ at its endpoint on $\odot P$.

▶ By the Tangent Line to Circle Theorem, $\overline{ST}$ is tangent to $\odot P$.

EXAMPLE 4 Finding the Radius of a Circle

In the diagram, point B is a point of tangency. Find the radius r of $\odot C$.

SOLUTION

You know from the Tangent Line to Circle Theorem that $\overline{AB} \perp \overline{BC}$, so $\triangle ABC$ is a right triangle. You can use the Pythagorean Theorem.

$AC^2 = BC^2 + AB^2$ Pythagorean Theorem
$(r + 50)^2 = r^2 + 80^2$ Substitute.
$r^2 + 100r + 2500 = r^2 + 6400$ Multiply.
$100r = 3900$ Subtract r^2 and 2500 from each side.
$r = 39$ Divide each side by 100.

▶ The radius is 39 feet.

Laurie's Notes — Teacher Actions

- The theorems relating to tangents can be investigated using dynamic geometry software. See the *Teaching Strategy* on page T-572.
- Have partners discuss strategies for proving each theorem and then share as a class.
- **Wait Time:** Pose Example 4, and give partners sufficient *Wait Time* to work on the problem. Students should recall how to square a binomial and perform the symbolic manipulations.
- ❓ "Does B have to be a point of tangency to solve this problem in this manner?" yes

> **CONSTRUCTION** Constructing a Tangent to a Circle

Given ⊙C and point A, construct a line tangent to ⊙C that passes through A. Use a compass and straightedge.

SOLUTION

Step 1	Step 2	Step 3
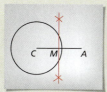	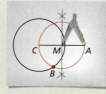	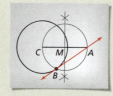
Find a midpoint Draw $\overline{AC}$. Construct the bisector of the segment and label the midpoint M.	**Draw a circle** Construct ⊙M with radius MA. Label one of the points where ⊙M intersects ⊙C as point B.	**Construct a tangent line** Draw $\overleftrightarrow{AB}$. It is a tangent to ⊙C that passes through A.

> **EXAMPLE 5** Using Properties of Tangents

$\overline{RS}$ is tangent to ⊙C at S, and $\overline{RT}$ is tangent to ⊙C at T. Find the value of x.

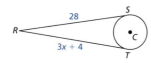

SOLUTION

RS = RT	External Tangent Congruence Theorem
28 = 3x + 4	Substitute.
8 = x	Solve for x.

▶ The value of x is 8.

Monitoring Progress Help in English and Spanish at *BigIdeasMath.com*

6. Is $\overline{DE}$ tangent to ⊙C?

7. $\overline{ST}$ is tangent to ⊙Q. Find the radius of ⊙Q.

8. Points M and N are points of tangency. Find the value(s) of x.

Section 10.1 Lines and Segments That Intersect Circles 577

English Language Learners

Organization

Have students complete a table similar to the one below to help them classify the lines, rays, and segments that intersect circles. Have them leave enough space to make a sketch to illustrate each term. Remind students that each figure may be named in several ways and that they should pay attention to whether the points used to name a figure are endpoints.

Term	Can be a ...	Intersects a circle ...
Chord	segment	in 2 points
Diameter	segment	in 2 points
Radius	segment	in 1 point
Secant	line	in 2 points
Tangent	line, ray, or segment	in 1 point

Extra Example 5

$\overline{JH}$ is tangent to ⊙L at H, and $\overline{JK}$ is tangent to ⊙L at K. Find the value of x.

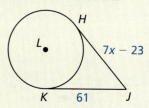

$x = 12$

MONITORING PROGRESS ANSWERS

6. yes

7. 7

8. ±3

Laurie's Notes — Teacher Actions

- **FYI:** The construction of a tangent from an external point is presented here with many of the properties of a tangent. The proof of why this is a valid construction requires the Measure of an Inscribed Angle Theorem in a later section. $\overline{CA}$ is a diameter and because the measure of inscribed ∠CBA is $\frac{1}{2}$ the measure of the 180° intercepted arc, $\overleftrightarrow{AB} \perp \overline{CB}$.
- **Think-Pair-Share:** Have students answer Questions 6–8, and then share and discuss as a class.

Closure

- **Exit Ticket:** Draw a figure like the one in Example 4. When CB = 8 centimeters and AB = 12 centimeters, what is AC? $4\sqrt{13} \approx 14.4$ centimeters

Assignment Guide and Homework Check

ASSIGNMENT
Basic: 1–4, 5–33 odd, 39, 42, 49, 50
Average: 1–4, 6–34 even, 37–40, 42, 45, 48–50
Advanced: 1–4, 6–14 even, 20–34 even, 37–42, 44–50

HOMEWORK CHECK
Basic: 7, 11, 15, 23, 29
Average: 20, 26, 32, 37, 40
Advanced: 20, 26, 40, 45, 48

ANSWERS

1. They both intersect the circle in two points; Chords are segments and secants are lines.
2. When the context is measure, it refers to length.
3. concentric circles
4. tangent; It is outside the circle.
5. $\odot C$
6. $\overline{AC}, \overline{CD}$
7. $\overline{BH}, \overline{AD}$
8. $\overline{AD}$
9. $\overleftrightarrow{KG}$
10. $\overleftrightarrow{GE}$, F
11. 4

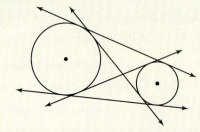

12. 0

13. 2

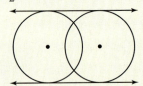

10.1 Exercises

Dynamic Solutions available at BigIdeasMath.com

Vocabulary and Core Concept Check

1. **WRITING** How are chords and secants alike? How are they different?
2. **WRITING** Explain how you can determine from the context whether the words *radius* and *diameter* are referring to segments or lengths.
3. **COMPLETE THE SENTENCE** Coplanar circles that have a common center are called _____.
4. **WHICH ONE DOESN'T BELONG?** Which segment does *not* belong with the other three? Explain your reasoning.

   chord   radius   tangent   diameter

Monitoring Progress and Modeling with Mathematics

In Exercises 5–10, use the diagram. *(See Example 1.)*

5. Name the circle.
6. Name two radii.
7. Name two chords.
8. Name a diameter.
9. Name a secant.
10. Name a tangent and a point of tangency.

In Exercises 11–14, copy the diagram. Tell how many common tangents the circles have and draw them. *(See Example 2.)*

11.
12.
13.
14.

In Exercises 15–18, tell whether the common tangent is *internal* or *external*.

15.
16.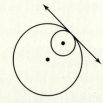
17.
18.

In Exercises 19–22, tell whether $\overline{AB}$ is tangent to $\odot C$. Explain your reasoning. *(See Example 3.)*

19.
20.
21.
22.

In Exercises 23–26, point B is a point of tangency. Find the radius r of $\odot C$. *(See Example 4.)*

23.
24.
25.
26.

578   Chapter 10   Circles

14. 1

15. external
16. internal
17. internal
18. external
19. yes; $\triangle ABC$ is a right triangle.
20. no; $\triangle ABC$ is not a right triangle.
21. no; $\triangle ABD$ is not a right triangle.
22. yes; $\triangle ABC$ is a right triangle.
23. 10
24. 3.75
25. 10.5
26. 16

CONSTRUCTION In Exercises 27 and 28, construct ⊙C with the given radius and point A outside of ⊙C. Then construct a line tangent to ⊙C that passes through A.

27. $r = 2$ in.
28. $r = 4.5$ cm

In Exercises 29–32, points B and D are points of tangency. Find the value(s) of x. *(See Example 5.)*

29.
30.
31.
32.

33. **ERROR ANALYSIS** Describe and correct the error in determining whether $\overline{XY}$ is tangent to ⊙Z.

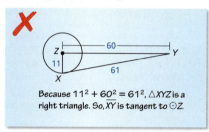

34. **ERROR ANALYSIS** Describe and correct the error in finding the radius of ⊙T.

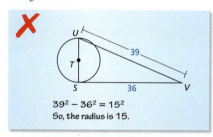

35. **ABSTRACT REASONING** For a point outside of a circle, how many lines exist tangent to the circle that pass through the point? How many such lines exist for a point on the circle? inside the circle? Explain your reasoning.

36. **CRITICAL THINKING** When will two lines tangent to the same circle not intersect? Justify your answer.

37. **USING STRUCTURE** Each side of quadrilateral TVWX is tangent to ⊙Y. Find the perimeter of the quadrilateral.

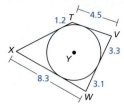

38. **LOGIC** In ⊙C, radii $\overline{CA}$ and $\overline{CB}$ are perpendicular. $\overrightarrow{BD}$ and $\overrightarrow{AD}$ are tangent to ⊙C.

   a. Sketch ⊙C, $\overline{CA}$, $\overline{CB}$, $\overrightarrow{BD}$, and $\overrightarrow{AD}$.
   b. What type of quadrilateral is CADB? Explain your reasoning.

39. **MAKING AN ARGUMENT** Two bike paths are tangent to an approximately circular pond. Your class is building a nature trail that begins at the intersection B of the bike paths and runs between the bike paths and over a bridge through the center P of the pond. Your classmate uses the Converse of the Angle Bisector Theorem to conclude that the trail must bisect the angle formed by the bike paths. Is your classmate correct? Explain your reasoning.

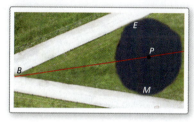

40. **MODELING WITH MATHEMATICS** A bicycle chain is pulled tightly so that $\overline{MN}$ is a common tangent of the gears. Find the distance between the centers of the gears.

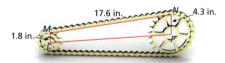

41. **WRITING** Explain why the diameter of a circle is the longest chord of the circle.

Section 10.1   Lines and Segments That Intersect Circles   579

Dynamic Teaching Tools
- Dynamic Assessment & Progress Monitoring Tool
- Interactive Whiteboard Lesson Library
- Dynamic Classroom with Dynamic Investigations

ANSWERS

27. Sample answer:

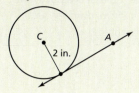

28. Sample answer:

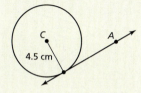

29. 5
30. 4
31. ±3
32. ±2
33. ∠Z is a right angle, not ∠YXZ; $\overline{XY}$ is not tangent to ⊙Z.
34. The diameter is 15; The radius is 7.5.
35. 2; 1; 0; Sample answer: There are two possible points of tangency from a point outside the circle, one from a point on the circle, and none from a point inside the circle.
36. when they are parallel; Sample answer: Two tangents perpendicular to the same diameter are parallel.
37. 25.6 units
38. a.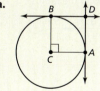

   b. square; $\overline{CA} \cong \overline{CB}, \overline{BD} \cong \overline{DA}$, $m\angle C = 90° = m\angle B = m\angle A$

39. yes; $\overline{PE}$ and $\overline{PM}$ are radii, so $\overline{PE} \cong \overline{PM}$.
40. about 17.78 in.
41. Sample answer: Every point is the same distance from the center, so the farthest two points can be from each other is opposite sides of the center.

ANSWERS
42–50. See Additional Answers.

Mini-Assessment

1. Tell whether the line, ray, or segment is best described as a *radius*, *chord*, *diameter*, *secant*, or *tangent* of ⊙P.

   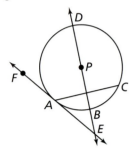

   a. $\overline{AE}$ tangent segment
   b. $\overline{BD}$ diameter
   c. $\overleftrightarrow{BD}$ secant
   d. $\overline{AC}$ chord

2. How many common external tangents do the circles have? How many common internal tangents?

   2 external; 0 internal

3. In the diagram, point S is a point of tangency. Find the radius r of ⊙Q.

   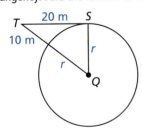

   15 meters

4. $\overline{RS}$ is tangent to ⊙C at S, and $\overline{RT}$ is tangent to ⊙C at T. Find the value of x.

   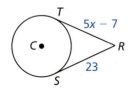

   $x = 6$

---

42. **HOW DO YOU SEE IT?** In the figure, $\overrightarrow{PA}$ is tangent to the dime, $\overrightarrow{PC}$ is tangent to the quarter, and $\overrightarrow{PB}$ is a common internal tangent. How do you know that $\overline{PA} \cong \overline{PB} \cong \overline{PC}$?

43. **PROOF** In the diagram, $\overline{RS}$ is a common internal tangent to ⊙A and ⊙B. Prove that $\dfrac{AC}{BC} = \dfrac{RC}{SC}$.

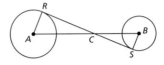

44. **THOUGHT PROVOKING** A polygon is *circumscribed* about a circle when every side of the polygon is tangent to the circle. In the diagram, quadrilateral ABCD is circumscribed about ⊙Q. Is it always true that $AB + CD = AD + BC$? Justify your answer.

45. **MATHEMATICAL CONNECTIONS** Find the values of x and y. Justify your answer.

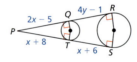

46. **PROVING A THEOREM** Prove the External Tangent Congruence Theorem.

Given $\overline{SR}$ and $\overline{ST}$ are tangent to ⊙P.
Prove $\overline{SR} \cong \overline{ST}$

47. **PROVING A THEOREM** Use the diagram to prove each part of the biconditional in the Tangent Line to Circle Theorem.

a. Prove indirectly that if a line is tangent to a circle, then it is perpendicular to a radius. (*Hint*: If you assume line m is not perpendicular to $\overline{QP}$, then the perpendicular segment from point Q to line m must intersect line m at some other point R.)

   Given   Line m is tangent to ⊙Q at point P.
   Prove $m \perp \overline{QP}$

b. Prove indirectly that if a line is perpendicular to a radius at its endpoint, then the line is tangent to the circle.

   Given $m \perp \overline{QP}$
   Prove   Line m is tangent to ⊙Q.

48. **REASONING** In the diagram, $AB = AC = 12$, $BC = 8$, and all three segments are tangent to ⊙P. What is the radius of ⊙P? Justify your answer.

Maintaining Mathematical Proficiency   Reviewing what you learned in previous grades and lessons

Determine whether the events are independent.   *(Section 5.2)*

49. A pouch contains three blue marbles and two red marbles. You randomly select one marble from the pouch, set it aside, then randomly select another marble from the pouch. Use a sample space to determine whether selecting a blue marble first and selecting a red marble second are independent events.

50. One-half of a spinner is blue and the other half is red. Use a sample space to determine whether randomly spinning blue and then red are independent events.

If students need help...	If students got it...
Resources by Chapter • Practice A and Practice B • Puzzle Time	**Resources by Chapter** • Enrichment and Extension • Cumulative Review
Student Journal • Practice	Start the *next* Section
Differentiating the Lesson Skills Review Handbook	

Laurie's Notes

Overview of Section 10.2

Introduction
- This is a short lesson on how to name and measure minor and major arcs in a circle. The definitions of congruent circles and congruent arcs then allow students to identify congruent arcs from given information.
- The lesson ends with proving that all circles are similar.

Formative Assessment Tips
- **Agreement Circles:** Students should be expected to defend and give rationale for their beliefs or ideas. This technique is a kinesthetic way to engage students in defending their beliefs about a statement made by the teacher.
- Ask students to stand in a large circle. Make a statement to the class about a geometric concept or relationship. If students agree with the statement, they take a few steps forward. If they disagree with the statement, they do nothing. Depending upon the number in each group, divide the students into smaller discussion groups so that each group has a mix of agree and disagree members. The expectation is that students will explain and defend why they have their particular belief.
- The technique can be used before formal instruction is given to assess students' conceptions about the content to be studied. It can also be used during the instruction to help students solidify their understanding of the content.
- Students are making public their belief when they need to step forward or remain still. The dialogue that goes on between the two sides is also important. Students can revise and adjust their thinking based upon new ideas that are shared by peers.
- Sample statements to make in this lesson.
  - Arcs that have the same measure are congruent.
  - Circles with the same radius are congruent.
  - All circles are congruent.
  - If two circles are congruent, then they are similar.
  - All circles are similar.
  - If two arcs are similar, then they are congruent.

Pacing Suggestion
- Once students have worked the exploration, continue with the formal lesson.

Dynamic Teaching Tools
Dynamic Assessment & Progress Monitoring Tool
Lesson Planning Tool
Interactive Whiteboard Lesson Library
Dynamic Classroom with Dynamic Investigations

What Your Students Will Learn

- Find arc measures of circles given the measure of central angles, and find the sum of the measures of two arcs.
- Identify congruent arcs of the same circle or congruent circles.
- Prove all circles are similar.

Laurie's Notes

Exploration

Motivate

- Have students write on scrap paper to answer. Say, "List three things that are circular." Give students 15 seconds and then ask, "How many of you have pizza on your list?" It is fairly likely that all students in the room will have pizza on their lists!
- ? "How many pieces are pizzas often cut into?" 6, 8, or 10
- ? "What is the measure of each angle at the tip of each slice when the pizza is cut into 6 equal slices?" 60° "8 equal slices?" 45° "10 equal slices?" 36°
- Explain to students that in this lesson they will work with the central angle of a circle (the tip of the pizza slice) and also with the circular arc (the crust).

Exploration Note

- The exploration begins with a definition of a central angle and the measure of a circular arc. Check students' understanding of the definitions and notation.
- If dynamic geometry software is not available, students could still perform the construction with a protractor and straightedge. Gathering results from classmates may be sufficient to make a conjecture about the problem posed.

Exploration 1

- This exploration is a nice review of trigonometric ratios.
- Students must first construct each sketch. To find the measure of each circular arc, they need to measure the central angle.
- ? "How did you verify your answer using trigonometry?" One way is to set up a tangent ratio, because you know the opposite and adjacent sides of an angle in a right triangle. For example, using the figure below, $\tan DAC = \dfrac{CD}{AD}$.

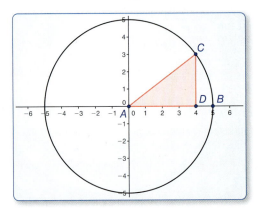

- **Reason Abstractly and Quantitatively:** In parts (c) and (d), students should offer other strategies for finding the measure of $\overarc{BC}$. For instance, in part (c) they may reason that the circular arc is the difference of the two circular arcs found in parts (a) and (b).

Communicate Your Answer

- For Question 3 there are different strategies that students could use to draw the circular arcs. Have students share and discuss the various methods.

Connecting to Next Step

- The exploration will help students make the connection between the measure of the central angle and the associated circular arc, which is the introduction to the formal lesson.

10.2 Finding Arc Measures

Essential Question How are circular arcs measured?

A **central angle** of a circle is an angle whose vertex is the center of the circle. A *circular arc* is a portion of a circle, as shown below. The measure of a circular arc is the measure of its central angle.

If $m\angle AOB < 180°$, then the circular arc is called a **minor arc** and is denoted by $\overparen{AB}$.

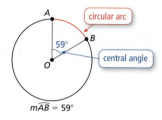

$m\overparen{AB} = 59°$

EXPLORATION 1 Measuring Circular Arcs

Work with a partner. Use dynamic geometry software to find the measure of $\overparen{BC}$. Verify your answers using trigonometry.

a. 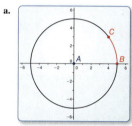   Points
A(0, 0)
B(5, 0)
C(4, 3)

b. 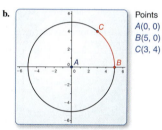   Points
A(0, 0)
B(5, 0)
C(3, 4)

c. 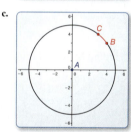   Points
A(0, 0)
B(4, 3)
C(3, 4)

d. 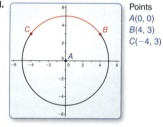   Points
A(0, 0)
B(4, 3)
C(−4, 3)

USING TOOLS STRATEGICALLY

To be proficient in math, you need to use technological tools to explore and deepen your understanding of concepts.

Communicate Your Answer

2. How are circular arcs measured?

3. Use dynamic geometry software to draw a circular arc with the given measure.
   a. 30°
   b. 45°
   c. 60°
   d. 90°

Section 10.2   Finding Arc Measures   581

Dynamic Teaching Tools

Dynamic Assessment & Progress Monitoring Tool
Lesson Planning Tool
Interactive Whiteboard Lesson Library
Dynamic Classroom with Dynamic Investigations

ANSWERS

1. a. 36.87°
   b. 53.13°
   c. 16.26°
   d. 106.26°

2. by their central angles

3. a.
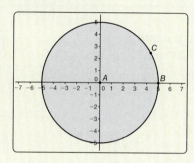

   b.
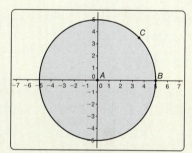

   c.
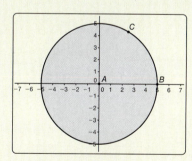

   d.
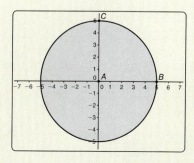

Section 10.2   581

Differentiated Instruction

Kinesthetic
Draw a large circle on the board with its center at about shoulder height. Have a student stand so his or her shoulders are at the center of the circle. Have the student use his or her arms to represent the sides of a central angle. For example, hold one arm straight overhead and one straight out to the side to represent a right angle. Point to the corresponding minor and major arcs for the angle. Repeat the activity for different angles.

Extra Example 1
Find the measure of each arc of $\odot C$, where $\overline{AB}$ is a diameter.

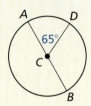

a. $\overset{\frown}{AD}$ 65°
b. $\overset{\frown}{DAB}$ 245°
c. $\overset{\frown}{BDA}$ 180°

10.2 Lesson

Core Vocabulary
central angle, p. 582
minor arc, p. 582
major arc, p. 582
semicircle, p. 582
measure of a minor arc, p. 582
measure of a major arc, p. 582
adjacent arcs, p. 583
congruent circles, p. 584
congruent arcs, p. 584
similar arcs, p. 585

STUDY TIP
The measure of a minor arc is less than 180°. The measure of a major arc is greater than 180°.

What You Will Learn
▶ Find arc measures.
▶ Identify congruent arcs.
▶ Prove circles are similar.

Finding Arc Measures
A **central angle** of a circle is an angle whose vertex is the center of the circle. In the diagram, $\angle ACB$ is a central angle of $\odot C$.

If $m\angle ACB$ is less than 180°, then the points on $\odot C$ that lie in the interior of $\angle ACB$ form a **minor arc** with endpoints A and B. The points on $\odot C$ that do not lie on the minor arc AB form a **major arc** with endpoints A and B. A **semicircle** is an arc with endpoints that are the endpoints of a diameter.

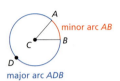

Minor arcs are named by their endpoints. The minor arc associated with $\angle ACB$ is named $\overset{\frown}{AB}$. Major arcs and semicircles are named by their endpoints and a point on the arc. The major arc associated with $\angle ACB$ can be named $\overset{\frown}{ADB}$.

Core Concept
Measuring Arcs
The **measure of a minor arc** is the measure of its central angle. The expression $m\overset{\frown}{AB}$ is read as "the measure of arc AB."

The measure of the entire circle is 360°. The **measure of a major arc** is the difference of 360° and the measure of the related minor arc. The measure of a semicircle is 180°.

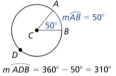

EXAMPLE 1   Finding Measures of Arcs
Find the measure of each arc of $\odot P$, where $\overline{RT}$ is a diameter.

a. $\overset{\frown}{RS}$
b. $\overset{\frown}{RTS}$
c. $\overset{\frown}{RST}$

SOLUTION

a. $\overset{\frown}{RS}$ is a minor arc, so $m\overset{\frown}{RS} = m\angle RPS = 110°$.

b. $\overset{\frown}{RTS}$ is a major arc, so $m\overset{\frown}{RTS} = 360° - 110° = 250°$.

c. $\overline{RT}$ is a diameter, so $\overset{\frown}{RST}$ is a semicircle, and $m\overset{\frown}{RST} = 180°$.

Laurie's Notes   Teacher Actions

• Discuss vocabulary: *minor arc*, *major arc*, and *semicircle*.
• **Turn and Talk:** "Why does a major arc require three letters to name?" With only two letters, it would be unclear whether the arc was a minor arc or a major arc.
• Write the *Core Concept*.
• **Think-Pair-Share:** Pose Example 1. Students should have little difficulty with these examples.
• "In any circle, is $m\overset{\frown}{ABC} = m\overset{\frown}{CBA}$? Explain." yes; It does not matter whether the major arc is named in a clockwise or counterclockwise direction.

Two arcs of the same circle are **adjacent arcs** when they intersect at exactly one point. You can add the measures of two adjacent arcs.

Postulate

Arc Addition Postulate

The measure of an arc formed by two adjacent arcs is the sum of the measures of the two arcs.

$m\widehat{ABC} = m\widehat{AB} + m\widehat{BC}$

EXAMPLE 2 Using the Arc Addition Postulate

Find the measure of each arc.

a. $\widehat{GE}$ b. $\widehat{GEF}$ c. $\widehat{GF}$

SOLUTION

a. $m\widehat{GE} = m\widehat{GH} + m\widehat{HE} = 40° + 80° = 120°$
b. $m\widehat{GEF} = m\widehat{GE} + m\widehat{EF} = 120° + 110° = 230°$
c. $m\widehat{GF} = 360° - m\widehat{GEF} = 360° - 230° = 130°$

EXAMPLE 3 Finding Measures of Arcs

A recent survey asked teenagers whether they would rather meet a famous musician, athlete, actor, inventor, or other person. The circle graph shows the results. Find the indicated arc measures.

a. $m\widehat{AC}$ b. $m\widehat{ACD}$
c. $m\widehat{ADC}$ d. $m\widehat{EBD}$

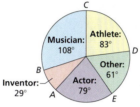

Whom Would You Rather Meet?

SOLUTION

a. $m\widehat{AC} = m\widehat{AB} + m\widehat{BC}$ b. $m\widehat{ACD} = m\widehat{AC} + m\widehat{CD}$
   $= 29° + 108°$ $= 137° + 83°$
   $= 137°$ $= 220°$

c. $m\widehat{ADC} = 360° - m\widehat{AC}$ d. $m\widehat{EBD} = 360° - m\widehat{ED}$
   $= 360° - 137°$ $= 360° - 61°$
   $= 223°$ $= 299°$

Monitoring Progress Help in English and Spanish at *BigIdeasMath.com*

Identify the given arc as a *major arc*, *minor arc*, or *semicircle*. Then find the measure of the arc.

1. $\widehat{TQ}$ 2. $\widehat{QRT}$ 3. $\widehat{TQR}$
4. $\widehat{QS}$ 5. $\widehat{TS}$ 6. $\widehat{RST}$

Section 10.2 Finding Arc Measures 583

English Language Learners

Visual Aid

Have students work in groups to make posters illustrating the core vocabulary terms used in this lesson. Include the terms *minor arc*, *major arc*, *semicircle*, *central angle*, *congruent circles*, *congruent arcs*, and *adjacent arcs*.

Extra Example 2

Find the measure of each arc.

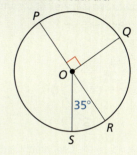

a. $\widehat{SQ}$ 125°
b. $\widehat{RPQ}$ 270°
c. $\widehat{PRS}$ 215°

Extra Example 3

A survey asked people how many minutes they spend brushing their teeth each morning. The circle graph shows the results. Find the indicated arc measures.

For How Long Did You Brush Your Teeth?

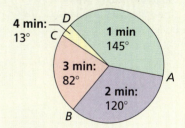

a. $m\widehat{ABC}$ 202°
b. $m\widehat{ACB}$ 240°
c. $m\widehat{BD}$ 95°
d. $m\widehat{CBD}$ 347°

MONITORING PROGRESS ANSWERS

1. minor arc; 120°
2. major arc; 240°
3. semicircle; 180°
4. minor arc; 160°
5. minor arc; 80°
6. semicircle; 180°

Section 10.2 583

Laurie's Notes Teacher Actions

- **Connection:** The Arc Addition Postulate is based on the Angle Addition Postulate and how circular arcs are measured.
- It can be confusing for students when many points appear on a circle and naming an arc such as $\widehat{GE}$ in Example 2, which does not include H. Explain that $\widehat{GE}$ and $\widehat{GHE}$ are the same arc. For simplicity and clarity, we use only the two letters in naming minor arc $\widehat{GE}$.
- Have partners work independently to solve Example 3, and then discuss as a class.

Extra Example 4
Tell whether the red arcs are congruent. Explain why or why not.

a.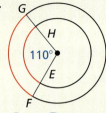

$\widehat{FG}$ and $\widehat{EH}$ are not congruent. They have the same measure, but they are arcs of circles that are not congruent.

b.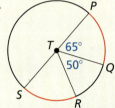

$\widehat{PQ} \cong \widehat{RS}$ by the Congruent Central Angles Theorem because they are arcs of the same circle and they have congruent central angles, $\angle PTQ \cong \angle RTS$.

c.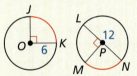

$\widehat{JK} \cong \widehat{MN}$ by the Congruent Central Angles Theorem because the arcs have congruent central angles of 90°. The circles are congruent because they both have radius 6.

Identifying Congruent Arcs

Two circles are **congruent circles** if and only if a rigid motion or a composition of rigid motions maps one circle onto the other. This statement is equivalent to the Congruent Circles Theorem below.

Theorem

Congruent Circles Theorem

Two circles are congruent circles if and only if they have the same radius.

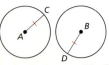

Proof Ex. 35, p. 588

$\odot A \cong \odot B$ if and only if $\overline{AC} \cong \overline{BD}$.

Two arcs are **congruent arcs** if and only if they have the same measure and they are arcs of the same circle or of congruent circles.

Theorem

Congruent Central Angles Theorem

In the same circle, or in congruent circles, two minor arcs are congruent if and only if their corresponding central angles are congruent.

Proof Ex. 37, p. 588

$\widehat{BC} \cong \widehat{DE}$ if and only if $\angle BAC \cong \angle DAE$.

EXAMPLE 4 Identifying Congruent Arcs

Tell whether the red arcs are congruent. Explain why or why not.

a.   b.   c.

STUDY TIP
The two circles in part (c) are congruent by the Congruent Circles Theorem because they have the same radius.

SOLUTION

a. $\widehat{CD} \cong \widehat{EF}$ by the Congruent Central Angles Theorem because they are arcs of the same circle and they have congruent central angles, $\angle CBD \cong \angle FBE$.

b. $\widehat{RS}$ and $\widehat{TU}$ have the same measure, but are not congruent because they are arcs of circles that are not congruent.

c. $\widehat{UV} \cong \widehat{YZ}$ by the Congruent Central Angles Theorem because they are arcs of congruent circles and they have congruent central angles, $\angle UTV \cong \angle YXZ$.

Laurie's Notes Teacher Actions

? **"How would you define congruent arcs?"** Students will likely say that two arcs are congruent if their central angles are congruent. They do not consider the size of the circle!
• An arc of 45° on noncongruent circles will have the same arc measure but not the same length. This naturally leads to the next question.
? **"How would you define congruent circles?"** Congruent circles have the same radius.
• Remind students that congruence could also be defined in terms of rigid motions. An alternate approach is described in the *Teaching Strategy* on page T-612.

Monitoring Progress Help in English and Spanish at *BigIdeasMath.com*

Tell whether the red arcs are congruent. Explain why or why not.

7.

8.

Proving Circles Are Similar

Theorem

Similar Circles Theorem

All circles are similar.

Proof p. 585; Ex. 33, p. 588

PROOF   **Similar Circles Theorem**

All circles are similar.

Given ⊙C with center C and radius r,
⊙D with center D and radius s

Prove ⊙$C \sim$ ⊙D

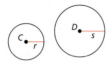

First, translate ⊙C so that point C maps to point D. The image of ⊙C is ⊙C' with center D. So, ⊙C' and ⊙D are concentric circles.

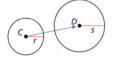

⊙C' is the set of all points that are r units from point D. Dilate ⊙C' using center of dilation D and scale factor $\dfrac{s}{r}$.

This dilation maps the set of all the points that are r units from point D to the set of all points that are $\dfrac{s}{r}(r) = s$ units from point D. ⊙D is the set of all points that are s units from point D. So, this dilation maps ⊙C' to ⊙D.

Because a similarity transformation maps ⊙C to ⊙D, ⊙$C \sim$ ⊙D.

Two arcs are **similar arcs** if and only if they have the same measure. All congruent arcs are similar, but not all similar arcs are congruent. For instance, in Example 4, the pairs of arcs in parts (a), (b), and (c) are similar but only the pairs of arcs in parts (a) and (c) are congruent.

Assignment Guide and Homework Check

ASSIGNMENT

Basic: 1, 2, 3–25 odd, 26, 27, 31, 36, 39–42

Average: 1, 2, 6–16 even, 17, 20–24 even, 25–27, 31, 32, 36, 39–42

Advanced: 1, 2, 6–18 even, 19–27, 30, 31, 34–42

HOMEWORK CHECK

Basic: 3, 13, 15, 17, 23

Average: 16, 17, 22, 24, 31

Advanced: 18, 22, 23, 35, 38

ANSWERS

1. congruent arcs
2. the one with diameter 6 in.; the others have diameter 12 in.
3. $\widehat{AB}$, 135°; $\widehat{ADB}$, 225°
4. $\widehat{EF}$, 68°; $\widehat{EGF}$, 292°
5. $\widehat{JL}$, 120°; $\widehat{JKL}$, 240°
6. $\widehat{MN}$, 170°; $\widehat{MPN}$, 190°
7. minor arc; 70°
8. minor arc; 65°
9. minor arc; 45°
10. minor arc; 70°
11. semicircle; 180°
12. semicircle; 180°
13. major arc; 290°
14. major arc; 315°
15. a. 132°
    b. 147°
    c. 200°
    d. 160°
16. a. 138°
    b. 180°
    c. 222°
    d. 138°
17. a. 103°
    b. 257°
    c. 196°
    d. 305°
    e. 79°
    f. 281°

10.2 Exercises

Dynamic Solutions available at *BigIdeasMath.com*

Vocabulary and Core Concept Check

1. **VOCABULARY** Copy and complete: If ∠ACB and ∠DCE are congruent central angles of ⊙C, then $\widehat{AB}$ and $\widehat{DE}$ are _____.

2. **WHICH ONE DOESN'T BELONG?** Which circle does *not* belong with the other three? Explain your reasoning.

Monitoring Progress and Modeling with Mathematics

In Exercises 3–6, name the red minor arc and find its measure. Then name the blue major arc and find its measure.

3.

4.

5.

6.

In Exercises 7–14, identify the given arc as a *major arc*, *minor arc*, or *semicircle*. Then find the measure of the arc. *(See Example 1.)*

7. $\widehat{BC}$
8. $\widehat{DC}$
9. $\widehat{ED}$
10. $\widehat{AE}$
11. $\widehat{EAB}$
12. $\widehat{ABC}$
13. $\widehat{BAC}$
14. $\widehat{EBD}$

In Exercises 15 and 16, find the measure of each arc. *(See Example 2.)*

15. a. $\widehat{JL}$
    b. $\widehat{KM}$
    c. $\widehat{JLM}$
    d. $\widehat{JM}$

16. a. $\widehat{RS}$
    b. $\widehat{QRS}$
    c. $\widehat{QST}$
    d. $\widehat{QT}$

17. **MODELING WITH MATHEMATICS** A recent survey asked high school students their favorite type of music. The results are shown in the circle graph. Find each indicated arc measure. *(See Example 3.)*

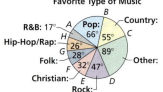

a. $m\widehat{AE}$
b. $m\widehat{ACE}$
c. $m\widehat{GDC}$
d. $m\widehat{BHC}$
e. $m\widehat{FD}$
f. $m\widehat{FBD}$

586   Chapter 10   Circles

18. **ABSTRACT REASONING** The circle graph shows the percentages of students enrolled in fall sports at a high school. Is it possible to find the measure of each minor arc? If so, find the measure of the arc for each category shown. If not, explain why it is not possible.

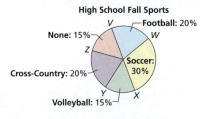

In Exercises 19–22, tell whether the red arcs are congruent. Explain why or why not. *(See Example 4.)*

19.   20.

21.

22.

MATHEMATICAL CONNECTIONS In Exercises 23 and 24, find the value of x. Then find the measure of the red arc.

23.   24.

25. **MAKING AN ARGUMENT** Your friend claims that any two arcs with the same measure are similar. Your cousin claims that any two arcs with the same measure are congruent. Who is correct? Explain.

26. **MAKING AN ARGUMENT** Your friend claims that there is not enough information given to find the value of x. Is your friend correct? Explain your reasoning.

27. **ERROR ANALYSIS** Describe and correct the error in naming the red arc.

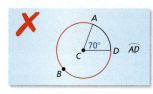

28. **ERROR ANALYSIS** Describe and correct the error in naming congruent arcs.

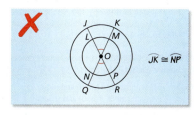

29. **ATTENDING TO PRECISION** Two diameters of $\odot P$ are $\overline{AB}$ and $\overline{CD}$. Find $m\widehat{ACD}$ and $m\widehat{AC}$ when $m\widehat{AD} = 20°$.

30. **REASONING** In $\odot R$, $m\widehat{AB} = 60°$, $m\widehat{BC} = 25°$, $m\widehat{CD} = 70°$, and $m\widehat{DE} = 20°$. Find two possible measures of $\widehat{AE}$.

31. **MODELING WITH MATHEMATICS** On a regulation dartboard, the outermost circle is divided into twenty congruent sections. What is the measure of each arc in this circle?

Section 10.2   Finding Arc Measures   587

ANSWERS

32. a. 15°
    b. 90°
    c. 3 A.M.

33. Translate ⊙A left a units so that point A maps to point O. The image of ⊙A is ⊙A' with center O, so ⊙A' and ⊙O are concentric circles. Dilate ⊙A' using center of dilation O and scale factor $\frac{r}{s}$, which maps the points s units from point O to the points $\frac{r}{s}(s) = r$ units from point O. So, this dilation maps ⊙A' to ⊙O. Because a similarity transformation maps ⊙A to ⊙O, ⊙O ~ ⊙A.

34. yes; Both radii are $\overline{CD}$.

35–38. See Additional Answers.

39. 15; yes
40. about 18.38; no
41. about 13.04; no
42. about 9.80; no

Mini-Assessment

Use ⊙O, where $\overline{AD}$ is a diameter.

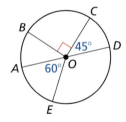

1. Name a semicircle. *Sample answer:* $\widehat{ABD}$, $\widehat{ACD}$ or $\widehat{AED}$

2. Find $m\widehat{AC}$. 135°

3. Find $m\widehat{DAE}$. 240°

4. Tell whether $\widehat{DB}$ and $\widehat{ED}$ are congruent. Explain why or why not. $\widehat{DB}$ and $\widehat{ED}$ are not congruent. They are arcs of the same circle, but $m\angle DOB = 45° + 90° = 135°$ and $m\angle EOD = 180° - 60° = 120°$. So, the corresponding central angles are not congruent.

5. Tell whether $\widehat{AB}$ and $\widehat{CD}$ are congruent. Explain why or why not. $\widehat{AB} \cong \widehat{CD}$ because they are arcs of the same circle with congruent central angles, $\angle AOB \cong \angle COD$.

---

588   Chapter 10

---

32. **MODELING WITH MATHEMATICS** You can use the time zone wheel to find the time in different locations across the world. For example, to find the time in Tokyo when it is 4 P.M. in San Francisco, rotate the small wheel until 4 P.M. and San Francisco line up, as shown. Then look at Tokyo to see that it is 9 A.M. there.

a. What is the arc measure between each time zone on the wheel?

b. What is the measure of the minor arc from the Tokyo zone to the Anchorage zone?

c. If two locations differ by 180° on the wheel, then it is 3 P.M. at one location when it is _____ at the other location.

33. **PROVING A THEOREM** Write a coordinate proof of the Similar Circles Theorem.

Given   ⊙O with center $O(0, 0)$ and radius r,
        ⊙A with center $A(a, 0)$ and radius s

Prove   ⊙O ~ ⊙A

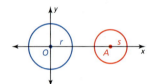

Maintaining Mathematical Proficiency   Reviewing what you learned in previous grades and lessons

Find the value of x. Tell whether the side lengths form a Pythagorean triple. *(Section 9.1)*

39. 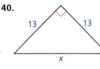   40. 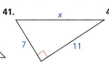   41. (7, 11, x triangle)   42.

588   Chapter 10   Circles

---

34. **ABSTRACT REASONING** Is there enough information to tell whether ⊙C ≅ ⊙D? Explain your reasoning.

35. **PROVING A THEOREM** Use the diagram on page 584 to prove each part of the biconditional in the Congruent Circles Theorem.

a. Given $\overline{AC} \cong \overline{BD}$
   Prove ⊙A ≅ ⊙B

b. Given ⊙A ≅ ⊙B
   Prove $\overline{AC} \cong \overline{BD}$

36. **HOW DO YOU SEE IT?** Are the circles on the target *similar* or *congruent*? Explain your reasoning.

37. **PROVING A THEOREM** Use the diagram to prove each part of the biconditional in the Congruent Central Angles Theorem.

a. Given $\angle BAC \cong \angle DAE$
   Prove $\widehat{BC} \cong \widehat{DE}$

b. Given $\widehat{BC} \cong \widehat{DE}$
   Prove $\angle BAC \cong \angle DAE$

38. **THOUGHT PROVOKING** Write a formula for the length of a circular arc. Justify your answer.

If students need help...	If students got it...
Resources by Chapter • Practice A and Practice B • Puzzle Time	Resources by Chapter • Enrichment and Extension • Cumulative Review
Student Journal • Practice	Start the *next* Section
Differentiating the Lesson Skills Review Handbook	

Laurie's Notes

Overview of Section 10.3

Introduction
- This lesson builds upon the language and circle concepts from the beginning of the chapter, and it also integrates prior skills such as working with perpendicular bisectors and the Pythagorean Theorem.
- The goal of the lesson is to determine whether two chords in a circle are congruent. If the chords are congruent, then there are other relationships that will be true.

Teaching Strategy
- Use the *Motivate* to launch the lesson. It is a classic application of the theorems in this lesson.
- Archeologists, forensic scientists, and tradesmen who have a portion of a circular object use the theorems in this lesson to determine something about the original size of the object.
- Make a handout with a sketch of the broken plate on it. As the closure for today's lesson, have students perform constructs to locate the center of the circle. Measure to determine the diameter of the circle.

Applications
- Finding a point that is equidistant from three (noncollinear) fixed points is fairly common. Where does a business locate a warehouse when it wants to be equidistant from its three stores? Where should three friends meet when you want everyone to travel the same distance from their house? Where do you locate a cell tower when you want to provide service to three locations?

Pacing Suggestion
- The formal lesson is fairly short. Through the explorations, students discover the relationship between the diameter of a circle and chords that intersect a diameter. The Perpendicular Chord Bisector Theorem and the Perpendicular Chord Bisector Converse will make sense to students.

Dynamic Teaching Tools
Dynamic Assessment & Progress Monitoring Tool
Lesson Planning Tool
Interactive Whiteboard Lesson Library
Dynamic Classroom with Dynamic Investigations

What Your Students Will Learn

- Use chords of circles to find lengths and arc measures in circles.

Laurie's Notes

Exploration

Motivate

- If possible, bring a piece of a broken dinner plate that has a portion of the edge remaining. If you have never broken a plate and saved a shattered piece, then you could instead tear a paper plate and make believe!
- If you are a storyteller, give details of the event that led up to the breaking of the plate. My (fictional) story was a dinner party at the new neighbor's house. I offered to help clear the table, and my heel caught on the carpet. The hostess's dress was hit with flying leftovers, and the plate that I was carrying broke. After apologizing profusely, I asked whether I could take part of the plate home. Then I display the broken piece of china.
- ❓ "What was the diameter of the plate?" *Give time for students to stop laughing and to consider the question. Students generally have a few ideas for estimating the diameter.*
- Explain to students that in this lesson they will discover a method for locating the center of a circle.

Exploration Note

- ❓ You might begin these explorations by asking something such as, "Is every chord a diameter?" *no* "Is every diameter a chord?" *yes*
- If dynamic geometry software is not available, students could still perform the constructions with a protractor and straightedge. Gathering results from classmates may be sufficient to make a conjecture about the problem posed.

Exploration 1

- This exploration serves to remind students that the diameter passes through the center of the circle—in this case, the origin.

Exploration 2

- **Construct Viable Arguments and Critique the Reasoning of Others:** Students need to use more than eyesight to state that the chord on the perpendicular bisector of chord $\overline{BC}$ passes through the center of the circle and therefore is a diameter.
- ❓ "How do you know the perpendicular bisector of chord $\overline{BC}$ passes through A, besides the fact that it looks like it does?" *One way students might prove the perpendicular bisector of chord $\overline{BC}$ passes through A is to construct a perpendicular through A to the chord.* Are the two lines the same? On most software, the equations of these lines are displayed and the equations will be the same.
- **Construct Viable Arguments and Critique the Reasoning of Others:** Have students state their conjectures and explain their reasoning. Discuss.

Exploration 3

- Be sure that students investigate for different length chords.
- Students should discover that if a diameter of a circle is perpendicular to a chord of the circle, then the diameter bisects the chord.
- **Construct Viable Arguments and Critique the Reasoning of Others:** Have students state their conjectures and explain their reasoning. Discuss.

Communicate Your Answer

- The conjectures made by students are two ways to show that a chord is a diameter of a circle.

Connecting to Next Step

- The explorations allow students to discover two of the theorems in today's lesson.

10.3 Using Chords

Essential Question What are two ways to determine when a chord is a diameter of a circle?

EXPLORATION 1 Drawing Diameters

Work with a partner. Use dynamic geometry software to construct a circle of radius 5 with center at the origin. Draw a diameter that has the given point as an endpoint. Explain how you know that the chord you drew is a diameter.

a. (4, 3)   **b.** (0, 5)   **c.** (−3, 4)   **d.** (−5, 0)

EXPLORATION 2 Writing a Conjecture about Chords

Work with a partner. Use dynamic geometry software to construct a chord $\overline{BC}$ of a circle A. Construct a chord on the perpendicular bisector of $\overline{BC}$. What do you notice? Change the original chord and the circle several times. Are your results always the same? Use your results to write a conjecture.

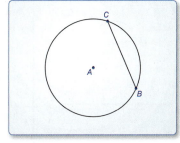

EXPLORATION 3 A Chord Perpendicular to a Diameter

Work with a partner. Use dynamic geometry software to construct a diameter $\overline{BC}$ of a circle A. Then construct a chord $\overline{DE}$ perpendicular to $\overline{BC}$ at point F. Find the lengths DF and EF. What do you notice? Change the chord perpendicular to $\overline{BC}$ and the circle several times. Do you always get the same results? Write a conjecture about a chord that is perpendicular to a diameter of a circle.

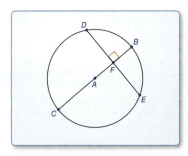

LOOKING FOR STRUCTURE

To be proficient in math, you need to look closely to discern a pattern or structure.

Communicate Your Answer

4. What are two ways to determine when a chord is a diameter of a circle?

---

2. Check students' work; The perpendicular bisector is a diameter; yes; A perpendicular bisector of a chord is a diameter of the circle.

3. Check students' work; $DF = EF$; yes; If a chord is perpendicular to a diameter of a circle, then the diameter is a perpendicular bisector of the chord.

4. when it is a perpendicular bisector of a chord or passes through the center of the circle

---

Dynamic Teaching Tools

Dynamic Assessment & Progress Monitoring Tool
Lesson Planning Tool
Interactive Whiteboard Lesson Library
Dynamic Classroom with Dynamic Investigations

ANSWERS

1. a.

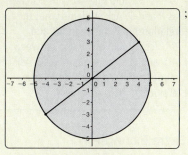

It passes through the center.

b.

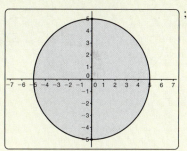

It passes through the center.

c.

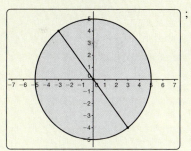

It passes through the center.

d.

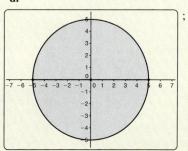

It passes through the center.

Differentiated Instruction

Kinesthetic
Have students draw a circle and a chord that is *not* a diameter on patty paper. Students should label the chord $\overline{AB}$, and then fold the paper to create the perpendicular bisector of $\overline{AB}$. Have students label the points where the fold intersects the circle as P and Q. Ask students what the relationship is between $\overset{\frown}{AP}$ and $\overset{\frown}{BP}$ **(congruent arcs)** and between $\overset{\frown}{AQ}$ and $\overset{\frown}{BQ}$ **(congruent arcs)**. Ask students what appears to be true about $\overline{PQ}$. **(It is a diameter.)**

Extra Example 1
In the diagram, $\odot P \cong \odot Q$, $\overline{FG} \cong \overline{JK}$, and $m\overset{\frown}{JK} = 120°$. Find $m\overset{\frown}{FG}$.

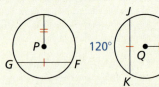

$m\overset{\frown}{FG} = 120°$

10.3 Lesson

What You Will Learn
▶ Use chords of circles to find lengths and arc measures.

Core Vocabulary
Previous
chord
arc
diameter

Using Chords of Circles
Recall that a *chord* is a segment with endpoints on a circle. Because its endpoints lie on the circle, any chord divides the circle into two arcs. A diameter divides a circle into two semicircles. Any other chord divides a circle into a minor arc and a major arc.

READING

If $\overset{\frown}{GD} \cong \overset{\frown}{GF}$, then the point G, and any line, segment, or ray that contains G, bisects $\overline{FD}$.

$\overline{EG}$ bisects $\overline{FD}$.

Theorems

Congruent Corresponding Chords Theorem
In the same circle, or in congruent circles, two minor arcs are congruent if and only if their corresponding chords are congruent.

Proof Ex. 19, p. 594

$\overset{\frown}{AB} \cong \overset{\frown}{CD}$ if and only if $\overline{AB} \cong \overline{CD}$.

Perpendicular Chord Bisector Theorem
If a diameter of a circle is perpendicular to a chord, then the diameter bisects the chord and its arc.

Proof Ex. 22, p. 594

If $\overline{EG}$ is a diameter and $\overline{EG} \perp \overline{DF}$, then $\overline{HD} \cong \overline{HF}$ and $\overset{\frown}{GD} \cong \overset{\frown}{GF}$.

Perpendicular Chord Bisector Converse
If one chord of a circle is a perpendicular bisector of another chord, then the first chord is a diameter.

Proof Ex. 23, p. 594

If $\overline{QS}$ is a perpendicular bisector of $\overline{TR}$, then $\overline{QS}$ is a diameter of the circle.

EXAMPLE 1 Using Congruent Chords to Find an Arc Measure

In the diagram, $\odot P \cong \odot Q$, $\overline{FG} \cong \overline{JK}$, and $m\overset{\frown}{JK} = 80°$. Find $m\overset{\frown}{FG}$.

SOLUTION
Because $\overline{FG}$ and $\overline{JK}$ are congruent chords in congruent circles, the corresponding minor arcs $\overset{\frown}{FG}$ and $\overset{\frown}{JK}$ are congruent by the Congruent Corresponding Chords Theorem.

▶ So, $m\overset{\frown}{FG} = m\overset{\frown}{JK} = 80°$.

Laurie's Notes — Teacher Actions

? "How would you know whether two chords are congruent?" Answers will vary.
- For each of the theorems, write the hypothesis and ask students to state the conclusion.
- The last two theorems follow from the explorations, so the conclusions should be obvious.
- **Turn and Talk:** "Discuss how you might prove each of these theorems." Allow quality *Wait Time* so that students have time to consider each theorem. Discuss proof outlines as a class.

EXAMPLE 2 Using a Diameter

a. Find HK.   b. Find $m\widehat{HK}$.

SOLUTION

a. Diameter $\overline{JL}$ is perpendicular to $\overline{HK}$. So, by the Perpendicular Chord Bisector Theorem, $\overline{JL}$ bisects $\overline{HK}$, and $HN = NK$.

▶ So, $HK = 2(NK) = 2(7) = 14$.

b. Diameter $\overline{JL}$ is perpendicular to $\overline{HK}$. So, by the Perpendicular Chord Bisector Theorem, $\overline{JL}$ bisects $\widehat{HK}$, and $m\widehat{HJ} = m\widehat{JK}$.

$m\widehat{HJ} = m\widehat{JK}$	Perpendicular Chord Bisector Theorem
$11x° = (70 + x)°$	Substitute.
$10x = 70$	Subtract x from each side.
$x = 7$	Divide each side by 10.

▶ So, $m\widehat{HJ} = m\widehat{JK} = (70 + x)° = (70 + 7)° = 77°$, and $m\widehat{HK} = 2(m\widehat{HJ}) = 2(77°) = 154°$.

EXAMPLE 3 Using Perpendicular Bisectors

Three bushes are arranged in a garden, as shown. Where should you place a sprinkler so that it is the same distance from each bush?

SOLUTION

Step 1

Label the bushes A, B, and C, as shown. Draw segments $\overline{AB}$ and $\overline{BC}$.

Step 2

Draw the perpendicular bisectors of $\overline{AB}$ and $\overline{BC}$. By the Perpendicular Chord Bisector Converse, these lie on diameters of the circle containing A, B, and C.

Step 3

Find the point where the perpendicular bisectors intersect. This is the center of the circle, which is equidistant from points A, B, and C.

Monitoring Progress Help in English and Spanish at *BigIdeasMath.com*

In Exercises 1 and 2, use the diagram of $\odot D$.

1. If $m\widehat{AB} = 110°$, find $m\widehat{BC}$.
2. If $m\widehat{AC} = 150°$, find $m\widehat{AB}$.

In Exercises 3 and 4, find the indicated length or arc measure.

3. CE
4. $m\widehat{CE}$

Section 10.3 Using Chords 591

Extra Example 2

a. Find KH. 28
b. Find $m\widehat{HLK}$. 196°

Extra Example 3
A telephone company plans to install a cell tower that is the same distance from the centers of three towns, labeled P, Q, and R. Where should the cell tower be placed?

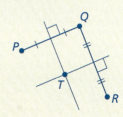

The tower should be placed at point T, where the perpendicular bisectors of $\overline{PQ}$ and $\overline{QR}$ intersect.

MONITORING PROGRESS ANSWERS

1. 110°
2. 105°
3. 10
4. 144°

Laurie's Notes — Teacher Actions

- **Think-Pair-Share:** Pose Example 2. Students should have little difficulty with either question.
- **Turn and Talk:** Sketch the garden and bushes in Example 3. "There are three bushes in the garden. Where should you place the sprinkler so that it is equidistant to all three bushes?"
- **Wait Time:** Solicit ideas from different partners.
- **? Model with Mathematics:** "Can you think of other real-life applications similar to Example 3?" See the *Applications* on page T-588.

Section 10.3   591

Extra Example 4
In the diagram, $EP = EQ = 12$, $CD = 5x + 7$, and $AB = 7x - 3$. Find the radius of $\odot E$.

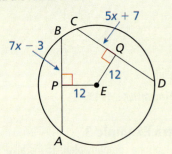

The radius is 20 units.

MONITORING PROGRESS ANSWER
5. 15

Theorem

Equidistant Chords Theorem

In the same circle, or in congruent circles, two chords are congruent if and only if they are equidistant from the center.

Proof Ex. 25, p. 594 $\overline{AB} \cong \overline{CD}$ if and only if $EF = EG$.

EXAMPLE 4   Using Congruent Chords to Find a Circle's Radius

In the diagram, $QR = ST = 16$, $CU = 2x$, and $CV = 5x - 9$. Find the radius of $\odot C$.

SOLUTION

Because $\overline{CQ}$ is a segment whose endpoints are the center and a point on the circle, it is a radius of $\odot C$. Because $\overline{CU} \perp \overline{QR}$, $\triangle QUC$ is a right triangle. Apply properties of chords to find the lengths of the legs of $\triangle QUC$.

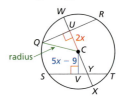

Step 1 Find CU.

Because $\overline{QR}$ and $\overline{ST}$ are congruent chords, $\overline{QR}$ and $\overline{ST}$ are equidistant from C by the Equidistant Chords Theorem. So, $CU = CV$.

$CU = CV$	Equidistant Chords Theorem
$2x = 5x - 9$	Substitute.
$x = 3$	Solve for x.

So, $CU = 2x = 2(3) = 6$.

Step 2 Find QU.

Because diameter $\overline{WX} \perp \overline{QR}$, $\overline{WX}$ bisects $\overline{QR}$ by the Perpendicular Chord Bisector Theorem.

So, $QU = \frac{1}{2}(16) = 8$.

Step 3 Find CQ.

Because the lengths of the legs are $CU = 6$ and $QU = 8$, $\triangle QUC$ is a right triangle with the Pythagorean triple 6, 8, 10. So, $CQ = 10$.

▶ So, the radius of $\odot C$ is 10 units.

Monitoring Progress Help in English and Spanish at *BigIdeasMath.com*

5. In the diagram, $JK = LM = 24$, $NP = 3x$, and $NQ = 7x - 12$. Find the radius of $\odot N$.

Laurie's Notes   Teacher Actions

- State the Equidistant Chords Theorem, and ask students to compare and contrast it with the Congruent Corresponding Chords Theorem. Both theorems are about congruent chords. One involves the intercepted minor arcs, while the other looks at distances to the center of a circle.

COMMON ERROR In Example 4, students often stop after they have solved for x. They need to finish the problem by finding the radius of the circle.

Closure

- Have students determine the diameter of the broken plate from the *Motivate*. See the *Teaching Strategy* on page T-588.

10.3 Exercises

Dynamic Solutions available at BigIdeasMath.com

Vocabulary and Core Concept Check

1. **WRITING** Describe what it means to bisect a chord.

2. **WRITING** Two chords of a circle are perpendicular and congruent. Does one of them have to be a diameter? Explain your reasoning.

Monitoring Progress and Modeling with Mathematics

In Exercises 3–6, find the measure of the red arc or chord in ⊙C. *(See Example 1.)*

3.

4.

5.

6.

In Exercises 7–10, find the value of x. *(See Example 2.)*

7.

8.

9.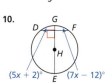

10.

11. **ERROR ANALYSIS** Describe and correct the error in reasoning.

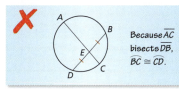

Because $\overline{AC}$ bisects $\overline{DB}$, $\overset{\frown}{BC} \cong \overset{\frown}{CD}$.

12. **PROBLEM SOLVING** In the cross section of the submarine shown, the control panels are parallel and the same length. Describe a method you can use to find the center of the cross section. Justify your method. *(See Example 3.)*

In Exercises 13 and 14, determine whether $\overline{AB}$ is a diameter of the circle. Explain your reasoning.

13.

14.

In Exercises 15 and 16, find the radius of ⊙Q. *(See Example 4.)*

15.

16.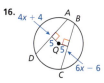

17. **PROBLEM SOLVING** An archaeologist finds part of a circular plate. What was the diameter of the plate to the nearest tenth of an inch? Justify your answer.

Section 10.3 Using Chords 593

Assignment Guide and Homework Check

ASSIGNMENT
Basic: 1, 2, 3–17 odd, 18, 26–28
Average: 1, 2, 6–10 even, 11, 12–22 even, 26–28
Advanced: 1, 2, 6, 10–22, 24, 26–28

HOMEWORK CHECK
Basic: 3, 7, 13, 15, 18
Average: 10, 14, 16, 18, 22
Advanced: 14, 16, 18, 22, 24

ANSWERS

1. Split the chord into two segments of equal length.
2. no; One chord does not necessarily bisect the other.
3. 75°
4. 5
5. 170°
6. 11
7. 8
8. 40°
9. 5
10. 7
11. $\overline{AC}$ and $\overline{DB}$ are not perpendicular; $\overset{\frown}{BC}$ is not congruent to $\overset{\frown}{CD}$.
12. *Sample answer:* Draw the perpendicular bisector of the control panels and find the midpoint; Because the control panels are parallel and congruent chords of the cross section, they are equal distances from the center, and their perpendicular bisectors form a diameter.
13. yes; The triangles are congruent, so $\overline{AB}$ is a perpendicular bisector of $\overline{CD}$.
14. no; $CE \neq ED$
15. 17
16. 13
17. about 13.9 in.; The perpendicular bisectors intersect at the center, so the right triangle with legs of 6 inches and 3.5 inches have a hypotenuse equal to the length of the radius.

Section 10.3 593

Dynamic Teaching Tools
- Dynamic Assessment & Progress Monitoring Tool
- Interactive Whiteboard Lesson Library
- Dynamic Classroom with Dynamic Investigations

ANSWERS
18–25. See Additional Answers.
26. yes; The diameter of the tire that is perpendicular to the ground is also perpendicular to $\overline{AB}$, so it bisects $\widehat{AB}$.
27. 259°
28. 122°

Mini-Assessment

The circle has center P.

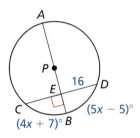

1. Find the value of x. 12
2. Find CD. 32
3. The circle has center O. Find AB.

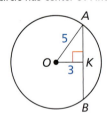

$AB = 8$

4. Q is the center of the circle. $CD = AB = 24$, $QM = x + 3$, and $QN = 2x + 1$. Find the radius of $\odot Q$.

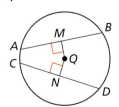

13 units

594   Chapter 10

---

18. **HOW DO YOU SEE IT?** What can you conclude from each diagram? Name a theorem that justifies your answer.

   a.   b.

   c.   d.

19. **PROVING A THEOREM** Use the diagram to prove each part of the biconditional in the Congruent Corresponding Chords Theorem.

   a. Given $\overline{AB}$ and $\overline{CD}$ are congruent chords.
      Prove $\widehat{AB} \cong \widehat{CD}$
   b. Given $\widehat{AB} \cong \widehat{CD}$
      Prove $\overline{AB} \cong \overline{CD}$

20. **MATHEMATICAL CONNECTIONS** In $\odot P$, all the arcs shown have integer measures. Show that x must be even.

21. **REASONING** In $\odot P$, the lengths of the parallel chords are 20, 16, and 12. Find $m\widehat{AB}$. Explain your reasoning.

22. **PROVING A THEOREM** Use congruent triangles to prove the Perpendicular Chord Bisector Theorem.
   Given $\overline{EG}$ is a diameter of $\odot L$.
   $\overline{EG} \perp \overline{DF}$
   Prove $\overline{DC} \cong \overline{FC}$, $\widehat{DG} \cong \widehat{FG}$

23. **PROVING A THEOREM** Write a proof of the Perpendicular Chord Bisector Converse.
   Given $\overline{QS}$ is a perpendicular bisector of $\overline{RT}$.
   Prove $\overline{QS}$ is a diameter of the circle L.

   (Hint: Plot the center L and draw $\triangle LPT$ and $\triangle LPR$.)

24. **THOUGHT PROVOKING** Consider two chords that intersect at point P. Do you think that $\dfrac{AP}{BP} = \dfrac{CP}{DP}$? Justify your answer.

25. **PROVING A THEOREM** Use the diagram with the Equidistant Chords Theorem on page 592 to prove both parts of the biconditional of this theorem.

26. **MAKING AN ARGUMENT** A car is designed so that the rear wheel is only partially visible below the body of the car. The bottom edge of the panel is parallel to the ground. Your friend claims that the point where the tire touches the ground bisects $\overline{AB}$. Is your friend correct? Explain your reasoning.

Maintaining Mathematical Proficiency
Reviewing what you learned in previous grades and lessons

Find the missing interior angle measure. *(Section 7.1)*

27. Quadrilateral $JKLM$ has angle measures $m\angle J = 32°$, $m\angle K = 25°$, and $m\angle L = 44°$. Find $m\angle M$.
28. Pentagon $PQRST$ has angle measures $m\angle P = 85°$, $m\angle Q = 134°$, $m\angle R = 97°$, and $m\angle S = 102°$. Find $m\angle T$.

594   Chapter 10   Circles

If students need help...	If students got it...
Resources by Chapter • Practice A and Practice B • Puzzle Time	Resources by Chapter • Enrichment and Extension • Cumulative Review
Student Journal • Practice	Start the *next* Section
Differentiating the Lesson Skills Review Handbook	

10.1–10.3 What Did You Learn?

Core Vocabulary

circle, p. 574
center, p. 574
radius, p. 574
chord, p. 574
diameter, p. 574
secant, p. 574
tangent, p. 574

point of tangency, p. 574
tangent circles, p. 575
concentric circles, p. 575
common tangent, p. 575
central angle, p. 582
minor arc, p. 582
major arc, p. 582

semicircle, p. 582
measure of a minor arc, p. 582
measure of a major arc, p. 582
adjacent arcs, p. 583
congruent circles, p. 584
congruent arcs, p. 584
similar arcs, p. 585

Core Concepts

Section 10.1
Lines and Segments That Intersect Circles, p. 574
Coplanar Circles and Common Tangents, p. 575

Tangent Line to Circle Theorem, p. 576
External Tangent Congruence Theorem, p. 576

Section 10.2
Measuring Arcs, p. 582
Arc Addition Postulate, p. 583
Congruent Circles Theorem, p. 584

Congruent Central Angles Theorem, p. 584
Similar Circles Theorem, p. 585

Section 10.3
Congruent Corresponding Chords Theorem, p. 590
Perpendicular Chord Bisector Theorem, p. 590

Perpendicular Chord Bisector Converse, p. 590
Equidistant Chords Theorem, p. 592

Mathematical Practices

1. Explain how separating quadrilateral *TVWX* into several segments helped you solve Exercise 37 on page 579.
2. In Exercise 30 on page 587, what two cases did you consider to reach your answers? Are there any other cases? Explain your reasoning.
3. Explain how you used inductive reasoning to solve Exercise 24 on page 594.

Keeping Your Mind Focused While Completing Homework

- Before doing homework, review the Concept boxes and Examples. Talk through the Examples out loud.
- Complete homework as though you were also preparing for a quiz. Memorize the different types of problems, formulas, rules, and so on.

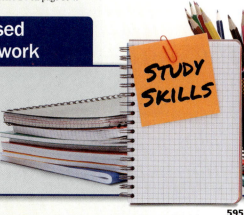

Dynamic Teaching Tools
Dynamic Assessment & Progress Monitoring Tool
Interactive Whiteboard Lesson Library
Dynamic Classroom with Dynamic Investigations

ANSWERS

1. *Sample answer:* The External Tangent Congruency Theorem could be used.
2. None of the arcs overlapped at more than one point, and point *E* was on $\overset{\frown}{CD}$; yes; *Sample answer:* Point *C* could be on $\overline{AB}$.
3. A general conclusion was reached from several specific cases.

ANSWERS
1. ⊙P
2. Sample answer: $\overline{PK}$
3. $\overline{NK}$
4. Sample answer: $\overline{JL}$
5. $\overleftrightarrow{QS}$
6. $\overleftrightarrow{QM}$
7. 8
8. 7
9. minor arc; 144°
10. minor arc; 43°
11. minor arc; 110°
12. semicircle; 180°
13. major arc; 216°
14. major arc; 317°
15. congruent; They are in the same circle and $m\widehat{JM} = m\widehat{KL}$.
16. not congruent; The circles are not congruent.
17. 100°
18. 17
19. a. 30°
    b. 150°
    c. Sample answer: 5:00

10.1–10.3 Quiz

In Exercises 1–6, use the diagram. *(Section 10.1)*

1. Name the circle.
2. Name a radius.
3. Name a diameter.
4. Name a chord.
5. Name a secant.
6. Name a tangent.

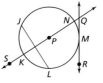

Find the value of x. *(Section 10.1)*

7.
8.

Identify the given arc as a *major arc*, *minor arc*, or *semicircle*. Then find the measure of the arc. *(Section 10.2)*

9. $\widehat{AE}$
10. $\widehat{BC}$
11. $\widehat{AC}$
12. $\widehat{ACD}$
13. $\widehat{ACE}$
14. $\widehat{BEC}$

Tell whether the red arcs are congruent. Explain why or why not. *(Section 10.2)*

15.
16.

17. Find the measure of the red arc in ⊙Q. *(Section 10.3)*

18. In the diagram, $AC = FD = 30$, $PG = x + 5$, and $PJ = 3x - 1$. Find the radius of ⊙P. *(Section 10.3)*

19. A circular clock can be divided into 12 congruent sections. *(Section 10.2)*

    a. Find the measure of each arc in this circle.
    b. Find the measure of the minor arc formed by the hour and minute hands when the time is 7:00.
    c. Find a time at which the hour and minute hands form an arc that is congruent to the arc in part (b).

Laurie's Notes

Overview of Section 10.4

Introduction
- This lesson introduces the language and properties of inscribed angles. The key relationship is that the measure of an inscribed angle is half the measure of the intercepted arc.
- This relationship is extended to inscribed angles that intercept the same arc and to inscribed angles that intercept a diameter of the circle.
- The lesson ends by looking at inscribed polygons.

Formative Assessment Tips
- **Always-Sometimes-Never True (AT-ST-NT):** This strategy is useful in assessing whether students overgeneralize or under generalize a particular concept.
- **Reason Abstractly and Quantitatively** and **Construct Viable Arguments and Critique the Reasoning of Others:** When answering, a student should be asked to justify his or her answer, and other students listening to the justification should critique the reasoning.
- *AT-ST-NT* statements help students practice the habit of checking validity when a statement (conjecture) is made. Are there different cases that need to be checked? Is there a counterexample that would show the conjecture to be false?
- Using *AT-ST-NT* statements encourages discourse. To develop these statements for a lesson, consider the common errors or misconceptions that students have relating to the goal(s) of the lesson. Allow private think time before students share their thinking with partners or the whole class.

Another Way
- There is a quick demonstration that you can do to help students visualize and understand the theorem about inscribed angles. The demonstration can be done on an interactive board, under a document camera, or on an overhead projector.
- Use or reproduce a 360° protractor. Draw an angle with sufficiently long rays. Place the angle with its vertex at the center of the protractor. Ask students to note the measure of the angle and the measure of the intercepted arc.
- As the angle is translated away from the center of the protractor, observe what happens to the measure of the intercepted arc.

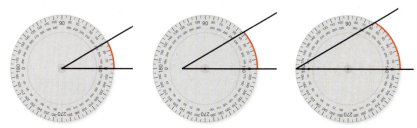

- As the angle is translated left in the image above, the angle measure remains the same and the measure of the intercepted arc is increasing.
- This same technique can be used in the next lesson to demonstrate other angle relationships in a circle.

Pacing Suggestion
- The explorations allow students to discover a relationship about inscribed angles and intercepted arcs, and the angles of an inscribed quadrilateral. Examples related to the theorems in the formal lesson should go quickly.

What Your Students Will Learn

- Use inscribed angles of circles to find angle and arc measures of circles.
- Use inscribed polygons (or the circumscribed circle of a polygon) to find angle measures in polygons.

Laurie's Notes

Exploration

Motivate
- Show students a diagram of several soccer players standing on a circle that contains the posts of the goal. Discuss with them whether any player has the greatest "kicking angle" for the goal. Tell students that in this lesson they will see why each player has the same "kicking angle."

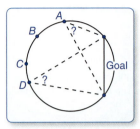

Discuss
- Discuss the vocabulary: inscribed angle and intercepted arc.
- If dynamic geometry software is not available, students could still perform the construction with a protractor and straightedge. Gathering results from classmates may be sufficient to make a conjecture about the problem posed.

Exploration 1
- Have students work with their partners to investigate inscribed angles. Be sure that students try different conditions such as those shown.

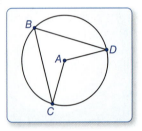 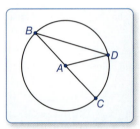 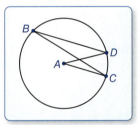

- As students make repeated trials, data should be recorded in a table.
- **Construct Viable Arguments and Critique the Reasoning of Others:** Have students state their conjectures and explain their reasoning. Discuss.

Exploration 2
- Before students begin, ask, "What is the sum of the angles of a quadrilateral?" 360°
  "Do you think this relationship will change if you inscribe the quadrilateral in a circle?"
  Hopefully, students will say *no*.
- With the vertices of the quadrilateral moving dynamically, students should recognize that the sum of the angles is still 360° and that opposite angles are supplementary.

Communicate Your Answer
- Students should summarize the conjectures made in each of the explorations.

Connecting to Next Step
- Each exploration is related to a theorem in today's lesson. Students will have a good sense of the theorem and perhaps how it might be proven.

10.4 Inscribed Angles and Polygons

Essential Question How are inscribed angles related to their intercepted arcs? How are the angles of an inscribed quadrilateral related to each other?

An **inscribed angle** is an angle whose vertex is on a circle and whose sides contain chords of the circle. An arc that lies between two lines, rays, or segments is called an **intercepted arc**. Recall that a polygon is an inscribed polygon when all its vertices lie on a circle.

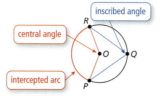

EXPLORATION 1   Inscribed Angles and Central Angles

Work with a partner. Use dynamic geometry software.

a. Construct an inscribed angle in a circle. Then construct the corresponding central angle.

b. Measure both angles. How is the inscribed angle related to its intercepted arc?

c. Repeat parts (a) and (b) several times. Record your results in a table. Write a conjecture about how an inscribed angle is related to its intercepted arc.

Sample

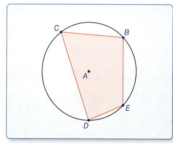

> **ATTENDING TO PRECISION**
> To be proficient in math, you need to communicate precisely with others.

EXPLORATION 2   A Quadrilateral with Inscribed Angles

Work with a partner. Use dynamic geometry software.

a. Construct a quadrilateral with each vertex on a circle.

b. Measure all four angles. What relationships do you notice?

c. Repeat parts (a) and (b) several times. Record your results in a table. Then write a conjecture that summarizes the data.

Communicate Your Answer

3. How are inscribed angles related to their intercepted arcs? How are the angles of an inscribed quadrilateral related to each other?

4. Quadrilateral *EFGH* is inscribed in $\odot C$, and $m\angle E = 80°$. What is $m\angle G$? Explain.

Section 10.4   Inscribed Angles and Polygons   597

English Language Learners

Group Activity
Form small groups of English learners and English speakers. Ask each group to collaborate to write a definition for one or two of the following terms: *inscribed angle, intercepted arc, subtend, inscribed polygon,* and *circumscribed circle*. Include a sketch for each term.

Extra Example 1
Find the indicated measure.

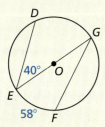

a. $m\widehat{DG}$ 80°
b. $m\angle G$ 29°

10.4 Lesson

Core Vocabulary
inscribed angle, *p. 598*
intercepted arc, *p. 598*
subtend, *p. 598*

Previous
inscribed polygon
circumscribed circle

What You Will Learn
▶ Use inscribed angles.
▶ Use inscribed polygons.

Using Inscribed Angles

Core Concept

Inscribed Angle and Intercepted Arc
An **inscribed angle** is an angle whose vertex is on a circle and whose sides contain chords of the circle. An arc that lies between two lines, rays, or segments is called an **intercepted arc**. If the endpoints of a chord or arc lie on the sides of an inscribed angle, then the chord or arc is said to **subtend** the angle.

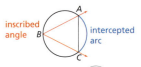

$\angle B$ intercepts $\widehat{AC}$.
$\widehat{AC}$ subtends $\angle B$.
$\overline{AC}$ subtends $\angle B$.

Theorem

Measure of an Inscribed Angle Theorem
The measure of an inscribed angle is one-half the measure of its intercepted arc.

$m\angle ADB = \frac{1}{2} m\widehat{AB}$

Proof Ex. 35, p. 604

The proof of the Measure of an Inscribed Angle Theorem involves three cases.

Case 1 Center *C* is on a side of the inscribed angle.

Case 2 Center *C* is inside the inscribed angle.

Case 3 Center *C* is outside the inscribed angle.

EXAMPLE 1 Using Inscribed Angles

Find the indicated measure.
a. $m\angle T$
b. $m\widehat{QR}$

SOLUTION

a. $m\angle T = \frac{1}{2} m\widehat{RS} = \frac{1}{2}(48°) = 24°$

b. $m\widehat{TQ} = 2 m\angle R = 2 \cdot 50° = 100°$
Because $\widehat{TQR}$ is a semicircle, $m\widehat{QR} = 180° - m\widehat{TQ} = 180° - 100° = 80°$.

Laurie's Notes | Teacher Actions

- Write the *Core Concept*, which should be familiar from the explorations.
- Introduce the Measure of an Inscribed Angle Theorem. It is helpful that students have explored this theorem dynamically with software. Even if students have explored this theorem with software, a quick visual demonstration resonates with them. See *Another Way* on page T-596.
- Have partners work independently to solve Example 1, and then discuss as a class.

EXAMPLE 2 Finding the Measure of an Intercepted Arc

Find $m\widehat{RS}$ and $m\angle STR$. What do you notice about $\angle STR$ and $\angle RUS$?

SOLUTION

From the Measure of an Inscribed Angle Theorem, you know that $m\widehat{RS} = 2m\angle RUS = 2(31°) = 62°$.

Also, $m\angle STR = \frac{1}{2}m\widehat{RS} = \frac{1}{2}(62°) = 31°$.

▶ So, $\angle STR \cong \angle RUS$.

Example 2 suggests the Inscribed Angles of a Circle Theorem.

Theorem

Inscribed Angles of a Circle Theorem

If two inscribed angles of a circle intercept the same arc, then the angles are congruent.

Proof Ex. 36, p. 604

$\angle ADB \cong \angle ACB$

EXAMPLE 3 Finding the Measure of an Angle

Given $m\angle E = 75°$, find $m\angle F$.

SOLUTION

Both $\angle E$ and $\angle F$ intercept $\widehat{GH}$. So, $\angle E \cong \angle F$ by the Inscribed Angles of a Circle Theorem.

▶ So, $m\angle F = m\angle E = 75°$.

Monitoring Progress 🔊 Help in English and Spanish at *BigIdeasMath.com*

Find the measure of the red arc or angle.

1.
2.
3.

Section 10.4 Inscribed Angles and Polygons 599

Extra Example 2

Find $m\widehat{HML}$ and $m\angle HJL$. What do you notice about $\angle HJL$ and $\angle LKH$?

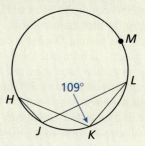

$m\widehat{HML} = 218°$ and $m\angle HJL = 109°$. So, $\angle HJL \cong \angle LKH$.

Extra Example 3

Given $m\angle C = 68°$, find $m\angle B$.

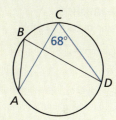

$m\angle B = 68°$

MONITORING PROGRESS ANSWERS

1. 45°
2. 76°
3. 72°

Laurie's Notes Teacher Actions

- **Teaching Tip:** Sometimes it is helpful to extend the rays forming the inscribed angle to help students visualize the angle.
- **Construct Viable Arguments and Critique the Reasoning of Others:** Write the Inscribed Angles of a Circle Theorem.
- **Turn and Talk:** "How could you prove this theorem?" Give sufficient *Wait Time*, and then solicit outlines for a proof.
- **Think-Pair-Share:** Have students answer Questions 1–3, and then share and discuss as a class.

Differentiated Instruction

Kinesthetic

Before introducing the Inscribed Right Triangle Theorem, have students explore this concept. Divide the class into groups. Each group should use a compass to draw two circles, and then in each circle draw an inscribed triangle in which one side is a diameter of the circle. The students in each group should compare their figures and make a conjecture about the measure of an angle inscribed in a semicircle. They should then use a protractor to verify their conjecture.

Extra Example 4

Find the value of each variable.

a.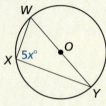

$x = 18$

b.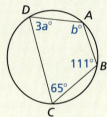

$a = 23, b = 115$

Using Inscribed Polygons

Recall that a polygon is an inscribed polygon when all its vertices lie on a circle. The circle that contains the vertices is a circumscribed circle.

Theorems

Inscribed Right Triangle Theorem

If a right triangle is inscribed in a circle, then the hypotenuse is a diameter of the circle. Conversely, if one side of an inscribed triangle is a diameter of the circle, then the triangle is a right triangle and the angle opposite the diameter is the right angle.

$m\angle ABC = 90°$ if and only if $\overline{AC}$ is a diameter of the circle.

Proof Ex. 37, p. 604

Inscribed Quadrilateral Theorem

A quadrilateral can be inscribed in a circle if and only if its opposite angles are supplementary.

$D, E, F,$ and G lie on $\odot C$ if and only if $m\angle D + m\angle F = m\angle E + m\angle G = 180°$.

Proof Ex. 38, p. 604; BigIdeasMath.com

EXAMPLE 4 Using Inscribed Polygons

Find the value of each variable.

a.

b.

SOLUTION

a. $\overline{AB}$ is a diameter. So, $\angle C$ is a right angle, and $m\angle C = 90°$ by the Inscribed Right Triangle Theorem.

$$2x° = 90°$$
$$x = 45$$

▶ The value of x is 45.

b. *DEFG* is inscribed in a circle, so opposite angles are supplementary by the Inscribed Quadrilateral Theorem.

$m\angle D + m\angle F = 180°$ $m\angle E + m\angle G = 180°$

$z + 80 = 180$ $120 + y = 180$

$z = 100$ $y = 60$

▶ The value of z is 100 and the value of y is 60.

600   Chapter 10   Circles

Laurie's Notes Teacher Actions

? Always-Sometimes-Never True: "Every triangle can be circumscribed. Explain." **always true** Students may recall the water sprinkler problem from Section 10.3.

- **Predict, Explain, Observe:** "Every quadrilateral can be circumscribed." Give groups time to investigate using dynamic geometry software. If they have completed Exploration 2 on page 597, they will already have a sense about this probe.

EXAMPLE 5 **Using a Circumscribed Circle**

Your camera has a 90° field of vision, and you want to photograph the front of a statue. You stand at a location in which the front of the statue is all that appears in your camera's field of vision, as shown. You want to change your location. Where else can you stand so that the front of the statue is all that appears in your camera's field of vision?

SOLUTION

From the Inscribed Right Triangle Theorem, you know that if a right triangle is inscribed in a circle, then the hypotenuse of the triangle is a diameter of the circle. So, draw the circle that has the front of the statue as a diameter.

▶ The statue fits perfectly within your camera's 90° field of vision from any point on the semicircle in front of the statue.

Monitoring Progress Help in English and Spanish at *BigIdeasMath.com*

Find the value of each variable.

4.

5.

6.

7.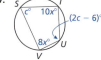

8. In Example 5, explain how to find locations where the left side of the statue is all that appears in your camera's field of vision.

Section 10.4   Inscribed Angles and Polygons   601

Extra Example 5
In Example 5, explain how to find locations where the right side of the statue is all that is seen in your camera's field of vision. Stand at any point on the semicircle whose diameter is the right side of the base of the statue.

MONITORING PROGRESS ANSWERS
4. $x = 90, y = 50$
5. $x = 17.5, y = 90$
6. $x = 98, y = 112$
7. $x = 10, c = 62$
8. Draw the circle that has the left side of the statue as a diameter.

Laurie's Notes — Teacher Actions

❓ "What does a 90° field of vision mean?" Answers will vary.
- **Teaching Tip:** To model what the 90° field of vision means, use a file folder opened to a 90° angle. Use a sharp object to poke a hole that acts as the viewfinder. Select some wide object in the room, such as the width of your whiteboard. Ask a student to model Example 5.
- Discuss related applications.

Closure
- **Exit Ticket:** Quadrilateral *ABCD* is inscribed in a circle. Let $m\angle A = 72°$ and $m\angle B = 113°$. Find $m\angle C$ and $m\angle D$. $m\angle C = 108°, m\angle D = 67°$

Section 10.4   601

Assignment Guide and Homework Check

ASSIGNMENT

Basic: 1, 2, 3–19 odd, 22, 29, 32, 41–44

Average: 1, 2, 6–16 even, 17, 18–32 even, 35, 41–44

Advanced: 1, 2, 8–16 even, 17, 18–38 even, 41–44

HOMEWORK CHECK

Basic: 5, 11, 15, 17

Average: 8, 12, 16, 20, 35

Advanced: 12, 16, 20, 34

ANSWERS

1. intercepted arc
2. Find $m\angle AGC$; 130°; 90°
3. 42°
4. 85°
5. 10°
6. 134°
7. 120°
8. 100°
9. $\angle ACB \cong \angle ADB$, $\angle DAC \cong \angle DBC$
10. $\angle WXZ \cong \angle WYZ$, $\angle YZX \cong \angle XWY$
11. 51°
12. 80°
13. $x = 100$, $y = 85$
14. $m = 120$, $k = 60$
15. $a = 20$, $b = 22$
16. $x = 30$, $y = 28$
17. The inscribed angle was not doubled; $m\angle BAC = 2(53°) = 106°$

10.4 Exercises

Dynamic Solutions available at *BigIdeasMath.com*

Vocabulary and Core Concept Check

1. **VOCABULARY** An arc that lies between two lines, rays, or segments is called a(n) _____.

2. **DIFFERENT WORDS, SAME QUESTION** Which is different? Find "both" answers.

   Find $m\angle ABC$.   Find $m\angle AGC$.

   Find $m\angle AEC$.   Find $m\angle ADC$.

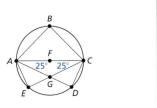

Monitoring Progress and Modeling with Mathematics

In Exercises 3–8, find the indicated measure. *(See Examples 1 and 2.)*

3. $m\angle A$

4. $m\angle G$

5. $m\angle N$

6. $m\overset{\frown}{RS}$

7. $m\overset{\frown}{VU}$

8. $m\overset{\frown}{WX}$

In Exercises 9 and 10, name two pairs of congruent angles.

9.

10.

In Exercises 11 and 12, find the measure of the red arc or angle. *(See Example 3.)*

11.

12.

In Exercises 13–16, find the value of each variable. *(See Example 4.)*

13.

14.

15.

16.

17. **ERROR ANALYSIS** Describe and correct the error in finding $m\overset{\frown}{BC}$.

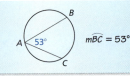

602   Chapter 10   Circles

18. **MODELING WITH MATHEMATICS** A *carpenter's square* is an L-shaped tool used to draw right angles. You need to cut a circular piece of wood into two semicircles. How can you use the carpenter's square to draw a diameter on the circular piece of wood? (*See Example 5.*)

MATHEMATICAL CONNECTIONS In Exercises 19–21, find the values of *x* and *y*. Then find the measures of the interior angles of the polygon.

19.

20.

21.

22. **MAKING AN ARGUMENT** Your friend claims that ∠PTQ ≅ ∠PSQ ≅ ∠PRQ. Is your friend correct? Explain your reasoning.

REASONING In Exercises 23–28, determine whether a quadrilateral of the given type can always be inscribed inside a circle. Explain your reasoning.

23. square
24. rectangle
25. parallelogram
26. kite
27. rhombus
28. isosceles trapezoid

29. **MODELING WITH MATHEMATICS** Three moons, A, B, and C, are in the same circular orbit 100,000 kilometers above the surface of a planet. The planet is 20,000 kilometers in diameter and m∠ABC = 90°. Draw a diagram of the situation. How far is moon A from moon C?

30. **MODELING WITH MATHEMATICS** At the movie theater, you want to choose a seat that has the best *viewing angle*, so that you can be close to the screen and still see the whole screen without moving your eyes. You previously decided that seat F7 has the best viewing angle, but this time someone else is already sitting there. Where else can you sit so that your seat has the same viewing angle as seat F7? Explain.

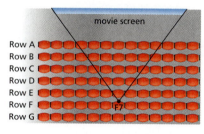

31. **WRITING** A right triangle is inscribed in a circle, and the radius of the circle is given. Explain how to find the length of the hypotenuse.

32. **HOW DO YOU SEE IT?** Let point *Y* represent your location on the soccer field below. What type of angle is ∠AYB if you stand anywhere on the circle except at point *A* or point *B*?

Section 10.4 Inscribed Angles and Polygons 603

ANSWERS

18. Place the right angle of the carpenter's square on the edge of the circle and connect the points where the sides intersect the edge of the circle.
19. $x = 25$, $y = 5$; 130°, 75°, 50°, 105°
20. $x = 9$, $y = 6$; 54°, 36°, 126°, 144°
21. $x = 30$, $y = 20$; 60°, 60°, 60°
22. yes; The angles intercept the same arc.
23. yes; Opposite angles are always supplementary.
24. yes; Opposite angles are always supplementary.
25. no; Opposite angles are not always supplementary.
26. no; Opposite angles are not always supplementary.
27. no; Opposite angles are not always supplementary.
28. yes; Opposite angles are always supplementary.
29.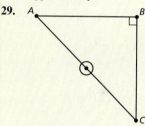

    220,000 km
30. any point on the circle circumscribed about the triangle that seat F7 makes with the movie screen; The angle intercepts the same arc.
31. double the radius
32. right angle

ANSWERS

33. Each diagonal splits the rectangle into two right triangles.
34. yes; Every triangle has a circumcenter.
35–37. See Additional Answers.
38. $m\widehat{EDG}$; Measure of an Inscribed Angle; $2m\angle F$; $m\angle D + m\angle F = 180°$; $m\angle E + m\angle G = 180°$
39. 2.4 units
40. a. $\dfrac{JH}{GJ} = \dfrac{GJ}{FJ}$; Geometric Mean (Altitude) Theorem
    b. 6 in.; 2 in.; $2\sqrt{3}$ in.; $4\sqrt{3}$ in.
41. $x = \dfrac{145}{3}$
42. $x = 126$
43. $x = 120$
44. $x = 180$

Mini-Assessment

1. Find $m\widehat{DF}$ and $m\widehat{ED}$.

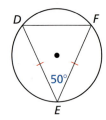

$m\widehat{DF} = 100°$, $m\widehat{ED} = 130°$

2. Find $m\widehat{CD}$.

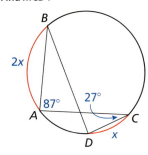

$m\widehat{CD} = 44°$

3. Find the value of each variable.

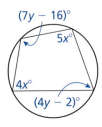

$x = 20$, $y = 18$

---

33. **WRITING** Explain why the diagonals of a rectangle inscribed in a circle are diameters of the circle.

34. **THOUGHT PROVOKING** The figure shows a circle that is circumscribed about $\triangle ABC$. Is it possible to circumscribe a circle about any triangle? Justify your answer.

35. **PROVING A THEOREM** If an angle is inscribed in $\odot Q$, the center Q can be on a side of the inscribed angle, inside the inscribed angle, or outside the inscribed angle. Prove each case of the Measure of an Inscribed Angle Theorem.

   a. **Case 1**
      Given $\angle ABC$ is inscribed in $\odot Q$.
      Let $m\angle B = x°$.
      Center Q lies on $\overline{BC}$.
      Prove $m\angle ABC = \frac{1}{2} m\widehat{AC}$
      (*Hint*: Show that $\triangle AQB$ is isosceles. Then write $m\widehat{AC}$ in terms of x.)

   b. **Case 2** Use the diagram and auxiliary line to write **Given** and **Prove** statements for Case 2. Then write a proof.

   c. **Case 3** Use the diagram and auxiliary line to write **Given** and **Prove** statements for Case 3. Then write a proof.

36. **PROVING A THEOREM** Write a paragraph proof of the Inscribed Angles of a Circle Theorem. First, draw a diagram and write **Given** and **Prove** statements.

37. **PROVING A THEOREM** The Inscribed Right Triangle Theorem is written as a conditional statement and its converse. Write a plan for proof for each statement.

38. **PROVING A THEOREM** Copy and complete the paragraph proof for one part of the Inscribed Quadrilateral Theorem.

   Given $\odot C$ with inscribed quadrilateral $DEFG$
   Prove $m\angle D + m\angle F = 180°$, $m\angle E + m\angle G = 180°$

   By the Arc Addition Postulate, $m\widehat{EFG} + \underline{\quad} = 360°$ and $m\widehat{FGD} + m\widehat{DEF} = 360°$. Using the _____ Theorem, $m\widehat{EDG} = 2m\angle F$, $m\widehat{EFG} = 2m\angle D$, $m\widehat{DEF} = 2m\angle G$, and $m\widehat{FGD} = 2m\angle E$. By the Substitution Property of Equality, $2m\angle D + \underline{\quad} = 360°$, so ___. Similarly, ___.

39. **CRITICAL THINKING** In the diagram, $\angle C$ is a right angle. If you draw the smallest possible circle through C tangent to $\overline{AB}$, the circle will intersect $\overline{AC}$ at J and $\overline{BC}$ at K. Find the exact length of $\overline{JK}$.

40. **CRITICAL THINKING** You are making a circular cutting board. To begin, you glue eight 1-inch boards together, as shown. Then you draw and cut a circle with an 8-inch diameter from the boards.

   a. $\overline{FH}$ is a diameter of the circular cutting board. Write a proportion relating GJ and JH. State a theorem to justify your answer.

   b. Find FJ, JH, and GJ. What is the length of the cutting board seam labeled $\overline{GK}$?

Maintaining Mathematical Proficiency *Reviewing what you learned in previous grades and lessons*

Solve the equation. Check your solution. *(Skills Review Handbook)*

41. $3x = 145$
42. $\frac{1}{2}x = 63$
43. $240 = 2x$
44. $75 = \frac{1}{2}(x - 30)$

If students need help...	If students got it...
Resources by Chapter • Practice A and Practice B • Puzzle Time	**Resources by Chapter** • Enrichment and Extension • Cumulative Review
Student Journal • Practice	Start the *next* Section
Differentiating the Lesson Skills Review Handbook	

Laurie's Notes

Overview of Section 10.5

Introduction
- This lesson extends the relationships found in the last lesson about angles and intercepted arcs of a circle. There are four additional theorems about angle relationships in a circle.
- The lesson ends with a real-life application about the northern lights, a topic used as the *Motivate* for the lesson.

Teaching Strategy
- On page T-596, another way was suggested for investigating angle relationships in a circle—in particular, when two lines intersect on a circle forming an inscribed angle. In this lesson, the two lines might intersect in the interior or the exterior of the circle.
- Students can become overwhelmed, believing that there are a lot of theorems (formulas) to remember. To help students organize and summarize their thinking, have them consider the three intersection possibilities described in the *Core Concept* on page 606.
- **Turn and Talk:** Have students note similarities and differences in the theorems relating to these three cases. Students should recognize the following:
  - When the lines intersect the circle, arcs are intercepted.
  - All of the formulas involve taking $\frac{1}{2}$ of a measure.
  - When the lines intersect inside the circle, you add arc measures.
  - When the lines intersect outside the circle, you subtract arc measures.
  - When the lines intersect on the circle, there is only one arc to consider.

Pacing Suggestion
- The explorations allow students to discover additional relationships about angles formed when two lines intersect with a circle. Examples related to the theorems in the formal lesson should go quickly.

Dynamic Teaching Tools

Dynamic Assessment & Progress Monitoring Tool
Lesson Planning Tool
Interactive Whiteboard Lesson Library
Dynamic Classroom with Dynamic Investigations

What Your Students Will Learn

- Find angle and arc measures formed by chords, tangents, and secants of circles.
- Use circumscribed angles to find angle and arc measures in circles.

Laurie's Notes

Exploration

Motivate
- Ask whether anyone can speak about aurora borealis, commonly known as the northern lights. Let students share what they know.
- **FYI:** Polar lights (aurora polaris) are a natural phenomenon found in both the northern and southern hemispheres. The polar lights are a natural light display in the sky, particularly in the high-latitude (Arctic and Antarctic) regions, caused by the collision of energetic charged particles with atoms in the high-altitude atmosphere.
- In this lesson, the connection to high altitudes is made.

Exploration Note
- If dynamic geometry software is not available, students could still perform the construction with a protractor and straightedge. Gathering results from classmates may be sufficient to make a conjecture about the problem posed.

Exploration 1
- Have students work with partners to explore the problem posed.
- Given the dynamic nature of the construction, students should click and drag on point C to observe any patterns.
- As students make repeated trials, data should be recorded in a table.
- **Construct Viable Arguments and Critique the Reasoning of Others:** Have students state their conjectures and explain their reasoning. Discuss.

Exploration 2
- This exploration is similar to the first.
- Students will likely ask which pair of vertical angles to measure, the obtuse pair or the acute pair—to which I generally respond, "Yes."
- As students make repeated trials, data should be recorded in a table.

Communicate Your Answer
- Listen for both conjectures made by students as a result of the two explorations.

Connecting to Next Step
- This lesson contains four theorems, two of which students have now explored using dynamic geometry software.

10.5 Angle Relationships in Circles

Essential Question When a chord intersects a tangent line or another chord, what relationships exist among the angles and arcs formed?

EXPLORATION 1 Angles Formed by a Chord and Tangent Line

Work with a partner. Use dynamic geometry software.

a. Construct a chord in a circle. At one of the endpoints of the chord, construct a tangent line to the circle.

b. Find the measures of the two angles formed by the chord and the tangent line.

c. Find the measures of the two circular arcs determined by the chord.

d. Repeat parts (a)–(c) several times. Record your results in a table. Then write a conjecture that summarizes the data.

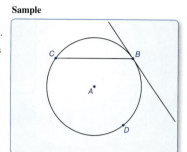

Sample

EXPLORATION 2 Angles Formed by Intersecting Chords

Work with a partner. Use dynamic geometry software.

a. Construct two chords that intersect inside a circle.

b. Find the measure of one of the angles formed by the intersecting chords.

c. Find the measures of the arcs intercepted by the angle in part (b) and its vertical angle. What do you observe?

d. Repeat parts (a)–(c) several times. Record your results in a table. Then write a conjecture that summarizes the data.

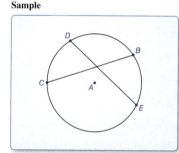

Sample

CONSTRUCTING VIABLE ARGUMENTS
To be proficient in math, you need to understand and use stated assumptions, definitions, and previously established results.

Communicate Your Answer

3. When a chord intersects a tangent line or another chord, what relationships exist among the angles and arcs formed?

4. Line m is tangent to the circle in the figure at the left. Find the measure of ∠1.

5. Two chords intersect inside a circle to form a pair of vertical angles with measures of 55°. Find the sum of the measures of the arcs intercepted by the two angles.

Section 10.5 Angle Relationships in Circles 605

English Language Learners

Notebook Development
Have students record in their notebooks all the theorems introduced in this lesson. For each theorem, include a sketch and an example of how to use the theorem to solve a problem.

Extra Example 1
Line *m* is tangent to the circle. Find the measure of the red angle or arc.

a.
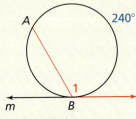
$m\angle 1 = 120°$

b.
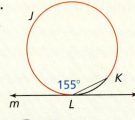
$m\widehat{KJL} = 310°$

MONITORING PROGRESS ANSWERS
1. 105°
2. 196°
3. 160°

10.5 Lesson

What You Will Learn
▶ Find angle and arc measures.
▶ Use circumscribed angles.

Core Vocabulary
circumscribed angle, *p. 608*
Previous
tangent
chord
secant

Finding Angle and Arc Measures

Theorem

Tangent and Intersected Chord Theorem
If a tangent and a chord intersect at a point on a circle, then the measure of each angle formed is one-half the measure of its intercepted arc.

Proof Ex. 33, p. 612

$m\angle 1 = \frac{1}{2}m\widehat{AB} \qquad m\angle 2 = \frac{1}{2}m\widehat{BCA}$

EXAMPLE 1 Finding Angle and Arc Measures

Line *m* is tangent to the circle. Find the measure of the red angle or arc.

a.   b.

SOLUTION

a. $m\angle 1 = \frac{1}{2}(130°) = 65°$ b. $m\widehat{KJL} = 2(125°) = 250°$

Monitoring Progress Help in English and Spanish at *BigIdeasMath.com*

Line *m* is tangent to the circle. Find the indicated measure.

1. $m\angle 1$ 2. $m\widehat{RST}$ 3. $m\widehat{XY}$

Core Concept

Intersecting Lines and Circles
If two nonparallel lines intersect a circle, there are three places where the lines can intersect.

on the circle   inside the circle   outside the circle

Laurie's Notes — Teacher Actions

- The first theorem in the lesson was explored on the previous page.
- ❓ "If $\overline{AB}$ is a diameter, what is $m\angle 1$ and $m\angle 2$ in the diagram?" They are each 90°.
- If $\overline{AB}$ is not a diameter, one of the intercepted arcs will be greater than 180° and one will be less than 180°.
- Have partners work independently to solve Example 1 and Questions 1–3. Discuss as a class.
- The *Core Concept* summarizes the three cases of intersecting lines and circles. See the *Teaching Strategy* on page T-604.

Theorems

Angles Inside the Circle Theorem

If two chords intersect *inside* a circle, then the measure of each angle is one-half the *sum* of the measures of the arcs intercepted by the angle and its vertical angle.

$m\angle 1 = \frac{1}{2}(m\widehat{DC} + m\widehat{AB})$,
$m\angle 2 = \frac{1}{2}(m\widehat{AD} + m\widehat{BC})$

Proof Ex. 35, p. 612

Angles Outside the Circle Theorem

If a tangent and a secant, two tangents, or two secants intersect *outside* a circle, then the measure of the angle formed is one-half the *difference* of the measures of the intercepted arcs.

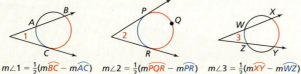

$m\angle 1 = \frac{1}{2}(m\widehat{BC} - m\widehat{AC})$ $m\angle 2 = \frac{1}{2}(m\widehat{PQR} - m\widehat{PR})$ $m\angle 3 = \frac{1}{2}(m\widehat{XY} - m\widehat{WZ})$

Proof Ex. 37, p. 612

EXAMPLE 2 Finding an Angle Measure

Find the value of *x*.

a.

b.

SOLUTION

a. The chords $\overline{JL}$ and $\overline{KM}$ intersect inside the circle. Use the Angles Inside the Circle Theorem.

$x° = \frac{1}{2}(m\widehat{JM} + m\widehat{LK})$

$x° = \frac{1}{2}(130° + 156°)$

$x = 143$

▶ So, the value of *x* is 143.

b. The tangent $\overrightarrow{CD}$ and the secant $\overrightarrow{CB}$ intersect outside the circle. Use the Angles Outside the Circle Theorem.

$m\angle BCD = \frac{1}{2}(m\widehat{AD} - m\widehat{BD})$

$x° = \frac{1}{2}(178° - 76°)$

$x = 51$

▶ So, the value of *x* is 51.

Monitoring Progress Help in English and Spanish at *BigIdeasMath.com*

Find the value of the variable.

4.

5.

Section 10.5 Angle Relationships in Circles 607

Differentiated Instruction

Visual

To help students see that the three cases of the Angles Outside the Circle Theorem are related, use a chalkboard compass or two yardsticks and a large circle drawn on the board. Use the sides of the compass to represent the sides of an angle whose vertex is outside the circle. Open and close the compass so the sides represent two tangents, then two secants, and then a tangent and a secant.

Extra Example 2

Find the value of *x*.

a.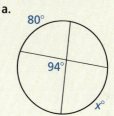

$x = 92$

b.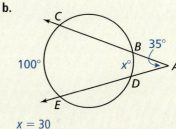

$x = 30$

MONITORING PROGRESS ANSWERS

4. 61
5. 104

Laurie's Notes Teacher Actions

- The first theorem on this page was explored on page 605. If time permits, have students explore the three cases of the theorem. An alternative is to demonstrate the theorem dynamically. An online search will reveal ready-made demonstrations.
- **Turn and Talk:** "How do you summarize the three cases for finding the angle measure of the angle formed by two intersecting lines and a circle?" See the *Teaching Strategy* on page T-604.

Section 10.5 607

Extra Example 3

Find the value of x.

a.

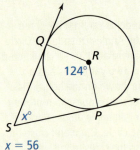

x = 56

b.

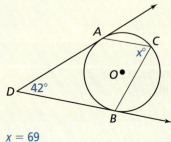

x = 69

Using Circumscribed Angles

Core Concept

Circumscribed Angle

A **circumscribed angle** is an angle whose sides are tangent to a circle.

circumscribed angle

Theorem

Circumscribed Angle Theorem

The measure of a circumscribed angle is equal to 180° minus the measure of the central angle that intercepts the same arc.

Proof Ex. 38, p. 612

$m\angle ADB = 180° - m\angle ACB$

EXAMPLE 3 Finding Angle Measures

Find the value of x.

a. 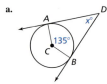   b.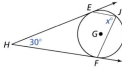

SOLUTION

a. Use the Circumscribed Angle Theorem to find $m\angle ADB$.

$m\angle ADB = 180° - m\angle ACB$	Circumscribed Angle Theorem
$x° = 180° - 135°$	Substitute.
$x = 45$	Subtract.

▶ So, the value of x is 45.

b. Use the Measure of an Inscribed Angle Theorem and the Circumscribed Angle Theorem to find $m\angle EJF$.

$m\angle EJF = \frac{1}{2}m\widehat{EF}$	Measure of an Inscribed Angle Theorem
$m\angle EJF = \frac{1}{2}m\angle EGF$	Definition of minor arc
$m\angle EJF = \frac{1}{2}(180° - m\angle EHF)$	Circumscribed Angle Theorem
$m\angle EJF = \frac{1}{2}(180° - 30°)$	Substitute.
$x = \frac{1}{2}(180 - 30)$	Substitute.
$x = 75$	Simplify.

▶ So, the value of x is 75.

Laurie's Notes — Teacher Actions

- Write the *Core Concept* and provide a sketch.
- State the Circumscribed Angle Theorem. Ask, "How might you prove this theorem?" Give *Wait Time*, and then solicit ideas from students. Write a brief outline of the proof.
- **Think-Alouds:** Pose Example 3 and say, "To solve part (a), I need to ...." Ask partner A to think aloud for partner B to hear the problem-solving process. When students have finished the problem, reverse roles for part (b). Discuss solutions as a class.

EXAMPLE 4 Modeling with Mathematics

The northern lights are bright flashes of colored light between 50 and 200 miles above Earth. A flash occurs 150 miles above Earth at point C. What is the measure of $\widehat{BD}$, the portion of Earth from which the flash is visible? (Earth's radius is approximately 4000 miles.)

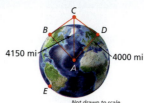

Not drawn to scale

SOLUTION

1. **Understand the Problem** You are given the approximate radius of Earth and the distance above Earth that the flash occurs. You need to find the measure of the arc that represents the portion of Earth from which the flash is visible.

2. **Make a Plan** Use properties of tangents, triangle congruence, and angles outside a circle to find the arc measure.

3. **Solve the Problem** Because $\overline{CB}$ and $\overline{CD}$ are tangents, $\overline{CB} \perp \overline{AB}$ and $\overline{CD} \perp \overline{AD}$ by the Tangent Line to Circle Theorem. Also, $\overline{BC} \cong \overline{DC}$ by the External Tangent Congruence Theorem, and $\overline{CA} \cong \overline{CA}$ by the Reflexive Property of Congruence. So, $\triangle ABC \cong \triangle ADC$ by the Hypotenuse-Leg Congruence Theorem. Because corresponding parts of congruent triangles are congruent, $\angle BCA \cong \angle DCA$. Solve right $\triangle CBA$ to find that $m\angle BCA \approx 74.5°$. So, $m\angle BCD \approx 2(74.5°) = 149°$.

> **COMMON ERROR**
> Because the value for $m\angle BCD$ is an approximation, use the symbol $\approx$ instead of $=$.

$m\angle BCD = 180° - m\angle BAD$	Circumscribed Angle Theorem
$m\angle BCD = 180° - m\widehat{BD}$	Definition of minor arc
$149° \approx 180° - m\widehat{BD}$	Substitute.
$31° \approx m\widehat{BD}$	Solve for $m\widehat{BD}$.

▶ The measure of the arc from which the flash is visible is about 31°.

4. **Look Back** You can use inverse trigonometric ratios to find $m\angle BAC$ and $m\angle DAC$.

$$m\angle BAC = \cos^{-1}\left(\frac{4000}{4150}\right) \approx 15.5°$$

$$m\angle DAC = \cos^{-1}\left(\frac{4000}{4150}\right) \approx 15.5°$$

So, $m\angle BAD \approx 15.5° + 15.5° = 31°$, and therefore $m\widehat{BD} \approx 31°$.

Monitoring Progress Help in English and Spanish at *BigIdeasMath.com*

Find the value of *x*.

6.

7.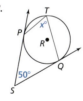

8. You are on top of Mount Rainier on a clear day. You are about 2.73 miles above sea level at point B. Find $m\widehat{CD}$, which represents the part of Earth that you can see.

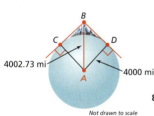

Not drawn to scale

Extra Example 4

Use the information in Example 4. A flash occurs 100 miles above Earth at point C. Find the measure of $\widehat{BD}$, the portion of Earth from which the flash is visible.

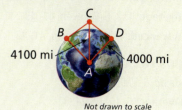

Not drawn to scale

about 25°

MONITORING PROGRESS ANSWERS

6. 60
7. 65
8. about 4°

Laurie's Notes Teacher Actions

- **Reason Abstractly and Quantitatively** and **Paired Verbal Fluency:** Have students pair up and follow the protocol described on page T-224. Ask partner B to explain what he or she knows about the problem and a strategy for solving. When it is partner A's turn, he or she should build on the reasoning of partner B or explain why partner B's strategy is faulty. Discuss solutions as a class when students have finished.

Closure

- **Point of Most Significance:** Ask students to identify, aloud or on paper, the most significant point (or part) in the lesson that aided their learning.

Assignment Guide and Homework Check

ASSIGNMENT

Basic: 1, 2, 3–23 odd, 30, 34, 41–43
Average: 1, 2, 6–34 even, 41–43
Advanced: 1, 2, 6–40 even, 41–43

HOMEWORK CHECK

Basic: 3, 7, 9, 11, 23
Average: 6, 10, 12, 24, 32
Advanced: 10, 12, 24, 32, 40

ANSWERS

1. outside
2. Find the measure of the central angle that intercepts the same arc and subtract it from 180°.
3. 130°
4. 234°
5. 130°
6. 110°
7. 115
8. 90
9. 56
10. 38
11. 40
12. 11
13. 34
14. 15
15. $\angle SUT$ is not a central angle; $m\angle SUT = \frac{1}{2}(m\widehat{QR} + m\widehat{ST}) = 41.5°$
16. The $\frac{1}{2}$ was left out of the equation; $m\angle 1 = \frac{1}{2}(122 - 70) = 26°$
17. 60°; Because the sum of the angles of a triangle always equals 180°, solve the equation $90 + 30 + x = 180$.
18. 60°; When a chord intersects a tangent line, the angle formed is half of the measure of the intercepted arc, which in this case is 120°.
19. 30°; Because the sum of the angles of a triangle always equals 180°, solve the equation $60 + 90 + x = 180$.
20. 90°; Because $\angle 2$ and $\angle 5$ sum to 90° and form a straight line with $\angle 4$, the angles are supplementary.
21. 30°; This angle is complementary to $\angle 2$, which is 60°.
22. 60°; This angle is congruent to $\angle 1$.

10.5 Exercises

Dynamic Solutions available at *BigIdeasMath.com*

Vocabulary and Core Concept Check

1. **COMPLETE THE SENTENCE** Points A, B, C, and D are on a circle, and $\overleftrightarrow{AB}$ intersects $\overleftrightarrow{CD}$ at point P. If $m\angle APC = \frac{1}{2}(m\widehat{BD} - m\widehat{AC})$, then point P is _____ the circle.

2. **WRITING** Explain how to find the measure of a circumscribed angle.

Monitoring Progress and Modeling with Mathematics

In Exercises 3–6, line t is tangent to the circle. Find the indicated measure. *(See Example 1.)*

3. $m\widehat{AB}$

4. $m\widehat{DEF}$

5. $m\angle 1$

6. $m\angle 3$

In Exercises 7–14, find the value of x. *(See Examples 2 and 3.)*

7.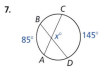

8.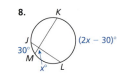

9.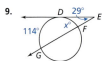

10.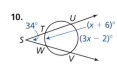

11.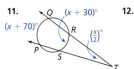

12.

13.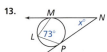

14.

ERROR ANALYSIS In Exercises 15 and 16, describe and correct the error in finding the angle measure.

15.

16.

In Exercises 17–22, find the indicated angle measure. Justify your answer.

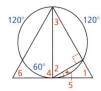

17. $m\angle 1$
18. $m\angle 2$
19. $m\angle 3$
20. $m\angle 4$
21. $m\angle 5$
22. $m\angle 6$

610 Chapter 10 Circles

23. **PROBLEM SOLVING** You are flying in a hot air balloon about 1.2 miles above the ground. Find the measure of the arc that represents the part of Earth you can see. The radius of Earth is about 4000 miles. *(See Example 4.)*

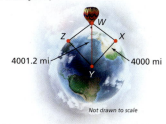

24. **PROBLEM SOLVING** You are watching fireworks over San Diego Bay S as you sail away in a boat. The highest point the fireworks reach F is about 0.2 mile above the bay. Your eyes E are about 0.01 mile above the water. At point B you can no longer see the fireworks because of the curvature of Earth. The radius of Earth is about 4000 miles, and $\overline{FE}$ is tangent to Earth at point T. Find $m\overset{\frown}{SB}$. Round your answer to the nearest tenth.

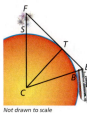

25. **MATHEMATICAL CONNECTIONS** In the diagram, $\overrightarrow{BA}$ is tangent to $\odot E$. Write an algebraic expression for $m\overset{\frown}{CD}$ in terms of x. Then find $m\overset{\frown}{CD}$.

26. **MATHEMATICAL CONNECTIONS** The circles in the diagram are concentric. Write an algebraic expression for c in terms of a and b.

27. **ABSTRACT REASONING** In the diagram, $\overrightarrow{PL}$ is tangent to the circle, and $\overline{KJ}$ is a diameter. What is the range of possible angle measures of $\angle LPJ$? Explain your reasoning.

28. **ABSTRACT REASONING** In the diagram, $\overline{AB}$ is any chord that is not a diameter of the circle. Line m is tangent to the circle at point A. What is the range of possible values of x? Explain your reasoning. (The diagram is not drawn to scale.)

29. **PROOF** In the diagram, $\overrightarrow{JL}$ and $\overrightarrow{NL}$ are secant lines that intersect at point L. Prove that $m\angle JPN > m\angle JLN$.

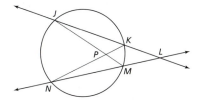

30. **MAKING AN ARGUMENT** Your friend claims that it is possible for a circumscribed angle to have the same measure as its intercepted arc. Is your friend correct? Explain your reasoning.

31. **REASONING** Points A and B are on a circle, and t is a tangent line containing A and another point C.

    a. Draw two diagrams that illustrate this situation.

    b. Write an equation for $m\overset{\frown}{AB}$ in terms of $m\angle BAC$ for each diagram.

    c. For what measure of $\angle BAC$ can you use either equation to find $m\overset{\frown}{AB}$? Explain.

32. **REASONING** $\triangle XYZ$ is an equilateral triangle inscribed in $\odot P$. $\overline{AB}$ is tangent to $\odot P$ at point X, $\overline{BC}$ is tangent to $\odot P$ at point Y, and $\overline{AC}$ is tangent to $\odot P$ at point Z. Draw a diagram that illustrates this situation. Then classify $\triangle ABC$ by its angles and sides. Justify your answer.

Section 10.5 Angle Relationships in Circles 611

ANSWERS

33–39. See Additional Answers.

40. $m\widehat{AB} = 85°$, $m\widehat{ED} = 75°$;
Sample answer: $m\angle DHC = 65°$, $m\widehat{AE} = 45°$, $m\widehat{AF} = 25°$, $m\angle J = 30°$, so $m\widehat{AB} = 85°$; $m\angle FGD = 90°$, $m\widehat{FD} = 95°$, and $m\widehat{ED} = 75°$.

41. $x = -4$, $x = 3$

42. $x = 6 \pm \sqrt{71}$

43. $x = -3$, $x = -1$

Mini-Assessment

1. $\overrightarrow{WX}$ is tangent to the circle. Find $m\widehat{XYZ}$.

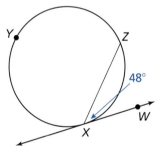

$m\widehat{XYZ} = 264°$

2. Find the value of x.

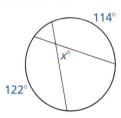

$x = 62$

3. Find the values of the variables.

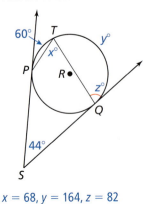

$x = 68$, $y = 164$, $z = 82$

---

33. **PROVING A THEOREM** To prove the Tangent and Intersected Chord Theorem, you must prove three cases.

    a. The diagram shows the case where $\overline{AB}$ contains the center of the circle. Use the Tangent Line to Circle Theorem to write a paragraph proof for this case.

    b. Draw a diagram and write a proof for the case where the center of the circle is in the interior of $\angle CAB$.

    c. Draw a diagram and write a proof for the case where the center of the circle is in the exterior of $\angle CAB$.

34. **HOW DO YOU SEE IT?** In the diagram, television cameras are positioned at A and B to record what happens on stage. The stage is an arc of $\odot A$. You would like the camera at B to have a 30° view of the stage. Should you move the camera closer or farther away? Explain your reasoning.

    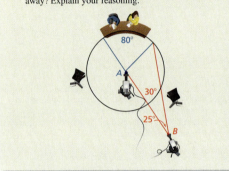

35. **PROVING A THEOREM** Write a proof of the Angles Inside the Circle Theorem.

    Given Chords $\overline{AC}$ and $\overline{BD}$ intersect inside a circle.

    Prove $m\angle 1 = \frac{1}{2}(m\widehat{DC} + m\widehat{AB})$

36. **THOUGHT PROVOKING** In the figure, $\overleftrightarrow{BP}$ and $\overleftrightarrow{CP}$ are tangent to the circle. Point A is any point on the major arc formed by the endpoints of the chord $\overline{BC}$. Label all congruent angles in the figure. Justify your reasoning.

    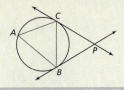

37. **PROVING A THEOREM** Use the diagram below to prove the Angles Outside the Circle Theorem for the case of a tangent and a secant. Then copy the diagrams for the other two cases on page 607 and draw appropriate auxiliary segments. Use your diagrams to prove each case.

38. **PROVING A THEOREM** Prove that the Circumscribed Angle Theorem follows from the Angles Outside the Circle Theorem.

In Exercises 39 and 40, find the indicated measure(s). Justify your answer.

39. Find $m\angle P$ when $m\widehat{WZY} = 200°$.

    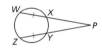

40. Find $m\widehat{AB}$ and $m\widehat{ED}$.

    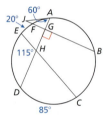

Maintaining Mathematical Proficiency
Reviewing what you learned in previous grades and lessons

Solve the equation using the Quadratic Formula. *(Section 4.5)*

41. $x^2 + x = 12$
42. $x^2 = 12x + 35$
43. $-3 = x^2 + 4x$

If students need help...	If students got it...
Resources by Chapter • Practice A and Practice B • Puzzle Time	**Resources by Chapter** • Enrichment and Extension • Cumulative Review
Student Journal • Practice	Start the *next* Section
Differentiating the Lesson Skills Review Handbook	

Laurie's Notes

Overview of Section 10.6

Introduction
- This lesson is about segment lengths in a circle when chords, secants, and tangents intersect a circle. Students may find the lesson difficult, in part because of the language involved in trying to describe the segments. Help students to focus on what the theorems state about the product of different segments.

Teaching Strategy
- This lesson is a good time to return to transformations and look at the proofs from a transformational approach versus synthetically.
- To model this, consider the Segments of Chords Theorem. It is possible to use transformations to show that the triangles are similar and therefore corresponding sides are proportional.
- Begin by constructing two chords and then drawing segments to form △DCF and △BEF (See Figure 1.)
- Construct the bisector of ∠CFE. (See Figure 2.)
- Reflect △DCF in the angle bisector. (See Figure 3.) At this point, students can dilate to map △DC'F' onto △BEF.

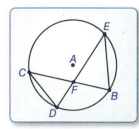
Figure 1

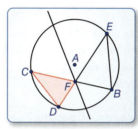
Figure 2

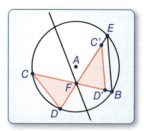
Figure 3

- You want students to give explanations of the constructions along the way. A common error is that students try to rotate △DCF about point F, and this approach does not work.

Pacing Suggestion
- The explorations allow students to make conjectures about segments that are formed when two chords intersect or two secants intersect. In the formal lesson, the discussion of the theorems should not take long.

What Your Students Will Learn

- Use segments of chords, tangents, and secants of circles to find lengths of line segments.

Laurie's Notes

Exploration

Motivate

- Pose the following problem: You are parasailing above the ocean and looking out at the horizon. If you know your height above the water and the radius of Earth, do you think you can calculate how far you can see?
- Students should recognize this as an application of the Pythagorean Theorem.
- Explain to students that they will look at an alternate way to solve this problem using a new theorem in this lesson.

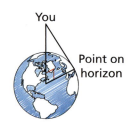

Exploration Note

- The two explorations about segment lengths in a circle are more directed than other explorations in this chapter. It is less likely that students would discover the relationship without a bit of guidance.
- Unlike other explorations, measurement with a ruler may not be accurate enough for students to observe a pattern and make a conjecture. Dynamic geometry software is recommended.

Exploration 1

- Have partners perform the construction and take the measurements as directed.
- Given the dynamic nature of the software, students should be able to click and drag on one of the points of the circle to investigate many cases.
- ? "What do you notice about the products of the segments of the chords?" The products are the same.
- **Construct Viable Arguments and Critique the Reasoning of Others:** Have students state their conjectures and explain their reasoning. Discuss.
- ? "Is your conjecture still true if one or both of the chords is a diameter?" yes
- ? "Do you have an idea as to how you could prove your conjecture?" Answers will vary.

Exploration 2

- Have partners perform the construction and take the measurements as directed.
- Given the dynamic nature of the software, students should be able to click and drag on one of the points of the circle to investigate many cases.
- ? "What do you notice about the products of the segments of the secants?" The products are the same.
- **Construct Viable Arguments and Critique the Reasoning of Others:** Have students state their conjectures and explain their reasoning. Discuss.
- ? "Is your conjecture still true if one of the secants passes through the center of the circle?" yes
- ? "Do you have an idea as to how you could prove your conjecture?" Answers will vary.

Communicate Your Answer

- Listen for both conjectures made by students as a result of the two explorations.

Connecting to Next Step

- This lesson contains three theorems, two of which students have now explored using dynamic geometry software.

10.6 Segment Relationships in Circles

Essential Question What relationships exist among the segments formed by two intersecting chords or among segments of two secants that intersect outside a circle?

EXPLORATION 1 Segments Formed by Two Intersecting Chords

Work with a partner. Use dynamic geometry software.

a. Construct two chords $\overline{BC}$ and $\overline{DE}$ that intersect in the interior of a circle at a point F.

b. Find the segment lengths BF, CF, DF, and EF and complete the table. What do you observe?

BF	CF	BF · CF

DF	EF	DF · EF

Sample

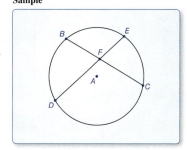

> **REASONING ABSTRACTLY**
> To be proficient in math, you need to make sense of quantities and their relationships in problem situations.

c. Repeat parts (a) and (b) several times. Write a conjecture about your results.

EXPLORATION 2 Secants Intersecting Outside a Circle

Work with a partner. Use dynamic geometry software.

a. Construct two secants $\overleftrightarrow{BC}$ and $\overleftrightarrow{BD}$ that intersect at a point B outside a circle, as shown.

b. Find the segment lengths BE, BC, BF, and BD, and complete the table. What do you observe?

BE	BC	BE · BC

BF	BD	BF · BD

Sample

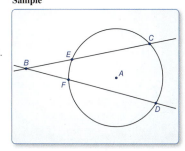

c. Repeat parts (a) and (b) several times. Write a conjecture about your results.

Communicate Your Answer

3. What relationships exist among the segments formed by two intersecting chords or among segments of two secants that intersect outside a circle?

4. Find the segment length AF in the figure at the left.

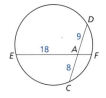

Section 10.6   Segment Relationships in Circles   613

English Language Learners
Graphic Organizer
Have students make a Concept Circle that shows how to solve problems using the Segments of Chords Theorem.

Extra Example 1

Find *AB* and *PQ*.

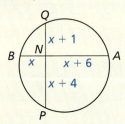

$AB = 14$, $PQ = 13$

MONITORING PROGRESS ANSWERS

1. 8
2. 5

10.6 Lesson

Core Vocabulary

segments of a chord, *p. 614*
tangent segment, *p. 615*
secant segment, *p. 615*
external segment, *p. 615*

What You Will Learn

▶ Use segments of chords, tangents, and secants.

Using Segments of Chords, Tangents, and Secants

When two chords intersect in the interior of a circle, each chord is divided into two segments that are called **segments of the chord**.

Theorem

Segments of Chords Theorem

If two chords intersect in the interior of a circle, then the product of the lengths of the segments of one chord is equal to the product of the lengths of the segments of the other chord.

Proof Ex. 19, p. 618

$EA \cdot EB = EC \cdot ED$

EXAMPLE 1 Using Segments of Chords

Find *ML* and *JK*.

SOLUTION

$NK \cdot NJ = NL \cdot NM$	Segments of Chords Theorem
$x \cdot (x + 4) = (x + 1) \cdot (x + 2)$	Substitute.
$x^2 + 4x = x^2 + 3x + 2$	Simplify.
$4x = 3x + 2$	Subtract x^2 from each side.
$x = 2$	Subtract $3x$ from each side.

Find *ML* and *JK* by substitution.

$ML = (x + 2) + (x + 1)$ $\qquad JK = x + (x + 4)$
$ = 2 + 2 + 2 + 1$ $\qquad = 2 + 2 + 4$
$ = 7$ $\qquad = 8$

▶ So, *ML* = 7 and *JK* = 8.

Monitoring Progress 🔊 Help in English and Spanish at *BigIdeasMath.com*

Find the value of *x*.

1.
2.

Laurie's Notes | Teacher Actions

- **Make Sense of Problems and Persevere in Solving Them:** The first theorem in the lesson was explored on the previous page.
  - ❓ State the theorem. "Do you have an idea as to how you could prove the theorem?" *Answers will vary.*
  - If there are no references to similar triangles, draw $\overline{AD}$ and $\overline{CB}$ and repeat the question. Give *Wait Time* for students to discuss the proof with their partners. They will recognize that the triangles are similar by the Angle-Angle Similarity Theorem.
- **Popsicle Sticks:** Example 1 is an opportunity for students to review equation-solving skills. Use *Popsicle Sticks* to solicit a solution.

 **Core Concept**

Tangent Segment and Secant Segment

A **tangent segment** is a segment that is tangent to a circle at an endpoint. A **secant segment** is a segment that contains a chord of a circle and has exactly one endpoint outside the circle. The part of a secant segment that is outside the circle is called an **external segment**.

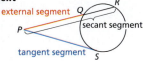

$\overline{PS}$ is a tangent segment.
$\overline{PR}$ is a secant segment.
$\overline{PQ}$ is the external segment of $\overline{PR}$.

Extra Example 2
Find the value of x.

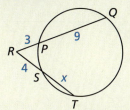

$x = 5$

MONITORING PROGRESS ANSWERS

3. 13
4. 3

 Theorem

Segments of Secants Theorem

If two secant segments share the same endpoint outside a circle, then the product of the lengths of one secant segment and its external segment equals the product of the lengths of the other secant segment and its external segment.

Proof Ex. 20, p. 618

$EA \cdot EB = EC \cdot ED$

EXAMPLE 2 Using Segments of Secants

Find the value of x.

SOLUTION

$RP \cdot RQ = RS \cdot RT$	Segments of Secants Theorem
$9 \cdot (11 + 9) = 10 \cdot (x + 10)$	Substitute.
$180 = 10x + 100$	Simplify.
$80 = 10x$	Subtract 100 from each side.
$8 = x$	Divide each side by 10.

▶ The value of x is 8.

Monitoring Progress Help in English and Spanish at *BigIdeasMath.com*

Find the value of x.

3.   4.

Section 10.6   Segment Relationships in Circles   615

Laurie's Notes — Teacher Actions

- Write the *Core Concept*. Color-coding the various segments is helpful for some students.
- State the theorem and ask, "Do you have an idea as to how you could prove the theorem?" Students will likely suspect similar triangles are involved. Give partners time to discuss how the theorem might be proven.
- **Thumbs Up:** Have students work independently to solve Example 2. Ask students to self-assess with *Thumbs Up*.

Extra Example 3
Find WX.

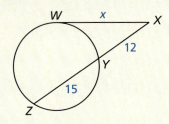

WX = 18

Extra Example 4
Find the radius of the circle.

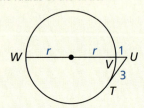

r = 4

MONITORING PROGRESS ANSWERS
5. 2
6. 4.8
7. 8
8. 36.75 ft

Theorem

Segments of Secants and Tangents Theorem

If a secant segment and a tangent segment share an endpoint outside a circle, then the product of the lengths of the secant segment and its external segment equals the square of the length of the tangent segment.

$EA^2 = EC \cdot ED$

Proof Exs. 21 and 22, p. 618

EXAMPLE 3 Using Segments of Secants and Tangents

Find RS.

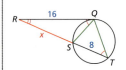

SOLUTION

$RQ^2 = RS \cdot RT$ Segments of Secants and Tangents Theorem

$16^2 = x \cdot (x + 8)$ Substitute.

$256 = x^2 + 8x$ Simplify.

$0 = x^2 + 8x - 256$ Write in standard form.

$x = \dfrac{-8 \pm \sqrt{8^2 - 4(1)(-256)}}{2(1)}$ Use Quadratic Formula.

$x = -4 \pm 4\sqrt{17}$ Simplify.

Use the positive solution because lengths cannot be negative.

▶ So, $x = -4 + 4\sqrt{17} \approx 12.49$, and $RS \approx 12.49$.

ANOTHER WAY

In Example 3, you can draw segments $\overline{QS}$ and $\overline{QT}$.

Because ∠RQS and ∠RTQ intercept the same arc, they are congruent. By the Reflexive Property of Congruence, ∠QRS ≅ ∠TRQ. So, △RSQ ~ △RQT by the AA Similarity Theorem. You can use this fact to write and solve a proportion to find x.

EXAMPLE 4 Finding the Radius of a Circle

Find the radius of the aquarium tank.

SOLUTION

$CB^2 = CE \cdot CD$ Segments of Secants and Tangents Theorem

$20^2 = 8 \cdot (2r + 8)$ Substitute.

$400 = 16r + 64$ Simplify.

$336 = 16r$ Subtract 64 from each side.

$21 = r$ Divide each side by 16.

▶ So, the radius of the tank is 21 feet.

Monitoring Progress Help in English and Spanish at *BigIdeasMath.com*

Find the value of x.

5.   6.   7.

8. **WHAT IF?** In Example 4, CB = 35 feet and CE = 14 feet. Find the radius of the tank.

Laurie's Notes — Teacher Actions

- Write the theorem and again have a brief discussion of the proof. Similar triangles are again involved.
- Work through Example 3 as shown. In addition, help students set up the alternate solution method shown in the margin.
- If time permits, have students return to the problem posed in the *Motivate*. Suggest that you are parasailing 100 feet in the air.

Closure
- **Muddiest Point:** Ask students to identify, aloud or on paper, the muddiest point(s) about the lesson. What was difficult to understand?

10.6 Exercises

Dynamic Solutions available at BigIdeasMath.com

Vocabulary and Core Concept Check

1. **VOCABULARY** The part of the secant segment that is outside the circle is called a(n) _____.

2. **WRITING** Explain the difference between a tangent segment and a secant segment.

Monitoring Progress and Modeling with Mathematics

In Exercises 3–6, find the value of x. (See Example 1.)

3.
4.

5.
6.

In Exercises 7–10, find the value of x. (See Example 2.)

7.
8.

9.
10.

In Exercises 11–14, find the value of x. (See Example 3.)

11.
12.

13.
14.

15. **ERROR ANALYSIS** Describe and correct the error in finding CD.

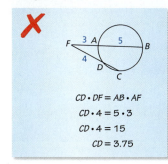

16. **MODELING WITH MATHEMATICS** The Cassini spacecraft is on a mission in orbit around Saturn until September 2017. Three of Saturn's moons, Tethys, Calypso, and Telesto, have nearly circular orbits of radius 295,000 kilometers. The diagram shows the positions of the moons and the spacecraft on one of Cassini's missions. Find the distance DB from Cassini to Tethys when $\overline{AD}$ is tangent to the circular orbit. (See Example 4.)

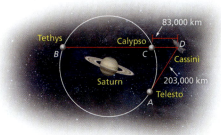

Section 10.6   Segment Relationships in Circles   617

Dynamic Teaching Tools
- Dynamic Assessment & Progress Monitoring Tool
- Interactive Whiteboard Lesson Library
- Dynamic Classroom with Dynamic Investigations

ANSWERS
17. about 124.5 ft
18. 4 cm/sec; It takes the sparkles 3 seconds to move from point C to point D. Because $CN = 12$ and the sparkles have 3 seconds to move from point C to point N, the sparkles need to move at a speed of 4 centimeters per second from point C to point N.
19–30. See Additional Answers.

Mini-Assessment

1. Find the value of x.

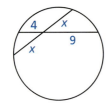

$x = 6$

2. Find the value of x.

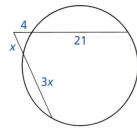

$x = 5$

3. Find BC and DC.

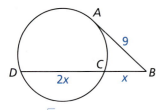

$BC = 3\sqrt{3} \approx 5.2$;
$DC = 6\sqrt{3} \approx 10.4$

4. The circle shows the cross-section of a pipe. Find the radius of the pipe.

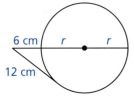

9 centimeters

---

17. **MODELING WITH MATHEMATICS** The circular stone mound in Ireland called Newgrange has a diameter of 250 feet. A passage 62 feet long leads toward the center of the mound. Find the perpendicular distance x from the end of the passage to either side of the mound.

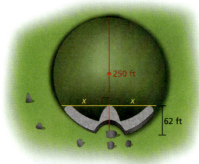

18. **MODELING WITH MATHEMATICS** You are designing an animated logo for your website. Sparkles leave point C and move to the outer circle along the segments shown so that all of the sparkles reach the outer circle at the same time. Sparkles travel from point C to point D at 2 centimeters per second. How fast should sparkles move from point C to point N? Explain.

19. **PROVING A THEOREM** Write a two-column proof of the Segments of Chords Theorem.

Plan for Proof Use the diagram from page 614. Draw $\overline{AC}$ and $\overline{DB}$. Show that $\triangle EAC$ and $\triangle EDB$ are similar. Use the fact that corresponding side lengths in similar triangles are proportional.

20. **PROVING A THEOREM** Prove the Segments of Secants Theorem. (*Hint*: Draw a diagram and add auxiliary line segments to form similar triangles.)

21. **PROVING A THEOREM** Use the Tangent Line to Circle Theorem to prove the Segments of Secants and Tangents Theorem for the special case when the secant segment contains the center of the circle.

22. **PROVING A THEOREM** Prove the Segments of Secants and Tangents Theorem. (*Hint*: Draw a diagram and add auxiliary line segments to form similar triangles.)

23. **WRITING EQUATIONS** In the diagram of the water well, AB, AD, and DE are known. Write an equation for BC using these three measurements.

24. **HOW DO YOU SEE IT?** Which two theorems would you need to use to find PQ? Explain your reasoning.

25. **CRITICAL THINKING** In the figure, $AB = 12$, $BC = 8$, $DE = 6$, $PD = 4$, and A is a point of tangency. Find the radius of $\odot P$.

26. **THOUGHT PROVOKING** Circumscribe a triangle about a circle. Then, using the points of tangency, inscribe a triangle in the circle. Must it be true that the two triangles are similar? Explain your reasoning.

Maintaining Mathematical Proficiency Reviewing what you learned in previous grades and lessons

Solve the equation by completing the square. (Section 4.4)

27. $x^2 + 4x = 45$
28. $x^2 - 2x - 1 = 8$
29. $2x^2 + 12x + 20 = 34$
30. $-4x^2 + 8x + 44 = 16$

618 Chapter 10 Circles

If students need help...	If students got it...
Resources by Chapter • Practice A and Practice B • Puzzle Time	Resources by Chapter • Enrichment and Extension • Cumulative Review
Student Journal • Practice	Start the *next* Section
Differentiating the Lesson Skills Review Handbook	

Laurie's Notes

Overview of Section 10.7

Introduction
- This last section of the chapter is about the standard equation of a circle. Students derive the equation in the exploration and then use the equation in the formal lesson. In addition to writing the equation, students will graph circles and solve systems of equations that include circles.
- The lesson includes a real-life application involving seismographs.

Teaching Strategy
- Students are always curious about how to graph a circle on the graphing calculator. Because the equation of a circle is not a function, the equation has to be written as a split rule function.
- For Example 5 in the lesson, there are three circles to graph. The first circle has center $(-2, 2.5)$ and radius 7. The equation is $(x + 2)^2 + (y - 2.5)^2 = 7^2$. Solve this equation for y.

$$(x + 2)^2 + (y - 2.5)^2 = 7^2$$
$$(y - 2.5)^2 = 49 - (x + 2)^2$$
$$y - 2.5 = \pm \sqrt{49 - (x + 2)^2}$$
$$y = \pm \sqrt{49 - (x + 2)^2} + 2.5$$

- Graph the two cases in Y1 and Y2. Use a square viewing window to make the graph circular.

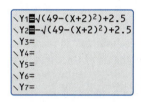 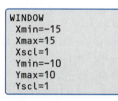 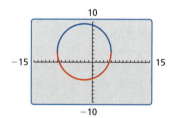

Pacing Suggestion
- Have students work through all three explorations and the questions that follow. You should be able to omit the *Core Concept* and Example 1.

What Your Students Will Learn

- Write and graph equations of circles in the coordinate plane.
- Write coordinate proofs involving circles.
- Solve real-life problems using graphs of circles.
- Solve nonlinear systems involving equations of circles and lines.

Laurie's Notes

Exploration

Motivate
- ❓ "What is a seismograph and what does it do?" Answers will vary.
- **FYI:** A rock thrown into a pond sends waves rippling out in all directions from the point of impact. Just as this impact sets waves in motion, an earthquake generates seismic waves that radiate out through the earth.
- Seismographs are instruments that detect and measure vibrations within the earth. Seismograph recordings are known as seismograms. The prefix *seismos* comes from the Greek, meaning "shock" or "earthquake." The devices are best known for studying earthquakes, although some types are used for underground surveying.

Exploration Note
- The approach in the first two explorations is to use the equation generated by the software to make a conjecture about the equation of a circle in standard form. In the third exploration, students derive the equation using the distance formula.

Exploration 1
- This first exploration will not take long to complete. Constructing concentric circles and examining the equations should be quick and easy for students at this point.
- ❓ **Look For and Express Regularity in Repeated Reasoning:** "What did you observe about the equations of the circles?" The equations were of the form $x^2 + y^2 = r^2$.
- ❓ "What would be the equation of a circle with center (0, 0) and radius 10?" $x^2 + y^2 = 100$

Exploration 2
- **Reason Abstractly and Quantitatively:** Students should recognize that in trying to see a pattern, they might want to use lattice points for the centers of the circle. Once they have a conjecture, they might want to test it with a random point.
- As you circulate, you might probe students about centers in all four quadrants if they have not constructed circles with centers in different quadrants or on either axis.
- ❓ **Look For and Express Regularity in Repeated Reasoning:** "What did you observe about the equations of the circles?" The equations were of the form $(x - h)^2 + (y - k)^2 = 4$.
- ❓ "What would be the equation of a circle with center (5, −2) and radius 2?" $(x - 5)^2 + (y + 2)^2 = 4$

Exploration 3
- **Construct Viable Arguments and Critique the Reasoning of Others:** This exploration is a great way to give evidence for the conjecture made in the last exploration, and it is nice to revisit an earlier skill.

Communicate Your Answer
- Listen for precise language as students state the equation of a circle with center (h, k) and radius r.

Connecting to Next Step
- Students should be ready to write an equation of a circle when given the radius and center.

10.7 Circles in the Coordinate Plane

Essential Question What is the equation of a circle with center (h, k) and radius r in the coordinate plane?

EXPLORATION 1 The Equation of a Circle with Center at the Origin

Work with a partner. Use dynamic geometry software to construct and determine the equations of circles centered at $(0, 0)$ in the coordinate plane, as described below.

Radius	Equation of circle
1	
2	

a. Complete the first two rows of the table for circles with the given radii. Complete the other rows for circles with radii of your choice.

b. Write an equation of a circle with center $(0, 0)$ and radius r.

EXPLORATION 2 The Equation of a Circle with Center (h, k)

Work with a partner. Use dynamic geometry software to construct and determine the equations of circles of radius 2 in the coordinate plane, as described below.

Center	Equation of circle
$(0, 0)$	
$(2, 0)$	

a. Complete the first two rows of the table for circles with the given centers. Complete the other rows for circles with centers of your choice.

b. Write an equation of a circle with center (h, k) and radius 2.

c. Write an equation of a circle with center (h, k) and radius r.

EXPLORATION 3 Deriving the Standard Equation of a Circle

Work with a partner. Consider a circle with radius r and center (h, k).

Write the Distance Formula to represent the distance d between a point (x, y) on the circle and the center (h, k) of the circle. Then square each side of the Distance Formula equation.

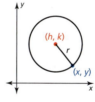

How does your result compare with the equation you wrote in part (c) of Exploration 2?

MAKING SENSE OF PROBLEMS
To be proficient in math, you need to explain correspondences between equations and graphs.

Communicate Your Answer

4. What is the equation of a circle with center (h, k) and radius r in the coordinate plane?

5. Write an equation of the circle with center $(4, -1)$ and radius 3.

ANSWERS

1. a. Sample answer:

Radius	Equation of Circle
1	$x^2 + y^2 = 1$
2	$x^2 + y^2 = 4$
3	$x^2 + y^2 = 9$
5	$x^2 + y^2 = 25$
6	$x^2 + y^2 = 36$
9	$x^2 + y^2 = 81$

   b. $x^2 + y^2 = r^2$

2. a. Sample answer:

Center	Equation of Circle
$(0, 0)$	$x^2 + y^2 = 4$
$(2, 0)$	$(x - 2)^2 + y^2 = 4$
$(0, 3)$	$x^2 + (y - 3)^2 = 4$
$(2, -3)$	$(x - 2)^2 + (y + 3)^2 = 4$
$(-1, 4)$	$(x + 1)^2 + (y - 4)^2 = 4$
$(-3, -6)$	$(x + 3)^2 + (y + 6)^2 = 4$

   b. $(x - h)^2 + (y - k)^2 = 4$

   c. $(x - h)^2 + (y - k)^2 = r^2$

3. $\sqrt{(x - h)^2 + (y - k)^2} = d$;
   $(x - h)^2 + (y - k)^2 = d^2$;
   If $d = r$, then the equations are the same.

4. $(x - h)^2 + (y - k)^2 = r^2$

5. $(x - 4)^2 + (y + 1)^2 = 9$

English Language Learners

Comprehension
Make sure students understand that the standard equation of a circle, $(x - h)^2 + (y - k)^2 = r^2$, contains variables and constants. Write the equation on the board. Then have students discuss which terms are variables (x, y) and which terms are constants (h, k, r). Be sure students understand that the constants determine the center (h, k) and the radius r of the circle. Show that for a given circle, the values of h, k, and r do not change, but the values of x and y change to represent the different points of the circle.

Extra Example 1
Write the standard equation of each circle.

a.
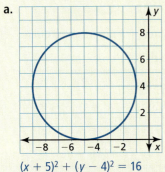
$(x + 5)^2 + (y - 4)^2 = 16$

b. a circle with center at the origin and radius 3.5 $x^2 + y^2 = 12.25$

MONITORING PROGRESS ANSWERS
1. $x^2 + y^2 = 6.25$
2. $(x + 2)^2 + (y - 5)^2 = 49$

10.7 Lesson

Core Vocabulary
standard equation of a circle, p. 620

Previous
completing the square

What You Will Learn
▶ Write and graph equations of circles.
▶ Write coordinate proofs involving circles.
▶ Solve real-life problems using graphs of circles.
▶ Solve nonlinear systems.

Writing and Graphing Equations of Circles

Let (x, y) represent any point on a circle with center at the origin and radius r. By the Pythagorean Theorem,

$$x^2 + y^2 = r^2.$$

This is the equation of a circle with center at the origin and radius r.

Core Concept

Standard Equation of a Circle

Let (x, y) represent any point on a circle with center (h, k) and radius r. By the Pythagorean Theorem,

$$(x - h)^2 + (y - k)^2 = r^2.$$

This is the **standard equation of a circle** with center (h, k) and radius r.

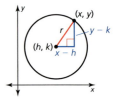

EXAMPLE 1 Writing the Standard Equation of a Circle

Write the standard equation of each circle.

a. the circle shown at the left

b. a circle with center $(0, -9)$ and radius 4.2

SOLUTION

a. The radius is 3, and the center is at the origin.

$(x - h)^2 + (y - k)^2 = r^2$	Standard equation of a circle
$(x - 0)^2 + (y - 0)^2 = 3^2$	Substitute.
$x^2 + y^2 = 9$	Simplify.

▶ The standard equation of the circle is $x^2 + y^2 = 9$.

b. The radius is 4.2, and the center is at $(0, -9)$.

$(x - h)^2 + (y - k)^2 = r^2$
$(x - 0)^2 + [y - (-9)]^2 = 4.2^2$
$x^2 + (y + 9)^2 = 17.64$

▶ The standard equation of the circle is $x^2 + (y + 9)^2 = 17.64$.

Monitoring Progress
Help in English and Spanish at *BigIdeasMath.com*

Write the standard equation of the circle with the given center and radius.

1. center: $(0, 0)$, radius: 2.5
2. center: $(-2, 5)$, radius: 7

Laurie's Notes Teacher Actions

- Write the Core Concept, the standard equation of a circle. Explain to students that it is desirable to leave the equation in this form versus squaring each binomial. Information about the graph is easy to see when it is left in standard form.

COMMON ERROR Students may use the incorrect sign. The equation is in the form of $(x - h)$. When h (or k) is a negative number and the subtraction is performed, it is written $(x + |h|)$.

EXAMPLE 2 **Writing the Standard Equation of a Circle**

The point $(-5, 6)$ is on a circle with center $(-1, 3)$. Write the standard equation of the circle.

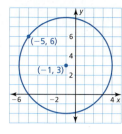

SOLUTION

To write the standard equation, you need to know the values of h, k, and r. To find r, find the distance between the center and the point $(-5, 6)$ on the circle.

$r = \sqrt{[-5 - (-1)]^2 + (6 - 3)^2}$ Distance Formula

$= \sqrt{(-4)^2 + 3^2}$ Simplify.

$= 5$ Simplify.

Substitute the values for the center and the radius into the standard equation of a circle.

$(x - h)^2 + (y - k)^2 = r^2$ Standard equation of a circle

$[x - (-1)]^2 + (y - 3)^2 = 5^2$ Substitute $(h, k) = (-1, 3)$ and $r = 5$.

$(x + 1)^2 + (y - 3)^2 = 25$ Simplify.

▶ The standard equation of the circle is $(x + 1)^2 + (y - 3)^2 = 25$.

CONNECTIONS TO ALGEBRA

Recall from Section 4.4 that to complete the square for the expression $x^2 + bx$, you add the square of half the coefficient of the term bx.

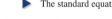

EXAMPLE 3 **Graphing a Circle**

The equation of a circle is $x^2 + y^2 - 8x + 4y - 16 = 0$. Find the center and the radius of the circle. Then graph the circle.

SOLUTION

You can write the equation in standard form by completing the square on the x-terms and the y-terms.

$x^2 + y^2 - 8x + 4y - 16 = 0$ Equation of circle

$x^2 - 8x + y^2 + 4y = 16$ Isolate constant. Group terms.

$x^2 - 8x + 16 + y^2 + 4y + 4 = 16 + 16 + 4$ Complete the square twice.

$(x - 4)^2 + (y + 2)^2 = 36$ Factor left side. Simplify right side.

$(x - 4)^2 + [y - (-2)]^2 = 6^2$ Rewrite the equation to find the center and the radius.

▶ The center is $(4, -2)$, and the radius is 6. Use a compass to graph the circle.

Monitoring Progress 🔊 Help in English and Spanish at *BigIdeasMath.com*

3. The point $(3, 4)$ is on a circle with center $(1, 4)$. Write the standard equation of the circle.

4. The equation of a circle is $x^2 + y^2 - 8x + 6y + 9 = 0$. Find the center and the radius of the circle. Then graph the circle.

Section 10.7   Circles in the Coordinate Plane   621

Extra Example 2

The point $(4, 1)$ is on a circle with center $(1, 4)$. Write the standard equation of the circle. $(x - 1)^2 + (y - 4)^2 = 18$

Extra Example 3

The equation of a circle is $x^2 + y^2 - 2x + 6y - 6 = 0$. Find the center and the radius of the circle. Then graph the circle. center $(1, -3)$; radius 4

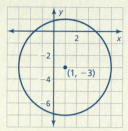

MONITORING PROGRESS ANSWERS

3. $(x - 1)^2 + (y - 4)^2 = 4$

4. center: $(4, -3)$, radius: 4

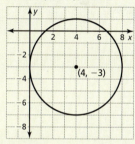

Laurie's Notes   Teacher Actions

- **?** Pose Example 2. "Will using any point on the circle result in the same standard equation for the circle?" yes
- **?** "What is the first step in solving this problem?" finding the radius of the circle
- Have students work with their partners to finish the example.
- Example 3 requires the technique of completing the square, twice.
- **Teaching Tip:** Write $(x^2 - 8x + \underline{?}) + (y^2 + 4y + \underline{?}) = 16$. "What needs to be in the blanks to make the quantity within each set of parentheses a perfect square trinomial?" 16 and 4
- The *Teaching Strategy* on page T-618 shows how to graph a circle in standard form on a graphing calculator.

Section 10.7   621

Extra Example 4

Prove or disprove that the point $(3, \sqrt{7})$ lies on the circle centered at the origin and containing the point (1, 4). The radius of the circle is $\sqrt{17} \approx 4.12$. The distance from $(3, \sqrt{7})$ to the origin is $\sqrt{16} = 4$. So, the point $(3, \sqrt{7})$ does not lie on the circle.

Extra Example 5

The epicenter of an earthquake is 10 miles away from $(-1, -3)$, 2 miles away from (5, 3), and 5 miles away from (2, 9). Find the coordinates of the epicenter.
about (5, 5)

MONITORING PROGRESS ANSWERS

5. The radius of the circle is 1.
$\sqrt{(1-0)^2 + (\sqrt{5}-0)^2} = \sqrt{6}$,
so $(1, \sqrt{5})$ does not lie on the circle.

6. Two circles can intersect at more than one point.

Writing Coordinate Proofs Involving Circles

EXAMPLE 4 Writing a Coordinate Proof Involving a Circle

Prove or disprove that the point $(\sqrt{2}, \sqrt{2})$ lies on the circle centered at the origin and containing the point (2, 0).

SOLUTION

The circle centered at the origin and containing the point (2, 0) has the following radius.

$r = \sqrt{(x-h)^2 + (y-k)^2} = \sqrt{(2-0)^2 + (0-0)^2} = 2$

So, a point lies on the circle if and only if the distance from that point to the origin is 2. The distance from $(\sqrt{2}, \sqrt{2})$ to (0, 0) is

$d = \sqrt{(\sqrt{2}-0)^2 + (\sqrt{2}-0)^2} = 2.$

▶ So, the point $(\sqrt{2}, \sqrt{2})$ lies on the circle centered at the origin and containing the point (2, 0).

Monitoring Progress Help in English and Spanish at BigIdeasMath.com

5. Prove or disprove that the point $(1, \sqrt{5})$ lies on the circle centered at the origin and containing the point (0, 1).

Solving Real-Life Problems

EXAMPLE 5 Using Graphs of Circles

The epicenter of an earthquake is the point on Earth's surface directly above the earthquake's origin. A seismograph can be used to determine the distance to the epicenter of an earthquake. Seismographs are needed in three different places to locate an earthquake's epicenter.

Use the seismograph readings from locations A, B, and C to find the epicenter of an earthquake.

- The epicenter is 7 miles away from $A(-2, 2.5)$.
- The epicenter is 4 miles away from $B(4, 6)$.
- The epicenter is 5 miles away from $C(3, -2.5)$.

SOLUTION

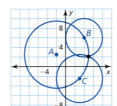

The set of all points equidistant from a given point is a circle, so the epicenter is located on each of the following circles.

⊙A with center $(-2, 2.5)$ and radius 7
⊙B with center (4, 6) and radius 4
⊙C with center $(3, -2.5)$ and radius 5

To find the epicenter, graph the circles on a coordinate plane where each unit corresponds to one mile. Find the point of intersection of the three circles.

▶ The epicenter is at about (5, 2).

Monitoring Progress Help in English and Spanish at BigIdeasMath.com

6. Why are three seismographs needed to locate an earthquake's epicenter?

Laurie's Notes — Teacher Actions

❓ Work through Example 4. "How else could you prove or disprove that $(\sqrt{2}, \sqrt{2})$ is on the circle?" See whether the coordinates satisfy the equation of the circle.

- **Use Appropriate Tools Strategically:** In Example 5, have students use a compass to construct the three circles on graph paper. Alternately, they can graph them using dynamic geometry software or a graphing calculator. The calculator technique is described in the *Teaching Strategy* on page T-618.

Solving Nonlinear Systems

In Section 4.8, you solved systems involving parabolas and lines. Now you can solve systems involving circles and lines.

EXAMPLE 6 Solving a Nonlinear System

Solve the system. $x^2 + y^2 = 1$ Equation 1

$y = \dfrac{x+1}{2}$ Equation 2

SOLUTION

Step 1 Equation 2 is already solved for y.

Step 2 Substitute $\dfrac{x+1}{2}$ for y in Equation 1 and solve for x.

$x^2 + y^2 = 1$	Equation 1
$x^2 + \left(\dfrac{x+1}{2}\right)^2 = 1$	Substitute $\dfrac{x+1}{2}$ for y.
$x^2 + \dfrac{(x+1)^2}{4} = 1$	Power of a Quotient Property
$4x^2 + (x+1)^2 = 4$	Multiply each side by 4.
$5x^2 + 2x - 3 = 0$	Write in standard form.
$x = \dfrac{-2 \pm \sqrt{2^2 - 4(5)(-3)}}{2(5)}$	Use the Quadratic Formula.
$x = \dfrac{-2 \pm \sqrt{64}}{10}$	Simplify.
$x = \dfrac{-2 \pm 8}{10}$	Evaluate the square root.
$x = \dfrac{-1 \pm 4}{5}$	Simplify.

So, $x = \dfrac{-1+4}{5} = \dfrac{3}{5}$ and $x = \dfrac{-1-4}{5} = -1$.

Step 3 Substitute $\dfrac{3}{5}$ and -1 for x in Equation 2 and solve for y.

$y = \dfrac{\frac{3}{5}+1}{2}$ Substitute for x in Equation 2. $y = \dfrac{-1+1}{2}$

$= \dfrac{4}{5}$ Simplify. $= 0$

 So, the solutions are $\left(\dfrac{3}{5}, \dfrac{4}{5}\right)$ and $(-1, 0)$.

Monitoring Progress *Help in English and Spanish at BigIdeasMath.com*

Solve the system.

7. $x^2 + y^2 = 1$
   $y = x$

8. $x^2 + y^2 = 2$
   $y = -x + 2$

9. $(x-1)^2 + y^2 = 4$
   $y = \dfrac{x+6}{2}$

Extra Example 6

Solve the system.
$x^2 + (y+2)^2 = 5$
$y = -2x + 3$
$(2, -1)$

MONITORING PROGRESS ANSWERS

7. $\left(\dfrac{\sqrt{2}}{2}, \dfrac{\sqrt{2}}{2}\right), \left(-\dfrac{\sqrt{2}}{2}, -\dfrac{\sqrt{2}}{2}\right)$
8. $(1, 1)$
9. no solution

Laurie's Notes   Teacher Actions

- ? "What does it mean to solve a system of equations?" *You want to find the ordered pair(s) that satisfy both equations.*
- ? "What are the intersection possibilities for a line and a circle?" *There could be 0, 1, or 2 points of intersection.*
- Pose Example 6 and give partners sufficient time to work through the problem. Some may use factoring versus the Quadratic Formula to solve for x.

Closure

- **Exit Ticket:** Write the equation of a circle with center $(-1, 3)$ and radius 5. Graph the equation. $(x+1)^2 + (y-3)^2 = 25$; Check students' graphs.

Assignment Guide and Homework Check

ASSIGNMENT

Basic: 1, 2, 3–51 odd, 58, 62, 68–73

Average: 1, 2, 10–58 even, 62, 68–73

Advanced: 1, 2, 20, 26–62 even, 63–73

HOMEWORK CHECK

Basic: 13, 25, 33, 39, 51

Average: 18, 22, 36, 40, 50

Advanced: 20, 26, 36, 40, 54

ANSWERS

1. $(x - h)^2 + (y - k)^2 = r^2$
2. The distance from the center to the point is the radius.
3. center: $(0, 0)$, radius: 4
4. center: $(0, 0)$, radius: 3
5. center: $(-3, 3)$, radius: 6
6. center: $(6, -6)$, radius: 3
7. center: $(-0.5, 0.5)$, radius: 2
8. center: $\left(\frac{1}{3}, \frac{4}{3}\right)$, radius: 2
9. $x^2 + y^2 = 4$
10. $x^2 + y^2 = 36$
11. $(x + 3)^2 + (y + 3)^2 = 81$
12. $(x - 6)^2 + (y - 3)^2 = 81$
13. $(x - 4)^2 + y^2 = 64$
14. $x^2 + (y + 4)^2 = 144$
15. $x^2 + y^2 = 49$
16. $(x - 4)^2 + (y - 1)^2 = 25$
17. $(x + 3)^2 + (y - 4)^2 = 1$
18. $(x - 3)^2 + (y + 5)^2 = 49$
19. $(x + 2)^2 + (y + 6)^2 = 1.21$
20. $(x + 10)^2 + (y + 8)^2 = 10.24$
21. $x^2 + y^2 = 36$
22. $(x - 1)^2 + (y - 2)^2 = 9$
23. $(x - 7)^2 + (y + 9)^2 = 25$
24. $(x + 4)^2 + (y - 3)^2 = 100$
25. $x^2 + y^2 = 58$
26. $(x + 1)^2 + (y + 3)^2 = 137$
27. The coordinates of the center should be subtracted;
$[x - (-3)]^2 + [y - (-5)]^2$
$= (x + 3)^2 + (y + 5)^2 = 9$

10.7 Exercises

Dynamic Solutions available at BigIdeasMath.com

Vocabulary and Core Concept Check

1. **VOCABULARY** What is the standard equation of a circle?
2. **WRITING** Explain why knowing the location of the center and one point on a circle is enough to graph the circle.

Monitoring Progress and Modeling with Mathematics

In Exercises 3–8, find the center and radius of the circle.

3.
4.
5.
6.
7. $(-0.5, 2.5)$ $(-0.5, -1.5)$
8. $\left(-\frac{5}{3}, \frac{4}{3}\right)$ $\left(\frac{7}{3}, \frac{4}{3}\right)$

In Exercises 9–20, write the standard equation of the circle. (See Example 1.)

9.
10.
11.
12.

13.
14.

15. a circle with center $(0, 0)$ and radius 7
16. a circle with center $(4, 1)$ and radius 5
17. a circle with center $(-3, 4)$ and radius 1
18. a circle with center $(3, -5)$ and radius 7
19. a circle with center $(-2, -6)$ and radius 1.1
20. a circle with center $(-10, -8)$ and radius 3.2

In Exercises 21–26, use the given information to write the standard equation of the circle. (See Example 2.)

21. The center is $(0, 0)$, and a point on the circle is $(0, 6)$.
22. The center is $(1, 2)$, and a point on the circle is $(4, 2)$.
23. The center is $(7, -9)$, and a point on the circle is $(3, -6)$.
24. The center is $(-4, 3)$, and a point on the circle is $(-10, -5)$.
25. The center is $(0, 0)$, and a point on the circle is $(3, -7)$.
26. The center is $(-1, -3)$, and a point on the circle is $(-5, 8)$.

27. **ERROR ANALYSIS** Describe and correct the error in writing the standard equation of the circle.

> ✗ The standard equation of the circle with center $(-3, -5)$ and radius 3 is $(x - 3)^2 + (y - 5)^2 = 9$.

624 Chapter 10 Circles

28. ERROR ANALYSIS Describe and correct the error in writing the standard equation of the circle.

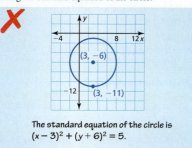

The standard equation of the circle is $(x − 3)^2 + (y + 6)^2 = 5$.

In Exercises 29–38, find the center and radius of the circle. Then graph the circle. *(See Example 3.)*

29. $x^2 + y^2 = 49$
30. $x^2 + y^2 = 64$
31. $(x + 5)^2 + (y − 3)^2 = 9$
32. $(x − 6)^2 + (y + 2)^2 = 36$
33. $x^2 + y^2 − 6x = 7$
34. $x^2 + y^2 + 4y = 32$
35. $x^2 + y^2 − 8x − 2y = −16$
36. $x^2 + y^2 + 4x + 12y = −15$
37. $x^2 + y^2 + 8x − 4y + 4 = 0$
38. $x^2 + y^2 + 12x − 8y + 16 = 0$

In Exercises 39–44, prove or disprove the statement. *(See Example 4.)*

39. The point (2, 3) lies on the circle centered at the origin with radius 8.
40. The point $(4, \sqrt{5})$ lies on the circle centered at the origin with radius 3.
41. The point $(\sqrt{6}, 2)$ lies on the circle centered at the origin and containing the point (3, −1).
42. The point $(\sqrt{7}, 5)$ lies on the circle centered at the origin and containing the point (5, 2).
43. The point (4, 4) lies on the circle centered at (1, 0) and containing the point (5, −3).
44. The point (2.4, 4) lies on the circle centered at (−3, 2) and containing the point (−11, 8).

45. MODELING WITH MATHEMATICS A city's commuter system has three zones. Zone 1 serves people living within 3 miles of the city's center. Zone 2 serves those between 3 and 7 miles from the center. Zone 3 serves those over 7 miles from the center. *(See Example 5.)*

a. Graph this situation on a coordinate plane where each unit corresponds to 1 mile. Locate the city's center at the origin.

b. Determine which zone serves people whose homes are represented by the points (3, 4), (6, 5), (1, 2), (0, 3), and (1, 6).

46. MODELING WITH MATHEMATICS Telecommunication towers can be used to transmit cellular phone calls. A graph with units measured in kilometers shows towers at points (0, 0), (0, 5), and (6, 3). These towers have a range of about 3 kilometers.

a. Sketch a graph and locate the towers. Are there any locations that may receive calls from more than one tower? Explain your reasoning.

b. The center of City A is located at (−2, 2.5), and the center of City B is located at (5, 4). Each city has a radius of 1.5 kilometers. Which city seems to have better cell phone coverage? Explain your reasoning.

In Exercises 47–56, solve the system. *(See Example 6.)*

47. $x^2 + y^2 = 4$
    $y = 1$
48. $x^2 + y^2 = 1$
    $y = -5$
49. $x^2 + y^2 = 64$
    $y = x - 10$
50. $x^2 + y^2 = 2$
    $y = x + 2$
51. $x^2 + y^2 = 16$
    $y = x + 16$
52. $x^2 + y^2 = 9$
    $y = x + 3$
53. $x^2 + y^2 = 1$
    $y = \dfrac{-x + 1}{2}$
54. $x^2 + y^2 = 1$
    $y = \dfrac{-5x + 13}{12}$
55. $(x - 3)^2 + (y + 1)^2 = 1$
    $y = 2x - 6$
56. $(x + 2)^2 + (y - 6)^2 = 25$
    $y = -2x - 23$

Section 10.7   Circles in the Coordinate Plane   625

ANSWERS

28. The radius should be squared; $(x − 3)^2 + (y + 6)^2 = 25$

29. center: (0, 0), radius: 7
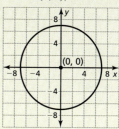

30. center: (0, 0), radius: 8
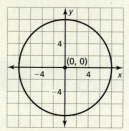

31. center: (−5, 3), radius: 3
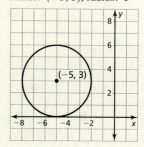

32. center: (6, −2), radius: 6
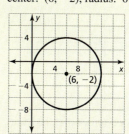

33. center: (3, 0), radius: 4
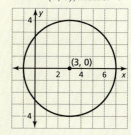

34. center: (0, −2), radius: 6
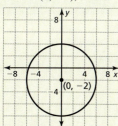

35. center: (4, 1), radius: 1

36-56. See Additional Answers.

Section 10.7   625

ANSWERS

57. a. Sample answer: Write the two numerators and the denominator of the coordinates.
b. Sample answer: $\left(\frac{3}{5}, \frac{4}{5}\right)$

58. a. C
b. D
c. B
d. A

59–67. See Additional Answers.

68. minor arc; 53°
69. minor arc; 90°
70. major arc; 270°
71. minor arc, 127°
72. semicircle; 180°
73. minor arc; 78°

Mini-Assessment

1. Write the standard equation of a circle with center $(-4, 0)$ and radius 1.5. $(x + 4)^2 + y^2 = 2.25$

2. The point $(3, 0)$ is on a circle with center $(0, 2)$. Write the standard equation of the circle. $x^2 + (y - 2)^2 = 13$

3. Find the center and the radius of the circle with equation $x^2 + y^2 - 4x + 2y + 1 = 0$. Then graph the circle. center $(2, -1)$; radius 2

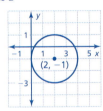

4. Prove or disprove that the point $(-2, 2\sqrt{3})$ lies on the circle centered at the origin with radius 4. The distance from $(-2, 2\sqrt{3})$ to the origin is $\sqrt{16} = 4$. So, the point $(-2, 2\sqrt{3})$ lies on the circle.

5. Solve the system.
$x^2 + y^2 = 25$
$y = \frac{3}{4}x$
$(4, 3), (-4, -3)$

626 Chapter 10

---

57. **REASONING** Consider the graph shown.

a. Explain how you can use the given points to form Pythagorean triples.

b. Identify another point on the circle that you can use to form the Pythagorean triple 3, 4, 5.

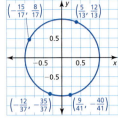

58. **HOW DO YOU SEE IT?** Match each graph with its equation.

a.   b.

c.   d.

A. $x^2 + (y + 3)^2 = 4$ B. $(x - 3)^2 + y^2 = 4$
C. $(x + 3)^2 + y^2 = 4$ D. $x^2 + (y - 3)^2 = 4$

59. **REASONING** Sketch the graph of the circle whose equation is $x^2 + y^2 = 16$. Then sketch the graph of the circle after the translation $(x, y) \to (x - 2, y - 4)$. What is the equation of the image? Make a conjecture about the equation of the image of a circle centered at the origin after a translation m units to the left and n units down.

Maintaining Mathematical Proficiency
Reviewing what you learned in previous grades and lessons

Identify the arc as a *major arc*, *minor arc*, or *semicircle*. Then find the measure of the arc. *(Section 10.2)*

68. $\overset{\frown}{RS}$
69. $\overset{\frown}{PR}$
70. $\overset{\frown}{PRT}$
71. $\overset{\frown}{ST}$
72. $\overset{\frown}{RST}$
73. $\overset{\frown}{QS}$

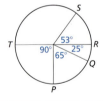

626   Chapter 10   Circles

---

60. **THOUGHT PROVOKING** A circle has center (h, k) and contains point (a, b). Write the equation of the line tangent to the circle at point (a, b).

61. **USING STRUCTURE** The vertices of $\triangle XYZ$ are $X(4, 5)$, $Y(4, 13)$, and $Z(8, 9)$. Find the equation of the circle circumscribed about $\triangle XYZ$. Justify your answer.

62. **MAKING AN ARGUMENT** Your friend claims that the equation of a circle passing through the points $(-1, 0)$ and $(1, 0)$ is $x^2 - 2yk + y^2 = 1$ with center $(0, k)$. Is your friend correct? Explain your reasoning.

MATHEMATICAL CONNECTIONS In Exercises 63–66, use the equations to determine whether the line is *a tangent*, *a secant*, *a secant that contains the diameter*, or *none of these*. Explain your reasoning.

63. Circle: $(x - 4)^2 + (y - 3)^2 = 9$
Line: $y = 6$

64. Circle: $(x + 2)^2 + (y - 2)^2 = 16$
Line: $y = 2x - 4$

65. Circle: $(x - 5)^2 + (y + 1)^2 = 4$
Line: $y = \frac{1}{5}x - 3$

66. Circle: $(x + 3)^2 + (y - 6)^2 = 25$
Line: $y = -\frac{4}{3}x + 2$

67. **REASONING** Four tangent circles are centered on the x-axis. The radius of $\odot A$ is twice the radius of $\odot O$. The radius of $\odot B$ is three times the radius of $\odot O$. The radius of $\odot C$ is four times the radius of $\odot O$. All circles have integer radii, and the point $(63, 16)$ is on $\odot C$. What is the equation of $\odot A$? Explain your reasoning.

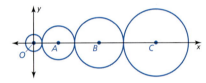

If students need help...	If students got it...
Resources by Chapter • Practice A and Practice B • Puzzle Time	**Resources by Chapter** • Enrichment and Extension • Cumulative Review
Student Journal • Practice	Start the *next* Section
Differentiating the Lesson Skills Review Handbook	

10.4–10.7 What Did You Learn?

Core Vocabulary

inscribed angle, p. 598
intercepted arc, p. 598
subtend, p. 598
circumscribed angle, p. 608
segments of a chord, p. 614
tangent segment, p. 615
secant segment, p. 615
external segment, p. 615
standard equation of a circle, p. 620

Core Concepts

Section 10.4
Inscribed Angle and Intercepted Arc, p. 598
Measure of an Inscribed Angle Theorem, p. 598
Inscribed Angles of a Circle Theorem, p. 599
Inscribed Right Triangle Theorem, p. 600
Inscribed Quadrilateral Theorem, p. 600

Section 10.5
Tangent and Intersected Chord Theorem, p. 606
Intersecting Lines and Circles, p. 606
Angles Inside the Circle Theorem, p. 607
Angles Outside the Circle Theorem, p. 607
Circumscribed Angle, p. 608
Circumscribed Angle Theorem, p. 608

Section 10.6
Segments of Chords Theorem, p. 614
Tangent Segment and Secant Segment, p. 615
Segments of Secants Theorem, p. 615
Segments of Secants and Tangents Theorem, p. 616

Section 10.7
Standard Equation of a Circle, p. 620
Writing Coordinate Proofs Involving Circles, p. 622
Solving Nonlinear Systems, p. 623

Mathematical Practices

1. What other tools could you use to complete the task in Exercise 18 on page 603?
2. You have a classmate who is confused about why two diagrams are needed in part (a) of Exercise 31 on page 611. Explain to your classmate why two diagrams are needed.

Performance Task:

Finding Locations

Scientists use seismographs to locate the epicenter of an earthquake. They use distance data taken from the instruments to create three circles. The intersection of these circles determines the precise location of the epicenter. This process, which is called trilateration, is also used by the Global Positioning System (GPS) to determine a person's location. How does it work?

To explore the answer to this question and more, check out the Performance Task and Real-Life STEM video at *BigIdeasMath.com*.

ANSWERS

1. radius
2. chord
3. tangent
4. diameter
5. secant
6. radius
7. internal
8. external
9. 2
10. 2
11. 12
12. tangent; $20^2 + 48^2 = 52^2$

10 Chapter Review

Dynamic Solutions available at *BigIdeasMath.com*

10.1 Lines and Segments That Intersect Circles (pp. 573–580)

In the diagram, $\overline{AB}$ is tangent to $\odot C$ at B and $\overline{AD}$ is tangent to $\odot C$ at D. Find the value of x.

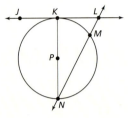

$AB = AD$ External Tangent Congruence Theorem
$2x + 5 = 33$ Substitute.
$x = 14$ Solve for x.

▶ The value of x is 14.

Tell whether the line, ray, or segment is best described as a *radius*, *chord*, *diameter*, *secant*, or *tangent* of $\odot P$.

1. $\overline{PK}$
2. $\overline{NM}$
3. $\overrightarrow{JL}$
4. $\overline{KN}$
5. $\overrightarrow{NL}$
6. $\overrightarrow{PN}$

Tell whether the common tangent is *internal* or *external*.

7.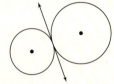
8.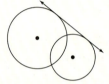

Points Y and Z are points of tangency. Find the value of the variable.

9.
10.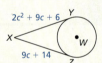
11.

12. Tell whether $\overline{AB}$ is tangent to $\odot C$. Explain.

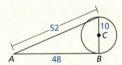

628 Chapter 10 Circles

10.2 Finding Arc Measures (pp. 581–588)

Find the measure of each arc of ⊙P, where $\overline{LN}$ is a diameter.

a. $\widehat{MN}$

▶ $\widehat{MN}$ is a minor arc, so $m\widehat{MN} = m\angle MPN = 120°$.

b. $\widehat{NLM}$

▶ $\widehat{NLM}$ is a major arc, so $m\widehat{NLM} = 360° - 120° = 240°$.

c. $\widehat{NML}$

▶ $\overline{NL}$ is a diameter, so $\widehat{NML}$ is a semicircle, and $m\widehat{NML} = 180°$.

Use the diagram above to find the measure of the indicated arc.

13. $\widehat{KL}$ 14. $\widehat{LM}$ 15. $\widehat{KM}$ 16. $\widehat{KN}$

Tell whether the red arcs are congruent. Explain why or why not.

17.

18.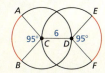

10.3 Using Chords (pp. 589–594)

In the diagram, $\odot A \cong \odot B$, $\overline{CD} \cong \overline{FE}$, and $m\widehat{FE} = 75°$. Find $m\widehat{CD}$.

Because $\overline{CD}$ and $\overline{FE}$ are congruent chords in congruent circles, the corresponding minor arcs $\widehat{CD}$ and $\widehat{FE}$ are congruent by the Congruent Corresponding Chords Theorem.

▶ So, $m\widehat{CD} = m\widehat{FE} = 75°$.

Find the measure of $\widehat{AB}$.

19.

20.

21.

22. In the diagram, $QN = QP = 10$, $JK = 4x$, and $LM = 6x - 24$. Find the radius of $\odot Q$.

ANSWERS

13. 100°
14. 60°
15. 160°
16. 80°
17. not congruent; The circles are not congruent.
18. congruent; The circles are congruent and $m\widehat{AB} = m\widehat{EF}$.
19. 61°
20. 65°
21. 91°
22. 26

ANSWERS
23. 80
24. $q = 100, r = 20$
25. 5
26. $y = 30, z = 10$
27. $m = 44, n = 39$
28. 28

10.4 Inscribed Angles and Polygons (pp. 597–604)

Find the value of each variable.

LMNP is inscribed in a circle, so opposite angles are supplementary by the Inscribed Quadrilateral Theorem.

$m\angle L + m\angle N = 180°$ $m\angle P + m\angle M = 180°$

$3a° + 3a° = 180°$ $b° + 50° = 180°$

$6a = 180$ $b = 130$

$a = 30$

▶ The value of *a* is 30, and the value of *b* is 130.

Find the value(s) of the variable(s).

23.
24.
25.
26.
27.
28.

10.5 Angle Relationships in Circles (pp. 605–612)

Find the value of *y*.

The tangent $\overrightarrow{RQ}$ and secant $\overrightarrow{RT}$ intersect outside the circle, so you can use the Angles Outside the Circle Theorem.

$y° = \frac{1}{2}(m\widehat{QT} - m\widehat{SQ})$ Angles Outside the Circle Theorem

$y° = \frac{1}{2}(190° - 60°)$ Substitute.

$y = 65$ Simplify.

▶ The value of *y* is 65.

Find the value of x.

29.
30.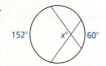
31.

32. Line ℓ is tangent to the circle. Find $m\widehat{XYZ}$.

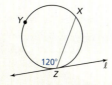

10.6 Segment Relationships in Circles (pp. 613–618)

Find the value of x.

The chords $\overline{EG}$ and $\overline{FH}$ intersect inside the circle, so you can use the Segments of Chords Theorem.

$EP \cdot PG = FP \cdot PH$ Segments of Chords Theorem
$x \cdot 2 = 3 \cdot 6$ Substitute.
$x = 9$ Simplify.

▶ The value of x is 9.

Find the value of x.

33.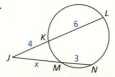
34.
35.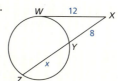

36. A local park has a circular ice skating rink. You are standing at point A, about 12 feet from the edge of the rink. The distance from you to a point of tangency on the rink is about 20 feet. Estimate the radius of the rink.

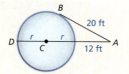

ANSWERS
29. 70
30. 106
31. 16
32. 240°
33. 5
34. 3
35. 10
36. about 10.7 ft

ANSWERS

37. $(x - 4)^2 + (y + 1)^2 = 9$
38. $(x - 8)^2 + (y - 6)^2 = 36$
39. $x^2 + y^2 = 16$
40. $x^2 + y^2 = 81$
41. $(x + 5)^2 + (y - 2)^2 = 1.69$
42. $(x + 7)^2 + (y - 6)^2 = 25$
43. center: $(6, -4)$, radius: 2

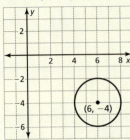

44. The radius of the circle is 5.
   $d = \sqrt{(0 - 4)^2 + (0 + 3)^2} = 5$, so $(4, -3)$ is on the circle.
45. no solution
46. $(0, -3), \left(\frac{4}{5}, -\frac{7}{5}\right)$

10.7 Circles in the Coordinate Plane (pp. 619–626)

Write the standard equation of the circle shown.

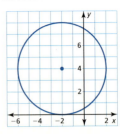

The radius is 4, and the center is $(-2, 4)$.

$(x - h)^2 + (y - k)^2 = r^2$ Standard equation of a circle
$[x - (-2)]^2 + (y - 4)^2 = 4^2$ Substitute.
$(x + 2)^2 + (y - 4)^2 = 16$ Simplify.

▶ The standard equation of the circle is $(x + 2)^2 + (y - 4)^2 = 16$.

Write the standard equation of the circle shown.

37. 38. 39.

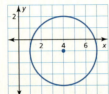 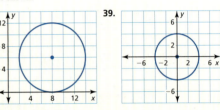

Write the standard equation of the circle with the given center and radius.

40. center: $(0, 0)$, radius: 9
41. center: $(-5, 2)$, radius: 1.3

42. The point $(-7, 1)$ is on a circle with center $(-7, 6)$. Write the standard equation of the circle.

43. The equation of a circle is $x^2 + y^2 - 12x + 8y + 48 = 0$. Find the center and the radius of the circle. Then graph the circle.

44. Prove or disprove that the point $(4, -3)$ lies on the circle centered at the origin and containing the point $(-5, 0)$.

Solve the system.

45. $x^2 + y^2 = 25$
    $y = -6$

46. $x^2 + (y + 2)^2 = 1$
    $y = 2x - 3$

632 Chapter 10 Circles

10 Chapter Test

Find the measure of each numbered angle in ⊙P. Justify your answer.

1.
2.
3.
4.

Use the diagram.

5. $AG = 2$, $GD = 9$, and $BG = 3$. Find GF.
6. $CF = 12$, $CB = 3$, and $CD = 9$. Find CE.
7. Solve the system. $x^2 + y^2 = 225$
   $y = x + 3$
8. Sketch a pentagon inscribed in a circle. Label the pentagon ABCDE. Describe the relationship between each pair of angles. Explain your reasoning.
   a. ∠CDE and ∠CAE
   b. ∠CBE and ∠CAE

Find the value of the variable. Justify your answer.

9.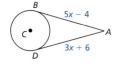
10.

11. Prove or disprove that the point $(2\sqrt{2}, -1)$ lies on the circle centered at $(0, 2)$ and containing the point $(-1, 4)$.

Prove the given statement.

12. $\widehat{ST} \cong \widehat{RQ}$
13. $\widehat{JM} \cong \widehat{LM}$
14. $\overline{DG} \cong \overline{FG}$

15. A bank of lighting hangs over a stage. Each light illuminates a circular region on the stage. A coordinate plane is used to arrange the lights, using a corner of the stage as the origin. The equation $(x - 13)^2 + (y - 4)^2 = 16$ represents the boundary of the region illuminated by one of the lights. Three actors stand at the points $A(11, 4)$, $B(8, 5)$, and $C(15, 5)$. Graph the given equation. Then determine which actors are illuminated by the light.

16. If a car goes around a turn too quickly, it can leave tracks that form an arc of a circle. By finding the radius of the circle, accident investigators can estimate the speed of the car.

    a. To find the radius, accident investigators choose points A and B on the tire marks. Then the investigators find the midpoint C of $\overline{AB}$. Use the diagram to find the radius r of the circle. Explain why this method works.
    b. The formula $S = 3.87\sqrt{fr}$ can be used to estimate a car's speed in miles per hour, where f is the *coefficient of friction* and r is the radius of the circle in feet. If $f = 0.7$, estimate the car's speed in part (a).

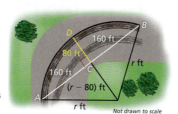

Not drawn to scale

10 Cumulative Assessment

ANSWERS

1. a. chord
   b. radius
   c. diameter
   d. chord
2. translation; $(x - 2, y + 2)$; 0; 3; 0; 3; 4; similarity; similar
3. $m\angle JPL = 90°$, $m\angle LPN = 90°$, and $\overline{PM}$ and $\overline{PK}$ are altitudes, so $\triangle JPL \sim \triangle PKL$ and $\triangle NPL \sim \triangle NMP$ by the Right Triangle Similarity Theorem. By the Transitive Property, $\triangle PKL \sim \triangle NMP$.
4. B
5. C

---

1. Classify each segment as specifically as possible.

   a. $\overline{BG}$ b. $\overline{CD}$

   c. $\overline{AD}$ d. $\overline{FE}$

2. Copy and complete the paragraph proof.

   Given Circle C with center $(2, 1)$ and radius 1,
   Circle D with center $(0, 3)$ and radius 4

   Prove Circle C is similar to Circle D.

   Map Circle C to Circle C' by using the _____ $(x, y) \rightarrow$ _____ so that Circle C' and Circle D have the same center at (__, __). Dilate Circle C' using a center of dilation (__, __) and a scale factor of ___. Because there is a _____ transformation that maps Circle C to Circle D, Circle C is _____ Circle D.

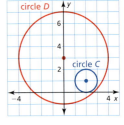

3. Use the diagram to write a proof.

   Given $\triangle JPL \cong \triangle NPL$
   $\overline{PK}$ is an altitude of $\triangle JPL$.
   $\overline{PM}$ is an altitude of $\triangle NPL$.

   Prove $\triangle PKL \sim \triangle NMP$

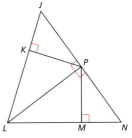

4. The equation of a circle is $x^2 + y^2 + 14x - 16y + 77 = 0$. What are the center and radius of the circle?

   Ⓐ center: $(14, -16)$, radius: 8.8

   Ⓑ center: $(-7, 8)$, radius: 6

   Ⓒ center: $(-14, 16)$, radius: 8.8

   Ⓓ center: $(7, -8)$, radius: 5.2

5. Which expression is *not* equivalent to the others?

   Ⓐ $\dfrac{x^{1/2} \cdot x^{5/4} \cdot \sqrt[4]{x}}{x^{-3} \cdot x^3}$

   Ⓑ $\dfrac{x^{-9} \cdot (x^{7/3})^3}{x^{-4}}$

   Ⓒ $\dfrac{x^5 \cdot x^3}{x^{-6}}$

   Ⓓ $\dfrac{\sqrt[3]{x} \cdot x^{-4/3}}{x^{-3}}$

634   Chapter 10   Circles

6. Which angles have the same measure as ∠ACB? Select all that apply.

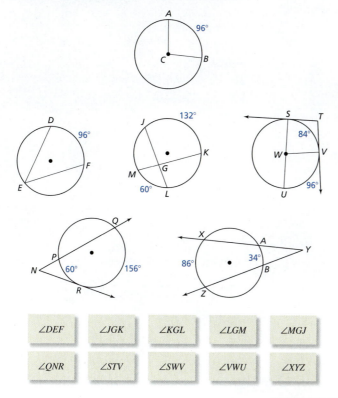

| ∠DEF | ∠JGK | ∠KGL | ∠LGM | ∠MGJ |

| ∠QNR | ∠STV | ∠SWV | ∠VWU | ∠XYZ |

7. A survey asked freshman and sophomore students about whether they think the new school mascot should be a tiger or a lion. The table shows the results of the survey.

   a. Complete the two-way table.
   b. What is the probability that a randomly selected student is a freshman and prefers a tiger?
   c. What is the probability that a randomly selected sophomore prefers a tiger?

		New Mascot		
		Tiger	Lion	Total
Class	Freshman			100
	Sophomore		48	
	Total	110		212

8. Your friend claims that the quadrilateral shown can be inscribed in a circle. Is your friend correct? Explain your reasoning.

ANSWERS

6. ∠JGK, ∠LGM, ∠STV, ∠VWU
7. a.

		New Mascot		
		Tiger	Lion	Total
Class	Freshman	46	54	100
	Sophomore	64	48	112
	Total	110	102	212

   b. about 0.22
   c. about 0.57
8. no; Opposite angles are not supplementary.

Chapter 11 Pacing Guide

Chapter Opener/Mathematical Practices	0.5 Day
Section 1	1.5 Days
Section 2	1 Day
Section 3	2 Days
Quiz	0.5 Day
Section 4	1.5 Days
Section 5	2 Days
Section 6	2 Days
Section 7	2 Days
Chapter Review/Chapter Tests	2 Days
Total Chapter 11	15 Days
Year-to-Date	158 Days

11 Circumference, Area, and Volume

- **11.1** Circumference and Arc Length
- **11.2** Areas of Circles and Sectors
- **11.3** Areas of Polygons
- **11.4** Volumes of Prisms and Cylinders
- **11.5** Volumes of Pyramids
- **11.6** Surface Areas and Volumes of Cones
- **11.7** Surface Areas and Volumes of Spheres

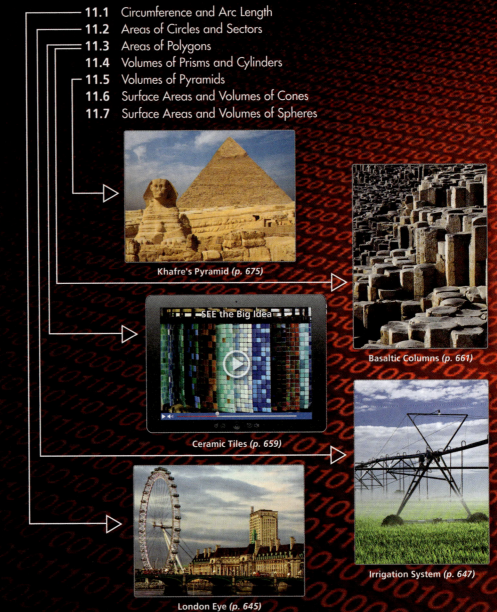

Khafre's Pyramid (p. 675)

Basaltic Columns (p. 661)

SEE the Big Idea

Ceramic Tiles (p. 659)

London Eye (p. 645)

Irrigation System (p. 647)

Laurie's Notes

Chapter Summary

- This last chapter on circumference, area, and volume finishes the study of measurement of solids that began in middle school.
- Students will come to this chapter with knowledge of many formulas for surface area and volume. These will be reviewed and a few new formulas added to the list.
- Different from middle school is that students now have a greater ability to solve equations. They also know the Pythagorean Theorem and trigonometry, so they are able to solve for measures that previously had to be told to them.
- In this chapter, students will do additional work with circles involving arc length and area of sectors. Students will also find the area of regular polygons.

What Your Students Have Learned

Middle School
- Find the areas of triangles.
- Use values of pi to estimate and calculate the circumference and area of circles.
- Solve problems involving surfaces areas and volumes of objects composed of prisms, pyramids, and cylinders.
- Solve problems involving volumes of cones, cylinders, and spheres.

Math I
- Solve multi-step linear equations in one variable that have variables on both sides using inverse operations.
- Use the properties of equality to justify steps in solving problems involving segment lengths and angle measures.
- Evaluate and simplify expressions with exponents.
- Classify polygons by the number of sides, and inscribe regular polygons in circles.
- Find perimeters and areas of polygons in the coordinate plane.

What Your Students Will Learn

Math II
- Use the formulas for circumference and arc length to find measures in circles.
- Measure angles in radians and convert between degree and radian measures.
- Find and use areas of sectors as fractional parts of the areas of circles.
- Find areas of rhombuses, kites, and regular polygons.
- Classify solids and name polyhedra.
- Find and use volumes of prisms, cylinders, pyramids, cones, spheres, and composite solids.
- Find surface areas of right cones and spheres.

Dynamic Teaching Tools
Dynamic Assessment & Progress Monitoring Tool
Lesson Planning Tool
Interactive Whiteboard Lesson Library
Dynamic Classroom with Dynamic Investigations
Real-Life STEM Videos

Scaffolding in the Classroom
Use vocabulary as a connector. When introducing vocabulary, ask students whether they are familiar with the word and what they think it means. Have them explain how they think the word relates to math. Students then find the definition and compare their original ideas to the actual meaning.

Questioning in the Classroom
One thing leads to another.
Try to build questions from the responses given. This requires students to listen to other students' responses.

Laurie's Notes

Maintaining Mathematical Proficiency

Finding Surface Area

- Remind students that the surface area of a prism is the sum of the areas of *all* the faces.
- Point out that the formula for the area A of a triangle with base b and corresponding height h is $A = \frac{1}{2}bh$.

COMMON ERROR Students may confuse lateral area and surface area. Referring to *total surface area* instead of *surface area* can help.

Finding a Missing Dimension

- Point out that the formula for the perimeter P of a rectangle $P = 2\ell + 2w$ can also be written as $P = 2(\ell + w)$.
- Remind students that after they substitute the known dimensions into the formula, they should solve using inverse operations, as they have done with other algebraic equations.

COMMON ERROR Students may confuse perimeter and area. Remind them that perimeter is measured in linear units and area is measured in square units.

Mathematical Practices (continued on page 638)

- The *Mathematical Practices* page focuses attention on how mathematics is learned—process versus content. This page demonstrates that a mathematically proficient student can create a coherent representation of a problem. The representation helps to make sense of the problem and/or aids in the solution of the problem.
- Use the *Mathematical Practices* page to help students develop mathematical habits of mind—how mathematics can be explored and how mathematics is thought about.

If students need help...	If students got it...
Student Journal • Maintaining Mathematical Proficiency	Game Closet at *BigIdeasMath.com*
Lesson Tutorials	Start the *next* Section
Skills Review Handbook	

Maintaining Mathematical Proficiency

Finding Surface Area

Example 1 Find the surface area of the prism.

$S = 2\ell w + 2\ell h + 2wh$ Write formula for surface area of a rectangular prism.

$= 2(2)(4) + 2(2)(6) + 2(4)(6)$ Substitute 2 for ℓ, 4 for w, and 6 for h.

$= 16 + 24 + 48$ Multiply.

$= 88$ Add.

▶ The surface area is 88 square inches.

Find the surface area of the prism.

1.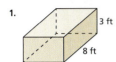
2.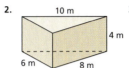
3.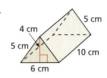

Finding a Missing Dimension

Example 2 A rectangle has a perimeter of 10 meters and a length of 3 meters. What is the width of the rectangle?

$P = 2\ell + 2w$ Write formula for perimeter of a rectangle.

$10 = 2(3) + 2w$ Substitute 10 for P and 3 for ℓ.

$10 = 6 + 2w$ Multiply 2 and 3.

$4 = 2w$ Subtract 6 from each side.

$2 = w$ Divide each side by 2.

▶ The width is 2 meters.

Find the missing dimension.

4. A rectangle has a perimeter of 28 inches and a width of 5 inches. What is the length of the rectangle?
5. A triangle has an area of 12 square centimeters and a height of 12 centimeters. What is the base of the triangle?
6. A rectangle has an area of 84 square feet and a width of 7 feet. What is the length of the rectangle?
7. **ABSTRACT REASONING** Write an equation for the surface area of a prism with a length, width, and height of x inches. What solid figure does the prism represent?

Dynamic Solutions available at *BigIdeasMath.com*

What Your Students Have Learned

- Find surface areas of prisms using formulas.
- Find missing dimensions of polygons using formulas.

ANSWERS

1. 158 ft^2
2. 144 m^2
3. 184 cm^2
4. 9 in.
5. 2 cm
6. 12 ft
7. $S = 6x^2$; cube

Vocabulary Review

Have students make Information Wheels for the following words.
- Surface area
- Area
- Perimeter

MONITORING PROGRESS ANSWERS

1.
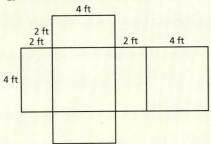

2.
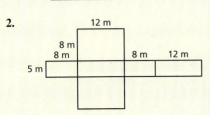

3.
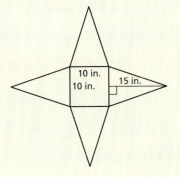

Mathematical Practices

Mathematically proficient students create valid representations of problems.

Creating a Valid Representation

Core Concept

Nets for Three-Dimensional Figures

A **net** for a three-dimensional figure is a two-dimensional pattern that can be folded to form the three-dimensional figure.

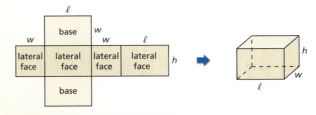

EXAMPLE 1 Drawing a Net for a Pyramid

Draw a net of the pyramid.

SOLUTION

The pyramid has a square base. Its four lateral faces are congruent isosceles triangles.

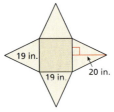

Monitoring Progress

Draw a net of the three-dimensional figure. Label the dimensions.

1. [rectangular prism 2 ft × 4 ft × 4 ft]
2. [rectangular prism 5 m × 8 m × 12 m]
3.

638 Chapter 11 Circumference, Area, and Volume

Laurie's Notes — Mathematical Practices (continued from page T-637)

- Students will be working with nets in this chapter, helping them make sense of how to find the surface area of three-dimensional figures. In the *Core Concept*, the net for a rectangular prism is shown. Students should be familiar with nets from middle school.
- Example 1 demonstrates how a net is drawn for a square pyramid. While there are other versions of a net that would fold to make the square pyramid, the one shown is easy to visualize folding into the pyramid.
- Give time for students to work through the *Monitoring Progress* questions. Be sure that students label the dimensions. Share nets with the whole class.

Laurie's Notes

Overview of Section 11.1

Introduction
- This lesson reintroduces the formula for the circumference of a circle. In the last chapter, students worked with measures of circular arcs. Now they will work with the length of a circular arc.
- The lesson ends with an introduction to radian measure.

Teaching Strategy
- The *Core Concept* on page 641 presents two methods for finding arc length.
- Reading the proportion, "In a circle, the ratio of the length of a given arc to the circumference is equal to the ratio of the measure of the arc to 360°." Note that the ratio on the left side is in linear units, while the ratio on the right side is in degrees.

$$\frac{\text{Arc length of } \widehat{AB}}{2\pi r} = \frac{m\widehat{AB}}{360°} \qquad \text{Note: } \frac{\text{linear units}}{\text{linear units}} = \frac{\text{degrees}}{\text{degrees}}$$

- Reading the equation that follows, "The length of a given arc is equal to a fractional amount of the circumference."

$$\text{Arc length of } \widehat{AB} = \frac{m\widehat{AB}}{360°} \cdot 2\pi r$$

- Structurally, the proportion and the equation are the same, which can be shown by multiplying both sides of the proportion by $2\pi r$. How students think about them is not the same! Students are thinking and reasoning proportionally with the former and thinking of a percentage or portion of a whole for the latter.
- Helping students understand each of these approaches is necessary.

Extensions
- The introduction to circumference can be investigated using dynamic geometry software. Inscribe a regular n-gon in a circle and find the perimeter. As n increases, the perimeter gets closer to the circumference.

Pacing Suggestion
- The explorations provide a brief introduction to arc length. When students have finished, transition to the formal lesson.

What Your Students Will Learn

- Use the formula for circumference to find radii, diameters, and circumferences of circles.
- Use arc lengths to find arc measures, radii, diameters, and circumferences of circles.
- Measure angles in radians and convert between degree and radian measures.

Laurie's Notes

Exploration

Motivate
- A winter sport near where I live that is growing exponentially is fat-tire biking. Given the abundance of mountain bike, cross country, and snowmobile trails and a growing group of summer mountain bike enthusiasts, fat-tire biking is a natural evolution.
- The tires average 4 inches wide versus 2 inches on a mountain bike or 1 inch on a road bike. The diameter of the tires varies.
- Explain to students that in this lesson they will solve problems about tires.

Exploration Note
- Students should be familiar with the formula for the circumference of a circle from middle school.

Exploration 1
- The fractional amount of the circumference is stated, so a simple calculation is needed.
- ? "How did you calculate the length of each arc?" Multiply the fractional part of the circle by $2\pi r$. Some students may also say πd for the circumference. It is also possible that some students will have set up a proportion to solve.
- ? "What type of unit is used to label your answer?" a linear unit
- ? **Attend to Precision:** "What is the exact answer for part (a)?" 8π units "What is an approximate answer for part (a)?" about 25.13 units

Exploration 2
- Students should recognize that this exploration is just an application of the problems found in the first exploration. The diameter is 25 inches.
- ? "What fractional part of the whole circumference do you need to find?" $\frac{1}{2}$ of the circumference
- ? **Attend to Precision:** "Should you leave your answer in terms of π? Explain." no; You need a decimal approximation to compare to the width of the rectangular box.

Communicate Your Answer

- Students will likely describe finding a fractional part of the circumference, or they will describe setting up a proportion. In either case, listen for the distinction between arc measure and arc length.

Connecting to Next Step
- Students have now found circumference and arc lengths. Today's lesson also includes applications and using arc length to find measures.

11.1 Circumference and Arc Length

Essential Question How can you find the length of a circular arc?

EXPLORATION 1 Finding the Length of a Circular Arc

Work with a partner. Find the length of each red circular arc.

a. entire circle

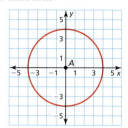

b. one-fourth of a circle

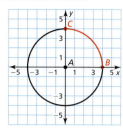

c. one-third of a circle

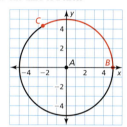

d. five-eighths of a circle

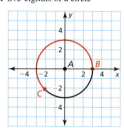

EXPLORATION 2 Using Arc Length

Work with a partner. The rider is attempting to stop with the front tire of the motorcycle in the painted rectangular box for a skills test. The front tire makes exactly one-half additional revolution before stopping. The diameter of the tire is 25 inches. Is the front tire still in contact with the painted box? Explain.

LOOKING FOR REGULARITY IN REPEATED REASONING

To be proficient in math, you need to notice if calculations are repeated and look both for general methods and for shortcuts.

Communicate Your Answer

3. How can you find the length of a circular arc?

4. A motorcycle tire has a diameter of 24 inches. Approximately how many inches does the motorcycle travel when its front tire makes three-fourths of a revolution?

Section 11.1 Circumference and Arc Length 639

Dynamic Teaching Tools

Dynamic Assessment & Progress Monitoring Tool
Lesson Planning Tool
Interactive Whiteboard Lesson Library
Dynamic Classroom with Dynamic Investigations

ANSWERS

1. a. $8\pi \approx 25.13$ units
   b. $2\pi \approx 6.28$ units
   c. $\frac{10}{3}\pi \approx 10.47$ units
   d. $\frac{15}{4}\pi \approx 11.78$ units

2. no; *Sample answer:* One-half revolution of the tire is about 3.27 feet.

3. Multiply the fraction of the circle the arc represents by the circumference of the circle.

4. about 56.55 in.

Section 11.1 639

> **English Language Learners**
>
> **Vocabulary**
> Some students may confuse *arc length* with *arc measure*. Point out that arc length is measured in linear units, while arc measure, like angle measure, is measured in degrees. Ask students whether two arcs with the same arc measure always have the same length? (no) Do two arcs with the same length always have the same arc measure? (no)

Extra Example 1
Find each indicated measure.
a. circumference of a circle with a radius of 11 inches
   The circumference is about 69.12 inches.
b. radius of a circle with a circumference of 4 millimeters
   The radius is about 0.64 millimeters.

MONITORING PROGRESS ANSWERS
1. about 15.71 in.
2. about 5.41 ft

11.1 Lesson

Core Vocabulary
circumference, p. 640
arc length, p. 641
radian, p. 643
Previous
circle
diameter
radius

What You Will Learn
- Use the formula for circumference.
- Use arc lengths to find measures.
- Solve real-life problems.
- Measure angles in radians.

Using the Formula for Circumference

The **circumference** of a circle is the distance around the circle. Consider a regular polygon inscribed in a circle. As the number of sides increases, the polygon approximates the circle and the ratio of the perimeter of the polygon to the diameter of the circle approaches $\pi \approx 3.14159\ldots$.

For all circles, the ratio of the circumference C to the diameter d is the same. This ratio is $\frac{C}{d} = \pi$. Solving for C yields the formula for the circumference of a circle, $C = \pi d$. Because $d = 2r$, you can also write the formula as $C = \pi(2r) = 2\pi r$.

> **Core Concept**
>
> **Circumference of a Circle**
> The circumference C of a circle is $C = \pi d$ or $C = 2\pi r$, where d is the diameter of the circle and r is the radius of the circle.
>
>
>
> $C = \pi d = 2\pi r$

EXAMPLE 1 Using the Formula for Circumference

Find each indicated measure.
a. circumference of a circle with a radius of 9 centimeters
b. radius of a circle with a circumference of 26 meters

ATTENDING TO PRECISION
You have sometimes used 3.14 to approximate the value of π. Throughout this chapter, you should use the π key on a calculator, then round to the hundredths place unless instructed otherwise.

SOLUTION

a. $C = 2\pi r$
   $= 2 \cdot \pi \cdot 9$
   $= 18\pi$
   ≈ 56.55
   ▶ The circumference is about 56.55 centimeters.

b. $C = 2\pi r$
   $26 = 2\pi r$
   $\dfrac{26}{2\pi} = r$
   $4.14 \approx r$
   ▶ The radius is about 4.14 meters.

Monitoring Progress Help in English and Spanish at *BigIdeasMath.com*

1. Find the circumference of a circle with a diameter of 5 inches.
2. Find the diameter of a circle with a circumference of 17 feet.

640   Chapter 11   Circumference, Area, and Volume

Laurie's Notes — Teacher Actions

- Students should be very comfortable using different forms of the circumference formula.
- ? "What type of number is π?" irrational
- ? "What does *irrational* mean?" It is a non-repeating, non-terminating decimal.
- Pose Example 1, and have partners work to solve each part. Discuss the margin note.
- Omit Questions 1 and 2 if you believe students are secure in working with the formula for circumference.

Using Arc Lengths to Find Measures

An **arc length** is a portion of the circumference of a circle. You can use the measure of the arc (in degrees) to find its length (in linear units).

Core Concept

Arc Length

In a circle, the ratio of the length of a given arc to the circumference is equal to the ratio of the measure of the arc to 360°.

$$\frac{\text{Arc length of } \widehat{AB}}{2\pi r} = \frac{m\widehat{AB}}{360°}, \text{ or}$$

$$\text{Arc length of } \widehat{AB} = \frac{m\widehat{AB}}{360°} \cdot 2\pi r$$

EXAMPLE 2 Using Arc Lengths to Find Measures

Find each indicated measure.

a. arc length of $\widehat{AB}$ b. circumference of $\odot Z$ c. $m\widehat{RS}$

SOLUTION

a. Arc length of $\widehat{AB} = \frac{60°}{360°} \cdot 2\pi(8)$

≈ 8.38 cm

b. $\frac{\text{Arc length of } \widehat{XY}}{C} = \frac{m\widehat{XY}}{360°}$

$\frac{4.19}{C} = \frac{40°}{360°}$

$\frac{4.19}{C} = \frac{1}{9}$

$37.71 \text{ in.} = C$

c. $\frac{\text{Arc length of } \widehat{RS}}{2\pi r} = \frac{m\widehat{RS}}{360°}$

$\frac{44}{2\pi(15.28)} = \frac{m\widehat{RS}}{360°}$

$360° \cdot \frac{44}{2\pi(15.28)} = m\widehat{RS}$

$165° \approx m\widehat{RS}$

Monitoring Progress Help in English and Spanish at BigIdeasMath.com

Find the indicated measure.

3. arc length of $\widehat{PQ}$ 4. circumference of $\odot N$ 5. radius of $\odot G$

Extra Example 2

Find each indicated measure.

a. arc length of $\widehat{PR}$

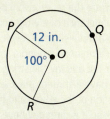

arc length of $\widehat{PR} \approx 20.94$ inches

b. circumference of $\odot P$

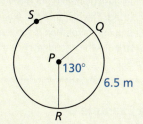

$C = 18$ meters

c. $m\widehat{JK}$

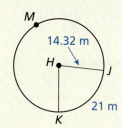

$m\widehat{JK} \approx 84°$

MONITORING PROGRESS ANSWERS

3. about 5.89 yd
4. 81.68 m
5. about 4.01 ft

Laurie's Notes — Teacher Actions

- **Attend to Precision:** Explain the difference between the length of an arc (linear units) and the measure of the arc (degrees).
- **Look For and Make Use of Structure:** To find arc length, set up a proportion (note the units) or find a fractional part of the circumference. Structurally, the proportion and the equation are the same—multiply both sides of the proportion by $2\pi r$ and you have the equation. Example 2 demonstrates these two approaches.
- **Make Sense of Problems and Persevere in Solving Them:** Students need to focus on what information is known, what they are solving for, and the approach they will use.

Differentiated Instruction

Kinesthetic
Demonstrate the relationship between the circumference of a wheel and the distance it travels in one revolution. Wrap a string once around a small wheel or tire and lay the string flat on a table. Have a student mark a point on the edge of the wheel and line up the mark so it is directly above one end of the string. Keep the string taut and roll the wheel along the string until the mark lies directly above the other end of the string. The distance the wheel traveled in one revolution is the length of the string—the circumference of the wheel.

Extra Example 3
The radius of a wheel on a toy truck is 4 inches. To the nearest foot, how far does the wheel travel when it makes 7 revolutions? **The wheel travels approximately 15 feet.**

Extra Example 4
A path is built around four congruent circular fields. The radius of each field is 100 feet. How long is the path? Round to the nearest hundred feet.

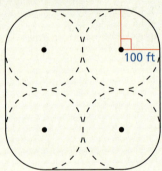

The path is about 1400 feet long.

MONITORING PROGRESS ANSWERS
6. about 68
7. about 445.4 m

Solving Real-Life Problems

EXAMPLE 3 Using Circumference to Find Distance Traveled

The dimensions of a car tire are shown. To the nearest foot, how far does the tire travel when it makes 15 revolutions?

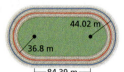

SOLUTION

Step 1 Find the diameter of the tire.
$$d = 15 + 2(5.5) = 26 \text{ in.}$$

Step 2 Find the circumference of the tire.
$$C = \pi d = \pi \cdot 26 = 26\pi \text{ in.}$$

Step 3 Find the distance the tire travels in 15 revolutions. In one revolution, the tire travels a distance equal to its circumference. In 15 revolutions, the tire travels a distance equal to 15 times its circumference.

$$\text{Distance traveled} = \text{Number of revolutions} \cdot \text{Circumference}$$
$$= 15 \cdot 26\pi \approx 1225.2 \text{ in.}$$

COMMON ERROR
Always pay attention to units. In Example 3, you need to convert units to get a correct answer.

Step 4 Use unit analysis. Change 1225.2 inches to feet.
$$1225.2 \text{ in.} \cdot \frac{1 \text{ ft}}{12 \text{ in.}} = 102.1 \text{ ft}$$

▶ The tire travels approximately 102 feet.

EXAMPLE 4 Using Arc Length to Find Distances

The curves at the ends of the track shown are 180° arcs of circles. The radius of the arc for a runner on the red path shown is 36.8 meters. About how far does this runner travel to go once around the track? Round to the nearest tenth of a meter.

SOLUTION

The path of the runner on the red path is made of two straight sections and two semicircles. To find the total distance, find the sum of the lengths of each part.

$$\text{Distance} = 2 \cdot \text{Length of each straight section} + 2 \cdot \text{Length of each semicircle}$$
$$= 2(84.39) + 2\left(\tfrac{1}{2} \cdot 2\pi \cdot 36.8\right)$$
$$\approx 400.0$$

▶ The runner on the red path travels about 400.0 meters.

Monitoring Progress Help in English and Spanish at *BigIdeasMath.com*

6. A car tire has a diameter of 28 inches. How many revolutions does the tire make while traveling 500 feet?

7. In Example 4, the radius of the arc for a runner on the blue path is 44.02 meters, as shown in the diagram. About how far does this runner travel to go once around the track? Round to the nearest tenth of a meter.

Laurie's Notes — Teacher Actions

- **Whiteboarding:** For the problems on this page, whiteboards are helpful to quickly share work.
- Students will recognize that Example 3 involves the circumference of the tire.
- Students need to pay attention to units.
- Have partners work independently to solve the problem as you circulate.
- ❓ "In Example 4, how do you find the distance around the track?" **The two circular arcs equal a circumference. Then add the two lengths of the straight section.**
- ❓ "Is the length of the circular arc the same for all runners?" **no**
- ❓ "Is the length of the straight section the same for all runners?" **yes**

Measuring Angles in Radians

Recall that in a circle, the ratio of the length of a given arc to the circumference is equal to the ratio of the measure of the arc to 360°. To see why, consider the diagram.

A circle of radius 1 has circumference 2π, so the arc length of $\widehat{CD}$ is $\dfrac{m\widehat{CD}}{360°} \cdot 2\pi$.

Recall that all circles are similar and corresponding lengths of similar figures are proportional. Because $m\widehat{AB} = m\widehat{CD}$, $\widehat{AB}$ and $\widehat{CD}$ are corresponding arcs. So, you can write the following proportion.

$$\dfrac{\text{Arc length of } \widehat{AB}}{\text{Arc length of } \widehat{CD}} = \dfrac{r}{1}$$

Arc length of $\widehat{AB} = r \cdot$ Arc length of $\widehat{CD}$

Arc length of $\widehat{AB} = r \cdot \dfrac{m\widehat{CD}}{360°} \cdot 2\pi$

This form of the equation shows that the arc length associated with a central angle is *proportional to the radius* of the circle. The constant of proportionality, $\dfrac{m\widehat{CD}}{360°} \cdot 2\pi$, is defined to be the **radian** measure of the central angle associated with the arc.

In a circle of radius 1, the radian measure of a given central angle can be thought of as the length of the arc associated with the angle. The radian measure of a complete circle (360°) is exactly 2π radians, because the circumference of a circle of radius 1 is exactly 2π. You can use this fact to convert from degree measure to radian measure and vice versa.

Core Concept

Converting between Degrees and Radians

Degrees to radians
Multiply degree measure by
$\dfrac{2\pi \text{ radians}}{360°}$, or $\dfrac{\pi \text{ radians}}{180°}$.

Radians to degrees
Multiply radian measure by
$\dfrac{360°}{2\pi \text{ radians}}$, or $\dfrac{180°}{\pi \text{ radians}}$.

EXAMPLE 5 Converting between Degree and Radian Measure

a. Convert 45° to radians.

b. Convert $\dfrac{3\pi}{2}$ radians to degrees.

SOLUTION

a. $45° \cdot \dfrac{\pi \text{ radians}}{180°} = \dfrac{\pi}{4}$ radian

 So, $45° = \dfrac{\pi}{4}$ radian.

b. $\dfrac{3\pi}{2}$ radians $\cdot \dfrac{180°}{\pi \text{ radians}} = 270°$

 So, $\dfrac{3\pi}{2}$ radians $= 270°$.

Monitoring Progress Help in English and Spanish at *BigIdeasMath.com*

8. Convert 15° to radians.

9. Convert $\dfrac{4\pi}{3}$ radians to degrees.

Extra Example 5

a. Convert 30° to radians. $\dfrac{\pi}{6}$ radian

b. Convert $\dfrac{3\pi}{8}$ radians to degrees. 67.5°

MONITORING PROGRESS ANSWERS

8. $\dfrac{\pi}{12}$ radian

9. 240°

Laurie's Notes   Teacher Actions

- Work slowly through the development of radian measure, being careful with language and symbols.
- Explain that the *radian* measure of a given central angle can be thought of as the length of the associated arc.
- Write the *Core Concept*, which states the conversion factors.
- Work through Example 5. It is helpful to finish with a circular model showing the angle and arc length marked for each problem.

Closure

- **Writing Prompt:** Explain how the radius, central angles, arc lengths, and circumference of a circle are related.

Assignment Guide and Homework Check

ASSIGNMENT

Basic: 1, 2, 3–23 odd, 29, 36, 37, 43, 44

Average: 1, 2–28 even, 29–31, 36–38, 43, 44

Advanced: 1, 2, 4–22 even, 25–33, 36–44

HOMEWORK CHECK

Basic: 3, 7, 13, 15, 21

Average: 6, 10, 14, 16, 22

Advanced: 8, 14, 16, 22, 31

ANSWERS

1. πd
2. Arc measure refers to the angle and arc length refers to the length.
3. about 37.70 in.
4. about 20.05 ft
5. 14 units
6. 5π in.
7. about 3.14 ft
8. about 50.02°
9. about 35.53 m
10. about 8.58 cm
11. The diameter was used as the radius; $C = \pi d = 9\pi$ in.
12. The arc measure should be divided by 360°; Arc length of $\widehat{GH} = \dfrac{m\widehat{GH}}{360°} \cdot 2\pi r = \dfrac{25}{12}\pi$ cm
13. 182 ft
14. 19
15. about 44.85 units
16. about 21.42 units

11.1 Exercises

Dynamic Solutions available at BigIdeasMath.com

Vocabulary and Core Concept Check

1. **COMPLETE THE SENTENCE** The circumference of a circle with diameter d is $C = $ _____.
2. **WRITING** Describe the difference between an arc measure and an arc length.

Monitoring Progress and Modeling with Mathematics

In Exercises 3–10, find the indicated measure. *(See Examples 1 and 2.)*

3. circumference of a circle with a radius of 6 inches
4. diameter of a circle with a circumference of 63 feet
5. radius of a circle with a circumference of 28π
6. exact circumference of a circle with a diameter of 5 inches
7. arc length of $\widehat{AB}$ 8. $m\widehat{DE}$

9. circumference of $\odot C$ 10. radius of $\odot R$

 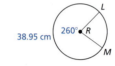

11. **ERROR ANALYSIS** Describe and correct the error in finding the circumference of $\odot C$.

12. **ERROR ANALYSIS** Describe and correct the error in finding the length of $\widehat{GH}$.

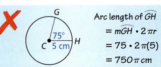

13. **PROBLEM SOLVING** A measuring wheel is used to calculate the length of a path. The diameter of the wheel is 8 inches. The wheel makes 87 complete revolutions along the length of the path. To the nearest foot, how long is the path? *(See Example 3.)*

14. **PROBLEM SOLVING** You ride your bicycle 40 meters. How many complete revolutions does the front wheel make?

In Exercises 15–18, find the perimeter of the shaded region. *(See Example 4.)*

15.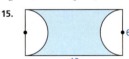

16.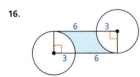

644   Chapter 11   Circumference, Area, and Volume

17.

18.

In Exercises 19–22, convert the angle measure. *(See Example 5.)*

19. Convert 70° to radians.

20. Convert 300° to radians.

21. Convert $\frac{11\pi}{12}$ radians to degrees.

22. Convert $\frac{\pi}{8}$ radian to degrees.

23. PROBLEM SOLVING The London Eye is a Ferris wheel in London, England, that travels at a speed of 0.26 meter per second. How many minutes does it take the London Eye to complete one full revolution?

24. PROBLEM SOLVING You are planning to plant a circular garden adjacent to one of the corners of a building, as shown. You can use up to 38 feet of fence to make a border around the garden. What radius (in feet) can the garden have? Choose all that apply. Explain your reasoning.

Ⓐ 7   Ⓑ 8   Ⓒ 9   Ⓓ 10

In Exercises 25 and 26, find the circumference of the circle with the given equation. Write the circumference in terms of π.

25. $x^2 + y^2 = 16$

26. $(x + 2)^2 + (y - 3)^2 = 9$

27. USING STRUCTURE A semicircle has endpoints $(-2, 5)$ and $(2, 8)$. Find the arc length of the semicircle.

28. REASONING $\overset{\frown}{EF}$ is an arc on a circle with radius r. Let $x°$ be the measure of $\overset{\frown}{EF}$. Describe the effect on the length of $\overset{\frown}{EF}$ if you (a) double the radius of the circle, and (b) double the measure of $\overset{\frown}{EF}$.

29. MAKING AN ARGUMENT Your friend claims that it is possible for two arcs with the same measure to have different arc lengths. Is your friend correct? Explain your reasoning.

30. PROBLEM SOLVING Over 2000 years ago, the Greek scholar Eratosthenes estimated Earth's circumference by assuming that the Sun's rays were parallel. He chose a day when the Sun shone straight down into a well in the city of Syene. At noon, he measured the angle the Sun's rays made with a vertical stick in the city of Alexandria. Eratosthenes assumed that the distance from Syene to Alexandria was equal to about 575 miles. Explain how Eratosthenes was able to use this information to estimate Earth's circumference. Then estimate Earth's circumference.

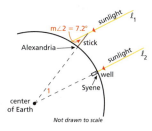

31. ANALYZING RELATIONSHIPS In $\odot C$, the ratio of the length of $\overset{\frown}{PQ}$ to the length of $\overset{\frown}{RS}$ is 2 to 1. What is the ratio of $m\angle PCQ$ to $m\angle RCS$?

Ⓐ 4 to 1   Ⓑ 2 to 1

Ⓒ 1 to 4   Ⓓ 1 to 2

32. ANALYZING RELATIONSHIPS A 45° arc in $\odot C$ and a 30° arc in $\odot P$ have the same length. What is the ratio of the radius r_1 of $\odot C$ to the radius r_2 of $\odot P$? Explain your reasoning.

ANSWERS

17. about 20.57 units

18. about 30.71 units

19. $\frac{7\pi}{18}$ radian

20. $\frac{5\pi}{3}$ radians

21. 165°

22. 22.5°

23. about 27.19 min

24. A, B; *Sample answer:* The maximum radius is about 8.06 feet.

25. 8π units

26. 6π units

27. about 7.85 units

28. a. The length doubles.
  b. The length doubles.

29. yes; *Sample answer:* The arc length also depends on the radius.

30. *Sample answer:* Angles 1 and 2 are alternate interior angles, so the arc measure is 7.2°; 28,750 mi

31. B

32. 2 to 3; *Sample answer:* $\frac{1}{4}\pi r_1 = \frac{1}{6}\pi r_2$ implies $\frac{r_1}{r_2} = \frac{2}{3}$.

ANSWERS

33. $2\frac{1}{3}$
34. a. about 62.83 in.
    b. about 26.66 cm
    c. about 16.32 in.
35. arc length of $\overset{\frown}{AB} = r\theta$; about 9.42 in.
36. They are the same; Sample answer: The radius of $\odot C$ is the same as the diameter of $\odot P$.
37. yes; Sample answer: The circumference of the red circle can be found using $2 = \frac{30°}{360°}C$. The circumference of the blue circle is double the circumference of the red circle.
38. a. $\frac{3\pi}{4}$ radian; Sample answer: The hour hand is halfway between 1 and 2, and the minute hand is on 6.
    b. $\frac{\pi}{24}$ radian; Sample answer: The hour hand is one-quarter way between 3 and 4, and the minute hand is on 3.
39. 28 units
40–44. See Additional Answers.

Mini-Assessment

Use the diagram.

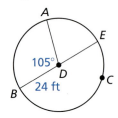

1. What is the circumference of $\odot D$? **about 150.80 feet**
2. What is the arc length of $\overset{\frown}{AB}$? **about 43.98 feet**
3. Children's bicycles are sized according to the diameter of their wheels. A bicycle recommended for six-year-olds has wheels whose diameters are 18 inches. How many revolutions does the wheel make when traveling 42 feet?
   about 9 revolutions
4. Convert 210° to radians.
   $\frac{7\pi}{6}$ **radians**
5. Convert $\frac{5\pi}{12}$ radians to degrees. **75°**

646   Chapter 11

---

33. **PROBLEM SOLVING** How many revolutions does the smaller gear complete during a single revolution of the larger gear?

34. **USING STRUCTURE** Find the circumference of each circle.
    a. a circle circumscribed about a right triangle whose legs are 12 inches and 16 inches long
    b. a circle circumscribed about a square with a side length of 6 centimeters
    c. a circle inscribed in an equilateral triangle with a side length of 9 inches

35. **REWRITING A FORMULA** Write a formula in terms of the measure θ of the central angle (in radians) that can be used to find the length of an arc of a circle. Then use this formula to find the length of an arc of a circle with a radius of 4 inches and a central angle of $\frac{3\pi}{4}$ radians.

36. **HOW DO YOU SEE IT?**
    Compare the circumference of $\odot P$ to the length of $\overline{DE}$. Explain your reasoning.

37. **MAKING AN ARGUMENT** In the diagram, the measure of the red shaded angle is 30°. The arc length a is 2. Your classmate claims that it is possible to find the circumference of the blue circle without finding the radius of either circle. Is your classmate correct? Explain your reasoning.

38. **MODELING WITH MATHEMATICS** What is the measure (in radians) of the angle formed by the hands of a clock at each time? Explain your reasoning.
    a. 1:30 P.M.   b. 3:15 P.M.

39. **MATHEMATICAL CONNECTIONS** The sum of the circumferences of circles A, B, and C is 63π. Find AC.

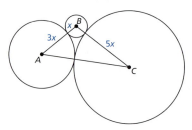

40. **THOUGHT PROVOKING** Is π a rational number? Compare the rational number $\frac{355}{113}$ to π. Find a different rational number that is even closer to π.

41. **PROOF** The circles in the diagram are concentric and $\overline{FG} \cong \overline{GH}$. Prove that $\overline{JK}$ and $\overline{NG}$ have the same length.

42. **REPEATED REASONING** $\overline{AB}$ is divided into four congruent segments, and semicircles with radius r are drawn.

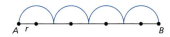

   a. What is the sum of the four arc lengths?
   b. What would the sum of the arc lengths be if $\overline{AB}$ was divided into 8 congruent segments? 16 congruent segments? n congruent segments? Explain your reasoning.

Maintaining Mathematical Proficiency Reviewing what you learned in previous grades and lessons

Find the area of the polygon with the given vertices. *(Skills Review Handbook)*

43. $X(2, 4)$, $Y(8, -1)$, $Z(2, -1)$
44. $L(-3, 1)$, $M(4, 1)$, $N(4, -5)$, $P(-3, -5)$

646   Chapter 11   Circumference, Area, and Volume

If students need help...	If students got it...
Resources by Chapter • Practice A and Practice B • Puzzle Time	Resources by Chapter • Enrichment and Extension • Cumulative Review
Student Journal • Practice	Start the *next* Section
Differentiating the Lesson Skills Review Handbook	

Laurie's Notes

Overview of Section 11.2

Introduction
- This lesson reintroduces the formula for the area of a circle. Students should be familiar with the formula from middle school.
- New in this lesson is how to find the area of a sector of a circle.
- Area of a circle is also used to solve a few geometric probability problems.

Formative Assessment Tips
- **Opposing Views:** This technique allows students to compare two or more solutions or ways of thinking about a problem.
- To be informative, students must present more than yes/no answers. Students must be able to explain their reasoning for classmates to hear. Once the different views are presented, all students should decide which view most closely reflects their own and give justification for their thinking. Students can respond in writing, or you can solicit viewpoints as a discussion prompt.
- **Construct Viable Arguments and Critique the Reasoning of Others:** Using this formative assessment technique allows you to check students' conceptual knowledge and their ability to construct a viable argument.

Extensions
- Exercise 28 on page 653 has a nice extension. Ask students to find the percent of the square that is shaded. Now imagine having 9 tangent circles drawn inside the same square. What percent of the square is shaded? Repeat for 16 and 25 tangent circles. Have students make observations.

Pacing Suggestion
- The explorations provide a brief introduction to areas of sectors. When students have finished, transition to the formal lesson. You might omit Example 1.

Dynamic Teaching Tools
Dynamic Assessment & Progress Monitoring Tool
Lesson Planning Tool
Interactive Whiteboard Lesson Library
Dynamic Classroom with Dynamic Investigations

What Your Students Will Learn

- Use the formula for the area of a circle to find radii, diameters, and areas of circles.
- Use the formula for the area of a circle to find geometric probabilities.
- Find areas of sectors as fractional parts of the areas of circles.
- Use areas of sectors to find areas of circles.

Laurie's Notes

Exploration

Motivate
- **Story Time:** When students enter, display this iconic symbol and ask students whether they are familiar with it. Tell them that you were such a fan of this video game that, in college, you had a rug in this shape in your dorm room. Everyone envied you!
- "How big was this rug you ask? I'll tell you at the end of class!"

Exploration Note
- Students should be familiar with the formula for the area of a circle from middle school.

Exploration 1
- The fractional amount of the area is stated, so a simple calculation is needed.
- ? "How did you calculate the area of each sector?" Multiply the fractional part of the circle by πr^2. It is also possible that students set up a proportion to solve.
- ? "What type of unit is used to label your answer?" a square unit
- ? **Attend to Precision:** "What is the exact answer for part (a)?" 36π square units
  "What is an approximate answer for part (a)?" about 113.10 square units

Exploration 2
- Students should recognize that this exploration is just an application of the problems found in the first exploration. The radius is 400 meters.
- ? "What fractional part of the whole area do you need to find?" $\frac{1}{3}$ of the area
- ? **Attend to Precision:** "Should you leave your answer in terms of π? Explain." no; You should use a decimal approximation.

Communicate Your Answer
- Students will likely describe finding a fractional part of the area, or they will describe setting up a proportion. In either case, listen for correct language.

Connecting to Next Step
- Students have now found the area of a circle and the areas of sectors. Applications of areas of sectors will be found in today's lesson.

11.2 Areas of Circles and Sectors

Essential Question How can you find the area of a sector of a circle?

EXPLORATION 1 Finding the Area of a Sector of a Circle

Work with a partner. A **sector of a circle** is the region bounded by two radii of the circle and their intercepted arc. Find the area of each shaded circle or sector of a circle.

a. entire circle

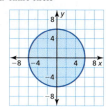

b. one-fourth of a circle

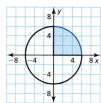

c. seven-eighths of a circle

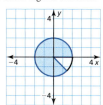

d. two-thirds of a circle

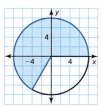

REASONING ABSTRACTLY
To be proficient in math, you need to explain to yourself the meaning of a problem and look for entry points to its solution.

EXPLORATION 2 Finding the Area of a Circular Sector

Work with a partner. A center pivot irrigation system consists of 400 meters of sprinkler equipment that rotates around a central pivot point at a rate of once every 3 days to irrigate a circular region with a diameter of 800 meters. Find the area of the sector that is irrigated by this system in one day.

Communicate Your Answer

3. How can you find the area of a sector of a circle?
4. In Exploration 2, find the area of the sector that is irrigated in 2 hours.

ANSWERS
1. a. about 113.10 square units
   b. about 28.27 square units
   c. about 11.00 square units
   d. about 134.04 square units
2. about 167,552 m^2
3. Multiply the fraction of the circle the sector represents by the area of the circle.
4. about 13,963 m^2

English Language Learners

Group Activity
Organize the students into small groups of English learners and English speakers. Give each group a problem similar to Example 1 and have each group solve it together. When finished, have students share their work at the board.

Extra Example 1
Find each indicated measure.
a. area of a circle with a radius of 8.5 inches The area of the circle is about 226.98 square inches.
b. diameter of a circle with an area of 153.94 square feet The radius is about 7 feet, so the diameter is about 14 feet.

MONITORING PROGRESS ANSWERS
1. about 63.62 m^2
2. about 7.50 ft

11.2 Lesson

What You Will Learn
- Use the formula for the area of a circle.
- Find areas of sectors.
- Use areas of sectors.

Core Vocabulary
geometric probability, *p. 649*
sector of a circle, *p. 650*
Previous
circle
radius
diameter
intercepted arc

Using the Formula for the Area of a Circle

You can divide a circle into congruent sections and rearrange the sections to form a figure that approximates a parallelogram. Increasing the number of congruent sections increases the figure's resemblance to a parallelogram.

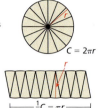

The base of the parallelogram that the figure approaches is half of the circumference, so $b = \frac{1}{2}C = \frac{1}{2}(2\pi r) = \pi r$. The height is the radius, so $h = r$. So, the area of the parallelogram is $A = bh = (\pi r)(r) = \pi r^2$.

Core Concept
Area of a Circle
The area of a circle is
$$A = \pi r^2$$
where r is the radius of the circle.

EXAMPLE 1   Using the Formula for the Area of a Circle

Find each indicated measure.
a. area of a circle with a radius of 2.5 centimeters
b. diameter of a circle with an area of 113.1 square centimeters

SOLUTION

a. $A = \pi r^2$ Formula for area of a circle
$= \pi \cdot (2.5)^2$ Substitute 2.5 for r.
$= 6.25\pi$ Simplify.
≈ 19.63 Use a calculator.

▶ The area of the circle is about 19.63 square centimeters.

b. $A = \pi r^2$ Formula for area of a circle
$113.1 = \pi r^2$ Substitute 113.1 for A.
$\dfrac{113.1}{\pi} = r^2$ Divide each side by π.
$6 \approx r$ Find the positive square root of each side.

▶ The radius is about 6 centimeters, so the diameter is about 12 centimeters.

Monitoring Progress Help in English and Spanish at *BigIdeasMath.com*

1. Find the area of a circle with a radius of 4.5 meters.
2. Find the radius of a circle with an area of 176.7 square feet.

Laurie's Notes   Teacher Actions

- Although students may be familiar with the area formula, they may not have seen its derivation. The model shown can be easily made out of heavyweight paper and used each year. There are also online simulations and applets that model this approach.
- Write the *Core Concept*.
- ❓ "What type of unit is used to measure area?"   a square unit
- **Think-Pair-Share:** Have students work independently to answer Example 1. Discuss as a class. You might omit Questions 1 and 2.

Probabilities found by calculating a ratio of two lengths, areas, or volumes are called **geometric probabilities**.

EXAMPLE 2 Using Area to Find Probability

You throw a dart at the board shown. Your dart is equally likely to hit any point inside the square board. Are you more likely to get 10 points or 0 points?

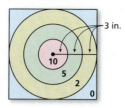

SOLUTION

The probability of getting 10 points is

$$P(10 \text{ points}) = \frac{\text{Area of smallest circle}}{\text{Area of entire board}}$$

$$= \frac{\pi \cdot 3^2}{18^2}$$

$$= \frac{9\pi}{324}$$

$$= \frac{\pi}{36}$$

$$\approx 0.0873.$$

The probability of getting 0 points is

$$P(0 \text{ points}) = \frac{\text{Area outside largest circle}}{\text{Area of entire board}}$$

$$= \frac{18^2 - (\pi \cdot 9^2)}{18^2}$$

$$= \frac{324 - 81\pi}{324}$$

$$= \frac{4 - \pi}{4}$$

$$\approx 0.215.$$

▶ You are more likely to get 0 points.

Monitoring Progress Help in English and Spanish at *BigIdeasMath.com*

Use the diagram in Example 2.

3. Are you more likely to get 10 points or 5 points?
4. Are you more likely to get 2 points or 0 points?
5. Are you more likely to get 5 or more points or 2 or less points?
6. Are you more likely to score points (10, 5, or 2) or get 0 points?

Section 11.2 Areas of Circles and Sectors 649

Extra Example 2

You throw a dart at the board shown. Your dart is equally likely to hit any point inside the square board. What is the probability that your dart lands in one of the two yellow regions in the smaller circle?

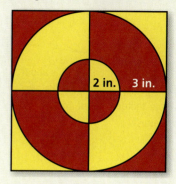

P(landing on one of the smaller yellow regions) $= \frac{2\pi}{100} \approx 0.063$

MONITORING PROGRESS ANSWERS

3. 5 points
4. 2 points
5. 2 or less points
6. score points

Laurie's Notes — Teacher Actions

- Pose Example 2 and give students time to consider how they might answer. Solicit ideas. Work through the problem.
- **Opposing Views:** Pose an extension by asking, "What if the target were enlarged so that the length of the radii were now 4, 8, and 12. Who believes you are more likely to score 10 points? 0 points?" Give students time to consider their answer. Some students will say 10 points, some will say 0 points, and some will be uncertain. Ask a few students from each viewpoint to share their reasoning. The class should decide that you are most likely going to score 0 points. What surprises students the most is that the probabilities do not change from the original question!
- **Think-Pair-Share:** Have students answer Questions 3-6, and then share and discuss as a class.

Extra Example 3
Find the areas of the sectors formed by ∠PSQ.

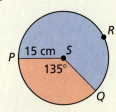

The area of the small sector is about 265.07 square centimeters. The area of the large sector is about 441.79 square centimeters.

MONITORING PROGRESS ANSWERS
7. about 205.25 ft^2
8. about 410.50 ft^2

Finding Areas of Sectors

A **sector of a circle** is the region bounded by two radii of the circle and their intercepted arc. In the diagram below, sector APB is bounded by $\overline{AP}$, $\overline{BP}$, and $\widehat{AB}$.

ANALYZING RELATIONSHIPS

The area of a sector is a fractional part of the area of a circle. The area of a sector formed by a 45° arc is $\frac{45°}{360°}$, or $\frac{1}{8}$ of the area of the circle.

Core Concept
Area of a Sector
The ratio of the area of a sector of a circle to the area of the whole circle (πr^2) is equal to the ratio of the measure of the intercepted arc to 360°.

$$\frac{\text{Area of sector } APB}{\pi r^2} = \frac{m\widehat{AB}}{360°}, \text{ or}$$

$$\text{Area of sector } APB = \frac{m\widehat{AB}}{360°} \cdot \pi r^2$$

EXAMPLE 3 Finding Areas of Sectors

Find the areas of the sectors formed by ∠UTV.

SOLUTION

Step 1 Find the measures of the minor and major arcs.
Because $m\angle UTV = 70°$, $m\widehat{UV} = 70°$ and $m\widehat{USV} = 360° - 70° = 290°$.

Step 2 Find the areas of the small and large sectors.

$\text{Area of small sector} = \frac{m\widehat{UV}}{360°} \cdot \pi r^2$ Formula for area of a sector

$= \frac{70°}{360°} \cdot \pi \cdot 8^2$ Substitute.

≈ 39.10 Use a calculator.

$\text{Area of large sector} = \frac{m\widehat{USV}}{360°} \cdot \pi r^2$ Formula for area of a sector

$= \frac{290°}{360°} \cdot \pi \cdot 8^2$ Substitute.

≈ 161.97 Use a calculator.

▶ The areas of the small and large sectors are about 39.10 square inches and about 161.97 square inches, respectively.

Monitoring Progress   Help in English and Spanish at *BigIdeasMath.com*

Find the indicated measure.

7. area of red sector
8. area of blue sector

650   Chapter 11   Circumference, Area, and Volume

Laurie's Notes — Teacher Actions

- **Connection:** Define *sector of a circle*. Explain that the area of the sector can be found using techniques similar to what was done in finding arc length.
- **Look For and Make Use of Structure:** Write the *Core Concept*. Structurally, the proportion and the equation are the same—multiply both sides of the proportion by πr^2 and you have the equation.
- Have students solve Example 3.
- ❓ "How could you check your two answers?" They should sum to the area of the circle.

Using Areas of Sectors

EXAMPLE 4 Using the Area of a Sector

Find the area of ⊙V.

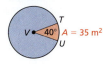

SOLUTION

Area of sector $TVU = \dfrac{m\widehat{TU}}{360°} \cdot$ Area of ⊙V   Formula for area of a sector

$35 = \dfrac{40°}{360°} \cdot$ Area of ⊙V   Substitute.

$315 =$ Area of ⊙V   Solve for area of ⊙V.

▶ The area of ⊙V is 315 square meters.

EXAMPLE 5 Finding the Area of a Region

A rectangular wall has an entrance cut into it. You want to paint the wall. To the nearest square foot, what is the area of the region you need to paint?

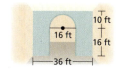

COMMON ERROR
Use the radius (8 feet), not the diameter (16 feet), when you calculate the area of the semicircle.

SOLUTION

The area you need to paint is the area of the rectangle minus the area of the entrance. The entrance can be divided into a semicircle and a square.

Area of wall = Area of rectangle − (Area of semicircle + Area of square)

$= 36(26) - \left[\dfrac{180°}{360°} \cdot (\pi \cdot 8^2) + 16^2 \right]$

$= 936 - (32\pi + 256)$

≈ 579.47

▶ The area you need to paint is about 579 square feet.

Monitoring Progress 🔊 Help in English and Spanish at *BigIdeasMath.com*

9. Find the area of ⊙H.

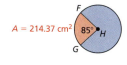

10. Find the area of the figure.

11. If you know the area and radius of a sector of a circle, can you find the measure of the intercepted arc? Explain.

Section 11.2 Areas of Circles and Sectors 651

Extra Example 4
Find the area of ⊙S.

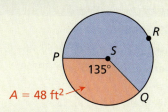

The area of ⊙S is 128 square feet.

Extra Example 5
A farmer has a field with the shape shown. Find the area of the shaded region to the nearest square meter.

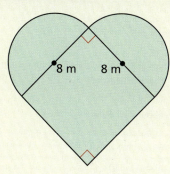

The area is about 114 square meters.

MONITORING PROGRESS ANSWERS

9. 907.92 cm²
10. about 43.74 m²
11. yes; Solve the formula for the measure of the intercepted arc.

Laurie's Notes Teacher Actions

❓ "In Example 4, what portion of the circle has an area of 35 square meters?" $\dfrac{40°}{360°} = \dfrac{1}{9}$
- Have students solve Example 4, being sure to label their answer.
- **Think-Alouds:** Pose Example 5 and say, "To solve this problem I need to …." Ask partner A to think aloud for partner B to hear the problem-solving process. When students have finished the problem, use *Popsicle Sticks* to solicit a response.

Closure
- **Exit Ticket:** Find the area of your teacher's college dorm room rug. The rug had a 6-foot diameter with a 60° sector removed. $\dfrac{15\pi}{2}$ or about 23.6 square feet

Assignment Guide and Homework Check

ASSIGNMENT

Basic: 1, 2, 3–31 odd, 34, 40, 46–49

Average: 1, 2, 8–38 even, 39, 40, 46–49

Advanced: 1, 2, 4, 10, 18, 22–46 even, 47–49

HOMEWORK CHECK

Basic: 5, 17, 21, 25, 31

Average: 8, 16, 20, 26, 30

Advanced: 10, 18, 22, 32, 34

ANSWERS

1. sector
2. yes; The sector area is proportional to the arc measure.
3. about 0.50 cm²
4. about 314.16 in.²
5. about 78.54 in.²
6. about 201.06 ft²
7. about 5.32 ft
8. about 11.00 in.
9. about 4.00 in.
10. 52 cm
11. about 0.09, or about 9%
12. about 0.21, or about 21%
13. about 0.02, or about 2%
14. about 0.56, or about 56%
15. about 0.65, or about 65%
16. about 0.67, or about 67%
17. about 0.79, or about 79%
18. about 0.44, or about 44%
19. about 52.36 in.²; about 261.80 in.²
20. about 177.88 cm²; about 437.87 cm²
21. about 937.31 m²; about 1525.70 m²
22. about 10.47 ft²; about 39.79 ft²
23. The diameter was substituted in the formula for area as the radius; $A = \pi(6)^2 \approx 113.10$ ft²
24. The angle measures should be on the same side of the proportion; $\frac{115}{360} = \frac{n}{255}$; $n \approx 81.46$ ft²

11.2 Exercises

Dynamic Solutions available at *BigIdeasMath.com*

Vocabulary and Core Concept Check

1. **VOCABULARY** A(n) _____ of a circle is the region bounded by two radii of the circle and their intercepted arc.

2. **WRITING** The arc measure of a sector in a given circle is doubled. Will the area of the sector also be doubled? Explain your reasoning.

Monitoring Progress and Modeling with Mathematics

In Exercises 3–10, find the indicated measure. (See Example 1.)

3. area of ⊙C

4. area of ⊙C

5. area of a circle with a radius of 5 inches

6. area of a circle with a diameter of 16 feet

7. radius of a circle with an area of 89 square feet

8. radius of a circle with an area of 380 square inches

9. diameter of a circle with an area of 12.6 square inches

10. diameter of a circle with an area of 676π square centimeters

In Exercises 11–18, you throw a dart at the board shown. Your dart is equally likely to hit any point inside the square board. Find the probability that your dart lands in the indicated region. (See Example 2.)

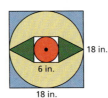

11. red
12. blue
13. white
14. yellow
15. yellow or red
16. yellow or green
17. not blue
18. not yellow

In Exercises 19–22, find the areas of the sectors formed by ∠DFE. (See Example 3.)

19.

20.

21.

22.

23. **ERROR ANALYSIS** Describe and correct the error in finding the area of the circle.

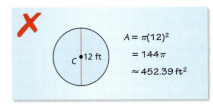

24. **ERROR ANALYSIS** Describe and correct the error in finding the area of sector XZY when the area of ⊙Z is 255 square feet.

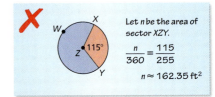

652   Chapter 11   Circumference, Area, and Volume

In Exercises 25 and 26, the area of the shaded sector is shown. Find the indicated measure. *(See Example 4.)*

25. area of ⊙M

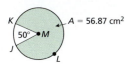

26. radius of ⊙M

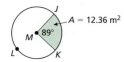

In Exercises 27–32, find the area of the shaded region. *(See Example 5.)*

27.   28.

29.   30.

31.   32.

33. **PROBLEM SOLVING** The diagram shows the shape of a putting green at a miniature golf course. One part of the green is a sector of a circle. Find the area of the putting green.

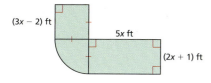

34. **MAKING AN ARGUMENT** Your friend claims that if the radius of a circle is doubled, then its area doubles. Is your friend correct? Explain your reasoning.

35. **MODELING WITH MATHEMATICS** The diagram shows the area of a lawn covered by a water sprinkler.

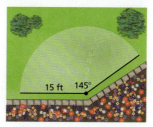

   a. What is the area of the lawn that is covered by the sprinkler?
   b. The water pressure is weakened so that the radius is 12 feet. What is the area of the lawn that will be covered?

36. **MODELING WITH MATHEMATICS** The diagram shows a projected beam of light from a lighthouse.

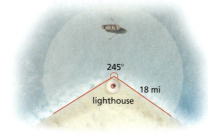

   a. What is the area of water that can be covered by the light from the lighthouse?
   b. What is the area of land that can be covered by the light from the lighthouse?

37. **ANALYZING RELATIONSHIPS** Look back at the Perimeters of Similar Polygons Theorem and the Areas of Similar Polygons Theorem in Section 8.3. How would you rewrite these theorems to apply to circles? Explain your reasoning.

38. **ANALYZING RELATIONSHIPS** A square is inscribed in a circle. The same square is also circumscribed about a smaller circle. Draw a diagram that represents this situation. Then find the ratio of the area of the larger circle to the area of the smaller circle.

Dynamic Teaching Tools
- Dynamic Assessment & Progress Monitoring Tool
- Interactive Whiteboard Lesson Library
- Dynamic Classroom with Dynamic Investigations

ANSWERS
25. about 66.04 cm^2
26. about 3.99 m
27. about 1696.46 m^2
28. about 85.84 in.^{2}
29. about 43.98 ft^2
30. about 301.59 cm^2
31. about 26.77 in.^{2}
32. about 7.63 m^2
33. about 192.48 ft^2
34. no; *Sample answer:* The area is proportional to the square of the radius.
35. a. about 285 ft^2
    b. about 182 ft^2
36. a. about 693 mi^2
    b. about 325 mi^2
37. *Sample answer:* change side lengths to radii and perimeter to circumference; Different terms need to be used because a circle is not a polygon.
38.

    2 to 1

ANSWERS

39. a. *Sample answer:* The total is 100%.

b. bus 234°; walk 90°; other 36°

How Students Get To School

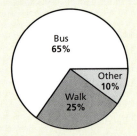

c. bus: about 8.17 in.²; walk: about 3.14 in.²; other: about 1.26 in.²

40–49. See Additional Answers.

Mini-Assessment

1. Find the area of a circle with radius 12 feet. about 452.39 square feet

2. Find the areas of the sectors formed by ∠LMN.

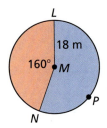

small sector: about 452.39 square meters; large sector: about 565.49 square meters

3. Find the area of ⊙E.

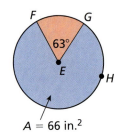

80 square inches

4. Find the area of the shaded region.

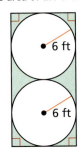

about 61.81 square feet

---

39. CONSTRUCTION The table shows how students get to school.

Method	Percent of students
bus	65%
walk	25%
other	10%

a. Explain why a circle graph is appropriate for the data.

b. You will represent each method by a sector of a circle graph. Find the central angle to use for each sector. Then construct the graph using a radius of 2 inches.

c. Find the area of each sector in your graph.

40. HOW DO YOU SEE IT? The outermost edges of the pattern shown form a square. If you know the dimensions of the outer square, is it possible to compute the total colored area? Explain.

41. ABSTRACT REASONING A circular pizza with a 12-inch diameter is enough for you and 2 friends. You want to buy pizzas for yourself and 7 friends. A 10-inch diameter pizza with one topping costs $6.99 and a 14-inch diameter pizza with one topping costs $12.99. How many 10-inch and 14-inch pizzas should you buy in each situation? Explain.

a. You want to spend as little money as possible.

b. You want to have three pizzas, each with a different topping, and spend as little money as possible.

c. You want to have as much of the thick outer crust as possible.

Maintaining Mathematical Proficiency
Reviewing what you learned in previous grades and lessons

Find the area of the figure. *(Skills Review Handbook)*

46.   **47.**   **48.**   **49.**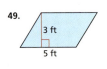

---

42. THOUGHT PROVOKING You know that the area of a circle is πr^2. Find the formula for the area of an *ellipse*, shown below.

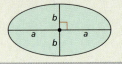

43. MULTIPLE REPRESENTATIONS Consider a circle with a radius of 3 inches.

a. Complete the table, where x is the measure of the arc and y is the area of the corresponding sector. Round your answers to the nearest tenth.

x	30°	60°	90°	120°	150°	180°
y						

b. Graph the data in the table.

c. Is the relationship between x and y linear? Explain.

d. If parts (a)–(c) were repeated using a circle with a radius of 5 inches, would the areas in the table change? Would your answer to part (c) change? Explain your reasoning.

44. CRITICAL THINKING Find the area between the three congruent tangent circles. The radius of each circle is 6 inches.

45. PROOF Semicircles with diameters equal to three sides of a right triangle are drawn, as shown. Prove that the sum of the areas of the two shaded crescents equals the area of the triangle.

Laurie's Notes

Overview of Section 11.3

Introduction
- Earlier in the book, students found the areas of triangles and special quadrilaterals. This lesson builds upon that knowledge and includes the area of kites, rhombuses, and regular polygons.
- To find the area of regular polygons, it is necessary to be able to find certain angle measures in the regular polygon along with the length of the apothem. Trigonometric ratios are used to find certain segment lengths that enable the area of the regular polygon to be found.

Teaching Strategy
- The explorations for this lesson help students to develop a strategy and understanding of how to find the area of any regular polygon. In the formal lesson, students will learn to find the length of the apothem using trigonometric ratios.
- If time permits, you could have students write a script or program for the dynamic geometry software they are familiar with that computes the area of regular polygons. Students should use a slider for the length of the side of the regular polygon and a second slider for the number of sides in the regular polygon.
- A quick online search will reveal that such work has already been done by others. Having students create the script deepens their understanding of the mathematics involved.

Pacing Suggestion
- The first exploration provides an opportunity for students to develop a conceptual understanding of how to find the area of a regular polygon. Transition to the formal lesson as soon as students have discussed the second exploration.

Dynamic Teaching Tools
Dynamic Assessment & Progress Monitoring Tool
Lesson Planning Tool
Interactive Whiteboard Lesson Library
Dynamic Classroom with Dynamic Investigations

What Your Students Will Learn

- Find areas of rhombuses and kites.
- Find angle measures in regular polygons inscribed in a circle.
- Find areas of regular polygons by dividing the polygons into congruent triangles.

Laurie's Notes

Exploration

Motivate
- When students enter the room, have the stop sign displayed. You could tell them a story about a colleague who failed to notice this sign last week, or you could simply ask them what information they would need in order to find the area of the sign.
- Explain to students that in this lesson they will learn how to find the area of a regular polygon.

Exploration Note
- It is important in this exploration that students not use the functionality of the software to find the area of the regular polygon. The intent is for students to develop a strategy for finding the area of each regular polygon, which will likely be similar to the formula presented in this lesson.

Exploration 1
- The definition of the apothem is given. Finding the length of the apothem for each regular polygon is something students will need to figure out. Circulate and ask probing questions without telling students how to do it.
- **Reason Abstractly and Quantitatively:** Have students describe their method for finding the length of the apothem. Expected approaches include the following:
  - Find the midpoint of the segment between (0, 0) and the midpoint of the top side when the number of sides is even.
  - Find the point of intersection of the angle bisectors.
  - Find the intersection point of the perpendicular bisectors of the sides.
  - Find the intersection point of the diagonals between opposite vertices when the number of sides is even.
- **Construct Viable Arguments and Critique the Reasoning of Others:** Have students describe their method for finding the area of the polygon. Peers should listen carefully to ensure that the method is valid.

Exploration 2
- Explain that it is okay to simply let the length of the apothem be denoted by the variable a. The way in which students represent or write the formula will likely vary.
- ❓ "Does your formula work for all polygons regardless of whether the number of sides is odd or even?" yes

Communicate Your Answer
- Verify that all students have the same area for Question 4, regardless of how the formula was written.

Connecting to Next Step
- Students should now have a good conceptual understanding of how to find the area of a regular polygon, one of the skills presented in this lesson.

11.3 Areas of Polygons

Essential Question How can you find the area of a regular polygon?

The **center of a regular polygon** is the center of its circumscribed circle.

The distance from the center to any side of a regular polygon is called the **apothem of a regular polygon**.

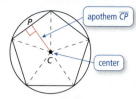

EXPLORATION 1 Finding the Area of a Regular Polygon

Work with a partner. Use dynamic geometry software to construct each regular polygon with side lengths of 4, as shown. Find the apothem and use it to find the area of the polygon. Describe the steps that you used.

a.
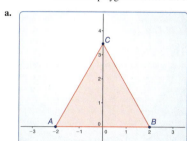

b.
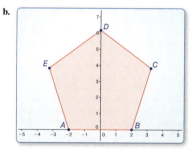

c.
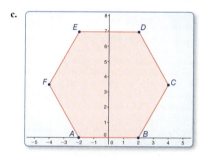

d.
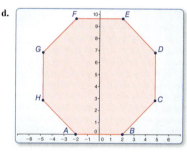

EXPLORATION 2 Writing a Formula for Area

REASONING ABSTRACTLY
To be proficient in math, you need to know and flexibly use different properties of operations and objects.

Work with a partner. Generalize the steps you used in Exploration 1 to develop a formula for the area of a regular polygon.

Communicate Your Answer

3. How can you find the area of a regular polygon?

4. Regular pentagon *ABCDE* has side lengths of 6 meters and an apothem of approximately 4.13 meters. Find the area of *ABCDE*.

Section 11.3 Areas of Polygons 655

Differentiated Instruction

Kinesthetic
Have students draw and cut a rhombus out of cardstock. Then have students cut along the diagonals of the rhombus to separate it into four congruent right triangles. Ask students to rearrange the four triangles in different ways to illustrate the formula for the area of a rhombus. For example, a student could arrange the triangles to form a rectangle with dimensions d_1 and $\frac{1}{2}d_2$.

Extra Example 1
Find the area of each rhombus or kite.

a.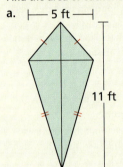

The area is 27.5 square feet.

b.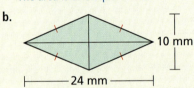

The area is 120 square millimeters.

MONITORING PROGRESS ANSWERS
1. 10 ft^2
2. 54 in.^2

11.3 Lesson

What You Will Learn
- Find areas of rhombuses and kites.
- Find angle measures in regular polygons.
- Find areas of regular polygons.

Core Vocabulary
center of a regular polygon, p. 657
radius of a regular polygon, p. 657
apothem of a regular polygon, p. 657
central angle of a regular polygon, p. 657

Previous
rhombus
kite

Finding Areas of Rhombuses and Kites

You can divide a rhombus or kite with diagonals d_1 and d_2 into two congruent triangles with base d_1, height $\frac{1}{2}d_2$, and area $\frac{1}{2}d_1\left(\frac{1}{2}d_2\right) = \frac{1}{4}d_1d_2$. So, the area of a rhombus or kite is $2\left(\frac{1}{4}d_1d_2\right) = \frac{1}{2}d_1d_2$.

 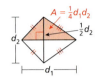

Core Concept

Area of a Rhombus or Kite

The area of a rhombus or kite with diagonals d_1 and d_2 is $\frac{1}{2}d_1d_2$.

EXAMPLE 1 Finding the Area of a Rhombus or Kite

Find the area of each rhombus or kite.

a.

b.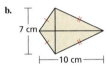

SOLUTION

a. $A = \frac{1}{2}d_1d_2$
$= \frac{1}{2}(6)(8)$
$= 24$

So, the area is 24 square meters.

b. $A = \frac{1}{2}d_1d_2$
$= \frac{1}{2}(10)(7)$
$= 35$

So, the area is 35 square centimeters.

Monitoring Progress Help in English and Spanish at BigIdeasMath.com

1. Find the area of a rhombus with diagonals $d_1 = 4$ feet and $d_2 = 5$ feet.
2. Find the area of a kite with diagonals $d_1 = 12$ inches and $d_2 = 9$ inches.

Laurie's Notes   Teacher Actions

? Make Sense of Problems and Persevere in Solving Them: "What are the attributes of the diagonals of a rhombus? A kite?" *Diagonals of a rhombus are perpendicular bisectors of one another. Diagonals of a kite are perpendicular.*

- Sketch the two quadrilaterals.
- "Develop a formula for the area of each that involves the diagonals." Given time and the suggestion that the formula involves the diagonals, many students can derive the formula.
- Have students work Example 1 with their partners. Discuss as a class.

Finding Angle Measures in Regular Polygons

The diagram shows a regular polygon inscribed in a circle. The **center of a regular polygon** and the **radius of a regular polygon** are the center and the radius of its circumscribed circle.

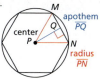

∠MPN is a central angle.

The distance from the center to any side of a regular polygon is called the **apothem of a regular polygon**. The apothem is the height to the base of an isosceles triangle that has two radii as legs. The word "apothem" refers to a segment as well as a length. For a given regular polygon, think of *an* apothem as a segment and *the* apothem as a length.

A **central angle of a regular polygon** is an angle formed by two radii drawn to consecutive vertices of the polygon. To find the measure of each central angle, divide 360° by the number of sides.

EXAMPLE 2 Finding Angle Measures in a Regular Polygon

In the diagram, *ABCDE* is a regular pentagon inscribed in ⊙*F*. Find each angle measure.

a. $m\angle AFB$ **b.** $m\angle AFG$ **c.** $m\angle GAF$

ANALYZING RELATIONSHIPS

$\overline{FG}$ is an altitude of an isosceles triangle, so it is also a median and angle bisector of the isosceles triangle.

SOLUTION

a. ∠AFB is a central angle, so $m\angle AFB = \dfrac{360°}{5} = 72°$.

b. $\overline{FG}$ is an apothem, which makes it an altitude of isosceles △AFB.
So, $\overline{FG}$ bisects ∠AFB and $m\angle AFG = \dfrac{1}{2}m\angle AFB = 36°$.

c. By the Triangle Sum Theorem, the sum of the angle measures of right △GAF is 180°.
$m\angle GAF = 180° - 90° - 36°$
$= 54°$
So, $m\angle GAF = 54°$.

Monitoring Progress Help in English and Spanish at *BigIdeasMath.com*

In the diagram, *WXYZ* is a square inscribed in ⊙*P*.

3. Identify the center, a radius, an apothem, and a central angle of the polygon.

4. Find $m\angle XPY$, $m\angle XPQ$, and $m\angle PXQ$.

Section 11.3 Areas of Polygons 657

English Language Learners

Build on Past Knowledge
Point out that the center, radius, and central angle of a regular polygon are the same as the center, radius, and central angle of its circumscribed circle. The only new term on this page is *apothem*.

Extra Example 2
In the diagram, polygon *ABCDEFGHJK* is a regular decagon inscribed in ⊙*P*. Find each angle measure.

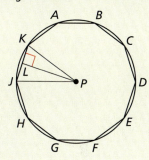

a. $m\angle KPJ$ 36°
b. $m\angle LPK$ 18°
c. $m\angle LJP$ 72°

MONITORING PROGRESS ANSWERS

3. center: *P*, radius: $\overline{PX}$ or $\overline{PY}$, apothem: $\overline{PQ}$, central angle: ∠XPY

4. $m\angle XPY = 90°$; $m\angle XPQ = 45°$; $m\angle PXQ = 45°$

Laurie's Notes Teacher Actions

- **Paired Verbal Fluency:** Have students pair up and follow the protocol described on page T-224. Ask students to share what they recall about polygons and regular polygons. If students completed the explorations, they should mention area and the apothem.
- **Whiteboarding:** Have partners work on Example 2 and display their results on whiteboards. Circulate and give sufficient *Wait Time*. Have students share their solutions.
- ❓ "Could the strategies you used for Example 2 be used to find the corresponding angles in any regular *n*-gon?" yes

Extra Example 3

A regular hexagon is inscribed in a circle with a diameter of 32 units. Find the area of the hexagon.

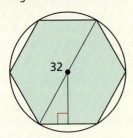

$A = \frac{1}{2}aP = \frac{1}{2}(8\sqrt{3})(6 \cdot 16)$, or about 665.11 square units

---

Finding Areas of Regular Polygons

You can find the area of any regular n-gon by dividing it into congruent triangles.

A = Area of one triangle • Number of triangles

$ = \left(\frac{1}{2} \cdot s \cdot a\right) \cdot n$ Base of triangle is s and height of triangle is a. Number of triangles is n.

$ = \frac{1}{2} \cdot a \cdot (n \cdot s)$ Commutative and Associative Properties of Multiplication

$ = \frac{1}{2} a \cdot P$ There are n congruent sides of length s, so perimeter P is $n \cdot s$.

> **READING DIAGRAMS**
> In this book, a point shown inside a regular polygon marks the center of the circle that can be circumscribed about the polygon.

🄲 Core Concept

Area of a Regular Polygon

The area of a regular n-gon with side length s is one-half the product of the apothem a and the perimeter P.

$A = \frac{1}{2}aP$, or $A = \frac{1}{2}a \cdot ns$

EXAMPLE 3 Finding the Area of a Regular Polygon

A regular nonagon is inscribed in a circle with a radius of 4 units. Find the area of the nonagon.

SOLUTION

The measure of central $\angle JLK$ is $\frac{360°}{9}$, or 40°. Apothem $\overline{LM}$ bisects the central angle, so $m\angle KLM$ is 20°. To find the lengths of the legs, use trigonometric ratios for right $\triangle KLM$.

$\sin 20° = \frac{MK}{LK}$ $\cos 20° = \frac{LM}{LK}$

$\sin 20° = \frac{MK}{4}$ $\cos 20° = \frac{LM}{4}$

$4 \sin 20° = MK$ $4 \cos 20° = LM$

The regular nonagon has side length $s = 2(MK) = 2(4 \sin 20°) = 8 \sin 20°$, and apothem $a = LM = 4 \cos 20°$.

▶ So, the area is $A = \frac{1}{2}a \cdot ns = \frac{1}{2}(4 \cos 20°) \cdot (9)(8 \sin 20°) \approx 46.3$ square units.

658   Chapter 11   Circumference, Area, and Volume

Laurie's Notes | Teacher Actions

- Sketch a portion of a regular n-gon.
- Derive the formula for the area of a regular polygon. The first step should be similar to what students wrote for Exploration 2 on page 655.
- Write both forms of the formula. Many students prefer $A = \frac{1}{2}ans$ because *ans* triggers "answer."
- **?** Pose Example 3. "What is needed to find the area?" the length of the apothem and the side length of the regular nonagon
- **Teaching Tip:** Sketch one isosceles triangle and label the given information. Hopefully students will think about trigonometric ratios!

EXAMPLE 4 Finding the Area of a Regular Polygon

You are decorating the top of a table by covering it with small ceramic tiles. The tabletop is a regular octagon with 15-inch sides and a radius of about 19.6 inches. What is the area you are covering?

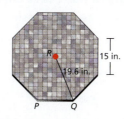

SOLUTION

Step 1 Find the perimeter P of the tabletop.
An octagon has 8 sides, so $P = 8(15) = 120$ inches.

Step 2 Find the apothem a. The apothem is height RS of $\triangle PQR$.
Because $\triangle PQR$ is isosceles, altitude $\overline{RS}$ bisects $\overline{QP}$.
So, $QS = \frac{1}{2}(QP) = \frac{1}{2}(15) = 7.5$ inches.

To find RS, use the Pythagorean Theorem for $\triangle RQS$.
$$a = RS = \sqrt{19.6^2 - 7.5^2} = \sqrt{327.91} \approx 18.108$$

Step 3 Find the area A of the tabletop.

$A = \frac{1}{2}aP$ Formula for area of a regular polygon

$= \frac{1}{2}(\sqrt{327.91})(120)$ Substitute.

≈ 1086.5 Simplify.

▶ The area you are covering with tiles is about 1086.5 square inches.

Monitoring Progress Help in English and Spanish at *BigIdeasMath.com*

Find the area of the regular polygon.

5.

6.

Extra Example 4

A mirror is in the shape of a regular nonagon with 6-inch sides. What is the area of the mirror?

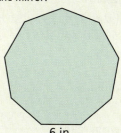

$A = \frac{1}{2}aP = \frac{1}{2}\left(\frac{3}{\tan 20°}\right)(9 \cdot 6)$, or about 222.5 square inches

MONITORING PROGRESS ANSWERS

5. about 151.57 square units
6. about 377.02 square units

Laurie's Notes Teacher Actions

- **Pass the Problem:** Explain this technique and pose Example 4. Give partners 1 to 2 minutes to discuss and begin a solution. Call "swap" and have partners continue to solve. Both groups debrief. Ask students what part of the problem was challenging. Was there a spot where they got stuck? What helped them make progress with the problem?

Closure

- **Exit Ticket:** Find the area of a stop sign when the side length is 10 inches.
about 482.8 square inches

Assignment Guide and Homework Check

ASSIGNMENT

Basic: 1, 2, 3–21 odd, 25–29 odd, 40, 42, 44, 53, 54

Average: 1, 2, 4–22 even, 26–36 even, 42, 44, 53, 54

Advanced: 1, 2, 4, 12, 16, 20, 24–32 even, 33–44, 50–54

HOMEWORK CHECK

Basic: 3, 15, 19, 27, 40

Average: 4, 16, 20, 28, 36

Advanced: 12, 20, 28, 34, 37

ANSWERS

1. Divide 360° by the number of sides.
2. Find the apothem of regular polygon ABCDE; 5.5; 6.8
3. 361 square units
4. 72 square units
5. 70 square units
6. 24 square units
7. P
8. Sample answer: ∠MPN
9. 5 units
10. 4.05 units
11. 36°
12. 20°
13. 15°
14. 51.4°
15. 45°
16. 22.5°
17. 67.5°
18. 135°
19. about 62.35 square units
20. about 289.24 square units
21. about 20.87 square units
22. about 127.31 square units
23. about 342.24 square units
24. about 90.82 square units
25. The side lengths were used instead of the diagonals; $A = \frac{1}{2}(8)(4) = 16$

11.3 Exercises

Dynamic Solutions available at *BigIdeasMath.com*

Vocabulary and Core Concept Check

1. **WRITING** Explain how to find the measure of a central angle of a regular polygon.

2. **DIFFERENT WORDS, SAME QUESTION** Which is different? Find "both" answers.

     Find the radius of ⊙F.   Find the apothem of polygon ABCDE.

   Find AF.   Find the radius of polygon ABCDE.

   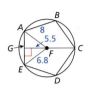

Monitoring Progress and Modeling with Mathematics

In Exercises 3–6, find the area of the kite or rhombus. (See Example 1.)

3.

4.

5. 6.

In Exercises 7–10, use the diagram.

7. Identify the center of polygon JKLMN.
8. Identify a central angle of polygon JKLMN.
9. What is the radius of polygon JKLMN?
10. What is the apothem of polygon JKLMN?

In Exercises 11–14, find the measure of a central angle of a regular polygon with the given number of sides. Round answers to the nearest tenth of a degree, if necessary.

11. 10 sides
12. 18 sides
13. 24 sides
14. 7 sides

In Exercises 15–18, find the given angle measure for regular octagon ABCDEFGH. (See Example 2.)

15. m∠GJH
16. m∠GJK
17. m∠KGJ
18. m∠EJH

In Exercises 19–24, find the area of the regular polygon. (See Examples 3 and 4.)

19.
20.
21.
22.

23. an octagon with a radius of 11 units
24. a pentagon with an apothem of 5 units

25. **ERROR ANALYSIS** Describe and correct the error in finding the area of the kite.

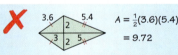

So, the area of the kite is 9.72 square units.

660   Chapter 11   Circumference, Area, and Volume

26. **ERROR ANALYSIS** Describe and correct the error in finding the area of the regular hexagon.

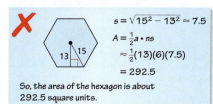

In Exercises 27–30, you throw a dart at the region shown. Your dart is equally likely to hit any point inside the region. Find the probability that your dart lands in the shaded region.

27.
28.
29.
30.

31. **MODELING WITH MATHEMATICS** Basaltic columns are geological formations that result from rapidly cooling lava. Giant's Causeway in Ireland contains many hexagonal basaltic columns. Suppose the top of one of the columns is in the shape of a regular hexagon with a radius of 8 inches. Find the area of the top of the column to the nearest square inch.

32. **MODELING WITH MATHEMATICS** A watch has a circular surface on a background that is a regular octagon. Find the area of the octagon. Then find the area of the silver border around the circular face.

CRITICAL THINKING In Exercises 33–35, tell whether the statement is *true* or *false*. Explain your reasoning.

33. The area of a regular *n*-gon of a fixed radius r increases as n increases.

34. The apothem of a regular polygon is always less than the radius.

35. The radius of a regular polygon is always less than the side length.

36. **REASONING** Predict which figure has the greatest area and which has the least area. Explain your reasoning. Check by finding the area of each figure.

    Ⓐ   Ⓑ
    Ⓒ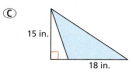

37. **USING EQUATIONS** Find the area of a regular pentagon inscribed in a circle whose equation is given by $(x - 4)^2 + (y + 2)^2 = 25$.

38. **REASONING** What happens to the area of a kite if you double the length of one of the diagonals? if you double the length of both diagonals? Justify your answer.

MATHEMATICAL CONNECTIONS In Exercises 39 and 40, write and solve an equation to find the indicated lengths. Round decimal answers to the nearest tenth.

39. The area of a kite is 324 square inches. One diagonal is twice as long as the other diagonal. Find the length of each diagonal.

40. One diagonal of a rhombus is four times the length of the other diagonal. The area of the rhombus is 98 square feet. Find the length of each diagonal.

41. **REASONING** The perimeter of a regular nonagon, or 9-gon, is 18 inches. Is this enough information to find the area? If so, find the area and explain your reasoning. If not, explain why not.

ANSWERS

42. no; *Sample answer:* A rhombus is not a regular polygon.

43. *Sample answer:* Let $QT = x$ and $TS = y$. The area of $PQRS$ is $\frac{1}{2}d_2x + \frac{1}{2}d_2y = \frac{1}{2}d_2(x+y) = \frac{1}{2}d_2d_1$.

44. *Sample answer:* Find the area of one of the triangles and multiply it by 6.

45. $A = \frac{1}{2}d^2$; $A = \frac{1}{2}d^2 = \frac{1}{2}(s^2 + s^2) = \frac{1}{2}(2s^2) = s^2$

46. The apothem is $\frac{s\sqrt{3}}{6}$, so the area is $A = \frac{1}{2}a \cdot ns = \frac{1}{4}s^2\sqrt{3}$.

47. about 6.47 cm   48. about 6.60 in.

49. about 92 square units

50. a circle; the circumference; the radius; *Sample answer:* The area of the circle is $\frac{1}{2}rC$; Substitute the expression into the formula for the area of a regular n-gon.

51. about 43 square units; *Sample answer:* $A = \frac{1}{2}aP$; There are fewer calculations.

52. $A = nr^2 \tan\left(\frac{180°}{n}\right) - nr^2 \sin\left(\frac{180°}{n}\right)\cos\left(\frac{180°}{n}\right)$

53–54. See Additional Answers.

Mini-Assessment

1. Find the area of the rhombus.

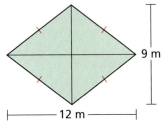

54 square meters

2. Find the measure of a central angle of a regular dodecagon (12-gon).   30°

3. Find the area of the regular pentagon.

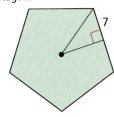

337.2 square units

662   Chapter 11

---

42. **MAKING AN ARGUMENT** Your friend claims that it is possible to find the area of any rhombus if you only know the perimeter of the rhombus. Is your friend correct? Explain your reasoning.

43. **PROOF** Prove that the area of any quadrilateral with perpendicular diagonals is $A = \frac{1}{2}d_1d_2$, where d_1 and d_2 are the lengths of the diagonals.

44. **HOW DO YOU SEE IT?** Explain how to find the area of the regular hexagon by dividing the hexagon into equilateral triangles.

45. **REWRITING A FORMULA** Rewrite the formula for the area of a rhombus for the special case of a square with side length s. Show that this is the same as the formula for the area of a square, $A = s^2$.

46. **REWRITING A FORMULA** Use the formula for the area of a regular polygon to show that the area of an equilateral triangle can be found by using the formula $A = \frac{1}{4}s^2\sqrt{3}$, where s is the side length.

47. **CRITICAL THINKING** The area of a regular pentagon is 72 square centimeters. Find the length of one side.

48. **CRITICAL THINKING** The area of a dodecagon, or 12-gon, is 140 square inches. Find the apothem of the polygon.

Maintaining Mathematical Proficiency   *Reviewing what you learned in previous grades and lessons*

Complete the two-way table. *(Section 5.3)*

53.

		Response		
		Yes	No	Total
Gender	Male		52	
	Female	48		110
	Total	65		

54.

		Response		
		Yes	No	Total
Class	Junior	65		105
	Senior		12	
	Total	116		

662   Chapter 11   Circumference, Area, and Volume

---

49. **USING STRUCTURE** In the figure, an equilateral triangle lies inside a square inside a regular pentagon inside a regular hexagon. Find the approximate area of the entire shaded region to the nearest whole number.

50. **THOUGHT PROVOKING** The area of a regular n-gon is given by $A = \frac{1}{2}aP$. As n approaches infinity, what does the n-gon approach? What does P approach? What does a approach? What can you conclude from your three answers? Explain your reasoning.

51. **COMPARING METHODS** Find the area of regular pentagon $ABCDE$ by using the formula $A = \frac{1}{2}aP$, or $A = \frac{1}{2}a \cdot ns$. Then find the area by adding the areas of smaller polygons. Check that both methods yield the same area. Which method do you prefer? Explain your reasoning.

52. **USING STRUCTURE** Two regular polygons both have n sides. One of the polygons is inscribed in, and the other is circumscribed about, a circle of radius r. Find the area between the two polygons in terms of n and r.

If students need help...	If students got it...
Resources by Chapter • Practice A and Practice B • Puzzle Time	Resources by Chapter • Enrichment and Extension • Cumulative Review
Student Journal • Practice	Start the *next* Section
Differentiating the Lesson Skills Review Handbook	

11.1–11.3 What Did You Learn?

Core Vocabulary

circumference, *p. 640*
arc length, *p. 641*
radian, *p. 643*
geometric probability, *p. 649*
sector of a circle, *p. 650*

center of a regular polygon, *p. 657*
radius of a regular polygon, *p. 657*
apothem of a regular polygon, *p. 657*
central angle of a regular polygon, *p. 657*

Core Concepts

Section 11.1
Circumference of a Circle, *p. 640*
Arc Length, *p. 641*

Converting between Degrees and Radians, *p. 643*

Section 11.2
Area of a Circle, *p. 648*

Area of a Sector, *p. 650*

Section 11.3
Area of a Rhombus or Kite, *p. 656*

Area of a Regular Polygon, *p. 658*

Mathematical Practices

1. In Exercise 13 on page 644, why does it matter how many revolutions the wheel makes?
2. Your friend is confused with Exercise 23 on page 652. What question(s) could you ask your friend to help them figure it out?
3. In Exercise 38 on page 661, write a proof to support your answer.

Dynamic Teaching Tools

Dynamic Assessment & Progress Monitoring Tool
Interactive Whiteboard Lesson Library
Dynamic Classroom with Dynamic Investigations

ANSWERS

1. *Sample answer:* Each rotation of the wheel increases the total distance measured.
2. *Sample answer:* Is 12 the radius of the circle?
3. *Sample answer:* The original area is $A = \frac{1}{2}d_1 d_2$. Doubling the length of one diagonal produces $\frac{1}{2}(2d_1)d_2 = 2\left(\frac{1}{2}d_1 d_2\right) = 2A$, and doubling the lengths of both diagonals produces $\frac{1}{2}(2d_1)(2d_2) = 4\left(\frac{1}{2}d_1 d_2\right) = 4A$.

Kinesthetic Learners

Incorporate physical activity.
- Act out a word problem as much as possible. Use props when you can.
- Solve a word problem on a large whiteboard. The physical action of writing is more kinesthetic when the writing is larger and you can move around while doing it.
- Make a review card.

STUDY SKILLS

11.1–11.3 Quiz

Find the indicated measure. *(Section 11.1)*

1. $m\widehat{EF}$

2. arc length of $\widehat{QS}$

3. circumference of $\odot N$

Convert the angle measure. *(Section 11.1)*

4. Convert 26° to radians.

5. Convert $\frac{5\pi}{9}$ radians to degrees.

Use the figure to find the indicated measure. *(Section 11.2)*

6. area of red sector

7. area of blue sector

In the diagram, *RSTUVWXY* is a regular octagon inscribed in ⊙*C*. *(Section 11.3)*

8. Identify the center, a radius, an apothem, and a central angle of the polygon.

9. Find $m\angle RCY$, $m\angle RCZ$, and $m\angle ZRC$.

10. The radius of the circle is 8 units. Find the area of the octagon.

11. The two white congruent circles just fit into the blue circle. What is the area of the blue region? *(Section 11.2)*

12. Find the area of each rhombus tile. Then find the area of the pattern. *(Section 11.3)*

ANSWERS

1. about 124.14°
2. about 5.79 cm
3. 60 in.
4. $\frac{13\pi}{90}$ radian
5. 100°
6. about 125.66 yd²
7. about 326.73 yd²
8. center *C*; radius $\overline{CY}$ or $\overline{CR}$; apothem $\overline{CZ}$; central angle $\angle YCR$
9. $m\angle RCY = 45°$; $m\angle RCZ = 22.5°$; $m\angle ZRC = 67.5°$
10. about 181 square units
11. about 56.55 m²
12. 89.49 mm²; 55.5 mm²; 4029.18 mm²

Laurie's Notes

Overview of Section 11.4

Introduction
- This is the first of four lessons on the volumes of solids. In this lesson, students will find the volumes of prisms and cylinders, solids that structurally are the same. Also reviewed are the volumes of similar solids.
- The formulas for prisms and cylinders were taught in middle school, and the only new content is Cavalieri's Principle.

Resources
- If you have a slinky, or a collection of interestingly shaped slinkies, use them to demonstrate Cavalieri's Principle.

Formative Assessment Tips
- **Example/Non-Example (student generated):** This technique asks students to generate examples and non-examples to demonstrate their understanding of a concept. To generate the examples and non-examples, students must have a deeper understanding of the concept than just memorization would yield.
- In generating the examples, students can also be asked why—meaning they should provide justification. Listen carefully to students' language as they provide the justification.
- Note that this technique is an instructional technique used by teachers to introduce concepts, such as making a list of expressions that are monomials and those that are not monomials. The difference here is that students are using their cognitive skills to generate the examples and non-examples.
- This technique can be used by individuals, partners, or small groups.
- Be sure to allow sufficient time so that students receive feedback from others and/or you about their examples and non-examples.

Applications
- **Note:** The *Exit Ticket* at the end of the lesson refers to the Leaning Tower of Pisa as a cylinder. There is actually a main cylinder and an upper cylinder (bell tower). Also, the interior and exterior diameters are not the same.

Pacing Suggestion
- Students will review some volume formulas in the explorations. You might be selective in the examples you do in the formal lesson.

Dynamic Teaching Tools
Dynamic Assessment & Progress Monitoring Tool
Lesson Planning Tool
Interactive Whiteboard Lesson Library
Dynamic Classroom with Dynamic Investigations

What Your Students Will Learn

- Classify solids and name polyhedra.
- Find volumes of prisms, cylinders, similar solids, and composite solids.

Laurie's Notes

Exploration

Motivate
- ❓ "What do you think the capacity is of one of the largest farm grain silos in the country?" Answers will vary.
- Share information about a silo located in Berks County, Pennsylvania.
  - It is equipped with an elevated floor that has a 12-foot clearance so that a feed truck can drive through and be quickly loaded.
  - It is equipped with an unloader, making the system capable of loading 50-plus tons of grain per hour and unloading 800 to 1200 pounds of grain per minute.
  - The silo has more than 94,000 cubic feet of storage capacity and should hold more than 4000 tons of corn silage.
- Share a picture of the silo, which you can find with an online search for "largest farm silo."
- Explain to students that in this lesson they will be finding volumes of prisms and cylinders.

Exploration Note
- **Model with Mathematics:** Students will benefit from having a model available; something as simple as a deck of cards will do.

Exploration 1
- **Common Misconception:** Not all students will be convinced that the volumes are the same for both stacks of paper. Let peers try to explain why the volumes remain the same. They will likely say something like, "The number of pieces of paper has not changed. None were added and none were taken away."
- **Attend to Precision** and **Construct Viable Arguments and Critique the Reasoning of Others:** Describing why the stacks have the same volume can be challenging for students. Push students to describe attributes of the stacks that go beyond referencing the fact that the stacks have the same number of pieces of paper.
- ❓ **Probing Question:** "Did the height of the stack change when it was twisted, and how do you know?" no; Both stacks have the same number of layers. "What do you know about each layer?" They each have the same area.
- **Construct Viable Arguments and Critique the Reasoning of Others:** Students should listen to one another's conjecture to see whether the reasoning is valid or not.

Exploration 2
- ❓ "How do you find the volume of a cylinder?" Multiply the area of the base times the height of the cylinder.
- Be prepared. Some students may still be uncertain that the volumes remained the same. In these two problems, the layers are not visible, but they are all circles!

Communicate Your Answer
- Question 4 may cause some students to change their mind about the volume remaining the same when the stack of paper is twisted.

Connecting to Next Step
- This exploration is an introduction to Cavalieri's Principle, which is introduced in this lesson.

11.4 Volumes of Prisms and Cylinders

Essential Question
How can you find the volume of a prism or cylinder that is not a right prism or right cylinder?

Recall that the volume V of a right prism or a right cylinder is equal to the product of the area of a base B and the height h.

$$V = Bh$$

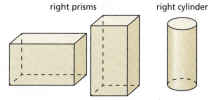
right prisms   right cylinder

EXPLORATION 1   Finding Volume

Work with a partner. Consider a stack of square papers that is in the form of a right prism.

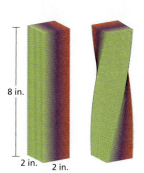

a. What is the volume of the prism?

b. When you twist the stack of papers, as shown at the right, do you change the volume? Explain your reasoning.

c. Write a carefully worded conjecture that describes the conclusion you reached in part (b).

d. Use your conjecture to find the volume of the twisted stack of papers.

ATTENDING TO PRECISION
To be proficient in math, you need to communicate precisely to others.

EXPLORATION 2   Finding Volume

Work with a partner. Use the conjecture you wrote in Exploration 1 to find the volume of the cylinder.

a. 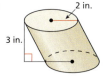   b.

Communicate Your Answer

3. How can you find the volume of a prism or cylinder that is not a right prism or right cylinder?

4. In Exploration 1, would the conjecture you wrote change if the papers in each stack were not squares? Explain your reasoning.

Section 11.4   Volumes of Prisms and Cylinders   665

Dynamic Teaching Tools
- Dynamic Assessment & Progress Monitoring Tool
- Lesson Planning Tool
- Interactive Whiteboard Lesson Library
- Dynamic Classroom with Dynamic Investigations

ANSWERS
1. a. 32 in.^{3}
   b. no; *Sample answer:* The amount of paper is the same.
   c. *Sample answer:* If two solids have the same height and the same cross-sectional area at every level, then they have the same volume.
   d. 32 in.^{3}
2. a. about 37.70 in.^{3}
   b. about 1178.10 cm^3
3. *Sample answer:* Multiply the area of a base by the vertical height.
4. no; *Sample answer:* Each piece of paper would still have the same area.

English Language Learners

Build on Past Knowledge
Remind students of their previous study of three-dimensional figures, including prisms, pyramids, cylinders, cones, and spheres. Before starting this lesson, have students describe each solid and how to name the solid. Have students draw an example of the solid on the board as they describe it.

Extra Example 1
Tell whether each solid is a polyhedron. If it is, name the polyhedron.

a.
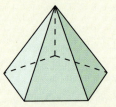
polyhedron; a pentagonal pyramid

b.

not a polyhedron

c.
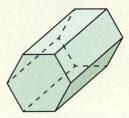
polyhedron; a hexagonal prism

11.4 Lesson

Core Vocabulary
polyhedron, *p. 666*
face, *p. 666*
edge, *p. 666*
vertex, *p. 666*
volume, *p. 667*
Cavalieri's Principle, *p. 667*
similar solids, *p. 669*

Previous
solid
prism
pyramid
cylinder
cone
sphere
base
composite solid

What You Will Learn
▶ Classify solids.
▶ Find volumes of prisms and cylinders.

Classifying Solids
A three-dimensional figure, or solid, is bounded by flat or curved surfaces that enclose a single region of space. A **polyhedron** is a solid that is bounded by polygons, called **faces**. An **edge** of a polyhedron is a line segment formed by the intersection of two faces. A **vertex** of a polyhedron is a point where three or more edges meet. The plural of polyhedron is *polyhedra* or *polyhedrons*.

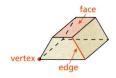

Core Concept
Types of Solids

Polyhedra	Not Polyhedra

prism   cylinder   cone

pyramid   sphere

Pentagonal prism

Bases are pentagons.

Triangular pyramid

Base is a triangle.

To name a prism or a pyramid, use the shape of the *base*. The two bases of a prism are congruent polygons in parallel planes. For example, the bases of a pentagonal prism are pentagons. The base of a pyramid is a polygon. For example, the base of a triangular pyramid is a triangle.

EXAMPLE 1 Classifying Solids

Tell whether each solid is a polyhedron. If it is, name the polyhedron.

a.   b.   c.

SOLUTION

a. The solid is formed by polygons, so it is a polyhedron. The two bases are congruent rectangles, so it is a rectangular prism.

b. The solid is formed by polygons, so it is a polyhedron. The base is a hexagon, so it is a hexagonal pyramid.

c. The cone has a curved surface, so it is not a polyhedron.

Laurie's Notes | Teacher Actions

- **Example/Non-Example:** Make two lists: polyhedra and not polyhedra. Give partners time to write a definition for polyhedra. Ask partners to sketch other examples of polyhedra or name objects in the room or other common objects that are examples. Use whiteboards to share and discuss results.
- Students should be familiar with certain vocabulary: vertex, face, edge, prism, pyramid, cylinder, cone, and sphere.
- ? "How do you name prisms and pyramids?" You name them by the base, such as pentagonal prism or triangular pyramid.

Monitoring Progress Help in English and Spanish at BigIdeasMath.com

Tell whether the solid is a polyhedron. If it is, name the polyhedron.

1.   2.   3.

Finding Volumes of Prisms and Cylinders

The **volume** of a solid is the number of cubic units contained in its interior. Volume is measured in cubic units, such as cubic centimeters (cm^3). **Cavalieri's Principle**, named after Bonaventura Cavalieri (1598–1647), states that if two solids have the same height and the same cross-sectional area at every level, then they have the same volume. The prisms below have equal heights h and equal cross-sectional areas B at every level. By Cavalieri's Principle, the prisms have the same volume.

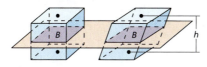

Core Concept

Volume of a Prism

The volume V of a prism is

$$V = Bh$$

where B is the area of a base and h is the height.

EXAMPLE 2   Finding Volumes of Prisms

Find the volume of each prism.

a. 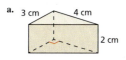   b.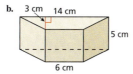

SOLUTION

a. The area of a base is $B = \frac{1}{2}(3)(4) = 6$ cm² and the height is $h = 2$ cm.

$$V = Bh = 6(2) = 12$$

▶ The volume is 12 cubic centimeters.

b. The area of a base is $B = \frac{1}{2}(3)(6 + 14) = 30$ cm² and the height is $h = 5$ cm.

$$V = Bh = 30(5) = 150$$

▶ The volume is 150 cubic centimeters.

Differentiated Instruction

Kinesthetic

You can use a stack of playing cards or notepaper to demonstrate Cavalieri's Principle as it applies to rectangular prisms. Use a stack of coins, circular game chips, or paper plates to demonstrate Cavalieri's Principle as it applies to cylinders.

Extra Example 2

Find the volume of each prism.

a.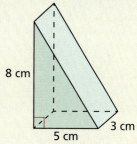

60 cubic centimeters

b.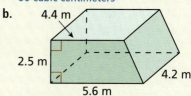

52.5 cubic meters

MONITORING PROGRESS ANSWERS

1. yes; square pyramid
2. no
3. yes; triangular prism

Laurie's Notes   Teacher Actions

- **Example/Non-Example:** Make groups of four students. To gain a sense of how students think about the concept of volume, ask them to make two lists: one that contains a contextual question about volume and one that does not. *Example:* How much gas does my car hold? *Non-Example:* How much wrapping paper do I need for that present? After sufficient time, have groups share. Give feedback as needed.
- Discuss Cavalieri's Principle and write the *Core Concept*. Omit Example 2 if you believe students are secure with finding the volume.

Extra Example 3
Find the volume of each cylinder.

a.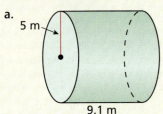

227.5π, or about 714.71 cubic meters

b.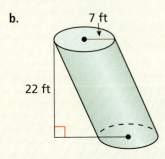

1078π, or about 3386.64 cubic feet

MONITORING PROGRESS ANSWERS
4. 180 m^3
5. about 2814.87 ft^3

Consider a cylinder with height h and base radius r and a rectangular prism with the same height that has a square base with sides of length $r\sqrt{\pi}$.

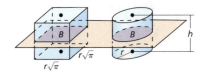

The cylinder and the prism have the same cross-sectional area, πr^2, at every level and the same height. By Cavalieri's Principle, the prism and the cylinder have the same volume. The volume of the prism is $V = Bh = \pi r^2 h$, so the volume of the cylinder is also $V = Bh = \pi r^2 h$.

Core Concept
Volume of a Cylinder

The volume V of a cylinder is

$$V = Bh = \pi r^2 h$$

where B is the area of a base, h is the height, and r is the radius of a base.

EXAMPLE 3 Finding Volumes of Cylinders

Find the volume of each cylinder.

a. 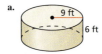   b.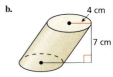

SOLUTION

a. The dimensions of the cylinder are $r = 9$ ft and $h = 6$ ft.

$$V = \pi r^2 h = \pi(9)^2(6) = 486\pi \approx 1526.81$$

▶ The volume is 486π, or about 1526.81 cubic feet.

b. The dimensions of the cylinder are $r = 4$ cm and $h = 7$ cm.

$$V = \pi r^2 h = \pi(4)^2(7) = 112\pi \approx 351.86$$

▶ The volume is 112π, or about 351.86 cubic centimeters.

Monitoring Progress Help in English and Spanish at *BigIdeasMath.com*

Find the volume of the solid.

4.    5.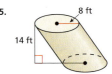

668   Chapter 11   Circumference, Area, and Volume

Laurie's Notes Teacher Actions

- Explain that Cavalieri's Principle does not apply to just rectangular prisms. Note the volume comparison of a prism and a cylinder.
- **Think-Alouds:** Pose Example 3(a) and say, "To solve this problem I need to ...." Ask partner A to think aloud for partner B to hear the problem-solving process. Switch roles and have partner B think aloud in solving part (b).
- Did students label the answers with correct units?
- **Think-Pair-Share:** Have students answer Questions 4 and 5, and then share and discuss as a class.

Core Concept

Similar Solids
Two solids of the same type with equal ratios of corresponding linear measures, such as heights or radii, are called **similar solids**. The ratio of the corresponding linear measures of two similar solids is called the *scale factor*. If two similar solids have a scale factor of k, then the ratio of their volumes is equal to k^3.

EXAMPLE 4 Finding the Volume of a Similar Solid

Cylinder A and cylinder B are similar. Find the volume of cylinder B.

SOLUTION

The scale factor is $k = \dfrac{\text{Radius of cylinder B}}{\text{Radius of cylinder A}}$

$= \dfrac{6}{3} = 2.$

Use the scale factor to find the volume of cylinder B.

$\dfrac{\text{Volume of cylinder B}}{\text{Volume of cylinder A}} = k^3$ The ratio of the volumes is k^3.

$\dfrac{\text{Volume of cylinder B}}{45\pi} = 2^3$ Substitute.

Volume of cylinder B $= 360\pi$ Solve for volume of cylinder B.

▶ The volume of cylinder B is 360π cubic centimeters.

COMMON ERROR
Be sure to write the ratio of the volumes in the same order you wrote the ratio of the radii.

Prism C, $V = 1536\ m^3$

Prism D, 3 m

Monitoring Progress 🔊 Help in English and Spanish at BigIdeasMath.com

6. Prism C and prism D are similar. Find the volume of prism D.

EXAMPLE 5 Finding the Volume of a Composite Solid

Find the volume of the concrete block.

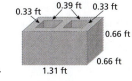

SOLUTION

To find the area of the base, subtract two times the area of the small rectangle from the large rectangle.

$B =$ Area of large rectangle $- 2 \cdot$ Area of small rectangle

$= 1.31(0.66) - 2(0.33)(0.39)$

$= 0.6072$

Using the formula for the volume of a prism, the volume is

$V = Bh = 0.6072(0.66) \approx 0.40.$

▶ The volume is about 0.40 cubic foot.

Monitoring Progress 🔊 Help in English and Spanish at BigIdeasMath.com

7. Find the volume of the composite solid.

English Language Learners

Build on Past Knowledge
Remind students of their study of similar polygons. Have students describe the characteristics of two similar triangles. Then connect students' understanding of similar polygons to the new concept of similar solids. Describe the characteristics of similar solids.

Extra Example 4
Square prism A and square prism B are similar. Each base edge of prism A is 4 inches, and each base edge of prism B is 6 inches. The volume of prism B is 135 cubic inches. Find the volume of prism A. *The volume of prism A is 40 cubic inches.*

Extra Example 5
Find the volume of the composite solid.

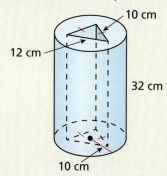

The volume is $3200\pi - 1536$, or about 8517.10 cubic centimeters.

MONITORING PROGRESS ANSWERS
6. $24\ m^3$
7. about $205.81\ ft^3$

Laurie's Notes — Teacher Actions

- Write the *Core Concept*, which students may recall from middle school.
- ❓ Pose Example 4. "What is the ratio of their linear measurements?" *The radii are in the ratio of 2:1.* "What is the ratio of their volumes?" *The volumes are in the ratio of $2^3:1^3$ or 8:1.*
- **Think-Pair-Share:** Have students answer Example 5, and then share and discuss as a class.

Closure
- **Exit Ticket:** Assume the Leaning Tower of Pisa is approximately a cylinder of height 55.9 meters and diameter of 15.5 meters. Find its volume. *about 10,547.9 cubic meters*

Assignment Guide and Homework Check

ASSIGNMENT

Basic: 1–6, 7–33 odd, 38, 41, 45–47

Average: 1–6, 8–34 even, 38, 40, 44–47

Advanced: 1, 2, 10, 14, 18, 22–40 even, 41–47

HOMEWORK CHECK

Basic: 7, 13, 17, 29, 31

Average: 10, 14, 18, 30, 32

Advanced: 14, 18, 24, 26, 34

ANSWERS

1. cubic units
2. the cone; The others are pyramids.
3. B
4. D
5. A
6. C
7. yes; pentagonal pyramid
8. yes; hexagonal prism
9. no
10. yes; trapezoidal prism
11. 6.3 cm³
12. 12 m³
13. 175 in.³
14. 924 m³
15. about 288.40 ft³
16. about 5528.22 cm³
17. about 628.32 ft³
18. about 1763.01 m³

11.4 Exercises

Dynamic Solutions available at *BigIdeasMath.com*

Vocabulary and Core Concept Check

1. **VOCABULARY** In what type of units is the volume of a solid measured?

2. **WHICH ONE DOESN'T BELONG?** Which solid does *not* belong with the other three? Explain your reasoning.

Monitoring Progress and Modeling with Mathematics

In Exercises 3–6, match the polyhedron with its name.

3.
4.
5.
6.

A. triangular prism
B. rectangular pyramid
C. hexagonal pyramid
D. pentagonal prism

In Exercises 7–10, tell whether the solid is a polyhedron. If it is, name the polyhedron. *(See Example 1.)*

7.
8.
9.
10.

In Exercises 11–14, find the volume of the prism. *(See Example 2.)*

11.
12.
13.
14.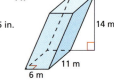

In Exercises 15–18, find the volume of the cylinder. *(See Example 3.)*

15.
16.
17.
18.

670 Chapter 11 Circumference, Area, and Volume

In Exercises 19 and 20, make a sketch of the solid and find its volume. Round your answer to the nearest hundredth.

19. A prism has a height of 11.2 centimeters and an equilateral triangle for a base, where each base edge is 8 centimeters.

20. A pentagonal prism has a height of 9 feet and each base edge is 3 feet.

21. ERROR ANALYSIS Describe and correct the error in identifying the solid.

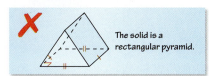

22. ERROR ANALYSIS Describe and correct the error in finding the volume of the cylinder.

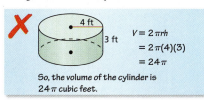

In Exercises 23–28, find the missing dimension of the prism or cylinder.

23. Volume = 560 ft³   **24.** Volume = 2700 yd³

25. Volume = 80 cm³   **26.** Volume = 72.66 in.³

27. Volume = 3000 ft³   **28.** Volume = 1696.5 m³

In Exercises 29 and 30, the solids are similar. Find the volume of solid B. *(See Example 4.)*

29.
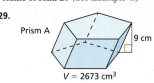
Prism A
V = 2673 cm³

Prism B

30.
Cylinder A   Cylinder B
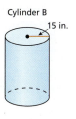
V = 4608π in.³

In Exercises 31–34, find the volume of the composite solid. *(See Example 5.)*

31. 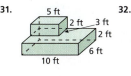   **32.**

33. 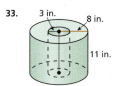   **34.**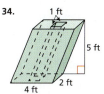

35. MODELING WITH MATHEMATICS The Great Blue Hole is a cylindrical trench located off the coast of Belize. It is approximately 1000 feet wide and 400 feet deep. About how many gallons of water does the Great Blue Hole contain? (1 ft³ ≈ 7.48 gallons)

Section 11.4   Volumes of Prisms and Cylinders   671

ANSWERS

19.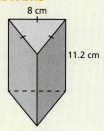

310.38 cm³

20.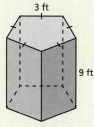

139.36 ft³

21. There are two parallel, congruent bases, so it is a prism, not a pyramid; The solid is a triangular prism.

22. The base circumference was used instead of the base area; $V = \pi r^2 h = 48\pi$ ft³

23. 10 ft
24. 15 yd
25. 4 cm
26. about 6.99 in.
27. about 11.04 ft
28. about 6.00 m
29. 99 cm³
30. 9000π in.³
31. 150 ft³
32. about 89.13 in.³
33. about 1900.66 in.³
34. 35 ft³
35. about 2,350,000,000 gal

ANSWERS

36. about 0.40 ft^3; *Sample answer:* subtracting the volumes; Using the formula for the volume of a prism is simpler.

37. *Sample answer:* The stacks have the same height and the rectangles have the same lengths, so the stacks have the same area.

38. They are the same; *Sample answer:* The stacks have the same height and cross-sectional area.

39. *Sample answer:*

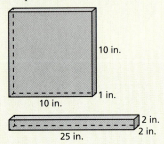

40. your cousin; The sides come together at a point.

41. no; The cross-sectional areas must be the same to use Cavalieri's Principle.

42. no; *Sample answer:* The bases are the same but the sides of the other faces change.

43–47. See Additional Answers.

Mini-Assessment

1. A cylinder has base radius 4 feet and height 8.5 feet. Find its volume. 136π, or about 427.26 cubic feet

2. The base of a triangular prism is an equilateral triangle with sides 20 inches long. The height of the prism is 8 inches. Find the volume of the prism. $800\sqrt{3}$, or about 1385.64 cubic inches

3. A cylindrical hole is drilled through a wooden block that is in the shape of a rectangular prism. Find the volume of the resulting solid.

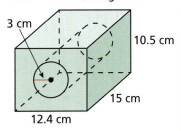

about 1528.88 cubic centimeters

672 Chapter 11

---

36. COMPARING METHODS The *Volume Addition Postulate* states that the volume of a solid is the sum of the volumes of all its nonoverlapping parts. Use this postulate to find the volume of the block of concrete in Example 5 by subtracting the volume of each hole from the volume of the large rectangular prism. Which method do you prefer? Explain your reasoning.

37. WRITING Both of the figures shown are made up of the same number of congruent rectangles. Explain how Cavalieri's Principle can be adapted to compare the areas of these figures.

38. HOW DO YOU SEE IT? Each stack of memo papers contains 500 equally-sized sheets of paper. Compare their volumes. Explain your reasoning.

39. OPEN-ENDED Sketch two rectangular prisms that have volumes of 100 cubic inches but different surface areas. Include dimensions in your sketches.

40. MAKING AN ARGUMENT Your friend says that the polyhedron shown is a triangular prism. Your cousin says that it is a triangular pyramid. Who is correct? Explain your reasoning.

41. MAKING AN ARGUMENT A prism and a cylinder have the same height and different cross-sectional areas. Your friend claims that the two solids have the same volume by Cavalieri's Principle. Is your friend correct? Explain your reasoning.

42. THOUGHT PROVOKING Cavalieri's Principle states that the two solids shown below have the same volume. Do they also have the same surface area? Explain your reasoning.

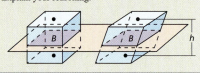

43. PROBLEM SOLVING A barn is in the shape of a pentagonal prism with the dimensions shown. The volume of the barn is 9072 cubic feet. Find the dimensions of each half of the roof.

44. PROBLEM SOLVING A wooden box is in the shape of a regular pentagonal prism. The sides, top, and bottom of the box are 1 centimeter thick. Approximate the volume of wood used to construct the box. Round your answer to the nearest tenth.

Maintaining Mathematical Proficiency
Reviewing what you learned in previous grades and lessons

Find the surface area of the regular pyramid. *(Skills Review Handbook)*

45.

46.

47.

672 Chapter 11 Circumference, Area, and Volume

If students need help...	If students got it...
Resources by Chapter • Practice A and Practice B • Puzzle Time	Resources by Chapter • Enrichment and Extension • Cumulative Review
Student Journal • Practice	Start the *next* Section
Differentiating the Lesson Skills Review Handbook	

Laurie's Notes

Overview of Section 11.5

Introduction
- This is the second of four lessons on the volumes of solids. In this lesson, students will find the volumes of pyramids.
- Also reviewed are volumes of similar solids.

Teaching Strategy
- In this lesson and the next, you want to draw attention to the structural similarities between a pyramid and a cone. Each has one base and a lateral portion. Help students recognize the similarities in the formulas for surface area and volume of each.

	Surface Area (words)	Surface Area (formula)	Volume (words)	Volume (formula)
Pyramid	area of base + area of lateral sides	$S = B + \frac{1}{2}P\ell$	$\frac{1}{3}$ area of base × height	$V = \frac{1}{3}Bh$
Cone	area of base + area of lateral side	$S = B + \pi r \ell$ $= \pi r^2 + \pi r \ell$	$\frac{1}{3}$ area of base × height	$V = \frac{1}{3}Bh$ $= \frac{1}{3}\pi r^2 h$

- In finding the lateral surface area of the cone, the formula could also be written as
$S = B + \frac{1}{2}C\ell = B + \frac{1}{2}(2\pi r)\ell = B + \pi r \ell$.
- Although cones are not introduced until the next lesson, begin to discuss the structure of the pyramid in the lesson.

Pacing Suggestion
- Now that students have reviewed the formula for the volume of a pyramid, you might begin the formal lesson with Example 2.

Dynamic Teaching Tools
Dynamic Assessment & Progress Monitoring Tool
Lesson Planning Tool
Interactive Whiteboard Lesson Library
Dynamic Classroom with Dynamic Investigations

What Your Students Will Learn

- Find volumes of pyramids.
- Use volumes of pyramids to find dimensions of pyramids and volumes of similar solids and composite solids.

Laurie's Notes

Exploration

Motivate
- Ask whether anyone has visited the famed Louvre Museum in Paris.
- The Louvre Pyramid is a large glass and metal pyramid, surrounded by three smaller pyramids, in the main courtyard. The large pyramid serves as the main entrance to the Louvre Museum.
- The Louvre Pyramid was designed by I. M. Pei, who also designed the Rock and Roll Hall of Fame in Cleveland, Ohio.
- Students will find the volume of the glass pyramid at the end of the lesson.

Exploration Note
- Students may recall the formula for the volume of a pyramid from middle school.

Exploration 1
- If you have a set of clear plastic solids to demonstrate this relationship, it would be more engaging than a static picture. Alternately, an online search will result in videos that demonstrate this relationship.
- ❓ "What do the images suggest in order to find the volume of a pyramid?" Multiply $\frac{1}{3}$ times the area of the base times the height.

Exploration 2
- This exploration is a good review of Section 11.3, where students found the areas of regular polygons.
- Circulate and listen to students' strategies for computing the volume.
- Ask for a volunteer to share his or her work.

Communicate Your Answer
- Students will likely guess that the relationship between cones and cylinders is the same as the relationship between pyramids and prisms.

Connecting to Next Step
- Students have discovered, or perhaps recalled, the formula for the volume of a pyramid. In the lesson, they will also solve problems about the volumes of similar pyramids.

11.5 Volumes of Pyramids

Essential Question How can you find the volume of a pyramid?

EXPLORATION 1 Finding the Volume of a Pyramid

Work with a partner. The pyramid and the prism have the same height and the same square base.

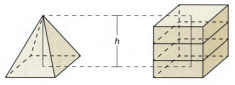

When the pyramid is filled with sand and poured into the prism, it takes three pyramids to fill the prism.

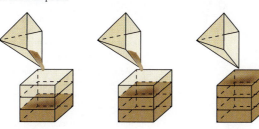

LOOKING FOR STRUCTURE

To be proficient in math, you need to look closely to discern a pattern or structure.

Use this information to write a formula for the volume V of a pyramid.

EXPLORATION 2 Finding the Volume of a Pyramid

Work with a partner. Use the formula you wrote in Exploration 1 to find the volume of the hexagonal pyramid.

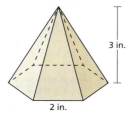

Communicate Your Answer

3. How can you find the volume of a pyramid?

4. In Section 11.6, you will study volumes of cones. How do you think you could use a method similar to the one presented in Exploration 1 to write a formula for the volume of a cone? Explain your reasoning.

Dynamic Teaching Tools

Dynamic Assessment & Progress Monitoring Tool
Lesson Planning Tool
Interactive Whiteboard Lesson Library
Dynamic Classroom with Dynamic Investigations

ANSWERS

1. $V = \frac{1}{3}Bh$
2. about 10.39 in.^{3}
3. Use the formula $V = \frac{1}{3}Bh$.
4. *Sample answer:* Use a cone and cylinder with the same height and circular base and determine how many cones of sand are needed to fill the cylinder.

English Language Learners

Notebook Development
Have students create a formula page in their notebooks for all the formulas they have studied so far in this chapter. Each formula should have a sketch and an example of how to use the formula.

Extra Example 1
Find the volume of the pyramid.

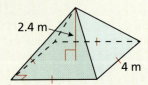

The volume is 12.8 cubic meters.

11.5 Lesson

Core Vocabulary
Previous
pyramid
composite solid

What You Will Learn
▶ Find volumes of pyramids.
▶ Use volumes of pyramids.

Finding Volumes of Pyramids

Consider a triangular prism with parallel, congruent bases △JKL and △MNP. You can divide this triangular prism into three triangular pyramids.

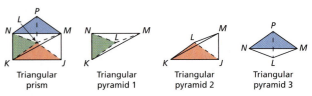

| Triangular prism | Triangular pyramid 1 | Triangular pyramid 2 | Triangular pyramid 3 |

Pyramid Q

You can combine triangular pyramids 1 and 2 to form a pyramid with a base that is a parallelogram, as shown at the left. Name this pyramid Q. Similarly, you can combine triangular pyramids 1 and 3 to form pyramid R with a base that is a parallelogram.

In pyramid Q, diagonal $\overline{KM}$ divides ▱JKNM into two congruent triangles, so the bases of triangular pyramids 1 and 2 are congruent. Similarly, you can divide any cross section parallel to ▱JKNM into two congruent triangles that are the cross sections of triangular pyramids 1 and 2.

Pyramid R

By Cavalieri's Principle, triangular pyramids 1 and 2 have the same volume. Similarly, using pyramid R, you can show that triangular pyramids 1 and 3 have the same volume. By the Transitive Property of Equality, triangular pyramids 2 and 3 have the same volume.

The volume of each pyramid must be one-third the volume of the prism, or $V = \frac{1}{3}Bh$. You can generalize this formula to say that the volume of any pyramid with any base is equal to $\frac{1}{3}$ the volume of a prism with the same base and height because you can divide any polygon into triangles and any pyramid into triangular pyramids.

🄲 Core Concept

Volume of a Pyramid
The volume V of a pyramid is
$$V = \frac{1}{3}Bh$$
where B is the area of the base and h is the height.

EXAMPLE 1 Finding the Volume of a Pyramid

Find the volume of the pyramid.

SOLUTION

$V = \frac{1}{3}Bh$ Formula for volume of a pyramid

$= \frac{1}{3}\left(\frac{1}{2} \cdot 4 \cdot 6\right)(9)$ Substitute.

$= 36$ Simplify.

▶ The volume is 36 cubic meters.

674 Chapter 11 Circumference, Area, and Volume

Laurie's Notes Teacher Actions

- The volume of pyramids was studied in middle school.
- See the *Teaching Strategy* on page T-672.
- The models described on this page can actually be constructed. The model, composed of three congruent pyramids that can be rearranged to make a prism, helps students make the connection to the $\frac{1}{3}$ in the formula.
- ❓ Write the *Core Concept*. "How do you find the area of the base of the pyramid?" It depends on what polygon it is!
- In Example 1, the height of the pyramid is 9 meters. The slant height is not 9 meters.

Monitoring Progress Help in English and Spanish at *BigIdeasMath.com*

Find the volume of the pyramid.

1.
2. (pyramid with height 20 cm and base 12 cm)

Using Volumes of Pyramids

EXAMPLE 2 Using the Volume of a Pyramid

Originally, Khafre's Pyramid had a height of about 144 meters and a volume of about 2,218,800 cubic meters. Find the side length of the square base.

SOLUTION

Khafre's Pyramid, Egypt

$V = \frac{1}{3}Bh$	Formula for volume of a pyramid
$2{,}218{,}800 \approx \frac{1}{3}x^2(144)$	Substitute.
$6{,}656{,}400 \approx 144x^2$	Multiply each side by 3.
$46{,}225 \approx x^2$	Divide each side by 144.
$215 \approx x$	Find the positive square root.

▶ Originally, the side length of the square base was about 215 meters.

EXAMPLE 3 Using the Volume of a Pyramid

Find the height of the triangular pyramid.

SOLUTION

The area of the base is $B = \frac{1}{2}(3)(4) = 6$ ft^2 and the volume is $V = 14$ ft^3.

$V = \frac{1}{3}Bh$	Formula for volume of a pyramid
$14 = \frac{1}{3}(6)h$	Substitute.
$7 = h$	Solve for *h*.

▶ The height is 7 feet.

Monitoring Progress Help in English and Spanish at *BigIdeasMath.com*

3. The volume of a square pyramid is 75 cubic meters and the height is 9 meters. Find the side length of the square base.

4. Find the height of the triangular pyramid at the left.

Section 11.5   Volumes of Pyramids   675

Differentiated Instruction

Auditory
Remind students that the base of a pyramid must be a polygon. Review the formulas for the areas of triangles and rectangles.

Extra Example 2
A square pyramid has a height of 12 centimeters and a volume of 64 cubic centimeters. Find the side length of the square base. **4 centimeters**

Extra Example 3
Find the height of the triangular pyramid.

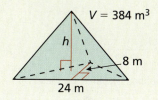

The height is 12 meters.

MONITORING PROGRESS ANSWERS

1. 400 cm^3
2. about 2494.15 cm^3
3. 5 m
4. 8 m

Laurie's Notes — Teacher Actions

- Question 2 will be more challenging due to the hexagonal base.
- **? Reason Abstractly and Quantitatively:** "What is known in Example 2?" *You know the volume of the square pyramid and its height.* "What are you trying to find?" *the length of an edge on the square base* Have students work independently to solve. Multiplying both sides by 3 clears the fraction.
- **Think-Pair-Share:** Have students answer Example 3, and then share and discuss as a class.

Extra Example 4
Square pyramid A and square pyramid B are similar. The height of pyramid A is 6 inches and the height of pyramid B is 15 inches. The volume of pyramid B is 312.5 cubic inches. Find the volume of pyramid A. *The volume of pyramid A is 20 cubic inches.*

Extra Example 5
Find the volume of the composite solid.

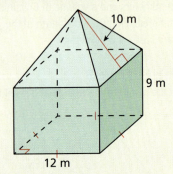

The volume is 1680 cubic meters.

MONITORING PROGRESS ANSWERS
5. 12 m^3
6. 96 ft^3

676   Chapter 11

EXAMPLE 4 Finding the Volume of a Similar Solid

Pyramid A and pyramid B are similar. Find the volume of pyramid B.

SOLUTION

The scale factor is $k = \dfrac{\text{Height of pyramid B}}{\text{Height of pyramid A}} = \dfrac{6}{8} = \dfrac{3}{4}$.

Use the scale factor to find the volume of pyramid B.

$\dfrac{\text{Volume of pyramid B}}{\text{Volume of pyramid A}} = k^3$ The ratio of the volumes is k^3.

$\dfrac{\text{Volume of pyramid B}}{96} = \left(\dfrac{3}{4}\right)^3$ Substitute.

Volume of pyramid B = 40.5   Solve for volume of pyramid B.

▶ The volume of pyramid B is 40.5 cubic meters.

Monitoring Progress 🔊 *Help in English and Spanish at BigIdeasMath.com*

5. Pyramid C and pyramid D are similar. Find the volume of pyramid D.

EXAMPLE 5 Finding the Volume of a Composite Solid

Find the volume of the composite solid.

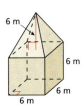

SOLUTION

$\begin{array}{c}\text{Volume of} \\ \text{solid}\end{array} = \begin{array}{c}\text{Volume of} \\ \text{cube}\end{array} + \begin{array}{c}\text{Volume of} \\ \text{pyramid}\end{array}$

$= s^3 + \dfrac{1}{3}Bh$ Write formulas.

$= 6^3 + \dfrac{1}{3}(6)^2 \cdot 6$ Substitute.

$= 216 + 72$ Simplify.

$= 288$ Add.

▶ The volume is 288 cubic meters.

Monitoring Progress 🔊 *Help in English and Spanish at BigIdeasMath.com*

6. Find the volume of the composite solid.

676   Chapter 11   Circumference, Area, and Volume

Laurie's Notes   **Teacher Actions**

- **Visitor Explanation:** Finish Example 4 and ask, "If a visitor entered the room right now, how would you explain what you are doing and why you are doing it?" Students should practice explanations with partners. Solicit oral responses or do a one-minute write.
- **COMMON ERROR** When computing the area of the base in Example 5, the $\dfrac{1}{3}$ is not squared; only the 6 is squared.

Closure
- **Exit Ticket:** Find the volume of the Louvre Pyramid. It reaches a height of about 72 feet, and the square base has side length of about 116 feet.   *about 322,944 cubic feet*

11.5 Exercises

Dynamic Solutions available at *BigIdeasMath.com*

Vocabulary and Core Concept Check

1. **VOCABULARY** Explain the difference between a triangular prism and a triangular pyramid.

2. **REASONING** A square pyramid and a cube have the same base and height. Compare the volume of the square pyramid to the volume of the cube.

Monitoring Progress and Modeling with Mathematics

In Exercises 3 and 4, find the volume of the pyramid. (*See Example 1.*)

3.

4.

In Exercises 5–8, find the indicated measure. (*See Example 2.*)

5. A pyramid with a square base has a volume of 120 cubic meters and a height of 10 meters. Find the side length of the square base.

6. A pyramid with a square base has a volume of 912 cubic feet and a height of 19 feet. Find the side length of the square base.

7. A pyramid with a rectangular base has a volume of 480 cubic inches and a height of 10 inches. The width of the rectangular base is 9 inches. Find the length of the rectangular base.

8. A pyramid with a rectangular base has a volume of 105 cubic centimeters and a height of 15 centimeters. The length of the rectangular base is 7 centimeters. Find the width of the rectangular base.

9. **ERROR ANALYSIS** Describe and correct the error in finding the volume of the pyramid.

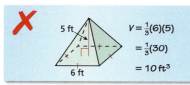

10. **OPEN-ENDED** Give an example of a pyramid and a prism that have the same base and the same volume. Explain your reasoning.

In Exercises 11–14, find the height of the pyramid. (*See Example 3.*)

11. Volume = 15 ft³

12. Volume = 224 in.³

13. Volume = 198 yd³

14. Volume = 392 cm³

In Exercises 15 and 16, the pyramids are similar. Find the volume of pyramid B. (*See Example 4.*)

15.
V = 256 ft³

16. Pyramid A   Pyramid B

V = 10 in.³

Section 11.5   Volumes of Pyramids   **677**

Assignment Guide and Homework Check

ASSIGNMENT
Basic: 1, 2, 3–19 odd, 22, 26–29
Average: 1, 2, 4, 6, 9, 10–20 even, 21, 22, 25–29
Advanced: 1, 2, 9–29

HOMEWORK CHECK
Basic: 3, 5, 7, 11, 17
Average: 6, 12, 16, 20, 21
Advanced: 12, 20, 21, 23, 24

ANSWERS

1. *Sample answer:* A triangular prism has two parallel bases that are triangles. A triangular pyramid has one base that is a triangle, and the other faces all intersect at a single point.

2. The volume of the square pyramid is $\frac{1}{3}$ of the volume of the cube.

3. 448 m³
4. 6 in.³
5. 6 m
6. 12 ft
7. 16 in.
8. 3 cm
9. One side length was used in the formula as the base area;
$V = \frac{1}{3}(6^2)(5) = 60$ ft³
10. *Sample answer:* A rectangular pyramid with a base area of 5 square meters and a height of 6 meters, and a rectangular prism with a base area of 5 square meters and a height of 2 meters; Both volumes are 10 cubic meters.
11. 5 ft
12. 7 in.
13. 12 yd
14. 24 cm
15. 4 ft³
16. 80 in.³

Dynamic Teaching Tools
- Dynamic Assessment & Progress Monitoring Tool
- Interactive Whiteboard Lesson Library
- Dynamic Classroom with Dynamic Investigations

ANSWERS
17. 72 in.^{3}
18. 666 cm^3
19. about 213.33 cm^3
20. 1440 in.^{3}
21. a. The volume doubles.
    b. The volume is 4 times greater.
    c. yes; Sample answer:
    Square pyramid: $V = \frac{1}{3}s^2h$
    Double height:
    $V = \frac{1}{3}s^2(2h) = 2\left(\frac{1}{3}s^2h\right)$
    Double side length of base:
    $V = \frac{1}{3}(2s)^2h = 4\left(\frac{1}{3}s^2h\right)$
22. Sample answer: The three pyramids have the same base and height as the prism, and the same volumes as each other, so each is $\frac{1}{3}$ the volume of the prism.
23. about 9.22 ft^3
24–29. See Additional Answers.

Mini-Assessment

1. Two similar pyramids have heights 8 feet and 5 feet. What is the ratio of their volumes? **512:125**

2. A square pyramid has height 13 cm. Each base edge is 6 cm long. Find the volume. **156 cubic centimeters**

3. The volume of the pyramid is 168 cubic inches. Find the length of the rectangular base.

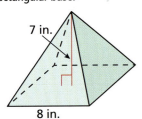

9 inches

4. The volume of a triangular pyramid with a right triangle base is 100 cubic feet. One leg of the base triangle is 4 feet. The height of the pyramid is 27 feet. Find the height of the base triangle. **about 5.56 feet**

678   Chapter 11

In Exercises 17–20, find the volume of the composite solid. *(See Example 5.)*

17.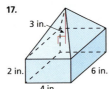

18.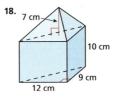

19.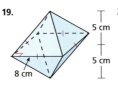

20.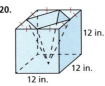

21. ABSTRACT REASONING A pyramid has a height of 8 feet and a square base with a side length of 6 feet.
   a. How does the volume of the pyramid change when the base stays the same and the height is doubled?
   b. How does the volume of the pyramid change when the height stays the same and the side length of the base is doubled?
   c. Are your answers to parts (a) and (b) true for any square pyramid? Explain your reasoning.

22. HOW DO YOU SEE IT? The cube shown is formed by three pyramids, each with the same square base and the same height. How could you use this to verify the formula for the volume of a pyramid?

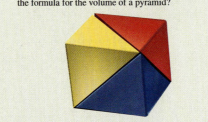

Maintaining Mathematical Proficiency Reviewing what you learned in previous grades and lessons

Find the value of x. Round your answer to the nearest tenth. *(Section 9.4 and Section 9.5)*

26.
27.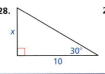
28.
29.

678   Chapter 11   Circumference, Area, and Volume

---

23. CRITICAL THINKING Find the volume of the regular pentagonal pyramid. Round your answer to the nearest hundredth. In the diagram, $m\angle ABC = 35°$.

24. THOUGHT PROVOKING A *frustum* of a pyramid is the part of the pyramid that lies between the base and a plane parallel to the base, as shown. Write a formula for the volume of the frustum of a square pyramid in terms of a, b, and h. (*Hint*: Consider the "missing" top of the pyramid and use similar triangles.)

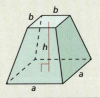

25. MODELING WITH MATHEMATICS Nautical deck prisms were used as a safe way to illuminate decks on ships. The deck prism shown here is composed of the following three solids: a regular hexagonal prism with an edge length of 3.5 inches and a height of 1.5 inches, a regular hexagonal prism with an edge length of 3.25 inches and a height of 0.25 inch, and a regular hexagonal pyramid with an edge length of 3 inches and a height of 3 inches. Find the volume of the deck prism.

If students need help...	If students got it...
Resources by Chapter • Practice A and Practice B • Puzzle Time	**Resources by Chapter** • Enrichment and Extension • Cumulative Review
Student Journal • Practice	Start the *next* Section
Differentiating the Lesson Skills Review Handbook	

Laurie's Notes

Dynamic Teaching Tools

- Dynamic Assessment & Progress Monitoring Tool
- Lesson Planning Tool
- Interactive Whiteboard Lesson Library
- Dynamic Classroom with Dynamic Investigations

Overview of Section 11.6

Introduction

- This is the third of four lessons on the volumes of solids. In this lesson, students will find the volume of a cone, which structurally is like a pyramid.
- A second topic presented in this lesson is the surface area of a cone.
- Students will also work with volumes of similar cones.

Teaching Strategy

- In the first exploration, students derive the formula for the surface area of a cone. Students begin with a circle of radius 3 inches. Then sectors of 60° are removed from the circle and a cone is formed. In the exploration, the students are gathering data about the radius, slant height, area of the base, lateral area, and total surface area.
- The one dimension that is not needed in the investigation is the height of the cone. As an extension to the exploration, have students solve for the height of each successive cone. Is there a pattern?
- Using a graphing calculator, store the base radius of the cone in L_1 and the slant height in L_2.
- Calculate the height of the cone $\left(L_3 = \sqrt{L_2{}^2 - L_1{}^2}\right)$ as shown in Figure 1, and store the height in L_3. (See Figure 2.)
- Figure 3 shows the graph of (L_1, L_3) or (radius of the base, height). As the radius increases, the height decreases.

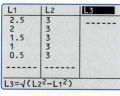

L1	L2	L3
2.5	3	------
2	3	
1.5	3	
1	3	
0.5	3	
------	------	

L3=√(L2²−L1²)

Figure 1

L1	L2	L3
2.5	3	1.6583
2	3	2.2361
1.5	3	2.5981
1	3	2.8284
0.5	3	2.958
------	------	------

L3(1)=1.658312395177

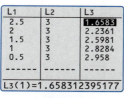

Figure 2

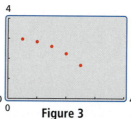

Figure 3

- Once students have the volume formula, they can also solve for the volume of the cone. Is there a pattern?
- Calculate the volume $\left(L_4 = (1/3)\pi L_1{}^2 L_3\right)$ as shown in Figure 4, and store the volume in L_4. (See Figure 5.)

L1	L2	L3	L4
2.5	3	1.6583	------
2	3	2.2361	
1.5	3	2.5981	
1	3	2.8284	
0.5	3	2.958	
------	------	------	

L4=(1/3)πL1²L3

Figure 4

L1	L2	L3	L4
2.5	3	1.6583	10.854
2	3	2.2361	9.3664
1.5	3	2.5981	6.1216
1	3	2.8284	2.9619
0.5	3	2.958	.77441
------	------	------	------

L4(1)=10.853629245932

Figure 5

- Figure 6 shows the graph of (L_1, L_4) or (radius of the base, volume). As the radius increases, the volume increases.

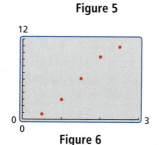

Figure 6

Pacing Suggestion

- Complete both explorations, the first of which takes a bit longer. Transition to the formal lesson by stating the *Core Concept*.

What Your Students Will Learn

- Find surface areas of right cones.
- Find volumes of cones.
- Use volumes of cones to find volumes of similar solids and composite solids.

Laurie's Notes

Exploration

Motivate
- "What does the net for a cone look like? Sketch it." Ask students to share their sketches. Many will have the lateral portion incorrect. Students will work with the net in the first exploration.

Exploration 1
- The first exploration focuses on the derivation of the formula for the surface area of a cone. It's cool that your students can derive a formula that generally is just given to students to use.
- Begin with a paper circle of radius 3 inches. A sector, $\frac{1}{6}$ of the original area of the circle, is removed each time. When taped, the remaining sector folds into a cone.
- Making a table to gather results will help facilitate this exploration.

Shape					
Radius of Base	$\frac{5}{2}$	2	$\frac{3}{2}$	1	$\frac{1}{2}$
Base Area	$\pi\left(\frac{5}{2}\right)^2$	$\pi(2)^2$	$\pi\left(\frac{3}{2}\right)^2$	$\pi(1)^2$	$\pi\left(\frac{1}{2}\right)^2$
Lateral Area	$\frac{5}{6}\pi(3)^2 = \frac{15}{2}\pi$	$\frac{2}{3}\pi(3)^2 = 6\pi$	$\frac{1}{2}\pi(3)^2 = \frac{9}{2}\pi$	$\frac{1}{3}\pi(3)^2 = 3\pi$	$\frac{1}{6}\pi(3)^2 = \frac{3}{2}\pi$
Total Area	$\frac{55}{4}\pi$	10π	$\frac{27}{4}\pi$	4π	$\frac{7}{4}\pi$

- **Observations:** The radius of the base is getting smaller. The radius of the lateral area is always 3 inches, and it is also the slant height of the cone when folded. To find the total area, you add the base area and the lateral area. Students will recognize that the area of the base is still πr^2. The pattern in the lateral area is that the coefficient of π is the radius of the base r times the slant height ℓ.
- So, the formula for the surface area of a cone is $\pi r^2 + r\ell\pi$.
- **Extension:** What is happening to the height of the cone? See the *Teaching Strategy* on page T-678.

Exploration 2
- If you have a set of clear plastic solids to demonstrate this relationship, it would be more engaging than a static picture. Alternately, an online search will result in videos that demonstrate this relationship.
- ❓ "What do the images suggest in order to find the volume of a cone?" Multiply $\frac{1}{3}$ times the area of the base times the height.

Communicate Your Answer
- Listen for a general understanding of how to find the surface area of a cone (lateral surface area + base area) even though the formula may not be fully developed.
- Knowing the volume of a pyramid helps students gain an understanding of the formula for the volume of a cone.

Connecting to Next Step
- Students will be finding both the surface area and the volume of a cone in the lesson.

11.6 Surface Areas and Volumes of Cones

Essential Question How can you find the surface area and the volume of a cone?

EXPLORATION 1   Finding the Surface Area of a Cone

Work with a partner. Construct a circle with a radius of 3 inches. Mark the circumference of the circle into six equal parts, and label the length of each part. Then cut out one sector of the circle and make a cone.

a. Explain why the base of the cone is a circle. What are the circumference and radius of the base?

b. What is the area of the original circle? What is the area with one sector missing?

c. Describe the surface area of the cone, including the base. Use your description to find the surface area.

EXPLORATION 2   Finding the Volume of a Cone

Work with a partner. The cone and the cylinder have the same height and the same circular base.

When the cone is filled with sand and poured into the cylinder, it takes three cones to fill the cylinder.

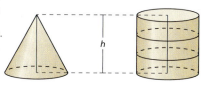

CONSTRUCTING VIABLE ARGUMENTS
To be proficient in math, you need to understand and use stated assumptions, definitions, and previously established results in constructing arguments.

Use this information to write a formula for the volume V of a cone.

Communicate Your Answer

3. How can you find the surface area and the volume of a cone?

4. In Exploration 1, cut another sector from the circle and make a cone. Find the radius of the base and the surface area of the cone. Repeat this three times, recording your results in a table. Describe the pattern.

Dynamic Teaching Tools

Dynamic Assessment & Progress Monitoring Tool
Lesson Planning Tool
Interactive Whiteboard Lesson Library
Dynamic Classroom with Dynamic Investigations

ANSWERS

1. a. *Sample answer:* The points on the edge of the base are all the same distance from the point on the same plane directly below the vertex of the cone; 5π in.; 2.5 in.
   b. 9π in.^{2}; 7.5π in.^{2}
   c. a circle with a radius of 2.5 in. and a sector which is $\frac{5}{6}$ of a circle with a radius of 3 in.; 13.75π in.^{2}
2. $V = \frac{1}{3}\pi r^2 h$
3. *Sample answer:* $S = \pi r^2 + \pi r \ell$; $V = \frac{1}{3}\pi r^2 h$
4. *Sample answer:* As the radius decreases, the total surface area decreases.

> **Differentiated Instruction**
>
> **Auditory**
> Ask students to describe the types of figures for which they are finding lateral area and surface area in this section. Make sure students understand that the formulas for surface area and lateral area apply to right cones only. Draw a non-right cone on the board and label its radius and height. Ask students to explain why the formula $\pi r \ell$ does not work to find the lateral area of this figure.

Extra Example 1
Find the surface area of the right cone.

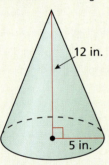

The surface area is 90π, or about 282.74 square inches.

MONITORING PROGRESS ANSWER
1. about 436.18 m²

11.6 Lesson

What You Will Learn

▶ Find surface areas of right cones.
▶ Find volumes of cones.
▶ Use volumes of cones.

Core Vocabulary
lateral surface of a cone, p. 680
Previous
cone
net
composite solid

Finding Surface Areas of Right Cones

Recall that a *circular cone*, or *cone*, has a circular *base* and a *vertex* that is not in the same plane as the base. The *altitude*, or *height*, is the perpendicular distance between the vertex and the base. In a *right cone*, the height meets the base at its center and the *slant height* is the distance between the vertex and a point on the base edge.

The **lateral surface of a cone** consists of all segments that connect the vertex with points on the base edge. When you cut along the slant height and lay the right cone flat, you get the net shown at the left. In the net, the circular base has an area of πr^2 and the lateral surface is a sector of a circle. You can find the area of this sector by using a proportion, as shown below.

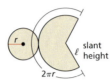

$$\frac{\text{Area of sector}}{\text{Area of circle}} = \frac{\text{Arc length}}{\text{Circumference of circle}} \quad \text{Set up proportion.}$$

$$\frac{\text{Area of sector}}{\pi \ell^2} = \frac{2\pi r}{2\pi \ell} \quad \text{Substitute.}$$

$$\text{Area of sector} = \pi \ell^2 \cdot \frac{2\pi r}{2\pi \ell} \quad \text{Multiply each side by } \pi \ell^2.$$

$$\text{Area of sector} = \pi r \ell \quad \text{Simplify.}$$

The surface area of a right cone is the sum of the base area and the lateral area, $\pi r \ell$.

Core Concept

Surface Area of a Right Cone
The surface area S of a right cone is

$$S = \pi r^2 + \pi r \ell$$

where r is the radius of the base and ℓ is the slant height.

EXAMPLE 1 Finding Surface Areas of Right Cones

Find the surface area of the right cone.

SOLUTION
$S = \pi r^2 + \pi r \ell = \pi \cdot 4^2 + \pi(4)(6) = 40\pi \approx 125.66$

▶ The surface area is 40π, or about 125.66 square inches.

Monitoring Progress Help in English and Spanish at *BigIdeasMath.com*

1. Find the surface area of the right cone.

680 Chapter 11 Circumference, Area, and Volume

Laurie's Notes | Teacher Actions

- The introduction to this lesson defines the vocabulary associated with a cone.
- The formula for the surface area of a cone is then derived using the schematic of the net shown in the margin. If students did not complete the first exploration, work through the derivation.
- **? Make Sense of Problems and Persevere in Solving Them:** "What is the relationship between the slant height ℓ, the radius r, and the height of the cone h?" $r^2 + h^2 = \ell^2$
- Have students work independently on Example 1 and Question 1, if needed.

Finding Volumes of Cones

Consider a cone with a regular polygon inscribed in the base. The pyramid with the same vertex as the cone has volume $V = \frac{1}{3}Bh$. As you increase the number of sides of the polygon, it approaches the base of the cone and the pyramid approaches the cone. The volume approaches $\frac{1}{3}\pi r^2 h$ as the base area B approaches πr^2.

🔵 Core Concept

Volume of a Cone

The volume V of a cone is

$$V = \frac{1}{3}Bh = \frac{1}{3}\pi r^2 h$$

where B is the area of the base, h is the height, and r is the radius of the base.

EXAMPLE 2 Finding the Volume of a Cone

Find the volume of the cone.

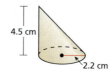

SOLUTION

$V = \frac{1}{3}\pi r^2 h$	Formula for volume of a cone
$= \frac{1}{3}\pi \cdot (2.2)^2 \cdot 4.5$	Substitute.
$= 7.26\pi$	Simplify.
≈ 22.81	Use a calculator.

▶ The volume is 7.26π, or about 22.81 cubic centimeters.

Monitoring Progress Help in English and Spanish at *BigIdeasMath.com*

Find the volume of the cone.

2.

3.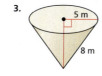

Extra Example 2

Find the volume of the cone.

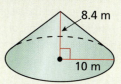

The volume is 280π, or about 879.65 cubic meters.

MONITORING PROGRESS ANSWERS

2. about 667.06 in.^{3}
3. about 163.49 m^3

Laurie's Notes   Teacher Actions

- Structurally, the cone is like a pyramid—a base and a lateral portion that converges to a vertex. It is not a surprise to students that the formula for the volume of a cone should be similar to the formula for the volume of a pyramid.
- Write the *Core Concept*. Note that the height of the cone is used in the formula, not the slant height.
- **Think-Pair-Share:** Have students answer Example 2, and then share and discuss as a class.

Extra Example 3
Cone A and cone B are similar. The height of cone A is 6 inches and the height of cone B is 2 inches The volume of cone B is 18π cubic inches. Find the volume of cone A. The volume of cone A is 486π cubic inches.

Extra Example 4
Find the volume of the composite solid.

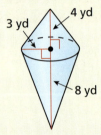

The volume is 36π, or about 113.10 cubic yards.

MONITORING PROGRESS ANSWERS
4. $6\pi \text{ cm}^3$
5. about 329.87 cm^3

Using Volumes of Cones

EXAMPLE 3 Finding the Volume of a Similar Solid

Cone A and cone B are similar. Find the volume of cone B.

Cone A   Cone B

SOLUTION

The scale factor is $k = \dfrac{\text{Radius of cone B}}{\text{Radius of cone A}} = \dfrac{9}{3} = 3$.

Use the scale factor to find the volume of cone B.

$\dfrac{\text{Volume of cone B}}{\text{Volume of cone A}} = k^3$ The ratio of the volumes is k^3.

$\dfrac{\text{Volume of cone B}}{15\pi} = 3^3$ Substitute.

Volume of cone B $= 405\pi$ Solve for volume of cone B.

▶ The volume of cone B is 405π cubic feet.

Monitoring Progress   Help in English and Spanish at *BigIdeasMath.com*

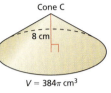

4. Cone C and cone D are similar. Find the volume of cone D.

EXAMPLE 4 Finding the Volume of a Composite Solid

Find the volume of the composite solid.

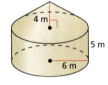

SOLUTION

Let h_1 be the height of the cylinder and let h_2 be the height of the cone.

$$\text{Volume of solid} = \text{Volume of cylinder} + \text{Volume of cone}$$

$= \pi r^2 h_1 + \dfrac{1}{3}\pi r^2 h_2$ Write formulas.

$= \pi \cdot 6^2 \cdot 5 + \dfrac{1}{3}\pi \cdot 6^2 \cdot 4$ Substitute.

$= 180\pi + 48\pi$ Simplify.

$= 228\pi$ Add.

≈ 716.28 Use a calculator.

▶ The volume is 228π, or about 716.28 cubic meters.

Monitoring Progress   Help in English and Spanish at *BigIdeasMath.com*

5. Find the volume of the composite solid.

682   Chapter 11   Circumference, Area, and Volume

Laurie's Notes | Teacher Actions

- **?** Pose Example 3. "What is the ratio of their linear measurements?" The radii are in the ratio of 3:1. "What is the ratio of their volumes?" The volumes are in the ratio of $3^3:1^3$ or 27:1.
- **?** Extension: "What is the ratio of their surface areas?" $3^2:1^2$ or 9:1
- **Turn and Talk:** "How can you find the volume of the composite figure in Example 4?" Discuss and have partners find the volume.

Closure
- **Exit Ticket:** Find the surface area and volume of a cone with a diameter of 8 centimeters and height of 3 centimeters. The surface area is 36π, or about 113.10 square centimeters. The volume is 16π, or about 50.27 cubic centimeters.

11.6 Exercises

Dynamic Solutions available at *BigIdeasMath.com*

Vocabulary and Core Concept Check

1. **WRITING** Describe the differences between pyramids and cones. Describe their similarities.

2. **COMPLETE THE SENTENCE** The volume of a cone with radius r and height h is $\frac{1}{3}$ the volume of a(n) _____ with radius r and height h.

Monitoring Progress and Modeling with Mathematics

In Exercises 3–6, find the surface area of the right cone. *(See Example 1.)*

3.

4.

5. A right cone has a radius of 9 inches and a height of 12 inches.

6. A right cone has a diameter of 11.2 feet and a height of 9.2 feet.

In Exercises 7–10, find the volume of the cone. *(See Example 2.)*

7.

8. (with 1 m, 2 m)

9. A cone has a diameter of 11.5 inches and a height of 15.2 inches.

10. A right cone has a radius of 3 feet and a slant height of 6 feet.

In Exercises 11 and 12, find the missing dimension(s).

11. Surface area = 75.4 cm²

12. Volume = 216π in.³
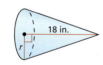

In Exercises 13 and 14, the cones are similar. Find the volume of cone B. *(See Example 3.)*

13. Cone A   Cone B
   $V = 32\pi$ ft³, 16 ft

14. Cone A   Cone B
   $V = 120\pi$ m³

In Exercises 15 and 16, find the volume of the composite solid. *(See Example 4.)*

15.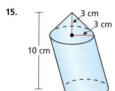

16.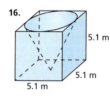

17. **ANALYZING RELATIONSHIPS** A cone has height h and a base with radius r. You want to change the cone so its volume is doubled. What is the new height if you change only the height? What is the new radius if you change only the radius? Explain.

Section 11.6 Surface Areas and Volumes of Cones 683

Assignment Guide and Homework Check

ASSIGNMENT

Basic: 1, 2, 3–15 odd, 18, 25, 27–30

Average: 1, 2, 6–18 even, 19–22, 25, 27–30

Advanced: 1, 2, 6, 8, 12–20 even, 21–30

HOMEWORK CHECK

Basic: 3, 7, 11, 13, 15

Average: 6, 12, 14, 16, 20

Advanced: 12, 14, 16, 20

ANSWERS

1. Sample answer: pyramids have a polygonal base, cones have a circular base; They both have sides that meet at a single vertex.
2. cylinder
3. about 603.19 in.²
4. about 219.44 cm²
5. about 678.58 in.²
6. about 288.00 ft²
7. about 1361.36 mm³
8. about 2.09 m³
9. about 526.27 in.³
10. about 48.97 ft³
11. $\ell \approx 5.00$ cm; $h \approx 4.00$ cm
12. 6 in.
13. 256π ft³
14. 7.68π m³
15. about 226.19 cm³
16. about 97.92 m³
17. $2h$; $r\sqrt{2}$; Sample answer: The original volume is $V = \frac{1}{3}\pi r^2 h$ and the new volume is $V = \frac{2}{3}\pi r^2 h$.

Dynamic Teaching Tools

Dynamic Assessment & Progress Monitoring Tool

Interactive Whiteboard Lesson Library

Dynamic Classroom with Dynamic Investigations

ANSWERS

18. **a.** 3; *Sample answer:* The volume of the cone-shaped container is $\frac{1}{3}$ the volume of the cylindrical container.

    b. the cylindrical container; *Sample answer:* Three cone-shaped containers cost $3.75.

19–30. See Additional Answers.

Mini-Assessment

1. Find the surface area of a right cone with height 9 meters and base radius 6 meters. $(36 + 18\sqrt{13})\pi$, or about 316.99 square meters

2. Find the volume of a cone with base diameter 30 centimeters and height 45 centimeters. 3375π, or about 10,603 cubic centimeters

3. An ice cream cone has a base radius of 2.1 centimeters and a volume of 56.4 cubic centimeters. Find the height of the cone.

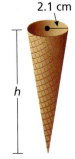

about 12.21 centimeters

4. Find the volume of the composite solid.

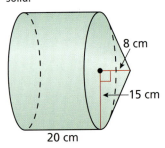

The volume is 5100π, or about 16,022 cubic meters.

684   Chapter 11

---

18. HOW DO YOU SEE IT A snack stand serves a small order of popcorn in a cone-shaped container and a large order of popcorn in a cylindrical container. Do not perform any calculations.

  a. How many small containers of popcorn do you have to buy to equal the amount of popcorn in a large container? Explain.

  b. Which container gives you more popcorn for your money? Explain.

In Exercises 19 and 20, find the volume of the right cone.

19.

20.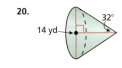

21. MODELING WITH MATHEMATICS
A cat eats half a cup of food, twice per day. Will the automatic pet feeder hold enough food for 10 days? Explain your reasoning. (1 cup ≈ 14.4 in.³)

22. MODELING WITH MATHEMATICS During a chemistry lab, you use a funnel to pour a solvent into a flask. The radius of the funnel is 5 centimeters and its height is 10 centimeters. You pour the solvent into the funnel at a rate of 80 milliliters per second and the solvent flows out of the funnel at a rate of 65 milliliters per second. How long will it be before the funnel overflows? (1 mL = 1 cm³)

Maintaining Mathematical Proficiency — Reviewing what you learned in previous grades and lessons

Find the indicated measure. *(Section 11.2)*

27. area of a circle with a radius of 7 feet

28. area of a circle with a diameter of 22 centimeters

29. diameter of a circle with an area of 256π square meters

30. radius of a circle with an area of 529π square inches

684   Chapter 11   Circumference, Area, and Volume

---

23. REASONING To make a paper drinking cup, start with a circular piece of paper that has a 3-inch radius, then follow the given steps. How does the surface area of the cup compare to the original paper circle? Find $m\angle ABC$.

24. THOUGHT PROVOKING A *frustum* of a cone is the part of the cone that lies between the base and a plane parallel to the base, as shown. Write a formula for the volume of the frustum of a cone in terms of a, b, and h. (*Hint*: Consider the "missing" top of the cone and use similar triangles.)

25. MAKING AN ARGUMENT In the figure, the two cylinders are congruent. The combined height of the two smaller cones equals the height of the larger cone. Your friend claims that this means the total volume of the two smaller cones is equal to the volume of the larger cone. Is your friend correct? Justify your answer.

26. MODELING WITH MATHEMATICS Water flows into a reservoir shaped like a right cone at a rate of 1.8 cubic meters per minute. The height and diameter of the reservoir are equal.

  a. Show that $V = \frac{\pi h^3}{12}$.

  b. Make a table showing the heights h of the water after 1, 2, 3, 4, and 5 minutes.

  c. Is there a linear relationship between the height of the water and time? Explain.

If students need help...	If students got it...
Resources by Chapter • Practice A and Practice B • Puzzle Time	Resources by Chapter • Enrichment and Extension • Cumulative Review
Student Journal • Practice	Start the *next* Section
Differentiating the Lesson Skills Review Handbook	

Laurie's Notes

Dynamic Teaching Tools
Dynamic Assessment & Progress Monitoring Tool
Lesson Planning Tool
Interactive Whiteboard Lesson Library
Dynamic Classroom with Dynamic Investigations

Overview of Section 11.7

Introduction
- This is the fourth and final lesson on the volumes of solids. In this lesson, students will find the volume of a sphere.
- A second topic presented in this lesson is the surface area of a sphere.

Teaching Strategy
- The sphere is the only solid that has just one dimension, the radius. As the year comes to a close and students are preparing to take Math III, have them discuss the formulas for surface area and volume of a sphere.
- **?** "What type of function is the surface area formula?" quadratic
- **?** "What type of function is the volume formula?" cubic
- **?** "What is the domain of each function?" the positive real numbers
- Use a graphing calculator to graph each function.

Pacing Suggestion
- Complete both explorations, the second of which takes a bit longer. Transition to the formal lesson by discussing the vocabulary of a sphere and stating the *Core Concept*.

What Your Students Will Learn

- Find surface areas of spheres and identify chords and great circles of spheres.
- Find volumes of spheres and composite solids.

Laurie's Notes

Exploration

Motivate
- Have students describe the surface of a half sphere, and ask them how they might calculate its surface area.
- Discuss with students that the surface area consists of the flat, circular base, for which they can use the area formula $A = \pi r^2$, and half a ball-shaped surface.
- Tell students that in this lesson they will learn and apply formulas for the surface area and volume of a sphere.

Exploration 1
- This first exploration is a quick, visual way for students to estimate the surface area of a sphere. Students should estimate that to find the surface area they need to multiply 4 times πr^2.

Exploration 2
- If you have a set of clear plastic solids to demonstrate this relationship, it would be more engaging than a static picture. Alternately, an online search will result in videos that demonstrate this relationship.
- **?** "What does the exploration suggest is the formula for the volume of a sphere?" Multiply $\frac{4}{3}$ times πr^3.

Communicate Your Answer
- Listen for a general understanding of how to find the surface area and volume of a sphere even though the formulas may not be fully developed.

Connecting to Next Step
- Students will be finding both the surface area and the volume of a sphere in the lesson.

11.7 Surface Areas and Volumes of Spheres

Essential Question How can you find the surface area and the volume of a sphere?

EXPLORATION 1 — Finding the Surface Area of a Sphere

Work with a partner. Remove the covering from a baseball or softball.

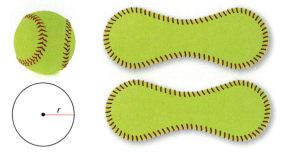

USING TOOLS STRATEGICALLY

To be proficient in math, you need to identify relevant external mathematical resources, such as content located on a website.

You will end up with two "figure 8" pieces of material, as shown above. From the amount of material it takes to cover the ball, what would you estimate the surface area S of the ball to be? Express your answer in terms of the radius r of the ball.

$S = $ _____   Surface area of a sphere

Use the Internet or some other resource to confirm that the formula you wrote for the surface area of a sphere is correct.

EXPLORATION 2 — Finding the Volume of a Sphere

Work with a partner. A cylinder is circumscribed about a sphere, as shown. Write a formula for the volume V of the cylinder in terms of the radius r.

$V = $ _____   Volume of cylinder

When half of the sphere (a *hemisphere*) is filled with sand and poured into the cylinder, it takes three hemispheres to fill the cylinder. Use this information to write a formula for the volume V of a sphere in terms of the radius r.

$V = $ _____   Volume of a sphere

Communicate Your Answer

3. How can you find the surface area and the volume of a sphere?

4. Use the results of Explorations 1 and 2 to find the surface area and the volume of a sphere with a radius of (a) 3 inches and (b) 2 centimeters.

Dynamic Teaching Tools
- Dynamic Assessment & Progress Monitoring Tool
- Lesson Planning Tool
- Interactive Whiteboard Lesson Library
- Dynamic Classroom with Dynamic Investigations

ANSWERS
1. $4\pi r^2$
2. $2\pi r^3$; $\frac{4}{3}\pi r^3$
3. Sample answer: $S = 4\pi r^2$, $V = \frac{4}{3}\pi r^3$
4. a. about 113.10 in.^{2}; about 113.10 in.^{3}
   b. about 50.27 cm^2; about 33.51 cm^3

Differentiated Instruction

Inclusion

After presenting the surface area and volume formulas for a sphere, challenge advanced students to derive formulas for the lateral area L, surface area S, and volume V of a hemisphere with radius r.
$L = 2\pi r^2$, $S = 3\pi r^2$, $V = \frac{2}{3}\pi r^3$

11.7 Lesson

What You Will Learn

▶ Find surface areas of spheres.
▶ Find volumes of spheres.

Core Vocabulary

chord of a sphere, p. 686
great circle, p. 686

Previous
sphere
center of a sphere
radius of a sphere
diameter of a sphere
hemisphere

Finding Surface Areas of Spheres

A *sphere* is the set of all points in space equidistant from a given point. This point is called the *center* of the sphere. A *radius* of a sphere is a segment from the center to a point on the sphere. A **chord of a sphere** is a segment whose endpoints are on the sphere. A *diameter* of a sphere is a chord that contains the center.

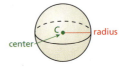

As with circles, the terms radius and diameter also represent distances, and the diameter is twice the radius.

If a plane intersects a sphere, then the intersection is either a single point or a circle. If the plane contains the center of the sphere, then the intersection is a **great circle** of the sphere. The circumference of a great circle is the circumference of the sphere. Every great circle of a sphere separates the sphere into two congruent halves called *hemispheres*.

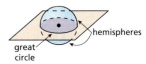

Core Concept

Surface Area of a Sphere

The surface area S of a sphere is

$$S = 4\pi r^2$$

where r is the radius of the sphere.

To understand the formula for the surface area of a sphere, think of a baseball. The surface area of a baseball is sewn from two congruent shapes, each of which resembles two joined circles.

So, the entire covering of the baseball consists of four circles, each with radius r. The area A of a circle with radius r is $A = \pi r^2$. So, the area of the covering can be approximated by $4\pi r^2$. This is the formula for the surface area of a sphere.

leather covering

686   Chapter 11   Circumference, Area, and Volume

Laurie's Notes | Teacher Actions

- **Teaching Tip:** If you have a globe, use it as a model as you discuss the vocabulary associated with a sphere. The globe is also useful in describing the role a great circle has in air travel. Traveling the path of a great circle is a shorter distance than traveling a latitude line.
- The baseball (or softball) cover has been the most visual way for my students to recall the formula for the surface area of a sphere.

> **EXAMPLE 1** Finding the Surface Areas of Spheres
>
> Find the surface area of each sphere.
>
> a. 8 in.
>
> b. 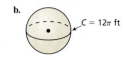 $C = 12\pi$ ft
>
> **SOLUTION**
>
> a. $S = 4\pi r^2$ Formula for surface area of a sphere
> $\quad = 4\pi(8)^2$ Substitute 8 for r.
> $\quad = 256\pi$ Simplify.
> $\quad \approx 804.25$ Use a calculator.
>
> ▶ The surface area is 256π, or about 804.25 square inches.
>
> b. The circumference of the sphere is 12π, so the radius of the sphere is $\frac{12\pi}{2\pi} = 6$ feet.
> $\quad S = 4\pi r^2$ Formula for surface area of a sphere
> $\quad\;\; = 4\pi(6)^2$ Substitute 6 for r.
> $\quad\;\; = 144\pi$ Simplify.
> $\quad\;\; \approx 452.39$ Use a calculator.
>
> ▶ The surface area is 144π, or about 452.39 square feet.
>
> **EXAMPLE 2** Finding the Diameter of a Sphere
>
> Find the diameter of the sphere.
>
> $S = 20.25\pi$ cm²
>
> **SOLUTION**
>
> $S = 4\pi r^2$ Formula for surface area of a sphere
> $20.25\pi = 4\pi r^2$ Substitute 20.25π for S.
> $5.0625 = r^2$ Divide each side by 4π.
> $2.25 = r$ Find the positive square root.
>
> ▶ The diameter is $2r = 2 \cdot 2.25 = 4.5$ centimeters.

COMMON ERROR
Be sure to multiply the value of r by 2 to find the diameter.

> **Monitoring Progress** 🔊 Help in English and Spanish at *BigIdeasMath.com*
>
> Find the surface area of the sphere.
>
> 1. 40 ft
>
> 2.  $C = 6\pi$ ft
>
> 3. Find the radius of the sphere.
> $S = 30\pi$ m²

English Language Learners

Class Activity
Have students work in groups to make posters describing planets in the solar system. Assign one planet to each group. Have them record significant data about the planet, including its diameter, surface area, volume, density, and mass.

Extra Example 1
Find the surface area of each sphere.

a. 3 ft

The surface area is 9π, or about 28.27 square feet.

b. 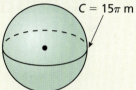 $C = 15\pi$ m

The surface area is 225π, or about 706.86 square meters.

Extra Example 2
Find the diameter of the sphere.

 $S = 144\pi$ cm²

The diameter is 12 centimeters.

MONITORING PROGRESS ANSWERS

1. about 5026.55 ft²
2. about 113.10 ft²
3. about 2.74 m

Laurie's Notes — Teacher Actions

❓ Reason Abstractly and Quantitatively: "If you know the circumference of a sphere, can you find its surface area? Explain." *yes; First solve for the radius using the formula for circumference, and then find the surface area.*

- Have students solve parts (a) and (b) in Example 1.
- **Turn and Talk:** "Describe how to find the diameter of a sphere when you know the surface area." Circulate and listen to strategies.
- Have students work independently to solve Example 2.

Extra Example 3
Find the volume of the sphere.

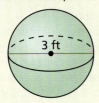

The volume is $\frac{9}{2}\pi$, or about 14.14 cubic feet.

Finding Volumes of Spheres

The figure shows a hemisphere and a cylinder with a cone removed. A plane parallel to their bases intersects the solids z units above their bases.

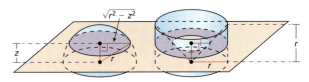

Using the AA Similarity Theorem, you can show that the radius of the cross section of the cone at height z is z. The area of the cross section formed by the plane is $\pi(r^2 - z^2)$ for both solids. Because the solids have the same height and the same cross-sectional area at every level, they have the same volume by Cavalieri's Principle.

$$V_{\text{hemisphere}} = V_{\text{cylinder}} - V_{\text{cone}}$$
$$= \pi r^2(r) - \tfrac{1}{3}\pi r^2(r)$$
$$= \tfrac{2}{3}\pi r^3$$

So, the volume of a sphere of radius r is

$$2 \cdot V_{\text{hemisphere}} = 2 \cdot \tfrac{2}{3}\pi r^3 = \tfrac{4}{3}\pi r^3.$$

> ### Core Concept
> **Volume of a Sphere**
> The volume V of a sphere is
> $$V = \tfrac{4}{3}\pi r^3$$
> where r is the radius of the sphere.
>
> $V = \tfrac{4}{3}\pi r^3$

EXAMPLE 3 Finding the Volume of a Sphere

Find the volume of the soccer ball.

4.5 in.

SOLUTION

$V = \tfrac{4}{3}\pi r^3$	Formula for volume of a sphere
$= \tfrac{4}{3}\pi(4.5)^3$	Substitute 4.5 for r.
$= 121.5\pi$	Simplify.
≈ 381.70	Use a calculator.

▶ The volume of the soccer ball is 121.5π, or about 381.70 cubic inches.

Laurie's Notes Teacher Actions

- The formula for the volume of a sphere is developed using the graphics shown. If students did not complete the second exploration, work through the derivation.
- **Teaching Tip:** Draw the cross sections showing the right triangles.
- Write the *Core Concept*. Students sometimes have difficulty with this formula because of the fraction.
- **? Reason Abstractly and Quantitatively:** "How do you find the volume of the soccer ball?" *Cube the radius and multiply by $\tfrac{4}{3}\pi$. Label the answer in cubic inches.*

EXAMPLE 4 Finding the Volume of a Sphere

The surface area of a sphere is 324π square centimeters. Find the volume of the sphere.

SOLUTION

Step 1 Use the surface area to find the radius.

$S = 4\pi r^2$ Formula for surface area of a sphere

$324\pi = 4\pi r^2$ Substitute 324π for S.

$81 = r^2$ Divide each side by 4π.

$9 = r$ Find the positive square root.

The radius is 9 centimeters.

Step 2 Use the radius to find the volume.

$V = \frac{4}{3}\pi r^3$ Formula for volume of a sphere

$= \frac{4}{3}\pi(9)^3$ Substitute 9 for r.

$= 972\pi$ Simplify.

≈ 3053.63 Use a calculator.

▶ The volume is 972π, or about 3053.63 cubic centimeters.

EXAMPLE 5 Finding the Volume of a Composite Solid

Find the volume of the composite solid.

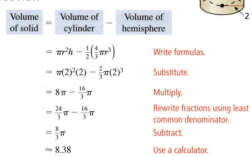

SOLUTION

Volume of solid = Volume of cylinder − Volume of hemisphere

$= \pi r^2 h - \frac{1}{2}\left(\frac{4}{3}\pi r^3\right)$ Write formulas.

$= \pi(2)^2(2) - \frac{2}{3}\pi(2)^3$ Substitute.

$= 8\pi - \frac{16}{3}\pi$ Multiply.

$= \frac{24}{3}\pi - \frac{16}{3}\pi$ Rewrite fractions using least common denominator.

$= \frac{8}{3}\pi$ Subtract.

≈ 8.38 Use a calculator.

▶ The volume is $\frac{8}{3}\pi$, or about 8.38 cubic inches.

Monitoring Progress Help in English and Spanish at BigIdeasMath.com

4. The radius of a sphere is 5 yards. Find the volume of the sphere.
5. The diameter of a sphere is 36 inches. Find the volume of the sphere.
6. The surface area of a sphere is 576π square centimeters. Find the volume of the sphere.
7. Find the volume of the composite solid at the left.

Extra Example 4
The surface area of a sphere is 676π square inches. Find the volume of the sphere.
The volume is about 2929.33π, or about 9202.77 cubic inches.

Extra Example 5
Find the volume of the composite solid.

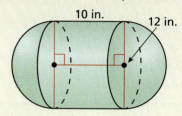

The volume is 648π, or about 2036 cubic inches.

MONITORING PROGRESS ANSWERS

4. about 523.60 yd^3
5. about 24,429.02 in.^{3}
6. about 7238.23 cm^3
7. about 7.33 m^3

Laurie's Notes — Teacher Actions

- **Pass the Problem:** Review this technique. Pose Example 4, and say that each part of the problem is to be solved. Give partners a minute of discussion time (if needed) before they begin. Call "swap" and have partners continue to solve. Have partners debrief.
- **Turn and Talk:** "How can you find the volume of the composite figure in Example 5?" Discuss and have partners find the volume.

Closure
- **3-2-1:** Hand out a 3-2-1 reflection sheet as described on page T-468.

Assignment Guide and Homework Check

ASSIGNMENT

Basic: 1, 2, 3–35 odd, 42, 48–51

Average: 1, 2, 6–30 even, 33–36, 42, 48–51

Advanced: 1, 2, 8–30 even, 33, 34–46 even, 48–51

HOMEWORK CHECK

Basic: 7, 11, 15, 23, 27

Average: 8, 20, 26, 34, 36

Advanced: 8, 20, 26, 36, 46

ANSWERS

1. The plane must contain the center of the sphere.
2. *Sample answer:* A hemisphere is one-half of a sphere.
3. about 201.06 ft^2
4. about 706.86 cm^2
5. about 1052.09 m^2
6. about 50.27 ft^2
7. 1 ft
8. 16 in.
9. 30 m
10. 14 cm
11. about 157.08 m^2
12. about 226.19 in.^{2}
13. about 2144.66 m^3
14. about 268.08 ft^3
15. about 5575.28 yd^3
16. about 1436.76 ft^3
17. about 4188.79 cm^3
18. about 179.59 in.^{3}
19. about 33.51 ft^3
20. about 5575.28 cm^3
21. The radius was squared instead of cubed; $V = \frac{4}{3}\pi(6)^3 \approx 904.78$ ft^3

11.7 Exercises

Dynamic Solutions available at *BigIdeasMath.com*

Vocabulary and Core Concept Check

1. **VOCABULARY** When a plane intersects a sphere, what must be true for the intersection to be a great circle?

2. **WRITING** Explain the difference between a sphere and a hemisphere.

Monitoring Progress and Modeling with Mathematics

In Exercises 3–6, find the surface area of the sphere. *(See Example 1.)*

3.

4.

5.

6.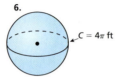

In Exercises 7–10, find the indicated measure. *(See Example 2.)*

7. Find the radius of a sphere with a surface area of 4π square feet.

8. Find the radius of a sphere with a surface area of 1024π square inches.

9. Find the diameter of a sphere with a surface area of 900π square meters.

10. Find the diameter of a sphere with a surface area of 196π square centimeters.

In Exercises 11 and 12, find the surface area of the hemisphere.

11.

12.

In Exercises 13–18, find the volume of the sphere. *(See Example 3.)*

13.

14.

15.

16.

17.

18.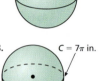

In Exercises 19 and 20, find the volume of the sphere with the given surface area. *(See Example 4.)*

19. Surface area = 16π ft^2

20. Surface area = 484π cm^2

21. **ERROR ANALYSIS** Describe and correct the error in finding the volume of the sphere.

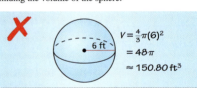

22. ERROR ANALYSIS Describe and correct the error in finding the volume of the sphere.

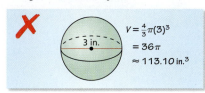

In Exercises 23–26, find the volume of the composite solid. *(See Example 5.)*

23.

24.

25.

26.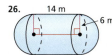

In Exercises 27–32, find the surface area and volume of the ball.

27. bowling ball
    $d = 8.5$ in.

28. basketball
    $C = 29.5$ in.

29. softball
    $C = 12$ in.

30. golf ball
    $d = 1.7$ in.

31. volleyball
    $C = 26$ in.

32. baseball
    $C = 9$ in.

33. MAKING AN ARGUMENT Your friend claims that if the radius of a sphere is doubled, then the surface area of the sphere will also be doubled. Is your friend correct? Explain your reasoning.

34. MODELING WITH MATHEMATICS The circumference of Earth is about 24,900 miles. Find the surface area of the Western Hemisphere of Earth.

35. MODELING WITH MATHEMATICS A silo has the dimensions shown. The top of the silo is a hemispherical shape. Find the volume of the silo.

36. MODELING WITH MATHEMATICS Three tennis balls are stored in a cylindrical container with a height of 8 inches and a radius of 1.43 inches. The circumference of a tennis ball is 8 inches.

a. Find the volume of a tennis ball.

b. Find the amount of space within the cylinder not taken up by the tennis balls.

37. ANALYZING RELATIONSHIPS Use the table shown for a sphere.

Radius	Surface area	Volume
3 in.	36π in.^{2}	36π in.^{3}
6 in.		
9 in.		
12 in.		

a. Copy and complete the table. Leave your answers in terms of π.

b. What happens to the surface area of the sphere when the radius is doubled? tripled? quadrupled?

c. What happens to the volume of the sphere when the radius is doubled? tripled? quadrupled?

38. MATHEMATICAL CONNECTIONS A sphere has a diameter of $4(x + 3)$ centimeters and a surface area of 784π square centimeters. Find the value of x.

ANSWERS

39. a. Earth: about 197.1 million mi²; moon: about 14.7 million mi²
    b. The surface area of the Earth is about 13.4 times greater than the surface area of the moon.
    c. about 137.9 million mi²
40. a. about 80.9 million mi²
    b. about 41%
41. about 50.27 in.²; *Sample answer:* The side length of the cube is the diameter of the sphere.
42. the hemisphere; *Sample answer:* The cone fits inside the hemisphere.
43. $V = \frac{1}{3}rS$
44. *Sample answer:* If θ is in radians, then $S = 2r^2\theta$.
45–51. See Additional Answers.

Mini-Assessment

1. Find the surface area and volume of a sphere with radius $\frac{3}{4}$ foot.
   $S = \frac{9}{4}\pi$, or about 7.07 square feet;
   $V = \frac{9}{16}\pi$, or about 1.77 cubic feet

2. The surface area of a sphere is 368.64π cubic centimeters. Find the diameter of the sphere.
   19.2 centimeters

3. Find the surface area and volume of the sphere.

   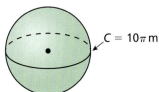
   $C = 10\pi$ m

   $S = 100\pi$, or about 314.16 square meters; $V = \frac{500}{3}\pi$, or about 523.60 cubic meters

4. Find the volume of the composite solid.

   6 ft

   $432\pi - 288\pi = 144\pi$, or about 452.39 cubic feet

692   Chapter 11

---

39. **MODELING WITH MATHEMATICS** The radius of Earth is about 3960 miles. The radius of the moon is about 1080 miles.
    a. Find the surface area of Earth and the moon.
    b. Compare the surface areas of Earth and the moon.
    c. About 70% of the surface of Earth is water. How many square miles of water are on Earth's surface?

40. **MODELING WITH MATHEMATICS** The Torrid Zone on Earth is the area between the Tropic of Cancer and the Tropic of Capricorn. The distance between these two tropics is about 3250 miles. You can estimate the distance as the height of a cylindrical belt around Earth at the equator.

    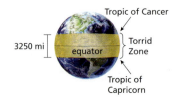

    a. Estimate the surface area of the Torrid Zone. (The radius of Earth is about 3960 miles.)
    b. A meteorite is equally likely to hit anywhere on Earth. Estimate the probability that a meteorite will land in the Torrid Zone.

41. **ABSTRACT REASONING** A sphere is inscribed in a cube with a volume of 64 cubic inches. What is the surface area of the sphere? Explain your reasoning.

42. **HOW DO YOU SEE IT?** The formula for the volume of a hemisphere and a cone are shown. If each solid has the same radius and $r = h$, which solid will have a greater volume? Explain your reasoning.

    $V = \frac{2}{3}\pi r^3$ $V = \frac{1}{3}\pi r^2 h$

43. **CRITICAL THINKING** Let V be the volume of a sphere, S be the surface area of the sphere, and r be the radius of the sphere. Write an equation for V in terms of r and S. $\left(Hint:\text{ Start with the ratio } \frac{V}{S}.\right)$

44. **THOUGHT PROVOKING** A *spherical lune* is the region between two great circles of a sphere. Find the formula for the area of a lune.

    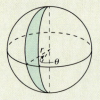

45. **CRITICAL THINKING** The volume of a right cylinder is the same as the volume of a sphere. The radius of the sphere is 1 inch. Give three possibilities for the dimensions of the cylinder.

46. **PROBLEM SOLVING** A *spherical cap* is a portion of a sphere cut off by a plane. The formula for the volume of a spherical cap is $V = \frac{\pi h}{6}(3a^2 + h^2)$, where a is the radius of the base of the cap and h is the height of the cap. Use the diagram and given information to find the volume of each spherical cap.
    a. $r = 5$ ft, $a = 4$ ft
    b. $r = 34$ cm, $a = 30$ cm
    c. $r = 13$ m, $h = 8$ m
    d. $r = 75$ in., $h = 54$ in.

47. **CRITICAL THINKING** A sphere with a radius of 2 inches is inscribed in a right cone with a height of 6 inches. Find the surface area and the volume of the cone.

Maintaining Mathematical Proficiency *Reviewing what you learned in previous grades and lessons*

Events A and B are disjoint. Find $P(A \text{ or } B)$. *(Section 5.4)*

48. $P(A) = 0.4, P(B) = 0.2$
49. $P(A) = 0.75, P(B) = 0.05$
50. $P(A) = \frac{1}{8}, P(B) = \frac{3}{4}$
51. $P(A) = \frac{2}{5}, P(B) = \frac{3}{10}$

692   Chapter 11   Circumference, Area, and Volume

If students need help...	If students got it...
Resources by Chapter • Practice A and Practice B • Puzzle Time	Resources by Chapter • Enrichment and Extension • Cumulative Review
Student Journal • Practice	Start the *next* Section
Differentiating the Lesson Skills Review Handbook	

11.4–11.7 What Did You Learn?

Core Vocabulary

polyhedron, *p. 666*
face, *p. 666*
edge, *p. 666*
vertex, *p. 666*
volume, *p. 667*
Cavalieri's Principle, *p. 667*
similar solids, *p. 669*
lateral surface of a cone, *p. 680*
chord of a sphere, *p. 686*
great circle, *p. 686*

Core Concepts

Section 11.4
Types of Solids, *p. 666*
Cavalieri's Principle, *p. 667*
Volume of a Prism, *p. 667*
Volume of a Cylinder, *p. 668*
Similar Solids, *p. 669*

Section 11.5
Volume of a Pyramid, *p. 674*

Section 11.6
Surface Area of a Right Cone, *p. 680*
Volume of a Cone, *p. 681*

Section 11.7
Surface Area of a Sphere, *p. 686*
Volume of a Sphere, *p. 688*

Mathematical Practices

1. In Exercise 15 on page 677, explain why the volume changed by a factor of $\frac{1}{64}$.
2. In Exercise 38 on page 691, explain the steps you used to find the value of x.

Performance Task:

Tabletop Tiling

A Penrose tiling has a variety of properties that make it attractive. How do the concepts of area and circumference contribute to the visual appeal of a tabletop tiled in this type of pattern?

To explore the answer to this question and more, check out the Performance Task and Real-Life STEM video at *BigIdeasMath.com*.

Dynamic Teaching Tools
Dynamic Assessment & Progress Monitoring Tool
Interactive Whiteboard Lesson Library
Dynamic Classroom with Dynamic Investigations

ANSWERS
1. *Sample answer:* The scale factor is $\frac{1}{4}$ and the ratio of the volumes is the scale factor cubed.
2. *Sample answer:* Substitute $r = 2x + 6$ into the surface area formula and set equal to 784π, then solve for x.

ANSWERS

1. about 30.00 ft
2. about 56.57 cm
3. about 26.09 in.
4. 218 ft
5. about 169.65 in.²
6. about 17.72 in.²
7. 173.166 ft²

11 Chapter Review

Dynamic Solutions available at *BigIdeasMath.com*

11.1 Circumference and Arc Length (pp. 639–646)

The arc length of $\widehat{QR}$ is 6.54 feet. Find the radius of $\odot P$.

$$\frac{\text{Arc length of } \widehat{QR}}{2\pi r} = \frac{m\widehat{QR}}{360°} \qquad \text{Formula for arc length}$$

$$\frac{6.54}{2\pi r} = \frac{75°}{360°} \qquad \text{Substitute.}$$

$$6.54(360) = 75(2\pi r) \qquad \text{Cross Products Property}$$

$$5.00 \approx r \qquad \text{Solve for } r.$$

▶ The radius of $\odot P$ is about 5 feet.

Find the indicated measure.

1. diameter of $\odot P$
2. circumference of $\odot F$
3. arc length of $\widehat{AB}$

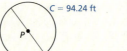

4. A mountain bike tire has a diameter of 26 inches. To the nearest foot, how far does the tire travel when it makes 32 revolutions?

11.2 Areas of Circles and Sectors (pp. 647–654)

Find the area of sector ADB.

$$\text{Area of sector } ADB = \frac{m\widehat{AB}}{360°} \cdot \pi r^2 \qquad \text{Formula for area of a sector}$$

$$= \frac{80°}{360°} \cdot \pi \cdot 10^2 \qquad \text{Substitute.}$$

$$\approx 69.81 \qquad \text{Use a calculator.}$$

▶ The area of sector ADB is about 69.81 square meters.

Find the area of the blue shaded region.

5.
6.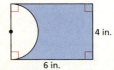
7.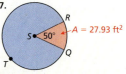

11.3 Areas of Polygons (pp. 655–662)

A regular hexagon is inscribed in ⊙H. Find (a) m∠EHG, and (b) the area of the hexagon.

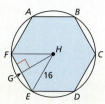

a. ∠FHE is a central angle, so $m\angle FHE = \dfrac{360°}{6} = 60°$.

Apothem $\overline{GH}$ bisects ∠FHE.

▶ So, $m\angle EHG = 30°$.

b. Because △EHG is a 30°-60°-90° triangle, $GE = \dfrac{1}{2} \cdot HE = 8$ and $GH = \sqrt{3} \cdot GE = 8\sqrt{3}$. So, $s = 2(GE) = 16$ and $a = GH = 8\sqrt{3}$.

$A = \dfrac{1}{2} a \cdot ns$ Formula for area of a regular polygon

$= \dfrac{1}{2}(8\sqrt{3})(6)(16)$ Substitute.

≈ 665.1 Simplify.

▶ The area is about 665.1 square units.

Find the area of the kite or rhombus.

8.
9.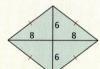
10.

Find the area of the regular polygon.

11.
12.
13.

14. A platter is in the shape of a regular octagon with an apothem of 6 inches. Find the area of the platter.

ANSWERS

8. 130 square units
9. 96 square units
10. 105 square units
11. about 201.20 square units
12. about 167.11 square units
13. about 37.30 square units
14. about 119.29 in.²

ANSWERS

15. yes; square prism
16. yes; heptagonal pyramid
17. no
18. 11.34 m³
19. about 100.53 mm³
20. about 27.53 yd³

11.4 Volumes of Prisms and Cylinders (pp. 665–672)

a. Tell whether the solid is a polyhedron. If it is, name the polyhedron.

▶ The solid is formed by polygons, so it is a polyhedron. The two bases are congruent octagons, so it is an octagonal prism.

b. Find the volume of the triangular prism.

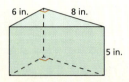

The area of a base is $B = \frac{1}{2}(6)(8) = 24$ in.² and the height is $h = 5$ in.

$V = Bh$ Formula for volume of a prism

$= 24(5)$ Substitute.

$= 120$ Simplify.

▶ The volume is 120 cubic inches.

Tell whether the solid is a polyhedron. If it is, name the polyhedron.

15.   16.   17.

Find the volume of the solid.

18. 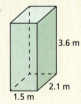   19. 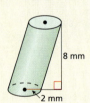   20.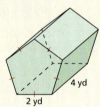

696   Chapter 11   Circumference, Area, and Volume

11.5 Volumes of Pyramids (pp. 673–678)

Find the volume of the pyramid.

$V = \frac{1}{3}Bh$ Formula for volume of a pyramid

$= \frac{1}{3}\left(\frac{1}{2} \cdot 5 \cdot 8\right)(12)$ Substitute.

$= 80$ Simplify.

▶ The volume is 80 cubic meters.

Find the volume of the pyramid.

21.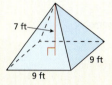

22.

23.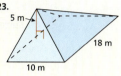

24. The volume of a square pyramid is 60 cubic inches and the height is 15 inches. Find the side length of the square base.

25. The volume of a square pyramid is 1024 cubic inches. The base has a side length of 16 inches. Find the height of the pyramid.

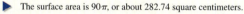

11.6 Surface Areas and Volumes of Cones (pp. 679–684)

Find the (a) surface area and (b) volume of the cone.

a. $S = \pi r^2 + \pi r \ell$ Formula for surface area of a cone

$= \pi \cdot 5^2 + \pi(5)(13)$ Substitute.

$= 90\pi$ Simplify.

≈ 282.74 Use a calculator.

▶ The surface area is 90π, or about 282.74 square centimeters.

b. $V = \frac{1}{3}\pi r^2 h$ Formula for volume of a cone

$= \frac{1}{3}\pi \cdot 5^2 \cdot 12$ Substitute.

$= 100\pi$ Simplify.

≈ 314.16 Use a calculator.

▶ The volume is 100π, or about 314.16 cubic centimeters.

ANSWERS
21. 189 ft³
22. 400 yd³
23. 300 m³
24. about 3.46 in.
25. 12 in.

ANSWERS

26. $S \approx 678.58$ cm^2; $V \approx 1017.88$ cm^3
27. $S \approx 2513.27$ cm^2; $V \approx 8042.48$ cm^3
28. $S \approx 439.82$ m^2; $V \approx 562.10$ m^3
29. 15 cm
30. $S \approx 615.75$ in.^{2}; $V \approx 1436.76$ in.^{3}
31. $S \approx 907.92$ ft^2; $V \approx 2572.44$ ft^3
32. $S \approx 2827.43$ ft^2; $V \approx 14{,}137.17$ ft^3
33. $S \approx 74.8$ million km^2; $V \approx 60.8$ billion km^3
34. about 272.55 m^3

Find the surface area and the volume of the cone.

26.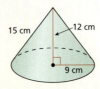
27.
28.

29. A cone with a diameter of 16 centimeters has a volume of 320π cubic centimeters. Find the height of the cone.

11.7 Surface Areas and Volumes of Spheres (pp. 685–692)

Find the (a) surface area and (b) volume of the sphere.

a. $S = 4\pi r^2$ Formula for surface area of a sphere
   $= 4\pi(18)^2$ Substitute 18 for r.
   $= 1296\pi$ Simplify.
   ≈ 4071.50 Use a calculator.

▶ The surface area is 1296π, or about 4071.50 square inches.

b. $V = \frac{4}{3}\pi r^3$ Formula for volume of a sphere
   $= \frac{4}{3}\pi(18)^3$ Substitute 18 for r.
   $= 7776\pi$ Simplify.
   $\approx 24{,}429.02$ Use a calculator.

▶ The volume is 7776π, or about 24,429.02 cubic inches.

Find the surface area and the volume of the sphere.

30.
31.
32.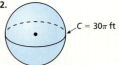

33. The shape of Mercury can be approximated by a sphere with a diameter of 4880 kilometers. Find the surface area and the volume of Mercury.

34. A solid is composed of a cube with a side length of 6 meters and a hemisphere with a diameter of 6 meters. Find the volume of the composite solid.

11 Chapter Test

Tell whether the solid is a polyhedron. If it is, name the polyhedron.

1.
2.
3.

Find the volume of the solid.

4.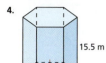
5.
6.
7.

Find the indicated measure.

8. circumference of ⊙F
9. $m\widehat{GH}$
10. area of shaded sector

11. Find the surface area of a right cone with a diameter of 10 feet and a height of 12 feet.

12. You have a funnel with the dimensions shown.
    a. Find the approximate volume of the funnel.
    b. You use the funnel to put oil in a car. Oil flows out of the funnel at a rate of 45 milliliters per second. How long will it take to empty the funnel when it is full of oil? (1 mL = 1 cm³)
    c. How long would it take to empty a funnel with a radius of 10 centimeters and a height of 6 centimeters if oil flows out of the funnel at a rate of 45 milliliters per second?
    d. Explain why you can claim that the time calculated in part (c) is greater than the time calculated in part (b) without doing any calculations.

13. A water bottle in the shape of a cylinder has a volume of 500 cubic centimeters. The diameter of a base is 7.5 centimeters. What is the height of the bottle? Justify your answer.

14. Find the area of a dodecagon (12 sides) with a side length of 9 inches.

15. In general, a cardboard fan with a greater area does a better job of moving air and cooling you. The fan shown is a sector of a cardboard circle. Another fan has a radius of 6 centimeters and an intercepted arc of 150°. Which fan does a better job of cooling you?

ANSWERS

1. no
2. yes; octagonal pyramid
3. yes; pentagonal prism
4. about 2577.29 m³
5. about 17.16 ft³
6. about 402.12 m³
7. $93\frac{1}{3}$ ft³
8. about 109.71 in.
9. about 74.27°
10. about 142.42 in.²
11. 90π ft² or about 282.74 ft²
12. a. about 376.99 cm³
    b. about 8.38 sec
    c. about 13.96 sec
    d. *Sample answer:* Changing the radius has a greater effect than changing the height.
13. about 11.32 cm; *Sample answer:* $500 = \pi(3.75)^3 h$, so $h \approx 11.32$.
14. about 906.89 in.²
15. the fan shown

ANSWERS
1. C
2. $\overline{PQ} \perp \overleftrightarrow{RS}$
3. a. about 4650 mm³
   b. about 75,267 mm³
4. $k = 4$

11 Cumulative Assessment

1. The directed line segment RS is shown. Point Q is located along $\overline{RS}$ so that the ratio of RQ to QS is 2 to 3. What are the coordinates of point Q?

   Ⓐ $Q(1.2, 3)$ Ⓑ $Q(4, 2)$
   Ⓒ $Q(2, 3)$ Ⓓ $Q(-6, 7)$

   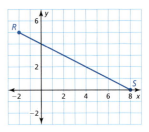

2. In the diagram, $\overleftrightarrow{RS}$ is tangent to $\odot P$ at Q and $\overline{PQ}$ is a radius of $\odot P$. What must be true about $\overleftrightarrow{RS}$ and $\overline{PQ}$? Select all that apply.

   $PQ = \frac{1}{2}RS$ $PQ = RS$ $\overline{PQ}$ is tangent to $\odot P$. $\overline{PQ} \perp \overleftrightarrow{RS}$

3. A crayon can be approximated by a composite solid made from a cylinder and a cone. A crayon box is a rectangular prism. The dimensions of a crayon and a crayon box containing 24 crayons are shown.

   a. Find the volume of a crayon.
   b. Find the amount of space within the crayon box not taken up by the crayons.

   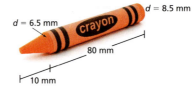

4. The graphs of $f(x) = |x|$ and $g(x) = |x + k|$ are shown. Find the value of k.

   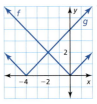

700   Chapter 11   Circumference, Area, and Volume

5. The top of the Washington Monument in Washington, D.C., is a square pyramid, called a *pyramidion*. What is the volume of the pyramidion?

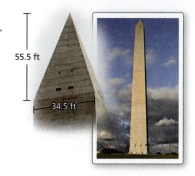

   Ⓐ 22,019.63 ft³

   Ⓑ 172,006.91 ft³

   Ⓒ 66,058.88 ft³

   Ⓓ 207,530.08 ft³

6. Prove or disprove that the point $(1, \sqrt{3})$ lies on the circle centered at the origin and containing the point (0, 2).

7. Enter the missing reasons in the proof of the Base Angles Theorem.

   Given $\overline{AB} \cong \overline{AC}$

   Prove $\angle B \cong \angle C$

   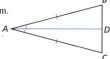

STATEMENTS	REASONS
1. Draw $\overline{AD}$, the angle bisector of $\angle CAB$.	1. Construction of angle bisector
2. $\angle CAD \cong \angle BAD$	2. _____
3. $\overline{AB} \cong \overline{AC}$	3. _____
4. $\overline{DA} \cong \overline{DA}$	4. _____
5. $\triangle ADB \cong \triangle ADC$	5. _____
6. $\angle B \cong \angle C$	6. _____

8. The diagram shows a square pyramid and a cone. Both solids have the same height, h, and the base of the cone has radius r. According to Cavalieri's Principle, the solids will have the same volume if the square base has sides of length ____.

   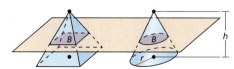

9. A triangle has vertices $X(-2, 2)$, $Y(1, 4)$, and $Z(2, -2)$. Your friend claims that a translation of $(x, y) \rightarrow (x + 2, y - 3)$ and a dilation by a scale factor of 3 will produce a similarity transformation. Do you support your friend's claim? Explain your reasoning.

Additional Answers

Chapter 1

Chapter 1 Maintaining Mathematical Proficiency

5.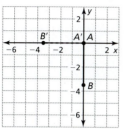

6.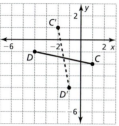

7.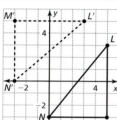

8.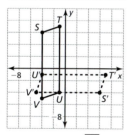

9. Sample answer: $\overline{AB}$ with endpoints $A(-1, 2)$ and $B(2, -1)$

10. The possible inputs for f are all real numbers and the output is a. The input for its reflection is a and the possible outputs are all real numbers.

1.1 Lesson Monitoring Progress

1.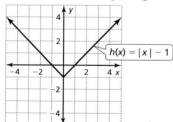

The graph of h is a vertical translation 1 unit down of the graph of f; domain: all real numbers; range: $y \geq -1$

2.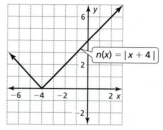

The graph of n is a horizontal translation 4 units left of the graph of f; domain: all real numbers; range: $y \geq 0$

3.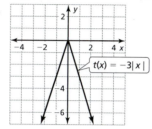

The graph of t is a vertical stretch of the graph of f by a factor of 3 and a reflection in the x-axis; domain: all real numbers; range: $y \leq 0$

4.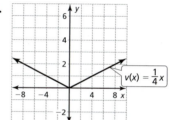

The graph of v is a vertical shrink of the graph of f by a factor of $\frac{1}{4}$; domain: all real numbers; range: $y \geq 0$

1.1 Monitoring Progress and Modeling with Mathematics

8.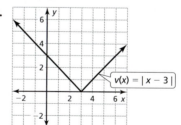

The graph of v is a horizontal translation 3 units right of the graph of f; domain: all real numbers; range: $y \geq 0$

9.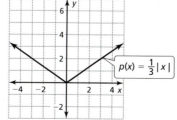

The graph of p is a vertical shrink of the graph of f by a factor of $\frac{1}{3}$; domain: all real numbers; range: $y \geq 0$

Additional Answers A1

10.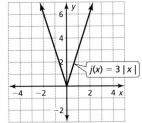
The graph of j is a vertical stretch of the graph of f by a factor of 3; domain: all real numbers; range: y ≥ 0

11.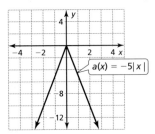
The graph of a is a vertical stretch of the graph of f by a factor of 5 and a reflection in the x-axis; domain: all real numbers; range: y ≤ 0

12.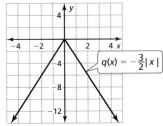
The graph of q is a vertical stretch of the graph of f by a factor of $\frac{3}{2}$ and a reflection in the x-axis; domain: all real numbers; range: y ≤ 0

13.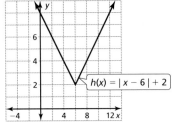
The graph of h is a vertical translation 2 units up of the graph of f.

14.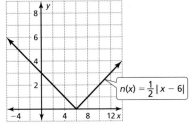
The graph of n is a vertical shrink of the graph of f by a factor of $\frac{1}{2}$.

15.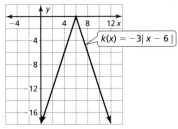
The graph of k is a vertical stretch of the graph of f by a factor of 3 and a reflection in the x-axis.

16.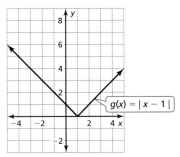
The graph of g is a horizontal translation 5 units left of the graph of f.

17.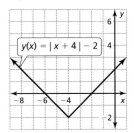
The graph of y is a horizontal translation 1 unit left of the graph of f.

18.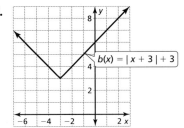
The graph of b is a vertical translation 5 units up of the graph of f.

19. The graph of g is a vertical translation 3 units down of the graph of f; k = −3

20. The graph of t is a horizontal translation 1 unit right of the graph of f; h = 1

21. The graph of p is a vertical stretch of the graph of f by a factor of 3 and a reflection in the x-axis; a = −3

22. The graph of w is a vertical shrink of the graph of f by a factor of 0.5; a = 0.5

23. $h(x) = |x| − 7$

24. $h(x) = |x + 10|$

25. $h(x) = \frac{1}{4}|x|$

26. $h(x) = −3|x|$

27.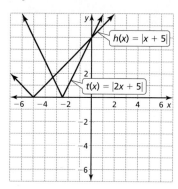

The graph of g is a horizontal shrink of the graph of f by a factor of $\frac{1}{3}$.

28.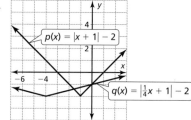

The graph of t is a horizontal shrink of the graph of h by a factor of $\frac{1}{2}$.

29.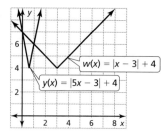

The graph of q is a horizontal stretch of the graph of p by a factor of 4.

30.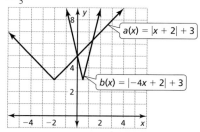

The graph of y is a horizontal shrink of the graph of w by a factor of $\frac{1}{5}$.

31.

The graph of b is a horizontal shrink of the graph of a by a factor of $\frac{1}{4}$ and a reflection in the y-axis.

32.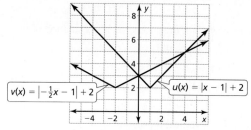

The graph of v is a horizontal stretch of the graph of u by a factor of 2 and a reflection in the y-axis.

38. The transformations are a horizontal translation 2 units right, then a horizontal shrink by a factor of $\frac{1}{2}$, then a vertical translation 3 units down.

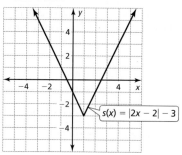

39. The transformations are a horizontal translation 1 unit left, then a reflection in the y-axis, then a vertical translation 5 units down.

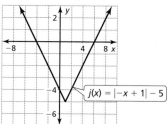

40. The transformations are a horizontal translation 1 unit left, then a horizontal stretch by a factor of 3, then a reflection in the y-axis, then a vertical translation 2 units up.

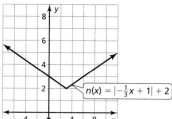

41. a.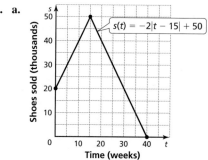

b. 50,000 pairs of shoes

42. **a.** $\left(\frac{5}{4}, 0\right)$
    b. yes; The location of the pocket (5, 5) is a solution of p.
43. **a.** $A'\left(-\frac{1}{2}, -2\right); B'(1, -5); C'(-4, -7)$
    b. $A'\left(\frac{5}{2}, 3\right); B'(4, 0); C'(-1, -2)$
    c. $A'\left(-\frac{1}{2}, -3\right); B'(1, 0); C'(-4, 2)$
    d. $A'\left(-\frac{1}{2}, 12\right); B'(1, 0); C'(-4, -8)$
44. **a.** $y = |x| + k$ is a vertical translation up of the graph of $y = |x|$ for positive values of k, and down for negative values of k.
    b. $y = |x - h|$ is a horizontal translation right of the graph of $y = |x|$ for positive values of h, and left for negative values of h.
    c. $y = a|x|$ is a vertical stretch or shrink by a factor of a of the graph of $y = |x|$ for positive values of a. It is both a vertical stretch or shrink by a factor of a and a reflection in the x-axis for negative values of a.
    d. $y = |ax|$ is a reflection in the y-axis of the graph of $y = |x|$ for negative values of a. $y = |ax|$ is a horizontal stretch of the graph of $y = |x|$ by a factor of $\frac{1}{a}$ when $0 < |a| < 1$, or a horizontal shrink by a factor of $\frac{1}{a}$ when $|a| > 1$.
45. The graph should have a horizontal shift right, not left.

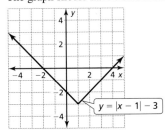

46. The graph should have a reflection in the x-axis.

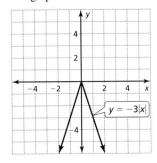

47. Sample answer: $f(x) = -|x| + 2$
48. Sample answer: $f(x) = -|x - 3| + 3$
49. The graph of p is a horizontal translation 6 units right of the graph of $f(x) = |x|$. The graph of q is a vertical translation 6 units down of the graph of $f(x) = |x|$.

51.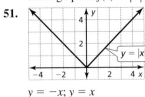

$y = -x; y = x$

52.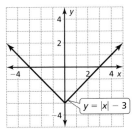

$y = x - 3; y = -x - 3$

53.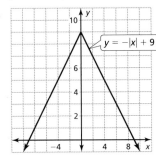

$y = x + 9; y = -x + 9$

54.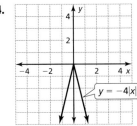

$y = 4x; y = -4x$

55.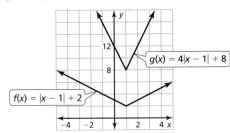

The graph of g is a vertical stretch by a factor of 4 of the graph of f.

56.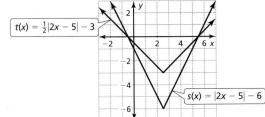

The graph of t is a vertical shrink by a factor of $\frac{1}{2}$ of the graph of s.

57.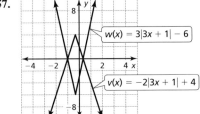

The graph of w is a vertical stretch of the graph of v by a factor of $\frac{3}{2}$ and a reflection in the x-axis.

58.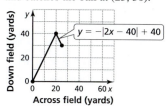

The graph of d is a vertical shrink of the graph of c by a factor of $\frac{1}{3}$ and a reflection in the x-axis.

59. The transformations are a vertical translation 4 units down, then a reflection in the x-axis, then a vertical shrink by a factor of $\frac{1}{2}$, then a horizontal translation 1 unit right; *Sample answer:* Determine the transformations from h to g, then perform the opposite transformations in reverse order.

60. *Sample answer:* $y = -|2x - 40| + 40$, $0 \leq x \leq 25$; The receiver runs a diagonal route that cuts back at an angle at the point (20, 40) and catches the ball at (25, 30).

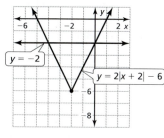

61.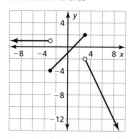

$x = 0$, $x = -4$

62. no; The form of a horizontal transformation of $y = f(x)$ is $y = f(x - h)$. So, if p is positive, the graph of $y = |x + p|$ is a negative horizontal translation.

63. $\left(\dfrac{h}{a}, k\right)$

1.1 Maintaining Mathematical Proficiency

64. $h = \dfrac{V}{\pi r^2}$

65. $w = \dfrac{P - 2\ell}{2}$ or $w = \dfrac{1}{2}P - \ell$

66. $y = 3x - 7$

67. $y = -2x + 6$

68. $y = -\dfrac{1}{2}x + \dfrac{5}{2}$

69. $y = -x + 1$

1.2 Monitoring Progress and Modeling with Mathematics

19.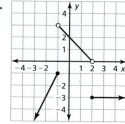

domain: all real numbers; range: $y \leq 2$

20.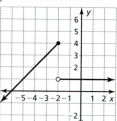

domain: all real numbers; range: $y \leq -1$ or $0 < y < 3$

21. The wrong expression was used to evaluate the function; $f(5) = 5 + 8 = 13$

22. The domain of $y = x + 6$ should be $x \leq 2$, and the domain of $y = 1$ should be $x > 2$.

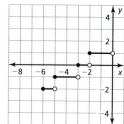

34.

domain: $-6 \leq x < 0$; range: $-2, -1, 0, 1$

35. $C(p) = \begin{cases} 180, & \text{if } 0 < p \leq 5 \\ 210, & \text{if } 5 < p \leq 6 \\ 240, & \text{if } 6 < p \leq 7 \\ 270, & \text{if } 7 < p \leq 8 \\ 300, & \text{if } 8 < p \leq 9 \end{cases}$

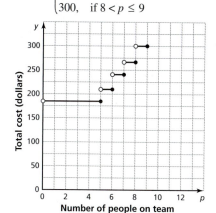

36. $f(x) = \begin{cases} 4, & \text{if } 0 < x \le 1 \\ 8, & \text{if } 1 < x \le 2 \\ 12, & \text{if } 2 < x \le 3 \\ 15, & \text{if } 3 < x \le 24 \end{cases}$

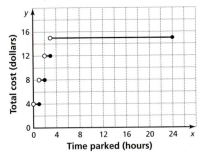

48. a. $f(x) = -\frac{2}{3}|x - 6| + 4$

b. $f(x) = \begin{cases} \frac{2}{3}x, & \text{if } x < 6 \\ -\frac{2}{3}x + 8, & \text{if } x \ge 6 \end{cases}$

49. a. -4

b. 6

50. a. The open circle at (2, 4) changes to a closed circle and the closed circle at (2, −3) changes to an open circle; The domain does not change, the range changes from $y < 4$ to $y \le 4$.

b. It does not change; The domain and range do not change; Because the value of each expression at $x = 1$ is 2, the two pieces of the function have the same endpoint. So, the graph does not change.

51.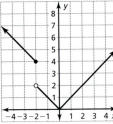

domain: all real numbers; range: $y \ge 0$

52. a. 100; At $x = 101$, $C < 10$.

b. yes; From the graph, you can make more than 550 copies with $40.

53.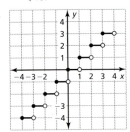

yes; yes; It is a function defined by two or more equations and it is defined by a constant value over each part of its domain.

54. The input value of 3 has two outputs; Either change ≤ 3 to < 3 in the first expression or change ≥ 3 to > 3 in the second expression.

55. a. $f(x) = \begin{cases} x, & \text{if } 0 \le x \le 2 \\ 2x - 2, & \text{if } 2 < x \le 8 \\ x + 6, & \text{if } 8 < x \le 9 \end{cases}$

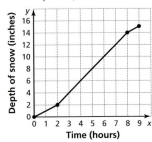

b. no; The total accumulation is 15 inches.

1.2 Maintaining Mathematical Proficiency

56. $-12 < r \le 13$

57. $t \le 4$ or $t \ge 18$

58. The transformation is a horizontal translation 5 units right.

59. The transformations are a vertical stretch by a factor of 2.5, then a reflection in the x-axis.

60. The transformations are a horizontal translation 1 unit left, then a vertical translation 6 units up.

61. The transformations are a horizontal translation 3 units left, then a vertical stretch by a factor of 4.

1.2 Mini-Assessment

4. $f(x) = \begin{cases} 60, & \text{if } 0 < x \le 1 \\ 100, & \text{if } 1 < x \le 2 \\ 140, & \text{if } 2 < x \le 3 \\ 180, & \text{if } 3 < x \le 4 \end{cases}$

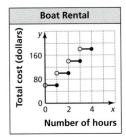

1.3 Monitoring Progress and Modeling with Mathematics

11. $x = 4y + 4$; 12

12. $x = \dfrac{3y - 12}{2}$; -3

13.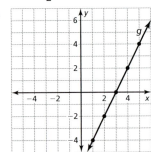

domain of inverse: all real numbers; range of inverse: all real numbers

14.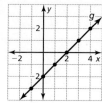

domain of inverse: all real numbers; range of inverse: all real numbers

15. $g(x) = \frac{1}{3}x$

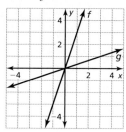

16. $g(x) = -\frac{1}{5}x$

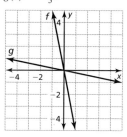

17. $g(x) = \frac{x+1}{4}$

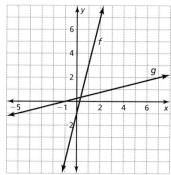

18. $g(x) = \frac{5-x}{2}$

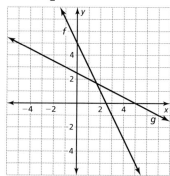

19. $g(x) = -\frac{x+2}{3}$

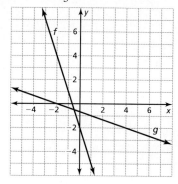

20. $g(x) = \frac{x-3}{2}$

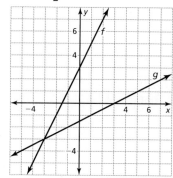

21. $g(x) = 3x - 24$

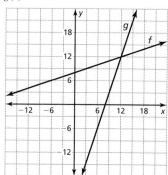

22. $g(x) = \frac{7-2x}{3}$

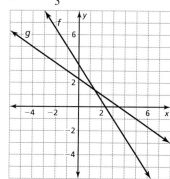

29. b.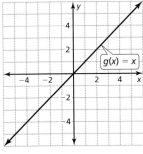
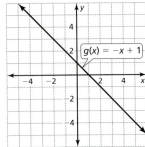
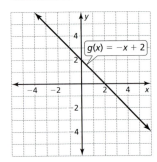

   c. The linear function $f(x) = mx + b$ is its own inverse when $m = -1$ and b is any real number or when $m = 1$ and $b = 0$.

30. *Sample answer:* $C = 2\pi r$, where C is the circumference of a circle and r is the radius of the circle. The inverse function is $r = \dfrac{C}{2\pi}$.

31. The inverse of $f(x) = mx + b$, where $b \neq 0$, is $g(x) = \dfrac{x - b}{m} = \dfrac{1}{m}x - \dfrac{b}{m}$, which is a linear function. The graph of the inverse function g has a slope of $\dfrac{1}{m}$ and a y-intercept of $-\dfrac{b}{m}$.

1.3 Maintaining Mathematical Proficiency

32. 248,832
33. -1331
34. $\dfrac{625}{1296}$
35. $\dfrac{1}{729}$
36. 8.4375×10^4
37. 9.44×10^{-3}
38. 4.007×10^8
39. 8.2×10^{-8}

1.1–1.3 Quiz

13. $h(x) = -2x$
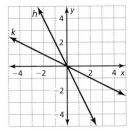

14. $h(x) = \tfrac{1}{3}x + \tfrac{5}{3}$
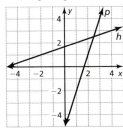

15. $h(x) = -5x + 30$
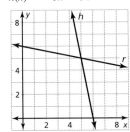

16. *Sample answer:* $y = |x|$, $y = -|x| + 5$
17. *Sample answer:* Cost of hiring a tutor for x hours
18. a. 56.5; It costs $56.50 to order 35 bracelets.
    b. $90; $225; $275
    c. *Sample answer:* Shipping costs

1.4 Monitoring Progress and Modeling with Mathematics

69. a. When $a > 1$ and $n < 0$, $a^n < a^{-n}$ because a^n will be less than 1 and a^{-n} will be greater than 1. When $a > 1$ and $n = 0$, $a^n = a^{-n} = 1$, because any number to the zero power is 1. When $a > 1$ and $n > 0$, $a^n > a^{-n}$ because a^n will be greater than 1 and a^{-n} will be less than 1.

    b. When $0 < a < 1$ and $n < 0$, $a^n > a^{-n}$, because a^n will be greater than 1 and a^{-n} will be less than 1. When $0 < a < 1$ and $n = 0$, $a^n = a^{-n} = 1$, because any number to the zero power is 1. When $0 < a < 1$ and $n > 0$, $a^n < a^{-n}$ because a^n will be less than 1 and a^{-n} will be greater than 1.

1.4 Maintaining Mathematical Proficiency

70. 5
71. -10
72. $\pm \tfrac{1}{8}$
73. natural number, whole number, integer, rational number, real number
74. rational number, real number
75. irrational number, real number

1.5 Monitoring Progress and Modeling with Mathematics

53. always; definition of rational exponent
54. sometimes; true when $x = 0$, $x = -1$, or $x = 1$, otherwise false
55. sometimes; false when $x = 0$ because division by 0 is undefined
56. sometimes: true when $x = 0$ or $x = 1$, otherwise false

1.5 Maintaining Mathematical Proficiency

57. $f(-3) = -16; f(0) = -10; f(8) = 6$
58. $w(-3) = 14; w(0) = -1; w(8) = -41$
59. $h(-3) = 16; h(0) = 13; h(8) = 5$
60. $g(-3) = -8; g(0) = 16; g(8) = 80$

1.6 Lesson Monitoring Progress

1. exponential growth

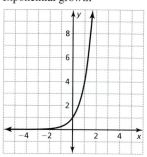

2. exponential decay

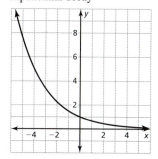

3. exponential decay

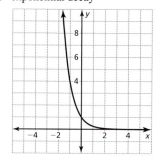

4. exponential growth

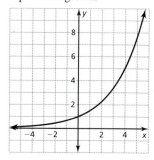

1.6 Monitoring Progress and Modeling with Mathematics

13. exponential growth

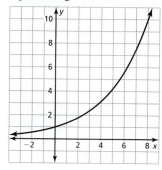

14. exponential decay

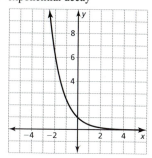

15. exponential growth

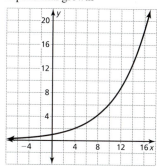

16. exponential decay

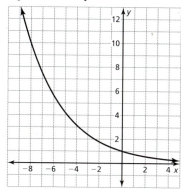

17. exponential decay

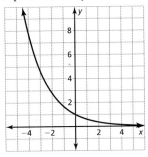

18. exponential growth

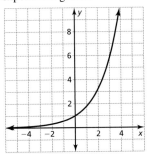

19. $b = 3$
20. $b = 5$
21. a. $y = 494(1.03)^t$; about 664; The population of Austin was about 664,000 at the end of the decade.
    b. about 6 years after the start of the decade
22. a. $y = 325(0.71)^t$; about 194; There are about 194 milligrams of ibuprofen in your blood after 1.5 hours.
    b. about 3.4 h
23. The percent decrease needs to be subtracted from 1 to produce the decay factor.
$y = \left(\begin{array}{c}\text{Initial}\\\text{amount}\end{array}\right)\left(\begin{array}{c}\text{Decay}\\\text{factor}\end{array}\right)^t$;
$y = 500(1 - 0.02)^t$; $y = 500(0.98)^t$
69. a. $y \approx 325(0.9943)^{60t} = 325(0.9943^{60})^t \approx 325(0.7097)^t$;
    $y \approx 325(0.843)^{2t} = 325(0.843^2)^t \approx 325(0.7106)^t$
    b. The initial amount of ibuprofen is 325 milligrams in each function. The percent decrease in the first function is about 0.57% each minute. The percent decrease in the second function is about 15.7% each half hour.
70. Account 1: about 1.0303; Account 2: about 1.0304; Account 3: about 1.0305; The quantity represents the annual growth rate of each account; $1030.34; $1030.42; $1030.45
71. a. $\dfrac{f(x+1)}{f(x)} = \dfrac{ab^{x+1}}{ab^x} = \dfrac{b^{x+1}}{b^x} = b^{(x+1)-x} = b^1 = b$
    b. The equation shows that when a value of the function is divided by the previous value, the answer is the constant b. Dividing the y-values in the table by the previous value does not always produce the same number.
72. Sample answer: $y = (1 - b)^x$

73. a. The decay factor is 0.9978. The percent decrease is 0.22%.
    b.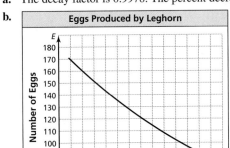
    c. about 134 eggs per year
    d. Replace $\dfrac{w}{52}$ with y, where y represents the age of the chicken in years.
74. $V = 1300(0.6782)^t$

1.6 Maintaining Mathematical Proficiency

75. $4x$
76. $-13y$
77. $5z + 9$
78. $-8w - 5$
79. $\dfrac{1}{x^7}$
80. x
81. $36x^2$
82. $16x^8$

Chapter 1 Review

19. $a = 210 - 1.25h$; 175
20. $\dfrac{1}{y^2}$
21. $\dfrac{1}{x^3}$
22. y^6
23. $\dfrac{25y^8}{4x^4}$

Chapter 1 Test

12. $h(x) = \dfrac{1}{3}x - \dfrac{5}{3}$

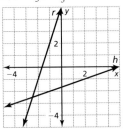

13. $h(x) = -\dfrac{4}{3}x - \dfrac{4}{9}$

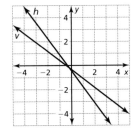

14. <

15. =

16. Sample answer: $y = \begin{cases} 0, & \text{if } x \leq -3 \\ x, & \text{if } -3 < x \leq 1 \\ 1, & \text{if } x > 1 \end{cases}$

17. a.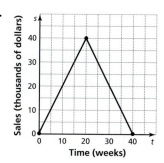

The transformations are a horizontal translation 20 units right, then a vertical stretch by a factor of 2, then a reflection in the x-axis, then a vertical translation 40 units up.

b. (20, 40); After 20 weeks, the sales reach a maximum of $40,000.

18. 1.5×10^4 sec; 15,000 sec

19. a. exponential decay
    b. 25% decrease
    c. in about 4.8 years

20. about 5.19%

Chapter 2

2.1 Monitoring Progress and Modeling with Mathematics

59. a. no; The product of two negative integers is always a positive integer.
    b. yes; The sum of two whole numbers is always a whole number.

60. a. $x^2 - 2x$; $x^2 - 12x$; $2x^2 - 14x$
    b. 520 ft²
    c. 1.3 gal; Divide the total area of the deck by the area that can be covered by 1 gallon of sealant to get the amount of sealant needed.

61. a. $19x^2 - 12x$
    b. $14,310

2.1 Maintaining Mathematical Proficiency

62. $5x + 4$
63. $34y - 34$
64. $22r - 1$

2.3 Monitoring Progress and Modeling with Mathematics

36. a. $(\pi x^2 - 12\pi x + 36\pi)$ mm²
    b. 4 times greater; The area increases from 4π square millimeters to 16π square millimeters, which is 4 times greater.

37. $(x + 11)(x - 11)$; $x^2 - 121$ fits the product side of the sum and difference pattern, so working backwards, a and b are the square roots of a^2 and b^2.

38. Use the Punnett Square which shows that $\frac{3}{4} = 75\%$ of the combinations result in green pods; Because both the G^2 and the Gy combinations result in green pods, add the coefficients of these terms in the polynomial, $0.25 + 0.5 = 0.75 = 75\%$.

39. $x^4 - 1$

40. $y^6 + 8y^3 + 16$
41. $4m^4 - 20m^2n^2 + 25n^4$
42. $r^6 - 36t^8$
43. no; $\left(4\frac{1}{3}\right)^2$ can be written as $\left(4 + \frac{1}{3}\right)^2$, however using the square of a binomial pattern results in $16 + \frac{8}{3} + \frac{1}{9}$, which is $18\frac{7}{9}$, not $16\frac{1}{9}$.

44. Sample answer: Reduce both sides by x feet; extend both sides by x feet; no; $(100 - x)^2 = (100 + x)^2$ only when $x = 0$.

45. $k = 64$

46. $x^3 + 3x^2 + 3x + 1$; $x^3 + 6x^2 + 12x + 8$; $a^3 + 3a^2b + 3ab^2 + b^3$

47. Sample answer: $a = 3, b = 4$

2.3 Maintaining Mathematical Proficiency

48. $6(2y - 3)$
49. $9(r + 3)$
50. $7(7s + 5t)$
51. $5(3x - 2y)$

2.4 Explorations

2.

x =	1	2	3	4	5	6
a. $(x - 1)(x - 2) = 0$	T	T	F	F	F	F
b. $(x - 2)(x - 3) = 0$	F	T	T	F	F	F
c. $(x - 3)(x - 4) = 0$	F	F	T	T	F	F
d. $(x - 4)(x - 5) = 0$	F	F	F	T	T	F
e. $(x - 5)(x - 6) = 0$	F	F	F	F	T	T
f. $(x - 6)(x - 1) = 0$	T	F	F	F	F	T

If the product of two values is 0, then at least one of the values must be 0. If $(x - a)$ is a factor of an equation, then $x = a$ is a solution.

2.4 Monitoring Progress and Modeling with Mathematics

45. no; Roots will occur if $x^2 + 3 = 0$ or $x^4 + 1 = 0$. However, solving these equations results in $x^2 = -3$ or $x^4 = -1$, and even powers of any number cannot be negative.

46. Sample answer: $(x - 1)^2(x - 2)(x - 3)$

47. a. $x = -y, x = \frac{1}{2}y$
    b. $x = \pm y, x = -4y$

48. $x = 7, x = 4$

2.4 Maintaining Mathematical Proficiency

49. 1, 10; 2, 5
50. 1, 18; 2, 9; 3, 6
51. 1, 30; 2, 15; 3, 10; 5, 6
52. 1, 48; 2, 24; 3, 16; 4, 12; 6, 8

2.5 Monitoring Progress and Modeling with Mathematics

45. $x^2 - 2x - 24 = 0$; Multiply $(x + 4)(x - 6)$.

46. a. Look for the x-intercepts and substitute them for p and q in $(x - p)(x - q)$.
    b. $(x - 2)(x + 3)$

47. a. $-x^2 + 38x$
    b. $-x^2 + 38x = 280$; 10 m

48. $(x + 2y)(x + 4y)$

49. $(r + 3s)(r + 4s)$
50. $(a - 2b)(a + 13b)$
51. $(x + 5y)(x - 7y)$

2.5 Maintaining Mathematical Proficiency
52. $p = 9$
53. $z = -17$
54. $c = -42$
55. $k = 0$

2.6 Monitoring Progress and Modeling with Mathematics
42. a.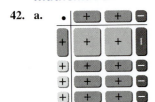

$2x^2 + 5x - 3 = (x + 3)(2x - 1)$

b.

$3x^2 - 2x - 1 = (x - 1)(3x + 1)$

43. 3.5 in.
44. 4 ft
45. $(k + 2j)(4k - j)$
46. $(2x - y)(3x + 4y)$
47. $-(a - 2b)(6a - 7b)$
48. $3m(2m + 5n)(3m - n)$

2.6 Maintaining Mathematical Proficiency
49.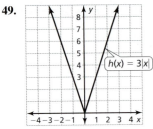

The graph of h is a vertical stretch of the graph of f by a factor of 3; domain: all real numbers; range: $y \geq 0$

50.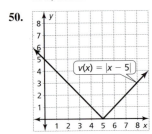

The graph of v is a horizontal translation 5 units right of the graph of f; domain: all real numbers; range: $y \geq 0$

51.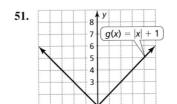

The graph of g is a vertical translation 1 unit up of the graph of f; domain: all real numbers; range: $y \geq 1$

52.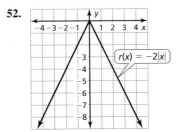

The graph of r is a vertical stretch of the graph of f by a factor of 2 and a reflection in the x-axis; domain: all real numbers; range: $y \leq 0$

53. -6
54. -2
55. 32
56. 9

2.7 Explorations
2.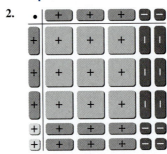

$9x^2 - 4;\ (3x + 2)(3x - 2)$

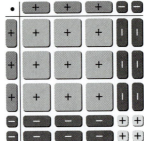

$9x^2 - 12x + 4;\ (3x - 2)^2$

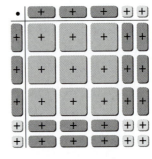

$9x^2 + 12x + 4;\ (3x + 2)^2$

2.7 Monitoring Progress and Modeling with Mathematics

46. a. the blue, yellow, and red regions; *Sample answer:*

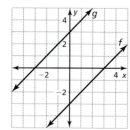

b. $(a - b)^2$ is represented by the yellow region. a^2 is the entire square. Subtracting $2ab$ requires removing 2 regions that each represent ab, which would be the blue and purple regions, and the red and purple regions. There is only one purple region, and subtracting $2ab$ required removing it twice, so one purple region needs to be replaced, which adds b^2.

47. a. $(9x^2 - 144)$ in.^{2}

b. 5 in.; Setting the polynomial in part (a) equal to 81 and solving results in $x = -5$ or $x = 5$. Length cannot be negative, so 5 is the solution.

48. a. $(x^3 + 16x)$ in.^{3}

b. $x = 3$; Setting the polynomial in part (a) equal to $25x$ and solving results in $x = 0$, $x = -3$, or $x = 3$. Because x is also a side length, and length cannot be negative or zero, 3 is the solution.

2.7 Maintaining Mathematical Proficiency

49. $2 \cdot 5^2$

50. $2^2 \cdot 11$

51. $5 \cdot 17$

52. $2^5 \cdot 3$

53. $g(x) = x + 3$

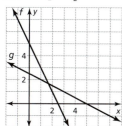

54. $g(x) = -\frac{1}{2}x + \frac{5}{2}$

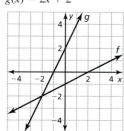

55. $g(x) = 2x + 2$

2.8 Monitoring Progress and Modeling with Mathematics

46. a. $(-h^3 + 5h^2 + 36h)$ in.^{3}

b. length: 3 in., width: 10 in., height: 6 in.; length: 4 in., width: 9 in., height: 5 in.

c. length: 4 in., width: 9 in., height: 5 in.; These dimensions result in a surface area of 202 square inches. The other possible dimensions result in a surface area of 216 square inches.

47. radius: 5, height: 8

48. $(x - 1)(x - 1)(x + 1)(x + 2)(x - 2)$

49. a. *Sample answer:* $w = 40$; When $w = 40$, factoring out $5x$ will leave a perfect square trinomial, so there will be 2 factors.

b. *Sample answer:* $w = 50$; When $w = 50$, factoring out $5x$ will leave a factorable trinomial that is not a perfect square, so there will be 3 factors.

2.8 Maintaining Mathematical Proficiency

50. $(2, -2)$

51. $(2, 3)$

52. $(4, 8)$

53. $(-6, -2)$

54. exponential growth

55. exponential decay

56. exponential growth

57. exponential growth

Chapter 3

Chapter 3 Mathematical Practices

6.

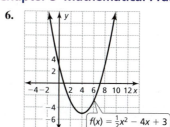

The graph opens up and the lowest point is at $(4, -5)$.

7.

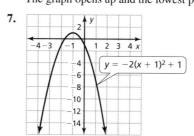

The graph opens down and the highest point is at $(-1, 1)$.

8.

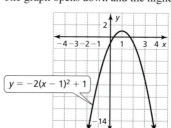

The graph opens down and the highest point is at $(1, 1)$.

9. *Sample answer:* They are all U-shaped; Some open up and some open down.

Additional Answers **A13**

3.1 Explorations

1. d.

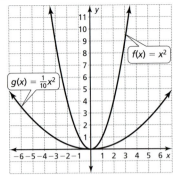

The graph of g is a vertical shrink by a factor of $\frac{1}{10}$ of the graph of f.

3.1 Lesson Monitoring Progress

3.

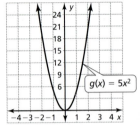

The graph of g is a vertical stretch by a factor of 5 of the graph of f.

4.

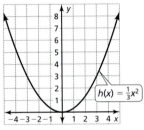

The graph of h is a vertical shrink by a factor of $\frac{1}{3}$ of the graph of f.

5.

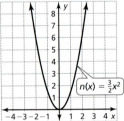

The graph of n is a vertical stretch by a factor of $\frac{3}{2}$ of the graph of f.

6.

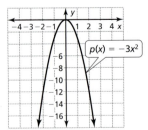

The graph of p is a vertical stretch by a factor of 3 and a reflection in the x-axis of the graph of f.

7.

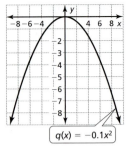

The graph of q is a vertical shrink by a factor of 0.1 and a reflection in the x-axis of the graph of f.

8.

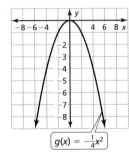

The graph of g is a vertical shrink by a factor of $\frac{1}{4}$ and a reflection in the x-axis of the graph of f.

3.1 Monitoring Progress and Modeling with Mathematics

9.

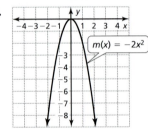

The graph of m is a vertical stretch by a factor of 2 and a reflection in the x-axis of the graph of f.

10.

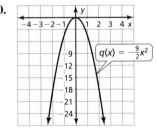

The graph of q is a vertical stretch by a factor of $\frac{9}{2}$ and a reflection in the x-axis of the graph of f.

11.

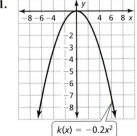

The graph of k is a vertical shrink by a factor of 0.2 and a reflection in the x-axis of the graph of f.

12.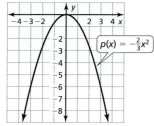

The graph of p is a vertical shrink by a factor of $\frac{2}{3}$ and a reflection in the x-axis of the graph of f.

13.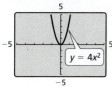

The graph of $y = 4x^2$ is a reflection in the x-axis of the graph of $y = -4x^2$.

14.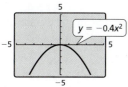

The graph of $y = -0.4x^2$ is a vertical shrink by a factor of $\frac{1}{10}$ of the graph of $y = -4x^2$.

15.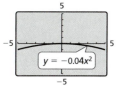

The graph of $y = -0.04x^2$ is a vertical shrink by a factor of $\frac{1}{100}$ of the graph of $y = -4x^2$.

16.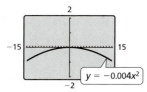

The graph of $y = -0.004x^2$ is a vertical shrink by a factor of $\frac{1}{1000}$ of the graph of $y = -4x^2$.

17. The graph of $y = 0.5x^2$ should be wider than the graph of $y = x^2$;

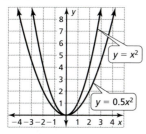

The graphs have the same vertex and the same axis of symmetry. The graph of $y = 0.5x^2$ is wider than the graph of $y = x^2$.

18. 300 feet high, 1000 feet wide

19. a. domain: $d \geq 0$, range: $z \geq 0$

b.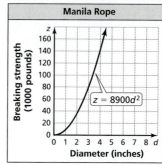

c. no; *Sample answer:* The relationship is quadratic, so a rope with 4 times the diameter will have $4^2 = 16$ times the breaking strength.

22. f is decreasing when $x < 0$. g is decreasing when $x > 0$.

23. f; $a = \frac{3}{4}$

24. yes; *Sample answer:* At an x-intercept, $y = 0$, so $ax^2 = 0$. By the definition of a quadratic function, $a \neq 0$, so by the Zero-Product Property, $x^2 = 0$, which means $x = 0$.

25. *Sample answer:* The vertex of a parabola that opens up is the minimum point, so its y-coordinate is the minimum value of y. The graph passes through $(6, -3)$, so 2 is not the minimum value of y.

26. sometimes; *Sample answer:* The graph of f will be narrower than the graph of g when $a > 1$, but it will be wider when $0 < a < 1$.

27. always; *Sample answer:* When $|a| > 1$, the graph of f will be a vertical stretch of the graph of g, so it will be narrower.

28. always; *Sample answer:* When $0 < |a| < 1$, the graph of f will be a vertical shrink of the graph of g, so it will be wider.

29. never; *Sample answer:* When $|a| > |d|$, the graph of f will be a vertical stretch of the graph of g, so it will be narrower, not wider.

30.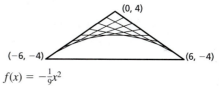

$f(x) = -\frac{1}{9}x^2$

31. a. 8 cm

b. no; *Sample answer:* A faster rotational speed would increase the depth. The diagram shown has a depth of 3.2 centimeters. A model of $y = 0.1x^2$ would only have a depth of 1.6 centimeters, so it would have a slower rotational speed.

3.1 Maintaining Mathematical Proficiency

32. 14

33. 3

34. -25

35. 11

3.1 Mini-Assessment

1. vertex: $(-2, -4)$, axis of symmetry: $x = -2$, domain: all real numbers, range: $y \geq -4$; when $x < -2$, y increases as x decreases, when $x > -2$, y increases as x increases.

2.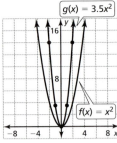

The graph of g is a vertical stretch by a factor of 3.5 of the graph of f.

3.2 Lesson Monitoring Progress

3.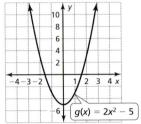

The graph of g is a vertical stretch by a factor of 2 and a vertical translation 5 units down of the graph of f.

4.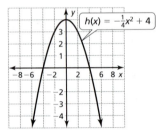

The graph of h is a vertical shrink by a factor of $\frac{1}{4}$, a reflection in the x-axis, and a vertical translation 4 units up of the graph of f.

5. **a.** The graph of g is a vertical translation 3 units up of the graph of f.

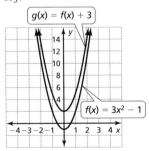

b. $g(x) = 3x^2 + 2$

3.2 Monitoring Progress and Modeling with Mathematics

8.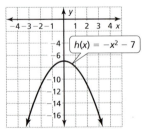

The graph of h is a reflection in the x-axis, and a vertical translation 7 units down of the graph of f.

9.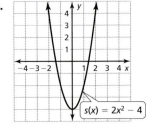

The graph of s is a vertical stretch by a factor of 2 and a vertical translation 4 units down of the graph of f.

10.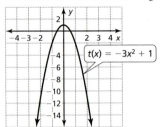

The graph of t is a vertical stretch by a factor of 3, a reflection in the x-axis, and a vertical translation 1 unit up of the graph of f.

11.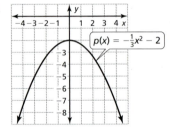

The graph of p is a vertical shrink by a factor of $\frac{1}{3}$, a reflection in the x-axis, and a vertical translation 2 units down of the graph of f.

12.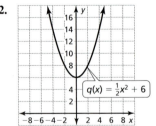

The graph of q is a vertical shrink by a factor of $\frac{1}{2}$ and a vertical translation 6 units up of the graph of f.

13. The graph of g is a vertical translation 2 units up of the graph of f.

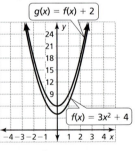

$g(x) = 3x^2 + 6$

A16 Additional Answers

14. The graph of g is a vertical translation 4 units down of the graph of f.

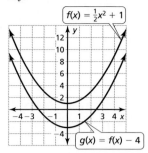

$g(x) = \frac{1}{2}x^2 - 3$

15. The graph of g is a vertical translation 3 units down of the graph of f.

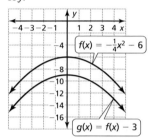

$g(x) = -\frac{1}{4}x^2 - 9$

16. The graph of g is a vertical translation 7 units up of the graph of f.

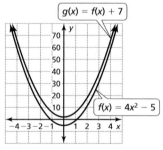

$g(x) = 4x^2 + 2$

17. The graph of $y = 3x^2 + 2$ is narrower, so it should be a stretch, not a shrink; The graph of $y = 3x^2 + 2$ is a vertical stretch by a factor of 3 and a translation 2 units up of the graph of $y = x^2$.

18. $g(x)$ is not graphed correctly;

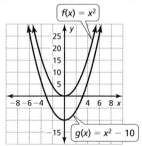

Both graphs open up and have the same axis of symmetry. However, the vertex of the graph of g, $(0, -10)$, is 10 units below the vertex of the graph of f, $(0, 0)$.

19. $x = 1, x = -1$
20. $x = 6, x = -6$
21. $x = 5, x = -5$
22. $x = 7, x = -7$
23. $x = 2, x = -2$
24. $x = 3, x = -3$
25. $x = \frac{1}{2}, x = -\frac{1}{2}$
26. $x = \frac{7}{2}, x = -\frac{7}{2}$

27. a. the height of the balloon after t seconds
    b. 3 sec
    c. When $k > 0$, the water balloon will take more than 3 seconds to hit the ground. When $k < 0$, the water balloon will take less than 3 seconds to hit the ground.

28. The x-intercept is 1.5. This means the apple hits the ground after 1.5 seconds. The y-intercept is 36. This means the apple fell from a height of 36 feet.

29. Sample answer:

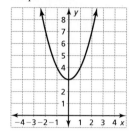

30. Sample answer:

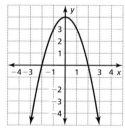

31. Sample answer:

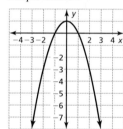

32. Sample answer:

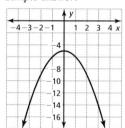

33. a. $T(x) = -44$; the distance between the balls
    b. 44 ft

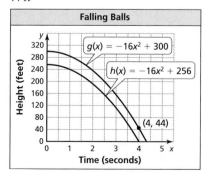

34. no; *Sample answer:* Changing *a* causes a vertical stretch or shrink, which does not change the vertex.

35. 12 ft by 12 ft

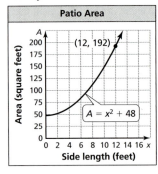

36. A

37. Graph the function and determine the *x*-intercepts; Set the function equal to 0, factor $-16t^2 + 400$, and apply the Zero-Product Property.

38. a. waterfall 1
   b. waterfall 2
   c. waterfall 1

39. a. $h = -16t^2 + 45$; $h = -16t^2 + 32$
   b. The graph of $h = -16t^2 + 32$ is a vertical translation 13 units down of the graph of $h = -16t^2 + 45$.

40. *Sample answer:* 8; $A = 2\left(\frac{1}{2}bh\right) = bh = 2(4) = 8$

41. (0, 5.8); *Sample answer:* The outer edges are located 40 feet from the center. Substituting this into $y = 0.012x^2$ indicates they are 19.2 feet above the *x*-axis. To be 25 feet above the *x*-axis, they must be vertically translated up 5.8 feet.

3.2 Maintaining Mathematical Proficiency

42. $-\frac{1}{3}$

43. $-\frac{3}{8}$

44. $\frac{7}{9}$

45. $\frac{5}{12}$

3.3 Lesson Monitoring Progress

6.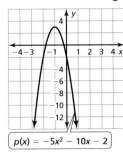

domain: all real numbers, range: $y \leq 3$

3.3 Monitoring Progress and Modeling with Mathematics

16.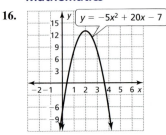

domain: all real numbers, range: $y \leq 13$

17.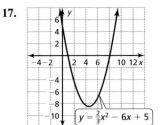

domain: all real numbers, range: $y \geq -\frac{17}{2}$

18.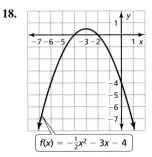

domain: all real numbers, range: $y \leq \frac{1}{2}$

19. There should be two negatives in the substitution, one from the formula and one because *b* is -12;
$x = -\dfrac{b}{2a} = -\dfrac{-12}{2(3)} = 2$; The axis of symmetry is $x = 2$.

20. The formula for the axis of symmetry has a negative sign;
The axis of symmetry is $x = -\dfrac{b}{2a} = -\dfrac{4}{2(1)} = -2$;
$f(-2) = (-2)^2 + 4(-2) + 3 = -1$; So, the vertex is $(-2, -1)$; The *y*-intercept is 3, so the points (0, 3) and $(-4, 3)$ lie on the graph.

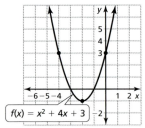

21. minimum value; -12

22. maximum value; 12

23. maximum value; -1

24. minimum value; $\frac{1}{2}$

25. maximum value; $66\frac{1}{2}$

26. minimum value; $-4\frac{1}{4}$

43. down; *Sample answer:* Because (3, 2) and (9, 2) have the same *y*-coordinate, any point with an *x*-coordinate between 3 and 9 lies on the part of the parabola between these two points that passes through the vertex. Because 7 is greater than 2, the vertex must be above these 2 points, so the parabola opens down.

44. the maximum value or minimum value of the function;
Sample answer: Because $-\dfrac{b}{2a}$ is the *x*-coordinate of the vertex, $f\left(-\dfrac{b}{2a}\right)$ is the *y*-coordinate of the vertex, which is also the maximum value or minimum value of the function.

45. $y = -2x^2 + 8x$

46. It is not possible to identify characteristics of parabola A with only two points; The vertex of parabola B is (3, −4) because 3 is halfway between 1 and 5, which are the x-coordinates of points that have the same y-coordinates. Parabola B opens up because −4, the y-coordinate of the vertex, is the minimum value.

47. 14 ft

48. *Sample answer:* 2; Sketch a graph of $y = x^2$ and a tangent line through (1, 1). Then select another point on the tangent line, (0, −1), and substitute these points into the slope formula to calculate a slope of 2.

49. $\dfrac{k^2}{8}$ ft²

3.3 Maintaining Mathematical Proficiency

50. The graph of q is a horizontal translation 6 units left of the graph of f.

51. The graph of h is a vertical shrink by a factor of 0.5 and a reflection in the x-axis of the graph of f.

52. The graph of g is a horizontal translation 2 units right and a vertical translation 5 units up of the graph of f.

53. The graph of p is a vertical stretch by a factor of 3 and a horizontal translation 1 unit left of the graph of f.

3.1–3.3 Quiz

9. The graph of g is a vertical translation 2 units up of the graph of f.

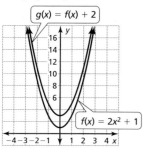

$g(x) = 2x^2 + 3$

10. The graph of g is a vertical translation 9 units down of the graph of f.

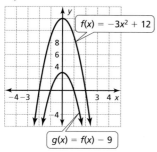

$g(x) = -3x^2 + 3$

11. The graph of g is a vertical translation 6 units down of the graph of f.

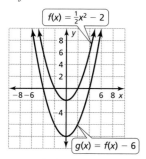

$g(x) = \frac{1}{2}x^2 - 8$

12. The graph of g is a vertical translation 1 unit up of the graph of f.

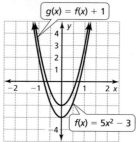

$g(x) = 5x^2 - 2$

13.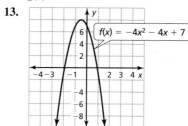

domain: all real numbers, range: $y \leq 8$

14.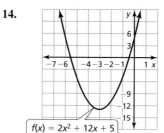

domain: all real numbers, range: $y \geq -13$

15.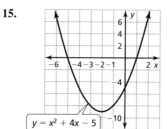

domain: all real numbers, range: $y \geq -9$

16.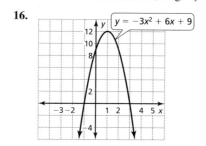

domain: all real numbers, range: $y \leq 12$

17. minimum value; −8
18. maximum value; 18
19. maximum value; 16
20. minimum value; −5
21. 2 sec
22. a. 1.25 sec
    b. the first pinecone; *Sample answer:* The second pinecone will take 1.5 seconds to fall, which is longer than the first.
23. domain: $0 \leq t \leq 2$, range: $0 \leq h \leq 18$; 18 ft

3.4 Lesson Monitoring Progress

4.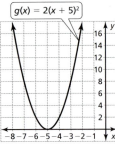

The graph of g is a horizontal translation 5 units left and a vertical stretch by a factor of 2 of the graph of f.

5.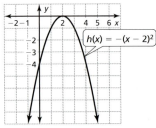

The graph of h is a horizontal translation 2 units right and a reflection in the x-axis of the graph of f.

7.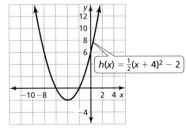

The graph of h is a vertical shrink by a factor of $\frac{1}{2}$, and a translation 4 units left and 2 units down of the graph of f.

8.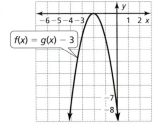

3.4 Monitoring Progress and Modeling with Mathematics

25.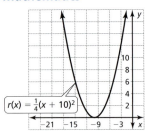

The graph of r is a horizontal translation 10 units left and a vertical shrink by a factor of $\frac{1}{4}$ of the graph of f.

26.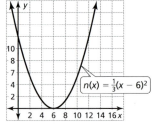

The graph of n is a horizontal translation 6 units right and a vertical shrink by a factor of $\frac{1}{3}$ of the graph of f.

27.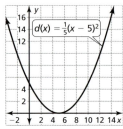

The graph of d is a horizontal translation 5 units right and a vertical shrink by a factor of $\frac{1}{5}$ of the graph of f.

28.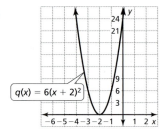

The graph of q is a horizontal translation 2 units left and a vertical stretch by a factor of 6 of the graph of f.

29. If $f(-x) = f(x)$, then the function is even; So, $f(x)$ is an even function.

30. h is the x-coordinate of the vertex, not the y-coordinate; Because $h = -8$, the vertex is $(-8, 0)$.

43.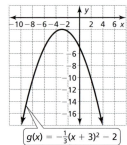

The graph of g is a vertical shrink by a factor of $\frac{1}{3}$, a reflection in the x-axis, and a translation 3 units left and 2 units down of the graph of f.

44.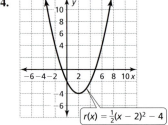

The graph of r is a vertical shrink by a factor of $\frac{1}{2}$, and a translation 2 units right and 4 units down of the graph of f.

49.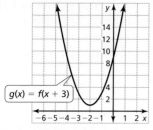

50.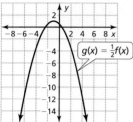

51.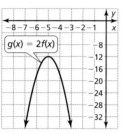

52.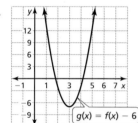

53.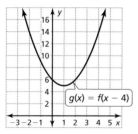

54.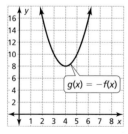

55. a.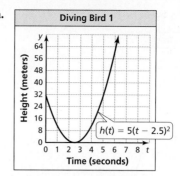

b.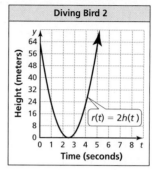

c. The graph of r is a vertical stretch by a factor of 2 of the graph of h; the second bird; *Sample answer:* Because $r(t)$ is twice $h(t)$, the second bird starts at a height twice as high as the first bird.

56. a.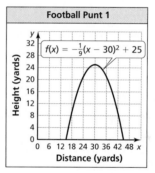

domain: $15 \le x \le 45$, range: $0 \le y \le 25$

b.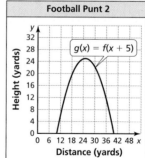

domain: $10 \le x \le 40$, range: $0 \le y \le 25$

c. The graph of g is a horizontal translation 5 units left of the graph of f; second possession; *Sample answer:* On the first possession, the ball is punted 15 yards from the kicker's goal line. On the second possession, the ball is punted 10 yards from the kicker's goal line.

63. $f(x) = -\frac{18}{125}(x - 25)^2 + 90$

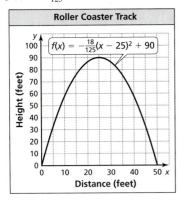

64. $f(x) = -\frac{1}{12}(x - 59)^2 + 300$

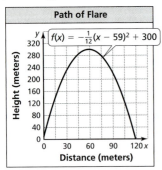

65. $y = 2(x - 2)^2 - 4$ **66.** $y = 3(x + 1)^2 - 4$

67. $f(x) = -5(x - 1)^2 + 8$ **68.** $f(x) = -(x + 2)^2 + 6$

69. no; *Sample answer:* The graph would not pass the Vertical Line Test.

70. $+$; $-$; *Sample answer:* The vertex is $(-2, -3)$, so in the vertex form $h = -2$ and $k = -3$.

71. The graph of h is a vertical translation 4 units up of the graph of f; $h(x) = -(x + 1)^2 + 2$

72. The graph of h is a horizontal translation 5 units right of the graph of f; $h(x) = 2(x - 6)^2 + 1$

73. The graph of h is a vertical stretch by a factor of 2 of the graph of f; $h(x) = 8(x - 2)^2 + 6$

74. The graph of h is a vertical shrink by a factor of $\frac{1}{3}$ of the graph of f; $h(x) = -\frac{1}{3}(x + 5)^2 - 2$

75. $y = (x - 2)^2 - 5$; $y = x^2 - 4x - 1$; *Sample answer:* The vertex, $(2, -5)$, can be quickly determined from the vertex form; The y-intercept, -1, can be quickly determined from the standard form.

76. a. true; *Sample answer:* If $g(x) = af(x)$, then $g(-x) = af(-x)$. Because f is even, $g(-x) = af(x) = g(x)$, so g is even.

b. true; *Sample answer:* If $g(x) = af(x)$, then $g(-x) = af(-x)$. Because f is odd, $g(-x) = a(-f(x)) = -af(x) = -g(x)$, so g is odd.

c. true; *Sample answer:* If $h(x) = f(x) + g(x)$, then $h(-x) = f(-x) + g(-x)$. Because f and g are even, $h(-x) = f(x) + g(x) = h(x)$, so h is even.

d. true; *Sample answer:* If $h(x) = f(x) + g(x)$, then $h(-x) = f(-x) + g(-x)$. Because f and g are odd, $h(-x) = -f(x) - g(x) = -(f(x) + g(x)) = -h(x)$, so h is odd.

e. false; *Sample answer:* A counterexample is illustrated by adding $f(x) = x^2 + 4$, an even function, to $g(x) = 3x$, an odd function. The sum $h(x) = x^2 + 3x + 4$ is neither even nor odd.

77. a. the second birdbath; *Sample answer:* The first birdbath has a depth of 4 inches and the second birdbath has a depth of 6 inches.

b. the first birdbath; *Sample answer:* The first birdbath has a width of 36 inches and the second birdbath has a width of 30 inches.

78. The graph of $y = 2x^2 + 8x + 8$ is a vertical stretch by a factor of 2 and a horizontal translation 2 units left of the graph of $y = x^2$; *Sample answer:* Factoring the right side of $y = 2x^2 + 8x + 8$ changes the equation to vertex form, $y = 2(x + 2)^2$, from which the transformations can be easily determined.

3.4 Mini-Assessment

4. $f(x) = -\frac{9}{25}(x - 10)^2 + 36$

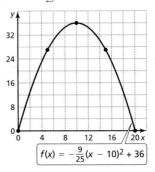

3.5 Lesson Monitoring Progress

3.

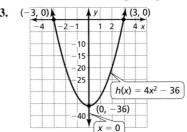

domain: all real numbers, range: $y \geq -36$

3.5 Monitoring Progress and Modeling with Mathematics

11.

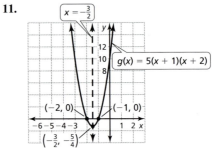

domain: all real numbers, range: $y \geq -\frac{5}{4}$

12.

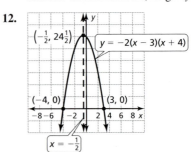

domain: all real numbers, range: $y \leq 24\frac{1}{2}$

A22   Additional Answers

13.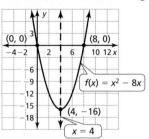
domain: all real numbers, range: $y \geq -9$

14.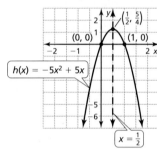
domain: all real numbers, range: $y \geq -16$

15.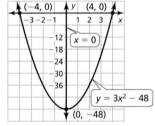
domain: all real numbers, range: $y \leq \frac{5}{4}$

16.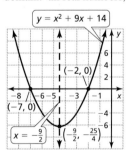
domain: all real numbers, range: $y \geq -48$

17.
domain: all real numbers, range: $y \geq -\frac{25}{4}$

18.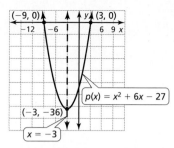
domain: all real numbers, range: $y \geq -36$

19.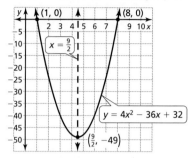
domain: all real numbers, range: $y \geq -49$

20.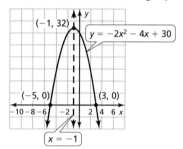
domain: all real numbers, range: $y \leq 32$

21. 2, 10
22. $-5, 1$
23. $-2, 2$
24. $-6, 6$
25. $-8, 3$
26. 4, 13
27. $-2, 7$
28. -1
29. D
30. F
31. C
32. B
33. A
34. E

61.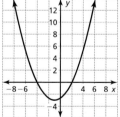

62. *Sample answer:*
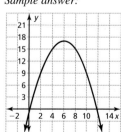

63. Sample answer:

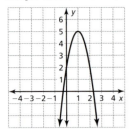

64. Sample answer:

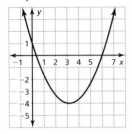

65. a. 4 ft
   b. $\frac{1}{2}$ ft
   c. $y = \frac{1}{6}(x^2 - 9)$

73. a. $f(x) = -3(x + 4)(x - 2)$
   b.

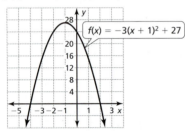

Sample answer: Plot the vertex $(-1, 27)$, which can be determined from the vertex form. Then plot the x-intercepts $(-4, 0)$ and $(2, 0)$, which can be determined from the intercept form. Draw a smooth curve through these points.

74. a. $g(x) = 2x(x - 2)$
   b.

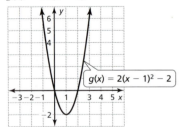

Sample answer: Plot the vertex $(1, -2)$, which can be determined from the vertex form. Then plot the x-intercepts $(0, 0)$ and $(2, 0)$, which can be determined from the intercept form. Draw a smooth curve through these points.

75. yes; Sample answer: Let p and q be real zeros of a quadratic function. When $p = q$, the function has exactly one real zero and can be written as $y = a(x - p)(x - p)$.

76. yes; Sample answer: In standard form, $f(x) = ax^2 + bx + c$, missing terms can be included by using 0 for b or c. In vertex form, $f(x) = a(x - h)^2 + k$, missing values can be included by using 0 for h or k.

77. $-k, 2k$

78. a.

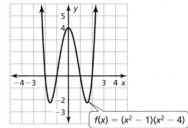

   b.

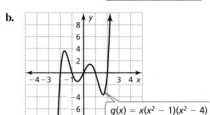

Sample answer: Factor the functions into intercept form. Plot the intercepts. Calculate the values for several additional points, then plot them to get an idea of the general shape of the curve. Draw a smooth curve through the plotted points.

79. Sample answer: $y = (x + 4)(x - 2)$, $y = 2(x + 4)(x - 2)$; The two given points are x-intercepts, so the graphs of any two quadratic functions having these two intercepts would intersect at these points.

80. Sample answer: $y = (x - 5)^2 + 2$, $y = -(x - 5)^2 + 10$; The two given points have the same y-coordinate, so the axis of symmetry of any parabola passing through them would be halfway between, which is $x = 5$. The vertex form can be used with $h = 5$ and any selected value of k, where $k \neq 6$, to find quadratic functions that pass through both points.

3.5 Maintaining Mathematical Proficiency

81. $\sqrt{2}$, or about 1.41
82. 10
83. $\sqrt{5}$, or about 2.24

3.6 Explorations

1. a.

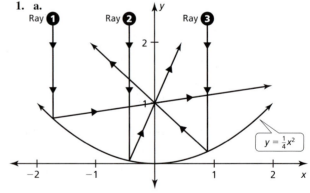

   b. All the reflected rays cross through the point $(0, 1)$.
   c. $(0, 1)$; This is the best place for the receiver because any ray entering the satellite dish will hit the receiver.

2.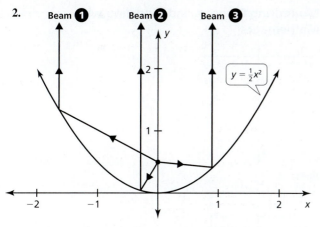

All beams leaving the parabola are parallel; yes; With a spotlight, all beams of light should be pointed at the same object.

3.6 Lesson Monitoring Progress

4. The focus is $\left(\frac{3}{2}, 0\right)$, the directrix is $x = -\frac{3}{2}$, and the axis of symmetry is the x-axis.

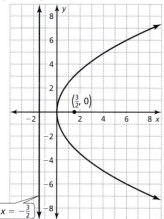

3.6 Monitoring Progress and Modeling with Mathematics

16. The focus is $(6, 0)$. The directrix is $x = -6$. The axis of symmetry is the x-axis.

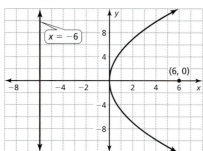

17. The focus is $(4, 0)$. The directrix is $x = -4$. The axis of symmetry is the x-axis.

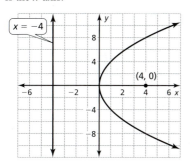

18. The focus is $(0, -12)$. The directrix is $y = 12$. The axis of symmetry is the y-axis.

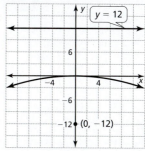

19. The focus is $\left(0, -\frac{1}{8}\right)$. The directrix is $y = \frac{1}{8}$. The axis of symmetry is the y-axis.

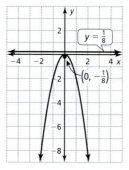

20. The focus is $\left(0, \frac{1}{32}\right)$. The directrix is $y = -\frac{1}{32}$. The axis of symmetry is the y-axis.

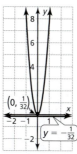

21. Instead of a vertical axis of symmetry, the graph should have a horizontal axis of symmetry.

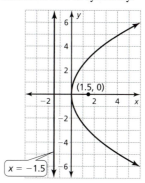

22. Because $p = -0.5$, the focus is $(-0.5, 0)$, and the directrix is $x = 0.5$.

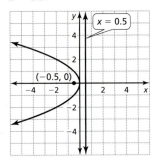

23. 9.5 in.; The receiver should be placed at the focus. The distance from the vertex to the focus is $p = \frac{38}{4} = 9.5$ in.

3.6 Maintaining Mathematical Proficiency

55. The inputs are equally spaced and the common ratio is 2; $f(0) = 0.25, f(n) = 2 \cdot f(n-1)$
56. The inputs are equally spaced and the common ratio is 3; $f(0) = 2, f(n) = 3 \cdot f(n-1)$
57. The inputs are equally spaced and the common ratio is $\frac{1}{2}$; $f(0) = 144, f(n) = \frac{1}{2} \cdot f(n-1)$
58. The inputs are equally spaced and the common ratio is $\frac{1}{5}$; $f(0) = 1250, f(n) = \frac{1}{5} \cdot f(n-1)$

3.7 Lesson Monitoring Progress

1.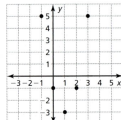

quadratic

2.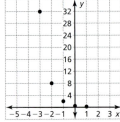

exponential

3.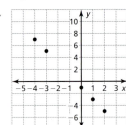

linear

3.7 Monitoring Progress and Modeling with Mathematics

13.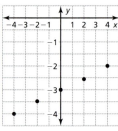

linear

14.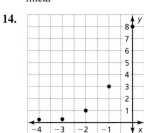

exponential

15. linear
16. exponential
17. exponential
18. quadratic
38. a. D; A parabola is the graph of a quadratic function, and opens down when the coefficient of x^2 is negative.
    b. C; A line is the graph of a linear function.
    c. B; The graph appears to represent an exponential function that has been translated down.
    d. A; A parabola is the graph of a quadratic function, and opens up when the coefficient of x^2 is positive.
39. The average rate of change of a linear function is constant because the dependent variable of a linear function increases by the same amount for each constant change in the independent variable. The average rate of change of a quadratic or exponential function is not constant because the dependent variable of a quadratic or exponential function changes by a different amount for each constant change in the independent variable.
40. quadratic; The second differences have a constant value of $9n - 5$.
41. Sample answer: $y = 1.5x^2$
42. no; Sample answer: There may not be enough points to clearly determine the shape of the graph.
43. no; The graph of an exponential growth function will always eventually have greater y-values than the graph of a quadratic function.
44. Sample answer: the width, the perimeter, the area; linear, linear, quadratic; $y = n$, $y = 4n$, $y = n^2$
45. Sample answer: $y = -0.414x^2 + 6.17x + 47.0$; Compare the linear, quadratic, and exponential models obtained using the regression features of a graphing utility; about $69.9 billion

3.7 Maintaining Mathematical Proficiency

46. 11
47. 5
48. 8
49. 3
50. $x^2 - 36$
51. $4y^2 - 25$
52. $16c^2 - 9d^2$
53. $9s^2 - 64t^2$

Chapter 3 Review

8.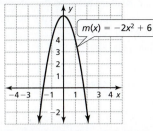

The graph of m is a vertical stretch by a factor of 2, a reflection in the x-axis, and a vertical translation 6 units up of the graph of f.

9.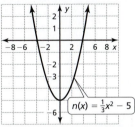

The graph of n is a vertical shrink by a factor of $\frac{1}{3}$ and a vertical translation 5 units down of the graph of f.

Chapter 3 Test

5.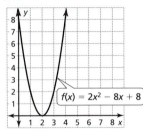

domain: all real numbers, range: $y \geq 0$

6.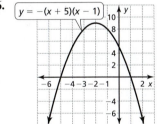

domain: all real numbers, range: $y \leq 9$

7.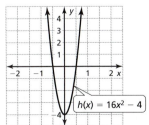

domain: all real numbers, range: $y \geq -4$

8. The focus is $\left(\frac{1}{8}, 0\right)$, the directrix is $x = -\frac{1}{8}$, and the axis of symmetry is $y = 0$.

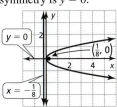

9. $y = -\frac{1}{28}(x + 3)^2 + 2$; Use the focus and vertex to solve for $p = -7$, and substitute p and (h, k) into $y = \frac{1}{4p}(x - h)^2 + k$.

10. $y = -\frac{1}{4}(x + 1)(x - 7)$; Substitute the given information into intercept form and use (5, 3) to solve for a.

11. $x = \frac{1}{16}y^2 + 2$; Use the directrix and the vertex to solve for $p = 4$ and substitute p and (h, k) into $x = \frac{1}{4p}(y - k)^2 + h$.

12. exponential; The y-values increase by a constant factor; $y = 8(2)^x$

13. quadratic; The second differences increase by a constant amount; $y = -2x^2$

14. a. 4.445 ft
    b. yes; Sample answer: After traveling a horizontal distance of 30 feet, the height is 3.6 feet, so the ball will be 0.6 feet above the net.

15. a. Sample answer: $a = 1, b = 0, c = 0$
    b. Sample answer: $a = 0, b = 1, c = 0$
    c. Sample answer: $a = 1, b = 2, c = 3$

16. 1, 3, 5; They are increasing by a constant amount.

Chapter 4

4.1 Monitoring Progress and Modeling with Mathematics

105. 377

106. a. $\left(\dfrac{1 + \sqrt{5}}{2}\right)^2 - \dfrac{1 + \sqrt{5}}{2} - 1 = 0$,

$\left(\dfrac{1 - \sqrt{5}}{2}\right)^2 - \dfrac{1 - \sqrt{5}}{2} - 1 = 0$

b. Sample answer: $DF = \dfrac{1 + \sqrt{5}}{2}$

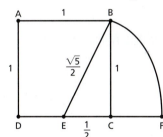

107. $\dfrac{2\sqrt[3]{x^2} - 2\sqrt[3]{x} + 2}{x + 1}$; Multiplying the numerator and denominator by $\sqrt[3]{x^2} - \sqrt[3]{x} + 1$ rationalizes the denominator.

4.1 Maintaining Mathematical Proficiency

108.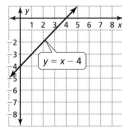

4

109.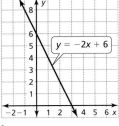
3

110.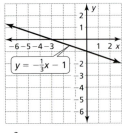
−3

111.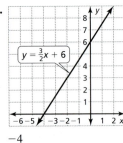
−4

112. $z^2 + 6z + 9$
113. $9a^2 − 30ab + 25b^2$
114. $x^2 − 64$
115. $16y^2 − 4$

4.2 Monitoring Progress and Modeling with Mathematics

51.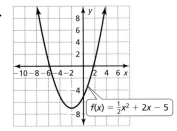
about 1.7, about −5.7

52.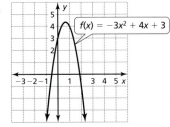
about 1.9, about −0.5

53. **a.** 118 ft, 198 ft, 246 ft, 262 ft, 246 ft, 198 ft, 118 ft, 6 ft
b. about 1.5 sec, about 6.5 sec
c. about 1.4 sec, about 6.6 sec

55. about 3.7 ft
56. about 6.2 m

57. Graph the function to determine which integers the solutions are between. Then make tables using x-values between the integers with an interval of 0.1. Look for a change of sign in the function values, then select the value closest to zero.

58. **a.** 2; The graphs intersect in two places.
b. The x-intercepts are 1 and −4.

59. Sample answer: Method 1; Only one graph needs to be drawn.

60. infinitely many; Sample answer:

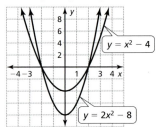

61. about 24.1 ft
62. yes; When $x = 57$, $y = 26.313$.
63. sometimes; $y = −2x^2 + 1$ has two x-intercepts, but $y = −2x^2 + (−1)$ has no x-intercepts.
64. always; The sign of y does not change.
65. never; The graph of $y = ax^2 + bx + c$ has at most two x-intercepts.

4.2 Maintaining Mathematical Proficiency

66. The inputs are equally spaced and the common ratio is $\frac{1}{6}$; $f(0) = 18, f(n) = \frac{1}{6} \cdot f(n − 1)$

67. The inputs are equally spaced and the common ratio is 4; $f(0) = 2, f(n) = 4 \cdot f(n − 1)$

4.3 Monitoring Progress and Modeling with Mathematics

42. $x = \frac{-b}{2a} \pm \sqrt{\frac{b^2 - 4ac}{4a^2}}$ or $x = \frac{-b \pm \sqrt{b^2 - 4ac}}{2a}$

43. $x = 6, x = −2$
44. **a.** $x = 14, x = −2$
b. $x = −3, x = −11$

4.3 Maintaining Mathematical Proficiency

45. $(x + 4)^2$ **46.** $(x − 2)^2$
47. $(x − 7)^2$ **48.** $(x + 9)^2$
49. $(x + 6)^2$ **50.** $(x − 11)^2$

4.4 Monitoring Progress and Modeling with Mathematics

68.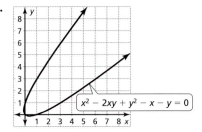
a parabola at an angle

71. yes; Substituting 23.50 for y in the model, the stock is worth $23.50 ten days and thirty days after the stock is purchased.

72. Factoring does not work for this equation, and graphing does not produce an exact solution.

73. length: 66 in., width: 6 in.

74. 0; If $c < -\left(\frac{b}{2}\right)^2$, then $\left(\frac{b}{2}\right)^2 + c < 0$, so there are no real solutions.

4.4 Maintaining Mathematical Proficiency

75. $y = \begin{cases} -x + 1, & \text{if } x < 0 \\ 1, & \text{if } x \geq 0 \end{cases}$

76. $y = \begin{cases} 0.5x - 1, & \text{if } x \leq 2 \\ 2x - 3, & \text{if } x > 2 \end{cases}$

77. $y = \begin{cases} 1, & \text{if } 0 \leq x < 2 \\ 3, & \text{if } 2 \leq x < 4 \\ 5, & \text{if } 4 \leq x < 6 \end{cases}$

78. $2\sqrt{3}$
79. $6\sqrt{2}$
80. $2\sqrt{5}$

4.5 Monitoring Progress and Modeling with Mathematics

39. $x \approx 0.74$, $x \approx -6.74$; *Sample answer:* $a = 1$ and b is even, so solve by completing the square.

40. $x \approx 2.24$, $x \approx -2.24$; *Sample answer:* The equation can be written in the form $x^2 = d$, so solve using square roots.

41. $x = -4$, $x = 3$; *Sample answer:* The equation is easily factorable, so solve by factoring.

42. $x \approx 3.73$, $x \approx 0.27$; *Sample answer:* $a = 1$ and b is even, so solve by completing the square.

43. $x \approx 2.19$, $x \approx -1.94$; *Sample answer:* The equation cannot be factored and $a \neq 1$, so solve using the Quadratic Formula.

44. $x = 1$, $x = -7$; *Sample answer:* The left side is a perfect square trinomial, so solve using square roots.

79. **a.** $(x + 1), (x - 6); x^2 - 5x - 6 = 0;$
    $x, (x - 2); x^2 - 2x = 0;$
    $\left(x + \frac{1}{2}\right), (x - 5); x^2 - \frac{9}{2}x - \frac{5}{2} = 0$

b.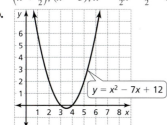

 3, 4

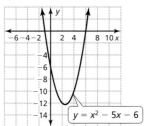

$-1, 6$

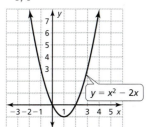

$0, 2$

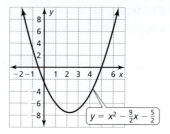

$-\frac{1}{2}, 5$

82. **a.** $k > -\frac{25}{64}$
    b. $k = -\frac{25}{64}$
    c. $k < -\frac{25}{64}$

4.5 Maintaining Mathematical Proficiency

83. $3x^2 - x + 2$
84. $x^3 + 4x^2 + 6$
85. $3x^2 - 3x + 1$
86. $-3x^3 + 7x^2 - 15x + 9$
87. $x^2 - 4$
88. $10x^3 - 2x^2 + 6x$
89. $-x^2 + 8x - 7$

4.6 Monitoring Progress and Modeling with Mathematics

63.

Powers of i	i^1	i^2	i^3	i^4	i^5	i^6	i^7	i^8	i^9	i^{10}	i^{11}	i^{12}
Simplified form	i	-1	$-i$	1	i	-1	$-i$	1	i	-1	$-i$	1

The results of i^n alternate in the pattern i, -1, $-i$, and 1.

4.7 Monitoring Progress and Modeling with Mathematics

22. *Sample answer:* $3x^2 - 6x + 6 = 0$
23. $\pm\sqrt{7}i$
24. $\pm 2\sqrt{2}$
25. $-4 \pm \sqrt{29}$
26. $\dfrac{-1 \pm 3i}{4}$
27. $5 \pm \sqrt{2}i$
28. $\dfrac{-9 \pm \sqrt{33}}{8}$
29. *Sample answer:* $a = 1$ and $c = 5$; $x^2 + 4x + 5 = 0$
30. *Sample answer:* $a = 2$ and $c = 4$; $2x^2 - 8x + 4 = 0$
31. *Sample answer:* $a = 5$ and $c = -5$; $5x^2 + 10x + 5 = 0$
32. *Sample answer:* $a = 4$ and $c = 5$; $4x^2 - 4x + 5 = 0$
38. functions f and g; function h; Functions f and g have real zeros because their graphs touch the x-axis. Function h has imaginary zeros because its graph does not touch the x-axis.
39. $-5x^2 + 8x - 12 = 0$
40. *Sample answer:* A basketball is pushed downward from a height of 3 feet at a speed of 6 feet per second; $h = -16t^2 - 6t + 3$; about 0.28 sec
41. **a.** $v_0 = 32\sqrt{10} \approx 101.19$ ft/sec
    b. about 3.2 sec; At $t = 2$, $h \approx 140$, so the rider is near the top and the graph is accurate.

Additional Answers **A29**

4.7 Maintaining Mathematical Proficiency

42. (4, 0); *Sample answer:* substitution because both equations are solved for y
43. (−4, 5); *Sample answer:* substitution because one equation is solved for x
44. (5, 3); *Sample answer:* elimination because one pair of like terms has the same coefficient
45. (−2, 7); *Sample answer:* substitution because one of the variables has a coefficient of 1
46. a. $x = 1$
    b. (1, 2)
47. a. $x = 0.25$
    b. (0.25, 2.875)
48. a. $x = -2$
    b. (−2, 3)

4.8 Monitoring Progress and Modeling with Mathematics

57. The possible number of solutions is 0, 1, or 2.

4.8 Maintaining Mathematical Proficiency

63.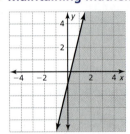

64.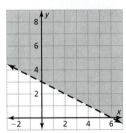

65.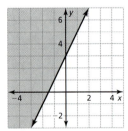

66.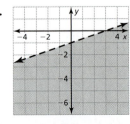

67.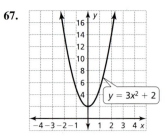

domain: all real numbers; range: $y \geq 2$

68.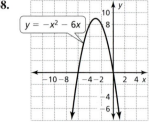

domain: all real numbers; range: $y \leq 9$

69.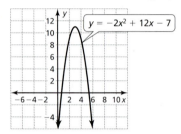

domain: all real numbers; range: $y \leq 11$

70.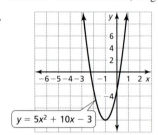

domain: all real numbers; range: $y \geq -8$

4.9 Lesson Monitoring Progress

1.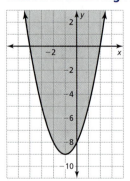

2.

3.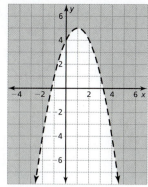

4.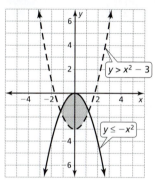

4.9 Monitoring Progress and Modeling with Mathematics

10.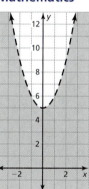

11.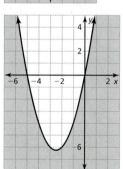

12.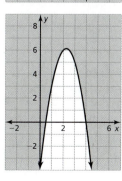

13.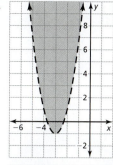

14.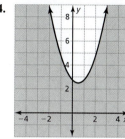

15. $y > f(x)$
16. $y > f(x)$
17. The graph should be solid, not dashed.

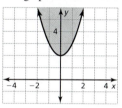

18. The wrong side of the parabola is shaded.

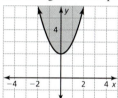

19. The solution represents weights that can be supported by shelves with various thicknesses.

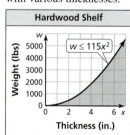

20. The solution represents weights that can be supported by wire ropes with various diameters.

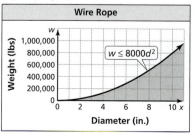

Additional Answers **A31**

21.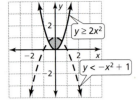

22.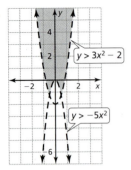

23.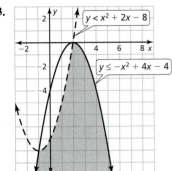

24.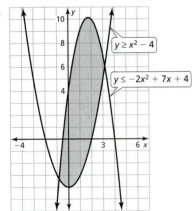

25.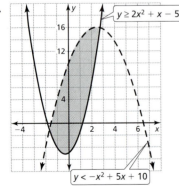

26.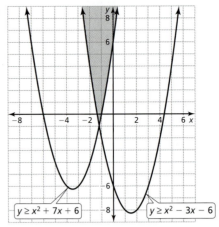

49. $0.00170x^2 + 0.145x + 2.35 > 10$, $0 \leq x \leq 40$; after about 37 days; Because $L(x)$ is a parabola, $L(x) = 10$ has two solutions. Because the x-value must be positive, the domain requires that the negative solution be rejected.

50. your friend; Any points with negative y-coordinates are solutions.

51. a. $\frac{32}{3} \approx 10.67$ square units

    b. $\frac{256}{3} \approx 85.33$ square units

52. Sample answer:

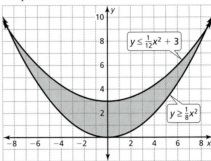

The intersection of $y \leq \frac{1}{12}x^2 + 3$ and $y \geq \frac{1}{8}x^2$ create a shape similar to a smile, which could be used by a company that sells toothpaste.

53. a. yes; The points on the parabola that are exactly 11 feet high are $(6, 11)$ and $(14, 11)$. Because these points are 8 feet apart, there is enough room for a 7-foot wide truck.

    b. 8 ft

    c. about 11.2 ft

4.9 Maintaining Mathematical Proficiency

54. maximum value; -1

55. maximum value; 25

56. minimum value; 0

57. minimum value; $-\frac{81}{4}$

58. $-7, 9$

59. $0, 4$

60. $2, 3$

Chapter 4 Review

56.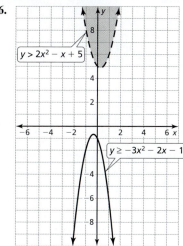

Chapter 5

5.1 Monitoring Progress and Modeling with Mathematics

30. $\frac{2}{3}$; $f(x) + c$ intersects the x-axis when c is 1, 2, 3, or 4.
31. $\frac{3}{400}$, or 0.75%; about 113; $(0.0075)15,000 = 112.5$
32. *Sample answer:* Box A contains three cards numbered 1, 2, and 3. Box B contains 2 cards numbered 1 and 2. One card is removed at random from each box. Find the probability that the product of the two numbers is at least 5; $\frac{1}{6}$

5.1 Maintaining Mathematical Proficiency

33. $3(x + 6)(x - 2)$
34. $(x - 1)(2x - 9)$
35. $(x + 3)(4x - 5)$
36. $x = -6, x = 6$
37. $x = -2, x = 2$
38. $x = -13, x = 5$
39. $x = \frac{7}{5}, x = \frac{13}{5}$

5.2 Monitoring Progress and Modeling with Mathematics

25. a. *Sample answer:* Put 20 pieces of paper with each of the 20 students' names in a hat and pick one; 5%
    b. *Sample answer:* Put 45 pieces of paper in a hat with each student's name appearing once for each hour the student worked. Pick one piece; about 8.9%
26. a. without
    b. with
27. yes; The chance that it will be rescheduled is $(0.7)(0.75) = 0.525$, which is greater than a 50% chance.
28. Event A represents rolling at least one 2. Event B represents the dice summing to 5; dependent; $P(A \text{ and } B) = \frac{2}{36}$ and $P(A)P(B) = \frac{11}{324}$
29. a. wins: 0%; loses: 1.99%; ties: 98.01%
    b. wins: 20.25%; loses: 30.25%; ties: 49.5%
    c. yes; Go for 2 points after the first touchdown, and then go for 1 point if they were successful the first time or 2 points if they were unsuccessful the first time; winning: 44.55%; losing: 30.25%
30. a. The occurrence of one event does not affect the occurrence of the other, so the probability of each event is the same whether or not the other event has occurred.
    b. yes; $P(A \text{ and } B) = P(A) \cdot P(B)$ and $P(A) = P(A \mid B)$.

5.2 Maintaining Mathematical Proficiency

31. $x = 0.2$
32. $x = 2$
33. $x = 0.15$

5.3 Explorations

1.

	Play an Instrument	Do Not Play an Instrument	Total
Speak a Foreign Language	16	30	46
Do Not Speak a Foreign Language	25	9	34
Total	41	39	80

   a. 41
   b. 46
   c. 16
   d. 9
   e. 25

3. *Sample answer:*

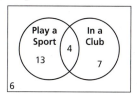

	Play a Sport	Do Not Play a Sport	Total
In a Club	4	7	11
Not in a Club	13	6	19
Total	17	13	30

Of the students in the class, 43.33% only play a sport, 13.33% play a sport and are in a club, 23.33% are only in a club, and 20% do neither activity. A total of 36.67% are in a club, and a total of 56.67% play a sport.

5.3 Lesson Monitoring Progress

3.

		Attendance	
		Attending	Not Attending
Class	Junior	0.353	0.634
	Senior	0.647	0.366

Given that a student is attending, the conditional relative frequency that he or she is a junior is about 35.3%. Given that a student is attending, the conditional relative frequency that he or she is a senior is about 64.7%. Given that a student is not attending, the conditional relative frequency that he or she is a junior is about 63.4%. Given that a student is not attending, the conditional relative frequency that he or she is a senior is about 36.6%.

4.

	Response	
Gender	In Favor	Against
Boy	0.636	0.364
Girl	0.808	0.192

Given that a student is a boy, the conditional relative frequency that he is in favor is about 63.6%. Given that a student is a girl, the conditional relative frequency that she is in favor is about 80.8%. Given that a student is a boy, the conditional relative frequency that he is against is about 36.4%. Given that a student is a girl, the conditional relative frequency that she is against is about 19.2%.

5.3 Monitoring Progress and Modeling with Mathematics

9.

	Gender		
Response	Male	Female	Total
Yes	0.376	0.430	0.806
No	0.111	0.083	0.194
Total	0.487	0.513	1

10.

	Vaccination		
Health	Received	Not Received	Total
Flu	0.1429	0.1518	0.2947
No Flu	0.2946	0.4107	0.7053
Total	0.4375	0.5625	1

19. *Sample answer:*

	Transportation to School			
Gender	Rides Bus	Walks	Car	Total
Male	6	9	4	19
Female	5	2	4	11
Total	11	11	8	30

	Transportation to School			
Gender	Rides Bus	Walks	Car	Total
Male	0.2	0.3	0.133	0.633
Female	0.167	0.067	0.133	0.367
Total	0.367	0.367	0.266	1

20. a. the parents surveyed that said no
   b. the total people that said yes
   c. the total people surveyed

21. Routine B is the best option, but your friend's reasoning of why is incorrect; Routine B is the best choice because there is a 66.7% chance of reaching the goal, which is higher than the chances of Routine A (62.5%) and Routine C (63.6%).

22.

	Preference		
Gender	Math	Science	Total
Male	93	57	150
Female	148	52	200
Total	241	109	350

23. a. about 0.438
   b. about 0.387

24. *Sample answer:* Venn diagrams show a visual representation of the data, and two-way tables organize the information into rows and columns; An advantage of a Venn diagram is that people who learn visually will easily understand them. A disadvantage is that as more categories are used, the Venn diagram becomes harder to draw and interpret. An advantage of a two-way table is that it is very easy to read and interpret, even with many categories. A disadvantage is that they are not as visual as Venn diagrams.

25. a. More of the current consumers prefer the leader, so they should improve the new snack before marketing it.
   b. More of the new consumers prefer the new snack than the leading snack, so there is no need to improve the snack.

26. *Sample answer:*

	Owns a Dog		
Gender	Yes	No	Total
Male	5	3	8
Female	7	5	12
Total	12	8	20

$$P(A \mid B) = \frac{P(B \mid A) \cdot P(A)}{P(B)}$$

$$P(\text{Male} \mid \text{yes}) = \frac{P(\text{yes} \mid \text{Male}) \cdot P(\text{Male})}{P(\text{yes})}$$

$$= \frac{\frac{5}{8} \cdot \frac{8}{20}}{\frac{12}{20}}$$

$$= \frac{5}{12}$$

5.3 Maintaining Mathematical Proficiency

27.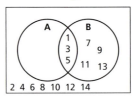

28.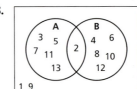

29.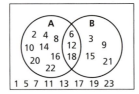

5.4 Monitoring Progress and Modeling with Mathematics

22. a. $P(A \text{ or } B \text{ or } C) = P(A) + P(B) + P(C)$

b. $P(A \text{ or } B \text{ or } C) = P(A) + P(B) + P(C) - P(A \text{ and } B)$
$- P(A \text{ and } C) - P(B \text{ and } C) + P(A \text{ and } B \text{ and } C)$

5.4 Maintaining Mathematical Proficiency

24.

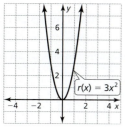

Both graphs open up and have the same vertex, (0, 0), and the same axis of symmetry, $x = 0$. The graph of r is narrower than the graph of f because the graph of r is a vertical stretch by a factor of 3 of the graph of f.

25.

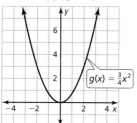

Both graphs open up and have the same vertex, (0, 0), and the same axis of symmetry, $x = 0$. The graph of g is wider than the graph of f because the graph of g is a vertical shrink by a factor of $\frac{3}{4}$ of the graph of f.

26.

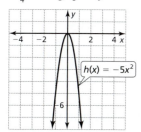

Both the graphs have the same vertex, (0, 0), and the same axis of symmetry, $x = 0$, but the graph of h opens down and is narrower than the graph of f, because the graph of h is a vertical stretch by a factor of 5 and a reflection in the x-axis of the graph of f.

5.5 Monitoring Progress and Modeling with Mathematics

42. a. $_nC_n = \dfrac{n!}{n!0!} = 1$

b. $_nC_{n-r} = \dfrac{n!}{(n-(n-r))!(n-r)!} = \dfrac{n!}{(r)!(n-r)!} = {_nC_r}$

c. $_nC_r + {_nC_{r-1}} = \dfrac{n!}{(n-r)!r!} + \dfrac{n!}{(n-r+1)!(r-1)!}$

$= \dfrac{n!(n-r+1) + n!r}{(n-r+1)!r!}$

$= \dfrac{n!n + n!}{(n-r+1)!r!}$

$= \dfrac{n!(n+1)}{(n-r+1)!r!}$

$= \dfrac{(n+1)!}{(n+1-r)!r!}$

$= {_{n+1}C_r}$

5.6 Monitoring Progress and Modeling with Mathematics

5.

w (value)	1	2
Outcomes	5	21
P(w)	$\dfrac{5}{26}$	$\dfrac{21}{26}$

Choosing a Letter

(Bar graph: P(w) vs Value; w=1 ≈ 0.2, w=2 ≈ 0.8)

6.

n (value)	1	2	3
Outcomes	10	90	900
P(n)	$\dfrac{1}{100}$	$\dfrac{9}{100}$	$\dfrac{9}{10}$

Number of Digits

(Bar graph: P(n) vs Number of digits)

14. a.

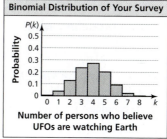

Binomial Distribution of Your Survey

Number of persons who believe UFOs are watching Earth

17. a. $P(0) \approx 0.099$, $P(1) \approx 0.271$, $P(2) \approx 0.319$, $P(3) \approx 0.208$, $P(4) \approx 0.081$, $P(5) \approx 0.019$, $P(6) \approx 0.0025$, $P(7) \approx 0.00014$

b.

x	0	1	2	3	4
P(x)	0.099	0.271	0.319	0.208	0.081

x	5	6	7
P(x)	0.019	0.0025	0.00014

c.

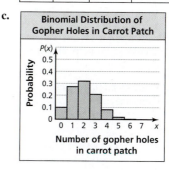

Binomial Distribution of Gopher Holes in Carrot Patch

Number of gopher holes in carrot patch

18. 0.2; 0.6
19. no; The data is skewed right, so the probability of failure is greater.
20. no; The probability of not choosing the coin 100 times is $\left(\frac{99}{100}\right)^{100} \approx 0.366$.
21. a. The statement is not valid, because having a male child and having a female child are independent events.
    b. 0.03125
    c.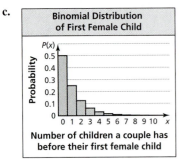
    skewed right
22. $p > 0.5$

5.6 Maintaining Mathematical Proficiency
23. FFF, FFM FMF, FMM, MMM, MMF, MFM, MFF
24. $D_1C_1, D_1C_2, D_1C_3, D_2C_1, D_2C_2, D_2C_3$

Chapter 5 Cumulative Assessment
2.
No zeros	$f(x) = 3x^2 + 4x + 2$	$f(x) = -6x^2 - 5$
One zero	$f(x) = 4x^2 - 8x + 4$	$f(x) = 7x^2$
Two zeros	$f(x) = -x^2 + 2x$	$f(x) = x^2 - 3x - 21$

Chapter 6

6.1 Monitoring Progress and Modeling with Mathematics
11. By the Distance Formula, $OB = \sqrt{a^2 + b^2}$, $BC = \sqrt{a^2 + b^2}$, $OD = \sqrt{a^2 + b^2}$, and $DC = \sqrt{a^2 + b^2}$. So, $\triangle OBC$ and $\triangle ODC$ are isosceles by definition.
12. By the Midpoint Formula, the coordinates of point X are $\left(\frac{r+q}{2}, \frac{s}{2}\right)$. By the Distance Formula, $XO = \sqrt{\frac{(r+q)^2 + s^2}{4}}$, $XZ = \sqrt{\frac{(r+q)^2 + s^2}{4}}$, $OW = a$ and $ZY = q$. By the definition of midpoint, $XY = XW$, so $\overline{XY} \cong \overline{XW}$. Because $XO = XZ$ and $OW = ZY, \overline{XO} \cong \overline{XZ}$ and $\overline{OW} \cong \overline{ZY}$. So, $\triangle XYZ \cong \triangle XWO$ by the SSS Congruence Theorem.
13. The wrong property is stated; $\overline{DE} \cong \overline{BE}$ by the Transitive Property of Congruence.
14. a.
STATEMENTS	REASONS
1. $\angle 2 \cong \angle 7$	1. Given
2. $\angle 7 \cong \angle 6$	2. Vertical Angles Congruence Theorem
3. $\angle 2 \cong \angle 6$	3. Transitive Property of Congruence
4. $m \parallel n$	4. Corresponding Angles Converse

b.
STATEMENTS	REASONS
1. $\angle 3$ and $\angle 5$ are supplementary	1. Given
2. $\angle 5$ and $\angle 6$ are supplementary	2. Linear Pair Postulate
3. $\angle 3 \cong \angle 6$	3. Congruent Supplements Theorem
4. $m \parallel n$	4. Alternate Interior Angles Converse

15. no; Sample answer: $\overrightarrow{BA}$ and $\overrightarrow{BE}$ are not opposite rays, so the angles are not vertical.

6.1 Mini-Assessment
3.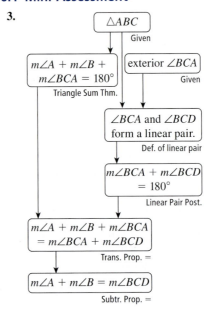

6.2 Lesson Monitoring Progress
1. 13.75
2. 27
3. 7.4

6.2 Monitoring Progress and Modeling with Mathematics
14. 16; $\overrightarrow{EG}$ is an angle bisector of $\angle FEH, \overline{FG} \perp \overline{EF}$, and $\overline{GH} \perp \overline{EH}$. So, by the Converse of the Angle Bisector Theorem, $FG = GH$. This means that $x + 11 = 3x + 1$, and the solution is $x = 5$. So, $FG = x + 11 = 5 + 11 = 16$.

33. a.

If $\overrightarrow{AD}$ bisects $\angle BAC$, then by definition of angle bisector, $\angle BAD \cong \angle CAD$. Also, because $\overline{DB} \perp \overline{AB}$ and $\overline{DC} \perp \overline{AC}$, by definition of perpendicular lines, $\angle ABD$ and $\angle ACD$ are right angles, and congruent to each other by the Right Angles Congruence Theorem. Also, $\overline{AD} \cong \overline{AD}$ by the Reflexive Property of Congruence. So, by the AAS Congruence Theorem, $\triangle ADB \cong \triangle ADC$. Because corresponding parts of congruent triangles are congruent, $DB = DC$. This means that point D is equidistant from each side of $\angle BAC$.

b.

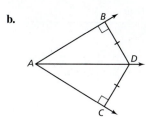

STATEMENTS	REASONS
1. $\overline{DC} \perp \overrightarrow{AC}, \overline{DB} \perp \overrightarrow{AB}, BD = CD$	1. Given
2. $\angle ABD$ and $\angle ACD$ are right angles.	2. Definition of perpendicular lines
3. $\triangle ABD$ and $\triangle ACD$ are right triangles.	3. Definition of a right triangle
4. $\overline{BD} \cong \overline{CD}$	4. Definition of congruent segments
5. $\overline{AD} \cong \overline{AD}$	5. Reflexive Property of Congruence
6. $\triangle ABD \cong \triangle ACD$	6. HL Congruence Theorem
7. $\angle BAD \cong \angle CAD$	7. Corresponding parts of congruent triangles are congruent.
8. $\overrightarrow{AD}$ bisects $\angle BAC$.	8. Definition of angle bisector

34. **a.** Roosevelt School; Because the corner of Main and 3rd is exactly 2 blocks of the same length from each hospital, and the two streets are perpendicular, 3rd Street is the perpendicular bisector of the segment that connects the two hospitals. Because Roosevelt School is on 3rd Street, it is the same distance from both hospitals by the Perpendicular Bisector Theorem.

   b. no; Because the corner of Maple and 2nd Street is approximately the midpoint of the segment that connects Wilson School to Roosevelt School, and 2nd Street is perpendicular to Maple, 2nd Street is the perpendicular bisector of the segment connecting Wilson and Roosevelt schools. By the contrapositive of the Converse of the Perpendicular Bisector Theorem, the Museum is not equidistant from the two schools because it is not on 2nd Street.

35. **a.** $y = x$
    b. $y = -x$
    c. $y = |x|$

36. no; In spherical geometry, all intersecting lines meet in two points, which are equidistant from each other because they are the two endpoints of a diameter of the circle.

37. Because $\overline{AD} \cong \overline{CD}$ and $\overline{AE} \cong \overline{CE}$, by the Converse of the Perpendicular Bisector Theorem, both points D and E are on the perpendicular bisector of $\overline{AC}$. So, $\overleftrightarrow{DE}$ is the perpendicular bisector of $\overline{AC}$. So, if $\overline{AB} \cong \overline{CB}$, then by the Converse of the Perpendicular Bisector Theorem, point B is also on $\overleftrightarrow{DE}$. So, points D, E, and B are collinear. Conversely, if points D, E, and B are collinear, then by the Perpendicular Bisector Theorem, point B is also on the perpendicular bisector of $\overline{AC}$. So, $\overline{AB} \cong \overline{CB}$.

38. **a.** Because $\overline{YW}$ is on plane P, and plane P is a perpendicular bisector of $\overline{XZ}$ at point Y, $\overline{YW}$ is a perpendicular bisector of $\overline{XZ}$ by definition of a plane perpendicular to a line. So, by the Perpendicular Bisector Theorem, $\overline{XW} \cong \overline{ZW}$.

    b. Because $\overline{YV}$ is on plane P, and plane P is a perpendicular bisector of $\overline{XZ}$ at point Y, $\overline{YV}$ is a perpendicular bisector of $\overline{XZ}$ by definition of a plane perpendicular to a line. So, by the Perpendicular Bisector Theorem, $\overline{XV} \cong \overline{ZV}$.

    c. First, $\overline{WV} \cong \overline{WV}$ by the Reflexive Property of Congruence. Then, because $\overline{XW} \cong \overline{ZW}$ and $\overline{XV} \cong \overline{ZV}$, $\triangle WVX \cong \triangle WVZ$ by the SSS Congruence Theorem. So, $\angle VXW \cong \angle VZW$ because corresponding parts of congruent triangles are congruent.

6.2 Maintaining Mathematical Proficiency

39. isosceles
40. scalene
41. equilateral
42. acute
43. right
44. obtuse

6.3 Monitoring Progress and Modeling with Mathematics

23. Sample answer:

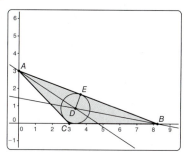

24. Sample answer:

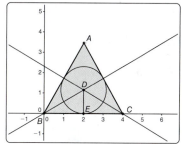

25. Because point G is the intersection of the angle bisectors, it is the incenter. But, because $\overline{GD}$ and $\overline{GF}$ are not necessarily perpendicular to a side of the triangle, there is not sufficient evidence to conclude that $\overline{GD}$ and $\overline{GF}$ are congruent; Point G is equidistant from the sides of the triangle.

26. Because point T is the intersection of the perpendicular bisectors, it is the circumcenter and is equidistant from the vertices of the triangle, not necessarily the sides; $TU = TW = TY$.

27. You could copy the positions of the three houses, and connect the points to draw a triangle. Then draw the three perpendicular bisectors of the triangle. The point where the perpendicular bisectors meet, the circumcenter, should be the location of the meeting place.

28. You should place the fountain at the incenter of the pond because the incenter is equidistant from the sides of the triangle.

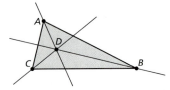

29. sometimes; If the scalene triangle is obtuse or right, then the circumcenter is outside or on the triangle, respectively. However, if the scalene triangle is acute, then the circumcenter is inside the triangle.

30. always; If the perpendicular bisector of one side of a triangle intersects the opposite vertex, then it divides the triangle into two congruent triangles. So, two sides of the original triangle are congruent because corresponding parts of congruent triangles are congruent.

31. sometimes; This only happens when the triangle is equilateral.

32. always; This is the Incenter Theorem.

42. b. (7, 7)

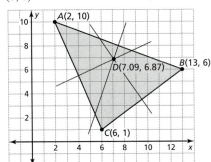

43. B

48. Sample answer:

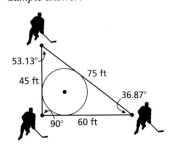

49. angle bisectors; about 2.83 in.

50. perpendicular bisectors; (10, 10); 10

51. $x = \dfrac{AB + AC - BC}{2}$ or $x = \dfrac{AB \cdot AC}{AB + AC + BC}$

6.3 Maintaining Mathematical Proficiency

52. $M(0, 5)$; $AB = 6$
53. $M(6, 3)$; $AB \approx 11.3$
54. $M(-0.5, -2)$; $AB \approx 10.8$
55. $M(-1, 7)$; $AB \approx 12.6$

56. $y = -\tfrac{1}{2}x + 9$

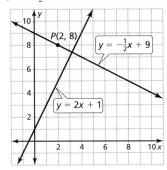

57. $x = 6$

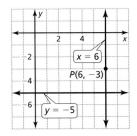

58. $y = \tfrac{3}{2}x + 6$

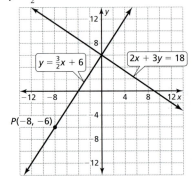

59. $y = \tfrac{1}{4}x + 2$

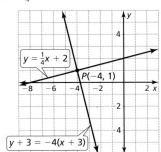

6.4 Monitoring Progress and Modeling with Mathematics

36. always; This is the definition of a centroid.

37. Both segments are perpendicular to a side of a triangle, and their point of intersection can fall either inside, on, or outside of the triangle. However, the altitude does not necessarily bisect the side, but the perpendicular bisector does. Also, the perpendicular bisector does not necessarily pass through the opposite vertex, but the altitude does.

38. All are segments that pass through the vertex of a triangle. A median connects a vertex with the midpoint of the opposite side. An altitude is perpendicular to the opposite side. An angle bisector bisects the angle through which it passes. The medians of a triangle intersect at a single point, and the same is true for the altitudes and angle bisectors of a triangle. Medians and angle bisectors always lie inside the triangle, but altitudes may be inside, on, or outside of the triangle.

39. 6.75 in.²; altitude

40. a. $EJ = 3KJ$
    b. $DK = 2KH$
    c. $FG = \frac{3}{2}FK$
    d. $KG = \frac{1}{3}FG$

41. $x = 2.5$
42. $x = 9$
43. $x = 4$
44. $x = 3$

45.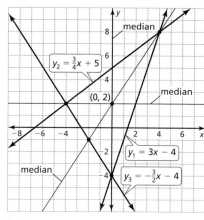
    (0, 2)

46. right triangle; The orthocenter of a right triangle is the vertex of the right angle.

51.

Sides $\overline{AB}$ and $\overline{BC}$ of equilateral $\triangle ABC$ are congruent. $\overline{AD} \cong \overline{CD}$ because $\overline{BD}$ is the median to $\overline{AC}$. Also, $\overline{BD} \cong \overline{BD}$ by the Reflexive Property of Congruence. So, $\triangle ABD \cong \triangle CBD$ by the SSS Congruence Theorem. $\angle ADB \cong \angle CDB$ and $\angle ABD \cong \angle CBD$ because corresponding parts of congruent triangles are congruent. Also, $\angle ADB$ and $\angle CDB$ are a linear pair. Because $\overline{BD}$ and $\overline{AC}$ intersect to form a linear pair of congruent angles, $\overline{BD} \perp \overline{AC}$. So, median $\overline{BD}$ is also an angle bisector, altitude, and perpendicular bisector of $\triangle ABC$.

52. Sample answer:

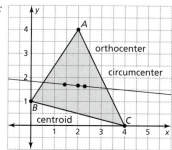

They are collinear.

53. Sample answer:

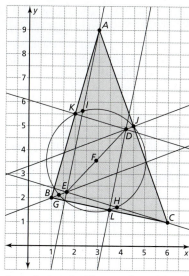

The circle passes through nine significant points of the triangle. They are the midpoints of the sides, the midpoints between each vertex and the orthocenter, and the points of intersection between the sides and the altitudes.

54. a.

STATEMENTS	REASONS
1. $\overline{LP}$ and $\overline{MQ}$ are medians of scalene $\triangle LMN$; $\overline{LP} \cong \overline{PR}$, $\overline{MQ} \cong \overline{QS}$	1. Given
2. $\overline{NP} \cong \overline{MP}$, $\overline{LQ} \cong \overline{NQ}$	2. Definition of median
3. $\angle LPM \cong \angle RPN$, $\angle MQL \cong \angle SQN$	3. Vertical Angles Congruence Theorem
4. $\triangle LPM \cong \triangle RPN$, $\triangle MQL \cong \triangle SQN$	4. SAS Congruence Theorem
5. $\overline{NR} \cong \overline{LM}$, $\overline{NS} \cong \overline{LM}$	5. Corresponding parts of congruent triangles are congruent.
6. $\overline{NS} \cong \overline{NR}$	6. Transitive Property of Congruence

b. It was shown in part (a) that $\triangle LPM \cong \triangle RPN$ and $\triangle MQL \cong \triangle SQN$. So, $\angle LMP \cong \angle RNP$ and $\angle MLQ \cong \angle SNQ$ because corresponding parts of congruent triangles are congruent. Then, $\overline{NS} \parallel \overline{LM}$ and $\overline{NR} \parallel \overline{LM}$ by the Alternate Interior Angles Converse.

c. Because $\overline{NS}$ and $\overline{NR}$ are both parallel to the same segment, $\overline{LM}$, they would have to be parallel to each other by the Transitive Property of Parallel Lines. However, because they intersect at point N, they cannot be parallel. So, they must be collinear.

6.5 Monitoring Progress and Modeling with Mathematics

22. The midpoint of $\overline{OC}$ is $F(p, 0)$. Because the slopes of $\overline{DF}$ and $\overline{BC}$ are the same $\left(-\dfrac{r}{p-q}\right)$, $\overline{DF} \parallel \overline{BC}$. $DF = \sqrt{p^2 - 2pq + q^2 + r^2}$ and $BC = 2\sqrt{p^2 - 2pq + q^2 + r^2}$. Because $\sqrt{p^2 - 2pq + q^2 + r^2} = \frac{1}{2}\left(2\sqrt{p^2 - 2pq + q^2 + r^2}\right)$, $DF = \frac{1}{2}BC$.

23. An eighth segment, $\overline{FG}$, would connect the midpoints of $\overline{DL}$ and $\overline{EN}$; $\overline{DE} \parallel \overline{LN} \parallel \overline{FG}$, $DE = \frac{3}{4}LN$, and $FG = \frac{7}{8}LN$; Because you are finding quarter segments and eighth segments, use $8p$, $8q$, and $8r$: $L(0, 0)$, $M(8q, 8r)$, and $N(8p, 0)$. Find the coordinates of X, Y, D, E, F, and G. $X(4q, 4r)$, $Y(4q + 4p, 4r)$, $D(2q, 2r)$, $E(2q + 6p, 2r)$, $F(q, r)$, and $G(q + 7p, r)$. The y-coordinates of D and E are the same, so $\overline{DE}$ has a slope of 0. The y-coordinates of F and G are also the same, so $\overline{FG}$ also has a slope of 0. $\overline{LM}$ is on the x-axis, so its slope is 0. Because their slopes are the same, $\overline{DE} \parallel \overline{LM} \parallel \overline{FG}$. Use the Ruler Postulate to find DE, FG, and LM. $DE = 6p$, $FG = 7p$, and $LN = 8p$. Because $6p = \frac{3}{4}(8p)$, $DE = \frac{3}{4}LN$. Because $7p = \frac{7}{8}(8p)$, $FG = \frac{7}{8}LN$.

24. Sample answer:

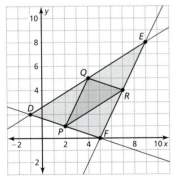

The crossbars on the ends of the swing set are midsegments.

25. a. 24 units
    b. 60 units
    c. 114 units

26. Two sides of the red triangle have a length of 4 tile widths. A yellow segment connects the midpoints, where there are two tile lengths on either side. The third red side has a length of 4 tile diagonals, and the other two yellow segments meet at the midpoint, where there are two tile diagonals on either side.

27. After graphing the midsegments, find the slope of each segment. Graph the line parallel to each midsegment passing through the opposite vertex. The intersections of these three lines will be the vertices of the original triangle: $(-1, 2)$, $(9, 8)$, and $(5, 0)$.

6.5 Maintaining Mathematical Proficiency

28. Sample answer: $-2 - (-7) = 5$, and $5 > -2$
29. Sample answer: An isosceles triangle whose sides are 5 centimeters, 5 centimeters, and 3 centimeters is not equilateral.

6.6 Explorations

1. c. Sample answer:

A(x, y)	B(x, y)	C(x, y)	AB	AC	BC
A(5, 1)	B(7, 4)	C(2, 4)	3.61	4.24	5
A(2, 4)	B(4, −2)	C(7, 6)	6.32	5.39	8.54
A(1, 0)	B(7, 0)	C(1, 7)	6	7	9.22

m∠A	m∠B	m∠C
78.69°	56.31°	45°
93.37°	38.99°	47.64°
90°	49.4°	40.6°

If one side of a triangle is longer than another side, then the angle opposite the longer side is larger than the angle opposite the shorter side. Similarly, if one angle of a triangle is larger than another angle, then the side opposite the larger angle is longer than the side opposite the smaller angle.

2. c. Sample answer:

A(x, y)	B(x, y)	C(x, y)	AB	AC + BC	AC
A(5, 1)	B(7, 4)	C(2, 4)	3.61	9.24	4.24
A(2, 4)	B(4, −2)	C(7, 6)	6.32	13.93	5.39
A(1, 0)	B(7, 0)	C(1, 7)	6	16.22	7
A(1, 0)	B(7, 0)	C(5, 1)	6	6.36	4.12

AB + BC	BC	AB + AC
8.61	5	7.85
14.86	8.54	11.71
15.22	9.22	13
8.24	2.24	10.12

The length of each side is less than the sum of the other two.

6.6 Monitoring Progress and Modeling with Mathematics

35. b. You could measure from distances that are closer together. In order to do this, you would have to use angle measures that are closer to 45°.

36. a. $x > 76$ km, $x < 1054$ km
    b. Because ∠2 is the smallest angle, the distance between Granite Peak and Fort Peck Lake must be the shortest side of the triangle. So, the second inequality becomes $x < 489$ kilometers.

37. ∠WXY, ∠Z, ∠YXZ, ∠WYX and ∠XYZ, ∠W; In △WXY, because $WY < WX < YX$, by the Triangle Longer Side Theorem, $m\angle WXY < m\angle WYX < m\angle W$. Similarly, in △XYZ, because $XY < YZ < XZ$, by the Triangle Longer Side Theorem, $m\angle Z < m\angle YXZ < m\angle XYZ$. Because $m\angle WYX = m\angle XYZ$ and ∠W is the only angle greater than either of them, you know that ∠W is the largest angle. Because △WXY has the largest angle and one of the congruent angles, the remaining angle, ∠WXY, is the smallest.

38. $EF < DF < DE$, $m\angle D < m\angle E < m\angle F$

42. The shortest route is along Washington Avenue. By the Triangle Inequality Theorem, the length of Washington Avenue must be shorter than the sum of the lengths of Eighth Street and View Street, as well as the sum of the lengths of Hill Street and Seventh Street.

43. It is given that $BC > AB$ and $BD = BA$. By the Base Angles Theorem, $m\angle 1 = m\angle 2$. By the Angle Addition Postulate, $m\angle BAC = m\angle 1 + m\angle 3$. So, $m\angle BAC > m\angle 1$. Substituting $m\angle 2$ for $m\angle 1$ produces $m\angle BAC > m\angle 2$. By the Exterior Angle Theorem, $m\angle 2 = m\angle 3 + m\angle C$. So, $m\angle 2 > m\angle C$. Finally, because $m\angle BAC > m\angle 2$ and $m\angle 2 > m\angle C$, you can conclude that $m\angle BAC > m\angle C$.

44. Because the sum of the lengths of the legs must be greater than the length of the base, the length of a leg must be greater than $\frac{1}{2}\ell$.

45. no; The sum of the other two sides would be 11 inches, which is less than 13 inches.

46. *Sample answer:*

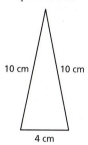

acute

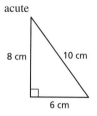

right

47.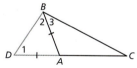

Assume $\overline{BC}$ is longer than or the same length as each of the other sides, $\overline{AB}$ and $\overline{AC}$. Then, $AB + BC > AC$ and $AC + BC > AB$. The proof for $AB + AC > BC$ follows.

STATEMENTS	REASONS
1. $\triangle ABC$	1. Given
2. Extend $\overline{AC}$ to D so that $\overline{AB} \cong \overline{AD}$.	2. Ruler Postulate
3. $AB = AD$	3. Definition of segment congruence
4. $AD + AC = DC$	4. Segment Addition Postulate
5. $\angle 1 \cong \angle 2$	5. Base Angles Theorem
6. $m\angle 1 = m\angle 2$	6. Definition of angle congruence
7. $m\angle DBC > m\angle 2$	7. Protractor Postulate
8. $m\angle DBC > m\angle 1$	8. Substitution Property
9. $DC > BC$	9. Triangle Larger Angle Theorem
10. $AD + AC > BC$	10. Substitution Property
11. $AB + AC > BC$	11. Substitution Property

48. The perimeter of $\triangle HGF$ must be greater than 4 and less than 24; Because of the Triangle Inequality Theorem, FG must be greater than 2 and less than 8, GH must be greater than 1 and less than 7, and FH must be greater than 1 and less than 9. So, the perimeter must be greater than $2 + 1 + 1 = 4$ and less than $8 + 7 + 9 = 24$.

49. Assume temporarily that another segment, $\overline{PA}$, where A is on plane M, is the shortest segment from P to plane M. By definition of the distance between a point and a plane, $\overline{PA} \perp$ plane M. This contradicts the given statement because there cannot be two different segments that share an endpoint and are both perpendicular to the same plane. So, the assumption is false, and because no other segment exists that is the shortest segment from P to plane M, it must be $\overline{PC}$ that is the shortest segment from P to plane M.

6.6 Maintaining Mathematical Proficiency

50. $\angle AEB$
51. $\angle ACD$
52. $\angle ADC$
53. $\angle CEB$

6.7 Explorations

1. f. *Sample answer:*

D	AC	BC	AB	BD	$m\angle ACB$	$m\angle BCD$
1. (4.75, 2.03)	2	3	3.61	2.68	90°	61.13°
2. (4.94, 2.5)	2	3	3.61	3.16	90°	75.6°
3. (5, 3)	2	3	3.61	3.61	90°	90°
4. (4.94, 3.5)	2	3	3.61	4	90°	104.45°
5. (3.85, 4.81)	2	3	3.61	4.89	90°	154.93°

6.7 Monitoring Progress and Modeling with Mathematics

18. The angle bisector of ∠FEG will also pass through incenter H. Then, $m\angle HEG + m\angle HFG + m\angle HGF = \frac{180°}{2} = 90°$, because they are each half of the measure of an angle of a triangle. By subtracting $m\angle HEG$ from each side, you can conclude that $m\angle HFG + m\angle HGF < 90°$. Also, $m\angle FHG + m\angle HFG + m\angle HGF = 180°$ by the Triangle Sum Theorem. So, $m\angle FHG > 90°$, which means that $m\angle FHG > m\angle HFG$ and $m\angle FHG > m\angle HGF$. So, $FG > FH$ and $FG > HG$.

19. Because $\overline{NR}$ is a median, $\overline{PR} \cong \overline{QR}$. $\overline{NR} \cong \overline{NR}$ by the Reflexive Property of Congruence. So, by the Converse of the Hinge Theorem, ∠NRQ > ∠NRP. Because ∠NRQ and ∠NRP form a linear pair, they are supplementary. So, ∠NRQ must be obtuse and ∠NRP must be acute.

20. $x > \frac{1}{2}$

21. $x > \frac{3}{2}$

22. $m\angle ACB < m\angle ACD$

23. △ABC is an obtuse triangle; If the altitudes intersect inside the triangle, then $m\angle BAC$ will always be less than $m\angle BDC$ because they both intercept the same segment, $\overline{CD}$. However, because $m\angle BAC > m\angle BDC$, ∠A must be obtuse, and the altitudes must intersect outside of the triangle.

24. The sum of the measures of the angles of a triangle in spherical geometry must be greater than 180°; The area of spherical $\triangle ABC = \frac{\pi r^2}{180°}(m\angle A + m\angle B + m\angle C - 180°)$, where r is the radius of the sphere.

Chapter 6 Test

12.

STATEMENTS	REASONS
1. △ABC is a right triangle.	1. Given
2. ∠C is a right angle.	2. Given (marked in diagram)
3. $m\angle C = 90°$	3. Definition of a right angle
4. $m\angle A + m\angle B + m\angle C = 180°$	4. Triangle Sum Theorem
5. $m\angle A + m\angle B + 90° = 180°$	5. Subtraction Property of Equality
6. $m\angle A + m\angle B = 90°$	6. Substitution Property of Equality
7. ∠A and ∠B are complementary.	7. Definition of complementary angles

15.

Chapter 7

7.1 Monitoring Progress and Modeling with Mathematics

47. $(n - 2) \cdot 180°$; When diagonals are drawn from the vertex of the concave angle as shown, the polygon is divided into $n - 2$ triangles whose interior angle measures have the same total as the sum of the interior angle measures of the original polygon.

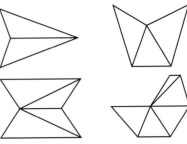

48. 90°; The base angles of △BPC are congruent exterior angles of the regular octagon, each with a measure of 45°. So, $m\angle BPC = 180° - 2(45°) = 90°$.

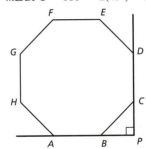

49.
   a. $h(n) = \frac{(n - 2) \cdot 180°}{n}$

   b. $h(9) = 140°$

   c. $n = 12$

   d.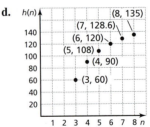

   The values of $h(n)$ increase by lesser amounts as n increases.

50. no; The interior angles are supplements of the adjacent exterior angles, and because the exterior angles have different values, the supplements will be different as well.

51. In a convex n-gon, the sum of the measures of the n interior angles is $(n - 2) \cdot 180°$ using the Polygon Interior Angles Theorem. Because each of the n interior angles forms a linear pair with its corresponding exterior angle, you know that the sum of the measures of the n interior and exterior angles is $180n°$. Subtracting the sum of the interior angle measures from the sum of the measures of the linear pairs gives you $180n° - [(n - 2) \cdot 180°] = 360°$.

52. 3; In order to have $\frac{540°}{180°} = 3$ more triangles formed by the diagonals, the new polygon will need 3 more sides.

7.1 Maintaining Mathematical Proficiency
53. $x = 101$
54. $x = 67$
55. $x = 16$
56. $x = 21$

7.2 Lesson Monitoring Progress

4.

STATEMENTS	REASONS
1. $ABCD$ and $GDEF$ are parallelograms.	1. Given
2. $\angle C$ and $\angle D$ are supplementary angles.	2. Parallelogram Consecutive Angles Theorem
3. $m\angle C + m\angle D = 180°$	3. Definition of supplementary angles
4. $\angle D \cong \angle F$	4. Parallelogram Opposite Angles Theorem
5. $m\angle D = m\angle F$	5. Definition of congruent angles
6. $m\angle C + m\angle F = 180°$	6. Substitution Property of Equality
7. $\angle C$ and $\angle F$ are supplementary angles.	7. Definition of supplementary angles

7.2 Monitoring Progress and Modeling with Mathematics

23.

STATEMENTS	REASONS
1. $ABCD$ and $CEFD$ are parallelograms.	1. Given
2. $\overline{AB} \cong \overline{DC}, \overline{DC} \cong \overline{FE}$	2. Parallelogram Opposite Sides Theorem
3. $\overline{AB} \cong \overline{FE}$	3. Transitive Property of Congruence

24.

STATEMENTS	REASONS
1. $ABCD$, $EBGF$, and $HJKD$ are parallelograms.	1. Given
2. $\angle 1 \cong \angle 2, \angle 3 \cong \angle 4, \angle 1 \cong \angle 4$	2. Parallelogram Opposite Angles Theorem
3. $\angle 2 \cong \angle 4$	3. Transitive Property of Congruence
4. $\angle 2 \cong \angle 3$	4. Transitive Property of Congruence

36. Sample answer:

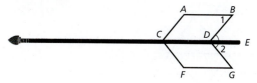

$m\angle 1 = m\angle G$ because corresponding parts of congruent figures are congruent. Note that $\overline{AB} \parallel \overleftrightarrow{CD} \parallel \overline{FG}$. So, $m\angle 1 = m\angle BDE$ and $m\angle GDE = m\angle G$ because they are pairs of alternate interior angles, and $m\angle 1 = m\angle GDE$ by the Transitive Property of Equality. Also, by the Angle Addition Postulate, $m\angle 2 = m\angle BDE + m\angle GDE$. By substituting, you get $m\angle 2 = m\angle 1 + m\angle 1 = 2m\angle 1$.

37.

STATEMENTS	REASONS
1. $ABCD$ is a parallelogram.	1. Given
2. $\overline{AB} \parallel \overline{DC}, \overline{BC} \parallel \overline{AD}$	2. Definition of parallelogram
3. $\angle BDA \cong \angle DBC$, $\angle DBA \cong \angle BDC$	3. Alternate Interior Angles Theorem
4. $\overline{BD} \cong \overline{BD}$	4. Reflexive Property of Congruence
5. $\triangle ABD \cong \triangle CDB$	5. ASA Congruence Theorem
6. $\angle A \cong \angle C, \angle B \cong \angle D$	6. Corresponding parts of congruent triangles are congruent.

38.

STATEMENTS	REASONS
1. $PQRS$ is a parallelogram.	1. Given
2. $\overline{QR} \parallel \overline{PS}$	2. Definition of parallelogram
3. $\angle Q$ and $\angle P$ are supplementary.	3. Consecutive Interior Angles Theorem
4. $x° + y° = 180°$	4. Definition of supplementary angles

39.

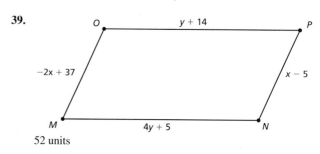

52 units

44. yes;

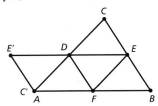

Any triangle, such as △ABC, can be partitioned into four congruent triangles by drawing the midsegment triangle, such as △DEF. Then, one triangle, such as △CDE, can be rotated 180° about a vertex, such as D, to create a parallelogram as shown.

45. 3; (4, 0), (−2, 4), (8, 8)

46.

STATEMENTS	REASONS
1. $\overline{EK}$ bisects ∠FEH and $\overline{FJ}$ bisects ∠EFG. EFGH is a parallelogram.	1. Given
2. m∠PEH = m∠PEF, m∠PFE = m∠PFG	2. Definition of angle bisector
3. m∠HEF = m∠PEH + m∠PEF, m∠EFG = m∠PFE + m∠PFG	3. Angle Addition Postulate
4. m∠HEF = m∠PEF + m∠PEF, m∠EFG = m∠PFE + m∠PFE	4. Substitution Property of Equality
5. m∠HEF = 2(m∠PEF), m∠EFG = 2(m∠PFE)	5. Distributive Property
6. m∠HEF + m∠EFG = 180°	6. Parallelogram Consecutive Angles Theorem
7. 2(m∠PEF) + 2(m∠PFE) = 180°	7. Substitution Property of Equality
8. 2(m∠PEF + m∠PFE) = 180°	8. Distributive Property
9. m∠PEF + m∠PFE = 90°	9. Division Property of Equality
10. m∠PEF + m∠PFE + m∠EPF = 180°	10. Triangle Sum Theorem
11. 90° + m∠EPF = 180°	11. Substitution Property of Equality
12. m∠EPF = 90°	12. Subtraction Property of Equality
13. ∠EPF is a right angle.	13. Definition of right angle
14. $\overline{EK} \perp \overline{FJ}$	14. Definition of perpendicular lines

47.

STATEMENTS	REASONS
1. $\overleftrightarrow{GH} \parallel \overleftrightarrow{JK} \parallel \overleftrightarrow{LM}$, $\overline{GJ} \cong \overline{JL}$	1. Given
2. Construct $\overline{PK}$ and $\overline{QM}$ such that $\overleftrightarrow{PK} \parallel \overleftrightarrow{GL} \parallel \overleftrightarrow{QM}$.	2. Construction
3. GPKJ and JQML are parallelograms.	3. Definition of parallelogram
4. ∠GHK ≅ ∠JKM, ∠PKQ ≅ ∠QML	4. Corresponding Angles Theorem
5. $\overline{GJ} \cong \overline{PK}$, $\overline{JL} \cong \overline{QM}$	5. Parallelogram Opposite Sides Theorem
6. $\overline{PK} \cong \overline{QM}$	6. Transitive Property of Congruence
7. ∠HPK ≅ ∠PKQ, ∠KQM ≅ ∠QML	7. Alternate Interior Angles Theorem
8. ∠HPK ≅ ∠QML	8. Transitive Property of Congruence
9. ∠HPK ≅ ∠KQM	9. Transitive Property of Congruence
10. △PHK ≅ △QKM	10. AAS Congruence Theorem
11. $\overline{HK} \cong \overline{KM}$	11. Corresponding sides of congruent triangles are congruent.

7.2 Maintaining Mathematical Proficiency

48. yes; Alternate Interior Angles Converse

49. yes; Alternate Exterior Angles Converse

50. no; By the Consecutive Interior Angles Converse, consecutive interior angles need to be supplementary for the lines to be parallel, and the consecutive interior angles are not supplementary.

7.3 Monitoring Progress and Modeling with Mathematics

18.

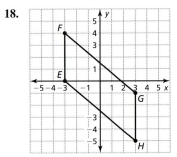

Because $EF = GH = 4$, $\overline{EF} \cong \overline{GH}$. Because both $\overline{EF}$ and $\overline{GH}$ are vertical lines with undefined slope, they are parallel. $\overline{EF}$ and $\overline{GH}$ are opposite sides that are both congruent and parallel. So, EFGH is a parallelogram by the Opposite Sides Parallel and Congruent Theorem.

19.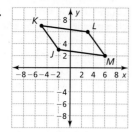

Because $JK = LM = 5$ and $KL = JM = \sqrt{65}$, $\overline{JK} \cong \overline{LM}$ and $\overline{KL} \cong \overline{JM}$. Because both pairs of opposite sides are congruent, quadrilateral $JKLM$ is a parallelogram by the Parallelogram Opposite Sides Converse.

20.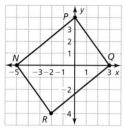

Because the slopes of $\overline{NP}$ and $\overline{QR}$ are both $\frac{4}{5}$, they are parallel. Because the slopes of $\overline{PQ}$ and $\overline{NR}$ are both $-\frac{4}{3}$, they are parallel. Because both pairs of opposite sides are parallel, $NPQR$ is a parallelogram by definition.

39.

STATEMENTS	REASONS
1. $\angle A \cong \angle C$, $\angle B \cong \angle D$	1. Given
2. Let $m\angle A = m\angle C = x°$ and $m\angle B = m\angle D = y°$	2. Definition of congruent angles
3. $m\angle A + m\angle B + m\angle C + m\angle D = x° + y° + x° + y° = 360°$	3. Corollary to the Polygon Interior Angles Theorem
4. $2(x°) + 2(y°) = 360°$	4. Simplify.
5. $2(x° + y°) = 360°$	5. Distributive Property
6. $x° + y° = 180°$	6. Division Property of Equality
7. $m\angle A + m\angle B = 180°$, $m\angle A + m\angle D = 180°$	7. Substitution Property of Equality
8. $\angle A$ and $\angle B$ are supplementary. $\angle A$ and $\angle D$ are supplementary.	8. Definition of supplementary angles
9. $\overline{BC} \parallel \overline{AD}$, $\overline{AB} \parallel \overline{DC}$	9. Consecutive Interior Angles Converse
10. $ABCD$ is a parallelogram.	10. Definition of parallelogram

40.

STATEMENTS	REASONS
1. $\overline{QR} \parallel \overline{PS}$, $\overline{QR} \cong \overline{PS}$	1. Given
2. $\angle SQR \cong \angle QSP$	2. Alternate Interior Angles Theorem
3. $\overline{QS} \cong \overline{QS}$	3. Reflexive Property of Congruence
4. $\triangle QRS \cong \triangle SPQ$	4. SAS Congruence Theorem
5. $\angle QSR \cong \angle SQP$	5. Corresponding parts of congruent triangles are congruent.
6. $\overline{QP} \parallel \overline{RS}$	6. Alternate Interior Angles Converse
7. $PQRS$ is a parallelogram.	7. Definition of parallelogram

41.

STATEMENTS	REASONS
1. Diagonals $\overline{JL}$ and $\overline{KM}$ bisect each other.	1. Given
2. $\overline{KP} \cong \overline{MP}$, $\overline{JP} \cong \overline{LP}$	2. Definition of segment bisector
3. $\angle KPL \cong \angle MPJ$	3. Reflexive Property of Congruence
4. $\triangle KPL \cong \triangle MPJ$	4. SAS Congruence Theorem
5. $\angle MKL \cong \angle KMJ$, $\overline{KL} \cong \overline{MJ}$	5. Corresponding parts of congruent triangles are congruent.
6. $\overline{KL} \parallel \overline{MJ}$	6. Alternate Interior Angles Converse
7. $JKLM$ is a parallelogram.	7. Opposite Sides Parallel and Congruent Theorem (Thm. 7.9)

42.

STATEMENTS	REASONS
1. $DEBF$ is a parallelogram, $AE = CF$	1. Given
2. $\overline{DE} \cong \overline{BF}, \overline{FD} \cong \overline{EB}$	2. Parallelogram Opposite Sides Theorem
3. $\angle DFB \cong \angle DEB$	3. Parallelogram Opposite Angles Theorem
4. $\angle AED$ and $\angle DEB$ form a linear pair. $\angle CFB$ and $\angle DFB$ form a linear pair.	4. Definition of linear pair
5. $\angle AED$ and $\angle DEB$ are supplementary. $\angle CFB$ and $\angle DFB$ are supplementary.	5. Linear Pair Postulate
6. $\angle AED \cong \angle CFB$	6. Congruent Supplements Theorem
7. $\overline{AE} \cong \overline{CF}$	7. Definition of congruent segments
8. $\triangle AED \cong \triangle CFB$	8. SAS Congruence Theorem
9. $\overline{AD} \cong \overline{CB}$	9. Corresponding parts of congruent triangles are congruent.
10. $AB = AE + EB$, $DC = CF + FD$	10. Segment Addition Postulate
11. $FD = EB$	11. Definition of congruent segments
12. $AB = CF + FD$	12. Substitution Property of Equality
13. $AB = DC$	13. Transitive Property of Equality
14. $\overline{AB} \cong \overline{DC}$	14. Definition of congruent segments
15. $ABCD$ is a parallelogram.	15. Parallelogram Opposite Sides Converse

45. 8; By the Parallelogram Opposite Sides Theorem, $\overline{AB} \cong \overline{CD}$. Also, $\angle ABE$ and $\angle CDF$ are congruent alternate interior angles of parallel segments $\overline{AB}$ and $\overline{CD}$. Then, you can use the Segment Addition Postulate, the Substitution Property of Equality, and the Reflexive Property of Congruence to show that $\overline{DF} \cong \overline{BE}$. So, $\triangle ABE \cong \triangle CDF$ by the SAS Congruence Theorem, which means that $AE = CF = 8$ because corresponding parts of congruent triangles are congruent.

46. Sample answer:

47. If every pair of consecutive angles of a quadrilateral is supplementary, then the quadrilateral is a parallelogram; In $ABCD$, you are given that $\angle A$ and $\angle B$ are supplementary, and $\angle B$ and $\angle C$ are supplementary. So, $m\angle A = m\angle C$. Also, $\angle B$ and $\angle C$ are supplementary, and $\angle C$ and $\angle D$ are supplementary. So, $m\angle B = m\angle D$. So, $ABCD$ is a parallelogram by the Parallelogram Opposite Angles Converse.

48. By the definition of a right angle, $m\angle A = 90°$. Because $ABCD$ is a parallelogram, and opposite angles of a parallelogram are congruent, $m\angle A = m\angle C = 90°$. Because consecutive angles of a parallelogram are supplementary, $\angle C$ and $\angle B$ are supplementary, and $\angle C$ and $\angle D$ are supplementary. So, $90° + m\angle B = 180°$ and $90° + m\angle D = 180°$. This gives you $m\angle B = m\angle D = 90°$. So, $\angle B$, $\angle C$, and $\angle D$ are right angles.

49. Given quadrilateral $ABCD$ with midpoints E, F, G, and H that are joined to form a quadrilateral, you can construct diagonal $\overline{BD}$. Then $\overline{FG}$ is a midsegment of $\triangle BCD$, and $\overline{EH}$ is a midsegment of $\triangle DAB$. So, by the Triangle Midsegment Theorem, $\overline{FG} \parallel \overline{BD}$, $FG = \frac{1}{2}BD$, $\overline{EH} \parallel \overline{BD}$, and $EH = \frac{1}{2}BD$.
So, by the Transitive Property of Parallel Lines, $\overline{EH} \parallel \overline{FG}$ and by the Transitive Property of Equality, $EH = FG$. Because one pair of opposite sides is both congruent and parallel, $EFGH$ is a parallelogram by the Opposite Sides Parallel and Congruent Theorem.

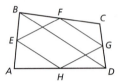

50. Based on the given information, $\overline{GH}$ is a midsegment of $\triangle EBC$, and $\overline{FJ}$ is a midsegment of $\triangle EAD$. So, by the Triangle Midsegment Theorem, $\overline{GH} \parallel \overline{BC}$, $GH = \frac{1}{2}BC$, $\overline{FJ} \parallel \overline{AD}$, and $FJ = \frac{1}{2}AD$. Also, by the Parallelogram Opposite Sides Theorem and the definition of a parallelogram, $\overline{BC}$ and $\overline{AD}$ are congruent and parallel. So, by the Transitive Property of Parallel Lines, $\overline{AD} \parallel \overline{FJ} \parallel \overline{GH} \parallel \overline{BC}$ and by the Transitive Property of Equality, $\frac{1}{2}BC = GH = FJ = \frac{1}{2}AD$. Because one pair of opposite sides is both congruent and parallel, $FGHJ$ is a parallelogram by the Opposite Sides Parallel and Congruent Theorem.

7.3 Maintaining Mathematical Proficiency

51. parallelogram
52. rectangle
53. square
54. rhombus

7.1–7.3 Quiz

22. a. $\overline{JK} \cong \overline{ML}$ by the Parallelogram Opposite Sides Theorem, $\overline{JM} \cong \overline{KL}$ by the Parallelogram Opposite Sides Theorem, $\angle J \cong \angle KLM$ by the Parallelogram Opposite Angles Theorem, $\angle M \cong \angle JKL$ by the Parallelogram Opposite Angles Theorem

b. Opposite Sides Parallel and Congruent Theorem

c. 3 ft, 123°, 57°, 57°; $ST = QR$ by the Parallelogram Opposite Sides Theorem, $m\angle QTS = m\angle QRS$ by the Parallelogram Opposite Angles Theorem, $m\angle TQR = m\angle TSR = 180° - 123°$ by the Parallelogram Consecutive Angles Theorem

7.4 Monitoring Progress and Modeling with Mathematics

17. always; All angles of a rectangle are congruent.

18. always; Opposite sides are congruent.

19. sometimes; Some rectangles are squares.

20. always; The diagonals of a rectangle are congruent.

21. sometimes; Some rectangles are squares.

22. sometimes; Some rectangles are squares.

23. no; All four angles are not congruent.
24. yes; Opposite sides are congruent and the angles are 90°.
25. 11
26. 66
27. 4
28. 8.4

73.

STATEMENTS	REASONS
1. $PQRS$ is a parallelogram. $\overline{PR}$ bisects $\angle SPQ$ and $\angle QRS$. $\overline{SQ}$ bisects $\angle PSR$ and $\angle RQP$.	1. Given
2. $\angle SRT \cong \angle QRT$, $\angle RQT \cong \angle RST$	2. Definition of angle bisector
3. $\overline{TR} \cong \overline{TR}$	3. Reflexive Property of Congruence
4. $\triangle QRT \cong \triangle SRT$	4. AAS Congruence Theorem
5. $\overline{QR} \cong \overline{SR}$	5. Corresponding parts of congruent triangles are congruent.
6. $\overline{QR} \cong \overline{PS}$, $\overline{PQ} \cong \overline{SR}$	6. Parallelogram Opposite Sides Theorem
7. $\overline{PS} \cong \overline{QR} \cong \overline{SR} \cong \overline{PQ}$	7. Transitive Property of Congruence
8. $PQRS$ is a rhombus.	8. Definition of rhombus

74.

STATEMENTS	REASONS
1. $WXYZ$ is a rhombus.	1. Given
2. $\overline{WX} \cong \overline{XY} \cong \overline{YZ} \cong \overline{WZ}$	2. Definition of a rhombus
3. $\overline{XV} \cong \overline{XV}, \overline{YV} \cong \overline{YV}, \overline{ZV} \cong \overline{ZV}$, and $\overline{WV} \cong \overline{WV}$	3. Reflexive Property of Congruence
4. $WXYZ$ is a parallelogram.	4. Definition of a rhombus
5. $\overline{XZ}$ bisects $\overline{WY}$. $\overline{WY}$ bisects $\overline{XZ}$.	5. Parallelogram Diagonals Theorem
6. $\overline{WV} \cong \overline{YV}, \overline{XV} \cong \overline{ZV}$	6. Definition of segment bisector
7. $\triangle WXV \cong \triangle YXV \cong \triangle YZV \cong \triangle WZV$	7. SSS Congruence Theorem
8. $\angle WXV \cong \angle YXV$, $\angle XYV \cong \angle ZYV$, $\angle YZV \cong \angle WZV$, $\angle ZWV \cong \angle XWV$	8. Corresponding parts of congruent triangles are congruent.
9. $\overline{WY}$ bisects $\angle ZWX$ and $\angle XYZ$. $\overline{ZX}$ bisects $\angle WZY$ and $\angle YXW$.	9. Definition of angle bisector

81. If a quadrilateral is a rhombus, then it has four congruent sides; If a quadrilateral has four congruent sides, then it is a rhombus; The conditional statement is true by the definition of rhombus. The converse is true because if a quadrilateral has four congruent sides, then both pairs of opposite sides are congruent. So, by the Parallelogram Opposite Sides Converse, it is a parallelogram with four congruent sides, which is the definition of a rhombus.

82. If a quadrilateral is a rectangle, then it has four right angles; If a quadrilateral has four right angles, then it is a rectangle; The conditional statement is true by definition of rectangle. The converse is true because if a quadrilateral has four right angles, then both pairs of opposite angles are congruent. So, by the Parallelogram Opposite Angles Converse, it is a parallelogram with four right angles, which is the definition of a rectangle.

83. If a quadrilateral is a square, then it is a rhombus and a rectangle; If a quadrilateral is a rhombus and a rectangle, then it is a square; The conditional statement is true because if a quadrilateral is a square, then by definition of a square, it has four congruent sides, which makes it a rhombus by the Rhombus Corollary, and it has four right angles, which makes it a rectangle by the Rectangle Corollary; The converse is true because if a quadrilateral is a rhombus and a rectangle, then by the Rhombus Corollary, it has four congruent sides, and by the Rectangle Corollary, it has four right angles. So, by the definition, it is a square.

85.

STATEMENTS	REASONS
1. $\triangle XYZ \cong \triangle XWZ$, $\angle XYW \cong \angle ZWY$	1. Given
2. $\angle YXZ \cong \angle WXZ$, $\angle YZX \cong \angle WZX$, $\overline{XY} \cong \overline{XW}$, $\overline{YZ} \cong \overline{WZ}$	2. Corresponding parts of congruent triangles are congruent.
3. $\overline{XZ}$ bisects $\angle WXY$ and $\angle WZY$.	3. Definition of angle bisector
4. $\angle XWY \cong \angle XYW$, $\angle WYZ \cong \angle ZWY$	4. Base Angles Theorem
5. $\angle XYW \cong \angle WYZ$, $\angle XWY \cong \angle ZWY$	5. Transitive Property of Congruence
6. $\overline{WY}$ bisects $\angle XWZ$ and $\angle XYZ$.	6. Definition of angle bisector
7. WXYZ is a rhombus.	7. Rhombus Opposite Angles Theorem

86.

STATEMENTS	REASONS
1. $\overline{BC} \cong \overline{AD}$, $\overline{BC} \perp \overline{DC}$, $\overline{AD} \perp \overline{DC}$	1. Given
2. $\overline{BC} \parallel \overline{AD}$	2. Lines Perpendicular to a Transversal Theorem
3. ABCD is a parallelogram.	3. Opposite Sides Parallel and Congruent Theorem
4. $m\angle DAB = m\angle BCD$, $m\angle ABC = m\angle ADC$	4. Parallelogram Opposite Angles Theorem
5. $m\angle BCD = m\angle ADC = 90°$	5. Definition of perpendicular lines
6. $m\angle DAB = m\angle BCD = m\angle ABC = m\angle ADC = 90°$	6. Transitive Property of Equality
7. ABCD has four right angles.	7. Definition of a right angle
8. ABCD is a rectangle.	8. Definition of a rectangle

87.

STATEMENTS	REASONS
1. PQRS is a rectangle.	1. Given
2. PQRS is a parallelogram.	2. Definition of a rectangle
3. $\overline{PS} \cong \overline{QR}$	3. Parallelogram Opposite Side Theorem
4. $\angle PQR$ and $\angle QPS$ are right angles.	4. Definition of a rectangle
5. $\angle PQR \cong \angle QPS$	5. Right Angles Congruence Theorem
6. $\overline{PQ} \cong \overline{PQ}$	6. Reflexive Property of Congruence
7. $\triangle PQR \cong \triangle QPS$	7. SAS Congruence Theorem
8. $\overline{PR} \cong \overline{SQ}$	8. Corresponding parts of congruent triangles are congruent.

88.

STATEMENTS	REASONS
1. PQRS is a parallelogram, $\overline{PR} \cong \overline{SQ}$	1. Given
2. $\overline{PS} \cong \overline{QR}$	2. Parallelogram Opposite Sides Theorem
3. $\overline{PQ} \cong \overline{PQ}$	3. Reflexive Property of Congruence
4. $\triangle PQR \cong \triangle QPS$	4. SSS Congruence Theorem
5. $\angle SPQ \cong \angle RQP$	5. Corresponding parts of congruent triangles are congruent.
6. $m\angle SPQ \cong m\angle RQP$	6. Definition of congruent angles
7. $m\angle SPQ + m\angle RQP = 180°$	7. Parallelogram Consecutive Angles Theorem
8. $2m\angle SPQ = 180°$ and $2m\angle RQP = 180°$	8. Substitution Property of Equality
9. $m\angle SPQ = 90°$ and $m\angle RQP = 90°$	9. Division Property of Equality
10. $m\angle RSP = 90°$ and $m\angle QRS = 90°$	10. Parallelogram Opposite Angles Theorem
11. $\angle SPQ, \angle RQP, \angle RSP,$ and $\angle QRS$ are right angles.	11. Definition of a right angle
12. PQRS is a rectangle.	12. Definition of a rectangle

7.5 Monitoring Progress and Modeling with Mathematics

35.

STATEMENTS	REASONS
1. $\overline{JL} \cong \overline{LN}$, $\overline{KM}$ is a midsegment of $\triangle JLN$.	1. Given
2. $\overline{KM} \parallel \overline{JN}$	2. Triangle Midsegment Theorem
3. $JKMN$ is a trapezoid.	3. Definition of trapezoid
4. $\angle LJN \cong \angle LNJ$	4. Base Angles Theorem
5. $JKMN$ is an isosceles trapezoid.	5. Isosceles Trapezoid Base Angles Converse

36.

STATEMENTS	REASONS
1. $ABCD$ is a kite, $\overline{AB} \cong \overline{CB}$, $\overline{AD} \cong \overline{CD}$	1. Given
2. $\overline{BD} \cong \overline{BD}$, $\overline{ED} \cong \overline{ED}$	2. Reflexive Property of Congruence
3. $\triangle BCD \cong \triangle BAD$	3. SSS Congruence Theorem
4. $\angle CDE \cong \angle ADE$	4. Corresponding parts of congruent triangles are congruent.
5. $\triangle CED \cong \triangle AED$	5. SAS Congruence Theorem
6. $\overline{CE} \cong \overline{AE}$	6. Corresponding parts of congruent triangles are congruent.

44.

STATEMENTS	REASONS
1. $QRST$ is an isosceles trapezoid.	1. Given
2. $\angle QTS \cong \angle RST$	2. Isosceles Trapezoid Base Angles Theorem
3. $\overline{QT} \cong \overline{RS}$	3. Definition of an isosceles trapezoid
4. $\overline{TS} \cong \overline{TS}$	4. Reflexive Property of Congruence
5. $\triangle QST \cong \triangle RTS$	5. SAS Congruence Theorem
6. $\angle TQS \cong \angle SRT$	6. Corresponding parts of congruent triangles are congruent.

47. Given kite $EFGH$ with $\overline{EF} \cong \overline{FG}$ and $\overline{EH} \cong \overline{GH}$, construct diagonal $\overline{FH}$, which is congruent to itself by the Reflexive Property of Congruence. So, $\triangle FGH \cong \triangle FEH$ by the SSS Congruence Theorem, and $\angle E \cong \angle G$ because corresponding parts of congruent triangles are congruent. Next, assume temporarily that $\angle F \cong \angle H$. Then $EFGH$ is a parallelogram by the Parallelogram Opposite Angles Converse, and opposite sides are congruent. However, this contradicts the definition of a kite, which says that opposite sides cannot be congruent. So, the assumption cannot be true and $\angle F$ is not congruent to $\angle H$.

50. no; A concave kite and a convex kite can have congruent corresponding sides and a pair of congruent corresponding angles, but the kites are not congruent.

51. a.

STATEMENTS	REASONS
1. $JKLM$ is an isosceles trapezoid, $\overline{KL} \parallel \overline{JM}$, $\overline{JK} \cong \overline{LM}$	1. Given
2. $\angle JKL \cong \angle MLK$	2. Isosceles Trapezoid Base Angles Theorem
3. $\overline{KL} \cong \overline{KL}$	3. Reflexive Property of Congruence
4. $\triangle JKL \cong \triangle MLK$	4. SAS Congruence Theorem
5. $\overline{JL} \cong \overline{KM}$	5. Corresponding parts of congruent triangles are congruent.

b. If the diagonals of a trapezoid are congruent, then the trapezoid is isosceles. Let $JKLM$ be a trapezoid, $\overline{KL} \parallel \overline{JM}$ and $\overline{JL} \cong \overline{KM}$. Construct line segments through K and L perpendicular to $\overline{JM}$ as shown below.

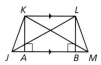

Because $\overline{KL} \parallel \overline{JM}$, $\angle AKL$ and $\angle KLB$ are right angles, so $KLBA$ is a rectangle and $\overline{AK} \cong \overline{BL}$. Then $\triangle JLB \cong \triangle MKA$ by the HL Congruence Theorem. So, $\angle LJB \cong \angle KMA$. $\overline{JM} \cong \overline{JM}$ by the Reflexive Property of Congruence. So, $\triangle KJM \cong \triangle LMJ$ by the SAS Congruence Theorem. Then $\angle KJM \cong \angle LMJ$, and the trapezoid is isosceles by the Isosceles Trapezoid Base Angles Converse.

52. rhombus

STATEMENTS	REASONS
1. $\overline{JK} \cong \overline{LM}$, E is the midpoint of $\overline{JL}$, F is the midpoint of $\overline{KL}$, G is the midpoint of $\overline{KM}$, H is the midpoint of $\overline{JM}$	1. Given
2. $\overline{EF}$ is a midsegment of $\triangle JKL$, $\overline{FG}$ is a midsegment of $\triangle KML$, $\overline{GH}$ is a midsegment of $\triangle JKM$, $\overline{EH}$ is a midsegment of $\triangle JML$	2. Definition of midsegment
3. $\overline{EF} \parallel \overline{JK}$, $\overline{FG} \parallel \overline{LM}$, $\overline{GH} \parallel \overline{JK}$, and $\overline{EH} \parallel \overline{LM}$	3. Triangle Midsegment Theorem
4. $\overline{EF} \parallel \overline{GH}$, $\overline{FG} \parallel \overline{EH}$	4. Transitive Property of Parallel Lines
5. EFGH is a parallelogram.	5. Definition of parallelogram
6. $EF = \frac{1}{2}JK$, $FG = \frac{1}{2}LM$, $GH = \frac{1}{2}JK$, $EH = \frac{1}{2}LM$	6. Trapezoid Midsegment Theorem
7. $JK = LM$	7. Definition of congruent segments
8. $FG = \frac{1}{2}JK$, $EH = \frac{1}{2}JK$	8. Substitution Property of Equality
9. $EF = FG = GH = EH$	9. Transitive Property of Equality
10. $\overline{EF} \cong \overline{FG} \cong \overline{GH} \cong \overline{EH}$	10. Definition of congruent segments
11. EFGH is a rhombus.	11. Definition of a rhombus

7.5 Maintaining Mathematical Proficiency

53.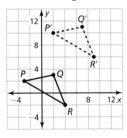

54.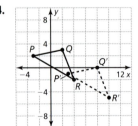

55.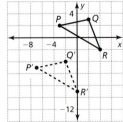

56. yes; The corresponding angles are congruent and the corresponding side lengths are proportional.

57. no; $\frac{20}{6} \neq \frac{24}{8}$, the side lengths are not proportional.

58. yes; The corresponding angles are congruent and the corresponding side lengths are proportional.

Chapter 7 Cumulative Assessment

7.

STATEMENTS	REASONS
1. $\overline{BC} \parallel \overline{AD}$, $\angle EBC \cong \angle ECB$, $\angle ABE \cong \angle DCE$	1. Given
2. ABCD is a trapezoid.	2. Definition of trapezoid
3. $m\angle EBC = m\angle ECB$, $m\angle ABE = m\angle DCE$	3. Definition of congruent angles
4. $m\angle ABE + m\angle EBC = m\angle ABC$, $m\angle DCE + m\angle ECB = m\angle DCB$	4. Angle Addition Postulate
5. $m\angle ABE + m\angle EBC = m\angle ABE + m\angle EBC$	5. Reflexive Property of Equality
6. $m\angle ABE + m\angle EBC = m\angle DCE + m\angle ECB$	6. Substitution Property of Equality
7. $m\angle ABC = m\angle DCB$	7. Transitive Property of Equality
8. $\angle ABC \cong \angle DCB$	8. Definition of congruent angles
9. ABCD is an isosceles trapezoid.	9. Isosceles Trapezoid Base Angles Converse

Chapter 8

8.1 Explorations

4. The difference between the x-value of each vertex of $\triangle A'B'C'$ and the x-value of the center of dilation is equal to k times the difference between its corresponding x-value of $\triangle ABC$ and the x-value of the center of dilation. The difference between the y-value of each vertex of $\triangle A'B'C'$ and the y-value of the center of dilation is equal to k times the difference between its corresponding y-value of $\triangle ABC$ and the y-value of the center of dilation. Each side of $\triangle A'B'C'$ is k times as long as its corresponding side of $\triangle ABC$. Each angle of $\triangle A'B'C'$ is congruent to its corresponding angle of $\triangle ABC$.

Sample answer:

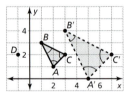

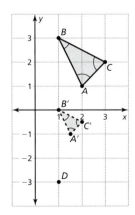

8.1 Lesson Monitoring Progress

3.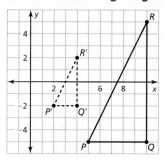

8.1 Monitoring Progress and Modeling with Mathematics

7.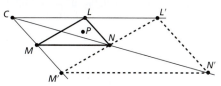
Not drawn to scale.

8.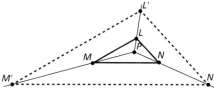
Not drawn to scale.

9.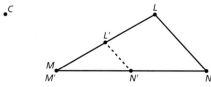
Not drawn to scale.

10.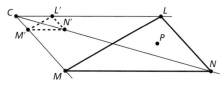
Not drawn to scale.

11.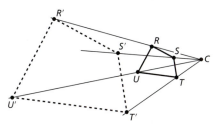
Not drawn to scale.

12.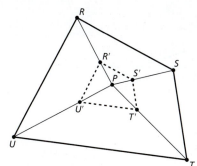
Not drawn to scale.

13.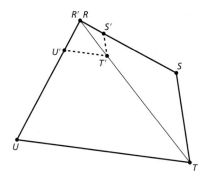
Not drawn to scale.

14.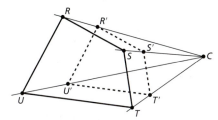
Not drawn to scale.

19.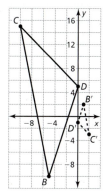

20.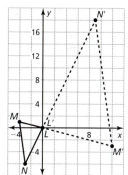

21.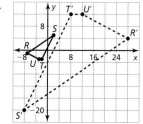

22.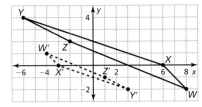

45.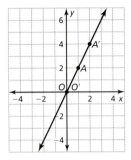

a. $O'A' = 2(OA)$
b. $\overleftrightarrow{O'A'}$ coincides with $\overleftrightarrow{OA}$.

46.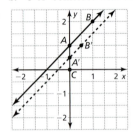

a. $A'B' = \frac{1}{2}(AB)$
b. $\overleftrightarrow{A'B'} \parallel \overleftrightarrow{AB}$ and the y-intercept of $\overleftrightarrow{A'B'}$ is half of the y-intercept of $\overleftrightarrow{AB}$.

47. $k = \frac{1}{16}$

48. no; It is true that dilating a figure with a scale factor of 1 will not enlarge or reduce the image nor will -1. However, by dilating with a scale factor that is negative, the image is rotated by 180°.

49.
a. $P = 24$ units, $A = 32$ square units
b.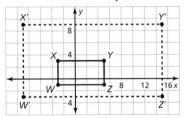

$P = 72$ units, $A = 288$ square units; The perimeter of the dilated rectangle is three times the perimeter of the original rectangle. The area of the dilated rectangle is nine times the area of the original rectangle.

c.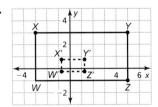

$P = 6$ units, $A = 2$ square units; The perimeter of the dilated rectangle is $\frac{1}{4}$ the perimeter of the original rectangle. The area of the dilated rectangle is $\frac{1}{16}$ the area of the original rectangle.

d. The perimeter changes by a factor of k. The area changes by a factor of k^2.

8.2 Explorations

2. c. Sample answer:

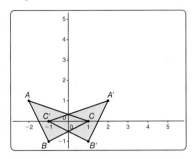

yes; Because the corresponding sides are congruent and the corresponding angles are congruent, the image is similar to the original triangle.

d. Sample answer:

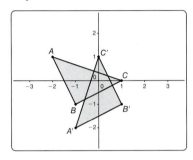

yes; Because the corresponding sides are congruent and the corresponding angles are congruent, the image is similar to the original triangle.

8.2 Extra Example 3

Translate square ABCD so that point A maps to point E. Because translations map segments to parallel segments and $\overline{AD} \parallel \overline{EH}$, the image of $\overline{AD}$ lies on $\overline{EH}$.

Because translations preserve length and angle measure, the image of ABCD, $EB'C'D'$, is a square with side length s. Because all the interior angles of a square are right angles, $\angle B'ED' \cong \angle FEH$. When $\overrightarrow{ED'}$ coincides with $\overrightarrow{EH}$, $\overrightarrow{EB'}$ coincides with $\overrightarrow{EF}$. So $\overline{EB'}$ lies on $\overline{EF}$. Next, dilate square $EB'C'D'$ using center of dilation E. Choose the scale factor to be the ratio of the side lengths of EFGH and $EB'C'D'$, which is $\frac{2s}{s} = 2$.

8.2 Lesson Monitoring Progress

5. Translate $\triangle JKL$ so that point L maps to point P. Because translations map segments to parallel segments and $\overline{LJ} \parallel \overline{PM}$, the image of $\overline{LJ}$ lies on $\overline{PM}$.

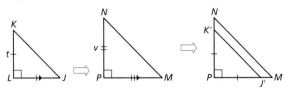

Because translations preserve side lengths and angle measures, the image of $\triangle JKL$, $\triangle J'K'P$, is a right isosceles triangle with leg length r. Because $\angle K'PJ'$ and $\angle NPM$ are right angles, they are congruent. When $\overrightarrow{PJ'}$ coincides with $\overrightarrow{PM}$, $\overrightarrow{PK'}$ coincides with $\overrightarrow{PN}$. So, $\overline{PK'}$ lies on $\overline{PN}$. Next, dilate $\triangle J'K'P$ using center of dilation P. Choose the scale factor to be the ratio of the side lengths of $\triangle MNP$ and $\triangle J'K'P$, which is $\frac{v}{t}$.

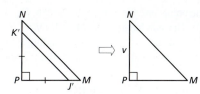

The dilation maps $\overline{PJ'}$ to $\overline{PM}$ and $\overline{PK'}$ to $\overline{PN}$ because the images of $\overline{PJ'}$ and $\overline{PK'}$ have side length $\frac{v}{t}(t) = v$ and the segments $\overline{PJ'}$ and $\overline{PK'}$ lie on lines passing through the center of dilation. So, the dilation maps K' to N and J' to M. A similarity transformation maps $\triangle JKL$ to $\triangle MNP$. So, $\triangle JKL$ is similar to $\triangle MNP$.

8.2 Monitoring Progress and Modeling with Mathematics

13. Reflect $\triangle ABC$ in $\overleftrightarrow{AB}$. Because reflections preserve side lengths and angle measures, the image of $\triangle ABC$, $\triangle ABC'$, is a right isosceles triangle with leg length j. Also because $\overleftrightarrow{AC} \perp \overleftrightarrow{BA}$, point C' is on $\overleftrightarrow{AC}$. So, $\overline{AC'}$ is parallel to $\overline{RT}$.

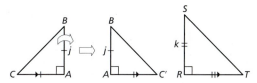

Then translate $\triangle ABC'$ so that point A maps to point R. Because translations map segments to parallel segments and $\overline{AC'} \parallel \overline{RT}$, the image of $\overline{AC'}$ lies on $\overline{RT}$.

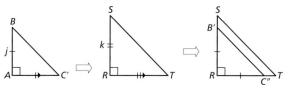

Because translations preserve side lengths and angle measures, the image of $\triangle ABC'$, $\triangle RB'C''$, is a right isosceles triangle with leg length j. Because $\angle B'RC''$ and $\angle SRT$ are right angles, they are congruent. When $\overrightarrow{RC''}$ coincides with $\overrightarrow{RT}$, $\overrightarrow{RB'}$ coincides with $\overrightarrow{RS}$. So, $\overline{RB'}$ lies on $\overline{RS}$. Next, dilate $\triangle RB'C''$ using center of dilation R. Choose the scale factor to be the ratio of the side lengths of $\triangle RST$ and $\triangle RB'C''$, which is $\frac{k}{j}$.

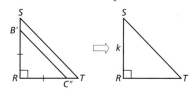

The dilation maps $\overline{RC''}$ to $\overline{RT}$ and $\overline{RB'}$ to $\overline{RS}$ because the images of $\overline{RC''}$ and $\overline{RB'}$ have side length $\frac{k}{j}(j) = k$ and the segments $\overline{RC''}$ and $\overline{RB'}$ lie on lines passing through the center of dilation. So, the dilation maps C'' to T and B' to S. A similarity transformation maps $\triangle ABC$ to $\triangle RST$. So, $\triangle ABC$ is similar to $\triangle RST$.

14. If necessary, rotate $\square JKLM$ about point M so that $\overleftrightarrow{ML'} \parallel \overleftrightarrow{TS}$.

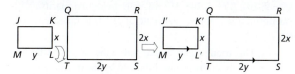

Note that because rotations are rigid motions, $\square J'K'L'M$ is congruent to $\square JKLM$. Translate $\square J'K'L'M$ so that point M maps to point T. Because translations map segments to parallel segments and $\overline{ML'} \parallel \overline{TS}$, the image of $\overline{ML'}$ lies on $\overline{TS}$.

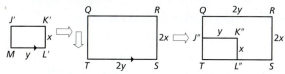

Because translations preserve side lengths and angle measures, the image of $\square J'K'L'M$, $\square J''K''L''T$, is a rectangle with side lengths x and y. Because all interior angles of a rectangle are right angles, $\angle J''TL''$ and $\angle QTS$ are congruent. When $\overrightarrow{TL''}$ coincides with $\overrightarrow{TS}$, $\overrightarrow{TJ''}$ coincides with $\overrightarrow{TQ}$. So, $\overline{TJ''}$ lies on $\overline{TQ}$. Next, dilate $\square J''K''L''T$ using center of dilation T. Choose the scale factor to be the ratio of the side lengths of $\square QRST$ and $\square J''K''L''T$, which is $\frac{2x}{x} = \frac{2y}{y} = 2$.

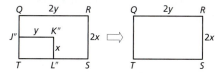

The dilation maps $\overline{TL''}$ to $\overline{TS}$ and $\overline{TJ''}$ to $\overline{TQ}$ because the images of $\overline{TL''}$ and $\overline{TJ''}$ have side lengths $y(2) = 2y$ and $x(2) = 2x$, and the segments $\overline{TL''}$ and $\overline{TJ''}$ lie on lines passing through the center of dilation. So, the dilation maps L'' to S and J'' to Q. The image of K'' lies $y(2) = 2y$ units to the right of the image of J'', and $x(2) = 2x$ units above the image of L''. So, the image of K'' is R. A similarity transformation maps $\square JKLM$ to $\square QRST$. So, $\square JKLM$ is similar to $\square QRST$.

16. Figure A is not similar to Figure B because the scale factor (A to B) of the shorter legs is $\frac{1}{2}$, and the scale factor (A to B) of the longer legs is $\frac{2}{3}$; Figure A is not similar to Figure B.

18. a. yes; The suns appear to be a dilation of one another.
    b. no; The hearts are about the same height but one is wider than the other.

19. Sample answer:

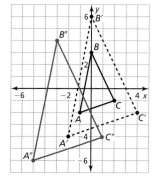

$\triangle A''B''C''$ can be mapped to $\triangle ABC$ by a translation 3 units right and 2 units up, followed by a dilation with center at the origin and a scale factor of $\frac{1}{2}$.

20. sometimes; As long as the center of dilation and the center of rotation are the same, rotations and dilations are commutative.

21. $J(-8, 0)$, $K(-8, 12)$, $L(-4, 12)$, $M(-4, 0)$; $J''(-9, -4)$, $K''(-9, 14)$, $L''(-3, 14)$, $M''(-3, -4)$; yes; A similarity transformation mapped quadrilateral $JKLM$ to quadrilateral $J''K''L''M''$.

22. **a.** yes; This triangle can be mapped to the larger one by a 180° rotation about the origin, followed by the translation $(x, y) \rightarrow (x + 5, y + 4)$, followed by a dilation with center $(1, 1)$ and a scale factor of 2. Because one can be mapped to the other by a similarity transformation, the triangles are similar.

   b. The triangle formed when the midpoints of a triangle are connected is always similar to the original triangle.

8.2 Maintaining Mathematical Proficiency

23. obtuse
24. straight
25. acute
26. right

8.3 Monitoring Progress and Modeling with Mathematics

46. always
47. sometimes
48. never
49. yes; All four angles of each rectangle will always be congruent right angles.
50. yes; The light, object, and image form similar triangles.
51. about 1116 mi

52.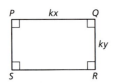

Let $KLMN$ and $PQRS$ be similar rectangles as shown. The ratio of corresponding side lengths is $\frac{KL}{PQ} = \frac{x}{kx} = \frac{1}{k}$. The perimeter of $KLMN$ is $2x + 2y$ and the perimeter of $PQRS$ is $2kx + 2ky$. So, the ratio of the perimeters is

$\frac{2x + 2y}{2kx + 2ky} = \frac{2x + 2y}{k(2x + 2y)} = \frac{1}{k}$. Because both ratios equal $\frac{1}{k}$,

the ratios are equal. So,

$\frac{KL + LM + MN + NK}{PQ + QR + RS + SP} = \frac{KL}{PQ} = \frac{LM}{QR} = \frac{MN}{RS} = \frac{NK}{SP}$.

53.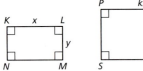

Let $KLMN$ and $PQRS$ be similar rectangles as shown. The ratio of corresponding side lengths is $\frac{KL}{PQ} = \frac{x}{kx} = \frac{1}{k}$. The area of $KLMN$ is xy and the area of $PQRS$ is $(kx)(ky) = k^2xy$.

So, the ratio of the areas is $\frac{xy}{k^2xy} = \frac{1}{k^2} = \left(\frac{1}{k}\right)^2$. Because the

ratio of corresponding side lengths is $\frac{1}{k}$, any pair of

corresponding side lengths can be substituted for $\frac{1}{k}$. So,

$\frac{\text{Area of } KLMN}{\text{Area of } PQRS} = \left(\frac{KL}{PQ}\right)^2 = \left(\frac{LM}{QR}\right)^2 = \left(\frac{MN}{RS}\right)^2 = \left(\frac{NK}{SP}\right)^2$.

54. no; Three angles determine a unique triangle on a sphere.

55. $x = \frac{1 + \sqrt{5}}{2}$; $x = \frac{1 + \sqrt{5}}{2}$ satisfies the proportion $\frac{1}{x} = \frac{x-1}{1}$.

56. The coordinates of the points are $A(-3, 0)$, $B(0, 4)$, $C(6, 0)$, $D(0, -8)$, and $O(0, 0)$. The side lengths are $OA = 3$, $OB = 4$, $AB = 5$, $OC = 6$, $OD = 8$, and $CD = 10$. Corresponding side lengths are proportional with a scale factor of 2, so the triangles are proportional.

8.3 Maintaining Mathematical Proficiency

57. $x = 63$
58. $x = 66$
59. $x = 120$
60. $x = 108$

8.4 Explorations

1. **a.** Check students' work.

   b, d, and e. Sample answer for columns 4, 5, and 6:

	1.	2.	3.	4.	5.	6.
$m\angle A, m\angle D$	106°	88°	40°	45°	90°	15°
$m\angle B, m\angle E$	31°	42°	65°	90°	60°	150°
$m\angle C$	43°	50°	75°	45°	30°	15°
$m\angle F$	43°	50°	75°	45°	30°	15°
AB	2	2	3	2	2	2
DE	4	3	5	6	4	3.16
BC	2.82	2.61	2	2	4	2
EF	5.64	3.91	3.33	6	8	3.16
AC	1.51	1.75	2.81	2.83	3.46	3.86
DF	3.02	2.62	4.69	8.49	6.93	6.11

   c. yes; Corresponding angles are congruent, and the corresponding side lengths are proportional.

   e. no

   f. Two triangles with two pairs of congruent corresponding angles are similar.

8.4 Monitoring Progress and Modeling with Mathematics

22. yes;

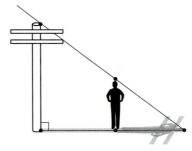

If you and the telephone pole are both standing at right angles with the ground, then two similar triangles are formed, and you can use a proportion to estimate the height.

23. yes; Corresponding angles are congruent.
24. yes; The angle measures could be 30°, 60°, and 90°.
25. no; $94° + 87° > 180°$
26. no; $m\angle X = 100°$ and $100° + 85° > 180°$
27. *Sample answer:* Because the triangles are similar, the ratios of the vertical sides to the horizontal sides are equal.
28. *Sample answer:* $\triangle KLM$ and $\triangle JLN$; The triangles are similar, so you can use the proportion $\frac{x}{x + 20} = \frac{100}{125}$.

29. The angle measures are 60°.
30. **a.** no; Not all rectangles are similar.
    b. no; Not all rectangles are similar.
31.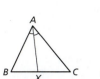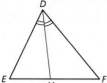

    Let △ABC ~ △DEF with a scale factor of k, and $\overline{AX}$ and $\overline{DY}$ be angle bisectors as shown. Then ∠C ≅ ∠F, m∠CAB = m∠FDE, 2m∠CAX = m∠CAB and 2m∠FDY = m∠FDE. By the Substitution Property of Equality, 2m∠CAX = 2m∠FDY, so m∠CAX = m∠FDY. Then △ACX ~ △DFY by the AA Similarity Theorem, and because corresponding side lengths are proportional, $\frac{AX}{DY} = \frac{AC}{DF} = k$.

32. Let △XYZ be any triangle such that a = XY and b = XZ, and the altitudes to $\overline{XY}$ and $\overline{XZ}$ are $\overline{AZ}$ and $\overline{BY}$, respectively. Then, ∠XAZ and ∠XBY are congruent right angles and ∠YXB ≅ ∠ZXA, so △BXY ~ △AXZ by the AA Similarity Theorem. Because corresponding side lengths of similar figures are proportional, $\frac{AZ}{BY} = \frac{XZ}{XY} = \frac{b}{a}$. Note: When m∠X = 90°, XY = BY and XZ = AZ, so the ratio is still $\frac{b}{a}$.

33. about 17.1 ft; △AED ~ △CEB, so $\frac{DE}{BE} = \frac{4}{3}$. △DEF ~ △DBC, so $\frac{EF}{30} = \frac{DE}{DB} = \frac{4}{7}$ and $EF = \frac{120}{7}$.

8.4 Maintaining Mathematical Proficiency
34. yes; Use the SAS Congruence Theorem.
35. yes; Use the SSS Congruence Theorem.
36. no; ∠PSQ is not necessarily a right angle.

8.5 Explorations
1. b, d, and g. *Sample answer for column 7:*

	1.	2.	3.	4.
AB	5	5	6	15
BC	8	8	8	20
AC	10	10	10	10
DE	10	15	9	12
EF	16	24	12	16
DF	20	30	15	8
m∠A	52.41°	52.41°	53.13°	104.48°
m∠B	97.9°	97.9°	90°	28.96°
m∠C	29.69°	29.69°	36.87°	46.57°
m∠D	52.41°	52.41°	53.13°	104.48°
m∠E	97.9°	97.9°	90°	28.96°
m∠F	29.69°	29.69°	36.87°	46.57°

	5.	6.	7.
AB	9	24	3
BC	12	18	5
AC	8	16	6
DE	12	8	9
EF	15	6	15
DF	10	8	18
m∠A	89.6°	48.59°	56.25°
m∠B	41.81°	41.81°	93.82°
m∠C	48.59°	89.6°	29.93°
m∠D	85.46°	44.05°	56.25°
m∠E	41.65°	67.98°	93.82°
m∠F	52.89°	67.98°	29.93°

8.5 Monitoring Progress and Modeling with Mathematics
15. no; The included angles are not congruent.
16. the pieces with side lengths of 5.25 inches and 7 inches
17. D; ∠M ≅ ∠M, so △MNP ~ △MRQ by the SAS Similarity Theorem.
18. not necessarily; The acute angles might not be congruent.

19.

STATEMENTS	REASONS
1. $\angle A \cong \angle D$, $\dfrac{AB}{DE} = \dfrac{AC}{DF}$	1. Given
2. Draw $\overline{PQ}$ so that P is on $\overline{AB}$, Q is on $\overline{AC}$, $\overline{PQ} \parallel \overline{BC}$, and $AP = DE$.	2. Parallel Postulate
3. $\angle APQ \cong \angle ABC$	3. Corresponding Angles Theorem
4. $\angle A \cong \angle A$	4. Reflexive Property of Congruence
5. $\triangle APQ \sim \triangle ABC$	5. AA Similarity Theorem
6. $\dfrac{AB}{AP} = \dfrac{AC}{AQ} = \dfrac{BC}{PQ}$	6. Corresponding sides of similar figures are proportional.
7. $\dfrac{AB}{DE} = \dfrac{AC}{AQ}$	7. Substitution Property of Equality
8. $AQ \cdot \dfrac{AB}{DE} = AC$, $DF \cdot \dfrac{AB}{DE} = AC$	8. Multiplication Property of Equality
9. $AQ = AC \cdot \dfrac{DE}{AB}$, $DF = AC \cdot \dfrac{DE}{AB}$	9. Multiplication Property of Equality
10. $AQ = DF$	10. Transitive Property of Equality
11. $\overline{AQ} \cong \overline{DF}$, $\overline{AP} \cong \overline{DE}$	11. Definition of congruent segments
12. $\triangle APQ \cong \triangle DEF$	12. SAS Congruence Theorem
13. $\overline{PQ} \cong \overline{EF}$	13. Corresponding parts of congruent triangles are congruent.
14. $PQ = EF$	14. Definition of congruent segments
15. $\dfrac{AB}{DE} = \dfrac{AC}{DF} = \dfrac{BC}{EF}$	15. Substitution Property of Equality
16. $\triangle ABC \sim \triangle DEF$	16. SSS Similarity Theorem

20. SAS Similarity Theorem
21. If two angles are congruent, then the triangles are similar by the AA Similarity Theorem.

22. a. no; Two right trapezoids can have proportional heights and lengths of one base and not be similar.
    b. yes; The fourth pair of sides must be proportional, and the angles must be congruent.
    c. no; The angles might not be congruent.
    d. no; Two kites can have proportional sides and one pair of congruent angles but not be similar.
23. no; no; The sum of the angle measures would not be 180°.
24. Sample answer:

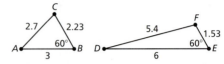

8.5 Maintaining Mathematical Proficiency
25. -1
26. $\dfrac{1}{3}$
27. 2

8.6 Monitoring Progress and Modeling with Mathematics

27.

STATEMENTS	REASONS
1. $\overline{QS} \parallel \overline{TU}$	1. Given
2. $\angle RQS \cong \angle RTU$, $\angle RSQ \cong \angle RUT$	2. Corresponding Angles Theorem
3. $\triangle RQS \sim \triangle RTU$	3. AA Similarity Theorem
4. $\dfrac{QR}{TR} = \dfrac{SR}{UR}$	4. Corresponding side lengths of similar figures are proportional.
5. $QR = QT + TR$, $SR = SU + UR$	5. Segment Addition Postulate
6. $\dfrac{QT + TR}{TR} = \dfrac{SU + UR}{UR}$	6. Substitution Property of Equality
7. $\dfrac{QT}{TR} + \dfrac{TR}{TR} = \dfrac{SU}{UR} + \dfrac{UR}{UR}$	7. Rewrite the proportion.
8. $\dfrac{QT}{TR} + 1 = \dfrac{SU}{UR} + 1$	8. Simplify.
9. $\dfrac{QT}{TR} = \dfrac{SU}{UR}$	9. Subtraction Property of Equality

28.

STATEMENTS	REASONS
1. $\dfrac{ZY}{YW} = \dfrac{ZX}{XV}$	1. Given
2. $\dfrac{YW}{ZY} = \dfrac{XV}{ZX}$	2. Rewrite the proportion.
3. $\dfrac{YW}{ZY} + 1 = \dfrac{XV}{ZX} + 1$	3. Addition Property of Equality
4. $\dfrac{YW}{ZY} + \dfrac{ZY}{ZY} = \dfrac{XV}{ZX} + \dfrac{ZX}{ZX}$	4. Substitution Property of Equality
5. $\dfrac{YW + ZY}{ZY} = \dfrac{XV + ZX}{ZX}$	5. Add fractions.
6. $ZW = YW + ZY$, $ZV = XV + ZX$	6. Segment Addition Postulate
7. $\dfrac{ZW}{ZY} = \dfrac{ZV}{ZX}$	7. Substitution Property of Equality
8. $\angle Z \cong \angle Z$	8. Reflexive Property of Congruence
9. $\triangle ZWV \sim \triangle ZYX$	9. SAS Similarity Theorem
10. $\angle ZYX \cong \angle ZWV$	10. Corresponding angles of similar triangles are congruent.
11. $\overline{YX} \parallel \overline{WV}$	11. Corresponding Angles Converse

35. Because $\overline{WX} \parallel \overline{ZA}$, $\angle XAZ \cong \angle YXW$ by the Corresponding Angles Theorem and $\angle WXZ \cong \angle XZA$ by the Alternate Interior Angles Theorem. So, by the Transitive Property of Congruence, $\angle XAZ \cong \angle XZA$. Then $\overline{XA} \cong \overline{XZ}$ by the Converse of the Base Angles Theorem, and by the Triangle Proportionality Theorem, $\dfrac{YW}{WZ} = \dfrac{XY}{XA}$. Because $XA = XZ$, $\dfrac{YW}{WZ} = \dfrac{XY}{XZ}$.

36. If a ray that passes through a vertex of a triangle divides the opposite side into segments whose lengths are proportional to the lengths of the other two sides, then the ray bisects the angle through which it passes; yes; The ray divides the triangle into proportional parts. So, the angles must be congruent.

37. The Triangle Midsegment Theorem is a specific case of the Triangle Proportionality Theorem when the segment parallel to one side of a triangle that connects the other two sides also happens to pass through the midpoints of those two sides.

38. no; You can use the ratio of AD and BE to find the rate.

39. •———————•
     x

40. By the Vertical Angles Congruence Theorem, $\angle NPA \cong \angle CPM$, $\angle BXP \cong \angle CXM$, and $\angle PZB \cong \angle NZA$. By the Alternate Interior Angles Theorem, $\angle M \cong \angle PAN$, $\angle M \cong \angle BPX$, and $\angle N \cong \angle ZPB$. So, $\triangle APN \sim \triangle MPC$, $\triangle CXM \sim \triangle BXP$, and $\triangle BZP \sim \triangle AZN$ by the AA Similarity Theorem. Then $\dfrac{CX}{XB} = \dfrac{CM}{BP}$, $\dfrac{AP}{PM} = \dfrac{AN}{CM}$, and $\dfrac{BZ}{ZA} = \dfrac{BP}{AN}$, and by the Triangle Proportionality Theorem $\dfrac{AY}{YC} = \dfrac{AP}{PM}$. So, $\dfrac{AY}{YC} = \dfrac{AN}{CM}$ by the Transitive Property of Equality. Then $\dfrac{AY}{YC} \cdot \dfrac{CX}{XB} \cdot \dfrac{BZ}{ZA} = \dfrac{AN}{CM} \cdot \dfrac{CM}{BP} \cdot \dfrac{BP}{AN} = 1$.

8.6 Maintaining Mathematical Proficiency

41. a, b
42. c
43. $x = \pm 11$
44. $x = \pm 3$
45. $x = \pm 7$

Chapter 9

9.1 Explorations

2. c.

STATEMENTS	REASONS
1. $\triangle ABC \sim \triangle ACD \sim \triangle CBD$	1. Given
2. $\dfrac{c}{b} = \dfrac{b}{c-d}, \dfrac{c}{a} = \dfrac{a}{d}$	2. Corresponding sides of similar figures are proportional.
3. $c(c - d) = b^2$, $cd = a^2$	3. Cross Products Property
4. $c^2 - cd = b^2$	4. Distributive Property
5. $c^2 - a^2 = b^2$	5. Substitution Property of Equality
6. $c^2 = a^2 + b^2$	6. Addition Property of Equality
7. $a^2 + b^2 = c^2$	7. Symmetric Property of Equality

9.1 Monitoring Progress and Modeling with Mathematics

38. Let the sides of the first triangle be represented by a_1, b_1, and c_1, and the sides of the second triangle be represented by a_2, b_2, and c_2. By using the Subtraction Property of Equality on $c^2 = a^2 + b^2$, you can say $b^2 = c^2 - a^2$. So, $b_1^2 = c_1^2 - a_1^2$ and $b_2^2 = c_2^2 - a_2^2$. We are given that $c_1 = c_2$, and one pair of legs are congruent. Because a and b are interchangeable in the Pythagorean Theorem, we can say that either $a_1 = a_2$ or $b_1 = b_2$. For simplicity, let's use $a_1 = a_2$. By the Substitution Property of Equality, we get $b_1^2 = c_2^2 - a_2^2$. So, by the Transitive Property, $b_1^2 = b_2^2$. Then, by taking the positive square root of each side, you get $b_1 = b_2$. So, the two right triangles are congruent by the SSS Congruence Theorem.

39.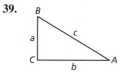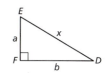

Let $\triangle ABC$ be any triangle so that the square of the length, c, of the longest side of the triangle is equal to the sum of the squares of the lengths, a and b, of the other two sides: $c^2 = a^2 + b^2$. Let $\triangle DEF$ be any right triangle with leg lengths of a and b. Let x represent the length of its hypotenuse. Because $\triangle DEF$ is a right triangle, by the Pythagorean Theorem, $a^2 + b^2 = x^2$. So, by the Transitive Property, $c^2 = x^2$. By taking the positive square root of each side, you get $c = x$. So, $\triangle ABC \cong \triangle DEF$ by the SSS Congruence Theorem.

40. no; Let $m = 2$ and $n = -1$. The resulting integers are -4, 3, and 5. The integers in a Pythagorean triple must be positive.

41. no; They can be part of a Pythagorean triple if 75 is the hypotenuse: $21^2 + 72^2 = 75^2$

42. given; Pythagorean Theorem; Substitution Property; definition of a right angle; Substitution Property; definition of acute angle; definition of acute triangle

43.

STATEMENTS	REASONS
1. In $\triangle ABC$, $c^2 > a^2 + b^2$, where c is the length of the longest side. $\triangle PQR$ has side lengths a, b, and x, where x is the length of the hypotenuse and $\angle R$ is a right angle.	1. Given
2. $a^2 + b^2 = x^2$	2. Pythagorean Theorem
3. $c^2 > x^2$	3. Substitution Property
4. $c > x$	4. Take the positive square root of each side.
5. $m\angle R = 90°$	5. Definition of a right angle
6. $m\angle C > m\angle R$	6. Converse of the Hinge Theorem
7. $m\angle C > 90°$	7. Substitution Property
8. $\angle C$ is an obtuse angle.	8. Definition of obtuse angle
9. $\triangle ABC$ is an obtuse triangle.	9. Definition of obtuse triangle

9.1 Maintaining Mathematical Proficiency

44. $\dfrac{7\sqrt{2}}{2}$

45. $\dfrac{14\sqrt{3}}{3}$

46. $4\sqrt{2}$

47. $4\sqrt{3}$

9.2 Monitoring Progress and Modeling with Mathematics

20. a. the triangle farthest to the left with legs that are each 1 unit

b. the third triangle from the left with legs that are 1 unit and $\sqrt{3}$ units, and a hypotenuse that is $\sqrt{4} = 2$ units

21. Given $\triangle JKL$, which is a 30°-60°-90° triangle, whose shorter leg, $\overline{KL}$, has length x, construct $\triangle JML$, which is congruent and adjacent to $\triangle JKL$. Because corresponding parts of congruent triangles are congruent, $LM = KL = x$, $m\angle M = m\angle K = 60°$, $m\angle MJL = m\angle KJL = 30°$, and $JM = JK$. Also, by the Angle Addition Postulate, $m\angle KJM = m\angle KJL + m\angle MJL$, and by substituting, $m\angle KJM = 30° + 30° = 60°$. So, $\triangle JKM$ has three 60° angles, which means that it is equiangular by definition, and by the Corollary to the Converse of the Base Angles Theorem, it is also equilateral. By the Segment Addition Postulate, $KM = KL + LM$, and by substituting, $KM = x + x = 2x$. So, by the definition of an equilateral triangle, $JM = JK = KM = 2x$. By the Pythagorean Theorem, $(JL)^2 + (KL)^2 = (JK)^2$. By substituting, you get $(JL)^2 + x^2 = (2x)^2$, which is equivalent to $(JL)^2 + x^2 = 4x^2$, when simplified. When the Subtraction Property of Equality is applied, you get $(JL)^2 = 4x^2 - x^2$, which is equivalent to $(JL)^2 = 3x^2$. By taking the positive square root of each side, $JL = x\sqrt{3}$. So, the hypotenuse of the 30°-60°-90° triangle, $\triangle JKL$, is twice as long as the shorter leg, and the longer leg is $\sqrt{3}$ times as long as the shorter leg.

22.

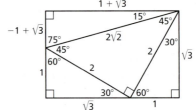

This rectangle contains two 30°-60°-90° triangles with side lengths of 1, 2, and $\sqrt{3}$ units, one 45°-45°-90° triangle with side lengths of 2, 2, and $2\sqrt{2}$ units, and one 15°-75°-90° triangle with side lengths of $1 + \sqrt{3}$, $-1 + \sqrt{3}$, and $2\sqrt{2}$ units.

23. *Sample answer:* Because all isosceles right triangles are 45°-45°-90° triangles, they are similar by the AA Similarity Theorem. Because both legs of an isosceles right triangle are congruent, the legs will always be proportional. So, 45°-45°-90° triangles are all similar by the SAS Similarity Theorem also.

24. no; The length of the legs of the triangles added are divided by $\sqrt{2}$ in each stage. The triangles that are added in stage 1, for example, have a hypotenuse that is 1 unit long. So, the legs must be $\dfrac{1}{\sqrt{2}} = \dfrac{\sqrt{2}}{2}$. So, the length of the legs in stage 8 will be $\dfrac{1}{(\sqrt{2})^8} = \dfrac{1}{16}$.

25. $T(1.5, 1.6)$

9.3 Monitoring Progress and Modeling with Mathematics

41.

STATEMENTS	REASONS
1. Draw $\triangle ABC$, $\angle BCA$ is a right angle.	1. Given
2. Draw a perpendicular segment (altitude) from C to $\overline{AB}$, and label the new point on $\overline{AB}$ as D.	2. Perpendicular Postulate
3. $\triangle ADC \sim \triangle CDB$	3. Right Triangle Similarity Theorem
4. $\dfrac{BD}{CD} = \dfrac{CD}{AD}$	4. Corresponding sides of similar figures are proportional.
5. $CD^2 = AD \cdot BD$	5. Cross Products Property

A58 Additional Answers

42.

STATEMENTS	REASONS
1. Draw $\triangle ABC$, $\angle BCA$ is a right angle.	1. Given
2. Draw a perpendicular segment (altitude) from C to $\overline{AB}$, and label the new point on $\overline{AB}$ as D.	2. Perpendicular Postulate
3. $\triangle ADC \sim \triangle CDB$	3. Right Triangle Similarity Theorem
4. $\dfrac{DB}{CB} = \dfrac{CB}{AB}$, $\dfrac{AD}{AC} = \dfrac{AC}{AB}$	4. Corresponding sides of similar figures are proportional.
5. $CB^2 = DB \cdot AB$, $AC^2 = AD \cdot AB$	5. Cross Products Property

43.

The two smaller triangles are congruent; Their corresponding sides lengths are represented by the same variables. So, they are congruent by the SSS Congruence Theorem.

45.

STATEMENTS	REASONS
1. $\triangle ABC$ is a right triangle. Altitude $\overline{CD}$ is drawn to hypotenuse $\overline{AB}$.	1. Given
2. $\angle BCA$ is a right angle.	2. Definition of right triangle
3. $\angle ADC$ and $\angle BDC$ are right angles.	3. Definition of perpendicular lines
4. $\angle BCA \cong \angle ADC \cong \angle BDC$	4. Right Angles Congruence Theorem
5. $\angle A$ and $\angle ACD$ are complementary. $\angle B$ and $\angle BCD$ are complementary.	5. Corollary to the Triangle Sum Theorem
6. $\angle ACD$ and $\angle BCD$ are complementary.	6. Definition of complementary angles
7. $\angle A \cong \angle BCD$, $\angle B \cong \angle ACD$	7. Congruent Complements Theorem
8. $\triangle CBD \sim \triangle ABC$, $\triangle ACD \sim \triangle ABC$, $\triangle CBD \sim \triangle ACD$	8. AA Similarity Theorem

9.3 Maintaining Mathematical Proficiency

46. $x = 65$

47. $x = 116$

48. $x = \dfrac{26}{3} \approx 8.7$

49. $x = \dfrac{23}{6} \approx 3.8$

9.4 Monitoring Progress and Modeling with Mathematics

20. 45°; greater than 45°; less than 45°; If the ratio of the legs is equal to 1, then the legs are congruent, and all isosceles right triangles are 45°-45°-90° triangles. If one acute angle measure of a right triangle is less than 45°, then the other acute angle is greater than 45°. So, the side opposite the smaller acute angle will be the shorter leg, which means the ratio will be less than 1, and the leg opposite the larger acute angle will be the longer leg, which means the ratio will be greater than 1.

21. no; The Sun's rays form a right triangle with the length of the awning and the height of the door. The tangent of the angle of elevation equals the height of the door divided by the length of the awning, so the length of the awning equals the quotient of the height of the door, 8 feet, and the tangent of the angle of elevation, 70°: $x = \dfrac{8}{\tan 70°} \approx 6.5$ ft

22. $\tan A = \dfrac{a}{b}$; $\tan B = \dfrac{b}{a}$; They are reciprocals of each other; complementary

23. You cannot find the tangent of a right angle, because each right angle has two adjacent legs, and the opposite side is the hypotenuse. So, you do not have an opposite leg and an adjacent leg. If a triangle has an obtuse angle, then it cannot be a right triangle, and the tangent ratio only works for right triangles.

24.

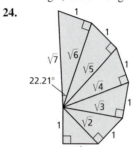

about 22.2°

25. **a.** about 33.4 ft

b. 3 students at each end; The triangle formed by the 60° angle has an opposite leg that is about 7.5 feet longer than the opposite leg of the triangle formed by the 50° angle. Because each student needs 2 feet of space, 3 more students can fit on each end with about 1.5 feet of space left over.

9.5 Monitoring Progress and Modeling with Mathematics

53. $\dfrac{\sin A}{\cos A} = \dfrac{\dfrac{\text{length of side opposite } A}{\text{length of hypotenuse}}}{\dfrac{\text{length of side adjacent to } A}{\text{length of hypotenuse}}} \cdot \dfrac{\text{length of hypotenuse}}{\text{length of hypotenuse}}$

$= \dfrac{\text{length of side opposite } A}{\text{length of side adjacent to } A}$

$= \tan A$

54. sine; You are given the length of the opposite leg.

55. a.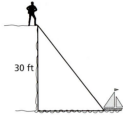

b.
Angle of depression	40°	50°	60°	70°	80°
Approximate length of line of sight (feet)	46.7	39.2	34.6	31.9	30.5

c. View of Sailboat from Cliff

(scatter plot: Angle of Depression (degrees) on x-axis, Approximate Length of Line of Sight (feet) on y-axis)

d. 60 ft

56. (a) represesents sin x and (b) represents cos x; sin 0 = 0 and cos 0 = 1. Substitute 0 for x in each series. Part (a) yields 0 and part (b) yields 1, so (a) represents sin x and (b) represents cos x; $\sin \frac{\pi}{6} \approx 0.5$, $\cos \frac{\pi}{6} \approx 0.87$

58. In $\triangle ABC$, $\sin C = \frac{h}{a}$. So, $h = a \sin C$. When you substitute this into the formula for the area of a triangle, you get $A = \frac{1}{2}bh = \frac{1}{2}b(a \sin C) = \frac{1}{2}ab \sin C$; about 9.0 square units

9.5 Maintaining Mathematical Proficiency

59. $x = 8$; yes
60. $x = 12\sqrt{2} \approx 17.0$; no
61. $x = 45$; yes
62. $x = 6\sqrt{2} \approx 8.5$; no

63.
		Gender		
		Male	Female	Total
Class	Sophomore	0.10	0.20	0.30
	Junior	0.12	0.18	0.30
	Senior	0.19	0.21	0.40
	Total	0.41	0.59	1

9.6 Monitoring Progress and Modeling with Mathematics

23. (triangle diagram: 4.76°, 96.4 in., 8 in., 96.1 in.)

24. a. about 32.5°
    b. *Sample answer:* The ratio is 8 inches to 15 inches; about 28.1°

25. about 36.9°; $PQ = 3$ centimeters and $PR = 4$ centimeters, so $m\angle R = \tan^{-1}\left(\frac{3}{4}\right) \approx 36.9°$.

26. no; *Sample answer:* Measures from a 30°-60°-90° triangle illustrate a counter example: $\tan^{-1}\sqrt{3} = 60°$, but $\frac{1}{\tan\sqrt{3}} \approx 33.1$.

27. $KM \approx 7.8$ ft, $JK \approx 11.9$ ft, $m\angle JKM = 49°$; $ML \approx 19.5$ ft, $m\angle MKL \approx 68.2°$, $m\angle L \approx 21.8°$

28. $TS \approx 9.1$ m, $TU \approx 8.2$ m, $m\angle UTS = 26°$; $TV \approx 14.1$ m, $TW \approx 10.9$ m, $m\angle V \approx 50.5°$, $m\angle VTW \approx 39.5°$

29. a. *Sample answer:* $\tan^{-1}\frac{3}{1}$; about 71.6°
    b. *Sample answer:* $\tan^{-1}\frac{4}{3}$; about 53.1°

30. a. $\sin^{-1}(\sin x) = x$; Inverse operations undo each other.
    b. $\tan(\tan^{-1} y) = y$; Inverse operations undo each other.
    c. $\cos(\cos^{-1} z) = z$; Inverse operations undo each other.

31. Because the sine is the ratio of the length of a leg to the length of the hypotenuse, and the hypotenuse is always longer than either of the legs, the sine cannot have a value greater than 1.

32. $AB = CD = 6$ cm, $BD = 2\sqrt{10}$ cm, $AD = BC = 2$ cm; $m\angle A = m\angle C = 90°$, $m\angle ADB = m\angle CBD \approx 71.6°$, $m\angle BDC = m\angle DBA \approx 18.4°$

9.6 Maintaining Mathematical Proficiency

33. $x = 8$
34. $x = 26$
35. $x = 2.46$
36. $x = 11.1125$

Chapter 9 Cumulative Assessment

6.
STATEMENTS	REASONS
1. $\overline{HE} \cong \overline{HG}$, $\overline{EG} \perp \overline{DF}$	1. Given
2. $\overline{HF} \cong \overline{HF}$, $\overline{DH} \cong \overline{DH}$	2. Reflexive Property of Congruence
3. $\angle EHF$, $\angle GHF$, $\angle GHD$, and $\angle EHD$ are right angles.	3. Definition of perpendicular lines
4. $\angle EHF \cong \angle GHF$, $\angle GHD \cong \angle EHD$	4. Right Angles Congruence Theorem
5. $\triangle EHF \cong \triangle GHF$, $\triangle EHD \cong \triangle GHD$	5. SAS Congruence Theorem
6. $\overline{FE} \cong \overline{FG}$, $\overline{DE} \cong \overline{DG}$	6. Corresponding parts of congruent triangles are congruent.

Chapter 10

10.1 Lesson Monitoring Progress

5. 0

10.1 Monitoring Progress and Modeling with Mathematics

42. $\overline{PA} \cong \overline{PB}$ and $\overline{PB} \cong \overline{PC}$ by the External Tangent Congruency Theorem.

43. $\angle ARC \cong \angle BSC$ and $\angle ACR \cong \angle BCS$, so $\triangle ARC \sim \triangle BSC$ by the AA Similarity Theorem. Because corresponding sides of similar figures are proportional, $\dfrac{AC}{BC} = \dfrac{RC}{SC}$.

44. yes; $\overline{AZ} = \overline{AW}, \overline{BW} = \overline{BX}, \overline{CX} = \overline{CY},$ and $\overline{DY} = \overline{DZ}$, so, $(AW + WB) + (CY + YD) = (AZ + DZ) + (BX + CX)$.

45. $x = 13, y = 5$; $2x - 5 = x + 8$ and $2x + 4y - 6 = 2x + 14$.

46. By the Tangent Line to Circles Theorem, $m\angle PRS = m\angle PTS = 90°$. $\overline{PT} \cong \overline{PR}$ and $\overline{PS} \cong \overline{PS}$, so $\triangle PTS \cong \triangle PRS$ by the HL Congruence Theorem. So, $\overline{SR} \cong \overline{ST}$.

47. a. Assume m is not perpendicular to $\overline{QP}$. The perpendicular segment from Q to m intersects m at some other point R. Then $QR < QP$, so R must be inside $\odot Q$, and m must be a secant line. This is a contradiction, so m must be perpendicular to $\overline{QP}$.

   b. Assume m is not tangent to $\odot Q$. Then m must intersect $\odot Q$ at a second point R. $\overline{QP}$ and $\overline{QR}$ are both radii of $\odot Q$, so $\overline{QP} \cong \overline{QR}$. Because $m \perp \overline{QP}$, $QP < QR$. This is a contradiction, so m must be tangent to $\odot Q$.

48. $2\sqrt{2}$; Sample answer: Use the given lengths and the Pythagorean Theorem to find $AE = 8\sqrt{2}$. So, $PF = 2\sqrt{2}$.

10.1 Maintaining Mathematical Proficiency

49. not independent

50. independent

10.2 Monitoring Progress and Modeling with Mathematics

35. a. Translate $\odot B$ so that point B maps to point A. The image of $\odot B$ is $\odot B'$ with center A. Because $\overline{AC} \cong \overline{BD}$, this translation maps $\odot B'$ to $\odot A$. A rigid motion maps $\odot B$ to $\odot A$, so $\odot A \cong \odot B$.

    b. Because $\odot A \cong \odot B$, the distance from the center of the circle to a point on the circle is the same for each circle. So, $\overline{AC} \cong \overline{BD}$.

36. similar; The circles each have different radii.

37. a. $m\overset{\frown}{BC} = m\angle BAC, m\overset{\frown}{DE} = m\angle DAE$ and $m\angle BAC = m\angle DAE$, so $m\overset{\frown}{BC} = m\overset{\frown}{DE}$. Because $\overset{\frown}{BC}$ and $\overset{\frown}{DE}$ are in the same circle, $\overset{\frown}{BC} \cong \overset{\frown}{DE}$.

    b. $m\overset{\frown}{BC} = m\angle BAC$ and $m\overset{\frown}{DE} = m\angle DAE$. Because $\overset{\frown}{BC} \cong \overset{\frown}{DE}, \angle BAC \cong \angle DAE$.

38. $s = \dfrac{\pi r \theta}{180}$; The circumference $= 2\pi r$, so the arc length $s = 2\pi r \cdot \dfrac{\theta}{360}$.

10.3 Monitoring Progress and Modeling with Mathematics

18. a. $\overline{AB}$ is a diameter; Perpendicular Chord Bisector Converse
    b. $\overline{AB} \cong \overline{CD}$; Congruent Corresponding Chords Theorem
    c. $\overline{JH}$ bisects $\overset{\frown}{FG}$ and $\overline{FG}$; Perpendicular Chord Bisector Theorem
    d. $\overline{NP} \cong \overline{LM}$; Equidistant Chords Theorem

19. a. Because $PA = PB = PC = PD$, $\triangle PDC \cong \triangle PAB$ by the SSS Congruence Theorem. So, $\angle DPC \cong \angle APB$ and $\overline{AB} \cong \overline{CD}$.

    b. $PA = PB = PC = PD$, and because $\overline{AB} \cong \overline{CD}$, $\angle DPC \cong \angle APB$. By the SAS Congruence Theorem, $\triangle PDC \cong \triangle PAB$, so $\overline{AB} \cong \overline{CD}$.

20. Sample answer: $\overline{AC} \cong \overline{BC}$, so $\overset{\frown}{AC} \cong \overset{\frown}{BC}$ and $m\overset{\frown}{AB} = 360 - 2m\overset{\frown}{AC}$. $m\overset{\frown}{AC}$ is an integer, so $2m\overset{\frown}{AC}$ is even and $360 - 2m\overset{\frown}{AC}$ is even.

21. about 16.26°; Sample answer: Draw radii $\overline{PA}$ and $\overline{PB}$, and the radius perpendicular to the two chords. Find the angles at P of the right triangles formed and subtract to find $m\angle APB$.

22. $\overline{LD} \cong \overline{LF}$ and $\overline{LC} \cong \overline{LC}$, so $\triangle LDC \cong \triangle LFC$ by the HL Congruence Theorem. Then $\overline{DC} \cong \overline{FC}$ and $\angle DLC \cong \angle FLC$. So, $\overline{DG} \cong \overline{FG}$.

23. $\overline{TP} \cong \overline{PR}, \overline{LP} \cong \overline{LP}$, and $\overline{LT} \cong \overline{LR}$, so $\triangle LPR \cong \triangle LPT$ by the SSS Congruence Theorem. Then $\angle LPT \cong \angle LPR$, so $m\angle LPT = m\angle LPR = 90°$. By definition, $\overline{LP}$ is a perpendicular bisector of $\overline{RT}$, so L lies on $\overline{QS}$. Because $\overline{QS}$ contains the center, $\overline{QS}$ is a diameter of $\odot L$.

24. yes; Sample answer: Draw several pairs of chords to see that the segment lengths are proportional.

25. If $\overline{AB} \cong \overline{CD}$, then $\overline{GC} \cong \overline{FA}$. Because $\overline{EC} \cong \overline{EA}$, $\triangle ECG \cong \triangle EAF$ by the HL Congruence Theorem, so $\overline{EF} \cong \overline{EG}$ and $EF = EG$. If $EF = EG$, then because $\overline{EC} \cong \overline{ED} \cong \overline{EA} \cong \overline{EB}$, $\triangle AEF \cong \triangle BEF \cong \triangle DEG \cong \triangle CEG$ by the HL Congruence Theorem. Then $\overline{AF} \cong \overline{BF} \cong \overline{DG} \cong \overline{CG}$, so $\overline{AB} \cong \overline{CD}$.

10.4 Monitoring Progress and Modeling with Mathematics

35. a. $\overline{QB} \cong \overline{QA}$, so $\triangle ABC$ is isosceles. By the Base Angles Theorem, $\angle QBA \cong \angle QAB$, so $m\angle BAQ = x°$. By the Exterior Angles Theorem, $m\angle AQC = 2x°$. Then $m\overset{\frown}{AC} = 2x°$, so $m\angle ABC = x° = \frac{1}{2}(2x)° = \frac{1}{2}m\overset{\frown}{AC}$.

    b. Given: $\angle ABC$ is inscribed in $\odot Q$. $\overline{DB}$ is a diameter; Prove: $m\angle ABC = \frac{1}{2}m\overset{\frown}{AC}$; By Case 1, proved in part (a), $m\angle ABD = \frac{1}{2}m\overset{\frown}{AD}$ and $m\angle CBD = \frac{1}{2}m\overset{\frown}{CD}$. By the Arc Addition Postulate, $m\overset{\frown}{AD} + m\overset{\frown}{CD} = m\overset{\frown}{AC}$. By the Angle Addition Postulate, $m\angle ABD + m\angle CBD = m\angle ABC$. Then $m\angle ABC = \frac{1}{2}m\overset{\frown}{AD} + \frac{1}{2}m\overset{\frown}{CD}$
    $= \frac{1}{2}(m\overset{\frown}{AD} + m\overset{\frown}{CD})$
    $= \frac{1}{2}m\overset{\frown}{AC}$.

    c. Given: $\angle ABC$ is inscribed in $\odot Q$. $\overline{DB}$ is a diameter; Prove: $m\angle ABC = \frac{1}{2}m\overset{\frown}{AC}$; By Case 1, proved in part (a), $m\angle DBA = \frac{1}{2}m\overset{\frown}{AD}$ and $m\angle DBC = \frac{1}{2}m\overset{\frown}{CD}$. By the Arc Addition Postulate, $m\overset{\frown}{AC} + m\overset{\frown}{CD} = m\overset{\frown}{AD}$, so $m\overset{\frown}{AC} = m\overset{\frown}{AD} - m\overset{\frown}{CD}$. By the Angle Addition Postulate, $m\angle DBC + m\angle ABC = m\angle DBA$, so $m\angle ABC = m\angle DBA - m\angle DBC$. Then $m\angle ABC = \frac{1}{2}m\overset{\frown}{AD} - \frac{1}{2}m\overset{\frown}{CD}$
    $= \frac{1}{2}(m\overset{\frown}{AD} - m\overset{\frown}{CD})$
    $= \frac{1}{2}m\overset{\frown}{AC}$.

36.

Given: $\angle ABC$ and $\angle ADC$ are inscribed angles intercepting $\widehat{AC}$;
Prove: $\angle ABC \cong \angle ADC$; By the Measure of an Inscribed Angle Theorem, $m\angle ABC = \frac{1}{2}m\widehat{AC}$ and $m\angle ADC = \frac{1}{2}m\widehat{AC}$. By the Transitive Property of Equality, $m\angle ABC = m\angle ADC$. So, $\angle ABC \cong \angle ADC$.

37. To prove the conditional, find the measure of the intercepted arc of the right angle and the definition of a semicircle to show the hypotenuse of the right triangle must be the diameter of the circle. To prove the converse, use the definition of a semicircle to find the measure of the angle opposite the diameter.

10.5 Monitoring Progress and Modeling with Mathematics

33. a. By the Tangent Line to Circle Theorem, $m\angle BAC$ is 90°, which is half the measure of the semicircular arc.

    b.

    By the Tangent Line to Circle Theorem, $m\angle CAD = 90°$. $m\angle DAB = \frac{1}{2}m\widehat{DB}$ and by part (a), $m\angle CAD = \frac{1}{2}m\widehat{AD}$. By the Angle Addition Postulate, $m\angle BAC = m\angle BAD + m\angle CAD$. So, $m\angle BAC = \frac{1}{2}m\widehat{DB} + \frac{1}{2}m\widehat{AD} = \frac{1}{2}(m\widehat{DB} + m\widehat{AD})$. By the Arc Addition Postulate, $m\widehat{DB} + m\widehat{AD} = m\widehat{ADB}$, so $m\angle BAC = \frac{1}{2}(m\widehat{ADB})$.

    c.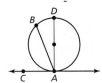

    By the Tangent Line to Circle Theorem, $m\angle CAD = 90°$. $m\angle DAB = \frac{1}{2}m\widehat{DB}$ and by part (a), $m\angle DAC = \frac{1}{2}m\widehat{ABD}$. By the Angle Addition Postulate, $m\angle BAC = m\angle DAC - m\angle DAB$. So, $m\angle BAC = \frac{1}{2}m\widehat{ABD} - \frac{1}{2}m\widehat{DB} = \frac{1}{2}(m\widehat{ABD} - m\widehat{DB})$. By the Arc Addition Postulate, $m\widehat{ABD} - m\widehat{DB} = m\widehat{AB}$, so $m\angle BAC = \frac{1}{2}(m\widehat{AB})$.

34. closer; The smaller arc needs to be 20°.

35.

STATEMENTS	REASONS
1. Chords $\overline{AC}$ and $\overline{BD}$ intersect.	1. Given
2. $m\angle ACB = \frac{1}{2}m\widehat{AB}$ and $m\angle DBC = \frac{1}{2}m\widehat{DC}$	2. Measure of an Inscribed Angle Theorem
3. $m\angle 1 = m\angle DBC + m\angle ACB$	3. Exterior Angle Theorem
4. $m\angle 1 = \frac{1}{2}m\widehat{DC} + \frac{1}{2}m\widehat{AB}$	4. Substitution Property of Equality
5. $m\angle 1 = \frac{1}{2}(m\widehat{DC} + m\widehat{AB})$	5. Distributive Property

36.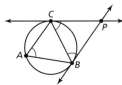

$\angle PCB$, $\angle PBC$, and $\angle CAB$ intercept $\widehat{BC}$.

37. By the Exterior Angle Theorem, $m\angle 2 = m\angle 1 + m\angle ABC$, so $m\angle 1 = m\angle 2 - m\angle ABC$. By the Tangent and Intersected Chord Theorem, $m\angle 2 = \frac{1}{2}m\widehat{BC}$ and by the Measure of an Inscribed Angle Theorem, $m\angle ABC = \frac{1}{2}m\widehat{AC}$. By the Substitution Property, $m\angle 1 = \frac{1}{2}m\widehat{BC} - \frac{1}{2}m\widehat{AC} = \frac{1}{2}(m\widehat{BC} - m\widehat{AC})$;

By the Exterior Angle Theorem, $m\angle 1 = m\angle 2 + m\angle 3$, so $m\angle 2 = m\angle 1 - m\angle 3$. By the Tangent and Intersected Chord Theorem, $m\angle 1 = \frac{1}{2}m\widehat{PQR}$ and $m\angle 3 = \frac{1}{2}m\widehat{PR}$. By the Substitution Property, $m\angle 2 = \frac{1}{2}m\widehat{PQR} - \frac{1}{2}m\widehat{PR} = \frac{1}{2}(m\widehat{PQR} - m\widehat{PR})$;

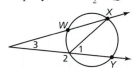

By the Exterior Angle Theorem, $m\angle 1 = m\angle 3 + m\angle WXZ$, so $m\angle 3 = m\angle 1 - m\angle WXZ$. By the Measure of an Inscribed Angle Theorem, $m\angle 1 = \frac{1}{2}m\widehat{XY}$ and $m\angle WXZ = \frac{1}{2}m\widehat{WZ}$. By the Substitution Property, $m\angle 3 = \frac{1}{2}m\widehat{XY} - \frac{1}{2}m\widehat{WZ} = \frac{1}{2}(m\widehat{XY} - m\widehat{WZ})$.

38.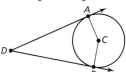

By the Angles Outside the Circle Theorem,
$m\angle ADB = \frac{1}{2}((360° - m\widehat{AB}) - m\widehat{AB})$
$= \frac{1}{2}(360° - 2m\widehat{AB})$
$= 180° - m\widehat{AB}$
$= 180° - m\angle ACB$.

39. 20°; Sample answer: $m\widehat{WY} = 160°$ and $m\widehat{WX} = m\widehat{ZY}$, so
$m\angle P = \frac{1}{2}(m\widehat{WZ} - m\widehat{XY})$
$= \frac{1}{2}((200° - m\widehat{ZY}) - (160° - m\widehat{WX}))$
$= \frac{1}{2}(40°)$.

10.6 Monitoring Progress and Modeling with Mathematics

19.

STATEMENTS	REASONS
1. $\overline{AB}$ and $\overline{CD}$ are chords intersecting in the interior of the circle.	1. Given
2. $\angle AEC \cong \angle DEB$	2. Vertical Angles Congruence Theorem
3. $\angle ACD \cong \angle ABD$	3. Inscribed Angles of a Circle Theorem
4. $\triangle AEC \sim \triangle DEB$	4. AA Similarity Theorem
5. $\dfrac{EA}{ED} = \dfrac{EC}{EB}$	5. Corresponding side lengths of similar triangles are proportional.
6. $EB \cdot EA = EC \cdot ED$	6. Cross Products Property

20.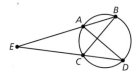

STATEMENTS	REASONS
1. $\overline{EB}$ and $\overline{ED}$ are secant segments.	1. Given
2. $\angle BEC \cong \angle DEA$	2. Reflexive Property of Congruence
3. $\angle ABC \cong \angle ADC$	3. Inscribed Angles of a Circle Theorem
4. $\triangle BCE \sim \triangle DAE$	4. AA Similarity Theorem
5. $\dfrac{EA}{EC} = \dfrac{ED}{EB}$	5. Corresponding side lengths of similar triangles are proportional.
6. $EA \cdot EB = EC \cdot ED$	6. Cross Products Property

21.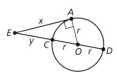

By the Tangent Line to Circle Theorem, $\angle EAO$ is a right angle, which makes $\triangle AEO$ a right triangle. By the Pythagorean Theorem, $(r + y)^2 = r^2 + x^2$. So, $r^2 + 2yr + y^2 = r^2 + x^2$. By the Subtraction Property of Equality, $2yr + y^2 = x^2$. Then $y(2r + y) = x^2$, so $EC \cdot ED = EA^2$.

22.

STATEMENTS	REASONS
1. $\overline{EA}$ is a tangent segment and $\overline{ED}$ is a secant segment.	1. Given
2. $\angle E \cong \angle E$	2. Reflexive Property of Congruence
3. $m\angle EAC = \frac{1}{2}m\widehat{AC}$	3. Tangent and Intersected Chord Theorem
4. $m\angle ADC = \frac{1}{2}m\widehat{AC}$	4. Measure of an Inscribed Angle Theorem
5. $m\angle EAC = m\angle ADC$	5. Transitive Property of Equality
6. $\angle EAC \cong \angle ADC$	6. Definition of congruence
7. $\triangle EAC \sim \triangle EDA$	7. AA Similarity Theorem
8. $\dfrac{EA}{ED} = \dfrac{EC}{EA}$	8. Corresponding side lengths of similar triangles are proportional.
9. $EA^2 = EC \cdot ED$	9. Cross Products Property

23. $BC = \dfrac{AD^2 + (AD)(DE) - AB^2}{AB}$

24. Sample answer: Segments of Secants and Tangents Theorem, Pythagorean Theorem; Use the equation $RQ^2 = RS \cdot RP$ to find RQ. Then use the Pythagorean Theorem to find PQ.

25. $2\sqrt{10}$

26. no; The side lengths are not proportional.

10.6 Maintaining Mathematical Proficiency

27. $x = -9, x = 5$

28. $x = 1 \pm \sqrt{10}$

29. $x = -7, x = 1$

30. $x = 1 \pm 2\sqrt{2}$

10.7 Monitoring Progress and Modeling with Mathematics

36. center: $(-2, -6)$, radius: 5

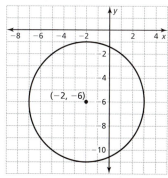

Additional Answers A63

37. center: $(-4, 2)$, radius: 4

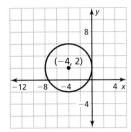

38. center: $(-6, 4)$, radius: 6

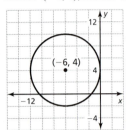

39. The radius of the circle is 8. $\sqrt{(2-0)^2 + (3-0)^2} = \sqrt{13}$, so $(2, 3)$ does not lie on the circle.
40. The radius of the circle is 3. $\sqrt{(4-0)^2 + (\sqrt{5}-0)^2} = \sqrt{21}$, so $(4, \sqrt{5})$ does not lie on the circle.
41. The radius of the circle is $\sqrt{10}$.
$\sqrt{(\sqrt{6}-0)^2 + (2-0)^2} = \sqrt{10}$, so $(\sqrt{6}, 2)$ does lie on the circle.
42. The radius of the circle is $\sqrt{29}$.
$\sqrt{(\sqrt{7}-0)^2 + (5-0)^2} = \sqrt{32}$, so $(\sqrt{7}, 5)$ does not lie on the circle.
43. The radius of the circle is 5. $\sqrt{(4-1)^2 + (4-0)^2} = 5$, so $(4, 4)$ does lie on the circle.
44. The radius of the circle is 10.
$\sqrt{(2.4+3)^2 + (4-2)^2} = \sqrt{33.16}$, so $(\sqrt{6}, 4)$ does not lie on the circle.
45. a.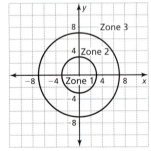

b. zone 2, zone 3, zone 1, zone 1, zone 2
46. a.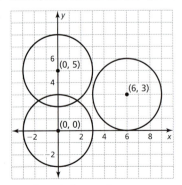

yes; Two circles overlap.

b. City B; Parts of City A do not have coverage.

47. $(-\sqrt{3}, 1), (\sqrt{3}, 1)$
48. no solution
49. $(5 - \sqrt{7}, -5 - \sqrt{7}), (5 + \sqrt{7}, -5 + \sqrt{7})$
50. $(-1, 1)$
51. no solution
52. $(-3, 0), (0, 3)$
53. $-\left(\frac{3}{5}, \frac{4}{5}\right), (1, 0)$
54. $\left(\frac{5}{13}, \frac{12}{13}\right)$
55. $(2.2, -1.6), (3, 0)$
56. no solution
59.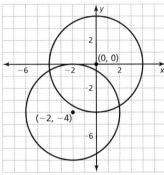

The equation of the image is $(x + 2)^2 + (y + 4)^2 = 16$; The equation of the image of a circle after a translation m units to the left and n units down is $(x + m)^2 + (y + n)^2 = r^2$.

60. $y - b = \dfrac{h - a}{b - k}(x - a)$
61. $(x - 4)^2 + (y - 9)^2 = 16$; $m\angle Z = 90°$, so $\overline{XY}$ is a diameter.
62. yes; The diameter perpendicularly bisects the chord from $(-1, 0)$ to $(1, 0)$, so the center is on the y-axis at $(0, k)$ and the radius is $k^2 + 1$.
63. tangent; The system has one solution.
64. none of these; The system has no solution.
65. secant; The system has two solutions, and $(5, -1)$ is not on the line.
66. secant that contains the diameter; The system has two solutions, and $(-3, 6)$ is on the line.
67. $(x - 15)^2 + y^2 = 100$; Sample answer: If r is the radius of $\odot O$, C is at $(15r, 0)$, and the distance to $(63, 16)$ is $4r$.

Chapter 11

11.1 Monitoring Progress and Modeling with Mathematics

40. no; Sample answer: The values are the same to 6 decimal places; $\dfrac{104{,}348}{33{,}215}$

41. Sample answer:

STATEMENTS	REASONS
1. $\overline{FG} \cong \overline{GH}$, $\angle JFK \cong \angle KLF$	1. Given
2. $FG = GH$	2. Definition of congruent segments
3. $FH = FG + GH$	3. Segment Addition Postulate
4. $FH = 2FG$	4. Substitution Property of Equality
5. $m\angle JFK = m\angle KFL$	5. Definition of congruent angles
6. $m\angle JFL = m\angle JFK + m\angle KFL$	6. Angle Addition Postulate
7. $m\angle JFL = 2m\angle JFK$	7. Substitution Property of Equality
8. $\angle NFG \cong \angle JFL$	8. Vertical Angles Congruence Theorem
9. $m\angle NFG = m\angle JFL$	9. Definition of congruent angles
10. $m\angle NFG = 2m\angle JFK$	10. Substitution Property of Equality
11. arc length of $\widehat{JK}$ $= \dfrac{m\angle JFK}{360°} \cdot 2\pi FH$, arc length of $\widehat{NG}$ $= \dfrac{m\angle NFG}{360°} \cdot 2\pi FG$	11. Formula for arc length
12. arc length of $\widehat{JK}$ $= \dfrac{m\angle JFK}{360°} \cdot 2\pi(2FG)$, arc length of $\widehat{NG}$ $= \dfrac{2m\angle JFK}{360°} \cdot 2\pi FG$	12. Substitution Property of Equality
13. arc length of $\widehat{NG}$ $=$ arc length of $\widehat{JK}$	13. Transitive Property of Equality

42. a. $4\pi r$
    b. $4\pi r$; $4\pi r$; $4\pi r$; The total length of the segments is the same.

11.1 Maintaining Mathematical Proficiency

43. 15 square units
44. 42 square units

11.2 Monitoring Progress and Modeling with Mathematics

40. yes; Sample answer: The side length of each of the small squares is $\frac{1}{8}$ the side length of the outer square, and each small square is either completely colored or has a quarter circle colored.

41. a. You should buy two 14-inch pizzas; Sample answer: The area is 98π square inches and the cost is $25.98.
    b. You should buy two 10-inch pizzas and one 14-inch pizza; Sample answer: Buying three 10-inch pizzas is the only cheaper option, and it would not be enough pizza.
    c. You should buy four 10-inch pizzas; Sample answer: The total circumference is 20π inches.

42. πab
43. a. 2.4 in.^{2}; 4.7 in.^{2}; 7.1 in.^{2}; 9.4 in.^{2}; 11.8 in.^{2}; 14.1 in.^{2}
    b.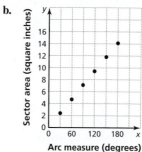
    c. yes; Sample answer: The rate of change is constant.
    d. yes; no; Sample answer: The rate of change will still be constant.
44. about 5.81 in.^{2}
45. Sample answer: Let $2a$ and $2b$ represent the lengths of the legs of the triangle. The areas of the semicircles are $\frac{1}{2}\pi a^2$, $\frac{1}{2}\pi b^2$, and $\frac{1}{2}\pi(a^2+b^2)$. $\frac{1}{2}\pi a^2 + \frac{1}{2}\pi b^2 = \frac{1}{2}\pi(a^2+b^2)$, and subtracting the areas of the unshaded regions from both sides leaves the area of the crescents on the left and the area of the triangle on the right.

11.2 Maintaining Mathematical Proficiency

46. 54 in.^{2}
47. 49 ft^2
48. 58.5 in.^{2}
49. 15 ft^2

11.3 Maintaining Mathematical Proficiency

53.

		Response		
		Yes	No	Total
Gender	Male	17	52	69
	Female	48	62	110
	Total	65	114	179

54.

		Response		
		Yes	No	Total
Gender	Junior	65	40	105
	Senior	51	12	63
	Total	116	52	168

11.4 Monitoring Progress and Modeling with Mathematics

43. 36 ft, 15 ft
44. 218.0 cm^3

11.4 Maintaining Mathematical Proficiency

45. 16 m^2
46. 406.3 cm^2
47. 680.4 in.^{2}

11.5 Monitoring Progress and Modeling with Mathematics

24. $V = \frac{1}{3}(a^2 + ab + b^2)h$
25. about 78 in.^{3}

11.5 Maintaining Mathematical Proficiency

26. 12.9
27. 12.6
28. 5.8
29. 16.0

11.6 Monitoring Progress and Modeling with Mathematics

19. about 3716.85 ft^3
20. about 574.82 yd^3
21. yes; *Sample answer:* The automatic pet feeder holds about 12 cups of food.
22. about 17.45 sec
23. It is half; about 60°
24. $V = \frac{1}{3}\pi h(a^2 + ab + b^2)$
25. yes; *Sample answer:* The base areas are the same and the total heights are the same.
26. a. Because $r = \frac{1}{2}h$, $V = \frac{1}{3}\pi r^2 h = \frac{1}{3}\pi\left(\frac{1}{2}h\right)^2 h = \frac{\pi h^3}{12}$.

   b.
Time (in minutes)	Height (in meters)
1	1.90
2	2.40
3	2.74
4	3.02
5	3.25

   c. no; *Sample answer:* The change in height is not constant.

11.6 Maintaining Mathematical Proficiency

27. about 153.94 ft^2
28. about 380.13 cm^2
29. 32 m
30. 23 in.

11.7 Monitoring Progress and Modeling with Mathematics

45. *Sample answer:* radius 1 in. and height $\frac{4}{3}$ in.; radius $\frac{1}{3}$ in. and height 12 in.; radius 2 in. and height $\frac{1}{3}$ in.
46. a. about 54.45 ft^3
    b. about 28,500.53 cm^3
    c. about 2077.64 m^3
    d. about 522,170.40 in.^{3}
47. $S \approx 113.10$ in.^{2}, $V \approx 75.40$ in.^{3}

11.7 Maintaining Mathematical Proficiency

48. 0.6
49. 0.8
50. $\frac{7}{8}$
51. $\frac{7}{10}$

English-Spanish Glossary

English | Spanish

A

absolute value function *(p. 4)* A function that contains an absolute value expression

adjacent arcs *(p. 583)* Arcs of a circle that have exactly one point in common

altitude of a triangle *(p. 363)* The perpendicular segment from a vertex of a triangle to the opposite side or to the line that contains the opposite side

angle of depression *(p. 551)* The angle that a downward line of sight makes with a horizontal line

angle of elevation *(p. 544)* The angle that an upward line of sight makes with a horizontal line

apothem of a regular polygon *(p. 657)* The distance from the center to any side of a regular polygon

arc length *(p. 641)* A portion of the circumference of a circle

asymptote *(p. 42)* A line that a graph approaches more and more closely

axis of symmetry *(p. 124)* The vertical line that divides a parabola into two symmetric parts

función de valor absoluto *(p. 4)* Una función que contiene una expresión de valor absoluto

arcos adyacentes *(p. 583)* Arcos de un círculo que tienen exactamente un punto en común

altitud de un triángulo *(p. 363)* El segmento perpendicular desde el vértice de un triángulo al lado opuesto o a la recta que contiene el lado opuesto

ángulo de depresión *(p. 551)* El ángulo formado entre una recta de vista descendente y una recta horizontal

ángulo de elevación *(p. 544)* El ángulo formado entre una recta de vista ascendente y una recta horizontal

apotema de un polígono regular *(p. 657)* La distancia desde el centro a cualquier lado de un polígono regular

longitud de arco *(p. 641)* Una porción de la circunferencia de un círculo

asíntota *(p. 42)* Una recta a la que una gráfica se acerca cada vez más

eje de simetría *(p. 124)* La línea vertical que divide una parábola en dos partes simétricas

B

base angles of a trapezoid *(p. 442)* Either pair of consecutive angles whose common side is a base of a trapezoid

bases of a trapezoid *(p. 442)* The parallel sides of a trapezoid

binomial *(p. 63)* A polynomial with two terms

binomial distribution *(p. 321)* A type of probability distribution that shows the probabilities of the outcomes of a binomial experiment

binomial experiment *(p. 321)* An experiment in which there are a fixed number of independent trials, exactly two possible outcomes for each trial, and the probability of success is the same for each trial

ángulos de la base de un trapecio *(p. 442)* Cualquier par de ángulos consecutivos cuyo lado común es la base de un trapezoide

bases de un trapecio *(p. 442)* Los lados paralelos de un trapezoide

binomio *(p. 63)* Un polinomio con dos términos

distribución del binomio *(p. 321)* Un tipo de distribución de probabilidades que muestra las probabilidades de los resultados posibles de un experimento del binomio

experimento del binomio *(p. 321)* Un experimento en el que hay un número fijo de pruebas independientes, exactamente dos resultados posibles para cada prueba, y la probabilidad de éxito es la misma para cada prueba

C

Cavalieri's Principle (p. 667) If two solids have the same height and the same cross-sectional area at every level, then they have the same volume.

center of a circle (p. 574) The point from which all points on a circle are equidistant

center of dilation (p. 462) The fixed point in a dilation

center of a regular polygon (p. 657) The center of a polygon's circumscribed circle

central angle of a circle (p. 582) An angle whose vertex is the center of a circle

central angle of a regular polygon (p. 657) An angle formed by two radii drawn to consecutive vertices of a polygon

centroid (p. 362) The point of concurrency of the three medians of a triangle

chord of a circle (p. 574) A segment whose endpoints are on a circle

chord of a sphere (p. 686) A segment whose endpoints are on a sphere

circle (p. 574) The set of all points in a plane that are equidistant from a given point

circumcenter (p. 352) The point of concurrency of the three perpendicular bisectors of a triangle

circumference (p. 640) The distance around a circle

circumscribed angle (p. 608) An angle whose sides are tangent to a circle

closed (p. 64) When an operation performed on any two numbers in the set results in a number that is also in the set

combination (p. 314) A selection of objects in which order is not important

common tangent (p. 575) A line or segment that is tangent to two coplanar circles

completing the square (p. 216) To add a constant c to an expression of the form $x^2 + bx$ so that $x^2 + bx + c$ is a perfect square trinomial

Principio de Cavalieri (p. 667) Si dos sólidos tienen la misma altura y la misma área transversal en todo nivel, entonces tienen el mismo volumen.

centro de un círculo (p. 574) El punto desde donde todos los puntos en un círculo son equidistantes

centro de dilatación (p. 462) El punto fijo en una dilatación

centro de un polígono regular (p. 657) El centro del círculo circunscrito de un polígono

ángulo central de un círculo (p. 582) Un ángulo cuyo vértice es el centro de un círculo

ángulo central de un polígono regular (p. 657) Un ángulo formado por dos radios extendidos a vértices consecutivos de un polígono

centroide (p. 362) El punto de concurrencia de las tres medianas de un triángulo

cuerda de un círculo (p. 574) Un segmento cuyos puntos extremos están en un círculo

cuerda de una esfera (p. 686) Un segmento cuyos puntos extremos están en una esfera

círculo (p. 574) El conjunto de todos los puntos en un plano que son equidistantes de un punto dado

circuncentro (p. 352) El punto de concurrencia de las tres bisectrices perpendiculares de un triángulo

circunferencia (p. 640) La distancia alrededor de un círculo

ángulo circunscrito (p. 608) Un ángulo cuyos lados son tangentes a un círculo

operación interna (p. 64) Cuando una operación efectuada en dos números cualesquiera del conjunto da como resultado un número que también está en el conjunto

combinación (p. 314) Una selección de objetos en la que el orden no es importante

tangente común (p. 575) Una recta o segmento que es tangente a dos círculos coplanarios

completando el cuadrado (p. 216) Agregar una constante c a una expresión de la forma $x^2 + bx$ para que $x^2 + bx + c$ sea un trinomio de cuadrado perfecto

complex conjugates *(p. 241)* Pairs of complex numbers of the forms $a + bi$ and $a - bi$, where $b \neq 0$

complex number *(p. 238)* A number written in the form $a + bi$, where a and b are real numbers

compound event *(p. 306)* The union or intersection of two events

concentric circles *(p. 575)* Coplanar circles that have a common center

concurrent *(p. 352)* Three or more lines, rays, or segments that intersect in the same point

conditional probability *(p. 289)* The probability that event B occurs given that event A has occurred, written as $P(B|A)$

conditional relative frequency *(p. 297)* The ratio of a joint relative frequency to the marginal relative frequency in a two-way table

congruent arcs *(p. 584)* Arcs that have the same measure and are of the same circle or of congruent circles

congruent circles *(p. 584)* Circles that can be mapped onto each other by a rigid motion or a composition of rigid motions

conjugates *(p. 194)* Binomials of the form $a\sqrt{b} + c\sqrt{d}$ and $a\sqrt{b} - c\sqrt{d}$, where a, b, c, and d are rational numbers

coordinate proof *(p. 339)* A style of proof that involves placing geometric figures in a coordinate plane

cosine *(p. 548)* For an acute angle of a right triangle, the ratio of the length of the leg adjacent to the acute angle to the length of the hypotenuse

counterexample *(p. 191)* An example that proves that a general statement is not true

---

D

---

decay factor *(p. 42)* The value of b in an exponential decay function of the form $y = ab^x$, where $a > 0$ and $0 < b < 1$

degree of a monomial *(p. 62)* The sum of the exponents of the variables in the monomial

degree of a polynomial *(p. 63)* The greatest degree of the terms in a polynomial

dependent events *(p. 289)* Two events in which the occurrence of one event does affect the occurrence of the other event

conjugados complejos *(p. 241)* Pares de números complejos de las formas $a + bi$ y $a - bi$, donde $b \neq 0$

número complejo *(p. 238)* Un número escrito en la forma $a + bi$, donde a y b son números reales

evento compuesto *(p. 306)* La unión o intersección de dos eventos

círculos concéntricos *(p. 575)* Círculos coplanarios que tienen un centro en común

concurrente *(p. 352)* Tres o más rectas, rayos o segmentos que se intersectan en el mismo punto

probabilidad condicional *(p. 289)* La probabilidad de que el evento B ocurra dado que el evento A ha ocurrido, escrito como $P(B|A)$

frecuencia relativa condicional *(p. 297)* La razón de una frecuencia relativa conjunta a la frecuencia relativa marginal en una tabla de doble entrada

arcos congruentes *(p. 584)* Arcos que tienen la misma medida y que son del mismo círculo o de círculos congruentes

círculos congruentes *(p. 584)* Círculos que pueden superponerse sobre sí mismos mediante un movimiento rígido o una composición de movimientos rígidos

conjugados *(p. 194)* Binomios de la forma $a\sqrt{b} + c\sqrt{d}$ y $a\sqrt{b} - c\sqrt{d}$, donde a, b, c y d son números racionales

prueba de coordenadas *(p. 339)* Un estilo de prueba que implica colocar figuras geométricas en un plano coordenado

coseno *(p. 548)* Para un ángulo agudo de un triángulo rectángulo, la razón de la longitud del cateto adyacente al ángulo agudo a la longitud de la hipotenusa

contraejemplo *(p. 191)* Un ejemplo que prueba que una afirmación general no es verdadera

factor de decaimiento *(p. 42)* El valor de b en una función de decaimiento exponencial de la forma $y = ab^x$, donde $a > 0$ y $0 < b < 1$

grado de un monomio *(p. 62)* La suma de los exponentes de las variables en el monomio

grado de un polinomio *(p. 63)* El grado mayor de los términos en un polinomio

eventos dependientes *(p. 289)* Dos eventos en los que la ocurrencia de un evento afecta la ocurrencia del otro evento

diagonal (p. 404) A segment that joins two nonconsecutive vertices of a polygon

diameter (p. 574) A chord that contains the center of a circle

dilation (p. 462) A transformation in which a figure is enlarged or reduced with respect to a fixed point

directed line segment (p. 502) A segment that represents moving from point A to point B is called the directed line segment AB.

directrix (p. 162) A fixed line perpendicular to the axis of symmetry, such that the set of all points (x, y) of the parabola are equidistant from the focus and the directrix

discriminant (p. 228) The expression $b^2 - 4ac$ in the Quadratic Formula

disjoint events (p. 306) Two events that have no outcomes in common

diagonal (p. 404) Un segmento que une dos vértices no consecutivos de un polígono

diámetro (p. 574) Una cuerda que contiene el centro de un círculo

dilatación (p. 462) Una transformación en la cual una figura se agranda o reduce con respecto a un punto fijo

segmento de línea dirigido (p. 502) Un segmento que representa el moverse del punto A al punto B se llama el segmento de línea dirigido AB.

directriz (p. 162) Una recta fija perpendicular al eje de simetría de modo tal, que el conjunto de todos los puntos (x, y) de la parábola sean equidistantes del foco y la directriz

discriminante (p. 228) La expresión $b^2 - 4ac$ en la Fórmula Cuadrática

eventos disjunto (p. 306) Dos eventos que no tienen resultados en común

edge (p. 666) A line segment formed by the intersection of two faces of a polyhedron

enlargement (p. 462) A dilation in which the scale factor is greater than 1

equiangular polygon (p. 405) A polygon in which all angles are congruent

equidistant (p. 344) A point is equidistant from two figures when it is the same distance from each figure.

equilateral polygon (p. 405) A polygon in which all sides are congruent

even function (p. 146) A function $y = f(x)$ is even when $f(-x) = f(x)$ for each x in the domain of f.

event (p. 280) A collection of one or more outcomes in a probability experiment

experimental probability (p. 282) The ratio of the number of successes, or favorable outcomes, to the number of trials in a probability experiment

exponential decay function (p. 42) A function of the form $y = a(1 - r)^t$, where $a > 0$ and $0 < r < 1$

exponential function (p. 42) A nonlinear function of the form $y = ab^x$, where $a \neq 0$, $b \neq 1$, and $b > 0$

borde (p. 666) Un segmento de línea formado por la intersección de dos caras de un poliedro

agrandamiento (p. 462) Una dilatación en donde el factor de escala es mayor que 1

polígono equiangular (p. 405) Un polígono en donde todos los ángulos son congruentes

equidistante (p. 344) Un punto es equidistante desde dos figuras cuando está a la misma distancia de cada figura.

polígono equilátero (p. 405) Un polígono en donde todos los lados son congruentes

par función (p. 146) Una función $y = f(x)$ es par cuando $f(-x) = f(x)$ para cada x en el dominio de f.

evento (p. 280) Una colección de uno o más resultados en un experimento de probabilidades

probabilidad experimental (p. 282) La razón del número de éxitos, o resultados favorables, con respecto al número de pruebas en un experimento de probabilidades

función de decaimiento exponencial (p. 42) Una función de la forma $y = a(1 - r)^t$, donde $a > 0$ y $0 < r < 1$

función exponencial (p. 42) Una función no lineal de la forma $y = ab^x$, donde $a \neq 0$, $b \neq 1$, y $b > 0$

exponential growth function (p. 42) A function of the form $y = a(1 + r)^t$, where $a > 0$ and $r > 0$

external segment (p. 615) The part of a secant segment that is outside the circle

función de crecimiento exponencial (p. 42) Una función de la forma $y = a(1 + r)^t$, donde $a > 0$ y $r > 0$

segmento externo (p. 615) La parte de un segmento secante que está fuera del círculo

face (p. 666) A flat surface of a polyhedron

cara (p. 666) Una superficie plana de un poliedro

factored completely (p. 108) A polynomial that is written as a product of unfactorable polynomials with integer coefficients

factorizado completamente (p. 108) Un polinomio que se escribe como un producto de polinomios no factorizables con coeficientes de números enteros

factored form (p. 82) A polynomial that is written as a product of factors

forma factorizada (p. 82) Un polinomio que se escribe como un producto de factores

factoring by grouping (p. 108) To use the Distributive Property to factor a polynomial with four terms

factorización por agrupamiento (p. 108) Uso de la propiedad distributiva para factorizar un polinomio con cuatro términos

flowchart proof (flow proof) (p. 338) A type of proof that uses boxes and arrows to show the flow of a logical argument

prueba de organigrama (prueba de flujo) (p. 338) Un tipo de prueba que usa casillas y flechas para mostrar el flujo de un argumento lógico

focus (p. 162) A fixed point in the interior of a parabola, such that the set of all points (x, y) of the parabola are equidistant from the focus and the directrix

foco (p. 162) Un punto fijo en el interior de una parábola, de tal forma que el conjunto de todos los puntos (x, y) de la parábola sean equidistantes del foco y la directriz

FOIL Method (p. 71) A shortcut for multiplying two binomials by finding the sum of the products of the first terms, outer terms, inner terms, and last terms

método FOIL o de multiplicación de binomios (p. 71) Un método rápido para multiplicar dos binomios encontrando la suma de los productos de los primeros términos, los segundos términos, los terceros términos y los últimos términos

geometric mean (p. 534) The positive number x that satisfies $\frac{a}{x} = \frac{x}{b}$
So, $x^2 = ab$ and $x = \sqrt{ab}$.

media geométrica (p. 534) El número positivo x que satisface $\frac{a}{x} = \frac{x}{b}$
Entonces, $x^2 = ab$ and $x = \sqrt{ab}$.

geometric probability (p. 649) A probability found by calculating a ratio of two lengths, areas, or volumes

probabilidad geométrica (p. 649) Una probabilidad hallada al calcular la razón de dos longitudes, áreas o volúmenes

great circle (p. 686) The intersection of a plane and a sphere such that the plane contains the center of the sphere

gran círculo (p. 686) La intersección de un plano y una esfera, de tal forma que el plano contiene el centro de la esfera

growth factor (p. 42) The value of b in an exponential growth function of the form $y = ab^x$, where $a > 0$ and $b > 1$

factor de crecimiento (p. 42) El valor de b en una función de crecimiento exponencial de la forma $y = ab^x$, donde $a > 0$ y $b > 1$

I

imaginary number *(p. 238)* A number written in the form $a + bi$, where a and b are real numbers and $b \neq 0$

imaginary unit *i* *(p. 238)* The square root of -1, denoted $i = \sqrt{-1}$

incenter *(p. 355)* The point of concurrency of the angle bisectors of a triangle

independent events *(p. 288)* Two events in which the occurrence of one event does not affect the occurrence of another event

index of a radical *(p. 36)* The value of n in the radical $\sqrt[n]{a}$

indirect proof *(p. 378)* A style of proof in which you temporarily assume that the desired conclusion is false, then reason logically to a contradiction

This proves that the original statement is true.

inscribed angle *(p. 598)* An angle whose vertex is on a circle and whose sides contain chords of the circle

intercept form *(p. 154)* A quadratic function written in the form $f(x) = a(x - p)(x - q)$, where $a \neq 0$

intercepted arc *(p. 598)* An arc that lies between two lines, rays, or segments

inverse cosine *(p. 558)* An inverse trigonometric ratio, abbreviated as $\cos^{-1}$

For acute angle A, if $\cos A = z$, then $\cos^{-1} z = m\angle A$.

inverse function *(p. 21)* Functions that undo each other

inverse relation *(p. 20)* When the input and output values of the original relation are switched

inverse sine *(p. 558)* An inverse trigonometric ratio, abbreviated as $\sin^{-1}$

For acute angle A, if $\sin A = y$, then $\sin^{-1} y = m\angle A$.

inverse tangent *(p. 558)* An inverse trigonometric ratio, abbreviated as $\tan^{-1}$

For acute angle A, if $\tan A = x$, then $\tan^{-1} x = m\angle A$.

isosceles trapezoid *(p. 442)* A trapezoid with congruent legs

número imaginario *(p. 238)* Un número escrito de la forma $a + bi$, donde a y b son números reales y $b \neq 0$

unidad imaginaria *i* *(p. 238)* La raíz cuadrada de -1, denotado $i = \sqrt{-1}$

incentro *(p. 355)* El punto de concurrencia de las bisectrices de los ángulos de un triángulo

eventos independientes *(p. 288)* Dos eventos en los que la ocurrencia de un evento no afecta la ocurrencia de otro evento

índice de un radical *(p. 36)* El valor de n en el radical $\sqrt[n]{a}$

prueba indirecta *(p. 378)* Un estilo de prueba en donde uno asume temporalmente que la conclusión deseada es falsa, luego se razona de forma lógica hasta llegar a una contradicción

Esto prueba que el enunciado original es vertdadero.

ángulo inscrito *(p. 598)* Un ángulo cuyo vértice está en un círculo y cuyos lados contienen cuerdas del círculo

forma intersección *(p. 154)* Una ecuación cuadrática escrita en la forma $f(x) = a(x - p)(x - q)$, donde $a \neq 0$

arco interceptado *(p. 598)* Un arco que descansa entre dos rectas, rayos o segmentos

coseno inverso *(p. 558)* Una razón trigonométrica inversa, abreviada como $\cos^{-1}$

Para un ángulo agudo A, si $\cos A = z$, entonces $\cos^{-1} z = m\angle A$.

función inversa *(p. 21)* Funciones que se anulan entre sí

relación inversa *(p. 20)* Cuando los valores y los resultados de la relación original se intercambian

seno inverso *(p. 558)* Una razón trigonométrica inversa, abreviada como $\sin^{-1}$

Para un ángulo agudo A, si $\sin A = y$, entonces $\sin^{-1} y = m\angle A$.

tangente inversa *(p. 558)* Una razón trigonométrica inversa, abreviada como $\tan^{-1}$

Para un ángulo agudo A, si $\tan A = x$, entonces $\tan^{-1} x = m\angle A$.

trapecio isósceles *(p. 442)* Un trapecio con catetos congruentes

joint frequency *(p. 296)* Each entry in a two-way table

joint relative frequency *(p. 297)* The ratio of a frequency that is not in the total row or the total column to the total number of values or observations in a two-way table

frecuencia conjunta *(p. 296)* Cada valor en una tabla de doble entrada

frecuencia relativa conjunta *(p. 297)* La razón de una frecuencia que no está en la hilera total o columna total del número total de valores u observaciones en una tabla de doble entrada

kite *(p. 445)* A quadrilateral that has two pairs of consecutive congruent sides, but opposite sides are not congruent

papalote *(p. 445)* Un cuadrilátero que tiene dos pares de lados congruentes consecutivos, pero los lados opuestos no son congruentes

lateral surface of a cone *(p. 680)* Consists of all segments that connect the vertex with points on the base edge of a cone

superficie lateral de un cono *(p. 680)* Consiste en todos los segmentos que conectan el vértice con puntos en el borde base de un cono

leading coefficient *(p. 63)* The coefficient of the first term of a polynomial written in standard form

coeficiente principal *(p. 63)* El coeficiente del primer término de un polinomio escrito en forma estándar

legs of a trapezoid *(p. 442)* The nonparallel sides of a trapezoid

catetos de un trapecio *(p. 442)* Los lados no paralelos de un trapezoide

like radicals *(p. 196)* Radicals with the same index and radicand

radicales semejantes *(p. 196)* Radicales con el mismo índice y radicando

major arc *(p. 582)* An arc with a measure greater than 180°

arco mayor *(p. 582)* Un arco con una medida mayor de 180°

marginal frequency *(p. 296)* The sums of the rows and columns in a two-way table

frecuencia marginal *(p. 296)* Las sumas de las hileras y columnas en una tabla de doble entrada

marginal relative frequency *(p. 297)* The sum of the joint relative frequencies in a row or a column in a two-way table

frecuencia relativa marginal *(p. 297)* La suma de las frecuencias relativas conjuntas en una hilera o columna en una tabla de doble entrada

maximum value *(p. 137)* The y-coordinate of the vertex of the graph of $f(x) = ax^2 + bx + c$, where $a < 0$

valor máximo *(p. 137)* La coordenada y del vértice del gráfico de $f(x) = ax^2 + bx + c$, donde $a < 0$

measure of a major arc *(p. 582)* The measure of a major arc's central angle

medida de arco mayor *(p. 582)* La medida del ángulo central de un arco mayor

measure of a minor arc *(p. 582)* The measure of a minor arc's central angle

medida de arco menor *(p. 582)* La medida del ángulo central de un arco menor

median of a triangle *(p. 362)* A segment from a vertex of a triangle to the midpoint of the opposite side

mediana de un triángulo *(p. 362)* Un segmento desde el vértice de un triángulo hasta el punto medio del lado opuesto

midsegment of a trapezoid *(p. 444)* The segment that connects the midpoints of the legs of a trapezoid

midsegment of a triangle *(p. 372)* A segment that connects the midpoints of two sides of a triangle

minimum value *(p. 137)* The y-coordinate of the vertex of the graph of $f(x) = ax^2 + bx + c$, where $a > 0$

minor arc *(p. 582)* An arc with a measure less than $180°$

monomial *(p. 62)* A number, a variable, or the product of a number and one or more variables with whole number exponents

mutually exclusive events *(p. 306)* Two events that have no outcomes in common

segmento medio de un trapezoide *(p. 444)* El segmento que conecta los puntos medios de los catetos de un trapezoide

segmento medio de un triángulo *(p. 372)* Un segmento que conecta los puntos medios de dos lados de un triángulo

valor mínimo *(p. 137)* La coordenada y del vértice del gráfico de $f(x) = ax^2 + bx + c$, donde $a > 0$

arco menor *(p. 582)* Un arco con una medida menor de $180°$

monomio *(p. 62)* Un número, una variable o el producto de un número y una o más variables con exponentes en números enteros

eventos mutuamente exclusivos *(p. 306)* Dos eventos que no tienen resultados en común

---

N

---

n factorial *(p. 312)* The product of the integers from 1 to n, for any positive integer n

nth root of a *(p. 36)* For an integer n greater than 1, if $b^n = a$, then b is an nth root of a.

factorial de n *(p. 312)* El producto de los números enteros de 1 a n, para cualquier número entero positivo n

raíz de orden n de a *(p. 36)* Para un número entero n mayor que 1, si $b^n = a$, entonces b es una raíz de orden n de a.

O

---

odd function *(p. 146)* A function $y = f(x)$ is odd when $f(-x) = -f(x)$ for each x in the domain of f.

odds *(p. 283)* A comparison of the number of favorable and unfavorable outcomes of an event when all outcomes are equally likely

orthocenter *(p. 363)* The point of concurrency of the lines containing the altitudes of a triangle

outcome *(p. 280)* The possible result of a probability experiment

overlapping events *(p. 306)* Two events that have one or more outcomes in common

impar funcíon *(p. 146)* Una función $y = f(x)$ es impar cuando $f(-x) = -f(x)$ para cada x en el dominio de f.

probabilidad de un evento *(p. 283)* Una comparación del número de resultados favorables y desfavorables de un evento cuando todos los resultados son igualmente probables

ortocentro *(p. 363)* El punto de concurrencia de las líneas que contienen las alturas de un triángulo

resultado *(p. 280)* El resultado posible de un experimento de probabilidad

eventos superpuestos *(p. 306)* Dos eventos que tienen uno o más resultados en común

P

parabola *(p. 124)* The U-shaped graph of a quadratic function

paragraph proof *(p. 337)* A style of proof that presents the statements and reasons as sentences in a paragraph, using words to explain the logical flow of an argument

parallelogram *(p. 412)* A quadrilateral with both pairs of opposite sides parallel

permutation *(p. 312)* An arrangement of objects in which order is important

piecewise function *(p. 12)* A function defined by two or more equations

point of concurrency *(p. 352)* The point of intersection of concurrent lines, rays, or segments

point of tangency *(p. 574)* The point at which a tangent line intersects a circle

polyhedron *(p. 666)* A solid that is bounded by polygons

polynomial *(p. 63)* A monomial or a sum of monomials

probability distribution *(p. 320)* A function that gives the probability of each possible value of a random variable

probability of an event *(p. 280)* A measure of the likelihood, or chance, that an event will occur

probability experiment *(p. 280)* An action, or trial, that has varying results

proof *(p. 336)* A logical argument that uses deductive reasoning to show that a statement is true

pure imaginary number *(p. 238)* A number written in the form $a + bi$, where $a = 0$ and $b \neq 0$

Pythagorean triple *(p. 518)* A set of three positive integers a, b, and c that satisfy the equation $c^2 = a^2 + b^2$

parábola *(p. 124)* El gráfico en forma de U de una función cuadrática

prueba en forma de párrafo *(p. 337)* Un estilo de prueba que presenta los enunciados y motivos como oraciones en un párrafo, usando palabras para explicar el flujo lógico de un argumento

paralelogramo *(p. 412)* Un cuadrilátero con ambos pares de lados opuestos paralelos

permutación *(p. 312)* Una disposición de objetos en la que el orden es importante

función por tramos *(p. 12)* Una función definida por dos o más ecuaciones

punto de concurrencia *(p. 352)* El punto de intersección de rectas, rayos o segmentos concurrentes

punto de tangencia *(p. 574)* El punto en donde una recta tangente intersecta a un círculo

poliedro *(p. 666)* Un sólido que está encerrado por polígonos

polinomio *(p. 63)* Un monomio o una suma de monomios

distribución de probabilidad *(p. 320)* Una función que da la probabilidad de cada valor posible de una variable aleatoria

probabilidad de un evento *(p. 280)* Una medida de la probabilidad o posibilidad de que ocurrirá un evento

experimento de probabilidad *(p. 280)* Una acción o prueba que tiene resultados variables

prueba *(p. 336)* Un argumento lógico que usa el razonamiento deductivo para mostrar que un enunciado es verdadero

número imaginario puro *(p. 238)* Un número escrito en la forma $a + bi$, donde $a = 0$ y $b \neq 0$

triple pitagórico *(p. 518)* Un conjunto de tres números enteros positivos a, b, y c que satisfacen la ecuación $c^2 = a^2 + b^2$

Q

quadratic equation *(p. 202)* A nonlinear equation that can be written in the standard form $ax^2 + bx + c = 0$, where $a \neq 0$

Quadratic Formula *(p. 226)* The real solutions of the quadratic equation $ax^2 + bx + c = 0$ are

$$x = \frac{-b \pm \sqrt{b^2 - 4ac}}{2a}$$

where $a \neq 0$ and $b^2 - 4ac \geq 0$.

quadratic function *(p. 124)* A nonlinear function that can be written in the standard form $y = ax^2 + bx + c$, where $a \neq 0$

quadratic inequality in one variable *(p. 262)* An inequality of the form $ax^2 + bx + c < 0$, $ax^2 + bx + c > 0$, $ax^2 + bx + c \leq 0$, or $ax^2 + bx + c \geq 0$, where a, b, and c are real numbers and $a \neq 0$

quadratic inequality in two variables *(p. 260)* An inequality of the form $y < ax^2 + bx + c$, $y > ax^2 + bx + c$, $y \leq ax^2 + bx + c$, or $y \geq ax^2 + bx + c$, where a, b, and c are real numbers and $a \neq 0$

ecuación cuadrática *(p. 202)* Una ecuación no lineal que puede escribirse en la forma estándar $ax^2 + bx + c = 0$, donde $a \neq 0$

fórmula cuadrática *(p. 226)* Las soluciones reales de la ecuación cuadrática $ax^2 + bx + c = 0$ son

$$x = \frac{-b \pm \sqrt{b^2 - 4ac}}{2a}$$

donde $a \neq 0$ y $b^2 - 4ac \geq 0$.

función cuadrática *(p. 124)* Una ecuación no lineal que puede escribirse en la forma estándar $y = ax^2 + bx + c$, donde $a \neq 0$

desigualdad cuadrática en una variable *(p. 262)* Una desigualdad de la forma $ax^2 + bx + c < 0$, $ax^2 + bx + c > 0$, $ax^2 + bx + c \leq 0$, o $ax^2 + bx + c \geq 0$, donde a, b, y c son números reales y $a \neq 0$

desigualdad cuadrática en dos variables *(p. 260)* Una desigualdad de la forma $y < ax^2 + bx + c$, $y > ax^2 + bx + c$, $y \leq ax^2 + bx + c$, o $y \geq ax^2 + bx + c$, donde a, b, y c son números reales y $a \neq 0$

R

radian *(p. 643)* A unit of measurement for angles

radical *(p. 36)* An expression of the form $\sqrt[n]{a}$

radical expression *(p. 192)* An expression that contains a radical

radius of a circle *(p. 574)* A segment whose endpoints are the center and any point on a circle

radius of a regular polygon *(p. 657)* The radius of a polygon's circumscribed circle

random variable *(p. 320)* A variable whose value is determined by the outcomes of a probability experiment

rationalizing the denominator *(p. 194)* To eliminate a radical from the denominator of a fraction by multiplying by an appropriate form of 1

rectangle *(p. 432)* A parallelogram with four right angles

recursive rule for an exponential function *(p. 44)* A rule that gives the initial value of the function $f(0)$, and a recursive equation that tells how a value $f(n)$ is related to a preceding value $f(n-1)$

radián *(p. 643)* Una unidad de medida para ángulos

radical *(p. 36)* Una expresión de la forma $\sqrt[n]{a}$

expresión radical *(p. 192)* Una expresión que contiene un radical

radio de un círculo *(p. 574)* Un segmento cuyos puntos extremos son el centro y cualquier punto en un círculo

radio de un polígono regular *(p. 657)* El radio del círculo circunscrito de un polígono

variable aleatoria *(p. 320)* Una variable cuyo valor está determinado por los resultados de un experimento de probabilidad

racionalización del denominador *(p. 194)* Eliminar un radical del denominador de una fracción multiplicando por una forma apropiada de 1

rectángulo *(p. 432)* Un paralelogramo con cuatro ángulos rectos

regla recursiva de una función exponencial *(p. 44)* Una regla que da el valor inicial de la función $f(0)$ y una ecuación recursiva que indica cómo se relaciona un valor $f(n)$ con el valor precedente $f(n-1)$

reduction (p. 462) A dilation in which the scale factor is greater than 0 and less than 1

regular polygon (p. 405) A convex polygon that is both equilateral and equiangular

repeated roots (p. 83) Two or more roots of an equation that are the same number

rhombus (p. 432) A parallelogram with four congruent sides

roots (p. 82) The solutions of a polynomial equation

reducción (p. 462) Una dilatación en donde el factor de escala es mayor que 0 y menor que 1

polígono regular (p. 405) Un polígono convexo que es tanto equilátero como equiángulo

raíces repetidas (p. 83) Dos o más raíces de una ecuación que son el mismo número

rombo (p. 432) Un paralelogramo con cuatro lados congruentes

raíces (p. 82) Las soluciones deuna ecuación de polinomios

S

sample space (p. 280) The set of all possible outcomes for an experiment

scale factor (p. 462) The ratio of the lengths of the corresponding sides of the image and the preimage of a dilation

secant (p. 574) A line that intersects a circle in two points

secant segment (p. 615) A segment that contains a chord of a circle and has exactly one endpoint outside the circle

sector of a circle (p. 650) The region bounded by two radii of the circle and their intercepted arc

segments of a chord (p. 614) The segments formed from two chords that intersect in the interior of a circle

semicircle (p. 582) An arc with endpoints that are the endpoints of a diameter

similar arcs (p. 585) Arcs that have the same measure

similar figures (p. 470) Geometric figures that have the same shape but not necessarily the same size

similar solids (p. 669) Two solids of the same type with equal ratios of corresponding linear measures

similarity transformation (p. 470) A dilation or a composition of rigid motions and dilations

simplest form of a radical (p. 192) An expression involving a radical with index n that has no radicands with perfect nth powers as factors other than 1, no radicands that contain fractions, and no radicals that appear in the denominator of a fraction

espacio de muestra (p. 280) El conjunto de todos los resultados posibles de un experimento

factor de escala (p. 462) La razón de las longitudes de los lados correspondientes de la imagen y la preimagen de una dilatación

secante (p. 574) Una recta que intersecta a un círculo en dos puntos

segmento de secante (p. 615) Un segmento que contiene una cuerda de un círculo y que tiene exactamente un punto extremo fuera del círculo

sector de un círculo (p. 650) La región encerrada por dos radios del círculo y su arco interceptado

segmentos de una cuerda (p. 614) Los segmentos formados a partir de dos cuerdas que se intersectan en el interior de un círculo

semicírculo (p. 582) Un arco con puntos extremos que son los puntos extremos de un diámetro

arcos similares (p. 585) Arcos que tienen la misma medida

figuras similares (p. 470) Figuras geométricas que tienen la misma forma pero no necesariamente el mismo tamaño

sólidos similares (p. 669) Dos sólidos del mismo tipo con razones iguales de medidas lineales correspondientes

transformación de similitud (p. 470) Una dilatación o composición de movimientos rígidos y dilataciones

mínima expresión de un radical (p. 192) Una expresión que conlleva un radical con índice n que no tiene radicandos con potencias perfectas de orden n como factores distintos a 1, que no tiene radicandos que contengan fracciones y que no tiene radicales que aparezcan en el denominador de una fracción

sine *(p. 548)* For an acute angle of a right triangle, the ratio of the length of the leg opposite the acute angle to the length of the hypotenuse

solve a right triangle *(p. 559)* To find all unknown side lengths and angle measures of a right triangle

square *(p. 432)* A parallelogram with four congruent sides and four right angles

standard equation of a circle *(p. 620)* $(x - h)^2 + (y - k)^2 = r^2$, where r is the radius and (h, k) is the center

standard form of a polynomial *(p. 63)* A polynomial in one variable written with the exponents of the terms decreasing from left to right

step function *(p. 14)* A piecewise function defined by a constant value over each part of its domain

subtend *(p. 598)* If the endpoints of a chord or arc lie on the sides of an inscribed angle, the chord or arc is said to subtend the angle.

system of nonlinear equations *(p. 252)* A system in which at least one of the equations is nonlinear

seno *(p. 548)* Para un ángulo agudo de un triángulo rectángulo, la razón de la longitud del cateto enfrente del ángulo agudo a la longitud de la hipotenusa

resolver un triángulo recto *(p. 559)* Para encontrar todas las longitudes de los lados y las medidas de los ángulos desconocidas de un triángulo recto

cuadrado *(p. 432)* Un paralelogramo con cuatro lados congruentes y cuatro ángulos rectos

ecuación estándar de un círculo *(p. 620)* $(x - h)^2 + (y - k)^2 = r^2$, donde r es el radio y (h, k) es el centro

forma estándar de un polinomio *(p. 63)* Un polinomio en una variable escrita con los exponentes de los términos decreciendo de izquierda a derecha

función escalón *(p. 14)* Una función por tramos definida por un valor constante sobre cada parte de su dominio

subtender *(p. 598)* Si los puntos extremos de una cuerda o arco descansan en los lados de un ángulo inscrito, se dice que la cuerda o arco subtiende el ángulo.

sistema de ecuaciones no lineales *(p. 252)* Un sistema en donde al menos una de las ecuaciones no es lineal

tangent *(p. 542)* For an acute angle of a right triangle, the ratio of the length of the leg opposite the acute angle to the length of the leg adjacent to the acute angle

tangent of a circle *(p. 574)* A line in the plane of a circle that intersects the circle at exactly one point

tangent circles *(p. 575)* Coplanar circles that intersect in one point

tangent segment *(p. 615)* A segment that is tangent to a circle at an endpoint

theoretical probability *(p. 281)* The ratio of the number of favorable outcomes to the total number of outcomes when all outcomes are equally likely

trapezoid *(p. 442)* A quadrilateral with exactly one pair of parallel sides

trigonometric ratio *(p. 542)* A ratio of the lengths of two sides in a right triangle

tangente *(p. 542)* Para un ángulo agudo de un triángulo rectángulo, la razón de la longitud del cateto enfrente del ángulo agudo a la longitud del cateto adyacente al ángulo agudo

tangente de un círculo *(p. 574)* Una recta en el plano de un círculo que intersecta el círculo en exactamente un punto

círculos tangentes *(p. 575)* Círculos coplanarios que se intersectan en un punto

segmento de tangente *(p. 615)* Un segmento que es tangente a un círculo en un punto extremo

probabilidad teórica *(p. 281)* La razón del número de resultados favorables con respecto al número total de resultados cuando todos los resultados son igualmente probables

trapecio *(p. 442)* Un cuadrilátero con exactamente un par de lados paralelos

razón trigonométrica *(p. 542)* Una razón de las longitudes de dos lados en un triángulo recto

trinomial *(p. 63)* A polynomial with three terms

two-column proof *(p. 336)* A type of proof that has numbered statements and corresponding reasons that show an argument in a logical order

two-way table *(p. 296)* A frequency table that displays data collected from one source that belong to two different categories

trinomio *(p. 63)* Un polinomio con tres términos

prueba de dos columnas *(p. 336)* Un tipo de prueba que tiene enunciados numerados y motivos correspondientes que muestran un argumento en un orden lógico

tabla de doble entrada *(p. 296)* Una tabla de frecuencia que muestra los datos recogidos de una fuente que pertenece a dos categorías distintas

---

V

vertex *(p. 4)* The point where a graph changes direction

vertex form of an absolute value function *(p. 6)* An absolute value function written in the form $f(x) = a|x - h| + k$, where $a \neq 0$

vertex form of a quadratic function *(p. 148)* A quadratic function written in the form $f(x) = a(x - h)^2 + k$, where $a \neq 0$

vertex of a parabola *(p. 124)* The lowest point on a parabola that opens up or the highest point on a parabola that opens down

vertex of a polyhedron *(p. 666)* A point of a polyhedron where three or more edges meet

volume *(p. 667)* The number of cubic units contained in the interior of a solid

vértice *(p. 4)* El punto en donde un gráfico cambia de dirección

forma de vértice de una función de valor absoluto *(p. 6)* Una función de valor absoluto escrita en la forma $f(x) = a|x - h| + k$, donde $a \neq 0$

forma de vértice de una función cuadrática *(p. 148)* Una función cuadrática escrita en la forma $f(x) = a(x - h)^2 + k$, donde $a \neq 0$

vértice de una parábola *(p. 124)* El punto más bajo de una parábola que se abre hacia arriba o el punto más alto de una parábola que se abre hacia abajo

vértice de un poliedro *(p. 666)* Un punto de un poliedro donde se encuentran tres o más bordes

volumen *(p. 667)* El número de unidades cúbicas contenidas en el interior de u sólido

---

Z

zero of a function *(p. 132)* An *x*-value of a function *f* for which $f(x) = 0$

Zero-Product Property *(p. 82)* If the product of two real numbers is 0, then at least one of the numbers is 0.

cero de una función *(p. 132)* Un valor *x* de una función *f* para el cual $f(x) = 0$

propiedad de producto cero *(p. 82)* Si el producto de dos números reales es 0, entonces al menos uno de los números es 0.

Index

A

30°-60°-90° right triangles
  finding sine and cosine of, 550
  finding tangent in, 543
  side lengths of, 525, 527
30°-60°-90° Triangle Theorem, 527
45°-45°-90° (isosceles) right triangles
  defined, 526
  finding sine and cosine of, 550
  side lengths of, 525–526
  in standard position, 516
45°-45°-90° Triangle Theorem, 526
AA, *See* Angle-Angle (AA) Similarity Theorem
Absolute value functions
  defined, 4
  graphing, 3–7, 52
    combining transformations, 7
    identifying graphs, 3
    stretching, shrinking, and reflecting, 5–6
    translating graphs, 4
  as piecewise functions, 15
  vertex form of, 6
  writing, 15
Acute triangles
  in circumscribed circle, 353
  classifying by Pythagorean inequalities, 521
  orthocenter of, 364
Addition
  of complex numbers, 239–240
  of polynomials, 61, 64–65, 114
  of radicals and square roots, 191, 196
Adjacent arcs, 583
Ailles rectangle, 530
Algebra, Fundamental Theorem of, 246
Algebra tiles
  multiplying monomials and binomials using, 69
  using, 60
  writing expressions modeled by, 60
Algebraic expressions
  modeling, 60
  simplifying, 59
Altitudes of cones, 680
Altitudes of triangles, 361–365, 395
  concept summary of, 365
  defined, 363
  isosceles triangles, 365
  properties of, 361
  using, 363–365

Angle(s)
  in circles
    angle relationships, 605–609, 630–631
    and arc measures, 606–607
    central angles, 581, 582, 597
    inscribed angles, 597–599
    measuring in radians, 643
  circumscribed, 608–609
  complementary, sine and cosine of, 548
  identifying, using inverse trigonometric ratios, 558
  measures of (*See* Angle measures)
  of polygons, 403–407, 452
Angle Bisector Theorem, 346
Angle Bisector Theorem, Converse of the, 346
Angle bisectors, 343–347, 393
  concept summary of, 365
  defined, 346
  finding angle measures using, 346
  points on, 343
  properties of, 351
  proportionality in triangle, 502
  using, 346–347
Angle measures
  in circles
    finding, 606–608
    in radians, 643
  of kites, 445
  of polygons, 403–407, 452
    exterior, 406–407
    interior, 404–406
    regular polygons, 403, 407, 657
  of rhombuses, 434
  of triangles
    finding, using trigonometric ratios, 558
    ordering, 380
    and side lengths, 377, 379–380
Angle of depression, 551
Angle of elevation, 544, 551
Angle-Angle (AA) Similarity Theorem, 486–487
  proof of, 486
  using, 486–487
Angles Inside the Circle Theorem, 607
Angles Outside the Circle Theorem, 607
Another Way
  graphing quadratic functions, 147
  negative exponents, 30

  probability, sample space and outcomes, 280
  quadratic inequalities, 263
  segments of secants and tangents, 616
  solving quadratic equations
    by factoring, 203, 210
    by rewriting formulas, 212
  solving right triangles, 559
  Table of Trigonometric Ratios, 558
  writing quadratic equations, 247
  zeros of functions, 204
Another Way
  inscribed angles, T-596
  nth roots, T-34
  permutations, 313
  quadratic functions, T-100
  simplifying radical expressions, T-244
  sketching piecewise functions, T-10
Apothem of regular polygon, 655, 657
Applications, T-60, T-244, T-376, T-460, T-588, T-664
Arc(s), circular, 581–585, 629
  adjacent, 583
  congruent, 584–585
  defined, 581
  intercepted, 597, 598
  lengths (*See* Arc lengths)
  major, 582
  measuring (*See* Arc measures)
  minor, 581, 582
  naming, 582
  similar, 585
Arc Addition Postulate, 583
Arc length(s), 639–643, 694
  defined, 641
  finding, 639
  using, 639
    to find measures, 641
    to find distances, 642
Arc measures, 581–585, 629
  of congruent arcs, 584–585
  finding, 582–583
    using angle relationships in circles, 606–607
    using arc lengths, 641
    using congruent chords, 590
  of intercepted arcs, 599
  of minor and major arcs, 582
  proving circles are similar, 585
Area(s), 636, *See also* Surface areas
  of circles, 647–651, 694

formula for, 648
sectors of circles, 647, 650–651, 694
after dilations, 460
of ellipses, 654
of kites, 656
of polygons, 655–659, 695
    formula for, 655, 658
    regular polygons, 655, 658–659
    similar polygons, 479
of regions, 651
of rhombuses, 656
of semicircles, 651
of trapezoids, 72
using to find probability, 649

Areas of Similar Polygons Theorem, 479

Arithmetic mean, compared to geometric mean, 531

Assignment Guide, *occurs at the beginning of each exercise set*

Asymptotes, 42

Average rate of change
comparing functions using, 173–174
defined, 173
using and interpreting, 173

Axis of symmetry
defined, 124
finding, 136–137
and standard equations of parabolas, 163–164

B

Basal area, 48
Base (of cones), 680
Base (of solids), 666
Base (of trapezoids), 442
Base angles (of trapezoids), 442
Bayes' Theorem, 302
Biconditional statements, 434
Binomial(s)
defined, 63
finding factors, 89, 95
multiplying, 69–72, 114, 571
    with trinomials, 72
    using FOIL Method, 71, 77
square of binomial pattern, 75–76
sum and difference pattern, 75, 77

Binomial distributions, 319–322, 328
constructing, 322
defined, 321
interpreting, 322

Binomial experiments, 321
Birthday problem, 318
Bisecting angles, *See* Angle bisector
Bisectors of triangles, 351–356, 394
angle (*See* Angle bisectors)
circumcenter of triangle, 352–354
circumscribing circle about triangle, 353–354
incenter of triangle, 355–356
inscribing circle within triangle, 356
perpendicular (*See* Perpendicular bisector(s))

C

Calculator, *See* Graphing calculator
Carpenter's square, 603
Cavalieri, Bonaventura, 667
Cavalieri's Principle, 667–668
Center of circle, 572, 574
Center of dilation, 462
Center of regular polygon, 655, 657
Center of sphere, 686
Central angle of circles
defined, 581, 582
and inscribed angles, 597
Central angle of regular polygon, 657
Centroid of triangle
defined, 362
finding, 362–363
Centroid Theorem, 362
Chapter Summary, T-xxxii, T-58, T-120, T-188, T-276, T-332, T-400, T-458, T-514, T-570, T-636
Chords of circles, 589–592, 629
defined, 574, 590
identifying, 574
intersections among, 605, 613–614
intersections with tangents on circles, 605
perpendicular to diameter, 589
using, 589–592, 629
Chords of spheres, 686
Circles, 570
analyzing relationships among, 572
angle relationships in, 605–609, 630–631
    finding angle and arc measures, 606–607
    using circumscribed angles, 608–609
arc measures of, 581–585, 629
areas of, 647–651, 694
center of, 572, 574
central angle of, 581, 582, 597
chords of (*See* Chords of circles)
circumference of, 639–643, 694
circumscribed about triangle, 353–354, 601
concentric, 575
congruent, 584
in coordinate plane, 619–623, 632
defined, 572, 574
diameters of, 574, 589, 591
drawing by using string, 573
equations of, 619–621
graphing, 621, 622
inscribed angles in, 597–599, 630
inscribed polygons in, 600–601, 630
lines and segments that intersect, 573–577, 606, 628
polygon circumscribed about, 580
proving similarity in, 585
radius of, 574, 592, 616
segment relationships in, 613–616, 631
tangent, 572, 575
writing coordinate proofs involving, 622

Circular arcs, *See* Arc(s), circular
Circular cones, 680, *See also* Cones
Circumcenter of triangles, 352–354
circumscribing circle about triangle, 353–354
defined, 352
finding, 354
Circumcenter Theorem, 352
Circumference of circles, 636
and arc length, 639–643, 694
defined, 640
finding distances using, 642
formula for, 640
Circumscribed Angle Theorem, 608
Circumscribed angles
defined, 608
using, 608–609
Circumscribed circle, 353–354
Closed set of numbers, 64
Closure, *occurs at the end of each lesson*
Coefficient of friction, 633
Coin flip, 279, 280, 311, 319
Combinations, 311–315, 328
counting, 314
defined, 314
finding probability using, 315
formula, 314–315
Common Error, T-1, 29, T-41, T-59, 77, 78, 84, T-107, 109, T-121, 125, T-129, 165, T-189, 195, T-259, T-277, 283, T-333, 345, 379, T-401, 421, 422, 442, 444, T-459, T-515, 543, T-571, 592, 620, T-637, 676
Common Errors
angle approximation, 609
angle symbol compared to less than symbol, 379
area of semicircle, 651

completing the square, 217
diameter of sphere, 687
falling objects, 132
geometric mean of right triangle, 535
indirect proofs, 379
negative signs, 137
paying attention to units, 642
polynomials, 64
probability
   and binomial distribution, 322
   overlapping events, 307
proportions, 533
ratio of volumes, 669

Common external tangents, 575
Common internal tangents, 575
Common Misconceptions, 30, T-279, 432, 435, 477, 479, 501, T-540, 542, T-665
Common tangents, 575
Compass, 354
Complementary angles, sine and cosine of, 548
Complements of events, 281–282
Completing the square
defined, 216
deriving Quadratic Formula by, 225
finding maximum and minimum values, 218–219
solving quadratic equations by, 215–220, 269, 571
   compared to other methods, 229
in standard equation of circle, 621

Complex conjugates
defined, 241
multiplying, 241

Complex numbers, 237–241, 270
adding and subtracting, 239–240
defined, 237, 238
equality of two, 239
imaginary unit i, 237, 238–239
multiplying, 240–241
standard form of, 238
subsets of set of, 237

Complex plane, 244
Composite solids, volumes of, 669, 676, 682, 689
Compound events, 306–307
Compound inequalities, writing, 333, 381
Concentric circles, 575
Concept Summaries
factoring polynomials completely, 109
factoring trinomials, relationships between signs, 92
number of solutions of quadratic equation, 203
segments, lines, rays, and points in triangles, 365
types of proofs, 339
ways to prove quadrilateral is parallelogram, 423

Concurrency, point of, 352
Concurrent lines, rays, or segments, 352
Conditional probabilities
comparing, 299
defined, 289
finding with conditional relative frequencies, 298
finding with tables, 291

Conditional relative frequencies, 297–298
Cones
defined, 680
frustum of, 684
lateral surfaces of, 680
surface areas of, 679–680, 697–698
volumes of, 679, 681–682, 697–698

Congruence transformations, 470
Congruent arcs
defined, 584
identifying, 584–585

Congruent Central Angles Theorem, 584
Congruent chords
using to find arc measures, 590
using to find radius of circle, 592

Congruent circles, 584
Congruent Circles Theorem, 584
Congruent Corresponding Chords Theorem, 590
Congruent Parts of Parallel Lines Corollary, 418
Conjectures
making, about trapezoids, 441
writing
   on chords, 589
   on permutations and combinations, 311
   on polynomial equations, 81
   on similar right triangles, 531

Conjugates
complex, 241
defined, 194
rationalizing the denominator with, 194

Connecting to Next Step, *occurs at the end of each exploration*
Connections to Algebra
completing the square, 621
compound inequalities, 381
Product Property of Square Roots, 519

Connections to Geometry, area of trapezoid, 72
Consecutive vertices, 404
Construction
binomial distribution, 322
centroid of triangle, 362
circumscribing circle about triangle, 354
dilation, 464
inscribing circle within triangle, 356
point along directed line segment, 503
probability distribution, 320
tangent to circle, 577

Contingency tables, *See* Two-way tables
Contradiction, Proof by, 378
Contrapositive of the Triangle Proportionality Theorem, 500
Converses of theorems
Converse of the Angle Bisector Theorem, 346
Converse of the Hinge Theorem, 386–387
Converse of the Perpendicular Bisector Theorem, 344
Converse of the Pythagorean Theorem, 520–521
Converse of the Triangle Proportionality Theorem, 500
Isosceles Trapezoid Base Angles Converse, 443
Parallelogram Diagonals Converse, 422
Perpendicular Chord Bisector Converse, 590

Convex polygons, 404
Coordinate plane
circles in, 619–623, 632
dilating figures in, 461, 463–464
dilating lines in, 461
parallelograms in, 415, 424, 436
trapezoids in
   identifying, 442
   midsegments of, 444
triangle midsegments in, 372

Coordinate proofs
defined, 339
involving circles, 622
writing, 339, 622

Coordinate rule for dilations, 463–464
Coplanar circles, 575
Corollaries
Congruent Parts of Parallel Lines Corollary, 418

Corollary to the Polygon Interior
    Angles Theorem, 405
Rectangle Corollary, 432
Rhombus Corollary, 432
Square Corollary, 432
Corresponding lengths, in similar
    polygons, 477
Corresponding parts, of similar
    polygons, 476
Cosine ratios, 547–552, 566
    of 45° and 30° angles, 550
    calculating, 547
    of complementary angles, 548
    defined, 548
    finding, 548, 550
    inverse, 558
    using, 548–549
Counterexamples, 191
Critical values, 262
Cube roots
    finding, 35
    using properties of, 193
Cylinders, volumes of, 665–669, 696
    formula for, 31, 668

D

Data
    choosing functions to model,
        170–171
    writing functions to model, 172
Decay, *See* Exponential decay
Decay factor, 42, 46, *See also*
        Exponential decay function
Deductive reasoning, 135
Degree of a monomial, 62
Degree of a polynomial, 63
Degrees, converting between radians
    and, 643
Denominators, rationalizing, 194–195
Dependent events, 287–291, 326
    comparing to independent events,
        290
    defined, 287, 289
    determination of, 287
    probability of, 289–290
Depreciation rate, 46
Depression, angle of, 551
Diagonal(s)
    of polygons, 404
    of special parallelograms,
        434–436
Diagonal lengths
    of parallelograms, 422
    of rectangles, 435
Diagrams, *See also* Reading Diagrams
    tree
        finding sample space using, 280

        reading, 311
    Venn, 295, 433
Diameters of circles
    chord perpendicular to, 589
    defined, 574
    drawing, 589
    identifying, 574
    using, 591
Diameters of spheres, 686–687
Die roll, 279, 282, 287, 320–321
Difference of two squares pattern,
    102
Differences, identifying functions with,
    171
Differentiated Instruction,
    Throughout, See for example:
    12, 64, 174, 239, 312, 423
Dilations, 461–465, 508
    centers of, 462
    comparing triangles after, 475
    constructing, 464
    coordinate rule for, 463–464
    defined, 462
    finding perimeter and area after, 460
    identifying, 462
    performing, in coordinate plane,
        461, 463–464
    scale factors of, 462
        finding, 465
        negative, 464
    and similarity, 469
Directed line segments
    constructing point along, 503
    defined, 502
    partitioning, 502–503
Directrix, 162–163
Discriminant, in Quadratic Formula
    defined, 228
    interpreting, 228–229
    using, 247–248
Discuss, 6, T-11, T-27, T-35, T-61,
    T-81, T-145, T-201, T-209,
    T-225, T-237, T-251, T-259,
    T-287, T-469, T-597
Disjoint events, 305–308, 327
    defined, 305–306
    finding probability of, 305–306
Distance, using circumference and arc
    length to find, 642
Distance Formula, writing equation of
    parabola using, 162–163
Distance from a point to a line,
    defined, 346
Distributive Property
    adding and subtracting radicals
        using, 196
    multiplying binomials using, 70

    multiplying complex numbers using,
        240–241
Division
    Property
        Power of a Quotient, 30
        Quotient of Powers, 29
    of radicals and square roots, 191
Dodecagons, 407
Domain of a function
    defined, 137
    restricting, 2
Dynamic geometry software
    drawing perpendicular bisector, 334
    drawing triangles, 516
Dynamic Teaching Tools, *occurs on*
    each Chapter Summary page,
    on each Section Overview
    page, on each Exploration
    page, on the second page of
    each exercise set, and on each
    What Did You Learn? page

E

Edges (of polyhedrons), 666
Elevation, angle of, 544, 551
Elimination, solving nonlinear systems
    by, 253–254
Ellipses, areas of, 654
English Language Learners,
    Throughout, See for example:
    20, 90, 137, 227, 297, 420
Enlargement, 462
Enrichment Strategies, *occurs in each*
    chapter opener, at the end of
    each exercise set, and in each
    chapter test
Equations
    approximating solutions of, 190
    of circles, 619–621
    graphing linear, 121
    linear systems of, solving by
        graphing, 189
    matching equivalent forms of, 81
    multi-step, using structure to solve,
        401
    nonlinear systems of (*See* Nonlinear
        systems of equations)
    polynomial, solving in factored
        form, 81–84, 115
    quadratic (*See* Quadratic equations)
    standard
        of circle, 619, 620–621
        of parabola, 163–164
    writing
        for circles, 620–621
        for functions, 11
        for parabolas, 162–165

for perpendicular bisectors, 347
for perpendicular lines, 333
Equiangular polygons, 405
Equidistant (point), 344
Equidistant Chords Theorem, 592
Equilateral polygons, 405
Even functions, 146
Events
complements of, 281–282
compound, 306–307
defined, 280
disjoint, 305–308, 327
independent and dependent, 287–291, 326
overlapping, 305–308, 327
probability of (*See* Probability)
Exactly two answers, 281
Experiment(s)
binomial, 321
probability experiments, 280
sample spaces of, 279–283
Experimental probabilities, 282–283
defined, 282
finding, 282–283, 287
Exploration Note, T-41, T-69, T-75, T-107, T-153, T-169, T-279, T-305, T-319, T-403, T-411, T-419, T-431, T-485, T-493, T-517, T-525, T-581, T-589, T-605, T-613, T-619, T-639, T-647, T-655, T-665, T-673
Exponential decay, 43
Exponential decay functions, 42–46, 54
defined, 42
graphing, 43
parent function for, 42
Exponential decay models, 43–44
Exponential functions
comparing to quadratic and linear functions, 169–174, 184
defined, 42
graphing, 41, 43
growth and decay functions, 42–46, 54
identifying, 170–171
using differences or ratios, 171
using graphs, 41, 170
modeling with, 43–44
properties of exponents, 27–31, 53
radicals and rational exponents, 35–38, 54
rewriting, 45
writing recursive rules for, 44–45
Exponential growth, 43
Exponential growth functions, 42–45, 54
defined, 42

graphing, 43
parent function for, 42
Exponential growth models, 43–44
Expressions
algebraic
modeling, 60
simplifying, 59
evaluating, 121
interpreting, 65
radical
defined, 192
simplest form of, 192, 526
Extensions, T-11, 65, T-81, T-107, 139, T-279, 353, 355, T-360, T-361, 362, T-370, T-371, 402, 412, 415, 421, 460, T-461, T-469, T-475, 518, T-531, 549, T-557, T-638, T-646, T-679, 682
Exterior Angle Inequality Theorem, 384
Exterior angle measures, of polygons, 406–407
External resources, using, 225, 685
External segments, 615
External Tangent Congruence Theorem, 576
Extra Examples, *occur in each lesson*

F

Faces (of polyhedrons), 666
Factored completely, 108
Factored form
defined, 81, 82
matching to standard forms, 107
of polynomial equations, 81–84
Factorial numbers, n, 312–313
Factoring
polynomials
$ax^2 + bx + c$, 95–98, 115
completely, 107–110, 116
defined, 90
difference of two squares, 102
by grouping, 108
perfect square trinomials, 103, 189
special products, 101–104, 116
using the GCF, 83–84
$x^2 + bx + c$, 89–92, 115
solving quadratic equations by, 203, 210–211
compared to other methods, 229
Factors and zeros, 156
Favorable outcomes, 281, 283
First differences, 171
Flow proofs, *See* Flowchart proofs
Flowchart proofs
concept summary of, 339

defined, 338
writing, 338
Focus of parabola, 161–165, 183
FOIL Method
multiplying binomials using, 71, 77
multiplying complex numbers using, 240–241
For Your Information, T-3, T-417
Formative Assessment Tips,
Agree-Disagree Statement, T-152
Agreement Circles, T-580
Always-Sometimes-Never True, T-596
Example/Nonexample (student generated), T-664
Exit Ticket, T-60
Fact-First Questioning, T-492
Give Me Five, T-546
I Used to Think . . . But Now I Know, T-244
Learning Goals Inventory, T-334
Look Back, T-384
Muddiest Point, T-376
No-Hands Questioning, T-236
Opposing Views, T-646
Paired Verbal Fluency, T-224
Partner Speaks, T-304
Pass the Problem, T-214
Point of Most Significance, T-74
Popsicle Sticks, T-40
Predict, Explain, Observe (P-E-O Probe), T-278
Response Cards, T-200
Think-Alouds, T-460
Think-Pair-Share, T-18
Three-Minute Pause, T-144
3-2-1, T-468
Thumbs Up, T-34
Turn and Talk, T-2
Visitor Explanation, T-294
Wait Time, T-250
Which One Doesn't Belong?, T-540
Whiteboarding, T-26
Writing Prompt, T-190
Formulas
circle
arc length, 641
area, 648
area of sectors, 650
circumference, 640
combinations, 314
cone
surface area, 680
volume, 681
cylinder, volume, 31, 668
Distance Formula, 162–163
falling objects, 65

for input of functions, 21
permutations, 313
polygon, area, 655, 658
prism, volume, 667
pyramid, volume, 674
Pythagorean Theorem, 518
Quadratic Formula (See Quadratic Formula)
rhombus or kite, area, 656
sphere
   radius, 38
   surface area, 686
   volume, 688
spherical cap, volume, 692
spherical lune, area, 692
trapezoid, area, 72
Frequencies, 296–298
  joint, 296
  marginal, 296
  relative, 297–298
Frequency tables, See Two-way tables
Frustum
  of cones, 684
  of pyramids, 678
Functions
  absolute value (See Absolute value functions)
  comparing, using average rates of change, 173–174
  domain of
    defined, 137
    restricting, 2
  even and odd, 146
  exponential (See Exponential functions)
  finding zeros of (See Zeros of functions)
  identifying, 170–171
    using differences or ratios, 171
    using graphs, 170
  inverses of, 19–22, 53
  linear (See Linear functions)
  parent, 42, 122, 124
  piecewise, 11–15, 52–53
  quadratic (See Quadratic functions)
  range of, 137
Fundamental Counting Principle, 312
Fundamental Theorem of Algebra, 246
FYI, T-27, T-75, 104, T-236, T-541, 577, T-605, T-619

G

GCF, See Greatest common factor (GCF)
General rules, 27

Geometric mean
  compared to arithmetic mean, 531
  defined, 531, 534
  using, 534–535
Geometric Mean (Altitude) Theorem, 534
Geometric Mean (Leg) Theorem, 534
Geometric probabilities, 649
Geometric relationships, proving, 335–339, 392
Geometry drawing software, See Dynamic geometry software
Golden ratio, 267, 484
Golden rectangle, 195
Graph(s)
  of absolute value functions, 3–7
    identifying, 3
    stretching, shrinking, and reflecting, 5–6
    transforming, 3–7
    translating, 4
  of exponential functions, characteristics of, 41
  identifying functions using, 170
  of linear functions, transforming, 1
  of parent quadratic functions, characteristics of, 124
  of systems of quadratic inequalities, 261
Graphing
  absolute value functions, 3–7, 52
  circles, 621, 622
  exponential functions, 41, 43
  linear equations, 121
  parabolas
    equations of, 163
    with horizontal axis of symmetry, 163
  piecewise functions, 13
  quadratic equations, solving by, 201–205, 268
    compared to other methods, 229
    with complex solutions, 245
  quadratic functions
    $f(x) = a(x - p)(x - q)$, 153–157, 182
    $f(x) = a(x - h)^2 + k$, 145–149, 181
    $f(x) = ax^2$, 123–126, 180
    $f(x) = ax^2 + bx + c$, 135–139, 181
    $f(x) = ax^2 + c$, 129–132, 180
    parent function, 122, 124
    using zeros, 156
  quadratic inequalities, 260–262, 272
    systems of, 261
    in two variables, 260–261
  step functions, 14

  systems of linear equations, 189
  systems of nonlinear equations, 252
Graphing calculator
  combinations, 314
  functions with restricted domain, 2
  intersect feature, 227, 255
  maximum feature, 139, 149
  permutations, 313
  π key, 640
  quadratic functions, 129, 145
  quadratic inequalities, 259
  table feature, 44, 255, 318
  tangent key, 541
  variables displayed, 263
  viewing windows, 255
  zero feature, 149
Great circle, 686
Greatest common factor (GCF)
  factoring out, 96
  factoring polynomials using, 83–84
  finding, 59, 83
Greatest common monomial factor, 83
Grouping, factoring by, 108
Growth, See Exponential growth
Growth factor, 42, See also Exponential growth function

H

Hemispheres, 685, 686
Hinge Theorem, 386–388
Hinge Theorem, Converse of, 386–387
Histograms
  analyzing, 319
  making, 277
Homework Check, occurs at the beginning of each exercise set
Horizontal axis of symmetry, of parabolas, 163–164
Horizontal shrinks, of absolute value functions, 6
Horizontal stretches, as nonrigid transformations, 465
Horizontal translations
  of absolute value functions, 4
  of quadratic functions, 130

I

Imaginary numbers
  defined, 238
  nth roots as, 36
  pure, 238
  and quadratic equations, 225
  in set of complex numbers, 237
Imaginary solutions, of quadratic equations, 245–248

Imaginary unit *i*, 238
  defined, 237, 238
  equality of two complex numbers, 239
  finding square roots of negative numbers, 238–239
Incenter of triangle
  defined, 355
  inscribing circle within triangle, 356
  using, 355–356
Incenter Theorem, 355
Independent events, 287–291, 326
  comparing to dependent events, 290
  defined, 287, 288
  determination of, 287, 288–289
  probability of, 288–289
Index of radical, 36
Indirect measurement, using geometric mean of right triangle, 535
Indirect proofs, 377–381, 396
  defined, 378
  writing, 378
Induction, 402
Inductive hypothesis, 402
Inequalities
  compound, writing, 333, 381
  in one triangle, 377–381, 396
  Pythagorean, 521
  quadratic, 259–263, 272
  in two triangles, 385–388, 396
Inscribed angles, 597–599, 630
  defined, 597, 598
  finding measure of, 599
  using, 598–599
Inscribed Angles of a Circle Theorem, 599
Inscribed polygons, 600–601, 630
  defined, 600
  using, 600–601
Inscribed Quadrilateral Theorem, 600
Inscribed Right Triangle Theorem, 600
Intercept form, of quadratic functions
  defined, 154
  using to find zeros of functions, 155–156
Intercepted arcs, 597, 598
Interior angle measures, of polygons, 404–406
  finding unknown, 405
  number of sides of polygon, 405
  sum of angle measures, 404
Intersecting lines
  and circles, 606
  examples of, 606

Intersection of events, 306–307
Introduction, *occurs at the beginning of each section*
Inverse cosine, 558
Inverse functions, 21
Inverse of the Triangle Proportionality Theorem, 500
Inverse relation, 20
Inverse sine, 558
Inverse tangent, 558
Inverse trigonometric ratios, 558–559
Inverses of functions, 19–22, 53
  finding algebraically, 22
  formula for input of function, 21
Isosceles right triangles
  defined, 526
  finding sine and cosine of, 550
  side lengths of, 525–526
  in standard position, 516
Isosceles trapezoid(s)
  defined, 442
  properties of, 443
Isosceles Trapezoid Base Angles Converse, 443
Isosceles Trapezoid Base Angles Theorem, 443
Isosceles Trapezoid Diagonals Theorem, 443
Isosceles triangles, *See also* Isosceles right triangles
  median and altitude of, 365

J

Joint frequencies, 296
Joint relative frequencies, 297–298

K

Kite(s)
  angle measures of, 445
  areas of, 656
  defined, 445
  properties of, 441, 445, 454
Kite Diagonals Theorem, 445
Kite Opposite Angles Theorem, 445

L

Lake frontage, 505
Lateral surfaces of cones, 680
Latus rectum, 168
Laurie's Notes, *occurs for each exploration and lesson*
Leading coefficient, 63
Legs (of right triangles)
  finding with sine and cosine ratios, 549

  finding with tangent ratio, 543
Legs (of trapezoids), 442
Like radicals, 196
Likelihoods, and probabilities, 278
Line(s)
  concurrent, 352
  in coordinate plane, dilation of, 461
  intersecting in circles, 606
  parallel
    identifying, 401
    proportionality with three lines, 501
  perpendicular
    identifying, 401
    writing equations of, 333
  that intersect circles, 573–577, 606, 628
  in triangles, 334
Line segments, *See* Segment(s), line
Linear equations
  graphing, 121
  systems of, solving by graphing, 189
Linear functions
  comparing to exponential and quadratic functions, 169–174, 184
  identifying, 170–171
    using differences or ratios, 171
    using graphs, 170
  transformations of graphs of, 1
Linear systems, solving by graphing, 189

M

Maintaining Mathematical Proficiency Notes, T-1, T-59, T-121, T-189, T-277, T-333, T-401, T-459, T-515, T-571, T-637
Major arcs
  defined, 582
  measuring, 582
Marginal frequencies, 296
Marginal relative frequencies, 297–298
Mathematical induction, 402
Mathematical Practices, *Throughout. See for example:* T-1, 2, T-3, 6, 13, 14, T-27, 31, T-35, 37, 38, T-41, 42, 45, T-59, 60, T-61, T-69, 72, T-75
Maximum values
  defined, 137
  of quadratic functions, finding, 137–139
    by completing the square, 218–219

Means
  arithmetic, 531
  geometric, 531, 534–535
Measure(s)
  of angles (*See* Angle measures)
  of arc (*See* Arc measures)
  indirect, 535
Measure of an Inscribed Angle Theorem, 598
Medians
  of trapezoids, 444
  of triangles, 361–365, 395
    concept summary of, 365
    defined, 362
    isosceles triangles, 365
    properties of, 361
    using, 362–363
Midsegment(s)
  of trapezoids, 444
  of triangles, 371–374, 395
    defined, 372
    using in coordinate plane, 372
    using Triangle Midsegment Theorem, 373–374
Midsegment triangles, 372
Mini-Assessment, *occurs at the end of each exercise set*
Minimum values
  defined, 137
  of quadratic functions, finding, 137–138
    by completing the square, 218
Minor arcs
  defined, 581
  measuring, 582
Moai, 567
Modeling
  algebraic expressions, with algebra tiles, 60
  choosing functions for, 170–171
  of dropped objects, 248
  with exponential functions, 43–44
  of launched or thrown objects, 248
  writing functions for, 172
Modeling with Mathematics
  circles, northern lights, 609
  polynomials
    arch of fireplace, 84
    falling objects, 65, 104
    Punnett square and gene combinations, 78
    terrarium as rectangular prism, 110
  probabilities and likelihoods, 278
  quadratic equations
    chalkboard on door, 220
    golden rectangle, 195
    perimeter and area of land, 263
    wolf breeding pairs, 227
  quadratic functions
    height of water balloons, 139
    water fountains, 149
  relationships within triangles, city streets, 374
  right triangles and trigonometry
    angle of depression and skiing on mountain, 551
    angle of elevation and height of tree, 544
    equilateral triangle road sign, 528
    roof height, 533
  similarity
    height of flagpole, 488
    swimming pool, 478
Monomials
  defined, 62
  finding degrees of, 62
  multiplying, 69
Motivate, *occurs on each exploration page*
Multiplication
  of complex numbers, 240–241
  of polynomials, 69–72, 114
    binomials, 69–72, 77, 571
    monomials, 69
  Property
    Power of a Power, 29, 36
    Power of a Product, 30
    Product of Powers, 29
  of radicals and square roots, 191, 196
Multi-step equations, using structure to solve, 401

N

***n* factorial,** 312–313
Negative exponents, 28
Negative numbers, square roots of, 238–239
Negative scale factors, 464
Nets for three-dimensional solids, 638
No solutions, quadratic equations with, 203, 245
Nonlinear systems of equations
  defined, 252
  solving, 251–255, 272
    algebraically, by elimination, 253–254
    algebraically, by substitution, 253–254
    approximating solutions, 254–255
    by graphing, 252
    involving circles, 623
Nonrigid transformations, 465

***n*th root of *a*,** 36
***n*th roots,** 35–37
Number of solutions, of quadratic equations, 203, 209, 228–229

O

Obtuse triangles
  in circumscribed circle, 353
  classifying by Pythagorean inequalities, 521
  orthocenter of, 364
Odd functions, 146
Odds
  defined, 283
  finding, 283
Ohms, 240
One, special properties of, 81
Operations, *See* Addition; Division; Multiplication; Subtraction
Opposite Sides Parallel and Congruent Theorem, 422
Orthocenter of triangle
  defined, 363
  finding, 364
  type of triangle, and location, 364
Outcomes
  defined, 280
  favorable, 281, 283
  odds of, 283
Overlapping events, 305–308, 327
  defined, 305, 306
  finding probability of, 305–307

P

Pacing Guide, xxxii, 58, 120, 188, 276, 332, 400, 458, 514, 570, 636
Pacing Suggestion, *occurs at the beginning of each section*
Parabolas
  defined, 124, 162
  directrix of, 162–163
  equations of
    graphing, 163
    standard, 163–164
    translated, 165
    using Distance Formula to write, 162–163
    writing, 162–165
  focus of, 161–165, 183
  graphing
    equations of, 163
    with horizontal axis of symmetry, 163
  latus rectum of, 168
  satellite dishes and spotlights, 161
  x-intercepts of, finding number of, 229

Parabolic reflectors, 165
Paragraph proofs
   concept summary of, 339
   defined, 337
   writing, 337
Parallel lines
   identifying, 401
   proportionality with three lines, 501
Parallel Postulate, 495
Parallelogram(s)
   in coordinate plane, 415, 424, 436
   defined, 412
   diagonal lengths of, 422
   properties of, 411–415, 452
   proving quadrilateral is parallelogram, 419–424, 453
   side lengths of, 421
   special, properties of, 431–436, 453–454
   writing two-column proof for, 414
Parallelogram Consecutive Angles Theorem, 413
Parallelogram Diagonals Converse, 422
Parallelogram Diagonals Theorem, 413
Parallelogram Opposite Angles Theorem, 412
Parallelogram Opposite Sides Theorem, 412
Parent functions
   for exponential decay functions, 42
   for exponential growth functions, 42
   for quadratic functions, 122, 124
Partitioning
   defined, 503
   of directed line segments, 502–503
Patterns
   difference of two squares pattern, 102
   in dilation, 460
   inverse functions, 21
   perfect square trinomial pattern, 103
   for similar figures, 460
   square of binomial pattern, 75–76
   sum and difference pattern, 75, 77
Percent, finding, 277
Perfect square trinomial pattern, 103
Perfect square trinomials
   completing the square, 216
   factoring, 103, 189
Performance Tasks
   Building a Roof Truss, 391
   Challenging the Rock Wall, 563
   Diamonds, 451
   Finding Locations, 627
   Flight Path of a Bird, 113
   Golden Ratio, 267
   Pool Maintenance, 507
   Pool Shots, 51
   Population Density, 693
   Risk Analysis, 325
   Solar Energy, 179
Perimeters
   after dilations, 460
   of similar polygons, 478
Perimeters of Similar Polygons Theorem, 478
Permutations, 311–315, 328
   counting, 312–313
   defined, 312
   finding probability using, 313
   formulas, 313
Perpendicular bisector(s), 343–347, 393
   concept summary of, 365
   defined, 344
   drawing, 334
   points on, 343
   properties of, 351
   using, 344–345
   using chords of circles, 590–591
   writing equations for, 347
Perpendicular Bisector Theorem, 344
Perpendicular Bisector Theorem, Converse of the, 344
Perpendicular Chord Bisector Converse, 590
Perpendicular Chord Bisector Theorem, 590
Perpendicular lines
   identifying, 401
   writing equations of, 333
Perspective drawing, 511
Pi (π), 640
Piecewise functions, 11–15, 52–53
   absolute value function as, 15
   defined, 12
   graphing and writing, 13
Point(s)
   on angle bisectors, 343
   on perpendicular bisectors, 343
Point of concurrency, 352
Point of tangency, 574
Polygon(s), 400, *See also* Quadrilaterals; Regular polygons; Similar polygons
   angle measures of, 403–407, 452
      exterior, 406–407
      interior, 404–406
   areas of, 655–659, 695
   circumscribed about circle, 580
   inscribed in circle, 600–601
   number of sides of, 405
Polygon Exterior Angles Theorem, 406
Polygon Interior Angles Theorem, 404
Polygon Interior Angles Theorem, Corollary to the, 405
Polyhedrons, 666–667
Polynomial(s), 58, *See also* Binomial(s); Monomials
   adding and subtracting, 61, 64–65, 114
   classifying, 63
   defined, 63
   degrees of, 63
   factoring
      $ax^2 + bx + c$, 95–98, 115
      completely, 107–110, 116
      defined, 90
      by grouping, 108
      special products, 101–104, 116
      using GCF, 83–84
      $x^2 + bx + c$, 89–92, 115
   multiplying, 69–72, 114
   special products of, 75–78, 101–104, 114, 116
   in standard form, 63, 107
Polynomial equations, solving in factored form, 81–84, 115
Postulates
   Arc Addition Postulate, 583
   Parallel Postulate, 495
   Volume Addition Postulate, 672
Power, defined, 29
Power of a Power Property, 29, 36
Power of a Product Property, 30
Power of a Quotient Property, 30
Precision, Attending to
   clear definitions, 237, 469
   communication, 665
   exactly two answers, 281
   rounding probabilities, 322
   rounding trigonometric ratios and lengths, 542
   standard position for right triangles, 516
   statement of problem, 356
   table of values and graphing, 156
   use π key on calculator, 640
Prisms
   surface area of, 637
   volumes of, 665–669, 696
Probability, 276
   binomial distributions, 319–322, 328
   of complements of events, 281–282
   of compound events, 306–307
   conditional, 289, 291, 299

of disjoint and overlapping events, 305–308, 327
of events
defined, 278, 280
and likelihoods, 278
experimental, 282–283, 287
frequencies, 296–298
geometric, 649
of independent and dependent events, 287–291, 326
odds, 283
permutations and combinations, 311–315, 328
sample spaces, 279–283, 326
theoretical, 280–282, 287
two-way tables, 295–299, 327
using area to find, 649
Probability distributions, 320–321
binomial, 319–322
constructing, 320
defined, 320
interpreting, 321
Probability experiments, 280
Problem Solving
analyzing givens, relationships, and goals, 251
conjecturing about solutions, 245
guess, check, and revise strategy, 190
trying special cases, 122
Product of Powers Property, 29
Product Property of Square Roots, 192, 519
Proofs
by Contradiction, 378
coordinate
defined, 339
involving circles, 622
writing, 339, 622
defined, 335, 336
flowchart or flow, 338, 339
indirect, 377–381
paragraph, 337
summary of types of, 339
two-column, 336, 414
writing reasons in, 335
writing steps in, 335
Proofs of theorems
Angle-Angle (AA) Similarity Theorem, 486
Circumcenter Theorem, 352
Converse of the Hinge Theorem, 387
Kite Diagonals Theorem, 445
Parallelogram Diagonals Theorem, 414
Parallelogram Opposite Sides Theorem, 412

Perpendicular Bisector Theorem, 344
Pythagorean Theorem, 517
Rhombus Diagonals Theorem, 434
Side-Side-Side (SSS) Similarity Theorem, 495
Similar Circles Theorem, 585
Triangle Larger Angle Theorem, 379
Triangle Midsegment Theorem, 373
Properties
Distributive Property, 70, 196, 240–241
Negative Exponents, 28
Power of a Power, 29, 36
Power of a Product, 30
Power of a Quotient, 30
Product of Powers, 29
Product Property of Square Roots, 192, 519
Quotient of Powers, 29
Quotient Property of Square Roots, 192–193
Zero Exponent, 28
Zero-Product, 81, 82–83
Proportion(s), solving, 515
Proportionality, 499–503, 510
finding relationships, 499
in partitioning directed line segments, 502–503
ratios forming, 459
statements of, 476
of three parallel lines, 501
with triangle angle bisector, 502
using theorems of, 500–502
Punnett square and gene combinations, 78
Pure imaginary numbers, 238
Pyramidion, 701
Pyramids
frustum of, 678
nets for, 638
slant height of, 569
volumes of, 673–676, 697
Pythagoras, 518
Pythagorean identities, 552, *See also* Trigonometric identities
Pythagorean Inequalities Theorem, 521
Pythagorean Theorem, 517–521, 564
classifying triangles using, 521
proving without words, 517
using, 518–519
using converse of, 520–521
Pythagorean Theorem, Converse of the, 520–521
Pythagorean triples, 518

Q

Quadratic equations, 188
complex numbers in, 237–241, 270
defined, 202
number of solutions of, 203, 209, 228–229
properties of radicals, 191–196, 268
solving
approximating solutions, 211–212
choosing method for, 229–230
by completing the square, 215–220, 269, 571
with complex solutions, 245–248, 271
by factoring, 203, 210
by graphing, 201–205, 268
with imaginary solutions, 245–248
nonlinear systems of equations, 251–255, 272
summary of methods for, 229
using Quadratic Formula, 225–230, 270
using square roots, 209–212, 269
Quadratic Formula
defined, 226
deriving, 225–226
interpreting the discriminant in, 228–229, 247–248
solving quadratic equations using, 225–230, 270
compared to other methods, 229
Quadratic functions, 120
characteristics of, 124–125
comparing to linear and exponential functions, 169–174, 184
defined, 124
even and odd functions, 146
focus of a parabola, 161–165, 183
graphing
$f(x) = a(x - p)(x - q)$, 153–157, 182
$f(x) = a(x - h)^2 + k$, 145–149, 181
$f(x) = ax^2$, 123–126, 180
$f(x) = ax^2 + bx + c$, 135–139, 181
$f(x) = ax^2 + c$, 129–132, 180
parent function, 122
using zeros, 153, 156
identifying, 170–171
using differences or ratios, 171
using graphs, 170
intercept form of, 154, 155–156
using to find zeros, 155–156
interpreting forms of, 219

standard form of, 124
vertex form of, 148, 218
Quadratic inequalities, 259–263, 272
graphing, 259–263
in one variable, 262–263
solving, 259–263
in two variables, 260–261
Quadrilaterals, 400, *See also* Kite(s); Parallelogram(s); Trapezoid(s)
with inscribed angles, 597
inscribed in circle, 600
proving quadrilateral is parallelogram, 419–424, 453
special
classifying, 433
identifying, 431, 446
Questioning in the Classroom, T-1, T-59, T-121, T-189, T-277, T-333, T-401, T-459, T-515, T-571, T-637
Quotient of Powers Property, 29
Quotient Property of Square Roots, 192–193

R

Radian measure
converting between degrees and, 643
defined, 643
Radians, 643
Radical(s)
adding and subtracting, 196
defined, 36
multiplying, 196
Product Property of Square Roots, 192
properties of, 191–196, 268, 515
Quotient Property of Square Roots, 192–193
roots and rational exponents, 35–38, 54
Radical expressions
defined, 192
simplest form of, 192, 526
Radicand, 37
Radius of circles
defined, 574
finding with congruent chords, 592
finding with segments, 616
identifying, 574
Radius of regular polygons, 657
Radius of spheres, 38, 686
Random variables, 320
Range of a function, 137
Ratio(s)
of areas of similar polygons, 479
golden, 267, 484
identifying functions using, 171

of perimeters of similar polygons, 478
proportions formed with, 459
trigonometric (*See* Trigonometric ratios)
Rational exponents, 37, 54
Rationalizing the denominator, 194–195
Rays
concurrent, 352
tangent, 574
in triangles, 334
Reactance, 240
Reading
abbreviations: sin, cos, hyp., 548
abbreviations: tan, opp., adj., 542
biconditionals, 434
bisector of circle arc, 590
centers of dilations, 463
circles, radius and diameter, 574
circum- prefix, 353
contradictions, 378
corresponding lengths, 477
inverse tangent, 558
*n*th roots, 36
odds, 283
raked stage, 560
scale factors of dilations, 462, 465
statements of proportionality, 476
trapezoid midsegments, 444
tree diagrams, 311
triangle altitudes, 364
triangle area formula, 363
triangle midsegments, 372
two-way tables, 296
Reading Diagrams
center of circle circumscribed about polygon, 658
center of dilations, 463
rely on marked information, 446
tree diagrams, 311
Real *n*th roots of *a*, 36
Real numbers, 237
Real-life problems
circles
circumference and distance traveled, 642
graphs of, earthquake and seismograph, 622
exponential functions
annual inflation rate, 38
beach ball radius, 38
car depreciation, 46
photons from jellyfish, 31
volume of cylinder, 31
polynomials
arch of fireplace, 84

falling distance and time, 65, 104
hockey trapezoidal region, 72
land area, 92
Punnett square and gene combinations, 78
terrarium as rectangular prism, 110
width of reserve, 98
probability
adults with pets, 283
diagnostic test for diabetes, 308
quadratic equations
complex numbers and electrical circuits, 240
dimensions of touch tank, 212
height of football, 205
height of rocket, 219
horizon distance, 195
quadratic inequality, rope climbing, 261
quadratic functions
falling object, 132
parabolic reflectors, 165
satellite dish, 126
water fountains, 149
relationships within triangles
biking, 388
circumcenter or incenter for lamppost placement, 356
distance in city, 353
soccer goal, 347
right triangles and trigonometry
angle of depression and skiing on mountain, 551
angle of elevation and height of tree, 544
equilateral triangle road sign, 528
height of ramp, 528
roof height, 533
skyscrapers and support beams, 519
solving right triangles and raked stage, 560
similarity
height of flagpole, 488
scale factor and length of image, 465
triangles and shoe rack, 501
Reasoning
deductive, 135
proving geometric relationships, 335–339
visual, of similar triangles, 487
Reasoning Abstractly
algebraic properties, 335
counterexamples, 191

different properties of objects and operations, 107, 517, 655
entry point into solutions, 647
quantities and relationships, 287, 613
recall meanings, 69
Reasoning Quantitatively
depreciation, 46
quantities and relationships, 123
Rectangle(s)
Ailles, 530
defined, 432
diagonal lengths in, 435
finding missing dimensions of, 637
properties of, 432–433
Rectangle Corollary, 432
Rectangle Diagonals Theorem, 435
Recursive rules
defined, 44
writing
for exponential functions, 44–45
for quadratic functions, 172
Reduction, 462
Reflections
of absolute value function, 5
of quadratic function, 125
Regions, areas of, 651
Regular polygons
angle measures in, 403
finding, 407, 657
apothem of, 655, 657
areas of, 655, 658–659
center of, 655, 657
central angle of, 657
defined, 405
radius of, 657
Relationships within triangles, 332
bisectors of triangles, 351–356, 394
indirect proof and inequalities in one triangle, 377–381, 396
inequalities in two triangles, 385–388, 396
medians and altitudes of triangles, 361–365, 395
perpendicular and angle bisectors, 343–347, 393
triangle midsegments, 371–374, 395
Relative frequencies
conditional, 297–298
finding, 297–298
joint, 297–298
marginal, 297–298
Remember
calculator viewing windows, 255
convex polygon, 404
decimal written as percent, 38

distance from point to line, 162
dodecagon, 407
domain and range, 137
$f(x)$ notation, 124
function values closest to 0, 254
functions/zeros, graphs/x-intercepts, equations/solutions, 205
functions/zeros and graphs/x-intercepts, 155
Fundamental Counting Principle, 312
inverse operations, 162
nonlinear systems, 253
polygon in coordinate plane, 415
power and exponents, 29
quadratic functions, stretch, shrink, and reflection, 125
quadratic functions, translations, 130
radical in simplest form, 526
radicand, 37
recursive rules, 44
scientific notation, 31
transformations of graphs, 5
translations of graphs, 4
Triangle Inequality Theorem, 521
triangle side lengths, 527
Repeated roots, 83
Resistance, 240
Resources, T-2, T-34, T-122, T-128, T-144, T-160, T-190, T-224, T-342, T-350, T-360, T-370, T-430, T-460, T-468, T-474, T-484, T-516, T-540, T-664
Reteaching Strategies, *occurs in each chapter opener, at the end of each exercise set, and in each chapter test*
Rhombus(es)
angle measures in, 434
areas of, 656
defined, 432
properties of, 432–433
Rhombus Corollary, 432
Rhombus Diagonals Theorem, 434
Rhombus Opposite Angles Theorem, 434
Right cones
defined, 680
surface areas of, 679–680
Right Triangle Similarity Theorem, 532
Right triangles, 514
in circumscribed circle, 353
classifying, 521
cosine ratios of, 547–552, 566
inscribed in circle, 600

isosceles (*See* Isosceles right triangles)
orthocenter of, 364
Pythagorean Theorem, 517–521, 564
similar, 531–535
identifying, 532
using geometric mean, 534–535
sine ratios of, 547–552, 566
solving, 557–560, 566
defined, 559
using inverse trigonometric ratios, 558–559
special (*See* Special right triangles)
standard position for, 516
tangent ratios of, 541–544, 565
verifying, 520
Rigid motions, 469
Riser, 562
Roots of polynomial equations
defined, 82
repeated, 83
Rules, *See also* Recursive rules
general, 27
for properties of exponents, 27

S

Sample spaces, 279–283, 326
defined, 279, 280
finding, 280
SAS, *See* Side-Angle-Side (SAS)
Scaffolding in the Classroom, T-xxxii, T-58, T-120, T-188, T-276, T-332, T-400, T-458, T-514, T-570, T-636
Scale factors
of dilations
defined, 462
finding, 465
negative, 464
of similar solids, defined, 669
Scientific notation, 31
Secant lines
defined, 574
identifying, 574
intersecting outside circles, 613
Secant segments, 615–616
Second differences, 171
Section Overview, *occurs at the beginning of each section*
Sectors of circles
areas of, 647, 650–651, 694
defined, 647, 650
Segment(s), line
of chords, secants, and tangents, 614–616
directed
constructing point along, 503

defined, 402
partitioning, 502–503
relationships in circles, 613–616, 631
tangent, 574
that intersect circles, 573–577, 628
of triangles, 334
finding length of, 500
Segments of chords, 614
Segments of Chords Theorem, 614
Segments of Secants and Tangents Theorem, 616
Segments of Secants Theorem, 615
Semicircles
areas of, 651
defined, 582
Series circuit, 240
Shrinks
horizontal, of absolute value functions, 6
vertical
of absolute value functions, 5
of quadratic functions, 125
Side(s), of polygons, finding number of, 405
Side lengths
of parallelograms, 421
of triangles, 377–381
and angle measures, 377, 379–380
ordering, 380
special right triangles, 525–527
Side-Angle-Side (SAS) Similarity Theorem, 496
Side-Side-Side (SSS) Similarity Theorem, 494–495
proof of, 495
using, 494–495
Similar arcs, 585
Similar circles, 585
Similar Circles Theorem, 585
Similar figures, See also Similar polygons; Similar triangles
defined, 470
identifying, 459
proving similarity, 472
Similar polygons, 475–480, 509, See also Similar triangles
areas of, 479
corresponding lengths in, 477
corresponding parts of, 476
determining polygons are similar, 480
perimeters of, 478
Similar solids
defined, 669
volumes of, 669, 676, 682
Similar triangles
comparing after dilation, 475

deciding triangles are similar, 493
identifying, 532
proving similarity
by AA, 485–488, 509
by SAS, 496, 510
by SSS, 494–495, 510
right triangles, 531–535, 565
using geometric mean, 534–535
Similarity, 458
after dilations, 469
proportionality theorems, 499–503, 510
and rigid motions, 469
and transformations, 469–472, 508
Similarity statements, 476
Similarity transformations, 470–472
Simplest form, of radical expressions, 192, 526
Sine ratios, 547–552, 566
of 45° and 30° angles, 550
calculating, 547
of complementary angles, 548
defined, 548
finding, 548, 550
inverse, 558
using, 548–549
Slant height (of pyramid), 569
Slant height (of right cone), 680
Solids, See Three-dimensional figures
Solutions
approximating, 190
imaginary, 245–248
to quadratic equations
with complex solutions, 245–248, 271
with no solutions, 203, 245
number of, 203, 209, 228–229
Special cases, trying, 122
Special parallelograms, properties of, 431–436, 453–454
Special quadrilaterals
classifying, 433
identifying, 431, 446
Special right triangles
finding sine and cosine in, 550
finding tangents in, 543
side lengths of, 525–528, 564
30°-60°-90° triangle, 525, 527
isosceles (45°-45°-90°), 525–526
solving, 557
Speeds, comparing, 169
Spheres
defined, 686
diameter of, 686–687
radius of, 38, 686
surface areas of, 685–687, 698
volumes of, 685, 688–689, 698
Spherical cap, 692
Spherical lune, 692

Spreadsheets, 81
Square(s)
defined, 432
properties of, 432–433
Square Corollary, 432
Square of binomial pattern, 75–76
Square roots
of negative numbers, 238–239
operations with, 191
Product Property of, 192, 519
Quotient Property of, 192–193
solving quadratic equations using, 209–212, 269
compared to other methods, 229
SSS, See Side-Side-Side (SSS)
Standard equations
of circle, 619, 620–621
of parabola, 163–164
Standard form
of complex numbers, 238
of polynomials
defined, 63
matching to factored form, 107
of quadratic functions, 124
Standard position, for right triangles, 516
Step functions, 14
Straightedge, 354
Stretches
horizontal, as nonrigid transformations, 465
vertical
of absolute value functions, 5
as nonrigid transformations, 465
of quadratic functions, 125
String, drawing circles using, 573
Structure
in corresponding parts of similar polygons, 476
discerning, in dilations, 460
solving multi-step equations using, 401
Study Skills
Analyzing Your Errors, Misreading Directions, 491
Form a Weekly Study Group, Set Up Rules, 539
Keeping a Positive Attitude, 235
Keeping Your Mind Focused during Class, 429
Keeping Your Mind Focused While Completing Homework, 595
Kinesthetic Learners, 663
Learning Visually, 143
Make Note Cards, 25
Making a Mental Cheat Sheet, 303
Preparing for a Test, 87
Rework Your Notes, 369

Substitution, solving nonlinear systems by, 253–254
Subtend, 598
Subtraction
  of complex numbers, 239–240
  of polynomials, 61, 64–65, 114
  of radicals and square roots, 191, 196
Success of trials, 282
Sum and difference pattern, 75, 77
Surface areas
  of cones, 679–680, 697–698
  of prisms, 637
  of spheres, 685–687, 698
Symmetric about the origin, 146
Symmetry, axis of, See Axis of symmetry
Systems of linear equations, solving by graphing, 189
Systems of nonlinear equations
  defined, 252
  solving, 251–255, 272
    approximating solutions, 254–255
    by elimination, 253–254
    by graphing, 252
    by substitution, 253–254
Systems of quadratic inequalities, graphing, 261

T

Table of Trigonometric Ratios, 543, 558
Tables
  finding conditional probabilities with, 291
  two-way (See Two-way tables)
  using, 81
Tangency, point of, 574
Tangent(s)
  common
    defined, 575
    drawing and identifying, 575
  constructing to circle, 577
  defined, 574
  identifying, 574–575
  intersection with chords on circle, 605
  using properties of, 576–577
Tangent and Intersected Chord Theorem, 606
Tangent circles, 572, 575
Tangent Line to Circle Theorem, 576
Tangent lines
  to circle, 574
  to curve, 142
Tangent ratios, 541–544, 565
  calculating, 541
  defined, 542

finding, 542–543
inverse, 558
using, 542–543
Tangent rays, 574
Tangent segments, 574, 615–616
Teaching Strategy
  compass construction/paper folding, T-342, T-360, T-370
  connection to multiplying whole numbers, T-68
  Draw a Diagram, T-430
  exterior angles of a polygon, T-402
  graphing calculator, T-10, T-134, T-168, T-286, T-310, T-318, T-678
  Proportional Triangles, T-498
  Selective Responses, T-3, T-258
  Rule of 3, T-122, T-128
  specific teaching strategy for lesson, T-80, T-94, T-100, T-106, T-160, T-208, T-350, T-440, T-474, T-484, T-516, T-524, T-530, T-556, T-572, T-588, T-604, T-612, T-618, T-638, T-654, T-672, T-684
  Transformational Proof, T-410
  transparencies, T-418
  using algebra tiles with polynomials, T-88
Teaching Tip, 5, 12, 62, 64, 71, 82, 90, 108, 110, T-161, 162, 194, 196, 210, 246, T-279, T-305, T-335, 363, 471, 494, 502, 532, 534, T-573, 599, 601, 621, 658, 686, 688
Theorems
  30°-60°-90° Triangle Theorem, 527
  45°-45°-90° Triangle Theorem, 526
  Angle Bisector Theorem, 346
  Angle-Angle (AA) Similarity Theorem, 486–487
  Angles Inside the Circle Theorem, 607
  Angles Outside the Circle Theorem, 607
  Areas of Similar Polygons, 479
  Bayes' Theorem, 302
  Centroid Theorem, 362
  Circumcenter Theorem, 352
  Circumscribed Angle Theorem, 608
  Congruent Central Angles Theorem, 584
  Congruent Circles Theorem, 584
  Congruent Corresponding Chords Theorem, 590
  Contrapositive of the Triangle Proportionality Theorem, 500

Converse of the Angle Bisector Theorem, 346
Converse of the Hinge Theorem, 386–387
Converse of the Perpendicular Bisector Theorem, 344
Converse of the Pythagorean Theorem, 520–521
Converse of the Triangle Proportionality Theorem, 500
Equidistant Chords Theorem, 592
Exterior Angle Inequality Theorem, 384
External Tangent Congruence Theorem, 576
Geometric Mean (Altitude) Theorem, 534–535
Geometric Mean (Leg) Theorem, 534–535
Hinge Theorem, 386–388
Incenter Theorem, 355
Inscribed Angles of a Circle Theorem, 599
Inscribed Quadrilateral Theorem, 600
Inscribed Right Triangle Theorem, 600
Inverse of the Triangle Proportionality Theorem, 500
Isosceles Trapezoid Base Angles Converse, 443
Isosceles Trapezoid Base Angles Theorem, 443
Isosceles Trapezoid Diagonals Theorem, 443
Kite Diagonals Theorem, 445
Kite Opposite Angles Theorem, 445
Measure of an Inscribed Angle Theorem, 598
Opposite Sides Parallel and Congruent Theorem, 422
Parallelogram Consecutive Angles Theorem, 413–414
Parallelogram Diagonals Converse, 422
Parallelogram Diagonals Theorem, 413–414
Parallelogram Opposite Angles Theorem, 412–413
Parallelogram Opposite Sides Theorem, 412–413
Perimeters of Similar Polygons, 478
Perpendicular Bisector Theorem, 344–345
Perpendicular Chord Bisector Converse, 590

Perpendicular Chord Bisector Theorem, 590
Polygon Exterior Angles Theorem, 406–407
Polygon Interior Angles Theorem, 404–405
Pythagorean Inequalities Theorem, 521
Pythagorean Theorem, 517–521
Rectangle Diagonals Theorem, 435
Rhombus Diagonals Theorem, 434
Rhombus Opposite Angles Theorem, 434
Right Triangle Similarity Theorem, 532
Segments of Chords Theorem, 614
Segments of Secants and Tangents Theorem, 616
Segments of Secants Theorem, 615
Side-Angle-Side (SAS) Similarity Theorem, 496
Side-Side-Side (SSS) Similarity Theorem, 494–495
Similar Circles Theorem, 585
Tangent and Intersected Chord Theorem, 606
Tangent Line to Circle Theorem, 576
Three Parallel Lines Theorem, 501
Trapezoid Midsegment Theorem, 444
Triangle Angle Bisector Theorem, 502
Triangle Inequality Theorem, 381, 521
Triangle Larger Angle Theorem, 379, 527
Triangle Longer Side Theorem, 379
Triangle Midsegment Theorem, 373–374
Triangle Proportionality Theorem, 500

Theoretical probabilities, 280–282
defined, 281
finding, 281, 287
Three Parallel Lines Theorem, 501
Three-dimensional figures (solids), 666–669
classifying, 666–667
naming, 666
nets for, 638
volumes of, 667–669
composite solids, 669, 676, 682, 689
similar solids, 669, 676, 682
Tools, *See* Dynamic geometry software
Transformations
congruence and, 470
of figures, dilations, 461–465
of graphs of absolute value functions, 3–7
combining transformations, 7
stretching, shrinking, and reflecting, 5–6
vertex form of, 6
of graphs of linear functions, 1
of graphs of parent quadratic functions, 124
similarity and, 469–472, 508
Translated parabolas, 165
Translations
horizontal
of absolute value functions, 4
of quadratic functions, 130
vertical
of absolute value functions, 4
of quadratic functions, 130–131
Trapezoid(s), 441–446
area of, 72
in coordinate plane, 442
defined, 442
isosceles, 442
making conjectures about, 441
midsegments of, 444
properties of, 441–446, 454
Trapezoid Midsegment Theorem, 444
Tread, 562
Tree diagrams
finding sample space using, 280
reading, 311
Trials of probability experiments, 282
Triangle(s)
altitudes of, 361–365, 395
angle measures of, 377, 379–380
angle-angle similarity, 486–487
bisectors of (*See* Bisectors of triangles)
centroid of, 362–363
circle circumscribed about, 353–354, 601
circumcenter of, 352–354
classifying by Pythagorean inequalities, 521
comparing measures in, 385–388
dilation of
comparing after, 475
in coordinate plane, 461
incenter of, 355–356
inequalities
in one triangle, 377–381, 396
in two triangles, 385–388, 396
lines, rays, and segments in, 334
medians of, 361–365, 395
midsegments of, 371–374, 395

relating sides and angles of, 377, 379–380
relationships within (*See* Relationships within triangles)
right (*See* Right triangles)
side lengths of (*See* side lengths)
similar (*See* Similar triangles)
Triangle Angle Bisector Theorem, 502
Triangle Inequality Theorem, 381, 521
Triangle Larger Angle Theorem, 379, 527
Triangle Longer Side Theorem, 379
Triangle Midsegment Theorem, 373–374
Triangle Proportionality Theorem, 500
Triangle Proportionality Theorem, Contrapositive of the, 500
Triangle Proportionality Theorem, Converse of the, 500
Triangle Proportionality Theorem, Inverse of the, 500
Trigonometric identities
defined, 552
finding trigonometric values, 552
using, 552
Trigonometric ratios, *See also* Cosine ratios; Sine ratios; Tangent ratios
defined, 542
inverse, solving right triangles using, 558–559
Table of, 543, 558
Trigonometry, 514, *See also* Right triangles
unit circle, 516
Trinomials
defined, 63
factoring $x^2 + bx + c$, 89–92, 115
multiplying with binomials, 72
Two-column proofs
concept summary of, 339
defined, 336
writing, 336
for parallelograms, 414
Two-way frequency tables, *See* Two-way tables
Two-way tables, 295–299, 327
defined, 295–296
finding conditional probabilities, 298–299
finding relative frequencies, 297–298
making, 295–296
and Venn diagrams, 295

Union of events, 306–307
Unit circle trigonometry, 516

Variables, random, 320
Venn diagrams
    classifying parallelograms, 433
    and two-way tables, 295
Vertex (vertices)
    in absolute value functions,
        defined, 4
    of cones, 680
    of parabolas
        defined, 162
        and standard equations of
            parabolas, 163–164
    of polygons, consecutive, 404
    of polyhedrons, defined, 666
    in quadratic functions
        comparing x-intercepts with, 135
        defined, 124
        finding, 136–137
Vertex form
    of absolute value functions, 6
    of quadratic functions, 148, 218
Vertical axis of symmetry, of
        parabolas, 163–164
Vertical shrinks
    of absolute value functions, 5
    of quadratic functions, 125
Vertical stretches
    of absolute value functions, 5
    as nonrigid transformations, 465
    of quadratic functions, 125
Vertical translations
    of absolute value functions, 4

    of quadratic functions, 130–131
Viewing angle, 603
Volume(s), 636
    of composite solids, 669, 676, 682, 689
    of cones, 679, 681–682, 697–698
    of cylinders, 665–669, 696
    defined, 667
    after dilations, 460
    of prisms, 665–669, 696
    of pyramids, 673–676, 697
    of similar solids, 669, 676, 682
    of spheres, 685, 688–689, 698
Volume Addition Postulate, 672
Volume formulas
    of cones, 681
    of cylinders, 31, 668
    of prisms, 667
    of pyramids, 674
    of spheres, 688
    of spherical caps, 692

What Your Students Have Learned,
    occurs in each chapter opener
    and on each Maintaining
    Mathematical Proficiency
    page
What Your Students Will Learn,
    occurs in each chapter opener
    and at the beginning of each
    exploration page
Wheel of Theodorus, 530
Writing
    absolute value functions, 15
    compound inequalities, 333, 381
    coordinate proofs, 339, 622
    equations for functions, 11

    equations of circles, 620–621
    exponential models, 43–44
    formula for area of polygons, 655
    formula for input of functions, 21
    indirect proofs, 378
    piecewise functions, 13
    recursive rules, 44–45, 172
    rewriting equations and formulas, 212
    rewriting exponential functions, 45
    rewriting trigonometric functions, 549
    rules for exponents, 27
    step functions, 14

x-intercepts
    comparing with vertex, 135
    finding, 129, 135
    of parabolas, finding number of, 229

Zero(s)
    graphing quadratic functions using, 153, 156
    special properties of, 81
Zero exponents, 28
Zero-Product Property, 81, 82–83
Zeros of functions
    defined, 132
    factors and, 156
    of quadratic functions, finding, 204, 246–247
        using intercept form, 155–156

Postulates

Integrated Mathematics I

Ruler Postulate (p. 388)
The points on a line can be matched one to one with the real numbers. The real number that corresponds to a point is the coordinate of the point. The distance between points A and B, written as AB, is the absolute value of the difference of the coordinates of A and B.

Segment Addition Postulate (p. 390)
If B is between A and C, then $AB + BC = AC$.
If $AB + BC = AC$, then B is between A and C.

Protractor Postulate (p. 415)
Consider $\overleftrightarrow{OB}$ and a point A on one side of $\overleftrightarrow{OB}$. The rays of the form $\overrightarrow{OA}$ can be matched one to one with the real numbers from 0 to 180. The measure of $\angle AOB$, which can be written as $m\angle AOB$, is equal to the absolute value of the difference between the real numbers matched with $\overrightarrow{OA}$ and $\overrightarrow{OB}$ on a protractor.

Angle Addition Postulate (p. 417)
If P is in the interior of $\angle RST$, then the measure of $\angle RST$ is equal to the sum of the measures of $\angle RSP$ and $\angle PST$.

Two Point Postulate (p. 460)
Through any two points, there exists exactly one line.

Line-Point Postulate (p. 460)
A line contains at least two points.

Line Intersection Postulate (p. 460)
If two lines intersect, then their intersection is exactly one point.

Three Point Postulate (p. 460)
Through any three noncollinear points, there exists exactly one plane.

Plane-Point Postulate (p. 460)
A plane contains at least three noncollinear points.

Plane-Line Postulate (p. 460)
If two points lie in a plane, then the line containing them lies in the plane.

Plane Intersection Postulate (p. 460)
If two planes intersect, then their intersection is a line.

Linear Pair Postulate (p. 480)
If two angles form a linear pair, then they are supplementary.

Parallel Postulate (p. 499)
If there is a line and a point not on the line, then there is exactly one line through the point parallel to the given line.

Perpendicular Postulate (p. 499)
If there is a line and a point not on the line, then there is exactly one line through the point perpendicular to the given line.

Translation Postulate (p. 546)
A translation is a rigid motion.

Reflection Postulate (p. 554)
A reflection is a rigid motion.

Rotation Postulate (p. 564)
A rotation is a rigid motion.

Integrated Mathematics II

Arc Addition Postulate (p. 583)
The measure of an arc formed by two adjacent arcs is the sum of the measures of the two arcs.

Theorems

Integrated Mathematics I

Properties of Segment Congruence (p. 471)
Segment congruence is reflexive, symmetric, and transitive.
Reflexive For any segment AB, $\overline{AB} \cong \overline{AB}$.
Symmetric If $\overline{AB} \cong \overline{CD}$, then $\overline{CD} \cong \overline{AB}$.
Transitive If $\overline{AB} \cong \overline{CD}$ and $\overline{CD} \cong \overline{EF}$, then $\overline{AB} \cong \overline{EF}$.

Properties of Angle Congruence (p. 471)
Angle congruence is reflexive, symmetric, and transitive.
Reflexive For any angle A, $\angle A \cong \angle A$.
Symmetric If $\angle A \cong \angle B$, then $\angle B \cong \angle A$.
Transitive If $\angle A \cong \angle B$ and $\angle B \cong \angle C$, then $\angle A \cong \angle C$.

Right Angles Congruence Theorem (p. 478)
All right angles are congruent.

Congruent Supplements Theorem (p. 479)
If two angles are supplementary to the same angle (or to congruent angles), then they are congruent.

Congruent Complements Theorem (p. 479)
If two angles are complementary to the same angle (or to congruent angles), then they are congruent.

Vertical Angles Congruence Theorem (p. 480)
Vertical angles are congruent.

Corresponding Angles Theorem (p. 504)
If two parallel lines are cut by a transversal, then the pairs of corresponding angles are congruent.

Alternate Interior Angles Theorem (p. 504)
If two parallel lines are cut by a transversal, then the pairs of alternate interior angles are congruent.

Alternate Exterior Angles Theorem (p. 504)
If two parallel lines are cut by a transversal, then the pairs of alternate exterior angles are congruent.

Consecutive Interior Angles Theorem (p. 504)
If two parallel lines are cut by a transversal, then the pairs of consecutive interior angles are supplementary.

Corresponding Angles Converse (p. 510)
If two lines are cut by a transversal so the corresponding angles are congruent, then the lines are parallel.

Alternate Interior Angles Converse (p. 511)
If two lines are cut by a transversal so the alternate interior angles are congruent, then the lines are parallel.

Alternate Exterior Angles Converse (p. 511)
If two lines are cut by a transversal so the alternate exterior angles are congruent, then the lines are parallel.

Consecutive Interior Angles Converse (p. 511)
If two lines are cut by a transversal so the consecutive interior angles are supplementary, then the lines are parallel.

Transitive Property of Parallel Lines (p. 513)
If two lines are parallel to the same line, then they are parallel to each other.

Linear Pair Perpendicular Theorem (p. 522)
If two lines intersect to form a linear pair of congruent angles, then the lines are perpendicular.

Perpendicular Transversal Theorem (p. 522)
In a plane, if a transversal is perpendicular to one of two parallel lines, then it is perpendicular to the other line.

Lines Perpendicular to a Transversal Theorem (p. 522)
In a plane, if two lines are perpendicular to the same line, then they are parallel to each other.

Slopes of Parallel Lines (p. 528)
In a coordinate plane, two distinct nonvertical lines are parallel if and only if they have the same slope. Any two vertical lines are parallel.

Slopes of Perpendicular Lines (p. 528)
In a coordinate plane, two nonvertical lines are perpendicular if and only if the product of their slopes is -1. Horizontal lines are perpendicular to vertical lines.

Composition Theorem (p. 546)
The composition of two (or more) rigid motions is a rigid motion.

Reflections in Parallel Lines Theorem (p. 572)
If lines k and m are parallel, then a reflection in line k followed by a reflection in line m is the same as a translation. If A'' is the image of A, then
1. $\overline{AA''}$ is perpendicular to k and m, and
2. $AA'' = 2d$, where d is the distance between k and m.

Reflections in Intersecting Lines Theorem (p. 573)

If lines k and m intersect at point P, then a reflection in line k followed by a reflection in line m is the same as a rotation about point P. The angle of rotation is $2x°$, where $x°$ is the measure of the acute or right angle formed by lines k and m.

Triangle Sum Theorem (p. 589)

The sum of the measures of the interior angles of a triangle is 180°.

Exterior Angle Theorem (p. 590)

The measure of an exterior angle of a triangle is equal to the sum of the measures of the two nonadjacent interior angles.

Corollary to the Triangle Sum Theorem (p. 591)

The acute angles of a right triangle are complementary.

Properties of Triangle Congruence (p. 597)

Triangle congruence is reflexive, symmetric, and transitive.
Reflexive For any triangle $\triangle ABC$, $\triangle ABC \cong \triangle ABC$.
Symmetric If $\triangle ABC \cong \triangle DEF$, then $\triangle DEF \cong \triangle ABC$.
Transitive If $\triangle ABC \cong \triangle DEF$ and $\triangle DEF \cong \triangle JKL$, then $\triangle ABC \cong \triangle JKL$.

Third Angles Theorem (p. 598)

If two angles of one triangle are congruent to two angles of another triangle, then the third angles are also congruent.

Side-Angle-Side (SAS) Congruence Theorem (p. 602)

If two sides and the included angle of one triangle are congruent to two sides and the included angle of a second triangle, then the two triangles are congruent.

Base Angles Theorem (p. 608)

If two sides of a triangle are congruent, then the angles opposite them are congruent.

Converse of the Base Angles Theorem (p. 608)

If two angles of a triangle are congruent, then the sides opposite them are congruent.

Corollary to the Base Angles Theorem (p. 609)

If a triangle is equilateral, then it is equiangular.

Corollary to the Converse of the Base Angles Theorem (p. 609)

If a triangle is equiangular, then it is equilateral.

Side-Side-Side (SSS) Congruence Theorem (p. 618)

If three sides of one triangle are congruent to three sides of a second triangle, then the two triangles are congruent.

Hypotenuse-Leg (HL) Congruence Theorem (p. 620)

If the hypotenuse and a leg of a right triangle are congruent to the hypotenuse and a leg of a second right triangle, then the two triangles are congruent.

Angle-Side-Angle (ASA) Congruence Theorem (p. 626)

If two angles and the included side of one triangle are congruent to two angles and the included side of a second triangle, then the two triangles are congruent.

Angle-Angle-Side (AAS) Congruence Theorem (p. 627)

If two angles and a non-included side of one triangle are congruent to two angles and the corresponding non-included side of a second triangle, then the two triangles are congruent.

Integrated Mathematics II

Perpendicular Bisector Theorem (p. 344)

In a plane, if a point lies on the perpendicular bisector of a segment, then it is equidistant from the endpoints of the segment.

Converse of the Perpendicular Bisector Theorem (p. 344)

In a plane, if a point is equidistant from the endpoints of a segment, then it lies on the perpendicular bisector of the segment.

Angle Bisector Theorem (p. 346)

If a point lies on the bisector of an angle, then it is equidistant from the two sides of the angle.

Converse of the Angle Bisector Theorem (p. 346)

If a point is in the interior of an angle and is equidistant from the two sides of the angle, then it lies on the bisector of the angle.

Circumcenter Theorem (p. 352)

The circumcenter of a triangle is equidistant from the vertices of the triangle.

Incenter Theorem (p. 355)

The incenter of a triangle is equidistant from the sides of the triangle.

Centroid Theorem (p. 362)
The centroid of a triangle is two-thirds of the distance from each vertex to the midpoint of the opposite side.

Triangle Midsegment Theorem (p. 373)
The segment connecting the midpoints of two sides of a triangle is parallel to the third side and is half as long as that side.

Triangle Longer Side Theorem (p. 379)
If one side of a triangle is longer than another side, then the angle opposite the longer side is larger than the angle opposite the shorter side.

Triangle Larger Angle Theorem (p. 379)
If one angle of a triangle is larger than another angle, then the side opposite the larger angle is longer than the side opposite the smaller angle.

Triangle Inequality Theorem (p. 381)
The sum of the lengths of any two sides of a triangle is greater than the length of the third side.

Hinge Theorem (p. 386)
If two sides of one triangle are congruent to two sides of another triangle, and the included angle of the first is larger than the included angle of the second, then the third side of the first is longer than the third side of the second.

Converse of the Hinge Theorem (p. 386)
If two sides of one triangle are congruent to two sides of another triangle, and the third side of the first is longer than the third side of the second, then the included angle of the first is larger than the included angle of the second.

Polygon Interior Angles Theorem (p. 404)
The sum of the measures of the interior angles of a convex n-gon is $(n-2) \cdot 180°$.

Corollary to the Polygon Interior Angles Theorem (p. 405)
The sum of the measures of the interior angles of a quadrilateral is 360°.

Polygon Exterior Angles Theorem (p. 406)
The sum of the measures of the exterior angles of a convex polygon, one angle at each vertex, is 360°.

Parallelogram Opposite Sides Theorem (p. 412)
If a quadrilateral is a parallelogram, then its opposite sides are congruent.

Parallelogram Opposite Angles Theorem (p. 412)
If a quadrilateral is a parallelogram, then its opposite angles are congruent.

Parallelogram Consecutive Angles Theorem (p. 413)
If a quadrilateral is a parallelogram, then its consecutive angles are supplementary.

Parallelogram Diagonals Theorem (p. 413)
If a quadrilateral is a parallelogram, then its diagonals bisect each other.

Parallelogram Opposite Sides Converse (p. 420)
If both pairs of opposite sides of a quadrilateral are congruent, then the quadrilateral is a parallelogram.

Parallelogram Opposite Angles Converse (p. 420)
If both pairs of opposite angles of a quadrilateral are congruent, then the quadrilateral is a parallelogram.

Opposite Sides Parallel and Congruent Theorem (p. 422)
If one pair of opposite sides of a quadrilateral are congruent and parallel, then the quadrilateral is a parallelogram.

Parallelogram Diagonals Converse (p. 422)
If the diagonals of a quadrilateral bisect each other, then the quadrilateral is a parallelogram.

Rhombus Corollary (p. 432)
A quadrilateral is a rhombus if and only if it has four congruent sides.

Rectangle Corollary (p. 432)
A quadrilateral is a rectangle if and only if it has four right angles.

Square Corollary (p. 432)
A quadrilateral is a square if and only if it is a rhombus and a rectangle.

Rhombus Diagonals Theorem (p. 434)
A parallelogram is a rhombus if and only if its diagonals are perpendicular.

Rhombus Opposite Angles Theorem (p. 434)
A parallelogram is a rhombus if and only if each diagonal bisects a pair of opposite angles.

Rectangle Diagonals Theorem (p. 435)
A parallelogram is a rectangle if and only if its diagonals are congruent.

Isosceles Trapezoid Base Angles Theorem (p. 443)
If a trapezoid is isosceles, then each pair of base angles is congruent.

Isosceles Trapezoid Base Angles Converse (p. 443)
If a trapezoid has a pair of congruent base angles, then it is an isosceles trapezoid.

Isosceles Trapezoid Diagonals Theorem (p. 443)
A trapezoid is isosceles if and only if its diagonals are congruent.

Trapezoid Midsegment Theorem (p. 444)
The midsegment of a trapezoid is parallel to each base, and its length is one-half the sum of the lengths of the bases.

Kite Diagonals Theorem (p. 445)
If a quadrilateral is a kite, then its diagonals are perpendicular.

Kite Opposite Angles Theorem (p. 445)
If a quadrilateral is a kite, then exactly one pair of opposite angles are congruent.

Perimeters of Similar Polygons (p. 478)
If two polygons are similar, then the ratio of their perimeters is equal to the ratios of their corresponding side lengths.

Areas of Similar Polygons (p. 479)
If two polygons are similar, then the ratio of their areas is equal to the squares of the ratios of their corresponding side lengths.

Angle-Angle (AA) Similarity Theorem (p. 486)
If two angles of one triangle are congruent to two angles of another triangle, then the two triangles are similar.

Side-Side-Side (SSS) Similarity Theorem (p. 494)
If the corresponding side lengths of two triangles are proportional, then the triangles are similar.

Side-Angle-Side (SAS) Similarity Theorem (p. 496)
If an angle of one triangle is congruent to an angle of a second triangle and the lengths of the sides including these angles are proportional, then the triangles are similar.

Triangle Proportionality Theorem (p. 500)
If a line parallel to one side of a triangle intersects the other two sides, then it divides the two sides proportionally.

Converse of the Triangle Proportionality Theorem (p. 500)
If a line divides two sides of a triangle proportionally, then it is parallel to the third side.

Three Parallel Lines Theorem (p. 501)
If three parallel lines intersect two transversals, then they divide the transversals proportionally.

Triangle Angle Bisector Theorem (p. 502)
If a ray bisects an angle of a triangle, then it divides the opposite side into segments whose lengths are proportional to the lengths of the other two sides.

Pythagorean Theorem (p. 518)
In a right triangle, the square of the length of the hypotenuse is equal to the sum of the squares of the lengths of the legs.

Converse of the Pythagorean Theorem (p. 520)
If the square of the length of the longest side of a triangle is equal to the sum of the squares of the lengths of the other two sides, then the triangle is a right triangle.

Pythagorean Inequalities Theorem (p. 521)
For any $\triangle ABC$, where c is the length of the longest side, the following statements are true.
If $c^2 < a^2 + b^2$, then $\triangle ABC$ is acute.
If $c^2 > a^2 + b^2$, then $\triangle ABC$ is obtuse.

45°-45°-90° Triangle Theorem (p. 526)
In a 45°-45°-90° triangle, the hypotenuse is $\sqrt{2}$ times as long as each leg.

30°-60°-90° Triangle Theorem (p. 527)
In a 30°-60°-90° triangle, the hypotenuse is twice as long as the shorter leg, and the longer leg is $\sqrt{3}$ times as long as the shorter leg.

Right Triangle Similarity Theorem (p. 532)
If the altitude is drawn to the hypotenuse of a right triangle, then the two triangles formed are similar to the original triangle and to each other.

Geometric Mean (Altitude) Theorem (p. 534)
In a right triangle, the altitude from the right angle to the hypotenuse divides the hypotenuse into two segments. The length of the altitude is the geometric mean of the lengths of the two segments of the hypotenuse.

Geometric Mean (Leg) Theorem (p. 534)
In a right triangle, the altitude from the right angle to the hypotenuse divides the hypotenuse into two segments. The length of each leg of the right triangle is the geometric mean of the lengths of the hypotenuse and the segment of the hypotenuse that is adjacent to the leg.

Tangent Line to Circle Theorem (p. 576)
In a plane, a line is tangent to a circle if and only if the line is perpendicular to a radius of the circle at its endpoint on the circle.

External Tangent Congruence Theorem (p. 576)
Tangent segments from a common external point are congruent.

Congruent Circles Theorem *(p. 584)*
Two circles are congruent circles if and only if they have the same radius.

Congruent Central Angles Theorem *(p. 584)*
In the same circle, or in congruent circles, two minor arcs are congruent if and only if their corresponding central angles are congruent.

Similar Circles Theorem *(p. 585)*
All circles are similar.

Congruent Corresponding Chords Theorem *(p. 590)*
In the same circle, or in congruent circles, two minor arcs are congruent if and only if their corresponding chords are congruent.

Perpendicular Chord Bisector Theorem *(p. 590)*
If a diameter of a circle is perpendicular to a chord, then the diameter bisects the chord and its arc.

Perpendicular Chord Bisector Converse *(p. 590)*
If one chord of a circle is a perpendicular bisector of another chord, then the first chord is a diameter.

Equidistant Chords Theorem *(p. 592)*
In the same circle, or in congruent circles, two chords are congruent if and only if they are equidistant from the center.

Measure of an Inscribed Angle Theorem *(p. 598)*
The measure of an inscribed angle is one-half the measure of its intercepted arc.

Inscribed Angles of a Circle Theorem *(p. 599)*
If two inscribed angles of a circle intercept the same arc, then the angles are congruent.

Inscribed Right Triangle Theorem *(p. 600)*
If a right triangle is inscribed in a circle, then the hypotenuse is a diameter of the circle. Conversely, if one side of an inscribed triangle is a diameter of the circle, then the triangle is a right triangle and the angle opposite the diameter is the right angle.

Inscribed Quadrilateral Theorem *(p. 600)*
A quadrilateral can be inscribed in a circle if and only if its opposite angles are supplementary.

Tangent and Intersected Chord Theorem *(p. 606)*
If a tangent and a chord intersect at a point on a circle, then the measure of each angle formed is one-half the measure of its intercepted arc.

Angles Inside the Circle Theorem *(p. 607)*
If two chords intersect inside a circle, then the measure of each angle is one-half the sum of the measures of the arcs intercepted by the angle and its vertical angle.

Angles Outside the Circle Theorem *(p. 607)*
If a tangent and a secant, two tangents, or two secants intersect outside a circle, then the measure of the angle formed is one-half the difference of the measures of the intercepted arcs.

Circumscribed Angle Theorem *(p. 608)*
The measure of a circumscribed angle is equal to 180° minus the measure of the central angle that intercepts the same arc.

Segments of Chords Theorem *(p. 614)*
If two chords intersect in the interior of a circle, then the product of the lengths of the segments of one chord is equal to the product of the lengths of the segments of the other chord.

Segments of Secants Theorem *(p. 615)*
If two secant segments share the same endpoint outside a circle, then the product of the lengths of one secant segment and its external segment equals the product of the lengths of the other secant segment and its external segment.

Segments of Secants and Tangents Theorem *(p. 616)*
If a secant segment and a tangent segment share an endpoint outside a circle, then the product of the lengths of the secant segment and its external segment equals the square of the length of the tangent segment.

Integrated Mathematics III

Law of Sines *(p. 487)*
The Law of Sines can be written in either of the following forms for $\triangle ABC$ with sides of length a, b, and c.

$$\frac{\sin A}{a} = \frac{\sin B}{b} = \frac{\sin C}{c}$$

$$\frac{a}{\sin A} = \frac{b}{\sin B} = \frac{c}{\sin C}$$

Law of Cosines *(p. 494)*
If $\triangle ABC$ has sides of length a, b, and c, then the following are true.
$a^2 = b^2 + c^2 - 2bc \cos A$
$b^2 = a^2 + c^2 - 2ac \cos B$
$c^2 = a^2 + b^2 - 2ab \cos C$

Reference

Properties

Properties of Equality

Let a, b, and c be real numbers.

Addition Property of Equality
If $a = b$, then $a + c = b + c$.

Multiplication Property of Equality
If $a = b$, then $a \cdot c = b \cdot c$, $c \neq 0$.

Reflexive Property of Equality
$a = a$

Transitive Property of Equality
If $a = b$ and $b = c$, then $a = c$.

Subtraction Property of Equality
If $a = b$, then $a - c = b - c$.

Division Property of Equality
If $a = b$, then $a \div c = b \div c$, $c \neq 0$.

Symmetric Property of Equality
If $a = b$, then $b = a$.

Substitution Property of Equality
If $a = b$, then a can be substituted for b (or b for a) in any equation or expression.

Properties of Inequality

Let a, b, and c be real numbers.

Addition Property of Inequality
If $a > b$, then $a + c > b + c$.
If $a < b$, then $a + c < b + c$.

Multiplication Property of Inequality ($c > 0$)
If $a > b$ and $c > 0$, then $ac > bc$.
If $a < b$ and $c > 0$, then $ac < bc$.

Multiplication Property of Inequality ($c < 0$)
If $a > b$ and $c < 0$, then $ac < bc$.
If $a < b$ and $c < 0$, then $ac > bc$.

Subtraction Property of Inequality
If $a > b$, then $a - c > b - c$.
If $a < b$, then $a - c < b - c$.

Division Property of Inequality ($c > 0$)
If $a > b$ and $c > 0$, then $\dfrac{a}{c} > \dfrac{b}{c}$.
If $a < b$ and $c > 0$, then $\dfrac{a}{c} < \dfrac{b}{c}$.

Division Property of Inequality ($c < 0$)
If $a > b$ and $c < 0$, then $\dfrac{a}{c} < \dfrac{b}{c}$.
If $a < b$ and $c < 0$, then $\dfrac{a}{c} > \dfrac{b}{c}$.

* The Properties of Inequality are also true for $\geq$ and $\leq$.

Properties of Exponents

Let a and b be real numbers and let m and n be rational numbers.

Zero Exponent
$a^0 = 1$, where $a \neq 0$

Negative Exponent
$a^{-n} = \dfrac{1}{a^n}$, where $a \neq 0$

Product of Powers Property
$a^m \cdot a^n = a^{m+n}$

Quotient of Powers Property
$\dfrac{a^m}{a^n} = a^{m-n}$, where $a \neq 0$

Power of a Power Property
$(a^m)^n = a^{mn}$

Power of a Product Property
$(ab)^m = a^m b^m$

Power of a Quotient Property
$\left(\dfrac{a}{b}\right)^m = \dfrac{a^m}{b^m}$, where $b \neq 0$

Rational Exponents
$a^{m/n} = (a^{1/n})^m = (\sqrt[n]{a})^m$

Rational Exponents
$a^{-m/n} = \dfrac{1}{a^{m/n}} = \dfrac{1}{(a^{1/n})^m} = \dfrac{1}{(\sqrt[n]{a})^m}$, where $a \neq 0$

Properties of Absolute Value

Let a and b be real numbers.

$|a| \geq 0$ $|-a| = |a|$ $|ab| = |a||b|$ $\left|\dfrac{a}{b}\right| = \dfrac{|a|}{|b|}$, $b \neq 0$

Properties of Radicals
Let a and b be real numbers and let n be an integer greater than 1.

Product Property of Square Roots
$\sqrt{ab} = \sqrt{a} \cdot \sqrt{b}$, where $a, b \geq 0$

Quotient Property of Square Roots
$\sqrt{\dfrac{a}{b}} = \dfrac{\sqrt{a}}{\sqrt{b}}$, where $a \geq 0$ and $b > 0$

Product Property of Radicals
$\sqrt[n]{ab} = \sqrt[n]{a} \cdot \sqrt[n]{b}$

Quotient Property of Radicals
$\sqrt[n]{\dfrac{a}{b}} = \dfrac{\sqrt[n]{a}}{\sqrt[n]{b}}$, where $b \neq 0$

Square Root of a Negative Number
1. If r is a positive real number, then $\sqrt{-r} = i\sqrt{r}$.
2. By the first property, it follows that $(i\sqrt{r})^2 = -r$.

Properties of Logarithms
Let b, m, and n be positive real numbers with $b \neq 1$.

Product Property
$\log_b mn = \log_b m + \log_b n$

Quotient Property
$\log_b \dfrac{m}{n} = \log_b m - \log_b n$

Power Property
$\log_b m^n = n \log_b m$

Properties of Segment and Angle Congruence

Reflexive Property of Congruence
For any segment AB, $\overline{AB} \cong \overline{AB}$.

For any angle A, $\angle A \cong \angle A$.

Symmetric Property of Congruence
If $\overline{AB} \cong \overline{CD}$, then $\overline{CD} \cong \overline{AB}$.

If $\angle A \cong \angle B$, then $\angle B \cong \angle A$.

Transitive Property of Congruence
If $\overline{AB} \cong \overline{CD}$ and $\overline{CD} \cong \overline{EF}$, then $\overline{AB} \cong \overline{EF}$.

If $\angle A \cong \angle B$ and $\angle B \cong \angle C$, then $\angle A \cong \angle C$.

Other Properties

Property of Equality for Exponential Equations
If $b > 0$ and $b \neq 1$, then $b^x = b^y$ if and only if $x = y$.

Zero-Product Property
If a and b are real numbers and $ab = 0$, then $a = 0$ or $b = 0$.

Property of Equality for Logarithmic Equations
If b, x, and y are positive real numbers with $b \neq 1$, then $\log_b x = \log_b y$ if and only if $x = y$.

Transitive Property of Parallel Lines
If $p \parallel q$ and $q \parallel r$, then $p \parallel r$.

Distributive Property
Sum
$a(b + c) = ab + ac$

Difference
$a(b - c) = ab - ac$

Triangle Inequalities

Triangle Inequality Theorem

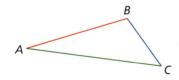

$AB + BC > AC$
$AC + BC > AB$
$AB + AC > BC$

Pythagorean Inequalities Theorem

If $c^2 < a^2 + b^2$, then $\triangle ABC$ is acute.

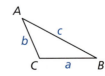

If $c^2 > a^2 + b^2$, then $\triangle ABC$ is obtuse.

Patterns

Square of a Binomial Pattern
$(a + b)^2 = a^2 + 2ab + b^2$
$(a - b)^2 = a^2 - 2ab + b^2$

Cube of a Binomial Pattern
$(a + b)^3 = a^3 + 3a^2b + 3ab^2 + b^3$
$(a - b)^3 = a^3 - 3a^2b + 3ab^2 - b^3$

Difference of Two Squares Pattern
$a^2 - b^2 = (a + b)(a - b)$

Sum of Two Cubes Pattern
$a^3 + b^3 = (a + b)(a^2 - ab + b^2)$

Sum and Difference Pattern
$(a + b)(a - b) = a^2 - b^2$

Completing the Square
$x^2 + bx + \left(\dfrac{b}{2}\right)^2 = \left(x + \dfrac{b}{2}\right)^2$

Perfect Square Trinomial Pattern
$a^2 + 2ab + b^2 = (a + b)^2$
$a^2 - 2ab + b^2 = (a - b)^2$

Difference of Two Cubes Pattern
$a^3 - b^3 = (a - b)(a^2 + ab + b^2)$

Algebra Theorems

The Remainder Theorem
If a polynomial $f(x)$ is divided by $x - k$, then the remainder is $r = f(k)$.

The Factor Theorem
A polynomial $f(x)$ has a factor $x - k$ if and only if $f(k) = 0$.

The Rational Root Theorem
If $f(x) = a_n x^n + \cdots + a_1 x + a_0$ has *integer* coefficients, then every rational solution of $f(x) = 0$ has the form
$\dfrac{p}{q} = \dfrac{\text{factor of constant term } a_0}{\text{factor of leading coefficient } a_n}$.

The Irrational Conjugates Theorem
Let f be a polynomial function with rational coefficients, and let a and b be rational numbers such that $\sqrt{b}$ is irrational. If $a + \sqrt{b}$ is a zero of f, then $a - \sqrt{b}$ is also a zero of f.

The Fundamental Theorem of Algebra

Theorem If $f(x)$ is a polynomial of degree n where $n > 0$, then the equation $f(x) = 0$ has at least one solution in the set of complex numbers.

Corollary If $f(x)$ is a polynomial of degree n where $n > 0$, then the equation $f(x) = 0$ has exactly n solutions provided each solution repeated twice is counted as two solutions, each solution repeated three times is counted as three solutions, and so on.

The Complex Conjugates Theorem
If f is a polynomial function with real coefficients, and $a + bi$ is an imaginary zero of f, then $a - bi$ is also a zero of f.

Descartes's Rule of Signs
Let $f(x) = a_n x^n + a_{n-1} x^{n-1} + \cdots + a_2 x^2 + a_1 x + a_0$ be a polynomial function with real coefficients.
- The number of positive real zeros of f is equal to the number of changes in sign of the coefficients of $f(x)$ or is less than this by an even number.
- The number of negative real zeros of f is equal to the number of changes in sign of the coefficients of $f(-x)$ or is less than this by an even number.

Formulas

Algebra

Slope
$$m = \frac{y_2 - y_1}{x_2 - x_1}$$

Slope-intercept form
$$y = mx + b$$

Point-slope form
$$y - y_1 = m(x - x_1)$$

Standard form of a linear equation
$Ax + By = C$, where A and B are not both 0

Vertex form of an absolute value function
$f(x) = a|x - h| + k$, where $a \neq 0$

Standard form of a quadratic function
$f(x) = ax^2 + bx + c$, where $a \neq 0$

Vertex form of a quadratic function
$f(x) = a(x - h)^2 + k$, where $a \neq 0$

Intercept form of a quadratic function
$f(x) = a(x - p)(x - q)$, where $a \neq 0$

Quadratic Formula
$$x = \frac{-b \pm \sqrt{b^2 - 4ac}}{2a}, \text{ where } a \neq 0$$

Standard equation of a circle
$(x - h)^2 + (y - k)^2 = r^2$

Standard form of a polynomial function
$f(x) = a_n x^n + a_{n-1} x^{n-1} + \cdots + a_1 x + a_0$

Exponential growth function
$y = a(1 + r)^t$, where $a > 0$ and $r > 0$
or
$y = ab^x$, where $a > 0$ and $b > 1$

Exponential decay function
$y = a(1 - r)^t$, where $a > 0$ and $0 < r < 1$
or
$y = ab^x$, where $a > 0$ and $0 < b < 1$

Logarithm of y with base b
$\log_b y = x$ if and only if $b^x = y$

Change-of-base formula
$\log_c a = \dfrac{\log_b a}{\log_b c}$, where a, b, and c are positive real numbers
with $b \neq 1$ and $c \neq 1$.

Explicit rule for an arithmetic sequence
$a_n = a_1 + (n - 1)d$

Explicit rule for a geometric sequence
$a_n = a_1 r^{n-1}$

Recursive equation for an arithmetic sequence
$a_n = a_{n-1} + d$

Recursive equation for a geometric sequence
$a_n = r \cdot a_{n-1}$

Sum of n terms of 1
$$\sum_{i=1}^{n} 1 = n$$

Sum of first n positive integers
$$\sum_{i=1}^{n} i = \frac{n(n + 1)}{2}$$

Sum of squares of first n positive integers
$$\sum_{i=1}^{n} i^2 = \frac{n(n + 1)(2n + 1)}{6}$$

Sum of first n terms of an arithmetic series
$$S_n = n\left(\frac{a_1 + a_n}{2}\right)$$

Sum of first n terms of a geometric series
$$S_n = a_1\left(\frac{1 - r^n}{1 - r}\right), \text{ where } r \neq 1$$

Sum of an infinite geometric series
$$S = \frac{a_1}{1 - r} \text{ provided } |r| < 1$$

Statistics

Sample mean
$$\bar{x} = \frac{\Sigma x}{n}$$

Standard deviation
$$\sigma = \sqrt{\frac{(x_1 - \bar{x})^2 + (x_2 - \bar{x})^2 + \cdots + (x_n - \bar{x})^2}{n}}$$

z-Score
$$z = \frac{x - \mu}{\sigma}$$

Margin of error for sample proportions
$$\pm \frac{1}{\sqrt{n}}$$

Probability and Combinatorics

Theoretical Probability = $\dfrac{\text{Number of favorable outcomes}}{\text{Total number of outcomes}}$

Experimental Probability = $\dfrac{\text{Number of successes}}{\text{Number of trials}}$

Probability of the complement of an event
$P(\overline{A}) = 1 - P(A)$

Probability of independent events
$P(A \text{ and } B) = P(A) \cdot P(B)$

Probability of dependent events
$P(A \text{ and } B) = P(A) \cdot P(B \mid A)$

Probability of compound events
$P(A \text{ or } B) = P(A) + P(B) - P(A \text{ and } B)$

Odds

Odds in favor = $\dfrac{\text{Number of favorable outcomes}}{\text{Number of unfavorable outcomes}}$

Odds against = $\dfrac{\text{Number of unfavorable outcomes}}{\text{Number of favorable outcomes}}$

Permutations
$${}_nP_r = \dfrac{n!}{(n-r)!}$$

Combinations
$${}_nC_r = \dfrac{n!}{(n-r)! \cdot r!}$$

Binomial experiments
$P(k \text{ successes}) = {}_nC_k p^k (1-p)^{n-k}$

The Binomial Theorem
$(a+b)^n = {}_nC_0 a^n b^0 + {}_nC_1 a^{n-1} b^1 + {}_nC_2 a^{n-2} b^2 + \cdots + {}_nC_n a^0 b^n$, where n is a positive integer.

Coordinate Geometry

Midpoint Formula
$$\left(\dfrac{x_1 + x_2}{2}, \dfrac{y_1 + y_2}{2}\right)$$

Distance Formula
$$d = \sqrt{(x_2 - x_1)^2 + (y_2 - y_1)^2}$$

Right Triangles

Pythagorean Theorem

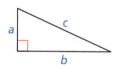

$a^2 + b^2 = c^2$

45°-45°-90° Triangles

hypotenuse = leg $\cdot \sqrt{2}$

30°-60°-90° Triangles

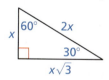

hypotenuse = shorter leg $\cdot$ 2
longer leg = shorter leg $\cdot \sqrt{3}$

Trigonometric Ratios

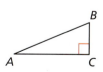

$\sin A = \dfrac{BC}{AB}$

$\cos A = \dfrac{AC}{AB}$

$\tan A = \dfrac{BC}{AC}$

$\sin^{-1} \dfrac{BC}{AB} = m\angle A$

$\cos^{-1} \dfrac{AC}{AB} = m\angle A$

$\tan^{-1} \dfrac{BC}{AC} = m\angle A$

Sine and cosine of complementary angles
Let A and B be complementary angles. Then the following statements are true.

$\sin A = \cos(90° - A) = \cos B$

$\sin B = \cos(90° - B) = \cos A$

$\cos A = \sin(90° - A) = \sin B$

$\cos B = \sin(90° - B) = \sin A$

Trigonometry

General definitions of trigonometric functions
Let θ be an angle in standard position, and let (x, y) be the point where the terminal side of θ intersects the circle $x^2 + y^2 = r^2$. The six trigonometric functions of θ are defined as shown.

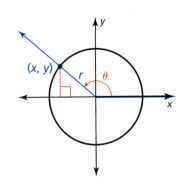

$$\sin \theta = \frac{y}{r} \qquad \cos \theta = \frac{x}{r} \qquad \tan \theta = \frac{y}{x}, x \neq 0$$

$$\csc \theta = \frac{r}{y}, y \neq 0 \qquad \sec \theta = \frac{r}{x}, x \neq 0 \qquad \cot \theta = \frac{x}{y}, y \neq 0$$

Conversion between degrees and radians
$180° = \pi$ radians

Arc length of a sector
$s = r\theta$

Area of a sector
$A = \frac{1}{2}r^2\theta$

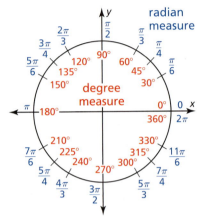

Reciprocal Identities

$$\sin \theta = \frac{1}{\csc \theta} \qquad \cos \theta = \frac{1}{\sec \theta} \qquad \tan \theta = \frac{1}{\cot \theta}$$

$$\csc \theta = \frac{1}{\sin \theta} \qquad \sec \theta = \frac{1}{\cos \theta} \qquad \cot \theta = \frac{1}{\tan \theta}$$

Tangent and Cotangent Identities

$$\tan \theta = \frac{\sin \theta}{\cos \theta} \qquad \cot \theta = \frac{\cos \theta}{\sin \theta}$$

Pythagorean Identities
$\sin^2 \theta + \cos^2 \theta = 1$
$1 + \tan^2 \theta = \sec^2 \theta$
$1 + \cot^2 \theta = \csc^2 \theta$

Negative Angle Identities
$\sin(-\theta) = -\sin \theta$
$\cos(-\theta) = \cos \theta$
$\tan(-\theta) = -\tan \theta$
$\csc(-\theta) = -\csc \theta$
$\sec(-\theta) = \sec \theta$
$\cot(-\theta) = -\cot \theta$

Cofunction Identites

$$\sin\left(\frac{\pi}{2} - \theta\right) = \cos \theta \qquad \csc\left(\frac{\pi}{2} - \theta\right) = \sec \theta$$

$$\cos\left(\frac{\pi}{2} - \theta\right) = \sin \theta \qquad \sec\left(\frac{\pi}{2} - \theta\right) = \csc \theta$$

$$\tan\left(\frac{\pi}{2} - \theta\right) = \cot \theta \qquad \cot\left(\frac{\pi}{2} - \theta\right) = \tan \theta$$

Sum Formulas
$\sin(a + b) = \sin a \cos b + \cos a \sin b$
$\cos(a + b) = \cos a \cos b - \sin a \sin b$

$$\tan(a + b) = \frac{\tan a + \tan b}{1 - \tan a \tan b}$$

Difference Formulas
$\sin(a - b) = \sin a \cos b - \cos a \sin b$
$\cos(a - b) = \cos a \cos b + \sin a \sin b$

$$\tan(a - b) = \frac{\tan a - \tan b}{1 + \tan a \tan b}$$

Any Triangle

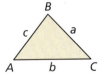

Area
Area $= \frac{1}{2}bc \sin A$
Area $= \frac{1}{2}ac \sin B$
Area $= \frac{1}{2}ab \sin C$

Law of Sines
$$\frac{\sin A}{a} = \frac{\sin B}{b} = \frac{\sin C}{c}$$

$$\frac{a}{\sin A} = \frac{b}{\sin B} = \frac{c}{\sin C}$$

Law of Cosines
$a^2 = b^2 + c^2 - 2bc \cos A$
$b^2 = a^2 + c^2 - 2ac \cos B$
$c^2 = a^2 + b^2 - 2ab \cos C$

Polygons

Triangle Sum Theorem

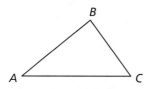

$m\angle A + m\angle B + m\angle C = 180°$

Exterior Angle Theorem

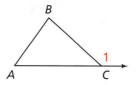

$m\angle 1 = m\angle A + m\angle B$

Triangle Midsegment Theorem

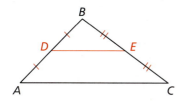

$\overline{DE} \parallel \overline{AC}$

$DE = \tfrac{1}{2} AC$

Trapezoid Midsegment Theorem

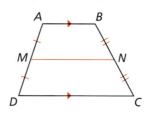

$\overline{MN} \parallel \overline{AB}$

$\overline{MN} \parallel \overline{DC}$

$MN = \tfrac{1}{2}(AB + CD)$

Polygon Interior Angles Theorem

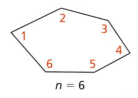

$m\angle 1 + m\angle 2 + \cdots + m\angle n = (n - 2) \cdot 180°$

Polygon Exterior Angles Theorem

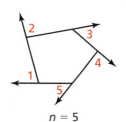

$m\angle 1 + m\angle 2 + \cdots + m\angle n = 360°$

Geometric Mean (Altitude) Theorem

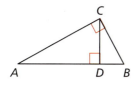

$CD^2 = AD \cdot BD$

Geometric Mean (Leg) Theorem

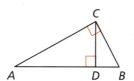

$CB^2 = DB \cdot AB$

$AC^2 = AD \cdot AB$

Circles

Arc length

Arc length of $\widehat{AB} = \dfrac{m\widehat{AB}}{360°} \cdot 2\pi r$

Area of a sector

Area of sector $APB = \dfrac{m\widehat{AB}}{360°} \cdot \pi r^2$

Central angles

$m\angle ACB = m\widehat{AB}$

Inscribed angles

$m\angle ADB = \tfrac{1}{2}m\widehat{AB}$

Tangent and intersected chord

$m\angle 1 = \tfrac{1}{2}m\widehat{AB}$

$m\angle 2 = \tfrac{1}{2}m\widehat{BCA}$

Angles and Segments of Circles

Two chords

$m\angle 1 = \tfrac{1}{2}(m\widehat{AC} + m\widehat{DB})$

$EA \cdot EB = EC \cdot ED$

Two secants

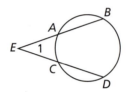

$m\angle 1 = \tfrac{1}{2}(m\widehat{BD} - m\widehat{AC})$

$EA \cdot EB = EC \cdot ED$

Tangent and secant

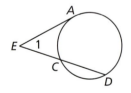

$m\angle 1 = \tfrac{1}{2}(m\widehat{AD} - m\widehat{AC})$

$EA^2 = EC \cdot ED$

Two tangents

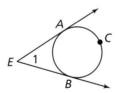

$m\angle 1 = \tfrac{1}{2}(m\widehat{ACB} - m\widehat{AB})$

$EA = EB$

Other Formulas

Simple Interest
$I = Prt$

Compound Interest
$A = P\left(1 + \dfrac{r}{n}\right)^{nt}$

Continuously Compounded Interest
$A = Pe^{rt}$

Distance
$d = rt$

Geometric mean
$x = \sqrt{a \cdot b}$

Density
$\text{Density} = \dfrac{\text{Mass}}{\text{Volume}}$

Similar polygons or similar solids with scale factor k
Ratio of perimeters $= k$
Ratio of areas $= k^2$
Ratio of volumes $= k^3$

Perimeter, Area, and Volume Formulas

Square

$P = 4s$
$A = s^2$

Rectangle

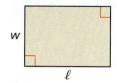

$P = 2\ell + 2w$
$A = \ell w$

Triangle

$P = a + b + c$
$A = \frac{1}{2}bh$

Circle

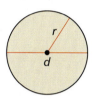

$C = \pi d$ or $C = 2\pi r$
$A = \pi r^2$

Parallelogram

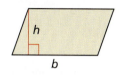

$A = bh$

Trapezoid

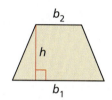

$A = \frac{1}{2}h(b_1 + b_2)$

Rhombus/Kite

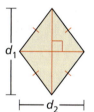 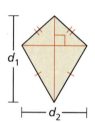

$A = \frac{1}{2}d_1 d_2$

Regular n-gon

$A = \frac{1}{2}aP$ or $A = \frac{1}{2}a \cdot ns$

Prism

$L = Ph$
$S = 2B + Ph$
$V = Bh$

Cylinder

$L = 2\pi rh$
$S = 2\pi r^2 + 2\pi rh$
$V = \pi r^2 h$

Pyramid

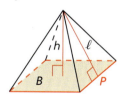

$L = \frac{1}{2}P\ell$
$S = B + \frac{1}{2}P\ell$
$V = \frac{1}{3}Bh$

Cone

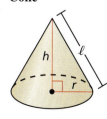

$L = \pi r \ell$
$S = \pi r^2 + \pi r \ell$
$V = \frac{1}{3}\pi r^2 h$

Sphere

$S = 4\pi r^2$
$V = \frac{4}{3}\pi r^3$

Conversions

U.S. Customary

1 foot = 12 inches
1 yard = 3 feet
1 mile = 5280 feet
1 mile = 1760 yards
1 acre = 43,560 square feet
1 cup = 8 fluid ounces
1 pint = 2 cups
1 quart = 2 pints
1 gallon = 4 quarts
1 gallon = 231 cubic inches
1 pound = 16 ounces
1 ton = 2000 pounds

U.S. Customary to Metric

1 inch = 2.54 centimeters
1 foot ≈ 0.3 meter
1 mile ≈ 1.61 kilometers
1 quart ≈ 0.95 liter
1 gallon ≈ 3.79 liters
1 cup ≈ 237 milliliters
1 pound ≈ 0.45 kilogram
1 ounce ≈ 28.3 grams
1 gallon ≈ 3785 cubic centimeters

Time

1 minute = 60 seconds
1 hour = 60 minutes
1 hour = 3600 seconds
1 year = 52 weeks

Temperature

$C = \frac{5}{9}(F - 32)$

$F = \frac{9}{5}C + 32$

Metric

1 centimeter = 10 millimeters
1 meter = 100 centimeters
1 kilometer = 1000 meters
1 liter = 1000 milliliters
1 kiloliter = 1000 liters
1 milliliter = 1 cubic centimeter
1 liter = 1000 cubic centimeters
1 cubic millimeter = 0.001 milliliter
1 gram = 1000 milligrams
1 kilogram = 1000 grams

Metric to U.S. Customary

1 centimeter ≈ 0.39 inch
1 meter ≈ 3.28 feet
1 meter ≈ 39.37 inches
1 kilometer ≈ 0.62 mile
1 liter ≈ 1.06 quarts
1 liter ≈ 0.26 gallon
1 kilogram ≈ 2.2 pounds
1 gram ≈ 0.035 ounce
1 cubic meter ≈ 264 gallons

Credits

Front Matter
i Rainbow-Pic/Shutterstock.com; **vii** ©iStockphoto.com/Christopher Futcher; **viii** Will Rodrigues/Shutterstock.com; **ix** anyaivanova/Shutterstock.com; **x** J.D.S/Shutterstock.com; **xi** ©iStockphoto.com/MinistryOfJoy; **xii** Nadiia Gerbish/Shutterstock.com; **xiii** sculpies/Shutterstock.com; **xiv** FreshPaint/Shutterstock.com; **xv** koh sze kiat/Shutterstock.com; **xvi** ©iStockphoto.com/Pamela Moore; **xvii** ©iStockphoto.com/-hakusan-; **xviii** Baloncici/Shutterstock.com; **xix** Monkey Business Images/Shutterstock.com; **xxiii** Tyler Olson/Shutterstock.com

Chapter 1
0 *top left* Olga Kovalenko/Shutterstock.com; *top right* kavram/Shutterstock.com; *center left* Karyna Che/Shutterstock.com; *bottom right* Pressmaster/Shutterstock.com; *bottom left* Will Rodrigues/Shutterstock.com; **9** Pakhnyushcha/Shutterstock.com; **14** Pressmaster/Shutterstock.com; **16** manish mansinh/Shutterstock.com; **25** Nattika/Shutterstock.com; **26** Pashin Georgiy/Shutterstock.com; **32** *top right* Vitaly Korovin/Shutterstock.com; *bottom right* Jiri Hera/Shutterstock.com; **33** *center left* Mark III Photonics/Shutterstock.com; *top right* NASA; **38** Glenda M. Powers/Shutterstock.com, siamionau pavel/Shutterstock.com; **40** *top right* ©iStockphoto.com/katkov, ©iStockphoto.com/Andrew Johnson; *center left* Dodecahedron by User:Cyp (*http://commons.wikimedia.org/wiki/File:Dodecahedron.jpg*), CC-BY-SA 3.0; **46** risteski goce/Shutterstock.com; **48** bornholm/Shutterstock.com, Masalski Maksim/Shutterstock.com; **49** *bottom left* Olga Kovalenko/Shutterstock.com; *top right* Tupungato/Shutterstock.com; *center right* U.S. Department of Energy; **50** Egon Zitter/Shutterstock.com **51** tuulijumala/Shutterstock.com, Aleksej Zhagunov/Shutterstock.com; **55** Africa Studio/Shutterstock.com

Chapter 2
58 *top left* anyaivanova/Shutterstock.com; *top right* jaroslava V/Shutterstock.com; *center left* luchschen/Shutterstock.com; *bottom right* Rudy Balasko/Shutterstock.com; *bottom left* Andresr/Shutterstock.com; **65** ©iStockphoto.com/edge69; **66** ecco/Shutterstock.com ; **67** *center left* Johanna Goodyear/Shutterstock.com; *exercise 51* ©iStockphoto.com/edge69; *exercise 52* Matt Antonino/Shutterstock.com; *exercise 53* ©iStockphoto.com/ lumpynoodles; **72** *left* d3images/Shutterstock.com; *right* Fejas/Shutterstock.com; **74** Andresr/Shutterstock.com; **78** Eric Isselee/Shutterstock.com, Erik Lam/Shutterstock.com; **80** Tamara Kulikova/Shutterstock.com; **86** Rudy Balasko/Shutterstock.com; **87** Nattika/Shutterstock.com; **88** *top right* Tuja/Shutterstock.com; *bottom right* ©iStockphoto.com/tirc83; **93** deer boy/Shutterstock.com; **94** *top left* luchschen/Shutterstock.com; *center left* cobalt88/Shutterstock.com; **98** jaroslava V/Shutterstock.com; **107** G Tipene/Shutterstock.com; **105** Mark Stout Photography/Shutterstock.com; **106** *top left* MarijaPiliponyte/Shutterstock.com; *center left* Sergiy Telesh/Shutterstock.com; **110** Dirk Ercken/Shutterstock.com; **111** Anna Breitenberger/Shutterstock.com; **113** tuulijumala/Shutterstock.com, HERGON/Shutterstock.com

Chapter 3
120 *top left* homydesign/Shutterstock.com; *top right* J.D.S/Shutterstock.com *center left* Susana Ortega/Shutterstock.com; *bottom right* MishAl/Shutterstock.com; *bottom left* Ron Zmiri/Shutterstock.com; **127** Charles Brutlag/Shutterstock.com; **128** Wolna/Shutterstock.com, ©iStockphoto.com/PLAINVIEW; **132** ©iStockphoto.com/edge69; **134** Ron Zmiri/Shutterstock.com; **141** *top left* James Thew/Shutterstock.com; *bottom right* James Hoenstine/Shutterstock.com; **143** Nattika/Shutterstock.com; **144** Rob Byron/Shutterstock.com; **149** dompr/Shutterstock.com; **151** Paul St. Clair/Shutterstock.com; **152** Susana Ortega/Shutterstock.com; **159** Serg64/Shutterstock.com; **167** inxti/Shutterstock.com; **176** Zarja/Shutterstock.com; **177** homydesign/Shutterstock.com; **179** tuulijumala/Shutterstock.com, J.D.S/Shutterstock.com

Chapter 4
188 *top left* Pavel L Photo and Video/Shutterstock.com; *top right* paul cowell/Shutterstock.com; *center left* rickyd/Shutterstock.com; *bottom right* NatalieJean/Shutterstock.com; *bottom left* ©iStockphoto.com/MinistryOfJoy; **195** *top left* Radovan Spurny/Shutterstock.com, Worldpics/Shutterstock.com, wong yu liang/Shutterstock.com; *center left* ©iStockphoto.com/MinistryOfJoy **198** Jon Le-Bon/Shutterstock.com; **199** Dan Gerber/Shutterstock.com; **205** Richard Paul Kane/Shutterstock.com; **207** *top left* Ben Haslam/Haslam Photography/Shutterstock.com; *bottom right* Don Williamson/Shutterstock.com; **208** Steve Cukrov/Shutterstock.com; **213** Zhukov Oleg/Shutterstock.com; **214** Baker Alhashki/Shutterstock.com; **221** Klaus Bösch Rainbow Vision Sandpicture; **223** *exercise 57* graphixmania/Shutterstock.com; *exercise 58* Petrovic Igor/Shutterstock.com; *center right* © Kaleb Timberlake | Dreamstime.com; **224** *top right* Ermek/Shutterstock.com; *center right* © South12th | Dreamstime.com; **231** © Kjersti Joergensen | Dreamstime.com; **232** *center left* ©iStockphoto.com/edhor; *top right* Vereshchagin Dmitry/Shutterstock.com; **233** bikeriderlondon/Shutterstock.com, Aspen Photo/Shutterstock.com, Ruth Peterkin/Shutterstock.com; **235** Nattika/Shutterstock.com; **236** ©iStockphoto.com/mikdam; **240** rickyd/Shutterstock.com; **248** *top left* © Matthew Apps | Dreamstime.com; *top center* muzsy/Shutterstock.com; *top right* © Vlue | Dreamstime.com; **250** *top left* paul cowell/Shutterstock.com; *center left* Haslam Photography/Shutterstock.com; **257** luigi nifosi/Shutterstock.com; **265** *bottom left* © Tammygaffney | Dreamstime.com; *top right* Dan Breckwoldt/Shutterstock.com; *center right* Pavel L Photo and Video/Shutterstock.com; **266** *center left* BlueRingMedia/Shutterstock.com; *center right* © Tatjana Keisa | Dreamstime.com; **267** tuulijumala/Shutterstock.com, barbar34/Shutterstock.com

Chapter 5
276 *top left* Margaret M Stewart/Shutterstock.com; *top right* Mikhail Pogosov/Shutterstock.com; *center left* Nadiia Gerbish/Shutterstock.com; *bottom right* Tyler Olson/Shutterstock.com; *bottom left* bikeriderlondon/Shutterstock.com; **279** *top* Sascha Burkard/Shutterstock.com; *center* Ruslan Gi/Shutterstock.com; *bottom right* cobalt88/Shutterstock.com, Sanchai Khudpin/Shutterstock.com; **286** ©iStockphoto.com/hugolacasse; **287** *top right* cTermit/Shutterstock.com; *top center* Picsfive/Shutterstock.com; **290** ©iStockphoto.com/andriikoval; **292** *center left* Marques/Shutterstock.com, Laurin Rinder/Shutterstock.com; *bottom left* oliveromg/Shutterstock.com; **294** *center left* cobalt88/Shutterstock.com; *center right* bikeriderlondon/Shutterstock.com; **299** Tyler Olson/Shutterstock.com; **300** *bottom left* Andrey Bayda/Shutterstock.com; *bottom right* Image Point Fr/Shutterstock.com; **303** Nattika/Shutterstock.com; **305** CrackerClips Stock Media/Shutterstock.com; **309** ©iStockphoto.com/Andy Cook; **310** *top left* Nadiia Gerbish/Shutterstock.com; *center right* bikeriderlondon/Shutterstock.com; **313** Mikhail Pogosov/Shutterstock.com; **315** *center top* Alan Bailey/Shutterstock.com; *center bottom* ©iStockphoto.com/zorani; **317** Tony Oshlick/Shutterstock.com; **318** Telnov Oleksii/Shutterstock.com; **319** Sascha Burkard/Shutterstock.com; **321** claudiofichera/Shutterstock.com; **323** Margaret M Stewart/Shutterstock.com; **324** ecco/Shutterstock.com; **325** tuulijumala/Shutterstock.com, MARGRIT HIRSCH/Shutterstock.com; **329** Christos Georghiou/Shutterstock.com

Chapter 6

332 *top left* Warren Goldswain /Shutterstock.com; *top right* puttsk/Shutterstock.com; *center left* sculpies/Shutterstock.com; *bottom right* Olena Mykhaylova/Shutterstock.com; *bottom left* Nikonaft/Shutterstock.com; **345** Nikonaft/Shutterstock.com; **347** sababa66/Shutterstock.com; **349** Tobias W/Shutterstock.com; **353** filip robert/Shutterstock.com, John T Takai/Shutterstock.com; **358** *top right* Matthew Cole/Shutterstock.com; *center right* artkamalov/Shutterstock.com, anton_novik/Shutterstock.com; **360** *center left* Olena Mykhaylova/Shutterstock.com; *top right* Lonely/Shutterstock.com; **367** Laschon Maximilian/Shutterstock.com; **369** Nattika/Shutterstock.com; **373** sculpies/Shutterstock.com; **374** Shumo4ka/Shutterstock.com; **375** Alexandra Lande/Shutterstock.com, Dan Kosmayer/Shutterstock.com; **376** clearviewstock/Shutterstock.com; **380** Dzm1try/Shutterstock.com; **386** ©iStockphoto.com/enchanted_glass; **388** Warren Goldswain /Shutterstock.com; **391** tuulijumala/Shutterstock.com, Brandon Bourdages/Shutterstock.com

Chapter 7

400 *top left* FreshPaint/Shutterstock.com; *top right* © Swinnerrr | Dreamstime.com; *center left* Kieputin on Linnanmäki Amusement Park, Helsinki, Finland, by Ppntori; *bottom right* nacroba/Shutterstock.com; *bottom left* Reprinted with permission of The Photo News.; **404** © Peter Spirer | Dreamstime.com; **409** Reprinted with permission of The Photo News.; **414** Marco Prati/Shutterstock.com; **421** ©iStockphoto.com/A-Digit; **422** Ann Paterson; **427** kokandr/Shutterstock.com; **429** Nattika/Shutterstock.com; **430** Matthew Cole/Shutterstock.com; **347** *exercise 9* Ksenia Palimski/Shutterstock.com; *exercise 10* Sundraw Photography/Shutterstock.com; *exercise 11* EvenEzer/Shutterstock.com; *exercise 12* donatas1205/Shutterstock.com; **439** © Swinnerrr | Dreamstime.com; **451** tuulijumala/Shutterstock.com, Korpithas/Shutterstock.com

Chapter 8

458 *top left* Sfocato/Shutterstock.com; *top right* © Paul Maguire | Dreamstime.com; *center left* koh sze kiat/Shutterstock.com; *bottom right* Roman Gorielov/Shutterstock.com; *bottom left* Odua Images/Shutterstock.com; **465** *top right* Odua Images/Shutterstock.com; *center right and bottom left* Henrik Larsson/Shutterstock.com; **467** *exercise 31* sarra22/Shutterstock.com; *exercise 32* irin-k/Shutterstock.com; *exercise 33* paulrommer/Shutterstock.com; *exercise 34* Evgeniy Ayupov/Shutterstock.com; *exercise 35 top left* Evgeniy Ayupov/Shutterstock.com; *exercise 35 top right* irin-k/Shutterstock.com; *exercise 35 bottom left* Evgeniy Ayupov/Shutterstock.com; *exercise 35 bottom right* Ambient Ideas/Shutterstock.com; **478** koh sze kiat/Shutterstock.com; **484** pilgrim.artworks/Shutterstock.com; **488** Rigucci/Shutterstock.com; **491** Nattika/Shutterstock.com; **492** Sarawut Padungkwan/Shutterstock.com; **496** bkp/Shutterstock.com; **498** Sfocato/Shutterstock.com; **507** tuulijumala/Shutterstock.com, bmnarak/Shutterstock.com; **511** *center right* egreggsphoto/Shutterstock.com; *bottom right* ssguy/Shutterstock.com

Chapter 9

514 *top left* V. J. Matthew/Shutterstock.com; *top right* ©iStockphoto.com/Sportstock; *center left* © Vacclav | Dreamstime.com; *bottom right* ©iStockphoto.com/Pamela Moore; *bottom left* robert paul van beets/Shutterstock.com; **523** *top left* robert paul van beets/Shutterstock.com; *center left* Naypong/Shutterstock.com; *center right* ©iStockphoto.com/susandaniels; **529** AND Inc/Shutterstock.com; **535** ©iStockphoto.com/Pamela Moore; **536** *exercise 9* © Jorg Hackemann | Dreamstime.com; *exercise 10* © Det-anan Sunonethong | Dreamstime.com; **537** © Alex Zarubin | Dreamstime.com, Kamenetskiy Konstantin/Shutterstock.com; **539** Nattika/Shutterstock.com; **545** JASON TENCH/Shutterstock.com; **551** *top left* ©iStockphoto.com/Sportstock; *top right* IPFL/Shutterstock.com; **554** ©iStockphoto.com/ollo; **555** robin2/Shutterstock.com, Vladimir Zadvinskii/Shutterstock.com, Natali Snailcat/Shutterstock.com; **563** tuulijumala/Shutterstock.com, Sorbis/Shutterstock.com; **567** *bottom right* Andrew McDonough/Shutterstock.com, zzveillust/Shutterstock.com; *bottom right* Alberto Loyo/Shutterstock.com, Bairachnyi Dmitry/Shutterstock.com; **569** Nestor Noci/Shutterstock.com

Chapter 10

570 *top left* ©iStockphoto.com/-hakusan-; *top right* © Aaron Rutten | Dreamstime.com; *center left* ©iStockphoto.com/falun; *bottom right* FikMik/Shutterstock.com; *bottom left* ©iStockphoto.com/martin_k; **573** ©iStockphoto.com/Rebirth3d; **579** kbgen/Shutterstock.com; **587** Markus Gann/Shutterstock.com; **593** *top right* TsuneoMP/Shutterstock.com, Kjpargeter/Shutterstock.com; *bottom right* M. Unal Ozmen/Shutterstock.com; **594** ©iStockphoto.com/falun; **595** Nattika/Shutterstock.com; **596** tale/Shutterstock.com; **603** © Vadim Yerofeyev | Dreamstime.com, Andrey Eremin/Shutterstock.com; **609** leonello calvetti/Shutterstock.com; **611** Sergey Nivens/Shutterstock.com, sonya etchison/Shutterstock.com; **617** OHishiapply/Shutterstock.com, Paul B. Moore/Shutterstock.com, BlueRingMedia/Shutterstock.com; **618** *center left* © Stbernarstudio | Dreamstime.com; *bottom left* Alhovik/Shutterstock.com; **622** ©iStockphoto.com/-hakusan-; **627** tuulijumala/Shutterstock.com, Condor 36/Shutterstock.com

Chapter 11

636 *top left* Pius Lee/Shutterstock.com; *top right* NCG/Shutterstock.com; *center left* Baloncici/Shutterstock.com; *bottom right* Elena Eliseeva/Shutterstock.com; *bottom left* maigi/Shutterstock.com; **639** Mauro Rodrigues/Shutterstock.com; **642** Gwoeii/Shutterstock.com; **644** *top right* ©iStockphoto.com/Prill Mediendesign & Fotografie; *center right* Gena73/Shutterstock.com; **645** Mikio Oba/Shutterstock.com; **464** ©iStockphoto.com/Charles Mann; **647** *bottom left* Elena Eliseeva/Shutterstock.com; *bottom right* Crop circles north of Umatilla, Oregon, USA by Sam Beebe (*http://commons.wikimedia.org/wiki/File:Crop_circles_north_of_Umatilla,_Oregon,_USA.jpg*), CC-BY-2.0; **654** Victoria Kalinina/Shutterstock.com; **659** *top left* Baloncici/Shutterstock.com; *top right* ©iStockphoto.com/Olgertas , keellla/Shutterstock.com; **661** *center left* NCG/Shutterstock.com; *bottom left* Laschon Maximilian/Shutterstock.com; **663** Nattika/Shutterstock.com; **671** Wata51/Shutterstock.com; **672** © Fallsview | Dreamstime.com; **675** Pius Lee/Dreamstime.com; **678** © Richard Lowthian | Dreamstime.com; **685** Mark Herreid/Shutterstock.com; **686** Dan Thornberg/Shutterstock.com; **691** *exercise 27* Nomad_Soul/Shutterstock.com; *exercise 28* Lightspring/Shutterstock.com; *exercise 29* Mark Herreid/Shutterstock.com; *exercise 30* Keattikorn/Shutterstock.com; *exercise 31* Lightspring/Shutterstock.com; *exercise 32* Dan Thornberg/Shutterstock.com; *top right* MaxyM/Shutterstock.com; **692** leonello calvetti/Shutterstock.com; **693** tuulijumala/Shutterstock.com, Peter Hermes Furian/Shutterstock.com; **699** viritphon/Shutterstock.com; **700** *bottom left* Lucie Lang/Shutterstock.com; *bottom right* Stephanie Frey/Shutterstock.com; **701** Orhan Cam/Shutterstock.com, pmphoto/Shutterstock.com